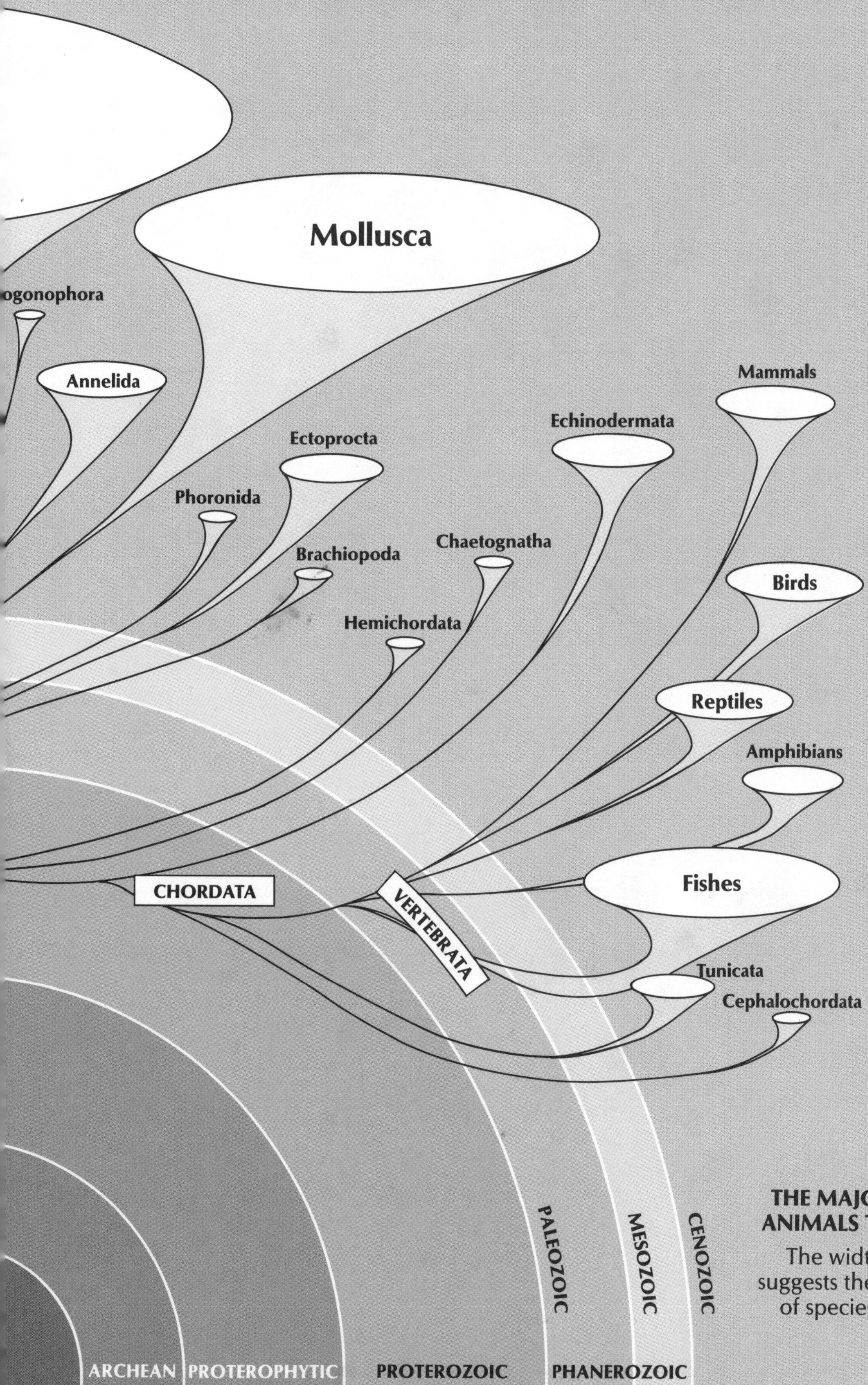

Mollusca

ogonophora

Annelida

Mammals

Echinodermata

Ectoprocta

Phoronida

Chaetognatha

Brachiopoda

Birds

Hemichordata

Reptiles

Amphibians

Fishes

CHORDATA

VERTEBRATA

Tunicata

Cephalochordata

PALEOZOIC

MESOZOIC

CENOZOIC

ARCHEAN PROTEROPHYTIC PROTEROZOIC PHANEROZOIC

THE MAJOR GROUPS OF ANIMALS THROUGH TIME

The width of each oval suggests the relative number of species in that group.

BIOLOGY
OF
ANIMALS

Cover and front matter artwork © by GUY COHELEACH
***Siberian Chase*, 1979**

Original text artwork by WILLIAM C. OBER, M.D., Crozet, Virginia

FIFTH EDITION

BIOLOGY OF ANIMALS

CLEVELAND P. HICKMAN, Jr.
Department of Biology, Washington and Lee University,
Lexington, Virginia

LARRY S. ROBERTS
Department of Biological Sciences, Texas Tech University,
Lubbock, Texas

FRANCES M. HICKMAN
Emeritus, Department of Zoology, DePauw University,
Greencastle, Indiana

with 949 illustrations including 791 in color

TIMES MIRROR/MOSBY COLLEGE PUBLISHING

ST. LOUIS • TORONTO • BOSTON • LOS ALTOS 1990

To
The Animals, Great and Small,
that enrich our lives and
teach us so much about ourselves

Developmental Editor: *Kathleen Scogna*
Project Manager: *Kathleen L. Teal*
Production Editor: *Teresa Breckwoldt*
Design: *Liz Fett*
Manuscript Editor: *Alice Fugate*
Production: *Jeanne Genz*

FIFTH EDITION

A division of The C.V. Mosby Company
11830 Westline Industrial Drive, St. Louis, MO 63146

Printed in the United States of America

Library of Congress Cataloging in Publication Data

Hickman, Cleveland P.
Biology of animals/Cleveland P. Hickman, Jr., Larry S. Roberts, Frances M. Hickman.—5th ed.
p. cm.
Includes bibliographical references.
ISBN 0-8016-5481-5
1. Zoology. I. Roberts, Larry S. II. Hickman, Frances Miller. III. Title.
QL47.2.H528 1990
591—dc20 89-5197
CIP

TS/VH/VH 9 8 7 6 5 4 3 2

Preface

Biology of Animals is a nonmajors and mixed-majors textbook designed for a one-semester course in general zoology. The book is organized into four parts, in which subject matter progresses from unifying general principles that govern the lives of all animals from paramecia to primates (Part One), to themes in animal form and function (Part Two), to an extensive, phylogenetically arranged survey of diversity in the animal world (Parts Three and Four). This was the successful organizational arrangement of the previous four editions of *Biology of Animals* and the preference of most instructors who have been contacted for their opinions. Each teacher will, of course, adapt the book to the particular needs of the class, selecting topics or rearranging the sequence as appropriate.

CHANGES IN THIS EDITION

Although *Biology of Animals* has been through four well-received editions, this fifth edition was extensively reworked and revised. In producing this edition, we benefited from having just revised this book's larger sibling, *Integrated Principles of Zoology,* now in its eighth edition. Many of the improvements to that text, such as strengthened topical presentations, many new illustrations, and pedagogical aids, were carried over into this fifth edition of *Biology of Animals.* In addition, we carefully examined all aspects of the fourth edition to improve teaching and learning aids, update information, and revise old illustrations or add new ones. This task was vastly assisted by the suggestions of seven professional zoologists who read and commented on the entire text. The revision is too extensive to describe in detail, but some of the more important changes are highlighted here.

Part One is devoted to the broad underpinnings of biology, dealing with the general principles of cell biology, metabolism, genetics, evolution, and ecology—introductory material that applies widely, often universally, to the animal world. Basic chemistry in Chapter 1 is limited to what is essential for the material that follows. Several sections were reworked and a new section added that explains what zoology is and why an understanding of zoology helps us to understand ourselves. Chapter 2 covers cell structure, cell division, and cellular metabolism. Several sections were extensively rewritten, including the fluid-mosaic model of the plasma membrane and the role of glycoproteins, the cytoskeleton, and junctional complexes. Descriptions of diffusion and osmosis were clarified. To help the student understand the role and action of enzymes in cellular metabolism, sections on energy and life, free energy, and activation energy were added. Chapter 3 is a presentation of both classical and molecular genetics and includes a description of protein synthesis. Several sections were rewritten, and an important addition covers molecular genetics of cancer.

Chapter 4, the centerpiece of Part One, is intended as a readable and nontechnical explanation of current evolutionary theory. Several sections were rewritten and many new illustrations added. Several changes were made in Chapter 5: the opening discussion of conditions for life on earth was substantially shortened, and several sections were partly rewritten, including those on terrestrial biomes, the food chain, niche concept, and population growth.

Part Two deals with form and function in animals. In Chapter 6 the topics of animal architecture and of animal integument, support, and movement—treated in separate chapters in the fourth edition—have been combined into a single chapter. Some topics, particularly the introductory treatment of development and the explanation of basic animal body plans, have been moved to Chapters 13 and 14 respectively, where they more effectively serve as essential background material for the survey chapters in Parts Three and Four. The remaining material on body organization in Chapter 6 was rewritten to be more comparative. Chapter 9 (Digestion and Nutrition) was also largely rewritten and reorganized to provide a much more comparative focus. Several sections in Chapter 8 (Homeostasis) and Chapter 10 (Nervous Coordination) were revised and reillustrated. Figures 8-5 through 8-8 are examples of how new, informative illustrations were designed to complement the text.

Several changes were made in Chapter 13 (Reproduction and Development) to provide a better coverage and at the same time to tighten the writing. Descriptions of cleavage patterns and early embryonic development, formerly in Chapter 6, were integrated into the unit on the developmental process. The description of fertilization was updated with a more accurate explanation of how polyspermy is blocked. Also rewritten in this chapter were the sections on parthenogenesis and the essay "What good is sex?" A new section was added on reproductive patterns, describing different breeding patterns and cycles, and adaptations for care and nourishment of the embryo.

Parts Three and Four present the evolutionary parade of animal diversity and adaptations. This unit begins with a brief history of the science of systematics and explanations of traditional and modern approaches to animal classification. Figure 14-5, a new illustration showing in branching pattern the major advances in architectural themes of animals, will be helpful in introducing the student to the diversity chapters that follow.

The remaining chapters in Part Three cover the panorama of the invertebrate animals. Each chapter received detailed attention, with updating and rewriting as needed. Numerous new figures have been added or replace old figures. Chapter 17 (Radiate Animals) and Chapter 18 (Acoelomate Animals) were extensively reorganized. The rather detailed treatments of structure and function of hydra and planaria were generalized to embrace all cnidarians and platyhelminths in the respective chapters, with the specific discussions of hydra and planaria being abbreviated accordingly. A short paragraph and an illustration of the phylum Loricifera were added to Chapter 19.

The vertebrates are the theme of Part Four. Much of Chapter 25 (Chordates: Vertebrate Beginnings) was rewritten. The remaining vertebrate chapters received minor updatings and clarifications, but as a whole not substantially changed. The classification of the mammals (Chapter 30) was abbreviated and its position moved forward in the chapter to parallel the placement of classifications in the other diversity chapters.

TEACHING AND LEARNING AIDS

Vocabulary Development

Key words are boldfaced and the derivations of generic names of animals are given where they first appear in the text. In addition, the derivation of many technical and zoological terms are provided in the text; in this way students gradually become familiarized with the more common roots that comprise many technical terms. An extensive glossary of more than 1000 terms provides pronunciation, derivation, and definition of each term.

Chapter Prologues

A new and distinctive feature with this edition is an opening prologue set off in a panel at the beginning of each chapter. The essays embrace a variety of themes and facts that relate in each instance to the subject of the chapter. Some present biological, particularly evolutionary, principles; others (especially those in the survey sections) illuminate distinguishing characteristics of the group treated in the chapter. Each is intended to present an important concept drawn from the chapter in an interesting manner that will facilitate learning by students, as well as engage their interest and pique their curiosity.

Marginal Notes

Marginal notes, which appear throughout the book, augment the text material and offer interesting sidelights without interrupting the narrative. We prepared many new marginal notes for this edition and revised many of the existing notes.

For Review

Each chapter ends with a concise summary, a list of review questions, and annotated selected references. The review questions enable the student to self-test his or her retention and understanding of the more important chapter material.

Art Program

The appearance and usefulness of this edition has been further enhanced with many new original drawings, all of them full color paintings, by William C. Ober. Bill Ober's artistic skills, knowledge of biology, and experience gained from an earlier career as a practicing physician have enriched this text through all its editions. Bill has been ably assisted for this edition by Claire Garrison.

SUPPLEMENTS

Instructor's Manual

The Instructor's Manual for this edition has been broadly revised and expanded. In addition to test questions, we have prepared a chapter outline, a commentary and lesson plan, and a listing of source materials for each chapter of the text. We trust this will be of particular value for first-time users of the text, but we think experienced teachers will also find much of value, particularly in the commentary and source material sections.

Laboratory Manual

The laboratory manual by Hickman and Hickman, *Laboratory Studies in Integrated Zoology*, now in its seventh edition, has been extensively rewritten and reillustrated. Although it was written for a year-long course in zoology, it can be adapted conveniently for semester or term courses by judicious selection of exercises. The large wall chart, "Chief taxonomic subdivisions and organ systems of animals," is especially appreciated by students.

ACKNOWLEDGMENTS

We wish to thank the zoologists engaged by Mosby to review the fourth edition of this text, and who in doing so suggested countless improvements that guided our approach to this revision. We are especially indebted to Peter Dalby, Clarion University, whose detailed line-by-line review of the entire text, combined with an extensive knowledge of zoology, transcended the usual expectations for a reviewer. We thank Kenneth L. Goodhue-McWilliams, California State University, Fullerton, who has generously and warmly shared his perceptive teaching philosophy with us and encouraged us in moving toward the elusive "holistic" text. It is doubtful that we have fully realized his vision for this revision, but his suggestions will be beneficial for years to come. Others to whom we owe special thanks for their insights are DuWayne C. Englert, Southern Illinois University at Carbondale; Standley E. Lewis, St. Cloud State University; Grover C. Miller, North Carolina State University; Lee F. Braithwaite, Brigham Young University; Jeffrey Pommerville, Texas A&M University. We have gained immeasurably from their insights. Our good friend Dean Foster, who has encouraged us with previous editions, read the new chapter prologues and applied his wit and vision to their improvement.

We feel fortunate to have had the continued support of the conscientious staff of Times Mirror/Mosby in the preparation of this fifth edition. We especially thank our editor David Brake, developmental editor Kathleen Scogna, production editor Teresa Breckwoldt, project manager Kathleen Teal, and manuscript editor Alice Fugate. Design and production were directed by Elizabeth Fett and Jean Genz respectively.

CLEVELAND P. HICKMAN, Jr.
LARRY S. ROBERTS
FRANCES M. HICKMAN

Contents

PART ONE Evolution of Animal Life

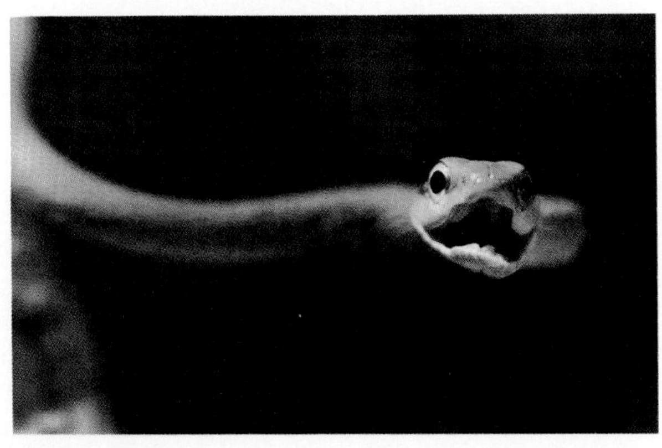

PART TWO Animal Form and Function

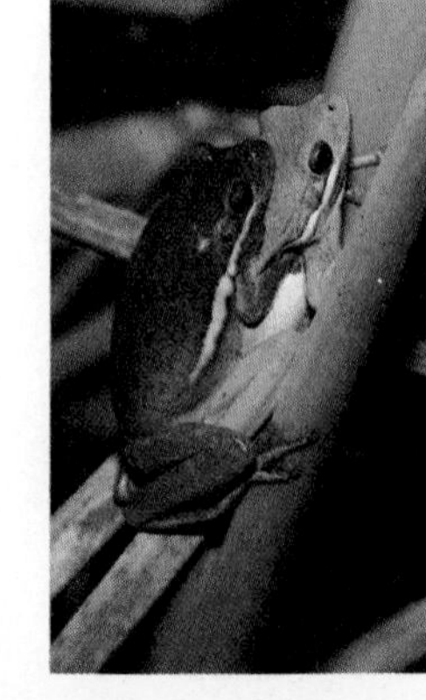

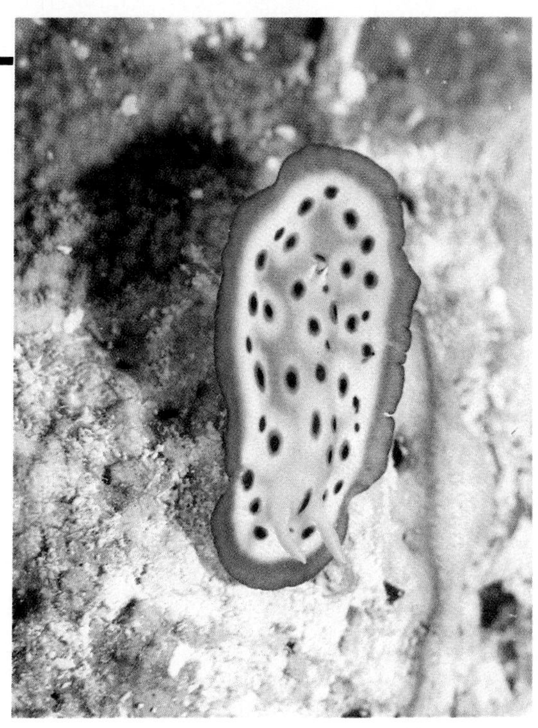

PART THREE The Invertebrate Animals

PART FOUR The Vertebrate Animals

PART ONE

Evolution of Animal Life

THAT MYSTERY OF MYSTERIES . . .

Cosmologists endeavor excitedly to understand the origins of the universe in toto and our own little solar system in particular. With backgrounds variously in physics, chemistry, astronomy, and mathematics, these scientists agree that the universe, including the planet Earth, is about 4.5 billion years old. Difficult as the question of cosmic origin is, more challenging still is the issue of the origin of life itself. Most intriguing of recent findings is the consistent picture of very early life on earth as revealed by the fossil record. The oldest fossils yet discovered are filamentous and sheath-enclosed colonial microorganisms from Western Australia that have been firmly dated radioactively at 3.3 to 3.5 billion years old. These are the earliest known life forms, among the first so far as we know to have closed themselves off from the rest of the universe as living, self-reproducing units. Responding then to the forces of natural selection, protocells such as these began to evolve, eventually yielding the web of life on earth today. Throughout time, life and environment have evolved together, each deeply marking the other. The primitive earth, with its reducing atmosphere of ammonia, methane, and water, was superbly fit for the prebiotic synthesis that led to life's beginnings. Yet, it was totally unsuited, indeed lethal, for the kinds of organisms that inhabit the earth today, just as early forms of life could not survive in our present environment.

On such matters did Charles Darwin ponder, that "mystery of mysteries" as he called it, the origins of life and species. It is a journey through time which, little by little, comes to light; the history of life forms that have been, and are being, evolved on this earth.▪

CHAPTER 1

Life

General Considerations, Basic Molecules, and Origins

Zoology (Gr. *zōon,* animal, + *logos,* discourse on, study of) is the scientific study of animals. It is commonly considered a subdivision of an even broader science, biology (Gr. *bios,* life, + *logos,* discourse on, study of), the study of all life. The panorama of animal life, how animals function, live, reproduce, and interact with their environment, is exciting, fascinating, and awe inspiring. A complete understanding of all phenomena included in zoology is beyond the ability of any single person, perhaps of all humanity, but the satisfaction of knowing as much as possible is worth the effort. In the chapters to follow, we hope to give you an introduction to this science and to share our excitement in the pursuit of it.

Why study zoology? One of the best reasons is curiosity: curiosity about animals, how they function, and how they relate to one another and to their environment. Understanding why things happen or how they have come to be as they are is satisfying to us. Study of zoology helps not only to understand animals in general, it helps us to understand ourselves, the kind of animals we call humans. Every educated person should have at least a modest grasp of human function to live better and more happily in the world. Certainly, a study of zoology helps us to understand current issues related to such phenomena as environmental destruction and pollution, extinction of species, genetic engineering, in vitro fertilization, the hazards of radiation, and many others. These are, of course, reasons why any person should have some knowledge of zoology, but some people want to study certain aspects in depth and contribute in a tangible way to human knowledge and to the solution of problems in the world. Many students take courses in zoology to help them prepare for a medical, veterinary, or other health-care career, either in clinical practice or as researchers on diseases of humans or other animals.

Some students may wish to become professional zoologists. Because no one person can be expert in all of zoology, zoologists usually specialize in one area and apply their research efforts to a narrow subspecialty within that area. Specialties in zoology can be classified in various ways according to the animal group studied. There are entomologists (who study insects), ornithologists (birds), mammalogists (mammals), ichthyologists (fishes), herpetologists (amphibians and reptiles), protozoologists (protozoa), acarologists (ticks and mites), arachnologists (spiders), malacologists (molluscs), and many others. But even these are broad groupings; an ichthyologist, for example, would not attempt to assimilate everything about all fishes but would specialize in one aspect of certain types—perhaps classification, behavior, or the biology of their populations. Specialties are also classified along functional lines, irrespective of the organisms studied. For example, a physiologist studies the life processes of organisms, an immunologist studies the immune system and its mechanisms, and a pathologist studies diseases and how the organism responds to them. Other specialties focus on levels of organization: molecular biologists concentrate on the molecular level, population biologists study population regulation and interaction, and an ecosystem ecologist might study all interactions of all the populations in an ecosystem and their relationships with nonliving components.

The Scientific Study of Animals

On March 19, 1981, the governor of Arkansas signed into law the Balanced Treatment for Creation-Science and Evolution-Science Act (Act 590 of 1981). A historic lawsuit was tried in December 1981 in the court of Judge William R. Overton, U.S. District Court, Eastern District of Arkansas. Plaintiffs included religious leaders and groups of several denominations. They contended that the law was a violation of the First Amendment of the U.S. Constitution, which prohibits "establishment of religion" by the government. On January 5, 1982, Judge Overton permanently enjoined the State of Arkansas from enforcing Act 590.

A basic understanding of zoology requires an understanding of what science is, what it is not, and how knowledge is gained by scientists. It is interesting that a succinct definition of science was rendered in a court of law. On the basis of testimony in the lawsuit, Judge William R. Overton stated the following essential characteristics of science:

1. *It is guided by natural law.* The pursuit of scientific knowledge must be guided by the physical and chemical laws that govern the state of existence and interactions of atoms, subatomic particles, molecules, and so on.

2. *It has to be explanatory by reference to natural law.* Scientific knowledge must explain what is observed by reference to natural law without requiring the intervention of any supernatural being or force. One may believe, as many scientists do, that the universe was brought into existence by the action of a supernatural being, but such a belief is neither within the realm of science nor contradictory to the tenets of science.

3. *It is testable against the empirical world.* We must be able to observe events in the real world, directly or indirectly, for them to have scientific value, and testing of hypotheses and theories must be accessible to our senses or to instruments that can measure the events.

4. *Its conclusions are tentative, that is, are not necessarily the final word.* If we draw a conclusion relative to some event, the conclusion is always subject to further verification. We must always be ready to discard or modify our conclusion if it is inconsistent with further observations.

5. *It is falsifiable.* If our conclusion is not compatible with further evidence, then that evidence has shown that the conclusion is false. As Judge Overton stated, "While anybody is free to approach a scientific inquiry in any fashion they choose, they cannot properly describe the methodology used as scientific, if they start with a conclusion and refuse to change it regardless of the evidence developed during the course of the investigation."

SCIENTIFIC METHOD

Induction in logic is the process of reasoning from the part to the whole, that is, generalizing from particular observations to apply a conclusion more widely. **Deduction** is reasoning from the general to the particular, that is, from given premises to their necessary conclusion.

The manner in which a scientist seeks to gain new knowledge or explain natural phenomena is known as the scientific method. It has sometimes been described as ordinary common sense raised to a higher level and applied systematically. The first step is **observation.** The scientist observes a series of events, often indirectly by means of instruments. Frequently, the events are "observed" by reading descriptions of the observations made by other scientists as recorded in the scientific literature. By use of instruments or literature, the scientist may observe otherwise inaccessible events, such as those too small to be seen or those that may have taken place many years previously (Figure 1-1). After considering the observations, the scientist seeks to generalize about them, that is, to make a statement of explanation about the observations, such as their cause, mechanism, and relationship to each other. This is called **inductive reasoning,** which is simply the mental process that allows us to form statements about information we perceive. The statement becomes the **hypothesis.** To have any scientific value, the hypothesis must then be tested. On the basis of the hypothesis, the scientist must deduce its consequences **(deductive reasoning)** to make a **prediction** about future observations. The scientist must say, "If my hypothesis is a valid explanation of past observations, then future observations ought to have certain characteristics." If the observations do have such characteristics, they constitute evidence in favor of the hypothesis and the hypothesis gains strength as an explanation of the events. In experimental science the "future observations" are in the form of experiments. An **experiment** is a manipulative process by which a prediction made on the basis of the hypothesis can be tested. A certain condition or manipulation is applied to an entity (such as a plant, an animal, a container of a substance, or a body of water), and the results are observed. The information acquired (commonly called **data,** singular **datum**) is critically examined, usually by analyzing the data via a branch of mathematics known as statistics. If the data are as predicted, the hypothesis is supported (not proved), but if the data are otherwise, the hypothesis is invalidated. Thus the data are evidence for or against the hypothesis. The condition or manipulation that has been applied and is being tested is referred to as the **experimental variable.** To have confidence that the experimental variable is responsible for the effect observed, another entity just like the first one must be subjected to all the conditions that prevailed during the experiment *except* the experimental variable. This part of the experiment is called the **control.** The difference observed between the experimental subject and the control is thus the effect of the experimental variable. In practice an experiment is almost never performed on a single entity, such as an individual plant or animal, but on a group of individuals; therefore there are an experimental group and a control group. Furthermore, the experiment is usually repeated a number of times. The more data obtained, the more confidence can be placed in the conclusion about the hypothesis.

We should emphasize that the inductive framing of a good hypothesis is a creative process and requires skill and talent. Hailman (1977) stated that "induction is as mysterious as the creation of a great painting or symphony, and like artists and composers, scientists differ in their creative abilities." Admiring colleagues may say of a scientist that the person "asks good questions."

If a hypothesis becomes supported by a great deal of data, and particularly if it is very powerful (that is, explains a wide variety of related phenomena), the hypothesis may attain the status of a **theory.** The student should understand that the meaning of the word "theory," when used by scientists, is not "speculation" as it is in ordinary English usage. The failure to make this distinction has been prominent in the creationism versus evolution controversy (Figure 1-2). The creationists have spoken of evolution as "only a theory," as if it were little better than a guess. In fact, the theory of evolution is supported by such massive evidence that most biologists view repudiation of evolution as tantamount to repudiation of reality. Nonetheless, evolution, along with other theories in science, has not been *proved* in a mathematical sense but is testable, tentative, and falsifiable. It has been tested for more than 120 years, and to date there is no scientific evidence that it is false; indeed, organic evolution is accepted as the cornerstone of biology. On the other hand, although much has been learned about the *mechanisms* of evolution, they continue to be explored and clarified.

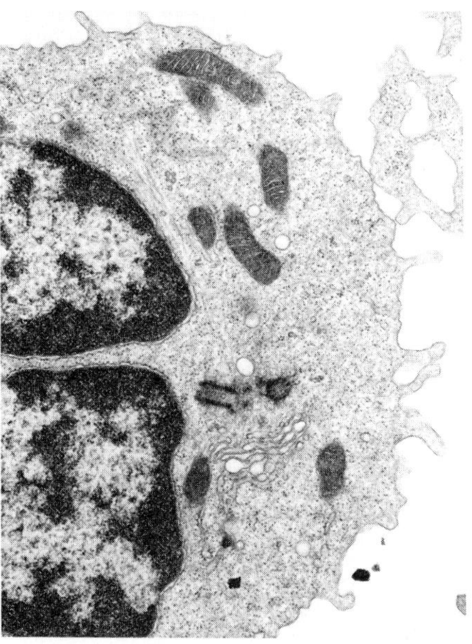

FIGURE 1-1

Scientific instruments are extensions of the scientist's senses. The electron microscope allows us to explore the structure of cells visually. This is a human white blood cell magnified 14,500 times.

Electron micrograph by K.G. Murti/Visuals Unlimited.

FIGURE 1-2

Modern evolutionary theory is strongly identified with Charles Robert Darwin who, with Alfred Russel Wallace, provided the first credible explanation of evolution—natural selection. This photograph of Darwin was taken in 1854 when he was 45 years old.

Courtesy American Museum of Natural History.

Some Basic Chemistry

Within recent years, it has been accepted as a first principle of biology that living systems and their constituents obey physical and chemical laws. Within the cells of any organism, the living substance is composed of a multitude of nonliving constituents: proteins, nucleic acids, fats, carbohydrates, waste metabolites, crystalline aggregates, pigments, and many others. Physical and chemical interactions of such substances account for the many processes essential to life, including digestion and absorption of nutrients, derivation of energy, removal of waste, communication of cells with each other, conduction of nerve impulses, and transmission of genetic information from one generation to the next. Because these phenomena will be discussed in later pages, we must present some basic information on chemistry and biochemistry here.

ELEMENTS AND ATOMS

All matter is composed of **elements,** which are substances that cannot be subdivided further by ordinary chemical reactions. Only 92 elements occur naturally. The elements are designated by one or two letters derived from their Latin or English names (Table 1-1). The elements are composed of discrete units called **atoms,** which are the smallest components into which an element can be subdivided by normal chemical means.

Table 1-1 Some of the most important elements in living organisms

Element	Symbol	Atomic number	Approximate atomic weight
Carbon	C	6	12
Oxygen	O	8	16
Hydrogen	H	1	1
Nitrogen	N	7	14
Phosphorus	P	15	31
Sodium	Na	11	23
Sulfur	S	16	32
Chlorine	Cl	17	35
Potassium	K	19	39
Calcium	Ca	20	40
Iron	Fe	26	56
Iodine	I	53	127

Subatomic particles

Each atom is composed of subatomic particles, of which there are three with which we need concern ourselves: protons, neutrons, and electrons. Every atom consists of a positively charged nucleus surrounded by a negatively charged system of electrons (Figures 1-3 and 1-4). The nucleus, containing most of the atom's mass, is made up of protons and neutrons clustered together in a very small volume. These two particles have about the same mass, each being about 2000 times heavier than an electron. The protons bear positive charges, and the neutrons are uncharged (neutral). Although the number of protons in the nucleus is the same as the number of electrons revolving around the nucleus, the number of neutrons may vary. For every positively charged proton in the nucleus, there is a negatively charged electron; the total charge of the atom is thus neutral.

The **atomic number** of an element is equal to the number of protons in the nucleus, whereas the **atomic weight** is nearly equal to the number of protons plus the number of neutrons (explanation of why atomic weight is not exactly equal to protons plus neutrons can be found in any introductory chemistry text). The mass of the electrons may be neglected.

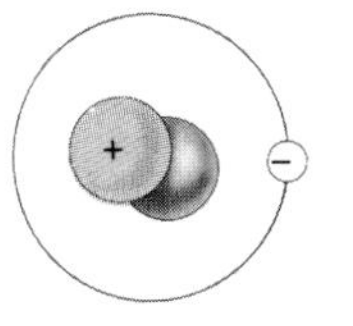
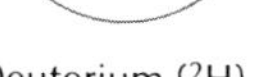

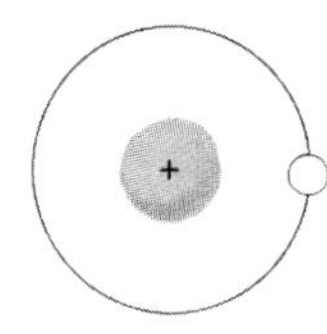

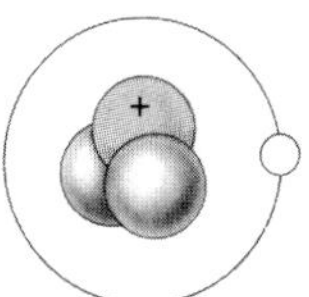

FIGURE 1-3

Three isotopes of hydrogen. Of the three isotopes, hydrogen 1 makes up about 99.98% of all hydrogen, and deuterium (heavy hydrogen) makes up about 0.02%. Tritium is radioactive and is found only in traces in water. Numbers indicate approximate atomic weights. Most elements are mixtures of isotopes. Some elements (for example, tin) have as many as 10 isotopes.

From Raven, P.H., and G.B. Johnson. 1989. Biology, ed. 2. St. Louis, Times Mirror/Mosby.

Isotopes

It is possible for two atoms of the same element to have the same number of protons in their nuclei but have a different number of neutrons. Such different forms, having the same number of protons but different atomic weights, are called **isotopes.** For example, the predominant form of hydrogen in nature has 1 proton and no neutron (abbreviated ^1H, the superscript number standing for the total protons and neutrons) (Figure 1-3). Another form (deuterium [^2H]) has 1 proton and 1 neutron. Tritium (^3H) has 1 proton and 2 neutrons. Some isotopes are unstable, undergoing a spontaneous disintegration with the emission of one or more of three types of particles, or rays: gamma rays (a form of electromagnetic radiation), beta rays (electrons), and alpha rays (positively charged helium nuclei stripped of their electrons). These unstable isotopes are said to be **radioactive.** Using radioisotopes, biologists are able to trace movements of elements and tagged compounds through organisms. Our present understanding of metabolic pathways in animals and plants is in large part the result of this powerful analytical tool. Among the commonly used radioisotopes are tritium, carbon 14 (^{14}C), and phosphorus 32 (^{32}P).

Electron "shells" of atoms

According to Niels Bohr's planetary model of the atom, the electrons revolve around the nucleus of an atom in circular orbits of precise energy and size. All of the orbits of any one energy and size comprise an electron shell. This simplified picture of the atom has been greatly modified by more recent experimental evidence; definite electron pathways are no longer hypothesized, and an electron shell is better understood as a thick region of space around the nucleus rather than a narrow shell of a particular radius out in space.

However, the old planetary model with the idea of electronic shells is still useful in interpreting chemical phenomena. The number of concentric shells required to contain an element's electrons varies with the element. Each shell can hold a maximum number of electrons. The first shell next to the atomic nucleus

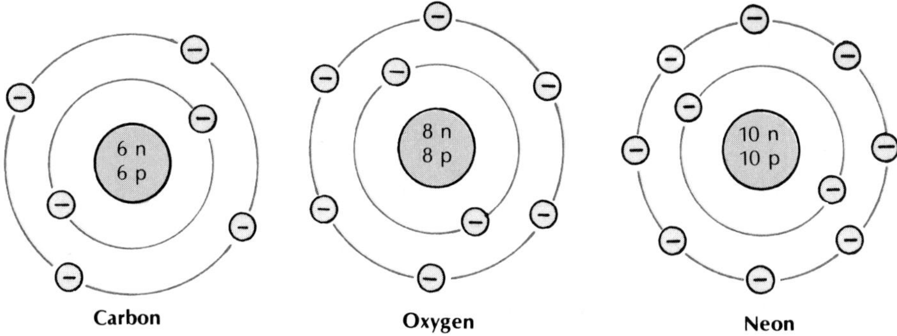

FIGURE 1-4

Electron shells of three common atoms. Since no atom can have more than 8 electrons in its outermost shell and 2 electrons in its innermost shell, neon is chemically inactive. However, the second shells of carbon and oxygen, with 4 and 6 electrons, respectively, are open so that these elements are electronically unstable and react chemically whenever appropriate atoms come into contact. Chemical properties of atoms are determined by their outermost electron shells.

can hold a maximum of 2 electrons, and the second shell can hold 8; other shells also have a maximum number, but no atom can have more than 8 electrons in its outermost shell. Inner shells are filled first, and if there are not enough electrons to fill all the shells, the outer shell is left incomplete. Hydrogen has 1 proton in its nucleus and 1 electron in its single orbit but no neutron. Since its shell can hold 2 electrons, it has an incomplete shell. Helium has 2 electrons in its single shell, and its nucleus is made up of 2 protons and 2 neutrons. Since the 2-electron arrangement in helium's shell is the maximum number for this shell, the shell is closed and precludes all chemical activity, that is, helium is **inert.** There is no known compound of helium. Neon is another inert gas because its outer shell contains 8 electrons, the maximum number (Figure 1-4). However, stable compounds of xenon (an inert gas) with fluorine and oxygen are formed under special conditions. Oxygen has an atomic number of 8. Its 8 electrons are arranged with 2 in the first shell and 6 in the second shell (Figure 1-4). It is active chemically, forming compounds with almost all elements except inert gases.

Molecules and compounds

Combination of the atoms of an element with each other or with those of other elements by chemical bonds creates **molecules.** The **molecular weight** of a substance is the total of all the atomic weights in one of its molecules. In a chemical formula, the symbol for an element stands for 1 atom of the element, with additional atoms indicated by appropriately placed numbers. Thus atmospheric nitrogen is N_2 (each molecule is composed of 2 atoms of nitrogen), and water is H_2O (2 atoms of hydrogen and 1 of oxygen in each molecule), and so on. When the molecules of a substance contain atoms of 2 or more different elements, the substance is known as a **compound.** Examples of other compounds are carbon dioxide (CO_2) and calcium carbonate ($CaCO_3$).

CHEMICAL BONDS

As noted above, atoms joined to each other by chemical bonds form molecules, and atoms of each element form molecules with atoms of other elements in particular ways, depending on the number of electrons in their outer orbits.

Ionic bonds

Every atom has a tendency to gain a stable configuration of electrons in its outer shell. The number of electrons in the outer shell varies from 0 to 8. With either 0 or 8 in this shell, the element is chemically inactive. When there are fewer than 8 electrons in the outer shell, the atom will tend to lose or gain electrons to create an outer shell of 8, which will result in a charged **ion** because the numbers of protons and electrons in the atom are no longer equal. Atoms with 1 to 3 electrons in the outer shell tend to lose them to other atoms and to become positively charged ions because of the excess protons in the nucleus (Figure 1-5). Atoms with 5 to 7 electrons in the outer orbit tend to gain electrons from other atoms and to become negatively charged ions because they have more electrons than protons. Positive and negative ions tend to unite since unlike electrostatic charges attract each other.

Let us examine how 2 atoms with incomplete outer shells, sodium and chlorine, can interact to fill their outer shells. Sodium, with 11 electrons, has 2 electrons in its first shell, 8 in its second shell, and only 1 in the third shell. The third shell is highly incomplete; if this third-shell electron were lost, the second shell would be the outermost shell and would produce a stable atom. Chlorine, with 17 electrons, has 2 in the first shell, 8 in the second, and 7 in the incomplete third shell. Chlorine must gain an electron to fill the outer shell and become a stable atom. Clearly, the transfer of the third-shell sodium electron to the incomplete chlorine third shell would yield simultaneous stability to both atoms.

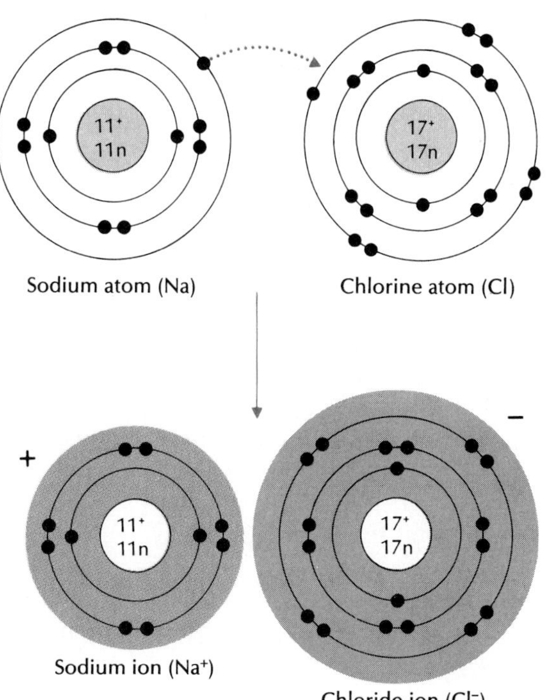

FIGURE 1-5

Ionic bond. When 1 atom of sodium and 1 of chlorine react to form a molecule, a single electron in the outer shell of sodium is transferred to the outer shell of chlorine. This causes the outer or second shell (third shell is empty) of sodium to have 8 electrons and also chlorine to have 8 electrons in its outer or third shell. The compound thus formed is sodium chloride (NaCl). By losing 1 electron, sodium becomes a positive ion, and by gaining 1 electron, chlorine (chloride) becomes a negative ion. This ionic bond is the strong electrostatic force acting between positively and negatively charged ions.

From Raven, P.H., and G.B. Johnson, 1989. Biology, ed. 2. St. Louis, Times Mirror/Mosby.

Sodium, now with 11 protons but only 10 electrons, becomes electropositive (Na^+). In gaining an electron from sodium, chlorine contains 18 electrons but only 17 protons and thus becomes an electronegative chloride ion (Cl^-). Since unlike charges attract, a strong electrostatic force, called an **ionic bond** (Figure 1-5), is formed. The ionic compound formed, sodium chloride (NaCl), can be represented in electron dot notation ("fly-speck formulas") as:

$$\mathrm{Na}\cdot + \cdot\ddot{\underset{\cdot\cdot}{\mathrm{Cl}}}\mathbf{:} \rightarrow \mathrm{Na}^+ + (\mathbf{:}\ddot{\underset{\cdot\cdot}{\mathrm{Cl}}}\mathbf{:})^-$$

The number of dots shows the number of electrons present in the original outer shell of the atom: 7 in the case of the neutral chlorine atom and 8 for the chloride ion; 1 in the case of the neutral sodium atom and none for the sodium ion. Note that the second shell of the sodium ion has now become its outer shell and contains the stable number of 8 electrons (see Figure 1-5).

If an element with 2 electrons in its outer shell, such as calcium, reacts with chlorine, it must give them both up, one to each of 2 chlorine atoms, and calcium becomes doubly positive, forming calcium chloride ($CaCl_2$):

$$\mathrm{Ca}\mathbf{:} + 2\cdot\ddot{\underset{\cdot\cdot}{\mathrm{Cl}}}\mathbf{:} \rightarrow \mathrm{Ca}^{2+} + (\mathbf{:}\ddot{\underset{\cdot\cdot}{\mathrm{Cl}}}\mathbf{:})^-$$

Processes that involve the **loss of electrons** are called **oxidation** reactions; those that involve the **gain of electrons** are **reduction** reactions. Since oxidation and reduction always occur simultaneously, each of these processes is really a "half-reaction." The entire reaction is called an **oxidation-reduction** reaction, or simply a **redox** reaction. The terminology is confusing because oxidation-reduction reactions involve electron transfers, rather than (necessarily) any reaction with oxygen. However, it is easier to learn the system than to try to change accepted usage.

Covalent bonds

Stability can also be achieved when 2 atoms share electrons. Let us again consider the chlorine atom, which, as we have seen, has an incomplete 7-electron outer shell. Stability is attained by gaining an electron. One way this can be

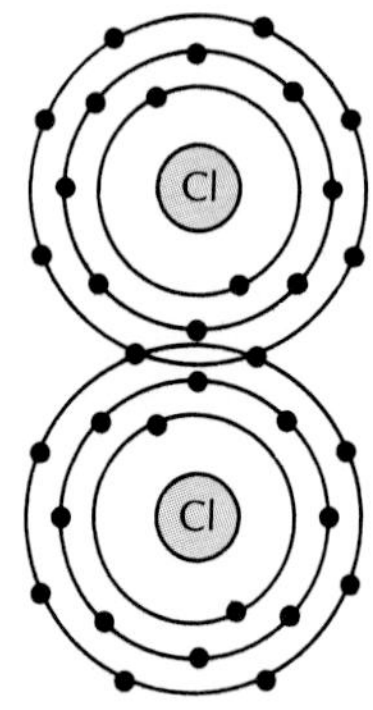

FIGURE 1-6

Covalent bond. Each chlorine atom has 7 electrons in its outer shell, and by sharing one pair of electrons, each atom acquires a complete outer shell of 8 electrons, thus forming a molecule of chlorine (Cl_2). Such a reaction is called a molecular reaction, and such bonds are called covalent bonds.

done is for 2 chlorine atoms to *share* one pair of electrons (Figure 1-6). To do this, the 2 chlorine atoms must *overlap* their third shells so that the electrons in these shells can now spread themselves over both atoms, thereby completing the filling of both shells. Many other elements can form **covalent** (or electron-pair) **bonds.** Examples are hydrogen (H_2)

$$\mathrm{H}\cdot + \mathrm{H}\cdot \rightarrow \mathrm{H}{:}\mathrm{H}$$

and oxygen (O_2)

$$\underset{\cdot\cdot}{\ddot{\mathrm{O}}}{:} + {:}\underset{\cdot\cdot}{\ddot{\mathrm{O}}} \rightarrow \underset{\cdot\cdot}{\ddot{\mathrm{O}}}{::}\underset{\cdot\cdot}{\ddot{\mathrm{O}}}$$

In this case oxygen must share two pairs of electrons to achieve stability. Each atom now has 8 electrons available to its outer shell, the stable number.

Covalent bonds are of great significance to living systems, since the major elements of living matter (carbon, oxygen, nitrogen, hydrogen) almost always share electrons in strong covalent bonds. The stability of these bonds is essential to the integrity of DNA and other macromolecules, which, if easily dissociated, would result in biological disorder.

The outer shell of carbon contains 4 electrons. This element is endowed with great potential for forming a variety of atomic configurations with itself and other molecules. It can, for example, share its electrons with hydrogen to form methane:

$$\cdot\underset{\cdot}{\dot{\mathrm{C}}}\cdot + 4\,\mathrm{H}\cdot \rightarrow \begin{array}{c} \mathrm{H} \\ \mathrm{H}{:}\underset{\cdot\cdot}{\ddot{\mathrm{C}}}{:}\mathrm{H} \\ \mathrm{H} \end{array}$$

Carbon now achieves stability with 8 electrons, and each hydrogen atom becomes stable with 2 electrons. Carbon can also bond with itself (and hydrogen) to form, for example, ethane (C_2H_6):

$$\begin{array}{c} \mathrm{H}\ \mathrm{H} \\ \mathrm{H}{:}\underset{\cdot\cdot}{\ddot{\mathrm{C}}}{:}\underset{\cdot\cdot}{\ddot{\mathrm{C}}}{:}\mathrm{H} \\ \mathrm{H}\ \mathrm{H} \end{array} \quad \text{or} \quad \begin{array}{ccccccc} & & \mathrm{H} & & \mathrm{H} & & \\ & & | & & | & & \\ \mathrm{H} & — & \mathrm{C} & — & \mathrm{C} & — & \mathrm{H} \\ & & | & & | & & \\ & & \mathrm{H} & & \mathrm{H} & & \end{array}$$

Carbon also forms covalent bonds with oxygen, resulting in carbon dioxide (CO_2):

$$\cdot\underset{\cdot}{\dot{\mathrm{C}}}\cdot + 2\,\underset{\cdot\cdot}{\ddot{\mathrm{O}}}{:} \rightarrow \underset{\cdot\cdot}{\ddot{\mathrm{O}}}{::}\mathrm{C}{::}\underset{\cdot\cdot}{\ddot{\mathrm{O}}}$$

This is a "double-bond" configuration usually written as O=C=O. Carbon can even form triple bonds as, for example, in acetylene (C_2H_2):

$$\mathrm{H}{:}\mathrm{C}{:::}\mathrm{C}{:}\mathrm{H} \text{ or } \mathrm{H—C{\equiv}C—H}$$

The significant aspect of each of these molecules is that each carbon gains a share in 4 electrons from atoms nearby, thus attaining the stability of 8 electrons. The sharing may occur between carbon and other elements or other carbon atoms, and in many instances the 8-electron stability is achieved by means of multiple bonds.

These examples only begin to illustrate the amazing versatility of carbon. It is a part of virtually all compounds comprising living substance, and without carbon, life as we know it would not exist.

Hydrogen bonds

Hydrogen bonds are described as "weak" bonds because they require relatively much less energy to break than ionic or covalent bonds. They do not form by transfer or sharing of electrons but result from unequal charge distribution on a molecule.

In all of the examples of covalent bonds described above, the electrons are shared equally between the atoms of the molecules. Atoms of some elements, however, have a greater electron-attracting power (electronegativity) than atoms of other elements, so that the electrons in the covalent bonds are shared *unequally.* This results in a charge distribution across a molecule composed of such elements so that one end of the molecule is more positive and one end (the end with the more electronegative element) is more negative. A molecule with this kind of charge separation is called a **polar molecule**. Water molecules are very polar; the electrons shared in the covalent bonds between the oxygen atom and the 2 hydrogen atoms are much more strongly attracted to the oxygen than the hydrogen (Figure 1-7). The bond angles of the 2 hydrogen atoms are not 180 degrees away from each other around the oxygen but form an angle of about 105 degrees (Figure 1-7). The side of the molecule away from the hydrogens is more negative, and the hydrogen side is more positive.

The electrostatic attraction between the electropositive area of one water molecule forms a hydrogen bond with the electronegative area of an adjacent molecule. The ability of water molecules to form hydrogen bonds with each other (Figure 1-8) accounts for many unusual properties of this unique substance. Hydrogen bonds are important in the formation and function of other biologically active substances, such as proteins and nucleic acids (p. 16).

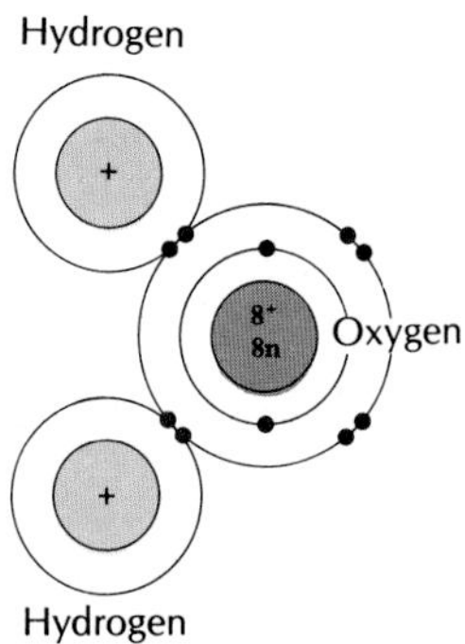

FIGURE 1-7

Molecular structure of water. The 2 hydrogen atoms bonded covalently to an oxygen atom are arranged at an angle of about 105 degrees. Since the electrical charge is not symmetrical, the molecule is polar with positively charged and negatively charged ends.

From Raven, P.H., and G.B. Johnson, 1989. Biology, ed. 2. St. Louis, Times Mirror/Mosby.

ACIDS, BASES, AND SALTS

The hydrogen ion (H^+) is one of the most important ions in living organisms. The hydrogen atom contains a single electron. When this electron is completely transferred to another atom (not just shared with another atom as in the covalent bonds with carbon), only the hydrogen nucleus with its positive proton remains. Any molecule that dissociates in solution and gives rise to a hydrogen ion is an **acid**. An acid is classified as strong or weak, depending on the extent to which the acid molecule is dissociated in solution. Examples of strong acids that dissociate completely in water are hydrochloric acid ($HCl \rightarrow H^+ + Cl^-$) and nitric acid ($HNO_3 \rightarrow H^+ + NO_3^-$). Weak acids, such as carbonic acid ($H_2CO_3 \rightleftharpoons H^+ + HCO_3^-$), dissociate only slightly. A solution of carbonic acid is mostly undissociated carbonic acid molecules with only a small number of bicarbonate (HCO_3^-) and hydrogen ions (H^+) present.

A **base** contains negative ions called hydroxide ions and may be defined as a molecule or ion that will accept a proton (hydrogen ion). Bases are produced when compounds containing them are dissolved in water. Sodium hydroxide (NaOH) is a strong base because it will dissociate completely in water into sodium (Na^+) and hydroxide (OH^-) ions. Among the characteristics of bases is their ability to combine with hydrogen ions, thus decreasing their concentration. Like acids, bases vary in the extent to which they dissociate in aqueous solutions into hydroxide ions.

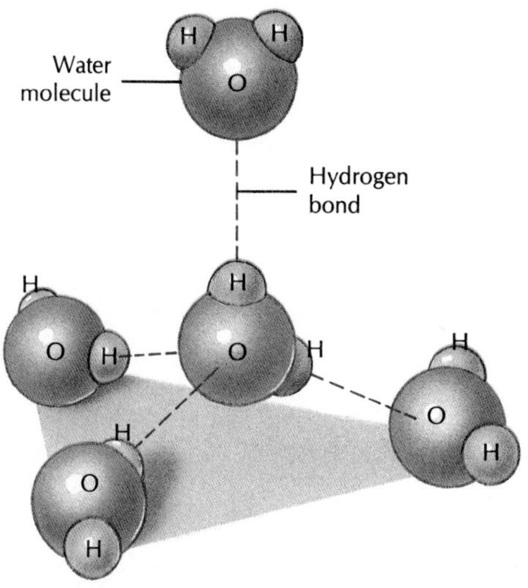

FIGURE 1-8

Geometry of water molecules. Each water molecule is linked by hydrogen bonds *(dashed lines)* to 4 other water molecules. If imaginary lines are used to connect the divergent oxygen atoms, a tetrahedron is obtained. In ice, the individual tetrahedrons associate to form an open lattice structure.

A **salt** is a compound resulting from the chemical interaction of an acid and a base. Common salt, sodium chloride (NaCl), is formed by the interaction of hydrochloric acid (HCl) and sodium hydroxide (NaOH). In water the HCl is dissociated into H^+ and Cl^- ions. The hydrogen and hydroxide ions combine to form water (HOH, commonly written as H_2O), and the sodium and chloride ions remain as a dissolved form of salt (Na^+Cl^-):

$$\underset{\textbf{Acid}}{H^+Cl^-} + \underset{\textbf{Base}}{Na^+OH^-} \rightarrow \underset{\textbf{Salt}}{Na^+Cl^-} + H_2O$$

Organic acids are usually characterized by having in their molecule the carboxyl group (—COOH). They are weak acids because a relatively small proportion of the H^+ reversibly dissociates from the carboxyl:

$$\begin{array}{c} R-C=O \\ | \\ O-H \end{array} \rightleftharpoons \begin{array}{c} R-C=O \\ | \\ O-^{-} \end{array} + H^+$$

R refers to an atomic grouping unique to the molecule. Some common organic acids are acetic, citric, formic, lactic, and oxalic. Many more of these will be encountered later in discussions of cellular metabolism.

Hydrogen ion concentration (pH)

Solutions are classified as acid, basic, or neutral according to the proportion of hydrogen (H^+) and hydroxide (OH^-) ions they possess. In acid solutions there is an excess of hydrogen ions; in alkaline, or basic, solutions the hydroxide ion is more common; and in neutral solutions both hydrogen and hydroxide ions are present in equal numbers.

To express the acidity or alkalinity of a substance, a logarithmic scale, a type of mathematical shorthand, is employed that uses the numbers 1 to 14. This is the pH, defined as:

$$pH = \log_{10}\frac{1}{[H^+]}$$

or

$$pH = -\log_{10}[H^+]$$

where $[H^+]$ represents hydrogen ion concentration

Thus pH is the negative logarithm of the hydrogen ion concentration. In other words, when the hydrogen ion concentration is expressed exponentially, pH is the exponent, but with the *opposite* sign; if $[H^+] = 10^{-2}$, then $pH = -(-2) = +2$. Unfortunately, pH can be a confusing concept because, as the $[H^+]$ decreases, the pH increases. Numbers below 7 indicate an acid range, and numbers above 7 indicate alkalinity. The number 7 indicates neutrality, that is, the presence of equal numbers of H^+ and OH^- ions. According to this logarithmic scale, a pH of 3 is 10 times more acid than one of 4; a pH of 9 is 10 times more alkaline than one of 8. The pH values of some common substances are orange juice, 2.0; tomato juice, 4.8; milk, 6.6; and blood, approximately 7.3.

ORGANIC MOLECULES

The term **organic compounds** was originally applied only to substances derived from living organisms, but it now encompasses carbon compounds in general. However, some common compounds containing carbon, such as CO_2 and $CaCO_3$, are inorganic. Hence, organic compounds are specifically those carbon compounds in which the principle bonds are carbon to carbon and carbon to hydrogen. The key property of carbon that makes it so suitable for life is that it can bond with other carbon atoms to form an almost limitless repertoire of large molecules. Many organic compounds also contain oxygen, nitrogen, sulfur, phosphorus, salts, and other elements.

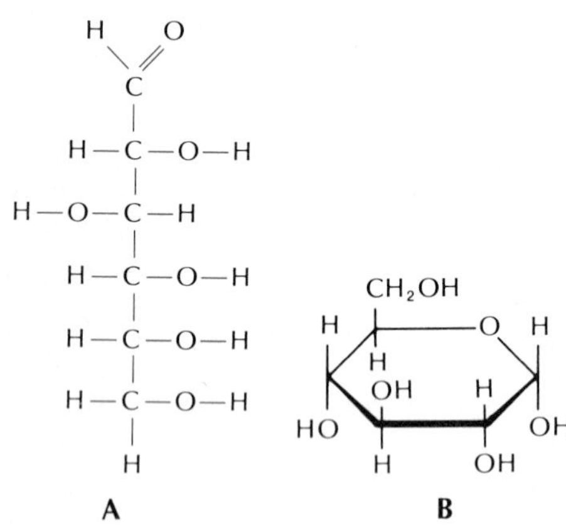

FIGURE 1-9

Two ways of depicting the structural formula of the simple sugar glucose. In **A** the carbon atoms are shown in open-chain form. When dissolved in water, glucose tends to assume a ring form as in **B.** In this ring model the carbon atoms located at each turn in the ring are usually not shown.

Carbohydrates: nature's most abundant organic substance

Carbohydrates are compounds of carbon, hydrogen, and oxygen. They are usually present in the ratio of 1 C:2 H:1 O and are grouped as H—C—OH. Familiar examples of carbohydrates are sugars, starches, and cellulose (the woody structure of plants). There is more cellulose on earth than all other organic materials combined. Glucose, a simple carbohydrate, is made synthetically from water and carbon dioxide by green plants, with the aid of the sun's energy. This process, called **photosynthesis**, is a reaction on which all life depends, for it is the starting point in the formation of food.

Carbohydrates are usually divided into the following three classes: (1) **monosaccharides**, or simple sugars (2) **disaccharides**, or double sugars; and (3) **polysaccharides**, or complex sugars. Simple sugars are most commonly composed of carbon chains containing 4 carbons (tetroses), 5 carbons (pentoses), or 6 carbons (hexoses). Simple sugars, such as glucose, galactose, and fructose, all contain a free sugar group,

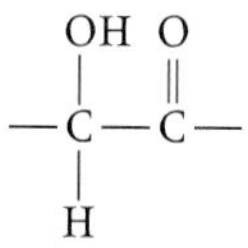

$$\begin{array}{c} OH\ \ O \\ |\ \ \ \ \| \\ -C-C- \\ | \\ H \end{array}$$

in which the double-bonded O may be attached to the terminal C of a chain or to a nonterminal C. The hexose **glucose** (also called dextrose) is the most important carbohydrate in the living world. Glucose is often shown as a straight chain (Figure 1-9, *A*), but in water tends to form a ring structure (Figure 1-9, *B*). Galactose and fructose are also biologically important hexoses, and ribose is a vital pentose (Figure 1-10).

Disaccharides are double sugars formed by the bonding of two simple sugars. An example is maltose (malt sugar), composed of 2 glucose molecules. As shown in Figure 1-11, the 2 glucose molecules are bonded together by the removal of a molecule of water. This **condensation reaction**, with the sharing of an oxygen atom by the two sugars, characterizes the formation of all disaccharides. (A reaction in the reverse direction, adding a molecule of water, is a **hydrolysis**, or **hydrolytic reaction**.) Two other common disaccharides are sucrose (ordinary cane, or table, sugar), formed by the linkage of glucose and fructose, and lactose (milk sugar), composed of glucose and galactose.

Polysaccharides are made up of many molecules of simple sugars (usually glucose) linked together in long chains. Such substances with many units linked together are referred to by the chemist as **polymers**. Their empirical formula is usually written $(C_6H_{10}O_5)_n$, where *n* stands for the unknown number of simple sugar molecules of which they are composed. Starch is the common storage form of sugar in most plants and is an important food constituent for animals. **Glycogen** is an important storage form for sugar in animals. It is found mainly in liver and muscle cells in vertebrates. When needed, glycogen is converted into glucose and is delivered by the blood to the tissues. Another polymer of glucose is **cellulose**, which is the principal structural carbohydrate of plants.

The main role of carbohydrates in cells is to serve as a source of chemical energy. Glucose is the most important of these energy carbohydrates. Some carbohydrates become basic components of protoplasmic structure, such as the pentoses that form constituent groups of nucleic acids and of nucleotides.

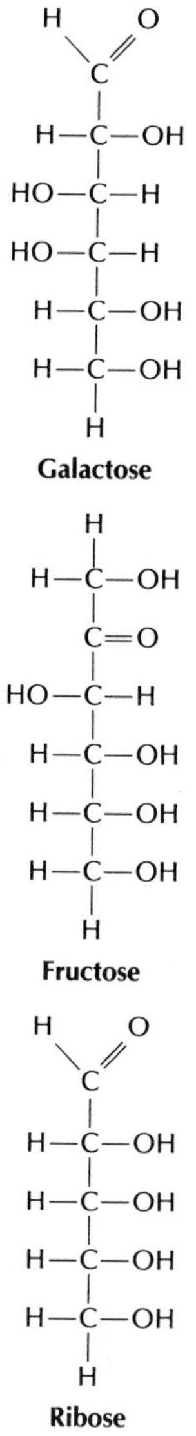

FIGURE 1-10

These three sugars are the most common monosaccharides. Galactose and ribose are aldehyde sugars; fructose is a ketone sugar.

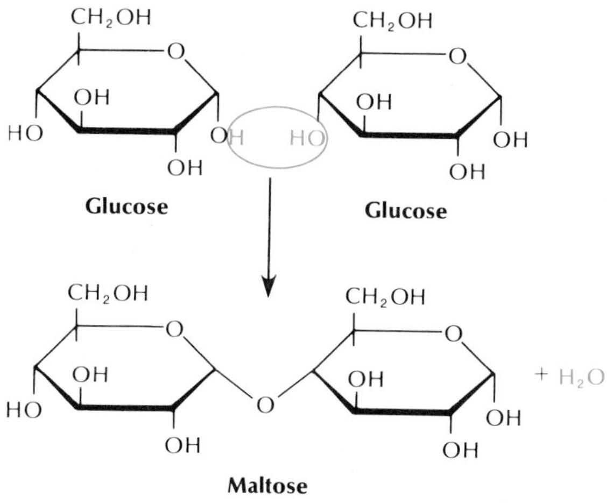

FIGURE 1-11

Formation of a double sugar (disaccharide maltose) from 2 glucose molecules with the removal of 1 molecule of water. This type of reaction is a condensation reaction. The reverse reaction, adding a molecule of water to maltose and forming 2 glucose molecules, is a hydrolytic reaction.

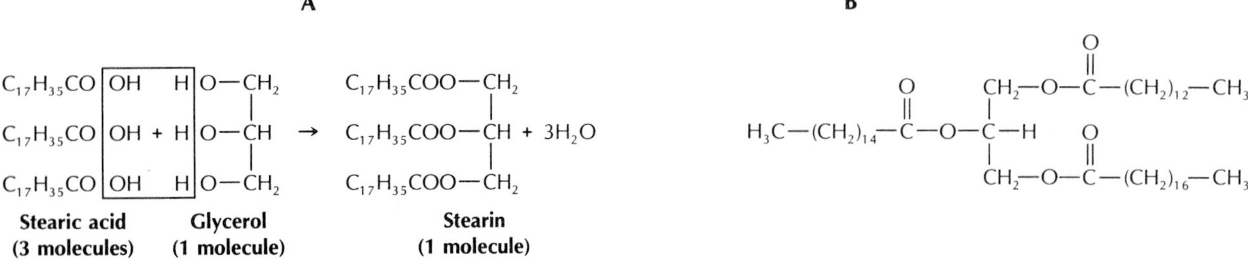

FIGURE 1-12

Neutral fats. **A,** Formation of a neutral fat, stearin, from 3 molecules of stearic acid (a fatty acid) and 1 molecule of glycerol. **B,** A neutral fat bearing three different fatty acids.

Lipids: fuel storage and building material

Lipids are fats and fatlike substances. They are composed of nonpolar molecules. This feature makes them virtually insoluble in water but soluble in organic solvents such as acetone and ether. Three principal groups of lipids are neutral fats, phospholipids, and steroids.

Neutral fats. The neutral, or "true" fats are major fuels of animals. Stored fat may be derived directly from dietary fat or indirectly from dietary carbohydrates that are converted to fat for storage. Fats are oxidized and released into the bloodstream as needed to meet tissue demands, especially for muscles.

In general, organic substances with a high proportion of hydroxyl (—OH) or other polar groups compared to the number of carbon atoms in the molecule are soluble in water, as are those with ionizable groups, such as acids. As the number of polar groups compared to the carbons decreases, solubility in water decreases, and solubility in nonpolar solvents increases.

Neutral fats are **triglycerides,** which are molecules consisting of glycerol and 3 molecules of fatty acids. Neutral fats are therefore esters, that is, a combination of an alcohol (glycerol) and an acid. Fatty acids in triglycerides are simply long-chain monocarboxylic acids; they vary in size but are commonly 14 to 24 carbons long. The production of a typical fat by a condensation reaction of glycerol and stearic acid is shown in Figure 1-12, *A*. In this reaction it can be seen that the 3 fatty acid molecules have bonded with the OH group of the glycerol to form stearin (a neutral fat), with the production of 3 molecules of water. A **monoglyceride** is a glycerol molecule with only 1 fatty acid, and a **diglyceride** has 2 fatty acids.

Most triglycerides contain two or three different fatty acids attached to glycerol, bearing ponderous names such as myristoyl stearoyl glycerol (Figure 1-12, *B*). The fatty acids in this triglyceride are **saturated;** that is, every carbon within the chain holds 2 hydrogen atoms. Saturated fats, more common in animals than in plants, are usually solid at room temperature. **Unsaturated** fatty acids, typical of plant oils, have 2 or more carbon atoms joined by double bonds; that is, the carbons are not "saturated" with hydrogen atoms and are able to form additional bonds with other atoms. Two common unsaturated fatty acids are oleic acid and linoleic acid (Figure 1-13). Plant fats such as peanut oil and corn oil tend to be liquid at room temperature. (See Table 1-2.)

Table 1-2 Some types of neutral fats

Animal
Lard
Butter
"Fat"
Plant
Corn oil
Peanut oil
Cottonseed oil
Soybean oil
Olive oil
Safflower oil
Palm oil*
Coconut oil*

*These plant oils are considered saturated.

Phospholipids. Unlike the the fats that are fuels and serve limited structural roles in the cell, phospholipids are important components of the molecular organization of cells, especially membranes. They resemble triglycerides in structure, except that 1 of the 3 fatty acids is replaced by phosphoric acid and an organic base containing nitrogen. An example is lecithin, an important phospholipid of nerve membrane (Figure 1-14). Because the phosphate group on phospholipids is charged and polar and therefore soluble in water and the remainder of the molecule is nonpolar, phospholipids can bridge two environments and bind water-soluble molecules such as proteins to water-insoluble materials.

Steroids. Steroids are complex alcohols, structurally unlike fats but having fatlike properties. The steroids are a large group of biologically important molecules, including cholesterol (Figure 1-15), vitamin D, many adrenocortical hormones, and the sex hormones.

$CH_3—(CH_2)_7—CH{=}CH—(CH_2)_7—COOH$

Oleic acid

$CH_3—(CH_2)_4—CH{=}CH—CH_2—CH{=}CH—(CH_2)_7—COOH$

Linoleic acid

FIGURE 1-13

Unsaturated fatty acids; oleic acid having one double bond and linoleic acid having two double bonds. The remainder of the hydrocarbon chains of both acids is saturated.

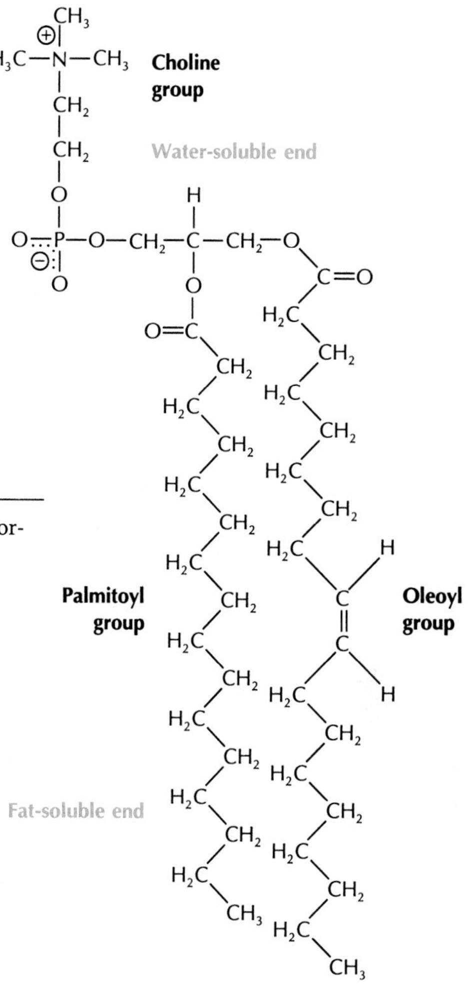

Choline group

Water-soluble end

Palmitoyl group

Oleoyl group

Fat-soluble end

FIGURE 1-14

Lecithin (phosphatidyl choline), an important phospholipid of nerve membranes.

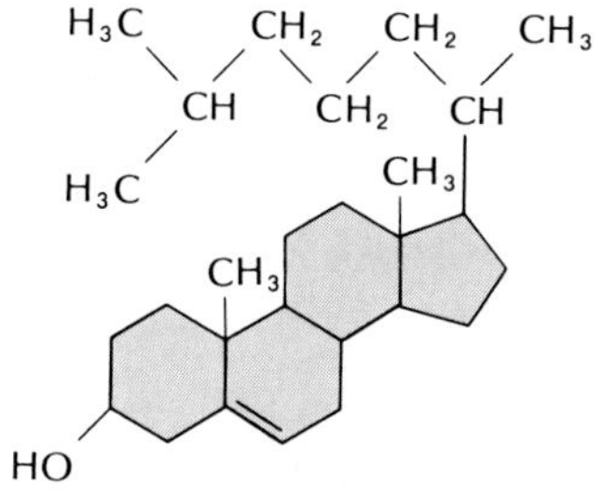

Cholesterol

FIGURE 1-15

Cholesterol, a steroid. All steroids have a basic skeleton of four rings (three 6-carbon rings and one 5-carbon ring) with various side groups attached.

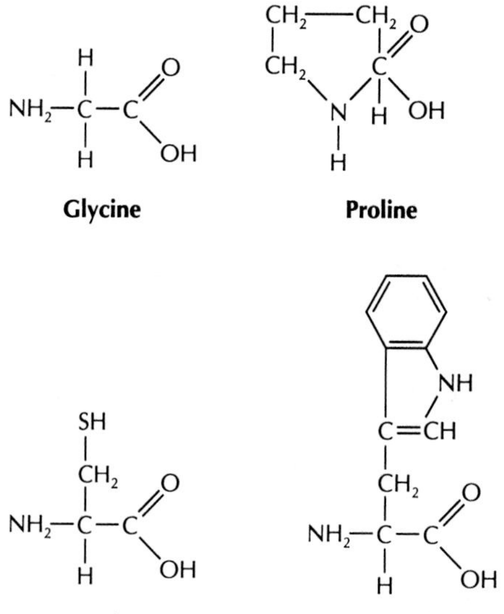

Glycine

Proline

Amino acids and proteins

Proteins are large, complex molecules composed of 20 commonly occurring amino acids (Figure 1-16). The amino acids are linked together by **peptide bonds** to form long, chainlike polymers. In the formation of a peptide bond, the carboxyl group of one amino acid is linked by a covalent bond to the amino group of another, with the elimination of water, as follows:

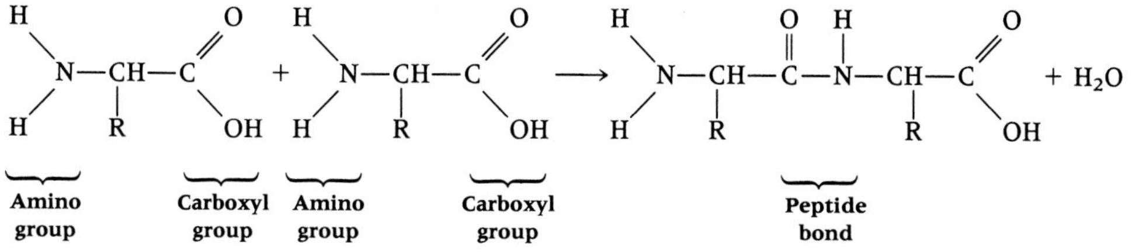

$$H_2N{-}CH(R){-}COOH + H_2N{-}CH(R){-}COOH \longrightarrow H_2N{-}CH(R){-}C(=O){-}NH{-}CH(R){-}COOH + H_2O$$

Amino group · Carboxyl group · Amino group · Carboxyl group · Peptide bond

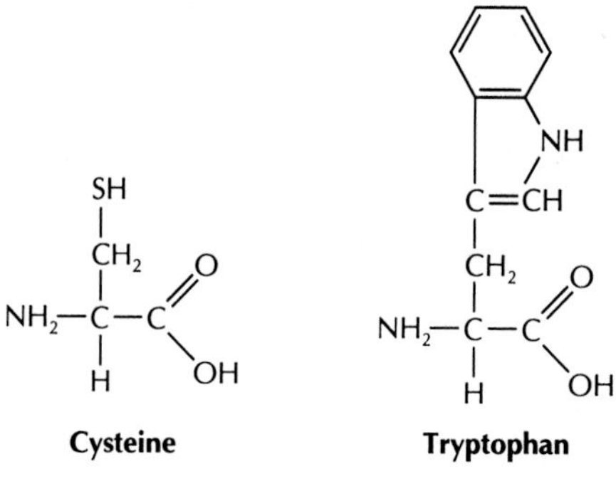

FIGURE 1-16

Four of the 20 naturally occurring amino acids.

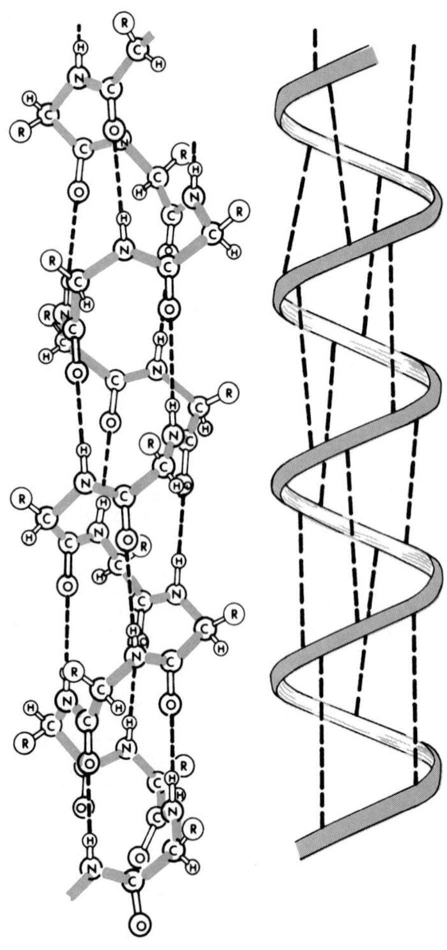

FIGURE 1-17

Alpha-helix pattern of a polypeptide chain. *Dashed lines,* Hydrogen bonds that stabilize adjacent turns of the helix. *R,* Amino acid side chains.

Modified from Green, D. 1956. Currents of biochemical research. New York, Interscience Publishers, Inc.

The combination of two amino acids by a peptide bond forms a dipeptide, and, as is evident, there is still a free amino group on one and a free carboxyl group on the other; therefore additional amino acids can be joined to both ends until a long chain in produced. A molecule consisting of many joined amino acids, but not as complex as a protein, is a **polypeptide.** The 20 different kinds of amino acids can be arranged in an enormous variety of sequences of up to hundreds of amino acid units; therefore, it is not difficult to account for the practically countless varieties of proteins among living organisms.

A **protein** is not just a long string of amino acids; it is a highly organized molecule. For convenience, biochemists have recognized four levels of protein organization called primary, secondary, tertiary, and quaternary.

The **primary structure** of a protein refers to the identity and sequence of amino acids making up the polypeptide chain. Because the bonds between the amino acids in the chain are characterized by a limited number of stable angles, certain recurrent structural patterns are assumed by the chain. This is called the **secondary structure,** and it is often that of an **alpha-helix,** that is, helical turns in a clockwise direction like a screw (Figure 1-17). The spirals of the chains are stabilized by hydrogen bonds, usually between a hydrogen atom of one amino acid and the peptide-bond oxygen of another amino acid in an adjacent turn of the helix.

The protein chain (primary structure) not only spirals into helical configurations (secondary structure), but also the helices themselves bend and fold, giving the protein its complex, yet stable, three-dimensional **tertiary structure** (Figure 1-18). The folded chains are stabilized by the interactions between side groups of amino acids. One of these interactions is the **disulfide bond,** a covalent bond between the sulfur atoms in pairs of cysteine (sis'tee-in) units that are brought together by folds in the polypeptide chain. Other attractive forces that help stabilize the tertiary structure of proteins are hydrogen bonds, ionic bonds, and lipophilic interactions.

The term **quaternary structure** describes those proteins which contain more than one polypeptide chain unit. For example, hemoglobin (the oxygen-carrying substance in blood) of higher vertebrates is composed of four polypeptide subunits nested together into a single protein molecule. The tertiary and quaternary structures of a protein are vitally important to its function, for example, as an enzyme, because if the shape is wrong, the enzyme will not fit the substrate (see p. 46).

Proteins as enzymes. Proteins perform many functions in living things. They serve as the structural framework of cells and form many cell components. However, the most important role of proteins by far is as **enzymes,** the biological catalysts required for almost every reaction in the body (p. 45).

Enzymes lower the energy required for specific reactions and enable life processes to proceed at moderate temperatures. They control the reactions by which food is digested, absorbed, and metabolized. They promote the synthesis of structural materials for growth, maintenance, and repair of the body. They determine the release of energy used in respiration, muscle contraction, physical and mental activities, and many other activities. Enzyme action is described in Chapter 2.

Nitrogen base derivatives

Many vital compounds in cells contain units of **nitrogenous bases,** which are ring compounds containing nitrogen as well as carbon (p. 77). Some of these, such as ATP, NAD, and FAD, play critical roles in energy metabolism (Chapter 2). Others are organized into **nucleic acids,** which are complex substances of high molecular weight.

The sequence of nitrogenous bases in nucleic acids encodes the genetic information necessary for all aspects of biological inheritance. They not only direct the synthesis of enzymes and other proteins but are also the only molecules that have the power (with the help of the right enzymes) to replicate themselves. The two kinds of nucleic acids in cells are **deoxyribonucleic acid (DNA)** and **ribonucleic acid (RNA)**. They are polymers of repeated units called **nucleotides**, each containing a sugar, nitrogenous base, and phosphate group. Because the structure of nucleic acids is crucial to the mechanism of inheritance and protein synthesis, this subject is discussed further in Chapter 3.

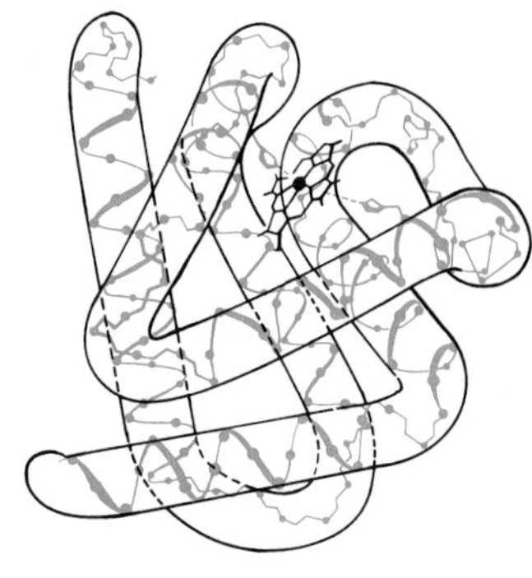

FIGURE 1-18

Three-dimensional tertiary structure of the protein myoglobin. Adjacent folds of the polypeptide chain are held together by disulfide bonds that form between pairs of cysteine molecules. In the upper center of the molecule is the heme group, which combines with oxygen.

From Neurath, H. 1964. The proteins, ed. 2, vol II. New York, Academic Press, Inc.

What is Life?

This is a very difficult question. Life can be defined only in terms of the characteristics we attribute to it. However, having introduced the basic molecules of life, and to preface our discussion of life's origins, let us briefly examine these characteristics.

Organization. Combinations of large molecules like proteins, fats, carbohydrates, and nucleic acids are highly organized into the complex and dynamic systems of physicochemical coordination found in living cells. Many of the large molecules are organized into the various structures of the cell, where thousands of chemical reactions take place.

Metabolism and growth. Metabolism is the collective name given the essential chemical processes that go on in living cells and organisms, including digestion, production of energy (respiration), and synthesis of molecules and structures. The assimilation of energy allows the organisms to grow, whether in size or division into one or more new cells.

Reproduction and inheritance. Whether a particular individual is able to reproduce, the ability must be present in the population of those organisms and of course is necessary for the continued existence of the population. During reproduction the progeny inherits genetic information in the form of DNA from its parent(s).

Irritability. Irritability is the ability to respond to stimuli in the environment. The stimulus and response may be very simple or quite complex.

Adaptability. An individual organism has the ability to adjust to changes in its environment, although the range of adjustment in particular cases may be quite narrow.

Interaction with the environment. Every organism interacts with both living and nonliving components in its environment. The evolutionary history of the organism has placed it in a specific environment that has determined the structural, functional, and behavioral properties of the organism.

Almost every criterion of life has its counterpart in the nonliving world. Some nonliving systems are even capable of a limited amount of metabolism or reproduction. However, only living things have combined these properties into unique structural and functional patterns.

Origin of Life

Considering the exquisite organization of living organisms and the complexity of the molecules and the reactions that result in the properties of life, how could life have originated from nonliving substances? To most biologists the question of life's beginnings is one of profound interest. Despite the complexity, the biologist is struck by a remarkable unity at the molecular and cellular level. All organisms, from humans to the small microbes that transcend the arbitrary boundary between life and nonlife, share two kinds of basic biomolecules—nucleic acid and protein. Except in some viruses, DNA is the material of inheritance. The sequence of nitrogenous bases in DNA provides the code for the amino acid sequence in proteins, and the code is the same throughout the living world. Of all

FIGURE 1-19

Louis Pasteur, holding in his left hand one of the swan-necked flasks used to demonstrate the absence of spontaneous generation.

Courtesy Parke-Davis.

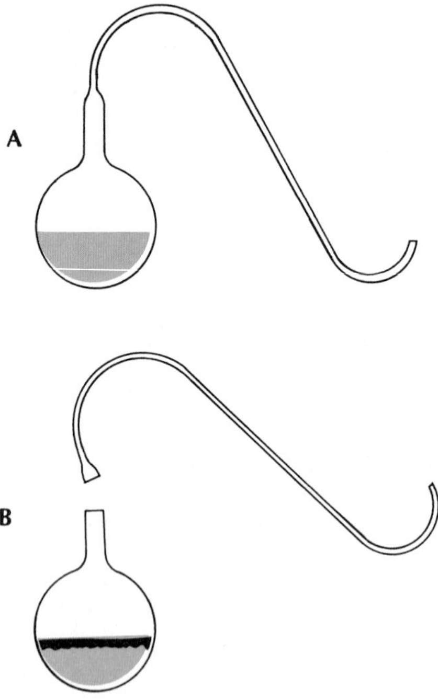

FIGURE 1-20

Louis Pasteur's swan-necked flask experiment. **A,** Sugared yeast water boiled in swan-neck flask remains sterile until neck is broken. **B,** Within 48 hours, flask is swarming with living microorganisms.

the possible amino acids, only 20 are normally found in proteins, regardless of source. Certain metabolic processes that convert foodstuffs into a usable form of energy consistently occur in a wide range of organisms, from the simplest to the most complex. These and many other examples of molecular and functional identity suggest that all life may have had a common beginning.

We must admit at the outset that we do not know how life on earth originated. However, in the last 30 years or so a multidisciplinary effort of scientists from several specialties has made it possible to construct a scenario in which simple living organisms evolved from nonliving constituents more than 3.5 billion years BP (before present). These studies are not attempts to prove or disprove any religious or philosophical belief, but rather they are endeavors to provide an intellectually satisfying account of how life could have arisen on earth by natural means. In the sections to follow, the student should note how scientific methodology was used to shed light on these questions.

HISTORICAL PERSPECTIVE

From ancient times it was commonly believed that life could arise by **spontaneous generation** from dead material, in addition to arising from parental organisms by reproduction **(biogenesis)**. Frogs appeared to arise from damp earth, mice from putrefied matter, insects from dew, maggots from decaying meat, and so on.

The question of spontaneous generation fell under the scrutiny of experimental science in the sixteenth and seventeenth centuries. However, the doctrine was too firmly entrenched to be disbelieved. It remained for the great French scientist Louis Pasteur in 1860 to silence all but the most stubborn proponents of spontaneous generation (Figure 1-19). The most famous of an elegant series of experiments involved the use of flasks with the necks drawn out into a long "swan neck" (Figure 1-20). It was known that microorganisms would appear "spontaneously" in nutrient broth left open to the air. Pasteur's hypothesis was that these microorganisms were carried to the broth by dust particles in the air. Broth was placed in the swan-neck flask and thoroughly boiled. His control was broth in a straight-neck flask similarly treated. Dust particles entering the swan-neck flask were trapped at the bottom of the neck; thus the broth in this flask remained sterile. The broth in the straight-neck flask was soon teeming with microorganisms, and the hypothesis was supported. As a control on the ability of the broth in the swan-neck flask to support growth, the neck was broken off, and colonies of microorganisms quickly began to grow in the flask.

Pasteur's experiments were so convincing that they ended further inquiry into the spontaneous origins of life for a long period. The rebirth of interest into the origins of life occurred in the 1920s. The Russian biochemist Alexander I. Oparin and the British biologist J.B.S. Haldane independently proposed that life originated on earth after an inconceivably long period of "abiogenic molecular evolution." They suggested that the simplest living units came into being gradually by the progressive assembly of organic molecules into more complex organic molecules. These molecules would react with each other to form living microorganisms. Although their proposals differed somewhat on the composition of the earth's early atmosphere, they agreed that the atmosphere lacked free oxygen. The Oparin-Haldane hypothesis greatly influenced theoretical speculation on the origins of life during the 1930s and 1940s.

MODERN EXPERIMENTATION

Finally in 1953 Stanley Miller, while a Ph.D. candidate at the University of Chicago, made the first attempt to simulate with laboratory apparatus the conditions believed to prevail on the primitive earth. His experiment was designed to test the Oparin-Haldane hypothesis by simulating conditions that would have prevailed on the earth, then determining whether biologically important molecules could be produced. Miller built an apparatus that would circulate a mixture of methane, hydrogen, ammonia, and water (representing the atmosphere of the early earth) past an electric spark (Figure 1-21). The spark represented lightning, an energy source to provide necessary energy for the chemical reactions. Water in the flask was boiled to produce steam that helped circulate the gases. The products formed in the electrical discharge were condensed in the condenser and collected in the U-tube and small flask (representing the ocean). The control was an apparatus containing the same materials but with no sparking.

After a week of continuous sparking, the water containing the products was analyzed. The results were surprising. Approximately 15% of the carbon that was originally in the "atmosphere" had been converted into organic compounds that collected in the "ocean." The most striking finding was that many compounds related to life were synthesized. These included four amino acids commonly found in proteins, urea, and several simple fatty acids. No such compounds were found in the control apparatus. Thus the hypothesis was supported.

Miller discovered that amino acids were not formed directly in the spark, but rather were produced by the condensation of certain reactive intermediates, especially hydrogen cyanide and formaldehyde reacting with ammonia. It was found that hydrogen cyanide would react with ammonia under prebiotic conditions to form adenine, one of the four bases found in nucleic acids and a component of adenosine triphosphate (ATP), the universal energy intermediate of living systems. Adenine is a purine, a complex molecule chemically (p. 77). The ease with which it was produced under prebiotic conditions suggests that it came to occupy a central position in biochemistry because it was abundant on the primitive earth. Other nucleic acid bases and sugars have been synthesized under conditions believed to have prevailed on the primitive earth.

Since Miller's original experiments were conducted it has become increasingly clear that the earth's early atmosphere was not as strongly reducing (conditions in which all compounds tend to be reduced, see p. 9) as thought previously, but intermediate or slightly reducing, although free oxygen was absent in either case. (Our present atmosphere is strongly oxidizing.) However, Miller and others have shown that amino acids and other organic compounds were produced when carbon dioxide was substituted for methane and molecular nitrogen for ammonia; that is, oxidized compounds were substituted for reduced compounds.

Need for concentration

The next stage in chemical evolution involved the condensation of amino acids, purines, pyrimidines, and sugars to yield larger molecules that resulted in proteins and nucleic acids. Such condensations do not occur easily in dilute solutions because the presence of excess water tends to drive reactions toward decomposition (hydrolysis). Prebiotic synthesis must have occurred in restricted regions where concentrations were higher, and modern experimentation has shown that any of a variety of mechanisms could have been effective.

Concentration may have occurred by evaporation in lakes, ponds, or tide pools. Dilute aqueous solutions also could have been concentrated by freezing, a technique employed to produce applejack from cider in the northern United States and Canada. As ice freezes, organic solutes are concentrated in the solution that separates from the pure ice. Although the Oparin-Haldane hypothesis suggested that the primitive ocean was a *warm* primordial soup, there is increasing evidence that prebiotic synthesis may have occurred in a cold rather than a warm ocean.

Another way prebiotic molecules might have been concentrated is by adsorption on the surface of clay and other minerals. Clay has the capacity to concen-

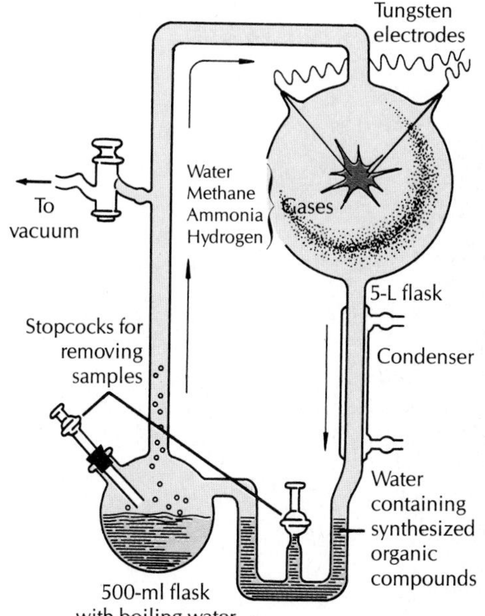

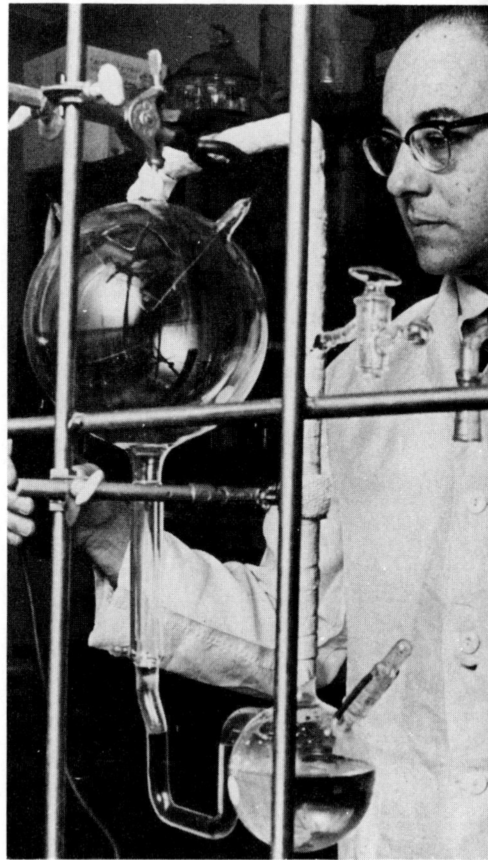

FIGURE 1-21

Stanley Miller and apparatus used in experiment on the synthesis of amino acids with an electric spark in a reducing atmosphere.

Photograph courtesy S.L. Miller.

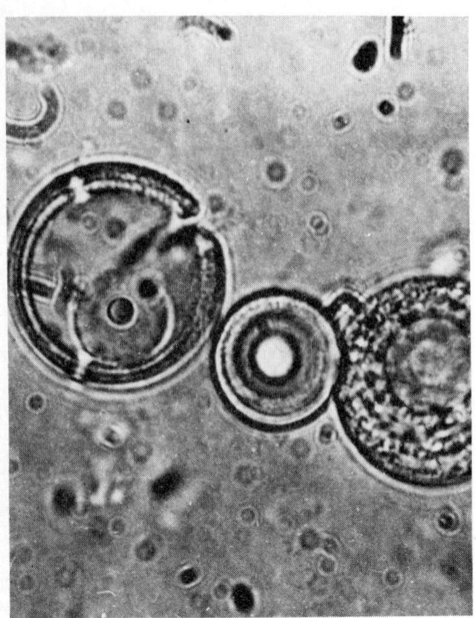

FIGURE 1-22

Electron micrograph of proteinoid microspheres. These proteinlike bodies can be produced in the laboratory from polyamino acids and may represent precellular forms. They have definite internal ultrastructure. (×1700.)

Courtesy S.W. Fox.

trate and condense large amounts of organic molecules from an aqueous solution.

Other possible mechanisms of concentration include **coacervate droplets** (L. *coacervare*, to assemble or cluster together), prepared by Oparin, and **proteinoid microspheres**, studied by the American scientist Sidney Fox. Coacervate droplets are formed when two or more polymers of opposite charge are mixed. Even though the coacervate droplets could be floating in the ocean or in a lake, they would provide a locally nonaqueous environment that was favorable for condensation reactions. Fox studied the thermal synthesis of polypeptides to form "proteinoids." When a mixture of all 20 amino acids is heated to 180°C in water, a good yield of a protein concentrate is obtained, and boiling the water causes the proteinoids to form enormous numbers of hard, minute spherules called microspheres. The proteinoid microspheres (Figure 1-22) possess certain characteristics of living systems.

ORIGIN OF LIVING SYSTEMS

The first living organisms were cells: autonomous membrane-bound units with a complex functional organization that permitted the essential activity of self-reproduction. The primitive chemical systems we have described lack this essential property. The principal problem in understanding the origin of life is explaining how primitive chemical systems could have become organized into living, autonomous, self-reproducing cells.

As we have seen, a lengthy chemical evolution on the primitive earth produced several molecular components of living forms. In a later stage of evolution, nucleic acids (DNA and RNA) began to behave as simple genetic systems that directed the synthesis of proteins, especially enzymes. However, this has led to a troublesome chicken-egg paradox: (1) How could nucleic acids have appeared without enzymes to synthesize them? (2) How could enzymes have evolved without nucleic acids to direct their synthesis? These questions are predicated on a long-accepted dogma that only proteins could act as enzymes. Startling new evidence has suggested that RNA in some instances may have enzymic activity! Therefore the earliest enzymes could have been RNA. Nevertheless, proteins have several important advantages over RNA as catalysts, and the first protocells with protein enzymes would have had a powerful selective advantage over those with only RNA.

Once this stage of organization was reached, natural selection (p. 103) began acting on these primitive self-replicating systems. This was a critical point. Before this stage, biogenesis was shaped by the favorable physical and chemical conditions on the primitive earth and by the nature of the reacting elements themselves. When self-replicating systems became responsive to changing conditions in the environment, their subsequent evolution became directed. The more rapidly replicating and more successful systems were favored and they replicated even faster. In short, the most efficient forms survived. From this evolved the genetic code and fully directed protein synthesis (pp. 82–83). The entity could be called a living organism.

ORIGIN OF METABOLISM

Living cells today are organized systems that possess complex and highly ordered sequences of enzyme-mediated reactions. How did such vastly complex metabolic schemes develop?

Organisms that depend on organic molecules they have not synthesized for their food supplies are known as **heterotrophs** (Gr. *heteros*, another + *trophos*, feeder), whereas those organisms that can synthesize organic molecules from inorganic sources using light or another source of energy are called **autotrophs** (Gr. *autos*, self + *trophos*, feeder) (Figure 1-23). The earliest microorganisms are sometimes referred to as **primary heterotrophs** because they existed before there were any autotrophs. They were probably anaerobic, bacteria-like organisms similar to modern *Clostridium*, and they obtained all their nutrients directly from the environment. Chemical and physical processes of the primordial earth had already supplied generous stores of nutrients in the prebiotic soup. There would be neither advantage nor need for the earliest organisms to synthesize their own compounds, as long as they were freely available from the environment.

Once the supply of a required compound was exhausted or became precarious (perhaps because of an increase in the numbers of organisms using it), those protocells able to convert a precursor to the required compound would have a tremendous advantage over those that lacked this ability. As this situation was repeated over and again, long reaction sequences could develop.

It is important to realize that an enzyme is normally required to catalyze each one of these reactions. So, when we say that early protocells developed a reaction sequence as we have described (A made from B, B from C, and so on), we are really assuming that the appropriate enzymes appeared to catalyze these reactions. The numerous enzymes of cellular metabolism appeared when cells became able to use proteins for catalytic functions and thereby gained a selective advantage. No planning was required; the results were achieved through natural selection.

FIGURE 1-23

Koala, a heterotroph, feeding on a eucalyptus tree, an autotroph. All heterotrophs depend for their nutrients directly or indirectly on autotrophs that capture the sun's energy to synthesize their own nutrients.

Photograph by C.P. Hickman, Jr.

Appearance of photosynthesis and oxidative metabolism

Eventually, almost all usable energy-rich nutrients of the prebiotic soup were consumed. This ushered in the next stage of biochemical evolution: the use of readily available solar radiation to provide metabolic energy. In an age of increasing scarcity of nutrient molecules, it is easy to appreciate what an advantage the autotrophs had over the primary heterotrophs.

In plant photosynthesis, water is the source of the electrons that are used to reduce carbon dioxide to sugars (that is, to add hydrogen), and molecular oxygen is liberated:

$$6\ CO_2 + 6\ H_2O \xrightarrow[\text{enzymes}]{\text{light}} C_6H_{12}O_6 + 6\ O_2$$

This is a summary of the many reactions now known to comprise the process of photosynthesis. Undoubtedly, these reactions did not appear all at once, and other reduced compounds, such as hydrogen sulfide (H_2S), were probably the early electron donors, rather than H_2O. As these reducing agents were used up, oxygen-evolving photosynthesis appeared. Ozone (a unique compound of 3 oxygen atoms) began to accumulate in the atmosphere, and, as a result, strong ultraviolet radiation was screened out.

At this important juncture, accumulating oxygen began to interfere with cellular metabolism, which up to this point had evolved under anaerobic conditions. As the atmosphere slowly changed from a somewhat reducing to a highly oxidizing one, a new and highly efficient kind of energy metabolism appeared: oxidative (aerobic) metabolism. By using the available oxygen as a terminal electron acceptor (see pp. 51–52) and oxidizing glucose completely to carbon dioxide and water, much of the bond energy stored by photosynthesis could be recovered. Most living forms became wholly dependent on oxidative metabolism.

PRECAMBRIAN LIFE

As depicted on the inside back cover of this book, the Precambrian period spanned the geological time before the beginning of the Cambrian period 600 million years ago. Thus about seven eighths of the age of the earth is encompassed by the Precambrian period. At the beginning of the Cambrian period, most of the major phyla of invertebrate animals made their appearance within a few million years. This has been called the "Cambrian explosion" because before this

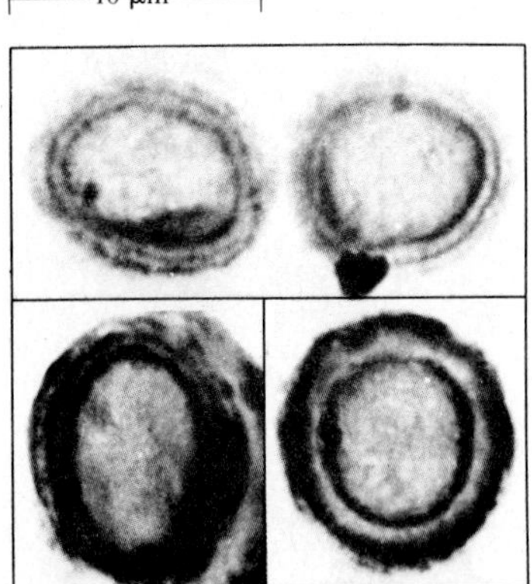

FIGURE 1-24

Fossil bacteria from the Bitter Springs Formation in Australia. These fossils are too small to be visualized except with an electron microscope. They are about 850 million years old.

From Schopf, J.W. 1971. J. Paleontol. **45**:925-960

time fossil deposits were apparently rare and almost devoid of anything more complex than single-celled algae. We now recognize that the apparent rarity of Precambrian fossils was because they escaped notice owing to their microscopic size (Figure 1-24). What were the forms of life that existed on earth before the burst of evolutionary activity in the early Cambrian world, and what organisms were responsible for the momentous change from a reducing to an oxidizing atmosphere?

Prokaryotes and the age of blue-green algae

The earliest bacteria-like organisms proliferated, giving rise to a great variety of bacterial forms, some of which were capable of photosynthesis. From these arose the oxygen-evolving **cyanobacteria** (blue-green algae) some 3 billion years ago.

Bacteria and cyanobacteria are called **prokaryotes**, meaning literally "before nucleus." They contain a single chromosome composed of a single, large molecule of DNA not located in a membrane-bound nucleus, but found in a nuclear region, or **nucleoid**. The DNA is associated with proteins different from those associated with eukaryote DNA, and prokaryotes lack membranous organelles such as mitochondria, plastids, Golgi apparatus, and endoplasmic reticulum (Chapter 2). They reproduce by fission or budding, never by true mitotic cell division.

Bacteria and especially cyanobacteria ruled the earth's oceans unchallenged for some 1.5 to 2 billion years. The cyanobacteria reached the zenith of their success approximately 1 billion years ago when filamentous forms produced great floating mats on the ocean surface. This long period of cyanobacteria dominance, encompassing approximately two thirds of the history of life, has been called with justification the "age of blue-green algae." Bacteria and cyanobacteria are so completely different from forms of life that evolved later that they are placed in a separate kingdom, Monera (p. 335).

The name "algae" is misleading because it suggests a relationship to the eukaryotic green algae, and many scientists prefer the alternative name "cyanobacteria" for the group. These were the organisms responsible for producing oxygen released into the atmosphere. Study of the biochemical reactions in present cyanobacteria suggests that they evolved in a time of fluctuating oxygen concentration. For example, although they can tolerate atmospheric concentrations of oxygen (21%), the optimum concentration for many of their metabolic reactions is only 10%.

Appearance of the eukaryotes

Approximately 1.5 billion years ago, after the accumulation of an oxygen-rich atmosphere, organisms with nuclei appeared. These **eukaryotes** (true nucleus) have cells with membrane-bound nuclei containing DNA complexed with certain characteristic proteins. Eukaryotes are generally larger than prokaryotes, contain much more DNA, and usually divide by some form of mitosis (p. 40). Within their cells are numerous membranous organelles, including mitochondria in which the enzymes for oxidative metabolism are packaged. Protozoa, fungi, algae (other than the blue-greens), higher plants, and multicellular animals are composed of eukaryotic cells.

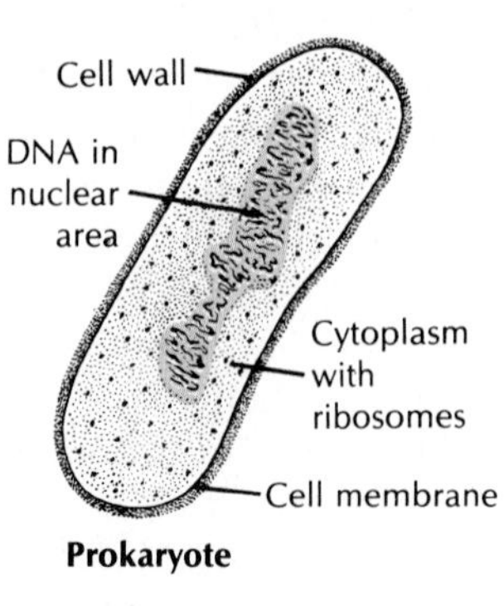

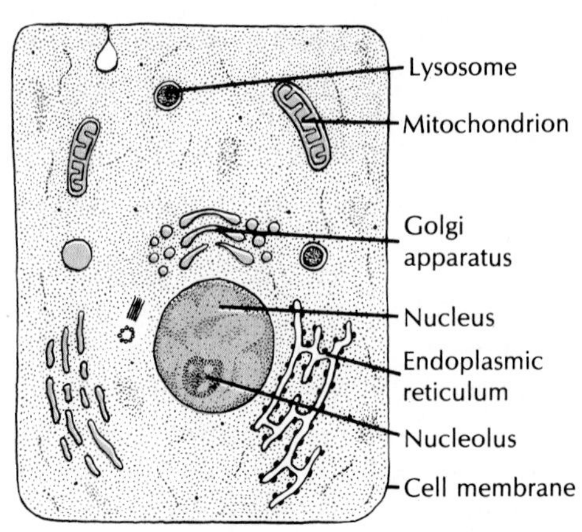

FIGURE 1-25

Comparison of prokaryote and eukaryote cells. The prokaryote cell is about one-tenth the size of the eukaryote cells.

Prokaryotes and eukaryotes are profoundly different from each other (Figure 1-25) and clearly represent a marked dichotomy in the evolution of life. The ascendancy of the eukaryotes resulted in a rapid decline in the dominance of cyanobacteria as the eukaryotes proliferated and fed on them.

Why were the eukaryotes immediately so successful? Probably because they developed an important process facilitating rapid evolution — sex. Sex promotes great genetic variability by mixing the genes of each two individuals that mate. By preserving favorable genetic variants, natural selection encourages rapid evolutionary change. Prokaryotes propagate effectively and efficiently, but their mechanisms for interchange of genes, which do occur in some cases, lack the systematic genetic recombination characteristic of sexual reproduction.

The organizational complexity of the eukaryotes compared to that of the prokaryotes is so much greater that it is difficult to visualize how a eukaryote could have arisen from any known prokaryote. It has been proposed by Margulis and others that eukaryotes did not, in fact, arise from any single prokaryote, but were derived from a symbiosis (life together) of two or more types. Mitochondria and plastids, for example, each contain their own complement of DNA (apart from the nucleus of the cell), which has some prokaryote characteristics. Mitochondria contain the enzymes of oxidative metabolism, and plastids (a plastid with chlorophyll is a chloroplast) carry out photosynthesis. It is easy to see how a host cell that was able to accommodate such guests in its cytoplasm would have had enormous competitive advantages.

Eukaryotes may have originated more than once. They were no doubt unicellular and many were photosynthetic autotrophs. Some of these lost their photosynthetic ability and became heterotrophs, feeding on the autotrophs and the prokaryotes. As the cyanobacteria were cropped, their dense filamentous mats began to thin, providing space for other species. Carnivores appeared and fed on the herbivores. Soon a balanced ecosystem of carnivores, herbivores, and primary producers appeared. This was ideal for evolutionary diversity. By freeing space, cropping herbivores encouraged a greater diversity of producers, which in turn promoted the evolution of new and more specialized croppers. An ecological pyramid appeared with carnivores at the top.

The burst of evolutionary activity that followed at the end of the Precambrian period and beginning of the Cambrian period was unprecedented; nothing approaching it has occurred since. Nearly all animal and plant phyla appeared and established themselves within a relatively brief period of a few million years. The eukaryotic cell made possible the richness and diversity of life on earth today.

It was originally believed that the asexual prokaryotes could only stamp out carbon copies of the parental cells, genetic change occurring solely when a mutation intervened. However, several processes of genetic recombination are now known in bacteria. For example, when certain viruses infect bacterial cells, the viruses may pick up a fragment of the bacterial DNA and transfer it to new cells during the course of subsequent infection *(transduction)*. Sometimes, DNA in the environment (released by cell death or other processes) can somehow penetrate the cell wall and membrane of other bacteria and be incorporated into their genetic complement *(transformation)*. Under certain circumstances, bacteria of differing "mating types" can actually form cytoplasmic bridges between each other and transfer segments of DNA *(conjugation)*. Conjugation often involves the transfer of *plasmids* between bacteria. Plasmids are small circles of DNA carried by bacteria in addition to their large, circular chromosome. They are about 1/1000 the size of the bacterial chromosome and carry only a few genes, but they are an important element in the new and exciting methods of "genetic engineering."

The hypothesis of symbiotic origin of eukaryotes described here is supported by studies of the amino acid sequence in certain proteins and base sequence in nucleic acids. Furthermore, there is some evidence that microtubules in eukaryotes were acquired symbiotically. *Microtubules* are tiny tubular structures characteristic of eukaryotes that are important in cell motility (as in cilia and flagella, for example) and mitotic and meiotic cell division. Margulis and co-workers found microtubules in certain spirochaete bacteria (which are prokaryotes) that were constituted of a protein (tubulin) apparently similar to that in microtubules of all eukaryotes.

■ Summary

Zoology is the scientific study of animals. Science is characterized by a particular approach to the acquisition of human knowledge. It is guided by, and is explanatory with reference to, natural law, and it is testable, tentative, and falsifiable. The scientific method is the manner in which scientific knowledge is gained, and it has proved to be a powerful tool. Its essential constituents are observing events, forming a hypothesis on the basis of the observations, testing the hypothesis, and drawing a conclusion on the basis of the tests. Tests may take the form of experiments. Every experiment must have two components: an experimental group and a control group. The control group is exactly like the experimental group and is treated exactly the same, except that the experimental variable is not applied to the control group. Results of the tests of the hypothesis are the data. A hypothesis for which there is a great deal of supporting data, particularly one that explains a very large number of observations, may be elevated to the status of a theory.

All matter is composed of elements, which are themselves composed of discrete units called atoms. Atoms consist of a positively charged nucleus of very small volume, surrounded by negatively charged electrons of almost no mass. The nucleus is made up of 1 or more postively charged protons (the number of protons is the atomic number) and unchanged neutrons. The total number of protons and neutrons is nearly equal to the atomic weight of the element. Forms of an element that differ in the number of neutrons in the nucleus are isotopes, and isotopes that spontaneously disintegrate are radioactive isotopes.

The innermost "shell" of an atom can hold only 2 electrons, and each outer shell can hold a maximum of 8 electrons. Inner shells are filled first, and atoms tend to complete their shells by gaining or losing electrons or by sharing electrons with other atoms.

Molecules are formed when 2 or more atoms are joined together by chemical bonds. When a molecule is composed of 2 or more different elements, it is a compound. Ionic bonds involve loss or gain of electrons. Loss of electrons is oxidation, and gain of electrons is reduction. Covalent bonds involve sharing of electrons. Hydrogen bonds are weak bonds resulting from unequal charge distribution in a covalent bond.

The acidity or alkalinity of solutions is expressed by the pH scale, defined as the negative logarithm of the hydrogen ion concentration.

Carbon is especially versatile in bonding with itself or with other atoms and is the only element capable of forming the variety of molecules found in living things. Carbohydrates are composed primarily of carbon, hydrogen, and oxygen grouped as H—C—OH. Sugars serve as immediate sources of energy in living systems. Monosaccharides, or simple sugars, may bond together to form disaccharides or polysaccharides, which serve as storage forms of sugar or perform structural roles. Lipids exist principally as fats, phospholipids, and steroids.

Proteins are large molecules composed of amino acids linked together by peptide bonds. Proteins have a primary, secondary, tertiary, and often, quarternary structure. Proteins perform many functions, especially as enzymes (biological catalysts).

Nucleic acids are polymers of nucleotide units, each composed of a sugar, a nitrogenous base, and a phosphate group. They contain the coding for inheritance and function in protein synthesis.

Life is characterized by a high degree of organization, metabolism and growth, reproduction and inheritance, irritability, adaptability, and interaction with the environment.

A remarkable uniformity in the chemical constituents of living things and many of the reactions that go on in their cells suggests that life on earth may have had a common origin. A.I. Oparin and J.B.S. Haldane proposed a long "abiogenic molecular evolution" on earth in which organic molecules slowly accumulated. The atmosphere of the primitive earth was somewhat reducing, and little or no free oxygen was present. The main energy source to power chemical reactions was probably lightning. Miller showed the plausibility of the Oparin-Haldane hypothesis by simple, but ingenious, experiments. Although the compounds could have accumulated in rather dilute solutions, condensation reactions could have occurred after concentration during freezing, adsorption onto clay particles, in coacervates, or in protein microspheres. When self-replicating systems became responsive to the forces in the environment, evolution proceeded more rapidly.

The first organisms were the primary heterotrophs, living on the energy stored up in molecules dissolved in the primordial soup. As such molecules were used up, autotrophs had a great selective advantage. Molecular oxygen began to accumulate in the atmosphere as an end product of photosynthesis, and finally the atmosphere became oxidizing. The organisms responsible for this were apparently cyanobacteria (blue-green algae). The cyanobacteria and true bacteria are prokaryotes, organisms that lack a membrane-bound nucleus and other membranous organelles in their cytoplasm. The eukaryotes originated after the atmosphere was oxidizing and apparently arose from symbiotic unions of two or more types of prokaryotes. Eukaryotes have most of their genetic material (DNA) borne in a membrane-bound nucleus and have mitochondria and other features. They include the eukaryotic algae, fungi, other plants, and all animals. Their evolutionary success is due in great degree to the variability conferred by sexual reproduction.

REVIEW QUESTIONS

1. Explain the difference between science and, for example, a religious belief.
2. Name some common, everyday event, and tell how you would investigate it by using the scientific method.
3. Distinguish among the following: element, atom, proton, neutron, electron.
4. How do isotopes of elements differ from each other?
5. Write the names of the elements for each of the following symbols: P, Cl, K, Fe, H, N, C, S, Ca, O, Na, I.
6. Distinguish among ionic bonds, covalent bonds, and hydrogen bonds.
7. In the reaction below, name the following: electron donor, electron acceptor.

$$K + Cl \rightarrow K^+ + Cl^-$$

8. Match the items in the left column with the most appropriate terms in the right column:

___$H_2^{2+}SO_4^{2-}$	a. Acid
___K^+OH^-	b. Base
___K^+Cl^-	c. Salt
___$CH_3{-}C{=}O$ with $O^-\ H^+$ on the C	
___$Ca^{2+}SO_4^{2-}$	

9. A solution with a pH of 6 is 10 times more acid than one with a pH of 7. Explain why this is so.
10. Name two simple carbohydrates, two storage carbohydrates, and a structural carbohydrate.
11. What are characteristic differences in molecular structure between lipids and carbohydrates?
12. Explain the difference between primary, secondary, tertiary, and quaternary structure of a protein.
13. What are the important nucleic acids in a cell, and of what units are they constructed?
14. Name and briefly explain the characteristics of life.
15. In regard to the experiments of Louis Pasteur and Stanley Miller described in this chapter, explain what constituted the following in each case: observations, induction, hypothesis, deduction, prediction, data, control.
16. Explain the significance of the Miller experiments.
17. What are several mechanisms by which organic molecules in the prebiotic ocean could have been concentrated so that further reactions could occur?
18. Distinguish among the following: primary heterotroph, autotroph, secondary heterotroph.
19. What is the origin of the oxygen in the present-day atmosphere, and what is its metabolic significance to most organisms living today?
20. Distinguish between prokaryotes and eukaryotes as completely as you can.

SELECTED REFERENCES

Cairns-Smith, A.G. 1985. The first organisms. Sci. Am. **252:**90–100 (June). *Argues that a genetic system based on nucleic acids was too complex for the earliest organisms to have evolved. Perhaps the earliest organisms depended on clay crystals with genetic properties.*

Cech, T.R. 1986. RNA as an enzyme. Sci. Am. **255:**64–75. *Recounts some of the recent investigations of RNA as a catalyst.*

Hailman, J.P. 1977. Optical signals. Animal communication and light. Bloomington, Indiana University Press. *The introduction has a concise statement of the scientific method and scientific epistemology.*

Margulis, L. 1982. Early life. Boston, Science Books International. *A simplified account of the development of life on earth. Well-illustrated; undergraduate level.*

Overton, W.R. 1982. Judgment, injunction, and memorandum opinion in the case of McLean vs Arkansas Board of Education. Science **215:**934–943. *Judge Overton's opinion is reprinted verbatim. It is highly recommended reading.*

Trachtman, P. 1984. Searching for the origins of life. Smithsonian **15**(3):42–51 (June). *Current research on the origin of life.*

Vidal, G. 1984. The oldest eukaryotic cells. Sci. Am. **250:**48–57 (Feb.). *Careful examination of microscopic fossils shows that eukaryotes evolved in the form of unicellular plankton about 1.4 billion years ago.*

Woese, C.R. 1984. The origin of life. Carolina Biology Readers, no. 13. Burlington, N.C., Carolina Biological Supply Co. *Thought-provoking account of the author's views; critique of traditional concepts, with focus on problems.*

SUNLIGHT AND CELLS

It is a remarkable fact that living forms, from amebas and unicellular algae to whales and giant redwood trees, are formed from a single type of building unit: the cell. All animals and plants are composed of cells and cell products. Thus the cell theory, or cell doctrine as it is more appropriately called, is another of the great unifying concepts of biology.

New cells come from division of preexisting cells, and the activity of a multicellular organism as a whole is the sum of the activities of its constituent cells and their interactions. The energy to support virtually all life activities flows from sunlight that is captured by green plants and transformed by the process of photosynthesis into chemical bond energy. Chemical bond energy is a form of potential energy that can be released when the bond is broken; the energy is used to perform electrical, mechanical, and osmotic tasks in the cell. Ultimately, all energy is dissipated, little by little, into heat. This is in accord with the second law of thermodynamics, which states that there is a tendency in nature to proceed toward a state of greater molecular disorder, or entropy. Thus the high degree of molecular organization in living cells is attained and maintained only as long as energy fuels the organization.▪

CHAPTER 2

The Cell as the Unit of Life

CELL CONCEPT

More than 300 years ago the English scientist and inventor Robert Hooke, using a primitive compound microscope, observed boxlike cavities in slices of cork and leaves. He called these compartments "little boxes or cells." In the years that followed Hooke's first demonstration of the remarkable powers of the microscope before the Royal Society of London in 1663, biologists gradually began to realize that cells were far more than simple containers filled with "juices."

Cells are the fabric of life. Even the most primitive cells are enormously complex structures that form the basic units of all living matter. All tissues and organs are composed of cells. In humans an estimated 40 trillion cells interact, each performing its specialized role in an organized community. In single-celled organisms, all the functions of life are performed within the confines of one microscopic package. There is no life without cells.

With the exception of some eggs, which are the largest cells (in volume) known, cells are small and mostly invisible to the unaided eye. Consequently, our understanding of cells is paralleled by technical advances in the resolving power of microscopes. The Dutch microscopist A. van Leeuwenhoek, using high-quality single lenses that he had made, sent letters to the Royal Society of London containing detailed descriptions of the numerous organisms he had observed (1673 to 1723). In the early nineteenth century, the improved design of the microscope permitted biologists to distinguish objects only 1 μm apart. This advance was quickly followed by new discoveries that laid the groundwork for the **modern cell theory**—a theory stating that all living organisms are composed of cells and that all cells are derived from preexisting cells.

In 1838 Matthias Schleiden, a German botanist, announced that all plant tissue was composed of cells. A year later one of his countrymen, Theodor Schwann, described animal cells as being similar to plant cells, which is an understanding that had been long delayed because the animal cell is bounded only by a nearly invisible plasma membrane and because it lacks the distinct cell wall characteristic of the plant cell. In 1858 Rudolf Virchow realized that

Units of measurement commonly used in microscopic study are micrometers, nanometers, and angstroms:
1 micrometer (μm) = 1/1,000,000 meter,
1 nanometer (nm) = 1/1,000,000,000 meter,
1 angstrom (Å) = 1/10,000,000,000 meter.
Thus 1 m = 10^3 mm = 10^6 μm = 10^9 nm = 10^{10} Å.

all cells came from preexisting cells; thus Schleiden, Schwann, and Virchow are credited with the unifying cell theory that ushered in a new era of productive exploration in cell biology.

In 1840 J. Purkinje introduced the term **protoplasm** to describe the cell contents. Protoplasm was at first believed to be a granular, gel-like mixture with special and elusive life properties of its own; the cell was thus viewed as a bag of thick soup containing a nucleus. Later the interior of the cell became increasingly visible as microscopes were improved and better tissue-sectioning and staining techniques were introduced. Rather than being a uniform granular soup, the cell interior is composed of numerous unique structures called **cell organelles**, each performing a specific function in the life of the cell. Today we realize that the components of a cell are so highly organized, structurally and functionally, that describing its contents as "protoplasm" is a bit like describing the contents of an automobile engine as "autoplasm."

HOW CELLS ARE STUDIED

Because cells are submicroscopic structures with complex and delicate internal organization, new technical approaches were required for the visual exploration of cell structure and function. The light microscope, with all its variations and modifications, has contributed more to biological investigation than any other instrument developed by humans. It has been a powerful exploratory tool for 300 years and continues to be so more than 45 years after the invention of the electron microscope. But, until the electron microscope was perfected, our concept of the cell was limited to that which could be seen with magnifications of 1000 to 2000 diameters. This is the practical limit of the light microscope. In addition to microscopy, modern biochemical techniques have contributed enormously to our understanding of cell structure and function.

The electron microscope, invented in Germany in approximately 1931 and developed in the 1940s, employs high voltages to drive a beam of electrons through a vacuum to visualize the object being studied. The wavelength of electrons is approximately 0.00001 that of ordinary white light, thus permitting far greater magnification (compare *A* and *B* of Figure 2-1). Electron microscopes can detect objects as small as 0.001 the size of objects discernible with the best light

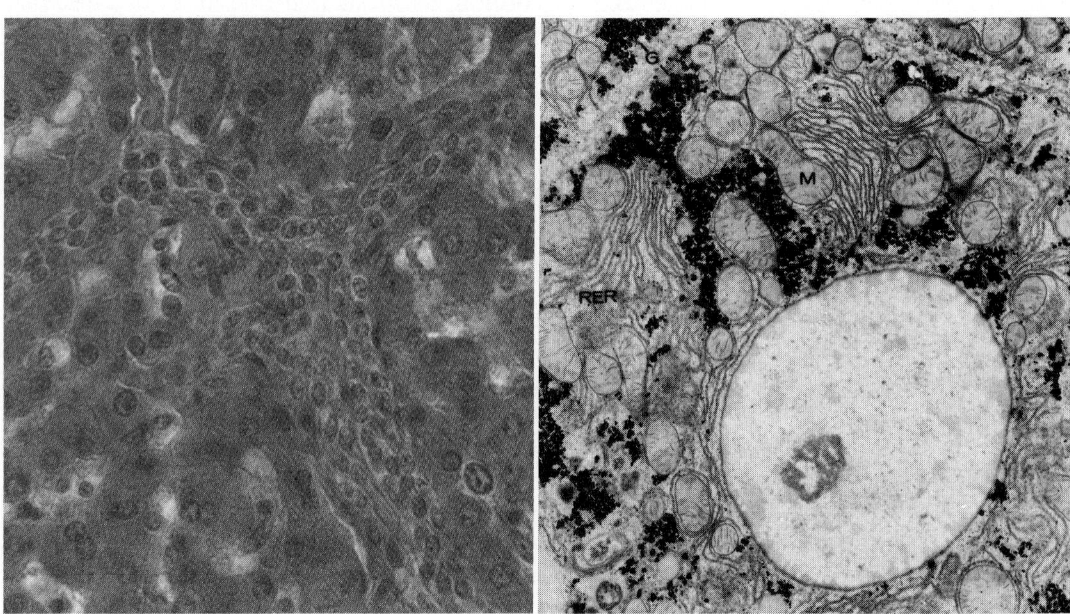

FIGURE 2-1

Liver cells. **A,** Magnified approximately 400 times through light microscope. Note prominently stained nucleus in each polyhedral cell. **B,** Portion of single liver cell, magnified approximately 5000 times by electron microscope. Single large nucleus dominates field; mitochondria *(M)*, rough endoplasmic reticulum *(RER)*, and glycogen granules *(G)* are also seen.

A, Photograph by J.D. Cunningham/Visuals Unlimited; **B,** from Morgan, C.R., and R.A. Jersild, Jr. 1970. Anat. Rec. **166**:575–586.

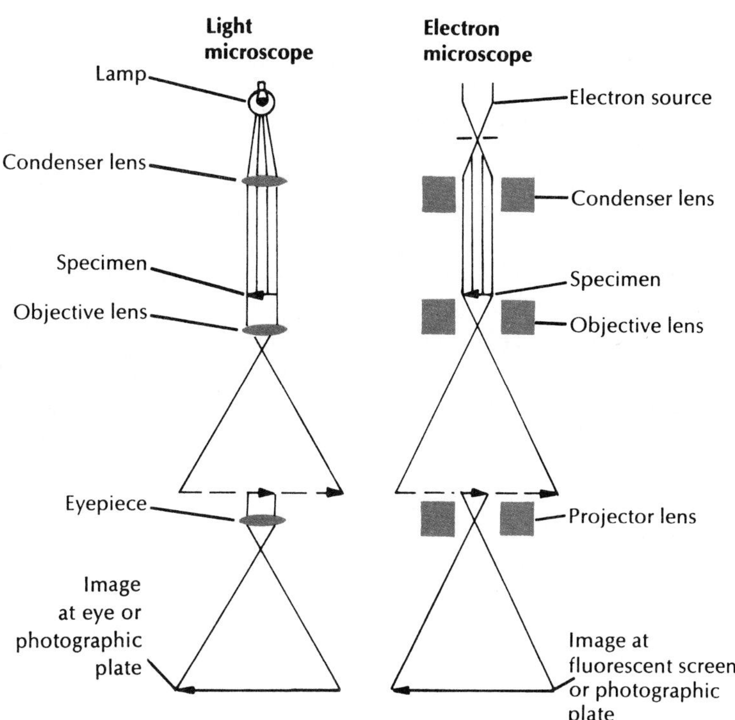

FIGURE 2-2

Comparison of optical paths of light and electron microscopes. Note that to facilitate comparison, schematic of light microscope has been inverted from its usual orientation with light source below and image above.

microscope. Even large molecules such as DNA and proteins can actually be seen with the electron microscope.

Ordinary light, as used with the first microscope, is replaced in the electron microscope with a beam of electrons emitted from a tungsten filament, the glass lenses of the light microscope are replaced with magnets for focusing the electron beam, and the human eye is replaced with a fluorescent screen or a camera (Figure 2-2). In preparation for viewing, specimens are cut into extremely thin sections and treated with "electron stains" (ions of elements such as osmium, lead, or uranium) to increase contrast between different structures. Image formation depends on differences in electron scattering, which in turn depend on the density of objects in the electron beam. Because the electrons pass through the specimen to the photographic plate, the instrument is called a **transmission electron microscope.**

In contrast, specimens prepared for the **scanning electron microscope** are not sectioned, and electrons do not pass through them. The whole specimen is bombarded with electrons, causing secondary electrons to be emitted. An apparent three-dimensional image is recorded in the photograph. Although the magnification capability of the scanning instrument is not as great as the transmission microscope, a great deal has been learned about the surface features of organisms and cells. Examples of scanning electron micrographs are shown on pp. 259 and 362.

Advances in the techniques of cell study were not limited to improvements in microscopes but have included new methods of tissue preparation, staining for microscopic study, and the great contributions of modern biochemistry. For example, the various organelles of cells have differing, characteristic densities. Cells can be broken up with most of the organelles remaining intact, then centrifuged in a density gradient (Figure 2-3), and relatively pure preparations of each organelle may be recovered. Thus the biochemical functions of the various organelles may be studied separately. The DNA and various types of RNA can be extracted and studied. Many enzymes can be purified and their characteristics determined. The use of radioactive isotopes has allowed elucidation of many

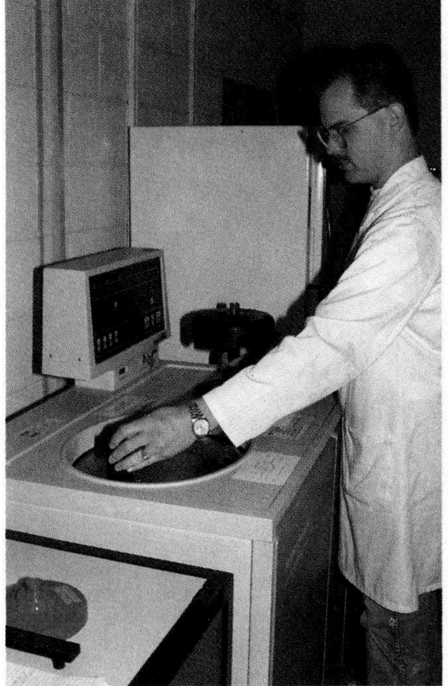

FIGURE 2-3

A rotor containing samples is being placed in an ultracentrifuge. Spinning at high speeds, such devices exert forces many thousands of times the force of gravity on the samples.

Photograph by L.S. Roberts.

metabolic reactions and pathways in the cell. Modern chromatographic techniques can separate chemically similar intermediates and products. Many more examples could be cited, and these have contributed enormously to our present understanding of cell structure and function.

Organization of the Cell

If we were to restrict our study of cells to fixed and sectioned tissues, we would be left with the erroneous impression that cells are static, quiescent, rigid structures. In fact, the cell interior is in a constant state of upheaval. Most cells are continually changing shape, pulsing and heaving; their organelles twist and regroup in a cytoplasm teeming with granules, fat globules, and vesicles of various sorts. This description is derived from studies with time-lapse photography of living cell cultures. If we could see the swift shuttling of molecular traffic through gates in the cell membrane and the metabolic energy transformations within cell organelles, we would have an even stronger impression of internal turmoil. However, the cell is anything but a bundle of disorganized activity. There is order and harmony in the cell's functioning that represents the elusive phenomenon we call life. Studying this dynamic miracle of evolution through the microscope, we

Table 2-1 Comparison of prokaryotic and eukaryotic cells

Characteristic	Prokaryotic cell	Eukaryotic cell
Cell size	Mostly small (1-10μm)	Mostly large (10-100μm)
Genetic system	DNA with some nonhistone protein; simple, circular chromosome in nucleoid; nucleoid not membrane bound	DNA complexed with histone and nonhistone proteins in complex chromosomes within nucleus with membranous envelope
Cell division	Direct by binary fission or budding; no mitosis	Some form of mitosis; centrioles in many; mitotic spindle present
Sexual system	Absent in most; highly modified if present	Present in most; male and female partners; gametes that fuse
Nutrition	Absorption by most; photosynthesis by some	Absorption, ingestion; photosynthesis by some
Energy metabolism	No mitochondria; oxidative enzymes bound to cell membrane, not packaged separately; great variation in metabolic pattern	Mitochondria in most; oxidative enzymes packaged therein; more unified pattern of oxidative metabolism
Organelles	None	Many
Intracellular movement	None	Cytoplasmic streaming, phagocytosis, pinocytosis
Flagella/cilia	Not with "9+2" microtubular pattern	With "9+2" microtubular pattern
Cell wall	Contains disaccharide chains cross-linked peptides	If present, not with disaccharide polymers linked with peptides

realize that as we gradually comprehend more and more about this unit of life and how it operates, we are gaining a greater understanding of the nature of life itself.

PROKARYOTIC AND EUKARYOTIC CELLS

The radically different cell plan of prokaryotes and eukaryotes was described in Chapter 1. A fundamental distinction, expressed in their names, is that prokaryotes lack the membrane-bound nucleus present in all eukaryotic cells. Other major differences are summarized in Table 2-1.

Prokaryotic organisms — bacteria and cyanobacteria — constitute the kingdom Monera. All other organisms are eukaryotes distributed among four kingdoms: the unicellular kingdom Protista (protozoa and nucleated algae) and three multicellular kingdoms, Plantae (green plants), Fungi (true fungi), and Animalia (multicellular animals). The kingdom classifications are discussed in Chapter 14. The following discussion is restricted to eukaryotic cells, of which all animals are composed.

COMPONENTS OF THE EUKARYOTIC CELL AND THEIR FUNCTION

If the inside of the cheek is gently scraped with a blunt instrument and if the scrapings are put on a slide in a drop of physiological salt solution and examined unstained with a microscope, living cells can be seen. The flat circular cells with small nuclei are the squamous epithelial cells that line the mouth region.

Flat epithelial cells are only one variety of many different shapes assumed by cells. Although many cells, because of surface tension forces, assume a spherical shape when freed from restraining influences, others retain their shape under most conditions because of their characteristic cytoskeleton, or framework of microtubules.

The eukaryotic cell is enclosed within a thin, differentially permeable **plasma membrane** (Figure 2-4). The most prominent organelle is the spherical or ovoid

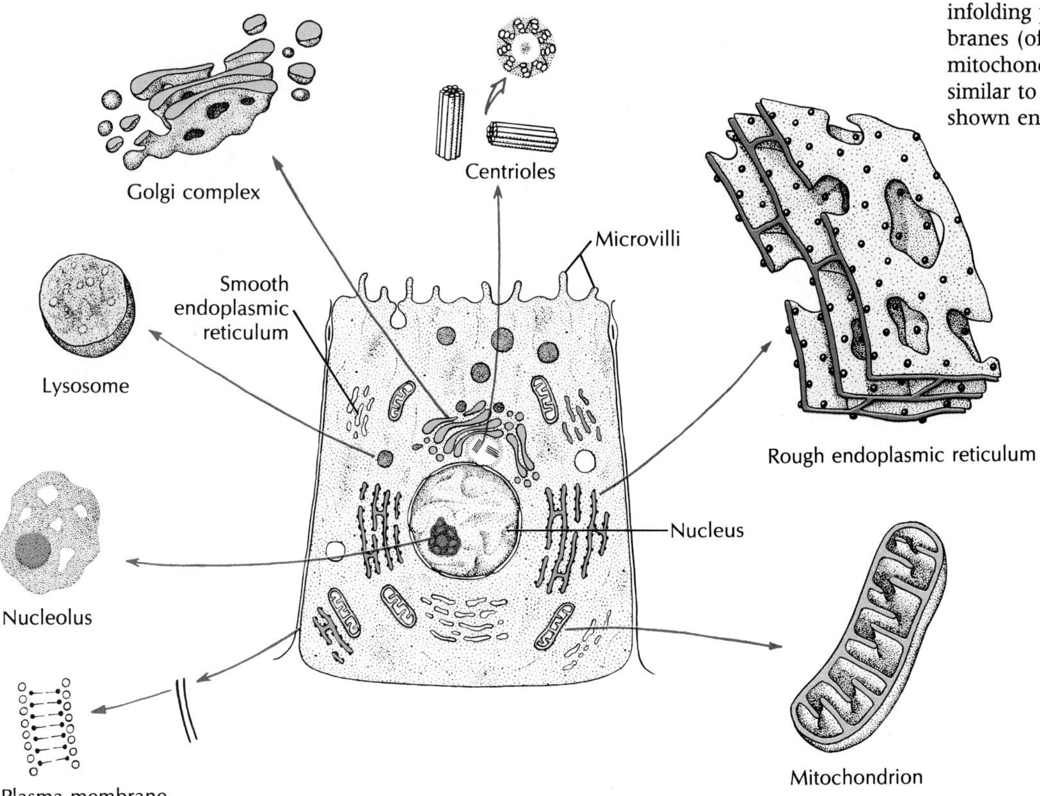

FIGURE 2-4

Generalized cell with principal organelles, as might be seen with the electron microscope. Each of the major organelles is shown enlarged. Membranes of organelles are believed to be continuous with, or derived from, the plasma membrane by an infolding process. Structure of other membranes (of nucleus, endoplasmic reticulum, mitochondria, and others) is probably similar to that of plasma membrane, shown enlarged at lower left.

nucleus, enclosed within its membranous envelope. The region outside the nucleus is regarded as cytoplasm. Within the cytoplasm are many organelles, such as mitochondria, Golgi complexes, centrioles, and endoplasmic reticulum. Plant cells typically contain **plastids**, some of which are photosynthetic organelles, and bear a cell wall containing cellulose outside the plasma membrane.

The **fluid-mosaic model** is the currently accepted concept of the plasma membrane. With the electron microscope, the cell membrane appears as two dark lines, each approximately 3 nm thick, at each side of a light zone. The entire membrane is 8 to 10 nm thick. This image is the result of a phospholipid bilayer, that is, two layers of phospholipid molecules, all oriented with their water-soluble ends toward the outside and their fat-soluble portions toward the inside of the membrane (Figure 2-5). An important characteristic of the phospholipid bilayer is that it is a liquid, giving the membrane flexibility and allowing the phospholipid molecules to move sideways freely within their own monolayer. Molecules of cholesterol are interspersed in the lipid portion of the bilayer (Figure 2-5). They make the membrane even less permeable and decrease its flexibility.

Glycoproteins (proteins with carbohydrates attached) are essential components of plasma membranes. Some of these proteins catalyze the transport of substances such as negatively charged ions across the membrane. Others act as specific receptors for various molecules or as highly specific markings. For example, the self-nonself recognition that enables the immune system to react to

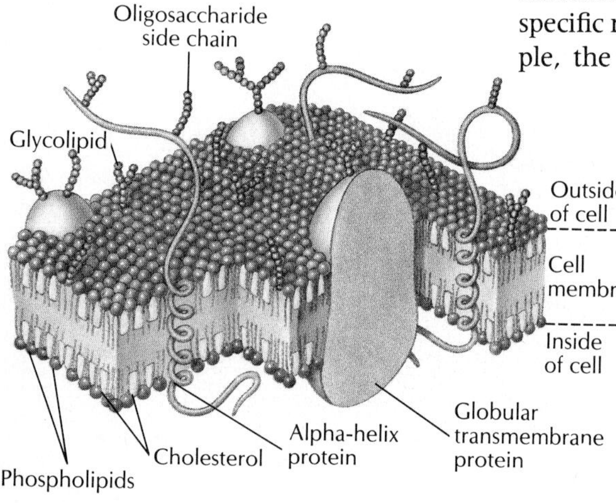

FIGURE 2-5

Fluid-mosaic model of a cell membrane.

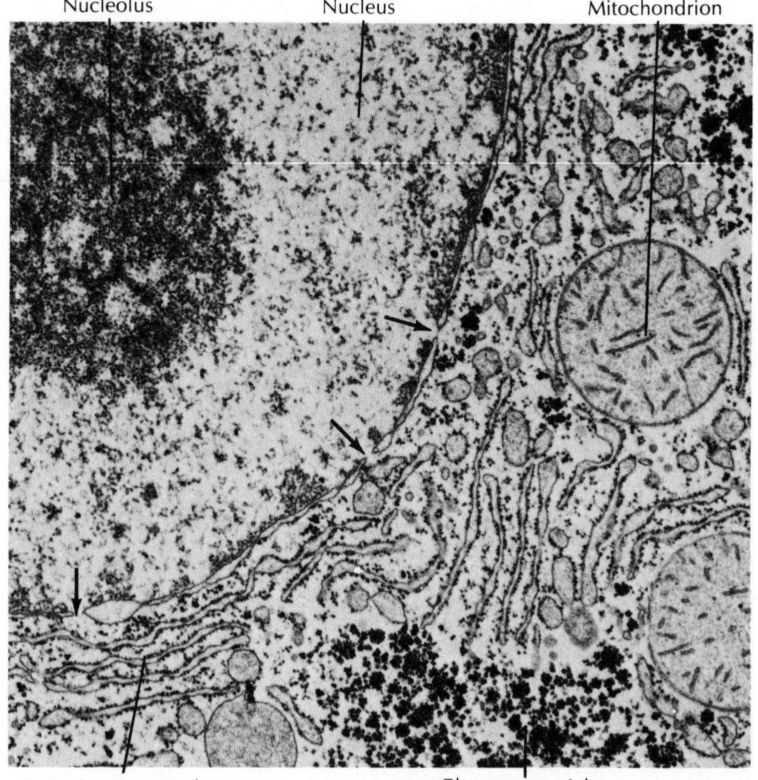

FIGURE 2-6

Electron micrograph of part of hepatic cell of rat showing portion of nucleus *(left)* and surrounding cytoplasm. Endoplasmic reticulum and mitochondria are visible in cytoplasm, and pores *(arrows)* can be seen in nuclear envelope. (×14,000; scale bar 1 μm.)

Courtesy G.E. Palade, The Rockefeller University, New York.

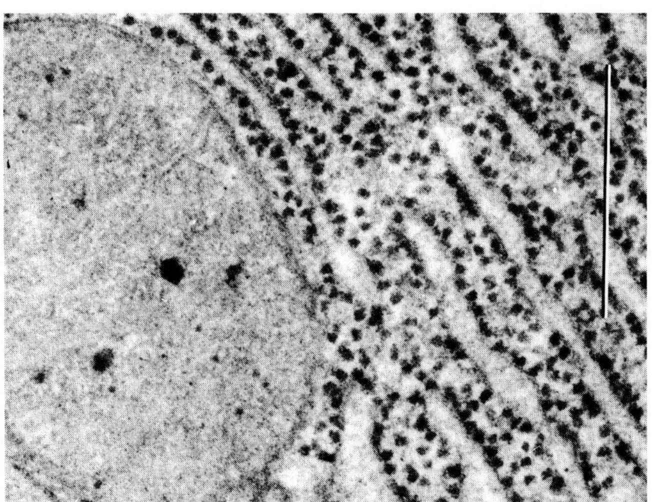

FIGURE 2-7

Electron micrograph of portion of pancreatic exocrine cell from guinea pig showing rough endoplasmic reticulum with ribosomes (small dark granules). Oval body *(left)* is mitochondrion. (×66,000; scale bar, 0.5 μm.)

Courtesy G.E. Palade.

invaders (Chapter 7) is based on proteins of this type. Some aggregations of protein molecules form pores through which small polar molecules may enter. Like the phospholipid molecules, most of the glycoproteins can move laterally in the membrane, although more slowly.

The nucleus is bounded by a double membrane of lipid bilayers, the **nuclear envelope** (Figure 2-6), which contains less cholesterol than the plasma membrane. The nucleus contains the **chromatin**, a complex of DNA, basic proteins called **histones**, and nonhistone protein. The chromatin carries the genetic information, that is, the code that results in most of the components characteristic of the cell after transcription and translation (see Chapter 3). **Nucleoli** are specialized parts of certain chromosomes that stain in a characteristically dark manner. They carry multiple copies of the DNA information to synthesize ribosomal RNA. After transcription from the DNA, the ribosomal RNA combines with protein to form a **ribosome**, detaches from the nucleolus, and passes to the cytoplasm through pores in the nuclear envelope.

The outer membrane of the nuclear envelope is continuous with extensive membranous elements in the cytoplasm called **endoplasmic reticulum (ER)** (Figures 2-4, 2-6, and 2-7). The space between the membranes of the nuclear envelope communicates with the channels **(cisternae)** in the ER. The ER is a complex of membranes that separates some of the products of the cell from the synthetic machinery that produces them, apparently functioning as routes for transport of proteins within the cell. The membranes of the ER may be lined on their outer surfaces with ribosomes and are thus designated **rough ER**, or they may lack the ribosomal lining and be called **smooth ER**. Smooth ER functions in the synthesis of lipids and phospholipids. Protein synthesized by the ribosomes on the rough ER enters the cisternae and from there is transported to the Golgi apparatus or complex. The **Golgi complex** (Figure 2-8) is composed of a stack of membranous vesicles that functions in storage, modification, and packaging of protein products, especially secretory products. The vesicles do not synthesize protein but may add complex carbohydrates to the molecules. Small vesicles of ER containing protein pinch off and then fuse with sacs on the "forming face" of the Golgi. After modification, the proteins bud off vesicles on the "maturing face" of the complex (Figure 2-8). The contents of some of these vesicles may be expelled to the outside of the cell, as secretory products destined to be exported from a glandular cell. Others may contain digestive enzymes that remain in the same cell that produces them. Such vesicles are called **lysosomes** (literally "loosening body," a body capable of causing lysis, or disintegration). The enzymes that they contain are involved in the breakdown of foreign material, including bac-

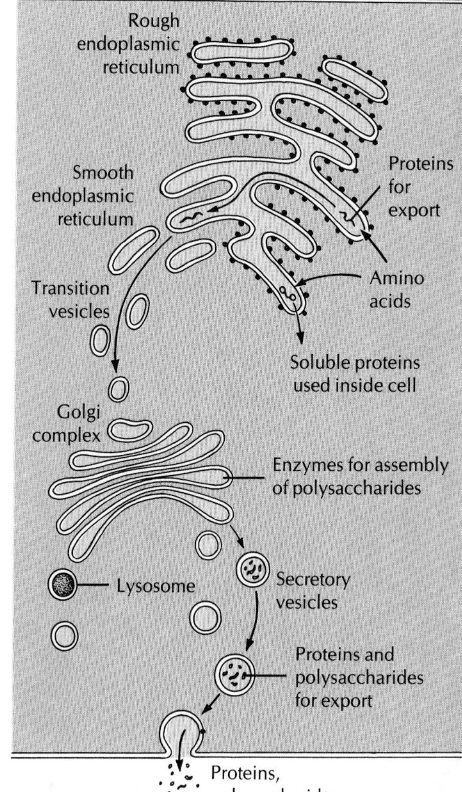

FIGURE 2-8

System for assembling, isolating, and secreting proteins for export from, and use inside, a eukaryotic cell.

teria engulfed by the cell. Lysosomes also are capable of breaking down injured or diseased cells and worn-out cellular components, since the enzymes they contain are so powerful that they kill the cell that formed them if the lysosome membrane ruptures. In normal cells the enzymes remain safely enclosed within the protective membrane.

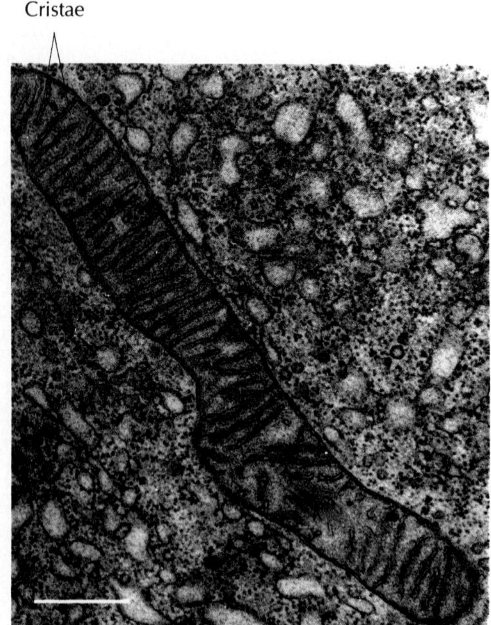

FIGURE 2-9

Electron micrograph of elongated mitochondrion in pancreatic exocrine cell of guinea pig. Many ribosomes and membrane-bound vesicles are present. (×30,000; scale bar, 0.5 μm.)

Courtesy G.E. Palade.

Mitochondria (Figure 2-9) (sing., **mitochondrion**) are conspicuous organelles present in nearly all eukaryotic cells. They are diverse in shape, size, and number; some are rodlike, and others are more or less spherical in shape. They may be scattered uniformly through the cytoplasm, or they may be localized near cell surfaces and other regions where there is unusual metabolic activity. The mitochondrion is composed of a double membrane. The outer membrane is smooth, whereas the inner membrane is folded into numerous platelike projections called **cristae** (Figures 2-4 and 2-9), which increase the internal surface area where chemical reactions take place as if in an assembly line. Mitochondria are often called "powerhouses of the cell" because enzymes located on the cristae carry out the energy-yielding steps of aerobic metabolism. ATP, the most important energy storage molecule of all cells, is produced in this organelle. Mitochondria are self-replicating. They have a tiny, circular chromosome, much like the chromosomes of prokaryotes except that it is much smaller. It contains DNA that specifies some, but not all, of the proteins of the mitochondrion. Mitochondria also have ribosomes and synthesize some of their own proteins.

Eukaryotic cells characteristically have a system of tubules and filaments that form the **cytoskeleton** (Figure 2-10). These provide support and maintain the form of the cell, and in many cells they provide a means of locomotion and translocation of organelles within the cell. **Microfilaments** are thin, linear structures, which electron microscopists first observed distinctly in muscle cells, where they are responsible for the ability of the cell to contract. They are made of a protein called **actin.** Several dozen other proteins are known that bind with actin and determine its configuration and behavior in particular cells. One of these is **myosin,** whose interaction with actin causes contraction in muscle and other cells. **Microtubules,** somewhat larger than microfilaments, are tubular structures composed of a protein called **tubulin.** They play a vital role in moving the chromosomes toward the daughter cells during cell division as will be seen later, and they are important in intracellular architecture, organization, and transport. In addition, microtubules form essential parts of the structures of cilia and flagella.

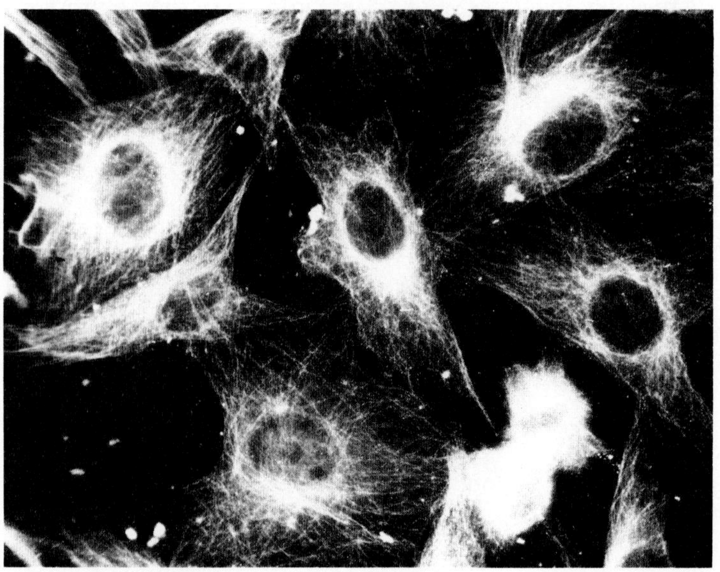

FIGURE 2-10

The microtubules in kidney cells of a baby hamster have been rendered visible by treating them with a preparation of fluorescent proteins that specifically bind to tubulin. The large dark bodies within the cells are the nuclei; the cell boundaries are not visible.

Photograph by K.G. Murti/Visuals Unlimited.

Microtubules radiate from a microtubule organizing center near the nucleus, sometimes called the **centrosome.** Within the centrosome are found a pair of **centrioles** (Figure 2-11), and the microtubules radiating from the centrioles form the **aster.** The centrioles are themselves composed of microtubules. Each centriole of a pair lies at right angles to the other and is a short cylinder of nine triplets of microtubules. They replicate before cell division. **Intermediate filaments** are larger than microfilaments but smaller than microtubules. There are five biochemically distinct types of intermediate filaments, and their composition and arrangement depend on the cell type in which they are found.

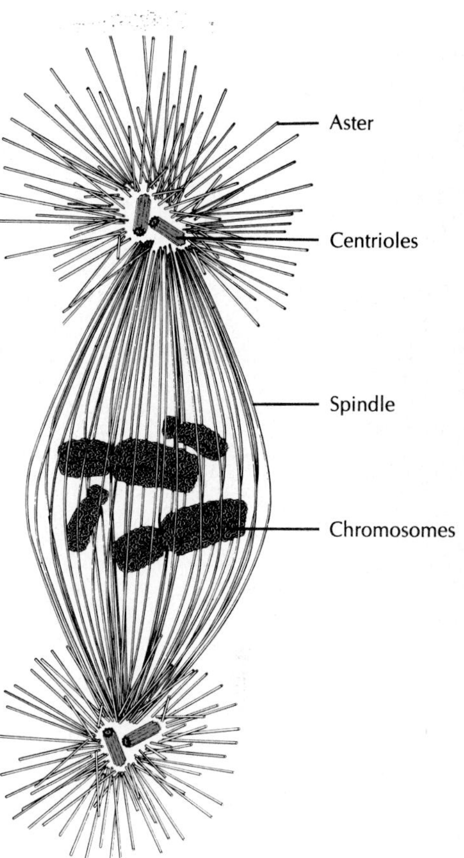

FIGURE 2-11

Fine structure of mitotic apparatus. At each pole of the spindles is a clear zone occupied by a pair of centrioles and surrounded by short microtubules. Other microtubules form spindle fibers, some of which extend from pole to pole while others attach to chromosomes.

From DuPraw, E.J. 1968. Cell and molecular biology. New York, Academic Press, Inc.

SURFACES OF CELLS AND THEIR SPECIALIZATIONS

The free surface of epithelial cells (cells that cover the surface of a structure or line a tube or cavity) sometimes bear either **cilia** or **flagella.** These are locomotory extensions of the cell surface that serve to sweep materials past the cell. In single-celled animals (many of the protozoa) and some primitive multicellular forms, they propel the entire animal through a liquid medium. Flagella provide the means of locomotion for the male reproductive cells of most animals and many plants.

Cilia occur in large numbers on each cell and are relatively short (5 to 10 μm). Flagella typically, although not always, occur singly or in a few numbers per cell and are long whiplike structures that may reach 150 μm in length. The distinction between cilia and flagella is that cilia propel water parallel to the surface to which the cilium is attached, whereas a flagellum propels water parallel to the main axis of the flagellum (Figure 6-15, page 164). Their internal structure is the same. With few exceptions, the internal structures of locomotory cilia and flagella are composed of a long cylinder of nine pairs of microtubules enclosing a central pair (see Figure 15-3, p. 351). At the base of each cilium or flagellum is found a **basal body (kinetosome),** which is identical in structure to a centriole. Basal bodies play an important role in forming these locomotory structures.

Cells covering the surface of a structure or cells packed together in a tissue may have specialized junctional complexes between them. Nearest the free surface, the two apposing cell membranes appear to fuse, forming a **tight junction** (Figure 2-12). Tight junctions function as seals to prevent the passage of molecules between cells from one side of a cell layer to another, because there is usually a space of about 20nm between the plasma membranes of adjacent cells. At various points small ellipsoid discs occur, just beneath the cell membrane in each cell. These appear to act as "spot-welds" and are called **desmosomes.** From each desmosome a tuft of intermediate filaments extends into the cytoplasm, and linker proteins extend through the plasma membrane into the intercellular space to bind the discs together. Desmosomes are not seals but appear to increase the strength of the tissue. **Gap junctions,** rather than serving as points of attachment, provide a means of intercellular communication. They have tiny canals between cells, so that the cytoplasm becomes continuous, and small molecules can pass from one cell to the other.

Another specialization of the cell surfaces is the interdigitations of confronted cell surfaces where the plasma membranes of the cells infold and interdigitate very much like a zipper. They are especially common in the epithelial cells of kidney tubules. The distal or apical boundaries of some epithelial cells, as seen with the electron microscope, show regularly arranged **microvilli.** They are small, fingerlike projections consisting of tubelike evaginations of the plasma membrane with a core of cytoplasm (Figure 2-12). They are seen clearly in the lining of the intestine where they greatly increase the absorptive and digestive surface. Such specializations appear as brush borders by the light microscope. The spaces between the microvilli are continuous with tubules of the ER, which may facilitate the movement of materials into the cells.

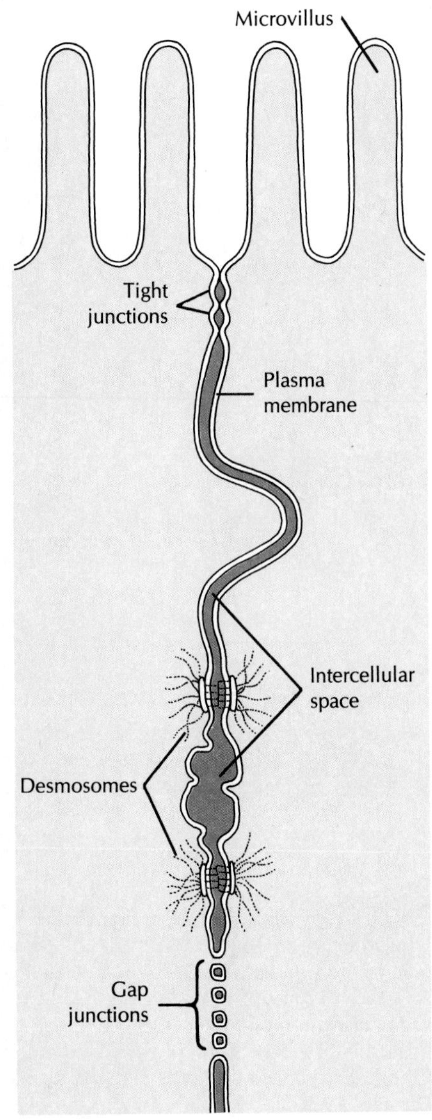

FIGURE 2-12

Two opposing plasma membranes forming the boundary between two epithelial cells. Various kinds of junctional complexes are found. The tight junction is a firm, adhesive band completely encircling the cell. Desmosomes are isolated "spot-welds" between cells. Gap junctions serve as sites of intercellular communication. Intercellular space may be greatly expanded in epithelial cells of some tissues.

MEMBRANE FUNCTION

The incredibly thin, yet sturdy, plasma membrane that encloses every cell is vitally important in maintaining cell integrity. Once believed to be a rather static entity that defined cell boundaries and kept cell contents from spilling out, the plasma membrane has proved to be a dynamic structure having remarkable activity and selectivity. It is a permeability barrier that separates the internal and external environment of the cell, regulates the vital flow of molecular traffic into and out of the cell, and provides many of the unique functional properties of specialized cells.

The fluid-mosaic model of membrane structure has already been described. The membrane appears to be remarkably restless and fluid, with proteins constantly moving and reorganizing their molecular configuration. Some of the proteins that extend through the membrane act as channels through which small molecules and ions such as sodium, chloride, and potassium are allowed to pass.

Membrane permeability

The plasma membrane acts as a gatekeeper for the entrance and exit of the many substances involved in cell metabolism. Some substances can pass through with ease, others enter slowly and with difficulty, and still others cannot enter at all. Because conditions outside the cell are different from and more variable than conditions within the cell, it is necessary that the passage of substances across the membrane be rigorously controlled.

We recognize three principal ways that a substance may traverse the cell membrane: (1) by **simple diffusion** along a concentration gradient; (2) by a **mediated-transport system,** in which the substance binds to a specific site that in some way assists it across the membrane; and (3) by **endocytosis,** in which the substance is enclosed within a vesicle that forms on and detaches from the surface, then enters the cell.

Simple diffusion and osmosis. A solution is a homogeneous phase consisting of a solvent and one or more solutes. The **solvent** is the component present in highest abundance, into which the other components, the **solutes,** are dissolved. If a membrane separates two solutions, one of which has more solute molecules than the other, a **concentration gradient** instantly exists between the solutions. More solute molecules strike the membrane from one side than from the other. Assuming the membrane is **permeable** to the solute, there is a net movement of solute toward the side having the lower concentration. The solute diffuses "downhill" across the membrane until concentrations on both sides are equal, and an equal number of solute particles are passing through the membrane in both directions.

Most cell membranes are **selectively permeable,** that is, permeable to water (the solvent) and permeable to some solutes but impermeable to others. In simple diffusion, it is this selectivity that regulates molecular traffic. As a rule, gases (such as oxygen and carbon dioxide), urea, and lipid-soluble solutes are the only solutes that can diffuse through biological membranes with any degree of freedom. This is because the lipid nature of membranes forms a natural barrier to most biologically important molecules that are not lipid soluble. Since many water-soluble molecules readily pass through membranes, such movements cannot be explained by simple diffusion. Instead sugars, as well as many electrolytes and macromolecules, are moved across membranes by carrier-mediated processes, described on pp. 37–39.

If a membrane is placed between two unequal concentrations of solutes to which the membrane is impermeable, water flows through the membrane from the more diluted (with respect to the solute) to the more concentrated solution. To understand why this happens, we must realize that the molecules of *water* are more concentrated on the side of the membrane where the solute molecules are less concentrated (more diluted). The water molecules move across the membrane down a concentration gradient from an area where the *water* molecules are more concentrated to an area on the other side of the membrane where they are less concentrated. This is **osmosis.**

Osmosis can be demonstrated by a simple experiment in which a selectively permeable membrane such as cellophane is tied tightly over the end of a funnel, making an **osmometer.** The funnel is filled with a sugar solution and placed in a beaker of pure water so that the water levels inside and outside the funnel are equal. In a short time the water level in the glass tube of the funnel rises, indicating that water is passing through the cellophane membrane into the sugar solution (Figure 2-13).

Inside the funnel are sugar molecules, as well as water molecules. In the beaker outside the funnel are only water molecules. Thus the concentration of water is less on the inside because some of the available space is occupied by the larger, nondiffusible sugar molecules. A concentration gradient is said to exist for water molecules in the system. Water diffuses from the region of greater concentration of water (pure water outside) to the region of lesser concentration (sugar solution inside).

As water enters the sugar solution, the fluid level in the funnel rises, creating a hydrostatic pressure because of gravity inside the osmometer. Eventually the hydrostatic pressure produced by the increasing weight of solution in the funnel pushes water molecules out as fast as they enter. The level in the funnel becomes stationary and the system is in equilibrium. The **osmotic pressure** of the solution is equal to the hydrostatic pressure necessary to prevent further net entry of water. Thus osmotic pressure is measured in units of hydrostatic pressure, for example, atmospheres or millimeters of mercury (mm Hg). Mammalian serum has an osmotic pressure of 7.75 atmospheres. If a cell is placed in a solution that has a higher osmotic pressure **(hyperosmotic, hypertonic),** the cell loses water and becomes small and wrinkled. If the cell is placed in a solution with a lower osmotic pressure **(hypoosmotic, hypotonic)** than that inside the cell, water enters the cell and it swells.

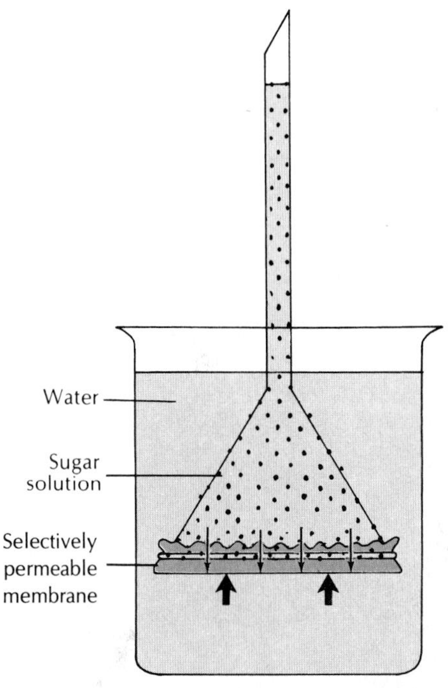

FIGURE 2-13

Simple membrane osmometer. The membrane covering the opening is permeable to water molecules but not sugar, hence water diffuses across the membrane *(large arrows)* from an area of higher concentration (in beaker) to an area of lower concentration (inside funnel). The fluid rises in the neck of the funnel until the hydrostatic pressure due to gravity causes a number of water molecules to pass out *(small arrows)* equal to those diffusing into the funnel. At that point, the osmotic pressure of the fluid in the funnel equals the hydrostatic pressure.

The concept of osmotic pressure is not without problems. A solution reveals an osmotic "pressure" only when it is separated from solvent (usually water in biology) by a selectively permeable membrane. It can be disconcerting to think of an isolated bottle of sugar solution or your blood serum as having osmotic "pressure" much as compressed gas in a bottle would have. Furthermore, the osmotic pressure is really the hydrostatic pressure that must be applied to a solution to keep it from gaining water. Consequently, biologists frequently use the term **osmotic potential** rather than osmotic pressure. However, since the term "osmotic pressure" is too firmly fixed in our vocabulary to be easily dislodged, it is necessary to understand the usage despite its potential confusion.

Mediated transport. We have seen that the cell membrane is an effective barrier to the free diffusion of most molecules of biological significance. Yet it is essential that such materials enter and leave the cell. Nutrients such as sugars and materials for growth such as amino acids must enter the cell, and the wastes of metabolism must leave. Such molecules are moved across the membrane by special mechanisms built into the structure of the membrane. This is called **mediated transport,** meaning that a specific transport mechanism mediates, or facilitates, transfer across the membrane barrier.

Experimental evidence suggests that mediated transport of sugars, amino acids, and other solutes involves the reversible combination with membrane proteins. Such proteins are called **carriers:** protein molecules positioned within the membrane and capable of shuttling from one membrane surface to the other. It is assumed that the carrier molecule captures a solute molecule to be transported, forming a solute-carrier complex. It moves or rotates to the opposite surface with its fare, where the solute detaches and leaves the membrane. The carrier moves back, again presenting its attachment site for the pickup and transport of another solute molecule. Protein carriers are usually quite specific, recognizing and transporting only a limited group of chemical substances or perhaps even a single substance.

At least two distinctly different kinds of carrier-mediated transport mechanisms are recognized: (1) **facilitated transport,** in which the carrier assists a molecule to diffuse through the membrane that it cannot otherwise penetrate, and without the expenditure of energy (ATP), and (2) **active transport,** in which energy (ATP) is supplied to the carrier systems to transport molecules in the direction opposite to the gradient (Figure 2-14). Facilitated transport therefore differs from active transport in that it sponsors movement in a downhill direction (in the direction of the concentration gradient) only and requires no metabolic energy to drive the carrier system.

In higher animals facilitated transport is important for the transport of glucose (blood sugar) into body cells that burn it as a principal energy source for the synthesis of ATP. The concentration of glucose is greater in the blood than in the cells that consume it, favoring inward diffusion, but glucose is a polar molecule that does not, by itself, penetrate the membrane rapidly enough to support the metabolism of many cells; the carrier system increases the inward flow of glucose.

In active transport, molecules are moved uphill against the forces of passive diffusion. Active transport always involves the expenditure of energy because materials are pumped against a concentration gradient. Among the most important active transport systems in all animals are those which maintain sodium and potassium gradients between cells and the surrounding extracellular fluid or external environment. Most animal cells require a high internal concentration of potassium for protein synthesis at the ribosome and for certain enzymatic func-

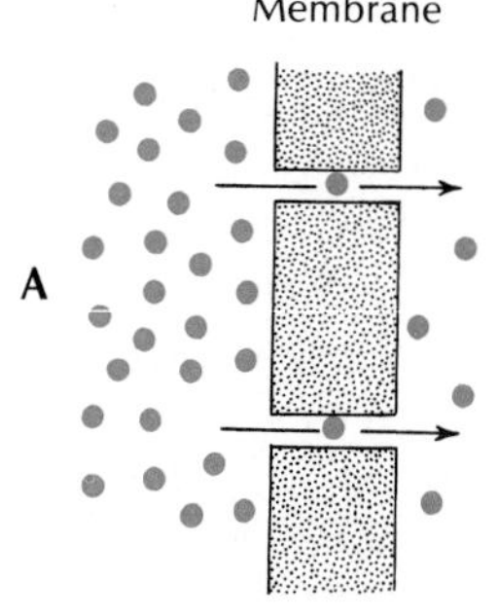

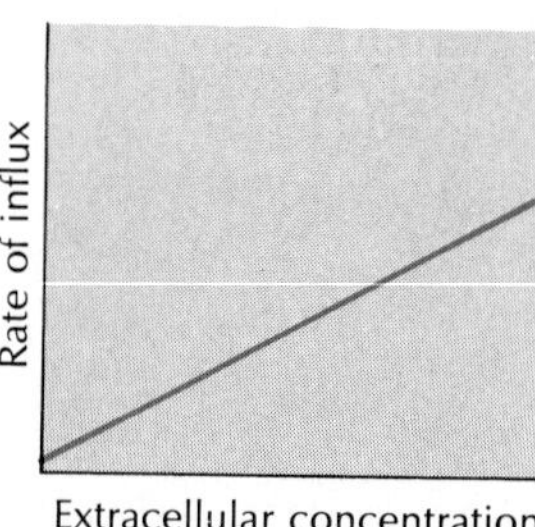

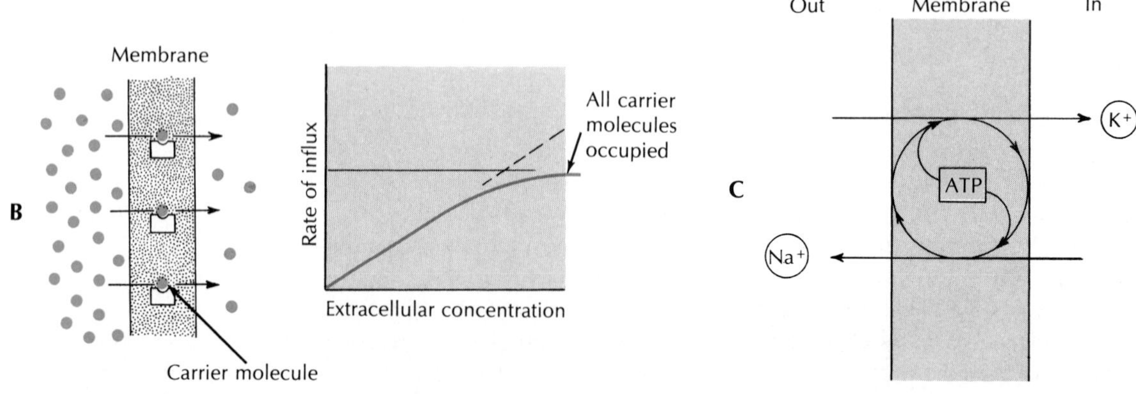

FIGURE 2-14

A, Diffusion through membrane channels. **B,** Carrier-mediated transport. Simple diffusion through membrane channels differs from carrier-mediated transport in that the latter shows saturation at high concentration. Simple diffusion does not exhibit saturation at high concentration. **C,** Sodium-potassium exchange pump, powered by bond energy of ATP, maintains the normal gradients of these ions across the cell membrane. Sodium is more concentrated outside the cell, and potassium is more concentrated inside.

tions. The potassium concentration may be 20 to 50 times greater inside the cell than outside. Sodium, on the other hand, may be 10 times more concentrated outside the cell than inside. Both of these electrolyte gradients are maintained by the active transport of potassium into and sodium out of the cell. It is known that in many cells the outward pumping of sodium is linked to the inward pumping of potassium; the same carrier molecule is used for both. As much as 10% to 40% of all the energy produced by the cell is used by the **sodium-potassium exchange pump.**

Endocytosis. The ingestion of solid or fluid material by cells was observed by microscopists nearly 100 years before phrases like "passive diffusion" and "mediated transport" were a part of the biologist's vocabulary. Endocytosis is a collective term that describes the processes of **phagocytosis, pinocytosis,** and **receptor-mediated endocytosis.**

Phagocytosis, which literally means "cell eating," is a common method of feeding among the protozoa and lower animals. It is also the way in which white blood cells (leukocytes) engulf cellular debris and uninvited microbes in the blood. By phagocytosis, the cell membrane forms a pocket that engulfs the solid material. The membrane-enclosed vesicle then detaches from the cell surface and moves into the cytoplasm where its contents are digested by intracellular enzymes (Figure 9-6, p. 225).

Pinocytosis, or "cell drinking," is similar to phagocytosis except that drops of fluid are sucked discontinuously through tubular channels into cells to form tiny vesicles (Figure 2-15). These may combine to form larger vacuoles. When a cell engages in pinocytosis, it brings within itself whatever molecules are contained in the fluid droplets. Thus it is nonspecific.

In contrast, receptor-mediated endocytosis is highly specific. Proteins of the plasma membrane specifically bind particular molecules (Figure 2-15), which may be present in the extracellular fluid in very low concentrations. The area of the cell surface that contains the receptors and bound particles then invaginates and is brought within the cell as in phagocytosis and pinocytosis.

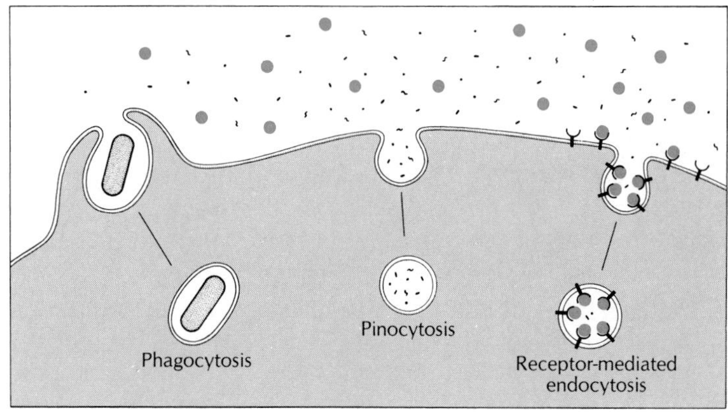

FIGURE 2-15

Three types of endocytosis. In phagocytosis the cell membrane binds to a large particle and extends to engulf it. In pinocytosis a small area of cell membrane invaginates to finally close off a droplet of surrounding medium. Receptor-mediated endocytosis is a mechanism for highly selective uptake of large molecules or particles. Binding of the ligand to the specific receptor on the surface membrane stimulates invagination of the receptor-ligand complex.

Cell Division

All cells of the body arise from the division of preexisting cells. All the cells found in most multicellular organisms have originated from the division of a single cell, the **zygote,** which is formed from the union (fertilization) of an **egg** and a **sperm.** Cell division provides the basis for one form of growth, both sexual and asexual reproduction, and transmission of hereditary qualities from one cell generation to another cell generation.

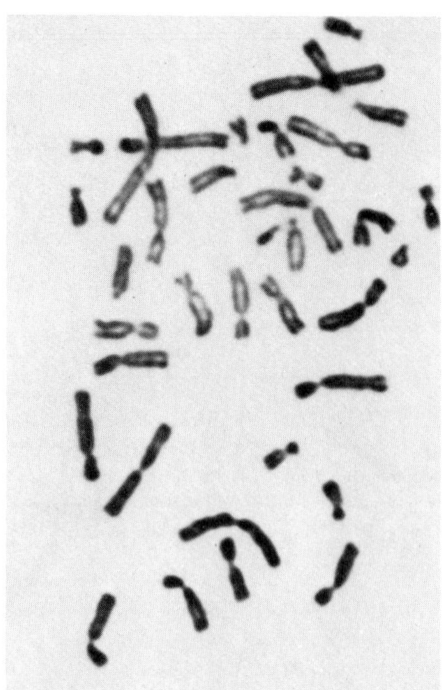

FIGURE 2-16

Human chromosomes.
Courtesy L.R. Emmons.

In the formation of **body cells (somatic cells)** for growth, maintenance, and repair, the nuclear division is referred to as **mitosis.** The cell divisions that occur in the formation of **germ cells** (ova and sperm) are called **meiotic divisions;** meiosis is described in Chapter 3.

STRUCTURE OF CHROMOSOMES

As mentioned earlier, DNA in the eukaryotic cell occurs in chromatin, a complex of DNA with histone and nonhistone protein. In fact, the chromatin is organized into a number of discrete bodies called **chromosomes** (''color bodies''), so named because they stain deeply with certain biological dyes when the chromosomes are condensed. In cells that are not dividing, the chromatin is loosely organized and dispersed, so that the individual chromosomes cannot be distinguished. Prior to division, the chromatin condenses, and the chromosomes can be recognized and their individual morphological characteristics determined. They are of varied lengths and shapes, some bent and some rodlike (Figure 2-16). Their number is constant for the species, and every body cell (but not the germ cells) has the same number of chormosomes regardless of the cell's function. A human being, for example, has 46 chromosomes in each body (somatic) cell.

During mitosis (nuclear division) the chromosomes shorten and become increasingly condensed and distinct, and each assumes a characteristic shape. At some point on the chromosome is a **centromere,** or constriction, to which are attached several spindle fibers that pull the chromosome toward the pole during mitosis.

Chromosomes always occur in pairs in somatic cells, or two of each kind. Of each pair, one comes from one parent and the other from the other parent. Thus there are 23 pairs in the human species. Each pair usually has certain characteristics of shape and form that aid in identification. A biparental organism begins with the union of two gametes, each of which furnishes a **haploid** set of chromosomes (23 in humans) to produce a somatic or **diploid** number of chromosomes (46 in humans). The chromosomes of a haploid set are also called a **genome.** Thus a fertilized egg consists of a paternal genome (chromosomes contributed by the father) and a maternal genome (chromosomes contributed by the mother). (The term ''genome'' is also sometimes used for all the chromosomes in a diploid nucleus.)

How is DNA packaged in chromosomes? Packaging is a formidable problem because, by some estimates, the DNA in a human cell is nearly 4 m in length, yet is packed into 46 chromosomes having an aggregate length of only 200 μm. The DNA must obviously be coiled or folded up in some way to fit into a small space. Yet the packaging must be arranged so that the genetic instructions are accessible for reading during the transcription process (the formation of messenger RNA from nuclear DNA, described in Chapter 3).

In 1974 it was discovered that the chromatin is composed of repeating subunits, called **nucleosomes** (Figure 2-17). Each nucleosome is a narrow ''spool'' or disc of histone proteins around which two turns of double-helical DNA are wound to form a superhelix (Figure 2-18). The nucleosomes are linked together by the continuous DNA strand much like beads on a string. This arrangement is believed to explain the knobby appearance of chromatin fibers as revealed by high-resolution electron micrographs.

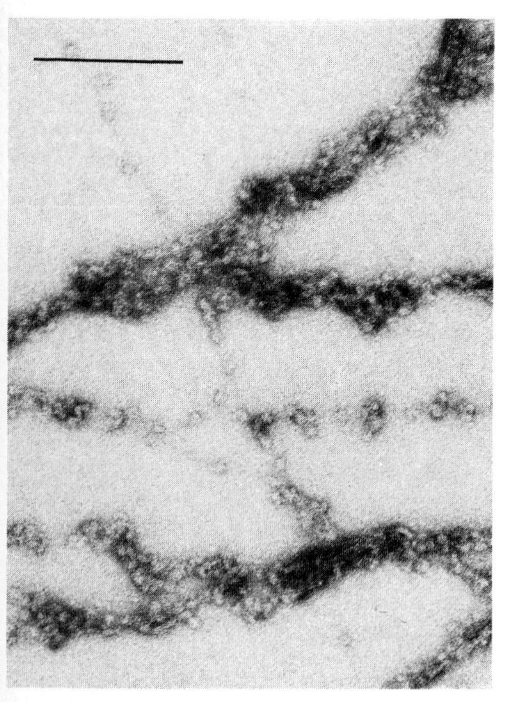

FIGURE 2-17

Chromatin prepared from chicken red blood cells. Some strands are stretched and demonstrate individual nucleosomes. (Original magnification ×100,000; scale bar, 0.2 μm.)
Courtesy C.L.F. Woodcock.

STAGES IN MITOSIS

There are two distinct phases of cell division: the division of the nuclear chromosomes **(mitosis)** and the division of the cytoplasm **(cytokinesis).** Mitosis, the division of the nucleus (that is, chromosomal segregation), is certainly the most obvious and complex part of cell division and that of greatest interest to the cytologist. Cytokinesis normally immediately follows mitosis, although there are occasions when the nucleus may divide a number of times without a corresponding division of the cytoplasm. In such a case the resulting mass of protoplasm containing many nuclei is referred to as a **multinucleate cell.** An example is the giant resorptive cell type of bone (osteoclast) that may contain 15 to 20 nuclei. Sometimes a multinucleate mass is formed by cell fusion rather than nuclear proliferation. This arrangement is called a **syncytium.** An example is vertebrate skeletal muscle, which is composed of multinucleate fibers that have arisen by the fusion of numerous embryonic cells.

The process of mitosis is arbitrarily divided for convenience into four successive stages or phases, although one stage merges into the next without sharp lines of transition. These phases are prophase, metaphase, anaphase, and telophase (Figure 2-19). When the cell is not actively dividing, it is in the "resting" stage, or **interphase.** However, the cell is not really "resting" at this stage because its metabolism is active and because the DNA content of the nucleus is being duplicated. Thus, when the cell begins "active" mitosis, it already has a double set of chromosomes.

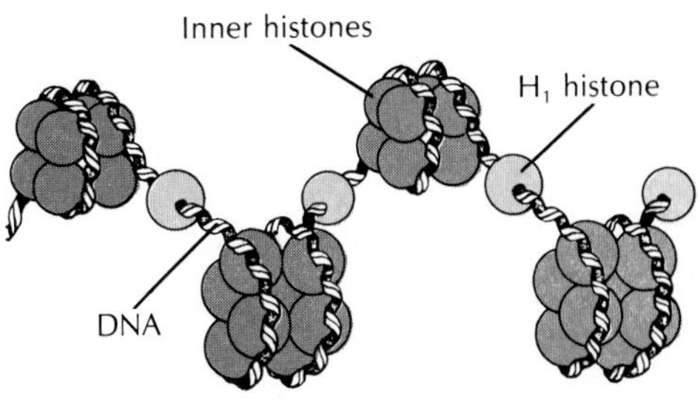

FIGURE 2-18

Current model of chromatin showing nucleosome organization. The DNA molecule winds one and three fourths turns around a group of eight histone molecules, two each of four different kinds. A fifth class of histones (H_1) is found in the spacer region.

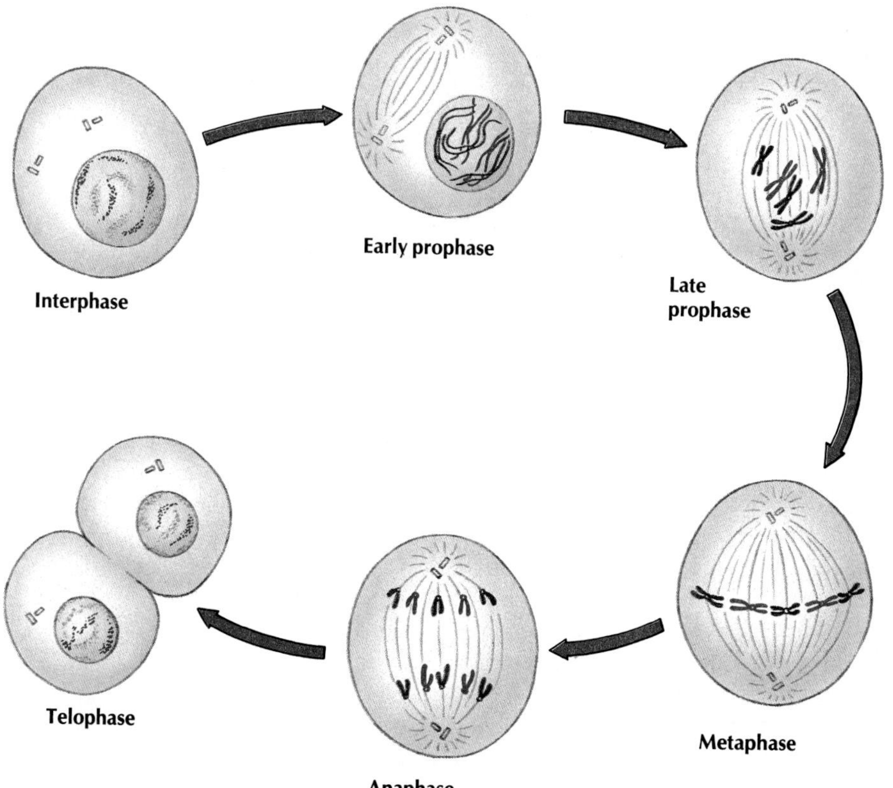

FIGURE 2-19

Stages of mitosis, showing division of a cell with two pairs of chromosomes. One chromosome of each pair is shown in red.

Prophase

Before the beginning of prophase, the centrioles replicate, and each pair of centrioles migrates toward opposite sides of the nucleus. At the same time, fine fibers (microtubules) known as **spindle fibers** appear between the two pairs of centrioles to form a football-shaped **spindle,** so named because of its resemblance to nineteenth-century wooden spindles used to twist thread together in spinning. Other microtubules radiate outward from each centriole to form **asters.** The entire structure is the **mitotic apparatus,** and it increases in size and prominence as the centrioles move farther apart (Figures 2-11 and 2-20).

FIGURE 2-20

Stages of mitosis in whitefish.

Courtesy of General Biological Supply House, Inc., Chicago.

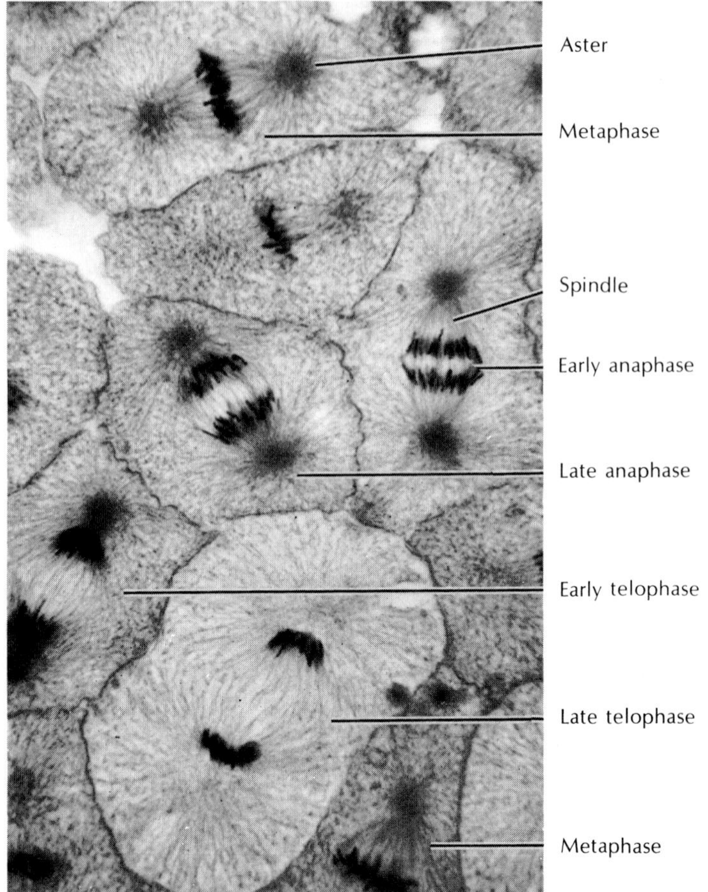

At the same time, the diffuse nuclear chromatin condenses to form visible chromosomes. These actually consist of two identical sister **chromatids** formed during a period of interphase. The sister chromatids are joined together at their **centromere,** which is the position on the chromsome where the spindle fiber attaches. During prophase the nuclear envelope fragments into small vesicles and disappears.

Metaphase

During metaphase the condensed sister chromatids rapidly migrate to the middle of the nuclear region to form a **metaphasic plate** (Figures 2-19 and 2-20). The centromeres line up precisely on the plate with the arms of the chromatids trailing off randomly in various directions. Spindle fibers attach to each centromere at early metaphase.

Anaphase

Microtubules, such as spindle fibers, consist of strands of tubulin subunits. One way that the spindle fibers could pull the chromosomes toward the poles is by losing tubulin subunits at the poles, dragging the chromosomes by their centromeres.

The two chromatids of each double chromosome thicken and separate. The single centromere that has held the two chromatids together now splits so that two independent chromosomes, each with its own centromere, are formed. The chromosomes part more, appearing to be pulled by the spindle fibers attached to the centromeres. The arms of each chromosome trail along behind as though the chromosome were being dragged through a resisting medium. However, the fibers maintain the same thickness throughout anaphase, neither thickening nor thinning as the chromosomes separate.

Telophase

When the daughter chromosomes reach their respective poles, telophase has begun. The daughter chromosomes are crowded together and stain intensely with histological stains. The spindle fibers disappear and the chromosomes lose their identity, reverting to the diffuse chromatin network characteristic of the interphase nucleus. The small vesicles of nuclear membrane begin to rejoin and reassemble the nuclear envelope prior to chromosome decondensation.

CYTOKINESIS: CELL DIVISION

During the final stages of nuclear division a cleavage furrow appears on the surface of the dividing cell and encircles it at the midline of the spindle. The cleavage furrow deepens and pinches the plasma membrane as though it were being tightened by an invisible rubber band. Microfilaments are present in the area of constriction, and actin and myosin filaments slide together to shorten the microfilaments in the same manner that muscle cells contract (p. 166). Finally, the infolding edges of the plasma membrane meet and fuse, completing cell division.

CELL CYCLE

The complete sequence of events resulting in mitosis, the actual cell division, and the events that follow are called the cell cycle. A cell prepares to divide before the actual division occurs. A human cell in tissue culture completes a cycle every 18 to 22 hours, yet division occupies only approximately 1 hour of this period.

The most important preparation—replication of DNA—occurs during the latter part of interphase, termed the S period (period of synthesis) (Figure 2-21). In humans, each chromosome contains approximately 175 million nucleotide pairs arranged in a double helix that makes one turn for every 10 nucleotide pairs. Accordingly, there are approximately 17.5 million turns in the DNA in each chromosome, all of which must somehow replicate and untangle during the S period.

The S period is preceded and succeeded by G_1 and G_2 periods, respectively (*G* stands for gap), when no DNA synthesis is occurring. It is believed that enzymes and substrates are being prepared during the G_1 period for the DNA replication that follows. During G_2, spindle and aster proteins are being synthesized. The energy demands of the cell are high during the G_2 period, and cellular metabolism continues throughout the cell cycle.

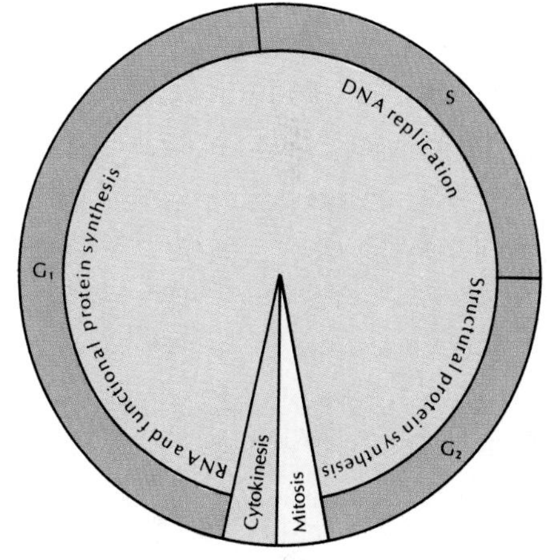

FIGURE 2-21

Cell cycle, showing relative duration of recognized periods. *S*, G_1, and G_2 are periods within interphase: *S*, synthesis of DNA; G_1, presynthetic period; G_2, postsynthetic period. Actual duration of the cycle and the different periods vary considerably in different cell types.

Cellular Metabolism

All cells must obtain energy, synthesize their own internal structure, control much of their own activity, and guard their boundaries. **Cellular metabolism** refers to the collective total of chemical processes that occur within living cells to accomplish these activities. Although the enormous number of reactions in their aggregate are extremely complex, the central metabolic routes through which matter and energy are channeled are not difficult to understand.

ENERGY AND LIFE

Energy is the capacity for performing work. It cannot be created or destroyed, but it can be changed in form. One form is potential energy, or energy of position, as opposed to kinetic energy, or energy of motion. Energy in the bonds of organic molecules is a kind of potential energy. Energy is required to form the bonds, and energy is released when they are broken, but until they are broken, the bond energy remains potential.

In living organisms the cells gradually liberate the chemical potential energy in organic molecules and couple the release to a variety of energy-consuming

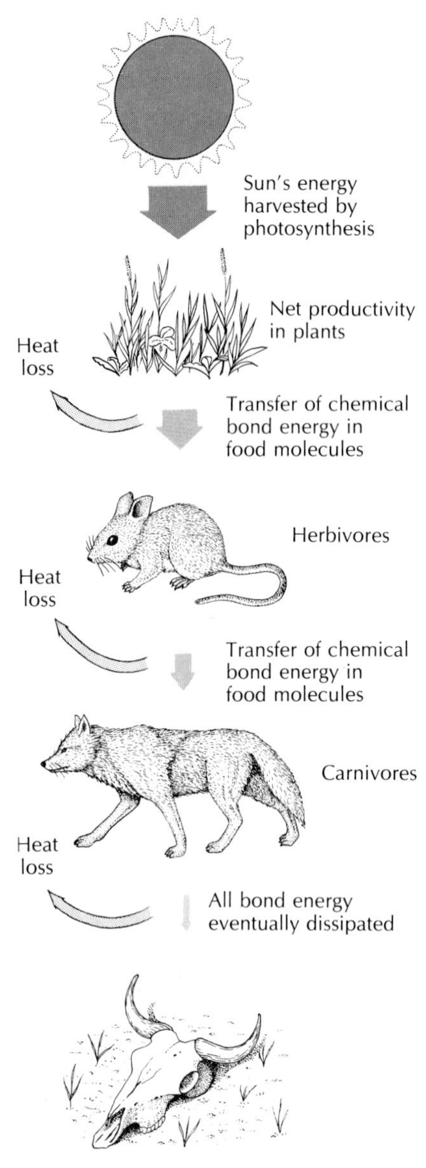

FIGURE 2-22

Solar energy sustains virtually all life on earth. With each energy transfer, however, much energy is lost as heat.

processes. As living organisms convert stored energy into other forms, and eventually into heat, less and less remains in reserve. How is it replenished? The ultimate source of energy for life on earth is the sun. The energy of sunlight is captured by green plants, which transform a portion of this energy into chemical bond energy (food energy). Herbivorous animals eat plants and convert a small part of the potential chemical energy into animal tissue and a large part into heat that is invisibly dissipated into space. Carnivorous animals eat the herbivores, and they in turn are eaten by other carnivores. At each step a large part of the energy is lost as heat (Figure 2-22).

At the cellular level, animals and plants use chemical energy stored in food to carry out hundreds of activities. Thus animals are totally dependent on plants, which fortunately accumulate enough energy to sustain both themselves and the animals that feed on them. Plant photosynthesis and respiration of all organisms are inextricably woven together in a vast cycle that is driven by the constant flow of energy from the sun.

Free energy

To describe the energy changes that take place in chemical reactions, biochemists use the concept of **free energy.** Free energy is simply the energy in a system available for doing work. In a molecule, free energy equals the energy present in chemical bonds minus the energy that cannot be used. The majority of reactions in cells release free energy and are said to be **exergonic** (Gr. *ex,* out, + *ergon,* work). Such reactions are spontaneous and always proceed "downhill" since free energy is lost from the system. Thus,

However, many important reactions in cells require the addition of free energy and are said to be **endergonic** (Gr. *endon,* within, + *ergon,* work). Such reactions have to "pushed uphill" because they end up with more energy than they started with:

C D → CD

Reactants Energy input **Products containing more free energy than reactants**

As we will see in a later section, ATP is the ubiquitous, energy-rich intermediate used by organisms to power important uphill reactions such as those required for active transport and cellular synthesis.

ROLE OF ENZYMES

Enzymes and activation energy

For any reaction to occur, even exergonic ones that tend to proceed spontaneously, the chemical bonds must first be destabilized. For example, if the reaction involves splitting a covalent bond, the atoms forming the bond must first be stretched apart to make them less stable. Some energy, termed the **activation energy**, must be supplied before the bond will be stressed enough to break. Only then will there be an overall loss of free energy and the formation of reaction products. Activation energy can be likened to the energy needed to push a cart over the crest of a hill before it will roll spontaneously down the other side, liberating its potential energy as the cart descends.

One way to activate chemical reactants is to raise the temperature. By increasing the rate of molecular collisions and pushing chemical bonds apart, heat can provide the activation energy to make a reaction proceed. But metabolic reactions must occur at biologically tolerable temperatures, temperatures that are too

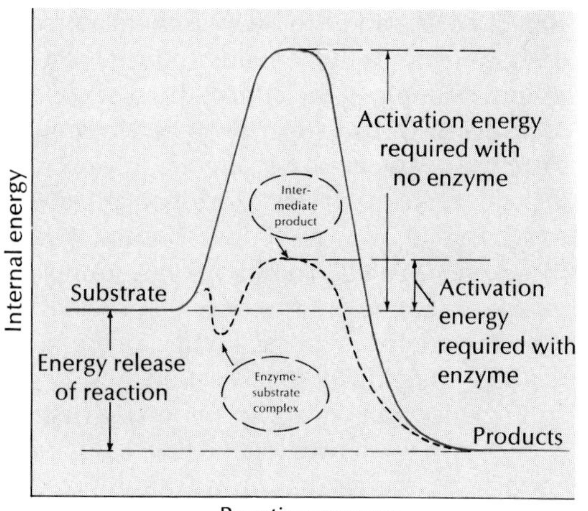

FIGURE 2-23

Energy changes during enzyme catalysis of a substrate. The overall reaction proceeds with a net release of energy. In the absence of an enzyme, substrate is stable because of the large amount of activation energy needed to disrupt strong chemical bonds. The enzyme reduces the energy barrier by forming a chemical intermediate with a much lower internal energy state.

low to allow reactions to proceed beyond extremely slow rates. Instead, living systems employ a different strategy to increase reaction rates: **catalysts.**

Catalysts are chemical substances that increase reaction rates without affecting the products of the reaction and without being altered or destroyed as a result of the reaction. A catalyst cannot make an energetically impossible reaction happen; it simply accelerates a reaction that would have proceeded at a very slow rate otherwise.

Enzymes are the catalysts of the living world; they reduce the amount of activation energy required for a reaction. In effect, an enzyme steers the reaction through one or more intermediate steps, each of which requires much less activation energy than that required for a single-step reaction (Fig. 2-23). Enzymes affect only the reaction rate. They do not in any way alter the free energy change of a reaction, nor do they change the proportions of reactants and products in a reaction.

Nature of enzymes

Enzymes are complex molecules varying in size from small, simple proteins with a molecular weight of 10,000 to highly complex molecules with molecular weights up to 1 million. Many enzymes are pure proteins—delicately folded and interlinked chains of amino acids. Other enzymes require the participation of small nonprotein groups called **cofactors** to perform their enzymatic function. In some cases these cofactors are metallic ions (such as ions of iron, copper, zinc, magnesium, potassium, and calcium) that form a functional part of the enzyme. Examples are carbonic anhydrase, which contains zinc; the cytochromes, which contain iron; and troponin (a muscle contraction enzyme), which contains calcium. Another class of cofactors, called **coenzymes,** is organic. All coenzymes contain groups derived from vitamins, compounds that must be supplied in the diet. All of the B complex vitamins are coenzyme compounds. Since animals have lost the ability to synthesize the vitamin components of coenzymes, it is obvious that a vitamin deficiency can be serious. However, unlike dietary fuels and nutrients that must be replaced, once burned or assembled into structural materials, vitamins are recovered in their original form and used repeatedly. Examples of coenzymes that contain vitamins are nicotinamide adenine dinucleotide (NAD), which contains nicotinic acid; coenzyme A, which contains pantothenic acid; and flavin adenine dinucleotide (FAD), which contains riboflavin.

Action of enzymes

An enzyme functions by combining in a highly specific way with its **substrate,** the molecule on which it acts. The enzyme bears an active site located within a

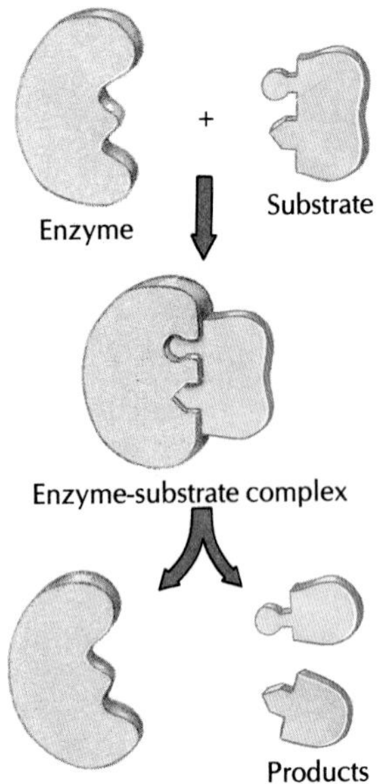

FIGURE 2-24

Interaction of substrate and enzyme. The enzyme changes shape when it binds the substrate, becoming complementary in shape only after the substrate is bound.

cleft or pocket and contains a unique molecular configuration. The active site has a flexible surface that enfolds and conforms to the substrate (Figure 2-24). This is a good example of the critical importance of the secondary and tertiary structure of proteins, those properties that determine their shapes, because if the tertiary structure of an enzyme is altered, its catalytic properties are destroyed. The binding of enzyme to substrate forms an **enzyme-substrate complex (ES complex)** in which the substrate is secured by weak bonds to several points in the active site of the enzyme. The ES complex is not strong and will quickly dissociate, but during this fleeting moment the enzyme provides a unique chemical environment that stresses certain chemical bonds in the substrate so that much less energy is required to complete the reaction.

Enzymes that engage in important main-line sequences—such as the crucial energy-providing reactions of the cell that go on constantly—seem to operate in enzyme sets rather than in isolation. For example, the conversion of glucose to carbon dioxide and water proceeds through 19 reactions, each requiring a specific enzyme. Main-line enzymes are found in relatively high concentrations in the cell, and they may implement quite complex and highly integrated enzymatic sequences. One enzyme carries out one step, then passes the product to another enzyme, which catalyzes another step, and so on. The reactions may be said to be coupled.

Specificity of enzymes

One of the most distinctive attributes of enzymes is their high specificity. This is a consequence of the exact molecular fit that is required between enzymes and substrates. Furthermore, an enzyme catalyzes only one reaction; unlike reactions carried out in the organic chemist's laboratory, no side reactions or by-products result. Specificity of both substrate and reaction is obviously essential to prevent a cell from being overwhelmed with useless by-products.

However, there is some variation in degree of specificity. Some enzymes, such as succinic dehydrogenase, will catalyze the oxidation (dehydrogenation) of one substrate only, succinic acid. Others, such as proteases (for example, pepsin and trypsin), will act on almost any protein, but each protease has its own particular point of attack in the protein. Usually an enzyme will take on one substrate molecule at a time, catalyze its chemical change, release the product, and then repeat the process with another substrate molecule. The enzyme may repeat this process billions of times until it is finally worn out (a few hours to several years) and is broken down by scavenger enzymes in the cell. Some enzymes undergo successive catalytic cycles at dizzying speeds of up to a million cycles per minute; most operate at slower rates.

Enzyme-catalyzed reactions

Enzyme-catalyzed reactions are reversible. This is signified by the double arrows between substrate and products.

$$\text{Fumaric acid} + H_2O \rightleftharpoons \text{Malic acid}$$

However, for various reasons the reactions catalyzed by most enzymes tend to go in one direction. For example, the proteolytic enzyme pepsin can degrade proteins into amino acids, but it cannot accelerate the rebuilding of amino acids into any significant amount of protein. The same is true of most enzymes that catalyze the hydrolysis of large molecules such as nucleic acids, polysaccharides, lipids, and proteins. There is usually one set of reactions and enzymes that break them down, but they must be resynthesized by a different set of reactions that are catalyzed by different enzymes. This apparent irreversibility exists because the chemical equilibrium usually favors the formation of the smaller degradation products.

The net **direction** of any chemical reaction depends on the relative energy contents of the substances involved. If there is little change in the chemical-bond energy of the substrate and products, the reaction is more easily reversible. However, if large quantities of energy are released as the reaction proceeds in one direction, more energy must be provided in some way to drive the reaction in the reverse direction. Thus many enzyme-catalyzed reactions are in practice irreversible, unless the reaction is coupled to another that makes energy available. In the cell, both reversible and irreversible reactions are combined in complex ways to make possible both synthesis and degradation.

CELLULAR ENERGY TRANSFER

Unlike the combusion of fuel in a fire, which is an explosive event with rapid release of energy as heat, metabolic oxidations are flameless and of low temperature. They proceed gradually, and the energy liberated is coupled to a great variety of energy-consuming reactions. Although metabolic energy exchanges proceed with impressive efficiency, heat is inevitably liberated. Heat can be put to some use, of course; the endothermic vertebrates (birds and mammals) use it to elevate and maintain a constant internal body temperature. But for the most part, heat is a useless commodity to a cell, since it is a nonspecific form of energy that cannot be captured and redistributed to power metabolic processes. There is actually only one way in which the oxidative release of energy is made available for use by cells: it is coupled to the production of high-energy phosphate bonds, usually in the form of **ATP (adenosine triphosphate)** by addition of inorganic phosphate **to ADP (adenosine diphosphate)** (Figure 2-25).

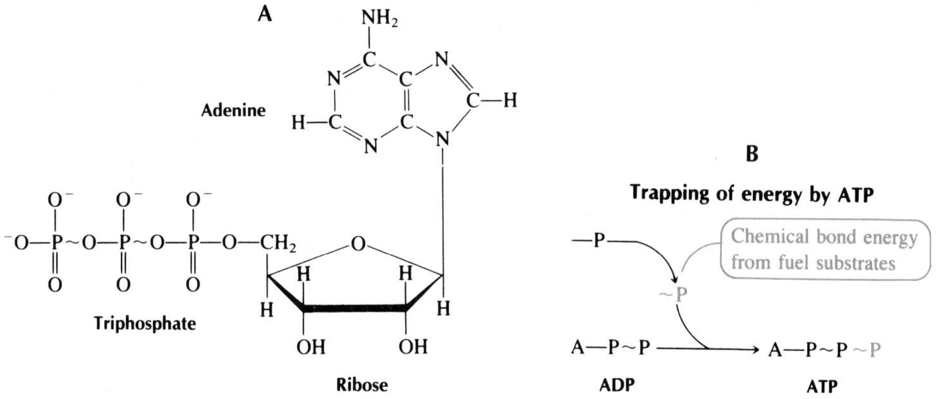

A

Adenine

NH_2

Triphosphate

Ribose

B

Trapping of energy by ATP

—P

Chemical bond energy from fuel substrates

~P

A—P~P → A—P~P~P

ADP ATP

FIGURE 2-25

A, Structure of ATP. **B,** ATP formation from ADP.

The ATP molecule consists of a purine (adenine), a 5-carbon sugar (ribose), and 3 molecules of phosphoric acid linked together by two pyrophosphate bonds to form a triphosphate group. The pyrophosphate bonds are called **high-energy bonds** because they are repositories of a great deal of chemical energy. This energy has been transferred to ATP from that released in breaking other bonds in the respiratory process. Respiration, by the stepwise oxidation of fuel substrates, redistributes bond energies so that a few high-energy bonds are created and stored in ATP. Obviously this energy is gained at the expense of fuel energy; the end products of cellular respiration, carbon dioxide and water, contain much less bond energy than do the fuel substrates (for example, glucose) that entered the oxidative pathway.

The high-energy pyrophosphate bonds of ADP and ATP are designated by the "tilde" symbol ~. Thus a low-energy phosphate bond is shown as —P, and a high-energy one as ~ P. ADP can be represented as A—P ~ P and ATP as A—P ~ P ~ P.

The amount of ATP produced in respiration depends on its rate of use. In other words, ATP is produced by one set of reactions and immediately consumed by another. It is not stored; rather fuel in the form of carbohydrates and fats is stored. ATP is formed as it is needed, primarily by oxidative processes in the mitochondria. Oxygen is not consumed unless ADP and phosphate molecules are available, and these do not become available until ATP is hydrolyzed by some energy-consuming process. *Energy metabolism is therefore mostly self-regulating.*

CELLULAR RESPIRATION

How electron transport is used to trap chemical bond energy

Having seen that ATP is the one common energy denominator by which all cellular machines are powered, we are in a position to ask how this energy is captured from fuel substrates. This question directs us to an important generalization: *all cells obtain their chemical energy requirements from oxidation-reduction reactions.* This means simply that in the degradation of fuel molecules, hydrogen atoms (electrons and protons) are passed from electron donors to electron acceptors with a release of energy. A portion of this energy is trapped and used to form the high-energy bonds in ATP. The release of energy during electron transfer and its conservation as ATP is the mainspring of cell activity and was a crucial evolutionary achievement.

Because they are so important, let us review what we mean by oxidation-reduction ("redox") reactions. In these reactions there is a transfer of electrons from an electron donor (the reducing agent) to an electron acceptor (the oxidizing agent). As soon as the electron donor loses its electrons, it becomes oxidized. As soon as the electron acceptor accepts electrons, it becomes reduced. In other words, a reducing agent becomes oxidized when it reduces another compound, and an oxidizing agent becomes reduced when it oxidizes another compound. Thus for every oxidation there must be a corresponding reduction.

In an oxidation-reduction reaction the electron donor and electron acceptor form a redox pair:

$$\underset{\text{(reducing agent)}}{\text{Electron donor}} \rightleftharpoons e^- + \underset{\text{(oxidizing agent)}}{\text{Electron acceptor}} + \text{Energy}$$

When electrons are accepted by the oxidizing agent, energy is liberated because the electrons move to a more stable position. In the cell, the electrons flow through a series of carriers. Each carrier is reduced by accepting electrons and then is reoxidized by passing electrons to the next carrier in the series. By transferring electrons stepwise in this manner, energy is gradually released, and a maximum yield of ATP is realized.

Aerobic versus anaerobic metabolism

Ultimately, the electrons are transferred to a **final electron acceptor**. The nature of this final acceptor is the key that determines the overall efficiency of cellular metabolism. The heterotrophs can be divided into two great groups: **aerobes**, those which use molecular oxygen as the final electron acceptor, and **anaerobes**, those which employ some other molecule as the final electron acceptor.

As we have discussed (Chapter 1), life originated in the absence of oxygen, and the abundance of atmospheric oxygen was produced after photosynthetic organisms evolved. Some strictly anaerobic organisms still exist and indeed play some important roles in specialized habitats. However, evolution has favored aerobic metabolism, not only because oxygen became available, but also because aerobic metabolism is vastly more efficient than anaerobic metabolism. In the absence of oxygen, only a very small fraction of the bond energy present in foodstuffs can be released. For example, when an anaerobic microorganism degrades glucose, the final electron acceptor (such as pyruvic acid) still contains most of the energy of the original glucose molecule. On the other hand, an aerobic organism, using oxygen as the final electron acceptor, can completely oxidize glucose to carbon dioxide and water. Almost 20 times as much energy is released when glucose is completely oxidized as when it is degraded only to the stage of lactic acid. An obvious advantage of aerobic metabolism is that a much smaller quantity of foodstuffs is required to maintain a given rate of metabolism.

General description of respiration

Aerobic metabolism is more familiarly known as true **cellular respiration**, defined as the oxidation of fuel molecules with molecular oxygen as the final electron acceptor. As mentioned previously, the oxidation of fuel molecules describes the *removal of electrons* and *not* the direct combination of molecular oxygen with fuel molecules. Let us look at this process in general before considering it in more detail.

Hans Krebs, the British biochemist who contributed so much to our understanding of respiration, described three stages in the complete oxidation of fuel molecules to carbon dioxide and water (Figure 2-26). In stage I, foodstuffs passing through the intestinal tract are digested into small molecules that can be absorbed into the circulation. There is no useful energy yield during digestion, which is discussed in Chapter 9. In stage II, most of the degraded foodstuffs are converted into two 3-carbon units (pyruvic acid). This occurs in the cytoplasm.

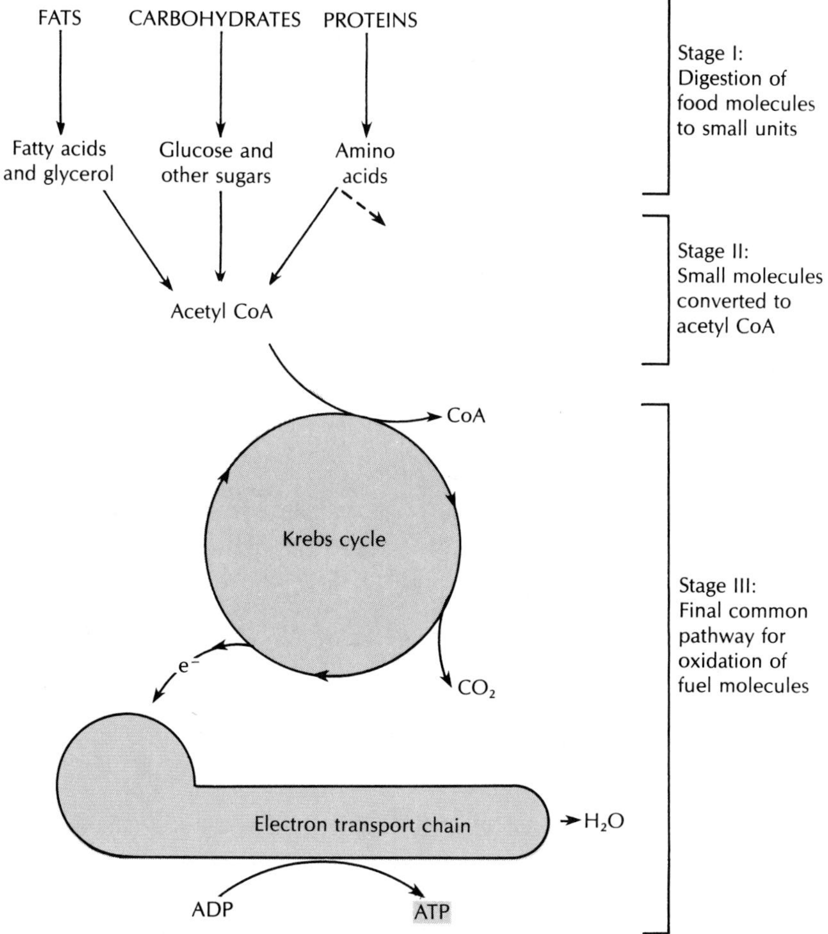

FIGURE 2-26

Overview of cellular respiration, showing the three stages in the complete oxidation of food molecules to carbon dioxide and water.

The pyruvic acid molecules then enter the mitochondria, where in another reaction they join with a coenzyme (coenzyme A) to form acetyl-CoA. Some ATP is generated in stage II, but the yield is small compared with that obtained in the final stage of respiration. In stage III the final oxidation of fuel molecules occurs, with a large yield of ATP. This stage takes place entirely in the mitochondria. Acetyl coenzyme A is channeled into the citric acid cycle where the acetyl group is completely oxidized to carbon dioxide. Electrons released from the acetyl groups are transferred to special carriers that pass them to electron acceptor compounds in the electron transport chain. At the end of the chain the electrons (and the protons accompanying them) are accepted by molecular oxygen to form water.

Glycolysis

We begin our journey through the stages of respiration with glycolysis, a nearly universal pathway in living organisms that converts glucose into pyruvic acid. In a series of reactions, glucose and other 6-carbon monosaccharides are split into 3-carbon fragments, **pyruvic acid** (Figure 2-27). A single oxidation occurs during glycolysis, and each molecule of glucose yields 2 molecules of ATP. In this pathway the carbohydrate molecule is phosphorylated twice by ATP, first to glucose-6-phosphate (not shown in Figure 2-27) and then to fructose-1,6-diphosphate. The fuel has now been "primed" with phosphate groups in these preparatory reactions and is sufficiently reactive to enable subsequent reactions to proceed. This is a kind of deficit financing that is required for an ultimate energy return many times greater than the original energy investment.

Fructose-1,6-diphosphate is next cleaved into two 3-carbon sugars, which undergo an oxidation, with the electrons being accepted by nicotinamide adenine dinucleotide (NAD), a derivative of the vitamin niacin. NAD serves as a carrier molecule to convey high-energy electrons to the final electron transport chain, where ATP will be produced.

The two 3-carbon sugars next undergo several reactions, ending with the formation of 2 molecules of pyruvic acid (Figure 2-27). In two of these steps, a molecule of ATP is produced. In other words, each 3-carbon sugar yields 2 ATP

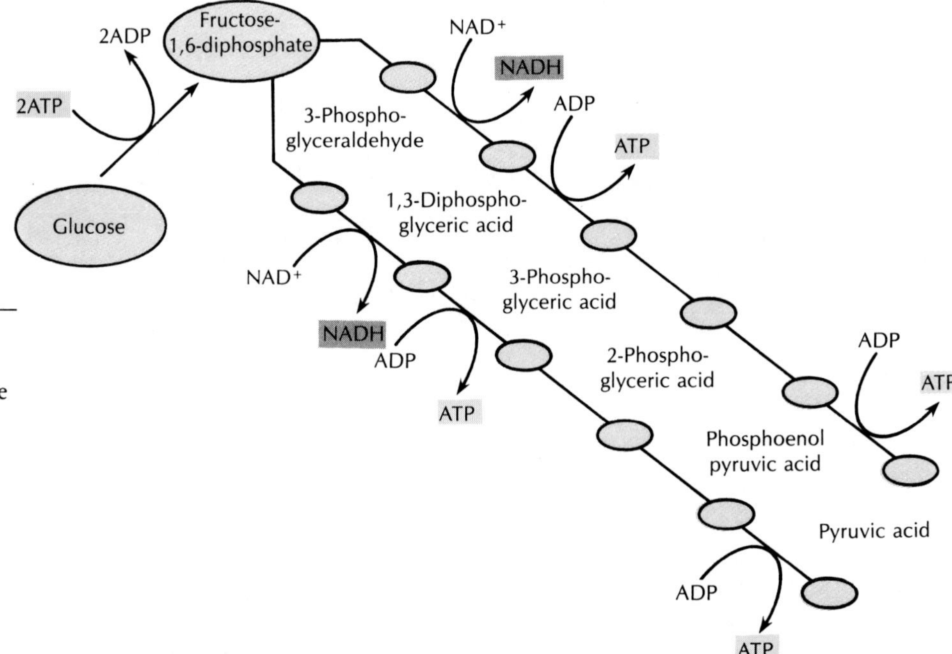

FIGURE 2-27

Glycolysis. Glucose is phosphorylated in two steps and raised to a higher energy level. High-energy fructose-1,6-diphosphate is split into triose phosphates that are oxidized exergonically to pyruvic acid, yielding ATP and NADH.

molecules, and since there are two 3-carbon sugars, 4 ATP molecules are generated. Recalling that 2 ATP molecules were used to prime the glucose initially, the net yield up to this point is 2 ATP molecules.

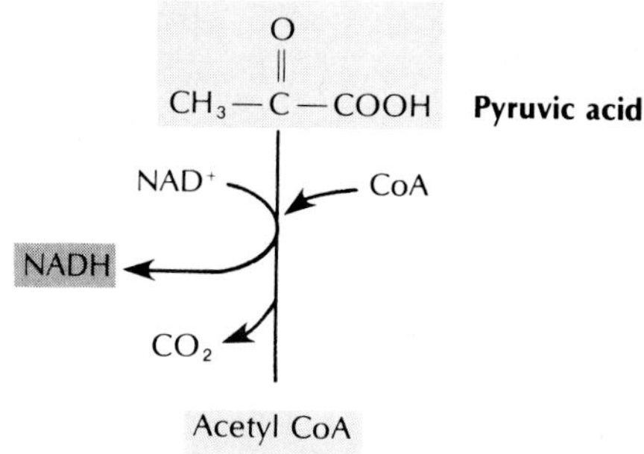

FIGURE 2-28

Formation of acetyl coenzyme A from pyruvic acid.

Acetyl coenzyme A: strategic intermediate in respiration

In aerobic metabolism the 2 molecules of pyruvic acid formed during glycolysis enter the mitochondrion. There, each molecule is oxidized, and one of the carbons is released as carbon dioxide (Figure 2-28). The 2-carbon residue condenses with **coenzyme A** to form **acetyl coenzyme A** (acetyl-CoA).

Acetyl coenzyme A is a critically important compound. Its oxidation in the citric acid cycle (below) provides energized electrons to generate ATP, and it is a crucial intermediate in lipid metabolism (p. 55).

Citric acid cycle: oxidation of acetyl coenzyme A

The degradation (oxidation) of the 2-carbon acetyl group of acetyl coenzyme A occurs in a cyclic sequence called the **citric acid cycle** (also called the tricarboxylic acid cycle [TCA cycle] and Krebs cycle) (Figure 2-29). The acetyl coenzyme A condenses with a 4-carbon acid (oxaloacetic acid), releasing the coenzyme A to react again with pyruvic acid. Through a series of reactions the 2 carbons from the acetyl group are released as carbon dioxide, and the oxaloacetic acid is regenerated. Electrons in the oxidations are transferred to NAD and to FAD (flavine adenine dinucleotide, another electron acceptor), and a pyrophosphate bond is generated in the form of guanosine triphosphate (GTP). This high-energy phosphate is readily transferred to ADP to form ATP.

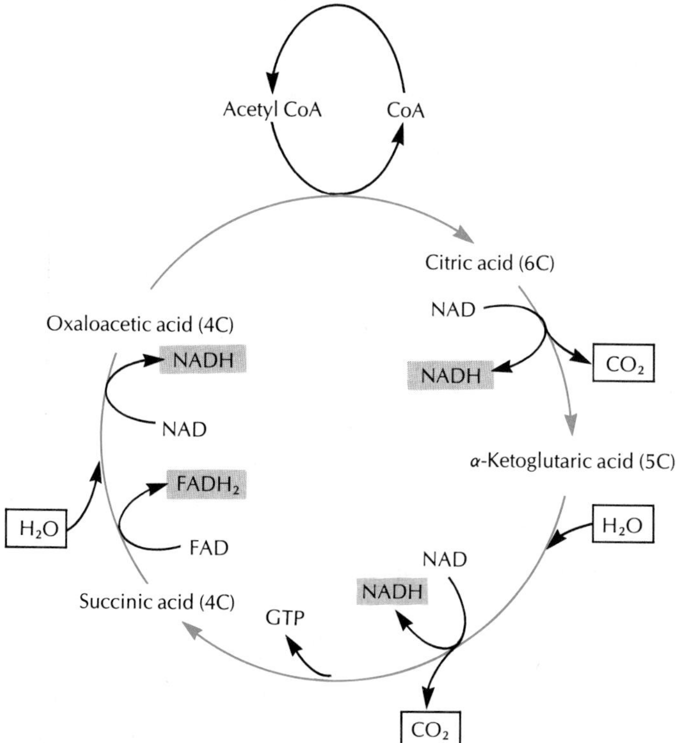

FIGURE 2-29

Krebs cycle in outline form showing the production of reduced NAD, reduced FAD, guanosine triphosphate (GTP), and carbon dioxide.

Electron transport chain

The transfer of electrons from reduced NAD and FAD to the final electron acceptor, molecular oxygen, is accomplished in an elaborate electron transport chain (Figure 2-30). The function of this chain is to permit the controlled release of free energy to drive the synthesis of ATP. At three points along the chain, ATP production occurs by the phosphorylation of ADP. This method of energy capture

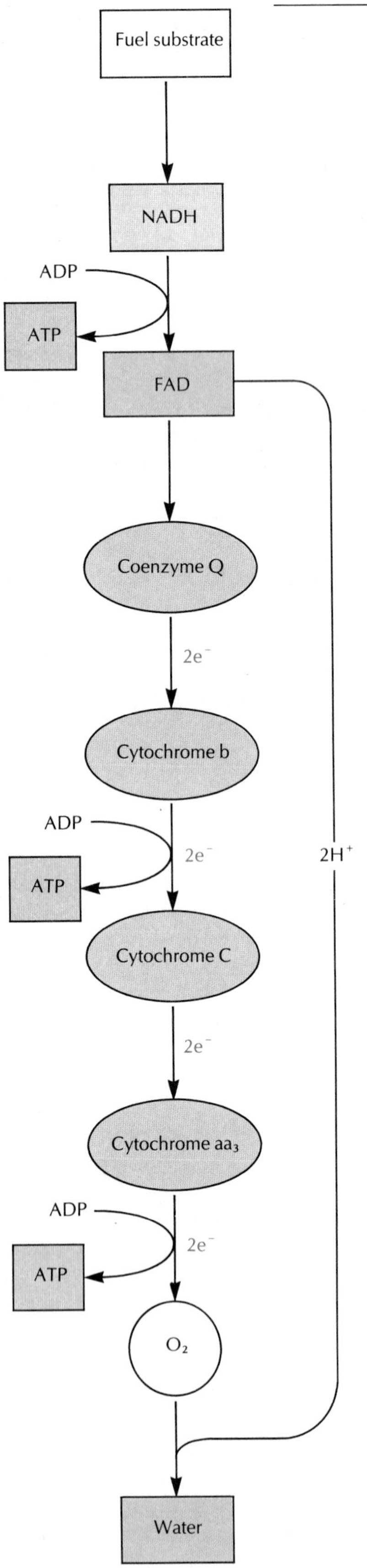

Fuel substrate
NADH
ADP
ATP
FAD
Coenzyme Q
$2e^-$
Cytochrome b
ADP
$2e^-$
ATP
$2H^+$
Cytochrome C
$2e^-$
Cytochrome aa_3
ADP
$2e^-$
ATP
O_2
Water

FIGURE 2-30

Electron transport chain. Electrons are transferred from one carrier to the next, terminating with molecular oxygen to form water. A carrier is reduced by accepting electrons and then is reoxidized by donating electrons to the next carrier. ATP is generated at three points in the chain. These electron carriers are located in the inner membrane of mitochondria.

is called oxidative phosphorylation because the formation of high-energy phosphate is coupled to oxygen consumption, and this depends on the demand for ATP by other metabolic activities within the cell. The actual *mechanism* of ATP formation by oxidative phosphorylation is not known with certainty. The most widely accepted explanation is the chemiosmotic coupling model (Hinkle and McCarty, 1978).

Oxidative phosphorylation, a complex process, would be unable to function efficiently, if at all, were the enzymes just floating freely in the cytoplasm of the cell. There is now abundant evidence that the oxidative enzymes and electron carriers are arranged in a highly ordered state on the inner membranes of the mitochondria.

The outer membrane of the mitochondrion is a smooth sac enclosing the inner membrane. On the cristae of the inner membrane there are enormous numbers of minute, stalked particles called submitochondrial particles (Figure 2-31). Within the inner membrane are the electron carriers of the respiratory chain.

Electrons are transported from NADH to oxygen through a series of electron carriers (Figure 2-30). The 2 electrons from NADH pass through an electron cascade, a series of discrete steps that permits the gradual harvesting of electron energy. By this means, the oxidation of one NADH yields 3 ATP molecules. The reduced FAD from the citric acid cycle enters the electron transport chain at a lower level and so yields only 2 ATP molecules.

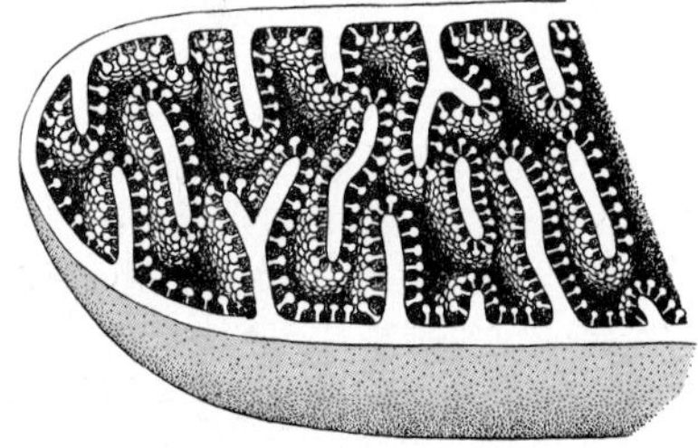

FIGURE 2-31

Representation of a section of mitochondrion as seen through a high-resolution electron microscope, showing the submitochondrial particles. The density of the particles is actually many times greater than depicted.

Efficiency of oxidative phosphorylation

We are now in a position to calculate the ATP yield from the complete oxidation of glucose (Figure 2-32). The overall reaction is:

$$\text{Glucose} + 2\ \text{ATP} + 36\ \text{ADP} + 36\ \text{P} + 6\ O_2 \rightarrow 6\ CO_2 + 2\ \text{ADP} + 36\ \text{ATP} + 6\ H_2O$$

ATP has been generated at several points along the way (Table 2-2). The cytoplasmic NADH generated in glycolysis requires a molecule of ATP to fuel transport of each molecule of NADH into the mitochondrion; therefore, each NADH from glycolysis results in only 2 ATP (total of 4), compared to the 3 ATP per NADH (total of 6) formed within the mitochondria. Accounting for the 2 ATP used in the priming reactions in glycolysis, the net yield may be as high as 36 molecules of ATP per molecule of glucose. The overall efficiency of aerobic oxidation of glucose is about 38%, comparing very favorably with human-designed energy conversion systems, which seldom exceed 5% to 10% efficiency.

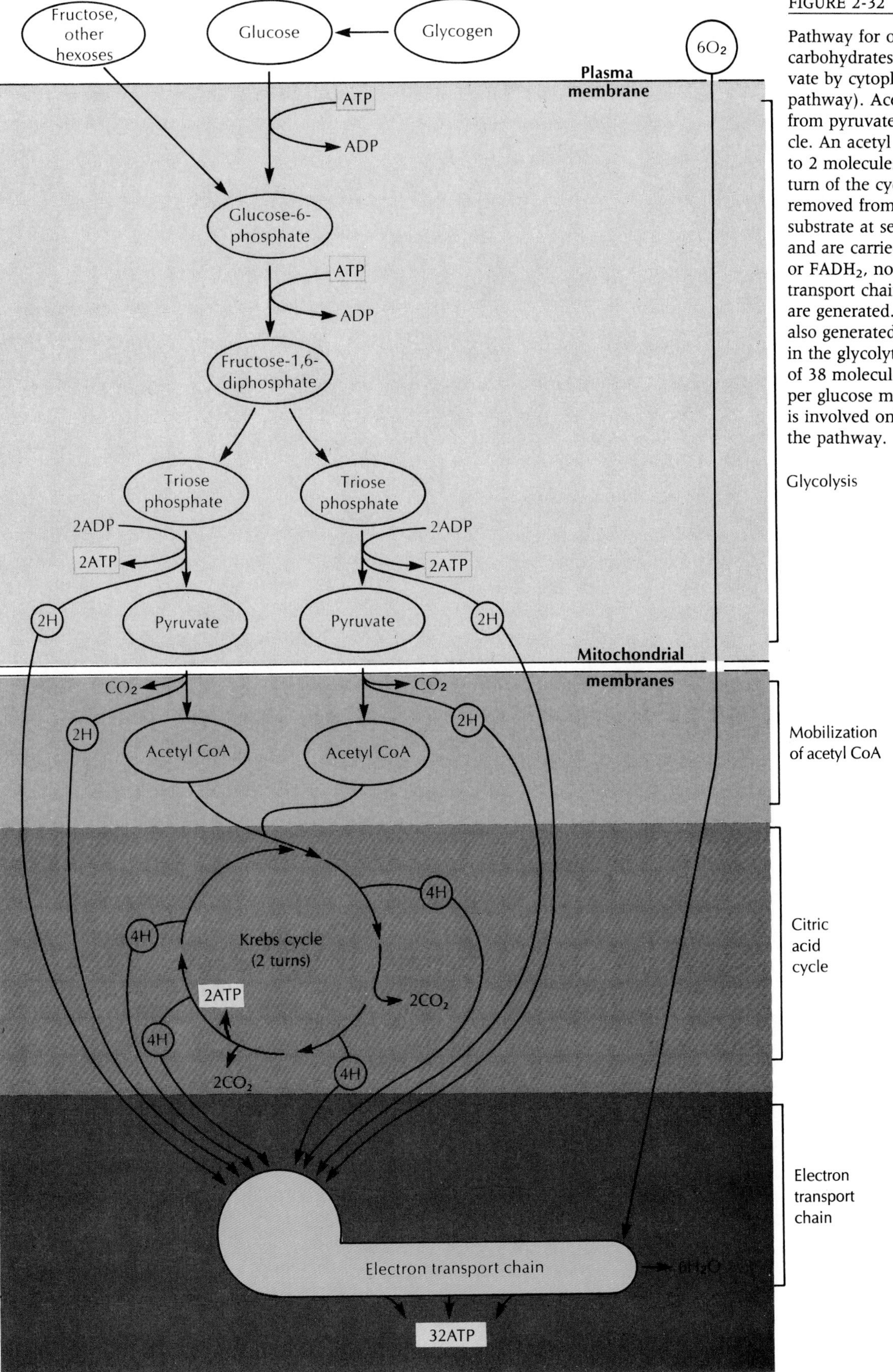

FIGURE 2-32

Pathway for oxidation of glucose and other carbohydrates. Glucose is degraded to pyruvate by cytoplasmic enzymes (glycolytic pathway). Acetyl coenzyme A is formed from pyruvate and is fed into the Krebs cycle. An acetyl group (2 carbons) is oxidized to 2 molecules of carbon dioxide with each turn of the cycle. Pairs of electrons (2H) are removed from the carbon skeleton of the substrate at several points in the pathway and are carried by oxidizing agents (NADH or $FADH_2$, not shown) to the electron transport chain where 32 molecules of ATP are generated. Four molecules of ATP are also generated by substrate phosphorylation in the glycolytic pathway, yielding a total of 38 molecules of ATP (36 molecules net) per glucose molecule. Molecular oxygen is involved only at the very end of the pathway.

Table 2-2 Calculation of total ATP molecules generated in respiration

ATP generated	Source
4	Directly in glycolysis
2	As GTP in citric acid cycle
4	From NADH in glycolysis
6	From NADH produced in pyruvic acid to acetyl coenzyme A reaction
4	From reduced FAD in citric acid cycle
18	From NADH produced in citric acid cycle
38 Total	
−2	Used in priming reactions in glycolysis
36 Net	

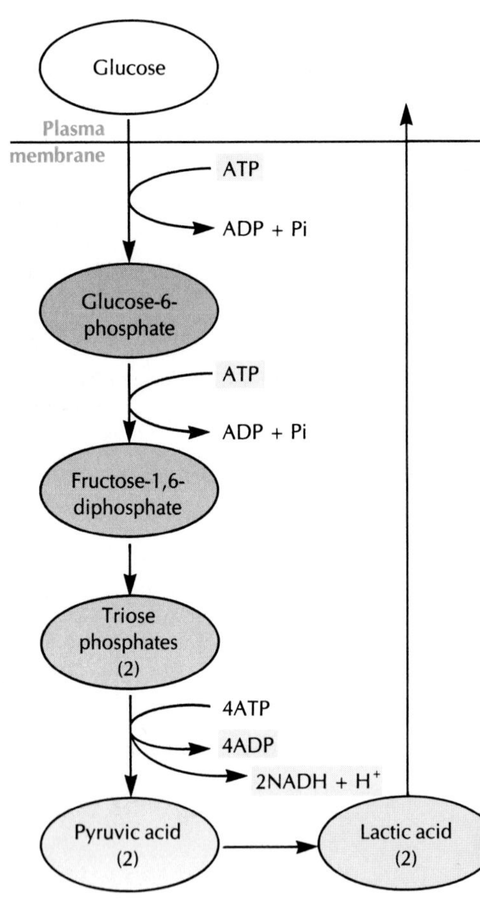

FIGURE 2-33

Anaerobic glycolysis, a process that proceeds in the absence of oxygen. Glucose is broken down to 2 molecules of pyruvic acid, generating 4 molecules of ATP and yielding 2, since 2 molecules of ATP are used to produce fructose-1,6-diphosphate. Pyruvic acid, the final electron acceptor for the hydrogen atoms and electrons released during pyruvic acid formation, is converted to lactic acid.

Anaerobic glycolysis: generating ATP without oxygen

Up to this point we have been describing aerobic cellular respiration. We will now consider how animals generate ATP without oxygen, that is, anaerobically.

Under anaerobic conditions, glucose and other 6-carbon sugars are first broken down stepwise to a pair of 3-carbon pyruvic acid molecules, yielding 2 molecules of ATP and 4 atoms of hydrogen (4 reducing equivalents, represented by 2 NADH + H^+). But in the absence of molecular oxygen, further oxidation of pyruvic acid cannot occur because the citric acid cycle and electron transport system cannot operate and cannot, therefore, provide a mechanism for reoxidation of the NADH produced in glycolysis. The problem is neatly solved in most animal cells by reducing pyruvic acid to lactic acid (Figure 2-33). Pyruvic acid becomes the final electron acceptor and lactic acid the end product of anaerobic glycolysis. In **alcoholic fermentation** (as in yeast, for example) the steps are identical to glycolysis down to pyruvic acid. One of its carbons is then released as carbon dioxide, and the resulting 2-carbon compound is reduced to ethanol, thus regenerating the NAD.

Anaerobic glycolysis is only one-eighteenth as efficient as the complete oxidation of glucose to carbon dioxide and water, but its key virtue is that it provides *some* high-energy phosphate in situations where oxygen is absent or in short supply. Many microorganisms live in places where oxygen is severely depleted, such as waterlogged soil, in mud of lake or sea bottom, or within a decaying carcass. Vertebrate skeletal muscle may rely heavily on glycolysis during short bursts of activity when contraction is too rapid and too powerful to be sustained by oxidative phosphorylation. The lactic acid that accumulates in the muscle diffuses out into the blood and is carried to the liver where it is metabolized.

Some animals rely heavily on anaerobic glycolysis during normal activities. For example, diving birds and mammals fall back on glycolysis almost entirely to give them the energy needed to sustain a long dive. Salmon would never reach their spawning grounds were it not for anaerobic glycolysis providing almost all the ATP used in the powerful muscular bursts needed to carry them up rapids and falls. Many parasitic animals have dispensed with oxidative phosphorylation entirely. They secrete relatively reduced end products of their energy metabolism, such as succinic acid, acetic acid, and propionic acid. These compounds are produced in mitochondrial reactions that derive several more molecules of ATP than does the cycle from glycolysis to lactic acid, although these sequences are still far less efficient than the classical electron transport system.

METABOLISM OF LIPIDS

The first step in the breakdown of a triglyceride is the hydrolysis of glycerol from the 3 fatty acid molecules (Figure 2-34). Glycerol, a 3-carbon carbohydrate, is phosphorylated and enters the glycolytic pathway.

The remainder of the triglyceride molecule consists of fatty acids. One of the abundant naturally occurring fatty acids is **stearic acid.**

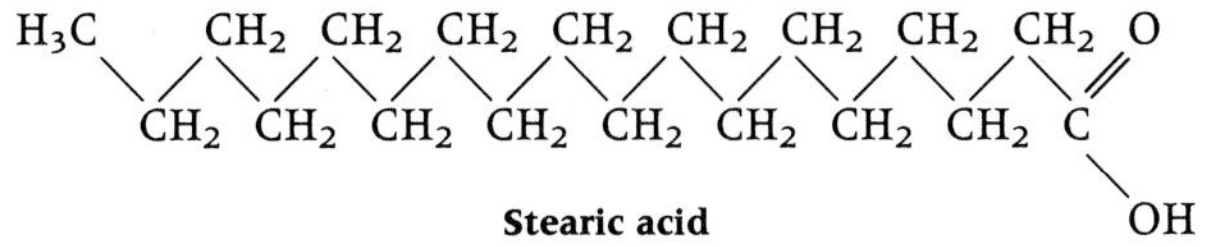

$$H_3C-CH_2-CH_2-CH_2-CH_2-CH_2-CH_2-CH_2-CH_2-CH_2-CH_2-CH_2-CH_2-CH_2-CH_2-CH_2-CH_2-C(=O)-OH$$

Stearic acid

The long hydrocarbon chain of a fatty acid is sliced up by oxidation, 2 carbons at a time; these are released from the end of the molecule as acetyl coenzyme A. Although 2 high-energy phosphate bonds are required to prime each 2-carbon fragment, energy is derived both from the reduction of NAD and FAD in the oxidations and from the acetyl group as it is degraded in the citric acid cycle. It can be calculated that the complete oxidation of 18-carbon stearic acid will net 146 ATP molecules. By comparison, 3 molecules of glucose (also totaling 18 carbons) yield 108 ATP molecules. Since there are 3 fatty acids in each triglyceride molecule, a total of 440 ATP molecules is formed. An additional 22 molecules of ATP are generated in the breakdown of glycerol, giving a grand total of 462 molecules of ATP. Little wonder that fat is considered the king of animal fuels! Fats are more concentrated fuels than carbohydrates because fats are almost pure hydrocarbons; they contain more hydrogen per carbon atom than sugars do, and it is the energized electrons of hydrogen that generate high-energy bonds, when they are carried through the mitochondrial electron transport system.

Fat stores are derived principally from surplus fats and carbohydrates in the diet. Acetyl coenzyme A is the source of carbon atoms used to build fatty acids. Since all major classes of organic molecules (carbohydrates, fats, and proteins) can be degraded to acetyl coenzyme A, all can be converted into stored fat. The biosynthetic pathway for fatty acids resembles a reversal of the catabolic pathway already described but requires an entirely different set of enzymes. From acetyl coenzyme A, the fatty acid chain is assembled 2 carbons at a time. Because fatty acids release energy when they are oxidized, they obviously require an input of energy for their synthesis. This is provided principally by electron energy from glucose degradation. Thus the total ATP derived from oxidation of a molecule of triglyceride is not as great as previously calculated, because varying amounts of energy are required for synthesis and storage.

Glycogen reserves (mainly in the liver) provide about 1 day's supply of glucose. Lipids and proteins are used when the carbohydrate reserves are diminishing.

Stored fats are the greatest reserve fuel in the body. Most of the usable fat resides in adipose tissue that is composed of specialized cells packed with globules of triglycerides. Adipose tissue is widely distributed in the abdominal cavity, in muscles, around deep blood vessels, and especially under the skin. Women average about 30% more fat than men, and this is responsible in no small measure for the curved contours of the female figure. However, its aesthetic contribution is strictly subsidiary to its principal function as an internal fuel depot. Indeed, humans can only too easily deposit large quantities of fat, generating personal unhappiness and hazards to health.

The physiological and psychological aspects of obesity are now being investigated by many researchers. There is increasing evidence that body fat deposition is regulated by a feeding control center located in the lateral and ventral regions of the hypothalamus, an area in the floor of the forebrain. The set point of this regulator determines the normal weight for the individual, which may be rather persistently maintained above or below what is considered normal for the human population. Thus, obesity is not always due to overindulgence and lack of self-control, despite popular notions to the contrary.

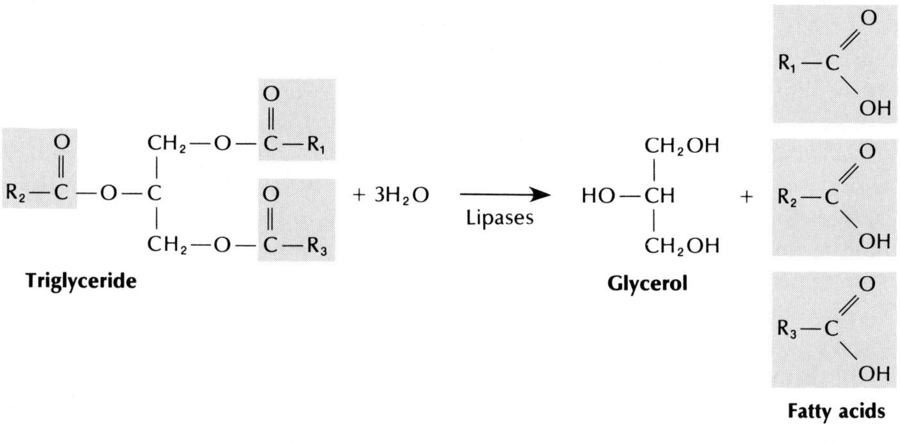

$$\underset{\textbf{Triglyceride}}{R_2-C(=O)-O-CH(CH_2-O-C(=O)-R_1)(CH_2-O-C(=O)-R_3)} + 3H_2O \xrightarrow{\text{Lipases}} \underset{\textbf{Glycerol}}{HO-CH(CH_2OH)(CH_2OH)} + \underset{\textbf{Fatty acids}}{R_1-C(=O)OH,\; R_2-C(=O)OH,\; R_3-C(=O)OH}$$

FIGURE 2-34

Hydrolysis of a triglyceride (neutral fat) by intracellular lipase. The R groups of each fatty acid represent a hydrocarbon chain.

METABOLISM OF PROTEINS

Since proteins are composed of amino acids, 20 kinds in all (p. 15), the central topic of our consideration is amino acid metabolism. Amino acid metabolism is complex. For one thing each of the 20 amino acids requires a separate pathway of biosynthesis and degradation. For another, amino acids are precursors to tissue proteins, enzymes, nucleic acids, and other nitrogenous constituents that form the very fabric of the cell. The central purpose of carbohydrate and fat oxidation is to provide energy needed to construct and maintain these vital macromolecules.

Let us begin with the **amino acid pool** in the blood and extracellular fluid from which the tissues draw their requirements. When animals eat proteins, these are digested in the gut, releasing the constituent amino acids, which are then absorbed (Figure 2-35). Tissue proteins also are hydrolyzed during normal growth, repair, and tissue restructuring; their amino acids join those derived from protein foodstuffs to enter the amino acid pool. A portion of the amino acid pool is used to rebuild tissue proteins, but most animals ingest a protein surplus. Since amino acids are not excreted as such in any significant amounts, they must be disposed of in some way. In fact, amino acids can be and are metabolized through oxidative pathways to yield high-energy phosphate. In short, excess proteins serve as fuel as do carbohydrates and fats. Their importance as fuel obviously depends on the nature of the diet. In carnivores that ingest a diet of almost pure protein and fat, nearly half of their high-energy phosphate is derived from amino acid oxidation.

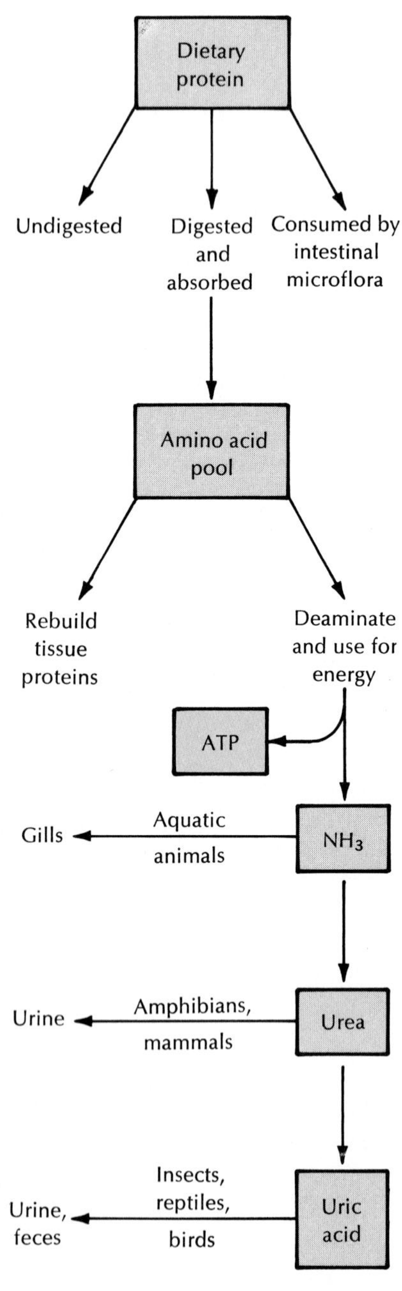

FIGURE 2-35

Fate of dietary protein.

Before entering the fuel depot, nitrogen must be removed from the amino acid molecule. This can be by **deamination** (the amino group splits off to form ammonia and a keto acid) or by **transamination** (the amino group is transferred to a keto acid to yield a new amino acid). Thus amino acid degradation yields two main products, ammonia and carbon skeletons, which are handled in different ways.

Once the nitrogen atoms are removed, the carbon skeletons of amino acids can be completely oxidized, usually by way of pyruvic acid or acetic acid. These residues then enter regular routes used by carbohydrate and fat metabolism.

The other product of amino acid degradation is ammonia. Ammonia is highly toxic because it reacts with α-ketoglutaric acid to form glutamic acid (an amino acid). Any accumulation of ammonia effectively removes this important intermediate from the citric acid cycle (see Figure 2-29) and inhibits respiration. Disposal of ammonia offers little problem to aquatic animals because it is soluble and readily diffuses into the surrounding medium through the respiratory surfaces. Terrestrial forms cannot get rid of ammonia so conveniently and must detoxify it by converting it to a relatively nontoxic compound. The two principal compounds formed are **urea** and **uric acid**, although a variety of other detoxified forms of ammonia are excreted by different invertebrate and vertebrate groups. Among vertebrates, amphibians and especially mammals produce urea. Reptiles and birds, as well as many terrestrial invertebrates, produce uric acid.

The key feature that seems to determine the choice of nitrogenous waste is the availability of water in the environment. When water is abundant, the chief nitrogenous waste is ammonia. When water is restricted, it is urea. And for animals living in truly arid habitats, it is uric acid. Uric acid is highly insoluble and easily precipitates from solution, allowing its removal in solid form. The embryos of birds and reptiles benefit greatly from excretion of nitrogenous waste as uric acid because the waste cannot be eliminated through their shells. During embryonic development, the harmless, solid uric acid is retained in one of the extraembryonic membranes. When the hatchling emerges into its new world, the accumulated uric acid, along with the shell and membranes that supported development, is discarded.

MANAGEMENT OF METABOLISM

The complex pattern of enzyme reactions that constitutes metabolism cannot be explained entirely in terms of physicochemical laws or chance happenings. Although some enzymes do indeed "flow with the tide," the activity of others is rigidly controlled. In the former case, suppose the purpose of an enzyme is to convert A to B. If B is removed by conversion into another compound, the enzyme will tend to restore the original ratio of B to A. Since many enzymes act reversibly, this can result, according to the metabolic situation prevailing, in synthesis or degradation. For example, an excess of an intermediate in the citric acid cycle would result in its contribution to glycogen synthesis; a depletion of such a metabolite would lead to glycogen breakdown. This automatic compensation (equilibration) is not, however, sufficient to explain all that actually takes place in an organism, as for example, what happens at branch points in a metabolic pathway.

Mechanisms exist for critically regulating enzymes in both *quantity* and *activity*. Enzyme induction in bacteria is an example of quantity regulation (p. 84). The genes leading to synthesis of the enzyme are switched on or off, depending on the presence or absence of a substrate molecule. In this way the *amount* of an enzyme can be controlled. It is a relatively slow process.

Mechanisms that alter activity of enzymes can quickly and finely adjust metabolic pathways to changing conditions in a cell. The presence or increase in concentration of some molecules can alter the shape (conformation) of particular enzymes, thus activating or inhibiting the enzyme (Figure 2-36). For example, phosphofructokinase, which catalyzes the phosphorylation of glucose-6-phosphate to fructose-1,6-diphosphate (Figure 2-32), is inhibited by high concentrations of ATP or citric acid. Their presence means that a sufficient amount of precursors has reached the citric acid cycle and additional glucose is not needed.

As well as being subject to alteration in physical shape, some enzymes exist in both an active and an inactive form. These may be chemically different. Enzymes that degrade glycogen (phosphorylase) and synthesize it (synthase) are examples. Conditions that activate the phosphorylase tend to inactivate the synthase and vice versa.

Many cases of enzyme regulation are known, but these selected examples must suffice to illustrate the importance of enzyme regulation in the integration of metabolism.

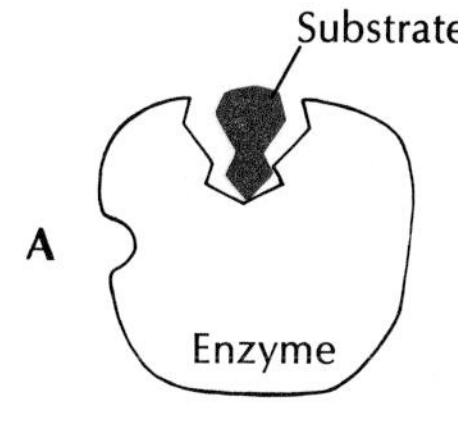

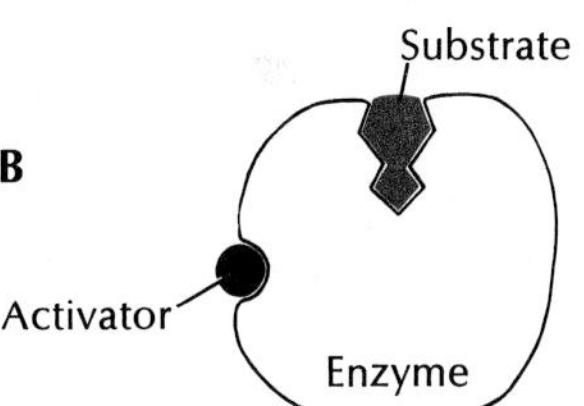

FIGURE 2-36

Enzyme regulation. **A,** The active site of an enzyme may only loosely fit its substrate in the absence of an activator. **B,** With the regulatory site of the enzyme occupied by an activator, the enzyme binds the substrate, and the site becomes catalytically active.

Summary

All living organisms are composed of one or more cells, which are the basic structural and functional units of life. Cells are studied by light microscopes, transmission and scanning electron microscopes, and biochemical methods. Cells are surrounded by a plasma membrane, and eukaryotic cells have a nucleus surrounded by a double membrane. Besides the chromatin, one or more nucleoli are usually found in the nucleus. The fluid-mosaic model of the membrane is a phospholipid bilayer with a mosaic of protein molecules as functional elements in the membrane. Outside the nuclear membrane is the cell cytoplasm, subdivided by a membranous network, the endoplasmic reticulum, which is often associated with ribosomes and probably functions in transport of materials within the cell. Among the organelles within the cell are the mitochondria, which contain the enzymes of oxidative energy metabolism. The Golgi complex functions in storage and packaging of proteins. Lysosomes, other membrane-bound vesicles, microfilaments, and microtubules are often found in the cytoplasm. Centrioles are short cylinders of nine triplet microtubules that help organize the mitotic spindle. Various specialized junctions occur between cells.

Substances can enter cells by diffusion, mediated transport, and endocytosis. Diffusion is the movement of molecules from an area of higher concentration to one of lower concentration, and osmosis is diffusion of water through a selectively permeable membrane due to osmotic pressure. Osmotic pressure is not the same as hydrostatic pressure but is defined in terms of equilibrium hydrostatic pressure. Solutes to which the membrane is not permeable require a carrier molecule to traverse it; carrier-mediated systems include facilitated transport (in the direction of a concentration gradient) and active transport (against a concentration gradient, which requires energy). Endocytosis includes bringing droplets (pinocytosis and receptor-mediated endocytosis) or particles (phagocytosis) into the cell.

The capacity to grow by cell multiplication is a fundamental characteristic of living systems. Ordinary somatic cells contain two of each kind of chromosome (hence are called diploid) and divide by mitosis. In mitosis the chromosomes replicate during interphase, and the replicated chromosomes (sister chromatids) are joined by a centromere. At the beginning of mitosis (prophase) the nuclear envelope disperses, and the chromosomes condense into recognizable bodies. At metaphase the replicated centrioles have moved to opposite poles of the cell, the spindle and asters of microtubules have formed, and the chromosomes are on the median plane of the cell. The centromeres divide, and one of each kind of chromosome is pulled toward the centriole by the attached spindle fiber (anaphase). At telophase the chromosomes are at the position of the new nucleus in each cell, and division of the cytoplasm (cytokinesis) begins. Mitosis itself is only a small part of the total cell cycle. In interphase, G_1, S, and G_2 periods are recognized, and the S period is the time when DNA is synthesized (the chromosomes are replicated).

Enzymes are pure proteins or proteins associated with nonprotein cofactors that vastly accelerate reaction rates in living systems. An enzyme does this by temporarily binding its reactant (substrate) onto an active site in a highly specific fit. In this configuration, internal activation energy barriers are lowered enough to disrupt and split the substrate, and the enzyme is restored to its original form.

Cells use the energy stored in chemical bonds of organic fuels by breaking the fuels down through a series of enzymatically controlled steps. This bond energy is transferred to ATP and packaged in the form of "high-energy" phosphate bonds. ATP is produced in cells as it is required to power various synthetic, secretory, or mechanical processes.

Glucose is an important source of energy for cells. In aerobic metabolism (respiration), the 6-carbon glucose is split into two 3-carbon molecules of pyruvate. Pyruvate is decarboxylated to form 2-carbon acetyl coenzyme A, a strategic intermediate that leads to the citric acid cycle. Acetyl coenzyme A can also be derived from fat breakdown. In the citric acid cycle, acetyl coenzyme A is oxidized in a series of reactions to carbon dioxide, yielding, in the course of the reactions, energized electrons that are passed to electron acceptor molecules (NAD and FAD). In the final stage, the energized electrons are passed along an electron transport chain consisting of a series of electron carriers located in the inner membranes of the mitochondrion. ATP is generated at three points along the chain as the electrons are passed from carrier to carrier and finally to oxygen. A net total of 36 molecules of ATP is generated from one molecule of glucose.

In the absence of oxygen (anaerobic glycolysis), glucose is degraded to two 3-carbon molecules of lactate, yielding two molecules of ATP. Although anaerobic glycolysis is vastly less efficient than respiration, it provides essential energy for muscle contraction when heavy energy expenditure outstrips the oxygen-delivery system of an animal; it also is the only source for microorganisms living in oxygen-free environments.

Triglycerides (neutral fats) are especially rich depots of metabolic energy because the fatty acids of which they are composed are highly reduced and anhydrous. Fatty acids are degraded by sequential removal of 2-carbon units, which enter the citric acid cycle through acetyl-CoA.

Amino acids in excess of requirements for synthesis of proteins and other bimolecules are used as fuel. They are degraded by deamination or transamination to yield ammonia and carbon skeletons. The latter enter the citric acid cycle to be oxidized. Ammonia is a highly toxic waste product that aquatic animals quickly dispose of through respiratory surfaces. Terrestrial animals, however, convert ammonia into much less toxic compounds, urea or uric acid, for disposal.

The integration of metabolic pathways is finely regulated by mechanisms that control both the amount and activity of enzymes. The quantity of some enzymes is regulated by certain molecules that switch on or off enzyme synthesis in the nucleus. Enzyme activity may be altered by the presence or absence of metabolites that cause conformational changes in enzymes and thus improve or diminish their effectiveness as catalysts.

REVIEW QUESTIONS

1. Explain the difference (in principle) between a light microscope and an electron microscope.
2. Give a one-sentence definition of each of the following: plasma membrane, chromatin, nucleus, nucleolus, rough endoplasmic reticulum (rough ER), Golgi complex, lysosomes, mitochondria, microfilaments, microtubules, centrioles, basal body (kinetosome), tight junction, gap junction, microvilli.
3. You place some red blood cells in a solution and observe that they swell and burst. You place some cells in another solution, and they shrink and become wrinkled. Explain what has happened in each case.
4. Explain the difference between osmotic and hydrostatic pressure.
5. Distinguish between two kinds of mediated transport.
6. Distinguish between three kinds of endocytosis.
7. Define the following: chromosome, nucleosome, haploid, diploid, centromere, genome, mitosis, cytokinesis, syncytium.
8. Name the stages of mitosis in order, and describe the behavior of the chromosomes at each stage.
9. When does the "S period" of a cell cycle occur, and what is happening at that period?
10. Briefly explain how enzymes are believed to work.
11. Why is the secondary and tertiary structure of proteins important in enzyme action?
12. How do you account for the specificity of enzymes?
13. Why is aerobic metabolism more efficient than anaerobic metabolism?
14. With respect to glycolysis, answer the following questions. What molecule does the pathway begin with? What are the products of the pathway? How many molecules of ATP are generated (gross and net)? Where in the cell does it occur?
15. Answer the questions in no. 14 with respect to the citric acid cycle.
16. Answer the questions in no. 14 with respect to the electron transport chain.
17. What is the importance of anaerobic glycolysis?
18. What is the significance of acetyl coenzyme A to lipid metabolism?
19. How are amino acids oxidized as energy sources?
20. Explain the relationship of ammonia, urea, and uric acid as nitrogenous wastes to the amount of water in an organism's environment.
21. Give three ways in which enzymes are regulated in cells.

SELECTED REFERENCES

Alberts, B., D. Bray, J. Lewis, M. Raff, K. Roberts, and J.D. Watson. 1983. Molecular biology of the cell. New York, Garland Publishing, Inc. *Chapter 2 presents an excellent summary of cellular metabolism.*

Darnell, J., H. Lodish, and D. Baltimore. 1986. Molecular cell biology. New York, Scientific American Books, W.H. Freeman & Co. *Up-to-date, thorough, and readable. Includes both cell biology and molecular biology. Advanced, but highly recommended.*

Hinkle, P.C., and R.E. McCarty. 1978. How cells make ATP. Sci. Am. **238:**104-123 (Mar.). *Describes the chemiosmotic theory of ATP formation.*

Rothman, J.E. 1985. The compartmental organization of the Golgi apparatus. Sci. Am. **253:**74-89 (Sept.). *Presents modern concepts of the Golgi; describes functional distinctions between compartments.*

Sheeler, P., and D.E. Bianchi. 1987. Cell biology: structure, biochemistry, and function, ed. 2. New York, John Wiley & Sons, Inc. *Well-written, well-illustrated introductory cell biology textbook.*

Stryer, L. 1988. Biochemistry, ed. 3. San Francisco, W.H. Freeman & Co., Publishers. *Clear explanations and good diagrams.*

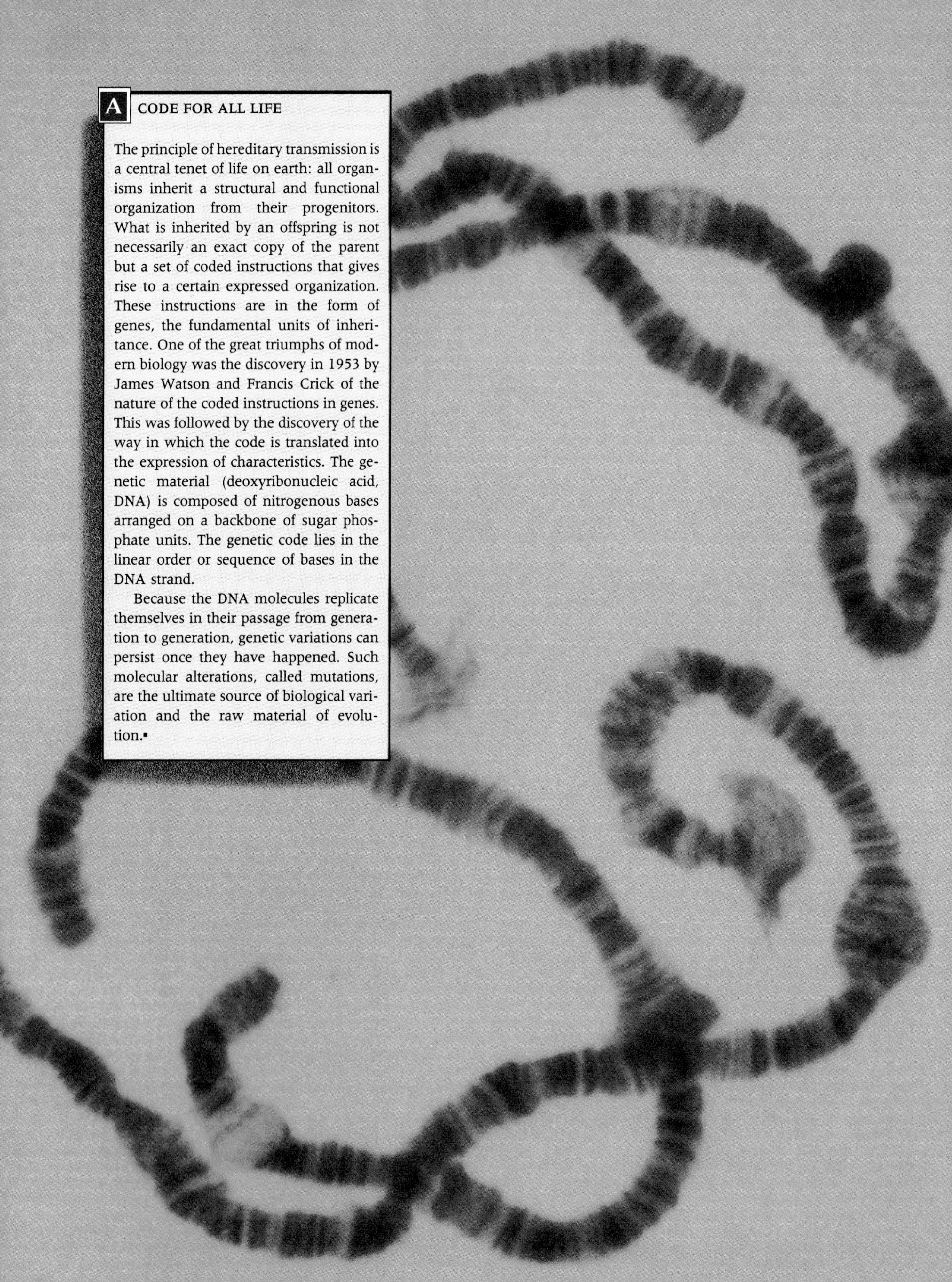

A CODE FOR ALL LIFE

The principle of hereditary transmission is a central tenet of life on earth: all organisms inherit a structural and functional organization from their progenitors. What is inherited by an offspring is not necessarily an exact copy of the parent but a set of coded instructions that gives rise to a certain expressed organization. These instructions are in the form of genes, the fundamental units of inheritance. One of the great triumphs of modern biology was the discovery in 1953 by James Watson and Francis Crick of the nature of the coded instructions in genes. This was followed by the discovery of the way in which the code is translated into the expression of characteristics. The genetic material (deoxyribonucleic acid, DNA) is composed of nitrogenous bases arranged on a backbone of sugar phosphate units. The genetic code lies in the linear order or sequence of bases in the DNA strand.

Because the DNA molecules replicate themselves in their passage from generation to generation, genetic variations can persist once they have happened. Such molecular alterations, called mutations, are the ultimate source of biological variation and the raw material of evolution.▪

CHAPTER 3

Genetic Basis of Evolution

A basic principle of modern evolutionary theory is that organisms attain their diversity of form, function, and behavior through hereditary modifications of preexisting lines of ancestors. It means that all known lineages of plants and animals are related by descent from simpler, common ancestral groups.

Heredity establishes the continuity of life forms. Although offspring and parents in a particular generation may look different, there is nonetheless a basic sameness that runs from generation to generation for any species of plant or animal. In other words, "like begets like." Yet children are not precise replicas of their parents. Some of their characteristics show resemblances to one or both parents, but they also demonstrate many traits not found in either parent. What is actually inherited by an offspring from its parents is a certain type of germinal organization **(genes)** that, under the influence of environmental factors, guides the orderly sequence of differentiation of the fertilized egg into a human being, bearing the unique physical characteristics as we see them. Each generation hands on to the next the instructions required for maintaining continuity of life.

The gene is the unit entity of inheritance, the germinal basis for every characteristic that appears in an organism. In a sense the primary function of an organism is to reproduce genes. The organism is a device, a vehicle for the transfer of genes from one generation to its descendants. It is a repository in which a portion of the gene "pool" of the population has been temporarily entrusted.

The study of what genes are and how they work is the science of genetics. It is the science that deals with the underlying causes of *resemblance*, as seen in the remarkable fidelity of reproduction, and of *variation*, which is the working material for organic evolution. Genetics has shown that all living forms use the same information storage, transfer, and translation system, and thus it has provided an explanation for both the stability of all life and its probable descent from a common ancestral form. This is one of the most important unifying concepts of biology.

Mendel's Investigations

The first man to formulate the cardinal principles of heredity was Gregor Johann Mendel (1822-1884), who was an Augustinian monk living in Brünn (Brno), Moravia. At that time Brünn was a part of Austria, but now it is in the central part of Czechoslovakia. While conducting breeding experiments in a small monastery garden from 1856 to 1864, he examined with great care the progeny of many thousands of plants. He worked out in elegant simplicity the laws governing the transmission of characters from parent to offspring. His discoveries, published in 1866, were of great potential significance, coming just after Darwin's publication of *The Origin of Species*. Yet these discoveries remained unappreciated and forgotten until 1900—some 35 years after the completion of the work and 16 years after Mendel's death.

Mendel's classic observations were based on the garden pea because it had been produced in pure strains by gardeners over a long period of time by careful selection. For example, some varieties were definitely dwarf and others were tall. A second reason for selecting peas was that they were self-fertilizing, but also

FIGURE 3-1

Seven experiments of Gregor Mendel based on his postulates. These are the results of monohybrid crosses for first and second generations.

Photograph from Historical Pictures Service, Chicago, Ill.

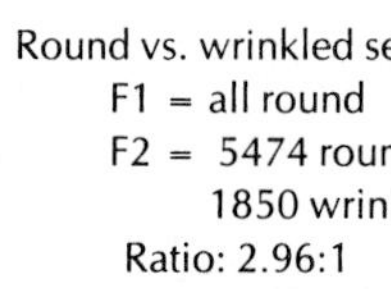

Round vs. wrinkled seeds
F1 = all round
F2 = 5474 round
1850 wrinkled
Ratio: 2.96:1

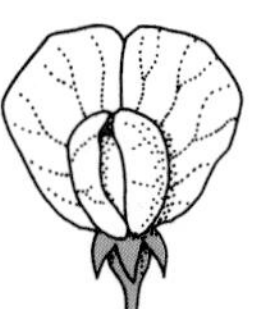

Purple vs. white flowers
F1 = all purple
F2 = 705 purple
224 white
Ratio: 3.15:1

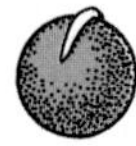

Yellow vs. green seeds
F1 = all yellow
F2 = 6022 yellow
2001 green
Ratio 3.01:1

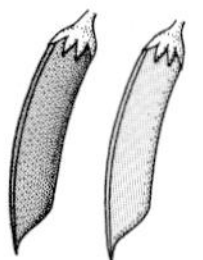

Green vs. yellow pods
F1 = all green
F2 = 428 green
152 yellow
Ratio 2.82:1

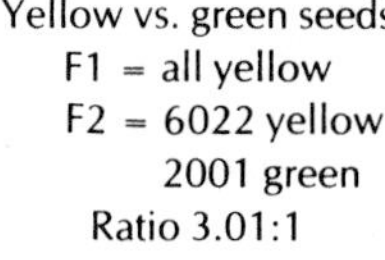
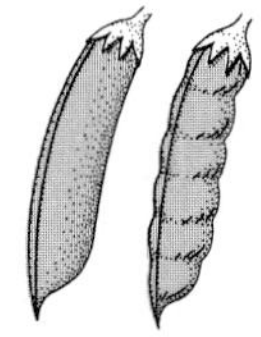

Inflated vs. constricted pods
F1 = all inflated
F2 = 882 inflated
299 constricted
Ratio 2.95: 1

Long vs. short stems
F1 = all long
F2 = 787 long
277 short
Ratio 2.84:1

Axial vs. terminal flowers
F1 = all axial
F2 = 651 axial
207 terminal
Ratio 3.14:1

capable of cross-fertilization. To simplify his problem he chose single characters and characters that were sharply contrasted. Mere qualitative and intermediate characters were carefully avoided. Mendel selected pairs of contrasting characters, such as tall plants, dwarf plants, smooth seeds, and wrinkled seeds (Figure 3-1).

Mendel crossed plants having one of these characters with others having the contrasting character. He did this by removing the stamens (male part, containing the pollen) from a flower to prevent self-fertilization and then placing on the stigma (female part of flower) of this flower pollen from the flower of the plant that had the contrasting character. He also prevented the experimental flowers from being pollinated from other sources such as wind and insects. When the cross-fertilized flower bore seeds, he noted the kind of plants (hybrids) that were produced from the planted seeds. Subsequently he crossed these hybrids among themselves to see what would happen.

Mendel knew nothing of the cytological basis of heredity, since chromosomes and genes were unknown to him. Although we can admire Mendel's power of intellect in his discovery of the principles of inheritance without knowledge of chromosomes, the principles are certainly easier to understand if we first examine chromosomal behavior, especially in meiosis.

Chromosomal Basis of Inheritance

In bisexual animals special **sex cells**, or **gametes** (ova and sperm), are responsible for providing the genetic information to the offspring. Scientific explanation of genetic principles required a study of germ cells and their behavior, which meant working backward from certain visible results of inheritance to the mechanism responsible for such results. The nuclei of sex cells were early suspected of furnishing the real answer to the mechanism. This applied especially to the chromosomes, since they appeared to be the only entities passed on in equal quantities from both parents to offspring.

When Mendel's laws were rediscovered in 1900, their parallelism with the cytological behavior of the chromosomes was obvious. Later experiments showed that the mechanism of heredity could be definitely assigned to the chromosomes. The next problem was to find out how chromosomes affected the hereditary pattern.

A giant stride in chromosomal genetics was made when the great American geneticist Thomas Hunt Morgan and his colleagues selected the fruit fly (*Drosophila* for their studies (1910-1920). It was cheaply and easily reared in bottles in the laboratory, fed on a simple medium of bananas and yeast. Most important, it produced a new generation every 10 days, enabling Morgan to proceed at least 25 times more rapidly than with organisms that take a year to mature, such as garden peas. Morgan's work led to the mapping of genes on chromosomes and founded the discipline of cytogenetics.

MEIOSIS: MATURATION DIVISION OF GAMETES

Every body cell contains *two* chromosomes bearing genes for the same set of characteristics, and the two members of each pair usually, but not always, have the same size and shape. The members of such a pair are called **homologous** chromosomes; one of each pair comes from the mother and the other from the father. Thus each cell normally has two genes coding for a given trait, one on each of the homologues. These may be alternative forms of the same gene, and, if so, they are **allelic genes**, or **alleles**. Sometimes only one of the alleles has an effect on the organism, although both are present in each cell, and either may be passed on to the progeny as a result of meiosis and subsequent fertilization.

Alleles are alternate forms of the same gene that have arisen by mutation of the DNA sequence. Like a baseball team with several pitchers, only one of whom can occupy the pitcher's mound at a time, only one allele can occupy a chromosomal locus. Alternate alleles for the locus may be on chromosomes of other individuals in the population of the species.

During an individual's growth, all the chromosomes of the mitotically dividing cells are replicated during the S period of each cell cycle (p. 43), so that each new cell contains the double set of chromosomes. In the reproductive organs, the gametes (germ cells) are formed by a kind of maturation division, called meiosis, which *separates* the homologous pairs of chromosomes. If it were not for this reductional division, the union of ovum (egg) and sperm would produce an individual with twice as many chromosomes as the parents. Continuation of this process in just a few generations could yield body cells with astronomical numbers of chromosomes.

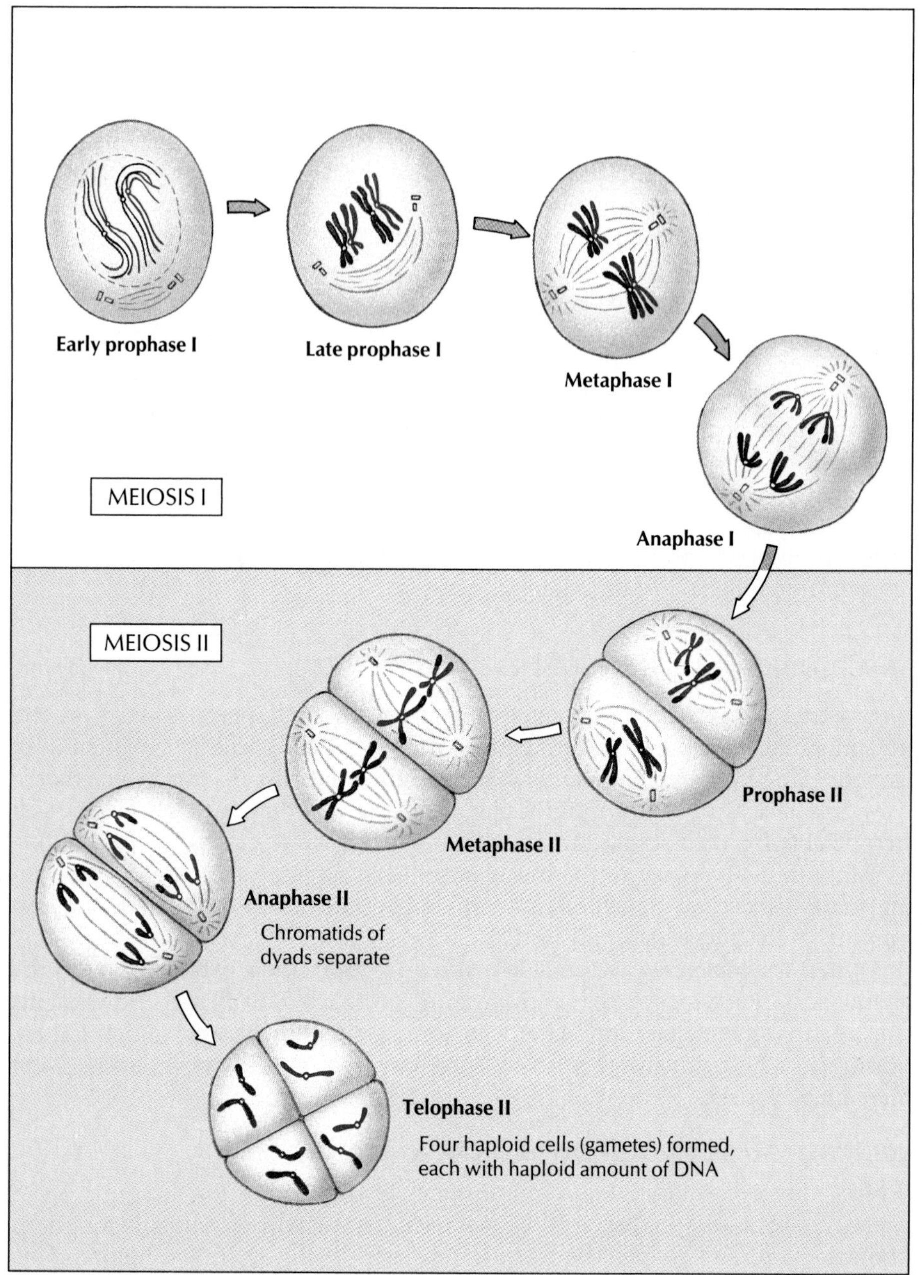

FIGURE 3-2

Meiosis in a sex cell with two pairs of chromosomes.

Meiosis consists of *two* nuclear divisions in which the chromosomes divide only once (Figure 3-2). The result is that mature gametes (ova and sperm) have only *one* member of each homologous chromosome pair, or a **haploid** (n) number of chromosomes. In humans the zygotes and all body cells normally have the **diploid** number (2n), or 46 chromosomes; the gametes have the haploid number (n), or 23.

Most of the unique features of meiosis occur during the prophase of the first meiotic division (Figure 3-2). The two members of each pair of homologous chromosomes come into side-by-side contact **(synapsis)** to form a **bivalent**. Each chromosome of the bivalent has already replicated to form two chromatids, each of which will become a new chromosome. The two chromatids are joined at one point, the centromere, so that each bivalent is made up of two pairs of chromatids, or *four* future chromosomes, and is thus called a **tetrad**. The position or location of any gene on a chromosome is the gene **locus** (pl., **loci**), and in synapsis all gene loci on a chromatid normally lie exactly opposite the corresponding loci on the homologous chromatid. Toward the end of the prophase, the chromosomes shorten and thicken and are ready to enter into the first meiotic division. In contrast to the mitosis, the centromeres holding the chromatids together *do not divide* at the beginning of anaphase. As a result, one of each pair of double-stranded chromosomes **(dyads)** is pulled toward each pole by the microtubules of the division spindle. Therefore at the end of the first meiotic division, the daughter cells contain *one* of *each* of the homologous chromosomes, so the total chromosome number has been reduced to the haploid. However, because the chromatids are still joined by the centromeres, each cell contains the 2n amount of DNA.

The second meiotic division more closely resembles the events in mitosis. The dyads are split at the beginning of anaphase by division of the centromeres, and single-stranded chromosomes move toward each pole. Thus by the end of the second meiotic division, the cells have the haploid number of chromosomes and the n amount of DNA. Each chromatid of the original tetrad exists in a separate nucleus. Four products are formed, each containing one complete haploid set of chromosomes and only one allele of each gene.

Speculation on how sex was determined in animals produced several incredible theories, for example, that the two testicles of the male contained different types of semen, one begetting males, the other females. It is not difficult to imagine the abuse and mutilation of domestic animals that occurred when attempts were made to alter the sex ratios of herds. Another theory asserted that sex of the offspring was determined by the more heavily sexed parent. An especially masculine father should produce sons, an effeminate father only daughters. Such ideas have lingered until recently.

CROSSING OVER

An important event often occurs when the chromatids of a bivalent exchange parts with the adjacent homologous (nonsister) chromatid. This phenomenon, called **crossing-over**, is shown in Figure 3-2. Crossing-over is important because the hereditary material is redistributed between homologous chromatids in one bivalent. The chromosomes exchange equivalent sections bearing allelic genes for the same traits, so each chromatid contains a full set of genes. But the genes are in new combinations.

While the chromosomes are in synapsis, a strand of one chromatid becomes joined with the homologous chromatid. As prophase continues, the homologues begin to move apart, revealing **chiasmata**, the connection points where crossing-over has occurred. There may be one or more chiasmata present in each bivalent, depending on the number of times the adjacent homologues have joined. When the chiasmata pull apart, the exchange is complete. Note that the resulting four gametes at the end of meiosis are all genetically different (Figure 3-2).

SEX DETERMINATION

Before the importance of chromosomes in heredity was realized in the early 1900s, how gender was determined was totally unknown. The first really scientific clue to the determination of sex came in 1902 when C. McClung observed that bugs (Hemiptera) produced two kinds of sperm in approximately equal numbers. One kind contained among its regular set of chromosomes a so-called accessory chromosome that was lacking in the other kind of sperm. Since all the eggs of these species had the same number of haploid chromosomes, half the sperm would have the same number of chromosomes as the eggs, and half of them would have one chromosome less. When an egg was fertilized by a spermatozoon carrying the accessory (sex) chromosome, the resulting offspring was a female; when fertilized by the spermatozoon without an accessory chromosome, the offspring was a male. There are therefore two kinds of chromosomes; **X chromosomes**, which determine sex (and sex-linked traits); and **autosomes**, which are all the other chromosomes. The particular type of sex determination just described is often called the XX-XO type, which indicates that the females have 2 X chromosomes and the males only 1 X chromosome (the O indicates absence of the chromosome). The XX-XO method of sex determination is depicted in Figure 3-3.

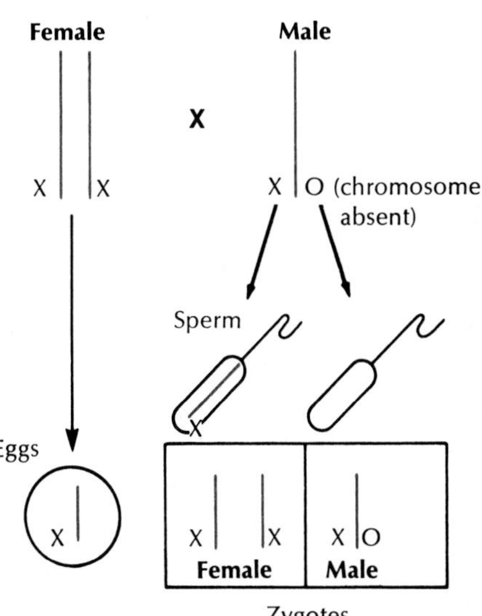

FIGURE 3-3

XX-XO sex determination.

We now know that not all animals with dioecious reproduction have their sex determined chromosomally. Several invertebrate examples are known; and the sex of alligators and some turtles is determined by the temperature at which the eggs are incubated. Some coral reef fishes actually change sex during their lifetimes, starting off as females, then becoming males.

Later, other types of sex determination were discovered. In humans and many other forms there are the same number of chromosomes in each sex; however, the sex chromosomes (XX) are alike in the female but unlike (XY) in the male. Hence the human egg contains 22 autosomes + 1 X chromosome; the sperm are of two kinds; half carry 22 autosomes + 1 X and half bear 22 autosomes + 1 Y. The Y chromosome is much smaller than the X and carries very little genetic information. At fertilization, when the 2 X chromosomes come together, the offspring are female; when X and Y come together, the offspring are male. The XX-XY kind of determination is shown in Figure 3-4.

A third type of sex determination is found in birds, moths, and butterflies in which the male has 2 X (or sometimes called ZZ) chromosomes and the female an X and Y (or ZW).

We should note that, in the case of X and Y chromosomes, the homologous chromosomes are unlike in size and shape. Therefore, they do not both carry the same genes. The genes of the X chromosome often do not have allelic counterparts on the diminutive Y chromosome. This fact is very important in sex-linked inheritance, which we shall discuss later.

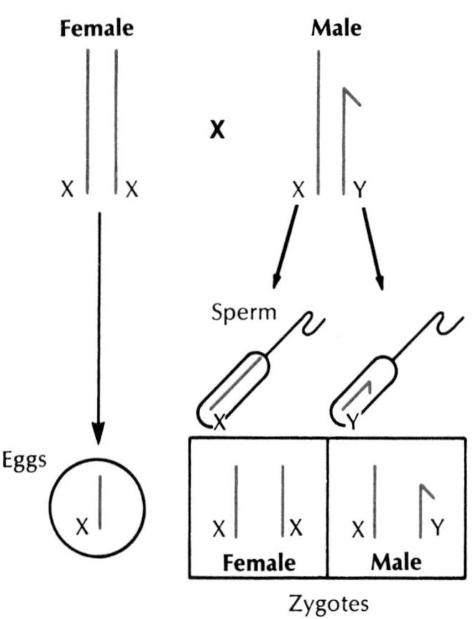

FIGURE 3-4

XX-XY sex determination. The Y chromosome is often represented by an inverted check mark.

Mendelian Laws of Inheritance

MENDEL'S FIRST LAW

Mendel's **law of segregation** states that *alternative alleles segregate independently of one another.* In one of Mendel's original experiments, he pollinated pure-line tall plants with the pollen of pure-line dwarf plants. He found that all the progeny of the first (F_1) generation were tall, just as tall as the tall parents of the cross. The reciprocal cross—dwarf plants pollinated with tall plants—gave the same result. This always happened no matter which way the cross was made. Obviously, this kind of inheritance was not a blending of two characteristics, since none of the progeny was of intermediate size.

Next Mendel selfed (self-fertilized) the tall F_1 plants and raised several hundred progeny, the second (F_2) generation. This time, *both* tall and dwarf plants appeared. Again, there was no blending (no plants of intermediate size), but the appearance of dwarf plants from all tall parental plants was surprising. The dwarf characteristic, present in the grandparents but not in the parents, had reappeared. When he counted the actual number of tall and dwarf plants in the F_2 generation, he discovered that there were almost exactly three times more tall plants than dwarf ones.

Mendel then repeated this experiment for the six other contrasting characters that he had chosen, and in every case he obtained ratios very close to 3:1 (Figure 3-1). At this point it must have been clear to Mendel that he was dealing with hereditary determinants for the contrasting characters that did not blend when brought together. Even though the dwarf characteristic disappeared in the F_1 generation, it reappeared fully expressed in the F_2 generation. He realized that the F_1 generation plants carried determinants (which he called "factors") of both tall and dwarf parents, even though only the tall characteristic was expressed in the F_1 generation.

In not reporting conflicting findings, which must surely have arisen as they do in any original research, Mendel has been accused of "cooking" his results. The chances are, however, that he carefully avoided ambiguous material to strengthen his central message, which we still regard as an exemplary achievement in experimental analysis.

Mendel called the tall factor **dominant** and the short **recessive**. Similarly, the other pairs of characters that he studied showed dominance and recessiveness. Whenever a dominant factor (gene) is present, the recessive one cannot produce an effect. The recessive factor will show up *only* when both factors are recessive, or in other words, a pure condition.

In representing his crosses, Mendel used letters as symbols; dominant characters were represented by capital letters, and for recessive characters he used the corresponding lowercase letters. Modern geneticists still follow this custom. Thus the factors for pure tall plants might be represented by *T/T*, the pure recessive by *t/t*, and the mix, or hybrid, of the two plants by *T/t*. The slash mark is to indicate that the genes are on homologous chromosomes. When the gametes unite in any fertilization, a **zygote** is formed. The zygote bears the complete genetic constitution of the organism. All the gametes produced by *T/T* must necessarily be *T*, whereas those produced by *t/t* must be *t*. Therefore a zygote produced by union of the two must be *T/t*, or a **heterozygote**. On the other hand, the pure tall plants *(T/T)* and pure dwarf plants *(t/t)* are **homozygotes**, meaning that the factors (genes) are alike on the homologous chromosomes. A cross involving only one pair of contrasting characters is called a monohybrid cross.

In the cross between tall and dwarf plants there are two types of *visible* characters—tall and dwarf. These are called **phenotypes**. On the basis of genetic formulas there are three *hereditary* types—*T/T*, *T/t*, and *t/t*. These are called **genotypes**. A genotype is a gene combination (*T/T*, *T/t*, or *t/t*), and the phenotype is the appearance of the organism (tall or dwarf).

In diagram form, one of Mendel's original crosses (tall plant and dwarf plant) could be represented as follows:

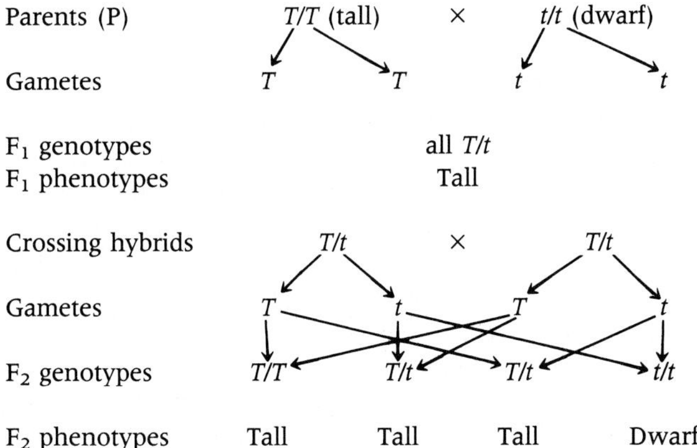

In other words, all possible combinations of F_1 gametes in the F_2 zygotes will yield a 3:1 phenotype ratio and a 1:2:1 genotype ratio. It is convenient in such crosses to use the checkerboard (Punnett square) method devised by Punnett for representing the various combinations resulting from a cross. In the F_2 cross the following scheme would apply:

Sperm \ **Ova**	½ *T*	½ *t*
½ *T*	¼ *T/T* (homozygous tall)	¼ *T/t* (hybrid tall)
½ *t*	¼ *T/t* (hybrid tall)	¼ *t/t* (homozygous dwarf)

Ratio: 3 tall to 1 dwarf.

The next step was an important one, because it enabled Mendel to test his hypothesis that every plant contained nonblending factors from both parents. He self-fertilized the plants in the F_2 generation; that is, the stigma of a flower was fertilized by the pollen of the same flower. The results showed that self-pollinated F_2 dwarf plants produced only dwarf plants, whereas one third of the F_2 tall plants produced tall and the other two thirds produced both tall and dwarf in the ratio of 3:1, just as the F_1 plants had done. Genotypes and phenotypes were as follows:

$$\begin{array}{ll} F_2 \text{ plants: Tall} & \left\{ \begin{array}{l} \tfrac{1}{4}\ T/T \xrightarrow{\text{selfed}} \text{all } T/T \text{ (homozygous tall)} \\ \tfrac{1}{2}\ T/t \xrightarrow{\text{selfed}} 1\ T/T{:}2\ T/t{:}1\ t/t \text{ (3 tall:1 dwarf)} \end{array} \right. \\ \text{Dwarf} & \tfrac{1}{4}\ t/t \xrightarrow{\text{selfed}} \text{all } t/t \text{ (homozygous dwarf)} \end{array}$$

This experiment showed that the dwarf plants were pure because they at all times gave rise to short plants when self-pollinated; the tall plants contained both pure tall and hybrid tall. It also demonstrated that, although the dwarf character disappeared in the F_1 plants, which were all tall, the character for dwarfness appeared in the F_2 plants.

Mendel reasoned that the factors for tallness and dwarfness were units that did not blend when they were together. The F_1 generation (the first generation of hybrids, or first filial generation) contained both of these units or factors, but when these plants formed their gametes, the factors separated so that each gamete had only one factor. In a pure plant both factors were alike; in a hybrid they were different. He concluded that individual germ cells were always pure with respect to a pair of contrasting factors, even though the gametes were formed from hybrids in which the contrasting characters were mixed.

This idea formed the basis for his first principle, the law of segregation, which states that *whenever two factors are brought together in a hybrid, they segregate into separate gametes that are produced by the hybrid*. Either one of the pair of genes of the parent passes with equal frequency to the gametes. We now understand that the factors segregate because there are two alleles for the trait, one on each chromosome of a homologous pair, but the gametes receive only one of each in meiosis. Thus, in current usage the law of segregation refers to the parting of homologous chromosomes during meiosis.

Mendel's great contribution was his quantitative approach to inheritance. This really marks the birth of genetics, since before Mendel, people believed that traits were blended like mixing together two colors of paint, a notion that unfortunately still lingers in the minds of many. If this were true, variability would be lost in hybridization between individuals. With particulate inheritance, on the other hand, different variations are retained and can be shuffled about and resorted like blocks.

Testcross

When one of the alleles is dominant, all individuals with the dominant gene are the same phenotype, whether they are homozygous or heterozygous. For instance, in Mendel's experiment with tall and dwarf characters, it is impossible to determine the genetic constitution of the tall plants of the F_2 generation by mere inspection of the tall plants. Three fourths of this generation are tall, but which of them are heterozygotes?

As Mendel reasoned, the test is to cross the questionable individuals with pure recessives. If the tall plant is homozygous, all the plants in such a testcross are tall, thus:

Parents *T/T* (tall) × *t/t* (dwarf)

Sperm \ **Ova**	*T*	*T*
t	*T/t* (hybrid tall)	*T/t* (hybrid tall)
t	*T/t* (hybrid tall)	*T/t* (hybrid tall)

All of the offspring are *T/t* (hybrid tall). If, on the other hand, the tall plant is heterozygous, half of the offspring are tall and half dwarf, thus:

Parents *T/t* (hybrid tall) × *t/t* (dwarf)

Sperm \ Ova	*T*	*t*
t	*T/t* (hybrid tall)	*t/t* (homozygous dwarf)
t	*T/t* (hybrid tall)	*t/t* (homozygous dwarf)

The **testcross** is often used in modern genetics for the analysis of the genetic constitution of the offspring, as well as for a quick way to make desirable homozygous stocks of animals and plants.

Intermediate inheritance

In some cases neither allele is completely dominant over the other, and the heterozygote phenotype shows characteristics either intermediate between or even quite distinct from those of the parents. This is called **intermediate, inheritance, incomplete dominance,** or **codominance**. In the four-o'clock flower (*Mirabilis*), homozygotes are red or white flowered, but heterozygotes have pink flowers. In a certain strain of chickens, a cross between those with black and splashed white feathers produces offspring that are not gray but a distinctive color called blue Andalusian (Figure 3-5). In each case, if the F_1s are crossed, the F_2s have a ratio of 1:2:1 in colors, or 1 red:2 pink:1 white and 1 black:2 blue:1 white, respectively. This can be illustrated for the chickens as follows:

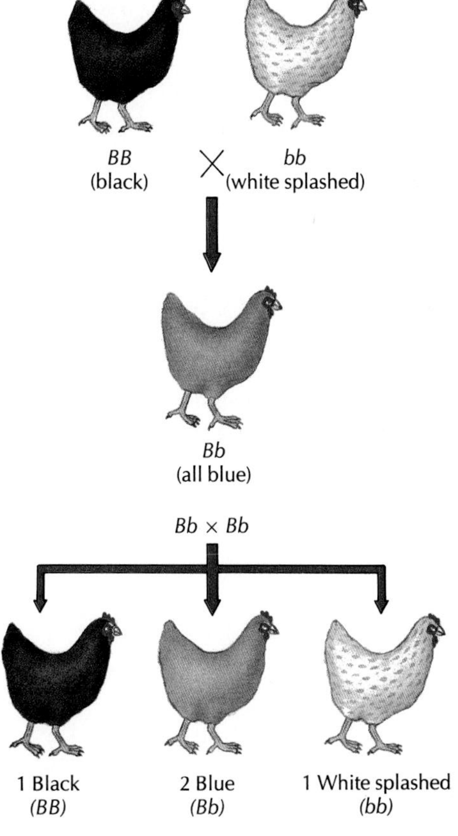

FIGURE 3-5

Cross between chickens with black and splashed white feathers. Black and white are homozygous; Andalusian blue is heterozygous.

	(black feathers)			(white feathers)	
Parents	*B/B*		×	*B′/B′*	
Gametes	all *B*			all *B′*	
F_1			*B/B′* (all blue)		
Crossing hydrids	*B/B′*		×	*B/B′*	
Gametes	*B,B′*			*B,B′*	
F_2 genotypes	*B/B*	*B/B′*		*B/B′*	*B′/B′*
F_2 phenotypes	Black	Blue		Blue	White

In this kind of a cross, the heterozygote is indeed a blending of both parental types. It is easy to see how observations of this type would encourage the notion of the blending concept of inheritance. However, in the cross of black and white chickens or red and white flowers, *only* the hybrid is a phenotypic blend; the homozygous strains breed true to the parental phenotypes.

When neither of the alleles is recessive, it is customary to represent both by capital letters and to distinguish them by the addition of a "prime" sign (*B′*) or by superscript letters, for example, B^b (equals black feathers) and B^w (equals white feathers).

MENDEL'S SECOND LAW

According to Mendel's **law of independent assortment,** *genes on different chromosomes assort independently during meiosis.* Thus the law deals with genes for two different characters that are borne on two different chromosomes. Mendel carried out experiments on peas that differed from each other by two or more genes, that is, experiments that involved two or more phenotypic characters.

Mendel had already established that tall plants were dominant to dwarf. He also noted that crosses between plants bearing yellow seeds and plants bearing green seeds produced plants with yellow seeds in the F_1 generation; therefore yellow was dominant to green. The next step was to make a cross between plants differing in these two characteristics. When a tall plant with yellow seeds (*T/T*

G/G) was crossed with a dwarf plant with green seeds (*t/t g/g*), the F_1 plants were tall and yellow as expected (*T/t G/g*).

Parents	*T/T G/G* (tall, yellow)	×	*t/t g/g* (dwarf, green)
Gametes	all *TG*		all *tg*
F_1		*T/t G/g* (tall, yellow)	

The F_1 hybrids were then crossed with each other, and the F_2 results are shown in Figure 3-6.

FIGURE 3-6

Punnett square showing dihybrid cross between tall plants with yellow seeds and dwarf plants with green seeds. The phenotypic ratio in the F_2 is 9 tall yellow : 3 tall green : 3 dwarf yellow : 1 dwarf green. Note that the 9:3:3:1 ratio only results when the original cross is between homozygous dominant and homozygous recessive.

Parents	*T/T G/G* (tall yellow)	×		*t/t g/g* (dwarf green)
Gametes	all *TG*			all *tg*
F_1		*T/t G/g* (hybrid tall, hybrid yellow)		
Crossing hybrids	*T/t G/g*	×		*T/t G/g*
Gametes	*TG, Tg, tG, tg*			*TG, Tg, tG, tg*
F_2		(see checkerboard)		

	TG	*Tg*	*tG*	*tg*
TG	*T/T G/G* Pure tall Pure yellow	*T/T G/g* Pure tall Hybrid yellow	*T/t G/G* Hybrid tall Pure yellow	*T/t G/g* Hybrid tall Hybrid yellow
Tg	*T/T G/g* Pure tall Hybrid yellow	*T/T g/g* Pure tall Pure green	*T/t G/g* Hybrid tall Hybrid yellow	*T/t g/g* Hybrid tall Pure green
tG	*T/t G/G* Hybrid tall Pure yellow	*T/t G/g* Hybrid tall Hybrid yellow	*t/t G/G* Pure dwarf Pure yellow	*t/t G/g* Pure dwarf Hybrid yellow
tg	*T/t G/g* Hybrid tall Hybrid yellow	*T/t g/g* Hybrid tall Pure green	*t/t G/g* Pure dwarf Hybrid yellow	*t/t g/g* Pure dwarf Pure green

Mendel already knew that a cross between two plants bearing a single pair of alleles of the genotype *T/t* would yield a 3 : 1 ratio. Similarly, a cross between two plants with the genotypes *G/g* would yield the same 3 : 1 ratio. If we examine *only* the tall and dwarf phenotypes in the dihybrid experiment, they total up to 12 tall and 4 dwarf, giving a ratio of 3 : 1. Likewise, a total of 12 plants have yellow seeds and 4 plants have green—again a 3 : 1 ratio. Thus the monohybrid ratio prevails for both traits when they are considered independently. The 9:3:3:1 ratio is nothing more than two 3 : 1 ratios combined.

$$3:1 \times 3:1 = 9:3:3:1$$

When one of the alleles is unknown, it can be designated by a dash (*T/-*). This designation can also be used when it is immaterial whether the genotype is heterozygote or homozygote, as when we total all of a certain phenotype. The dash could be either *T* or *t*.

The F_2 genotypes and phenotypes are as follows:

Genotypes		Phenotypes
1 *T/T G/G* 2 *T/t G/G* 2 *T/T G/g* 4 *T/t G/g*	9 *T/- G/-*	9 tall yellow
1 *T/T g/g* 2 *T/t g/g*	3 *T/- g/g*	3 tall green
1 *t/t G/G* 2 *t/t G/g*	3 *t/t G/-*	3 dwarf yellow
1 *t/t g/g*	3 *t/t g/g*	1 dwarf green

The results of this experiment show that the segregation of alleles for plant height is entirely independent of the segregation of alleles for seed color. Neither has any influence on the other. This is the basis for Mendel's second law, the law of independent assortment, which states that *alleles of different pairs of genes on different chromosomes segregate independently of one another*. The reason is that during meiosis, the member of any pair of homologous chromosomes received by a gamete is independent of the other chromosomes it receives. Of course, if the genes were on the same chromosome, they would assort together (be linked) unless crossing-over occurred. Linked genes are discussed on p. 74.

One way to estimate numbers of progeny from a cross with a given genotype or phenotype is to construct a Punnett square and count them up. With a monohybrid cross, this is very easy; with a dihybrid cross, a Punnett square is rather laborious; and with a trihybrid cross, it is very tedious. We can make such estimates much more easily by taking advantage of simple probability calculations. The basic assumption is that all genotypes of gametes of one sex have an equal chance of uniting with all the genotypes of gametes of the other sex, in proportion to the numbers of each present. This is generally true when the sample size is large enough, and the actual numbers observed come close to those predicted by the laws of probability.

We may define probability as follows:

$$\text{Probability (p)} = \frac{\text{Number of times an event happens}}{\text{Total number of trials or possibilities for the event to happen}}$$

For example, the probability (p) of a coin falling heads when tossed is ½ because the coin has two sides. The probability of rolling a three on a die is ⅙ because the die has six sides.

The probability of independent events occurring together (ordered events) involves the **product rule**, which is simply the product of their individual probabilities. When two coins are tossed together, the probability of getting two heads is $\frac{1}{2} \times \frac{1}{2} = \frac{1}{4}$, or 1 chance in 4. Or, the probability of rolling two threes simultaneously with two dice as follows:

$$\text{p (two threes)} = \frac{1}{6} \times \frac{1}{6} = \frac{1}{36}$$

Note, however, that a small sample size may give a result quite different from that predicted. Thus if we tossed the coin three times and it fell heads each time, we would not be much surprised. But, if we tossed the coin 1000 times, and the number of times it fell heads diverged very much from 500, we would strongly suspect that there was something wrong with the coin or with the way we were tossing it.

We can use the product rule to predict the ratios of inheritance in a monohybrid or dihybrid (or larger) crosses, if the genes sort independently in the gametes (as they did in all of Mendel's experiments). In other words, the mechanism of placing *T* into a gamete is independent of the mechanism of putting *t* into a gamete. Therefore, in a monohybrid cross the probability that a sperm carries the dominant is ½ and the same applies to an ovum. In a dihybrid cross involving *T/t* and *G/g*, the same thing applies; the probability of any gene appearing in a gamete is ½. Now we can apply the product rule to an F_1 plant *T/t G/g* to determine the frequency of each kind of gamete:

$$\text{p of gamete being } TG = \frac{1}{2} \times \frac{1}{2} = \frac{1}{4}$$
$$\text{p of gamete being } Tg = \frac{1}{2} \times \frac{1}{2} = \frac{1}{4}$$
$$\text{p of gamete being } tG = \frac{1}{2} \times \frac{1}{2} = \frac{1}{4}$$
$$\text{p of gamete being } tg = \frac{1}{2} \times \frac{1}{2} = \frac{1}{4}$$

From this point the probabilities for the genotype in each box of the Punnett square on p. 70 can be easily derived as follows:

p of plant being $T/T\ G/G = \frac{1}{4} \times \frac{1}{4} = \frac{1}{16}$
p of plant being $T/T\ G/g = \frac{1}{4} \times \frac{1}{4} = \frac{1}{16}$

and so on for all 16 boxes.

Collecting all similar phenotypes together we get:

Tall, yellow seed $\frac{9}{16} = 9$
Tall, green seed $\frac{3}{16} = 3$
Dwarf, yellow seed $\frac{3}{16} = 3$
Dwarf, green seed $\frac{1}{16} = 1$

Thus we have the 9:3:3:1 ratio, derived by the product rule. In fact, one quickly learns by experience how to determine the ratios of phenotypes without using either the Punnett squares or the product rule. In a dihybrid (9:3:3:1 ratio), for instance, those phenotypes which make up the dominance of both genes are $\frac{9}{16}$ of the whole F_2 generation; each of the $\frac{3}{16}$ phenotypes consists of one dominant and one recessive; and the $\frac{1}{16}$ phenotype consists of two recessives.

The F_2 ratios in any cross involving more than one pair of contrasting pairs can be found by combining the ratios in the cross of one pair of alleles. Thus the number of genotypes is $(3)^n$ and the proportion of phenotypes $(3:1)^n$ when one allele is dominant and the other recessive. For example, let us suppose that in a cross of two pairs of alleles the phenotypes are in the ratio of $(3:1)^2$, or 9:3:3:1. The genotypes in such a cross are $(3)^2$, or 9. If three pairs of characters are involved (trihybrid cross), the proportions, or ratios, of the phenotypes are then $(3:1)^3$, or 27:9:9:9:3:3:3:1. The genotypes are $(3)^3$, or 27. Obviously, as the number of gene pairs increases, the number of phenotypes and genotypes rises steeply.

MULTIPLE ALLELES

Earlier we defined alleles as the alternate forms of a gene. Whereas an individual can have no more than two alleles at a given locus (one each on each chromosome of the homologous pair), many more dissimilar alleles can exist in the population. An example is the set of multiple alleles that affect coat color in rabbits. The different alleles are C (normal color), c^{ch} (chinchilla color), c^h (Himalayan color), and c (albino). The four alleles fall into a dominance series with C dominant over all the others. The dominant allele is always written to the left and the recessive to the right:

C/c^h = normal color
c^{ch}/c^h = chinchilla color
c^h/c = Himalayan color

Multiple alleles arise through mutations at the gene locus over long periods of time. Any gene may mutate (p. 86) if given time and thus can give rise to slightly different alleles at the same locus.

SEX-LINKED INHERITANCE

It is known that the inheritance of some characters depends on the sex of the parent carrying the gene and the sex of the offspring. An example is red-green color blindness in humans, in which red and green colors are indistinguishable to varying degrees. Color-blind men greatly outnumber color-blind women. When color blindness does appear in women, their fathers are color blind. Furthermore, if a woman with normal vision who is a carrier of color blindness (a **carrier** has the gene but is phenotypically normal) bears sons, half of them are likely to be color blind, regardless of whether the father had normal or affected vision. How are these observations explained?

The color-blindness defect is a recessive trait, the gene for which is carried on

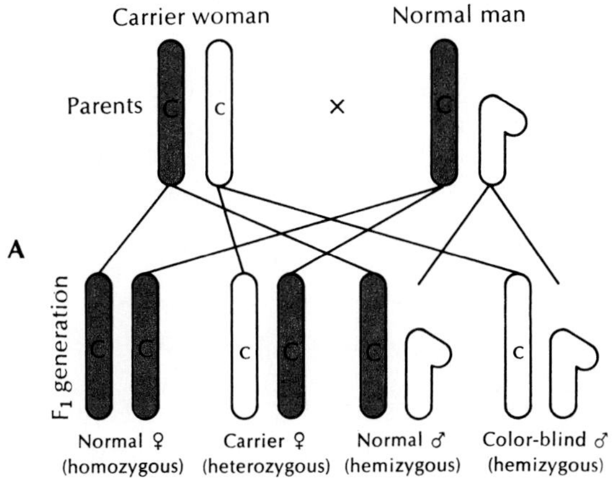

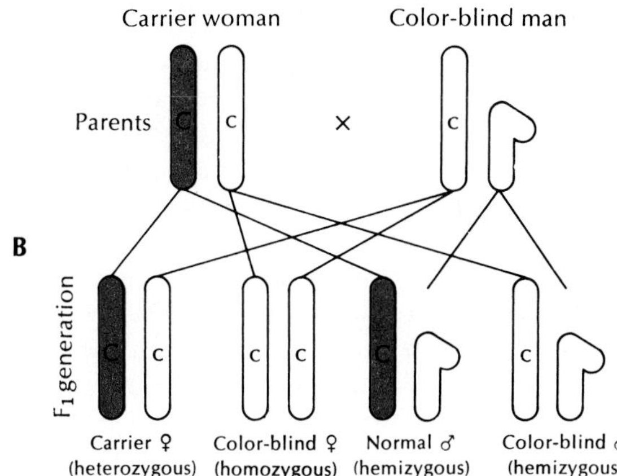

FIGURE 3-7

Sex-linked inheritance of red-green color blindness in humans. (*Hemizygous* refers to the condition in which only one allele is present, such as genes on the X chromosomes in males of humans.) **A,** Carrier mother and normal father produce color blindness in one half of their sons but in none of their daughters. **B,** Half of both sons and daughters of carrier mother and color-blind father are color blind.

the X chromosome. It is visibly expressed either when both genes are defective in the female or when only one defective gene is present in the male. The inheritance pattern is shown in Figure 3-7. When the mother is a carrier and the father is normal, half of the sons but none of the daughters are color blind. However, if the father is color blind and the mother is a carrier, half of the sons *and* half of the daughters are color blind (on the average and in a large sample). It is easy to understand then why the defect is much more prevalent in males: a single sex-linked recessive gene in the male has a visible effect. What would be the outcome of a mating between a homozygous normal woman and a color-blind man?

Another example of a sex-linked character was discovered by Morgan in *Drosophila*. The normal eye color of this fly is red, but mutations for white eyes do occur. The genes for eye color are known to be carried in the X chromosome. If a white-eyed male and a red-eyed female are crossed, all the F_1 offspring are red eyed because this trait is dominant (Figure 3-8). If these F_1 offspring are interbred, all the females of F_2 have red eyes, half the males have red eyes, and the other half have white eyes. No white-eyed females are found in this generation; only the males have the recessive character (white eyes). The gene for being white eyed is recessive and should appear in a homozygous condition. However, since the male has only one X chromosome (the Y does not carry a gene for eye color), white eyes appear whenever the X chromosome carries the gene for this trait. If the reciprocal cross is made in which the females are white eyed and the males red eyed, all the F_1 females are red eyed and all the males are white eyed (Figure 3-9). If these F_1 offspring are interbred, the F_2 generation shows equal numbers of red-eyed and white-eyed males and females.

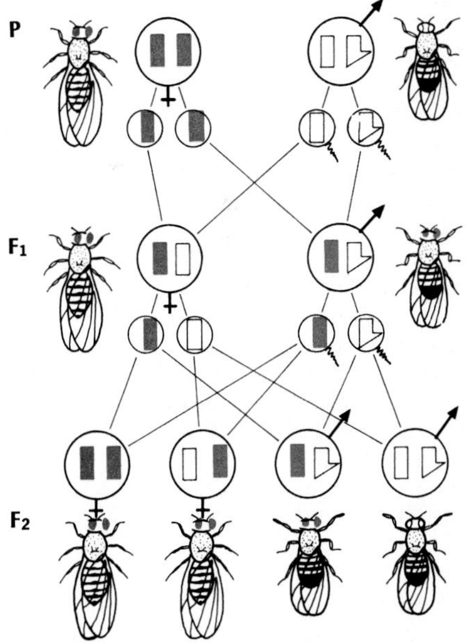

FIGURE 3-8

Sex-linked inheritance of eye color in fruit fly *(Drosophila)*. Genes for eye color are carried on X chromosome; Y carries no genes for eye color. Normal red is dominant to white. Homozygous red-eyed female mated with white-eyed male gives all red-eyed in F_1. F_2 ratios from F_1 cross are one homozygous red-eyed female and one heterozygous red-eyed female to one red-eyed male and one white-eyed male.

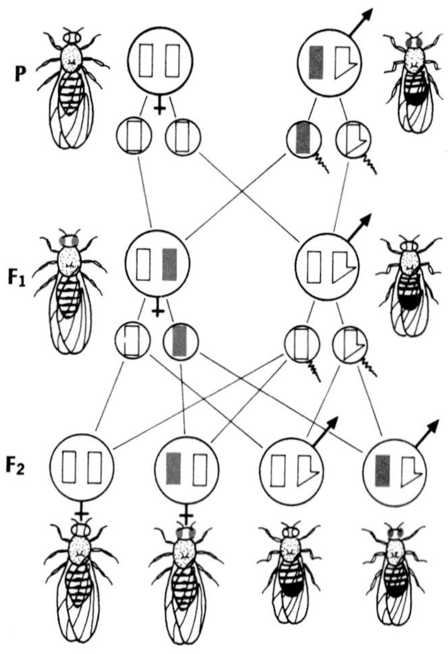

FIGURE 3-9

Reciprocal cross of Fig. 3-8 (homozygous white-eyed female with red-eyed male) gives white-eyed males and red-eyed females in F_1. F_2 shows equal numbers of red-eyed and white-eyed females and red-eyed and white-eyed males.

Geneticists commonly use the word "linkage" in two somewhat different senses. Sex-linkage refers to inheritance of a trait on the sex chromosomes, and thus its phenotypic expression depends on the sex of the organism and the factors already discussed. Autosomal linkage, or simply, linkage, refers to inheritance of the genes on a given autosomal chromosome. Letters used to represent such genes are normally written without a slash mark between them, indicating that they are on the same chromosome. For example, *AB/ab* shows that genes *A* and *B* are on the homologous chromosome. Interestingly, Mendel studied seven characteristics of garden peas, which assorted independently because they were on seven different chromosomes. If he had studied eight characteristics, he would not have found independent assortment in two of the traits because garden peas have only seven pairs of homologous chromosomes.

AUTOSOMAL LINKAGE AND CROSSING-OVER

Linkage

Since Mendel's laws were rediscovered in 1900, it became apparent that, contrary to Mendel's second law, not all factors segregate independently. Indeed, many traits are inherited together. Since the number of chromosomes in any organism is relatively small compared to the number of traits, each chromosome must contain many genes. All genes present on a chromosome are said to be **linked.** Linkage simply means that the genes are on the same chromosome, and all genes present on homologous chromosomes belong to the same linkage groups. Therefore there should be as many linkage groups as there are chromosome pairs.

In *Drosophila,* in which this principle has been worked out most extensively, there are four linkage groups that correspond to the four pairs of chromosomes found in these fruits flies. Usually, small chromosomes have small linkage groups, and large chromosomes have large groups.

Let us see how the Mendelian ratios can be altered by linkage, as illustrated by one of Morgan's experiments on *Drosophila.* In fruit fly genetics, the normal allele of any gene is called the **wild type,** since that allele is the most widespread in the wild state. It is usually dominant over its sister alleles, which are considered mutations of the normal wild allele. Morgan made a cross between wild-type fruit flies with normal bodies and wings and flies bearing two recessive mutant characters of black bodies and vestigial (small, very undeveloped) wings.

As expected, the F_1 generation of this cross was phenotypically the wild type, confirming that the alleles for black bodies and vestigial wings were recessive.

Morgan then made a testcross to learn more about the genotype of the F_1 generation. This was done by breeding back the F_1 hybrid generation *(AB/ab)* to the double-recessive flies of the parental generation *(ab/ab).* With independent assortment we should expect four different phenotypes in approximately equal numbers:

Phenotype	Expected ratio	Numbers obtained
Wild type	1	586 (46%)
Normal body, vestigial wing	1	106 (8%)
Black body, normal wing	1	111 (9%)
Black body, vestigial wing	1	465 (37%)

Instead Morgan obtained an excess of **parental** types (wild type and double-recessive type) and a deficiency of the two gene recombinations (called **recombinant** types). In a testcross such as this, linkage is indicated if the proportion of parental types exceeds 50%. This cross yielded 83% parental types and 17% recombinant types. Morgan concluded that the wild type and black-vestigial type had entered the dihybrid cross together and stayed together, or were linked.

Crossing-over

Linkage, however, is usually not complete. In the experiment just described, the parental forms totaled 83% rather than 100% as expected if linkage were complete. The fact that some recombinant types appear means that the linked genes have indeed separated, in this experiment 17% of the time. Separation of genes located on the same chromosome occurs because of **crossing-over.**

As described earlier, during the protracted prophase of the first meiotic division, chromatids of homologous chromosomes come to lie alongside each other, so that the locus of each gene is exactly opposite the same locus in the chromatid of the homologous chromosome. Sometimes the chromatids break and exchange equivalent portions; genes "cross over" from one chromosome to its homologue, and vice versa (Figure 3-10). Breaks and exchanges occur at corresponding points on nonsister chromatids. (Breaks and exchanges also occur between sister

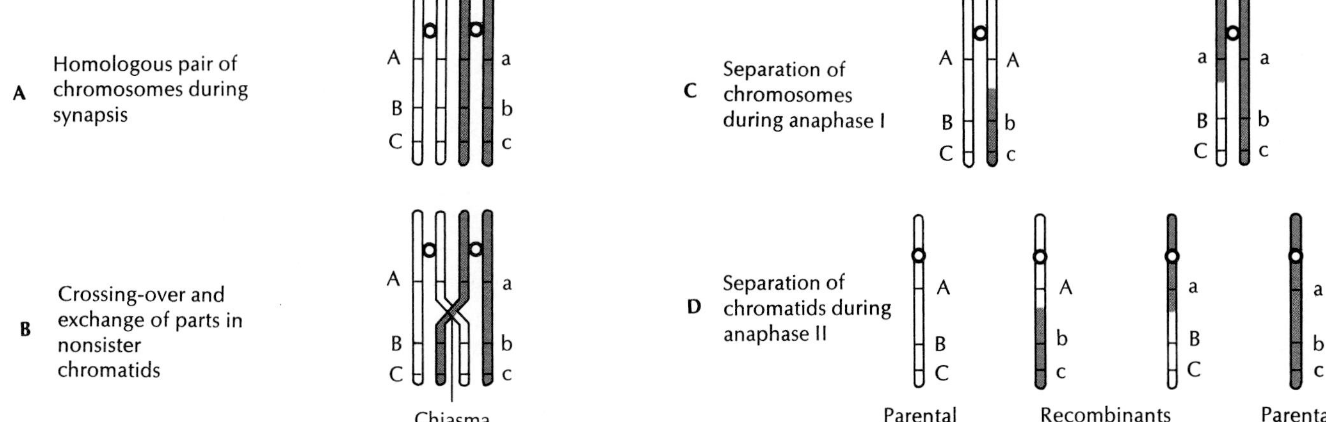

FIGURE 3-10

Crossing-over during meiosis. Nonsister chromatids exchange portions, so that none of the resulting gametes is genetically the same as any other. Gene *A* is farther from gene *B* than *B* is from *C;* therefore gene *A* is more frequently separated from *B* in crossing-over than *B* is from *C.*

chromatids but usually have no genetic significance because sister chromatids are identical.) Crossing-over then is a means for exchanging genes between homologous chromosomes and as such greatly increases the amount of genetic recombination. The frequency of crossing-over varies with the species, but usually at least one and often several cross-overs occur each time chromosomes pair in humans.

Because the frequency of recombination is proportional to the distance between loci, the comparative linear position of each locus can be determined. Laborious genetic experiments over many years have produced gene maps that indicate the position of more than 500 genes distributed on the four *Drosophila* chromosomes.

CHROMOSOMAL ABERRATIONS

Structural and numerical deviations from the norm that affect many genes at once are called chromosomal aberrations. They are sometimes called chromosomal mutations, but most cytogeneticists prefer to use the term "mutation" to refer to qualitative changes within a gene; gene mutations will be discussed on p. 86.

Despite the incredible precision of meiosis, chromosomal aberrations do occur, and they are more common than one might think. They are responsible for great economic benefit in agriculture. Unfortunately, they are also responsible for many human genetic malformations. It is estimated that five out of every 1000 humans are born with *serious* genetic defects, attributable to chromosomal anomalies. An even greater number of embryos with chromosomal defects are aborted spontaneously, far more than ever reach term.

Changes in chromosome numbers are called **euploidy** when there is the addition or deletion of complete haploid sets and **aneuploidy** when a single chromosome is added or subtracted from a diploid set. The most common kind of euploidy is **polyploidy,** the carrying of one or more additional sets of chromosomes. Such aberrations are much more common in plants than animals. Animals are much less tolerant of chromosomal aberrations because sex determination requires a delicate balance between the numbers of sex chromosomes and autosomes. Many domestic plant species are polyploid (cotton, wheat, apples, oats, tobacco, and others), and perhaps over 40% of flowering plants are believed to have originated in this manner. Horticulturists favor polyploids and often try to develop them because they have more intensely colored flowers and more vigorous vegetative growth.

Aneuploidy is usually caused by nondisjunctional separation of chromosomes during meiosis. If a pair of homologous chromosomes fails to separate during the

A *syndrome* is a group of symptoms associated with a particular disease or abnormality, although every symptom is not necessarily shown by every patient with the condition. An English physician, John Langdon Down, described the syndrome in 1866 that we now know is caused by trisomy 21. Because of Down's belief that the facial features of affected individuals were mongoloid in appearance, the condition has been known as mongolism. The resemblances are superficial, however, and the currently accepted names are trisomy 21 and Down syndrome. Among the numerous characteristics of the condition, the most disabling is severe mental retardation. This, as well as other conditions caused by chromosomal aberrations and several other birth defects, can be diagnosed *prenatally* by a procedure involving *amniocentesis*. The physician inserts a hypodermic needle through the abdominal wall of the mother and into the fluids surrounding the fetus (*not into* the fetus) and withdraws some of the fluid, which contains some fetal cells. The cells are grown in culture, their chromosomes are examined, and other tests done. If a severe birth defect is found, the mother has the option of having an abortion performed. As an extra "bonus," the sex of the fetus is learned after amniocentesis. How?

first meiotic division or sister chromatids fail to separate during the second meiotic division, both members go to one pole and none to the other. This results in one gamete having $n - 1$ number of chromosomes and another having $n + 1$ number of chromosomes. If the $n - 1$ gamete is fertilized by a normal n gamete, the result is a **monosomic** animal. Survival is rare because the lack of one chromosome gives an uneven balance of genetic instructions. **Trisomy,** the result of the fusion of a normal n gamete and an $n + 1$ gamete, is much more common, and several kinds of trisomic conditions are known in humans. Perhaps the most familiar is **trisomy 21,** or **Down syndrome.** As the name indicates, it involves an extra chromosome 21 combined with the chromosome pair 21, and it is caused by nondisjunction of that pair during meiosis. It occurs spontaneously, and there seldom is any family history of the abnormality. However, the risk of its appearance rises dramatically with increasing age of the mother; it occurs 40 times as often in pregnancies of women over 40 years old than among women between the ages of 20 and 30.

Structural aberrations involve whole sets of genes within a chromosome. A portion of a chromosome may be reversed, placing the linear arrangement of genes in reverse order **(inversion);** nonhomologous chromosomes may exchange sections **(translocation);** entire blocks of genes may be lost **(deletion);** or an extra section of chromosome may attach to a normal chromosome **(duplication).** These are all structural changes that usually do not produce phenotypic changes. Duplications, although rare, are important for evolution because they supply additonal genetic information that may enable new functions.

■ Gene Theory

GENE CONCEPT

The term "gene" (Gr. *genos,* descent) was coined by W. Johannsen in 1909 to refer to the hereditary factors of Mendel. Both cytological and genetic studies showed that genes, although of unknown chemical nature, were the fundamental units of inheritance. They were regarded as indivisible units of the chromosomes on which they were located. Studies with multiple mutant alleles demonstrated that alleles are in fact divisible by recombination; that is, *portions* of a gene are separable, and they have a fine structure. Furthermore, parts of many genes in eukaryotes are separated by sections of DNA that do not specify a part of the finished product **(introns).**

As the chief functional unit of genetic material, genes determine the basic architecture of every cell, the nature and life of the cell, the specific protein syntheses, the enzyme formation, the self-reproduction of the cell, and, directly or indirectly, the entire metabolic function of the cell. Because of their ability to mutate, to be assorted and shuffled around in different combinations, genes have become the basis for our modern interpretation of evolution. Genes are molecular patterns that can maintain their identities for many generations, can be self-duplicated in each generation, and can control processes by allowing their specificities to be copied.

One gene—one enzyme hypothesis

Since genes act to produce different phenotypes, we may infer that their action follows the scheme: gene → gene product → phenotypic expression. Furthermore, we may suspect that the gene product is usually a protein, because proteins, acting as enzymes, antibodies, hormones, and structural elements throughout the body, are the single most important group of biomolecules.

The first clear, well-documented study to link genes and enzymes was carried out on the common bread mold *Neurospora* by Beadle and Tatum in the early 1940s. This organism was ideally suited to a study of gene function for several reasons: these molds are much simpler to handle than fruit flies, they grow readily in well-defined chemical media, and they are haploid organisms that are consequently unencumbered with dominance relationships. Furthermore, mutations were readily induced by irradiation with ultraviolet light. Ultraviolet light–induced mutants, grown and tested in specific nutrient media, were found to have single-gene mutations that were inherited in accord with Mendelian principles of segregation. Each mutant strain was shown to be defective in one enzyme, which prevented that strain from synthesizing one or more complex molecules. Putting it another way, the ability to synthesize a particular molecule was controlled by a single gene.

From these experiments Beadle and Tatum set forth an important and exciting formulation: **one gene produces one enzyme.** For this work they were awarded the Nobel Prize in 1958. The new hypothesis was soon validated by the research of others who studied other biosynthetic pathways. Hundreds of inherited disorders, including dozens of human hereditary diseases, are caused by single mutant genes that result in the loss of a specific essential enzyme. We now know that a particular protein may be made of several chains of amino acids (polypeptides), each of which may be specified by a different gene, and not all proteins specified by genes are enzymes (for example, antibodies and hormones). Furthermore, genes direct the synthesis of various kinds of RNA. Therefore a gene now may be defined as a **nucleic acid sequence (usually DNA) that encodes a functional polypeptide or RNA sequence.**

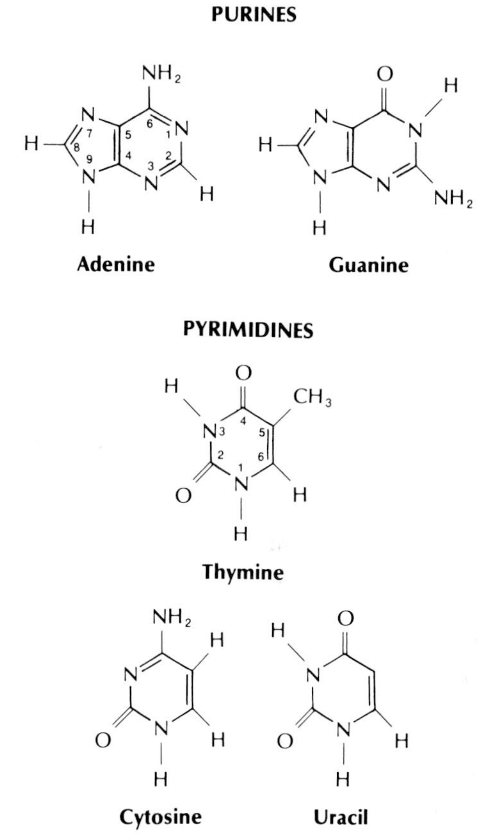

Storage and Transfer of Genetic Information

NUCLEIC ACIDS: MOLECULAR BASIS OF INHERITANCE

It has been known for many years that genes and chromosomes are made up chiefly of nucleoproteins; that is, they are macromolecules composed of **nucleic acids,** histones, and nonhistone proteins. To review briefly, nucleic acids are polymers of **nucleotides,** which are in turn composed of a **purine** or **pyrimidine** nitrogenous base (Figure 3-11), a sugar **(ribose** or **deoxyribose)** (Figure 3-12), and a phosphate group. The sugar, phosphate group, and nitrogenous base are linked as shown in this generalized scheme for a nucleotide.

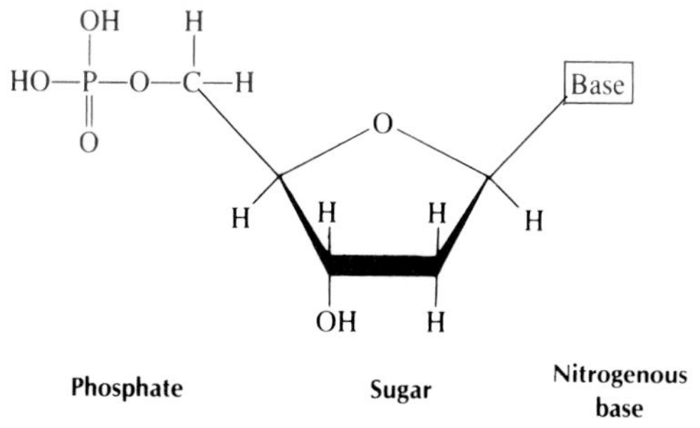

FIGURE 3-11

Purines and pyrimidines of DNA and RNA.

The sugar in DNA is a pentose (5-carbon) sugar called deoxyribose. In this representation, a carbon atom lies at each of the four corners of the pentagon (labeled 1 to 4) (Figure 3-12). Ribose, found in RNA, has the same formula except that there is a hydroxyl group (—OH) and a hydrogen on the number 2 carbon, rather than two hydrogens.

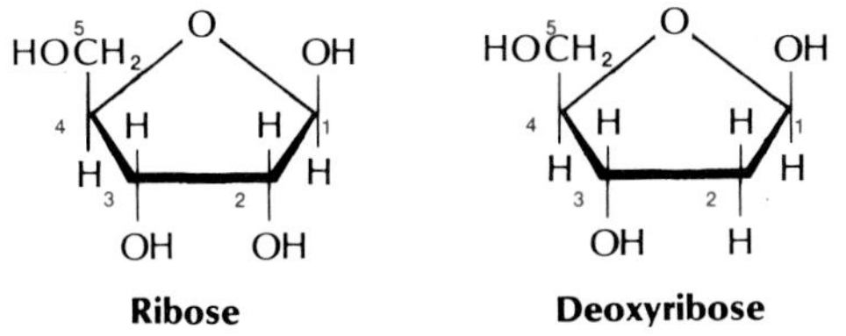

Of the nitrogenous bases in DNA, two of them are compounds composed of 9-membered double rings, classified as purines. They are **adenine** and **guanine** (Figure 3-11). As before, carbon atoms lie at each corner of the ring not occupied by some other atom, in this case nitrogen. The other two nitrogenous bases belong to a different class of organic compounds called pyrimidines, consisting of 6-membered rings. The two pyrimidines found in DNA are **thymine** and **cyto-**

FIGURE 3-12

Ribose and deoxyribose, the pentose sugars of nucleic acids. A carbon atom lies in each of the four corners of the pentagon (labeled 1 to 4). Ribose has a hydroxyl group (—OH) and a hydrogen on the number 2 carbon; deoxyribose has 2 hydrogens at this position.

FIGURE 3-13

Structure of polynucleotides, as in DNA. The chain is built of a backbone of phosphoric acid and deoxyribose sugar molecules. Each sugar holds a nitrogenous base side arm. Shown from top to bottom are adenine, guanine, thymine, and cytosine.

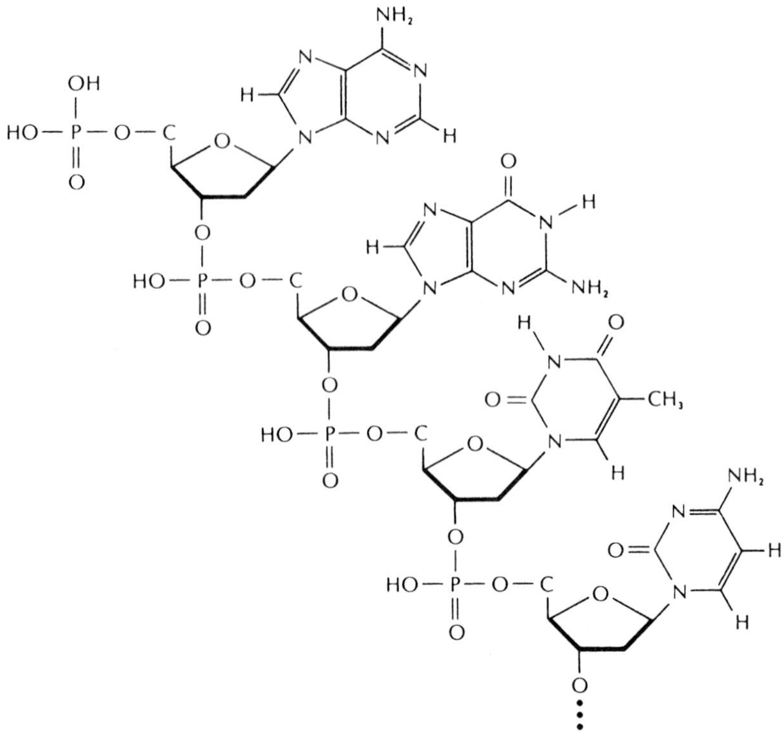

sine. In RNA, there is no thymine, but instead the similar pyrimidine **uracil** is found.

In DNA the backbone of the molecule is built of phosphoric acid and deoxyribose; to this backbone are attached the nitrogenous bases (Figure 3-13). However, one of the most interesting and important discoveries about the nucleic acids is that DNA is not a single polynucleotide chain; rather it consists of *two* complementary chains that are precisely cross-linked by specific hydrogen bonding between purine and pyrimidine bases. It was found that the number of adenines is equal to the number of thymines, and the number of guanines equals the number of cytosines. This fact suggests a pairing of bases: adenine with thymine (AT) and guanine with cytosine (GC) (Figure 3-14). The larger adenine (a purine) always attaches to the smaller thymine (a pyrimidine) by two hydrogen bonds; the larger guanine (a purine) always attaches to the smaller cytosine (a pyrimidine) by three hydrogen bonds, forming a ladder-shaped structure (Figure 3-15). The upright portions are the sugar-phosphate backbones, and the connecting rungs are the paired nitrogenous bases, AT or GC. However, the ladder is twisted into a **double helix** with approximately 10 base pairs for each complete turn of the helix (Figure 3-16).

FIGURE 3-14

Positions of hydrogen bonds between thymine and adenine and between cytosine and guanine in DNA.

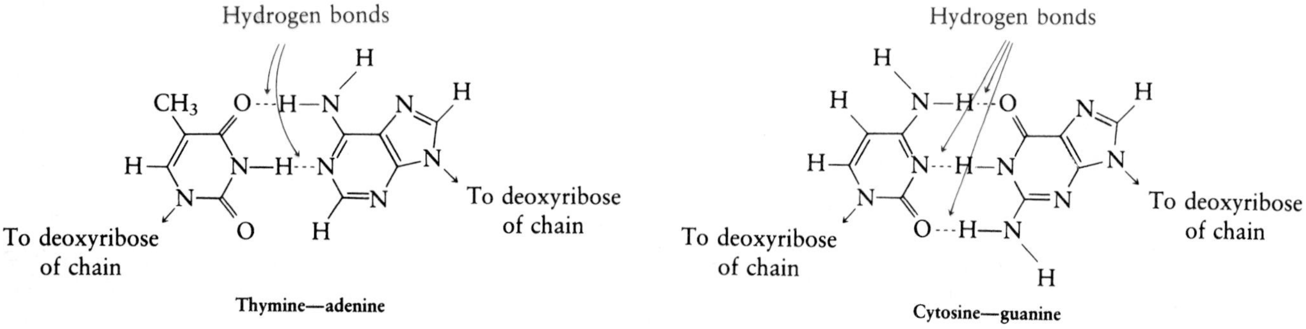

Hydrogen bonds

To deoxyribose of chain

To deoxyribose of chain

Thymine—adenine

Hydrogen bonds

To deoxyribose of chain

To deoxyribose of chain

Cytosine—guanine

FIGURE 3-15

DNA, showing how the complementary pairing of bases between the sugar-phosphate "backbones" keeps the double helix at a constant diameter for the entire length of the molecule. Dotted lines represent the three hydrogen bonds between each cytosine and guanine and the two hydrogen bonds between each adenine and thymine.

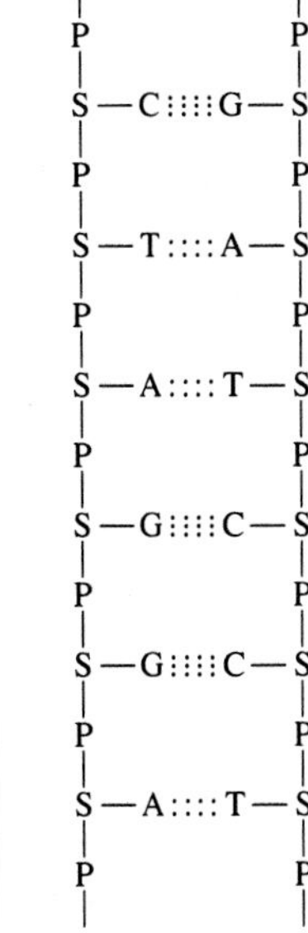

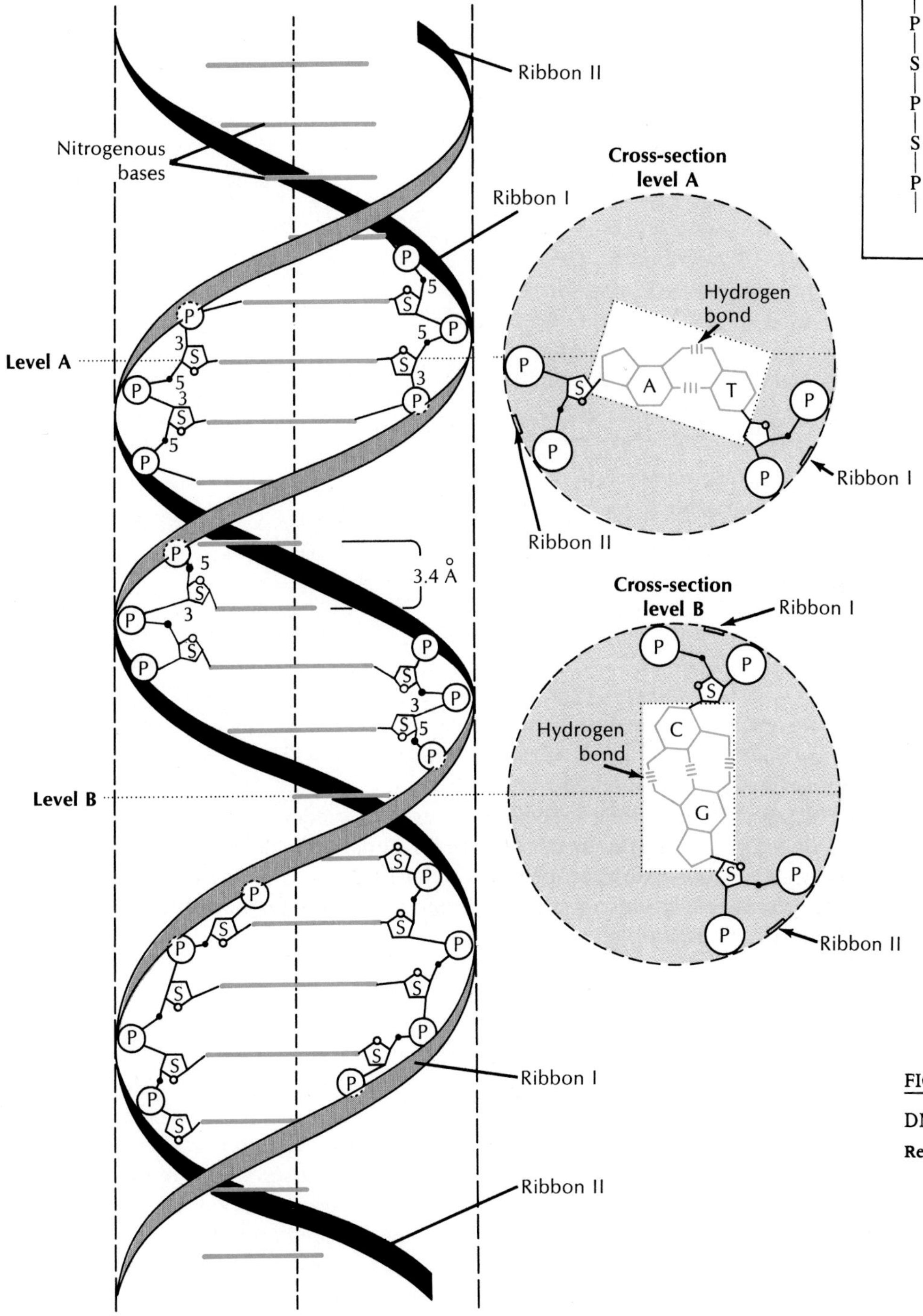

FIGURE 3-16

DNA molecule.

Redrawn from Etkin, W. 1973. BioScience 23:653.

The determination of the structure of DNA has been widely acclaimed as the single most important biological discovery of this century. It was based on the x-ray diffraction studies of Maurice H.F. Wilkins and Rosalind Franklin and on the ingenious proposals of Francis H.C. Crick and James D. Watson published in 1953. Watson, Crick, and Wilkins were later awarded the Nobel Prize for Medicine and Physiology for their momentous work.

RNA is very similar to DNA in structure except that it consists of a *single* polynucleotide chain, has ribose instead of deoxyribose, and has uracil instead of thymine. The three kinds of RNA (ribosomal, transfer, and messenger) are described below.

Every time a cell divides, the structure of DNA must be precisely copied in the daughter cells. This is called **replication** (Figure 3-17). During replication, the two strands of the double helix unwind, and each separated strand serves as a **template** against which a complementary strand is synthesized. That is, an enzyme (DNA polymerase) assembles a new strand of polynucleotides with a thymine group going next to the adenine group in the template strand, a guanine group next to the cytosine group, and so on.

FIGURE 3-17

Replication of DNA. The parent strands of DNA part, and DNA polymerase synthesizes daughter strands using the base sequence of parent strands as a template.

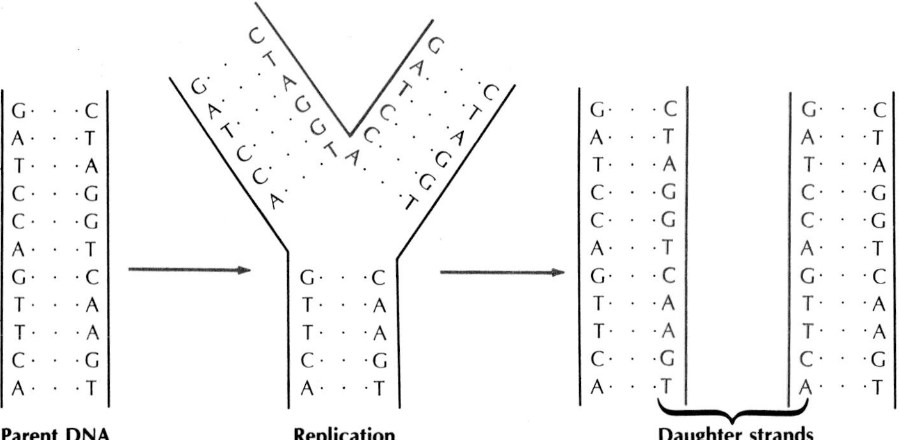

DNA coding by base sequence

Since DNA is the genetic material and is composed of a linear sequence of base pairs, an obvious extension of the Watson-Crick model is that the sequence of base pairs in DNA codes for, and is colinear with, the sequence of amino acids in a protein. The coding hypothesis had to account for the way a string of four different bases—a four-letter alphabet—could dictate the sequence of 20 different amino acids.

In the coding procedure, obviously there cannot be a 1:1 correlation between four bases and 20 amino acids. If the coding unit (often called a word) consists of two bases, only 16 words (4^2) can be formed, which cannot account for 20 amino acids. Therefore the protein code must consist of at least three bases or three letters because 64 possible words (4^3) can be formed by four bases when taken as triplets. This means that there could be a considerable redundancy of triplets, since DNA codes for just 20 amino acids. Later work by Crick confirmed that nearly all of the amino acids are specified by more than one triplet code. The triplet codes for each amino acid are shown for messenger RNA in Table 3-1.

Table 3-1 The genetic code: amino acids specified by codons of messenger RNA

First letter	Second letter: U	Second letter: C	Second letter: A	Second letter: G	Third letter
U	UUU, UUC: Phenylalanine UUA, UUG: Leucine	UCU, UCC, UCA, UCG: Serine	UAU, UAC: Tyrosine UAA, UAG: End chain	UGU, UGC: Cysteine UGA: End chain UGG: Tryptophane	U C A G
C	CUU, CUC, CUA, CUG: Leucine	CCU, CCC, CCA, CCG: Proline	CAU, CAC: Histidine CAA, CAG: Glutamine	CGU, CGC, CGA, CGG: Arginine	U C A G
A	AUU, AUC, AUA: Isoleucine AUG: Methionine*	ACU, ACC, ACA, ACG: Threonine	AAU, AAC: Asparagine AAA, AAG: Lysine	AGU, AGC: Serine AGA, AGG: Arginine	U C A G
G	GUU, GUC, GUA, GUG: Valine	GCU, GCC, GCA, GCG: Alanine	GAU, GAC: Aspartic acid GAA, GAG: Glutamic acid	GGU, GGC, GGA, GGG: Glycine	U C A G

*Also, begin chain.

TRANSCRIPTION AND THE ROLE OF MESSENGER RNA

Information is coded in DNA, but DNA does not participate directly in protein synthesis. It is obvious that an intermediary is required. This intermediary is another nucleic acid called **messenger RNA (mRNA)**. The triplet codes in DNA are **transcribed** into mRNA, with uracil substituting for thymine, and are now called **codons** (Table 3-1).

In addition to mRNA, the other types of RNA (ribosomal, transfer) are also transcribed directly from DNA, with each of the many mRNAs being determined by a gene. In this process of making a complementary copy of one strand or gene of DNA in the formation of RNA, an enzyme, **RNA polymerase**, is needed. In eukaryotes each type of RNA is transcribed by a specific type of RNA polymerase. The RNA contains a sequence of bases that complements the bases in one of the two DNA strands, just as the DNA strands complement each other. Thus, A in the coding DNA strand is replaced by U in RNA; C is replaced by G, G is replaced by C, and T is replaced by A. Only one of the two chains is used as the template for RNA synthesis because only one bears the triplet coding for AUG, in the initiation codon in RNA (Table 3-1). The reason why only one strand of the double-stranded DNA is a "coding strand" is that mRNA otherwise would always be formed in complementary pairs, and enzymes also would be synthesized in complementary pairs. In other words, two different enzymes would be produced for every DNA coding sequence instead of one. This would certainly lead to metabolic chaos.

Only one strand of DNA serves as the coding strand in all DNA except that found in plasmids (see p. 85). Messenger RNA can be transcribed from both DNA strands in one region of plasmid DNA, and this is the only known example of proteins being encoded in both DNA strands.

Genes on the DNA of prokaryotes are coded on a continuous stretch of DNA, which is transcribed into mRNA and then translated (see the following section). It was assumed that this was also the case for eukaryote genes until the surprising discovery a few years ago that stretches of DNA are transcribed in the nucleus into mRNA that does not encode the finished product. In other words, *the mRNA that is translated into protein in the cytoplasm is not the same mRNA that is transcribed*

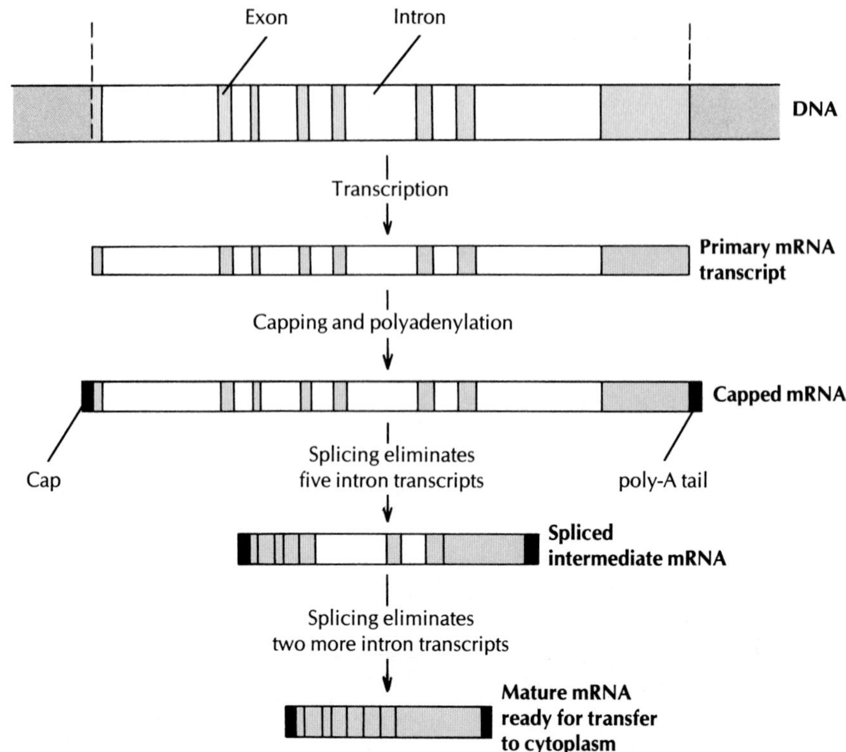

FIGURE 3-18

Transcription and maturation of ovalbumin gene of chicken. The entire gene of 7700 base pairs is transcribed to form the primary mRNA, then the cap of methyl guanine and the polyadenylate tail are added. After the introns are spliced out, the mature mRNA is transferred to the cytoplasm.

Redrawn from Chambon, P. 1981. Sci. Am. **244**:60-71 (May).

in the nucleus; pieces of the nuclear mRNA have been spliced out (Figure 3-18). It was thus discovered that many genes are split, interrupted by sequences of bases that do not code for the final product, and the mRNA transcribed from them must be edited or "matured" before translation in the cytoplasm. For example, a gene for the protein collagen in chickens is about 40,000 base pairs long, while the mature mRNA translated is only 5000 base pairs long. The intervening segments of DNA are now known as **introns**, while those that code for part of the mature RNA and are translated into gene products are called **exons**. Before the mRNA leaves the nucleus, the introns are spliced out and a methylated guanine **cap** is added at one end, while a tail of adenine nucleotides **(poly-*A*)** is added at the other (Figure 3-18). The cap and the poly-*A* tail are characteristic of mRNA molecules.

Even more recently it was found that the belief that introns are noncoding regions is not necessarily true. At least in some mitochondrial mRNAs, the introns may code for proteins, and in one case for a ribosomal component. In most of those studied so far, the proteins coded by the introns are "maturases," proteins that play some role in the splicing of the introns from which they came or introns from different genes. There are base sequences in some introns that are complementary to other base sequences in the intron, suggesting that the intron could fold so that the complementary sequences would pair. This may be necessary to control proper alignment of intron boundaries before splicing. Most surprising of all has been the discovery that, at least in some cases, the RNA can "self-catalyze" the excision of introns. The ends of the intron join; the intron thus becomes a small circle of RNA, and the exons are spliced together. This process does not fit the classical definition of an enzyme or other catalyst since the molecule itself is changed in the reaction.

TRANSLATION: FINAL STAGE IN INFORMATION TRANSFER

The **translation** process takes place on ribosomes, granular structures composed of protein and **ribosomal RNA (rRNA)**. Ribosomal RNA is composed of a large

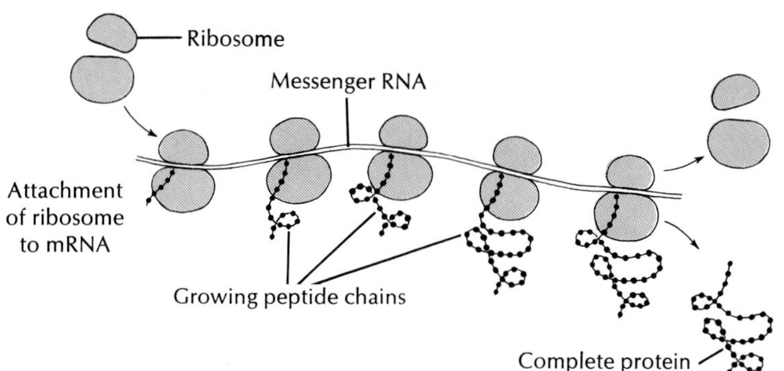

FIGURE 3-19

How the protein chain is formed. As ribosomes move along messenger RNA, the amino acids are added stepwise to form the polypeptide chain.

and a small subunit, and the small subunit comes to lie in a depression of the large subunit to form the functional ribosome (Figure 3-19). The mRNA molecules fix themselves to the ribosomes to form a messenger RNA–ribosome complex. Since only a short section of mRNA molecule is in contact with a single ribosome, the mRNA usually fixes itself to several ribosomes at once. The entire complex, called a **polyribosome** or **polysome**, allows several molecules of the same kind of protein to be synthesized at once, one on each ribosome of the polysome (Figure 3-19).

The assembly of proteins on the mRNA-ribosome complex requires the action of another kind of RNA called **transfer RNA (tRNA)**. The tRNAs are surprisingly large molecules that are folded in a complicated way in the form of a cloverleaf (Figure 3-20). The tRNA molecules collect the free amino acids from the cytoplasm and deliver them to the polysome, where they are assembled into a protein. There is a special tRNA molecule for every amino acid. Furthermore each tRNA is accompanied by a special **activating enzyme**. These enzymes are necessary to sort out and attach the correct amino acid to a site on the end of each tRNA by a process called **loading**.

On the end of the tRNA opposite the amino acid site, a special sequence of three bases (the **anticodon**) is exposed in just the right way to form base pairs with complementary bases (the codon) in the mRNA. The anticodon of the tRNA is the key to the correct sequencing of amino acids in the protein being assembled.

For example, alanine is assembled into a protein when it is signaled by the codon GCG in an mRNA. The translation is accomplished by alanine tRNA in which the anticodon is CGC. The alanine tRNA is first loaded with alanine by its activating enzyme. The loaded alanine tRNA enters the ribosome where it fits precisely into the right place on the mRNA stand. Then the next loaded tRNA specified by the mRNA code (glycine tRNA, for example) enters the ribosome and attaches itself beside the alanine tRNA. The two amino acids are united with a peptide bond (with the energy from a molecule of guanosine triphosphate), and the alanine tRNA falls off. The process continues stepwise as the protein chain is built (Figure 3-21). A protein of 500 amino acids can be assembled in less than 30 seconds.

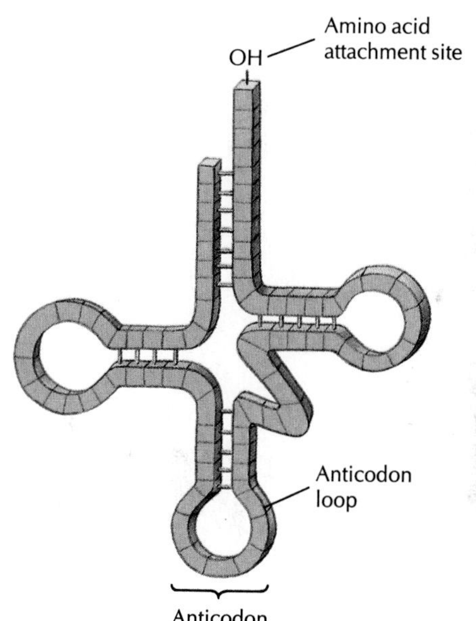

FIGURE 3-20

Structure of a tRNA molecule. The anticodon loop bears bases complementary to those in the mRNA codon. The other two loops function in binding to the ribosomes in protein synthesis. The amino acid is added to the free single-stranded —OH end by activating enzyme.

REGULATION OF GENE FUNCTION

In chapter 13 we will see how the orderly differentiation of an organism from fertilized ovum to adult requires the involvement of genetic material at every stage of development. Developmental biologists have provided convincing evidence that every cell in a developing embryo is genetically equivalent. Thus it is clear that as tissues differentiate (change developmentally), they use only a part of the genetic instruction present in every cell. Certain genes express themselves only at certain times and not at others. Indeed, there is reason to believe that in a

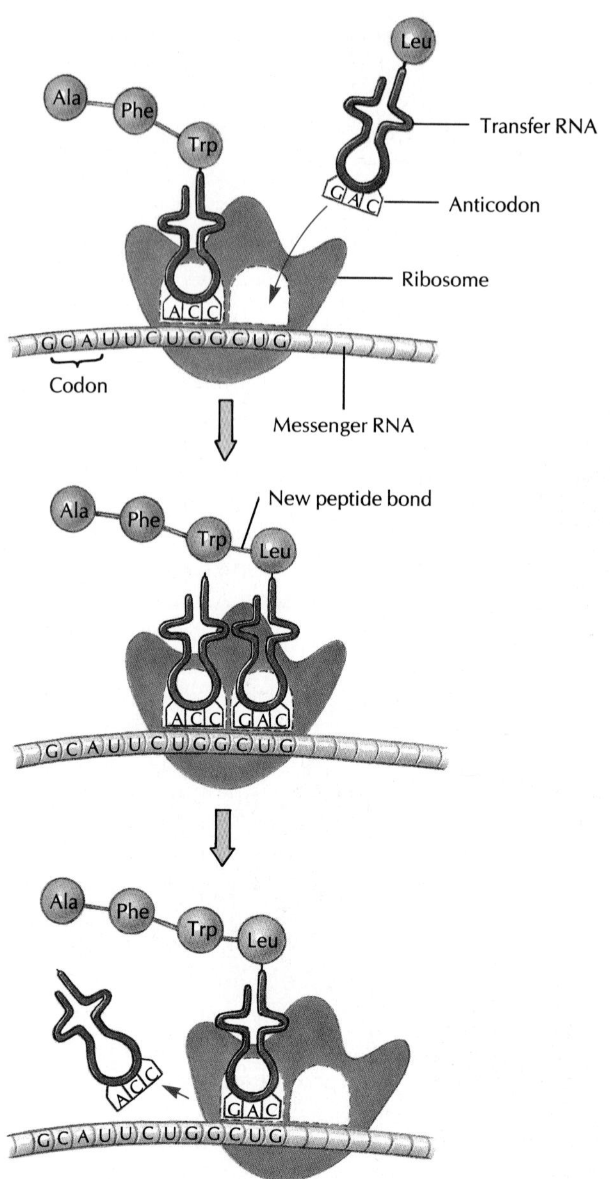

FIGURE 3-21

Formation of polypeptide chain on messenger RNA. As ribosome moves down messenger RNA molecule, transfer RNA molecules with attached amino acids enter ribosome *(top)*. Amino acids are joined together into polypeptide chain, and transfer RNA molecules leave ribosome *(bottom)*.

A

```
          NH2
          |
          C         CH3
        //4 \      /
      N 3    5 C
      |        ||
      C 2    6 C
    //  \ 1  /   \
   O      N        H
          |
          H
```

B

```
—P—S—P—S—P—S—P—S—
   |     |     |     |
   A     C—me G     T
   :     :     :     :
   T     G     C—me A
   |     |     |     |
—P—S—P—S—P—S—P—S—
```

FIGURE 3-22

Some genes in eukaryotes are apparently turned off by the methylation of some cytosine residues in the chain. **A,** Structure of 5-methyl cytosine. **B,** Cytosine residues next to guanine are those that are methylated in a strand, thus allowing both strands to be symmetrically methylated.

particular cell or tissue, most of the genes are inactive at any given moment. The problem in development is to explain how, if every cell has a full gene complement, certain genes are "turned on" and produce proteins that are required for a particular developmental stage while the other genes remain silent.

Actually, gene regulation is necessary throughout an organism's existence. The cellular enzyme systems that control all functional processes obviously require genetic regulation because enzymes have powerful effects even in minute amounts. Enzyme synthesis must be responsive to the influences of supply and demand.

The discoveries of two French scientists, J. Monod and F. Jacob, contributed greatly to our understanding of gene regulation. Using bacteria, they described the **operon model.** Knowledge of gene regulation in prokaryotes is important in recombinant DNA technology, but we now know that the operon concept has little if any applicability to gene regulation in eukaryotes.

Gene regulation in eukaryotes

A number of different phenomena in eukaryote cells are known that can serve as control points, and the following are a few examples.

Gene rearrangement. Genes coding for many, but not all, proteins are split; that is, they are not on continuous stretches of DNA. In the differentiation of lymphocytes, whose descendants can synthesize an enormous variety of different proteins called antibodies (p. 177), the genes coding for these proteins are actually *rearranged* during development. The different proteins result from subsequent transcription and translation.

DNA modification. An important mechanism for turning genes off appears to be methylation of cytosine residues in the 5′ position, that is, adding a methyl group (CH_3—) to the carbon in the 5′ position in the cytosine ring (Figure 3-22, *A*). This usually happens when the cytosine is next to a guanine residue; thus, the bases in the complementary DNA strand would also be a cytosine and a guanine (Figure 3-22, *B*). When the DNA is replicated, an enzyme recognizes the CG sequence and quickly methylates the daughter strand, maintaining the gene in an inactive state.

Transcriptional control. Some hormones act by entering the cell and affecting gene transcription directly. Progesterone, for example, enters the cells of the chicken oviduct and binds with a receptor protein, and this complex binds with the chromatin in the nucleus. The effect is to turn on genes that code for specific cellular products; that is, molecules of mRNA are produced that result in synthesis of egg albumin and other substances.

In a recent report the gene for fetal hemoglobin was turned on in an adult by treatment with a drug that causes demethylation of DNA. Because the fetal hemoglobin is functionally adequate, treatment with the drug results in a dramatic improvement in the clinical condition of persons with sickle cell anemia and beta thalassemia, which are caused by a defect in the gene for one of the adult hemoglobin chains.

METHODS IN MOLECULAR GENETICS

Progress in our understanding of genetic mechanisms on the molecular level, as discussed in the last few pages, has been almost breathtaking in the last few years. We can expect many more discoveries in the near future. This progress has been due largely to the effectiveness of many biochemical techniques now used in molecular biology. We have space to describe only a few briefly.

One of the most important tools in this technology is a series of enzymes called **restriction endonucleases.** Each of these enzymes, derived from bacteria, cleaves double-stranded DNA at particular sites determined by the particular base sequences at that point. Many of these endonucleases cut the DNA strands so that one has several bases projecting farther than the other strand (Figure 3-23), leaving what are called "sticky ends." When these DNA fragments are mixed with others that have been cleaved by the same endonuclease, and in the presence of DNA ligase, they tend to anneal (join) by the rules of complementary base pairing.

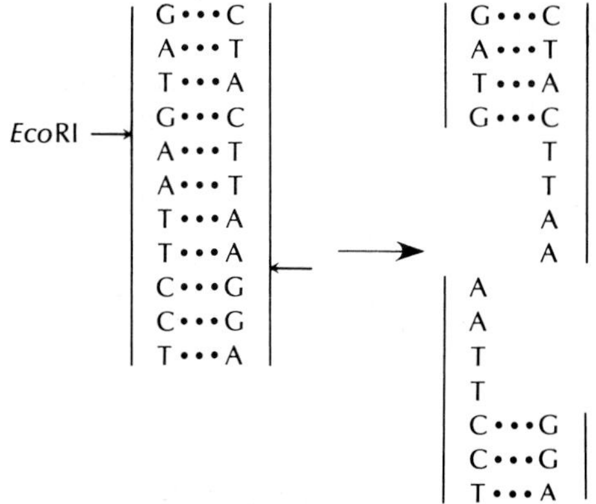

FIGURE 3-23

Action of restriction endonuclease, *Eco*RI. Such enzymes recognize specific base sequences that are palindromic (a palindrome is a word spelled the same backward and forward). *Eco*RI leaves "sticky ends," which are easily matched to other DNA fragments cleaved by the same enzyme and linked by DNA ligase.

If the DNA annealed after cleavage by the endonuclease is from two different sources, for example, a plasmid (see marginal note) and a mammal, the product is **recombinant DNA.** To make use of the recombinant DNA, the modified plasmid must be cloned in bacteria. The bacteria are treated with dilute calcium chloride to make them more susceptible to taking up the recombinant DNA, but the plasmids do not enter most of the cells present. The bacterial cells that have taken up the recombinant DNA can be identified if the plasmid has a marker, for example, resistance to an antibiotic. Then, only the bacteria that will grow in the presence of the antibiotic are those that have absorbed the recombinant DNA. Some viruses have also been used as carriers for recombinant DNA. The plasmids and viruses retain the ability to replicate in the bacterial cells; therefore the mammalian gene is amplified.

Besides their chromosomes, most prokaryote and at least some eukaryote cells have small circles of double-stranded DNA called *plasmids.* Though comprising only 1% to 3% of the bacterial genome, they may carry important genetic information, for example, resistance to an antibiotic. Plastids in plant cells (for example, chloroplasts) and mitochondria, found in most eukaryotic cells, are self-replicating and have their own complement of DNA in the form of small circles, reminiscent of plasmids. The DNA of mitochondria and plastids codes for some of their proteins, and some of their proteins are specified by nuclear genes.

Clearly, for this technique to be useful, the cloned bacteria must manufacture (and the mammalian gene in the recombinant DNA must code for) a product of some interest, such as a hormone. To find and isolate that gene from an entire genome is a formidable task, but it has been repeatedly accomplished in recent years.

The genetic complement of mammalian cells has also been artificially manipulated. Techniques include somatic cell hybridization (to produce monoclonal antibodies) and microinjection of genes into the nuclei of cells in culture.

Sources of Phenotypic Variation

The creative force of evolution is biological variation. Without variability among individuals, there could be no continued adaptation to a changing environment and no evolution.

There are actually several sources of variability, some of which we have already described. The independent assortment of chromosomes during meiosis is a random process that creates new chromosomal recombinations in the gametes. In addition, chromosomal crossing-over during meiosis allows recombination of linked genes between homologous chromosomes, further increasing variability. The presence of long noncoding sequences (introns) eliminates the need for unlikely precision in crossing-over, and some investigators have suggested that the evolutionary flowering of the eukaryotes as compared with the prokaryotes may be due, in part, to introns. In their evolution the prokaryotes seem to have sacrificed flexibility for efficiency.

The random fusion of gametes from both parents produces still another source of variation. Thus sexual reproduction multiplies variation and provides the diversity and plasticity necessary for a species to survive environmental change. Sexual reproduction with its sequence of gene segregation and recombination, generation after generation, is, as the geneticist T. Dobzhansky has said, the "master adaptation which makes all other evolutionary adaptations more readily accessible."

Although sexual reproduction reshuffles and amplifies whatever genetic diversity exists in the population, there must be ways to generate *new* genetic material. This happens through gene mutations and, sometimes, through chromosomal aberrations.

GENE MUTATIONS

There is a story that George Bernard Shaw once received a letter from a famous actress who suggested that they conceive a perfect child who would combine her beauty and his brains. He declined the offer, pointing out that the child could just as well inherit her brains and his beauty. Shaw was correct; the fusion of parental gametes is not only random, it is unpredictable.

Gene mutations are chemicophysical changes in genes resulting in an alteration of the original character. Although the actual mutation cannot be visually detected under the microscope, all gene mutations are believed to be changes of nucleotides in chromosomal DNA. A mutation may result in a codon substitution as, for example, in the condition in humans known as **sickle cell anemia**. Homozygotes with sickle cell trait usually die before the age of 30 because the ability of their red blood cells to carry oxygen is greatly impaired, a result of the substitution of only a single amino acid in the amino acid sequence of their hemoglobin. Other mutations may involve the deletion of one or more bases from a codon or the insertion of additional bases into the DNA chain. The transcription of mRNA will thus be shifted, leading to codons that specify incorrect amino acids.

Mutations are random because they are unpredictable and unrelated to the needs of the organism. Once a gene is mutated, it faithfully reproduces its new self just as it did before it was mutated. *Many, perhaps nearly all, mutant genes are actually harmful because they replace adaptive genes that have evolved and served the organism through its long evolution.* Sometimes, however, mutations are advantageous. These are of great significance to evolution because they furnish new possibilities on which natural selection works. Natural selection determines which new genes merit survival; the environment imposes a screening process that passes the fit and eliminates the unfit.

When an allele of a gene is mutated to the new allele, it tends to be recessive and its effects are normally masked by its partner allele. Only in the homozygous condition can such mutant genes be expressed. Thus a population carries a reservoir of mutant recessive genes, some of which are lethal but seldom expressed. Inbreeding encourages the formation of homozygotes and increases the probability of recessive mutants appearing.

Most mutations are destined for a brief existence. There are cases, however, in which mutations may be harmful or neutral under one set of environmental conditions and helpful under a different set. Should the environment change, there could be a new adaptation beneficial to the species. The earth's changing environment has provided numerous opportunities for new gene combinations and mutations, as evidenced by the great diversity of animal life today.

Frequency of mutations

Although whether or not a mutation occurs is random, different mutation rates prevail at different loci. Some *kinds* of mutations are more likely to occur than others, and individual genes differ considerably in length. A long gene (more base pairs) is more likely to have a mutation than a short gene. Nevertheless, it is possible to estimate average spontaneous rates for different organisms and traits.

Relatively speaking, genes are extremely stable. In the well-studied fruit fly *Drosophila* there is approximately one detectable mutation per 10,000 loci (rate of 0.01% per locus per generation). The rate for humans is one per 10,000 to one per 100,000 loci per generation. If we accept the latter, more conservative figure, then a single normal allele is expected to go through 100,000 generations before it is mutated. However, since human chromosomes contain 100,000 loci, every person carries approximately one new mutation. Similarly, each ovum or spermatozoon produced contains, on the average, one mutant allele.

Since most mutations are deleterious, these statistics are anything but cheerful. Fortunately, most mutant genes are recessive and are not detected by natural selection until by chance they have increased enough in frequency for homozygotes to be produced. At that point most are eliminated, since the zygote or individual carrying the mutants does not reproduce.

All animals carry large numbers of lethal and semilethal alleles in their gene pools. In human populations this has been called the "genetic load," or load of hidden mutations. By genetic recombination these continue to surface to produce mutant individuals less "fit" (in the Darwinian sense) than the "wild-type" genotype.

Molecular Genetics of Cancer

The fundamental defect in cancer cells is that they proliferate in an unrestrained manner **(neoplastic growth)**. The mechanism that controls the rate of division of normal cells has somehow broken down, and the cancer cells multiply much more rapidly, invading other tissues in the body. Cancer cells originate from normal cells that lose their constraint on division and become dedifferentiated (less specialized) in some degree. Thus there are many kinds of cancer, depending on the original founder cells of the tumor. In recent years mounting evidence has indicated that the change in many cancerous cells, perhaps all, has a genetic basis, and investigation of the genetic damage that causes cancer is now a major thrust of cancer research.

ONCOGENES AND PROTO-ONCOGENES

Some 40 genes have been identified in cells that, under certain circumstances of activation, are associated with neoplastic growth. They have been named **oncogenes** and are derived from normal cellular genes. In their benign, inactivated state these genes are often referred to as **proto-oncogenes**. In a number of cases the proteins encoded by oncogenes are known, but just how these proteins may account for the neoplastic phenotype remains unclear.

Because each somatic cell in vertebrates carries the entire genome, all the potential oncogenes are present in all cells. The same oncogene can be activated in founder cells of different tissue types, and these will result in different types of cancers.

Several mechanisms are now known that explain how proto-oncogenes can become oncogenes. For example, cellular genes can become oncogenes through the action of certain **retroviruses**. Retroviruses are viruses in which the genome is single-stranded RNA, and this is transcribed into DNA in the host cell by the enzyme reverse transcriptase. The viral DNA is then integrated into the host DNA, and the cell machinery is taken over to produce viral gene products. Insertion of viral DNA into host DNA is mutagenic and may damage cellular genes, as well as bring cellular genes under the influence of powerful regulatory elements of the viral genome. These events may enhance the production of gene product by releasing the proto-oncogene from its normal controls or by increasing transcription of the proto-oncogene by a viral enhancer.

Mutations have been implicated in the conversion of some proto-oncogenes to oncogenes. At lease one case is known in which a change in a single base (a thymine replacing a guanine) in a 5000 nucleotide gene is responsible for the transformation. This mechanism helps account for the fact that certain chemicals and ionizing radiation, which are known to cause mutations in cells, are also powerful carcinogens (cancer-causing agents).

Summary

In bisexual animals the genetic material is distributed to the offspring in the gametes (ova and sperm), produced in the process of meiosis. Each somatic cell in an organism has two chromosomes of each kind (homologous chromosomes) and is thus diploid. Meiosis separates the homologous chromosomes, so that each gamete has half the somatic chromosome number (haploid). In the first meiotic division, the centromeres do not divide, and each daughter cell receives one of each of the replicated homologous chromosomes with the chromatids still attached to the centromere. At the beginning of the first meiotic division, the replicated homologous chromosomes come to lie alongside each other (synapsis), forming a bivalent. The gene loci on one set of chromatids lie opposite the corresponding loci on the homologous chromatids. Portions of the adjacent chromatids can exchange with the nonsister chromatids (crossing-over) to produce new genetic combinations. At the second meiotic division, the centromeres divide, completing the reduction in chromosome number and amount of DNA. The diploid number is restored when the male and female gametes fuse to form the zygote. Gender is determined in most animals by the sex chromosomes; in humans, fruit flies, and many other animals, females have two X chromosomes, and males have an X and a Y.

Genes may be dominant, recessive, or intermediate; the recessive genes in the heterozygous genotype will not be expressed in the phenotype, but require the homozygous condition for overt expression. In a monohybrid cross involving a dominant gene and its recessive allele (both parents homozygous), the F_1 generation will all be heterozygous, whereas the F_2 genotypes will occur in a 1:2:1 ratio, and the phenotypes in a 3:1 ratio. This demonstrates Mendel's law of segregation. Heterozygotes in intermediate inheritance show phenotypes intermediate between the homozygous phenotypes, or sometimes they show a different phenotype altogether, with corresponding alterations in the phenotypic ratios. Dihybrid crosses (in which the alleles for the two traits are carried on separate pairs of homologous chromosomes) demonstrate Mendel's law of independent assortment, and the phenotypic ratios will be 9:3:3:1 with dominant and recessive characters. The ratios for monohybrid and dihybrid crosses can be determined by construction of a Punnett square, but the laws of probability allow calculation of the ratios in crosses of two or more characters more easily.

A gene on the X (sex) chromosome shows sex-linked inheritance and will produce an effect in the male, even if it is recessive, because the Y chromosome does not carry a corresponding allele. All genes on a given autosomal chromosome are linked and do not assort independently unless crossing-over occurs. Crossing-over increases the amount of genetic recombination in a population.

Occasionally, nondisjunction of one of the chromosomes in meiosis occurs, and one of the gametes ends up with one chromosome too many and the other with n − 1 chromosomes. Resulting zygotes usually do not survive; humans with 2n + 1 chromosomes may live, but they are born with serious defects, such as Down syndrome.

One gene most commonly controls the production of one protein or polypeptide (one gene-one polypeptide hypothesis), but the various types of RNA are also encoded on the genes.

The nucleic acids in the cell are DNA and RNA, which are large polymers of nucleotides composed of a nitrogenous base, pentose sugar, and phosphate group. The nitrogenous bases in DNA are adenine (A), guanine (G), thymine (T), and cytosine (C), and those in RNA are the same except that uracil (U) is substituted for thymine. DNA is a double-stranded, helical molecule in which the bases extend toward each other from the sugar-phosphate backbone: A always pairs with T and G with C. Thus the strands are antiparallel and complementary, being held in place by hydrogen bonds between the paired bases. In DNA replication the strands part, and the enzyme DNA polymerase synthesizes a new strand along each parent strand, using the parent strand as a template.

The sequence of the bases in DNA is a code for the amino acid sequence in the ultimate product protein. Each three bases (triplet) specify a particular amino acid.

Proteins are synthesized by transcription of the information coded into DNA into the base sequence of a molecule of messenger RNA (mRNA), which functions in concert with ribosomes (containing ribosomal RNA [rRNA] and protein) and transfer RNA (tRNA). Ribosomes attach to the strand of mRNA and move along it, assembling the amino acid sequence of the protein. Each amino acid is brought into position for assembly by a molecule of tRNA, which itself bears a base sequence (anticodon) complementary to the respective codons of the mRNA. In eukaryotes the sequences of bases in DNA coding for amino acids in a protein (exons) are interrupted by intervening sequences (introns). The introns are spliced out of the primary mRNA before it leaves the nucleus, and the protein is synthesized in the cytoplasm.

Genes, and the synthesis of the products for which they are responsible, must be regulated: turned on or off in response to varying environmental conditions or cell differentiation. Gene regulation in eukaryotes is complex, and a number of possible mechanisms are known. One inactivation mechanism is the methylation of cytosine residues that lie next to guanine residues in the DNA strand, thus preventing transcription. In some cells genes are activated by the direct action of a hormone in the nucleus.

Modern methods in molecular genetics have made spectacular advances possible. Restriction endonucleases cleave DNA at specific base sequences, and such cleaved DNA from different sources can be rejoined to form recombinant DNA. Combining mammalian with plasmid or viral DNA, a mammalian gene can be introduced into bacterial cells, which then multiply and express the mammalian gene.

A mutation is a physicochemical alteration in the bases of the DNA that changes the effect of the gene. Although rare and usually detrimental to the survival and reproduction of the organism, mutations are occasionally beneficial and provide new genetic material on which natural selection can work.

Many types of cancer (neoplastic growth) are associated with genetic damage in which certain normal genes (proto-oncogenes) become abnormally activated to oncogenes. Mechanisms for transforming proto-oncogenes into oncogenes include the action of retroviruses and mutation.

REVIEW QUESTIONS

1. What is the relationship between homologous chromosomes and alleles?
2. Describe or diagram the sequence of events in meiosis (both divisions).
3. What are the designations of the sex chromosomes in males of bugs, humans, and butterflies? How are these designations descriptive of the chromosomes?
4. Define the following: dominant, recessive, zygote, heterozygote, homozygote, phenotype, genotype, monohybrid cross, dihybrid cross.
5. Diagram by Punnett square a cross between individuals with the following genotypes: *A/a* × *A/a*, *A/a B/b* × *A/a B/b*.
6. Concisely state Mendel's law of segregation and his law of independent assortment.
7. Assuming brown eyes *(B)* are dominant over blue eyes *(b)*, determine the genotypes of all the following individuals. The blue-eyed son of two brown-eyed parents marries a brown-eyed daughter whose mother was brown eyed and whose father was blue eyed. Their child is blue eyed.
8. Recall that red color *(R)* in four-o'clock flowers is incompletely dominant over white *(R')*. In the following crosses, give the genotypes of the gametes produced by each parent and the flower color of the offspring: *R/R'* × *R/R'*, *R'/R'* × *R/R'*, *R/R* × *R/R'*, *R/R* × *R'/R'*.
9. A brown mouse is mated with two female black mice. When each female has produced several litters of young, the first female has had 48 black and the second female has had 14 black and 11 brown young. Can you deduce the pattern of inheritance of coat color and the genotypes of the parents?
10. Rough coat *(R)* is dominant over smooth coat *(r)* in guinea pigs, and black coat *(B)* is dominant over white *(b)*. If a homozygous rough black is mated with a homozygous smooth white, give the appearance of each of the following: F_1; F_2; offspring of F_1 mated with smooth, white parent; offspring of F_1 mated with rough, black parent.
11. Assume right-handedness *(R)* dominates over left-handedness *(r)* in humans, and that brown eyes *(B)* are dominant over blue *(b)*. A right-handed, blue-eyed man marries a right-handed, brown-eyed woman. Their two children are right handed, blue eyed and left handed, brown eyed. The man marries again, and this time the woman is right handed and brown eyed. They have 10 children, all right handed and brown eyed. What are the genotypes of the man and his two wives?
12. In *Drosophila*, red eyes (W) are dominant to white (w) and the characteristic is on the X chromosome. Vestigal wings *(v)* are recessive to normal *(V)*. What will be the appearance of the following crosses: X^W/X^w V/v × X^w/Y v/v, X^w/X^w V/v × X^W/Y V/v.
13. Assume that color blindness is a recessive character on the X chromosome. A man and woman with normal vision have the following offspring: daughter with normal vision who has one color-blind and one normal son; daughter with normal vision who has six normal sons; and a color-blind son who has a daughter with normal vision. What are the probable genotypes of all the individuals?
14. Assume an individual has the genotype *AB/ab*. With a 20% crossing-over rate between the two loci on this chromosome, what will the genotype of the gametes be? This person's spouse has a genotype *ab/ab*. What will the genotypes of their offspring be?
15. Distinguish the following: euploidy, aneuploidy, and polyploidy; monosomy and trisomy.
16. Name the purines and pyrimidines in DNA and tell which pair with each other in the double helix. What are the purines and pyrimidines in RNA and to what are they complementary in DNA?
17. Explain how DNA is replicated.
18. Why is it not possible for a codon to consist of only 2 bases?
19. Explain the transcription and processing of mRNA in the nucleus.
20. Explain the role of mRNA, tRNA, and rRNA in protein synthesis.
21. What are two ways that genes can be regulated in eukaryotes?
22. In modern molecular genetics, what is recombinant DNA, and how is it prepared?
23. Name three sources of phenotypic variation.
24. Distinguish between proto-oncogene and oncogene. What are two mechanisms whereby a proto-oncogene can become an oncogene?

SELECTED REFERENCES

Bishop, J.M. 1987. The molecular genetics of cancer. Science **235**:401-408. *A concise summary of the evidence to the present on oncogenes and their products.*

Cech, T.R. 1986. RNA as an enzyme. Sci. Am. **255**:64-75 (Nov.). *Evidence is accumulating that RNA can catalyze its own replication. The first nucleic acids may have duplicated themselves with the assistance of proteins.*

Darnell, J.E., Jr. 1983. The processing of RNA. Sci. Am. **249**:90-100 (Oct.). *Describes removal of introns, capping, and polyadenylation of RNA.*

Jenkins, J.B. 1979. Genetics, ed. 2. Boston, Houghton Mifflin Co. *Comprehensive introductory text.*

Koshland, D.E., Jr. 1987. Frontiers in recombinant DNA. Science **236**:1157. *This is a lead editorial in an issue of* Science *that contains a number of articles on exciting applications of recombinant DNA research.*

Motulsky, A.G. 1983. Impact of genetic manipulation on society and medicine. Science **219**:135-140. *A summary of ethical considerations raised by genetic engineering. We must improve biological education for everyone so that we can make informed decisions.*

Watson, J.D. 1969. The double helix. New York, Atheneum Publishers. *An enlightening and entertaining history of events leading to an understanding of the genetic code.*

THE DARWINIAN HERITAGE

Organic evolution, the great unifying theme of biology, explains the diversity of modern organisms, their characteristics, and their distribution as the historical outcome of change from previously existing forms. Evolutionary theory is strongly identified with Charles Robert Darwin, who with Alfred Russel Wallace presented the first credible explanation of the occurrence of evolution, the principle of natural selection. The principle is founded on the observed facts that no two organisms are exactly alike; that some variations, at least, are inheritable; that all groups tend to overproduce their kind; and that because more individuals are reproduced than can survive, there is a struggle for existence among them. In this struggle, according to Darwin, organisms possessing variations that make them better adapted to the environment will survive. If, because of their survival, they have more offspring than do those with less favorable variations, those with the "winning" genes will appear in greater proportion in the next generation. This is called microevolution.

Much of the current debate among evolutionists is over the rates at which evolutionary novelties appear. Is macroevolution, defined as evolution above the species level and acting over long periods of geologic time, merely the compounding of microevolutionary events? It would be rash to argue that our current neo-Darwinian theory is complete. Still, the rich discoveries that have been made in the last 40 years have served mainly to elaborate and strengthen the synthetic theory that emerged in the 1940s. Natural selection, supplemented by genetic drift, remains the centerpiece of evolutionary theory.▪

CHAPTER 4

Evolution of Animal Diversity

The wonder of the earth is not perhaps that it harbors life, although in the vastness of the universe earth is the only place we know of where life exists. It is that life on earth is so enormously diverse and pervasive. Even from space the ubiquity of life is evident in the kaleidoscopic patterns of grassland, forests, lakes, and croplands stretching across the earth's surface, their colors changing with the seasons.

Each ecosystem is biologically and physically distinct. Each is occupied by a great variety of organisms, living interdependently and manifesting every conceivable kind of adaptation to their surroundings. Nutrients are withdrawn and again released; energy captured by plants flows through the system bringing order out of disorder in apparent defiance of the second law of thermodynamics; organisms die and are replaced by offspring, a renewal that threads its descent faithfully through generations of ancestors. It is a drama of incalculable complexity that has been and is now unfolding before our eyes.

Yet despite the evident permanence of the natural world, change characterizes all things on earth and in the universe. Countless kinds of animals and plants have flourished and disappeared, leaving behind an imperfect fossil record of their former existence.

The earth itself bears its own record of change, transformed as it is by processes that have occurred over a vast span of time and still are at work today. These changes are irreversible, and for life they appear in the broad sense to be directional and progressive, as primitive organisms of the young earth have yielded to the advanced, complex creatures of the present. We call this historical process of organismal change due to environmental pressures **organic evolution.** Because every feature of life as we know it today is a product of the evolutionary process, biologists consider organic evolution to be the keystone of all biological knowledge.

The great contribution of Charles Robert Darwin and Alfred Russel Wallace (Figure 4-1) was that they simultaneously provided the first credible explanation for evolutionary change, the **theory of natural selection**. The principal alternative hypothesis to account for the diversity of life on earth is that of **special creation**, which holds that all life forms were divinely created much as we find them today. However, creationism rests on premises that do

A

B

FIGURE 4-1

Founders of theory of natural selection. **A,** Charles Robert Darwin (1809-1882), as he appeared in 1881, the year before his death. **B,** Alfred Russel Wallace (1823-1913) in 1895. Darwin and Wallace independently came to the same theory. A letter and essay from Wallace written to Darwin in 1858 spurred Darwin into writing *The Origin of the Species,* published in 1859.

A, Courtesy American Museum of Natural History; **B,** courtesy British Museum (Natural History), London.

not appear to be susceptible to factual verification and thus must be accepted on faith. The study of biology, however, must account for the physical evidence that life has evolved and changed over time. Students should be aware that the extensive evidence for evolution and acceptance of the fact that evolution has occurred are not necessarily incompatible with religious concepts of creation. Indeed, evolution can be viewed as a creative process operating over long periods of time. How matter and life may have appeared in our universe is not at issue in evolutionary theory. It is regrettable, therefore, that the recent reemergence of rigid antievolutionary views, in the guise of "creation science," should portray creation by a supernatural being and evolution as irreconcilable.

In our treatment of the principle of organic evolution in this chapter, we consider Darwin's theory of natural selection and the evidence for evolution, the concept of species and how they change with time, the emergence of neo-Darwinism, and the question of how new species arise. The chapter closes with a brief illustrated synopsis of human evolution.

Development of the Idea of Organic Evolution

Evolution is no longer a subject for debate among biologists; as a process it is known and accepted, even though the forces that determine the course of evolution are not fully understood. Yet, prior to the eighteenth century, much of the speculation on the origin of species rested on myth and superstition rather than on observation. Nevertheless, long before this time, there were those who were thinking about and attempting to interpret the order of nature.

Some of the early Greek philosophers, notably Xenophanes, Empedocles, and Aristotle, developed the germ of the idea of change and natural selection within the restrictions of belief in spontaneous generation. They recognized fossils as evidence of a former life that they believed had been destroyed by some natural catastrophe. Living in a spirit of intellectual inquiry, the Greeks failed to establish an evolutionary concept, probably because of their limited experience with the natural world.

With the gradual decline of ancient science, beginning well before the rise of Christianity, debate on evolution nearly ended. The opportunity for fresh thinking became even more restricted as the biblical account of the earth's creation became accepted as a tenet of faith. The year 4004 BC was fixed by Archbishop James Ussher (mid-seventeenth century) as the time of true creation of life. In this atmosphere evolutionary views were considered rebellious and heretical.

Still, some speculation continued. The French naturalist Buffon (1707-1788) stressed the influence of environment on the modifications of animal type. He also extended the age of earth to 70,000 years from 6000 years.

LAMARCKISM: THE FIRST SCIENTIFIC EXPLANATION OF EVOLUTION

The first complete explanation of evolution was authored by Jean Baptiste de Lamarck (1744-1829) (Figure 4-2), a pioneering French biologist. Lamarck, who published his concept in 1809, the year of Darwin's birth, was the first biologist to make a convincing case for the idea that fossils were the remains of extinct animals and not, as some argued, stones molded in chance imitations of life. Lamarck's idea of **inheritance of acquired characteristics** was engagingly simple: organisms, by striving to adapt to their environment, acquire adaptations during their lives that are passed on by heredity to their offspring. The giraffe, according to Lamarck, developed its long neck by constantly stretching for food. Genetic variation then arose preferentially in this direction, and the trait of long necks became common in the population. Snakes lost their legs, Lamarck believed, because legs were a handicap in moving through dense vegetation. A man who developed his muscles by exercise was supposed to pass his strength on to his sons. Lamarck's attitude toward adaptation was straightforward: an organism could get any adaptation it needed. Despite the uncluttered appeal and efficiency of Lamarck's concept, it was effectively disproved in this century by geneticists who could find no mechanism that would make such a process possible. The unveiling of the genetic code showed that DNA transcription and translation are a one-way process with no known way for the environment to feed information back into the DNA of the cells of the individual so that a new trait could be coded in that individual's gametes.

CHARLES LYELL AND THE DOCTRINE OF UNIFORMITARIANISM

While eighteenth- and early nineteenth-century zoologists wrestled with conflicting concepts of organic evolution, geologists were marshaling sound evidence for the physical evolution of the earth's crust. The geologist Sir Charles Lyell (1797-1875) (Figure 4-3) in his *Principles of Geology* (1830-1833) stated that the changes in the earth's surface were the result of ongoing natural forces operating with uniformity over great periods of time. The ideas that processes occurring now are the same as those in the past came to be called the doctrine of uniformitarianism. As one writer of the time stated, "no vestige of a beginning—no prospect of an end."

Lyell was able to show that such forces, acting over long periods of time, could account for all observed changes, including the formation of fossil-bearing rocks. His familiarity with fossils, with the natural history of contemporary marine and freshwater animals, and with sedimentary rocks led him to conclude that the earth's age must be reckoned in millions of years rather than in thousands.

Charles Darwin (1809-1882) was thus not the first to propose the idea of evolution nor even to suggest a mechanism for its action. His predecessors had already established an intellectual climate that made a hypothesis of evolutionary mechanism possible, if not inevitable. Darwin himself later acknowledged that the theory of natural selection had been suggested by others before him. But so forcefully did Darwin present his ideas and his array of carefully collected scientific data that no one since has been able to challenge his preeminence in this field. Today the fact of organic evolution can be denied only by abandoning reason. As the noted English biologist Sir Julian Huxley wrote, "Charles Darwin effected the greatest of all revolutions in human thought, greater than Einstein's or Freud's or even Newton's, by simultaneously establishing the fact and discovering the mechanism of organic evolution."*

FIGURE 4-2

Jean Baptiste de Lamarck (1744-1829), French naturalist who developed the first reasoned account of evolution. The scientific establishment of his time was contemptuous of his "use-and-disuse" theory of evolution. Yet, in France today, where Darwinism was accepted only grudgingly, Lamarck is regarded as the father of evolution. This engraving was made in 1821.

Courtesy of British Museum (Natural History), London.

■ Darwin's Great Voyage of Discovery

"After having been twice driven back by heavy south-western gales, Her Majesty's ship *Beagle,* a ten-gun brig, under the command of Captain Fitz Roy, R.N., sailed from Devonport on the 27th of December, 1831." Thus began Charles Darwin's account of the historic 5-year voyage of the *Beagle* around the world (Figure 4-4). Darwin, not quite 23 years old, had been invited to serve as naturalist without pay on the *Beagle,* a small vessel only 90 feet in length carrying a party of 74 persons, which was about to depart on a second extensive surveying voyage to South America and the Pacific. It was the beginning of one of the most important voyages of the nineteenth century.

During the 5-year voyage (1831-1836) (Figure 4-5), Darwin endured almost constant seasickness and the erratic companionship of the authoritarian Captain Fitz Roy. But his youthful physical strength and early training as a naturalist

FIGURE 4-3

Sir Charles Lyell (1797-1875), English geologist and friend of Darwin. His book *Principles of Geology* greatly influenced Darwin during Darwin's formative period. The photograph was made about 1856.

Courtesy British Museum (Natural History), London.

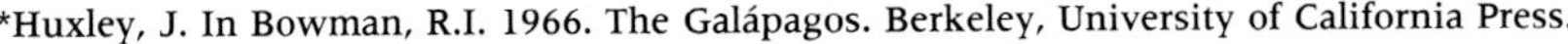

*Huxley, J. In Bowman, R.I. 1966. The Galápagos. Berkeley, University of California Press.

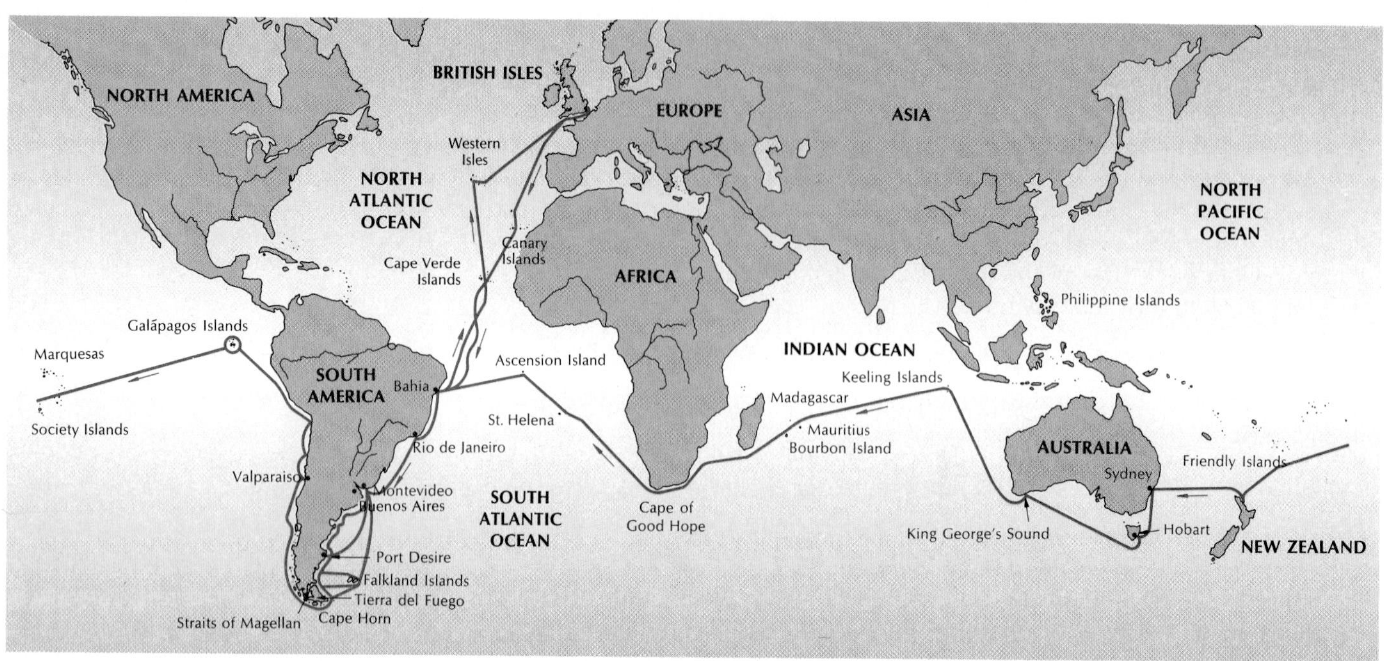

FIGURE 4-4

Five-year voyage of H.M.S. *Beagle*.

Modified from Moorhead, A. 1969. Darwin and the *Beagle*. New York, Harper & Row, Publishers, Inc.

FIGURE 4-5

Charles Darwin and H.M.S. *Beagle*.
A, Darwin in 1840, 4 years after the *Beagle* returned to England, and a year after his marriage to his cousin, Emma Wedgwood.
B, A replica of H.M.S. *Beagle* sails in the Straits of Magellan. The reconstruction was built for a BBC television series on Darwin's voyage.

A, From a watercolor by George Richmond, Bridgeman/Art Resource. **B,** photograph by Christopher Ralling.

equipped him for his work. The *Beagle* made many stops along the harbors and coasts of South America and adjacent regions. Darwin made extensive collections and observations on the fauna and flora of these regions. He unearthed numerous fossils of animals long since extinct and noted the resemblance between fossils of the South American pampas and the known fossils of North America. In the Andes he encountered seashells embedded in rocks at 13,000 feet above sea level. He experienced a severe earthquake and watched the mountain torrents that relentlessly wore away the earth—observations which strengthened his conviction that natural forces were responsible for the geological features of the earth. Finally, he realized that he was witnessing the results of evolution.

In mid-September of 1835, the *Beagle* arrived at the Galápagos Islands, a volcanic archipelago straddling the equator 600 miles west of the coast of Ecua-

FIGURE 4-6

The Galápagos Islands, viewed from the rim of a volcano.

Photograph by C.P. Hickman, Jr.

dor (Figure 4-6). The fame of the islands stems from their infinite strangeness. They are unlike any other islands on earth. Some visitors today are struck with an impression of awe and wonder; others with a sense of depression and dejection. Circled by capricious currents, surrounded by shores of twisted lava, bearing skeletal brushwood baked by the equatorial sun, almost devoid of lush tropical vegetation, inhabited by strange reptiles and by convicts stranded by the Ecuadorian government, the islands indeed had few admirers among mariners. By the middle of the seventeenth century, the islands were already known to the Spaniards as "Las Islas Galápagos"—the tortoise islands. The giant tortoises, used for food first by buccaneers and later by American and British whalers, sealers, and ships of war, were the islands' principal attraction. At the time of Darwin's visit, the tortoises were already being heavily exploited.

During the *Beagle's* 5-week visit to the Galápagos, Darwin began to develop his views of the evolution of life on earth. His original observations on the giant tortoises, marine iguanas that feed on seaweed, and family of drab ground finches that had evolved into several species, each with distinct beak and feeding adaptations, all contributed to the turning point in Darwin's thinking. Writing in his journal after leaving the Galápagos, Darwin noted that, ". . . one is astonished at the amount of creative force . . . displayed on these small, barren, and rocky islands." Later he commented that ". . . the zoology of the Archipelago . . . will be well worth examining; for such facts would undermine the stability of species."

Darwin was struck by the fact that, although the Galápagos Islands in the Pacific and the Cape Verde Islands in the Atlantic Ocean (visited earlier in this voyage of the *Beagle*) were similar in climate and topography, their fauna and flora were altogether different. He recognized that Galápagos plants and animals were related to those of the South American mainland, yet differed from them sometimes in curious ways. Each island often contained a unique species of a particular group of animals that was nonetheless related to forms on other islands. In short, Galápagos life must have originated in continental South America and then undergone modification in the various environmental conditions of the separate islands. He concluded that living forms were neither divinely created nor immutable; they were, in fact, the products of evolution. Although Darwin devoted only a few pages to Galápagos animals and plants in his monumental *The Origin of Species,* published more than two decades after visiting the islands, his observations on the unique character of the animals and plants were, in his own words, the "origin of all my views."

FIGURE 4-7

Darwin's study at Down House in Kent, England, preserved today much as it was when Darwin wrote *The Origin of Species*. Darwin maintained a set routine of work, kept meticulous records, and enjoyed experimentation. Although a famous scientist in his time, Darwin remained a mysterious figure to all but his family and a close circle of friends. His success is ascribed to many things, among them his ability to weave together a theoretical structure from a mountain of empirical observations and data. His intellectual development was much influenced by his scientific friends and nourished by his devoted family, an unfailing wellspring of love, respect, and support.

Photograph by C.P. Hickman, Jr.

On October 2, 1836, the *Beagle* landed in England. Most of Darwin's extensive collections had long since preceded him to England, as had most of his notebooks and diaries kept during the 5-year cruise. Three years after the *Beagle's* return to England, Darwin's journal was published. It was an instant success and required two additional printings within the first year. In later versions, Darwin made extensive changes and abbreviated the original ponderous title typical of nineteenth-century books to simply *The Voyage of the Beagle.* The fascinating account of his observations written in a simple, appealing style has made the book one of the most lasting and popular travel books of all time.

Curiously, the main product of Darwin's voyage, his theory of evolution, did not appear in print for more than 20 years after the *Beagle's* return. In 1838 he "happened to read for amusement" a gloomy essay on populations by the English economist T.R. Malthus (1766-1834), who stated that animal and plant populations, including human populations, tend to increase beyond the capacity of the environment to support them. Darwin had already been gathering information on the artificial selection of animals under domestication by humans. After reading Malthus' article, Darwin realized that a process of selection in nature, a "struggle for existence" because of overpopulation, could be a powerful force for evolution of wild species.

He allowed the idea to develop in his own mind until it was presented in an essay in 1844—still unpublished. Finally in 1856 he began to pull together his voluminous data into a work on the origin of species. He expected to write four volumes, a "very big" book, "as perfect as I can make it." However, his plans were to take an unexpected turn.

In 1858, he received a manuscript from Alfred Russel Wallace (1823-1913), an English naturalist in Malaya with whom he had been corresponding. Darwin was stunned to find that in a few pages, Wallace summarized the main points of the natural selection hypothesis that Darwin had been working on for two decades. Rather than withhold his own work in favor of Wallace as he was inclined to do, Darwin was persuaded by two close friends, the geologist Lyell and the botanist Hooker, to publish his views in a brief statement that would appear together with Wallace's papers in the *Journal of the Linnaean Society.* Portions of both papers were read before an unimpressed audience on July 1, 1858.

For the next year, Darwin worked urgently to prepare an "abstract" of the planned four-volume work. This was published in November, 1859, with the title *On the Origin of Species by Means of Natural Selection, or the Preservation of Favoured Races in the Struggle for Life.* The 1250 copies of the first printing were sold the first day! The book instantly generated a storm that has never completely abated. His views were to have extraordinary consequences on scientific and religious beliefs and remain among the greatest intellectual achievements of all time.

"Whenever I have found that I have blundered, or that my work has been imperfect, and when I have been contemptuously criticised, and even when I have been overpraised, so that I have felt mortified, it has been my greatest comfort to say hundreds of times to myself that 'I have worked as hard and as well as I could, and no man can do more than this.' "

Charles Darwin in his autobiography, 1876.

Once Darwin's excessive caution had been swept away by the publication of *The Origin of Species,* he entered an incredibly productive period of evolutionary thinking for the next 23 years, producing book after book. He died April 19, 1882, and was buried in Westminster Abbey. The little *Beagle* had already disappeared, having been retired in 1870 and presumably broken up for scrap.

Evidences for Evolution

RECONSTRUCTING THE PAST

Fossils and their formation

The strongest and most direct evidence for evolution is the fossil record of the past, even though fossils and the sediments containing them provide dim and imperfect views of an ancient life (Figure 4-8). The fossil record is like a moving picture film with most of the frames destroyed. Yet, incomplete as the record is, biologists rely on the discoveries of new fossils and the continued study of existing fossils to interpret phylogeny and relationships of both plant and animal life. It would be difficult indeed to make sense out of the evolutionary patterns or classification of organisms without the support of the fossil record. The documentary evidence of evolution as a general process, progressive changes in life from one geological era to another, past distribution of lands and seas, and environmental conditions of the past (paleoecology), all depend on what fossils teach us.

A fossil may be defined as the remains of past life uncovered from the crust of the earth. It refers to complete remains (for example, mammoths and insects embedded in fossil resin [amber]), actual hard parts (teeth and bones), and petrified skeletal parts that are infiltrated with silica or other minerals by water seepage (ostracoderms and molluscs), but also to molds, casts, impressions, and fossil excrement (coprolites).

The fossil record is biased because preservation is selective. Vertebrate skeletal parts and invertebrates with shells and other hard structures have left the best record (Figure 4-9). It is unlikely that jellyfish, worms, caterpillars, and such fossilize, but now and then a rare chance discovery such as the Burgess shale deposits of British Columbia and the Precambrian fossil bed of South Australia reveals an enormous amount of information about soft-bodied organisms.

A common method of fossil formation is the quick burial of animals under waterborne sediments. Rapid burial is usually important because it slows or prevents decomposition by oxidation, solution, bacterial action, and scattering of remains by scavengers.

Most fossils are laid down in deposits that become stratified. If left undisturbed, which is rare, the older strata are the deeper ones; however, the layers are usually tilted or folded or show faults (cracks). Often old deposits exposed by erosion are later covered with new deposits in a different plane. Sedimentary rock such as limestone may be exposed to tremendous pressures or heat during mountain building and may be metamorphosed into rocks such as crystalline quartzite, slate, or marble. Fossils are often destroyed during these processes.

FIGURE 4-8

Stalked crinoids from 85-million-year-old Cretaceous rocks. The fossil record of these echinoderms shows that they reached their peak millions of years earlier and began a slow decline to the present.

Photograph by A.J. Copley/Visuals Unlimited.

Estimating geological time

Long before the age of the earth was known, geologists began dividing its history into a table of succeeding events, using as a basis the accessible deposits and correlations from sedimentary rock. Time was divided into eons, eras, periods, and epochs. These are shown on the end paper inside the back cover of this book. Time during the last eon (Phanerozoic) is expressed in eras (Cenozoic), periods (Tertiary), epochs (Paleocene), and sometimes into smaller divisions of an epoch.

Until recently scientists had no way to measure the absolute passage of time. Relative dating was used, a method that relied on the sequence of layers of rock: the oldest at the bottom and youngest at the top. In the late 1940s, radiometric dating methods were developed for determining the absolute age in years of rock formations. Several independent methods are now used, all based on the radioactive decay of naturally occurring elements into other elements. The "radioactive clocks" proceed independent of pressure and temperature changes and therefore are not affected by often violent earth-building activities.

Especially important is potassium-argon dating, which can be used for comparatively young rocks (less than 1 million years old) as well as for the oldest rocks on earth. Potassium-40 (^{40}K) decays to argon-40 (^{40}A) with a half-life of 1.3 billion years. This means that half the original ^{40}K will decay in 1.3 billion years, and half the remaining ^{40}K will be gone at the end of the next 1.3 billion years. This continues until all the radioactive ^{40}K atoms are gone. To measure the

FIGURE 4-9

Fish fossil from rocks of the Green River Formation, Wyoming. Such fish swam here during the Eocene epoch of the Tertiary, approximately 55 million years ago.

Photograph by Ken Lucas/Biological Photo Service.

The more well-known carbon-14 (^{14}C) dating method is of little help in estimating the age of geological formations because the short half-life of ^{14}C restricts its use to quite recent events (less than about 40,000 years). It is especially useful, however, for archeological studies. This method is based on the production of radioactive ^{14}C (half-life of approximately 5570 years) in the upper atmosphere by bombardment of nitrogen-14 (^{14}N) with cosmic radiation. The radioactive ^{14}C enters the tissue of living animals and plants, and an equilibrium is established between atmospheric ^{14}C and and ^{14}N in the organism. At death, ^{14}C exchange with the atmosphere stops. In 5570 years only half of the original ^{14}C remains in the preserved fossil. Its age is found by comparing the ^{14}C content of the fossil with that of living organisms.

age of the rock, one calculates the ratio of remaining ^{40}K atoms to the amount of ^{40}K originally there (the remaining ^{40}K atoms plus the ^{40}A into which they have decayed).

There are many uncertainties in radioactive dating, and the techniques require sophisticated instrumentation and great care in methodology. Not all rocks or fossils can be dated. But if two or more different methods provide concordant answers, the age can be accepted as reliable, since it is most improbable that different isotopes would leach out of the sample at the same rate. In fact, in many thousands of analyses, rocks have been dated by three or more independent clocks and the ages determined have been in close agreement.

As far as the fossil record is concerned, the recorded history of life begins near the base of the Cambrian period of the Paleozoic era approximately 600 million years BP. The period before the Cambrian is called the Precambrian era. Although the Precambrian era occupies 85% of all geological time, it is a puzzling era because of the lack of macrofossils. It has received much less attention than later eras, partly because of the absence of fossils and partly because oil, which provides the commercial incentive for much geological work, seldom exists in Precambrian formations. There are, however, evidences for life in the Precambrian era: well-preserved bacteria and algae, as well as casts of jellyfish, sponge spicules, soft corals, segmented flatworms, and worm trails. Most, but not all, are microfossils.

Major extinctions

Extinction is a natural event in evolution. Virtually all (more than 99%) of the estimated 4 billion species of plants and animals that have ever lived on earth are extinct. Extinction has been caused by many things: global or regional climatic changes that adversely affect vegetation and habitat, or invasion of exotic animals that are able to outcompete the endemic fauna. Introduction of pathogens into a population may also cause extinction. Today habitat destruction is the principal cause of a serious rapid rise in the extinction rate of both animals and plants.

After the Cambrian explosion of biological diversity (see p. 21), five major mass extinctions stand out above the normal, or "background," extinction rate (Figure 4-10). The most cataclysmic of these extinction episodes happened about

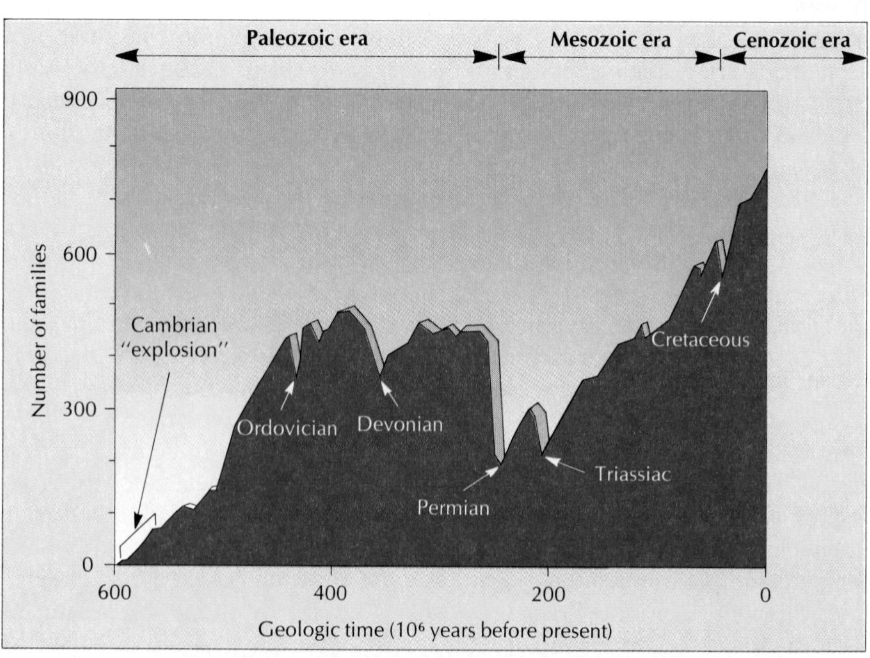

FIGURE 4-10

Changes in numbers of families of marine vertebrates and invertebrates through time from the Cambrian period to present as determined from the fossil record. Sharp drops represent the five major extinctions of skeletonized marine animals. Note that despite extinctions, the overall number of marine families has increased to the present.

From Raup, D.M. and J.J. Sepkoski, Jr. 1982. "Mass extinctions in the marine fossil record," Science **215**:1502-4.

225 million years ago when fully half of the families of shallow-water marine invertebrates and more than 90% of marine invertebrate species disappeared. This was the **Permian extinction.**

Other great extinctions of marine life occurred at the end of the Ordovician, Devonian, Triassic, and Cretaceous periods. The **Cretaceous extinction,** which occurred about 65 million years ago, marked the end of the dinosaurs as well as numerous marine invertebrates and many small reptile groups. The many terrestrial niches that opened at this time allowed the mammals to flourish and radiate.

What caused these extinctions? Many, perhaps most, geologists may now agree that the Cretaceous extinction coincided with the impact on earth of a huge extraterrestrial body—an asteroid or comet. There is much less agreement as to the cause of the other four mass extinctions. Whatever their cause, the great extinctions removed many life forms forever, including many "experimental" body plans organized in ways not seen in any living animal. Each extinction diminished diversity, exposed new niches for the survivors, and launched fresh evolutionary trends.

CLASSIFICATION AND HOMOLOGY

Biologists long before Darwin discovered that, despite the great diversity of life, it was possible to arrange living things into a logical system of classification. The "natural system of classification" that originated with Aristotle and later was formalized by Linnaeus (p. 329) is based on the degree of similarity of morphological characters.

For example, the domestic cat is clearly related to the wild cats and jungle cats of the Old World, and all are included in the same genus, *Felis.* This genus shares an obvious relationship with the "big cats" (lions, tigers, leopards, and jaguars of the genus *Panthera*), as well as lynxes and bobcats (genus *Lynx*); so all are grouped into a set of higher rank, the family Felidae. All members of this family share a number of traits that set them apart from all other families, such as a round head, 30 teeth, and feet with retractile claws. At the same time, the cat family is clearly related as a group to a number of other families (for example, the dog, bear, raccoon, weasel and otter, and hyena families). All of these bear a distinctive anatomy and way of life that characterize them as flesh eaters; they are grouped together in the order Carnivora.

Thus animals can be grouped together with respect to certain combinations of traits. Linnaeus had introduced his classification system partly as a means for cataloging nature and partly to reveal the Creator's purpose, since he believed that each group carried a set of similar features that corresponded to God's plan. Darwin, however, recognized an alternative explanation: the marvelous similarities that allowed plants and animals to be classified into neat hierarchies existed because of a "blood" relationship, caused by descent and divergence from a common ancestor. This became one of the strongest arguments for his theory.

One of the most important tools used by Linnaeus and his successors to search out natural relationships was the concept of homology. **Homology** (Gr. *homologia,* agreement) is the name given to structures that, because of common ancestry, share a fundamental structure and relative position. For example, the human arm, front leg of a frog, wing of a bat or bird, and flipper of a whale are homologous because they are built of the same bones, even though they look different and function differently (Figure 4-11). Each forelimb has about the same number of bones arranged in a similar way. Common structures, constructed of the same building blocks, must reflect common descent from an ancestor having these structures.

Embryological homologies remind us of our kinship with distant vertebrate ancestors. Gill arches, which in fish develop into respiratory organs, appear in the embryos of all vertebrates (Figure 4-12). In the adults of terrestrial vertebrates the

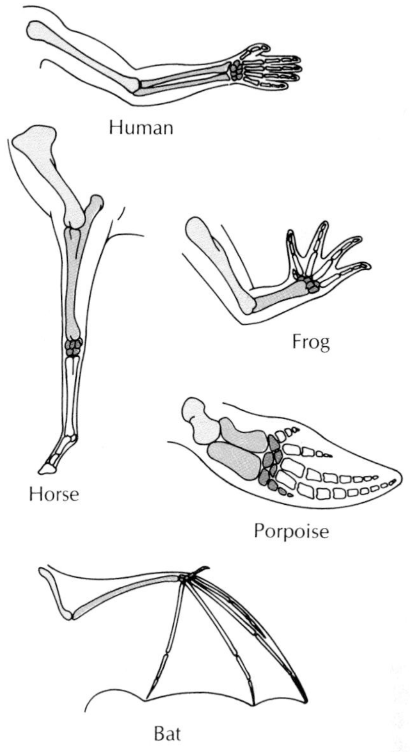

FIGURE 4-11

Forelimbs of five vertebrates to show skeletal homologies. *Yellow,* humerus; *blue,* radius and ulna; *pink,* wrist; *white,* phalanges. Most generalized or primitive limb is that of humans—the feature that has been a primary factor in human evolution because of its wide adaptability. Various types of limbs have been structurally modified for adaptations to particular functions.

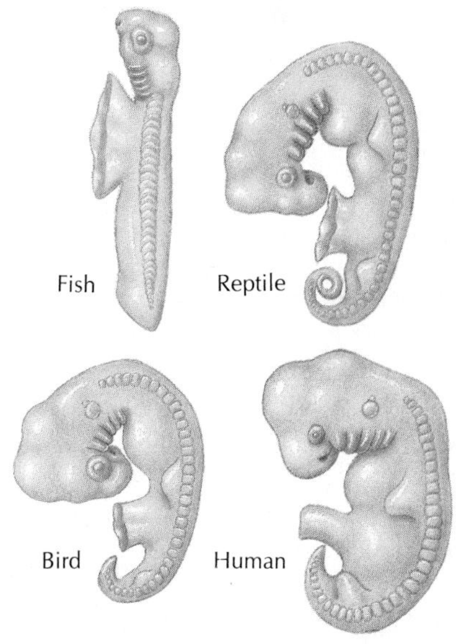

FIGURE 4-12

Comparison of gill arches of different vertebrate embryos. All are shown separated from yolk sac. Note the remarkable similarity of the four embryos at this early stage in development.

gill arches disappear altogether or become modified beyond recognition. Such persistence of history in our embryonic development occurs because early embryonic sequences are not easily modified. Embryological similarities, like anatomical homologies, attest to evolution.

As the evolutionary theory gained acceptance, the unveiling of homologies received new impetus because it was seen as a key to understanding the course of evolution. The example of the pentadactyl limb tells us that all these vertebrates are related by common ancestry and should be grouped together as four-footed, five-toed vertebrates. But, if we want to know if bats are more closely related to frogs or humans, the pentadactyl limb cannot help us because all three forms in question have it. We would discover, however, that bats and humans are warm-blooded, have hair, suckle their young, and possess other homologous features that relate them, whereas the frog has none of these things. The frog, on the other hand, possesses its own set of characteristics that relate it to salamanders and other amphibians. Working in this way, looking for more and more restricted homologies, we can characterize groups and subgroups, arriving at a classification that represents natural and true ancestry.

BIOCHEMICAL HOMOLOGIES

In the late 1950s rapid advancements in molecular biology suddenly opened an entirely new way of comparing relationships of animals. Before this, zoologists could be confident about relationships only through the large number of anatomical features animals shared. Now, molecular information can be used not only to verify common ancestry and homology but also to determine *exact degree* of relationship between two animals.

The genetic code with its dictionary of 64 DNA words of three letters each is universal for all living things, no matter how much they may differ in external appearance (see p. 80). This fact alone is virtual proof that all organisms have descended from a single common ancestor, since the chance of the same arbitrary dictionary arising twice in evolution is unimaginably small. The DNA dictionary cannot itself be used to determine distance of relationships because all living things use the same dictionary. But the protein molecules translated from DNA *can* be used to work out degrees of cousinship. The amino acid chains in protein molecules are like sentences constructed from the DNA dictionary. Such molecular sentences can be compared just as one can compare fossil shells, skulls, and limb bones. But unlike the fossil record, which is often poor or missing altogether, every living organism carries in its cells a molecular record of its ancestry.

Over long periods of time, most of the changes that have occurred at the molecular level are random mutations that have appeared at a constant rate, unaffected by natural selection. In other words, the gradual accumulation of word substitutions in the molecular sentence is proportional to the lapse of absolute time. This "molecular clock," as it is called, has armed taxonomists with a powerful tool that enables them both to determine degrees of relationship and to actually date the points of divergence in evolutionary history.

For example, it has been possible to trace the evolution of cytochrome c, a protein composed of 104 amino acid units, to the origin of eukaryotic cells more than 1 billion years ago. As different mutations accumulated, the amino acid sequences of the molecule in descendant species became different. Just how much any two species differ corresponds with the distance that they are separated on the phylogenetic tree. Elaborate family trees constructed entirely from amino acid sequence diversion agree remarkably well with those obtained by classical morphology and embryology. The study of molecular evolution is a very active field of research today. Unlike anatomical structures, which arise from the complex interactions of genes over lengthy periods of embryological development, protein molecules are the *immediate* products of DNA, which are the fundamental molecules of evolutionary change. Such studies carry us closer to an understanding of the genetic variation that underlies all evolutionary change.

Natural Selection

DARWIN'S THEORY OF NATURAL SELECTION

The theory of natural selection, often expressed in the phrase "survival of the fittest," was the centerpiece of Darwin's thoughts on the evolutionary process. It was the first plausible alternative to Lamarck's idea of evolution by way of inheritance of acquired characteristics. Darwin's theory was nearly as simple as Lamarck's (although less obvious) but was based on observable facts and supported by an abundance of evidence.

1. *All organisms show variation.* No two individuals are exactly alike. They differ in size, color, physiology, behavior, and other ways. Darwin realized that much of this variation is inherited even though he did not understand how. The mechanism of genetic mutation was to be discovered many years later. Darwin pointed out that humans have taken advantage of inherited variation to produce useful new breeds and races of livestock and plants. He believed that, if selection is possible under human control, it can also be produced by agencies operating in nature. He reasoned that *natural* selection can have the same effect as *artificial* selection.

2. *In nature all organisms produce more offspring than can survive.* In every generation the young are more numerous than the parents. Darwin calculated that, even in a slow-breeding species such as the elephant, a single pair breeding from age 30 to 90 and having only 6 young in this span could give rise to 19 million elephants in 750 years. Why, he asked, are there not more elephants?

3. *Accordingly, there is a struggle for survival.* If more individuals are born than can possibly survive, there must be a severe struggle for existence among them. Darwin wrote in *The Origin of Species,* ". . . it is the doctrine of Malthus applied with manifold force of the whole animal and vegetable kingdoms." Competition with other species, as well as members of the same species, for food, space, and shelter is commonplace. Other elements affecting survival include climate, predation, pathogens, and a host of other factors.

4. *Some individuals of a species have a better chance for survival than others.* Because individuals of a species differ, some are favored in the struggle for survival; other are handicapped and eliminated.

5. *The result is natural selection.* Out of the struggle for existence there results the survival of the fittest. Under natural selection, individuals bearing favorable variations survive and have a chance to breed and transmit their characteristics to their offspring, which themselves are more likely to successfully reproduce. The less fit die without reproducing (or they reproduce less). Natural selection is simply the differential survival and reproduction of favored variants. The process continues with succeeding generations so that organisms gradually become better adapted to their environment. Should the environment change, there must also be a change in those characters that have survival value, or the species is eliminated. Successful reproduction is what really counts in natural selection. The "most fit," then, refers to those individuals that have the most successful reproduction.

6. *Through natural selection new species originate.* The differential reproduction of variants can gradually transform species and result in the long-term "improvement" of types. According to Darwin, when different parts of an animal or plant population are confronted with slightly different environments, each diverges from the other. In time they differ enough from each other to form separate species. In this way two or more species may arise from a single ancestral species.

The popular phrase "survival of the fittest" was not originated by Darwin but was coined a few years earlier by the British philosopher Herbert Spencer, who anticipated some of Darwin's principles of evolution. Unfortunately the phrase later came to be coupled with unbridled aggression and violence in a bloody, competitive world. In fact, natural selection operates through many other characteristics of living things. The fittest animal may be the most helpful, the most caring, or the most loving. Fighting prowess is only one of several means toward successful reproductive advantage.

The use of the word "theory" to describe evolutionary science has caused many people to believe that the principle carries little more weight than an idea or a conjecture, since we often use the word in common speech (for example, "I have a theory about what really happened there"). When creationists charge that evolution is "just a theory," this may be what they have in mind. Scientists, on the other hand, use "theory" to refer to a body of generalizations and principles developed within a field of inquiry, such as mathematics or physics or biology. Thus we have the theory of numbers, theory of gravitation, and theory of evolution. Such theories are built from sets of facts, not conjectures and speculations (see p. 5). It is also important to emphasize that evolution as a process is a fact, and the only argument among scientists today is with the details of how it operates, not whether it operates.

Through adaptation to a changed environment, a group of animals may also diverge enough from their ancestors to become a different species.

Gregor Mendel published his theories on inheritance in 1868, 14 years before Darwin died. Darwin presumably never read the work (although it was found in Darwin's extensive library after his death), and, if there were others who knew of it, they failed to bring it to his attention. Had Mendel written Darwin about his results, it is possible that Darwin would have modified his theory. However, it is also possible that Darwin could not have seen the relationship of the hereditary mechanisms to gradual changes and continuous variations that represent the hub of Darwinian evolution. It took other scientists years to establish the relationship of genetics to Darwin's natural selection theory.

APPRAISAL OF DARWIN'S THEORY

Darwin supported his theory of evolution by an overwhelming amount of evidence that had never before been accumulated. He convincingly demonstrated evolution and then provided a logical explanation for it. Through the well-chosen body of evidence so lucidly expounded in *The Origin of Species,* he made evolution understandable. Nevertheless, despite its elegance, it is hardly surprising that parts of his theory have been modernized because of more recent biological knowledge. When Darwin wrote *The Origin of Species,* nothing was known about the inheritance of variation. It was not even known that sexual reproduction involved the combination of a single sperm with a single egg. Thus one always marvels that Darwin's deductions and conclusions were so sound in view of the scientific ignorance that prevailed at the time.

The most serious weakness of Darwin's theory was his failure to identify correctly the mechanism of inheritance. Darwin saw heredity as a blending phenomenon in which the characteristics of the parents melded together in the offspring. We now understand from our knowledge of genetics that there are always new variations arising in a species which can also play a role in speciation events. Thus the first step in modernizing Darwin's theory was the demonstration that the operative units in inheritance are self-reproducing, nonblending genes. Next there was the discovery of chromosomes in which genes were found to be precisely located in linkage groups. Biologists could then understand how the ultimate source of variation — mutations — could be inherited in the usual Mendelian fashion.

EMERGENCE OF NEO-DARWINISM: THE SYNTHETIC THEORY

In the 1930s a new breed of geneticists began to reevaluate Darwin's theory from a different perspective. These were population geneticists, scientists who studied variation in natural populations of animals and plants and who had a sound knowledge of statistics and mathematics, disciplines that earlier geneticists woefully lacked. Gradually, a new "synthetic theory" emerged that brought within its orbit several previously separate fields of evolutionary biology: population genetics, paleontology, systematics, and, later, animal behavior. Population geneticists showed that populations evolve through shifting gene frequencies produced by natural selection and, to a lesser extent, from genetic drift (to be discussed later). Paleontologists found that the evidence of the fossil record was consistent with the new synthetic theory. Systematists, too, embraced the new theory. Later, it became evident that the gradual evolution of populations (microevolution) could, over long periods of time, adequately explain major evolutionary events, such as the appearance of whole new taxonomic groups (macroevolution).

Central to the synthetic theory, then, was the recognition that evolution is a change in the genetic composition of populations. Such changes occur when the population is exposed to an environmental challenge — some change in the abiotic (physical) or biotic environment. Interactions between population and environment have become the dominant theme of the modern synthetic theory of evolution.

What happens when populations are faced with a drastic change in their environment? They may respond in any one of three ways. First, they may become extinct. This is the fate of most. Or, following a second course, they may somehow successfully meet all environmental challenges to which they are exposed, remaining mostly unchanged. Examples are the tuatara of New Zealand

(p. 604), the modern lungfishes, and oysters—all forms that closely resemble their ancestors of millions of years ago.

The third response to environmental challenge is the evolution, through mutations, of new adaptations that enable a population to invade a *new* ecological habitat. These are the winners in an evolutionary sense, because they generate new species.

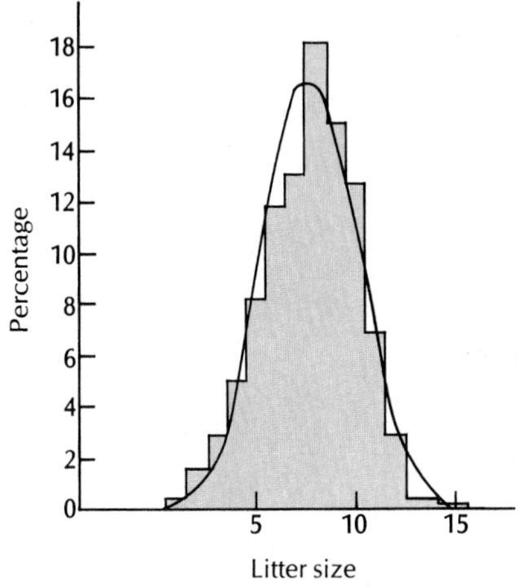

FIGURE 4-13

Frequency distribution of litter size in mice, showing how variation in litter size tends to conform to a bell-shaped, or normal, curve.

Adapted from Falconer, D.S. 1981. Introduction to quantitative genetics, ed. 2, London, Longman Group Ltd.

Microevolution: Genetic Change within Populations and Species

Microevolution is the change of gene frequencies within single, natural populations. These are the small evolutionary events that are going on all the time. The evolutionary process is usually much too slow to be observable in a single lifetime. Nevertheless, within individual populations it is often possible to observe shifts in gene frequencies that produce new adaptive traits. Occasionally we can witness the emergence of new species. We have stressed that evolution is the result of interaction between populations and their environments. How these interactions cause evolution is the subject of this section.

GENETIC VARIATION WITHIN SPECIES

Biological variation exists within all natural populations. We can verify this fact simply by measuring a single character in several dozen specimens selected from a local population: tail length in mice, length of a leg segment in grasshoppers, number of scales along the side of a trout, number of peas in a pod, and height of adult males of the human species. When the values are graphed with respect to frequency distribution, they tend to approximate a normal, or bell-shaped, probability curve (Figure 4-13). Most individuals fall near the mean, fewer fall above or below the mean, and the extremes make up the "tails" of the frequency curve with increasing rarity. The larger the population sample, the closer the frequencies resemble a normal curve.

As pointed out in Chapter 3, the ultimate source of all genetic variation is the mutation of genes. The *immediate* source of genetic variation, however, is not mutations because a mutation is a relatively rare event. Rather it is the recombination of existing alleles by sexual reproduction.

All of the alleles of all genes possessed by members of a population are referred to collectively as the **gene pool.** The interbreeding population is the visible manifestation of the gene pool, which thus has continuity through successive generations. The gene pool of large populations must be enormous, because at observed mutation rates many mutant alleles can be expected at all gene loci. Let us suppose that two alleles are present at a chromosomal locus: *A* and *a*. Two kinds of sex cells would be produced, and among individuals in the population there are three possible genotypes: *A/A*, *A/a*, and *a/a*. When two pairs of alleles are involved, 2^2 or four kinds of sex cells can be produced (see the Punnett square on p. 70), and within the population there are nine possible genotypes. With larger numbers of loci and pairs of alleles, the potential number of different kinds of sex cells rises rapidly, that is, 2^3, 2^4, 2^5, and so on.

Most animals have thousands of gene loci. If the number of gene loci in an individual were 10,000 (about average for many mammal species), and 10% (1000) of these loci were heterozygous, the individual could produce a staggering 2^{1000} genetically different kinds of sex cells, a number larger than all the atoms in the universe! So, with just two alleles existing for 10% of the genome, there is obviously copious material on which natural selection can act. Even if no new mutations occurred, the shuffling of existing genes would provide enough variation to fuel natural selection for a long time.

In fact, the amount of potential variation is even greater than our example suggests. After protein electrophoresis (a method used to distinguish different forms of the same enzyme [allozymes]) was introduced in the 1960s, geneticists were surprised to discover that at least 30% of the roughly 10,000 loci in *Drosophila* are polymorphic, with two to six alleles at each locus. These alleles appear at such high frequencies that the average fly is likely to be heterozygous at about 12% of its loci. But this is not the whole story because genes exert different influences in the presence of other genes. Gene A may act differently in the presence of gene B from what it does in the presence of gene C. The diversity produced by this interaction and the addition of new mutations now and then add to the complexity of population genetics.

Variation and natural selection

What does variation signify for evolution? As already stated, genetic variation produced in whatever manner furnishes the material on which natural selection works to produce evolution. Natural selection does this by favoring beneficial variations and eliminating those which are harmful to the organism. Selective advantages of this type represent a very slow process, but on a geological time scale they can bring about striking evolutionary changes represented by the various taxonomic units (species, genera, and so on), adaptive radiation groups, and the various kinds of adaptations.

The fact must be stressed, however, that natural selection works on the whole animal and not on a single hereditary characteristic. The organism that possesses the most beneficial combination of characteristics or "hand of cards" is going to be selected over one not so favored. This concept may help to explain some of those puzzling instances in which an animal may have certain characteristics of no advantage or that are actually harmful, but in the overall picture it has a winning combination. Thus in a population, pools of variations are created on which natural selection can work to produce evolutionary change.

GENETIC EQUILIBRIUM

In the human population, brown eyes are dominant to blue, curly hair is dominant to straight, and a Roman nose is dominant to a straight nose. Why hasn't the dominant gene gradually supplanted the recessive one in each instance so that we are all brown eyed, curly haired, and Roman nosed? It is a common belief that a character dependent on a dominant gene increases in proportion because of its dominance. This is not the case, because there is a tendency in *large* populations for genes to remain in equilibrium generation after generation. Strange as it may seem, a dominant gene does not change in frequency with respect to its recessive allele(s). Dominance describes the *phenotypic effect* of an allele, not its abundance.

The principle is based on a theorem that is the foundation of the entire genetic basis of evolution, the **Hardy-Weinberg equilibrium** (see box). According to this theorem, gene frequencies and genotype ratios in large biparental populations reach an equilibrium in one generation and *remain constant* thereafter *unless* disturbed by new mutations, natural selection, or genetic drift (chance). Such disturbances are the source materials of evolution and are described in the next section.

A rare gene, according to this principle, does not disappear merely because it is rare. That is why certain rare traits, such as albinism and cystic fibrosis, persist for many generations. Variation is retained even though evolutionary processes are not in active operation. Whatever changes occur in a population—gene flow from other populations, mutations, natural selection, and migration—involve the establishment of a new equilibrium with respect to the gene pool, and this new balance is maintained until upset by disturbing factors.

For example, albinism in humans is caused by a recessive allele a. Only one person in 20,000 is an albino, and this individual must be homozygous (a/a) for the recessive allele. Obviously there are many carriers in the population with normal pigmentation, that is, people who are heterozygous (A/a) for albinism. What is their frequency? A convenient way to calculate the frequencies of genes in a population is with the binomial expansion of $(p + q)^2$ (see box). We will let p represent the frequency of the dominant allele A, and q the frequency of the recessive allele a for albinism.

Assuming that mating is random (a questionable assumption for albinos, but one that we will accept for our example), genotype distribution will be $p^2 = A/A$, $2pq = A/a$, and $q^2 = a/a$. Only the frequency of genotype a/a is known with certainty, 1/20,000; therefore:

Hardy-Weinberg equilibrium: why gene frequencies do not change

The Hardy-Weinberg law is a logical consequence of Mendel's first law of segregation and expresses the tendency toward equilibrium inherent in Mendelian heredity.

Let us select for our example a population having a single locus bearing just two alleles *T* and *t*. The phenotypic expression of this gene might be, for example, the ability to taste a chemical compound called phenylthiocarbamide. Individuals in the population will be of three genotypes for this locus, *T/T*, *T/t* (both tasters), and *t/t* (nontasters). In a sample of 100 individuals, let us suppose we have determined that there are 20 of *T/T* genotype, 40 of *T/t* genotype, and 40 of *t/t* genotype. We could then set up a table showing the allelic frequencies as follows (remember that every individual has two loci for the gene and thus two alleles):

Genotype	Number of individuals	Number of *T* alleles	Number of *t* alleles
T/T	20	40	
T/t	40	40	40
t/t	40	—	80
TOTAL	100	80	120

Of the 200 alleles, the proportion of the *T* allele is 80/200 = 0.4 (40%); and the proportion of the *t* allele is 120/200 = 0.6 (60%). It is customary in presenting this equilibrium to use "p" and "q" to represent the two allele frequencies. The dominant gene is represented by p, the recessive by q. Thus:

$$p = \text{frequency of } T = 0.4$$
$$q = \text{frequency of } t = 0.6$$
$$\text{Therefore } p + q = 1$$

Having calculated allele frequencies in the sample, let us determine whether these frequencies will change spontaneously in a new generation of a population. Assuming the mating is random (and this is important; all mating combinations of genotypes must be equally probable), each individual will contribute an equal number of gametes to the "common pool" from which the next generation is formed. This being the case, the frequencies of gametes in the "pool" will be proportional to the allele frequencies in the sample. That is, 40% of the gametes will be *T*, and 60% will be *t* (ratio of 0.4:0.6). Both ova and sperm will, of course, show the same frequencies. The next generation is formed as follows:

Sperm \ Ova	*T* = 0.4	*t* = 0.6
T = 0.4	*T/T* = 0.16	*T/t* = 0.24
t = 0.6	*T/t* = 0.24	*t/t* = 0.36

Collecting the genotypes, we have:

$$\text{frequency of } T/T = 0.16$$
$$\text{frequency of } T/t = 0.48$$
$$\text{frequency of } t/t = 0.36$$

Next, we determine the values of p and q from the randomly mated populations. From the table above, we see that the frequency of *T* will be the sum of genotypes *T/T*, which is 0.16, and one-half of the genotype *T/t*, which is 0.24:

$$T(p) = \frac{0.16 + \frac{1}{2}(0.48)}{1} = 0.4$$

Similarly, the frequency of *t* will be the sum of genotypes *t/t*, which is 0.36, and one-half the genotype *T/t*, which is 0.24:

$$t(q) = \frac{0.36 + \frac{1}{2}(0.48)}{1} = 0.6$$

The new generation bears exactly the same gene frequencies as the parent population! Note that there has been no increase in the frequency of the dominant gene *T*. Thus *in a freely interbreeding, sexually reproducing population, the frequency of each allele remains constant generation after generation.* The more mathematically minded reader will recognize that the genotype frequencies *T/T*, *T/t*, and *t/t* are actually a binomial expansion of $(p + q)^2$:

$$(p + q)^2 = p^2 + 2pq + q^2 = 1$$

$$q^2 = 1/20{,}000$$
$$q = \sqrt{1/20{,}000} = \frac{1}{141}$$

Since the frequencies $p + q$ equal 1, we can determine frequency p by subtracting frequency q from 1:

$$p = 1 - q = \frac{141}{141} - \frac{1}{141} = \frac{140}{141}$$

The frequency of carriers is as follows: Recall that we stated that *A/a* is represented by $2pq$. Using the values for p and q just computed, $2pq$ becomes:

$$A/a = 2pq = 2 \times \frac{140}{141} \times \frac{1}{141} = \frac{1}{70}$$

Eugenics is the study that deals with the improvement of hereditary qualities in humans by social control of mating and reproduction. While the genetic argument against eugenics cited in the text is compelling in itself, a eugenics program lingered in the United States (and elsewhere) until finally dispatched following Adoph Hitler's attempt to "purify" races in Europe by sterilization.

One person in every 70 is a carrier, or *A/a*! Even though a recessive trait may be rare, it is amazing how common a recessive allele may be in a population. There is a message here for anyone proposing a eugenics program designed to free a population of a "bad" gene. It is practically impossible. Since only the homozygous recessives reveal the phenotype to be artificially selected against (by sterilization, for example), the gene continues to surface from the heterozygous carriers. For a recessive allele present in two of every 100 persons (but homozygous in only one in 10,000 persons), it would require 50 generations of complete selection against the homozygotes just to reduce its frequency to one in 100 persons.

PROCESSES OF EVOLUTION: HOW GENETIC EQUILIBRIUM IS UPSET

The Hardy-Weinberg law shows that populations do not change—and thus do not evolve—as long as certain conditions are met: (1) very large population, (2) random mating, (3) no mutations, (4) no selection, and (5) no migration. We already know that these conditions cannot all prevail indefinitely and that the Hardy-Weinberg equilibrium can be disturbed by mutations, genetic drift, natural selection, and migration. Earlier (p. 86) we discussed mutations, the ultimate source of variability in all populations. We will now look at other factors that cause gene frequencies to change and thus lead to evolution.

Thomas Henry Huxley (1825-1895), one of England's greatest zoologists, on first reading the convincing evidence of natural selection in Darwin's *The Origin of Species* is said to have exclaimed, "How extremely stupid not to have thought of that!" He became Darwin's foremost advocate and engaged in often bitter debates with Darwin's critics. Darwin, who disliked publicly defending his own work, was glad to leave such encounters to his "bulldog," as Huxley called himself.

Courtesy British Museum (Natural History), London.

Natural selection

We have already stated that natural selection is the principal guiding force in evolution, and we have alluded to its mechanism. We will now more firmly define what it does. *Natural selection guides evolution by sorting out new adaptive combinations from a population gene pool derived from mutations, recombination, and other sources of genetic variation acting over many generations.*

Let us examine this definition. Sexually reproducing populations are composed of many individuals, and each individual has a different genotype. These differences have arisen by mutations, genetic recombination, and perhaps genetic drift, nonrandom mating, or migration. Suppose that one of these genotypes is better than any of the others, that is, the animal bearing it is *better adapted* to the environment than are other animals having different genotypes. What basis have we for concluding that a particular genotype is better, that is, has greater **adaptive value** than others? The basis is reproductive success. Most organisms contribute progeny to the next generation. The genotype that contributes the most progeny, and therefore the most genotypes to the next generation, has the greatest adaptive value and chance of success. If this differential reproduction continues systematically generation after generation, the genotype will increase in frequency while others decline in frequency. Natural selection, then, produces differences in gene pools from one generation to the next. Changing gene frequencies is the whole basis of evolution.

We have stressed, and we will stress once more, that natural selection works on the whole organism—the entire phenotype—and not on individual genes. The winning genotype may well contain genes that have detrimental effects on the individual but are preserved because other genes are strongly beneficial.

Natural selection at work

There are many examples that show how natural selection alters populations in nature. Sometimes selection can proceed very rapidly as, for example, in the development of high resistance to insecticides by insects, especially flies and mosquitoes. Doses that at first killed almost all pests later were ineffective in controlling them. As more insects are exposed to insecticides, those most sensitive are killed, leaving more space and less competition for resistant strains to multiply. Thus, as a result of selection, mutations bestowing high resistance, but previously rare in the population, increased in frequency.

Perhaps the most famous instance of rapid selection is that of **industrial melanism** (dark pigmentation) in the peppered moth of England (Figure 4-14). Before 1850 the peppered moth was always white with black speckling in the wings and body. In 1849 a mutant black form of the species appeared. It became increasingly common, reaching frequencies of 98% in Manchester and other heavily industrialized areas by 1900. The peppered moth, like most moths, is active at night. It rests during the day in exposed places, depending on its cryptic coloration for protection. The mottled pattern of the normal white form blends perfectly with the lichen-covered tree trunks. With increasing industrialization, the soot and grime from thousands of chimneys killed the lichens and darkened the bark of trees for miles around centers such as Manchester. (This part of England was known as the "Black Country.") Against a dark background the white moth is conspicuous to predatory birds, whereas the mutant black form is camouflaged. The result was rapid natural selection: the easier-to-see white form was preferentially selected by birds, whereas the melanic form was subject to far less predation. Selection pressure thus tended to eliminate the white form while favoring the genes that contributed to black wings and body.

FIGURE 4-14

Light and melanic forms of the peppered moth *Biston betularia* on a soot-covered oak near industrial Birmingham, England. The dark melanic form blends with the sooty bark but the light form is plainly visible to predators.

Another way to describe the selective quality of the genes that determine color is in terms of their **fitness.** Fitness, or adaptive value, describes the way individuals differ in their reproductive success because of differences in the genes they carry. In this case the single dominant mutation for the black form of the peppered moth increased its fitness in England's "Black Country." This same trait, however, confers low fitness in nonpolluted countryside areas. On the other hand, genes for the white form confer high fitness in nonpolluted countryside areas but low fitness near industrial centers.

Of special interest is the rapidity of the change. Rather than requiring thousands of years, the change occurred in only about 50 years. White moths still survived in nonpolluted areas of England, and with the recent institution of pollution-control programs the white forms are reappearing in the woods around cities (Figure 4-15). Industrial melanism is a dramatic example of rapid directional selection caused by a changing environment and leading to shifting gene frequencies.

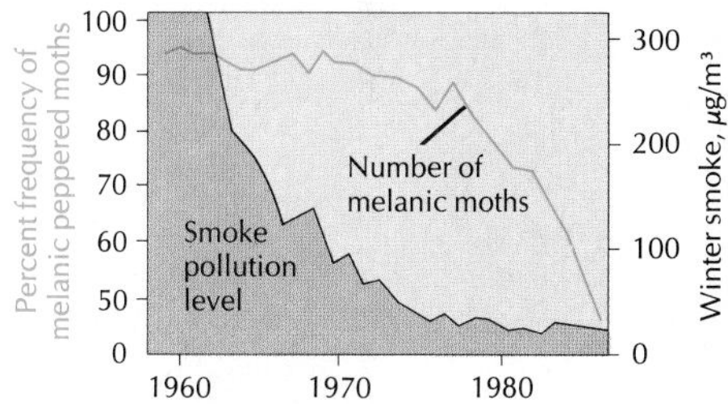

FIGURE 4-15

"Evolution in Reverse." Decline in the frequency of the melanic (black) form of the peppered moth with falling air pollution in industrial areas of England. The frequency of the melanic form still exceeded 90% in 1960 while smoke and sulfur dioxide emissions were still high. Later, as emissions fell and light-colored foliose lichens began to grow again on the tree trunks, the melanic form became more conspicuous to predators. By 1986, only 50% of the moths were still of the melanic form, the rest having been replaced by the light form.

From Brakefield, P.M., 1987, "Industrial melanism: do we have the answers?", Trends in Ecology and Evolution **2**:117-122.

Genetic drift

Genetic drift refers to accidental changes in gene frequency that may occur when a few individuals at random become isolated from a large population. This could happen when a small number of individuals (called **founders**) migrate to a remote habitat. These few will carry with them only an incomplete sample of the gene pool of the parent population; alleles not carried will be lost. Thus the founders are certain to be different from the parent population. The smaller the number of colonizers, the greater are random changes in gene frequencies that can lead to new species.

However, small populations that descend from founders may be less able to cope with a new environment because the loss by genetic drift of some genes present in the ancestral gene pool has reduced their "fitness," or capacity to adapt. This may be one reason why small populations on islands often become extinct. But despite the limited genetic diversity of founder populations, they are sometimes phenomenally successful. All of the billions of starlings in North America are the descendants of a few birds introduced into New York City in

During the 1960s, genetic drift was deemphasized as a factor of much importance in evolution because it could operate only in small populations. However, evolutionists now realize that most breeding populations of animals *are* small. A natural barrier such as a stream, or a ravine between two hilltops, may effectively separate a breeding population into two separate and independent units of selection.

1890 (p. 625). Most of the (unfortunately) successful introductions of birds and mammals into North America, New Zealand and Australia began from a handful of individuals. Once a founder population becomes established it is then guided in its evolution by natural selection.

■ Evolution of New Species

We have described how populations change over periods of time because of genetic variation. Mutations are the fundamental source of genetic variation, further enhanced by sexual interchange and recombination. Natural selection acting on these variations is the force that produces evolutionary change and maintains the adaptive well-being of populations. In this discussion we consider how new species arise. The process of species multiplication, that is, the division of a parent species into several daughter species, is called **speciation.**

SPECIES CONCEPT

Although a definition of species is needed before we can discuss species formation, biologists do not agree on a single rigid definition that applies in all cases. A traditional but now obsolete concept of species is as static and immutable units subject only to minor and unimportant variations. According to this view, a species is a group within which the individuals closely resemble one another but are clearly distinguishable from those of any other group. The flaw in this definition is the fact of biological variation. Many species are made up of individuals that can be arranged into completely intergrading series. Sometimes the gaps between species can be detected only with the greatest difficulty, and it becomes almost impossible to decide where two species are to be separated. Yet species are realities in nature. We observe and collect them, study them, sample their populations, and describe their behavior.

The difficulty with the traditional system is that classification is based on morphology alone (size, color, length of various parts, and so on). Other criteria can be and currently are being used such as behavioral differences, reproductive characteristics, genetic composition, and ecological niche. The single property that maintains the integrity of a species more than any other property, especially among biparental organisms, is *interbreeding.* The members of a species can interbreed freely with each other, produce fertile offspring, and share a common genetic pool. Interbreeding of different species usually is either physically or behaviorally impossible, or it produces sterile offspring. There are, of course, exceptions, but they are rare. Species are thus usually considered genetically closed systems, whereas **races** within a species are open systems and can exchange genes. **A species therefore may be defined as a group of organisms of interbreeding natural populations that is reproductively isolated from other groups and that shares a common gene pool** (Dobzhansky, 1951; Mayr, 1970).

HOW SPECIES BECOME REPRODUCTIVELY ISOLATED

The definition of a species states that gene pools of different species are **reproductively isolated** from each other. With the exception of occasional hybridization, gene flow between species does not occur. Reproductive isolation, then, ensures the independence of a new species. No new species could maintain its integrity if it freely interbred with other species. If it did interbreed with another species, and the offspring from these matings were just as viable and fertile as those of matings within the species, the two species would simply amalgamate. Reproductive isolation, then, is the *result* of successful speciation. How does it happen?

There appears to be only one certain way for barriers to be built between animal species: **geographical isolation.** Unless some individuals of a population can be segregated for many generations from the parent population, new adaptive variations that may arise may become lost through interbreeding with the parent population. Geographical isolation permits a unique sample of the population gene pool to be "pinched off," or segregated, so that diversification can occur.

Geographical isolation is not the same thing as reproductive isolation. Geographical isolation refers to the spatial separation of two populations. It prevents gene exchange and is a precondition for speciation. Reproductive isolation is the result of speciation, and refers to the various physical, physiological, ecological, and behavioral barriers that prevent interbreeding with any other species.

Allopatric speciation

Allopatric ("in another land") populations of species are those that occupy different geographical areas and do not interbreed because of barriers that separate these areas. Speciation that occurs as a result of geographical isolation is known as allopatric speciation.

Let us imagine a single interbreeding population of mammals that is split into two isolated populations by some geographical change — the uplifting of a mountain barrier, the sinking and flooding of a geological fault, or a climatic change that creates a hostile ecological barrier such as a desert. Gene flow between the two isolated populations is no longer possible. The populations are almost certainly different from each other even when first isolated, since just by chance alone one population contains alleles not present in the other.

With the passage of time genetic recombination accentuates the difference, especially if the populations are small. Mutations that appear in the separated populations are certain to be different. Furthermore, the climatic conditions of the separated regions are different; one may be warm and moist, the other cool and dry. Natural selection acting on the isolated gene pools favors those mutations and recombinations that best adapt the populations to their respective environments, unique food supplies, and new predator-prey relationships.

The two populations continue to diverge morphologically, physiologically, and behaviorally until distinct geographical races are formed. After some indefinite time span, perhaps only a few thousand years, but more likely a period of hundreds of thousands or even millions of years, evolutionary diversification progresses until the two races are reproductively isolated. If the barrier separating them fails (erosion of mountains, shifting of river flow, reestablishment of a hospitable ecological bridge), the two populations again intermix. However, they may no longer interbreed because fertility between them is impaired by their accumulated gene differences. They are now distinct species.

Speciation under these conditions is by no means inevitable. It happens only under the most ideal conditions: small population (and gene pool) size, fortunate mutations, and the right combination of selection pressures. The chances are good, in fact, that one or both populations isolated in this way become extinct. But over the vast span of time that organic evolution has had to work, the special conditions required for speciation have been repeated numerous times, indeed millions of times.

Adaptive radiation

Many times in the earth's history, a single parental population has given rise not just to one or two new species, but to an entire family of species. The multiplication of related species, each with their unique specializations that fit them for particular ecological niches, is called **adaptive radiation.**

Young islands are especially ideal habitats for rapid evolution. If they are formed by volcanoes that rose from a platform on the ocean floor, as were the Galápagos Islands, they are at first devoid of life. In time they are colonized by plants and animals from the continent or from other islands. These newcomers encounter an especially productive situation for evolution because of the abundance of new opportunities.

On the crowded mainland almost every ecological niche is already occupied. (An ecological niche is an animal's unique way of life in an environment; see Chapter 5.) Every animal is specialized and adapted for a particular way of life, and all food sources are exploited by one species or another. Although variations continue to appear within a population, those offspring in each generation most like their parents are most likely to survive, because they are best suited for that particular set of environmental conditions. Competition between different species is too keen on the mainland, where evolution has been proceeding for a very long time, for many new evolutionary experiments to succeed.

On a young island, however, new arrivals find new ecological opportunities and few or no competitors, predators, pathogens, and parasites. Those surviving the sea or air voyage and the landfall may be able to become established, multiply, and spread out. This is step one in the allopatric model of finch speciation shown in Figure 4-16. In step two of the model, immigrants from the first island disperse to other islands to form new populations. They are in a new habitat that in all probability differs ecologically from the first island. New variations may serve some of the immigrants in establishing themselves in new niches. Offspring that differ slightly from their parents may find that these differences enable them to exploit alternate food sources as yet unused by other animals. They thus flourish and produce offspring, some of which bear similar characteristics. With each passing generation these animals become increasingly successful at this alternative way of life, at the same time becoming increasingly different from their ancestors. In step three of the allopatric model (Figure 4-16), members of the derived populations of finches recolonize the original island and become established as a new species. This process may be, indeed must have been, repeated many times on these and other islands in the archipelago to eventually form the 13 species of Galápagos finches.

On archipelagoes, such as the Galápagos Islands, isolation plays an extremely important evolutionary role. Not only is the entire archipelago isolated from the continent, but each island is geographically isolated from the others by the sea, the most inhospitable environment for land animals. Moreover, each island is different from every other to a greater or lesser extent in its physical, climatic, and biotic characteristics. As colonizing animals find their way to the different islands, they are presented with an environmental challenge that is unique for each island. Moreover, each new arrival carries with it only a small fraction, a biased sample, of the population's gene pool. This further stimulates diversification, already encouraged by isolation, new ecological opportunities, and lack of competition. Archipelagoes, more than any other place on the earth's surface, offer the raw materials and opportunities for rapid evolutionary changes of great magnitude.

Woodpecker finch, one of the 13 species of the Galápagos Islands finches, using a slender twig as tool for feeding. This finch worked for about 15 minutes before spearing and removing a wood roach from a break in the tree.

Photograph by C.P. Hickman, Jr.

Darwin's finches

Let us consider the evolution of the family of 13 famous Galápagos finch species (Figure 4-17). Their fame rests not on their beauty—they are, in fact, inconspicuous, dull colored, and rather unmusical in song—but on the enormous impact they had in molding Darwin's theory of natural selection. Darwin noticed that the Galápagos finches (the name "Darwin's finches" was popularized in the 1940s by the British ornithologist David Lack) are clearly related to each other, but that each species differs from the others in several respects, especially in size and shape of the beak and in feeding habits. Darwin reasoned that, if the finches were specially created, it would require the strangest kind of coincidence for 13 similar finches to be created on the Galápagos Islands and nowhere else.

Darwin's finches are an excellent example of how new forms may originate from a single colonizing event. Lacking competition from other land birds, they underwent adaptive radiation, evolving along various lines to occupy available niches that on the mainland would have been denied to them. They thus assumed the characteristics of mainland families as diverse and unfinchlike as warblers and woodpeckers.

A fourteenth Darwin's finch is found on isolated Cocos Island, far to the north of the Galápagos archipelago. It is similar in appearance to the Galápagos finches

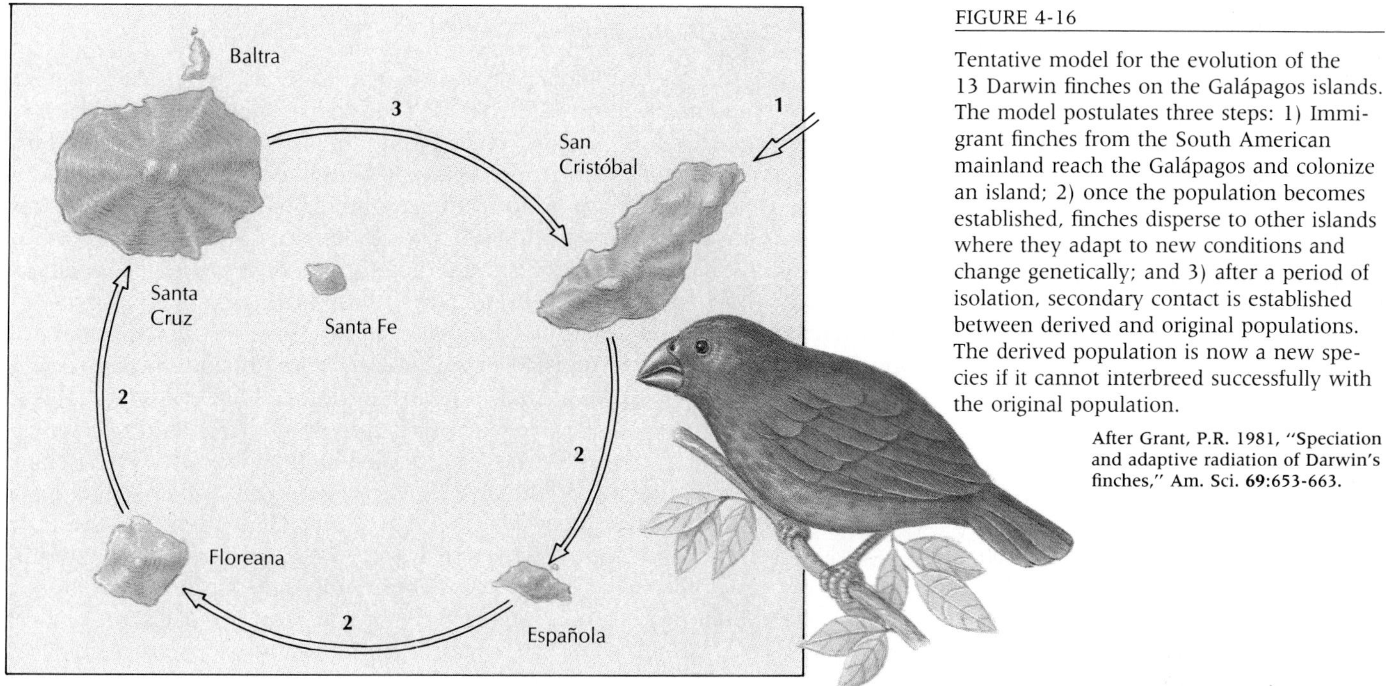

FIGURE 4-16

Tentative model for the evolution of the 13 Darwin finches on the Galápagos islands. The model postulates three steps: 1) Immigrant finches from the South American mainland reach the Galápagos and colonize an island; 2) once the population becomes established, finches disperse to other islands where they adapt to new conditions and change genetically; and 3) after a period of isolation, secondary contact is established between derived and original populations. The derived population is now a new species if it cannot interbreed successfully with the original population.

After Grant, P.R. 1981, "Speciation and adaptive radiation of Darwin's finches," Am. Sci. **69**:653-663.

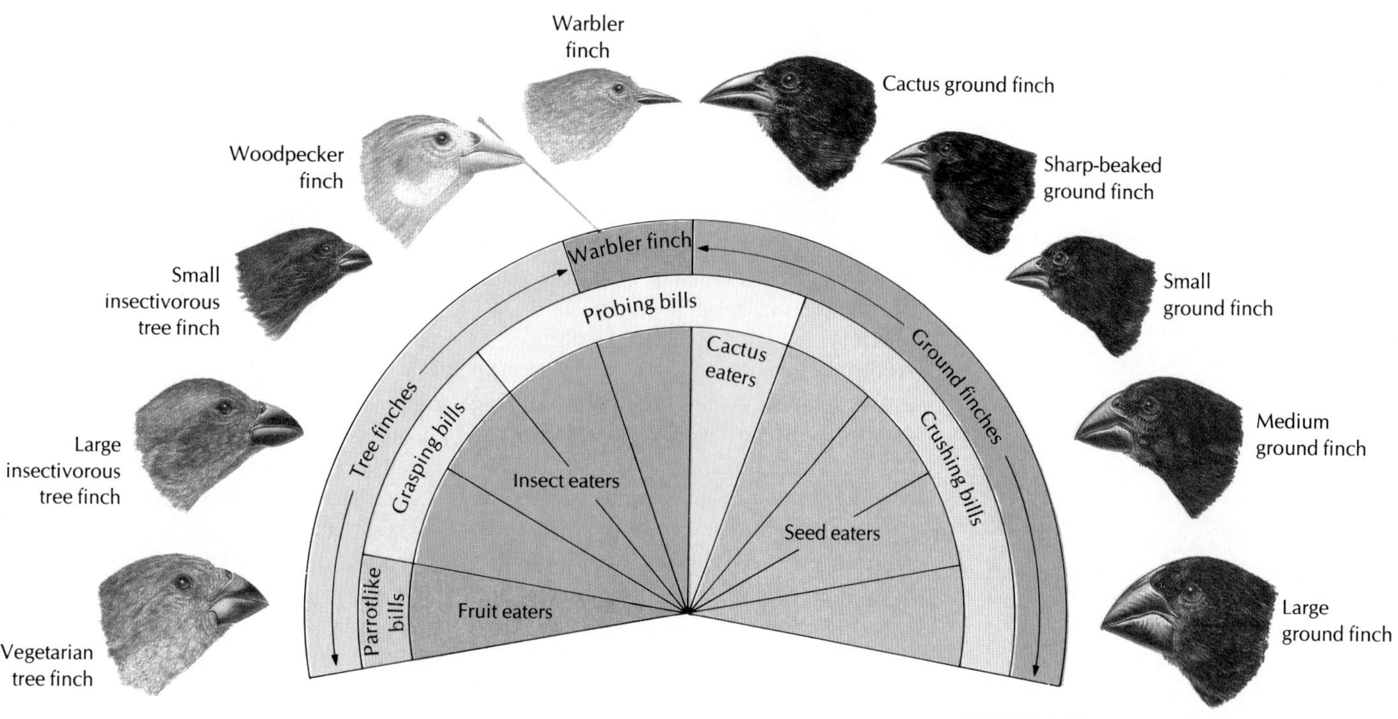

FIGURE 4-17

Adaptive radiation in ten species of Darwin's finches of Santa Cruz, one of the Galápagos Islands, showing differences in bills and feeding habits. All are believed to have descended from a single common ancestral finch from the South American continent.

From Raven, P.H., and G.B. Johnson, 1989. Biology, ed. 2. St. Louis, Times Mirror/Mosby.

and almost certainly has descended from the same ancestral stock. Since Cocos Island is a single small island with no opportunities for isolation of finch populations, there has been no species multiplication.

The Cocos Island example emphasizes the fundamental requirement for geographical isolation in speciation. In its absence, there is no other reasonable way for gene pools of animals to become separated long enough for genomic differences to accumulate.

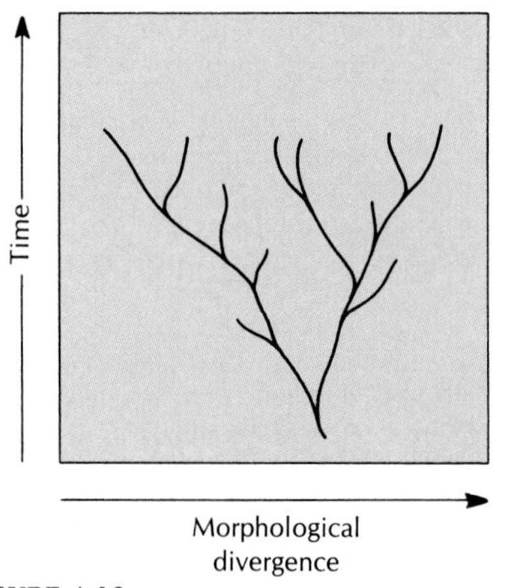

FIGURE 4-18

The gradualist model of evolutionary change in morphology, viewed as proceeding more or less steadily through time. Speciation events are represented as bifurcations in the phylogeny.

Macroevolution: Major Evolutionary Events

We have seen that the gradual accumulation of small changes in gene frequency within a population, leading to the formation of new species by classical Darwinian natural selection, is called microevolution. **Macroevolution** is the term used to describe large-scale events in organic evolution operating over millions of years. The appearance of major groups (for example, molluscs, land vertebrates, and birds) represents completely new designs for living. Can historical trends of this magnitude be explained by shifting gene frequencies within populations (that is, by microevolution) operating over vast time scales? The orthodox answer is yes, and the supporting evidence comes especially from molecular and physiological studies showing that organisms share homologies at many levels (DNA, RNA, blood proteins, cellular organizations, metabolic pathways, and so on).

Nevertheless, many evolutionists are troubled by the idea that evolution has always progressed at a stately pace, with lineages appearing and changing gradually over vast periods of time (Figure 4-18). More often than not, the fossil record suggests that evolution moves in leaps. Many species remain virtually unchanged for millions of years, then suddenly disappear to be replaced by a quite different, but related, form. Moreover, most major groups of animals appear abruptly in the fossil record, fully formed, and with no fossils yet discovered that form a transition from their parent groups. Thus it has seldom been possible to piece together ancestor-dependent sequences from the fossil record that show gradual, smooth transitions between species.

There are two principal explanations for the absence of transitional forms in the fossil record. The traditional explanation is that evolution has always progressed at the same steady pace, and that "gaps" in the record are imperfections because transitions occur by microevolution in restricted localities and in small populations, under circumstances not favorable to fossil formation. In this view, fossils appear in the record only after a major group has become widely established by migration. The alternative view, which appears to be gaining favor among paleontologists, is that the abruptness of the fossil record is a true record of macroevolutionary "spurts."

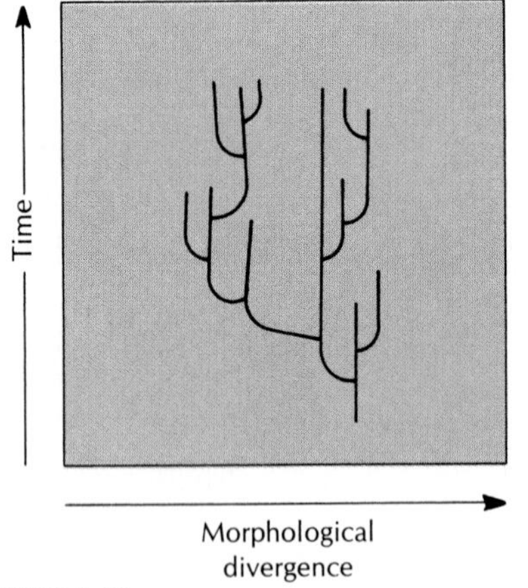

FIGURE 4-19

The punctuationist model sees evolutionary change as being concentrated in relatively rapid bursts *(straight lateral lines)* followed by prolonged periods of little or no change *(vertical lines).*

How can such spurts in evolution, as suggested in Figure 4-19, be explained? About 50 years ago the geneticist Richard Goldschmidt proposed that major jumps in evolution could occur in a single generation by macromutations. This is termed **evolution by saltation** (from the Latin *saltus,* meaning "a leap"). According to this idea, the distinctive features of higher taxa suddenly appeared: the first bat might have originated as the offspring of an early shrew-like mammal, or the first bird might have hatched, complete with feathers, from the egg of a reptile. Aside from the obvious dilemma such a macromutant would have in finding a mate to perpetuate itself, there is an almost insurmountable embryological problem: the genetic instructions for the correct sequence of development would have to be changed at many points. Thus the saltation hypothesis, while appealing in its directness, has never been endorsed by evolutionists.

There is a widely publicized alternative explanation for spurts in evolution, one that has received a wide measure of acceptance by evolutionists. It is called **punctuated equilibrium** (Eldredge and Gould, 1972). In this view, new species emerge rather suddenly, then remain mostly unchanged until they become extinct. In other words, long periods of species stasis — nonchange — are abruptly "punctuated" by the appearance of a better-adapted daughter species. This speciation event often coincides with, or is followed by, the extinction of the parent species. It has been observed that most species survive for 5 to 10 million years before becoming extinct, and during this time they apparently change very little. A speciation event that occurs in a small, isolated population over perhaps

50,000 years represents 1% or less of the species' life span. Fifty thousand years is a "geological instant" in the earth's history but represents thousands of generations in the evolving species. We know that gradual Darwinian selection can accomplish dramatic genetic changes much more rapidly than this. The hypothesis of punctuated equilibrium does not require macromutations and must not be confused with saltation. Many evolutionists believe that macroevolutionary leaps are simply the compounding of microevolutionary events: the gradual, though rapid, shift in characters under the influence of genetic drift and natural selection.

■ Summary

Organic evolution explains the diversity of living organisms, their characteristics, and their distribution as the historical outcomes of gradual, continuous change from previously existing forms. Evolutionary theory is strongly identified with Charles Robert Darwin who, with Alfred Russel Wallace, presented the first credible explanation for evolutionary change, the principle of natural selection. An alternative nineteenth-century scientific explanation of evolution, Jean Baptiste de Lamarck's concept of inheritance of acquired characteristics, was disproved early in this century by geneticists. Darwin derived most of the material later used to construct his theory when, as a young man, he made a 5-year voyage around the world on H.M.S. *Beagle*.

Evidences of organic evolution come from several sources. Fossils, the preserved remains of long-dead organisms, form an incomplete record of the progression of ancient life through geological time. The age of fossils can be reliably estimated because the geological formations in which they lie can be dated by measuring the radioactive decay of naturally occurring elements.

The similarities that allow organisms, both living and extinct, to be classified into neat hierarchies, are also evidence of evolution because of the presence of homologies: structures that, because of common ancestry, share a fundamental structural plan and relative position. Common ancestry is also revealed by molecular evolution, which involves comparing homologous proteins and other molecules in different species.

Natural selection is the guiding force of evolution. The principle is founded on the observed facts that no two organisms are exactly alike, that at least some variations are inheritable, that all species tend to overproduce their kind, and that more individuals are reproduced than can survive. Therefore survival and reproduction of some variants are favored over others; that is, organisms possessing those adaptations making them better adapted to the environment will survive and produce more offspring than those with less favorable variations. Thus their inheritable characteristics will appear in greater proportion in the next generation. The essence of natural selection is reproductive success.

Darwin's theory of natural selection was gradually modified in this century when the mechanisms of heredity were established by geneticists. The modern synthesis, a new view of evolution that emerged in the 1930s, emphasized natural selection by gradual shifting of gene frequencies within populations. This is also known, especially today, as microevolution.

Mutations are the ultimate source of all new variation. However, sexual reproduction, in which the genotypes of two parents are mixed in the offspring, is the immediate source of variation among individuals in a population.

In very large, interbreeding populations, gene frequencies tend to remain stable as predicted by the Hardy-Weinberg equilibrium. Such populations will not evolve unless disturbed by mutations, natural selection, or genetic drift. Genetic drift is the change in gene frequencies in small populations as a result of chance alone, in the absence of natural selection.

Human evolution

1

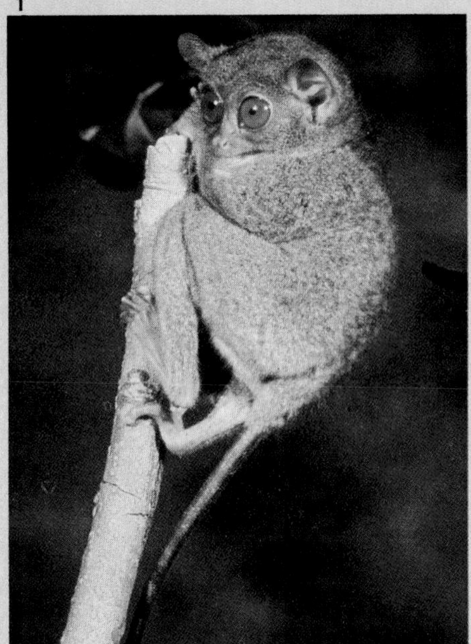

The publication of Charles Darwin's book *The Origin of Species* in 1859 precipitated an intellectual revolution in people's views of human origins. At the time that Darwin's views were first debated, few human fossils had been unearthed, but the current accumulation of fossil evidence has strongly vindicated Darwin's belief that humans descended from primate ancestors.

Humans are primates, a generalized group of mammals having an arboreal ancestry and characterized by having grasping hands with flat nails instead of claws and forward-facing eyes capable of stereoscopic vision. The commentaries that follow are matched to the numbered illustrations.

1. These adaptations, together with a large cerebral cortex that was essential for the precise timing, judgment of distance, and alertness required for an arboreal habitat, are evident in this tarsier of Mindanao Island. It is a **prosimian,** a group that also includes lemurs, lorises, and bush babies. Most are nocturnal in habit.
2. Forty million years ago the more advanced **anthropoids** emerged in Asia. Unlike the prosimians, all are active by day and have color vision. One group, the New World monkeys of South and Central America, represented here by the red howler, have prehensile tails.
3. A second group, the Old World monkeys, do not use their tails as prehensile organs. They are four footed in terrestrial locomotion, walking on their palms and soles. Most are arboreal, but the short-tailed baboons, such as this olive baboon, show a tendency for living on the ground.
4. A third group, the apes and humans, evolved separately from prosimian ancestors. About 25 million years ago, a group of large, arboreal fruit-eating apes called the dryopithecines appeared. On the ground they moved on all four legs as do their descendants, the modern pongids, or great (anthropoid) apes, such as this gorilla. The other pongids are the gibbon, orangutan, and chimpanzee.
5. The earliest known hominid was *Australopithecus afarensis* (the "southern African ape"), represented by this reconstructed skull. It appeared in the fossil record about 4 million years ago.
6. A particularly dramatic and important find was a 40% complete skeleton of *Australopithecus afarensis,* discovered in Ethiopia and dated at 2.9 million years BP. Lucy, as the skeleton was called, is depicted here beside a modern human being. Lucy was in life only about 3½

2

3

4

5

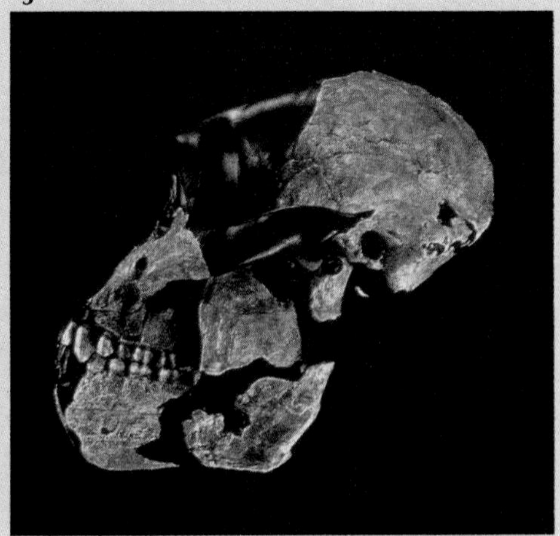

feet tall and had a chimpanzee-like face and a brain capacity of 400 cc, only one third that of modern humans. *A. afarensis* is thought by many paleoanthropologists to be ancestral to all human and humanlike forms that followed.

7. Two quite separate lines of hominids coexisted 2 million years ago. One line is represented by several fossils, which include a stocky ape man known as *Australopithecus robustus,* first discovered by Mary Leakey in Tanzania.
8. Coexisting with the robust ape man was a more delicate but larger-brained hominid, *Homo habilis.* The most dramatic find was this fossil discovered by Richard Leakey in Kenya, known as "1470 man," shown here with skull of a modern human, *Homo sapiens. H. habilis* was completely bipedal, fully erect, and had a cranial capacity of 775 cc. They used stone tools and were the first true humans.
9. *Homo erectus* appeared about 1½ million years ago, probably a descendant of *Homo habilis.* There is evidence that *Homo erectus,* with a cranial capacity of 1000 cc, had a rather complex social structure. They occupied much of the Old World, cooked animals that they had killed and butchered with stone and bone tools, and used caves and possibly crude wooden dwellings for shelter.
10. After the disappearance of *Homo erectus* about 300,000 years ago, subsequent human evolution threaded a complex course. Modern humans' immediate predecessors were the Neanderthal *Homo sapiens,* represented here as partial and complete restorations of the head. They emerged about 130,000 years ago. With a cranial capacity of 1300 cc (well within the range of modern brain size of 1000 to 2000 cc), they were proficient hunters and tool-users. The Neanderthals were replaced, and quite possibly exterminated, by the Cro-Magnons that appeared rather suddenly about 30,000 years ago. Their culture was far superior to that of the Neanderthals. Implement crafting improved rapidly, and human culture became enriched by aesthetics, artistry, and sophisticated language.

1 and **2,** Courtesy Zoological Society of San Diego; **3,** Photograph by C.P. Hickman, Jr.; **4,** Photograph by Joe McDonald/VU; **5** and **7,** Courtesy of Institute of Humans Origins: **8,** Photograph by Frank Awbrey/VU; **9,** Photograph by J.C. Cunningham/VU; **10,** Courtesy of American Museum of Natural History.

6

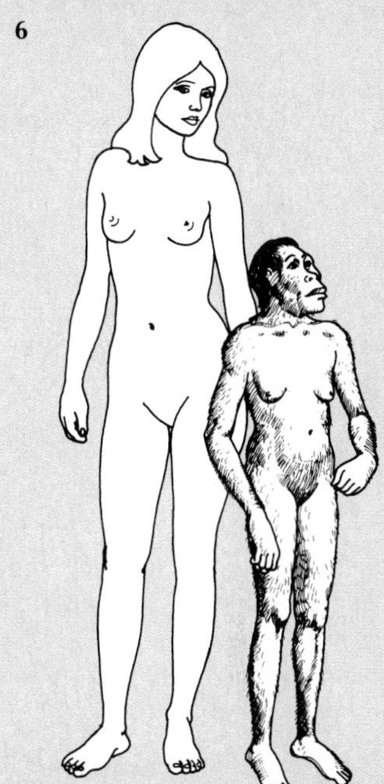

7

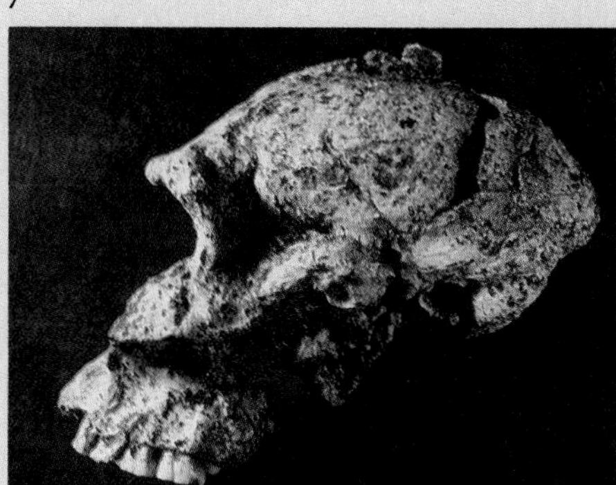

8

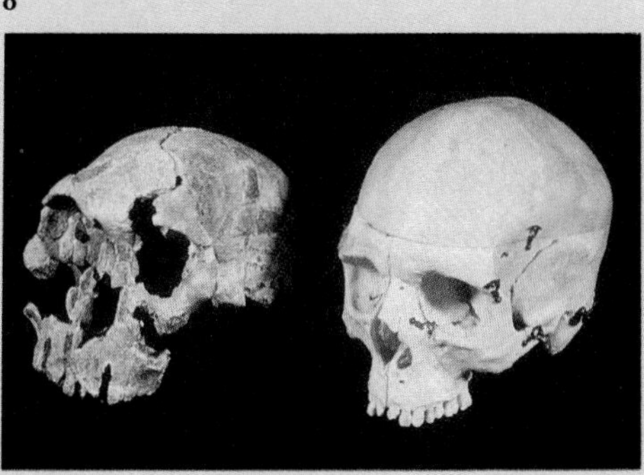

9

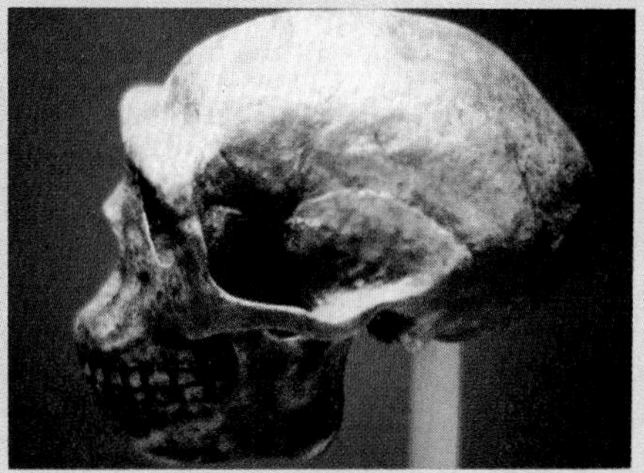

10

Natural selection, on the other hand, guides evolution by sorting out the better adapted organisms from a population composed of individuals with different genotypes. The spread of the mutant melanic form of the peppered moth in industrial England is an example of natural selection in action.

A species is defined as a natural, interbreeding population of organisms sharing a common gene pool that is reproductively isolated from other groups. The evolution of new species (speciation) may occur when some individuals become geographically isolated from the parent population. This is called allopatric speciation. Young islands, with their new ecological opportunities and absence of competitors for colonizing forms, are ideal habitats for adaptive radiation: multiplication of related species. The 13 finch species of the Galápagos Islands illustrate the crucial importance of isolation in speciation.

Large-scale events in life's history and the origin of new evolutionary trends are called macroevolution. While the traditional gradualistic model views evolution as proceeding smoothly through time, the more recently proposed punctuated equilibrium model suggests that evolution moves in leaps, with long periods of stasis being punctuated at intervals by bursts of rapid speciation. The fossil record appears to support this view.

Primates, a generalized group of mammals of arboreal ancestry, radiated into two major lines of descent: the prosimians (lemurs, lorises, and tarsiers) and the anthropoids (monkeys, apes, and hominids). The earliest hominids appeared about 4 million years ago; these were the bipedal australopithecines. These gave rise to, and coexisted with, the species *Homo habilis,* the first user of stone tools. *Homo erectus* appeared about 1.5 million years ago and was eventually replaced by *Homo sapiens* some 300,000 years ago.

REVIEW QUESTIONS

1. Briefly summarize Lamarck's concept of the mechanism of evolution. What is wrong with this concept?
2. To compress the earth's geological history into the nineteenth-century belief in a 6000-year-old earth, it was supposed that the history of the earth had been periodically interrupted by great cataclysms causing wholesale extinction and massive changes in the earth's surface. Why, then, was the publication of Charles Lyell's doctrine of uniformitarianism considered a turning point in geological thought? What was the significance of Lyell's work to Darwin?
3. Explain why the *Beagle's* visit to the Galápagos Islands in 1835 was so important to Darwin's thinking.
4. What was the key idea contained in Malthus' essay on populations that was to help Darwin formulate his theory of evolution?
5. Compare the roles of Darwin and Alfred Russel Wallace in developing the theory of natural selection.
6. Explain how each of the following serves as evidence that evolution has occurred: fossils; different ages of sequential layers of rock as determined by radioactive clocks; natural system of classification; biochemical homologies. How would a creationist explain these observations?
7. Provide a summary of Darwin's theory of natural selection. How does natural selection work?
8. If Darwin were alive today, what changes do you think he would make in this theory?
9. Creationists deride the idea of evolution, saying that it is "merely a theory." How would you answer this charge?
10. Explain why the following statement is or is not true: "Random mutations in genes are more likely to be favorable when the environment changes."
11. Distinguish between immediate and ultimate sources of variation in sexually reproducing populations.
12. Can natural selection act on a population if no new mutations occur? Why or why not?
13. It is a common belief that because some alleles are dominant and others are recessive, the dominants will eventually replace (drive out) all the recessives. How does the Hardy-Weinberg equilibrium answer this notion?
14. Assume you are sampling a trait in animal populations that is controlled by a single gene pair *A* and *a,* and that you can distinguish all three phenotypes *AA, Aa,* and *aa* (intermediate inheritance). Here are the results:

Population	*AA*	*Aa*	*aa*	Total
I	300	500	200	1000
II	400	400	200	1000

Calculate the *expected* distribution of phenotypes in each population. Is population I in equilibrium? Is population II in equilibrium?

15. If after studying a population for a trait determined by a single gene pair you find that the population is not in equilibrium, what possible reasons could there be for the lack of equilibrium?
16. Explain why genetic drift is more likely to occur in small populations.
17. What is the difference between genetic drift and natural selection? Between natural selection and adaptation?
18. Within a sexually reproducing population, what is the basis for concluding that a particular genotype makes an animal better adapted to its environment than other animals having different genotypes?
19. Explain how industrial melanism of the peppered moth has evolved under the influence of natural selection.
20. What is the difference between natural selection and speciation?
21. Offer a biological definition of species.
22. What is the difference between a "race" and a species?
23. Distinguish between reproductive isolation and geographical isolation.
24. How does allopatric speciation occur?
25. What is the evolutionary lesson provided by the Darwin's finches on the Galápagos Islands?
26. Distinguish between microevolution and macroevolution. Do they operate by separate evolutionary mechanisms?
27. What are the two principal explanations for the presence of "gaps" (absence of transitional forms) in the fossil record?
28. Is human evolution subject to the same forces as the evolution of other animals? Why or why not?

SELECTED REFERENCES

Berry, R.J., and A. Hallam. (eds). 1987. The encyclopedia of animal evolution. New York, Facts on File Publications. *Liberally illustrated treatment for the intelligent layman, with good treatments of early life on earth, the fossil evidence, and mechanisms of evolution.*

Darwin, C. 1859. On the origin of species by means of natural selection, or the preservation of favoured races in the struggle for life. London, John Murray. *There were five subsequent editions by the author.*

Dawkins, R. 1986. The blind watchmaker. New York, W.W. Norton & Company. *Through felicitous use of metaphor and analogy, Dawkins explains the origin of complex adaptations, coevolutionary "arms races," sexual selection, and more. A home computer is used to show how natural selection can produce order out of randomness.*

Eldredge, N. 1985. Time frames: the rethinking of Darwinian evolution and the theory of punctuated equilibria. New York, Simon and Schuster. *Clear account of background to, and evidence for, the punctuated equilibrium hypothesis.*

Eldredge, N., and S.J. Gould. 1972. Punctuated equilibria: an alternative to phyletic gradualism. In T.J.M. Schopf (ed.). Models in paleobiology. San Francisco, Freeman, Cooper & Co.

Futuyma, D.J. 1986. Evolutionary biology, ed. 2. Sunderland, Mass. Sinauer Associates, Inc. *Readable, broad coverage of evolutionary facts and principles; one of the best intermediate level texts. The second edition includes a new chapter on molecular evolution.*

Gould, S.J. 1977. Ever since Darwin: reflections in natural history. New York, W.W. Norton. *Fascinating collection of essays written with humor and cool logic.*

Irvine, W. 1955. Apes, angels, and Victorians: the story of Darwin, Huxley, and evolution. New York, McGraw-Hill Book Co. *Masterly description of two central figures in the nineteenth century Darwinian revolution, told with skill and wit.*

Lewin, R. 1987. Bones of contention: controversies in the search for human origins. New York, Simon and Schuster. *Lucid description of the great hominid fossil finds and the personalities of the scientists involved.*

Lowenstein, J.M. 1985. Molecular approaches to the identification of species. Am. Sci. **73**:541–547. *How newly developed immunological methods are used to unravel taxonomic relationships and dates of evolutionary divergence.*

Miller, J., and B. Van Loon. 1982. Darwin for beginners. New York, Pantheon Books, Inc. *Cleverly illustrated and witty presentation of the major concepts of evolution, Darwin's life and thought, and the controversy Darwin generated.*

Minkoff, E.C. 1983. Evolutionary biology. Reading, Mass. Addison-Wesley Publishing Company. *Well-organized coverage of paleontology, microevolution, and macroevolution.*

Patterson, C. 1978. Evolution. London, Routledge & Kegan Paul Ltd. *Lucid, succinct explanation of evolutionary theory. Highly recommended.*

EVERY SPECIES HAS ITS NICHE

The lavish richness of the earth's biomass is organized into a hierarchy of interacting units: the individual organism, the population, the community, and finally the ecosystem, that most bewilderingly complex of all natural systems. Central to ecological study is the habitat, the spatial location where an animal lives. What an animal does in its habitat, its profession as it were, is its niche: how it gets its food, how it arranges for its reproductive perpetuity—in short, how it survives and stays "fit" in the Darwinian sense. The niche for each population was originally defined by natural selection through competition with populations of other species for food, space, and other resources. Once the niche is defined, the owner excludes all others from it. This fact illustrates the competitive exclusion principle: no two species can live together indefinitely in the same niche. Thus a species and its niche become a unique way of life conditioned to avoid competition with the other ways of life around it. Peaceful coexistence fashioned by natural selection, not endless struggle, is the rule in the Darwinian world.▪

CHAPTER 5

The Earth's Environment

ONDITIONS FOR LIFE ON EARTH

All life is confined to a thin veneer of the earth called the **biosphere.** Viewers of the first remarkable photographs of earth taken from the Apollo spacecraft, revealing a beautiful blue and white globe lying against the limitless backdrop of space, were struck and perhaps humbled by our isolation and insignificance in the enormity of the universe. The phrase "spaceship earth" became a part of the vocabulary, and it helped people to realize that all the resources we will ever have for sustaining life are restricted to a thin layer of land and sea and a narrow veil of atmosphere above it. We could better appreciate just how thin the biosphere is if we could shrink the earth and all of its dimensions to a 1 m sphere. We would no longer perceive vertical dimensions on the earth's surface. The highest mountains would fail to penetrate a thin coat of paint applied to our shrunken earth; a fingernail's scratch on the surface would exceed the depth of the ocean's deepest trenches.

In a universe of billions of stars, our earth is a minor planet circling an ordinary star. Thousands of other stars are like our sun with planetary systems that conceivably could support life. Yet, of all these, our planet is the only one that we *know* supports life. Until proven otherwise, the earth is unique, a true wonder of an infinite universe.

What makes earth an especially fit environment for life? Most biologists would agree that foremost is the presence of liquid water on the earth's surface. Water, with its many extraordinary physical properties, not only provided the medium for the origin of life but also bestowed on earth a moderate climate suitable for the continued evolution of life. There are many other properties of earth that contribute to its fitness. Among these are a steady supply of light and heat from an unfailing sun; a suitable range of temperature for life, neither too hot nor too cold; a supply of the major and minor elements required by living matter; and a gravity force strong enough to hold an extensive gaseous atmosphere but not so strong that more than a trace of free hydrogen remains.

The many properties that make the earth wonderfully suitable for life were first recognized and examined in detail by Lawrence J. Henderson (1878-1942) in his book *The Fitness of the Environment*, published in 1913. His perceptive grasp of reciprocity between organism and environment has become a principle that underlies all ecological science.

Life itself has greatly influenced the earth's environment. The primitive earth of 3.5 billion years ago, barren, stormy, and volcanic with a reducing atmosphere of ammonia, methane, and water, was wonderfully fit for the prebiotic syntheses that led to life's beginnings. Yet, it was totally unsuited, indeed lethal, for the kinds of living organisms that inhabit the earth today, just as early forms of life could not survive in our present environment. The appearance of free oxygen in the atmosphere, produced largely if not almost entirely by living organisms, is an example of the reciprocity between organism and environment. Although oxygen was at first poisonous to early forms of life, its gradual accumulation over the ages from photosynthesis forced protective biochemical alterations to appear that led eventually to complete dependence on oxygen for life.

The organism and the environment share a reciprocal relationship. In this continuous interchange, both are altered in the process, and a favorable relationship is preserved. The environment of earth, with its living and nonliving components, is not a static entity but has undergone an evolution in every way as dramatic as the evolution of the animal kingdom. It is still changing today, more rapidly than ever before, under humanity's heavy-handed influence.

The earth's biosphere and the organisms in it have evolved together. As living organisms adapt and evolve, they act on and produce changes in their environment; in so doing they must themselves change.

FIGURE 5-1

Some of the major biomes of North America.

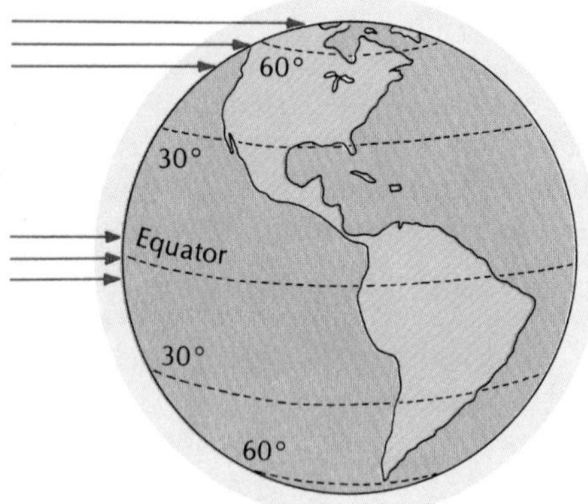

FIGURE 5-2

The earth's climate is determined by differential solar radiation between the poles and equator. Solar energy is spread across a much larger surface area at the poles than is an equivalent amount of energy at the equator.

■ Distribution of Life on Earth

TERRESTRIAL ENVIRONMENT: BIOMES

Botanists long ago recognized that the terrestrial environment of the earth could be divided into large units having a distinctive vegetation: forests, prairies, deserts, and so on. Animal distribution has always been more difficult to map because plant and animal distributions do not exactly coincide. Eventually zoogeographers accepted plant distribution as the basic biotic units which were called **biomes.** Each biome was identified by its dominant plant formation (Figure 5-1), but, since animals depend on plants, each biome supports a characteristic fauna.

Each biome is distinctive, but its borders are not. Anyone who has traveled across the United States knows that plant communities grade into one another over broad areas. The indistinct boundaries where dominant plants of adjacent biomes are mixed together form an almost continuous gradient called an **ecocline.** Thus, biomes are in some sense abstractions, a convenient way for us to organize our concepts about different life communities. Nevertheless, anyone can distinguish a grassland, deciduous forest, coniferous forest, or shrub desert by the dominant plants in each. And we can make reasonable assumptions about the kinds of animals that live in each biome.

Biome distinctiveness is determined mainly by climate, the characteristic pattern of rainfall and temperature of each region and the solar radiation it receives. Global variation in climate arises from the uneven heating of the atmosphere by the sun. Because of the low angle of the sun's rays striking the polar regions, atmospheric heating is much less there than at the equator. (Fig. 5-2).

The life on earth also influences differential heating by affecting the composition of the earth's atmosphere. The composition of the atmosphere is 78% nitrogen, 21% oxygen, 1% argon, 0.03% carbon dioxide, and variable amounts of water vapor. It is these last two components that are most important in determining climate. Carbon dioxide is released into the atmosphere by respiration and decay and removed by photosynthesis. Water vapor is added to the atmosphere mainly by plants. The importance of carbon dioxide and water vapor

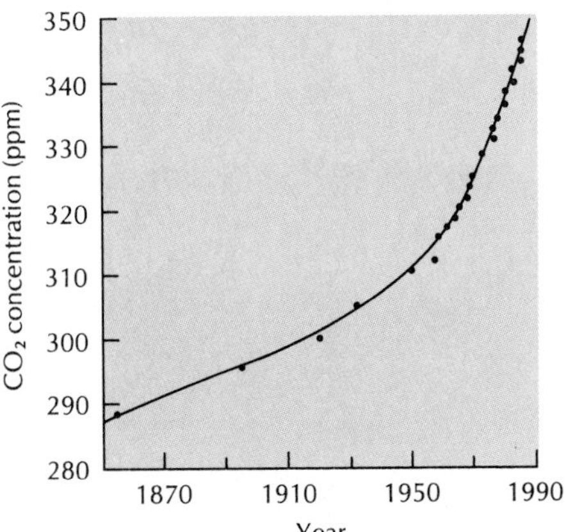

FIGURE 5-3

Rise in global atmospheric carbon dioxide for the past 140 years. Data points before 1958 come from analysis of air trapped in bubbles in glacial ice from sites around the world.

Adapted from Houghton, R.A. and G.M. Woodwell, 1989, Sci. Am., **260**: 36-44 (April).

The concern over the long-range effects of increasing atmospheric carbon dioxide, primarily from burning of fossil fuel, stems not from mere conjecture. Atmospheric carbon dioxide increased from an estimated 275 parts per million (ppm) before the Industrial Revolution to an average 345 ppm in 1985 and is increasing at a rate of 4.6% every 10 years. (Figure 5-3). It is expected to exceed 600 ppm in the next century. In the past century the global temperature has increased 0.4° C, and it will have increased 3° C (5.4° F) when the carbon dioxide has doubled in the next century, a warming of unprecedented magnitude. This is expected to shift climatic zones profoundly, creating drought conditions in the American wheatlands and wetter conditions along the coasts. Melting of the earth's ice sheets is another possibility, with flooding of coastal lowlands throughout the world. Whatever the outocme, a fascinating, unplanned global experiment is now under way.

relates to their role in producing an atmospheric thermal blanket over the earth, called the **greenhouse effect.** Carbon dioxide, water vapor, clouds, and, to a minor extent, ozone, absorb the sun's heat energy and keep the earth's surface at a higher temperature than it would be were there no atmosphere. This absorbed solar energy powers the earth's weather machinery and produces the broad climatic patterns that determine the distinctiveness of each biome.

The following is a selection of major biomes found in North America, but the same biomes comprised of somewhat different species are found on other continents where similar conditions exist.

Temperate deciduous forest

The temperate deciduous forest encompasses several forest types that change gradually from the Northeast to the South in North America. Deciduous, broad-leaved trees such as oak, maple, and beech that shed their leaves in winter predominate. Seasonal aspects are better defined in this biome than in any other. Animal populations in deciduous forests respond to seasonal change in various ways. Some, such as the insect-eating warblers, migrate. Others, such as the woodchuck, hibernate away the winter months. Others use available food (for example, deer) or stored food supplies (for example, squirrels). Hunting and habitat loss have eliminated virtually all the large carnivores that once roamed the eastern forests, such as mountain lions, bobcats, and wolves. Insect and invertebrate communities are abundant in the litter of the forest floor.

Coniferous forest

The coniferous forest is dominated by evergreens—pine, fir, spruce, and cedar—which are adapted to withstand freezing. The conical trees with their flexible branches shed snow loads easily. This area is the boreal (northern) forest, often referred to as **taiga** (a Russian word, pronounced "tie-ga"). In the central region of North America, the taiga merges into lake forest, largely destroyed by logging. The large southern pine forests occupy much of the southeastern United States.

Mammals of the coniferous forests are moose (Figure 5-4), snowshoe hare, several furbearers (including lynx), and a variety of rodents. They are adapted physiologically and behaviorally for long, cold, snowy winters. Some common birds are chickadees, nuthatches, warblers, and jays. Southern coniferous forests lack many of the mammals found in the North, but they have more reptiles and amphibians.

FIGURE 5-4

A moose browses on dwarf birch in the coniferous forest biome.
Photograph by C.P. Hickman, Jr.

Each month an area of undisturbed tropical forest the size of El Salvador is being converted to other uses. Unlike the forest clearing in the temperate zones, which made possible sustained, productive agriculture, tropical soils quickly become depleted, forcing farmers to move on to clear more forest. Another pressure on tropical forests is multinational timber companies that own or cut large tracts of timber for furniture of developed countries. Cattle ranchers also clear huge forest tracts to raise beef that is later sold to North America's fast-food chains. As one specialist remarked, "Everyone's hand is on the chain saw."

Tropical forest

Tropical forests are characterized by high rainfall, high humidity, relatively high and constant temperatures, and little seasonal variation in day length. In sharp contrast to temperate deciduous forests, dominated as they are by a relatively few tree species, tropical forests contain thousands of species, none of which is dominant. The variety of plant species is matched by the enormous numbers of animal species, with both plant and animal species stratified vertically (Figure 5-5). No other biome can match the tropical forests in its incredible variety of species. Food webs (see p. 131) are intricate and notoriously difficult for ecologists to unravel. The soil has low fertility, which may seem paradoxical in such a luxuriant biome. Nutrients released by decomposition are rapidly reabsorbed by plants, leaving no reservoir of humus. When the plants are removed for agriculture, the soil rapidly becomes a hard, red, bricklike crust called laterite.

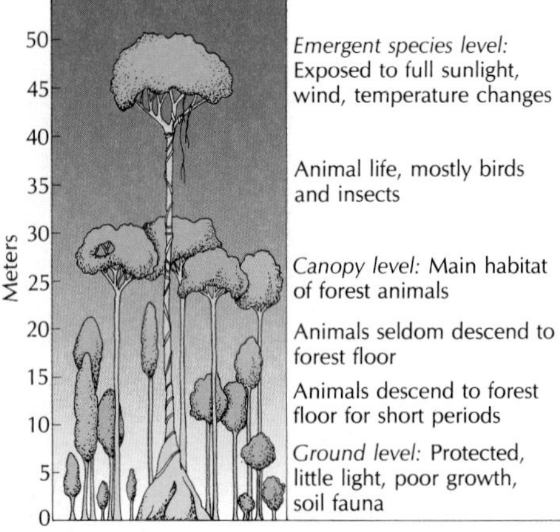

FIGURE 5-5

Profile of tropical forest, showing stratification of animal and plant life. The animal biomass is small compared with the biomass of the trees.

Grassland

One of the most extensive prairie biomes in the world is in North America, but the original plant and animal components have been almost completely destroyed by humans. Virtually all the major native grasses have been replaced by cultivated grains in croplands and by alien species of grasses in grazing lands. Rainfall is too scant to support trees but is sufficient to prevent the formation of deserts. Very few of the once dominant herbivore, bison, survive, but jackrabbits, prairie dogs, ground squirrels, and pronghorn antelope (Figure 5-6) remain. Mammalian predators include coyotes, ferrets, and badgers, although only coyotes are common.

Tundra

Tundra is characteristic of severe, cold climatic regions, especially the treeless Arctic regions and high mountaintops. The growing season is very short, and the soil may remain frozen for most of the year. Arctic tundra is covered with bogs, ponds, and a spongy mat of grasses (mosquitoes are abundant), whereas high-altitude tundras may be covered only with lichens and grasses. Despite the short growing season, the vegetation of dwarf woody plants, grasses, and lichens may be quite dense. Characteristic animals of the Arctic tundra are the lemming, caribou (Figure 5-7), arctic fox, arctic hare, ptarmigan, and (during the summer) many migratory waterfowl and shorebirds.

Desert

Deserts are extremely arid regions where rainfall is low (less than 25 cm a year), and water evaporation is high. Desert plants, such as various shrubs and cacti, have reduced foliage and other adaptations for conserving water. Many large desert animals have developed remarkable anatomical and physiological adaptations for keeping cool and conserving water (p. 215). Most smaller animals avoid the most severe conditions by living in burrows or adopting nocturnal habits. Mammals found there include the peccary, jackrabbit, kangaroo rat, and ground squirrel. Typical birds are the roadrunner, cactus wren, and burrowing owl. Reptiles are numerous. Arthropods include a great variety of insects and arachnids.

FIGURE 5-6

A pronghorn antelope quietly chews its cud after a morning of grazing on the grasslands of Montana. Pronghorn and bison were once the dominant ungulates of the North American prairie.

Photograph by C.P. Hickman, Jr.

AQUATIC ENVIRONMENTS

Because water is essential to life, aquatic environments are extremely important to the ecosphere and interact strongly with terrestrial environments. Of the earth's surface water, 99.9% is found in the oceans, but freshwater environments are nonetheless very important. They include running-water, or **lotic** habitats, and the standing-water, or **lentic** habitats. Lotic habitats follow a gradient from mountain brooks to streams and rivers. Brooks and streams with a very high velocity of water flow are high in dissolved oxygen due to their turbulence. Animals that live here are adapted to maintain their position in the current. They have organs of attachment, such as suckers and modified appendages, streamlined body shapes, or stages adapted for creeping under stones. Energy input is chiefly in the form of organic detritus washed in from adjacent terrestrial areas. More slowly moving rivers have less dissolved oxygen and more floating algae and other plants. Their fauna is tolerant of lower oxygen concentration.

Lentic habitats, such as ponds and lakes, tend to have still lower oxygen concentration, particularly in the deeper areas. Animals living on the bottom or on submerged vegetation (**benthos**) include snails and mussels, crustaceans, and

FIGURE 5-7

A large male caribou on the Alaskan tundra. The gregarious caribou travel in large herds, feeding in summer on grasses, dwarf willow, and birch, but in winter almost exlusively on lichen.

Photograph by C.P. Hickman, Jr.

Life without the sun

Animal community of giant pogonophoran tubeworms, some reaching 2m in length, grows in dense profusion near a Galápagos Rift thermal vent, photographed at 2800 m (about 9000 feet) from the deep submersible *Alvin.* Also visible in the photograph are mussels and crabs.

Until recently it was believed that all animals depended directly or indirectly on primary production from solar energy. However, in 1977 and 1979, dense communities of animals were discovered living on the sea floor adjacent to vents of hot water issuing from rifts (Galápagos Rift and East Pacific Rise) where tectonic plates on the sea floor are slowly spreading apart. These communities (see figure) included several species of molluscs, some crabs, polychaete worms, enteropneusts (acorn worms), and giant pogonophoran worms. The seawater above and immediately around vents is 7° to 23° C where it is heated by basaltic intrusions, whereas the surroundng normal seawater is 2° C.

It was discovered that the producers in the vent communities are chemoautotrophic bacteria that derive energy from the oxidation of the large amounts of hydrogen sulfide in the vent water and fix carbon dioxide into organic carbon. Some of the animals in the vent communities, for example, the bivalve molluscs, are filter feeders that ingest the bacteria. Others, such as the giant pogonophoran tube-worms, which lack mouths and digestive tracts, harbor colonies of symbiotic bacteria in their tissues and use the organic carbon that these bacteria synthesize.

Thus the deep-sea vent communities are self-contained, closed systems that depend entirely on energy issuing from the earth's interior. Each is a miniature ecosystem, altogether separate from other known ecosystems, all of which depend on solar energy and photosynthesis.

Photograph by D. Foster, Woods Hole Oceanographic Institute/Visuals Unlimited.

a wide variety of insects. Many swimming forms, called **nekton,** are found in lakes and larger ponds. Depending on the nutrients available, there may be a large contingent of small floating or feebly swimming plants and animals (**plankton**).

Acid rain, caused mainly by the industrial emissions of sulfur and nitrogen oxides from the burning of fossil fuels, has led to serious deterioration of water quality in lakes, ponds, and rivers of southern Scandinavia, central Europe, and eastern North America. These emissions interact with atmospheric moisture to yield high concentrations of hydrogen, sulfate, and nitrate ions. Acid rains having a pH below 4 are not uncommon. Most sensitive to acid rain are recently glaciated areas having low buffering capacity. The acidity of Maine's lakes increased eightfold between 1937 and 1974; hundreds of lakes in the Adirondacks are now sterile, or are approaching sterility, and are lost to practical and recreational purposes.

Oceans cover over 71% of the earth's surface to an average depth of 3.75 km (2.3 miles). Because water absorbs light, all production by plants is limited to the relatively shallow depths to which light penetrates (**photic zone,** 150 m or less). With a few notable exceptions (see box above), all life in deeper water (**aphotic zone**) must be supported by the light "rain" of organic particles produced in the photic zone.

The life of the ocean is divided into regions, or provinces, each with its own distinctive life forms (Figure 5-8): (1) **littoral,** or intertidal zone, where sea and land meet; (2) **neritic,** or shallow water zone, surrounding the continents and extending out to a depth of about 200 m; (3) **upwelling,** where deep nutrient-rich water is carried up into the photic zone; (4) **pelagic,** the vast open sea; and (5) **coral reefs,** built by carbonate-secreting organisms where tropical landmasses protrude into the photic zone (p. 392). **Estuaries,** where rivers flow into the sea, are very productive because of the load of nutrients brought to them by the river waters. Thus the most productive marine environments are reefs, estuaries, and zones of upwelling.

About 98% of the species of animals and plants in the sea live on the bottom (**benthic**), and only 2% live freely in the open ocean (**pelagic**). Of the benthic forms, most occur in the intertidal zone or shallow depths of the oceans. Less than 1% live in the deep ocean below 2000 m.

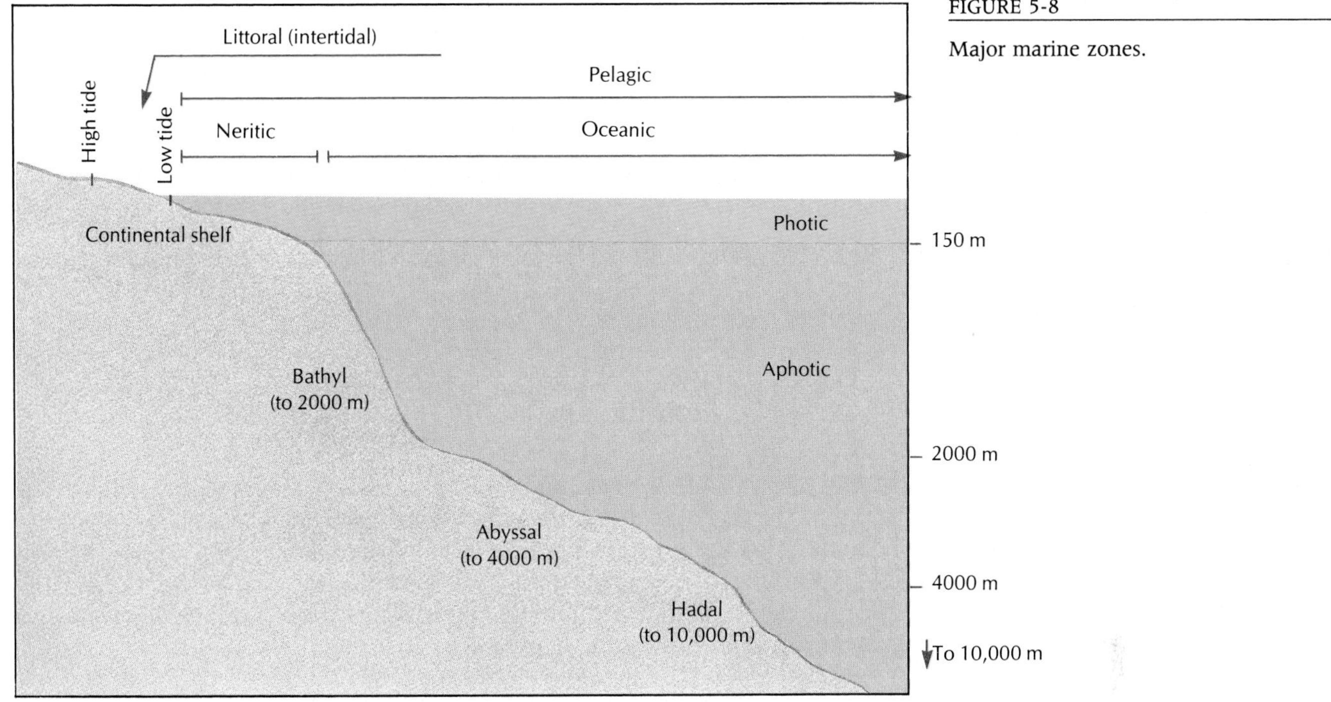

FIGURE 5-8

Major marine zones.

ANIMAL DISTRIBUTION: ZOOGEOGRAPHY

The study of zoogeography tries to explain why animals are found where they are, their patterns of dispersal, and the factors responsible for their dispersal. It is not always easy to explain why animals are distributed as they are, since similar habitats on separate continents may be occupied by quite different kinds of animals. A particular species may be absent from a region that supports similar animals for any of three reasons: (1) there may be barriers that prevent it from getting there; or (2) having gotten there, it may be unable to adapt to the new habitat or compete successfully with resident species and may become extinct; or (3) having once arrived and adapted, it has subsequently evolved into a distinct species. The history of an animal species or its ancestor must be known before one can understand why it is where it is. The fossil record is often essential, as well as knowledge of geological history. Geological change has been responsible for much of the alteration in animal (and plant) distribution and has been a powerful influence in shaping organic evolution.

Major faunal realms

The nineteenth-century naturalists recognized that it was possible to divide the terrestrial world into several distinct realms of animal distribution, separated from one another by topographical and climatic barriers. Philip L. Sclater (1858) first proposed a region system based on bird families. Later (1876), Alfred Russel Wallace, codiscoverer with Charles Darwin of the principle of natural selection, modified Sclater's pattern and applied it to vertebrates in general. Wallace recognized six land realms based on animal ranges. Although later scientists slightly modified his biogeographical boundaries, the Wallacean regions are the ones we recognize today (Figure 5-9). Great oceanic barriers serve as longitudinal dividers between the Nearctic and Palaearctic realms in the north and the Neotropic and Australian realms in the south. The great northern land mass is divided by a subtropic, warm temperate dry belt into Palaearctic, Ethiopian, and Oriental

FIGURE 5-9

The six major faunal realms of the earth.

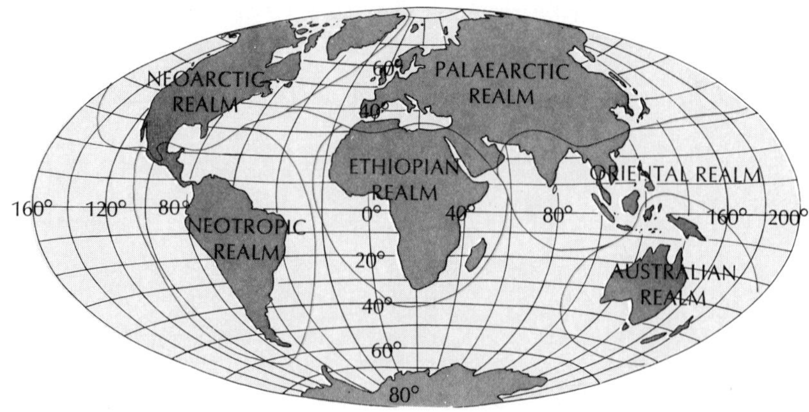

realms. These realms are better recognized by the differences between them than by faunal similarities within them. As with the biomes described earlier, the boundaries between realms are transition zones that are often poorly defined, especially if the barrier between them is largely climatic. Or they may be clear and distinct if the separation is a mountain range or salt water; to land animals, even a narrow saltwater strait is an effective barrier.

Analysis of the Environment

The biosphere is confined largely to the interfaces between land, air, and water. It does not penetrate very far below the land surface, high into the atmosphere, or abundantly into the depths of the ocean. Living things within the biosphere may be examined at several different levels of organization.

The most inclusive level of organization is the **ecosystem** (Figure 5-10). An ecosystem is a complex, largely self-sustaining natural system of which living organisms are a part, together with the nonliving components. All the interactions that bind the living (**biotic**) and nonliving (**abiotic**) components together are included in the ecosystem. A particular ecosystem is usually defined by the investigator studying it, and it may be quite large, such as a grassland, forest, lake, or even an ocean, or it may be more restricted, such as a riverbank or treehole. Whatever the size or the biological structures it contains, certain characteristics of any ecosystem can usually be described. The sun's energy is fixed by plants and then transferred to consumers and decomposers. Nutrients are cycled and recycled through the various living components of the ecosystem. No ecosystem is ever completely closed, however. There is always some flow of resources and organisms into and out of an ecosystem.

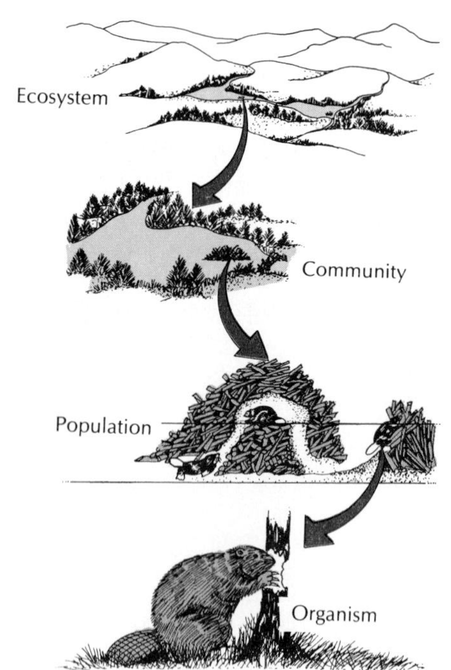

FIGURE 5-10

Relationships between ecosystem, community, population, and organism.

The next level of organization is the **community,** an assemblage of species sharing the same environment and having a certain distinctive unity (Figure 5-10). Communities comprise the living elements of an ecosystem. Like ecosystems, communities may be large or small, ranging from the coniferous forest community that may span a continent to the inhabitants of a rotting log community or the community of microorganisms living in a human's large intestine. The elements of a community are closely interdependent.

The **population,** the next lower level of organization, is an interbreeding group of organisms of the *same species* sharing a particular space. Every community is composed of several populations, including those of plants, animals, and microorganisms. Energy and nutrients flow through a population. Its size is regulated by its relationships to other populations in the community and by the abiotic characteristics of the ecosystem in which it is found. Because members of a population are interbreeding, they share a **gene pool** and thus are a distinct genetic unit.

The lowest level of organization within the biosphere, in ecological terms, is the **organism** itself. It is the living expression of the species. Each organism responds to its biotic and abiotic environment. If the ecologist is to understand why animals are distributed as they are, he or she must examine the varied mechanisms that animals use to compensate for environmental stresses and alterations.

Ecologists have, in fact, become increasingly interested in the physiological and behavioral mechanisms of animals. Both are intrinsic to the animal-habitat interrelationship. For example, the success of certain warm-blooded species under extreme temperature conditions such as in the Arctic or in a desert depends on near-perfect balance between heat production and heat loss and between appropriate insulation and special heat exchanges. Other species succeed in these situations by escaping the most extreme conditions by migration, hibernation, or torpidity. Insects, fishes, and other poikilotherms (animals having variable body temperature) compensate for temperature change by altering biochemical and cellular processes involving enzymes, lipid organization, and the neuroendocrine system. Thus the physiological capacities with which the animal is endowed permit it to live under changing and often adverse environmental conditions. Physiological studies are necessary to answer the "how" questions of ecology.

The animal's behavioral responses are also part of the animal-habitat interaction and of interest to the ecologist. Behavior is modified in obtaining food, finding shelter, escaping enemies and unfavorable environments, finding a mate, courting, and caring for the young. Genetically determined mechanisms such as behavioral repertoires are acted on by natural selection. Those which improve adaptability to the environment assist in survival and the evolution of the species.

The term **ecology,** coined in the last century by the German zoologist Ernst Haeckel, is derived from the Greek *oikos,* meaning "house" or "place to live." Haeckel called ecology the "relation of the animal to its organic as well as inorganic environment." Although we no longer restrict ecology to animals alone, Haeckel's definition is still basically sound.

In the last several years, the word *ecology* has come into popular misuse as a synonym for *environment,* which often makes biologists wince. As people concerned about the environment, we can be environmentalists; a person engaged in the scientific study of the relationship of organisms and their environment is an ecologist. He or she is usually an environmentalist too, but environment is not the same as ecology.

The term **environment** is frequently used in reference to the organism's immediate surroundings, but it is not always clear whether it is meant to include the living, as well as the nonliving, surroundings. To the ecologist it certainly includes both. Ultimately the environment consists of everything in the universe external to the organism. Since ecology is really biology of the environment, it obviously is a broad and far-flung field of study.

The ecologist may choose to focus on any level of organization within the biosphere. **Ecosystem analysis** is largely interdisciplinary, incorporating physics, chemistry, and other sciences to assist in the comprehension of the role of the environment in determining the distribution and abundance of organisms. **Community ecology** is similar but of more restricted scope; it is possible to focus on the interactions of a few species and to study energy transfers in detail. **Population biology** stresses genetics, evolution, seasonal changes, and other phenomena affecting population dynamics. Some ecologists study the organism itself (**organismic biology**) to see how it responds to the environment, hour by hour, day by day; such studies have become physiological and behavioral. All contribute to ecological understanding. None stands alone.

ECOSYSTEM ECOLOGY

The abiotic component of an ecosystem can be characterized by its physical parameters such as temperature, moisture, light, and altitude and by its chemical features, which include various essential nutrients. These characteristics determine the basic nature of the ecosystem.

The biotic component — the populations of plants, animals, and microorganisms that form the communities of the ecosystem — may be categorized into producers, consumers, and decomposers. The **producers,** mostly green plants, are autotrophs that use the energy of the sun to synthesize sugars from carbon dioxide and water by photosynthesis. This energy is made available to the **consumers** and **decomposers.** They are the heterotrophs that exploit the self-nourishing autotrophs by converting organic compounds of plants into compounds required for their own growth and activity. In this discussion we consider the flow of energy through the ecosystem, which involves the concepts of productivity and the food chain, biogeochemical cycling, and limiting factors within the environment.

Solar radiation and photosynthesis

Almost all life depends on the energy of the sun. The sun releases energy produced by the nuclear transmutation of hydrogen to helium. Solar radiation received at the earth's surface extends from wavelengths of approximately 280 to 13,500 nm (nanometers; 1000 nm = 1 micrometer). Ultraviolet radiation with wavelengths of less than 280 nm is cut off sharply by the ozone layer in the upper atmosphere (Figure 5-11). Radiation with wavelengths greater than 760 nm is long-wave infrared radiation that heats the atmosphere, warms the earth, and produces currents of air and water. The most important part of solar radiation lies between wavelengths of 310 and 760 nm; this is the portion that we call **visible light** because of its effect on the human retina. It is also the range that drives all important photobiological processes, including photosynthesis, photochemical effects, phototropism (orientation of plants toward light), and animal vision.

The flow of energy through the ecosystem begins with **photosynthesis.** Light striking a green plant is absorbed by chlorophyll, sending low-energy electrons

FIGURE 5-11

Narrowing of spectrum of sunlight by atmospheric absorption in seawater.

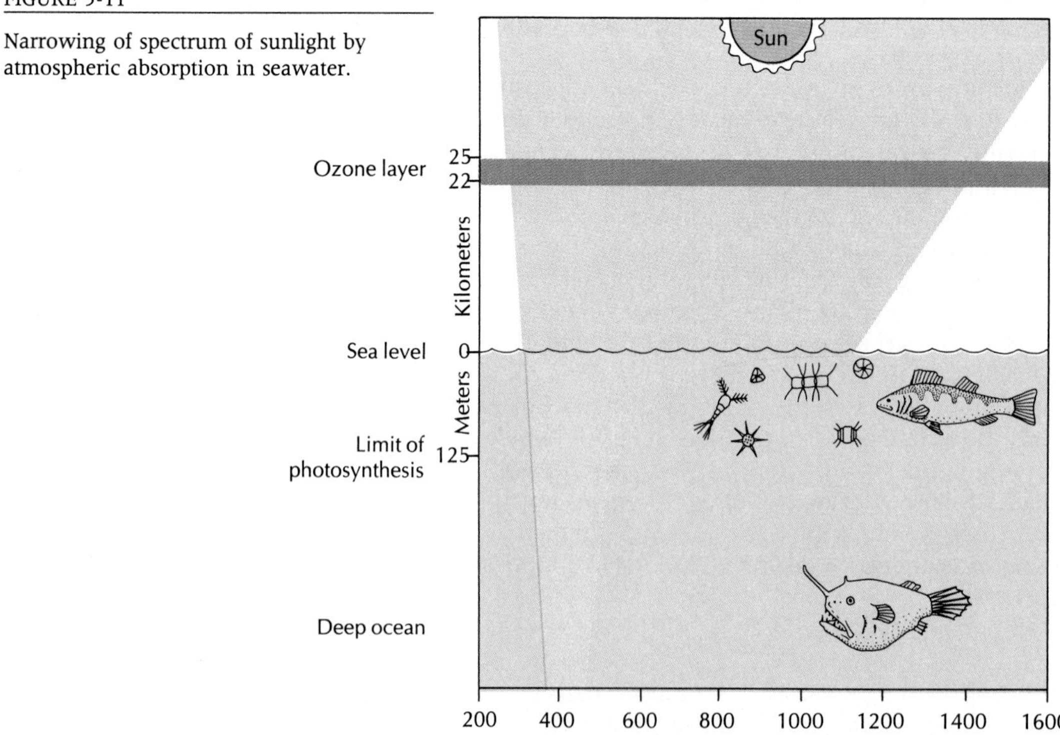

into a higher energy level. These excited electrons drop back to a ground state in approximately 10^{-7} seconds, but in this brief interval their energy is channeled into a sequence of energy-yielding reactions. Part of the energy is used to synthesize ATP; the remainder causes the reduction of pyridine nucleotides (NADP). Both ATP and reduced NADP are then used to synthesize sugars from carbon dioxide and water.

Other biochemical reactions in the plant can take this sugar (glucose), and with the addition of inorganic nutrients provided by the root system, provide all the other myriad organic molecules (proteins, nucleic acids, lipids, other carbohydrates) necessary for life. Photosynthesis in the individual leaf begins at low light intensity and increases until it reaches a maximum. The leaf achieves its highest rate of photosynthesis at only approximately one-tenth the intensity of full sunlight. During brief periods of very active growth, plants may store a maximum of nearly one-fifth the available light energy.

Production and the food chain

The energy accumulated by plants in photosynthesis is called **production.** Because it is the first step in the input of energy into the ecosystem, the *rate* of energy storage (that is, energy storage per unit of time) by plants is known as **primary productivity.** The total rate of energy storage, the **gross productivity,** is not entirely available for growth because plants also use energy for maintenance and reproduction. When this energy consumption, or plant **respiration,** is subtracted from the gross productivity, the **net primary productivity** remains. Plant growth results in the accumulation of plant **biomass.** Biomass is expressed as the *weight* of dry organic matter per unit of area and thus differs from productivity, which is the *rate* at which organic matter is formed by photosynthesis.

The level of productivity of different ecosystems depends on the availability of nutrients and the limitations of temperature levels and moisture availability. Highly productive ecosystems are intertidal communities, flood-plain forest, swamps and marshes, estuaries, coral reefs, and certain crop ecosystems (for example, rice and sugar). Most productive of all are intertidal kelp and sea palm communities in which net production can exceed 14,000 grams (nearly 31 lbs) of dry organic matter per square meter per year! Less productive (1000 to 2000 $g/m^2/yr$) are most temperate forests, most agricultural crops, lakes and streams, and grasslands. Least productive (70 to 200 $g/m^2/yr$) ecosystems are tundra and alpine regions, deserts, and the open ocean—the latter a surprise to many people who mistakenly view the ocean as a rich and unharvested source of food for all. Extreme desert, rock, and ice regions have virtually zero productivity.

The way in which the net primary productivity of plants in a community supports all the rest of life in the community is best seen in the **food chain.** Plants are eaten by consumers, which are themselves consumed by other consumers, and so on in a series of steps. Food chains are descriptions of the way energy flows through the ecosystem. A diagram of a food chain shows arrows leading from one species to another, meaning that the first species is food for the second (see left side of Figure 5-12). But the first may be food for several other organisms as well. Seldom, in fact, does one organism live exclusively on another. Although some food chains are simple and short, for example, the one in which the whale feeds mainly on plankton, it is more common for several food chains to be interwoven into a complex **food web** (Figure 5-12).

Food chains tend to follow a pattern. Green plants, the base of the food chain, are eaten by grazing **herbivores,** which can convert the energy stored in plants into animal tissue. This is the base of the **grazing food chain.** Herbivores may be eaten by small **carnivores** and these by large carnivores. There may be two or three, sometimes even four, levels of carnivores. At the end of the chain are the **top carnivores** that, lacking predators, die and decompose, replenishing the soil

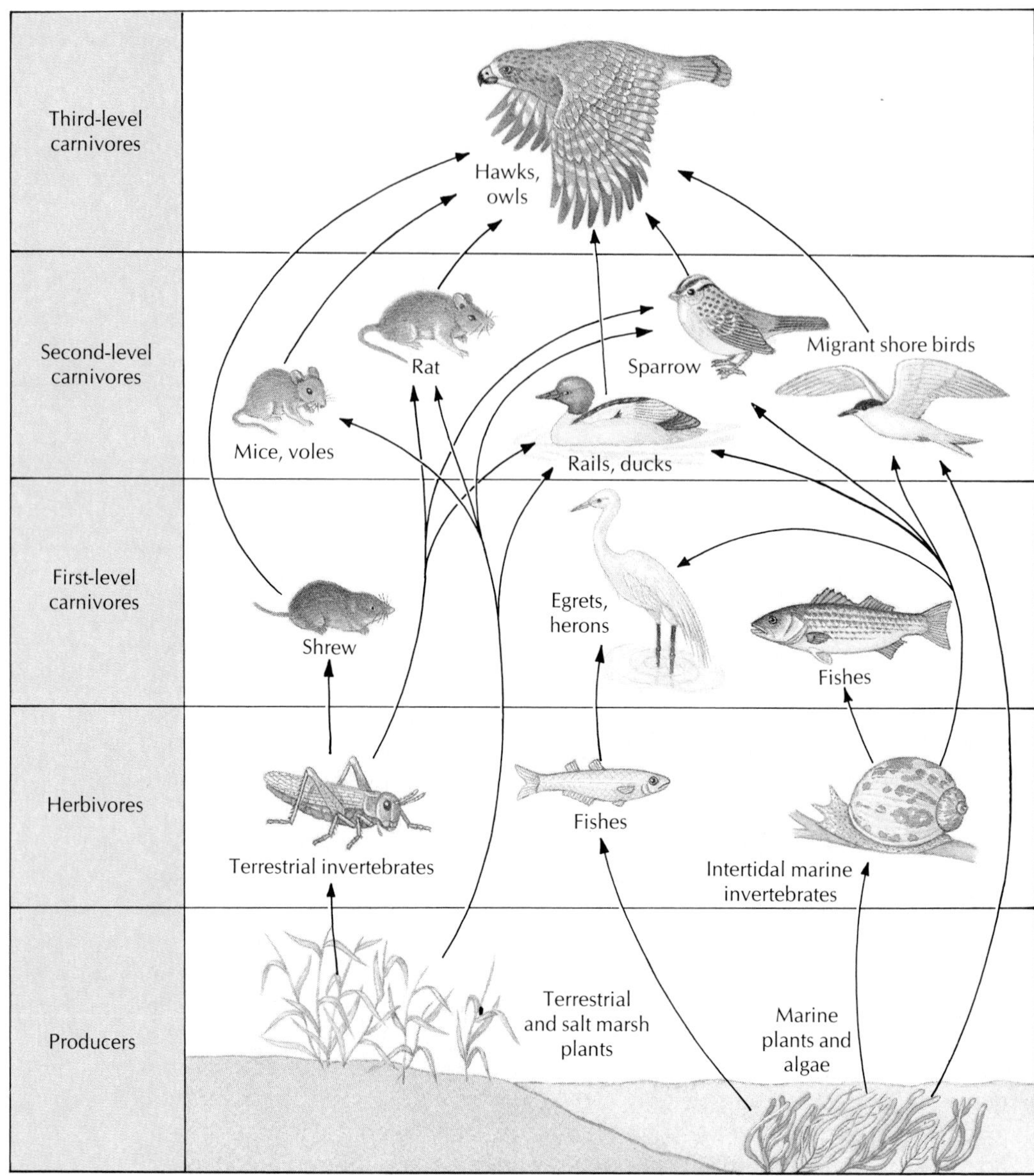

FIGURE 5-12

Midwinter food web in *Salicornia* salt marsh of San Francisco Bay area.

with nutrients for plants that start the chain. At each level, beginning with the plants, there are parasites, which would qualify as herbivores or carnivores, depending on whether they are parasites of plants or animals. They usually deprive their hosts of only insignificant amounts of energy, but other detrimental effects may be significant. Sometimes the parasites have parasites.

There are numerous examples of grazing food chains. In the forest, for instance, many small insects (primary consumers) feed on plants (producers). A smaller number of spiders and carnivorous insects (secondary consumers) prey on small insects; still fewer birds (tertiary consumers) live on the spiders and carnivorous insects; and finally one or two hawks or owls (quaternary or top consumers) prey on the birds.

The decomposers have traditionally been considered the final step in the grazing food chain, since they reduce organic matter into nutrients that become available again to the producers. Ecologists now recognize that the decomposers comprise their own distinct **detritus food chain**, consisting of **detritus feeders**, such as earthworms, mites, millipeds, crabs, aquatic worms, and molluscs, and **micro-**

organisms, such as bacteria and fungi. Dead organic matter, such as fallen leaves or dead animals, is decomposed and used by fungi, bacteria, and protozoa. Detritus feeders then eat the microorganisms as well as much of the dead organic matter directly. They are in turn eaten by small carnivores; the detritus food chain thus leads up into grazing food chains and it is very important in recycling nutrients into grazing food chains.

Trophic levels

Each step in the food chain may be considered a **trophic level** (meaning feeding level, from the Greek *trophe,* food). At each transfer to the next trophic level, 80% to 90% of the available energy is lost as heat because most of the fuel obtained in food is burned to do the work of staying alive. Only the 10% to 20% of food calories remaining can be used to build bodies. This is why the number of trophic levels is limited to four or five. It also explains why so few big carnivores can be supported by a given biomass of plants. There is a limit to how far the food chain can be stretched.

"R.F.D.2."

Courtesy Steve Stinson and Roanoke Times and World-News.

Humans, who occupy a position at the end of the chain, may eat the grain that fixes the sun's energy; this very short chain represents an efficient use of the potential energy. Humans also may eat beef from animals that eat grass that fixes the sun's energy; the addition of a trophic level decreases the available energy by an order of 10. In other words, it requires 10 times as much plant biomass to feed humans as meat eaters as to feed humans as grain eaters. Let us consider the person who eats the bass that eats the sunfish that eats the zooplankton that eats the phytoplankton that fixes the sun's energy. The tenfold loss of energy occurring at each trophic level in this five-step chain explains why bass do not form a very large part of the human diet. In this particular food chain, for a person to gain a pound by eating bass, the pond must produce 5 tons of phytoplankton biomass.

These figures need to be considered as we look to the sea for food. The productivity of the oceans is, in fact, very low and largely limited to regions of upwelling, estuaries, marshes, and reefs. Such areas occupy only a small part of the ocean. The rest is a watery desert.

Ocean fisheries supply 18% of the world's protein, but most of this is used to supplement livestock and poultry feed. If we remember the rule of 10-to-1 loss in energy with each transfer of material between trophic levels, then the use of fish as food for livestock rather than as food for humans is a poor use of a valuable resource in a protein-deficient world. Of the fishes that we do eat, the preference is for species such as flounder, tuna, and halibut, which are three or four levels up the food chain. Every 125 g of tuna requires 1 metric ton of phytoplankton food. If humans are to derive greater benefit from the oceans as a food source in the future, we must eat more of the fishes that are at lower trophic levels.

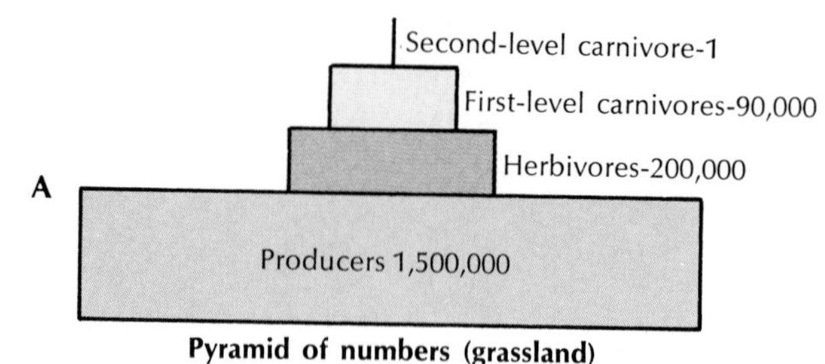

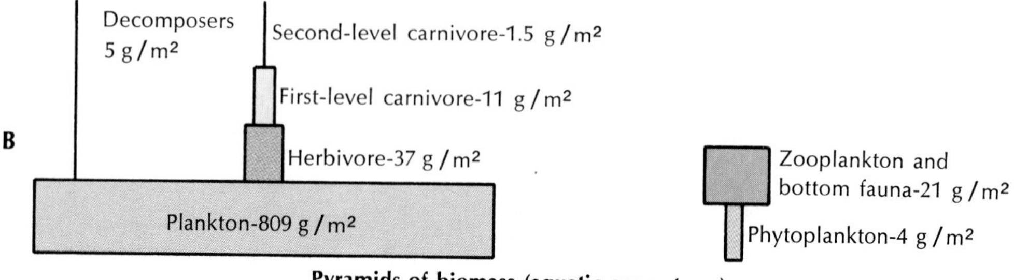

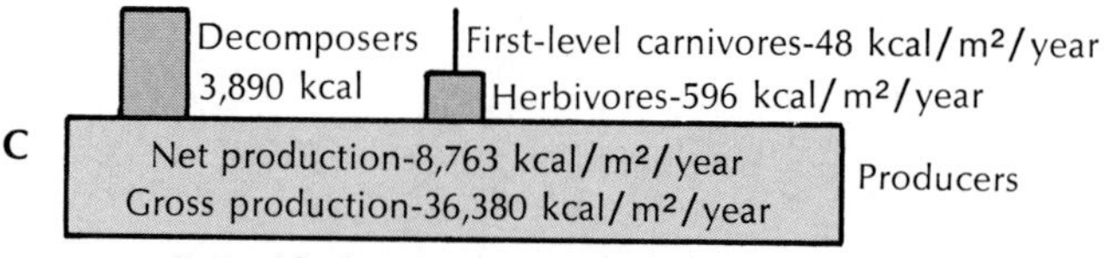

FIGURE 5-13

Ecological pyramids of numbers, biomass, and energy. Pyramids are generalized, since area within each trophic level is not scaled proportionally to quantitative differences in units given.

Redrawn from Smith, R.L. 1980. Ecology and field biology, ed. 3. New York, Harper & Row, Publishers, Inc.; and Odum, E.P. 1971. Fundamentals of ecology, ed. 3. Philadelphia, W.B. Saunders Co.

The concepts of food chains and ecological pyramids were invented and first explained in 1923 by Charles Elton, a young Oxford University ecologist. Working for a summer on a treeless arctic island, Elton watched the arctic foxes as they moved about, noting what they ate and, in turn, what their prey had eaten. As a result he was able to trace the complex cycling of nitrogen in food throughout the animal community. Elton realized that life in a food chain comes in discrete sizes, because each form had evolved to be much bigger than the thing it eats. He thus explained the common observation that large animals are rare while small animals are common.

When we examine the food chain in terms of biomass at each level, it is apparent that we can construct **ecological pyramids** of numbers or of biomass. A pyramid of numbers (Figure 5-13, *A*), also known as an **Eltonian pyramid** (after Charles Elton, who first devised the scheme), depicts the numbers of individual organisms that are transferred between each trophic level. Although providing a vivid impression of the great difference in numbers of organisms involved in each step, a pyramid of numbers does not indicate the actual weight of organisms at each level.

More instructive are pyramids of biomass (Figure 5-13, *B*), which depict the total bulk of organisms. Such pyramids usually slope upward because mass and energy are lost at each transfer. However, in some aquatic ecosystems in which the producers are the algae that have short life spans and rapid turnover rates, the pyramid is inverted. This happens because the algae can tolerate heavy exploitation by the zooplankton consumers. Therefore the base of the pyramid is smaller than the biomass it supports.

A third type of pyramid is the pyramid of energy, which shows rate of energy flow between levels (Figure 5-13, *C*). An energy pyramid is never inverted because less energy is transferred from each level than was put into it. A pyramid of energy gives the best overall picture of community structure because each level reveals its true importance in the community regardless of its biomass.

Nutrient cycles

All of the elements essential for life are derived from the environment, where they are present in the air, soil, rocks, and water. When plants and animals die and their bodies decay, or when organic substances are burned or oxidized, the elements and inorganic compounds essential for life processes (nutrients) are released and returned to the environment. Nutrients flow in a perpetual cycle

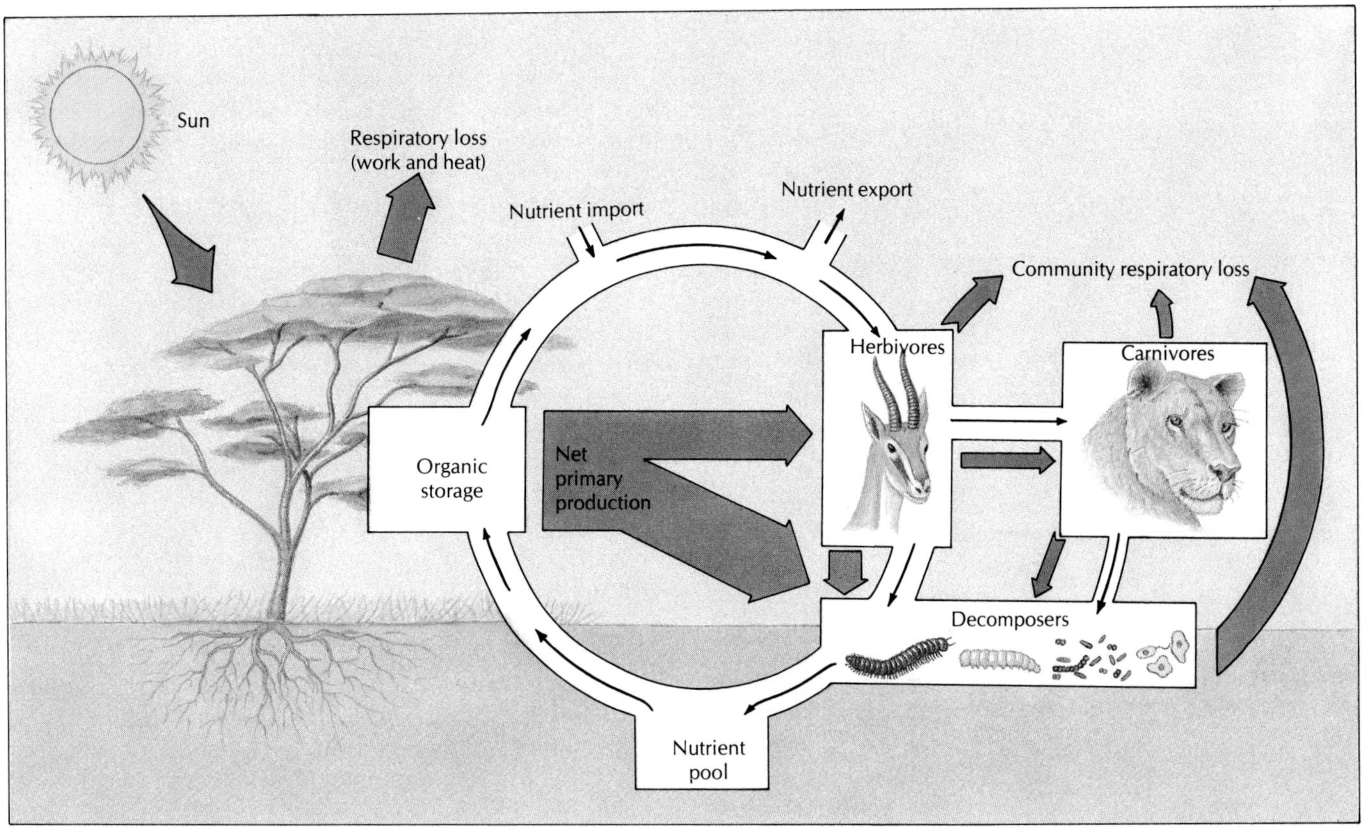

FIGURE 5-14

Nutrient cycles and energy flow in terrestrial ecosystem. Note that nutrients are recycled, whereas energy flow *(red)* is one way.

between the biotic and abiotic components of the ecosystem. Nutrient cycles are often called **biogeochemical cycles** because they involve exchanges between living organisms (bio-) and the rocks, air, and water of the earth's crust (geo-).

Nutrient and energy cycles are closely interrelated, since both influence the abundance of organisms in an ecosystem. However, unlike nutrients, which recirculate, energy flow follows one direction; it does not follow a cycle because it is lost as heat as it is used. The continuous input of energy from the sun keeps nutrients flowing and the ecosystem functioning (Figure 5-14). Among the most important biogeochemical cycles are those of carbon and nitrogen.

Carbon cycle. Carbon is the basic constituent of organic compounds and living tissue and is required by plants for the fixation of energy by photosynthesis. Carbon circulates between carbon dioxide (CO_2) gas in the atmosphere and living organisms through assimilation and respiration; it is also withdrawn into long-term reserves of fossil fuel deposits (humus and peat and finally coal and oil)(Figure 5-15).

The cycling of carbon parallels and is linked to the flow of energy that begins with the fixation of energy during photosynthetic production. Plants photosynthesize glucose, a six-carbon compound, from CO_2 that is withdrawn from the atmosphere; they then use this sugar to build higher carbohydrates, especially cellulose, the structural carbohydrate of plants. Plants require 1.6 kg of CO_2 from the atmosphere for each kilogram of cellulose produced. The concentration of CO_2 in the atmosphere today is only 0.03%.

Two aspects of the low concentration of CO_2 require emphasis. The first is that CO_2 availability limits energy fixation by plants. Physiologists have shown that, if the atmospheric CO_2 is increased by 10%, plant photosynthesis increases 5% to 8%.

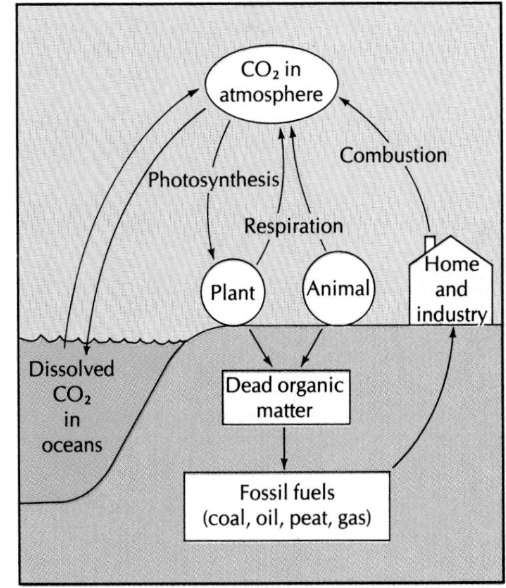

FIGURE 5-15

Carbon cycle showing circulation of carbon as CO_2 gas between living components of environment and long-term storage as fossil fuels.

The second point is that there has been a tremendous increase in the consumption of fossil fuels by humans in the last two decades. More than 10 billion tons of CO_2 enters the atmosphere each year from industrial and agricultural activities; of this 75% comes from burning fossil fuels and 25% from the release of CO_2 from soil because of frequent plowing (a surprising and unappreciated effect of present agricultural practice). There is concern that even small increases in CO_2, which traps radiated heat, may increase the temperature of the earth's biosphere (see marginal note, p. 123).

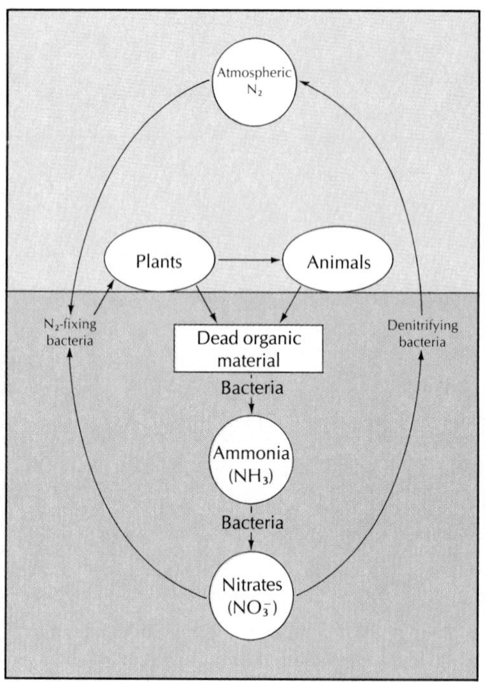

FIGURE 5-16

Nitrogen cycle showing circulation of nitrogen between organisms and through environment *(brown).* Microorganisms responsible for key conversions are indicated in circles and boxes.

Nitrogen cycle. Like carbon, nitrogen (N_2) is also a basic and essential constituent of living material, particularly in proteins and nucleic acids. Despite the high concentration of nitrogen in the atmosphere (79% of air), it is almost totally unavailable to life forms in its gaseous state (N_2). The most important contribution of the nitrogen cycle is converting N_2 into a chemical form that living organisms can use. This conversion is called **nitrogen fixation.** Some atmospheric N_2 is fixed by lightning, which produces ammonia and nitrates that are carried to earth by rain and snow. But at least 10 times as much N_2 is biologically fixed by bacteria and cyanobacteria (Figure 5-16). These **nitrogen fixers** are the essential link between the accessible inorganic nitrogen pool and all other organisms.

Most important in terrestrial systems are bacteria associated with legumes (members of the pea-clover-alfalfa family) whose nitrogenous contribution to soil enrichment is well known. The old agricultural practice of allowing fields to lie fallow (without crops) for a season every few years enabled N_2 fixers to replenish the element in a usable form. Nitrogen-fixing bacteria produce nodules on the roots of legumes within which N_2 is converted to ammonia and nitrates, which the plants can use to synthesize protein. Plant proteins are transferred to consumers that build their own proteins from the amino acids supplied. Plants' and animals' waste products (urea and excreta) and their ultimate decomposition provide for the return of organic nitrogen to the substrate. Then decomposers break down proteins, freeing ammonia, which is used by bacteria in a series of steps to form nitrate (NO_3^-). Nitrate may be used once again by plants or may be degraded to inorganic nitrogen by denitrifying bacteria and returned to the atmosphere. Nitrates also are carried by runoff to streams and lakes and eventually to the sea. A cycle similar to this terrestrial cycle occurs in aquatic ecosystems except that there is a steady loss of nitrogen to deep-sea sediments.

The nitrogen cycle is a near-perfect self-regulating cycle in which losses in one phase are balanced by gains in another. There is little absolute change in nitrogen in the biosphere as a whole. However, human activities have caused steady losses of soil nitrogen by slowing natural addition of organic nitrogen through current agricultural practice and by the harvesting of timber. The latter causes an especially heavy outflow, from both timber removal and soil disturbance.

COMMUNITIES

Communities represent the most tangible concept in ecology. They comprise the *biotic* portion of the ecosystem; each consists of a certain combination of species that forms a functional unit. Although communities are sometimes difficult to define because the assemblages of species within similar communities are not always the same, communities do exist, and they all possess a number of attributes. It is beyond the scope of this book to discuss all the aspects of community form and function. In this discussion we examine three important principles that operate in community organization.

Ecological dominance and keystone species

Biological communities are typically dominated by a single species or limited group of species that greatly influences the nature of the local environment. All other species in that community must adapt to conditions created by the dominants. Communities often are named for the dominant species, for example, black spruce forest, beech-maple forest, oyster community, and coral reef. It is not always easy to specify just what constitutes a dominant species. It may be the most numerous, the largest, or the most productive, or it may in some other manner exert the greatest influence on the rest of the community. In a woodland, a tree obviously means more to the community than a poison ivy plant (even though a casual visitor may carry away a more lasting impression of the poison ivy).

Dominant species are often so because they occupy space that might otherwise be occupied by other species. When a dominant species is eliminated for some reason, the community changes. The American chestnut once dominated large regions of the eastern United States, where it comprised more than 40% of the overstory trees in climax deciduous forests. After the invasion in 1900 of the chestnut blight (a fungus), which was apparently brought into New York City on nursery stock from Asia, the chestnut was eliminated from its entire range within a few years (Figure 5-17). While the disappearance of a tree of enormous commercial value was a great loss to the human community, the forest community quickly accommodated to the change. The chestnut position was replaced by oak, hickory, beech, poplar, and red maple, and chestnut-oak forests became oak and oak-hickory forests.

For the American chestnut, "dominant" did not mean "essential." The basic structure of eastern hardwood forests was not significantly altered. Sometimes, however, the disappearance of a single species can transform the fundamental framework of a community. Such species are called **keystone species.** For example, in 1983, a mysterious epidemic swept through the Caribbean populations of the sea urchin *Diadema antillarum,* destroying more than 95% of the animals. The immediate effect was on the algal community which, no longer grazed by the urchins, grew from a thin mat to a thick canopy of different algal composition. Both productivity and diversity on the coral reefs declined. *D. antillarum* is clearly a keystone predator species for Caribbean coral reef communities. On the west coast, the sea star *Pisaster ochraceous* is a keystone species. Sea stars are a major predator of the mussel *Mytilus californianus.* When the sea stars were experimen-

FIGURE 5-17

American chestnut tree *Castanea dentata,* and its original distribution in North America, showing the spread of chestnut blight following its accidental introduction into New York City in 1900.

Map courtesy Agricultural Research Service.

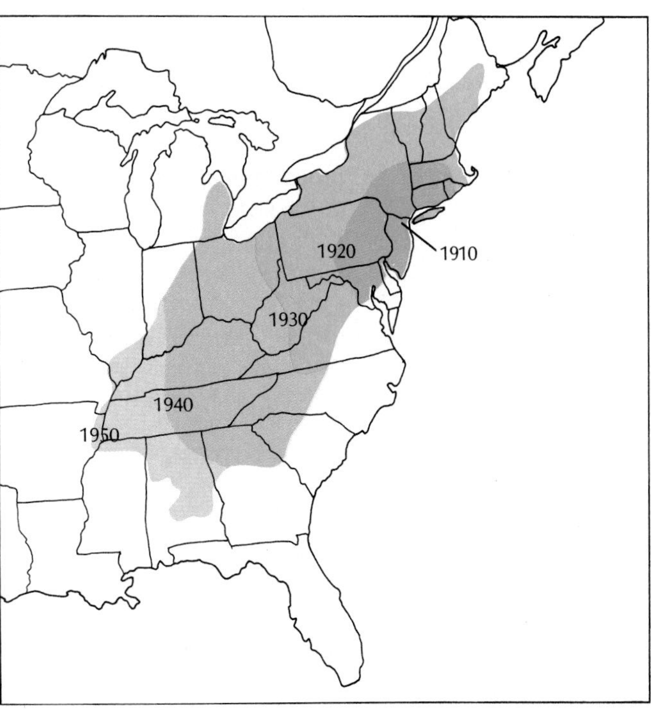

tally removed from a patch of Washington state coastline, the mussels expanded, taking over all the space previously occupied by some 25 species of other invertebrates and algae. Keystone species have been identified in other communities. Not all communities include keystone species, but for those that do keystone species are critical to community structure.

Ecological niche concept

An animal's position in the environment is characterized by more than the habitat in which we expect to find it; it also has a "profession," a special role in life that distinguishes it from all other species. This is its **niche,** which Elton defined as the animal's place in the biotic environment, that is, what it does and its relation to its food and to its enemies. The niche concept has now been broadened to include an animal's position in time and space as well. Moreover, an animal's niche is defined by and is a property of its community. Two populations of the same species in different communities will have different niches because they are interacting with different assemblages of other animals in the two communities.

It is an accepted rule that no two species can occupy the same niche at the same time and place: this is the competitive exclusion principle that we spoke of in the prologue to this chapter. For if they did, they would be in direct competition for exactly the same food and space. Should this happen, one species would have to diverge into a different niche, or move to a different habitat, or face extinction. Closely related, sympatric species might be expected to have similar niches and be in danger of direct competition.

One way closely related species coexist is by partitioning the resources among them. For example, in his classic study of Darwin's finches on the Galápagos Islands, the English ornithologist David Lack noticed that bill sizes of these birds depended on whether they occurred together on the same island (FIgure 5-18). On the islands Daphne and Hermanos, where *Geospiza fuliginosa* and *G. fortis* occur separately and are therefore not in competition with each other, beak sizes are nearly identical. But on the island Santa Cruz, where both *G. fuliginosa* and *G. fortis* coexist, their bill sizes do not overlap. This suggests resource partitioning, since bill size determines the size of seeds selected for food. Recent work by the American ornithologist Peter Grant has confirmed what Lack suspected: *G. fuliginosa* with its smaller bill selects smaller seeds than does *G. fortis* with its larger bill. This is an example of the work of evolution in the past. Where the two species coexisted, competition between them forced displacement of the bill sizes so that competition ceased. The absence of competition that we see today has been fittingly described as "the ghost of competition past."

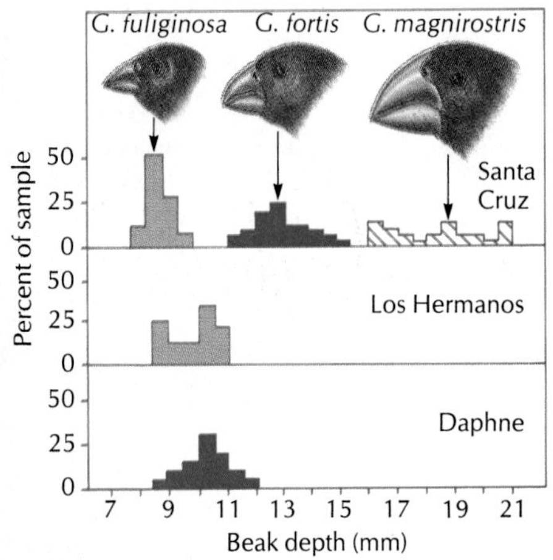

FIGURE 5-18

Displacement of beak sizes in Darwin's finches from the Galápagos Islands. Beak depths are given for the ground finches *Geospiza fuliginosa* and *G. fortis* where they occur together (sympatric) on Santa Cruz Island and where they occur alone on the islands Daphne and Hermosa. *G. magnirostris* is another large ground finch that lives on Santa Cruz.

After Lack, D. 1947, "Darwin's finches," Cambridge, Cambridge University Press.

The niche concept is a fundamental one to the biologist because it explains why animals avoid endless struggles with other animals. A population and its niche are reflections of the same thing: a unique way of life. A population of a species is master of its own niche and thus is not in direct competition with similar species, even though competition was used to decide the boundaries of the niche in the first place. This concept was discussed in Chapter 4 when we considered how speciation occurs.

POPULATIONS

As defined earlier, a population is a group of organisms belonging to the same species that share a particular space. Whether the population is gray squirrels or deer in an eastern woods or bluegill sunfish in a farm fishpond, it bears a number of attributes unique to the group. A population shares a common gene pool; it has a certain density, birth rate, death rate, age ratio, and reproductive potential; and it grows and differentiates much like the individual organisms of which it is composed.

Population interactions

Populations in a community may affect each other in various ways, although in certain cases there may be no apparent interaction (**neutral** effect). One interaction, **competition,** has already been mentioned. Although it is true that two populations cannot occupy the same niche at the same time and place, some degree of competition occurs in natural populations. Because natural selection favors decreased competition, this often leads to niche diversification. Another adaptation may be that the two populations use the same resource, but at different times of the year. One population may be excluded by a superior competitor from a habitat that it could otherwise occupy. When two such populations occupy adjacent habitats, competition may be intense along the border between them.

Most people are aware that lions, tigers, and wolves are predators, but the world of invertebrates is replete with predaceous animals, too. These range from protozoa, jellyfish and their relatives, and various worms to predaceous insects, starfish, and many others.

Another interaction is the **predator-prey** relationship, in which individuals of one population kill and feed on individuals in the prey population. Often a population is preyed on by more than one species of predator, and predators commonly have more than one prey species. Predators are large relative to their prey, and they usually consume several to many prey animals in their lifetime. A conceptually related interaction in the **parasite-host** relationship. Both parasites and predators feed at the expense of their host-prey. However, parasites are small relative to their hosts, and parasites have only one host—at least at a given stage in their life cycle. In most cases, parasites do not kill their hosts, since if the host dies, the parasite perishes along with it. Most parasites are **symbiotic,** that is, they live in (**endoparasites**) or on (**ectoparasites**) the body of their host. Organisms such as mosquitoes and biting flies are intermediate between parasites and predators as we have defined them; they are sometimes referred to as **micropredators** or **intermittent parasites.**

Another type of usually symbiotic interaction is **commensalism,** in which the commensal enjoys some benefit from the relationship but does not harm the host. A classic example of commensalism is the association of pilot fishes and remoras with sharks. These fish get the "crumbs" left over when the host makes its kill, but we now know that some remoras feed on ectoparasites of the sharks. This is an example of commensalism grading into **mutualism,** an interaction in which both species benefit. An example of mutualism is that of the termite and the protozoa living in its gut. The protozoa can digest the wood eaten by the termite, although the termite cannot, and the termite lives on the waste products of the protozoan metabolism. In return, the protozoa gain a congenial place to live and a food supply. Mutualism has been recognized in recent years as much more prevalent in the Animal Kingdom than previously believed (Figure 5-19). Many cases of mutualistic interactions are symbiotic, but many are not. Some investigators prefer to distinguish **protocooperation** from mutualism. Protocooperation describes mutually beneficial interactions that are not physiologically necessary to the survival of the partners.

FIGURE 5-19

Among the many examples of mutualism that abound in nature is the whistling thorn acacia of the African savannah and the ants that make their homes in the acacia's swollen galls. The acacia provides both protection for the ants' larvae *(lower photograph of opened gall)* and honey-like secretions used by the ants as food. In turn, the ants protect the tree from herbivores by swarming out as soon as the tree is touched. Giraffes, however, which love the tender acacia leaves, seem immune to the ants' fiery bites.

Photographs by C.P. Hickman, Jr.

Population growth

Living organisms have reproductive potential beyond that required for replacement of their numbers, some far in excess of such a requirement. Some female insects lay thousands of eggs, field mice can produce as many as 17 litters of four to seven young each year, and a single female codfish may spawn 6 million eggs per season. It has been calculated that a bacterium dividing three times per hour would produce a colony a foot deep over the entire earth in 1½ days, and the colony would be over our heads 1 hour later. Unrestricted geometric (exponential) growth can only happen in an environment with bounteous resources and no competition. This kind of growth, shown by the steeply rising curve in Figure 5-20, is called the **intrinsic growth rate** of a population and is referred to by the symbol r. Of course, the growth potential of a population can

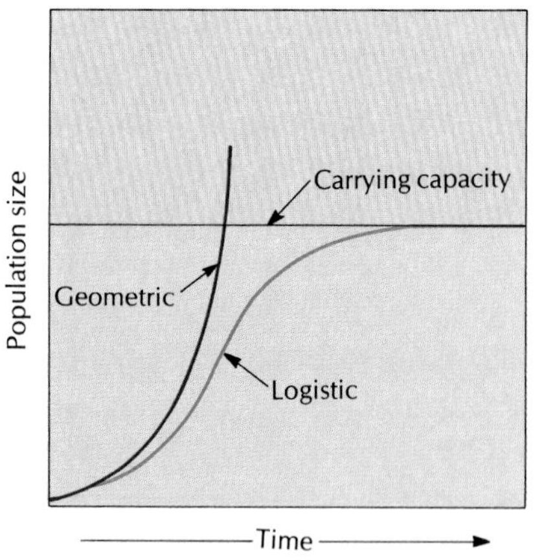

FIGURE 5-20

Population growth, showing geometric (exponential) growth of a species in an unlimited environment, and logistic growth in a limited environment.

Exponential and logistic growth

The sigmoid growth curve (Figure 5-20) can be described by a simple model called the logistic equation. The slope at any point on the growth curve is the *growth rate*, that is, how rapidly the population size is changing with time. If N represents the number of organisms and t the time, we can, in the language of calculus, express growth as an *instantaneous rate:*

dN/dt = the rate of change in the number of organisms per time at a particular instant

When populations are growing in an environment of unlimited resources (unlimited food and space, and no competition from other organisms), growth is limited only by the inherent capacity of the population to reproduce itself. Under these ideal conditions growth is expressed by the symbol r, which is defined as the instantaneous rate of population per capita. The index r is actually the difference between the birth rate and death rate per individual in the population at any instant. The growth rate of the population as a whole is then:

$$\frac{dN}{dt} = rN$$

This expression describes the rapid, exponential growth illustrated by the early upward-curving portion of the sigmoid growth curve (Figure 5-20). But the growth rate for populations in the real world slows as the upper population limit is approached, and eventually stops altogether. At this point N has reached its maximum density; that is, the space being studied has become "saturated" with animals. This limit is called the carrying capacity of the environment and is expressed by the symbol K. The sigmoid population growth curve can now be described by the logistic equation which is written as follows:

$$\frac{dN}{dt} = rN\left(\frac{K-N}{K}\right)$$

This equation states that the rate of population increase per unit of time (dN/dt) = rate of population growth per capita (r) × population size (N) × unutilized freedom for population growth ($[K-N]/K$). One can see from the equation that when the population approaches the carrying capacity, that is, when $K-N$ approaches 0, dN/dt also approaches 0 and the curve will flatten. It is not uncommon for populations to overshoot the carrying capacity of the environment so that N exceeds K. When this happens, the population runs out of some resource (usually food or shelter). The rate of population growth term dN/dt becomes negative and the population must decline.

never proceed unchecked. We are in no danger of being buried in bacteria, or codfish, or any other organism because as the population becomes more crowded, individuals compete with each other for limited resources, especially food and space. Eventually the population ceases to grow larger. Most populations tend to reach a certain density and then fluctuate about that level. What keeps populations in check? How is the "balance of nature" explained? Or is nature ever really "in balance"?

If we plot the numbers of individuals on a graph against units of time, the population describes a growth history that is sigmoid in shape (Figure 5-20). In the beginning, with ample food and space, the starting population breeds and grows as fast as its reproductive potential allows. We see this kind of growth—the intrinsic growth rate—as early summer plankton blooms appear in lakes, insect pests erupt, and weeds grow rapidly in old fields. But, as resources become depleted and the upper population limit is approached, growth slows and finally stops altogether. This limit, the maximum density the environment can support, is called the **carrying capacity** of the environment and is represented by the symbol *K*. The simple sigmoid curve has been demonstrated repeatedly in the laboratory with bacteria, protozoa, and some insects. The mathematical form of the growth curve, described by the **logistic equation,** is explained in the box on p. 140.

When populations reach the carrying capacity of the environment, they may remain at this level for a time, or they may fluctuate above and below the upper limit. For example, when sheep were introduced on the island of Tasmania around 1800, their growth was represented by a sigmoid curve with a small overshoot, followed by mild oscillations around a final population size of 1,700,000 sheep (Figure 5-21, *A*). A similar pattern but with larger fluctuations was recorded for a population of ring-necked pheasants introduced on an island in Ontario, Canada (Figure 5-21, *B*). Some populations oscillate cyclically with large year-to-year changes in abundance. One of the best-known examples is the cyclic oscillation in the abundance of lynx and snowshoe hare in which peaks of abundance alternate with spectacular "crashes" (Figure 30-22). One important reason for such oscillations is that the carrying capacity of the environment itself changes. Nature is seldom "in balance," at least not for very long, because of weather fluctuations, spread of diseases, changes in predation patterns, and other disturbances.

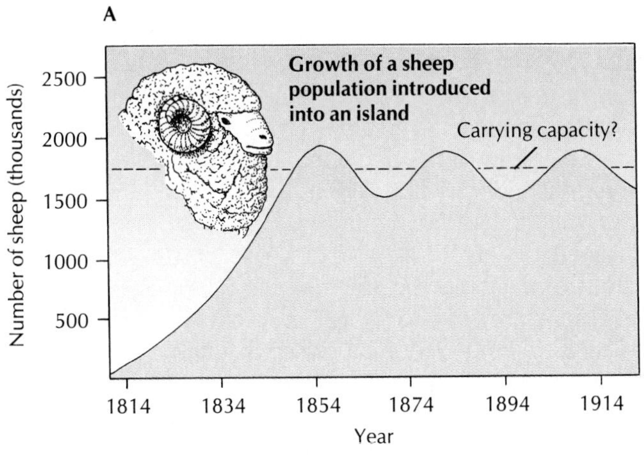

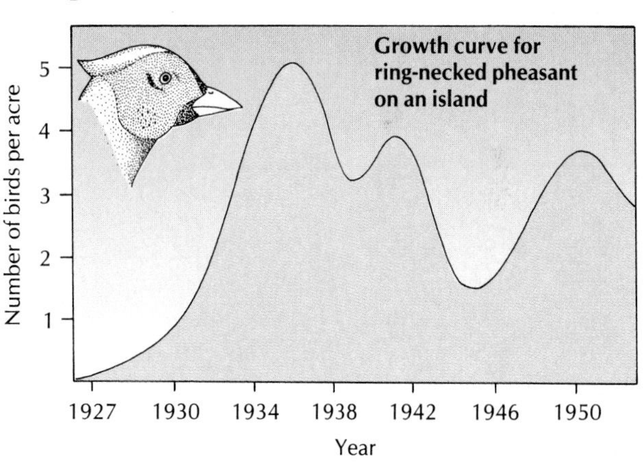

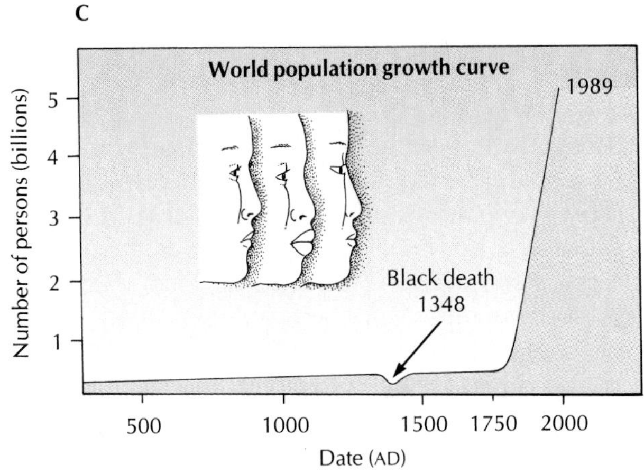

FIGURE 5-21

Growth curves for sheep **(A),** ring-necked pheasant **(B),** and world human populations throughout history **(C).** Note that while sheep population on a island is stable because of human control of the population, the ring-necked pheasant population oscillates greatly, probably because of large changes in carrying capacity. Where would you place the carrying capacity for the human population?

The growth of the human population was slow for a very long period of time. For most of their evolutionary history, humans were hunters and gatherers who depended on and were limited by the natural productivity of the environment. With the development of agriculture, the carrying capacity of the environment increased, and the population grew steadily from 5 million around 8000 BC, when agriculture was introduced, to 16 million around 4000 BC. Despite terrible famines, disease, and war, which took their toll, the population reached 500 million by 1650. With the coming of the Industrial Revolution in Europe and England in the eighteenth century, followed by a medical revolution, discovery of new lands for colonization, and better agriculture practices, the carrying capacity of the earth for humans increased dramatically. The population doubled to 1 billion around 1850. It doubled again to 2 billion by 1930, to 4 billion in 1976, and passed 5 billion in 1987 (Figure 5-21, *C*). Thus growth has been exponential and remains high, with more than 80 million people being added to the global population each year.

Recent surveys provide hope that the world population growth is slackening. Between 1970 and 1983 the annual growth rate decreased from 1.9% to 1.7%. At 1.7%, it will take 41 years for the world population to double rather than 36 years at the higher annual growth rate figure. The decrease is credited to better family planning programs, but increasing deaths from starvation and disease also are contributing factors.

Unlike other animals that can do little to increase the carrying capacity of their environment, humans have done so repeatedly. Unfortunately, when the carrying capacity is increased, humans respond as would any animal by increasing their population. Because the earth's resources are finite, the time will inevitably come when the carrying capacity can be extended no further. The question is whether humans will be able to anticipate this limit and check population growth in time or whether we will overshoot the resources and experience a crash. Indeed, for millions of Third World people time has already run out. The rapid growth of population is not something that thinking people view with equanimity.

How population growth is curbed

What determines the number of animals in a natural population? For a laboratory culture of animals the answer seems fairly clear. They grow until they reach the carrying capacity of the environment, at which point growth is suppressed by competition for the limited resources of space and food. Thus the forces that limit growth arise from within the population; they are **density-dependent** mechanisms caused by crowding. Food and space limitations are commonly observed density-dependent factors but others, such as disease and predation, may operate as well.

Transmission of infectious **disease** may be much higher under crowded conditions, so that much higher mortality due to parasitism occurs than would prevail at lower population densities. Epidemics may sweep through crowded populations as did the terrible bubonic plague (the "Black Death") through the crowded, dirty cities of Europe in the fourteenth century. **Predation** increases when prey becomes abundant and easy to catch, and larger populations of prey are often followed by an increase in the predator populations. **Shelter** for suitable nest sites and as refuge from bad weather becomes less available with crowding.

The continuous competition for food and space to live brings forth yet another density-dependent force: **stress**. Although the physiology of stress is not thoroughly understood, there is ample evidence that when certain natural populations become crowded, such as populations of lemmings, voles, and snowshoe hares, a neuroendocrine imbalance involving the pituitary and adrenal glands appears. Growth is suppressed. Reproduction fails and individuals become irritable and aggressive. Under such conditions the population crashes.

Overcrowding may also cause **emigration** (movement away) from the birth area. Overcrowded mice increase their locomotor activity and begin to explore new areas. In the case of lemmings, overpopulation produces the famous mass "marches" recorded at intervals in Scandinavia. When songbird populations

Table 5-1 Comparison of *K*- and *r*-Selected animals

Strategy	*K*-Selected species	*r*-Selected species
Population	At carrying capacity	Below capacity
Environment	Constant or predictable	Unstable
Intraspecific competition	Keen	Lax
Development	Slow	Fast
Body size	Large	Small
Investment per offspring	Large	Small
Number of offspring	Few	Many
Niche	Specialists	Generalists

From Gould, J.L. 1982. Ethology. New York, W.W. Norton & Co.

become overcrowded the less successful become "vagabond" birds that are forced out into less preferred habitats. In fact, emigration is one major force resulting in the colonization of new habitats. The rapid dispersal of the European starling throughout the United States and southern Canada, following its introduction into New York City in 1890, is an excellent example (p. 625).

Not all of the forces that limit growth are density dependent. An extreme change in the weather or unusually cold, hot, wet, or dry weather is a density-independent hazard of varying severity to animal populations. Local insect populations are sometimes pushed to the point of extinction by a severe winter. Hurricanes and volcanic eruptions may destroy entire populations. Hailstorms have been known to kill most of the young of a wading bird population. Prairie grass and forest fires, droughts, and floods all take their toll on animal populations.

Ecologists have distinguished between organisms whose populations are controlled more by density-dependent factors and those controlled by density-independent factors. Organisms with a range of adaptations to survive density-dependent control are called *K*-selected (from the *K* term in the logistic equation), and those with adaptations for density-independent control are *r*-selected (from the *r* term) (Table 5-1). In natural populations of various species, there is a continuum from highly *r*-selected to highly *K*-selected, and the terms are relative, that is, species A may be *r*-selected with respect to species B and *K*-selected compared to species C. Nevertheless, certain characteristics of each can be recognized. Organisms that are *r*-selected generally have very high reproductive rates, high (often catastrophic) mortality, short life, and population sizes variable in time, usually well below the carrying capacity of the environment, whereas *K*-selected organisms have the opposite characteristics. *K*-selected populations tend to comprise larger, long-lived individuals able to compete in crowded circumstances (*N* close to *K*), not making a heavy commitment to reproduction because that would reduce the chance of individual survival. Populations that are *r*-selected can rapidly exploit an unstable environment before better competitors arrive, producing large numbers of young that disperse quickly and increase the chances that at least a few will find another favorable place to grow. All such characteristics are genetically determined and are the products of relentless natural selection over the millenia.

K-selected species are often known as *K*-strategists, and *r*-selected organisms as *r*-strategists. For many *r*-strategists the carrying capacity of the environment for the adults is rarely reached because the mortality of the progeny is so high. This is true for many marine organisms that produce enormous numbers of offspring. Many parasites are *r*-strategists; unless the hosts are crowded, transmission to a new host may be hazardous, and the reproductive capacity of the parasite is high to offset the loss. Examples of *r*-strategists among plants would include the annual weeds that quickly colonize disturbed ground, reproduce, and disperse their seed before perennial species can mature, shading and crowding out the annuals, thus altering and stabilizing the environment.

Thus population numbers are influenced by factors generated within the population and by forces from without, and usually no single mechanism can account fully for how growth is curbed in a given population. Natural populations are controlled by density-dependent and density-independent forces. Humans are the greatest force of all. By altering animal habitats we change the balances that set the old limits to animal numbers. Our activities can increase or exterminate whole populations of animals.

■ Summary

The biosphere is the thin blanket of life surrounding the earth. The presence of life on earth is possible because numerous conditions for life are fulfilled on this planet. These include a steady supply of energy from the sun, the presence of water, a suitable range of temperatures, the correct proportion of major and minor elements, and the screening of lethal ultraviolet radiation by atmospheric ozone.

The earth environment and living organisms have evolved together, each deeply marked by the other.

The earth's terrestrial environment is composed of biomes that bear a distinctive array of plant life and associated animal life. Biome distinctiveness is determined mainly by climate, and climate in turn is influenced by earth's biota.

Freshwater environments include lotic habitats (streams and rivers) and lentic habitats (ponds and lakes). The biota of each is adapted to the conditions found there, such as water movement and level of dissolved oxygen. Animals living on the bottom are benthic, those swimming are nekton, and floating organisms are plankton. Plant production in the oceans is limited to the photic zone, and life in the aphotic zone is supported by particles sinking out of the photic zone.

Zoogeography is the study of animal distribution on earth. Several great faunal realms are recognized by distinctive animal associations. These are the Nearctic, Neotropical, Palearctic, Ethiopian, Oriental, and Australian realms.

The most inclusive level of environmental organization is the ecosystem, which includes interacting biotic and abiotic components. The biotic components are communities of various species of organisms, each species being composed of a population, and each population is the sum of the individual organisms that share a gene pool.

Energy flows into the ecosystem as sunlight, and life depends on the conversion of that energy into organic compounds by the fixation of CO_2 in green plants. Food chains of producers (plants) and consumers (herbivores and carnivores) may be recognized, but in nature, complex food webs exist. Ecological pyramids can be constructed to show the decrease in numbers of individuals, biomass, and energy at each trophic level. Nutrients, such as carbon, nitrogen, and other elements, are cycled and recycled in the ecosystem.

Each community is usually dominated by one or a few species that greatly influence the nature of the local environment. Each species has its own niche, or the role it plays in the ecosystem.

Populations in the community may have no interaction (neutral), or they may interact as competitors, predators and prey, parasites and hosts, commensals, mutuals, or protocooperators. Populations introduced to an environment where there is an upper limit to resources usually grow in a manner that can be described by a sigmoid curve; at first the growth is exponential, when food and space are not limiting, and then the curve levels off as the carrying capacity of the environment is reached. The curve can be expressed mathematically as the logistic equation. This equation takes into account the intrinsic rate of increase of the population *(r)* and the carrying capacity of the environment *(K)*. Species whose populations approach K and are regulated by density-dependent factors are K-selected, and species that generally do not approach K and are regulated by density-independent factors are r-selected. Populations that are r-selected are opportunists, quickly moving into an unstable environment and producing large numbers of progeny before better competitors arrive. Density-dependent mechanisms of population control include competition, disease, and predation; density-independent mechanisms include extreme weather, abrupt environmental change, and hazardous conditions for the progeny with accompanying high mortality.

REVIEW QUESTIONS

1. The American biochemist Lawrence J. Henderson argued that fitness is a mutual relationship between the organism and the environment. Can you think of examples of how the interchange between environment and animal life have produced changes in both?
2. What is the origin of oxygen on earth? What would happen to the earth's supply of oxygen if photosynthesis were suddenly to cease?
3. What effect is the burning of fossil fuels having on the atmosphere?
4. What is a biome? Briefly describe six examples of biomes.
5. What are the most productive marine environments and why are they so productive?
6. Although most of the open ocean is nutrient-poor, some areas have high productivity. Why?
7. What geographical boundaries or barriers divide the six faunal realms recognized by Alfred Russel Wallace?
8. What is the distinction between *ecology* and *environment*? Ask several nonbiologist friends how they would define *ecology*.
9. Discuss the distinction between ecosystem, community, and population.
10. What happens to solar energy reaching the earth?
11. Recently discovered animal communities surrounding deep-sea thermal vents are said to exist in total independence of solar energy. How can this be possible?
12. Define *production* as the word is used in ecology. What is the distinction between production, primary productivity, gross productivity, and net primary productivity?
13. What is a food chain? How does a food chain differ from a food web?
14. Distinguish between producers, herbivores, carnivores, and detritus feeders in a food chain.
15. What is a trophic level and how does it relate to a food chain?
16. How is it possible to have an inverted pyramid of biomass in which the consumers have a greater biomass than the producers? Can you think of an example of an inverted pyramid of *numbers* in which there are, for example, more herbivores than the plants upon which they feed?
17. What pools of CO_2 are involved in the carbon cycle? What is the significance of the carbon cycle?
18. What biological and nonbiological processes are involved in nitrogen fixation?
19. Species dominance among plants has sometimes been called "space capture." Is this an acceptable description?
20. Define the niche concept. How is direct competition avoided in sympatric species?
21. Define predation. How does predator-prey relationship differ from the parasite-host relationship?
22. Contrast exponential and logistic growth of a population. Under what conditions might you expect a population to exhibit exponential growth? Why cannot exponential growth be perpetuated indefinitely?
23. What do you think the carrying capacity of earth for the human population would be if everyone on earth enjoyed the standard of living of the highly developed nations of North America and Europe?
24. Population growth may be hindered by either density-dependent or density-independent mechanisms. Define and contrast these two mechanisms. Offer examples of how the human population growth might be curbed by either agent.
25. What are the characteristics of a K-selected species? Of an r-selected species? Are humans K-strategists or r-strategists (refer to marginal note on p. 143)?

SELECTED REFERENCES

Colinvaux, P. 1986. Ecology. New York, John Wiley & Sons. *Comprehensive college textbook.*

Cox, C.B., and P.D. Moore. 1985. Biogeography: an ecological and evolutionary approach, ed 4. Boston, Blackwell Scientific Publications. *Highly readable account with strong ecological emphasis.*

Ehrlich, P.R. 1986. The machinery of nature. New York, Simon and Schuster. *A leading American ecologist's personal and often earnest account of insights to be drawn from evolutionary ecology, and the inevitable fearful cost humans will have to pay if the lessons of ecology continue to be ignored.*

Ehrlich, P.R. and J. Roughgarden. 1987. The science of ecology. New York, Macmillan Publishing Company. *Comprehensive, with more emphasis on evolution and behavioral ecology than is usually found in introductory treatments.*

Krebs, C.J. 1988. The message of ecology. New York, Harper & Row, Publishers. *Intended as supplementary reading for college ecology courses, this paperback provides numerous fascinating examples to explain ecology's "message."*

Moore, P.D. (ed.). 1987. The encyclopedia of animal ecology. New York, Facts on File Publications. *Though ecological principles are sparsely treated in this book, the world's major ecosystems are surveyed with extensive text and beautifully illustrated with photographs, paintings, and helpful diagrams.*

PART TWO

Animal Form and Function

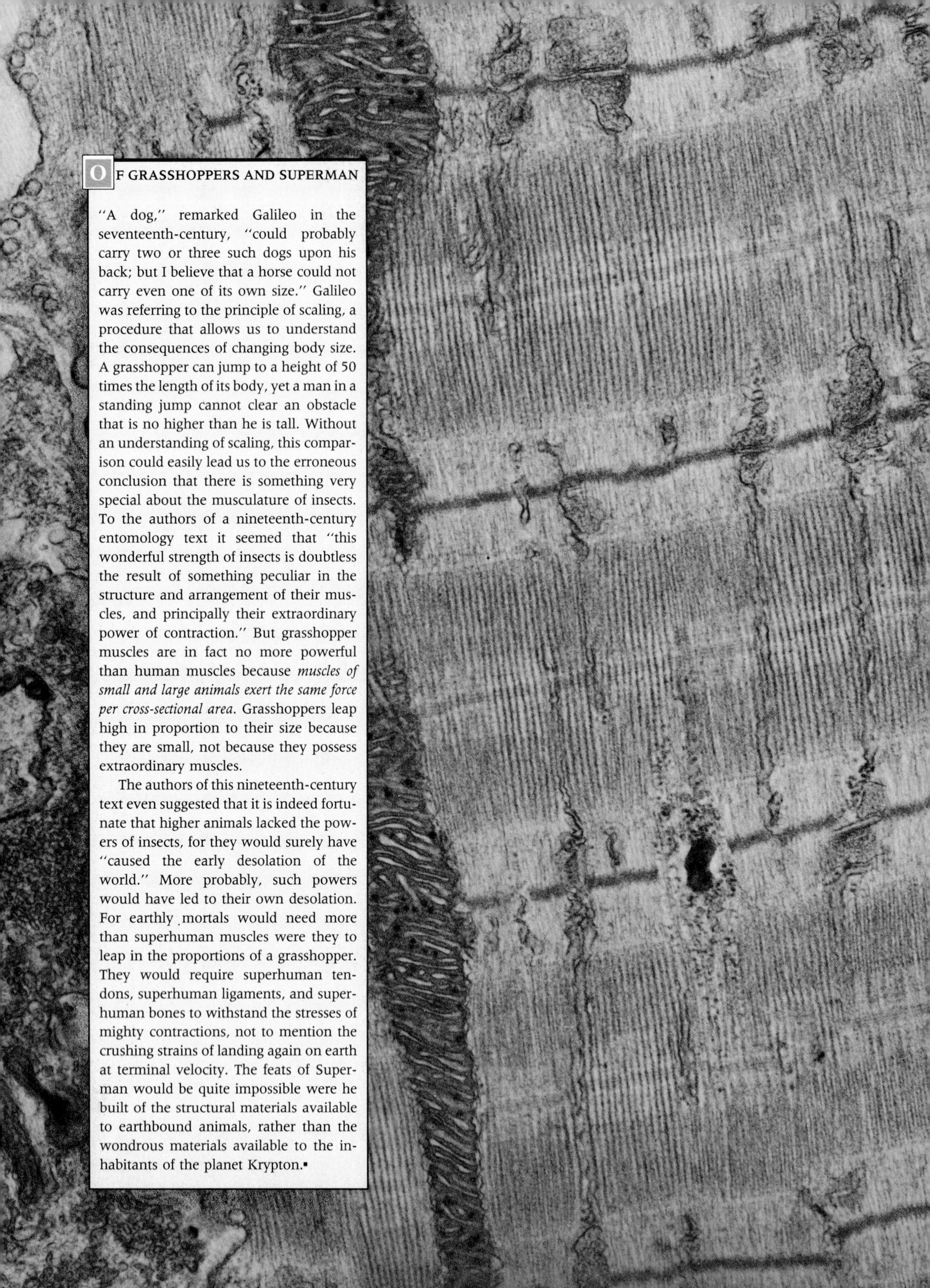

OF GRASSHOPPERS AND SUPERMAN

"A dog," remarked Galileo in the seventeenth-century, "could probably carry two or three such dogs upon his back; but I believe that a horse could not carry even one of its own size." Galileo was referring to the principle of scaling, a procedure that allows us to understand the consequences of changing body size. A grasshopper can jump to a height of 50 times the length of its body, yet a man in a standing jump cannot clear an obstacle that is no higher than he is tall. Without an understanding of scaling, this comparison could easily lead us to the erroneous conclusion that there is something very special about the musculature of insects. To the authors of a nineteenth-century entomology text it seemed that "this wonderful strength of insects is doubtless the result of something peculiar in the structure and arrangement of their muscles, and principally their extraordinary power of contraction." But grasshopper muscles are in fact no more powerful than human muscles because *muscles of small and large animals exert the same force per cross-sectional area*. Grasshoppers leap high in proportion to their size because they are small, not because they possess extraordinary muscles.

The authors of this nineteenth-century text even suggested that it is indeed fortunate that higher animals lacked the powers of insects, for they would surely have "caused the early desolation of the world." More probably, such powers would have led to their own desolation. For earthly mortals would need more than superhuman muscles were they to leap in the proportions of a grasshopper. They would require superhuman tendons, superhuman ligaments, and superhuman bones to withstand the stresses of mighty contractions, not to mention the crushing strains of landing again on earth at terminal velocity. The feats of Superman would be quite impossible were he built of the structural materials available to earthbound animals, rather than the wondrous materials available to the inhabitants of the planet Krypton.▪

CHAPTER 6

Animal Architecture

Body Organization, Support, and Movement

The English satirist Samuel Butler proclaimed that the human body was merely "a pair of pincers set over a bellows and a stewpan and the whole thing fixed upon stilts." While human attitudes toward the human body are distinctly ambivalent, most people less cynical than Butler would agree that the body is a triumph of intricate, living architecture. Less obvious, perhaps, is that the architecture of humans and most other animals conforms to the same well-defined plan. The basic uniformity of biological organization derives from the common ancestry of animals and from their basic cellular construction. Despite the vast differences of structural complexity of animals ranging from the simplest protozoa to humans, all share an intrinsic material design and fundamental functional plan.

We begin our study of animal form and function with the levels of organization in animal complexity. Then we will consider the body's basic structural framework: tissues, which are the "bricks and mortar" that comprise organs and systems; the integument, the body's outer enclosure; and the musculoskeletal system, a triumvirate of bone, muscle, and connective tissue organized to support the body, protect its internal organs, and give it mobility.

■ Levels of Organization in Animal Complexity

Increasing complexity of organization is the most evident feature of animal phylogeny. The simplest animals are the unicellular protozoa and, being small, have much less scope for complexity than do larger creatures. Size then is an important determinant of complexity. Nevertheless, unicellular forms are complete organisms that carry on all the basic life functions of more complex animals. Within the confines of their cell they often show remarkable organization and division of labor, such as skeletal elements, locomotor devices, fibrils, and beginnings of sensory structures. The metazoa, or multicellular animals, on the other hand, have a much greater capacity for internal specialization. Different cells can take on different functions, becoming increasingly specialized and increasingly efficient in carrying out particular tasks. Consequently, the metazoan cell is not the equivalent of a protozoan cell; it is only a specialized part of the whole organism and is incapable of independent existence.

We can recognize five **levels** of organization. Each level is more complex than the one before, and, as a general rule, it is a more advanced and more recent evolutionary product.

1. *Protoplasmic level of organization.* Protoplasmic organization is found in protozoa and other unicellular organisms. All life functions are confined within the boundaries of a single cell, the fundamental unit of life. Within the cell, the living substance is differentiated into organelles capable of carrying on specialized functions.

2. *Cellular level of organization.* Cellular organization is an aggregation of cells that are functionally differentiated. A division of labor is evident, so that some cells are concerned with, for example, reproduction, others with nutrition. Such cells have little tendency to become organized into tissues (a tissue is a group of similar cells organized to perform a common function). Some protozoan colonial forms having somatic and reproductive cells might be placed at the cellular level of organization. Many authorities also place sponges at this level.

3. *Cell-tissue level of organization.* A step beyond the preceding is the aggregation of similar cells into definite patterns or layers, thus becoming a tissue. Sponges are considered by some authorities to belong to this level, although the jellyfish and their relatives (Cnidaria) are usually identified as having the beginning of the tissue plan. Both groups are still largely of the cellular level of organization because most of the cells are scattered and not organized into tissues. An excellent example of a tissue in cnidarians is the **nerve net**, in which the nerve cells and their processes form a definite tissue structure, with the function of coordination.

4. *Tissue-organ level of organization.* The aggregtion of tissues into organs is a further step in advancement. Organs are usually made up of more than one kind of tissue and have a more specialized function than tissues. The first appearance of this level is in the flatworms (Platyhelminthes), in which there are a number of well-defined organs such as eyespots, digestive tract, and reproductive organs. In fact, the reproductive organs are well organized into a reproductive system.

5. *Organ-system level of organization.* When organs work together to perform some function we have the highest level of organization — the organ system. The systems are associated with the basic body functions — circulation, respiration, digestion, and the others. Typical of all the higher forms, this type of organization is first seen in the nemertean worms in which a complex digestive system complete with a mouth and anal opening is present.

■ Size and Complexity

It is evident from the preceding discussion that as animals become more complex, they tend to become larger. An ameba can be small because it can move without muscles, digest food without a gut, excrete its wastes without a kidney, coordinate its activities without a brain, and breathe without gills. More advanced multicellular animals have specialized tissues and organs for these functions, and consequently they must be larger to accommodate their increased complexity.

But if the single-celled protozoa are so successful, why has there been a tendency in animal lineages for increasingly larger multicellular forms to evolve? Certainly a prime selective force was the availability on earth of scores of habitats suitable for animals larger than a single cell. Small organisms, successful as they were (and are), long ago occupied all the available small ecological niches, and the only way a new species could succeed was to displace an organism from an existing niche or adapt to a new and larger one. Since single cells, excepting eggs, cannot become very large, multicellularity became a reasonably simple and highly adaptive path toward increased body size.

Once embarked on this path and as organisms became bulkier, increasing complexity was inevitable. As animals grow larger, the body surface (length^2) increases much more slowly than body volume (length^3) making it increasingly difficult for activities served from the surface to provision the mass of cells within. Systems had to evolve to supply nutrients and gases and to serve as an avenue for waste removal. Cilia and flagella are inadequate for moving anything larger than a flatworm, so muscles for locomotion appeared. And muscles in turn required some kind of skeletal support to work against.

If in becoming larger, animals lose the advantage of architectural simplicity, what is to be gained by being big? Perhaps the most obvious advantage is that in the predator-prey contest, predators are almost always larger than their prey. Exceptions are few and are usually related to an especially aggressive or social pack behavior on the part of the predator that compensates for small size. Since the predator must first catch its prey, whereas the prey must escape its predator if it is to survive, both predator and prey will experience selective pressures that encourage an evolutionary drift toward larger size.

Another advantage of large body size that is usually overlooked is that large animals can move themselves about at a lesser energy cost than small animals. This becomes clear when we look at the cost of running for a given group of animals (for example, mammals) of various body sizes (Figure 6-1). A large mammal consumes more oxygen than a small mammal, of course, but the cost of moving 1 g of its body over a given distance is much less for the large animal than for a small one. This makes sense when we compare the relative ease with which a horse can run a mile with the herculean task that running the same distance would present to a mouse!

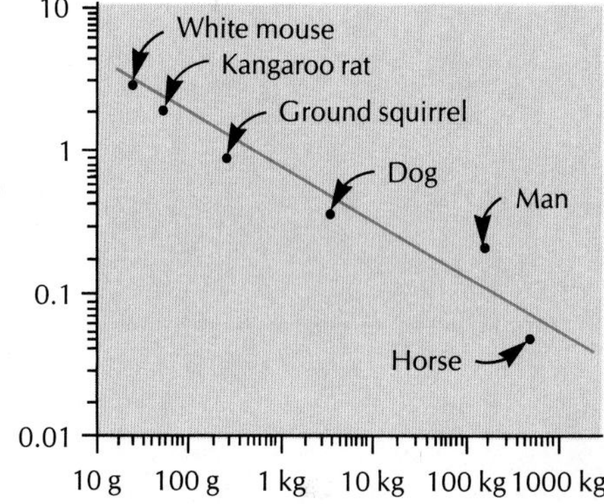

FIGURE 6-1

Net cost of running for mammals of various sizes. Each point represents the cost (measured in rate of oxygen consumption) of moving 1 g of body over 1 km. The cost decreases with increasing body size.

Modified from Schmidt-Nielsen, K. 1972. How animals work. New York, Cambridge University Press.

Other advantages to large size have been cited by biologists. Nevertheless, every animal, whether large or small, is a success in its ecological niche. In fact, the most abundant groups of animals on earth today are those with short generation times and, consequently, small body size. To these countless creatures, "small is beautiful."

Organization of the Body

The body of a multicellular animal consists of three elements: (1) cells, (2) body fluids, and (3) extracellular structural elements.

The animal body is separated from the outside world by a continuous outer layer of cells. This is the body's frontier that protects the controlled internal environment from the varying external environment. Within this protective wrapping are the cells of the body's interior. These are of great variety and serve numerous functions, for example, blood cells, muscle cells for movement, supportive cells, nervous and other cells for communication, and cells concerned with internal defense. All these will be treated in the chapters that follow.

Body fluids permeate all tissues and spaces in the body but are naturally separated into certain fluid "compartments." In all metazoa, the two major fluid compartments are the **intracellular space,** the cytoplasm within the body's cells, and the **extracellular space,** outside the cells. In animals with closed vascular systems (such as the vertebrates), the extracellular space can be further subdivided into the **blood plasma** (the fluid portion of the blood outside the blood cells; blood cells are really part of the intracellular compartments) and **interstitial fluid.** The interstitial (or intercellular) fluid is the tissue fluid that occupies the spaces surrounding the cells. Unlike the vertebrates, many invertebrate groups have open blood systems with no true separation of blood plasma from interstitial fluid. However, all metazoan invertebrates share with the vertebrates the basic subdivision of body fluids between the intracellular and extracellular compartments. These relationships will be explored further in Chapter 7.

The term *intercellular,* meaning "between cells," should not be confused with the term *intracellular,* meaning "within cells."

If we were to remove all the specialized cells and body fluids from the interior of the body, we would be left with the third element of the animal body: extracellular structural elements. This is the supportive material of the organism such as connective tissue, cartilage, and bone. It provides mechanical stability, protection, and a depot of materials for exchange and serves as a medium for extracellular reactions.

Two types of extracellular materials are recognized: **formed** and **amorphous** (meaning without definite form). One type of formed element that is extremely plentiful in the body is **collagen,** a white, fibrous protein material with great tensile strength. Collagen is in fact the most abundant protein in the animal kingdom, found in animal bodies wherever both flexibility and resistance to stretching are required, such as in skin, tendons and ligaments, and surrounding the eyeball. It is also found abundantly in cartilage and bone, but these tissues have special mechanisms to prevent the collagen from crumpling. **Elastic fibers** are another type of formed element. Unlike collagen, elastic fibers can be stretched but will spring back to their original length when the tension is released. Elastic fibers are found where springiness is required, such as in the walls of arteries and in the lungs. The amorphous extracellular substance is a jellylike, structureless material often called "ground substance." It is composed of complex, hydrated mucopolysaccharides that provide watery tissues with resistance to the free flow of water, thus preventing awkward shifts of body water when animals change positions.

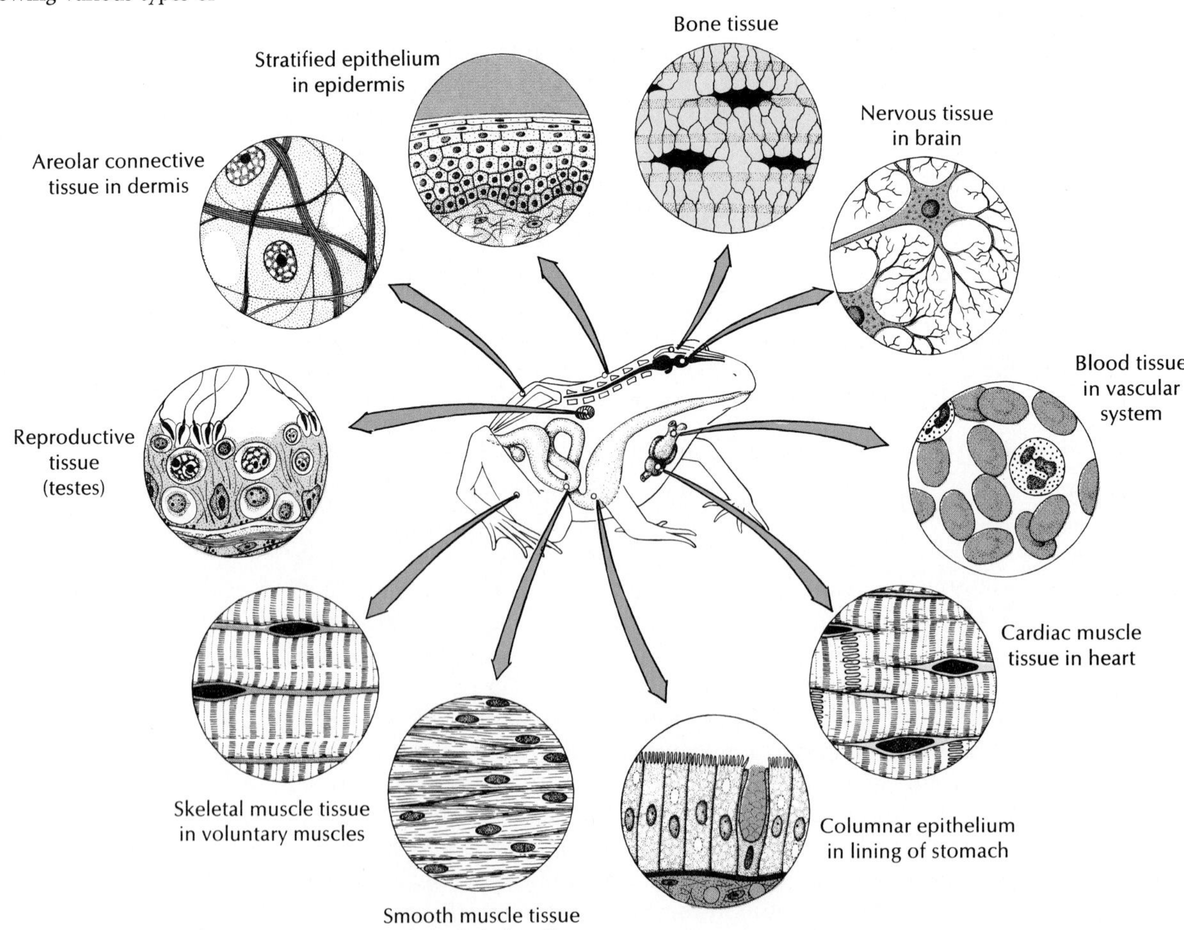

FIGURE 6-2

Diagram of frog showing various types of tissue.

TISSUES

A **tissue** is a group of similar cells (together with associated cell products) specialized for the performance of a common function. The study of tissues is called **histology.** All cells in metazoan animals take part in the formation of tissues (Figure 6-2). Sometimes the cells of a tissue may be of several kinds, and some tissues have a great many intercellular materials.

During embryonic development, the germ layers become differentiated into four kinds of tissues. These are epithelial, connective (including vascular), muscular, and nervous tissues. This is a surprisingly short list of basic tissue types that are able to meet the diverse requirements of animal life.

Epithelial tissue

An **epithelium** is a sheet of cells that covers an external or internal surface. On the outside of the body, the epithelium forms a protective covering. Inside, the epithelium lines all the organs of the body cavity, as well as ducts and passageways through which various materials and secretions move. On many surfaces the epithelial cells are often modified into glands that produce lubricating mucus or specialized products such as hormones or enzymes.

Epithelia are classified on the basis of cell form and number of cell layers. A **simple epithelium** is one layer thick (Figure 6-3), and its cells may be **squamous** (flat), as in the cheek cells of the mouth; **cuboidal** (short prisms), as in glands and ducts; or **columnar** (tall), as in the intestinal tract of most animals. In addition to simple epithelia that are present in nearly all metazoa, vertebrate animals often have epithelia arranged in several layers. This is called **stratified epithelium,** found for example in skin, sweat glands, and the urethra (Figure 6-4). Some stratified epithelia can change the number of their cell layers by being stretched out (**transitional,** such as the urinary bladder). Others have cells of different heights and give the appearance of stratified epithelia (**pseudostratified,** such as the trachea). Many of the above types of epithelia may be ciliated at their free surfaces (such as the oviduct).

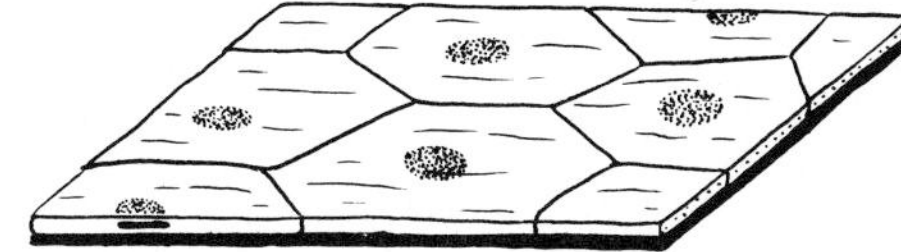

Simple squamous

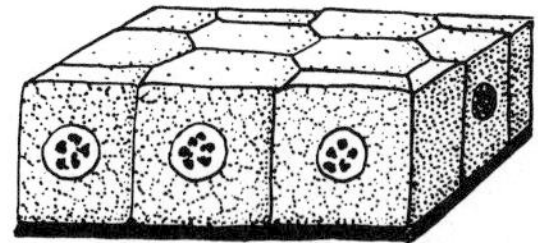

Simple cuboidal

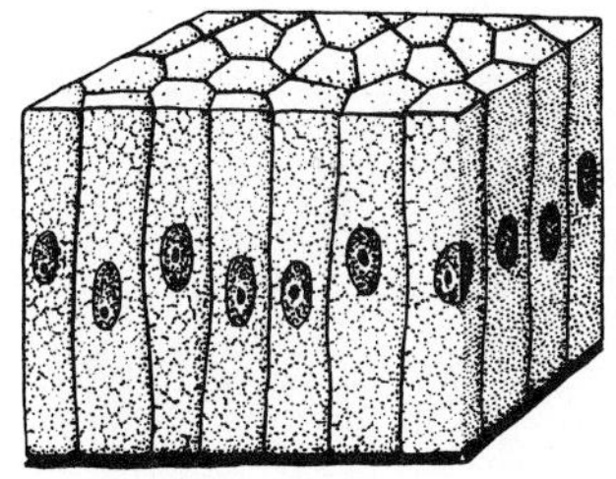

Simple columnar

FIGURE 6-3

Types of simple epithelium.

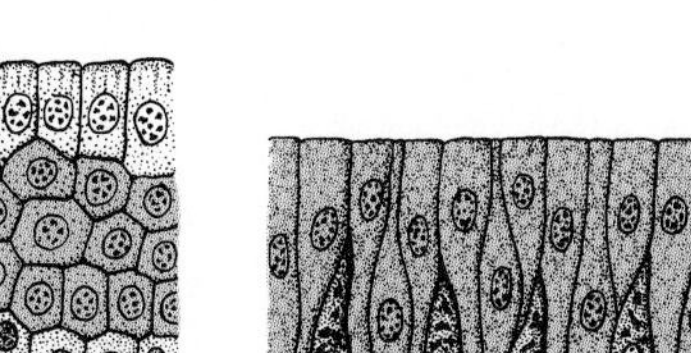

Stratified columnar

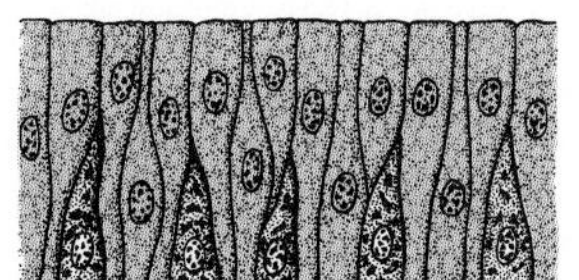

Pseudostratified

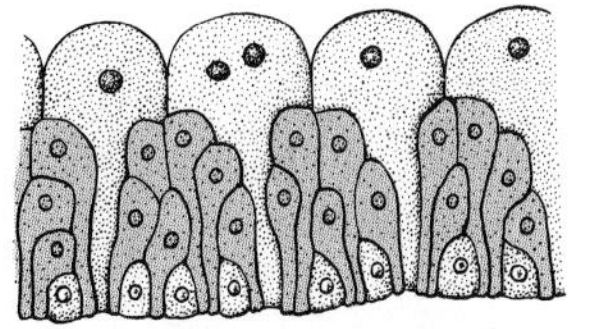

Transitional (relaxed)

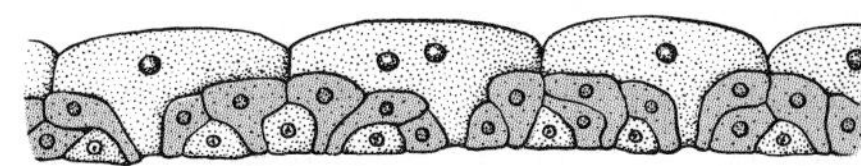

Transitional (stretched)

FIGURE 6-4

Types of stratified and transitional epithelial tissue.

Connective tissue

Connective tissues bind together and support all other structures. They are so common that the removal of all other tissues from the body would still leave the complete form of the body clearly apparent. Connective tissue is made up of relatively few **cells** and a great deal of formed materials such as **fibers** and ground substance (**matrix**) in which the fibers are embedded. There are three types of fibers: white or collagenous (the most common type), yellow or elastic, and branching or reticular. Connective tissue may be classified in various ways, but all the types fall under either **loose connective tissue** (reticular, areolar, adipose) or **dense connective tissue** (sheaths, ligaments, tendons, cartilage, bone) (Figures 6-5 and 6-6). The connective tissue of invertebrates, as in vertebrates, consists of cells, fibers, and ground substance, but it is not as elaborately developed.

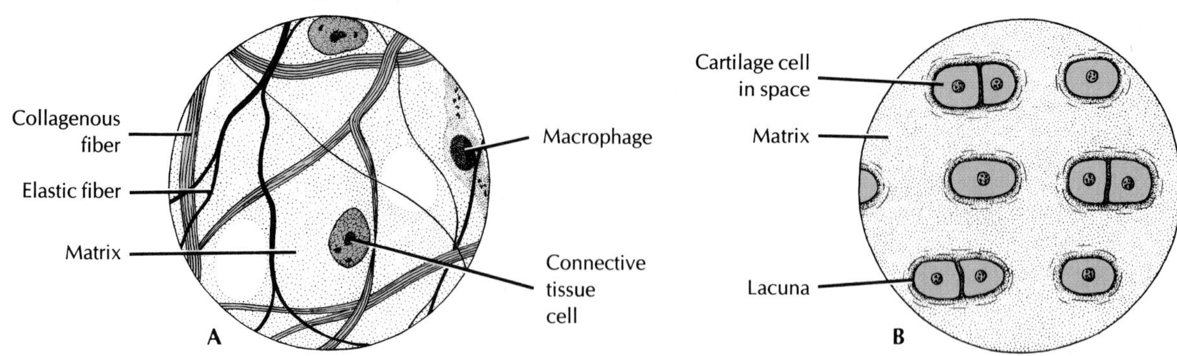

FIGURE 6-5

A, Areolar, a type of loose connective tissue. **B,** Hyaline cartilage, most common form of cartilage in body and a type of dense connective tissue.

Vascular tissue is considered by many histologists to be a type of fluid connective tissue. Vascular tissue includes blood, lymph, and tissue fluid. In vertebrates, blood is composed of **white blood cells** (leukocytes), **red blood cells** (erythrocytes), and **platelets,** all suspended in a liquid matrix, the **plasma.** Traveling through blood vessels, the blood carries to the tissue cells the materials necessary for the life processes. **Lymph** and **tissue fluids** arise from blood by filtration and serve in the exchange between cells and blood.

Erythrocytes are rare among invertebrates because when respiratory pigments such as hemoglobin are present, they usually circulate as giant molecules in the plasma. Most invertebrates have phagocytic blood cells, often called amebocytes, that correspond to the white blood cells of vertebrates.

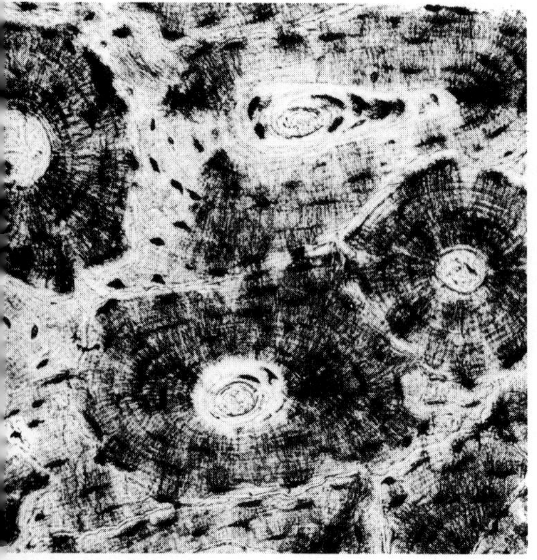

FIGURE 6-6

Section of bone, a type of dense connective tissue, showing several cylindrical osteons (haversian systems) typical of bone. (×180.)

Muscular tissue

Muscle is the most common tissue in the body of most animals. It is made up of elongated cells often called fibers, specialized for contraction. It originates (with few exceptions) from the mesoderm, and its unit is the cell or **muscle fiber.** Structurally, muscles are either **smooth** (fibers unstriped) or **striated** (fibers cross-striped) (Figure 6-7).

Smooth muscle is the most common type of invertebrate muscle, serving as body wall musculature, lining ducts and sphincters, and functioning in other contractile activities. Smooth muscle fibers are often present in epithelial cells **(epitheliomuscular cells),** which combine the functions of motility with those of digestion or protection. In vertebrates, smooth muscle lines the walls of blood vessels and surrounds internal organs such as the intestine and uterus. It is often called involuntary muscle, since its contraction is generally not consciously controlled.

Striated muscle appears in several invertebrate groups but is most highly developed in the arthropods. Insects have only striated muscle, found in the walls of the heart, intestine, and other internal organs and, with the exoskeleton to which it is attached, forms the locomotory apparatus. Two types of striated muscle are found in vertebrates. **Skeletal muscle,** composed of extraordinarily long fibers, moves the skeleton and is responsible for posture. It is voluntary muscle innervated by spinal and some cranial nerves. **Cardiac muscle,** the involuntary muscle of the vertebrate heart, is composed of shorter fibers arranged in sheets and is controlled by the autonomic nervous system.

FIGURE 6-7

Three kinds of vertebrate muscle fibers, as they appear when viewed with a light microscope.

Skeletal

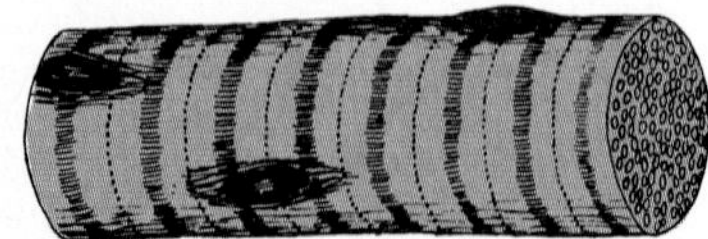

Cardiac

Smooth

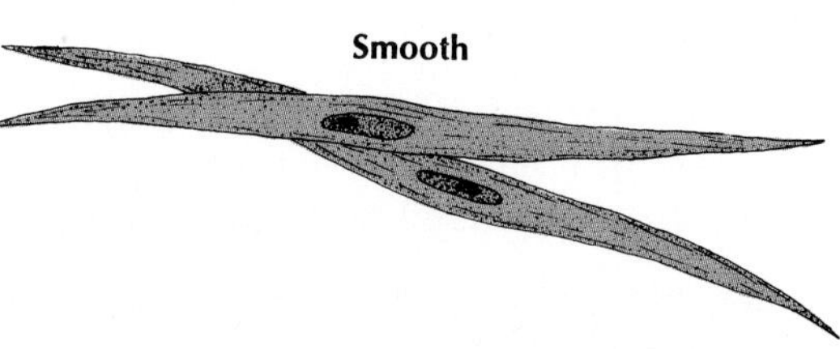

Nervous tissue

Nervous tissue is specialized for the reception of stimuli and the conduction of impulses. The structural and functional unit of the nervous system is the **neuron** (Figure 6-8), a nerve cell made up of a cell body containing the nucleus and its processes or fibers. These processes are of two types. **Dendrites,** often profusely branched extensions of the cell body, are the cell's receptive apparatus, often receiving information from several different sources. The **axon** of a nerve cell is a long, single fiber that typically carries information away from the cell body. In most animals the cell bodies of neurons are restricted to the central nervous system and ganglia (groups of cell bodies located outside the central nervous system), but the processes may extend long distances through the body. Neurons are arranged in chains, and the point of contact between neurons is a **synapse.** In vertebrates, some fibers are wrapped with an insulating myelin sheath. Insulating sheaths occur in some invertebrates, especially arthropods, but most invertebrate nerve fibers are naked.

Sensory neurons are concerned with conducting impulses from sensory **receptors** in the skin or sense organs to nerve centers (brain or spinal cord). **Motor neurons** carry impulses from the nerve centers to muscles or glands (**effectors**) that are thus stimulated to respond. **Association neurons** form various connections between other neurons.

ORGANS AND SYSTEMS

An **organ** is a group of tissues organized into a larger functional unit. In higher forms an organ may have most or all of the four basic tissue types. For example, the heart (Figure 6-9) has epithelial tissue for covering and lining, connective tissue for framework, muscular walls for contraction, nervous elements for coordination, and vascular tissue (a type of connective tissue) for transportation. Usually one tissue carries the burden of the organ's chief function, as muscle does in the heart; the other tissues perform supporting roles.

Organs are, in turn, associated in groups to form **systems,** with each system concerned with one of the basic functions. The higher metazoa have 11 organ systems: skeletal, muscular, integumentary, digestive, respiratory, circulatory,

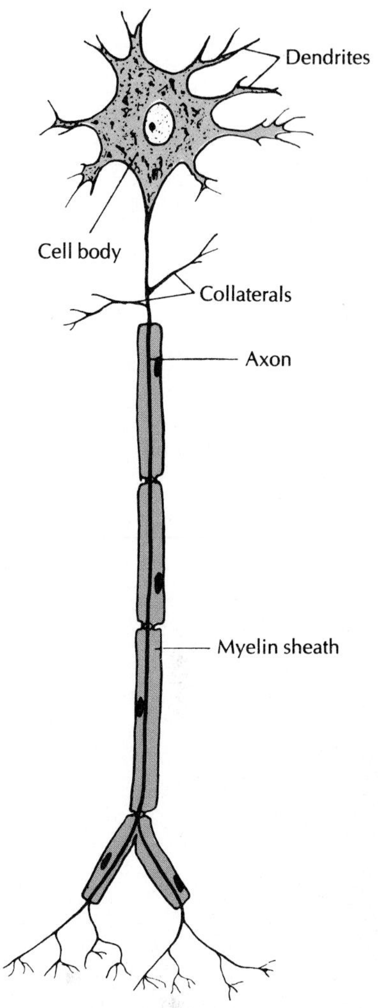

FIGURE 6-8

Neuron, showing the cell body and its extensions.

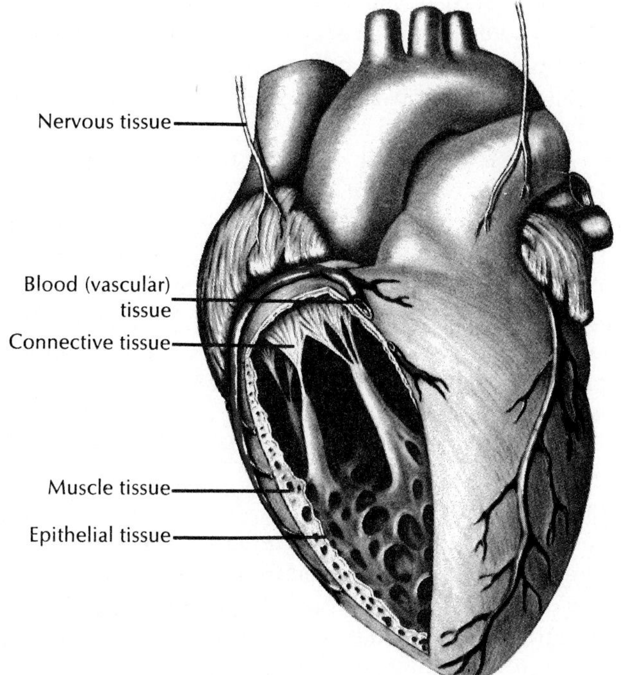

FIGURE 6-9

Heart, showing various types of tissue in its structure.

excretory, nervous, special sensory, endocrine, and reproductive. However, all living organisms perform the same basic functions. The need for procuring and using food and for movement, protection, perception, and reproduction are equally basic to an ameba, a clam, an insect, or a human being. Obviously, because of differences in size, structure, and environment, each must meet these problems in a different manner.

Integument among Various Groups of Animals

The integument is the outer covering of the body, a protective wrapping that includes the skin and all structures derived from or associated with the skin, such as hair, setae, scales, feathers, and horns. In most animals it is tough and pliable, providing mechanical protection against abrasion and puncture and forming an effective barrier against the invasion of bacteria. It may provide moisture proofing against fluid loss or gain. The skin helps protect the underlying cells against the damaging action of the ultraviolet rays of the sun. But, in addition to being a protective cover, the skin serves a variety of important regulatory functions. For example, in warm-blooded (homeothermic) animals, it is vitally concerned with temperature regulation, since most of the body's heat is lost through the skin; it contains mechanisms that cool the body when it is too hot and slow heat loss when the body is too cold. The skin contains sensory receptors that provide essential information about the immediate environment. It has excretory functions and in some forms respiratory functions as well. Through skin pigmentation the organism can make itself more or less conspicuous. Skin secretions can make the animal sexually attractive or repugnant or provide olfactory cues that influence behavioral interactions between individuals.

INVERTEBRATE INTEGUMENT

Many protozoa have only the delicate plasma membrane for external coverings; others, such as *Paramecium,* have developed a more rigid pellicle. Most multicellular invertebrates, however, have more complex tissue coverings. The principal covering is a single-layered **epidermis.** Some invertebrates have added a secreted noncellular **cuticle** over the epidermis for additional protection.

The molluscan epidermis is delicate and soft and contains mucous and other glands, some of which secrete the calcium carbonate of the shell. Cephalopod molluscs (squids and octopuses) have developed a more complex integument, consisting of cuticle, simple epidermis, layer of connective tissue, layer of reflecting cells (iridocytes), and thicker layer of connective tissue.

Arthropods have the most complex of invertebrate integuments, providing not only protection but also skeletal support. The development of a firm **exoskeleton** and jointed appendages suitable for the attachment of muscles has been a key feature in the extraordinary evolutionary success of this phylum, the largest of animal groups. The arthropod integument consists of a single-layered **epidermis** (also called more precisely **hypodermis**), which secretes a complex cuticle of two zones (Figure 6-10, *A*). The inner zone, the **procuticle,** is composed of protein and chitin (a polysaccharide) laid down in layers (lamellae) much like the veneers of plywood. The outer zone of cuticle, lying on the external surface above the procuticle, is the thin **epicuticle.** The epicuticle is a nonchitinous complex of proteins and lipids that provides a protective moisture-proofing barrier to the integument.

The arthropod cuticle may remain as a tough but soft and flexible layer, as it is, for example, in many microcrustaceans and insect larvae, or it may be hardened by either of two ways. In the decapod crustaceans, for example, crabs and lobsters, the cuticle is stiffened by **calcification,** the deposition of calcium carbonate. In insects hardening occurs when the protein molecules bond together with sta-

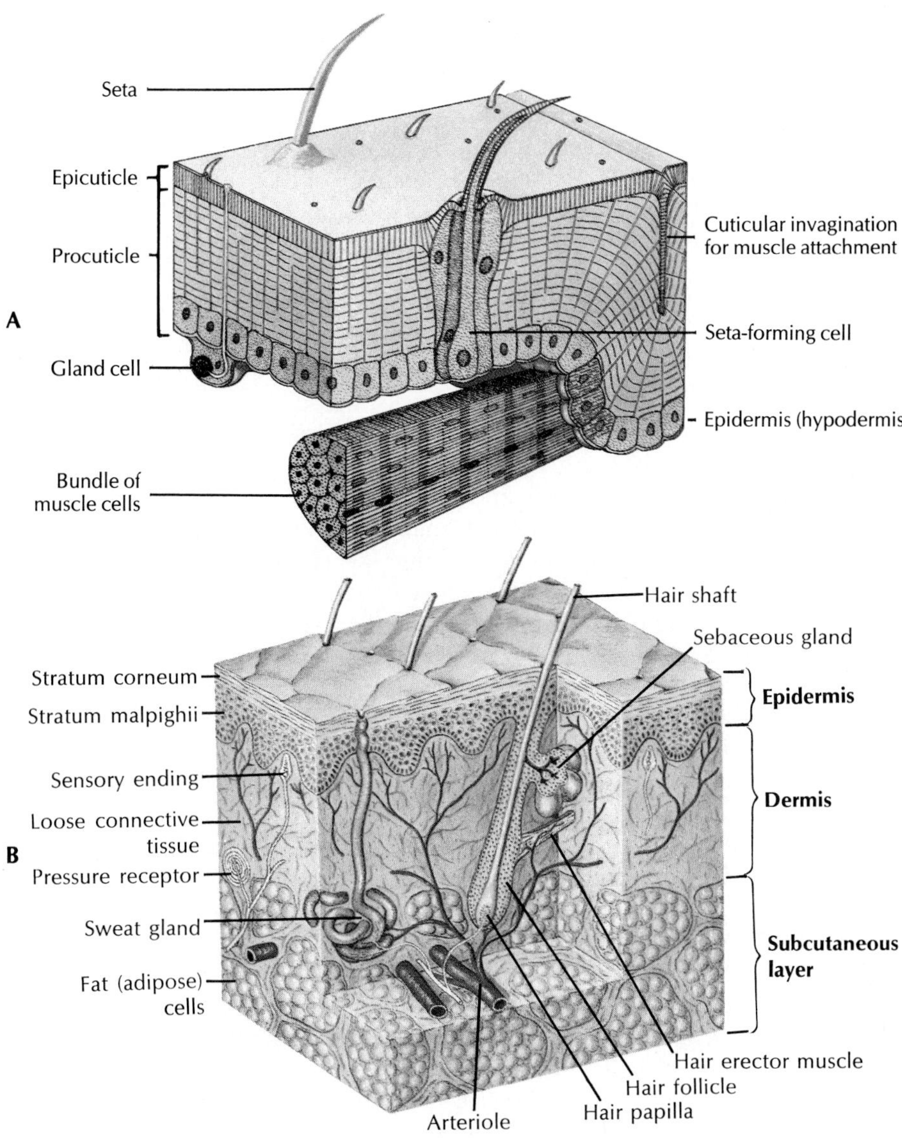

FIGURE 6-10

A, Structure of insect integument. This reconstruction shows a block of integument drawn at a point where the cuticle invaginates to provide exoskeletal muscle attachment. **B,** Structure of human skin.

bilizing cross-linkages, within and between the adjacent lamellae of the procuticle. The result of the process, called **sclerotization,** is the formation of a highly resistant and insoluble protein, **sclerotin.** Arthropod cuticle is one of the toughest materials synthesized by animals; it is strongly resistant to pressure and tearing and can withstand boiling in concentrated alkali, yet it is light, having a specific mass of only 1.3 (1.3 times the weight of water).

When arthropods molt, the epidermal cells first divide by mitosis. Enzymes secreted by the epidermis digest most of the procuticle; the digested materials are then absorbed and consequently not lost to the body. Then in the space beneath the old cuticle a new epicuticle and procuticle are formed. After the old cuticle is shed, the new cuticle is thickened and calcified or sclerotized.

VERTEBRATE INTEGUMENT

The basic plan of the vertebrate integument, as exemplified by human skin (Figure 6-10, *B*), includes a thin, outer stratified epithelial layer, the **epidermis,** derived from ectoderm and an inner, thicker layer, the **dermis,** or true skin, which is of mesodermal origin.

Although the epidermis is thin and appears simple in structure, it gives rise to most derivatives of the integument, such as hair, feathers, claws, and hooves. The

The reptiles were the first to exploit the adaptive possibilities of the remarkably tough protein keratin. The reptilian epidermal scale that develops from keratin is a much lighter and more flexible structure than the bony, dermal scale of fishes, yet it provides excellent protection from abrasion and desiccation. Scales may be overlapping structures, as in snakes and some lizards, or develop into plates as in turtles and crocodilians. Birds dedicated keratin to new uses. Feathers, beaks, and claws, as well as scales, are all epidermal structures composed of dense keratin. Mammals continued to capitalize on keratin's virtues by turning it into hair, hooves, claws, and nails. As a result of its keratin content, hair is by far the strongest material in the body. It has a tensile strength comparable to that of rolled aluminum and is nearly twice as strong, weight for weight, as the strongest bone.

dermis contains blood vessels, collagenous fibers, nerves, pigment cells, fat cells, and connective tissue cells called fibroblasts. These elements support, cushion, and nourish its overlying partner, which is devoid of blood vessels.

The epidermis is a stratified squamous epithelium (p. 153) consisting usually of several layers of cells. The basal part is made up of cells that undergo frequent mitosis to renew the layers that lie above. As the outer layers of cells are displaced upward by new generations of cells beneath, an exceedingly tough, fibrous protein called **keratin** accumulates in the interior of the cells. Gradually, keratin replaces all metabolically active cytoplasm. The cell dies and is eventually shed, lifeless and scalelike. Such is the origin of dandruff as well as a significant fraction of household dust. This process is called **keratinization,** and the cell, thus transformed, is said to be cornified. Cornified cells, highly resistant to abrasion and water diffusion, comprise the outermost **stratum corneum.** This epidermal layer becomes especially thick in areas exposed to persistent pressure or friction, such as calluses and the human palms and soles.

The **dermis,** as already mentioned, mainly serves a supportive role for the epidermis. Nevertheless, true bony structures, where they occur in the integument, are always dermal derivatives. Heavy bony plates were common in primitive ostracoderm and placoderm fishes of the Paleozoic era and persist in some living fishes such as sturgeons. Scales of contemporary fishes are bony dermal structures that have evolved from the bony armor of the Paleozoic fishes but are much smaller and more flexible. Although of dermal origin, fish scales are intimately associated with the thin, overlying epidermis; in some species the scales protrude through the epidermis, but typically the epidermis forms a continuous sheath that is reflected under the overlapping scales (Figure 6-11). Dermal bone also forms some of the flat bones of the skull and gives rise to antlers, which are outgrowths of dermal frontal bone.

Injurious effects of sunlight

The familiar vulnerability of the human skin to sunburn reminds us of the potentially damaging effects of ultraviolet radiation on cells. Many animals, such as protozoa and flatworms, are damaged or killed by ultraviolet radiation if exposed to the sun in shallow water. Most land animals are protected from such damage by the screening action of special body coverings, for example, the cuticle of arthropods, the scales of reptiles, and the feathers and fur of birds and mammals. Humans, however, lack such special coverings and must depend on thickening of the epidermis (stratum corneum) and on epidermal pigmentation for protection. Most ultraviolet radiation is absorbed in the epidermis but about 10% penetrates into the dermis. Damaged cells in both the epidermis and dermis release histamine and other vasodilator substances that cause blood vessel enlargement in the dermis and the characteristic red coloration of sunburn. Light skins suntan through the formation in the deeper epidermis of the pigment **melanin** and by "pigment darkening," that is, the photooxidative blackening of bleached pigment already present in the epidermis. Regrettably, Caucasians pay dearly for their sun worship. Excessive exposure to the sun accelerates aging and wrinkling of the skin and is directly responsible for approximately 400,000 cases of skin cancer each year in the United States alone.

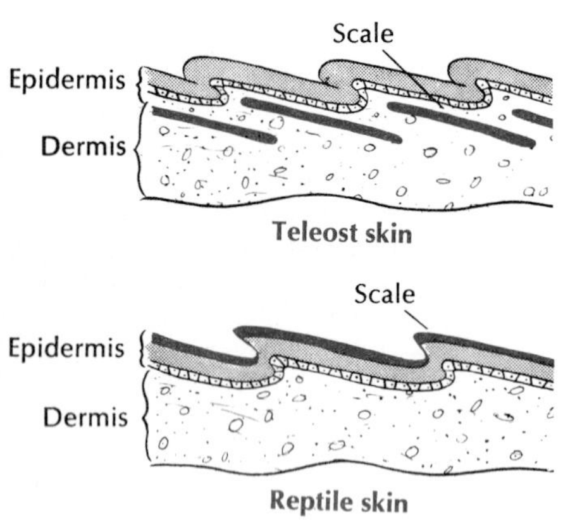

FIGURE 6-11

Integument of bony fishes and reptiles. Bony (teleost) fishes have bony scales from dermis, and reptiles have horny scales from epidermis. Dermal scales of fishes are retained throughout life. Since a new growth ring is added to each scale each year, fishery biologists use scales to tell the age of fishes. Epidermal scales of reptiles are shed periodically.

■ Skeletal Systems

Skeletons are supportive systems that provide rigidity to the body, surfaces for muscle attachment, and protection for vulnerable body organs. The familiar bone of the vertebrate skeleton is only one of several kinds of supportive and connective tissues serving various binding and weight-bearing functions, which are described in this discussion.

HYDROSTATIC SKELETONS

Not all skeletons are rigid; many invertebrate groups use their body fluids as an internal hydrostatic skeleton. The muscles in the body wall of the earthworm, for example, have no firm base for attachment but develop muscular force by contracting against the coelomic fluids, which are enclosed within a limited space and are incompressible, much like the hydraulic brake system of an automobile.

Alternate contractions of the circular and longitudinal muscles of the body wall enable the worm to thin and thicken, setting up backward-moving waves of motion that propel the animal forward (Figure 6-12). Earthworms and other annelids are helped by the septa that separate the body into more or less independent compartments. An obvious advantage is that if a worm is punctured or even cut into pieces, each part can still develop pressure and move. Worms that lack internal compartments, for example, the lugworm *Arenicola,* are rendered helpless if the body fluid is lost through a wound.

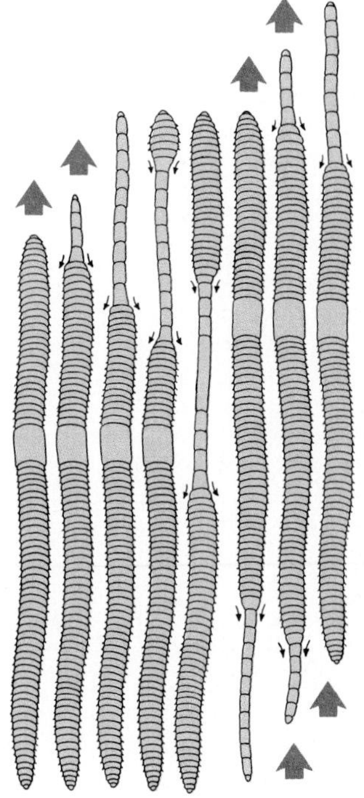

FIGURE 6-12

How an earthworm moves forward. When circular muscles contract, longitudinal muscles are stretched by internal fluid pressure and the worm elongates. Then, by alternate contraction of longitudinal and circular muscles, a wave of contraction passes from anterior to posterior. Bristle-like setae are extended to anchor the animal and prevent slippage.

RIGID SKELETONS

Rigid skeletons differ from hydrostatic skeletons in one fundamental way: rigid skeletons consist of rigid elements, usually jointed, to which muscles can attach. Muscles can only contract; to be lengthened they must be extended by the pull of an antagonistic set of muscles or other forces. Rigid skeletons provide the anchor points required by opposing sets of muscles, such as flexors and extensors.

There are two principal types of rigid skeletons: the **exoskeleton,** typical of molluscs and arthropods, and the **endoskeleton,** characteristic of echinoderms and vertebrates. The invertebrate exoskeleton may be mainly protective, but it may also perform a vital role in locomotion. An exoskeleton may take the form of a shell, as in molluscs, or a calcareous, proteinaceous, or chitinous plate, as in arthropods. It may be rigid, as in molluscs, or jointed and movable, as in arthropods. Unlike the endoskeleton, which grows with the animal, the exoskeleton is often a limiting coat of armor that must be periodically shed (molted) to make way for an enlarged replacement. Some invertebrate exoskeletons, such as the shells of snails and bivalves, grow with the animal.

The vertebrate endoskeleton is formed inside the body and is composed of bone and cartilage, which are forms of dense connective tissue. Bone not only supports and protects but is also the major body reservoir for calcium and phosphorus. In higher vertebrates the red blood cells and certain white blood cells are formed in the bone marrow.

The arthropod-type exoskeleton is perhaps a better arrangement for small animals than a vertebrate-type endoskeleton because a hollow cylindrical tube can support much more weight without collapsing than can a solid cylindrical rod of the same material and weight. Arthropods can thus enjoy both protection and structural support from their exoskeleton. But for larger animals the hollow cylinder would be completely impractical. If made thick enough to support the body weight, it would be too heavy to lift; but if kept thin and light, it would be extremely sensitive to buckling or shattering on impact. Finally, can you imagine the sad plight of a large animal when it shed its exoskeleton to molt?

Notochord and cartilage

The **notochord** (see Figure 25-1, p. 540), the semirigid axial rod of the most primitive chordates (protochordates) and vertebrate larvae and embryos, is also a primitive vertebrate supportive tissue. The notochord is composed of large, vacuolated cells and is surrounded by layers of elastic and fibrous sheaths. It is a stiffening device, preserving body shape during locomotion. Except in the most primitive chordates, for example lampreys, the notochord is surrounded or replaced by the backbone during embryonic development.

Cartilage is the major skeletal element of primitive vertebrates. The jawless fishes (like lampreys) and the sharks, skates, and rays (elasmobranchs) have purely cartilaginous skeletons, which strangely enough is a degenerative feature, since their Paleozoic ancestors had bony skeletons. Higher vertebrates have principally bony skeletons as adults with some cartilage interspersed. Cartilage is a soft, pliable, characteristically deep-lying tissue. Unlike most connective tissues, which are quite variable in form, cartilage is basically the same wherever it is found. The usual form, **hyaline cartilage** (Figure 6-5, *B*) has a clear, glossy appearance. It is composed of cartilage cells (**chondrocytes**) surrounded by firm complex protein gel interlaced with a meshwork of collagenous fibers. Blood vessels are virtually absent, making it almost impossible for a torn cartilage to heal. In addition to forming the cartilaginous skeleton of the primitive vertebrates and that of all vertebrate embryos, hyaline cartilage makes up the articulating surfaces of many bone joints of higher adult vertebrates and the supporting tracheal, laryngeal, and bronchial rings.

Bone

Bone is a living tissue that differs from other connective and supportive tissues by having significant deposits of inorganic calcium salts laid down in an extracellular matrix. Its structural organization is such that bone has nearly the tensile strength of cast iron, yet is only one third as heavy.

Bone is never formed in vacant space but is always laid down by replacement in areas occupied by some form of connecting tissue. Most bones develop from cartilage and are called **endochondral** ("within-cartilage") or **replacement bones.** The embryonic cartilage is gradually eroded, leaving it extensively honeycombed; bone-forming cells then invade these areas and begin depositing calcium salts around the strandlike remnants of the cartilage. A second type of bone is **membranous bone,** which develops directly from sheets of embryonic cells. In higher vertebrates, membranous bone is restricted to bones of the face and cranium; the remainder of the skeleton is endochondral bone. Whatever its embryonic origin, once bone is fully formed there is no difference in the histological structure; endochondral and membranous bone look the same.

Fully formed bone, however, may vary in density. **Spongy** (or cancellous) bone consists of an open, interlacing framework of bony tissue, oriented to give maximum strength under the normal stresses and strains that the bone receives. All bone develops first as spongy bone, but some bones, through further deposition of bone salts, become **compact.** Compact bone is dense, appearing solid to the unaided eye. Both spongy and compact bone are found in the typical long bones of the body (Figure 6-13).

Microscopic structure of bone. Compact bone is composed of a calcified bone matrix arranged in concentric rings. The rings contain cavities (**lacunae**) filled with bone cells (**osteocytes**) that are interconnected by many minute passages (**canaliculi**). These serve to distribute nutrients throughout the bone. This entire organization of lacunae and canaliculi is arranged into an elongated cyl-

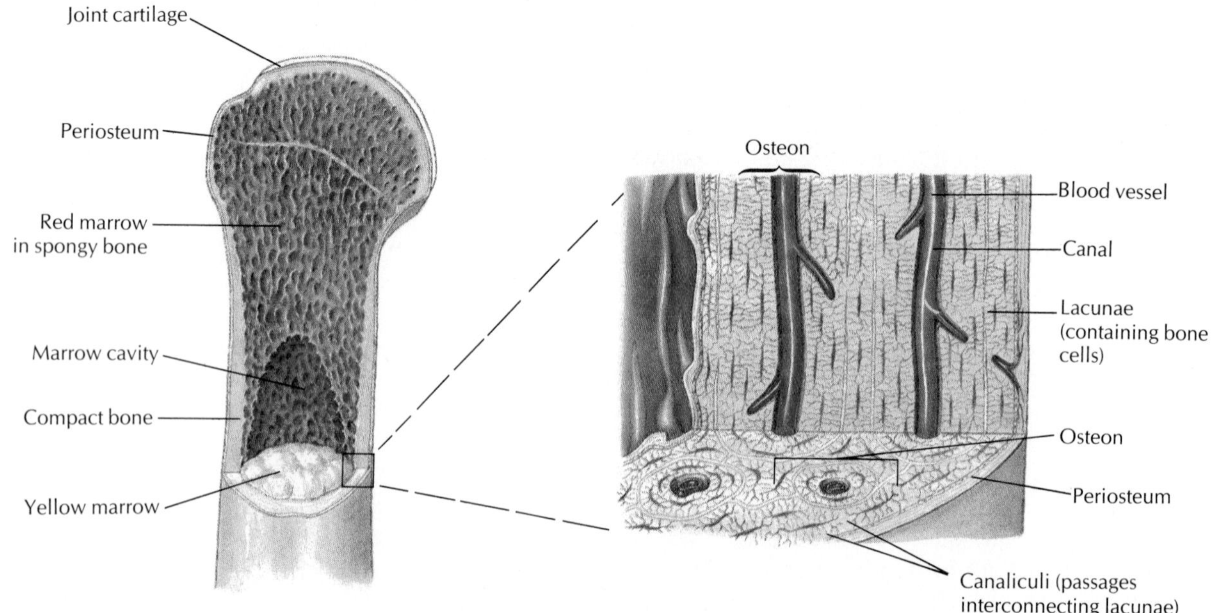

FIGURE 6-13

Structure of bone, showing the appearance of spongy and compact bone. The enlarged section shows how bone cells and the dense calcified matrix are arranged into units called osteons. Bone cells are entrapped within cell-like lacunae but receive nutrients from the circulatory system via tiny canaliculi that interlace the calcified matrix. Bone cells are known as osteoblasts when they are building bone, but, in mature bone shown here, they become resting osteocytes. Bone is covered with compact connective tissue called periosteum.

inder called an **osteon** (also called **haversian canals**) (Figure 6-13). Bone consists of bundles of osteons cemented together and interconnected with blood vessels and nerves. Because of blood vessels and nerves throughout bone, it is living tissue, although nonliving "ground substance" predominates. As a result of its living state, bone breaks can heal and bone diseases can be as painful as any other tissue disease.

Bone growth is a complex restructuring process, involving both its destruction internally by bone-resorbing cells (**osteoclasts**) and its deposition externally by bone-building cells (**osteoblasts**). Both processes occur simultaneously so that the marrow cavity inside grows larger by bone resorption while new bone is laid down outside by bone deposition. Bone growth responds to several hormones, in particular **parathyroid hormone** from the parathyroid gland, which stimulates bone resorption, and **calcitonin** from the thyroid gland, which inhibits bone resorption. These two hormones, together with a derivative of vitamin D, are responsible for maintaining a constant level of calcium in the blood (p. 273).

Following menopause, a woman loses 5% to 6% of her bone mass annually, often leading to the disease osteoporosis and increasing the risk of bone fractures. Dietary supplementation with calcium has been advocated to prevent such losses, but even large doses of calcium alone have little effect in slowing demineralization unless accompanied by therapy with the female sex hormone estrogen (because ovarian production of estrogen drops significantly after menopause). Among animals, only humans, especially females, are troubled with osteoporosis, perhaps a consequence of the long postreproductive life of the human species.

PLAN OF THE VERTEBRATE SKELETON

The vertebrate skeleton is composed of two main divisions: the **axial skeleton,** which includes the skull, vertebral column, sternum, and ribs, and the **appendicular skeleton,** which includes the limbs (or fins or wings) and the pectoral and pelvic girdles (Figure 6-14). Not surprisingly, the skeleton has undergone extensive remodeling in the course of vertebrate evolution. The move from water

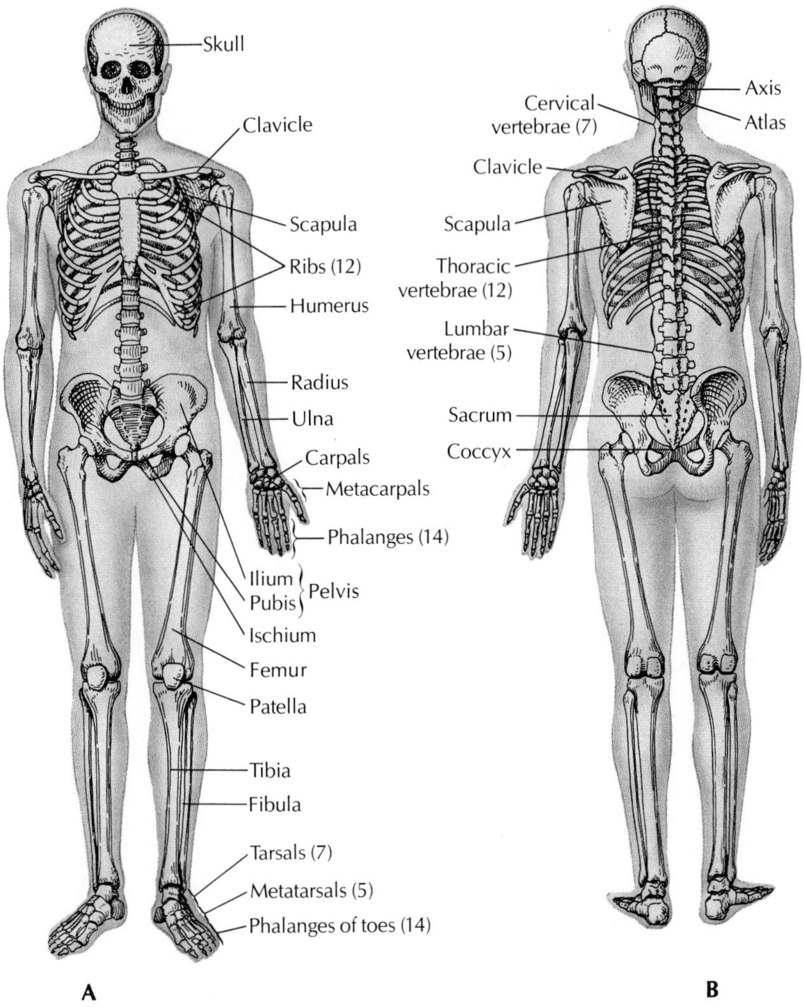

FIGURE 6-14

Human skeleton. **A,** Ventral view. **B,** Dorsal view. Numbers in parentheses indicate number of bones in that unit. In comparison with other mammals, the human skeleton is a patchwork of primitive and specialized parts. Erect posture, brought about by specialized changes in legs and pelvis, enabled the primitive arrangement of arms and hands (arboreal adaptation of human ancestors) to be used for manipulation of tools. Development of the skull and brain followed as a consequence of the premium natural selection put on dexterity and ability to appraise the environment.

to land forced dramatic changes in body form. With cephalization, that is, the concentration of brain, sense organs, and food-gathering and respiratory apparatus in the head, the skull became the most intricate portion of the skeleton. The lower vertebrates have a larger number of skull bones than the more advanced vertebrates. Some fish have 180 skull bones (a source of frustration to paleontologists); amphibians and reptiles, 50 to 95; and mammals, 35 or fewer. Humans have 29, the result of the loss of a number of bones and the fusion of others.

The vertebral column is the main stiffening axis of the postcranial skeleton. In fishes it serves much the same function as the notochord from which it is partially derived; that is, it provides points for muscle attachment and prevents telescoping of the body during muscle contraction. Since fish musculature is similar throughout the trunk and tail, fish vertebrae are differentiated only into trunk and caudal vertebrae.

With the evolution of the amphibious and terrestrial tetrapods (four-legged vertebrates), and no longer buoyed by the aquatic environment, the vertebral column became structurally adapted to withstand new regional stresses transmitted to the column by the two pairs of appendages. In the higher tetrapods, the vertebrae are differentiated into **cervical** (neck), **thoracic** (chest), **lumbar** (back), **sacral** (pelvic), and **caudal** (tail) vertebrae. In birds and also in humans the caudal vertebrae are reduced in number and size, and the sacral vertebrae are fused. The number of vertebrae varies among the different animals. The pythons seem to lead the list with over 400. In humans (Figure 6-14) there are 33 in the child, but in the adult 5 are fused to form the **sacrum** and 4 to form the **coccyx.** Besides the sacrum and coccyx, humans have 7 cervical, 12 thoracic, and 5 lumbar vertebrae. The number of cervical vertebrae (7) is constant in nearly all mammals, whether the neck is short as in dolphins, or long as in giraffes.

The first two cervical vertebrae, the **atlas** and the **axis,** are modified to support the skull and permit pivotal movements. The atlas bears the globe of the head much as the mythological Atlas bore the earth on his shoulders. The axis, the second vertebra, permits the head to turn from side to side.

Ribs are long or short skeletal structures that articulate medially with vertebrae and extend into the body wall. Primitive forms have a pair of ribs for every vertebra; they serve as stiffening elements in the connective tissue septa that separate the muscle segments and thus improve the effectiveness of the muscle contractions. Many fishes have both dorsal and ventral ribs, and some have numerous riblike intermuscular bones as well—all of which increase the difficulty and reduce the pleasure of our eating certain kinds of fish. Higher vertebrates have a reduced number of ribs, and some, such as frogs and toads, have no ribs at all. Others, such as elasmobranchs and some amphibians, have very short ribs. Primates except humans have 13 pairs of ribs; humans have 12 pairs, although approximately 1 person in 20 has a thirteenth pair. In mammals the ribs together form the thoracic basket, which supports the chest wall and prevents collapse of the lungs.

Turning now to the appendicular skeleton, we find that most vertebrates, fishes included, have paired appendages. All fishes except the agnathans have thin pectoral and pelvic fins that are supported by the pectoral and pelvic girdles, respectively. Forms above the fishes (except caecilians, snakes, and limbless lizards) have two pairs of **pentadactyl** (five-toed) limbs, also supported by girdles. The pentadactyl limb is similar in all tetrapods, alive and extinct; even when highly modified for various modes of life, the elements are rather easily homologized (p. 101).

Modifications of the basic pentadactyl limb for life in different environments involve the distal elements much more frequently than the proximal, and it is far more common for bones to be lost or fused than for new ones to be added. Horses and their relatives developed a foot structure for fleetness by elongation of the third toe. In effect, a horse stands on its third fingernail (hoof), much like a ballet dancer standing on the tips of the toes. The bird wing is a good example of distal modification. The bird embryo bears 13 distinct wrist and hand bones (carpals and metacarpals), which are reduced to 3 in the adult. Most of the finger bones (phalanges) are lost, leaving 4 bones in 3 digits (see p. 613). The proximal bones (humerus, radius, and ulna), however, are little modified in the bird wing.

In nearly all tetrapods the pelvic girdle is firmly attached to the axial skeleton, since the greatest locomotory forces transmitted to the body come from the hind limbs. The pectoral girdle, however, is much more loosely attached to the axial skeleton, providing the forelimbs with greater freedom for manipulative movements.

In humans the pectoral girdle is made up of 2 scapulae and 2 clavicles; the arm is made up of the humerus, ulna, radius, 8 carpals, 5 metacarpals, and 14 phalanges. The pelvic girdle consists of 3 fused bones — ilium, ischium, and pubis; the leg is made up of the femur, patella, tibia, fibula, 7 tarsals, 5 metatarsals, and 14 phalanges. Each bone of the leg has its counterpart in the arm with the exception of the patella. This correspondence in structure between anterior and posterior parts of the same individual is an example of **serial homology.**

■ Animal Movement

Movement is a distinctive characteristic of animals. Plants may show movement, but this usually results from changes in turgor pressure or growth rather than from specialized contractile proteins as in animals. Animal movement occurs in many forms in animal tissues, ranging from barely discernible streaming of cytoplasm, the swelling of mitochondria, and movement of chromosomes on the mitotic spindle during cell division, to sudden movements of powerful striated muscles of vertebrates. It has become evident that virtually all animal movement depends on a single fundamental mechanism: **contractile proteins,** which can change their form to elongate or contract. This contractile machinery is always composed of ultrafine fibrils — fine filaments, striated fibrils, or tubular fibrils (microtubules) — arranged to contract when powered by **ATP.** By far the most important protein contractile system is the **actomyosin system,** composed of two proteins, **actin** and **myosin.** This is an almost universal biomechanical system found from protozoa to vertebrates; it performs a long list of diverse functional roles. In this discussion we examine the three principal kinds of animal movement: ameboid, ciliary, and muscular.

AMEBOID MOVEMENT

Ameboid movement is a form of movement especially characteristic of amebas and other protozoa; it is also found in many wandering cells of higher animals, such as white blood cells, certain undifferentiated embryonic tissue (mesenchyme), and numerous other mobile cells that move through the tissue spaces. Ameboid cells change their shape by sending out and withdrawing **pseudopodia** (false feet) from any point on the cell surface. Such cells are surrounded by a delicate, highly flexible **plasma membrane,** also called **plasmalemma** (see Figure 15-1, p. 350). Beneath this lies a nongranular layer, the gel-like **ectoplasm,** which encloses the more liquid **endoplasm.**

Optical studies of an ameba in movement suggest that the outer gel-like layer of ectoplasm surrounds a rather fluid core of endoplasm. As the pseudopod extends, the inner endoplasm fountains out to the periphery, changing from the sol (liquid) to the gel state. The newly formed ectoplasm then slips posteriorly under the plasma membrane and is changed to endoplasm at the rear to begin another cycle. Although no completely satisfactory analysis exists, it seems certain that ameboid movement as just described is based on the same fundamental contractile system that powers vertebrate muscles: an actomyosin mechanism driven by ATP.

CILIARY AND FLAGELLARY MOVEMENT

Cilia are minute hairlike motile processes that extend from the surfaces of the cells of many animals. They are a particularly distinctive feature of ciliate protozoa, but, except for the nematodes in which motile cilia are absent and the arthropods in which they are rare, cilia are found in all major groups of animals. Cilia perform many roles either in moving small animals such as protozoa through their aquatic environment or in propelling fluids and materials across the epithelial surfaces of larger animals.

Cilia are of remarkably uniform diameter (0.2 to 0.5 μm) wherever they are found. The electron microscope has shown that each cilium contains a peripheral circle of nine double microtubules and an additional two microtubules in the center (see Figure 15-3, p. 351). (Exceptions to the 9 + 2 arrangement have been noted; sperm tails of flatworms have but one central microtubule.) A **flagellum** is a whiplike structure longer than a cilium and usually present singly or in small numbers at one end of a cell. They are found in flagellate protozoa, in animal spermatozoa, and in sponges. The main difference between a cilium and a flagellum is in their beating pattern rather than in their structure, since both look alike internally. A flagellum beats symmetrically with snakelike undulations so that the water is propelled parallel to the long axis of the flagellum. A cilium, in contrast, beats asymmetrically with a fast power stroke in one direction followed by a slow recovery during which the cilium bends as it returns to its original position. The water is propelled in direction parallel to the ciliated surface (Figure 6-15).

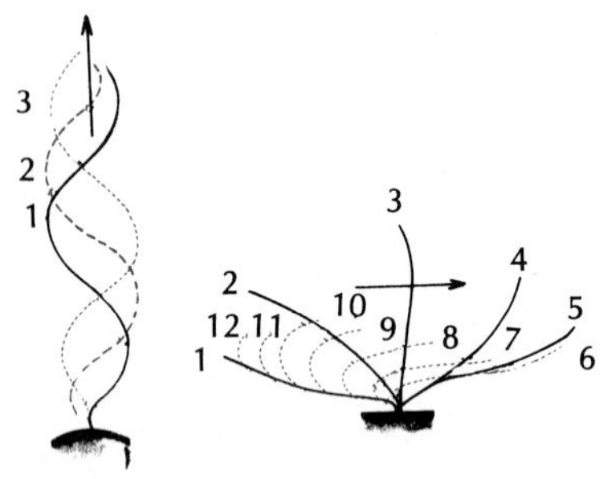

FIGURE 6-15

Flagellum beats in wavelike undulations, propelling water parallel to the main axis of the flagellum. Cilium propels water in direction parallel to the cell surface.

Modified from Sleigh, M.A. 1962. The biology of cilia and flagella. Copyright by Pergamon Press (Oxford).

According to the currently favored hypothesis of ciliary movement, the microtubules behave as "sliding filaments" that move past one another much like the sliding filaments of vertebrate skeletal muscle that is described in the next discussion. During ciliary flexion, microtubules on the concave side slide outward past microtubules on the convex side to increase curvature of the cilium; during the recovery stroke microtubules on the opposite side slide outward to bring the cilium back to its starting position. For such a system to work, the microtubules must be interconnected by molecular bridges, which, in fact, can be seen with the electron microscope.

MUSCULAR MOVEMENT

Contractile tissue is most highly developed in muscle cells, often called muscle **fibers.** Although muscle fibers themselves can only do work by contraction and cannot actively lengthen, they can be arranged in so many different configurations and combinations that almost any movement is possible.

The color of vertebrate muscle—red and white meat of chicken and fish, for example—is related to the speed of muscle contraction. Red muscle contains mostly slowly contracting fibers used for sustained locomotion, has a good blood supply and abundant mitochondria, and functions by aerobic metabolism. White meat, for example the bulk of fish muscle, contains rapidly contracting fibers and is designed for short bursts of vigorous activity; it works by anaerobic metabolism. Color is not always a reliable indicator of function, however. Many muscles contain both red and white fibers (e.g., pigeon breast muscle), and some, such as the golden flesh of salmon, are related to the fish's diet and have little to do with speed of contraction.

Types of vertebrate muscle

As discussed earlier in this chapter (p. 154), vertebrate muscle is broadly classified on the basis of the appearance of muscle cells (fibers) when viewed with a light microscope. **Striated muscle** appears transversely striped (striated), with alternating dark and light bands (Figure 6-16, *A*). We can recognize two types of striated muscle: **skeletal** and **cardiac muscle.** A third kind of vertebrate muscle is **smooth** (or visceral) **muscle,** which lacks the characteristic alternating bands of the striated type (Figure 6-16).

Skeletal muscle is typically organized into sturdy, compact bundles or bands. It is called skeletal muscle because it is attached to skeletal elements and is responsible for movements of the trunk, appendages, respiratory organs, eyes, mouthparts, and so on. Skeletal muscle fibers are extremely long (centimeters in length rather than micrometers), cylindrical, multinucleate cells that may reach from one end of the muscle to the other. They are packed into bundles called **fascicles** (L. *fasciculus,* small bundle), which are enclosed by tough connective tissue. The fascicles are in turn grouped into a discrete **muscle** surrounded by a thin connective tissue layer. Most skeletal muscles taper at their ends, where they connect by connective tissue bindings, called **tendons,** to bones. Other muscles, such as the ventral abdominal muscles, are flattened sheets.

In most fishes, amphibians, and to some extent reptiles, there is a segmented organization of muscles alternating with the vertebrae. Skeletal muscles of higher vertebrates, by splitting, fusion, and shifting, have developed into specialized muscles best suited for manipulating the jointed appendages that have evolved for locomotion on land. Skeletal muscle contracts powerfully and quickly but fatigues more rapidly than does smooth muscle. Skeletal muscle is sometimes called **voluntary muscle** because it is under conscious cerebral control.

Smooth muscle lacks the striations typical of skeletal muscle (Figure 6-16, *B*). The cells are long, tapering strands, each containing a single nucleus. Smooth muscle cells are organized into sheets of muscle circling the walls of the alimentary canal, blood vessels, respiratory passages, and urinary and genital ducts. Smooth muscle is typically slow acting and can maintain prolonged contractions with very little energy expenditure. Its contractions are involuntary and unconscious. The principal functions of smooth muscles are to push the material in a tube, such as the intestine, along its way by active contractions or to regulate the diameter of a tube, such as a blood vessel, by sustained contraction.

Cardiac muscle, the seemingly tireless muscle of the vertebrate heart, combines certain characteristics of both skeletal and smooth muscle (Figure 6-16, *C*). It is fast acting and striated like skeletal muscle, but contraction is under involuntary control like smooth muscle. Actually the nerves serving the heart can only speed up or slow down the rate of contraction; the heartbeat originates within specialized cardiac muscle, and the heart continues to beat even after all nerves are severed. Until recently, cardiac muscle was believed to be **syncytial** (a tissue with many nuclei not separated into discrete cells by cell membranes) with branching, interconnected fibers. Histologists, their understanding vastly increased by the electron microscope, now consider cardiac muscle to be composed of closely opposed but separate, uninucleate cell fibers.

Types of invertebrate muscle

Smooth and striated muscles are also characteristic of invertebrate animals, but there are many variations of both types and even instances in which the structural and functional features of vertebrate smooth and striated muscle are combined. Striated muscle appears in invertebrate groups as diverse as the primitive cnidarians and the advanced arthropods. The thickest muscle fibers known, approximately 3 mm in diameter and 6 cm long and easily seen with the unaided eye, are those of giant barnacles and of Alaska king crabs living along the Pacific coast of North America. These cells are so large that they can easily be penetrated with electrodes or micropipettes for physiological studies and are understandably popular with muscle physiologists.

It is not possible in this short space to describe adequately the tremendous diversity of muscle structure and function in the vast assemblage of invertebrates. We will mention only two functional extremes.

Bivalve molluscan muscles contain fibers of two types. One kind is striated muscle that can contract rapidly, enabling the bivalve to snap shut its valves when disturbed. Scallops use these "fast" muscle fibers to swim in their awkward manner. The second muscle type is smooth muscle, capable of slow, long-lasting contractions. Using these fibers, a bivalve can keep its valves tightly shut for hours or even days. Obviously these are no ordinary muscle fibers! It has been discovered that such adductor muscles use very little metabolic energy and receive remarkably few nerve impulses to maintain the activated state. The contracted state has been likened to a "catch mechanism" involving some kind of stable cross-linkage between the contractile proteins within the fiber. However, despite considerable research, there is still much uncertainty about how this adductor mechanism works.

A

B

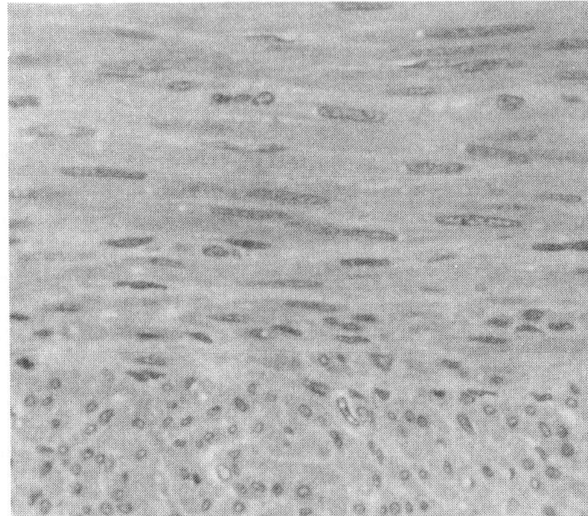

C

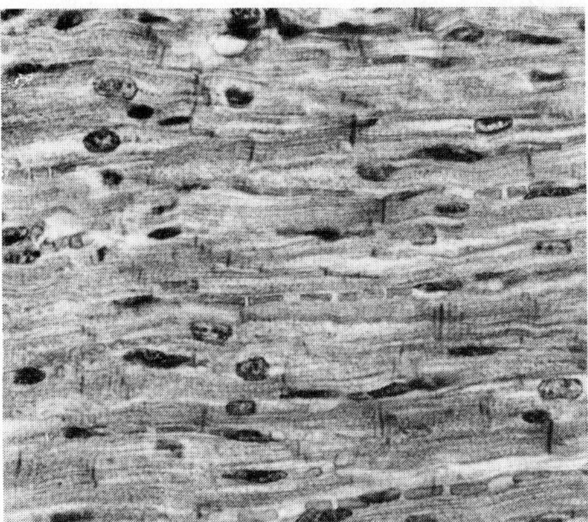

FIGURE 6-16

Photomicrographs of types of vertebrate muscle. **A,** Skeletal muscle (human) showing several striated fibers (cells) lying side by side. **B,** Smooth muscle (human) showing absence of striations. Note elongate nuclei in the long fibers. **C,** Cardiac muscle (monkey). Note the vertical bars, called intercalated discs, joining separate fibers end to end.

A, Courtesy J.W. Bamberger, B and C, Courtesy Carolina Biological Supply Co.

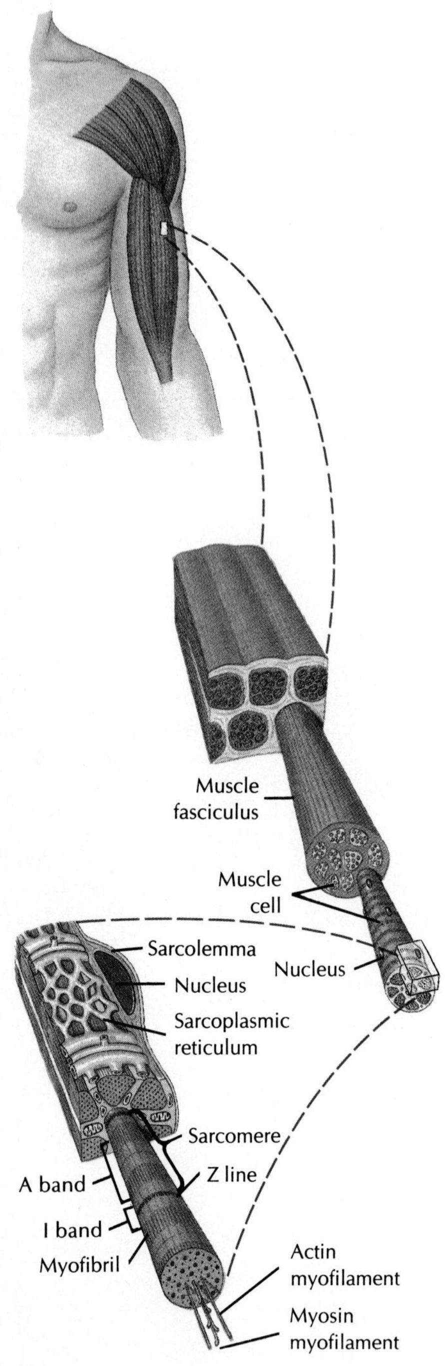

FIGURE 6-17

Organization of skeletal muscle from gross to molecular level. A skeletal muscle *(top)* is composed of thousands of multinucleated muscle fibers *(center)* with alternating A and I bands. Each muscle fiber is composed of thousands of myofibrils *(bottom)*. Each myofibril contains numerous thick (myosin) and thin (actin) filaments which interact to slide past each other during contraction to shorten the muscle. The sarcoplasmic reticulum is a network of tubules surrounding the myofibrils that serves as a communication system for carrying a depolarization to the filaments within the muscle fiber.

Insect flight muscles are virtually the functional antithesis of the slow, holding muscles of bivalves. The wings of some of the small flies operate at frequencies greater than 1000 beats per second. The so-called **fibrillar muscle,** which contracts at these incredible frequencies—far greater than even the most active of vertebrate muscles—shows unique characteristics. It has very limited extensibility; that is, the wing leverage system is arranged so that the muscles shorten only slightly during each downbeat of the wings. Furthermore, the muscles and wings operate as a rapidly oscillating system in an elastic thorax (Figure 22-33, p. 490). Since the muscles rebound elastically and are activated by stretch during flight, they receive impulses only periodically rather than one impulse per contraction; one reinforcement impulse for every 20 or 30 contractions is enough to keep the system active.

Structure of striated muscle

As we earlier pointed out, striated muscle is so named because of the periodic bands, plainly visible under the light microscope, that pass across the widths of the muscle cells. Each cell, or **fiber,** contains numerous **fibrils** (also called **myofibrils**) packed together and invested by the cell membrane, the **sarcolemma** (Figure 6-17). Also present in each fiber are several hundred nuclei usually located along the edge of the fiber, numerous mitochondria, a network of tubules called the **sarcoplasmic reticulum** (to be discussed later), and other cell inclusions typical of any living cell. Most of the fiber, however, is packed with the unique fibrils each 1 to 2 μm in diameter.

The characteristic banding of the muscle fiber represents the fine structure of the fibrils that make up the fiber. In the resting fiber are alternating light-staining and dark-staining bands called the **I bands** and **A bands,** respectively (Figure 6-17). The functional contractile unit of the fibril, the **sarcomere,** extends between successive **Z** lines. The fibril is actually an aggregate of much smaller parallel units called **filaments** (or **myofilaments).** These are of two kinds—thick ones, 11 nm in diameter composed of the protein **myosin,** and thin ones, 5 nm in diameter composed of the protein **actin** (Figure 6-17). These are the actual contractile proteins of muscle. The thick myosin filaments are confined to the A-band region. The thin actin filaments are located mainly in the light I bands but extend some distance into the the A band as well. In the relaxed muscle, they do not quite meet in the center of the A band. The **Z** line is a dense protein different from either actin or myosin, which serves as the attachment plane for the thin filaments and keeps them in register. These relationships are diagramed in Figure 6-17.

Sliding filament model of muscle contraction

In the 1950s the English physiologists A.F. Huxley and H.E. Huxley independently proposed the **sliding filament model** to explain striated muscle contraction. According to this model, the thick and thin filaments become linked together by molecular cross bridges, which act as levers to pull the filaments past each other. During contraction, the cross bridges on the thick filaments swing rapidly back and forth, alternately attaching to and releasing from special receptor sites on the thin filaments, and drawing the thin filaments past the thick in a kind of ratchet action. As contraction continues, the **Z** lines are pulled closer together (Figure 6-18). Thus the sarcomere shortens. Because all of the sarcomere units shorten together, the muscle contracts.

At the time the sliding filament model was proposed, the nature of the mechanism that pulled the thin (actin) filaments past the thick (myosin) was largely a mystery. In the intervening years, however, there have been important improvements in biochemical techniques and in the technology of electron microscopy and x-ray diffraction; these advances have permitted much better visualization of the complex architecture of the contractile machinery.

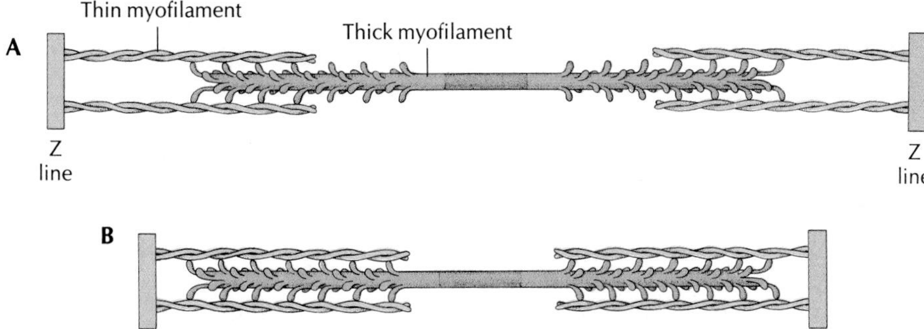

FIGURE 6-18

Sliding myofilament model, showing how thick and thin myofilaments interact during contraction. **A,** Muscle relaxed. **B,** Muscle contracted.

Each myosin molecule, shown in Figure 6-19, *A,* is composed of two linked polypeptide chains, each having a club-shaped "head." Lined up as they are in a bundle to form a thick myofilament (Figure 6-19, *B*), the double heads of each myosin molecule face outward from the center of the filament. These heads act as the molecular cross bridges that interact with the thin filaments during contraction.

The thin myofilaments are more complex because they are composed of three different proteins. The backbone of the thin filament is a double strand of the protein actin, twisted into a double helix. Surrounding the actin filament are two thin strands of another protein, **tropomyosin,** that lie near the grooves between the actin strands. Each tropomyosin strand is itself a double helix as shown in Figure 6-19, *C.*

The third protein of the thin filament is **troponin,** a complex of three globular proteins located at intervals along the filament. Troponin is a calcium-dependent switch that acts as the control point in the contraction process.

For shortening to occur, the cross bridges must attach, swivel, detach, and reattach at a point farther along the thin filament. The attach-pull-release cycle occurs in a series of steps (Figure 6-20). First, myosin binds with and then splits a molecule of ATP. The release of bond energy from ATP activates the myosin head, which attaches to the adjacent actin strand and swings 45 degrees, at the same

Human muscle tissue develops before birth. A newborn child's complement of skeletal muscle fibers is all that it will ever have. But while a weight-lifter and a young girl have a similar number of muscle fibers, the weight-lifter may have several times her strength because repeated high-intensity, short-duration exercise has induced the synthesis of additional actin and myosin filaments. Each fiber has hypertrophied, becoming larger and stronger. Endurance exercise such as long-distance running produces a very different response. Fibers do not become greatly stronger but develop more mitochondria and more myoglobin and become adapted for a high rate of oxidative phosphorylation. These changes, together with the development of more capillaries serving the fibers, lead to increased capacity for long-duration activity.

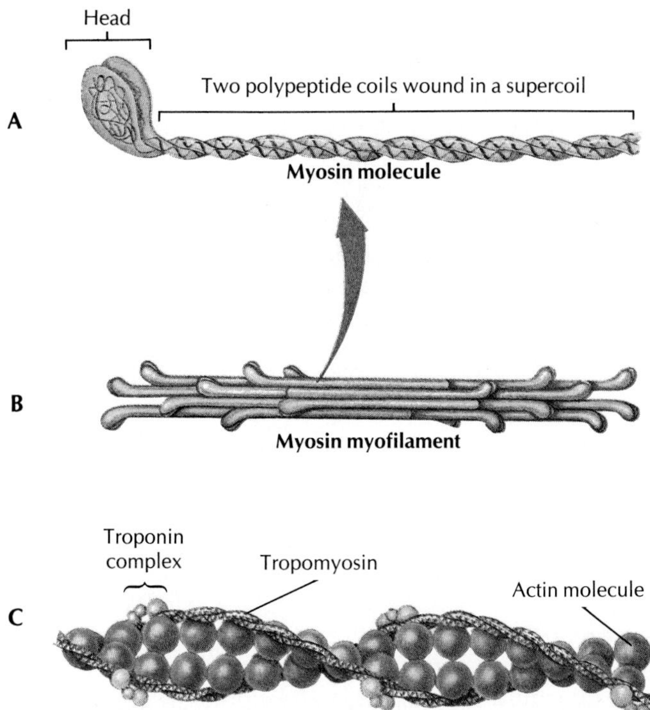

FIGURE 6-19

Molecular structure of thick and thin myofilaments of skeletal muscle. **A,** The myosin molecule is composed of two polypeptides coiled together and expanded at their ends into a globular head. **B,** The thick myofilament is composed of a bundle of myosin molecules with the globular heads extended outward. **C,** The thin myofilament consists of a double strand of actin surrounded by two tropomyosin strands. A globular protein complex, troponin, occurs in pairs at every seventh actin unit. Troponin is a calcium-dependent switch that controls the interaction between actin and myosin.

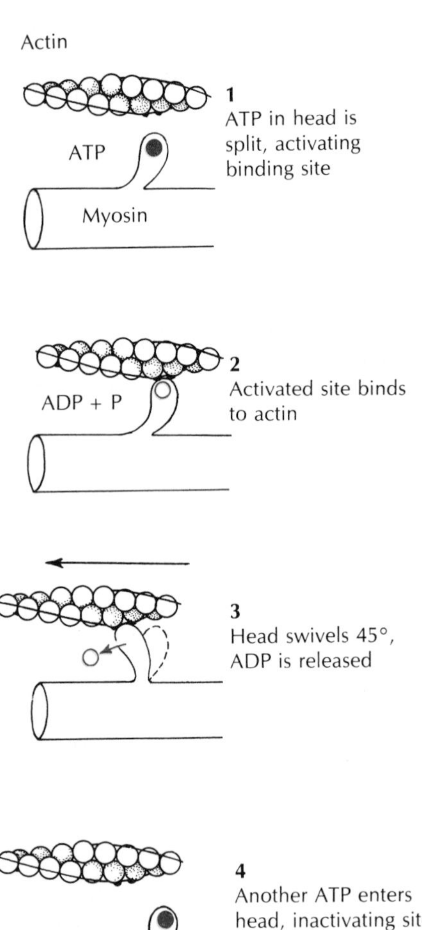

FIGURE 6-20

Successive steps in the attach-pull-release cycle that pulls actin past myosin and shortens the myofilament length.

time releasing the ADP molecule. This is the power stroke that pulls the actin filament a distance of about 10 nm, and it comes to an end when another ATP molecule binds to the myosin head, inactivating the site. Thus each cycle requires the expenditure of energy in the form of ATP.

As long as the muscle is stimulated and new ATP becomes available, the attach-pull-release cycle can repeat again and again, 50 to 100 times per second, pulling the thick and thin myofilaments past each other. While the distance each sarcomere can shorten is very small, this distance is multiplied by the thousands of sarcomeres lying end to end in a muscle fiber. Thus a strongly contracted muscle may shorten as much as one-third its resting length.

Stimulation of contraction

To contract, skeletal muscle must of course be stimulated. If the nerve supply to a muscle is severed, the muscle **atrophies,** or wastes away. Skeletal muscle fibers are grouped into **motor units,** each of which is a small number of muscle fibers innervated by a single motor neuron. As the nerve fiber approaches the muscle fibers within a unit, it splays out into many terminal branches. Each branch attaches to a muscle fiber at a special area called the **myoneural junction** (Figure 6-21). At the junction is a tiny gap, or **synaptic cleft,** that thinly separates a nerve fiber from muscle fiber. In the vicinity of the junction, the neuron stores a chemical, **acetylcholine,** in minute vesicles known as **synaptic vesicles.** Acetylcholine is released when a nerve impulse reaches the junction. The substance is a chemical mediator that diffuses across the narrow junction and acts on the muscle fiber membrane to generate an electrical depolarization. The depolarization spreads rapidly through the muscle fiber, causing it to contract. Thus the synapse is a special chemical bridge that couples together the electrical activities of nerve and muscle fibers.

For a long time physiologists were puzzled as to how the depolarization at the myoneural junction could spread quickly enough through the fiber to cause simultaneous contraction of all the densely packed filaments within. Then it was discovered that vertebrate skeletal muscle contains an elaborate communication system that performs just this function. It consists of infoldings of the sarcolemma, called **junctional folds,** which aid in the depolarization reaching the endoplasmic reticulum (called the **sarcoplasmic reticulum** in muscle). This system of fluid-filled channels runs parallel to the myofilaments and communicates with the sarcolemma that surrounds the fiber (Figure 6-21). The system is ideally arranged for speeding the electrical depolarization from the myoneural junction to the myofilament within.

How does the electrical depolarization activate the contraction process? In the resting, unstimulated muscle, shortening does not occur because the thin tropomyosin strands lie in a position on the actin filament that prevents the myosin heads from attaching to actin. When the muscle is stimulated and the electrical depolarization arrives at the sarcoplasmic reticulum surrounding the fibrils, calcium ions are released. Some of the calcium binds to the control protein troponin. Troponin immediately undergoes conformation changes that allow tropomyosin to move out of the blocking position, and the attach-pull-release cycle is set in motion. Shortening will continue as long as nerve impulses arrive at the neuromuscular junction and free calcium remains available around the myofilament. But when stimulation stops, the calcium is quickly pumped back onto the sarcoplasmic reticulum. Troponin resumes its original shape, tropomyosin moves back into its blocking position on actin, and the muscle relaxes.

Energy for contraction

ATP is the immediate source of energy for muscle, but the amount present will sustain contraction for only a fraction of a second. However, vertebrate muscle

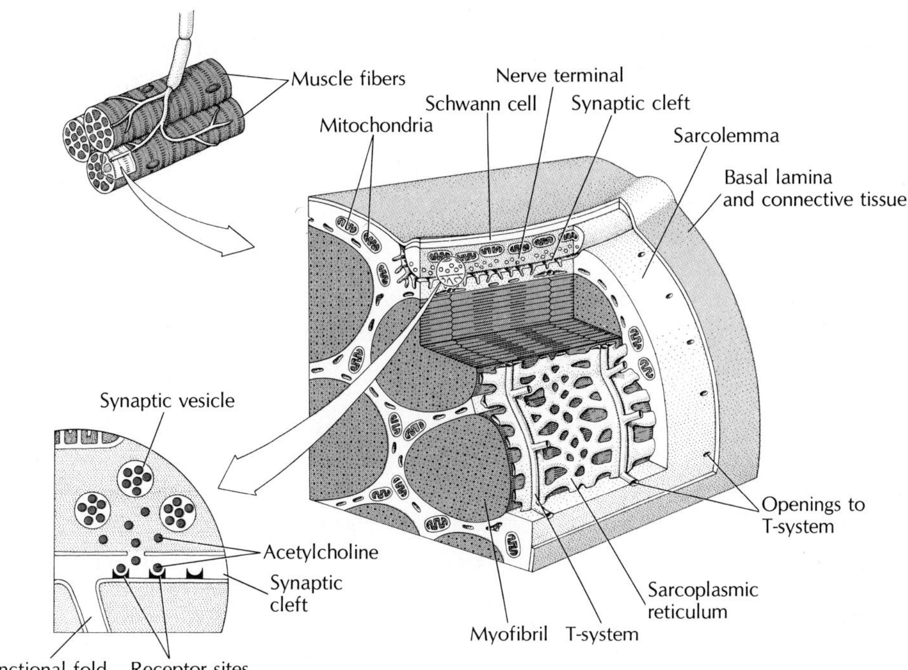

FIGURE 6-21

Section of vertebrate skeletal muscle showing nerve-muscle synapse (myoneural junction), sarcoplasmic reticulum, and connecting transverse tubules (T system). Arrival of a nerve impulse at the synapse triggers the release of acetylcholine into synaptic cleft *(inset at left)*. The binding of transmitter molecules to receptors generates membrane depolarization. This spreads across the sarcolemma, into the T system, and to the sarcoplasmic reticulum where the sudden release of calcium sets in motion the contractile machinery of the myofibril.

contains a much larger reservoir of high-energy phosphate, **creatine phosphate.** This compound contains even more free bond energy than ATP (p. 47) and thus can readily transfer its bond energy to ADP to form ATP.

$$\text{Creatine phosphate} + \text{ADP} \rightleftharpoons \text{ATP} + \text{Creatine}$$

The reserves of creatine phosphate are soon depleted in rapidly contracting muscle and must be restored by the oxidation of carbohydrate. The major store of carbohydrate in muscle is glycogen. In fact, about three fourths of all the glycogen in the body is stored in muscle (most of the rest is stored in the liver). Glycogen can be readily converted into glucose-6-phosphate, the first stage of glycolysis that leads into mitochondrial respiration and the generation of ATP (p. 50).

If muscular contraction is not too vigorous or too prolonged, glucose can be completely oxidized to carbon dioxide and water by **aerobic metabolism.** During prolonged or heavy exercise, however, the blood flow to the muscles, although greatly increased above the resting level, is insufficient to supply oxygen as rapidly as required for the complete oxidation of glucose. When this happens, the contractile machinery receives its energy largely by **anaerobic glycolysis,** a process that does not require oxygen (p. 54). The ability to take advantage of this anaerobic pathway, although not nearly as efficient as the aerobic one, is of great importance; without it, all forms of heavy muscular exertion would be impossible.

During anaerobic glycolysis in animals, glucose is degraded to lactic acid with the release of energy. This is used to resynthesize creatine phosphate, which in turn passes the energy to ADP for the resynthesis of ATP. Lactic acid accumulates in the muscle and diffuses rapidly into the general circulation. If the muscular exertion continues, the buildup of lactic acid causes enzyme inhibition and fatigue. Thus, the anaerobic pathway is a self-limiting one, since continued heavy exertion leads to exhaustion. The muscles incur an **oxygen debt** because the accumulated lactic acid must be oxidized by extra oxygen. After the period of exertion, oxygen consumption remains elevated until all of the lactic acid has been oxidized or resynthesized to glycogen.

Creatine phosphate, the high-energy compound of vertebrates that regenerates ATP during muscle contraction, is replaced in most invertebrates by arginine phosphate. Neither creatine phosphate nor arginine phosphate is stored in quantities sufficient to support contractions for more than a few seconds; sustained contractions depend ultimately on the oxidation of fuels: carbohydrates and fats.

Summary

From the relatively simple organisms that make up the beginnings of life on earth, animal evolution has progressed through a history of ever more intricately organized forms. Adaptive modifications for different niche requirements within a phyletic line of ancestry, however, have been constrained by the ancestral pattern of body architecture.

Organelles become integrated into cells, cells into tissues, tissues into organs, and organs into systems. Whereas a unicellular animal carries out all life functions within the confines of a single cell, an advanced multicellular animal is an organization of subordinate units that are united at successive levels. One correlate of increased body complexity is an increase in body size, which offers certain advantages such as more effective predation, reduced energy cost of locomotion, and improved homeostasis.

The metazoan animal body consists of cells, most of which are functionally specialized; body fluids, divided into the intracellular and extracellular fluid compartments; and the extracellular structural elements, which are fibrous or formless elements that serve various structural functions in the extracellular space. The cells of metazoans develop into various tissues made up of similar cells performing common functions. The basic tissue types are nervous, connective, epithelial, and muscular. Tissues are organized into larger functional units called organs, and organs are associated to form systems.

An animal is wrapped in a protective covering, the integument, which may be as simple as the delicate plasma membrane of a protozoan or as complex as the skin of a mammal. The arthropod exoskeleton is the most complex of invertebrate integuments, consisting of a two-layered cuticle secreted by a single-layered epidermis. It may be hardened by calcification or sclerotization and must be molted at intervals to permit body growth. Vertebrate integument consists of two layers: the epidermis, which gives rise to various derivatives such as hair, feathers, and claws; and the dermis, which supports and nourishes the epidermis. It also is the origin of bony derivatives such as fish scales and the antlers of deer.

Skeletons are supportive systems that may be hydrostatic or rigid. The hydrostatic skeletons of several soft-walled invertebrate groups depend on body wall muscles that contract against a noncompressible internal fluid of constant volume. Rigid skeletons have evolved with attached muscles that act with the supportive skeleton to produce movement. In higher animals, two forms of skeleton have appeared. Arthropods have an external skeleton, which must be periodically shed to make way for an enlarged replacement. The vertebrates developed an internal skeleton, a framework formed of cartilage or bone, that can grow with the animal, while, in the case of bone, additionally serving as a reservoir of calcium and phosphate.

Animal movement, whether in the form of cytoplasmic streaming, ameboid movement, or the contraction of an organized muscle mass, depends on specialized contractile proteins. The most important of these is the actomyosin system, which in higher animals is usually organized into elongated thick and thin filaments that slide past one another during contraction. When a muscle is stimulated, an electrical depolarization is conducted into the muscle fibers through the sarcoplasmic reticulum, causing the release of calcium. Calcium binds to a protein troponin complex associated with the thin actin filament. This causes tropomyosin to shift out of its blocking position and allows the myosin heads to cross-bridge with the actin filament. Powered by ATP, the myosin heads swivel back and forth to pull the thick and thin filaments past each other. Phosphate bond energy for contraction is supplied by carbohydrate fuels through a storage intermediate, creatine phosphate.

REVIEW QUESTIONS

1. Name the five levels of organization in animal complexity and explain how each successive level is more advanced than the one preceding it.
2. Can you suggest why, during the evolution of separate animal lineages, there has been a tendency for the maximum body size to increase? Why should it be inevitable that complexity tends to increase along with body size?
3. Body fluids of both vertebrates and invertebrates are separated into fluid "compartments." Name these compartments and explain how compartmentalization may differ in animals having open and closed circulatory systems.
4. What are the major kinds of extracellular structural elements in the metazoan body?
5. What are the four major tissue types in the body of a metazoan? For what function(s) is each tissue type specialized?
6. What is the difference between tissues, organs, and systems?
7. Describe the structure of the arthropod integument, and explain the difference in the way the cuticle is hardened in crustaceans and in insects.
8. Distinguish between epidermis and dermis in vertebrate integument, and describe the structural derivatives of these two layers.
9. Explain how human skin develops protection against the damaging effects of the ultraviolet rays of the sun.
10. Explain what a hydrostatic skeleton is; tell how it helps in movement; and name one invertebrate group in which hydrostatic skeletons are important in locomotion.
11. What is hyaline cartilage? Compare its distribution and function in lower and higher vertebrates.
12. What is the difference between endochondral and membranous bone? Between spongy and compact bone?
13. Discuss the role of osteoclasts, osteoblasts, parathyroid hormone, and calcitonin in bone growth.
14. Name the major skeletal components included in the axial and in the appendicular skeleton.
15. Describe the interaction of endoplasm and ectoplasm in ameboid movement.
16. Compare the structure and function of a cilium with that of a flagellum.
17. Describe the structural and functional features that distinguish each of the three types of vertebrate muscle.
18. What functional features set molluscan smooth muscle and insect fibrillar muscle apart from any known vertebrate muscle?
19. Explain how skeletal muscle shortens according to the sliding filament hypothesis.
20. Describe in sequence the events in muscle stimulation, explaining the role of each of the following: motor units, myoneural junction, acetylcholine, sarcoplasmic reticulum, calcium, troponin, tropomyosin.
21. Describe the immediate and reserve sources of energy for muscle contraction. Under what circumstances is an oxygen debt incurred during muscle contraction?

SELECTED REFERENCES

Alexander, R.M. 1982. Locomotion in animals. New York, Chapman and Hall. *Concise, fully comparative treatment. Introduced with a discussion of "sources of power" followed by treatment of mechanisms and energetics of locomotion on land, in water, and in the air. Undergraduate level.*

Caplan, A.J. 1984. Cartilage. Sci. Am. **251**:84-94 (Oct.). *Structure, aging, and development of vertebrate cartilage.*

Hadley, N.F. 1986. The arthropod cuticle. Sci. Am. **255**:104-112 (July). *Properties of this complex covering that accounts for much of the adaptive success of arthropods.*

Kessel, R.G., and R.H. Kardon. 1979. Tissues and organs: a text-atlas of scanning electron microscopy. San Francisco, W.H. Freeman & Co., Publishers. *Collection of excellent scanning electron micrographs with text.*

McMahon, T.A. 1984. Muscles, reflexes, and locomotion. Princeton, N.J., Princeton University Press. *Comprehensive, ranging from basic muscle mechanics to coordinated motion. Though sprinkled with mathematical models, the text is lucid throughout.*

McMahon, T.A., and J.T. Bonner. 1983. On size and life. New York, Scientific American Books, Inc. *Wide ranging, artfully written, and colorfully produced introductory book on the biology and physics of size.*

Nadel, E.R. 1985. Physiological adaptations to aerobic training. Am. Sci. **73(4)**:334-343 (July-Aug.). *The studies reported here on energy conversion in muscle were crucial to the training of a pilot for the Daedalus project, the successful world record 119-km flight of a human-powered aircraft in April, 1988.* (reported in Am.Sci. **76**, July-Aug., 1988.)

Shipman, P., A. Walker, and D. Bichell. 1985. The human skeleton. Cambridge, Mass., Harvard University Press. *Comprehensive view of the human skeleton.*

Spearman, R.I.C. 1973. The integument: a textbook of skin biology. Cambridge, Mass., Cambridge University Press. *Comparative treatment, embracing both invertebrates and vertebrates.*

Welsch, U., and V. Storch. 1976. Comparative animal cytology and histology. London, Sidgwick & Jackson. *Modern comparative histology with good treatment of invertebrates.*

WILLIAM HARVEY'S DISCOVERY

Ceaselessly, during a human life, the heart pumps blood through the arteries, capillaries, and veins: about 5 liters per minute, until by the end of a normal life the heart has contracted some 2.5 billion times and pumped 300,000 tons of blood. When the heart stops its contractions, life also ends.

The crucial importance of the heart and its contractions for human life has been known since antiquity, probably almost as long as humans have existed. However, the circuit flow of blood, the notion that the heart pumps blood into arteries through the circulation and receives it back in veins, only became known a few hundred years ago. The first correct description of blood flow by the English physician William Harvey initially received vigorous opposition when published in 1628. Centuries before, the Greek anatomist Galen had taught that air enters the heart from the windpipe and that blood was able to pass from one ventricle to the other through "pores" in the interventricular septum. He also believed that blood first flowed out of the heart into all vessels, then returned—a kind of ebb and flow of blood. Even though there was almost nothing correct about this concept, it was still doggedly trusted at the time of Harvey's publication. Harvey's conclusions were based on sound experimental evidence. He used a variety of animals for his experiments and chided human anatomists, saying that if only they had acquainted themselves with the anatomy of the lower vertebrates, they would have understood the blood's circuit. By tying ligatures on arteries, he noticed that the region between the heart and ligature swelled up. When veins were tied off, the swelling occurred beyond the ligature. When blood vessels were cut, blood flowed in arteries from the cut end nearest the heart; the reverse happened in veins. By means of such experiments, Harvey discovered the correct scheme of blood circulation, even though he could not see the capillaries that connected the arterial and venous flows.▪

CHAPTER 7

Internal Fluids

Immunity, Circulation, and Respiration

Single-celled organisms live a contact existence with their environment. Nutrients and oxygen are obtained and wastes are released directly across the cell surface. These animals are so small that no special internal transport system, beyond the normal streaming movements of the cytoplasm, is required. Even some primitive multicellular forms, such as sponges, cnidarians, and flatworms, have such a simple internal organization and low rate of metabolism that no circulatory system is needed. Most of the more advanced multicellular organisms, because of their size, activity, and complexity, require a specialized circulatory, or vascular, system to transport nutrients and respiratory gases to and from all tissues of the body. In addition to serving these primary transport needs, circulatory systems have acquired additional functions; hormones are moved about, finding their way to target organs where they assist the nervous system to integrate body function. Water, electrolytes, and the many other constituents of the body fluids are distributed and exchanged between different organs and tissues. An effective response to disease and injury is vastly accelerated by an efficient circulatory system. Warm-blooded birds and mammals depend heavily on the blood circulation to conserve or dissipate heat as required for the maintenance of constant body temperature.

Internal Fluid Environment

The body fluid of a single-celled animal is the cellular cytoplasm, a liquid-gel substance in which the various membrane systems and organelles of the cell are suspended. In multicellular animals the body fluids are divided into two main phases, the **intracellular** and the **extracellular.** The intracellular phase (also called intracellular fluid) is the collective fluid inside all the body's cells. The extracellular phase (or fluid) is the fluid outside and surrounding the cells (see Figure 7-2, *A*). Thus the cells, the sites of the body's crucial metabolic activities, are bathed by their own aqueous environment, the extracellular fluid that buffers them from the often harsh physical and chemical changes occurring outside the body. The importance of the extracellular fluid was first emphasized by the great French physiologist Claude Bernard (Figure 7-1). In

FIGURE 7-1

French physiologist Claude Bernard (1813-1878), one of the most influential of nineteenth century physiologists. Bernard believed in the constancy of the *milieu intérieur* ("internal environment"), which is the extracellular fluid bathing the cells. He pointed out that it is through the *milieu intérieur* that foods and wastes and gases are exchanged and through which chemical messengers are distributed. He wrote, "The living organism does not really exist in the external environment (the outside air or water) but in the liquid *milieu intérieur* . . . that bathes the tissue elements."

From Fulton, J.F., and L.G. Wilson. 1966. Selected readings in the history of physiology. Springfield, Ill., Charles C Thomas, Publisher.

animals having closed circulatory systems (vertebrates, annelids, and a few other invertebrate groups; see p. 181) the extracellular fluid is further subdivided into **blood plasma** and **interstitial (intercellular) fluid** (Figure 7-2, *A*). The blood plasma is contained within the blood vessels, whereas the interstitial fluid, or tissue fluid as it is sometimes called, occupies the space immediately around the cells. Nutrients and gases passing between the vascular plasma and the cells must traverse this narrow fluid separation. The interstitial fluid is constantly formed from the plasma by filtration through the capillary walls.

COMPOSITION OF THE BODY FLUIDS

All these fluid spaces—plasma, interstitial, and intracellular—differ from each other in solute composition, but all have one feature in common—they are mostly water. Despite their firm appearance, animals are 70% to 90% water. Humans, for example, are approximately 70% water by weight. Of this, 50% is cell water, 15% is interstitial fluid water, and the remaining 5% is in the blood plasma. As Figure 7-2, *A*, shows, it is the plasma space that serves as the pathway of exchange between the cells of the body and the outside world. This exchange of respiratory gases, nutrients, and wastes is accomplished by specialized organs (kidney, lung, gill, alimentary canal), as well as by the integument.

The body fluids contain many inorganic and organic substances in solution. Principal among these are the inorganic electrolytes and proteins. Figure 7-2, *B*, shows that **sodium, chloride,** and **bicarbonate ions** are the chief extracellular electrolytes, whereas **potassium, magnesium,** and **phosphate ions** and **proteins** are the major intracellular electrolytes. These differences are dramatic; they are always maintained despite the continuous flow of materials into and out of the

FIGURE 7-2

Fluid compartments of body. **A,** All body cells can be represented as belonging to a single large fluid compartment that is completely surrounded and protected by extracellular fluid *(milieu intérieur)*. This fluid is further subdivided into plasma and interstitial fluid. All exchanges with the environment occur across the plasma compartment. **B,** Electrolyte composition of extracellular and intracellular fluids. Total equivalent concentration of each major constituent is shown. Equal amounts of anions (negatively charged ions) and cations (positively charged ions) are in each fluid compartment. Note that sodium and chloride, major plasma electrolytes, are virtually absent from intracellular fluid (actually they are present in low concentration). Note the much higher concentration of protein inside the cells.

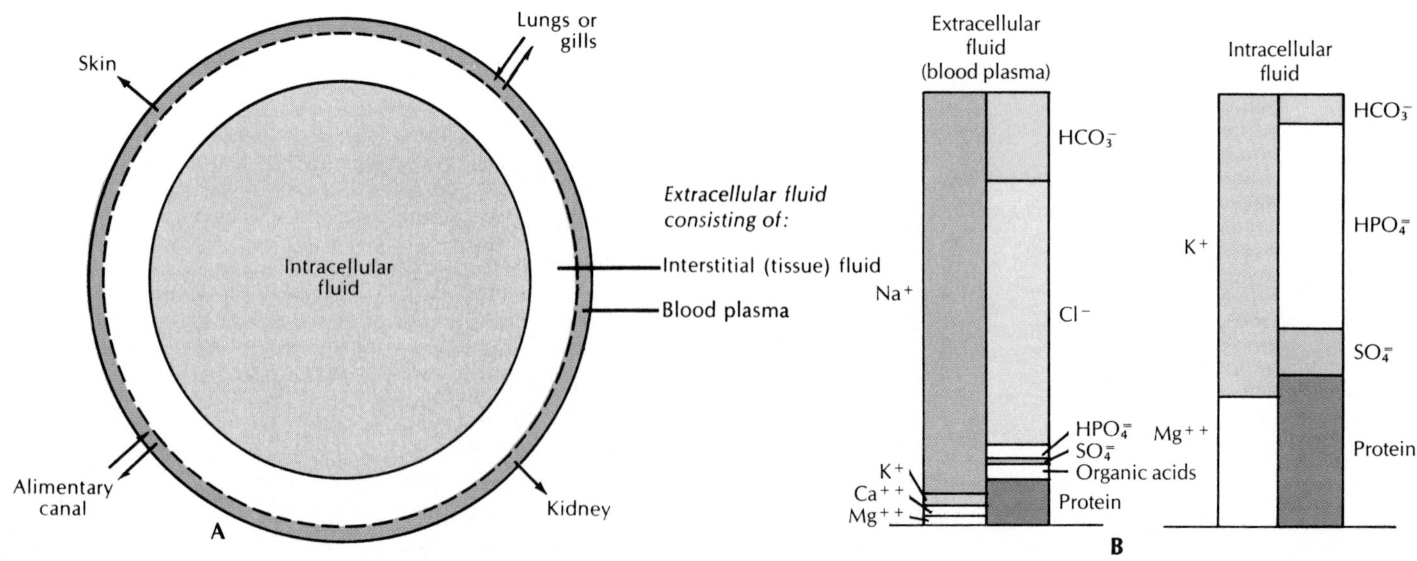

cells of the body. The two subdivisions of the extracellular fluid—plasma and interstitial fluid—have similar compositions except that the plasma has more proteins, which are mostly too large to filter through the capillary wall into the interstitial fluid.

Composition of blood

Among the lower invertebrates that lack a circulatory system (such as flatworms and cnidarians) it is not possible to distinguish a true "blood." These forms possess a clear, watery tissue fluid containing some primitive phagocytic cells, a little protein, and a mixture of salts similar to seawater. The "blood" of higher invertebrates with open circulatory systems is more complex and is often referred to as hemolymph (Gr. *haimo,* blood, + L. *lympha,* water). Invertebrates with closed circulatory systems, on the other hand, maintain a clear separation between blood contained within blood vessels and tissue (interstitial) fluid surrounding the vessels.

In vertebrates, blood is a complex liquid tissue composed of plasma and formed elements, mostly corpuscles, suspended in the plasma. When the red blood corpuscles and other formed elements are separated from the fluid components by centrifugation, the blood is found to be approximately 55% plasma and 45% formed elements.

The composition of mammalian blood is as follows:

Plasma

1. Water 90%
2. Dissolved solids, consisting of the plasma proteins (albumin, globulins, fibrinogen), glucose, amino acids, electrolytes, various enzymes, antibodies, hormones, metabolic wastes, and traces of many other organic and inorganic materials
3. Dissolved gases, especially oxygen, carbon dioxide, and nitrogen

Formed elements (Figure 7-3)

1. Red blood cells (erythrocytes), containing hemoglobin for the transport of oxygen and carbon dioxide
2. White blood cells (leukocytes), serving as scavengers and as defensive cells
3. Cell fragments (platelets in mammals) or cells (thrombocytes in other vertebrates) that function in blood coagulation

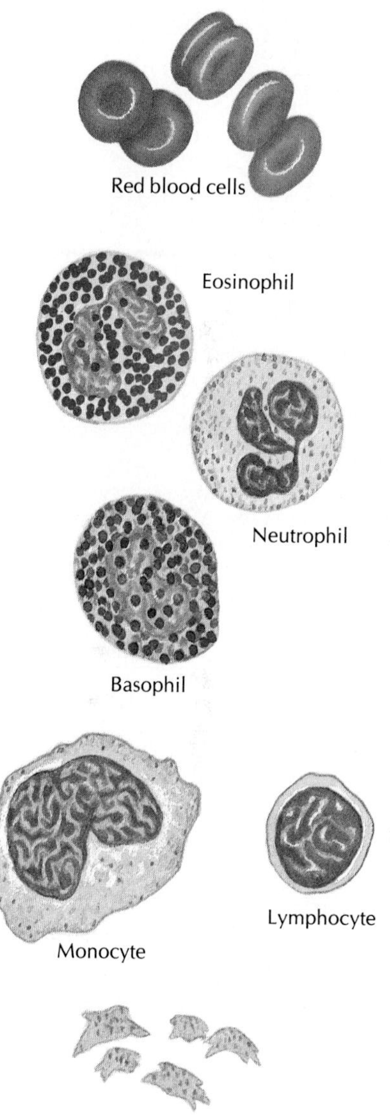

FIGURE 7-3

Formed elements of human blood. Hemoglobin-containing red blood cells of humans and other mammals lack nuclei, but those of all lower vertebrates have nuclei. Various leukocytes provide a wandering system of protection for the body. Platelets participate in the body's clotting mechanisms.

The plasma proteins are a diverse group of large and small proteins that perform numerous functions. The major protein groups are (1) **albumins,** the most abundant group, constituting 60% of the total, which help to keep the plasma in osmotic equilibrium with the cells of the body; (2) **globulins,** a diverse group of high-molecular weight proteins (35% of total) that includes immunoglobulins and various metal-binding proteins; and (3) **fibrinogen,** a very large protein that functions in blood coagulation. Blood **serum** is plasma minus the proteins involved in clot formation (see below).

Red blood cells, or **erythrocytes,** are present in enormous numbers in the blood, approximately 5.4 billion per milliliter of blood in an adult man and 4.8 billion in adult women. They are formed continuously from large nucleated **erythroblasts** in the red bone marrow. Here hemoglobin is synthesized and the cells divide several times. In mammals the nucleus shrinks during development to a small remnant and eventually disappears altogether. Many other characteristics of a typical cell also are lost: ribosomes, mitochondria, and most enzyme systems. What is left is a biconcave disc consisting of a baglike membrane packed with about 280 million molecules of the blood-transporting pigment **hemoglobin.** Approximately 33% of the erythrocyte by weight is hemoglobin. The biconcave shape (Figure 7-3) is a mammalian innovation that provides a larger surface for gas diffusion than would a flat or spherical shape. All other vertebrates have nucleated erythrocytes that are usually ellipsoidal in shape.

Red bone marrow is red because blood cells are being formed there. It is found in the spongy bone in the proximal ends of the humerus and femur (p. 160) and a few other bones in adults. Yellow bone marrow lies in the interior of the long bones and is a site of fatty deposition.

The erythrocyte enters the circulation for an average life span of approximately 4 months. During this time it may journey 700 miles, squeezing repeatedly through the capillaries, which are sometimes so narrow that the erythrocyte must bend to get through. At last it fragments and is quickly engulfed by large scavenger cells called **macrophages** located in the liver, bone marrow, and spleen. The iron from the hemoglobin is salvaged to be used again; the rest of the heme is converted to **bilirubin,** a bile pigment. It is estimated that 10 million erythrocytes are produced and another 10 million destroyed every second in the human body.

The white blood cells, or **leukocytes,** form a wandering system of protection for the body. In adults they number only approximately 7.5 million per milliliter of blood, a ratio of 1 white cell to 700 red cells. There are several kinds of white blood cells: **granulocytes** (subdivided into neutrophils, basophils, and eosinophils), and nongranulocytes, the **lymphocytes** and **monocytes** (Figure 7-3). The role of the leukocytes in the body's defense mechanisms will be discussed later.

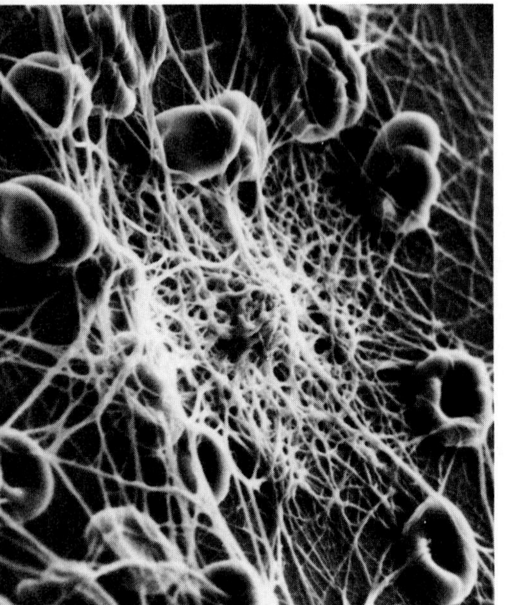

FIGURE 7-4

Human red blood cells entrapped in fibrin clot. Clotting is initiated after tissue damage by the disintegration of platelets in the blood, resulting in a complex series of intravascular reactions that end with the conversion of a plasma protein, fibrinogen, into long, tough, insoluble polymers of fibrin. Fibrin and entangled erythrocytes form the blood clot, which arrests bleeding. An aggregation of platelets probably underlies the raised mass of fibrin in center.

Courtesy N.F. Rodman.

Hemostasis: prevention of blood loss

It is essential that animals have ways of preventing the rapid loss of body fluids after an injury. Since blood is flowing and is under considerable hydrostatic pressure, it is especially vulnerable to hemorrhagic loss.

When a vessel is damaged, smooth muscle in the wall contracts, which causes the vessel lumen to narrow, sometimes so strongly that blood flow is completely stopped. This is a primitive but highly effective means of preventing hemorrhage used by invertebrates and vertebrates alike. Beyond this first defense against blood loss, all vertebrates, as well as some of the larger, active invertebrates with high blood pressures, have special cellular elements and proteins in the blood that are capable of forming plugs, or clots, at the injury site.

In higher vertebrates **blood coagulation** is the dominant hemostatic defense. Blood clots form as a tangled network of fibers from one of the plasma proteins, **fibrinogen.** The transformation of fibrinogen into a **fibrin** meshwork (Figure 7-4) that entangles blood cells to form a gel-like clot is catalyzed by the enzyme thrombin. Thrombin is normally present in the blood in an inactive form called **prothrombin,** which must be activated for coagulation to occur.

In this process, the blood platelets (Figure 7-3) play a vital role. Platelets are formed in the red bone marrow from certain large cells that regularly pinch off bits of their cytoplasm; thus they are fragments of cells. There are 150,000 to 300,000 platelets per cubic millimeter of blood. When the normally smooth inner surface of a blood vessel is disrupted, either by a break or by deposits of a cholesterol-lipid material, the platelets rapidly adhere to the surface and release **thromboplastin** and other clotting factors. These factors, along with factors released from damaged tissue and with calcium ions, initiate the conversion of prothrombin to the active thrombin. The stages in the formation of fibrin are summarized in Figure 7-5.

The catalytic sequence in this scheme is unexpectedly complex, involving a series of plasma protein factors, each normally inactive until activated by a previous factor in the sequence. The sequence behaves like a "cascade" with each reactant in the sequence leading to a large increase in the amount of the next reactant. At least 13 different plasma coagulation factors have been recognized. A deficiency of a single factor can delay or prevent the clotting process. Why has such a complex clotting mechanism evolved? Probably it is necessary to provide a fail-safe system capable of responding to any kind of internal or external hemorrhage that might occur and yet a system that cannot be activated into forming dangerous intravascular clots unless injury has occurred.

Hemophilia is one of the best known cases of sex-linked inheritance in humans. Actually two different loci on the X chromosome are involved. Classical hemophilia (hemophilia A) accounts for about 80% of persons with the condition, and the remainder are due to Christmas disease (hemophilia B). The allele at each locus results in the deficiency of a different platelet factor.

Several kinds of clotting abnormalities in humans are known. One of these, hemophilia, is a condition characterized by the failure of the blood to clot, so that even insignificant wounds can cause continuous severe bleeding. It is due to a rare mutation (about 1 in 10,000) on the X sex chromosome, resulting in an inherited lack of one of the platelet factors in males and in homozygous females. Called the "disease of kings," it once ran through several interrelated royal families of Europe, apparently having originated from a mutation in one of Queen Victoria's parents.

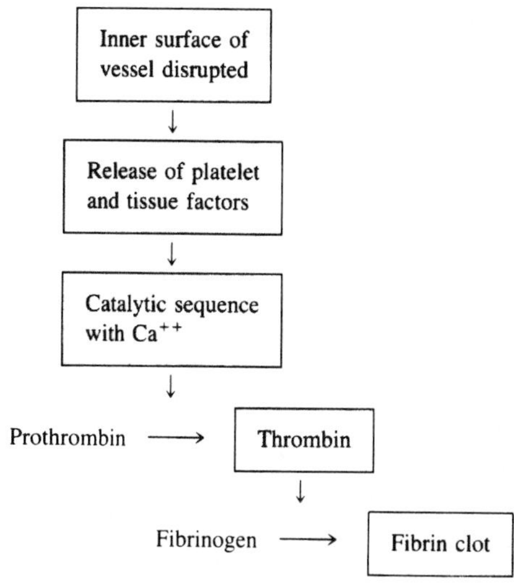

FIGURE 7-5

Stages in the formation of fibrin.

Defense Mechanisms of the Body

PHAGOCYTOSIS

Most animals have one or more mechanisms to protect themselves against invasion of a foreign body or infectious agent. These may be coincidental attributes of certain structures (a tough skin or high stomach acidity, for example), or they may be characteristics evolved as adaptations for defense. For defense against an invader, the invader first of all must be recognized. The cells in an animal must "know" when a substance does not belong in that animal; they must recognize "nonself." **Phagocytosis** illustrates nonself recognition, and it is found in almost all metazoa and is a feeding mechanism in protozoa. A cell that has this ability is a **phagocyte.** Phagocytosis is a process of engulfment of the invading particle within an invagination of the phagocyte's cell membrane (Figure 7-6). The invagination becomes pinched off, and the particle is thereby enclosed in an intracellular vacuole. Lysosomes empty digestive enzymes into the vacuole to destroy the particle. In metazoan invertebrates the cells performing this function are known as **amebocytes** (or another name, depending on the group of animals). If the particle is too large for phagocytosis, the amebocytes may gather around it and wall it off. In humans there are **fixed** and **mobile** phagocytes. The fixed phagocytes taken together form the **reticuloendothelial system** (**RE system**) and are found in the liver, spleen, lymph nodes, and other tissues. The RE system filters out and destroys particles and spent red blood cells from the blood that passes throughout the organs where the cells are located. The mobile phagocytes circulate in the blood and include the granulocytic leukocytes, especially neutrophils, and the monocytes, which become phagocytes (Figure 7-3). When monocytes move into the tissue from the blood, they differentiate into active phagocytes called **macrophages.**

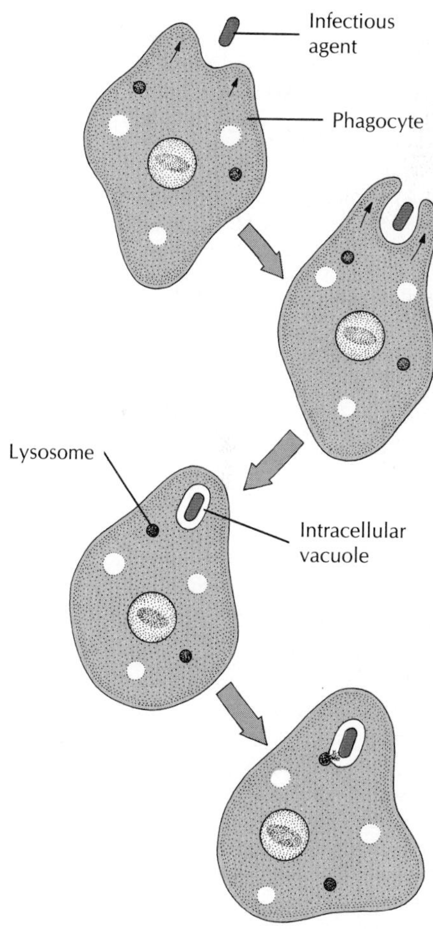

FIGURE 7-6

Phagocytosis. By pseudopodial movement, the phagocyte engulfs the particle. Lysosomes join with the vacuole containing the agent, pouring their contents (digestive enzymes or lysozymes) into the vacuole to destroy the particle.

ACQUIRED IMMUNE RESPONSE IN VERTEBRATES

Vertebrates have a specialized system of nonself recognition that results in increased resistance to a *specific* foreign substance or invader after repeated exposures. A substance that simulates an immune response is called an **antigen** (Gr. *anti,* against, + *genos,* birth). Antigens may be any of a variety of substances that have a molecular weight greater than 3000, most commonly proteins, and are usually (but not always) foreign to the host. There are two arms of the immune response, known as **humoral** (Figure 7-7) and **cellular.** Humoral immunity is based on **antibodies,** which are dissolved in and circulate in the blood, whereas cellular immunity is associated with cell surfaces. Although the two types interact, humoral immunity seems to be more important in a variety of bacterial infections, while the cellular response is of particular importance in tissue rejection reactions and a variety of viral, fungal, and parasitic infections. Antibodies are protein molecules of a class called **immunoglobulins,** which have a specific site that binds to a particular antigen.

Basis of self and nonself recognition

Nonself recognition is very specific; if tissue from one individual is transplanted into another individual of the same species, the graft will grow for a time

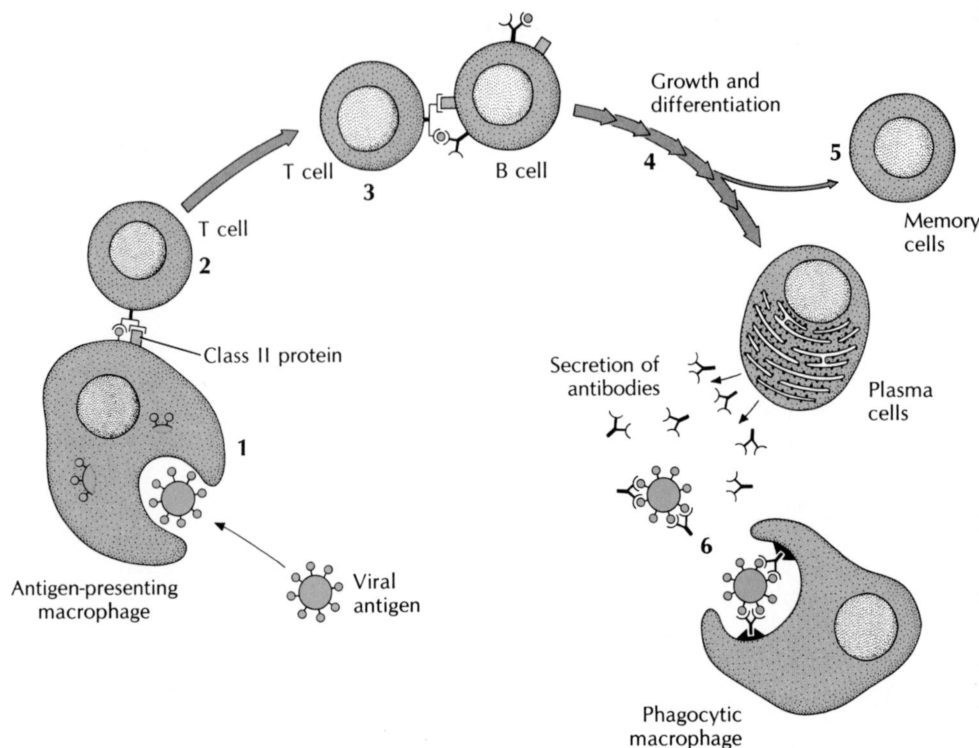

FIGURE 7-7

Stimulation of a humoral immune response by an antigen. *1,* Macrophage consumes antigen, partially digests it, and displays portions on its surface, along with class II MHC protein. *2,* T helper cell recognizes antigen and class II protein on macrophage and is activated. *3,* T helper then activates B cell, which carries same antigen and class II protein on its surface. *4,* Activated B cell multiplies, finally producing many plasma cells which secrete antibody. *5,* Some of B cell progeny become memory cells. *6,* Antibody produced by plasma cells binds to antigen and stimulates macrophages to consume antigen (opsonization).

The capability of an immune response develops over a period of time in the early development of the organism. All substances present at the time the capacity develops are recognized as self in later life. Unfortunately, the system of self and nonself recognition sometimes breaks down, and an animal may begin to produce antibodies against some part of its own body. This leads to one of several known *autoimmune diseases,* such as rheumatoid arthritis, multiple sclerosis, and myasthenia gravis.

A major problem of immunology is understanding how the mammalian genome could contain the information needed to produce at least a million different antibodies. The recently discovered answer is that antibody genes occur in pieces, rather than as continuous stretches of DNA, and that the antigen-recognizing sites on the antibody molecules are pieced together from information supplied by separate DNA sequences, which can be shuffled to increase vastly the diversity of gene products. The immense repertoire of antibodies is achieved in part by complex gene rearrangements (the exact mechanism of DNA switching is not yet clear) and in part by frequent somatic mutations that produce additional variation in antibody structure.

and then die as immunity against it arises. Without immunosuppression, tissue grafts can only grow successfully if they are between identical twins or between individuals of highly inbred strains of animals. It has been found in recent years that the molecular basis for this nonself recognition depends on certain proteins imbedded in the cell surface. These proteins are coded by certain genes, now known as the **major histocompatibility complex (MHC),** because they were discovered in tissue graft experiments. The MHC proteins are among the most variable known, and unrelated individuals almost always have different genes. There are two types of MHC proteins: class I and class II. Class I proteins are found on the surface of virtually all cells, whereas class II MHC proteins are found only on certain cells participating in the immune response, such as certain lymphocytes and macrophages.

Generation of a humoral response

The immune response is due primarily to the type of leukocytes called **lymphocytes.** There are two broad categories: **T lymphocytes (T cells)** and **B lymphocytes (B cells).** B cells have antibody molecules on their surface and give rise to cells that actively secrete antibodies into the blood. T cells have surface receptors that bind antigens, but the receptors are somewhat different in structure from antibodies. There are a vast array of B cells, each bearing on its surface molecules of antibody that will bind with one particular antigen, even though that antigen may never have been present in the body previously. There is probably an equal number of different T cells with receptors for specific antigens.

When an antigen is introduced into the body, it binds to a specific antibody on the surface of the appropriate B cell, but this is usually insufficient to activate the B cell to multiply. Some of the antigen is taken up by macrophages that partially digest it and then incorporate portions of the antigen into their own cell surface (Figure 7-7). Fragments of the antigen on the surfaces of the macrophages are recognized by a subset of T cells called **T helper cells (T_H),** in conjunction with the class II MHC protein of the macrophage surface. Both the class II protein and the antigen must be present; neither is effective alone. The T_H cells then activate the B cell that has the same processed antigen and a class II MHC protein on its surface. The B cell multiplies rapidly and produces many **plasma cells,** which secrete large quantities of antibody for a period of time, then die. Thus if the amount of antibody (**titer**) is measured soon after the antigen is injected, little or none can be detected. The titer rises rapidly as the plasma cells secrete antibody, then it may decrease somewhat as they die and the antibody is degraded (Figure 7-8). However, if another dose of antigen (the **challenge**) is given, there is no lag, and the antibody titer rises quickly to a higher level than after the first dose. This is the **secondary response,** and it occurs because some of the activated B cells gave rise to long-lived **memory cells.** There are many more memory cells present in the body than the original B cell with the appropriate antibody on its surface, and they rapidly multiply to produce additional plasma cells.

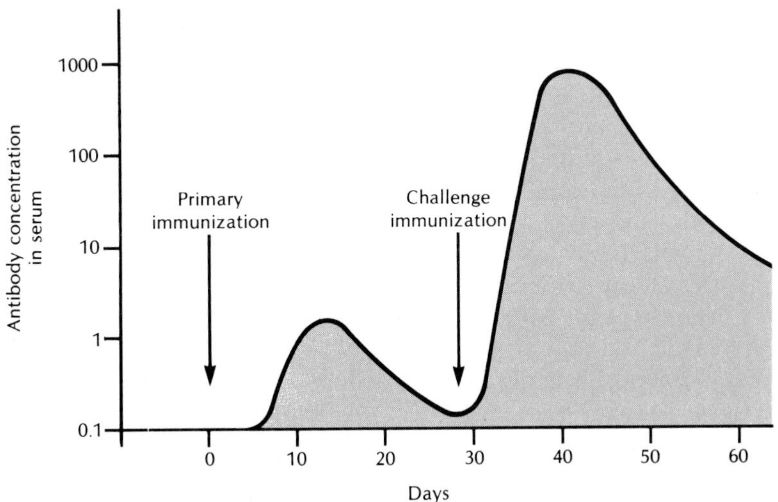

FIGURE 7-8

Typical immunoglobulin response after primary and challenge immunizations. The secondary response is a result of the large numbers of memory cells produced after the primary B-cell activation.

There are several ways known in which antibodies act to kill or render harmless an invader. In some cases the antibody bound to the surface of an invading cell acts in concert with certain enzymes and activators in the blood, collectively called **complement,** to lyse or break up the foreign cell. Commonly, an antibody bound to the surface of a particle facilitates engulfment by phagocytes, a phenomenon called **opsonization.**

Types of T cells. In addition to the T helper cells, three other T cell types are known. **T inducer cells** trigger the maturation of T lymphocytes from precursors into functionally distinct cells. **Cytotoxic T cells** help defend against viruses by killing cells that have been infected with a virus. They recognize class I MHC protein on the cell surface of infected cells in conjunction with viral protein that was left on the surface when the virus invaded the cell. **T suppressor cells** turn off the T_H cells and dampen the immune response; they are very important in regulating the immune response.

Certain biochemical markers on the surface of the T cells allow classification of the four types into two: T inducer and T helper cells are known as T4, and T suppressor and cytotoxic T cells are known as T8. T4 cells recognize class II MHC proteins in association with antigenic proteins on other cell surfaces, and T8 cells recognize class I MHC proteins along with foreign proteins.

AIDS (acquired immune deficiency syndrome) is an extremely serious disease which cripples the ability to mount an immune response. The first case was recognized in 1981, and by early 1989 over 86,000 cases and 40,000 deaths had been reported in the United States alone. AIDS patients are continuously plagued with infections by agents that cause insignificant problems in persons with normal immune responses. Although a variety of cell types are infected, the disease virus preferentially invades and destroys T4 lymphocytes. Normally T4 cells make up 60% to 80% of the T-cell population; in AIDS victims they can become too rare to be detected.

The cell-mediated response

Some immune responses involve little (if any) antibody but depend on the action of cells only. In cell-mediated immunity the antigen is also processed by

Only a few years ago transplantation of organs from one person to another seemed impossible. Then physicians began to transplant kidneys and depress the immune response in the recipient. It was very difficult to immunosuppress the recipient enough that the new organ would not be rejected and at the same time not leave the patient defenseless against infection. Cyclosporine, a new drug derived from a fungus, was discovered and now not only kidneys but also hearts, lungs, and livers can be transplanted. Cyclosporine inhibits the T cell growth factor and affects cytotoxic T cells more than T helpers. It has no effect on other white cells or on healing mechanisms, so that a patient can still mount an immune response but not reject the transplant. However, the patient must continue to take cyclosporine, because if the drug is stopped, the body will recognize the transplanted organ as foreign and reject it.

Certain cells in the dermis and other tissues have a type of antibody bound to their surface membranes. When the specific antigen binds to that antibody, the cells quickly secrete histamine and other substances. Some people become sensitized to certain antigens. When they are exposed to these antigens, excessive amounts of histamine are secreted, producing the symptoms of *allergy*. In some instances, the conditions may be so severe that life-threatening consequences result. People with allergies take medicine to counter the effects of excessive histamine (antihistamines).

macrophages, but only T cells respond. The T cells with the specific receptors for the particular antigen are activated and proliferate. They produce a variety of soluble effector molecules such as **interferons.** Interferons are a family of 20 to 25 proteins that act on other cells to increase their resistance to viral infection.

Like humoral immunity, cell-mediated immunity shows a secondary response as a result of large numbers of memory T cells that were produced from the original activation. For example, a second tissue graft (challenge) between the same donor and host will be rejected much more quickly than the first.

Inflammation

Inflammation is a vital process in mobilizing the body's defenses against an invading organism and in repairing the damage thereafter. Although inflammation is a nonspecific process, it is greatly influenced by the body's prior immunizing experience with the invader. Also, a more noxious foreign substance will produce more intense inflammation. Tissue damage initiates release of pharmacologically active substances (such as histamine) from certain cells in the area. These substances increase the diameter of nearby small blood vessels and also increase their permeability. Thus redness and warmth result from the increased amount of blood in the area (hence "inflammation"), and more proteins and fluid escape into the tissue, causing swelling. This swelling and change in permeability of the capillaries allows antibodies and leukocytes to move from the capillaries and easily reach the invader. The first phagocytic line of defense is the neutrophils, which may last only a few days, then macrophages (either fixed or differentiated from monocytes) become predominant. "Pus" of an infection is formed principally from exuded tissue fluid and spent phagocytes.

We are unaware of the majority of minute invasions that are always occurring. Most are efficiently disposed of and heal leaving little or no trace. However, if tissue damage has been more severe, fibrous connective tissue (scar tissue) will be deposited in the area.

Blood group antigens

ABO blood types. Blood differs chemically from person to person, and when two different (incompatible) blood types are mixed, **agglutination** (clumping together) of erythrocytes results. The basis of these chemical differences is naturally occurring antigens on the membranes of red blood cells. The best known of these inherited immune systems is the ABO blood group. The antigens A and B are inherited as dominant genes. Thus, as shown in Table 7-1, an individual with, for example, genes *A/A* or *A/O* develops A antigen (blood type A). The presence of a B gene produces B antigens (blood type B), and for the genotype *A/B* both A and B antigens develop on the erythrocytes (blood type AB).

There is an odd feature about the ABO system. Normally we would expect that a type A individual would develop antibodies against type B blood only if B cells were introduced into the body. In fact, type A persons always have anti-B antibodies in their blood, even without the prior exposure to type B blood. Similarly, type B individuals carry anti-A antibodies. Type AB blood has neither anti-A nor anti-B antibodies (since if it did, it would destroy its own blood cells), and type O blood has both anti-A and anti-B antibodies.

We see then that the blood group names identify their *antigen* content. Persons with type O blood are called universal donors because, lacking antigens, their blood can be infused into a person with any blood type. Even though it contains anti-A and anti-B antibodies, these are so diluted during transfusion that they do not react with A or B antigens in a recipient's blood. In practice, however, clinicians insist on matching blood types to prevent any possibility of incompatibility.

Table 7-1 Major blood groups

Blood type	Genotype	Antigens on red blood cells	Antibodies in serum	Can give blood to	Can receive blood from	Frequency in United States (%)		
						Whites	Blacks	Asians
O	*O/O*	None	Anti-A and anti-B	All	O	45	48	31
A	*A/A, A/O*	A	Anti-B	A, AB	O, A	41	27	25
B	*B/B, B/O*	B	Anti-A	B, AB	O, B	10	21	34
AB	*A/B*	AB	None	AB	All	4	4	10

Rh factor. Karl Landsteiner, an Austrian—later American—physician discovered the ABO blood groups in 1900. In 1940, 10 years after receiving the Nobel Prize, he made still another famous discovery. This was a blood group called the Rh factor, named after the Rhesus monkey, in which it was first found. Approximately 85% of white individuals in the United States have the factor (positive) and the other 15% do not (negative). Rh-positive and Rh-negative bloods are incompatible; shock and even death may follow their mixing when Rh-positive blood is introduced into an Rh-negative person who has been sensitized by an earlier transfusion of Rh-positive blood. Rh incompatibility accounts for a peculiar and often fatal form of anemia of newborn infants called **erythroblastosis fetalis.** If an Rh-negative mother has an Rh-positive baby (father is Rh-positive) she can become immunized by the fetal blood during the birth process. Anti-Rh antibodies can seep across the placenta during a subsequent pregnancy and agglutinate the fetal blood.

The genetics of the Rh factor are very much more complicated than it was believed when the factor was first discovered. Some authorities believe that three genes located close together on the same chromosome are involved, whereas others adhere to a system of one gene with many alleles. In 1968 a revision of the single gene concept listed 37 alleles necessary to account for the phenotypes then known. Furthermore, the frequency of the various alleles varies greatly between whites, Asians, and blacks.

Circulation

The circulatory system of vertebrates is made up of a system of tubes, the **blood vessels,** and a propulsive organ, the **heart.** This is a **closed circulatory system** because the circulating medium, the **blood,** is confined to vessels throughout its journey from the heart to the tissues and back again. Many invertebrates have an **open circulatory system;** the blood is pumped from the heart into blood vessels that open into tissue spaces. The blood circulates freely in direct contact with the tissues and then reenters open blood vessels to be propelled forward again. In invertebrates having open circulatory systems, there is no clear separation of the extracellular fluid into plasma and interstitial fluids, as there must be in closed systems. Closed systems are more suitable for large and active animals because the blood can be moved rapidly to the tissues needing it. In addition, flow to various organs can be readjusted to meet changing needs by varying the diameters of the blood vessels.

The closed circulatory system of vertebrates works along with the **lymphatic system.** This is a fluid "pickup" system. It re-collects fluid (lymph) that has been squeezed out through the walls of the capillaries and returns it to the blood circulation. In a sense "closed" circulatory systems are not absolutely closed because fluid is constantly leaking out into the tissue spaces. However, this leakage is but a small fraction of the total blood flow.

Erythroblastosis fetalis can now be prevented by giving an Rh-negative mother anti-Rh antibodies just after the birth of her first child. These antibodies remain long enough to neutralize any Rh-positive fetal blood cells that may have entered her circulation, thus preventing her own antibody machinery from being stimulated to produce the Rh-positive antibodies. Active, permanent immunity is blocked. If the mother has already developed an immunity, however, the baby may be saved by an immediate, massive transfusion of blood free of antibodies.

PLAN OF THE CIRCULATORY SYSTEM

All vertebrate vascular systems have certain features in common. A **heart** pumps the blood into **arteries** that branch and narrow into **aterioles** and then into a vast system of **capillaries.** Blood leaving the capillaries enters **venules** and then **veins** that return the blood to the heart. Figure 7-9 compares the circulatory systems of fish and mammals. The principal differences in circulation involve the heart in the transformation from gill to lung breathing.

The fish heart contains two main chambers, the **atrium** (or **auricle**) and the **ventricle.** Although there are also two subsidiary chambers, the **sinus venosus** and **conus arteriosus** (Figure 7-9), we still refer to the fish heart as a "two-chambered" heart. Blood makes a single circuit through the fish's vascular system; it is pumped from the heart to the gills, where it is oxygenated, and then flows into the dorsal aorta to be distributed to the body organs. After passing through the capillaries of the body organs and musculature, it returns by veins to the heart. In this circuit the heart must provide sufficient pressure to push the blood through two sequential capillary systems, one in the gills, and the other in the organ tissues. The principal disadvantage of the single-circuit system is that the gill capillaries offer so much resistance to blood flow that the pressure drops considerably before entering the dorsal aorta. This system can never provide high and continuous blood pressure to the body organs.

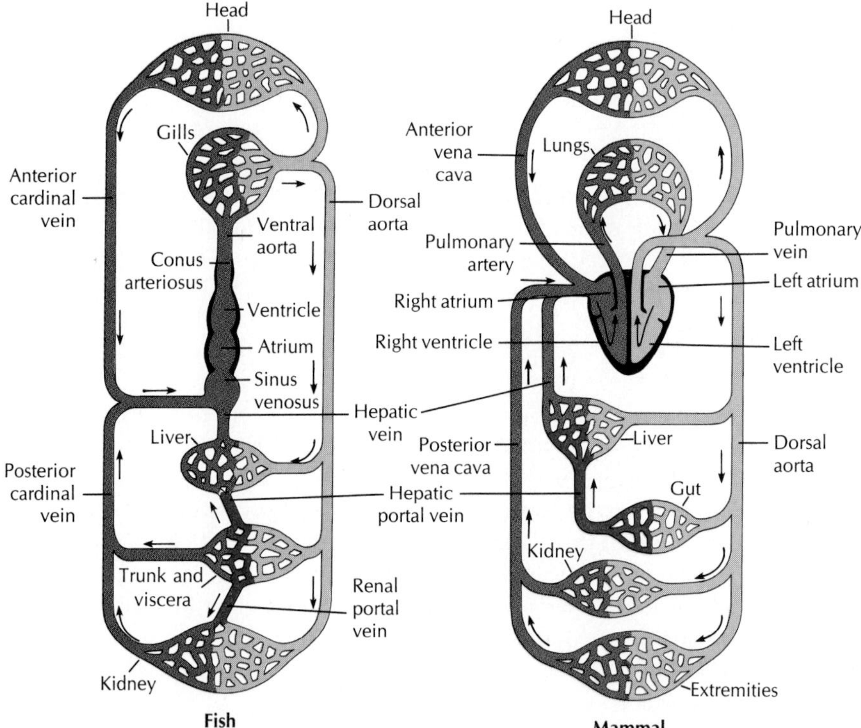

FIGURE 7-9

Plan of circulatory system of fish *(left)* and mammal *(right)*. *Pink,* oxygenated blood. *Dark red,* Deoxygenated blood. *Bright red,* Portal systems (blood collected from one bed of capillaries flows into another capillary bed).

The evolution of land forms with lungs and their need for highly efficient blood delivery resulted in the introduction of a double circulation. In the amphibians the atrium became partially (in some) or completely separated by a partition into two atria (Figure 7-10). Blood low in oxygen from the body is received into the right atrium, while oxygenated blood from the lungs comes from the left atrium. A partition partially separated the right and left portions of the ventricles, further increasing the efficiency of the heart. Separation of the ventricles is nearly complete in some reptiles (crocodilians) and is completely separate in birds and mammals. Thus one **systemic circuit** with its own pump provides oxygenated blood to the capillary beds of the body organs; another **pulmonary circuit** with its own pump sends deoxygenated blood to the lungs. Rather than actually developing two separate hearts, the existing two-chambered heart was divided down

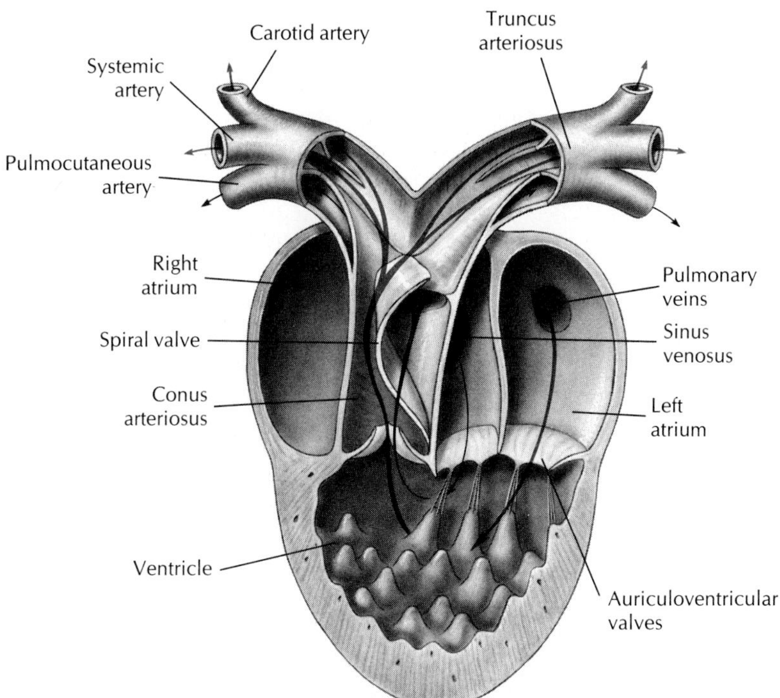

FIGURE 7-10

Frog heart.

the center into four chambers—really two two-chambered hearts lying side by side.

Such a great change in the vertebrate circulatory plan, involving not only the heart but the attendant plumbing as well, took many millions of years to evolve. The partial division of the atrium and ventricle began with the ancestors of present-day lungfishes. The course of the blood through this double circuit is shown in the right diagram of Figure 7-9.

Heart

The mammalian heart (Figure 7-11) is a muscular organ located in the thorax and covered by a tough, fibrous sac, the **pericardium.** As we have seen, the higher vertebrates have a four-chambered heart. Each half consists of a thinner-walled atrium and a thicker-walled ventricle. Heart (cardiac) muscle is a unique type of muscle found nowhere else in the body. It resembles striated muscle, but the cells are branched, and the dense end-to-end attachments between the cells are called **intercalated discs** (Figure 6-16, *C*, p. 165).

There are two sets of valves. **Atrioventricular valves (A-V valves)** separate the cavities of the atrium and ventricle in each half of the heart. The right A-V valve is called the **tricuspid valve** because it has three points, and the left A-V valve is the **bicuspid valve,** with two points. They permit blood to flow from atrium to ventricle but prevent backflow. Where the great arteries, the **pulmonary** from the right ventricle and the **aorta** from the left ventricle, leave the heart, **semilunar valves** prevent backflow into the ventricles.

The contraction of the heart is called **systole** (sis'to-lee), and the relaxation, **diastole** (dy-as'to-lee). The rate of the heartbeat depends on age, sex, and especially exercise. Exercise may increase the **cardiac output** (volume of blood forced from either ventricle each minute) more than fivefold. Both the heart **rate** and the **stroke volume** increase. Heart rates among vertebrates vary with the general level of metabolism and the body size. The cold-blooded codfish has a heart rate of approximately 30 beats per minute; a warm-blooded rabbit of about the same weight has a rate of 200 beats per minute. Small animals have higher heart rates

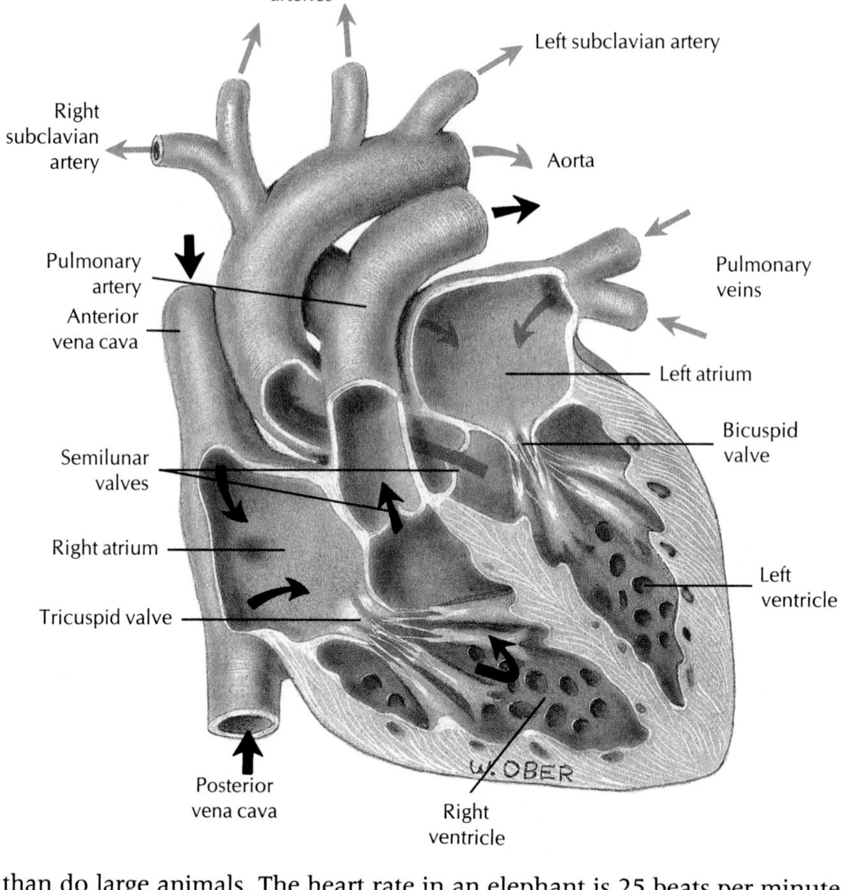

FIGURE 7-11

Human heart. Deoxygenated blood *(black arrows)* enters the right heart and is pumped to the lungs through the pulmonary arteries. Oxygenated blood *(red arrows)* returning from the lungs enters the left heart from the pulmonary veins and is pumped to the body. The left ventricular wall is thicker than the right ventricle, which needs only a minimal amount of muscular force to pump blood into the nearby lungs.

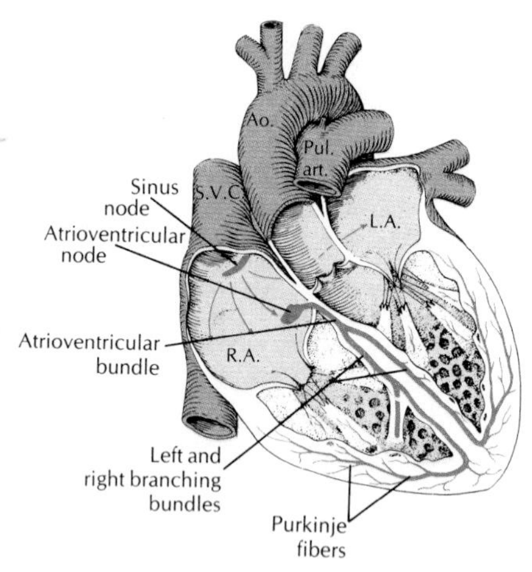

FIGURE 7-12

Neuromuscular mechanisms controlling heartbeat. Arrows indicate spread of excitation from the sinus node, across the atria, to the atrioventricular node. Wave of excitation is then conducted very rapidly to ventricular muscle over the specialized conducting bundles and Purkinje fiber system.

than do large animals. The heart rate in an elephant is 25 beats per minute, in a human 70 per minute, in a cat 125 per minute, in a mouse 400 per minute, and in the tiny 4 g shrew, the smallest mammal, the heart rate approaches a prodigious 800 beats per minute. We must marvel that the shrew's heart can sustain such a frantic pace throughout this animal's life, brief as it is.

Excitation of the heart. The heartbeat originates in a specialized muscle tissue, called the **sinus node,** located in the right atrium near the entrance of the caval veins (Figure 7-12). This tissue serves as the **pacemaker** of the heart. The contraction spreads across the two atria to the **atrioventricular (A-V) node,** where the impulse is held for less than a second as blood moves from the atria to the ventricles. The electrical impulse then penetrates the insulative connective tissue separating the atria and ventricles from each other via the **A-V bundles** into left and right halves, one for each ventricle. At this point the electrical activity is conducted very rapidly through the bundles to the apex of the ventricle and then through specialized fibers (**Purkinje fibers**), which continue up the walls of the ventricles. This arrangement allows the contraction to begin at the apex or "tip" of the ventricles and spread upward to squeeze out the blood in the most efficient way; it also ensures that both ventricles contract simultaneously.

Although the vertebrate heart can beat spontaneously—and the excised fish or amphibian heart does beat for hours in a balanced salt solution—the heart rate is normally under nervous control. The **control (cardiac) center** is located in the medulla and sends out two sets of nerves. Impulses sent along one set, the **vagus** nerves, apply a brake action to the heart rate, and impulses sent along the other set, the **accelerator** nerves, speed it up. Both sets of nerves terminate in the sinoatrial node, thus guiding the activity of the pacemaker.

The cardiac center in turn receives sensory information about a variety of stimuli. Pressure receptors (sensitive to blood pressure) and chemical receptors (sensitive to carbon dioxide and pH) are located at strategic points in the vascular system. This information is used by the cardiac center to increase or reduce the heart rate and cardiac output in response to activity or changes in body position. The heart is thus controlled by a series of feedback mechanisms that keep its activity constantly attuned to body needs.

Coronary circulation. It is no surprise that an organ as active as the heart needs a very good blood supply of its own. The heart muscle of the frog and other amphibians is so thoroughly channeled with spaces between the muscle fibers that sufficient oxygenated blood is squeezed through by the heart's own pumping action. In birds and mammals, however, the heart muscle is very thick and has such a high rate of metabolism that it must have its own vascular supply, the **coronary circulation.** The coronary arteries break up into an extensive capillary network surrounding the muscle fibers and provide them with oxygen and nutrients. Heart muscle has an extremely high oxygen demand, removing 80% of the oxygen from the blood, in contrast to most other body tissues, which remove only approximately 30%.

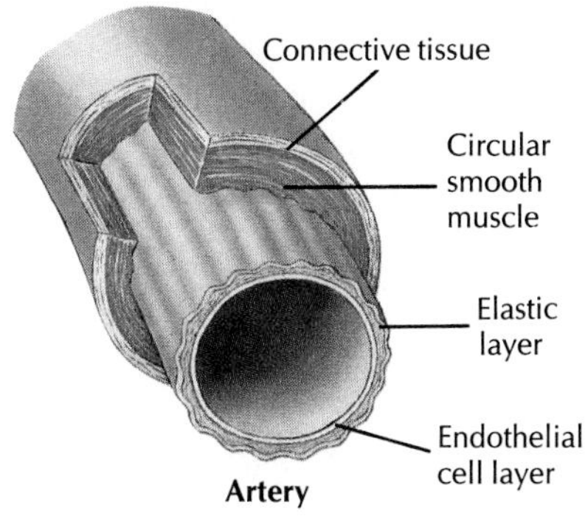

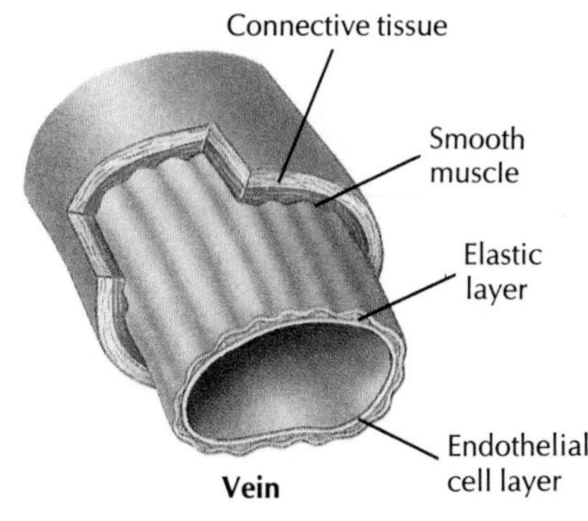

FIGURE 7-13

Artery and vein, showing layers. Note greater thickness of the muscular layer in the artery.

Arteries

All vessels leaving the heart are called arteries whether they carry oxygenated blood (aorta) or deoxygenated blood (pulmonary artery). To withstand high, pounding pressures, arteries are invested with layers of both elastic and tough inelastic connective tissue fibers (Figure 7-13). The elasticity of the arteries allows them to yield to the surge of blood leaving the heart during systole and then to squeeze down on the fluid column during diastole. This smooths out the blood pressure. Thus the normal arterial pressure in humans varies only between a high of 120 mm Hg (systole) and a low of 80 mm Hg (diastole) (usually expressed as 120/80 or 120 over 80), rather than dropping to zero during diastole as we might expect in a fluid system with an intermittent pump.

As the arteries branch and narrow into **arterioles,** the walls become mostly smooth muscle. Contraction of this muscle narrows the arterioles and reduces the flow of blood. The arterioles thus control the blood flow to body organs, diverting it to where it is needed most. The blood must be pumped with a hydrostatic pressure sufficient to overcome the resistance of the narrow passages through which the blood must flow. Consequently, large animals tend to have higher blood pressure than do small animals.

Blood pressure was first measured in 1733 by Stephen Hales, an English clergyman with unusual inventiveness and curiosity. He tied his mare, which was "to have been killed as unfit for service," on her back and exposed the femoral artery. This he cannulated with a brass tube, connecting it to a tall glass tube with the windpipe of a goose. The use of the windpipe was both imaginative and practical; it gave the apparatus flexibility "to avoid inconveniences that might arise if the mare struggled." The blood rose 8 feet in the glass tube and bobbed up and down with the systolic and diastolic beats of the heart. The weight of the 8-foot column of blood was equal to the blood pressure. We now express this as the height of a column of mercury, which is 13.6 times heavier than water. Hales' figures, expressed in millimeters of mercury, indicate that he measured a blood pressure of 180 to 200 mm Hg, about normal for a horse.

Today, blood pressure in humans is most commonly and easily measured with an instrument called a **sphygmomanometer.** Air is used to inflate a cuff on the upper arm to a pressure sufficient to close the arteries in the arm. As air is slowly released from the cuff, a person with a stethoscope held over the brachial

Thickening and loss of elasticity in the arteries is known as *arteriosclerosis.* When the arteriosclerosis is caused by fatty deposits of cholesterol in the artery walls, the condition is *atherosclerosis.* Such irregularities in the walls of blood vessels often cause blood to clot around them, forming a *thrombus.* When a bit of the thrombus breaks off and is carried by the blood to lodge elsewhere, it is an *embolus.* If the embolus blocks one of the coronary arteries, the person has a heart attack (a "coronary"). The portion of the heart muscle served by the branch of the coronary artery that is blocked is starved for oxygen. It may be replaced by scar tissue if the person survives.

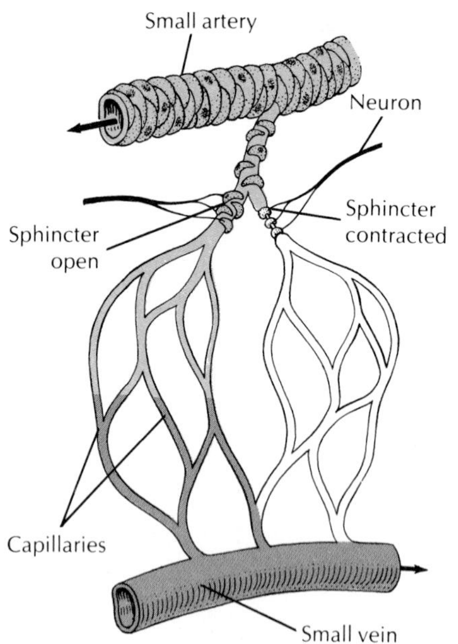

FIGURE 7-14

Capillary bed. Precapillary sphincters control blood flow through the capillaries.

artery (in the crook of the elbow) can hear the first spurts of blood through the artery when the pressure in the cuff allows the artery to open slightly. This is equivalent to the systolic pressure. As the pressure in the cuff is decreased, the sound heard with the stethoscope finally disappears when the blood is running smoothly through the artery. The pressure at which the sound disappears is the diastolic pressure.

Capillaries

The Italian Marcello Malpighi was the first to describe the capillaries in 1661, thus confirming the existence of the minute links between the arterial and venous systems that Harvey knew must be there but could not see. Malpighi studied the capillaries of the living frog's lung, which incidentally is still one of the simplest and most vivid preparations for demonstrating capillary blood flow.

The capillaries are present in enormous numbers, forming extensive networks in nearly all tissues. In muscle there are more than 2000 per square millimeter (1,250,000 per square inch), but not all are open at once. Indeed, perhaps less than 1% are open in resting skeletal muscle. But when the muscle is active, all the capillaries may open to bring oxygen and nutrients to the working muscle fibers and to carry away metabolic wastes (Figure 7-14).

Capillaries are extremely narrow, averaging less than 10 μm in diameter in mammals, which is hardly any wider than the red blood cells that must pass through them. Their walls are formed of a single layer of thin **endothelial** cells, held together by a delicate basement membrane and connective tissue fibers. Capillaries have a built-in leakiness that allows water and most dissolved substances except proteins to filter through into the interstitial space.

The actual situation is a bit more complicated because there is a small hydrostatic pressure in the interstitial fluid, and a small amount of protein does leak through the capillary wall. The protein tends to accumulate at the venule end of the capillary, building up a small osmotic pressure there. Although actual calculation of the pressure differences must take into account the interstitial fluid hydrostatic and osmotic pressures, the principle of the capillary fluid shift is as we have presented it.

Because of the low permeability of the capillaries to protein, the solute concentration in the blood is higher than in the interstitial fluid; therefore, there is a net osmotic pressure difference, due to the protein, of about 25 mm Hg in mammals. This osmotic force tends to move water into the capillary. The hydrostatic pressure of the blood at the arteriole end of the capillaries is about 40 mm Hg in humans. Thus there is a **net filtration pressure** (the hydrostatic pressure, which tends to force fluid out, less the osmotic pressure, which tends to draw water back in) of about 15 mm Hg at the arteriole end of the capillaries (Figure 7-15). Water and dissolved materials are forced out of the capillaries and circulate through the tissue space. As the blood proceeds through the narrow capillary, the blood pressure decreases steadily to perhaps 15 mm Hg. At this point the hydrostatic pressure is less than the osmotic pressure due to the plasma proteins, still approximately 25 mm Hg, and water is drawn back into the capillaries.

Thus it is the balance between hydrostatic pressure and protein osmotic pressure that determines the direction of capillary fluid shift. Normally, water is forced out of the capillary at the arteriole end, where hydrostatic pressure exceeds osmotic pressure, and drawn back into the capillary at the venule end, where osmotic pressure exceeds hydrostatic pressure. Any fluid left behind is picked up and removed by the **lymph capillaries** of the lymphatic system and is eventually returned to the circulatory system via larger lymph vessels. (This will be discussed later.)

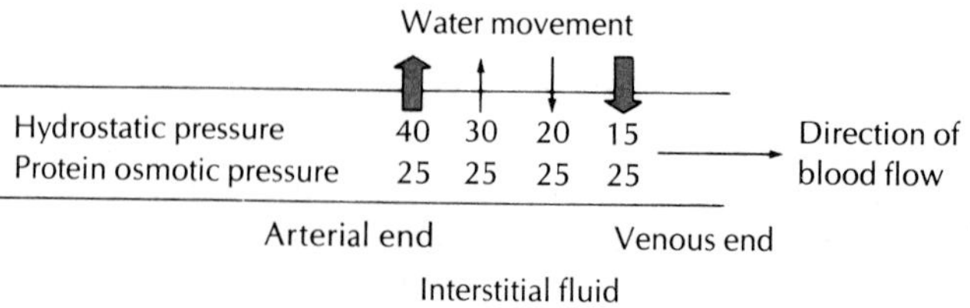

FIGURE 7-15

Fluid movement across the wall of a capillary. At the arterial end of the capillary, hydrostatic (blood) pressure exceeds protein osmotic pressure contributed by plasma proteins, and a plasma filtrate (shown as "water movement") is forced out. At the venous end, protein osmotic pressure exceeds the hydrostatic pressure, and fluid is drawn back in. In this way plasma nutrients are carried out into the interstitial space where they can enter cells, and metabolic end products from the cells are drawn back into the plasma and carried away.

Veins

The venules and veins into which the capillary blood drains for its return journey to the heart are thinner walled, less elastic, and of considerably larger diameter than their corresponding arteries and arterioles (Figure 7-13). Blood pressure in the venous system is low, from approximately 10 mm Hg, where capillaries drain into venules, to approximately zero in the right atrium. Because pressure is so low, the venous return gets assists from valves in the veins, body muscles surrounding the veins, and the rhythmical action of the lungs. If it were not for these mechanisms, the blood might pool in the lower extremities of a standing animal—a very real problem for people who must stand for long periods. Veins that lift blood from the extremities to the heart contain valves that divide the long column of blood into segments. When the skeletal muscles contract, as in even slight activity, the veins are squeezed, and the blood within them moves toward the heart because the valves within the veins keep the blood from slipping back. The well-known risk of fainting while standing at stiff attention in hot weather can usually be prevented by deliberately pumping the leg muscles. The negative pressure created in the thorax by the inspiratory movement of the lungs also speeds the venous return by sucking the blood up the large vena cava into the heart.

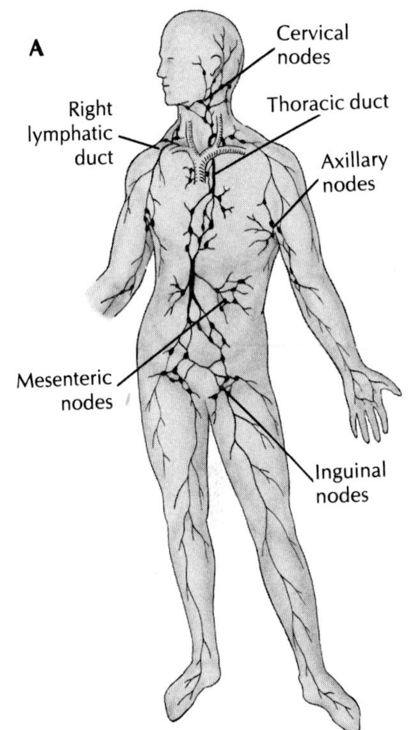

Lymphatic system

The lymphatic system is an extensive network of thin-walled vessels that is separate from the circulatory system. The system arises as blind-ended lymph capillaries in most tissues of the body. These unite to form larger and larger lymph vessels, which finally drain into veins in the lower neck (Figure 7-16).

The lymphatic system is an accessory drainage system for the body. As we have seen, the blood pressure in the arteriole end of the capillaries forces a plasma filtrate through the capillary walls and into the interstitial space. This tissue fluid bathing the cells is a clear, nearly colorless liquid. Tissue fluid and plasma are nearly identical except that tissue fluid contains very little protein, which was screened out as the plasma was squeezed through the capillary walls. Most of the tissue fluid returns to the vascular system at the venous end of the capillaries by the capillary fluid-shift mechanism described earlier. Usually, however, outflow from the capillaries slightly exceeds backflow. This difference is gathered up and returned to the circulatory system by lymphatic vessels. Tissue fluid is referred to as **lymph** as soon as it enters the lymph vessels. The rate of lymph flow is very low, a minute fraction of the blood flow.

Located at strategic intervals along the lymph vessels are **lymph nodes** (Figure 7-16) that have several defense-related functions. They are effective filters that remove foreign particles, especially bacteria, that might otherwise enter the general circulation. They are also centers for the storage, well-being, and multiplication of lymphocytes and plasma cells that produce antibodies—essential components of the body's defense mechanisms.

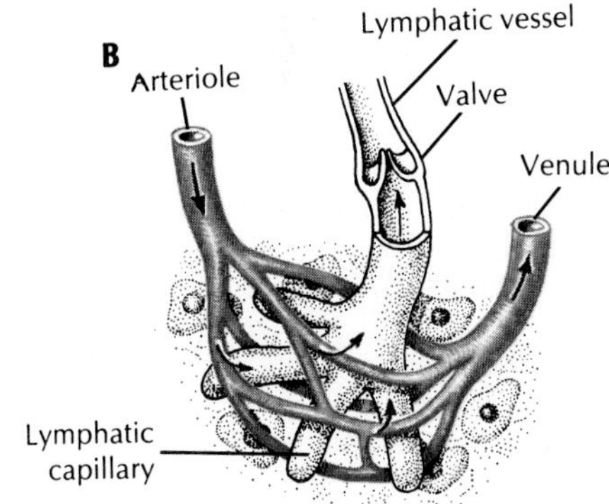

FIGURE 7-16

Human lymphatic system, showing major vessels, **A,** and a detail of the blood and lymphatic capillaries, **B.**

Respiration

The energy bound up in food is released by oxidative processes, usually with oxygen as the terminal electron acceptor. As oxygen is used by the body cells, carbon dioxide is produced; this process of gas exchange is called **respiration.** Small aquatic animals such as the one-celled protozoa obtain what oxygen they need by direct diffusion from the environment. Carbon dioxide is also lost by diffusion to the environment. Such a simple solution to the problem of gas exchange is really only possible for very small animals (less than 1 mm in diameter) that have a large surface relative to their volume or those having very low rates of metabolism.

As animals became larger and evolved a waterproof covering, specialized devices such as lungs and gills developed that greatly increased the effective surface for gas exchange. But, because gases diffuse so slowly through protoplasm, a circulatory system was necessary to distribute the gases to and from the deep tissues of the body. Even these adaptations were inadequate for advanced animals with their high rates of cellular respiration. The solubility of oxygen in the blood plasma is so low that plasma alone could not carry enough to support metabolic demands. With the evolution of special oxygen-transporting blood proteins such as hemoglobin, the oxygen-carrying capacity of the blood was greatly increased. Thus what began as a simple and easily satisfied requirement resulted in the evolution of several complex and essential respiratory and circulatory adaptations.

PROBLEMS OF AQUATIC AND AERIAL BREATHING

How an animal respires is largely determined by the nature of its environment. The two great arenas of animal evolution—water and land—are vastly different in their physical characteristics. The most obvious difference is that air contains far more oxygen—at least 20 times more—than does water.

Water at 5° C fully saturated with air contains approximately 9 ml of oxygen per liter, compared with air, which contains 210 ml of oxygen per 1 liter (21%). The solubility of oxygen in water decreases as the temperature rises. For example, water at 15° C contains approximately 7 ml of oxygen per liter, and at 35° C, only 5 ml of oxygen per liter. The relatively low concentrations of oxygen dissolved in water is the greatest respiratory problem facing aquatic animals. Unfortunately, it is not the only one. Oxygen diffuses much more slowly in water than in air, and water is much denser and more viscous than air. All of this means that successful aquatic animals must have evolved very efficient ways of removing oxygen from water. Yet even the most advanced fishes with highly efficient gills and pumping mechanisms may use as much as 20% of their energy just extracting oxygen from water. By comparison, a mammal uses only 1% to 2% of its resting metabolism to breathe.

It is essential that respiratory surfaces be thin and always kept wet to allow diffusion of gases between the environment and the underlying circulation (diffusion through a dry surface would be far too slow). This is hardly a problem for aquatic animals, immersed as they are in water, but it is a very real problem for air breathers. To keep the respiratory membranes moist and protected from injury, air breathers have in general developed invaginations of the body surface and then added pumping mechanisms to move air in and out. The lung is the best example of a successful solution to breathing on land. In general **evaginations** of the body surface, such as gills, are most suitable for aquatic respiration; **invaginations,** such as lungs, are best for air breathing. We can now consider the specific kinds of respiratory organs employed by animals.

Cutaneous respiration

Protozoa, sponges, cnidarians, and many worms respire by direct diffusion of gases between the organism and the environment. We have noted that this kind of **cutaneous respiration** is not adequate when the mass of living protoplasm exceeds approximately 1 mm in diameter. But, by greatly increasing the surface of the body relative to the mass, many multicellular animals respire in this way. Cutaneous respiration frequently supplements gill or lung breathing in larger animals such as amphibians and fishes. For example, an eel can exchange 60% of its oxygen and carbon dioxide through its highly vascular skin. During their winter hibernation, frogs and even turtles exchange all their respiratory gases through the skin while submerged in ponds or springs.

Gills

Gills of various types are more effective respiratory devices for life in water. Gills may be simple **external** extensions of the body surface, such as the **dermal papulae** of sea stars (p. 525) or the **branchial tufts** of marine worms (p. 456) and aquatic amphibians (p. 584). Most efficient are the **internal gills** of fishes (p. 571) and arthropods. Fish gills are thin filamentous structures, richly supplied with blood vessels arranged so that blood **flow** is opposite to the flow of water across the gills. This arrangement, called **countercurrent flow** (p. 571), provides for the greatest possible extraction of oxygen from water. Water flows over the gills in a steady stream, pulled and pushed by an efficient, two-valved, branchial pump (Figure 7-17). Gill ventilation is often assisted by the fish's forward movement through the water.

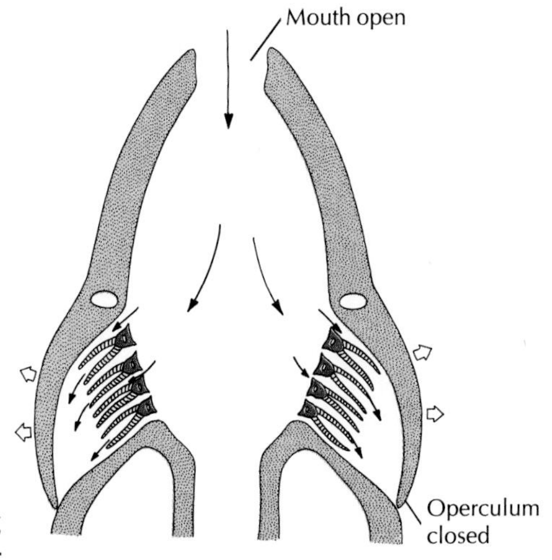

Operculum expanding, pulling water across gills

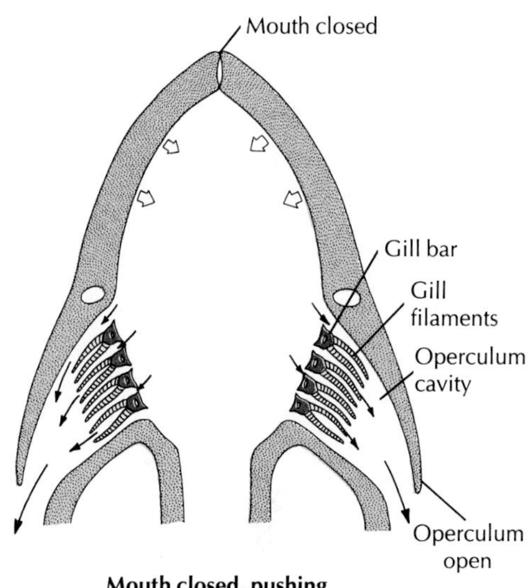

Mouth closed, pushing water across gills

FIGURE 7-17

How a fish ventilates its gills. Through the action of two skeletal muscle pumps, one in the mouth cavity, the other in the opercular cavity, water is drawn into the mouth, passes over the gills, and exits through the gill covers (opercular clefts).

Lungs

Gills are unsuitable for life in air because, when removed from the buoying water medium, the gill filaments collapse and stick together; a fish out of water rapidly asphyxiates despite the abundance of oxygen around it. Consequently air-breathing vertebrates possess lungs, highly vascularized internal cavities. Lungs of a sort are found in certain invertebrates (pulmonate snails, scorpions, some spiders, some small crustaceans), but these structures cannot be very efficiently ventilated.

Lungs that can be ventilated more efficiently are characteristic of the terrestrial vertebrates. The most primitive vertebrate lungs are those of lungfishes (Dipneusti), which use them to supplement, or even replace, gill respiration during periods of drought. Although of simple construction, the lungfish lung is supplied with a capillary network in its largely unfurrowed walls, a tubelike connection to the pharynx, and a primitive ventilating system for moving air in and out of the lung.

Amphibians also have simple baglike lungs, whereas in higher forms the inner surface area is vastly increased by numerous lobulations and folds (Figure 7-18). This increase is greatest in the mammalian lung, which is complexly divided into many millions of small sacs (**alveoli**), each veiled by a rich vascular network. It has been estimated that human lungs have a total surface area of from 50 to 90 m^2—50 times the area of the skin surface—and contain 1000 km of capillaries.

Moving air into and out of lungs was an evolutionary design problem that has been, of course, solved. However, we wonder whether an imaginative biological engineer, if given the proper resources, couldn't come up with a better design. Unlike the efficient one-way flow of water across fish gills, air must enter and exit a lung through the same channel. Furthermore, a tube of some length—the bronchi, trachea, and mouth cavity—connects the lungs to the outside. This is a "dead-air space" containing a volume of air that shuttles back and forth with each breath, adding to the difficulty of properly ventilating the lungs. In fact, lung ventilation is so inefficient that in normal breathing only approximately one sixth of the air in the lungs is replenished with each inspiration.

One group of vertebrates, the birds, vastly improved lung efficiency by adding an extensive system of air sacs (Figure 7-18 and p. 616) that serve as air reservoirs during ventilation. On inspiration, some 75% of the incoming air bypasses the lungs to enter the air sacs (gas exchange does not occur here). At expiration some of this fresh air passes directly through the lung passages and eventually into the one-cell thick **air capillaries** where gas exchange takes place. Thus the air capillaries receive nearly fresh air during both inspiration and expiration (Figure 29-10, p. 616). The beautifully designed bird lung is the result of selective pressures during the evolution of flight and its high metabolic demands.

FIGURE 7-18

Internal structures of lungs among vertebrate groups. Generally, the evolutionary trend has been from simple sacs with little exchange surface between blood and air spaces to complex, lobulated structures, each with complex divisions and extensive exchange surfaces.

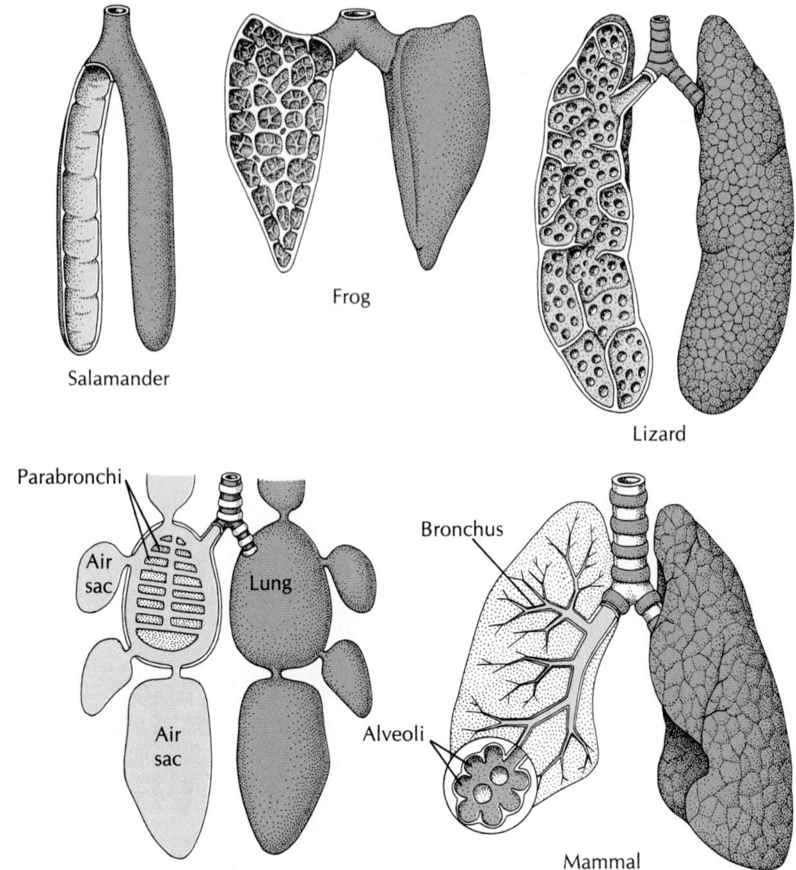

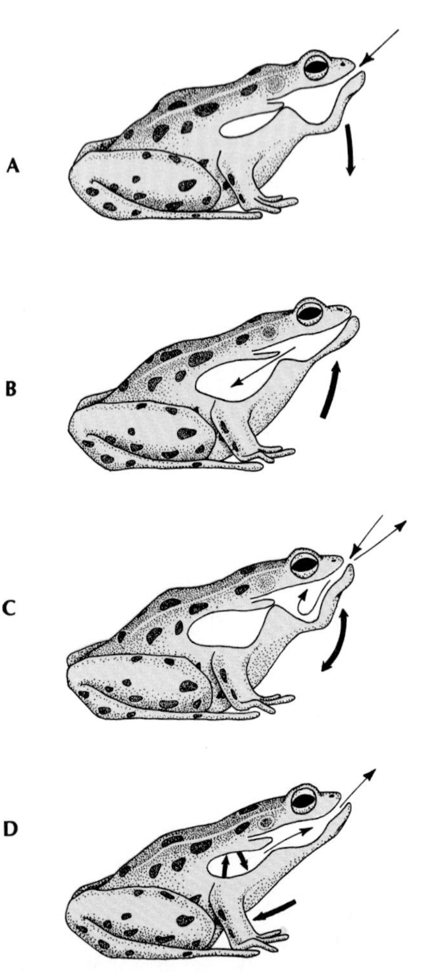

FIGURE 7-19

Breathing in the frog. The frog, a positive-pressure breather, fills its lungs by forcing air into them. **A,** Floor of mouth is lowered, drawing air in through the nostrils. **B,** With nostrils closed and glottis open, the frog forces air into lungs by elevating the floor of mouth. **C,** Mouth cavity is ventilated rhythmically for a period. **D,** Lungs are emptied by contraction of body wall musculature and by elastic recoil of lungs.

Modified from Gordon, M.S., and others. 1968. Animal function: principles and adaptations. New York, Macmillan, Inc.

Frogs force air into the lungs by first lowering the floor of the mouth to draw air into the mouth through the **external nares** (nostrils); then, by closing the nares and raising the floor of the mouth, air is driven into the lungs. Much of the time, however, frogs rhythmically ventilate only the mouth cavity, which serves as a kind of auxiliary "lung" (Figure 7-19). Amphibians therefore employ a **positive pressure** action to push air into their lungs, unlike most reptiles, birds, and mammals, which breathe by sucking air into the lungs (**negative pressure** action) by expansion of the rib cage.

Tracheal systems

Insects and certain other terrestrial arthropods (centipedes, millipedes, and some spiders) have a highly specialized type of respiratory system; in many respects it is the simplest, most direct, and most efficient respiratory system found in active animals. It consists of a system of tubes (**tracheae**) that branch repeatedly and extend to all parts of the body (p. 492). The smallest end channels are fluid-filled **tracheoles,** less than 1 μm in diameter, that sink into the plasma membranes of the body cells. Air enters the tracheal system through valvelike openings (**spiracles**). Carbon dioxide diffuses out in the opposite direction. Some insects can ventilate the tracheal system with body movements; the familiar telescoping movement of the bee abdomen is an example. The tracheal system is simple because blood is not needed to transport the respiratory gases; the cells have a direct pipeline to the outside.

RESPIRATION IN HUMANS

In mammals the respiratory system is made up of the following: the nostrils (external nares); the **nasal chamber,** lined with mucus-secreting epithelium; the

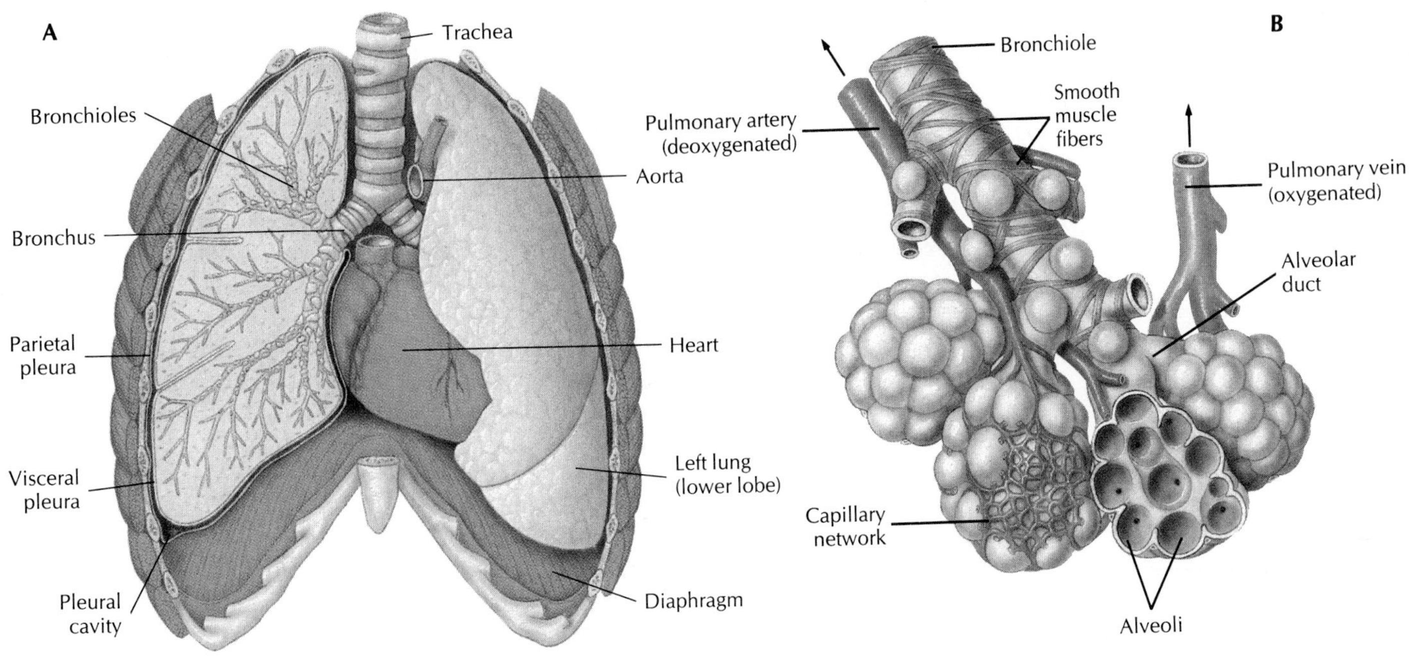

FIGURE 7-20

A, Lungs of human with right lung shown in section. **B,** Terminal portion of bronchiole showing air sacs with their blood supply. Arrows show direction of blood flow.

internal nares, the nasal openings which connect to the **pharynx** where the pathways of digestion and respiration cross; the **epiglottis,** a flap that folds over the **glottis** (the opening to the larynx) to prevent food from going the wrong way in swallowing; the **larynx,** or voice box; the **trachea,** or windpipe; and two **bronchi,** one to each lung (Figure 7-20). Within the lungs each bronchus divides and subdivides into small tubes (**bronchioles**) that lead via **alveolar ducts** to the air sacs (**alveoli**) (Figure 7-21). The single-layered endothelial walls of the alveoli are thin and moist to facilitate the exchange of gases between the air sacs and the adjacent blood capillaries. Air passageways are lined with both mucus-secreting and ciliated epithelial cells, which play an important role in conditioning the air before it reaches the alveoli. There are partial cartilage rings in the walls of the tracheae, bronchi, and even some of the bronchioles to prevent those structures from collapsing.

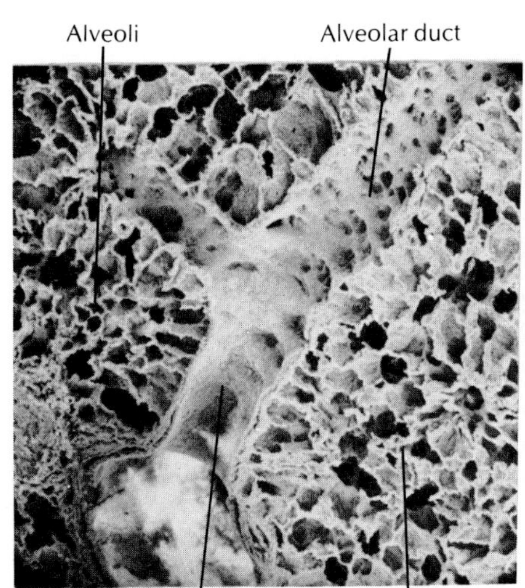

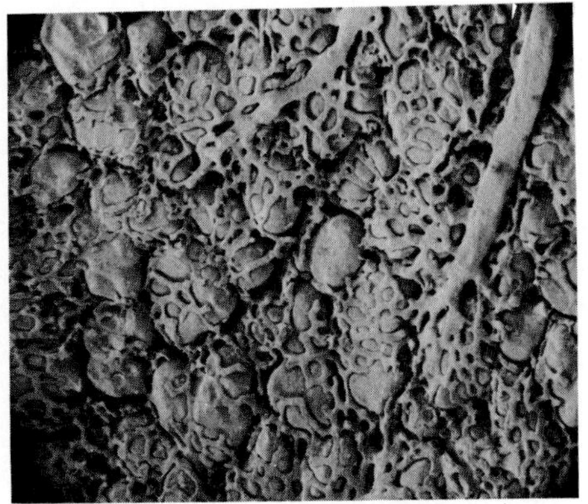

FIGURE 7-21

A, Scanning electron micrograph of mammalian lung. (×35.) **B,** Resin cast of lung, showing the extensive capillary network surrounding each alveolus. (×150.)

A, Courtesy E.E. Morrison; **B,** from Tissues and organs: a text-atlas of scanning electron microscopy, by Richard G. Kessel and Randy H. Kardon. W.H. Freeman and Co., Publishers. Copyright © 1979.

In its passage to the air sacs the air undergoes three important changes: (1) it is filtered free from most dust and other foreign substances, (2) it is warmed to body temperature, and (3) it is saturated with moisture.

The lungs consist of a great deal of elastic connective tissue and some muscle. They are covered by a thin layer of tough epithelium known as the **visceral pleura.** A similar layer, the **parietal pleura,** lines the inner surface of the walls of the chest (Figure 7-20). The two layers of the pleura are in contact and slide over one another as the lungs expand and contract. The "space" between the pleura, called the **pleural cavity,** contains a partial vacuum, which helps keep the lungs expanded to fill the pleural cavity. Therefore no real pleural space exists; the two pleura rub together, lubricated by tissue fluid (lymph). The chest cavity is bounded by the spine, ribs, and breastbone, and floored by the **diaphragm,** a dome-shaped, muscular partition between the chest cavity and abdomen. The diaphragm is found only in mammals.

Mechanism of breathing

The chest cavity is an air-tight chamber. In **inspiration** the ribs are pulled upward, the diaphragm is contracted and flattened, and the chest cavity is enlarged. The resultant increase in volume of the chest cavity causes the air pressure in the lungs to fall below atmospheric pressure: air rushes in through the air passageways to equalize the pressure. **Expiration** is a less active process than inspiration. When the muscles relax, the ribs and diaphragm return to their original position, and the chest cavity size decreases, the elastic lungs deflate, and air exits.

Coordination of breathing

It is well known that swimmers can remain submerged much longer if they vigorously hyperventilate first to blow off carbon dioxide from the lungs. This delays the overpowering urge to surface and breathe. The practice is dangerous because blood oxygen is depleted just as rapidly as without prior hyperventilation, and the swimmer may lose consciousness when the oxygen supply to the brain drops below a critical point. Several documented drownings among swimmers attempting long underwater swimming records have been caused by this practice.

Respiration must adjust itself to changing requirements of the body for oxygen. Breathing is normally involuntary and automatic but can come under voluntary control. Normal, quiet breathing is regulated by neurons centered in the medulla of the brain. These neurons spontaneously produce rhythmical bursts that lead to contraction of the diaphragm and the intercostal muscles between the ribs. The rhythm and depth of breathing are precisely regulated by the amount of carbon dioxide in the blood. Exercise raises the carbon dioxide level, and the breathing rate increases. Actually, the effects of carbon dioxide are due to the increase in blood hydrogen ion concentration (p. 194). The increase in hydrogen ions stimulates the respiratory center in the medulla, leading to increased rate and depth of respiration.

Composition of inspired, expired, and alveolar airs

The composition of expired and alveolar airs is not identical. Air in the alveoli contains less oxygen and more carbon dioxide than does the air that leaves the lungs. Inspired air has the composition of atmospheric air. Expired air is really a mixture of alveolar and inspired airs. The composition of the three kinds of air is shown in Table 7-2.

The water given off in expired air depends on the relative humidity of the external air and the activity of the person. At ordinary room temperature and with a relative humidity of approximately 50%, an individual in performing light work loses approximately 350 ml of water from the lungs each day.

Gaseous exchange in lungs

The diffusion of gases takes place in accordance with the laws of physical diffusion; that is, the gases pass from regions of higher concentration to those of lower concentration. The partial pressure of a gas refers to that pressure which the gas exerts in a mixture of gases, in other words, the part of the total pressure due to that component. If the atmospheric pressure at sea level is equivalent to

Table 7-2 Composition of respired air

	Inspired air (vol %)	Expired air (vol %)	Alveolar air (vol %)
Oxygen	20.96	16	14.0
Carbon dioxide	0.04	4	5.5
Nitrogen	79.00	80	80.5

760 mm Hg, the partial pressure of oxygen is 21% (percentage of oxygen in air) of 760, or 159 mm Hg in dry air. The partial pressure of oxygen in the lung alveoli is greater (100 mm Hg pressure) than it is in venous blood of lung capillaries (40 mm Hg pressure) (Figure 7-22). Oxygen then naturally diffuses into the capillaries. In a similar manner the carbon dioxide in the blood of the lung capillaries has a higher concentration (46 mm Hg) than has this same gas in the lung alveoli (40 mm Hg), so that carbon dioxide diffuses from the blood in the alveoli.

In the tissues respiratory gases also move according to their concentration gradients (Figure 7-22). The concentration of oxygen in the blood (100 mm Hg pressure) is greater than in the tissues (0 to 30 mm Hg pressure), and the carbon dioxide concentration in the tissues (45 to 68 mm Hg pressure) is greater than that in blood (40 mm Hg pressure). The gases in each case go from a high to a low concentration.

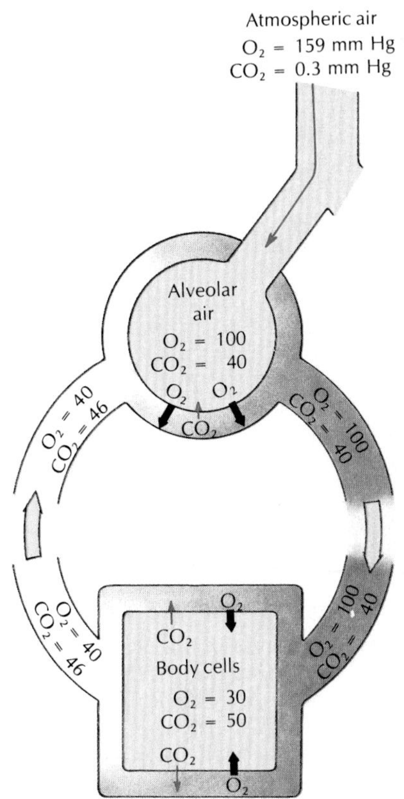

FIGURE 7-22

Exchange of respiratory gases in lungs and tissue cells. Numbers present partial pressures in millimeters of mercury (mm Hg).

Transport of gases in blood

In some invertebrates the respiratory gases are simply carried dissolved in the body fluids. However, the solubility of oxygen is so low in water that it is adequate only for animals with low rates of metabolism. For example, only approximately 1% of a human's oxygen requirement can be transported in this way. Consequently in many invertebrates and in the vertebrates, nearly all of the oxygen and a significant amount of carbon dioxide are transported by special colored proteins, or **respiratory pigments,** in the blood. In most animals (all vertebrates) these respiratory pigments are packaged into blood cells. This is necessary because, if this amount of respiratory pigment were free in blood, the blood would have the viscosity of syrup and would hardly flow through the blood vessels.

The most widespread respiratory pigment in the animal kingdom is **hemoglobin,** a red, iron-containing protein present in all vertebrates and many invertebrates. Each molecule of hemoglobin is made up of 5% **heme,** an iron-containing compound giving the red color to blood, and 95% **globin,** a colorless protein. The heme portion of the hemoglobin has a great affinity for oxygen; each gram of hemoglobin (there are approximately 15 g of hemoglobin in each 100 ml of human blood) can carry a maximum of approximately 1.3 ml of oxygen; each 100 ml of fully oxygenated blood contains approximately 20 ml of oxygen. Of course, for hemoglobin to be of value to the body it must hold oxygen in a loose, reversible chemical combination so that it can be released to the tissues. The actual amount of oxygen with which hemoglobin can combine depends on small structural changes in the hemoglobin molecule. These structural changes result from variation in the oxygen partial pressure surrounding the blood cells. When the oxygen concentration is high, as it is in the capillaries of the lung alveoli, the hemoglobin can actually combine with more oxygen than it can when the oxygen concentration is low, as it is in the systemic capillaries in the body tissues. Thus when the blood enters regions of low oxygen partial pressure, it is forced to

Although hemoglobin is the only vertebrate respiratory pigment, several other respiratory pigments are known among the invertebrates. *Hemocyanin,* a blue, copper-containing protein, is present in the crustaceans and most molluscs. Among other pigments is *chlorocruorin* (klora-croó-o-rin), a green-colored, iron-containing pigment found in four families of polychaete tube worms. Its structure and oxygen-carrying capacity are very similar to those of hemoglobin, but it is carried free in the plasma rather than being enclosed in blood corpuscles. *Hemerythrin* is a red pigment found in some polychaete worms. Although it contains iron, this metal is not present in a heme group (despite the name of the pigment!), and its oxygen-carrying capacity is poor.

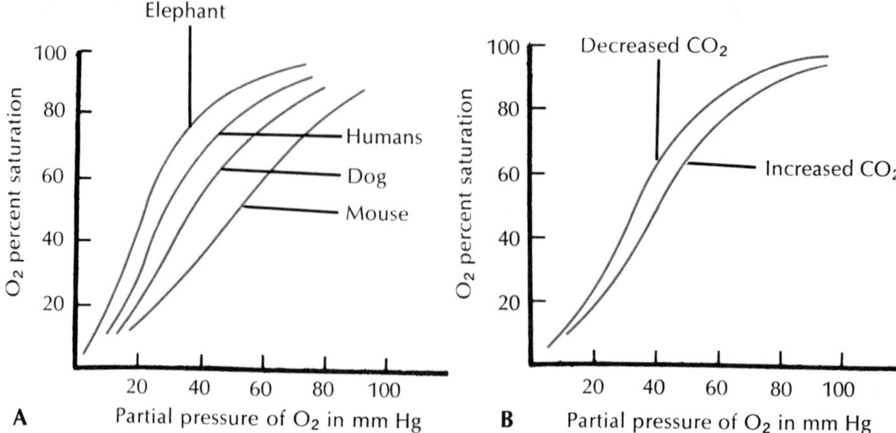

FIGURE 7-23

Oxygen dissociation curves. Curves show how amount of oxygen that can bind to hemoglobin is related to oxygen partial pressure. **A,** Small animals have blood that gives up oxygen more readily than does the blood of large animals. **B,** Hemoglobin is also sensitive to carbon dioxide partial pressure. As carbon dioxide enters blood from the tissues, it shifts the curve to the right, decreasing affinity of hemoglobin for oxygen. Thus the hemoglobin unloads more oxygen in the tissues where carbon dioxide concentration is higher.

Sickle cell anemia is an incurable, inherited condition (p. 86) in which a single amino acid in normal hemoglobin (HbA) is substituted. The ability of sickle cell hemoglobin (HbS) to carry oxygen is severely impaired, and the erythrocytes tend to crumple during periods of oxygen stress (for example, during exercise). Capillaries become clogged with the misshapen red cells; the affected area is very painful, and the tissue may die. About 1 in 10 black Americans carry the trait (heterozygous). Heterozygotes do not have sickle cell anemia and live normal lives, but if both parents are heterozygous, they have a 25% chance of producing offspring with the disease.

give up or unload its oxygen because of its structural change under these conditions. The relationship of carrying capacity to surrounding oxygen concentration is shown by oxygen dissociation curves (Figure 7-23). As these curves show, the lower the surrounding oxygen tension, the greater the quantity of oxygen released. This is an important characteristic because it allows more oxygen to be released to those tissues which need it most (have the lowest partial pressure of oxygen). More oxygen is loaded in the lungs and more unloaded in the tissues than if the combination and release depended on concentration gradients alone.

Another characteristic facilitating the release of oxygen to the tissues is the sensitivity of oxyhemoglobin (hemoglobin with bound oxygen) to carbon dioxide. Carbon dioxide shifts the oxygen dissociation curve to the right (Figure 7-23, *B*), a phenomenon that has been called the **Bohr effect** after the Danish scientist who first described it. Therefore as carbon dioxide enters the blood from the respiring tissues, it encourages the release of additional oxygen from the hemoglobin. The opposite event occurs in the lungs; as carbon dioxide diffuses from the venous blood into the alveolar space, the oxygen dissociation curve shifts back to the left, allowing more oxygen to be loaded onto the hemoglobin.

Transport of carbon dioxide by the blood. The same blood that transports oxygen to the tissues from the lungs must carry carbon dioxide back to the lungs on its return trip. However, unlike oxygen that is transported almost exclusively in combination with hemoglobin, carbon dioxide is transported in three major forms.

1. Most of the carbon dioxide, approximately 67%, is converted in the red blood cells into bicarbonate and hydrogen ions by undergoing the following series of reactions:

$$CO_2 + H_2O \rightleftharpoons \underset{\textbf{Carbonic acid}}{H_2CO_3}$$

This reaction would normally proceed very slowly, but an enzyme in the red blood cells, **carbonic anhydrase,** catalyzes the reaction, enabling it to proceed almost instantly. As soon as carbonic acid forms, it instantly and almost completely ionizes as follows:

$$\underset{\textbf{Carbonic acid}}{H_2CO_3} \rightleftharpoons \underset{\textbf{Bicarbonate ion}}{HCO_3^-} + \underset{\textbf{Hydrogen ion}}{H^+}$$

The hydrogen ion is buffered by several buffer systems in the blood, thus preventing a severe decrease in blood pH. The bicarbonate ion remains in solution in the plasma and red blood cell water since, unlike carbon dioxide, bicarbonate is extremely soluble.

2. Another fraction of the carbon dioxide, approximately 25%, combines reversibly with hemoglobin. Carbon dioxide does not combine with the heme group but with amino groups of several amino acids to form a carbaminohemoglobin. It is carried to the lungs and released as the hemoglobin loads with oxygen.

3. A third small fraction of the carbon dioxide, approximately 8%, is carried as the physically dissolved gas in the plasma and red blood cells.

Summary

The fluid in the body, whether intracellular, plasma, or interstitial, is mostly water, but it has many substances dissolved in it, including electrolytes and proteins. Mammalian blood consists of the fluid plasma and the formed elements, including red and white blood cells and platelets. The plasma has many dissolved solids, as well as dissolved gases. Mammalian red blood cells lose their nucleus during their development and contain the oxygen-carrying pigment, hemoglobin. White blood cells are important defensive elements. Platelets are vital in the process of clotting, necessary to prevent excess blood loss when the blood vessel is damaged. They release a series of factors that activate prothrombin to thrombin, an enzyme that causes fibrinogen to be changed to the gel form, fibrin.

Phagocytosis is one of the most important defense mechanisms in most animals. Specific immune responses of vertebrates are stimulated by antigens. The two arms of the immune response are humoral (mediated by antibodies) and cellular (associated with cell surfaces). The basis of self-nonself recognition lies in the genes of the major histocompatibility complex (MHC), which control production of certain proteins in cell surfaces. A humoral response is stimulated when an antigen is taken up by a macrophage, the macrophage presents parts of the antigen to a T lymphocytes (T helper cells), the T helpers stimulate specific B lymphocytes to proliferate, and finally plasma cells are produced that manufacture antibody to the antigen. Some of the activated B cells become memory cells, which mediate the secondary response. Among other actions, antibodies help destroy invading substances by opsonization and by lysis via complement. Other types of T cells are T inducers, cytotoxic T cells, and T suppressors. Cellular immunity depends on T cells that secrete substances such as interferons. Inflammation is a nonspecific process that is greatly influenced by the body's prior immunizing experience to a particular antigen. People have genetically determined antigens in the surfaces of their red blood cells (ABO blood groups and others); blood types must be compatible in transfusions, or the transfused blood will be agglutinated by antibodies in the recipient.

In a closed circulatory system, the heart pumps blood into arteries, then into arterioles of smaller diameter, through the bed of fine capillaries, through venules, and finally through the veins, which lead back to the heart. In fishes, which have a two-chambered heart with a single atrium and a single ventricle, the blood is pumped to the gills and then directly to the systemic capillaries throughout the body without first returning to the heart. The four-chambered heart of birds and mammals is more efficient than the two-chambered heart of fishes because the blood is pumped through the capillary bed of the lungs by one ventricle, then returned to the heart and pumped through the systemic circulation by another ventricle. One-way flow of blood during the heart's contraction (systole) and relaxation (diastole) is assured by valves between the atria and the ventricles and between the ventricles and the pulmonary arteries and the aorta. Although the heart can beat spontaneously, its rate is controlled by nerves from the central nervous system. The heart muscle uses a great deal of oxygen and has a well-developed coronary blood circulation. The walls of arteries are thicker than those of veins, and the connective tissue in the walls of arteries allows them to expand during systole and contract during diastole. Normal arterial blood pressure (hydrostatic) of humans in systole is 120 mm Hg and in diastole, 80 mm Hg. It decreases to about 40 mm Hg at the arteriolar end of the capillaries, about 15 mm Hg at the venule end of the capillaries, 10 mm Hg in the veins, and finally to near zero at the right atrium. Because the capillary walls are permeable to water, and there is a net osmotic pressure due to proteins in the plasma, water enters the surrounding tissue at the arteriole end and reenters the blood at the venule end of the capillaries. Not all the fluid reenters the blood at this point, but the remainder is collected by the lymphatic system and is returned to the blood through the thoracic duct.

Very small animals can depend on diffusion between the external environment and their tissues or cytoplasm for transport of respiratory gases, but larger animals require specialized organs, such as gills, tracheae, or lungs, for this function. Gills and lungs provide an increased surface area for exchange of respiratory gases between the blood and the environment. Since simple solution of gases in the blood may be insufficient for respiratory needs, many animals have special respiratory pigments and other mechanisms to help transport oxygen and carbon dioxide. The pigment in vertebrates, hemoglobin, undergoes small structural changes that enable it to combine with more oxygen at higher oxygen concentrations than it can at lower concentrations. This makes it possible for hemoglobin to load up more oxygen in the lungs and unload more in the tissues than if the process depended on concentration gradients alone. Carbon dioxide is carried from the tissues to the lungs in the blood as the bicarbonate ion, in combination with hemoglobin, and as the dissolved gas.

REVIEW QUESTIONS

1. Name the chief intracellular electrolytes and the chief extracellular electrolytes.
2. What is the fate of spent erythrocytes in the body?
3. Outline or briefly describe the sequence of events that leads to blood coagulation.
4. Phagocytosis is an important defense mechanism in most animals. How are phagocytes classified? Name two kinds of cells that are phagocytic.
5. Distinguish between T cells and B cells.
6. Outline the sequence of events in a humoral immune response from the introduction of antigen to the production of antibody.
7. Define the following: plasma cell, secondary response, memory cell, complement, opsonization.
8. Give the genotypes of each of the following blood types: A, B, O, AB. What happens when a person with type A gives blood to a person with type B? With type AB? With type O?
9. Trace the flow of the blood through the heart and pulmonary circulation, naming each of the chambers, valves, arteries, and veins through which the blood passes.
10. Explain why there is a net filtration pressure at the arteriole end of capillaries in humans, and explain the movement of fluid through the walls of the capillaries.
11. What is an advantage of a fish's gills for breathing in water and a disadvantage for breathing on land?
12. Trace the route of inspired air in humans from the nostrils to the smallest chamber of the lungs.
13. Explain the mechanism of breathing and how breathing is controlled.
14. Explain how oxygen is carried in the blood, including specifically the role of hemoglobin. Answer the same question but with regard to carbon dioxide transport.

SELECTED REFERENCES

Bellanti, J.A. 1985. Immunology III. Philadelphia, W.B. Saunders Co. *A good introductory immunology textbook.*

Gallo, R.C. and L. Montagnier. 1988. AIDS in 1988. Sci. Am. **259:**40-48 (Oct.). *Authors are the American and French scientists who independently discovered the AIDS virus. Lead article in an issue of Scientific American entirely devoted to current status and knowledge of AIDS.*

Lawn, R.M., and G.A. Vehar. 1986. The molecular genetics of hemophilia. Sci. Am. **254:**48-54 (Mar.). *The gene coding for the factor in the clotting cascade in the most common form of hemophilia has been isolated and cloned with recombinant DNA techniques. Prospects are good for a safe (virus-free), abundant source of the factor for treating hemophiliacs.*

Marrack, P., and J. Kappler. 1986. The T cell and its receptor. Sci. Am. **254:**36-45 (Feb.). *The molecule on the surface of T cells that recognizes antigen along with MHC protein has been identified.*

Perutz, M.F. 1978. Hemoglobin structure and respiratory transport. Sci. Am. **240:**92-125 (Dec.). *Hemoglobin transports oxygen and carbon dioxide between the lungs and tissues by clicking back and forth between two structures. Perutz and J.C. Kendrew won the Nobel Prize in 1962 for discovering the structure of hemoglobin.*

Randall, D.J., W.W. Burggren, A.P. Farrell, and M.S. Haswell. 1981. The evolution of air breathing in vertebrates. Cambridge, England, Cambridge University Press. *Traces the physiology of air breathing from aquatic ancestors.*

Robinson, T.F., S.M. Factor, and E.H. Sonnenblick. 1986. The heart as a suction pump. Sci. Am. **254:**84-91 (June). *Suggests that filling of heart in diastole is aided by elastic recoil of energy from systole.*

Tonegawa, S. 1985. The molecules of the immune system. Sci. Am. **253:**122-131 (Oct.) *Excellent review of structure and function of antibody molecules, T-cell surface receptors, and the genetics of antibody diversity. Tonegawa won the Nobel Prize in 1987 for his work.*

Zucker, M.B. 1980. The functioning of the blood platelets. Sci. Am. **242:**86-103 (June). *The small blood elements that act to stop blood flow from a wound also perform complex roles in health and disease.*

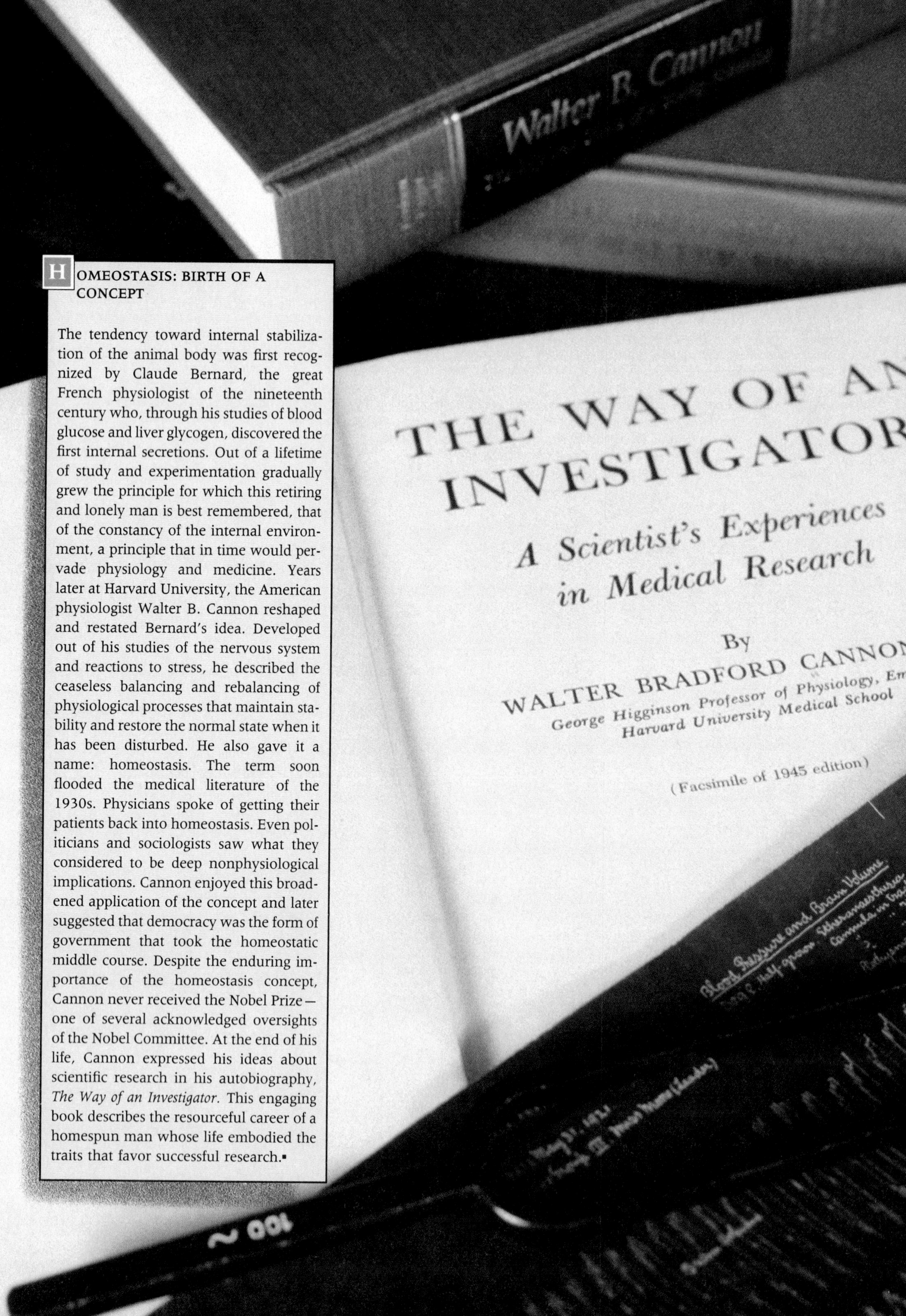

HOMEOSTASIS: BIRTH OF A CONCEPT

The tendency toward internal stabilization of the animal body was first recognized by Claude Bernard, the great French physiologist of the nineteenth century who, through his studies of blood glucose and liver glycogen, discovered the first internal secretions. Out of a lifetime of study and experimentation gradually grew the principle for which this retiring and lonely man is best remembered, that of the constancy of the internal environment, a principle that in time would pervade physiology and medicine. Years later at Harvard University, the American physiologist Walter B. Cannon reshaped and restated Bernard's idea. Developed out of his studies of the nervous system and reactions to stress, he described the ceaseless balancing and rebalancing of physiological processes that maintain stability and restore the normal state when it has been disturbed. He also gave it a name: homeostasis. The term soon flooded the medical literature of the 1930s. Physicians spoke of getting their patients back into homeostasis. Even politicians and sociologists saw what they considered to be deep nonphysiological implications. Cannon enjoyed this broadened application of the concept and later suggested that democracy was the form of government that took the homeostatic middle course. Despite the enduring importance of the homeostasis concept, Cannon never received the Nobel Prize—one of several acknowledged oversights of the Nobel Committee. At the end of his life, Cannon expressed his ideas about scientific research in his autobiography, *The Way of an Investigator.* This engaging book describes the resourceful career of a homespun man whose life embodied the traits that favor successful research.▪

CHAPTER 8

Homeostasis

Osmotic Regulation, Excretion, and Temperature Regulation

At the beginning of the preceding chapter we described the double-layered environment of the body's cells: the extracellular fluid, which immediately surrounds the cells, and the external environment of the outside world.

The life-supporting metabolic activities that occur within the body's cells can proceed only as long as they are bathed by a protective extracellular fluid of relatively constant composition and protected from extremes in environmental temperature. Yet an animal's world is seldom constant, varying not only in temperature but in the nutrients and other materials necessary for life. The animal itself requires a steady supply of these materials for its metabolic activities, which in turn produce heat and a continuous flow of products and wastes. Thus there are many elements within and without the animal that threaten to throw the protected cellular system out of balance.

This means that the fixity of body composition that is required for life must be maintained despite the continuous shifting of components within the system. This kind of stability is referred to as a dynamic **steady state,** or **homeostasis** ("same state"), the term coined by W.B. Cannon (Figure 8-1) and described in the prologue to this chapter.

Homeostasis is maintained by the coordinated activities of numerous body systems, such as the circulatory system, nervous system, endocrine system, and especially the organs that serve as sites of exchange with the external environment, which include the kidneys, lungs or gills, digestive tract, and skin. Through these organs oxygen, foodstuffs, minerals, and other constituents of the body fluids enter, water is exchanged, heat is lost, and metabolic wastes are eliminated.

We will look first at the problems of controlling the internal fluid environment of animals living in aquatic habitats. Next we will briefly examine how these problems are solved by terrestrial animals and consider the function of the organs that regulate the internal state. Finally we will look at the different ways animals solve the problem of living in a world of changing temperatures.

FIGURE 8-1

Walter Bradford Cannon (1871-1945), Harvard professor of physiology who coined the term *homeostasis* and developed the concept originated by French physiologist Claude Bernard (Figure 7-1).

From Fulton, J.F., and L.G. Wilson. 1966. Selected readings in the history of physiology. Springfield, Ill., Charles C Thomas, Publisher.

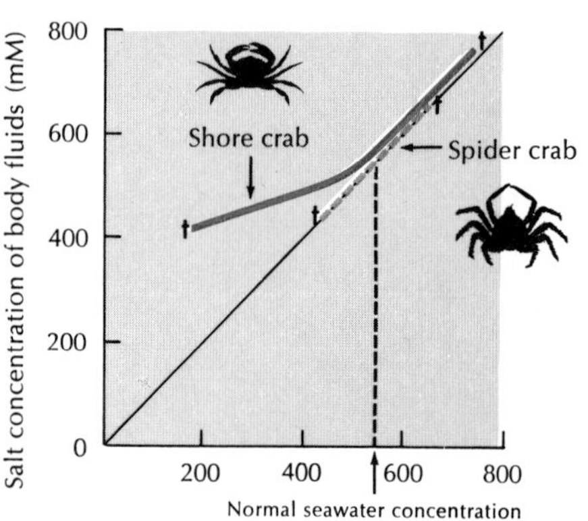

FIGURE 8-2

Salt concentration of body fluids of two crabs as affected by variations in the seawater concentration. The 45-degree line represents equal concentration between body fluids and seawater. Since the spider crab cannot regulate its body-fluid salt concentration, it conforms to whatever changes happen in the external seawater environment. The shore crab, however, can regulate osmotic concentration of its body fluids to some degree because in dilute seawater the shore crab can hold its body-fluid concentration above seawater concentration. For example, when seawater is 200 mM (millimolar), the shore crab's body fluid concentration is approximately 430 mM. Crosses at ends of lines indicate tolerance limits for each species.

Osmotic Regulation

HOW MARINE INVERTEBRATES MAINTAIN SALT AND WATER BALANCE

Most marine invertebrates are in osmotic equilibrium with their seawater environment. They have body surfaces that are permeable to salts and water so that their body fluid concentration rises or falls in conformity with changes in concentrations of seawater. Because such animals are incapable of regulating their body fluid osmotic pressure, they are referred to as **osmotic conformers.** Invertebrates living in the open sea are seldom exposed to osmotic fluctuations because the ocean is a highly stable environment. Oceanic invertebrates have, in fact, very limited abilities to withstand osmotic change. If they should be exposed to dilute seawater, they die quickly because their body cells cannot tolerate dilution and are helpless to prevent it. The animals are restricted to living in a narrow salinity range and are said to be **stenohaline** (Gr. *stenos,* narrow, + *hals,* salt). An example is the marine spider crab, represented in Figure 8-2.

Conditions along the coasts and in estuaries and river mouths are much less constant than those of the open ocean. Here animals must be able to withstand large and often abrupt salinity changes as the tides ebb and flow and mix with fresh water draining from rivers. These animals are referred to as **euryhaline** (Gr. *eurys,* broad, + *hals,* salt), meaning that they can survive a wide range of salinity changes, mainly because they demonstrate varying powers of **osmotic regulation.** For example, the brackish-water shore crab can resist body fluid dilution by dilute (brackish) seawater (Figure 8-2). Although the body fluid concentration falls, it does so less rapidly than the fall in seawater concentration. This crab is a **hyperosmotic regulator** because in a dilute environment it can maintain the concentration of its blood above that of the surrounding water.

What is the advantage of hyperosmotic regulation over osmotic conformity, and how is this regulation accomplished? The advantage is that by regulating against excessive dilution, thus protecting the body cells from extreme changes, these crabs can successfully live in the physically unstable but biologically rich coastal environment. Their powers of regulation are limited, however, since if the water is highly diluted, their regulation fails and they die.

To understand how the brackish-water shore crab and other coastal invertebrates achieve hyperosmotic regulation, let us examine the problems they face. First, the salt concentration of the internal fluids is greater than in the dilute seawater outside. This causes a steady osmotic influx of water. As with the membrane osmometer placed in a sugar solution (p. 37), water diffuses inward because it is more concentrated outside than inside. The shore crab is not nearly as permeable as a membrane osmometer—most of its shelled body surface is, in fact, almost impermeable to water—but the thin respiratory surfaces of the gills are highly permeable. Obviously the crab cannot insulate its gills with an impermeable hide and still breathe. The problem is solved by removing the excess water through the action of the kidney, known as the antennal gland, located in the crab's head.

The second problem is salt loss. Again, because the animal is saltier than its environment, it cannot avoid loss of ions by outward diffusion across the gills. Salt is also lost in the urine. The problem is solved by special salt-secreting cells in the gills that can actively remove ions from the dilute seawater and move them into the blood, thus maintaining the internal osmotic concentration. This is an **active transport** process that requires energy because ions must be transported against a concentration gradient, that is, from a lower salt concentration (in the dilute seawater) to an already higher one (in the blood). Active transport is described and illustrated on p. 38.

INVASION OF FRESH WATER

Some 400 million years ago, during the Silurian and Lower Devonian periods, the major groups of jawed fishes began to penetrate into brackish-water estuaries and then gradually into freshwater rivers. Before them lay a new, unexploited habitat already stocked with food in the form of insects and other invertebrates, which had preceded them into fresh water. However, the advantages of this new habitat were traded off for a tough physiological challenge: the necessity of developing effective osmotic regulation.

Freshwater animals must keep the salt concentration of their body fluids higher than that of the water. Water enters their bodies osmotically, and salt is lost by diffusion outward. Their problems are similar to those of the brackish-water crab, but more severe and unremitting. Fresh water is much more dilute than are coastal estuaries, and there is no retreat, no salty sanctuary into which the freshwater animal can retire for osmotic relief. It must and has become a permanent and highly efficient hyperosmotic regulator.

The scaled and mucus-covered body surface of a fish is about as waterproof as any flexible surface can be. In addition, freshwater fishes have several defenses against the problems of water gain and salt loss (see Figure 8-4). First, water that inevitably enters by osmosis across the gills is pumped out by the kidney, which is capable of forming a very dilute urine. Second, special salt-absorbing cells located in the gills move salt ions, principally sodium and chloride (present in small quantities even in fresh water), from the water to the blood. This, together with salt present in the fish's food, replaces diffusive salt loss. These mechanisms are so efficient that a freshwater fish devotes only a small part of its total energy expenditure to keeping itself in osmotic balance.

Crayfishes, aquatic insect larvae, mussels, and other freshwater animals are also hyperosmotic regulators and face the same hazards as freshwater fishes; they tend to gain too much water and lose too much salt. Not surprisingly, all of these forms solved these problems in the same direct way that fishes did. They excrete the excess water as urine, and they actively absorb salt from the water by some salt-transporting mechanism on the body surface.

Amphibians, when they are living in water, also must compensate for salt loss by actively absorbing salt from the water (Figure 8-3). They use their skin for this purpose. Physiologists learned some years ago that pieces of frog skin continue to transport sodium and chloride actively for hours when removed and placed in a specially balanced salt solution. Fortunately for biologists, but unfortunately for frogs, these animals are so easily collected and maintained in the laboratory that frog skin became a favorite membrane system for studies of ion-transport phenomena.

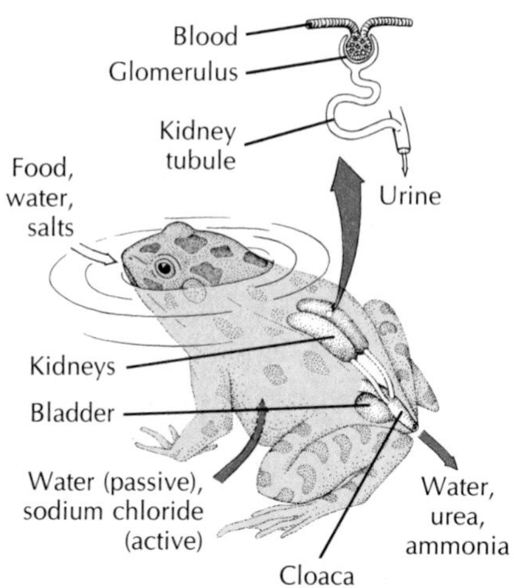

FIGURE 8-3

Water and solute exchange in a frog. Water enters the highly permeable skin and is excreted by the kidney. The skin also actively transports ions (sodium chloride) from the environment. The kidney forms a dilute urine by reabsorbing sodium chloride. Urine flows into the urinary bladder, where, during temporary storage, most of the remaining sodium chloride is removed and returned to the blood.

Modified from Webster, D., and M. Webster. 1974. Comparative vertebrate morphology. New York, Academic Press, Inc.

RETURN OF FISHES TO THE SEA

The marine bony fishes maintain the salt concentration of their body fluids at approximately one third that of seawater (body fluids = 0.3 to 0.4 gram moles per liter [M]; seawater = 1 M). They are **hypoosmotic regulators** because their body fluids are substantially more dilute than their seawater environment. Bony fishes living in the oceans today are descendants of earlier freshwater bony fishes that moved back into the sea during the Triassic period approximately 200 million years ago. The return to their ancestral sea was probably prompted by unfavorable climatic conditions on land and the deterioration of freshwater habitats, but we can only guess at the reasons. During the many millions of years that the freshwater fishes were adapting themselves so well to their environment, they established a body fluid concentration equivalent to approximately one third that of seawater, thus setting the pattern for all the vertebrates that were to evolve later, whether aquatic, terrestrial, or aerial. The ionic composition of vertebrate

When we express seawater or body fluid concentration in molarity, we are saying that the osmotic strength is equivalent to the molar concentration of an ideal solute having the same osmotic strength. In fact, seawater and animal body fluids are not ideal solutions because they contain electrolytes that dissociate in solution. A 1 M solution of sodium chloride (which dissociates in solution) has a much greater osmotic strenth than a 1 M solution of glucose, an ideal solute that does not dissociate in solution. Consequently, biologists usually express the osmotic strength of a biological solution in osmolarity rather than in molarity. A 1 osmolar solution exerts the same osmotic pressure as a 1 M solution of a nonelectrolyte.

body fluid is remarkably similar to that of dilute seawater too, a fact that is undoubtedly related to their marine heritage.

When some of the freshwater bony fishes of the Triassic period ventured back to the sea, they encountered a new set of problems. Having a much lower internal osmotic concentration than the seawater around them, they lost water and gained salt. Indeed the marine bony fish literally risks drying out, much like a desert mammal deprived of water.

Marine bony fishes, like their freshwater counterparts, have evolved an approximate set of compensation mechanisms (Figure 8-4). To offset water loss, the marine fish drinks seawater. Although this behavior brings needed water into the body, it is unfortunately accompanied by a great deal of unneeded salt. The latter is expelled in two ways: (1) the major sea salts (sodium, chloride, and potassium) are pumped out of the body by special salt-secretory cells located in the gills, and (2) the divalent ions remaining in the intestine, in particular magnesium, sulfate, and calcium, are voided with the feces or excreted by the kidney. In this roundabout way, marine fishes rid themselves of the excess sea salts they have drunk, resulting in a net gain of water, which replaces the water lost by osmosis. Samuel Taylor Coleridge's ancient mariner, surrounded by "water, water, everywhere, nor any drop to drink" would undoubtedly have been tormented even more had he known of the marine fishes' simple solution for thirst. A marine fish carefully regulates the amount of seawater it drinks, consuming only enough to replace water loss and no more.

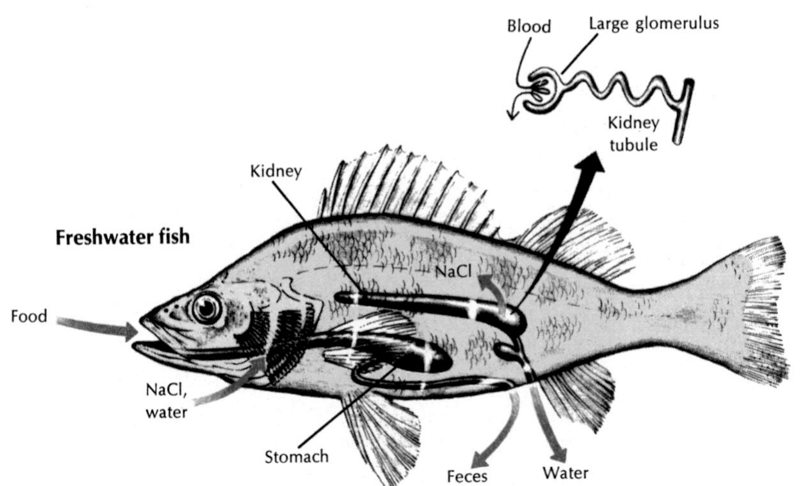

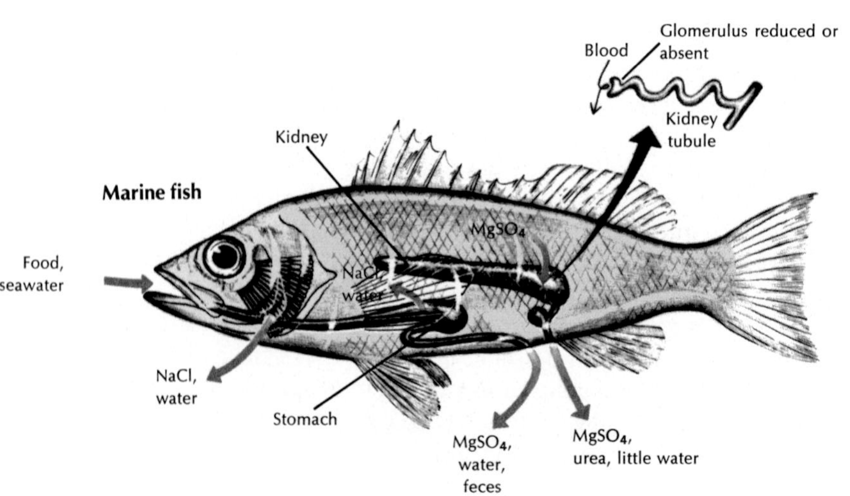

FIGURE 8-4

Osmotic regulation in freshwater and marine bony fishes. A freshwater fish maintains osmotic and ionic balance in its dilute environment by actively absorbing sodium chloride across the gills (some salt enters with food). To flush out excess water that constantly enters the body, the glomerular kidney produces a dilute urine by reabsorbing sodium chloride. A marine fish must drink seawater to replace water lost osmotically to its salty environment. Sodium chloride and water are absorbed from the stomach. Excess sodium chloride is secreted outward by the gills. Divalent sea salts, mostly magnesium sulfate, are eliminated with feces and secreted by the tubular kidney.

Modified from Webster, D., and M. Webster. 1974. Comparative vertebrate morphology. New York, Academic Press, Inc.

Table 8-1 Water balance in a human and a kangaroo rat, a desert rodent

	Human (%)	Kangaroo rat (%)
Gains		
Drinking	48	0
Free water in food	40	10
Metabolic water	12	90
Losses		
Urine	60	25
Evaporation (lungs and skin)	34	70
Feces	6	5

Some data from Schmidt-Nielsen, K. 1972. How animals work. New York, Cambridge University Press.

The high concentration of urea in the blood of sharks and their kin—more than 100 times as high as in mammals—could not be tolerated by most other vertebrates or invertebrates. In these animals, such high concentrations of urea disrupt the peptide bonds of proteins, thus altering protein configuration. Sharks have adapted biochemically to the presence of the urea that permeates all their body fluids, even penetrating freely into the cells. So accommodated are these fishes to urea that their tissues cannot function without it, and the heart will stop beating in its absence.

The cartilaginous fishes (sharks and rays) solve their water balance problems in a completely different way. This primitive group is almost totally marine. The salt composition of shark's blood is similar to that of the bony fishes, but the blood also carries a large content of organic compounds, especially **urea** and **trimethylamine oxide.** Urea is, of course, a metabolic waste that most animals quickly excrete. The shark kidney, however, conserves urea, causing it to accumulate in the blood. Added to the usual blood electrolytes, these nitrogenous compounds raise the blood osmotic pressure to exceed slightly that of seawater. In this way the sharks and their kin turn an otherwise useless waste material into an asset, eliminating the osmotic problem encountered by the marine bony fishes.

HOW TERRESTRIAL ANIMALS MAINTAIN SALT AND WATER BALANCE

The problems of living in an aquatic environment seem small indeed compared with the problems of life on land. Since animal bodies are mostly water, all metabolic activities proceed in water, and life itself was conceived in water, it would seem that animals were meant to stay in water. Yet many animals, like the plants preceding them, moved onto land, carrying their watery composition with them. Once on land, the terrestrial animals continued their adaptive radiation, solving the threat of desiccation, until they became abundant even in some of the most arid parts of the earth.

Terrestrial animals lose water by evaporation from the lungs and body surface, excretion in the urine, and elimination in the feces. Such losses are replaced by water in the food, drinking water if it is available, and forming **metabolic water** in the cells by oxidation of foodstuffs, especially carbohydrates. Certain insects—for example, desert roaches, certain ticks and mites, and the mealworm—are able to absorb water vapor directly from atmospheric air. In some desert rodents, metabolic water may constitute most of the animal's water gain.

Particularly revealing is a comparison (Table 8-1) of water balance in the human being, a nondesert mammal that drinks water, with that of the kangaroo rat, a desert rodent that may drink no water at all.

The kangaroo rat gains all of its water from its food (90% as metabolic water derived from the oxidation of foodstuffs, 10% as free moisture in the food). Even though humans eat foods with a much higher water content than the dry seeds that comprise much of the kangaroo rat's diet, people must still drink half their total water requirement.

Given ample water to drink, human beings can tolerate extremely high temperatures while preventing a rise in body temperature. Our ability to keep cool by evaporation was impressively demonstrated more than 200 years ago by a British scientist who remained for 45 minutes in a room heated to 260° F (126° C). A steak he carried in with him was thoroughly cooked but he remained uninjured and cool. Sweating rates may exceed 3 liters of water per hour under such conditions and cannot be long tolerated unless the lost water is replaced by drinking. Without water, a human continues to sweat unabated until the water deficit exceeds 10% of the body weight, when collapse occurs. With a water deficit of 12% a human is unable to swallow even if offered water, and death occurs when the water deficit reaches about 15% to 20%. Few people can survive more than a day or two in a desert without water. Thus people are not physiologically well adapted for desert climates, but prosper there nonetheless by virtue of their technological culture.

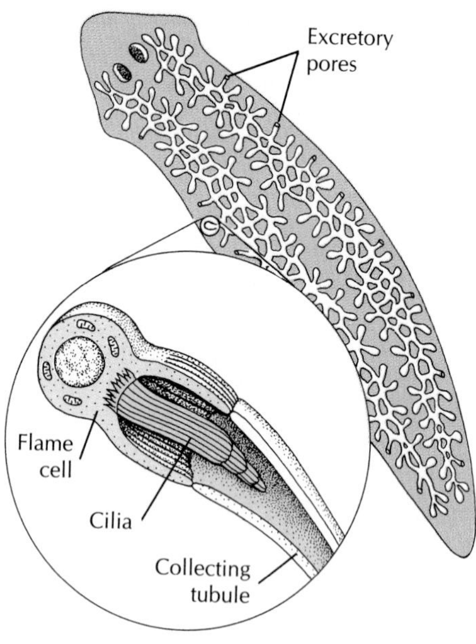

FIGURE 8-5

Flame cell system of a flatworm. Body fluids collected by flame cells (protonephridia) are passed down a system of ducts to excretory pores on the body surface.

The excretion of wastes presents a special problem in water conservation. The strategies that have evolved among terrestrial animals to excrete toxic ammonia, the primary end-product of protein catabolism, without losing water, are described in Chapter 2 (p. 56).

Marine birds and turtles have evolved a unique situation for excreting the large loads of salt eaten with their food. Located above each eye is a special **salt gland** capable of excreting a highly concentrated solution of sodium chloride—up to twice the concentration of seawater. In birds the salt solution runs out the nares (see p. 617). Marine lizards and turtles, like Alice in Wonderland's Mock Turtle, shed their salt gland secretion as salty tears. Salt glands are important accessory organs of salt excretion to these animals because their kidney cannot produce a concentrated urine, as can the mammalian kidney.

Invertebrate Excretory Structures

In such a variety of groups as make up the invertebrates, it is hardly surprising that there is a great variety of morphological structures serving as excretory organs. Many protozoa and some freshwater sponges have special excretory organelles called contractile vacuoles. The more advanced invertebrates have excretory organs that are basically tubular structures that form urine by first producing an ultrafiltrate or fluid secretion of the blood. This enters the proximal end of the tubule and is modified continuously as it flows down the tuble. The final product is urine.

CONTRACTILE VACUOLE

The tiny spherical intracellular vacuole of protozoa and freshwater sponges is not a true excretory organ, since ammonia and other nitrogenous wastes of metabolism readily leave the cell by direct diffusion across the cell membrane into the surrounding water. The contractile vacuole is really an organ of water balance. Because the cytoplasm of freshwater protozoa is considerably saltier than their freshwater environment, they tend to draw water into themselves by osmosis. In *Amoeba proteus* this excess water collects in numerous tiny vesicles surrounding the single thin membrane of the contractile vacuole (see Figure 15-5, p. 353). These vesicles then fuse with the vacuolar membrane, emptying their contents (a weak salt solution) into the contractile vacuole. This grows larger as water accumulates within it. Finally the vacuole is emptied through a pore on the surface, and the cycle is rhythmically repeated. Although the mechanism for filling is not fully understood, it is noteworthy that a layer of mitochondria surrounds the contractile vacuole. These are believed to provide energy for reabsorbing salts from the tiny vesicles as they are formed. This mechanism would create a hypoosmotic solution for excretion, while conserving valuable salts.

In *Paramecium* and many other ciliates, the contractile vacuole is a more complex structure, having a pore leading to the outside and several ampullae and canals radiating out from the contractile vacuole (see Figure 15-17, p. 361).

Contractile vacuoles are common in freshwater protozoa, sponges, and radiate animals (such as hydra), but rare or absent in marine forms of these groups that are isosmotic with seawater and consequently neither lose nor gain too much water.

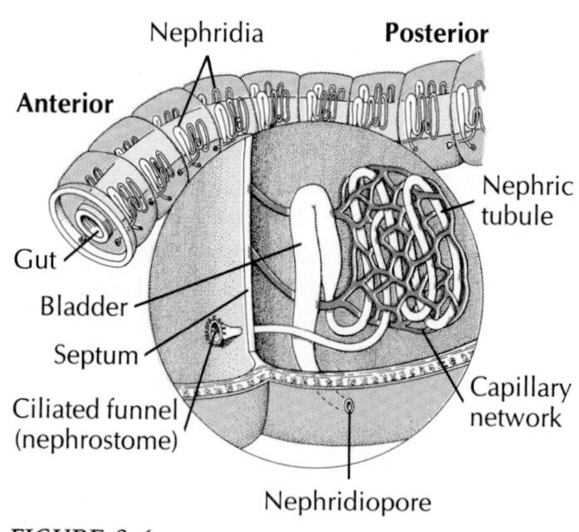

FIGURE 8-6

Excretory system of an earthworm. Each segment has a pair of large nephridia suspended in a fluid-filled coelom. Each nephridium occupies two segments because the ciliated funnel (nephrostome) drains the segment anterior to the segment containing the rest of the nephridium.

NEPHRIDIUM

The most common type of invertebrate excretory organ is the nephridium, a tubular structure designed to maintain the appropriate osmotic balance. One of the simplest arrangements is the flame cell system (or **protonephridium**) of the acoelomate flatworms and pseudocoelomate roundworms and their relatives.

In planaria and other flatworms the protonephridial system takes the form of two highly branched duct systems distributed throughout the body (Figure 8-5). Fluid enters the system through specialized "flame cells," moves slowly into and down the tubules, and is excreted through pores that open at intervals on the body surface. The rhythmical beat of the ciliary tuft, which resembles a tiny flickering flame, creates a negative pressure that draws fluid through delicate interdigitations between the flame cell and the tubule cell and drives it into the tubular portion of the system. In the tubule, certain molecules and ions are recovered by reabsorption, leaving wastes behind to be expelled. Nitrogenous wastes (mainly ammonia) diffuse across the body surface of the animal.

Note that the flame cell system, with its extensive branching throughout the body, is very unlike the condensed kidneys of vertebrates and the more advanced invertebrates. This is necessary in primitive invertebrates lacking a circulatory system which, in more advanced animals, carries the wastes to tubules that are gathered together into a compact excretory organ.

The protonephridium just described is a **closed** system. The tubules are closed on the inner end, and urine is formed from a fluid that must first enter the tubules by being transported across flame cells. A more advanced type of nephridium is the **open,** or "true," nephridium (**metanephridium**) that is found in several of the eucoelomate phyla such as the annelids (Figure 8-6), the molluscs, and several smaller phyla. The metanephridium is more advanced than the protonephridium in two important ways. First, the tubule is open at *both* ends, allowing fluid to be swept into the tubule through a ciliated funnel-like opening (**nephrostome**). Second, the metanephridium is surrounded by a network of blood vessels that help form urine by reabsorbing salts and other valuable materials from the tubular fluid. Despite these improvements, the basic process of urine formation is the same in protonephridia and metanephridia: fluid flows continuously through a tubule while materials are added here and taken away there, until urine is formed. We will see that the advanced kidneys of vertebrates operate in basically the same way.

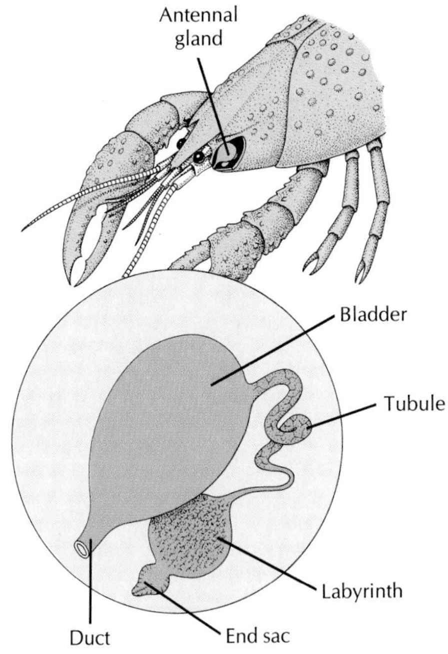

FIGURE 8-7

Antennal glands of a crayfish. These are filtration kidneys, that is, a filtrate of the blood is formed in the end sac. The filtrate is converted into urine as it passes down the tubule toward the bladder.

ARTHROPOD KIDNEYS

The **antennal glands** of crustaceans form a single, paired tubular structure located in the ventral part of the head (Figure 8-7). These excretory devices are an advanced design of the basic nephridial organ. However, they lack open nephrostomes. Instead, a protein-free filtrate of the blood (ultrafiltrate) is formed in the end sac by the hydrostatic pressure of the blood. In the tubular portion of the gland, the filtrate is modified by the selective reabsorption of certain salts and the active secretion of others. Thus crustaceans have excretory organs that are basically vertebrate-like in the functional sequence of urine formation.

Insects and spiders have a unique excretory system consisting of **malpighian tubules** that operate in conjunction with specialized glands in the wall of the rectum (Figure 8-8). The thin, elastic, blind malpighian tubules are closed and lack an arterial supply. Consequently urine formation cannot be initiated by blood ultrafiltration as in the crustaceans and vertebrates. Instead, salts, largely potassium, are actively secreted into the tubules from the hemolyph. This primary secretion of ions creates an osmotic drag that pulls water, solutes, and nitrogenous wastes, especially uric acid, into the tubule. Uric acid enters the upper end of the tubule as a soluble potassium urate, then precipitates out as insoluble uric acid in the proximal end of the tubule. Once the formative urine drains into the rectum, most of the water and potassium are reabsorbed by specialized rectal glands, leaving behind uric acid and other nitrogenous wastes that are disposed of in the feces. This unique excretory system is ideally suited for life in dry environments and has contributed to the great success of insects on land.

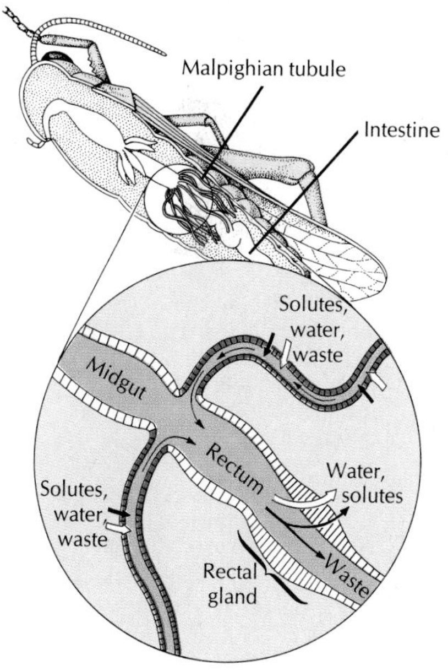

FIGURE 8-8

Malpighian tubules of insects. Malpighian tubules are located at the juncture of the midgut and hindgut (rectum). Solutes, especially potassium, are actively secreted into the tubules from the surrounding arthropod hemolymph. Water and wastes follow. This fluid drains into the rectum, where solutes and water are actively reabsorbed, leaving wastes to be excreted.

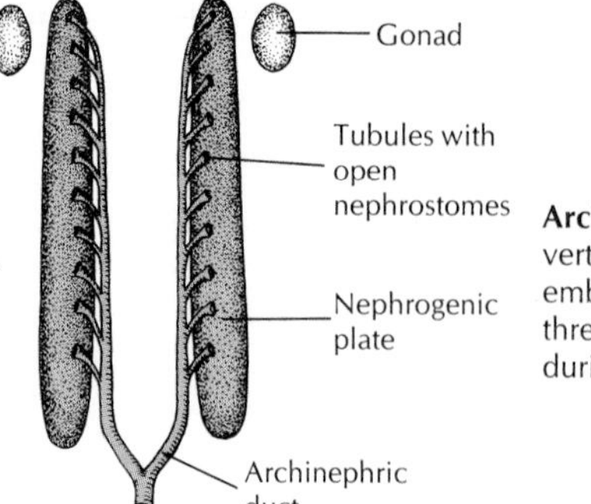

Archinephros: Ancestral prototype of vertebrate kidney and kidney found in embryo of hagfish; from this prototype, three successive kidneys evolved during vertebrate evolution

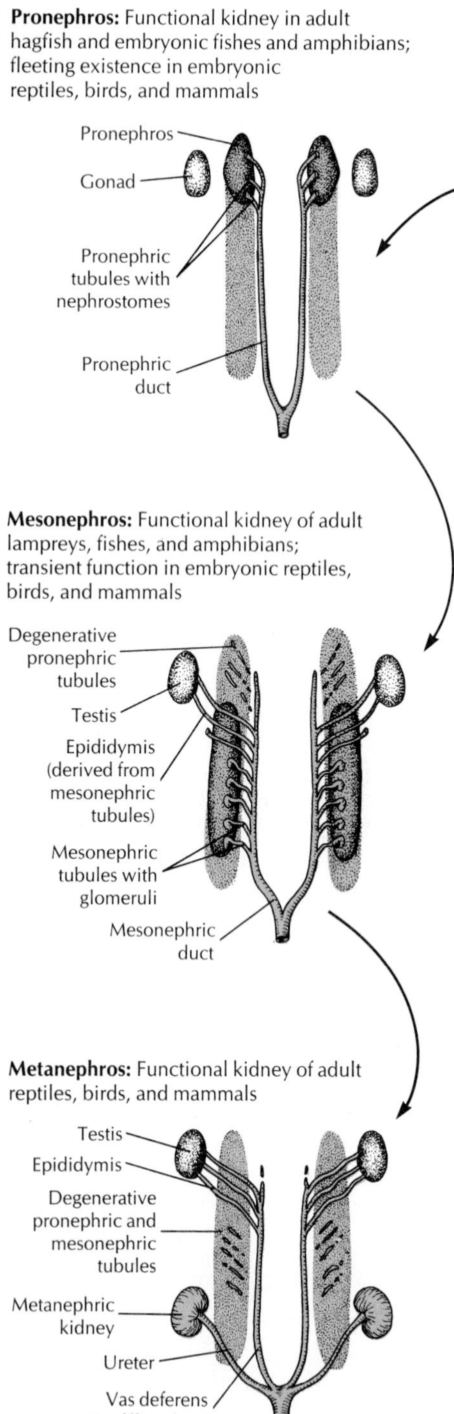

FIGURE 8-9

Evolution of male vertebrate kidney from archinephric prototype. *Pink,* Functional structures, *Tan,* Degenerative or undeveloped parts.

Vertebrate Kidneys

ANCESTRY AND EMBRYOLOGY

Reconstructing the evolution of the vertebrate kidney is admittedly difficult because soft organs such as kidneys are seldom preserved in fossils. Fortunately, a reasonably accurate record has persisted in the embryological development of the kidneys of living forms. Embryology is highly conservative, and developmental sequences, once fixed in the genes, are not easily modified. From this record it is believed that the kidney of the earliest vertebrates extended the length of the coelomic cavity and was made up of segmentally arranged tubules, each resembling an invertebrate nephridium. Each tubule opened at one end into the coelom by a nephrostome and at the other end into a common **archinephric duct.** This ancestral kidney has been called an **archinephros** ("first kidney"), and a segmental kidney very similar to the archinephros is found in the embryos of hagfishes and caecilians (Figure 8-9). Almost from the beginning, the reproductive system, which develops beside the excretory system from the same segmental blocks of trunk mesoderm, used the nephric ducts as a convenient conducting system for reproductive products. Thus, even though the two systems have nothing functionally in common, they are closely associated in their use of common ducts.

Kidneys of living vertebrates developed from this primitive plan. During embryonic development of higher vertebrates, there is a succession of three developmental stages of kidneys: **pronephros, mesonephros,** and **metanephros.** In all vertebrate embryos, the pronephros is the first and most primitive kidney to appear. As its name implies, it is located anteriorly in the body and becomes part of the persistent kidney in adult hagfishes. In all other vertebrates it degenerates during development and is replaced by a more centrally located mesonephros. The mesonephros is the functional kidney of embryonic amniotes (reptiles, birds, and mammals) and contributes to the adult kidney (called an opisthonephros) of fishes and amphibians.

The most advanced vertebrate kidney, the metanephros of the amniotes, is distinguished in several ways from the pronephros and metanephros. It is more caudally located and it is a more compact structure containing a very large number of nephric tubules. It is drained by a new duct, the **ureter,** which developed when the old archinephric duct was relinquished to the reproductive system of the male for sperm transport. Thus the three successive kidney types—pronephros, mesonephros, metanephros—succeed each other embryologically, and to some extent phylogenetically as well. Each type gives rise to a more functionally advanced adult kidney, each more compact and located more caudally than its predecessor.

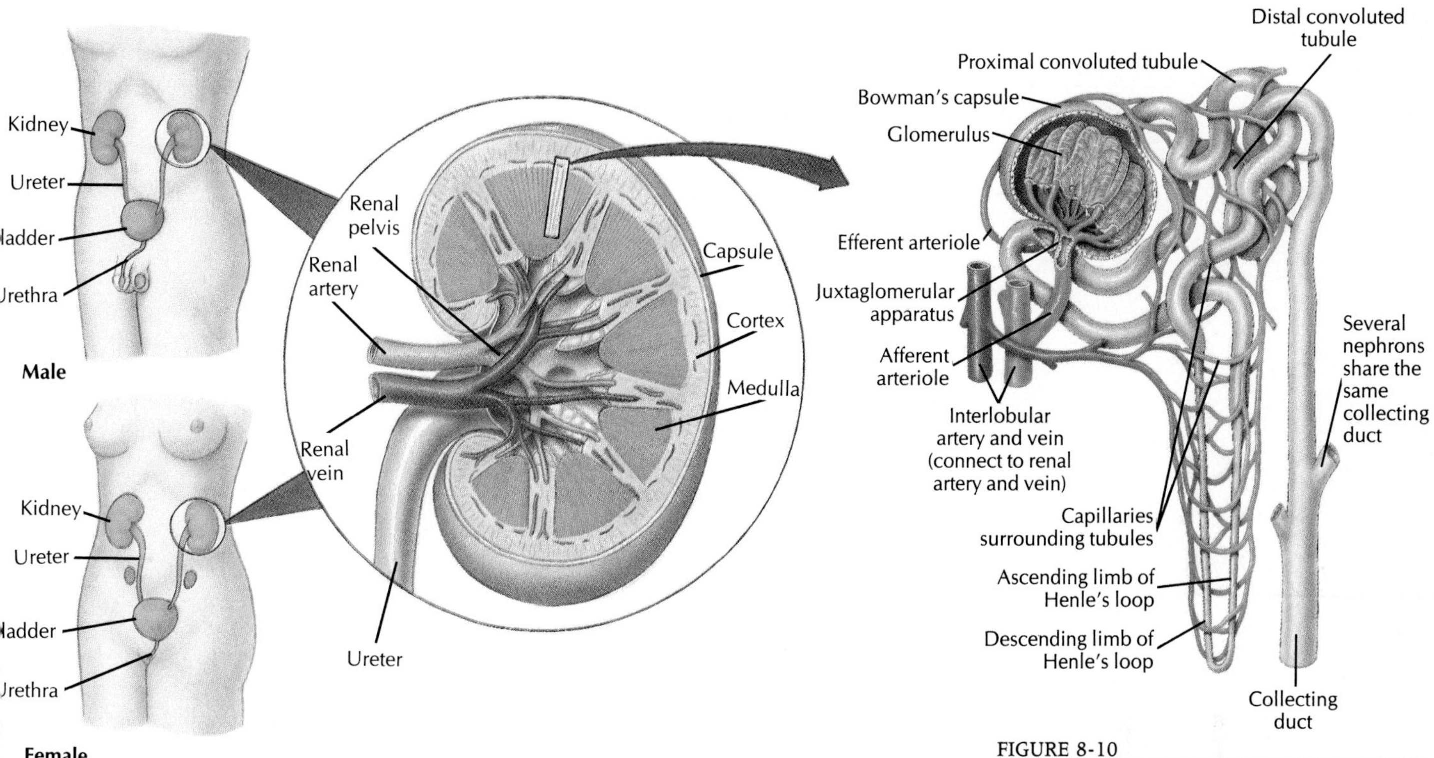

FIGURE 8-10

Urinary system of humans, with enlargements showing detail of the kidney and a single nephron.

VERTEBRATE KIDNEY FUNCTION

The kidneys of humans and other vertebrates play a critical role in the body's economy. Their failure means death; in this respect they are neither more nor less important than are the heart, lungs, and liver. The kidney is a part of many interlocking mechanisms that maintain homeostasis—constancy of the internal environment. However, the kidney's share in this regulatory council is an especially large one. It must and does individually monitor and regulate most of the major constituents of the blood and several minor constituents as well. In addition, it silently labors to remove a variety of potentially harmful substances that animals deliberately or unconsciously eat, drink, or inhale.

Perhaps even more remarkable is the way in which the kidney does its job. These small organs, which in human beings weigh less than 0.5% of the body weight, receive nearly 25% of the total cardiac output, amounting to approximately 2000 liters of blood per day! The vast blood flow is channeled to approximately 2 million **nephrons,** which comprise the bulk of the two human kidneys. Each nephron is a tiny excretory unit consisting of a pressure filter (**glomerulus**) and a long **nephric tubule.** Urine formation begins in the glomerulus where an ultrafiltrate of the blood is squeezed into the nephric tubule by the hydrostatic blood pressure. The ultrafiltrate then flows steadily down the twisted tubule. During its travel some substances are added to and others are subtracted from the ultrafiltrate. The final product of this process is urine.

All mammalian kidneys are paired structures that lie embedded in fat, anchored against the dorsal abdominal wall. The two **ureters,** 25 to 30 cm (10 to 12 inches) long in humans, extend from the **renal pelvis** to the dorsal surface of the **urinary bladder.** Urine is discharged from the bladder by way of the single **urethra** (Figure 8-10). In the male the urethra is the terminal portion of the reproductive system as well as of the excretory system. In the female the urethra is solely excretory in function, opening to the outside just anterior to the vagina.

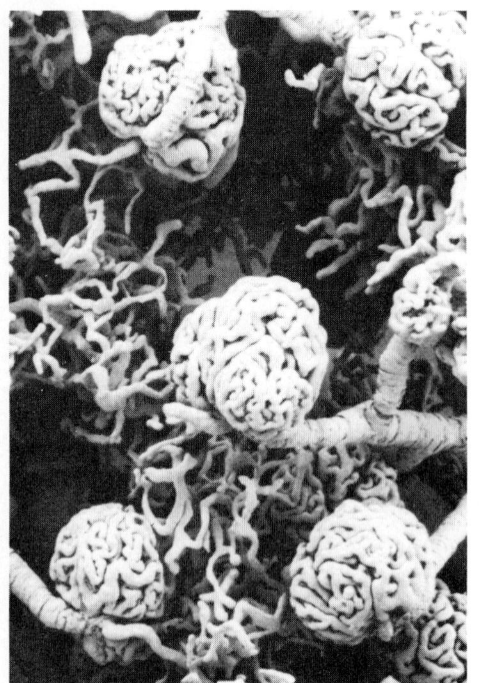

FIGURE 8-11

Scanning electron micrograph of a cast of the microcirculation of the mammalian kidney, showing several glomeruli and associated blood vessels. The Bowman's capsule, which normally surrounds each glomerulus, has been digested away in preparing the cast.

From Tissues and Organs: a text-atlas of scanning electron microscopy, by Richard G. Kessel and Randy H. Kardon, W.H. Freeman and Company. Copyright 1979.

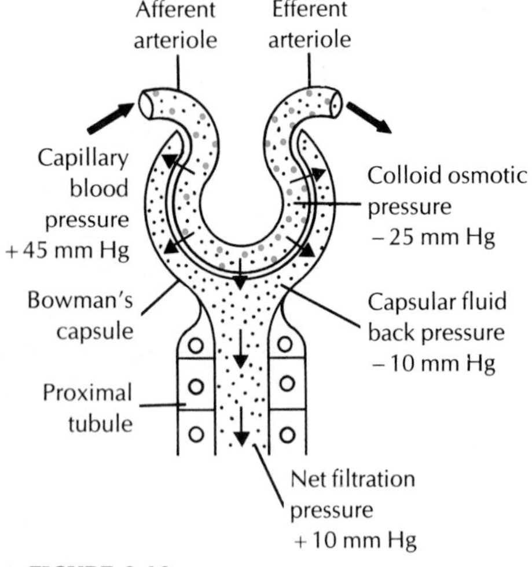

FIGURE 8-12

Pressure determining the net filtration pressure in a glomerulus. The net filtration pressure is the blood hydrostatic pressure, less the osmotic pressure contributed by the plasma proteins, less the capsular hydrostatic pressure.

Since each of the thousands of nephrons in the kidney forms urine independently, each is in a way a tiny, self-contained kidney that produces a minuscule amount of urine—perhaps only a few nanoliters per hour. This amount, multiplied by the number of nephrons in the kidney, produces the total urine flow. The kidney is an "in parallel" system of independent units. However, these "independent" nephrons actually work together to create large osmotic gradients in the kidney medulla. This makes it possible for the mammalian kidney to concentrate urine whose salt concentration is well above that of the blood.

As previously indicated, the nephron, with its pressure filter and tubule, is intimately associated with the blood circulation (Figure 8-10). Blood from the aorta is delivered to the kidney via the large **renal artery,** which breaks up into a branching system of smaller arteries. The arterial blood flows to each nephron through an **afferent arteriole** to the **glomerulus** (glo-mer'yoo-lus), which is a tuft of blood capillaries enclosed within a thin, cuplike **Bowman's capsule** (Figure 8-11). Blood leaves the glomerulus via the **efferent arteriole.** This vessel immediately breaks up again into an extensive system of capillaries, the **peritubular capillaries,** which completely surrounds the nephric tubules. Finally the blood from these many capillaries is collected by veins that unite to form the **renal vein.** This vein returns the blood to the vena cava.

The nephric tubule consists of several segments. The first segment, the **proximal convoluted tubule,** leads into a long, thin-walled, hairpin loop called the **loop of Henle** (Figure 8-10). This loop drops deep into the medulla of the kidney and then returns to the cortex to join the third segment, the **distal convoluted tubule.** The collecting duct empties into the kidney **pelvis,** a cavity that collects the urine before it passes into the **ureter,** on its way to the **urinary bladder** (Figure 8-10).

Glomerular filtration

Let us now return to the glomerulus, where the process of urine formation begins. The glomerulus acts as a specialized mechanical filter in which a protein-free filtrate resembling plasma is driven by the blood pressure across the capillary walls and into the fluid-filled space of Bowman's capsule. As shown in Figure 8-12, the net filtrate pressure is the difference between the blood pressure in the glomerular capillaries, believed to be approximately 45 mm Hg, and the opposing colloid osmotic and hydrostatic back pressures. Most important of these negative pressures is the colloid (protein) osmotic pressure, which is created because the proteins are too large to pass the glomerular membrane. The unequal distribution of protein causes the water concentration of the plasma to be less than the water concentration of the ultrafiltrate in Bowman's capsule. The osmotic gradient created, approximately 25 mm Hg, opposes filtration. Although small, a net filtrate pressure of 10 mm Hg is sufficient to force the ultrafiltrate that is formed down the nephric tubule.

Tubular modification of the formative urine

The ultrafiltrate that enters this complex tubular system must undergo extensive modification before it becomes urine. Approximately 200 liters of filtrate is formed each day by the average person's kidneys. Obviously the loss of this volume of body water, not to mention the many other valuable materials present in the filtrate, cannot be tolerated. How does tubular action convert the plasma filtrate into urine?

Two general processes are involved, **tubular reabsorption** and **tubular secretion.** Since the nephric tubules are at all points in close contact with the peritubular capillaries, materials can be transferred from the tubular lumen to the capillary blood plasma (tubular reabsorption) or from the blood plasma to the tubular lumen (tubular secretion).

Tubular reabsorption. The plasma contains a great variety of ions and molecules. With the exception of the plasma proteins, which are too large to pass through the glomerular filter, all the plasma components are filtered and most are reabsorbed. Some vital materials, such as glucose and amino acids, are completely reabsorbed. Others, such as sodium, chloride, and most other minerals, undergo variable reabsorption. That is, some are strongly reabsorbed and others weakly reabsorbed, depending on the body's need to conserve each mineral. Much of this reabsorption is by **active transport,** in which cellular energy is used to transport materials from the tubular fluid, across the cell, and into the peritubular blood that returns them to the general circulation.

For most substances there is an upper limit to the amount of substance that can be reabsorbed. This upper limit is termed the **transport maximum** for that substance. For example, glucose is normally completely reabsorbed by the kidney because the transport maximum for the glucose reabsorptive mechanism is poised well above the amount of glucose normally present in the plasma filtrate (Figure 8-13).

Unlike glucose, most of the mineral ions are excreted in the urine in variable amounts. Their excretion is regulated. The reabsorption of sodium, the dominant cation in the plasma, illustrates the flexibility of the reabsorption process. Approximately 600 g of sodium is filtered by the human kidneys every 24 hours. Nearly all of this is reabsorbed, but the exact amount is precisely matched to sodium intake. With a normal sodium intake of 4 g per day, the kidney excretes 4 g and reabsorbs 596 g each day. A person on a low-salt diet of 0.3 g of sodium per day still maintains salt balance because only 0.3 g escapes reabsorption. But with a very high salt intake, much above 10 g a day, the kidney cannot excrete sodium as fast as it enters. The unexcreted sodium chloride holds additional water in the body fluids, and the person begins to gain weight. (The salt intake of the average American is about 6 to 18 g a day, approximately 20 times more than the body needs, and three times more than is considerd acceptable for those predisposed to high blood pressure.)

It may seem odd that the kidney can excrete only 10 g of sodium chloride per day when approximately 600 g is filtered. It would appear that more sodium could be excreted by simply allowing more to escape tubular reabsorption. Unfortunately for salt lovers, the reabsorption of salt is not completely flexible. Some 80% to 85% of the salt and water filtered is reabsorbed in the proximal convoluted tubule; this is an **obligatory reabsorption** because it is governed entirely by a physical process (the osmotic pressure of the solutes) and cannot be controlled physiologically. In the distal convoluted tubule, however, sodium reabsorption is contolled by **aldosterone,** a steroid hormone from the adrenal gland. This is called **faculative reabsorption,** meaning that the reabsorption can be adjusted physiologically according to need. We can say the proximal reab-

In the disease diabetes mellitus ("sweet running through"), glucose rises to abnormally high concentrations in the blood plasma (hyperglycemia) because the hormone insulin, which enables the body cells to take up glucose, is deficient. As the blood glucose rises above a normal level of about 100 mg/100 ml of plasma, the concentration of glucose in the filtrate also rises, and more glucose must be reabsorbed by the proximal tubule. Eventually a point is reached (about 300 mg/100 ml of plasma) at which the reabsorptive capacity of the tubular cells is saturated. This is the transport maximum for glucose. Should the plasma glucose continue to rise, glucose spills over into the urine. In untreated diabetes, the victim's urine tastes sweet, thirst is unrelenting, and the body wastes away despite a large food intake. In England, the disease for centuries was appropriately called the "pissing evil."

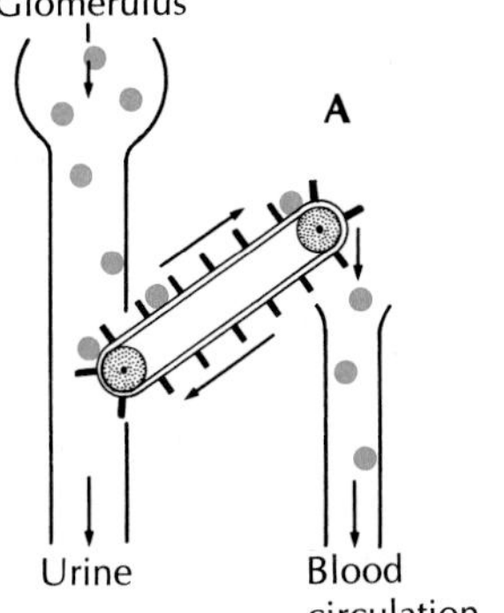

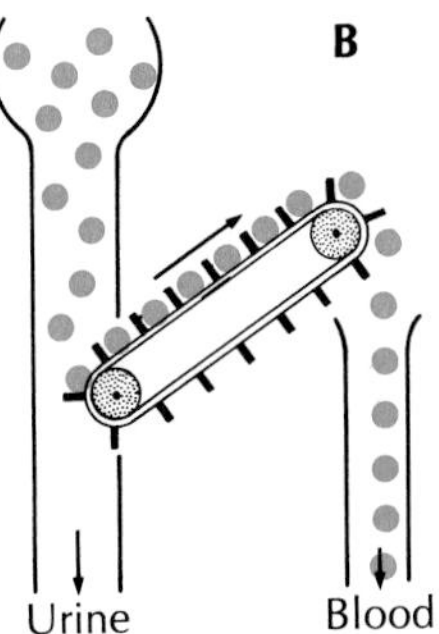

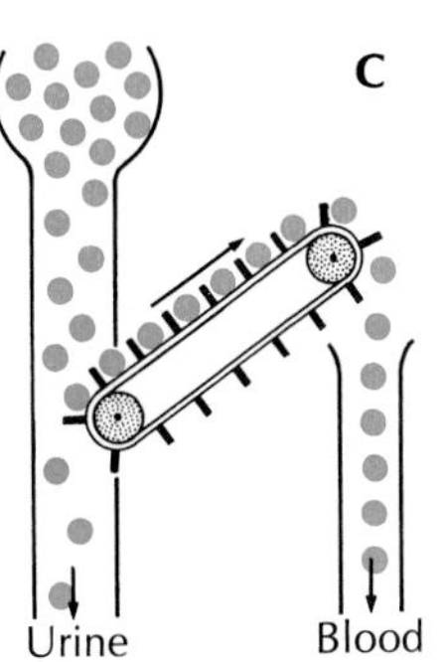

FIGURE 8-13

The mechanism for the tubular reabsorption of glucose can be likened to a conveyor belt running at constant speed. **A,** When the concentration of glucose in the filtrate is low, all is reabsorbed. **B,** When the glucose concentration in the filtrate has reached the transport maximum, all carrier sites for glucose are occupied. If the glucose rises further, **C,** as in the disease diabetes mellitus, some glucose escapes the carriers and appears in the urine.

Adapted from Pitts, R.F. 1963. Physiology of the kidney and body fluids. Chicago, Year Book Medical Publishers.

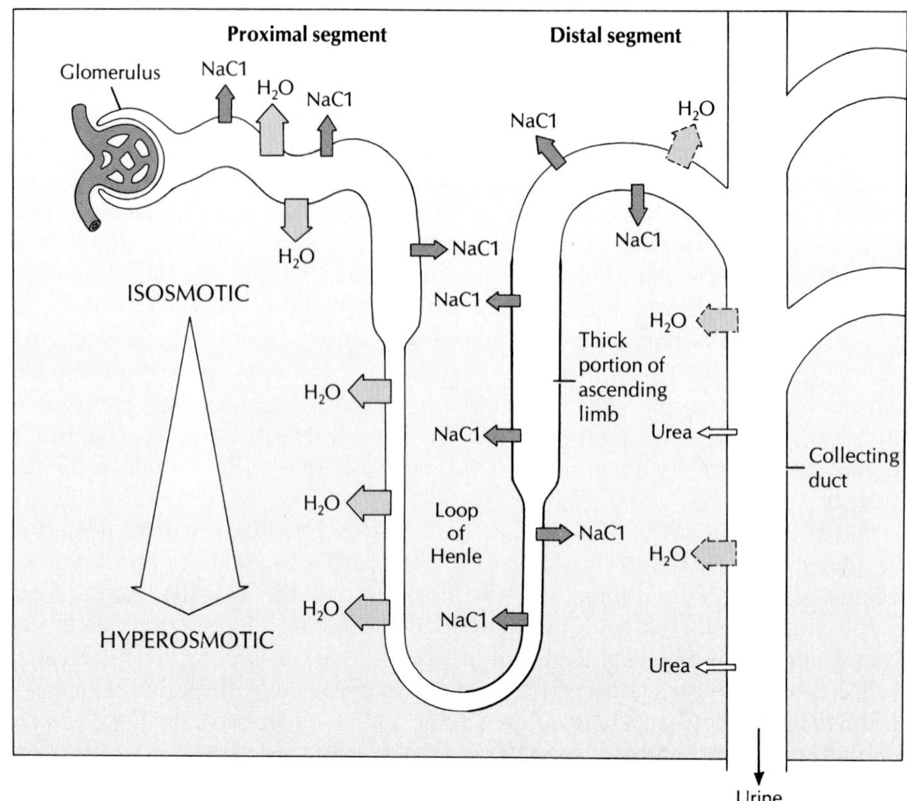

FIGURE 8-14

Mechanism of urine concentration in mammals. Sodium and chloride are pumped from the ascending limb of the loop of Henle, and water is withdrawn passively from the descending limb. Sodium chloride and urea reabsorbed from the collecting duct raise the osmotic concentration in the kidney medulla, creating an osmotic gradient for the controlled reabsorption of water from the collecting duct.

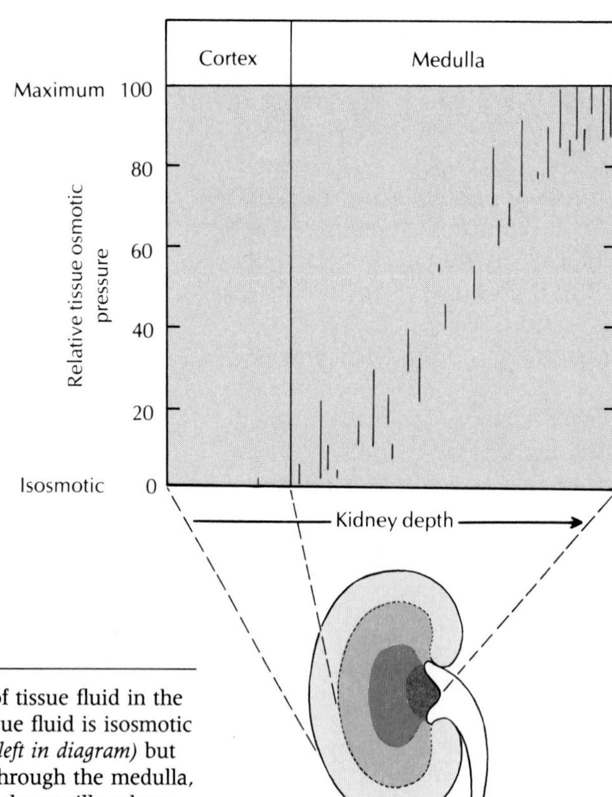

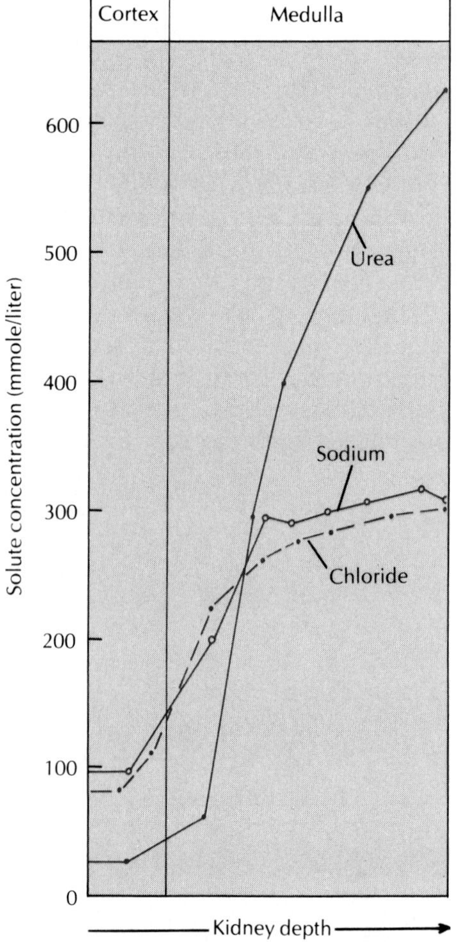

FIGURE 8-15

Osmotic concentration of tissue fluid in the mammalian kidney. Tissue fluid is isosmotic in the kidney cortex *(to left in diagram)* but increases continuously through the medulla, reaching a maximum at the papilla where the urine drains into the ureter.

***Left,* From Wirz, H., B. Hargitay, and W. Kuhn. 1951. Helv. Physiol. Acta 9:196-207. *Right,* After Ullrich, K.J., K. Kramer, and J. Boylan. 1961. Prog. Cardiovasc. Dis. 3:395-431.**

sorption is involuntary and distal reabsorption is voluntary, although, of course, we are not aware of the adjustments the kidney is performing on our behalf. The flexibility of distal reabsorption varies considerably in different animals: it is restricted in humans but very broad in many rodents. These differences have appeared because selective pressures during evolution have resulted in rodents adapted for dry environments. They must conserve water and at the same time excrete considerable sodium. Humans, however, were not designed to accommodate the large salt appetites many have. Our closest relatives, the great apes, are vegetarians with an average salt intake of less than 0.5 g a day.

Tubular secretion. In addition to reabsorbing large amounts of materials from the plasma filtrate, the kidney tubules are able to secrete certain substances into the tubular fluid. This process, which is the reverse of tubular reabsorption, enables the kidney to build up the urine concentrations of materials to be excreted, such as hydrogen and potassium ions, drugs, and various foreign organic materials. The distal convoluted tubule is the site of most tubular secretion.

In the kidneys of bony marine fishes, reptiles, and birds, tubular secretion is a much more highly developed process than it is in mammalian kidneys. Marine bony fishes actively secrete large amounts of magnesium and sulfate, which are by-products of their mode of osmotic regulation. Reptiles and birds excrete uric acid instead of urea as their major nitrogenous waste. The material is actively secreted by the tubular epithelium. Since uric acid is nearly insoluble, it forms crystals in the urine and requires little water for excretion. Thus the excretion of uric acid is an important adaptation for water conservation.

Water excretion

The osmotic pressure of the blood is closely regulated by the kidney. When fluid intake is high, the kidney excretes a dilute urine, saving salts and excreting water. When fluid intake is low, the kidney conserves water by forming a concentrated urine. A dehydrated person can concentrate urine to approximately four times blood osmotic concentration.

The capacity of the kidney of mammals and some birds to produce a concentrated urine involves the loop of Henle, the long hairpin loop between the proximal and distal convoluted tubules that extends into the renal medulla. The loops of Henle constitute a **countercurrent multiplier system.** Flow is in opposite directions in the two limbs, hence the name "countercurrent."

The functional characteristics of this system are as follows. Chloride is actively transported out of the thick portion of the ascending limb and into the surrounding tissue fluid (Figure 8-14). As the interstitium surrounding the loop becomes more concentrated with solute, water is passively withdrawn from the descending limb (which, unlike the ascending limb, is permeable to water but impermeable to sodium and chloride). The tubular fluid in the hairpin, now more concentrated, moves up the ascending limb, where still more sodium chloride is pumped out. In this way the effect of active ion transport in the ascending limb is multiplied as more water is withdrawn from the descending limb and more concentrated fluid is presented to the ascending limb ion pump. Also contributing significantly to tissue fluid concentration at the bottom of the hairpin loop is urea, which is reabsorbed from the collecting duct that lies parallel to the loop (Figures 8-14 and 8-15).

The final adjustment of urine concentration does not occur in the loops of Henle but in the collecting ducts. Formative urine that enters the distal tubule from the loop of Henle is dilute (because of active salt withdrawal) and is diluted still more by the active reabsorption of more sodium chloride in the distal tubule. The formative urine, low in solutes but carrying urea, now flows down into the collecting duct. Because of the high concentration of solutes surrounding the collecting duct, water is withdrawn from the urine. As the urine becomes more concentrated, urea also diffuses out and adds to the high osmotic pressure in the kidney medulla (Figure 8-15). The amount of water saved and the final concentration of the urine depend on the permeability of the walls of the collecting duct. This is controlled by the **antidiuretic hormone** (ADH, or vasopressin), which is released by the posterior pituitary gland (neurohypophysis), The release of this hormone is governed in turn by special receptors in the brain that constantly sense the osmotic pressure of the blood. When the blood osmotic pressure increases, as during dehydration, more ADH is released from the pituitary gland. ADH increases the permeability of the collecting duct, probably by expanding the size of pores in the walls of the duct. Then, as the fluid in the collecting duct passes through the hyperosmotic region of the kidney medulla, water diffuses through the pores into the surrounding interstitial fluid and is carried away by the blood circulation (Figure 8-14). The urine loses water and becomes more concentrated. Given this sequence of events for dehydration, it is not difficult to anticipate how the system responds to overhydration: the pituitary stops releasing ADH, the pores in the collecting duct walls close, and a large volume of dilute urine is excreted.

The varying ability of different mammals to form a concentrated urine is closely correlated with the length of the loops of Henle. The beaver, which has no need to conserve water in its aquatic environment, has short loops and can concentrate its urine to only approximately twice that of the blood plasma. Humans, with relatively longer loops, can concentrate urine 4.2 times that of the blood. As we would anticipate, desert mammals have much greater urine concentrating powers. The camel can produce a urine 8 times the plasma concentration, the gerbil 14 times, and the Australian hopping mouse 22 times. In this creature, the greatest urine concentrator of all, the loops of Henle extend to the tip of a long renal papilla that pushes out into the mouth of the ureter.

■ Temperature Regulation

We have seen that a fundamental problem facing an animal is keeping its internal environment in a state that permits normal cell function. Biochemical activities are sensitive to the chemical environment, and our discussion thus far has examined how the chemical environment is stabilized. Biochemical reactions are also extremely sensitive to temperature. The rate of most enzyme reactions doubles—some even triple—with every 10° C rise in temperature. Temperature therefore is a severe constraint for animals, all of which seek biochemical stability. When body temperatures drop too low, metabolic processes are slowed, reducing the amount of energy the animal can muster for activity and reproduction. If the body temperature rises too high, metabolic reactions become imbalanced, and enzymatic activity is impaired or even destroyed. Thus animals can succeed over only a restricted range of temperature, usually within the limits of 0° to 40° C. Animals must either find a habitat where they do not have to contend with temperature extremes, or they must develop the means of stabilizing their metabolism independent of temperature extremes.

ECTOTHERMY AND ENDOTHERMY

The terms "cold-blooded" and "warm-blooded" have long been used almost universally to divide animals into two groups: invertebrates and lower vertebrates that feel cold to the touch, and those, like humans, other mammals, and birds, that do not. It is true that the body temperature of mammals and birds is usually (though not always) warmer than the air temperature, but a "cold-blooded" animal is not necessarily cold. Tropical fishes, and insects and reptiles basking in the sun, may have body temperatures equaling or surpassing those of mammals. Moreover, many "warm-blooded" mammals hibernate, allowing their body temperature to approach the freezing point of water. Thus the terms "warm-blooded" and "cold-blooded" are hopelessly subjective and nonspecific but are so firmly entrenched in our vocabulary that most biologists find it easier to accept the usage than try to change people.

The terms **poikilothermic** (variable body temperature) and **homeothermic** (constant body temperature) are frequently used by zoologists as alternatives to "cold-blooded" and "warm-blooded," respectively. These terms are more precise and more informative, but they still offer difficulties. For example, deep sea fishes live in an environment having no perceptible temperature change. Even though their body temperature is absolutely stable, day in and day out, few would argue that deep sea fishes are homeotherms. Furthermore, among the homeothermic birds and mammals there are many that allow their body temperature to change between day and night, or, as with hibernators, between seasons.

Physiologists prefer yet another way to describe body temperatures, one that reflects the fact that an animal's body temperature is a balance between heat gain and heat loss. All animals produce heat from cellular metabolism, but in most the heat is conducted away as fast as it is produced. In these animals, the **ectotherms** — and the overwhelming majority of animals belong to this group — the body temperature is determined solely by the environment. Alternatively, there are some animals that are able to generate enough body heat to elevate their own body temperature to a high but stable level. Since the source of their body heat is internal, they are called **endotherms.** These, the favored few in the animal kingdom, are the birds and mammals, as well as a few reptiles and fast-swimming fishes and certain insects, that are at least partially endothermic. Endothermy allows birds and mammals to stabilize their internal temperature so that biochemical processes and nervous function can proceed at steady high levels of activity. Endotherms can thus remain active in winter and exploit habitats denied to ectotherms.

HOW ECTOTHERMS ACHIEVE TEMPERATURE INDEPENDENCE

Behavioral adjustments

Ectotherms, although having no control over their body temperature, are not helpless slaves to temperature. First, ectotherms often have the option of seeking out areas in the environment where the temperature is favorable to their activities. Second, some ectotherms, such as desert lizards, exploit hour-to-hour changes in solar radiation to keep their body temperature relatively constant (Figure 8-16). In the early morning they emerge from their burrows and bask in

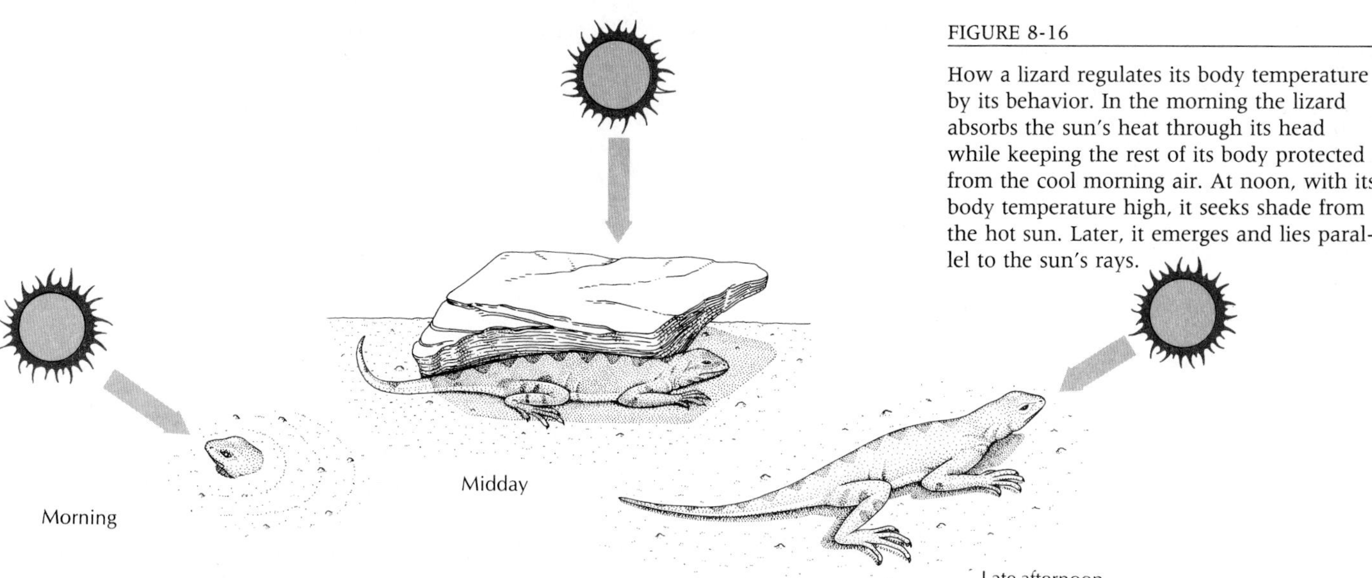

FIGURE 8-16

How a lizard regulates its body temperature by its behavior. In the morning the lizard absorbs the sun's heat through its head while keeping the rest of its body protected from the cool morning air. At noon, with its body temperature high, it seeks shade from the hot sun. Later, it emerges and lies parallel to the sun's rays.

the sun with their bodies flattened to absorb heat. As the day warms they turn to face the sun, to reduce the body area exposed, and raise their bodies from the hot substrate. In the hottest part of the day they may retreat to their burrows. Later they emerge to bask as the sun sinks lower and the air temperature drops.

These behavioral patterns help to maintain a relatively steady body temperature of 36° to 39° C while the air temperature is varying between 29° and 44° C. Some lizards can tolerate intense midday heat without shelter. The desert iguana of the southwestern United States prefers a body temperature of 42° C when active and can tolerate a rise to 47° C, a temperature that is lethal to all birds and mammals and most other lizards. The term "cold-blooded" clearly does not apply to these animals!

Metabolic adjustments

Even without the help of the behavioral adjustments just described, most ectotherms can adjust their metabolic rates to the prevailing temperature such that the intensity of metabolism remains mostly unchanged. This is called **temperature compensation** and involves complex biochemical and cellular adjustments. These adjustments enable a fish or a salamander, for example, to benefit from the same level of activity in both warm and cold environments. Thus whereas endotherms achieve metabolic homeostasis by regulating their body temperature, ectotherms accomplish much the same by directly regulating their metabolism. This also is a form of homeostasis.

TEMPERATURE REGULATION IN ENDOTHERMS

Most mammals have body temperatures between 36° and 38° C, somewhat lower than those of birds, which range between 40° and 42° C. This constant temperature is maintained by a delicate balance between heat production and heat loss—not a simple matter when these animals are constantly alternating between periods of rest and bursts of activity.

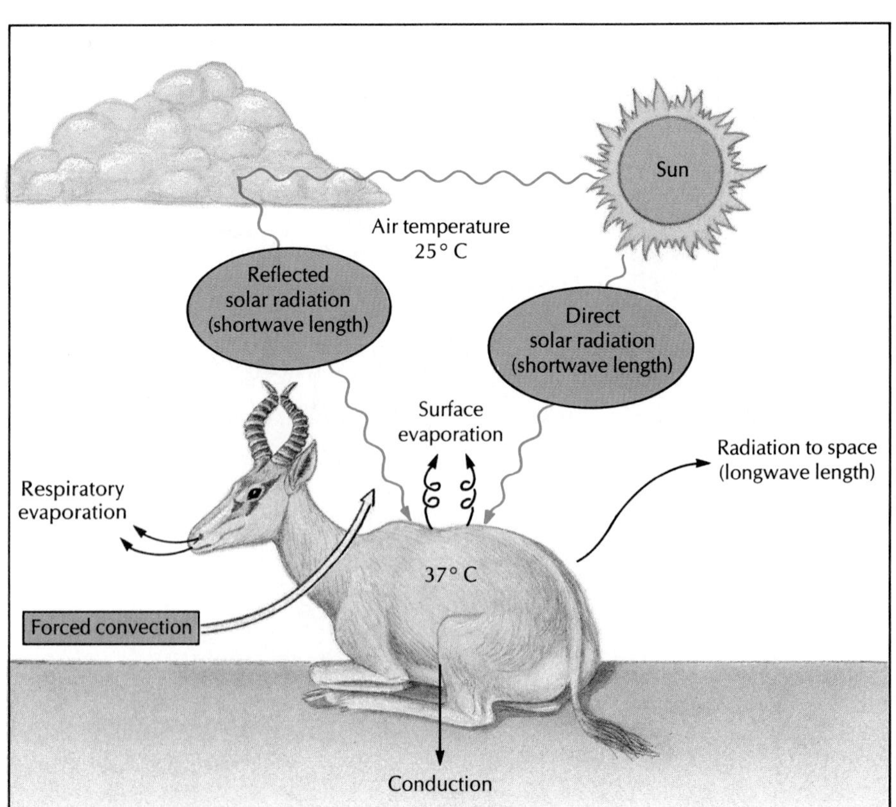

FIGURE 8-17

Exchange of heat between the animal and its environment on a warm day. Red arrows indicate sources of net heat gain by the animal (all radiation); black arrows are avenues of net heat loss (evaporative cooling, conduction to the ground, longwave radiation into space, and forced convection by the wind). If the air and ground temperatures were warmer than the animal, the arrows for forced convection, conduction, and radiation from space would be reversed. Then the animal could lose heat only by evaporative cooling.

Heat is produced by the animal's metabolism, which includes the oxidation of foodstuffs, basal cellular metabolism, and muscular contraction. Because much of an endotherm's daily caloric intake is required to generate body heat, especially in cold weather, the endotherm must eat more food than an ectotherm of the same size. Heat is lost by radiation, conduction, and convection (air movement) to a cooler environment and by evaporation of water (Figure 8-17). A bird or mammal can control both processes of heat production and heat loss within rather wide limits. If the animal becomes too cool, it can increase heat production by increasing muscular activity (exercise or shivering) and by decreasing heat loss by increasing its insulation. If it becomes too warm, it decreases heat production and increases heat loss. We will examine these processes in the following examples.

Adaptations for hot environments

Despite the harsh conditions of deserts — intense heat during the day, cold at night, and scarcity of water, vegetation, and cover — many kinds of animals live there successfully. The smaller desert mammals are mostly **fossorial** (living mainly in the ground) or **nocturnal,** that is, active at night. The lower temperature and high humidity of burrows help reduce water loss by evaporation. As explained earlier in this chapter (p. 203) some desert animals such as the kangaroo rat and the American desert ground squirrels can, if necessary, derive all the water they need from their dry food, thus drinking no water at all. Such animals produce a highly concentrated urine and form nearly solid feces.

The large desert ungulates (hooved mammals that chew their cud) obviously cannot escape the desert heat by living in burrows. Animals such as camels and desert antelopes (gazelle, oryx, and eland) possess a number of adaptations for coping with heat and dehydration. Those of the eland are shown in Figure 8-18. The mechanisms for controlling water loss and preventing overheating are closely linked together. The glossy, pallid color of the fur reflects direct sunlight, and the fur itself is an excellent insulation that works to keep heat out. Heat is lost by convection and conduction from the underside of the eland where the pelage is very thin. Fat tissue of the eland, an essential food reserve, is concentrated in a single hump on the back, instead of being uniformly distributed under the skin where it would impair heat loss by radiation. The eland avoids evaporative water

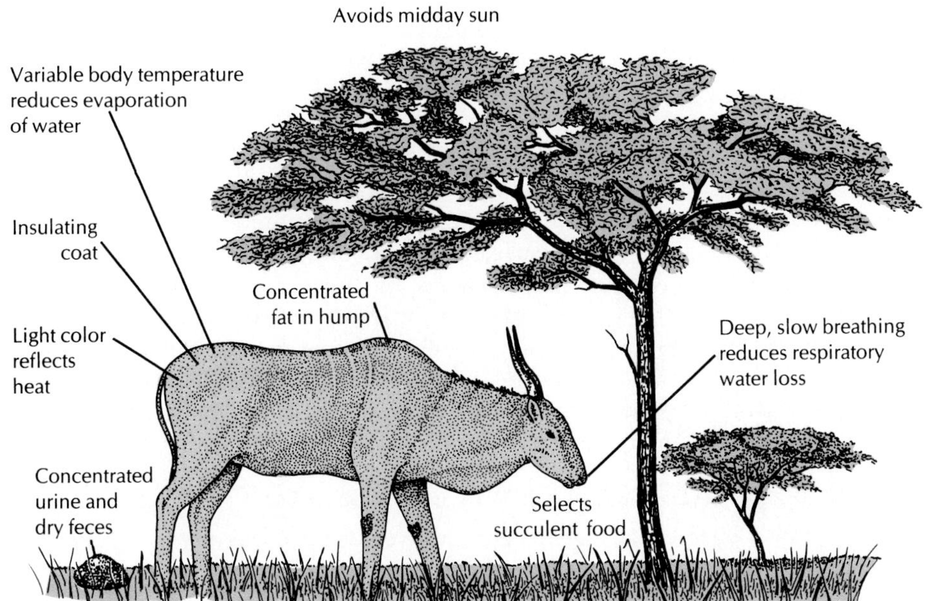

FIGURE 8-18

Physiological and behavioral adaptations of the common eland for maintaining heat balance in the hot, arid savannah of central Africa.

loss—the only means an animal has for cooling itself when the environmental temperature is higher than that of the body—by permitting its body temperature to decrease during the cool night and then increase slowly during the day as the body stores heat. Only when the body temperature reaches 41° C must the eland prevent further rise through **evaporative cooling** by sweating and panting. Water is also conserved by means of concentrated urine and dry feces. All of these adaptations are also found developed to a similar or even greater degree in camels, the most perfectly adapted of all large desert mammals.

Adaptations for cold environments

In cold environments mammals and birds use two major mechanisms to maintain homeothermy: (1) **decreased conductance,** that is, reduction of heat loss by increasing the effectiveness of the insulation and (2) **increased heat production.**

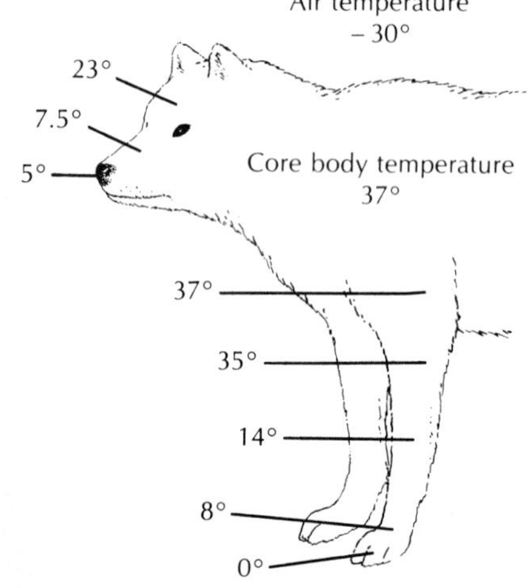

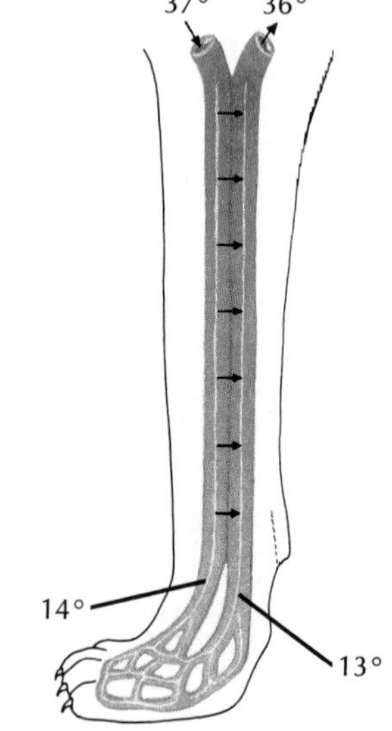

FIGURE 8-19

Countercurrent heat exchange in the leg of an arctic wolf. The upper diagram shows how the extremities cool when the animal is exposed to low air temperatures. The lower diagram depicts a portion of the front leg artery and vein, showing how heat is exchanged between outflowing arterial and inflowing venous blood. Heat is thus shunted back into body core and conserved.

The excellent insulation of the thick pelage of arctic animals is familiar. In all mammals living in cold regions of the earth fur thickness increases in winter, sometimes by as much as 50%. The thick underhair is the major insulating layer, whereas the longer and more visible guard hair serves as protection against wear and for protective coloration.

But the body extremities (legs, tail, ears, nose) of arctic mammals cannot be insulated as well as can the main part of the body. To prevent these parts from becoming major avenues of heat loss, they are allowed to cool to low temperatures, often approaching the freezing point. As warm arterial blood passes into a leg, for example, heat is shunted directly from artery to vein and carried back to the core of the body (Figure 8-19). This device prevents the loss of valuable body heat through the poorly insulated distal regions of the leg. A consequence of this **peripheral heat exchange system** is that the legs and feet must operate at low temperatures. The temperatures of the feet of the arctic fox and barren-ground caribou are just above the freezing point; in fact, the temperature may be below 0° C in the footpads and hooves. To keep feet supple and flexible at such low temperatures, fats in the extremities have very low melting points, perhaps 30° C lower than ordinary body fats.

In severely cold conditions all mammals can produce more heat by **augmented muscular activity** through exercise or shivering. We are all familiar with the effectiveness of both activities. A person can increase heat production as much as eighteenfold by violent shivering when maximally stressed by cold. Another source of heat is the increased oxidation of foodstuffs, especially brown fat stores (brown fat is described on p. 232). This mechanism is called **nonshivering thermogenesis.**

Small mammals the size of lemmings, moles, and mice meet the challenge of cold environments in a different way. Small mammals are not as well insulated as large mammals because there is an obvious practical limit to how much pelage a mouse, for example, can carry before it becomes an immobile bundle of fur. Consequently these forms have successfully exploited the excellent insulating qualities of snow by living under it in runways on the forest floor, where incidentally their food is also located. In this **subnivean environment** the temperature seldom drops below −5° C even though the air temperature above may fall to −50° C. The snow insulation decreases thermal conductance from small mammals in the same way that pelage does for large mammals. Living beneath the snow is really a type of avoidance response to cold.

ADAPTIVE HYPOTHERMIA IN BIRDS AND MAMMALS

Endothermy is energetically expensive. Whereas an ectotherm can survive for weeks in a cold environment without eating, an endotherm must always have energy resources to supply its high metabolic rate. The problem is especially acute

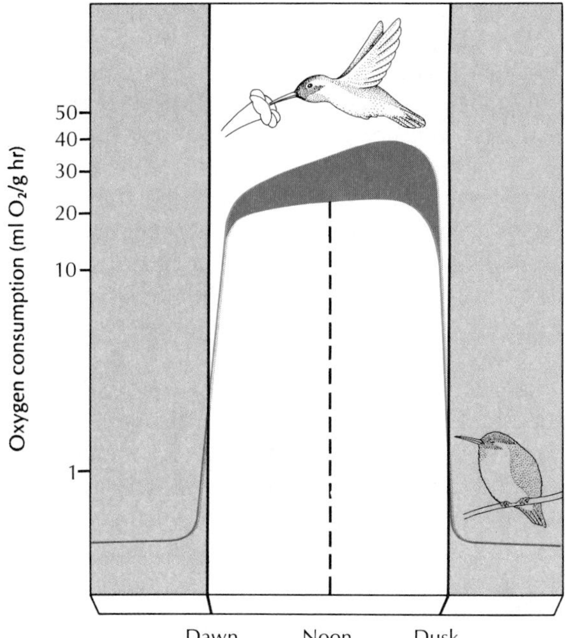

FIGURE 8-20

Daily torpor in hummingbirds. Body temperature and oxygen consumption are high when hummingbirds are active during the day but drop to one-twentieth these levels when they enter torpor in the evening. Torpor vastly lowers demands on the bird's limited energy reserves.

Adapted from Lasiewski, R.C. 1963. Physiol. Zool. **36**:122-140.

for small birds and mammals which, because of their intense metabolism, may have to consume food each day equal to their own body weight to maintain homeothermy. It is not surprising then that a few small birds and mammals have evolved ways to abandon homeothermy for periods ranging from a few hours to several months, allowing their body temperature to fall until it equals or remains just above the surrounding air temperature.

Some very small birds and mammals, such as hummingbirds and bats, maintain normal high body temperatures when active but allow their body temperature to drop profoundly when inactive and asleep (Figure 8-20). This is called **daily torpor,** an adaptive hypothermia that provides enormous energy savings to small endotherms that are never more than a few hours away from starvation at normal body temperatures.

Many small and medium-sized mammals in northern temperate regions solve the problem of winter scarcity of food and low temperature by entering a prolonged and controlled state of dormancy: **hibernation.** True hibernators, such as ground squirrels, jumping mice, marmots, and woodchucks (Figure 8-21), prepare for hibernation by building up large amounts of body fat. Entry into hibernation is gradual. After a series of "test drops" during which body temperature decreases a few degrees and then returns to normal, the animal cools to wihin a degree or less of the temperature of its surroundings. Metabolism decreases to a fraction of normal. In the ground squirrel, for example, the respiratory rate decreases from a normal rate of 200 per minute to 4 or 5 per minute, and the heart rate from 150 to 5 beats per minute. During arousal, the hibernator both shivers violently and employs nonshivering thermogenesis to produce heat.

Some mammals, such as bears, badgers, raccoons, skunks, and opossums, enter a state of prolonged sleep in winter with little or no decrease in body temperature. This is not true hibernation. Bears of the northern forest den-up for several months. Heart rate may decrease from 40 to 10 beats per minutes, but body temperature remains normal and the bear is awakened if sufficiently disturbed. One intrepid but reckless biologist narrowly escaped injury when he crawled into a den and attempted to measure the bear's rectal temperature with a thermometer!

FIGURE 8-21

Hibernating woodchuck *Marmota monax* (order Rodentia) in den exposed by road-building work sleeps on, unaware of the intrusion. Woodchucks begin hibernation in late September while the weather is still warm and may sleep 6 months. The animal is rigid and decidedly cold to the touch. Breathing is imperceptible, as slow as one breath every 5 minutes. Although it appears to be dead, it will awaken if the den temperature drops dangerously low.

Photograph by L.L. Rue III.

Summary

Throughout life, matter and energy pass through the body, producing perturbations of the internal physiological state. Homeostasis, the ability of an organism to maintain internal stability despite such perturbations, is a characteristic of all living systems. Homeostasis involves the coordinated activity of several physiological and biochemical mechanisms, and it is possible to relate major advances in animal evolution to increasing internal independence from the consequences of environmental change. In this chapter, we have examined two aspects of homeostasis: (1) the varying ability of animals to stabilize the osmotic and chemical composition of the blood, and (2) the capacity of animals to free themselves from the constraints of temperature change.

Most marine invertebrates must either depend on the stability of the ocean to which they conform osmotically, or they must be able to tolerate wide fluctuations in environmental salinity. Some of the latter show limited powers of osmotic regulation, that is, the capacity to resist internal osmotic change, through the evolution of specialized regulatory organs. All animals living in fresh water are hyperosmotic to their environment and have developed mechanisms for recovering salt from the environment and pumping out excess water that enters the body osmotically.

All vertebrate animals except the primitive hagfishes show excellent osmotic homeostasis. The marine bony fishes maintain their body fluids at a level distinctly hypoosmotic to their environment by drinking seawater and physiologically distilling it. The elasmobranchs (sharks and their kin) have adopted a strategy of near-osmotic conformity by retaining urea in the blood.

The kidney is the most important organ for regulating the chemical and osmotic composition of the blood. In all metazoa the kidney is some variation on a basic theme: a tubular structure that forms urine by introducing a fluid secretion or ultrafiltrate of the blood into a tubule in which it is selectively modified to form urine. The terrestrial vertebrates have especially sophisticated kidneys, since they must be able to regulate closely the water content of the blood by balancing off gains and expenditures. The basic excretory unit is the nephron, composed of a glomerulus in which an ultrafiltrate of the blood is formed, and a long nephric tubule in which the formative urine is selectively modified by the tubular epithelium. Water, salts, and other valuable materials are passed by reabsorption to the peritubular circulation, and certain wastes are passed by secretion from the circulation to the tubular urine. All mammals and some birds can produce a concentrated urine by means of a countercurrent multiplier system localized in the loops of Henle, a specialization not found in lower vertebrates.

Temperature has a profound effect on the rate of biochemical reactions and, consequently, on the metabolism and activity of all animals. Animals may be classified according to whether body temperature is variable (poikilothermic) or stable (homeothermic), or by the source of body heat, whether external (ectothermic) or internal (endothermic).

Ectotherms partially free themselves from thermal constraints by seeking out habitats with favorable temperatures, or by behavioral thermoregulation, or by adjusting their metabolism to the prevailing temperature through biochemical alterations.

The endothermic birds and mammals differ from ectotherms in having a much higher rate of metabolic heat production and a much lower rate of heat conductance from the body. They maintain constant body temperature by balancing heat production with heat loss.

Small mammals in hot environments for the most part escape intense heat and reduce evaporative water loss by burrowing. Large mammals employ several strategies for dealing with direct exposure to heat, including reflective insulation, heat storage by the body, and evaporative cooling.

Endotherms in cold environments maintain their body temperature by decreasing heat loss with thickened pelage or plumage and peripheral cooling and by increasing heat production through shivering or nonshivering thermogenesis.

Adaptive hypothermia is a strategy used by small mammals and birds to blunt energy demands during periods of inactivity (daily torpor) or periods of prolonged cold and minimal food availability (hibernation).

REVIEW QUESTIONS

1. Define homeostasis, and explain why stability of body fluids is considered a *dynamic* steady state.
2. Distinguish between the following pairs of terms: osmotic conformity and osmotic regulation, stenohaline and euryhaline, hyperosmotic and hypoosmotic.
3. Explain why marine bony fish, but not freshwater bony fish, must drink sea water to maintain osmotic balance.
4. Most marine invertebrates are osmotic conformers. How does their body fluid composition differ from that of the cartilaginous sharks and rays, which are also in near-osmotic equilibrium with their environment?
5. What strategy does the kangaroo rat use that allows it to exist in the desert without drinking any water?
6. In what animals would you expect to find a salt gland? What is its function?
7. Relate the function of the contractile vacuole to the following experimental observations: to expel an amount of fluid equal in volume to the volume of the animal required 4 to 53 minutes for some freshwater protozoa and between 2 and 5 hours for some marine species.
8. How does a protonephridium differ structurally and functionally from a true nephridium (metanephridium)? In what ways are they similar?
9. Explain why it is possible to reconstruct the evolution of the vertebrate kidney with the expectation of some accuracy even though we have no fossil record of kidney evolution.
10. In what ways does the nephridium of, for example, an earthworm parallel the human nephron in structure and function?
11. Describe what happens during the following stages in urine formation in the mammalian nephron: filtration, tubular reabsorption, tubular secretion.
12. Explain how the cycling of sodium chloride between the descending and ascending limbs of the loop of Henle in the mammalian kidney and the special permeability characteristics of these tubules produce high tissue fluid osmotic concentrations in the kidney medulla.
13. Explain how the antidiuretic hormone (ADH) controls the excretion of water in the mammalian kidney.
14. Define the following terms and comment on the limitations (if any) of each in describing the thermal relationships of animals: poikilothermy, homeothermy, ectothermy, endothermy.
15. Defend the statement: "Both ectotherms and endotherms achieve metabolic homeostasis in unstable temperature environments, but they do so by employing completely different physiological strategies."
16. Large mammals live successfully in deserts and in the arctic. Describe the different adaptations mammals use to maintain homeothermy in each environment.
17. Explain why it is advantageous for certain small birds and mammals to abandon homeothermy during brief or extended periods of their lives.

SELECTED REFERENCES

Beeuwkes, R. 1982. Renal countercurrent mechanisms, or how to get something for (almost) nothing. In C.R. Taylor et al. (eds.). A companion to animal physiology. New York, Cambridge University Press. *Clear explanation of the countercurrent multiplier mechanism of the mammalian kidney.*

Cossins, A.R., and K. Bowler. 1987. Temperature biology of animals. London, Chapman and Hall. *Comprehensive treatment of both ectotherms and endotherms.*

Hardy, R.N. 1983. Homeostasis, ed. 2. The Institute of Biology's Studies in Biology no. 63, London, Edward Arnold. *Introduces the history of the homeostasis concept; temperature and osmotic regulation are treated in the final chapter.*

Rankin, J.C., and J. Davenport. 1981. Animal osmoregulation. New York, John Wiley and Sons, Inc. *Concise and selective treatment.*

Riegel, J.A. 1972. Comparative physiology of renal excretion. New York, Hafner Publishing Co. *Excellent survey of excretory systems, both vertebrate and invertebrate.*

Smith, H.W. 1953. From fish to philosopher. Boston, Little, Brown & Co. *Classic account of vertebrate kidney evolution.*

A CONSUMING CORNUCOPIA

Sir Walter Raleigh observed that the difference between a rich man and a poor man is that the former eats when he pleases while the latter eats when he can get it. In today's crowded world, with 83 million people added each year to the world's population of 5 billion, the separation between the well-fed affluent and the hungry and malnourished poor reminds us that time has not diminished the shrewdness of Sir Raleigh's remark. Unlike the affluent, for whom food acquisition requires only the selection of prepackaged foods at a well-stocked supermarket, the world's poor can appreciate that for them, as for the rest of the animal kingdom, procuring food is fundamental to survival. For most animals, eating is the main business of living.

Potential food is everywhere and little remains unexploited. Animals bite, chew, nibble, crush, graze, browse, shred, rasp, filter, engulf, enmesh, suck, and soak up foods of incredible variety. What and how an animal eats profoundly affect an animal's feeding specialization, its behavior, its physiology, and its internal and external anatomy—in short, both its body form and its role in the web of life. The endless evolutionary jostling between predator and prey has provided compromise adaptations for eating and adaptations for avoiding being eaten.

By whatever means food may be secured, there is far less variation among animals in the subsequent digestive simplification of foods. Vertebrates and invertebrates alike utilize similar digestive enzymes. Even more uniform are the final biochemical pathways for nutrient utilization and energy transformation. The nourishment of animals is like a cornucopia in which the food flows in rather than out. A great diversity of foods procured by countless feeding adaptations streams into the mouth of the horn, is simplified, and finally applied to the common purpose of survival and reproduction.▪

CHAPTER 9

Digestion and Nutrition

All organisms require energy to maintain their highly ordered and complex structure. This energy is chemical bond energy that is released by transforming complex compounds acquired from the organism's environment into simpler ones.

The ultimate source of energy for life on earth is the sun. Sunlight is captured by chlorophyll molecules in green plants, which transform a portion of this energy into chemical bond energy (food energy). Green plants are **autotrophic** organisms; they require only inorganic compounds absorbed from their surroundings to provide the raw material for synthesis and growth. Most autotrophic organisms are the chlorophyll-bearing **phototrophs,** although some, the chemosynthetic bacteria, are **chemotrophs;** they gain energy from inorganic chemical reactions.

Almost all animals are **heterotrophic organisms** that depend on already synthesized organic compounds of plants and other animals to obtain the materials they will use for growth, maintenance, and reproduction of their kind. Since the food of animals, normally the complex tissues of other organisms, is usually too bulky to be absorbed directly by the body cells, it must be broken down, or digested, into soluble molecules that are small enough to be used.

Animals may be divided into a number of categories on the basis of dietary habits. **Herbivorous** animals feed mainly on plant life. **Carnivorous** animals feed mainly on herbivores and other carnivores. **Omnivorous** forms eat both plants and animals. Other categories are sometimes distinguished, such as **insectivorous** animals, which are those animals subsisting chiefly on insects.

The ingestion of foods and their simplification by digestion are only initial steps in nutrition. Foods reduced by digestion to soluble, molecular form are **absorbed** into the circulatory system and are **transported** to the tissues of the body. There they are **assimilated** into the structure of the cells. Oxygen is also transported by the blood to the tissues, where food products are **oxidized,** or burned to yield energy and heat. Much food is not immediately used but is **stored** for future use. Then the wastes produced by oxidation must be

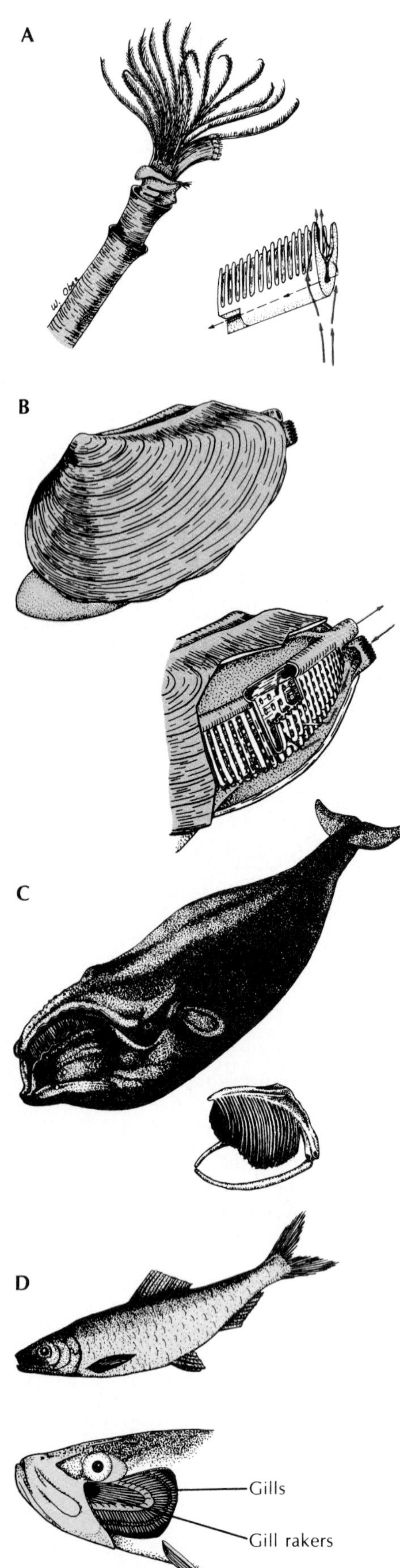

excreted. Food products unsuitable for digestion are **egested** in the form of feces. The sum total of all these processes is called **metabolism** (p. 43).

In this chapter we will first examine the feeding adaptations of animals. Next we will discuss digestion and absorption of foodstuffs. We will close with a consideration of nutritional requirements of animals.

■ Feeding Mechanisms

Only a few animals can absorb nutrients directly from their external environment. Some blood parasites and intestinal parasites may derive all their nourishment as primary organic molecules by surface absorption; some aquatic invertebrates may soak up parts of their nutritional needs directly from the water. Most animals, however, must work for their meals and the specializations that have evolved for food procurement are almost as numerous as are the species of animals. In this brief discussion we consider some of the major food-gathering devices.

FEEDING ON PARTICULATE MATTER

Drifting microscopic particles are found in the upper hundred meters of the ocean. Most of this multitude is **plankton,** plant and animal microorganisms too small to do anything but drift with the ocean's currents. The rest is organic debris, the disintegrating remains of dead plants and animals. Although this oceanic swarm of plankton forms a rich life domain, it is unevenly distributed. The heaviest plankton growth occurs in estuaries and in areas of upwelling, where there is an abundant nutrient supply. It is preyed on by numerous large animals, invertebrates and vertebrates, using a variety of feeding mechanisms.

One of the most important, successful, and widely employed feeding methods that has evolved is **filter feeding** (Figure 9-1). The majority of filter feeders use ciliated surfaces to produce currents that draw drifting food particles into their mouths. Most filter-feeding invertebrates, such as the tube-dwelling polychaete worms, bivalve molluscs, hemichordates, and most of the protochordates, entrap the particulate food on mucous sheets that convey the food into the digestive tract. Others, such as the fairy shrimps, water fleas, and barnacles, use sweeping movements of their setae-fringed legs to create water currents and entrap food, which is transferred to the mouth.

FIGURE 9-1

Some filter feeders and their feeding mechanisms. **A,** The marine fan worm (class Polychaeta, phylum Annelida) has a crown of tentacles. Numerous cilia on the edges of the tentacles draw water *(solid arrows)* between pinnules where food particles are entrapped in mucus; particles are then carried down a "gutter" in the center of the tentacle to the mouth *(broken arrows)*. **B,** Bivalve molluscs (class Bivalvia, phylum Mollusca) use their gills as feeding devices, as well as for respiration. Water currents created by cilia on the gills carry food particles into the inhalant siphon and between slits in the gills where they are entangled in a mucous sheet covering the gill surface. Particles are then transported by ciliated food grooves to the mouth (not shown). Arrows indicate direction of water movement. **C,** Whalebone whales (class Mammalia, phylum Chordata) filter out plankton, principally large crustaceans called "krill," with whalebone, or baleen. Water enters the swimming whale's open mouth by the force of the animal's forward motion and is strained out through the more than 300 horny baleen plates that hang down like a curtain from the roof of the mouth. Krill and other plankton caught in the baleen are periodically wiped off with the huge tongue and swallowed. **D,** Herring and other filter-feeding fishes (class Osteichthyes, phylum Chordata) use gill rakers which project forward from the gill arches into the pharyngeal cavity to strain off plankton. Herring swim almost constantly, forcing water and suspended food into the mouth; food is strained out by the gill rakers, and the water passes through the gill openings.

Filter-feeding has frequently evolved as a secondary modification among representatives of groups that are primarily selective feeders. Examples are many of the microcrustaceans, fishes such as herring, menhaden, and basking sharks, certain birds such as the flamingo, and the largest of all animals, the baleen (whalebone) whales. The vital importance of one component of the plankton, the diatoms, in supporting a great pyramid of filter-feeding animals is stressed by N.J. Berrill*:

> A humpback whale . . . needs a ton of herring in its stomach to feel comfortably full—as many as five thousand individual fish. Each herring, in turn, may well have 6000 or 7000 small crustaceans in its own stomach, each of which contains as many as 130,000 diatoms. In other words, some 400 billion yellow-green diatoms sustain a single medium-sized whale for a few hours at most.

Another type of particulate feeding exploits the deposits of organic detritus that accumulate on and in the substratum; this is called **deposit feeding.** Some deposit feeders, such as many annelids and some of the hemichordates, simply pass the substrate through their bodies, taking from it whatever provides nourishment. Others, such as the scaphopod molluscs, certain primitive bivalve molluscs, and some of the sedentary and tube-dwelling polychaete worms, use appendages to gather up organic deposits some distance from the body and move them toward the mouth (Figure 9-2).

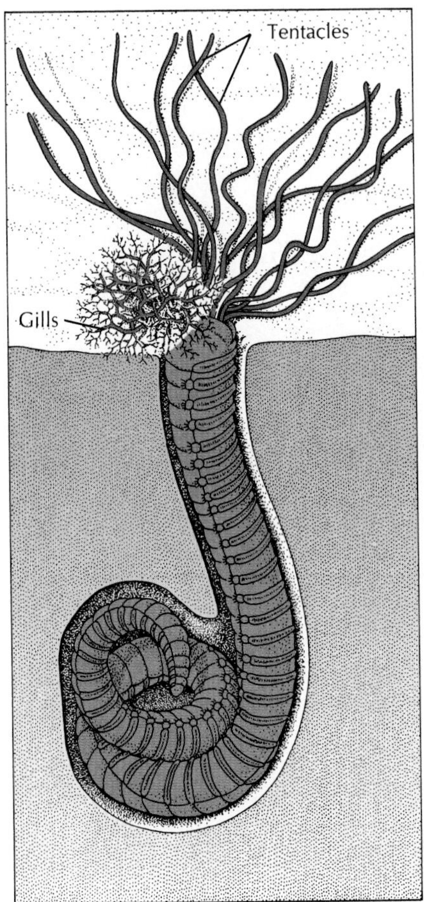

FIGURE 9-2

The annelid *Amphitrite* is a deposit feeder that lives in a mucus-lined burrow and extends long feeding tentacles in all directions across the surface. Food trapped on mucus is conveyed along the tentacles to the mouth.

FEEDING ON FOOD MASSES

Some of the most interesting animal adaptations are those which have evolved for procuring and manipulating solid food. Such adaptations and the animals bearing them are partly shaped by what the animal eats.

Predators must be able to locate, capture, hold, and swallow prey. Most carnivorous animals simply seize the food and swallow it intact, although some employ toxins that paralyze or kill the prey at the time of capture. While no true teeth appear among the invertebrates, many invertebrates have beaks or tooth-like structures for biting and holding, for example *Nereis* (p. 455). The teeth of lower vertebrates—those of fish, amphibians, and reptiles—are used principally to grip the prey and prevent its escape until it can be swallowed whole. Snakes and some fishes can swallow enormous meals. This, together with the absence of limbs, is associated with some striking feeding adaptations in these groups: recurved teeth for seizing and holding prey, and distensible jaws and stomachs to accommodate their large and infrequent meals (Figure 9-3). Birds lack teeth, but the bills are often provided with serrated edges or the upper bill is hooked for seizing and tearing apart prey.

*Berrill, N.J. 1958. You and the universe. New York, Dodd, Mead & Co.

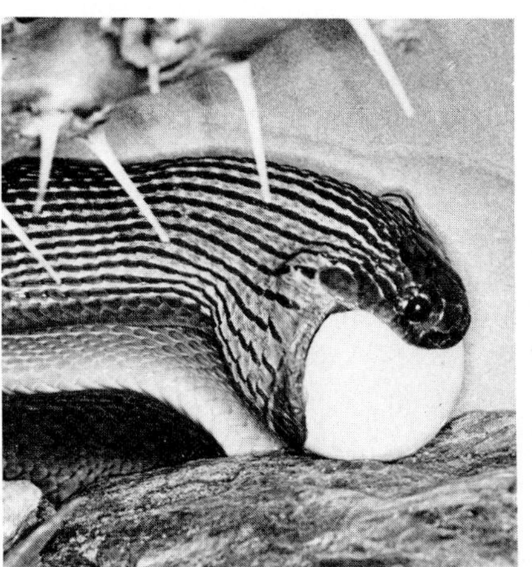

FIGURE 9-3

This African egg-eating snake, *Dasypeltis*, subsists entirely on hard-shelled birds' eggs, which it swallows whole. Its special adaptations are reduced size and number of teeth, enormously expansible jaw elements provided with elastic ligaments, and teeth-like vertebral spurs that puncture the shell. Shortly after the second photograph was taken, the snake punctured and collapsed the egg, swallowed its contents, and regurgitated the crushed shell.

Photographs courtesy of Carl Gans.

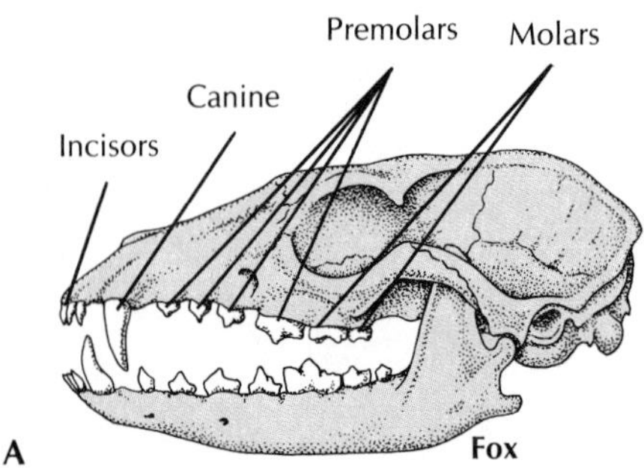

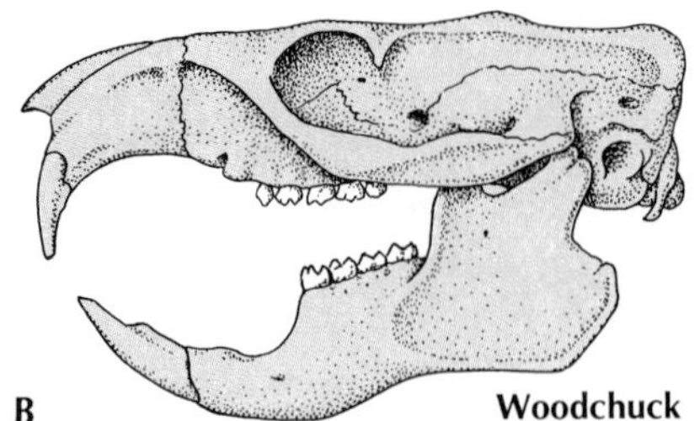

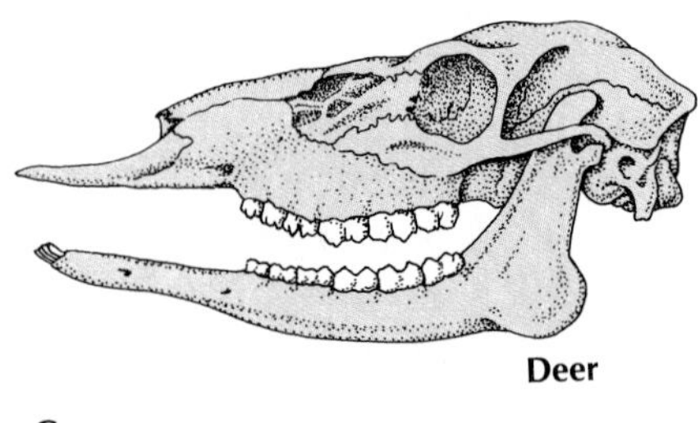

FIGURE 9-4

Mammalian dentition. **A,** Teeth of gray fox, a carnivore, showing the four types of teeth. **B,** Woodchuck, a rodent, has chisel-like incisors that continue to grow throughout life to replace wear. **C,** White-tailed deer, a browsing ungulate, with flat molars bearing complex ridges suited for grinding.

Many invertebrates are able to reduce food size by shredding devices (such as the shredding mouthparts of many crustaceans) or by tearing (such as the beak-like jaws of the cephalopod molluscs). Insects have three pairs of appendages on their heads that serve variously as jaws, chitinous teeth, chisels, tongues, or sucking tubes.

True mastication, that is, the chewing of food as opposed to tearing or crushing, is found only among the mammals. Mammals characteristically have four different types of teeth, each adapted for specific functions (Figure 9-4, *A*). **Incisors** are for nipping, cutting, and stripping; **canines** are designed for seizing, piercing, and tearing; **premolars** and **molars,** at the back of the jaw, are for grinding and crushing. This basic pattern is often greatly modified in animals having specialized food habits (Figure 9-4, *B* and *C*). Herbivores have suppressed canines but well-developed molars with enameled ridges for grinding. The well-developed, self-sharpening incisors of rodents grow throughout life and must be worn away by gnawing to keep pace with growth. Some teeth have become so highly modified that they are no longer useful for biting or chewing food. An elephant's tusk is a modified upper incisor used for defense, attack, and rooting (Figure 9-5), and the male wild boar has modified canines that are used as weapons.

FIGURE 9-5

An African elephant loosening soil from a salt lick with its tusk. Elephants use their powerful modified incisors in many ways in the search for food and water: plowing up the ground for roots, prying apart branches to reach the edible tissues, and drilling into dry riverbeds for water.

Photograph by C.P. Hickman, Jr.

Herbivorous, or plant-eating, animals have evolved special devices for crushing and cutting plant material. Certain invertebrates have scraping mouthparts, such as the radula of snails. Insects such as locusts have grinding and cutting mandibles; herbivorous mammals such as horses and cattle use wide, corrugated molars for grinding. All these mechanisms disrupt the tough cellulose cell wall to accelerate its digestion by intestinal microorganisms, as well as to release the cell contents for direct enzymatic breakdown.

FEEDING ON FLUIDS

Fluid feeding is especially characteristic of parasites, but it is certainly practiced among free-living forms as well. Some internal parasites (endoparasites) simply absorb the nutrients surrounding them, unwittingly provided by the host, whereas others bite and rasp off host tissue, suck blood, and feed on the contents of the host's intestine. External parasites (ectoparasites) such as leeches, lampreys, parasitic crustaceans, and insects use a variety of efficient piercing and sucking mouthparts to feed on blood or other body fluid. Unfortunately for humans and other warm-blooded animals, the ubiquitous mosquito excels in its bloodsucking habit. Alighting gently, the mosquito sets about puncturing its prey with an array of six needlelike mouthparts (Figure 22-35, *B*, p. 491). One of these is used to inject an anticoagulant saliva (responsible for the irritating itch that follows the "bite" and serving as a vector for microorganisms causing malaria, yellow fever, encephalitis, and other diseases); another mouthpart is a channel through which the blood is sucked. It is of little comfort that only the female of the species dines on blood (the male feeds on nectar). Far less annoying to people are the free-living butterflies, moths, and aphids that suck up plant fluids with long, tubelike mouthparts.

Digestion

In the process of digestion, which means literally "carrying asunder," organic foods are mechanically and chemically broken down into small units for absorption. Even though food solids consist principally of carbohydrates, proteins, and fats, the very components that make up the body of the consumer, these components must nevertheless be reduced to their simplest molecular units before they can be used. Each animal reassembles some of these digested and absorbed units into organic compounds of the animal's own unique pattern. Cannibals enjoy no special metabolic benefit from eating their own kind; they digest their victims just as thoroughly as they do food of another species.

In the protozoa and sponges, digestion is entirely **intracellular** (Figure 9-6). The food particle is enclosed within a food vacuole by phagocytosis (p. 177). Digestive enzymes are added and the products of digestion, the simple sugars, amino acids, and other molecules, are absorbed into the cell cytoplasm where they may be used directly or, in the case of multicellular animals, transferred to other cells. The inevitable food wastes are simply extruded from the cell.

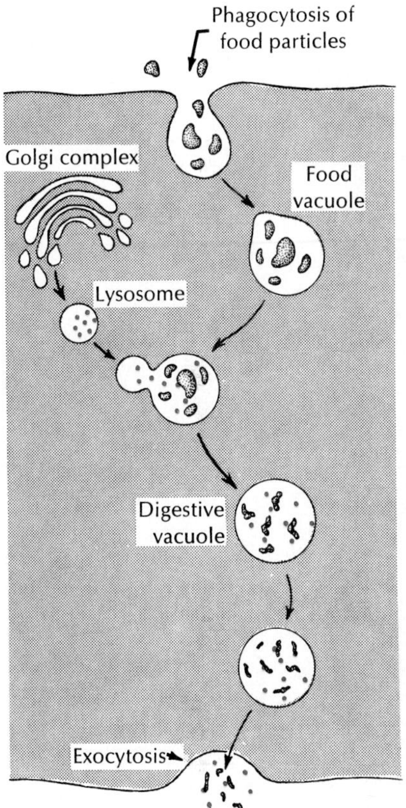

FIGURE 9-6

Intracellular digestion. Lysosomes containing digestive enzymes (lysozymes) are produced within the cell by the Golgi complex. Lysosomes fuse with food vacuoles and release enzymes that digest the enclosed food. Usable products of digestion are absorbed into the cytoplasm, and indigestible wastes are expelled.

There are important limitations to intracellular digestion, however. Only particles small enough to be phagocytized can be accepted, and every cell must be capable of secreting all of the necessary enzymes and of absorbing the products into the cytoplasm. These limitations were probably responsible for shaping the evolution of an **alimentary system** in which **extracellular** digestion of large food masses could take place. Extracellular digestion permits an efficient division of labor in which certain cells lining the **lumen** (cavity) of the alimentary canal are specialized for forming various digestive secretions while others become largely, or entirely, absorptive in function. Many of the lower metazoans, such as the radiates, turbellarian flatworms, and ribbon worms (nemerteans), practice both intracellular and extracellular digestion. With the evolution of greater body complexity and the appearance of complete mouth-to-anus alimentary systems, extracellular digestion became emphasized, together with increasing regional specialization of the digestive tract. In the most advanced animals, for example arthropods and vertebrates, digestion is almost entirely extracellular. The ingested food is exposed to various mechanical, chemical, and bacterial treatments and to different acidic and alkaline phases. Digestive juices are added at appropriate stages as the food passes through the alimentary canal.

ACTION OF DIGESTIVE ENZYMES

Mechanical processes of cutting and grinding by teeth and muscular mixing by the intestinal tract are important in digestion. However, the reduction of foods to small absorbable units relies principally on chemical breakdown by enzymes, discussed in Chapter 2 (pp. 44-47).

It must be stated that, although digestive enzymes are probably the best known and most studied of all enzymes, they represent but a small fraction of the numerous, perhaps thousands, of enzymes that ultimately regulate all processes in the body. The digestive enzymes are **hydrolytic** enzymes or **hydrolases,** so called because food molecules are split by the process of **hydrolysis,** that is, the breaking of a chemical bond by adding the components of water across it:

$$R—R + H_2O \xrightarrow[\text{enzyme}]{\text{Digestive}} R—OH + H—R$$

In this general enzymatic reaction R—R represents a food molecule that is split into two products, R—OH and R—H. Usually these reaction products must

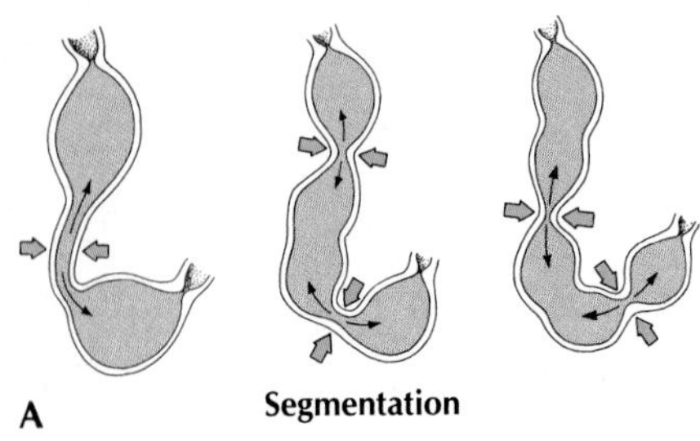

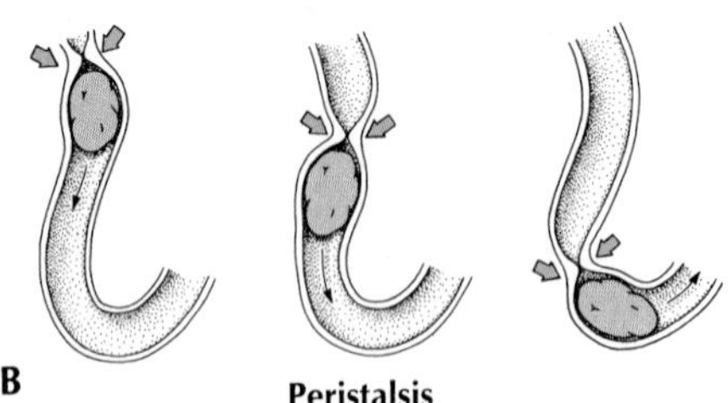

FIGURE 9-7

Movement of intestinal contents by segmentation and peristalsis. **A,** Segmentation movements of food showing how constrictions squeeze the food back and forth, mixing it with enzymes. The sequential mixing movements occur at about 1-second intervals. **B,** Peristalsis movement, showing how food is propelled forward by a traveling wave of contraction.

in turn be split repeatedly before the original molecule has been reduced to its numerous subunits. Proteins, for example, are composed of hundreds, or even thousands, of interlinked amino acids, which must be completely separated before the individual amino acids can be absorbed. Similarly, carbohydrates must be reduced to simple sugars. Fats (lipids) are reduced to molecules of glycerol and fatty acids, although some fats, unlike proteins and carbohydrates, may be absorbed without being completely hydrolyzed first. These are specific enzymes for each class of organic compounds. These enzymes are located in various regions of the alimentary canal in a sort of "enzyme chain," in which one enzyme may complete what another has started; the product moves along posteriorly for still further hydrolysis.

MOTILITY IN THE ALIMENTARY CANAL

Movement of food through the digestive tract is by **cilia** or by specialized **musculature,** and often by both. Movement is usually by cilia in the acoelomate and pseudocoelomate metazoa that lack the mesodermally derived gut musculature of the true coelomates, as well as in the eucoelomate bivalve molluscs in which the coelom is weakly developed. In animals with well-developed coeloms the gut usually has two opposing layers of muscle—a longitudinal layer, in which the smooth muscle fibers run parallel with the length of the gut, and a circular layer, in which the muscle fibers embrace the circumference of the gut. This arrangement is ideal for mixing and propelling foods. The most characteristic gut movement is **segmentation,** the alternate constriction of rings and smooth muscle of the intestine that constantly divide and squeeze the contents back and forth (Figure 9-7, *A*). Walter B. Cannon of homeostasis fame (p. 198), while still a medical student at Harvard in 1900, was the first to use x rays to watch segmentation in experimental animals that had been fed suspensions of barium sulfate. Segmentation serves to mix food but does not move it down the gut. Another kind of muscular action, called **peristalsis,** sweeps the food down the gut with waves of circular muscle contraction (Figure 9-7, *B*).

Organization of the Alimentary Canal

The metazoan alimentary canal can be divided into five major regions: (1) receiving, (2) conduction and storage, (3) grinding and early digestion, (4) terminal digestion and absorption, and (5) water absorption and concentration of solids. Food progresses from one region to the next, allowing digestion to proceed in sequential stages. A generalized vertebrate alimentary canal is illustrated in Figure 9-8, and that of an insect in Figure 9-9.

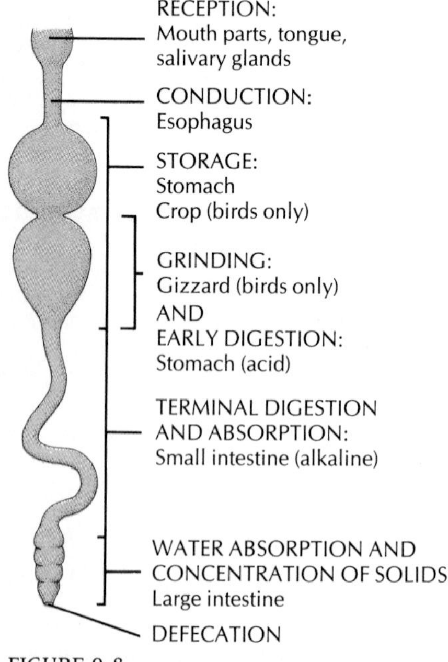

FIGURE 9-8

Generalized vertebrate digestive tract showing the functions of regions.

RECEIVING REGION

The first region of the alimentary canal consists of devices for feeding and swallowing. These include the mouthparts (for example, mandibles, jaws, teeth, radula, bills), the **buccal cavity,** and muscular **pharynx.** Most metazoans other than the filter feeders have **salivary glands** (buccal glands) that produce lubricating secretions containing mucus to assist swallowing. Salivary glands often have other specialized functions such as secretion of toxic enzymes to quiet struggling prey and secretion of salivary enzymes to begin digestion. The salivary secretion of the leech, for example, is a complex mixture containing several enzymes that dissolve tissues surrounding the bite and prevent blood coagulation.

Salivary **amylase,** a carbohydrate-splitting enzyme that begins the hydrolysis of plant and animal starches, is found only in certain herbivorous molluscs, some insects, and in the primate mammals including humans. Starches are long polymers of glucose. Salivary amylase does not completely hydrolyze starch, but breaks it down mostly into two-glucose fragments called **maltose.** Some free glucose, as well as longer fragments of starch, is also produced. When the food mass (bolus) is swallowed, salivary amylase continues to act for some time, digesting perhaps half of the starch before the enzyme is inactivated by the acid environment of the stomach. Further starch digestion resumes beyond the stomach in the intestine.

The tongue is a vertebrate innovation that assists in food manipulation and swallowing. In humans, swallowing begins with the tongue pushing the moistened food bolus toward the pharynx. The nasal cavity is reflexively closed by raising the soft palate. As the food slides into the pharynx, the epiglottis is tipped down over the windpipe, nearly closing it. Some particles of food may enter the opening of the windpipe, but are prevented from going further by contraction of laryngeal muscles. Once in the esophagus, the bolus is forced smoothly toward the stomach by peristaltic contraction of the esophageal muscles.

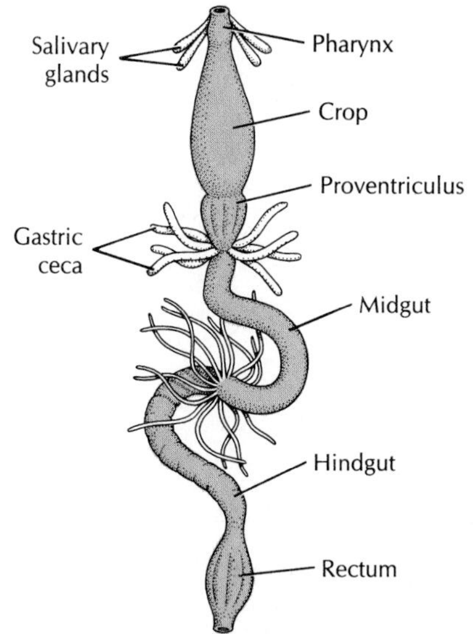

FIGURE 9-9

Insect digestive system (cockroach). The proventriculus is a gizzard containing chitinous teeth for grinding food.

CONDUCTION AND STORAGE REGION

The **esophagus** of vertebrates and many invertebrates serves to transfer food to the digestive region. In many animals the esophagus is expanded into a **crop** (Figure 9-8), used for food storage prior to digestion. Among vertebrates, only birds have a crop. This serves to store and soften food (grain, for example) before it passes to the stomach, or to allow mild fermentation of the food before it is regurgitated to feed nestlings.

REGION OF GRINDING AND EARLY DIGESTION

In most vertebrates and some invertebrates the **stomach** provides for initial digestion as well as for storage and mixing the food with digestive juices. Mechanical breakdown of food, especially plant food with its tough cellulose cell walls, is often continued in herbivorous animals by grinding and crushing devices in the stomach. The muscular **gizzard** of terrestrial oligochaete worms, many arthropods, and birds is assisted by stones and grit swallowed along with the food (annelids and birds) or by hardened linings (for example, the chitinous teeth of the insect proventriculus [Figure 9-9] and the calcareous teeth of the gastric mill of crustaceans).

Digestive diverticula—blind tubules or pouches arising from the main passage—often supplement the stomach of many invertebrates. They are usually lined with a multipurpose epithelium having cells specialized for secreting mucus or digestive enzymes, or having absorption or storage functions. Examples include the ceca of polychaete annelids, the digestive glands of bivalve molluscs, the hepatopancreas of crustaceans, and the pyloric ceca of sea stars.

Herbivorous vertebrates have evolved several strategies for exploiting cellulose-splitting microorganisms to get the most out of plant food. Despite its abundance on earth, the woody cellulose that encloses plant cells cannot be broken down by any of the enzymes that metazoan animals produce. Once plant food is ground up, however, and under anaerobic conditions of the gut, microorganisms ferment cellulose, producing fatty acids and sugars which the herbivore can use. While the ultimate fermentation machine is the multichambered stomach of the cud-chewing ruminants described on p. 639, many other animals harbor microorganisms in other parts of the gut, such as the intestine proper or the cecum.

The stomach of carnivorous and omnivorous vertebrates is typically a U-shaped muscular tube provided with glands that produce a proteolytic enzyme and a strong acid, the latter an adaptation that probably arose for killing prey and checking bacterial activity. When food reaches the stomach, the **cardiac sphincter** opens reflexively to allow food to enter, then closes to prevent regurgitation back into the esophagus. In humans, gentle peristaltic waves pass over the filled stomach at the rate of approximately three each minute; churning is most vigorous at the intestinal end where food is steadily released into the duodenum, the first region of the intestine. Approximately 2 liters of **gastric juice** is secreted each

The stomach contains both a strong acid, having a concentration some 4 million times that found in the blood, and a powerful proteolytic enzyme. Thus it seems remarkable that the stomach mucosa is not digested by its own secretions. That it is not is a result of another gastric secretion, mucin, a highly viscous organic compound that coats and protects the mucosa from both chemical and mechanical injury. We should note that despite the popular misconception of an "acid stomach" being unhealthy, a notion carefully nourished in media advertising, stomach acidity is normal and essential. Sometimes, however, the protective mucus coating fails, allowing the gastric juices to begin digesting the stomach. The result is a peptic ulcer.

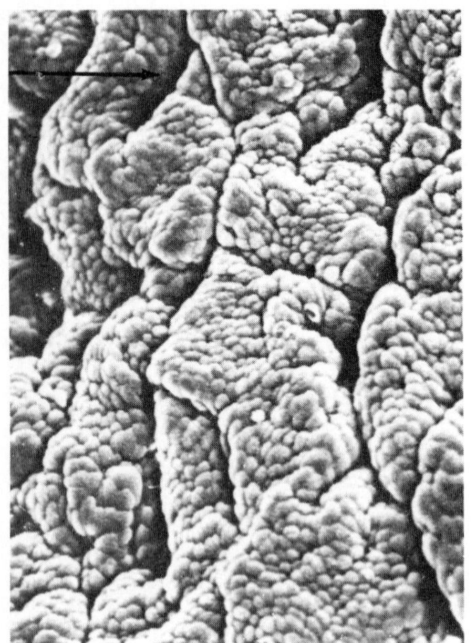

FIGURE 9-10

Scanning electron micrograph of the folded epithelium of the human stomach. Epithelial cells, resembling cobblestone paving, cover the stomach surface and line gastric glands. The arrow indicates the opening of one gland.

From Pfeiffer, C.J. 1970. J. Ultrastruct. Res. **33**:252.

FIGURE 9-11

Dr. William Beaumont at Fort Mackinac, Michigan Territory, collecting gastric juice from Alexis St. Martin.

From Myer, J.S. 1919. Life and letters of Dr. William Beaumont, St. Louis, The C.V. Mosby Co.

day by deep, tubular glands in the stomach wall (Figure 9-10). Two types of cells line these glands, **chief cells** which secrete **pepsin,** and **parietal cells,** which secrete **hydrochloric acid.** Pepsin is a **protease** (protein-splitting enzyme) that acts only in an acid medium—pH 1.6 to 2.4. It is a highly specific enzyme that splits large proteins by preferentially breaking down certain peptide bonds scattered along the peptide chain of the protein molecule. Although pepsin, because of its specificity, cannot completely degrade proteins, it effectively breaks them up into a number of small polypeptides. Protein digestion is completed in the intestine by other proteases that can together split all peptide bonds.

Rennin is a milk-curdling enzyme with only weak proteolytic activity found in the stomach of ruminant mammals. By clotting and precipitating the milk proteins it apparently slows the movement of milk through the stomach. Rennin extracted from the stomachs of calves is used in cheese making. Human infants, lacking rennin, digest milk proteins with acidic pepsin, just as adults do.

The secretion of the gastric juices is intermittent. Although a small volume of gastric juice is secreted continuously, even during prolonged periods of starvation, secretion is normally increased by the sight and smell of food, the presence of food in the stomach, and emotional states such as anxiety and anger.

The most unique and classic investigation in the field of digestion was made by U.S. Army surgeon William Beaumont during the years 1825 to 1833. His subject was a young, hard-living French Canadian voyageur, named Alexis St. Martin, who in 1822 had accidentally shot himself in the abdomen with a musket, the blast "blowing off integuments and muscles of the size of a man's hand, fracturing and carrying away the anterior half of the sixth rib, fracturing the fifth, lacerating the lower portion of the left lobe of the lungs, the diaphragm, and perforating the stomach." Miraculously the wound healed, but a permanent opening, or fistula, was formed that permitted Beaumont to see directly into the stomach (Figure 9-11). St. Martin became a permanent, although temperamental, patient in Beaumont's care, which included food and housing. Over a period of 8 years, Beaumont was able to observe and record how the lining of the stomach changed under different psychological and physiological conditions, how foods changed during digestion, the effect of emotional states on stomach motility, and many other facts about the digestive process of his famous patient.

REGION OF TERMINAL DIGESTION AND ABSORPTION: THE INTESTINE

The importance of the intestine varies widely among animal groups. In those invertebrates with extensive digestive diverticula in which food is broken and down and phagocytized, the intestine may serve only as a pathway for conducting wastes out of the body. In other invertebrates with simple stomachs, and in all vertebrates, the intestine is equipped for both digestion and absorption.

Devices for increasing the internal surface area of the intestine are highly developed in vertebrates but generally absent among invertebrates. Perhaps the most direct way to increase the absorptive surface of the gut is to increase its length. Coiling of the intestine is common among all vertebrate groups and reaches its highest development in mammals, in which the intestine's length may exceed eight times the total length of the body. While a coiled intestine is rare among invertebrates, other strategies for increasing surface area sometimes occur. For example the **typhlosole** of terrestrial oligochaete worms (Figure 21-10, p. 459), an inward folding of the dorsal intestinal wall that runs the full length of the intestine, effectively increases the internal surface area of the gut in a narrow body lacking space for a coiled intestine.

The more primitive fishes (lampreys and sharks) have longitudinal or spiral folds in their intestines. Higher vertebrates have developed elaborate folds and small fingerlike projections (**villi**) (Figure 9-12). Also the electron microscope reveals that each cell lining the intestinal cavity is bordered by hundreds of short, delicate processes called **microvilli** (Figure 9-13, *C* and *D*). These processes, together with larger villi and intestinal folds, may increase the internal surface of the intestine more than a million times compared to a smooth cylinder of the same diameter. The absorption of food molecules is enormously facilitated as a result.

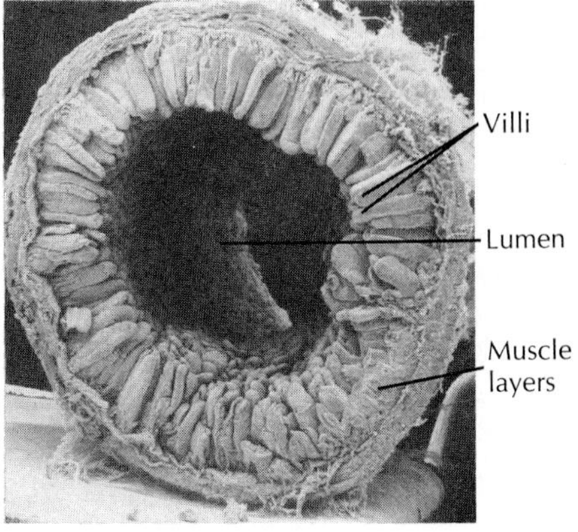

FIGURE 9-12

Scanning electron micrograph of a rat intestine showing the numerous fingerlike villi that project into the lumen and vastly increase the effective absorptive and secretory surface of the intestine. (×21.)

From Tissues and organs: a text-atlas of scanning electron microscopy, by Richard G. Kessel and Randy H. Kardon. W.H. Freeman & Co. Copyright 1979.

Digestion in the vertebrate small intestine

Food is released into the small intestine through the **pyloric sphincter,** which relaxes at intervals to allow entry of acidic stomach contents into the initial segment of the small intestine, the **duodenum.** Two secretions are poured into this region: **pancreatic juice** and **bile** (Figure 9-14). Both of these secretions have a high bicarbonate content, especially the pancreatic juice, which effectively neutralizes the gastric acid, raising the pH of the liquefied food mass, now called **chyme,** from 1.5 to 7 as it enters the duodenum. This change in pH is essential because all the intestinal enzymes are effective only in a neutral or slightly alkaline medium.

Pancreatic enzymes. Approximately 2 liters of pancreatic juice are secreted each day. The pancreatic juice contains several enzymes of major importance in digestion (Figure 9-14). Two powerful proteases, **trypsin** and **chymotrypsin,** continue the enzymatic digestion of proteins begun by pepsin, which is now inactivated by the alkalinity of the intestine. Trypsin and chymotrypsin, like pepsin, are highly specific proteases that split apart peptide bonds deep inside the protein molecule. The hydrolysis of the peptide linkage may be shown as:

$$\cdots-\mathrm{NH}-\mathrm{CH}(R^1)-\mathrm{C}(=\mathrm{O})-\mathrm{NH}-\mathrm{CH}(R^2)-\mathrm{C}(=\mathrm{O})-\cdots \;\xrightarrow{\mathrm{HOH}}\; \cdots-\mathrm{NH}-\mathrm{CH}(R^1)-\mathrm{C}(=\mathrm{O})\mathrm{OH} + \mathrm{H}\mathrm{N}\mathrm{H}-\mathrm{CH}(R^2)-\mathrm{C}(=\mathrm{O})-\cdots$$

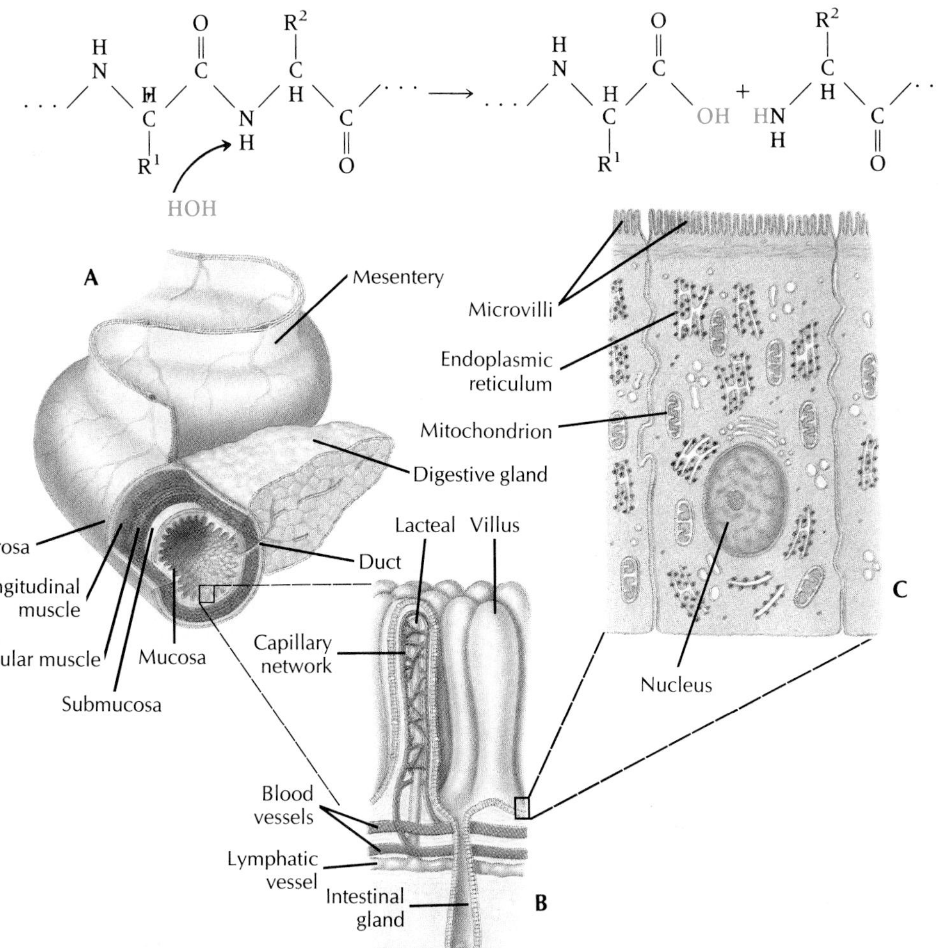

FIGURE 9-13

Organization of the vertebrate digestive tract, showing (**A**) the successive layers of mucosa, submucosa, muscle, and enveloping serosa; an enzyme-secreting digestive gland (e.g., pancreas); and the thin mesentery that positions the intestine within the body cavity. **B,** Portion of mucosal lining of intestine, showing fingerlike villi. **C,** Section of single mucosal lining cell. **D,** Microvilli on surface of mucosal cell, rat intestine. (×16,400.)

D, Electron micrograph courtesy J.D. Berlin.

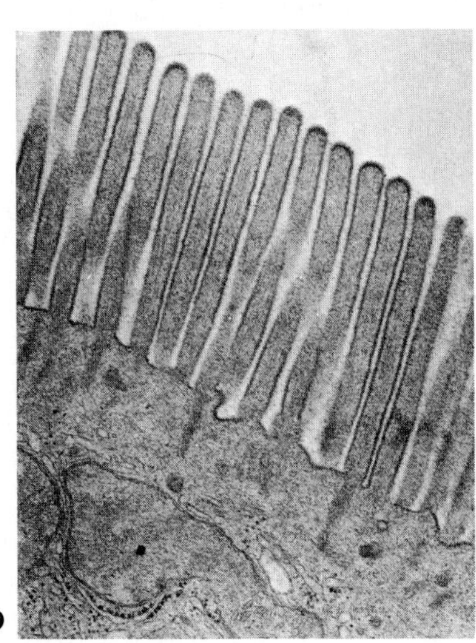

FIGURE 9-14

Secretions of mammalian alimentary canal with the principal components and the pH of each secretion.

Adapted from Eckert, R. and D. Randall. 1983. Animal physiology, ed. 2. San Francisco, W.H. Freeman & Co.

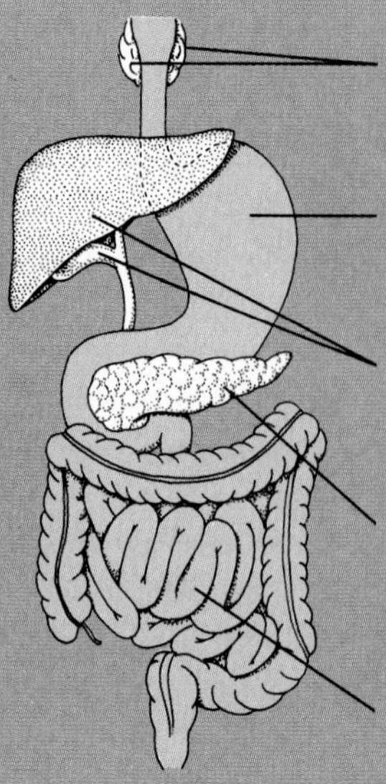

REGION	SECRETION	pH	COMPOSITION
Salivary glands	Saliva	6.5	Amylase Bicarbonate
Stomach	Gastric juice	1.5	Pepsin HCl Rennin in ruminant mammals
Liver and gallbladder	Bile	7-8	Bile salts and pigments Cholesterol
Pancreas	Pancreatic juice	7-8	Trypsin, Chymotrypsin, Carboxypeptidase, Lipase, Amylase, Nucleases Bicarbonate
Small intestine	Membrane enzymes	7-8	Aminopeptidase Maltase Lactase Sucrase Alkaline Phosphatase

Cells of the intestinal mucosa, like those of the stomach mucosa, are subjected to considerable wear and tear and are undergoing constant replacement. Cells deep in the crypt between adjacent villi divide rapidly and migrate up the villus. In mammals the cells reach the tip of the villus in about 2 days. There they are shed, along with their membrane enzymes, into the lumen at the rate of some 17 billion a day along the length of the human intestine. Before they are shed, however, these cells differentiate into absorptive cells that transport nutrients into the network of blood and lymph vessels, once digestion is complete.

Pancreatic juice also contains **carboxypeptidase,** which splits amino acids off the carboxyl ends of polypeptides; **pancreatic lipase,** which hydrolyzes fats into fatty acids and glycerol; **pancreatic amylase,** which is a starch-splitting enzyme identical to salivary amylase in its action; and **nucleases,** which degrade RNA and DNA to nucleotides.

Membrane enzymes. The cells lining the intestine have digestive enzymes embedded in their surface membrane that continue the digestion of carbohydrates, proteins, and phosphate compounds (Figure 9-14). Until recently it was thought that these enzymes were secreted into the intestinal lumen as intestinal "juice". Now it is known that the enzymes are actually a part of the microvillus membrane where, as attached glycoproteins, they accomplish much of the work of digestion. Among the membrane digestive enzymes is **aminopeptidase.** It splits terminal amino acids from the amino end of short peptides; its action is similar to that of the pancreatic enzyme carboxypeptidase. Several **disaccharidases** (enzymes that split 12-carbon sugar molecules into 6-carbon units) are also part of the microvillus membrane (Figure 9-14). These include **maltase,** which splits maltose into two molecules of glucose; **sucrase,** which splits sucrose to fructose and glucose; and **lactase,** which breaks down lactose (milk sugar) into glucose and galactose. Also present is **alkaline phosphatase,** which attacks a variety of phosphate compounds.

Bile is secreted by the cells of the **liver** into the **bile duct,** which drains into the upper intestine (duodenum). Between meals the bile is collected in the **gallbladder,** an expansible storage sac that releases the bile when stimulated by the presence of fatty food in the duodenum. Bile contains no enzymes. It is made up of water, bile salts, and pigments. The bile salts are essential for the complete absorption of fats, which, because of their tendency to remain in large, water-resistant globules, are especially resistant to enzymatic digestion. **Bile salts** reduce the surface tension of fats, so that they are broken up into small droplets by the churning movements of the intestine. This greatly increases the total surface exposure of fat particles, giving the fat-splitting lipases a chance to reduce them. The golden yellow color of bile is produced by the **bile pigments** that are breakdown products of hemoglobin from worn-out red blood cells. The bile pigments also give the feces its characteristic color.

The great versatility of the liver should be emphasized. Bile production is only one of the liver's many functions; it is a storehouse for glycogen, production center for the plasma proteins, site of protein synthesis and detoxification of protein wastes, and destruction of worn-out red blood cells, center for the metabolism of fat, amino acids, and carbohydrates, and many others.

Absorption

Little food is absorbed in the stomach because digestion is still incomplete and because of the limited surface exposure. Some materials, however, such as drugs and alcohol, are absorbed in part there, which explains their rapid action. Most digested foodstuffs are absorbed from the small intestine where the numerous finger-shaped **villi** provide an enormous surface area through which materials can pass from the intestinal lumen into the circulation.

Carbohydrates are absorbed almost exclusively as simple sugars (for example, glucose, fructose, and galactose) because the intestine is virtually impermeable to polysaccharides. Proteins, too, are absorbed principally as their amino acid subunits, although it is believed that very small amounts of small proteins or protein fragments may sometimes be absorbed. Simple sugars and amino acids are transferred across the intestinal epithelium by both passive and active processes.

Immediately after a meal these materials are in such high concentration in the gut that they readily diffuse into the blood, where their concentration is initially lower. However, if absorption were passive only, we would expect transfer to cease as soon as the concentrations of a substance became equal on both sides of the intestinal epithelium. This would permit valuable nutrients to be lost in the feces. In fact, very little is lost because passive transfer is supplemented by an **active transport** mechanism (p. 38) located in the epithelial cells that picks up the food molecules and transfers them into the blood. Materials are thus moved *against* their concentration gradient, a process that requires the expenditure of energy. Although not all food products are actively transported, those which are, such as glucose, galactose, and most of the amino acids, are handled by transport mechanisms that are specific for each kind of molecule.

As already described, fat droplets are emulsified by bile salts and then digested by pancreatic lipase. Triglycerides are thus broken down into fatty acids and monoglycerides, which are absorbed by simple diffusion. However, free fatty acids never enter the blood. Instead, during their passage through the intestinal epithelial cells, the fatty acids are resynthesized into triglycerides that pass out of the cells and into the **lacteals** (Figure 9-13, *B*). From the lacteals, the fat droplets enter the lymph system (Figure 7-16, p. 187) and eventually get into the blood by way of the thoracic duct. After a fatty meal, even a peanut butter sandwich, the presence of numerous fat droplets in the blood imparts a milky appearance to the blood plasma.

REGION OF WATER ABSORPTION AND CONCENTRATION OF SOLIDS

In the large intestine the indigestible remnants of digestion are consolidated by the reabsorption of water to form solid or semisolid feces for removal from the body by **defecation.** The reabsorption of water is especially significant for insects—particularly those living in dry environments—which must (and do) conserve nearly all water entering the rectum. Specialized **rectal glands** absorb water and ions as needed, leaving behind fecal pellets that are almost completely dry. In reptiles and birds, which also produce nearly dry feces, most of the water is reabsorbed in the cloaca. A white paste-like feces is formed containing both indigestible food wastes and uric acid.

In humans, the colon contains enormous numbers of bacteria that enter the sterile colon of the newborn infant. The bacteria break down organic wastes in the feces and provide some nutritional benefit by synthesizing certain vitamins (vitamin K and small quantities of some of the B vitamins), which are absorbed by the body.

Although milk is the universal food of newborn mammals and one of the most complete human foods, most adult humans cannot digest milk because they are deficient in lactase, the enzyme that hydrolyzes lactose (milk sugar). Lactose intolerance is genetically determined. It is characterized by abdominal bloating, cramps, flatulence, and watery diarrhea, all appearing within 30 to 90 minutes after ingesting milk or its unfermented byproducts. (Fermented dairy products, such as yogurt and cheese, create no intolerance problems.)

Northern Europeans and their descendants, which include the majority of North American whites, are most tolerant of milk. Many other ethnic groups are generally intolerant to lactose, including the Japanese, Chinese, Jews in Israel, Eskimos, South American Indians, and most African blacks. Only about 30% of North American blacks are tolerant; those who are tolerant are mostly descendants of slaves brought from east and central Africa where dairying is traditional, and tolerance to lactose is high.

THE SUPERBOWEL

I think that I shall never see
A tract more alimentary.
A tube whose velvet villi sway
Absorbing food along the way.
Whose surface folded and striate
Does rapidly regenerate.

A magic carpet whose fuzzy nap
Minuscule molecules entrap.
Then, microvilli with enzymes replete
The last hydrolyses complete.

A tunnel studded with protection
Against abrasion and infection
(Goblets their mucus spill
While lymphoid cells the microbes kill).

To top things off, it should be noted,
This Grand Canal is sugar coated!

From Fruhman, G.J. 1973. Perspect. Biol. Med. 17:66.

Brown fat, the tissue that mediates nonshivering thermogenesis, is brown because it is packed with mitochondria containing large quantities of iron-bearing cytochrome molecules. In ordinary body cells, ATP is generated by the flow of electrons down the respiratory chain (p. 51). This ATP is then used to power various cellular processes. In brown fat cells heat is generated instead of ATP. Thermogenesis is activated by the sympathetic nervous system, which responds to signals from the hypothalamus.

Regulation of Food Intake

Most animals unconsciously adjust food intake to balance energy expenditure. If energy expenditure is increased by, for example, increased physical activity, food intake is increased accordingly. Most vertebrates, from fish to mammals, eat for calories rather than bulk because, if the diet is diluted with fiber, they respond by eating more. Similarly, intake is adjusted downward following a period of several days when caloric intake is too high.

Food intake is regulated in large part by a "hunger" center located in the hypothalamus region of the brain. Blood sugar level has an important influence on this center because hunger pangs coincide with decreasing levels of blood glucose. While most animals seem able to stabilize their weight at normal levels with ease, many humans cannot. It is becoming clear that many obese people do not in fact eat more food than thin people. Rather they have a reduced capacity to burn off excess calories by "nonshivering thermogenesis." The defect has been traced to **brown fat,** a diffuse tissue located in adults in the chest, upper back, and near the kidneys (newborn mammals, including human infants, have much more brown fat than adults). In normal people an increased caloric intake induces brown fat tissue to dissipate the excess energy as heat. This is referred to as "diet-induced thermogenesis." The capacity is diminished in people tending toward obesity because they have less brown fat or because their brown fat does not respond to hypothalamic signals as it should. There are other reasons for obesity in addition to the fact that many people simply eat too much. Fat stores are supervised by the hypothalamus, which is set at a point that may be higher or lower than the population norm. A high setpoint can be lowered somewhat by exercise, but as dieters are painfully aware, the body defends its fat stores with remarkable efficiency.

Nutritional Requirements

The food of animals must include **carbohydrates, proteins, fats, water, mineral salts,** and **vitamins.** Carbohydrates and fats are required as fuels for energy demands of the body and for the synthesis of various substances and structures. Proteins (actually the amino acids of which they are composed) are needed for the synthesis of the body's specific proteins and other nitrogen-containing compounds. Water is required as the solvent for the body's chemistry and as a major component of all the body fluids. Inorganic salts are required as the anions and cations of body fluids and tissues and form important structural and physiological components throughout the body.

A vitamin is a relatively simple organic compound that is not a carbohydrate, fat, protein, or mineral and that is required in very small amounts in the diet for some specific cellular function. Vitamins are not sources of energy but are often associated with the activity of important enzymes that have vital metabolic roles. Plants and many microorganisms synthesize all the organic compounds they need; animals, however, have lost certain synthetic abilities during their long evolution and depend ultimately on plants to supply these compounds. Vitamins therefore represent synthetic gaps in the metabolic machinery of animals.

Vitamins are usually classified as fat soluble (soluble in fat solvents such as ether) or water soluble. The water-soluble ones include the B complex and vitamin C. (Table 9-1) The family of B vitamins, so grouped because the original B vitamin was subsequently found to consist of several distinct molecules, tends to be found together in nature. Almost all animals, vertebrate and invertebrate, require the B vitamins; they are "universal" vitamins. The dietary need for vitamin C and the fat-soluble vitamins A, D, E, and K tends to be restricted to the vertebrates, although some are required by certain invertebrates. Even within groups of close relationship, vitamin requirements are relative, not absolute. A

Table 9-1 Human nutrient requirements

Amino acids	
Phenylalanine	Methionine
Lysine	Cystine
Isoleucine	Tryptophan
Leucine	Threonine
Valine	
Polyunsaturated fatty acids	
Arachidonic	
Linoleic	
Linolenic	
Water-soluble vitamins	
Thiamine (B_1)	Folacin
Riboflavin (B_2)	Vitamin B_{12}
Niacin	Biotin
Pyridoxine (B_6)	Choline
Pantothenic acid	Ascorbic acid (C)
Fat-soluble vitamins	
A, D, E, and K	
Minerals	
Calcium	Silicon
Phosphorus	Vanadium
Sulfur	Tin
Potassium	Nickel
Chlorine	Selenium
Sodium	Manganese
Magnesium	Iodine
Iron	Molybdenum
Fluorine	Chromium
Zinc	Cobalt
Copper	

Modified from Scrimshaw, N.S., and V.R. Young. 1976. Sci. Am. **235**:50-64 (Sept.).

rabbit does not require vitamin C, but guinea pigs and humans do. Some songbirds require vitamin A, but others do not.

The recognition years ago that many human diseases and those of domesticated animals were caused by or associated with dietary deficiencies led biologists to search for specific nutrients that would prevent such diseases. These studies eventually yielded a list of "**essential" nutrients** for human beings and other animal species studied. The essential nutrients are those which are needed for normal growth and maintenance and *must* be supplied in the diet. In other words, it is "essential" that these nutrients be in the diet because the animal cannot synthesize them from other dietary constituents. Nearly 30 organic compounds (amino acids and vitamins) and 21 elements have been established as essential (Table 9-1). If we consider that the body contains thousands of different organic compounds, the list in Table 9-1 is remarkably short. Animal cells have marvelous powers of synthesis, enabling them to build compounds of enormous variety and complexity from a small, select group of raw materials.

In the average diet of Americans and Canadians, approximately 50% of the total calories (energy content) comes from carbohydrates and 40% comes from lipids. Proteins, essential as they are for structural needs, supply only a little more than 10% of the total calories of the average North American's diet. Carbohydrates are widely consumed because they are more abundant and cheaper than proteins or lipids. Actually humans and many other animals can subsist on diets devoid of carbohydrates, provided sufficient total calories and essential nutrients are present. Eskimos, before the decline of their native culture, lived on a diet that was high in fat and protein and very low in carbohydrate.

Atherosclerosis (Gr. *atheroma,* tumor containing gruellike matter, + *sclerosis,* to harden) is a degenerative disease in which fatty substances are deposited in the lining of arteries, resulting in narrowing of the passage and eventually hardening and loss of elasticity.

Lipids are needed principally to provide energy. However, at least three fatty acids are essential for humans because they cannot be synthesized. Much interest and research have been devoted to lipids in our diets because of the association between fatty diets and the disease **atherosclerosis.** The matter is complex, but evidence suggests that atherosclerosis may occur when the diet is high in saturated lipids (lipids with no double bonds in the carbon chains of the fatty acids) but low in polyunsaturated lipids (two or more double bonds in the carbon chains).

Proteins are expensive foods and limited in the diet. Proteins, of course, are not themselves the essential nutrients but rather contain essential amino acids. Of the 20 amino acids commonly found in proteins, 9 and possibly 11 are essential to humans (Table 9-1). The rest can be synthesized. Generally, animal proteins have more of the essential amino acids than do proteins of plant origin. All 9 of the essential amino acids must be present simultaneously in the diet for protein synthesis. If one or more is missing, the use of the other amino acids will be reduced proportionately; they cannot be stored and are broken down for energy. Thus heavy reliance on a single plant source will inevitably lead to protein deficiency. This problem can be corrected if two kinds of plant proteins having complementary strengths in essential amino acids are ingested together. For example, a balanced protein diet can by prepared by mixing wheat flour, which is poor only in lysine, with a legume (peas or beans), which is a good source of lysine but deficient in methionine and cystine. Each plant complements the other by having adequate amounts of those amino acids that are deficient in the other.

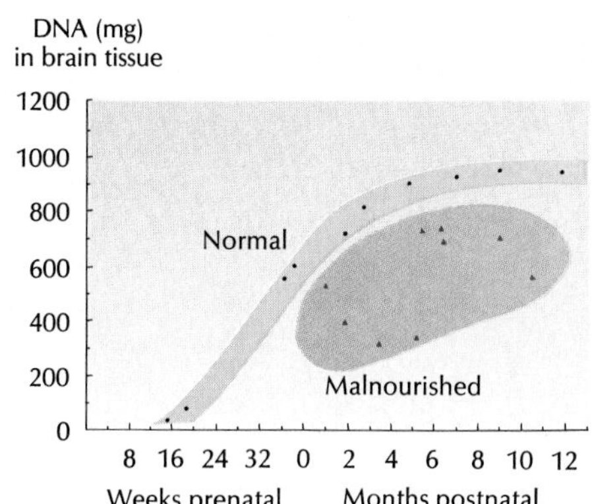

FIGURE 9-15

Effect of early malnutrition on cell number (measured as total DNA content) in the human brain. This graph shows that malnourished infants *(oval)* have far fewer brain cells than do normal infants *(growth curve).*

From Winick, M. 1976. Malnutrition and brain development. New York, Oxford University Press.

Because animal proteins are rich in essential amino acids, they are in great demand in all countries. North Americans eat far more animal proteins than do Asians and Africans; on the average a North American eats 66 g of animal protein a day, supplemented by milk, eggs, cereals, and legumes. In the Middle East, the individual consumption of protein is 14 g, in Africa 11 g, and in Asia 8 g.

Undernourishment and malnourishment rank as two of the world's oldest problems and remain the major health problems today, afflicting an eighth of the human population. Growing children and pregnant and lactating women are

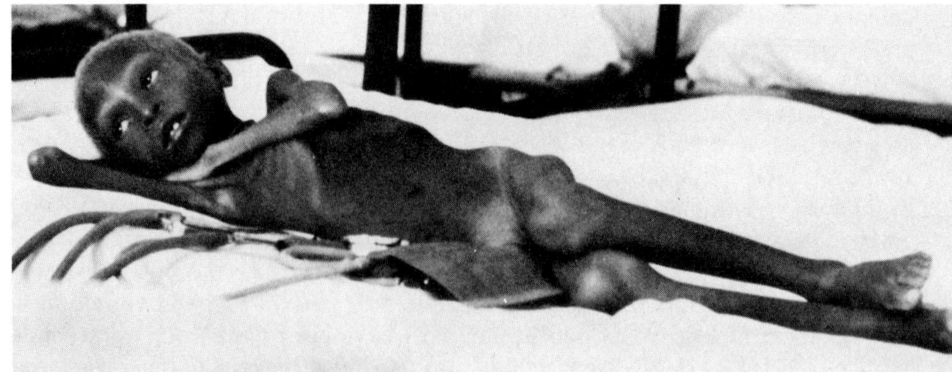

FIGURE 9-16

Biafran refugee child suffering severe malnutrition.

From Hosp. Tribune **8**:1, 1974.

Two different types of severe food deficiency are recognized: marasmus, general undernourishment from a diet low in both calories and protein, and kwashiorkor, protein malnourishment from a diet adequate in calories but deficient in protein. Marasmus (Gr. *marasmos,* a wasting away) is common in infants weaned too early and placed on low-calorie–low-protein diets; these children are listless, and their bodies waste away. Kwashiorkor is a West African word describing a disease a child gets when displaced from the breast by a newborn sibling. This disease is characterized by retarded growth, anemia, weak muscles, a bloated body with typical pot belly, acute diarrhea, susceptibility to infection, and high mortality. Ten million children the world over are seriously undernourished or malnourished, and 1 million children die of hunger each year in India alone.

especially vulnerable to the devastating effects of malnutrition. Cell proliferation and growth in the human brain are most rapid in the terminal months of pregnancy and the first year after birth. Adequate protein for neuron development is a requirement during this critical time to prevent neurological dysfunction. The brains of children who die of protein malnutrition during the first year of life have 15% to 20% fewer brain cells than those of normal children (Figure 9-15). Malnourished children who survive this period suffer permanent brain damage and cannot be helped by later corrective treatment (Figure 9-16).

The major cause of the world's precarious food situation is recent rapid population growth. The world population was 2 billion in 1930, reached 3 billion in 1960, passed 5 billion in 1987, and is expected to exceed 6 billion by the turn of the century. Each year, 83 million people are added. The equivalent of the total United States population is added to the world every 30 months. Thus the search for new ways to increase food production and distribution takes on a desperate urgency. (See Figure 9-17.)

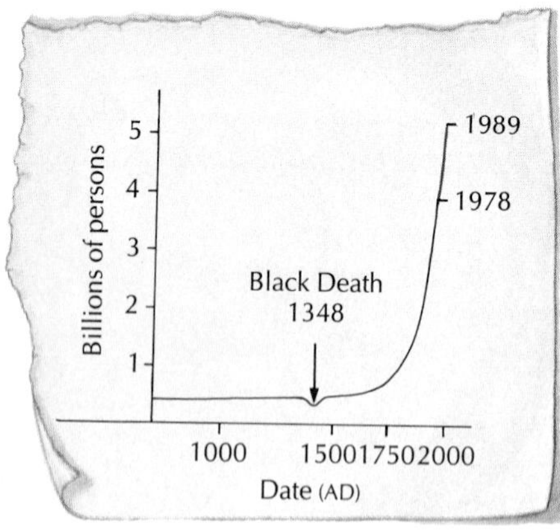

FIGURE 9-17

Portion of a graph for human population growth since 1300 AD, as it appeared in the 1978 edition of this book when the earth's population had passed 4 billion 2 years earlier, and updated to show the 1989 figure of approximately 5.2 billion.

■ Summary

Autotrophic organisms (mostly green plants), using inorganic compounds as raw materials, capture the energy of sunlight through photosynthesis and produce complex organic molecules. Heterotrophic organisms (bacteria, fungi, and animals) use the organic compounds synthesized by plants, and the chemical bond energy stored therein for their own nutritional and energy needs.

A large group of animals with very different levels of complexity feed by filtering out minute organisms and other particulate matter from the water. Others feed on organic detritus deposited in the substrate. Selective feeders, on the other hand, have evolved mechanisms for manipulating larger food masses, including various devices for seizing, scraping, boring, tearing, biting, and chewing. Fluid feeding is characteristic of endoparasites, which may absorb food across the general body surface, and of ectoparasites, herbivores, and predators that have developed specialized mouthparts for piercing and sucking.

Digestion is the process of breaking down food mechanically and chemically into molecular subunits for absorption. Digestion is intracellular in the protozoa and sponges. In more complex metazoans it is supplemented, and finally replaced entirely, by extracellular digestion, which takes place in sequential stages in a tubular cavity, the alimentary canal. Food is received in the mouth and mixed with lubricating saliva and then passed down the esophagus to regions where the food may be stored (crop), or ground (gizzard), or acidified and subjected to early digestion (vertebrate stomach). Among vertebrates, most digestion occurs in the small intestine. Enzymes from the pancreas and membrane enzymes embedded in the intestinal mucosal cells hydrolyze proteins, carbohydrates, fats, nucleic acids, and various phosphate compounds. The liver secretes bile, containing salts that emulsify fats. Once foodstuffs are digested, they are absorbed as molecular subunits (monosaccharides, amino acids, and fatty acids) into the blood or lymph vessels of the villi of the small intestine. The large intestine (colon) serves mainly to absorb water and minerals from the food wastes as they pass through. It also contains symbiotic bacteria that produce certain vitamins.

Most animals balance food intake with energy expenditure. Food intake is regulated primarily by a hunger center located in the hypothalamus. In mammals, should caloric intake exceed energy requirements, the excess calories normally are dissipated as heat in specialized brown fat tissue. A deficiency in this response is one cause of human obesity.

All animals require a balanced diet containing both fuels (mainly carbohydrates and lipids) and structural and functional components (proteins, minerals, and vitamins). For every multicellular animal, certain amino acids, lipids, vitamins, and minerals are "essential" dietary factors that cannot be produced by the animal's own synthetic machinery. Animal proteins are better-balanced sources of amino acids than are plant proteins, which tend to lack one or more essential amino acids. Undernourishment (marasmus) and protein malnourishment (kwashiorkor) are among the world's major health problems, afflicting millions of people.

REVIEW QUESTIONS

1. Distinguish between the following pairs of terms: autotrophic and heterotrophic; phototrophic and chemotrophic; herbivores and carnivores; omnivores and insectivores; anabolic and catabolic.
2. Filter feeding is one of the most important methods of feeding among animals. Explain what its characteristics, advantages, and limitations are, and name three different groups of animals that are filter feeders.
3. An animal's feeding adaptations are an integral part of an animal's life-style and usually shape the appearance of the animal itself. Discuss the contrasting feeding adaptations of carnivores and herbivores.
4. Explain how food is propelled through the digestive tract.
5. Compare intracellular with extracellular digestion and explain the advantages of the latter over intracellular digestion.
6. What structural modifications vastly increase the internal surface area of the intestine, and why is this large surface area important?
7. Trace the digestion and final absorption of a carbohydrate (starch) in the vertebrate gut, naming the carbohydrate-splitting enzymes, where they are found, the breakdown products of starch digestion, and in what form they are finally absorbed.
8. As in no. 7, trace the digestion and final absorption of a protein.
9. Explain how fats are emulsified, digested, and absorbed in the vertebrate gut.
10. Explain the phrase "diet-induced thermogenesis," and relate it to the problem of obesity in some people.
11. Name the basic classes of foods that serve mainly as (1) fuels and (2) structural and functional components.
12. Define a vitamin. What are the water-soluble and the fat-soluble vitamins?
13. Explain what is meant by the term "essential nutrients."
14. Explain the difference between saturated and unsaturated lipids, and comment on the current interest in these compounds as they relate to human health.
15. What is meant by "protein complementarity" among plant foods?

SELECTED REFERENCES

Carr, D.E. 1971. The deadly feast of life. Garden City, N.Y., Doubleday & Company. *What and how animals eat told with insight and wit.*

Doyle, J. 1985. Altered harvest: agriculture, genetics, and the fate of the world's food supply. New York, Viking Penguin Inc. *Examines the politics of the agricultural revolution and the environmental and biological costs of the American food production system.*

Jennings, J.B. 1973. Feeding, digestion and assimilation in animals, ed. 2. New York, St. Martin's Press, Inc. *A general, comparative approach. Excellent account of feeding mechanisms in animals.*

Lloyd, L.W., B.E. McDonald, and E.B. Crampton. 1978. Fundamentals of nutrition, ed. 2. San Francisco, W.H. Freeman & Co., Publishers. *Integrated approach to what nutrients are and how they are metabolized.*

Magee, D.F., and A.F. Dalley, II. 1987. Digestion and the structure and function of the gut. Basel, Switz., S. Karger AG. *Comprehensive treatment of mammalian (mostly human) digestion.*

Moog, F. 1981. The lining of the small intestine. Sci. Am. **245:**154-176 (Nov.). *Describes how the mucosal cells actively process foods.*

Owen, J. 1980. Feeding strategy. Chicago, University of Chicago Press. *Well-written and generously illustrated book from the series "Survival in the Wild."*

THE PRIVATE WORLD OF THE SENSES

By any measure, we humans enjoy a rich sensory world. We are continually bathed by information from the senses of vision, hearing, taste, olfaction, and touch. These classic five senses are supplemented by sensory inputs of cold, warmth, vibration, and pain, as well as numerous internal sensory receptors that operate silently and automatically to help keep our interior domain working smoothly. It is our senses that provide us with our impression of the environment. Yet our world, the world our senses perceive, is uniquely human. We share this exclusive world with no other animal, nor can we venture into the sensory world of any other animal except as an abstraction through our imagination.

The idea that each animal enjoys an unshared sensory world was first conceived by Jakob von Uexküll, a seldom-cited German biologist of the early part of this century. Von Uexküll asks us to try to enter the world of a tick through our imagination, supplemented by what we know of tick biology. It is a world of temperature, of light and dark, and of a single odor, that of butyric acid, a chemical common to all mammals. Insensible to all other stimuli, the tick clambers up a blade of grass to wait, for years if necessary, for the cues that will betray the presence of her prey. And later, swollen with blood, she drops to the earth, lays her eggs, and dies. The tick's impoverished sensory world, devoid of sensory luxuries and fine-tuned by natural selection for the world she will encounter, has ensured her single goal, reproduction.

A bird and a bat may share for a moment precisely the same environment. The worlds of their perceptions, however, are vastly different, structured by the limitations of the sensory windows each employs and by the brain that garners and processes what it needs for survival. For one it is a world dominated by vision; for the other, echolocation. The world of each is alien to the other, just as their worlds are to us.▪

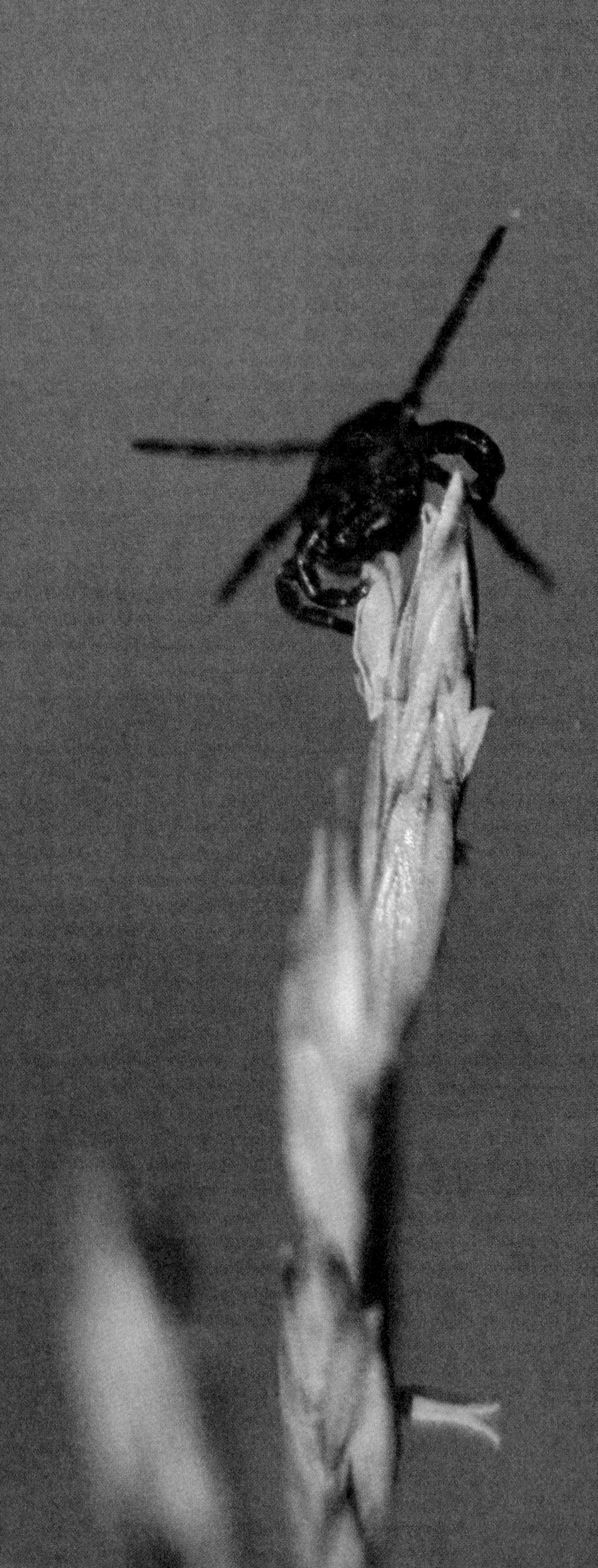

CHAPTER 10

Nervous Coordination

Nervous System and Sense Organs

The nervous system originated in a fundamental property of protoplasm — irritability. Each cell responds to stimulation in a manner characteristic of that type of cell. But certain cells have become highly specialized for receiving stimuli and for conducting impulses to various parts of the body. Through evolutionary changes, these cells have become organized into a vast communications network, and the most complex of all body systems. Indeed, the human brain is unparalleled in complexity among structures known to humans. The endocrine system is also part of the body's communication network and interacts continuously with the nervous system in the control of body function. Functionally, the nervous system differs from the endocrine system in its capacity to monitor external as well as internal changes and to respond immediately to such changes. Nervous responses are measured in milliseconds, whereas the fastest endocrine responses are measured in seconds.

The evolution of the nervous system has been correlated with the development of bilateral symmetry and cephalization. Along with this development, animals acquired specialized receptors for sensing the external environment and nervous centers for processing this information. The basic plan of the nervous system is to code sensory information and transmit it to regions of the central nervous system, where it is processed into appropriate action. This action may be any of several types, such as simple reflexes, automatic behavior patterns, conscious perception, or learning processes.

In this chapter we begin by examining neurons in some detail since they are the central functional elements of nervous systems. Then we explore the evolution of nervous systems, focusing at some length on nervous integration in the vertebrates. The final section of the chapter is devoted to the senses.

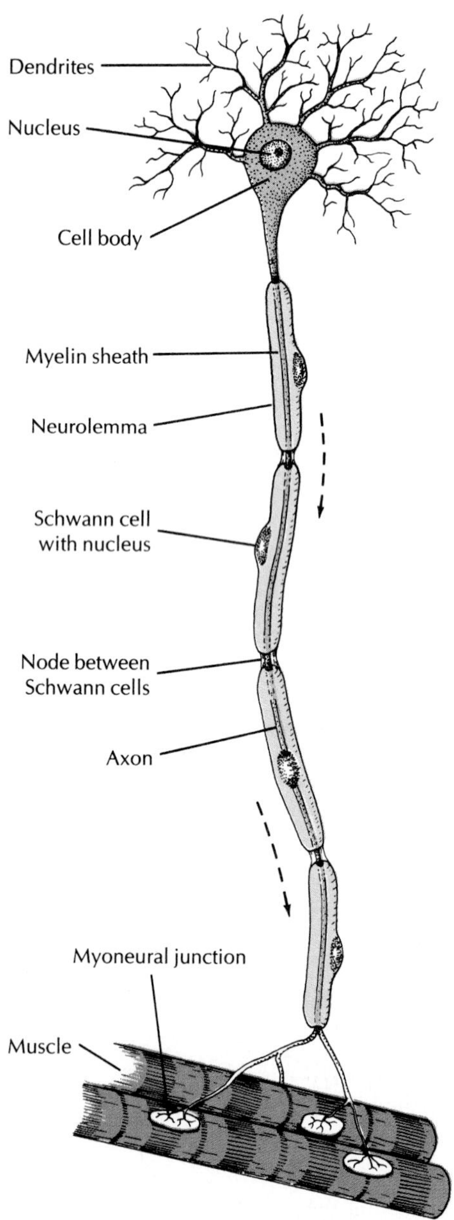

FIGURE 10-1

Structure of a motor (efferent) neuron.

The Neuron: Functional Unit of the Nervous System

The neuron is a cell body with all its processes. Although neurons assume many shapes, depending on their function and location, a typical kind is shown diagrammatically in Figure 10-1. From the nucleated body extend **processes** of two types. All but the simplest nerve cells have one or more cytoplasmic **dendrites.** As the name dendrite suggests (Gr., *dendron,* tree), these are often profusely branched. They are the nerve cell's receptive apparatus, often receiving information from several different sources at once. Some of these inputs are excitatory, others inhibitory.

From the nucleated cell body extends a single **axon** (Gr., *axon,* axle), often a long fiber (they can be meters in length in a large mammal), relatively uniform in diameter, that typically carries impulses away from the cell body. In vertebrates and some higher invertebrates, the axon is usually covered with an insulating sheath.

Neurons are commonly classified as **afferent,** or sensory; **efferent,** or motor; and **interneurons,** which are neither sensory nor motor but connect neurons with other neurons. Afferent and efferent neurons lie mostly outside the central nervous system (brain and nerve cord) while interneurons, which in humans comprise 99% of all nerve cells in the body, lie entirely within the central nervous system. Afferent neurons are connected to **receptors,** which function to convert some environmental stimulus into nerve impulses, which are carried by the afferent neurons into the central nervous system. Here the impulses may be perceived as conscious sensation. Impulses also move to efferent neurons, which carry them out by the peripheral system to **effectors,** such as muscles or glands.

In higher vertebrates, nerve processes (usually axons) are usually bundled together in a well-formed wrapping of connective tissue to form a **nerve** (Figure 10-2). The cell bodies of these nerve processes are located either in the central nervous system or in ganglia, which are discrete bundles of nerve cell bodies located outside the central nervous system.

Surrounding the neurons are nonnervous **neuroglial cells** (often simply called "glial" cells) that have a special relationship to the nerve cells. Glial cells are extremely numerous in the vertebrate brain, where they outnumber nerve cells 10 to 1 and may make up almost half the volume of the brain. Some glial cells form intimate insulating sheaths of lipid-containing **myelin** around nerve fibers. Vertebrate peripheral nerves are often enclosed by myelin laid down in

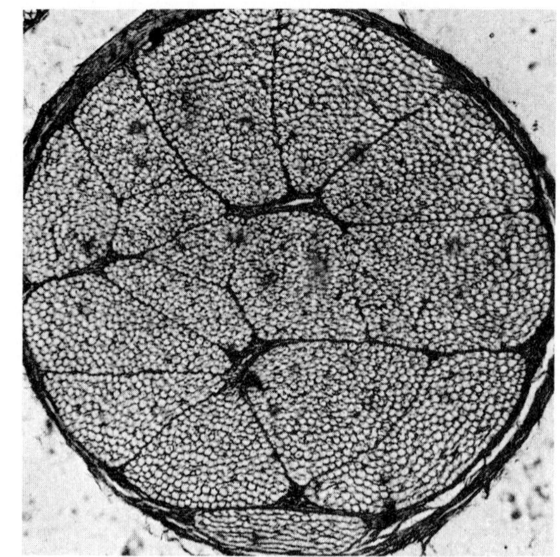

FIGURE 10-2

Cross section of nerve showing cut ends of nerve processes *(small white circles).* Such a trunk may contain thousands of both afferent and efferent fibers.

concentric rings by special glial cells called **Schwann cells** (Figure 10-3). These form the **myelin sheath,** which insulates the axon and greatly facilitates the transmission of impulses in these fibers. The functional roles of other glial cells remain to be clarified. Since many glial cells lie between blood vessels and nerve cells, they may nourish the nerve cells metabolically. It is certain that glial cells serve in the regenerative processes that follow injury and disease.

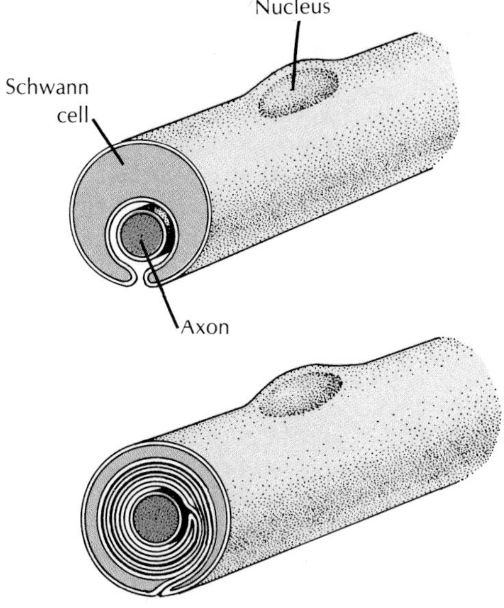

FIGURE 10-3

Development of the myelin sheath. The Schwann cell grows around the axon, and then rotates around it, enclosing the axon in a tight, multilayered, insulating myelin sheath.

NATURE OF THE NERVE IMPULSE

The nerve impulse is the chemical-electrical message of nerves, the common functional denominator of all nervous system activity. Despite the incredible complexity of the nervous system of advanced animals, nerve impulses are basically alike in all nerves and in all animals. An impulse is an "all-or-none" phenomenon; either the fiber is conducting an impulse, or it is not. Because all impulses are alike, the only way a nerve fiber can vary its effect on the tissue it innervates is by changing the frequency of impulse conduction. Frequency change is the language of a nerve fiber. A fiber may conduct no impulses at all or very few per second up to a maximum approaching 1000 per second. The higher the frequency (or rate) of conduction, the greater is the level of excitation.

The resting potential

To understand what happens when an impulse is conducted down a fiber, we need to know something about the resting, undisturbed fiber. Nerve cell membranes, like all cell membranes, have special permeability properties that create ionic imbalances. The interstitial fluid surrounding nerve cells contains relatively high concentrations of sodium (Na^+) and chloride (Cl^-) ions, but a low concentration of potassium ions (K^+). Inside the neuron, the ratio is reversed: the K^+ concentration is high but the Na^+ and Cl^- concentrations are low (Figure 10-4; see also Figure 7-2, *B,* p. 174). These differences are pronounced; there is approximately 10 times more Na^+ outside than in and 25 to 30 times more K^+ inside than out.

When at rest, the nerve cell membrane is selectively permeable to K^+, which can pass the membrane through passive ion-specific channels. The permeability to Na^+ and Cl^- is nearly zero because these channels are closed in the resting membrane. Potassium ions tend to diffuse outward through the membrane, following the concentration gradient. But since chloride ions cannot follow, each potassium ion that leaves the axon gives a positive charge to the outside of the membrane. The continued diffusion outward creates a charged membrane which is positive outside and negative inside. Very quickly the positive charge outside reaches a level that prevents any more K^+ from diffusing out of the axon. Now the resting membrane is at equilibrium, with a **resting membrane potential** that exactly balances the concentration gradient that forces the K^+ out. The resting potential is usually -70 mV (millivolts), with the inside of the membrane negative to the outside.

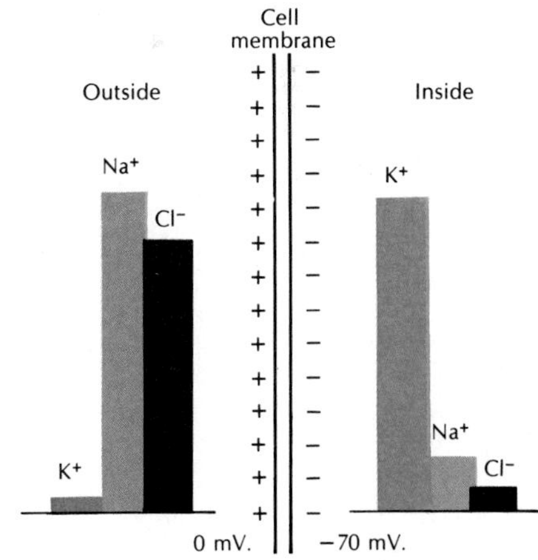

FIGURE 10-4

Ionic composition inside and outside a resting nerve cell. An active sodium pump located in the cell membrane drives sodium to the outside, keeping its concentration low inside. Potassium concentration is high inside, and although the membrane is "leaky" to potassium, this ion is held inside by the repelling positive charge outside the membrane.

The action potential

The nerve impulse is a rapidly moving change in electrical potential called the **action potential** (Figure 10-5). It is a very rapid and brief depolarization of the nerve fiber membrane; in fact, not only is the resting potential abolished, but in most nerve fibers the polarity actually reverses for an instant so that the outside becomes negative as compared to the inside. Then, as the action potential moves ahead, the membrane returns to its normal resting potential ready to conduct another impulse. The entire event occupies approximately a millisecond. Perhaps the most significant property of the nerve impulse is that it is self-propagating; that is, once started the impulse moves ahead automatically, much like the burning of a fuse.

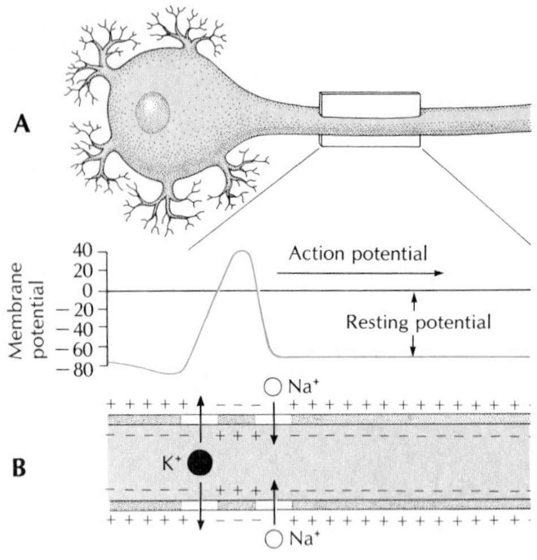

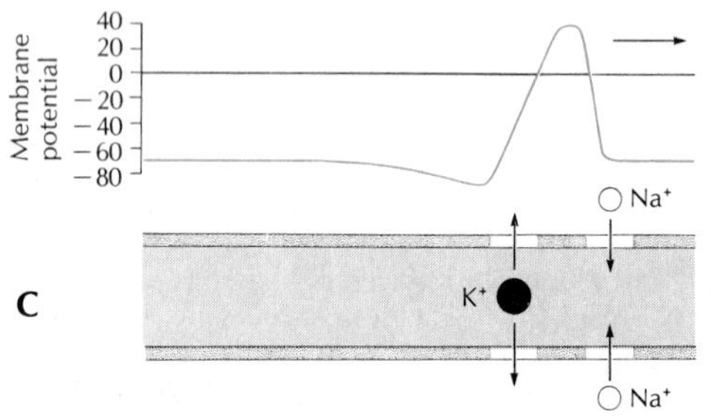

FIGURE 10-5

Conduction of the action potential of a nerve impulse. The impulse originates in cell body of the neuron (**A**) and moves toward the right. **B** and **C** show the electrical event and associated changes in localized membrane permeability to sodium and potassium. The position of the action potential in **C** is shown about 4 milliseconds after **B.** When the impulse arrives at a point, sodium gates are opened, allowing sodium ions to rush in. Sodium inflow reverses the membrane polarity, making the inner surface of the axon positive and the outside negative. Sodium gates then close and potassium gates open. Potassium ions can now penetrate the membrane and restore the normal resting potential. (Although the impulse is moving from left to right, it may be easier to follow the time course of events by reading the graphs from right to left.)

What causes the reversal of polarity in the cell membrane during passage of an action potential? We have seen that the resting potential depends on the high membrane permeability (leakiness) to K^+, some 50 to 70 times greater than the permeability to Na^+. When the action potential arrives at a given point, Na^+ channels suddenly open, permitting a flood of Na^+ to diffuse into the axon from the outside. Actually only a very minute amount of Na^+ moves across the membrane—less than one millionth of the Na^+ outside—but this sudden rush of positive ions wipes out the local membrane resting potential. The membrane is **depolarized,** creating an electrical "hole." Potassium ions, finding their electrical barrier gone, begin to move out. Then, as the action potential passes on, the membrane quickly regains its resting properties. It becomes once again practically impermeable to Na^+, and the outward movement of K^+ is checked.

The rising phase of the action potential is associated with the rapid influx (inward movement) of Na^+ (Figure 10-5). When the action potential reaches its peak, the Na^+ permeability is restored to normal, and K^+ permeability briefly increases above the resting level. This causes the action potential to decrease rapidly toward the resting membrane level.

The sodium pump

The resting cell membrane has a very low permeability to Na^+. Nevertheless some Na^+ leaks across, even in the resting condition. When the axon is active, Na^+ flows inward with each passing impulse. If not removed, the accumulation of Na^+ inside the axon would cause the resting potential of the fiber to decay. This is prevented by **sodium pumps,** each a complex of protein subunits embedded in the plasma membrane of the axon. Each sodium pump uses the energy stored in ATP to transport sodium from the inside to the outside of the membrane. The sodium pump in nerve axons, as in many other cell membranes, also moves K^+ into the axon while it is moving Na^+ out. Thus, it is an exchange pump that helps to restore the ion gradients of both Na^+ and K^+.

■ Synapses: Junction Points between Nerves

When an action potential passes down an axon to its terminal, it must cross a small gap, the **synapse** (Gr., *synapsis,* contact, union), separating it from another neuron or an effector organ. Two distinct kinds of synapses are known: electrical and chemical.

Electrical synapses, though much less common than chemical synapses, have been demonstrated in several invertebrate groups and are probably rather common in the nervous systems of higher vertebrates. Electrical synapses are points where ionic currents flow directly across a narrow **gap junction** from one neuron to another. Electrical synapses show no time lag and consequently are important for escape reactions.

Much more complex than electrical synapses are **chemical synapses.** These contain packets of specialized chemicals called **neurotransmitters.** Neurons bringing impulses toward chemical synapses are called **presynaptic neurons;** those carrying impulses away are **postsynaptic neurons.** At the synapse, the membranes are separated by a narrow gap, the synaptic cleft, having a width of approximately 20 nm.

The axon of most neurons divides at its end into many branches, each of which bears a synaptic knob that sits on the dendrites or cell body of the next neuron (Figure 10-6). The axon terminations of several neurons may almost cover a nerve cell body and its dendrites with thousands of synapses. Because a single impulse coming down a nerve axon splays out into the many branches and synaptic endings on the next nerve cell, many impulses therefore converge at the cell body at one instant.

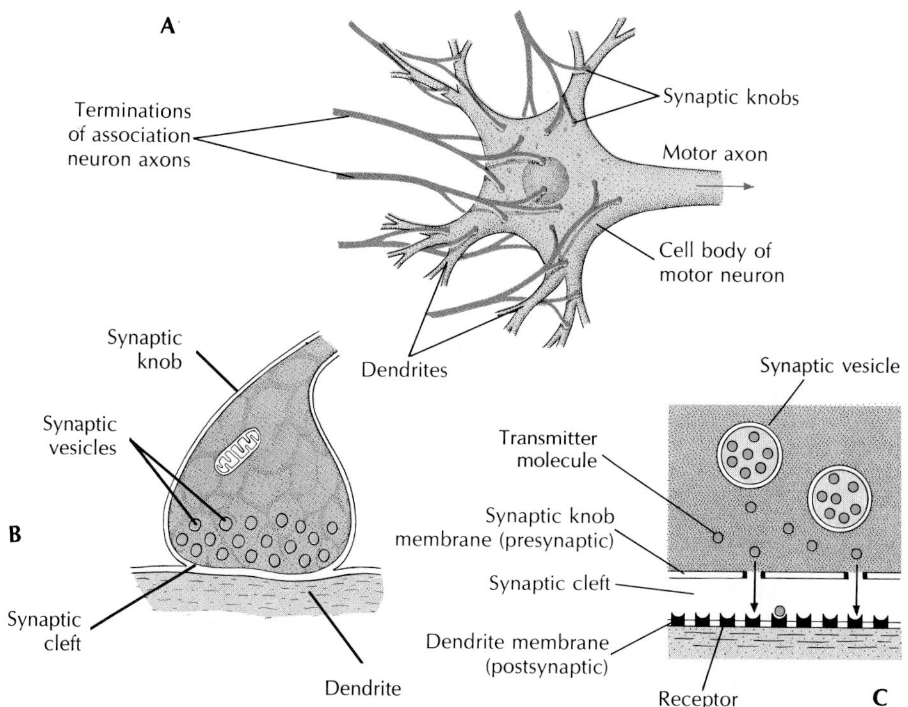

FIGURE 10-6

Transmission of impulses across nerve synapses. **A,** The cell body of a motor nerve is shown with the termination of interneurons. Each termination ends in a synaptic knob; hundreds of synaptic knobs may be on a single nerve cell body and its dendrites. **B,** A synaptic knob enlarged 60 times more than in **A.** An impulse traveling down the axon causes protein channels in the presynaptic membrane to open, releasing neurotransmitter molecules into the cleft. **C,** A synaptic cleft as it might appear under a high-resolution electron microscope. Neurotransmitter molecules move rapidly across the gap to bind briefly with receptor molecules in the postsynaptic membrane. This produces a potential change in the postsynaptic membrane.

The 20 nm fluid-filled gap between presynaptic and postsynaptic membranes prevents impulses from spreading directly to the postsynaptic neuron. Instead the synaptic knobs secrete a specific transmitter, usually acetylcholine, that communicates chemically with the postsynaptic cell. Inside the synaptic knobs are numerous tiny vesicles, each containing several thousand molecules of acetylcholine. Additional acetylcholine is also present in the cytoplasm of the synapse. Current evidence suggests that when an impulse arrives at the terminal knob a sequence of events as portrayed in Figure 10-7 occurs. The action potential opens protein channels in the presynaptic membrane, allowing a jet of acetylcholine to diffuse across the gap in a fraction of a millisecond and bind briefly to receptor molecules on the postsynaptic membrane. This creates a voltage change in the postsynaptic membrane. Whether the voltage change is large enough to trigger

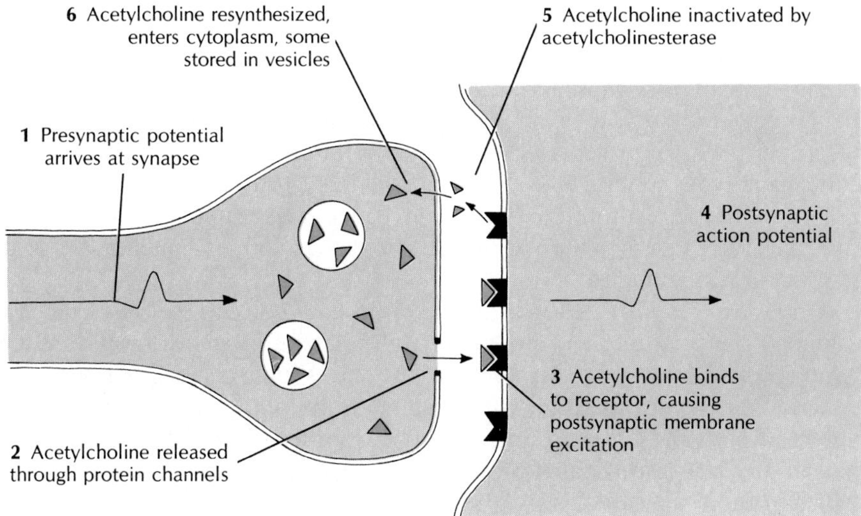

FIGURE 10-7

Sequence of events in synaptic transmission excitation and chemistry.

an action potential depends on how many acetylcholine molecules are released and how many channels are opened. The acetylcholine is rapidly destroyed by the enzyme **acetylcholinesterase.** This is important because, if not inactivated in this way, the transmitter would continue to stimulate indefinitely. The organophosphate insecticides and certain military nerve gases are poisonous for precisely this reason; that is, they inhibit acetylcholinesterase. The final step in the sequence is the resynthesis of the acetylcholine and its storage in vesicles, ready to respond to another impulse.

Several different chemical neurotransmitters have been identified in both vertebrate and invertebrate nervous systems. Some, such as acetylcholine and norepinephrine, depolarize the postsynaptic membrane; these are **excitatory synapses.** Other neurotransmitters such as γ-aminobutyric acid (GABA) hyperpolarize the postsynaptic membrane, which tends to stabilize the membrane against depolarization. These are **inhibitory synapses.** Most nerve cells in the central nervous system have both excitatory and inhibitory synapses among the hundreds of synaptic knobs on the dendrites and cell body of each nerve cell.

Whether or not an action potential will be fired depends on how the nerve cell responds to the balance of excitatory and inhibitory information it receives. If many excitatory synapses are received at one time, they may reduce the membrane voltage enough in the postsynaptic membrane to fire an impulse. Inhibitory impulses, however, stabilize the postsynaptic membrane, making it less likely that an action potential will fire. Thus it is the net effect of all excitatory and inhibitory inputs that determines whether the postsynaptic cell will fire or remain silent. The synapse is of great functional importance because it is a crucial part of the decision-making equipment of the central nervous system. In the synapse information is modulated from one nerve to the next.

■ Evolution of the Nervous System

INVERTEBRATES: THE PROGRESSION TOWARD CENTRALIZED NERVOUS SYSTEMS

The various metazoan phyla reveal a progressive increase in nervous system complexity that probably reflects in a general way the stages in the evolution of the nervous system. The simplest pattern of invertebrate nervous system is the **nerve net** of cnidarians (Figure 10-8, *A*), which, although primitive, is nonetheless a quantum leap in complexity beyond the protozoa, which lack any organized nervous network. The nerve net forms an extensive network that is found in and under the epidermis over all the body. An impulse starting in one part of this net is conducted in all directions, since synapses in the system do not restrict transmission to one-way movement, as they do in higher animals (see Figure 17-4, p. 379). There are no differentiated sensory, motor, or connector components in the strict meaning of those terms. Branches of the nerve net connect to receptors in the epidermis and to epithelial cells that have contractile properties. Most responses tend to be generalized, yet many are astonishingly complex for so simple a nervous system. It is interesting that this type of nervous system is found among higher animals in the form of nerve plexuses located, for example, in the intestinal wall, where they govern generalized intestinal movements such as peristalsis and segmentation.

Flatworms are provided with two anterior **ganglia** of nerve cells from which two main nerve trunks run posteriorly, with lateral branches extending to the various parts of the body (Figure 10-8, *B*). This is the true beginning of a differentiation into a **peripheral nervous system** (a communications network extending to all parts of the body) and a **central nervous system,** which coordinates everything. It is also the first appearance of the **linear** type of nervous system, which is more developed in higher invertebrates. Higher invertebrates have a

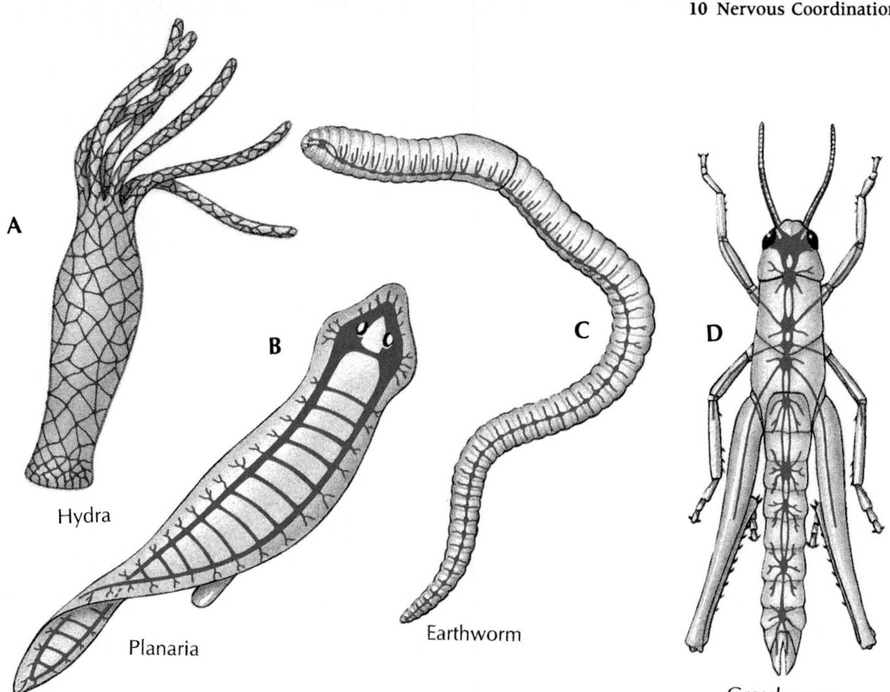

FIGURE 10-8

Invertebrate nervous systems. **A,** The nerve net of radiates, the simplest neural organization. **B,** The flatworm system, the simplest linear-type nervous system of two nerve trunks connected to a complex neuronal network. **C,** The annelid nervous system, organized into a bilobed brain and ventral cord with segmental ganglia. **D,** Arthropod nervous system with large ganglia and with more advanced sense organs.

more centralized nervous system, with the two longitudinal nerve cords fused (although still recognizable) and many ganglia present. The well-developed nervous system of annelids consists of a bilobed brain, a double nerve cord with ganglia, and distinctive **afferent** (sensory) and **efferent** (motor) neurons (Figure 10-8, *C*). The segmental ganglia serve as relay stations for coordinating regional activity.

The basic plan of the mollusc nervous system is a series of three pairs of well-defined ganglia, but in the most advanced molluscs (cephalopods such as octopus and squid), the ganglia have burgeoned into textured nervous centers of great complexity. Sense organs, too, are highly developed. With these advancements, the complexity of cephalopod behavior far outstrips that of any other invertebrate.

The basic plan of the arthropod nervous system (Figure 10-8, *D*) resembles that of annelids, but the ganglia are larger and sense organs are much better developed. Social behavior is often elaborate, particularly in the hymenopteran insects (bees, wasps, and ants), and most arthropods are capable of considerable manipulation of their environment. Despite the complexity of much insect behavior, insects are nevertheless reflex-bound animals incapable of higher learning behavior due in large part to their small size. More complex behavior had to await the evolution of vertebrates.

VERTEBRATES: THE SUCCESS OF ENCEPHALIZATION

The basic plan of the vertebrate nervous system is a hollow, *dorsal* nerve cord terminating anteriorly in a large ganglionic mass, the brain. This pattern is in pronounced contrast to the nerve cord of bilateral invertebrates, which is solid and ventral to the alimentary canal. Several trends in nervous system evolution are peculiar to the vertebrates, but of these certainly the most important is the continued process of **encephalization,** the great elaboration of the functional capacities of the brain.

The spinal cord

The brain and spinal cord compose the central nervous system. During early embryonic development, the spinal cord and brain begin as an ectodermal **neural groove,** which by folding and enlarging becomes a long, hollow **neural tube.** The cephalic end enlarges into the brain vesicles, and the rest becomes the spinal

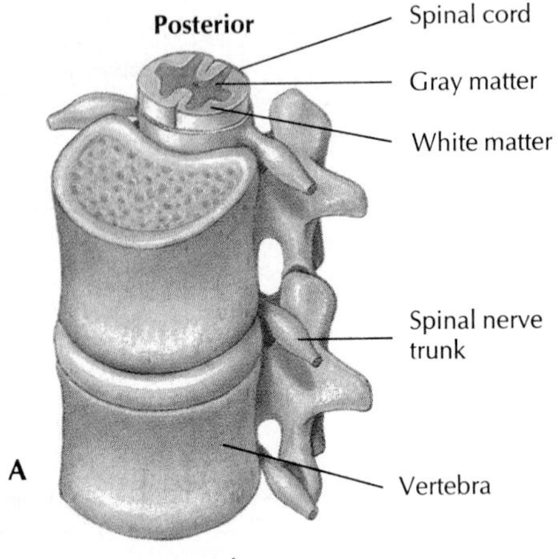

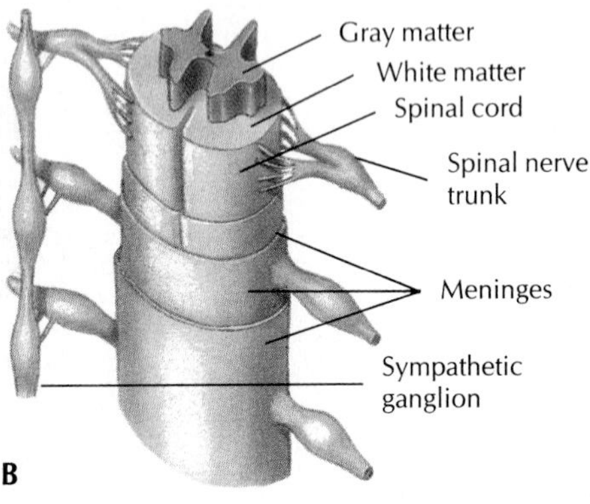

FIGURE 10-9

Human spinal cord and its protection. **A,** Two vertebrae showing the position of the spinal cord and emerging spinal nerve trunks. The cord is wrapped by three layers of membrane (meninges) between two of which lies a protective bath of cerebrospinal fluid. **B,** Spinal cord in section showing spinal nerves and sympathetic trunk.

cord. Unlike any invertebrate nerve cord, the segmental nerves of the spinal cord (31 pairs in humans) are separated into **dorsal sensory roots** and **ventral motor roots.** The sensory nerve cell bodies are gathered together into dorsal root (spinal) ganglia. Both dorsal (sensory) and ventral (motor) roots meet some distance beyond the spinal cord to form a mixed spinal nerve (Figure 10-9).

The spinal cord is enclosed by the vertebral canal and is additionally wrapped in three layers of membranes called **meninges** (men-in'jeez; Gr., *meningos,* membrane). In cross section the cord shows two zones (Figure 10-9, *B*). The inner zone of **gray matter,** resembling in shape the wings of a butterfly, contains the cell bodies of motor neurons and interconnecting association neurons (described below). The outer zone of **white matter** contains nerve bundles of axons and dendrites linking different levels of the cord with each other and with the brain.

The reflex arc. Many neurons work in groups called **reflex arcs** (Figure 10-10). There must be at least two neurons in a reflex arc, but usually there are more. The parts of a typical reflex arc consist of (1) a **receptor,** a sense organ in the skin, muscle, or other organ; (2) an **afferent** or sensory neuron, which carries impulses toward the central nervous system; (3) the **central nervous system,** where synaptic connections are made between the sensory neurons and the association neurons; (4) the **efferent** or motor neuron, which makes a synaptic connection with the association neuron and carries impulses out from the central nervous system; and (5) the **effector,** by which the animal responds to environmental changes. Examples of effectors are muscles, glands, ciliated cells, nematocysts of the radiate animals, electric organs of fish, and certain pigmented cells called chromatophores.

A reflex arc at its simplest consists of only two neurons—a sensory (afferent) neuron and a motor (efferent) neuron. Usually, however, association neurons are interposed (Figure 10-10). Association neurons may connect afferent and efferent neurons on the same side of the spinal cord, or on opposite sides, or connect them on different levels of the spinal cord, either on the same or opposite sides. In almost any reflex act a number of reflex arcs are involved. For instance, a single afferent neuron may make synaptic connections with many efferent neurons. In a similar way an efferent neuron may receive impulses from many afferent neurons.

A **reflex act** is the response to a stimulus acting over a reflex arc. It is **involuntary,** meaning that it is not under the control of the will. Many vital processes of the body, such as control of breathing, heartbeat, diameter of blood vessels, and sweat gland secretion are reflex acts. Some reflex acts are inherited and innate; others are acquired through learning processes.

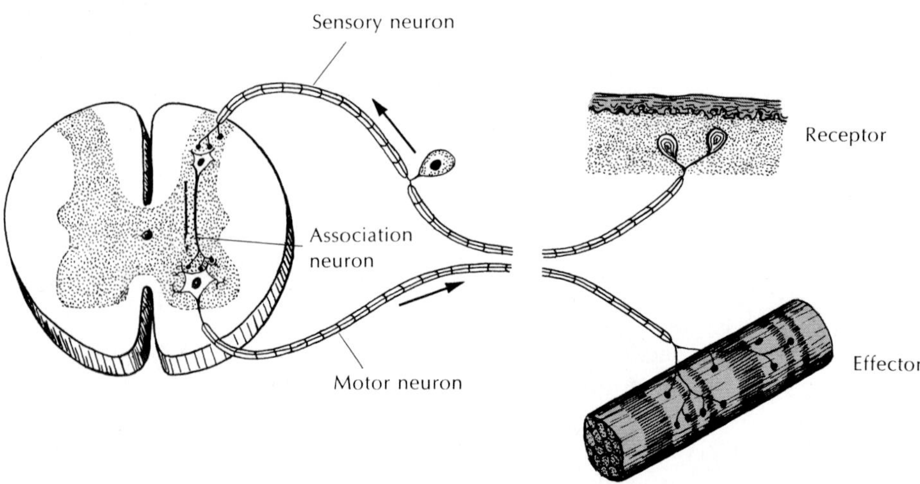

FIGURE 10-10

Reflex arc. Impulse generated in the receptor is conducted over an afferent (sensory) nerve to the spinal cord, relayed by an association neuron to an efferent (motor) nerve cell body and by the efferent axon to an effector.

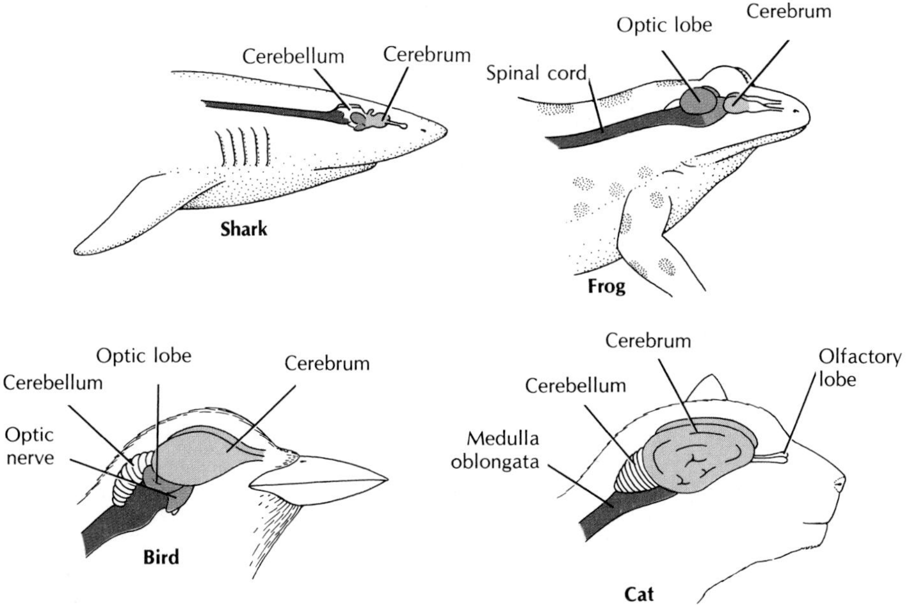

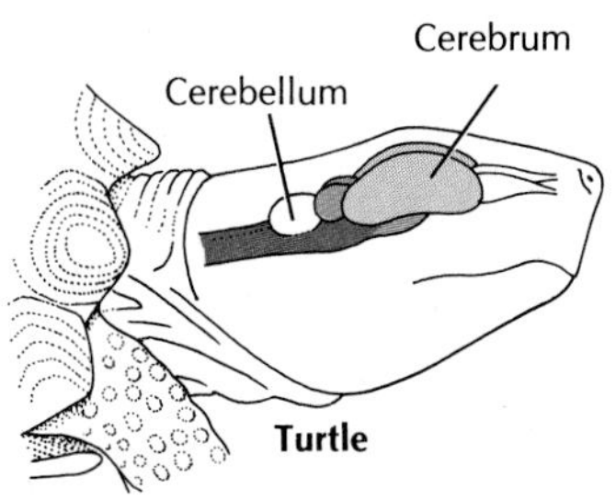

FIGURE 10-11

Evolution of the vertebrate brain. Note the progressive increase in size of the cerebrum. The cerebellum, concerned with equilibrium and motor coordination, is largest in animals whose balance and precise motor movements are well developed (fishes, birds, and mammals).

The brain

Unlike the spinal cord, which has changed little in structure during vertebrate evolution, the brain has changed dramatically. From the primitive linear brain of fishes and amphibians, it has expanded into the deeply fissured, enormously complex brain of mammals (Figure 10-11). It reaches its highest level in the human brain, which contains some 10 to 15 billion nerve cells, each connected, on the average, to about 1000 others. The ratio between the weight of the brain and that of the spinal cord affords a fair criterion of an animal's intelligence. In fish and amphibians this ratio is approximately 1:1; in humans the ratio is 55:1—in other words, the brain is 55 times heavier than the spinal cord. Although the human brain is not the largest (the sperm whale's brain is seven times heavier) nor the most convoluted (that of the porpoise is even more wrinkled), it is by all odds the best. This "great ravelled knot," as the British physiologist Sir Charles Sherrington called the human brain, in fact may be so complex that it will never be able to understand its own function.

The brain is made up of both white and gray matter, with the gray matter on the outside (in contrast to the spinal cord in which the gray matter is inside). The gray matter of the brain is mostly in the convoluted **cortex** of the cerebrum. In the deeper white matter of the cerebrum, myelinated bundles of nerve fibers connect the cortex with lower centers of the brain and spinal cord or connect one part of the cortex with another. Also in deeper portions of the brain are clusters of nerve cell bodies (nuclei) that provide synaptic junctions between the neurons of higher centers and those of lower centers.

The brain of the early vertebrate fishes had three principal divisions: a forebrain, the **prosencephalon;** a midbrain, the **mesencephalon;** and a hindbrain, the **rhombencephalon** (Figure 10-12 and Table 10-1). Each of the three parts was concerned with one or more of the special senses: the forebrain with the sense of smell, the midbrain with vision, and the hindbrain with hearing and balance. These primitive but very fundamental concerns of the brain have been in some instances amplified and in others reduced or overshadowed during continued evolution as sensory priorities were shaped by the animal's habitat and way of life.

Hindbrain. The **medulla,** the most posterior division of the brain, is really a conical continuation of the spinal cord. The medulla, together with the more

Although the large size of their brain undoubtedly makes humans the wisest of animals, it is apparent that they can do without much of it and still remain wise. Brain scans of persons with hydrocephalus (enlargement of the head as a result of pressure disturbances that cause the brain ventricles to enlarge many times their normal size) show that although many of them are functionally disabled, others are nearly normal. The cranium of one person with hydrocephalus was nearly filled with cerebrospinal fluid and the only remaining cerebral cortex was a thin layer of tissue, 1 mm thick, pressed against the cranium. Yet this young man, with only 5% of his brain, had achieved first-class honors in mathematics at a British university and was socially normal. This and other similarly dramatic observations suggest that there is enormous redundancy and spare capacity in corticocerebral function. It also suggests that the deep structures of the brain, which are relatively spared in hydrocephalus, may carry out functions once believed to be performed solely by the cortex.

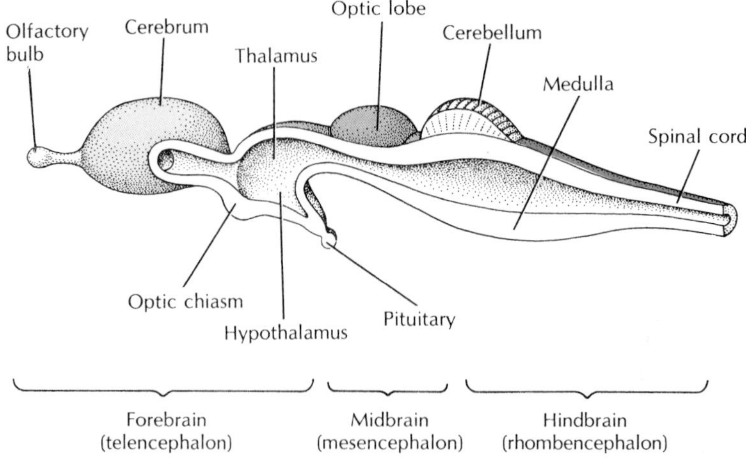

FIGURE 10-12

Brain of a fish, showing the basic organization of the vertebrate brain.

Table 10-1 Divisions of the vertebrate brain

Embryonic vesicle	Main components in adults
Forebrain (prosencephalon)	Olfactory bulb and tracts Cerebral cortex Basal ganglia Limbic system Thalamus Hypothalamus Pituitary gland
Midbrain (mesencephalon)	Optic lobes
Hindbrain (rhombencephalon)	Cerebellum Pons Medulla

anterior midbrain, constitutes the "brainstem," an area that controls numerous vital and largely subconscious activities such as heartbeat, respiration, vascular tone, and swallowing. Although it is small and mostly hidden from view by the much enlarged "higher" centers, it is, in fact, the most vital brain area. Whereas damage to higher centers may result in severely debilitating loss of sensory or motor function or of higher mental processes (e.g., learning, thinking, memory), damage to the brainstem usually is fatal.

The **pons,** between the medulla and the midbrain, is made up of a thick bundle of fibers that carry impulses from one side of the cerebellum to the other.

The **cerebellum,** lying above the medulla, is concerned with equilibrium, posture, and movement (Figure 10-13). Its development is directly correlated with the animal's mode of locomotion, agility of limb movement, and balance. It is usually weakly developed in amphibians and reptiles, which are relatively clumsy forms that stick close to the ground, and is well developed in the more agile bony fishes. It reaches its apogee in birds and mammals in which it is greatly expanded and folded. The cerebellum does not initiate movement but operates as a precision error-control center, or servomechanism, that programs a movement initiated somewhere else, such as the motor cortex. Primates and especially humans, who possess a manual dexterity far surpassing that of other animals, have the most complex cerebellum of all, since hand and finger movements may involve the simultaneous contraction and relaxation of hundreds of individual muscles.

Midbrain. The midbrain consists mainly of the optic lobes, which contain nuclei that serve as centers for visual and auditory reflexes. The midbrain has undergone little evolutionary change in size among vertebrates but has changed its function. It mediates the most complex behavior of fishes and amphibians. Such integrative functions were gradually assumed by the forebrain in higher vertebrates, and in mammals the midbrain is mainly a reflex center for eye muscles and a relay and analysis center for auditory information.

Forebrain. Just anterior to the midbrain lie the **thalamus** and **hypothalamus,** the most posterior elements of the forebrain. The egg-shaped thalamus is a major relay station that analyzes and passes sensory information to higher brain centers. In the hypothalamus, an area the size of a marble lying below the thalamus, are several "housekeeping" centers that regulate body temperature, water balance, appetite, and thirst—all functions concerned with the maintenance of internal constancy (homeostasis). Neurosecretory cells located in the hypothalamus produce several pituitary-regulating neurohormones (described in Chapter

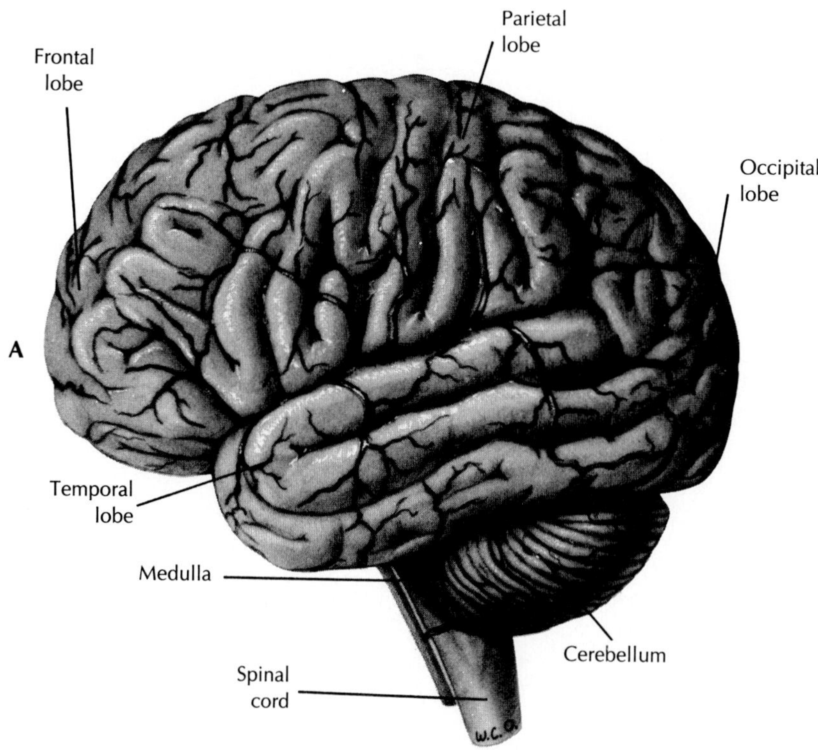

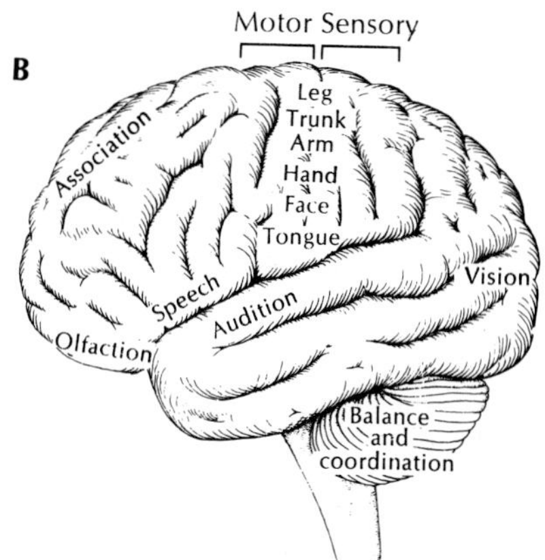

FIGURE 10-13

Human brain. **A,** External view of the left side of the brain showing lobes of the cerebrum, cerebellum, and brainstem. **B,** Localization of function on the left cerebral cortex and cerebellum.

11). The hypothalamus also contains centers for diverse emotions such as pleasure, aggression, rage, punishment, and sexual arousal.

The anterior portion of the forebrain, the **cerebrum** (Figure 10-13), can be divided into two anatomically distinct areas, the **paleocortex** and **neocortex.** Originally concerned with smell, it became well developed in the advanced fishes and early terrestrial vertebrates, which depend on this special sense. In mammals and especially in primates the paleocortex is a deep-lying area called the rhinencephalon ("nose brain"), but in humans it actually has little to do with the sense of smell. Better known as the **limbic system,** it has acquired a number of ill-defined functions concerned with mating, memory, and emotional control and species-specific behavior patterns.

The limbic system derives its name from its position: a ring-shaped belt bordering (L. *limbus,* edge, border) the neocortex.

Although a late arrival in vertebrate evolution, the neocortex completely overshadows the paleocortex and has become so expanded that it envelops much of the forebrain and all of the midbrain (Figure 10-13). Almost all the integrative activities primitively assigned to the midbrain were transferred to the neocortex or cerebral cortex as it is usually called.

Thus in mammals, and especially in human beings, there is a primitive and an advanced brain that mediate quite separate functions. The deep primitive brain, all of the brain but the cerebral cortex, governs the numerous vital functions that are removed from conscious control: respiration, blood pressure, heart rate, hunger, thirst, temperature balance, salt balance, sexual drive, and basic (sometimes irrational) emotions. It is also a complex endocrine gland that regulates the body's subservient endocrine system. This primitive brain is the *unconscious* mind. The other brain is the "new" brain, the cerebral cortex, wherein are seated intellect, reason, and integration of sensory information. This governing brain is the *conscious* mind.

The brain, of course, operates as a whole, and both the primitive and the advanced brains are intimately interconnected. Unconscious disturbances are communicated to the conscious brain, and the conscious brain may have powerful effects on the unconscious. Memory appears to transcend all brain levels

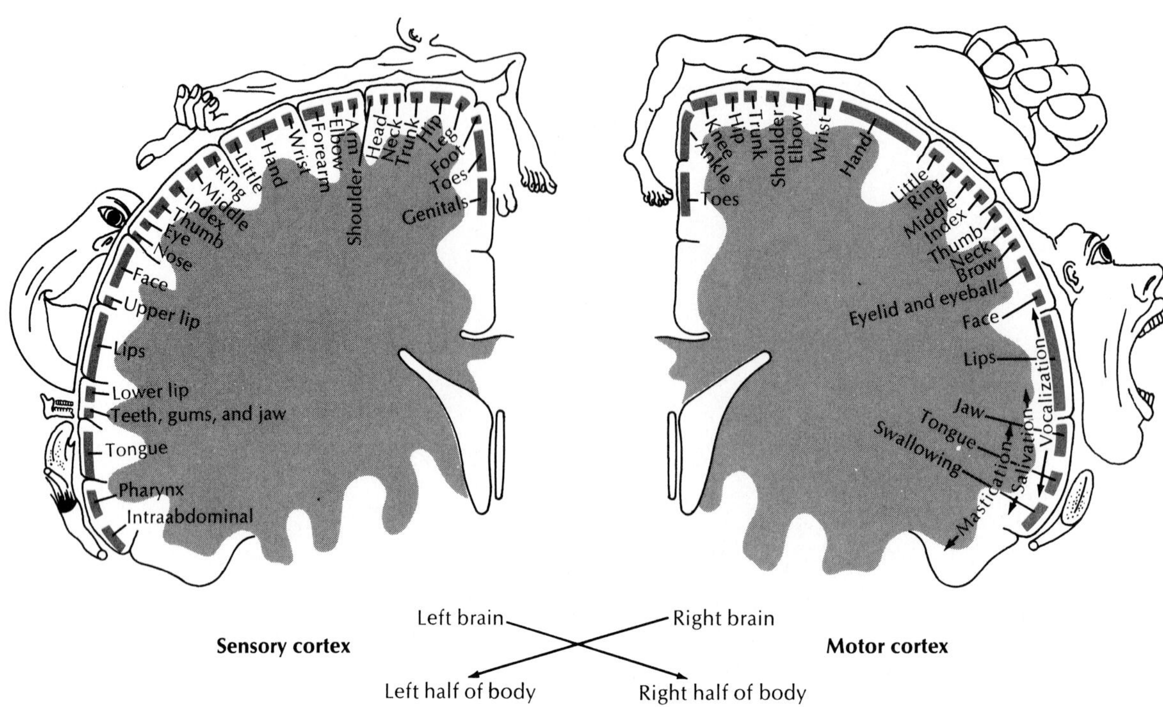

FIGURE 10-14

Arrangement of sensory and motor cortices. The location of sensory terminations from different parts of the body is shown at left; the sensory pattern is repeated as a mirror image on the right side. The origin of descending motor pathways to different parts of the body is shown at right. This pattern is repeated on the left brain. The motor cortex lies in front of the sensory, so the two are not superimposed. Ascending and descending tracts from each hemisphere connect to the opposite side of the body.

rather than being a property of any particular part of the brain as was once believed.

It has been possible to localize function in the cerebrum by direct stimulation of exposed brains of people and experimental animals, postmortem examination of persons suffering from various lesions, and surgical removal of specific brain areas in experimental animals. The cortex contains discrete motor and sensory areas (Figure 10-13 and 10-14) as well as large "silent" regions, called **association areas,** concerned with memory, judgment, reasoning, and other integrative functions. These regions are not directly connected to sense organs or muscles.

Hemispheric specialization. The cerebral cortex is incompletely divided into two **hemispheres** by a deep longitudinal fissure. The right and left hemispheres are bridged through the **corpus callosum,** a neural mass lying between and connecting the hemispheres. Through the corpus callosum the two hemispheres are able to transfer information and coordinate mental activities.

Until recently it was believed that one hemisphere, almost always the left, becomes functionally dominant over the other during childhood. This concept is now recognized as misleading. It is now known that the left and right brain hemispheres are specialized for entirely different functions: the left brain hemisphere (controlling the right side of the body) for language development, mathematical and learning capabilities, and sequential thought processes; and the right brain hemisphere (controlling the left side of the body) for spatial, musical, artistic, intuitive, and perceptual activities. It has long been known that even extensive damage to the right hemisphere may cause varying degrees of left-sided paralysis but has little effect on intellect. Conversely, damage to the left hemisphere usually has disastrous effects on intellect. Since these differences in brain symmetry and function exist at birth, they appear to be inborn rather than the result of developmental or environmental effects as previously believed.

Hemispheric specialization has long been considered a unique human trait, but was recently discovered in the brains of songbirds in which one side of the brain is specialized for song production.

The peripheral nervous system

The peripheral nervous system consists of nerve cells or extensions of nerve cells that lie *outside* the central nervous system. It is a communications system for the conduction of sensory and motor information between the brain and all parts of the body. The peripheral nervous system can be broadly subdivided into afferent and efferent components. As shown in the margin, the efferent system is much more complex, consisting of a somatic nervous system and an autonomic nervous system.

Peripheral nervous system
A. Afferent system (sensory)
B. Efferent system (motor)
 1. Somatic nervous system
 2. Autonomic nervous system
  a. Sympathetic nervous system
  b. Parasympathetic nervous system

Afferent system

Afferent (sensory) neurons carry signals from receptors in the periphery of the body to the central nervous system. The afferent neuron of a reflex arc (Figure 10-10) is representative of all afferent pathways in the peripheral nervous system. One long nerve process extends from the cell body in the dorsal root ganglion just outside the spinal cord to innervate receptors; another process passes from the cell body into the central nervous system where it connects with other neurons.

Efferent system

SOMATIC NERVOUS SYSTEM. Nerve fibers of the somatic division of the peripheral nervous system pass from the brain of the spinal cord to skeletal muscle fibers. These are called motor neurons because they control muscle movement. They release acetylcholine as the transmitter substance at the nerve endings. Motor neurons in the reflex arc (Figure 10-10) are representative of the functional position of the somatic nervous system. The somatic nervous system is a "voluntary" system because the movement of skeletal muscles is normally under cerebral control.

AUTONOMIC NERVOUS SYSTEM. The autonomic nerves govern the involuntary functions of the body that do not ordinarily affect consciousness. The cerebrum has no direct control over autonomic nerves as it has over the somatic nervous system. Thus we cannot by act of will stimulate or inhibit their action (although some people, yogis for example, manage to control such involuntary processes as heart rate, and others may be conditioned to control involuntary responses by biofeedback [instrument conditioning]). Autonomic nerves control the movements of the alimentary canal and heart, the contraction of the smooth muscle of the blood vessels, urinary bladder, iris of the eye, and others, plus the secretions of various glands.

Autonomic nerves originate in the brain or spinal cord as do the nerves of the somatic nervous system, but unlike the latter, the autonomic fibers consist of not one but two motor neurons. They synapse once after leaving the cord and before arriving at the effector organ. These synapses are located outside the spinal cord in ganglia. Fibers passing from the cord to the ganglia are called preganglionic autonomic fibers; those passing from the ganglia to the effector organs are called postganglionic fibers. These relationships are illustrated in Figure 10-15.

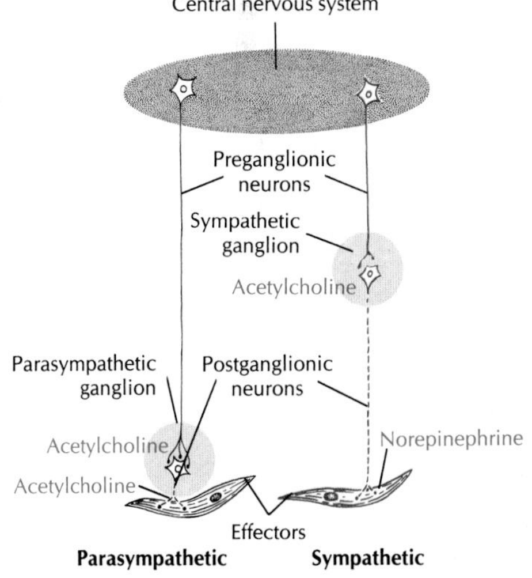

FIGURE 10-15

Arrangement of the preganglionic and postganglionic neurons of the sympathetic and parasympathetic divisions of the autonomic nervous system, and their neurotransmitters.

Subdivisions of the autonomic system are the **parasympathetic** and the **sympathetic** systems. Most organs in the body are innervated by both sympathetic and parasympathetic fibers, and their actions are antagonistic (Figure 10-16). If one fiber speeds up an activity, the other slows it down. However, neither kind of nerve is exclusively excitatory or inhibitory. For example, parasympathetic fibers inhibit heartbeat but excite peristaltic movements of the intestine; sympathetic fibers increase heartbeat but slow down peristaltic movement.

The **parasympathetic** system consists of motor neurons, some of which emerge from the brainstem by certain cranial nerves and others of which emerge from the sacral (pelvic) region of the spinal cord (Figure 10-16). Parasympathetic fibers *excite* the stomach and intestine, urinary bladder, bronchi, constrictor muscles of the iris, salivary glands, and coronary arteries. They *inhibit* the heart, intestinal sphincters, and sphincter of the urinary bladder.

In the **sympathetic** division the nerve cell bodies of all the preganglionic fibers are located in the thoracic and upper lumbar areas of the spinal cord. Their fibers pass out through the ventral roots of the spinal nerves, separate from these, and

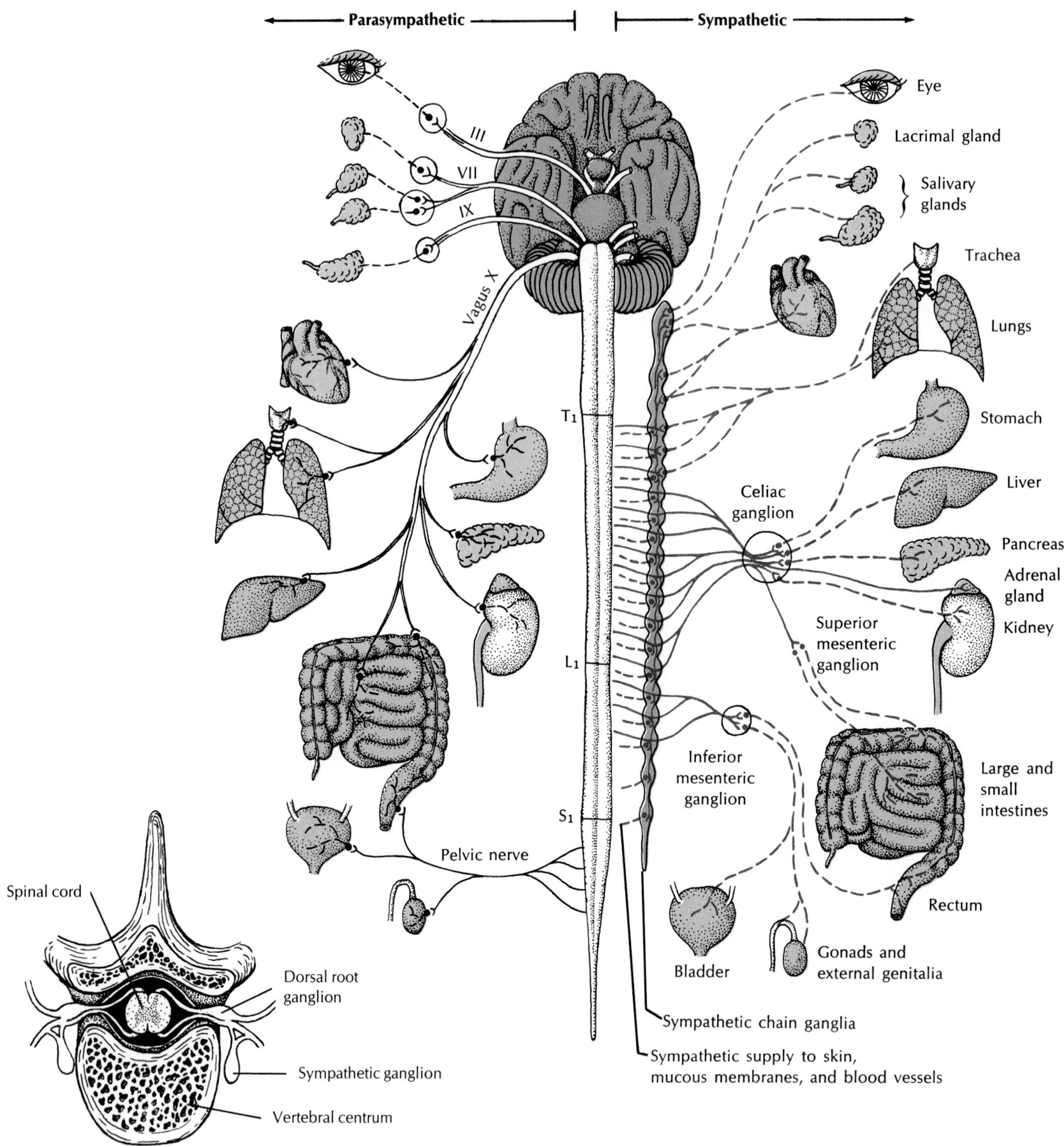

FIGURE 10-16

Autonomic nervous system in humans. *Left,* Parasympathetic division *(black)*. *Right,* Sympathetic division *(red)*. *Lower left,* Cross section of vertebral column shows position of sympathetic ganglia.

go to the sympathetic ganglia (Figure 10-16, *lower left*), which are paired and form a chain on each side of the spinal column. From these ganglia some of the postganglionic fibers run through spinal nerves to the limbs and body wall, where they innervate the blood vessels of the skin, the smooth muscles of the hair follicles, the sweat glands, etc.; other fibers run to the abdominal organs. Sympathetic fibers *excite* the heart, blood vessels, sphincters of the intestines, urinary bladder, dilator muscles of the iris, and others. They *inhibit* the stomach, intestine, and bronchial muscles and coronary arterioles. The importance of these responses in emergency reactions (fight, flight, fear, and rage) will be described in the next chapter (p. 275).

All preganglionic fibers, whether sympathetic or parasympathetic, release acetylcholine at the synapse with the postganglionic cells. The terminations of the parasympathetic and sympathetic nervous systems release different types of chemical transmitter substances (Figure 10-15). The parasympathetic postganglionic fibers release **acetylcholine** at their endings, whereas the sympathetic fibers release **norepinephrine** (also called noradrenaline). This difference is another important characteristic distinguishing the two parts of the autonomic nervous system.

Sense Organs

Animals require a constant inflow of information from the environment to regulate their lives. Sense organs are specialized receptors designed for detecting environmental status and change. An animal's sense organs are its first level of environmental perception; they are data input channels for the brain.

A **stimulus** is some form of energy—electrical, mechanical, chemical, or radiant. The task of the sense organ is to transform the energy form of the stimulus it receives into nerve impulses, the common language of the nervous system. In a very real sense, then, sense organs are biological transducers. A microphone, for example, is a transducer that converts mechanical (sound) energy into electrical energy. Like the microphone that is sensitive only to sound, sense organs are, as a rule, specific for one kind of stimulus energy. Thus eyes respond only to light, ears to sound, pressure receptors to pressure, and chemoreceptors to chemical molecules. But again, all of these different forms of energy are converted into nerve impulses.

Since all nerve impulses are qualitatively alike, how do animals perceive and distinguish the different **sensations** of varying stimuli? The answer is that the real perception of sensation is done in localized regions of the brain, where each sense organ has its own hookup. Impulses arriving at a particular sensory area of the brain can be interpreted in only one way. This is why pressure on the eye causes us to see "stars" or other visual patterns; the mechanical distortion of the eye initiates impulses in the optic nerve fibers that are perceived as light sensations. Although such an operation probably could never be done, a deliberate surgical switching of optic and auditory nerves would cause the recipient to literally see thunder and hear lightning!

CLASSIFICATION OF RECEPTORS

Receptors are traditionally classified on the basis of their location. Those near the external surface, called **exteroceptors,** keep the animal informed about the external environment. Internal parts of the body are provided with **interoceptors,** which pick up stimuli from the internal organs. Muscles, tendons, and joints have **proprioceptors,** which are sensitive to changes in the tension of muscles and provide the organism with a sense of body position.

Another way of classifying receptors is based on the form of energy to which the receptors respond, such as **chemical, mechanical, photo-,** or **thermal.**

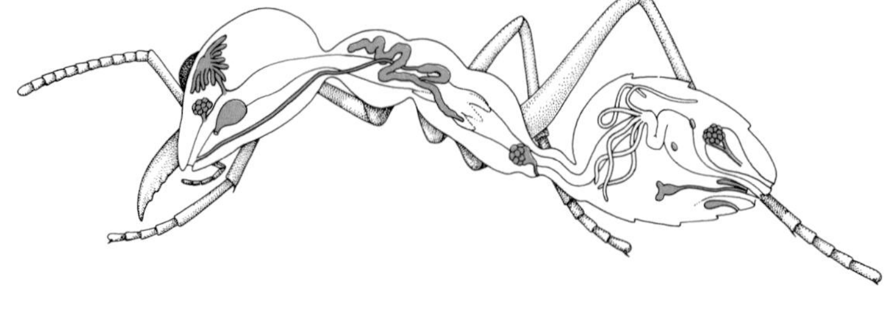

FIGURE 10-17

Pheromone-producing glands of an ant.
After Wilson, E.O., and W.M. Bossert, 1963. Chemical communication among animals. Recent Prog. Hormone Res. 19:673-716.

CHEMORECEPTION

Chemoreception is the most primitive and most universal sense in the animal kingdom. It probably guides the behavior of animals more than any other sense. The most primitive animals, protozoa, use **contact chemical receptors** to locate food and adequately oxygenated water and to avoid harmful substances. These receptors elicit an orientation behavior, called **chemotaxis,** toward or away from the chemical source. More advanced animals have specialized **distance chemical receptors.** These are often developed to a truly amazing degree of sensitivity. Distance chemoreception, usually referred to as a sense of smell or olfactory sense, guides feeding behavior, location and selection of sexual mates, territorial and trail marking, and alarm reactions of numerous animals.

The social insects and some other animals, including mammals, produce species-specific compounds, called **pheromones,** which constitute a highly developed chemical language. Pheromones are a diverse group of organic compounds released by an animal that affect the physiology or behavior of another individual of the same species. Ants, for example, are walking batteries of glands (Figure 10-17) that produce numerous chemical signals. These include releaser pheromones, such as alarm and trail pheromones, and primer pheromones, which alter the endocrine and reproductive systems of different castes in the colony. Insects have a variety of chemoreceptors on the body surface for sensing specific pheromones, as well as other, nonspecific odors.

In all vertebrates and in insects as well, the senses of **taste** and **smell** are clearly distinguishable. Although there are similarities between taste and smell receptors, in general the sense of taste is more restricted in response and is less sensitive than the sense of smell. Taste and smell centers are also located in different parts of the brain.

In higher forms, **taste buds** are found in the mouth cavity and especially on the tongue (Figure 10-18), where they provide a means for judging foods before they are swallowed. A taste bud consists of a cluster of several receptor cells surrounded by supporting cells and is provided with a small external pore through which the slender tips of the sensory cells project. Chemicals being tasted apparently combine with specific receptor sites on the microvilli of the receptor cells. Because they are subject to the wear and tear of abrasive and spicy foods, taste buds have a short life of only about 5 days and are continually being replaced.

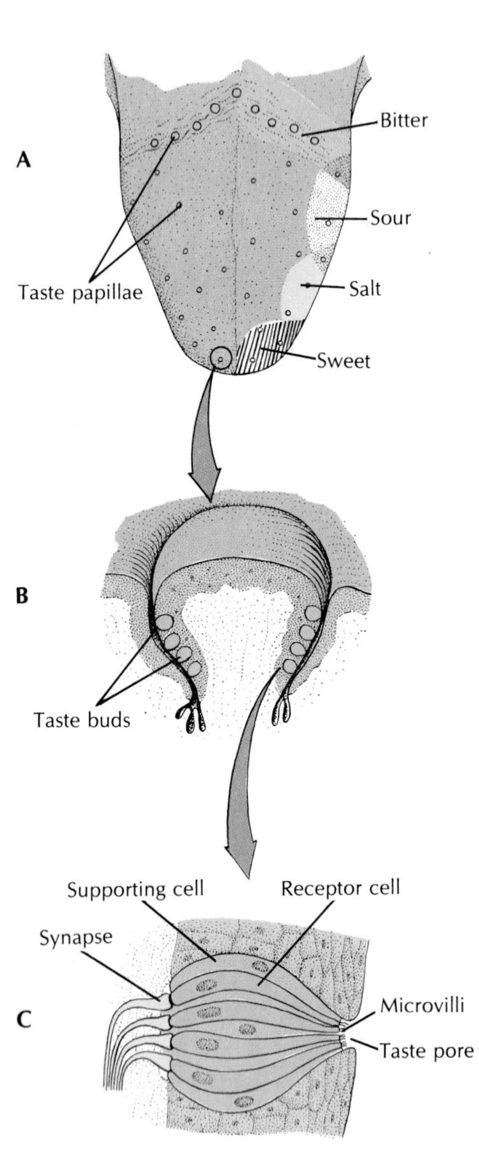

FIGURE 10-18

Taste receptors. **A,** Surface of human tongue showing regions of maximum sensitivity to the four primary taste sensations. **B,** Position of taste buds on a taste papilla. **C,** Structure of a taste bud.

The four basic taste sensations possessed by humans — sour, salt, bitter, and sweet — are each attributable to a different kind of taste bud. The tastes for salt and sweet are found mainly on the tip of the tongue, bitter at the base of the tongue, and sour along the sides of the tongue. Of these, the bitter taste is by far the most sensitive, since it serves as an early warning system against potentially dangerous foods, many of which are bitter.

The sense of smell is more complex than taste, and the basic nature of odor reception is still not known. The olfactory endings are located in a special epithelium covered with a sheet of mucus, positioned deep in the nasal cavity (Figure 10-19). There are perhaps 20 million olfactory receptors in the human nose, each ending in several projecting, cilia-like filaments. It is believed that the terminal cilia are the receptor sites, but they are so small that it has been difficult to obtain electrophysiological recordings from single cells. Even humans, although a species not renowned for detecting smells, can discriminate perhaps 10,000 different odors. The human nose can detect 1/25 of one-millionth of 1 mg of mercaptan, the odoriferous principle of the skunk. Even so, our olfactory abilities compare poorly to those of other mammals that rely on olfaction for their very survival. A dog explores new surroundings with its nose in the same way we do with our eyes. And a dog's nose is justifiably renowned; with some odorous sources a dog's nose is at least a million times more sensitive than ours. Dogs are assisted in their proficiency by having a nose located close to the ground where odors from passing creatures tend to linger.

Since the flavor of food depends a great deal on odors reaching the olfactory epithelium through the throat passage, taste and smell are easily confused. All the various "tastes" other than the four basic ones (sweet, sour, bitter, salt) are really the results of the flavors reaching the sense of smell in this manner. Food loses its appeal during a common cold because a stuffy nose blocks off odors rising from the mouth cavity.

Smells are distinctive, identifiable, and memorable, and yet nearly impossible to describe to another person except in vague, nonspecific terms. This difficulty of defining odor qualities objectively is another reason why olfactory research has lagged far behind other areas of sensory physiology. Of the many hypotheses that have been proposed to explain olfaction, a favored one suggests that a scent molecule fits into a corresponding receptor site, much like a key into a lock. This interaction somehow alters membrane permeability and results in depolarization in the receptor cell, which triggers a nerve impulse.

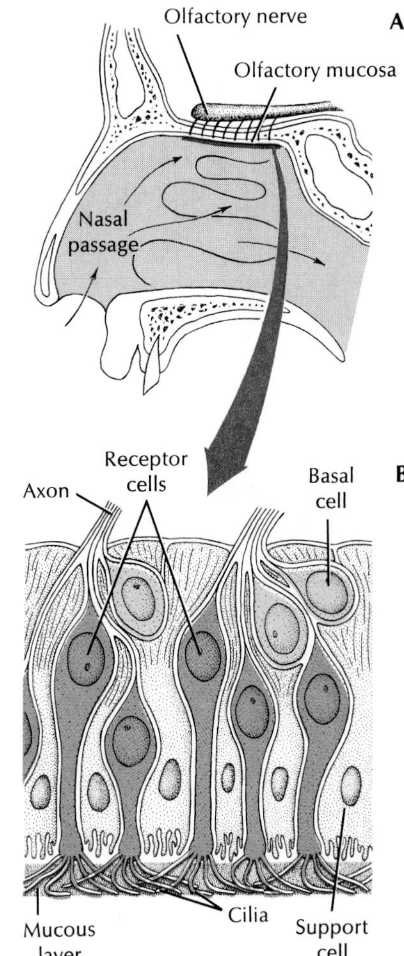

FIGURE 10-19

Human olfactory epithelium. **A,** The epithelium is a patch of tissue positioned in the roof of the nasal cavity. **B,** It is composed of supporting cells, basal cells, and olfactory receptor cells with cilia protruding from their free ends.

MECHANORECEPTION

Mechanoreceptors are sensitive to quantitative forces such as touch, pressure, stretching, sound, vibration, and gravity—in short, they respond to motion. Animals require a steady flow of information from mechanoreceptors in order to interact with their environment, feed themselves, maintain normal posture, and to walk, swim, or fly.

Touch and pain

Invertebrates have touch receptors of many kinds, especially the insects, which are especially well endowed with sensitive tactile hairs sensitive not only to touch but also to vibrations. The superficial touch receptors of vertebrates are distributed all over the body but tend to be concentrated in areas especially important in exploring and interpreting the environment. In most vertebrates these areas are on the face and limb extremities. Of the more than half a million separate sensitive spots on the human body surface, most are found on the lips, tongue, and fingertips (Figure 10-14). Many touch receptors are bare nerve-fiber terminals, but there is an assortment of other kinds of receptors of varying shapes and sizes. Each hair follicle is crowded with receptors that are sensitive to touch.

Pain receptors are relatively unspecialized nerve fiber endings that respond to a variety of stimuli signaling possible or real tissue damage. It is still uncertain whether pain fibers respond directly to injury or indirectly to some substance such as histamine, which is released by damaged cells.

Just as pain is a sign of danger, sensory pleasure is a sign of a stimulus useful to the subject. Pleasure depends on the internal state of the animal and is judged with reference to homeostasis and some physiological set point.

Pain is a distress call from the body signaling some noxious stimulus or internal disorder. Although there is no cortical pain center, discrete areas have been located in the brainstem where pain messages from the periphery terminate. These areas contain two kinds of recently discovered small peptides, called endorphins and enkephalins, that have morphinelike or opiumlike activity. When released, they bind with specific opiate receptors in the midbrain. They are the body's own analgesics.

Hearing

The ear is a specialized receptor for detecting sound waves in the surrounding environment. Because sound communication and reception are an integral part of the lives of higher vertebrates, we may be surprised to discover that most invertebrates inhabit a silent world. Only certain arthropod groups—crustaceans, spiders, and insects—have developed true sound-receptor organs. Even among the insects, only the locusts, cicadas, crickets, grasshoppers, and most moths possess ears, and these are of simple design: a pair of air pockets, each enclosed by a tympanic membrane that passes sound vibrations to sensory cells. Despite their spartan construction, insect ears are beautifully designed to detect the sound of a potential mate or a rival male.

Especially interesting are the ultrasonic detectors of certain nocturnal moths. These have evolved specifically to detect approaching bats and thus lessen the moth's chance of becoming a bat's evening meal (echolocation in bats is described on p. 641). Each moth ear possesses just two receptors (Figure 10-20). One of these, known as the A_1 receptor, will respond to the ultrasonic cries of a bat that is still too far away to detect the moth. As the bat approaches and its cries increase in intensity, the receptor fires more rapidly, informing the moth that the bat is coming nearer. Since the moth has two ears, its nervous system can determine the bat's position by comparing firing rates from the two ears. The moth's strategy is to fly away before the bat detects it. But if the bat continues its approach, the second (A_2) receptor in each ear, which responds only to high-intensity sounds, will fire. The moth responds immediately with an evasive maneuver, usually making a power dive to a bush or the ground where it is safe because the bat cannot distinguish the moth's echo from those of the surroundings.

In its evolution, the vertebrate ear originated as a balance organ, the **labyrinth.** In all jawed vertebrates, from fishes to humans, the labyrinth has a similar structure, consisting of two small chambers called the saccule and the utricle, and three semicircular canals. In fish the base of the saccule is extended into a tiny flap that, in the course of evolution, became enlarged and elaborated into the **cochlea,** the hearing receptor of tetrapod vertebrates.

The structure of the human ear is illustrated in Figure 10-21. The outer, or external, ear collects the sound waves and funnels them through the auditory canal to the **tympanic membrane** (eardrum) lying next to the middle ear. The middle ear is an air-filled chamber containing a remarkable chain of three tiny bones, or ossicles, known as the **malleus** (hammer), **incus** (anvil), and **stapes** (stirrup). These bones conduct the sound waves across the middle ear (Figure 10-21, *B*). The bridge of bones is so arranged that the force of sound waves pushing against the tympanic membrane is amplified as much as 90 times where the stapes contacts the oval window of the inner ear. Muscles attached to the middle ear bones contract when the ear receives very loud noises, thus protecting

FIGURE 10-20

Ear of a moth used to detect approaching bats. See text for explanation.

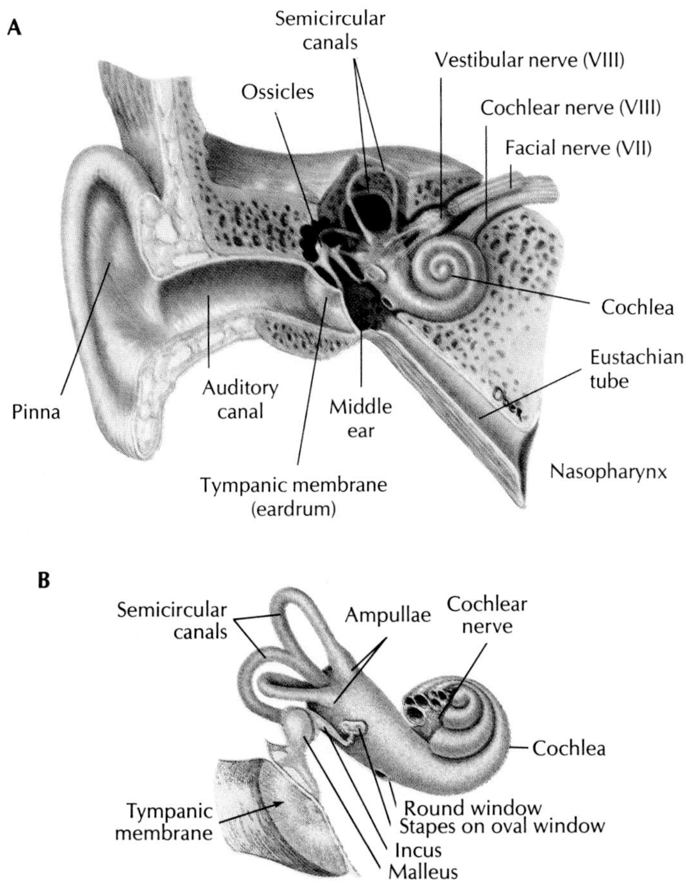

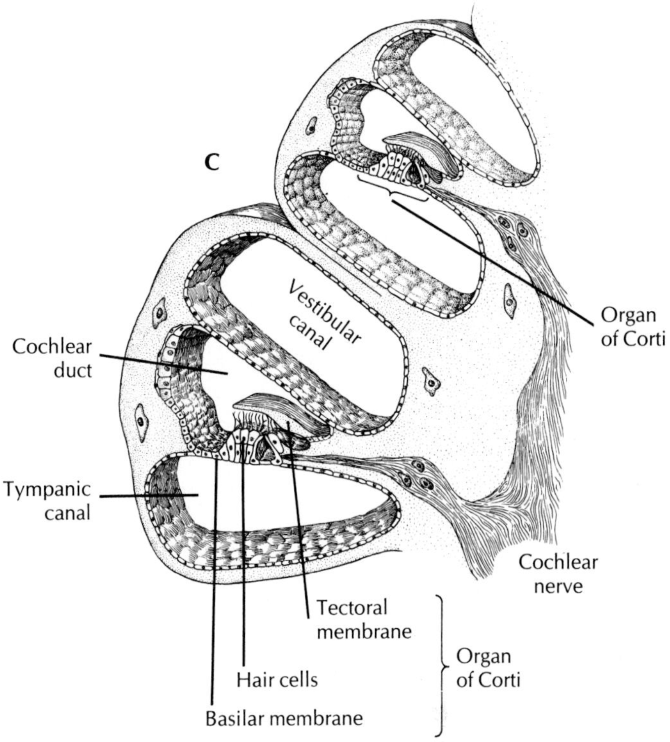

FIGURE 10-21

Human ear. **A,** Longitudinal section showing external, middle, and inner ear. **B,** Enlargement of middle ear and inner ear. The cochlea of the inner ear has been opened to show the arrangement of canals within. **C,** Enlarged cross section of cochlea showing the organ of Corti.

the inner ear from damage. However, these muscles cannot contract quickly enough to protect the inner ear from the damaging effects of a sudden blast, nor can they indefinitely protect the ear from sustained loud sounds such as highly amplified music. The middle ear connects with the pharynx by means of the eustachian tube, which acts as a safety device to equalize pressure on both sides of the tympanic membrane.

Within the inner ear is the organ of hearing, the **cochlea,** which is coiled like a snail's shell, making two and one half turns in humans (Figure 10-21). The cochlea is divided longitudinally into three tubular canals running parallel with one another. This relationship is indicated in Figure 10-22, in which the cochlea is shown stretched out. These canals become progressively smaller from the base of the cochlea to the apex. One of these canals is called the **vestibular canal;** its base is closed by the oval window. The **tympanic canal,** which is in communi-

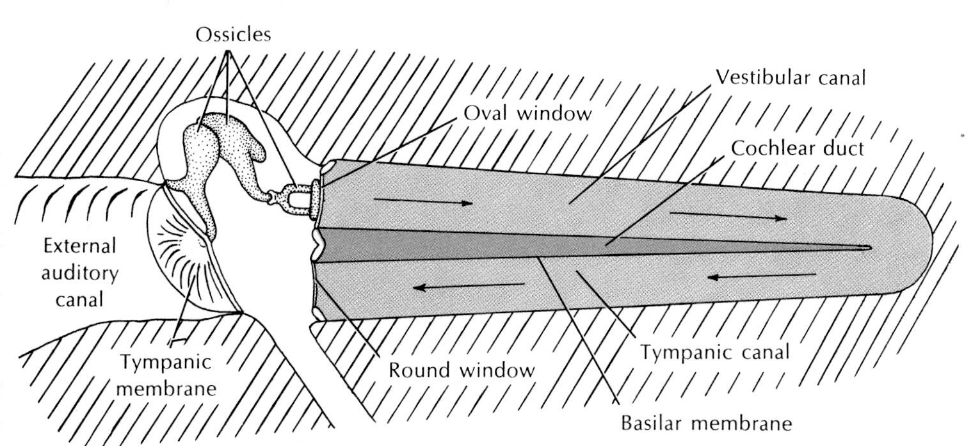

FIGURE 10-22

Mammalian ear as it would appear with the cochlea stretched out. Sound waves transmitted to the oval window produce vibration waves that travel down the basilar membrane. High-frequency vibrations cause the membrane to resonate at the end near the oval window before dying out; low-frequency tones travel farther down the basilar membrane.

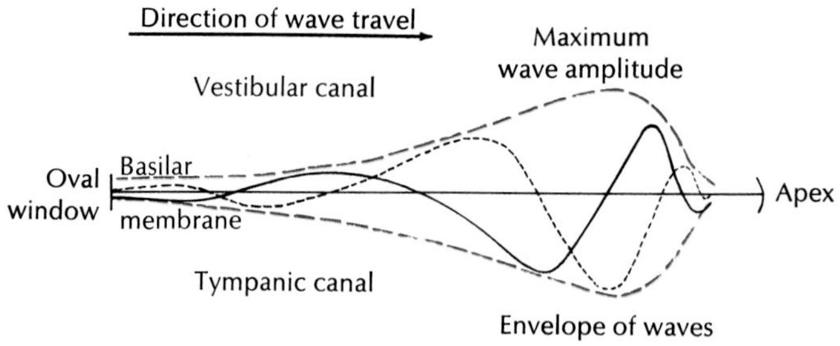

FIGURE 10-23

Traveling waves along the basilar membrane. The oval window is at left, and the cochlear apex at right. The two wave formations (*solid* and *dashed black lines*) occur at separate instants of time. The curves in color represent the extreme displacements of the membrane because of traveling waves.

cation with the vestibular canal at the tip of the cochlea, has its base closed by the round window. Between these two canals is the **cochlear duct,** which contains the organ of Corti, the actual sensory apparatus. Within the organ of Corti are fine rows of hair cells that run lengthwise from the base to the tip of the cochlea. There are at least 24,000 of these hair cells in the human ear, each cell with many hair-like extensions, resembling cilia, that project into the endolymph of the cochlear canal and each cell connected with neurons of the auditory nerve. The hair cells rest on the **basilar membrane,** which separates the tympanic canal and cochlear duct, and they are covered over by the **tectorial membrane** found directly above them.

Sounds with frequencies above 20,000 Hz (Hertz, cycles per second) are usually considered ultrasonic, since this is the practical limit perceived by humans with unimpaired hearing. Most vertebrates have much lower high-frequency limits: fish about 6000 Hz, amphibians 3500 Hz, and reptiles 5000 Hz. Only some mammals and a few birds among vertebrates can hear ultrasonic sounds. Some birds' limits extend to 29,000 Hz, although most end around 18,000 Hz. The limit for dogs is 44,000 Hz, for rats 72,000 Hz, and for bats 115,000 Hz. For a very large mammal such as an elephant, the high-frequency limit is 10,000 Hz. High-frequency limits are roughly related to body size, but strangely, even more to the distance between the ears. Thus mammals with the smallest heads, and therefore having close-set ears, hear the highest sounds!

When a sound wave strikes the ear, its energy is transmitted through the chain of bones of the middle ear to the oval window, which oscillates back and forth, driving the fluid of the vestibular and tympanic canals before it. Because these fluids are noncompressible, an inward movement of the oval window produces a corresponding outward movement of the round window. The fluid oscillations also cause the basilar membrane with its hair cells to vibrate simultaneously.

According to the **place hypothesis of pitch discrimination** formulated by von Békésy, different areas of the basilar membrane respond to different frequencies; that is, for every sound frequency, there is a specific "place" on the basilar membrane where the hair cells respond to that frequency. The initial displacement of the basilar membrane starts a wave traveling down the membrane, much as flipping a rope at one end starts a wave moving down the rope (Figure 10-23). The displacement wave increases in amplitude as it moves from the oval window toward the apex of the cochlea, reaching a maximum at the region of the basilar membrane where the natural frequency of the membrane corresponds to the sound frequency. Here, the membrane vibrates with such ease that the energy of the traveling wave is completely dissipated. Hair cells in that region are stimulated and the impulses conveyed to the fibers of the auditory nerve. Those impulses that are carried by certain fibers of the auditory nerve are interpreted by the hearing center as particular tones. The **loudness** of a tone depends on the

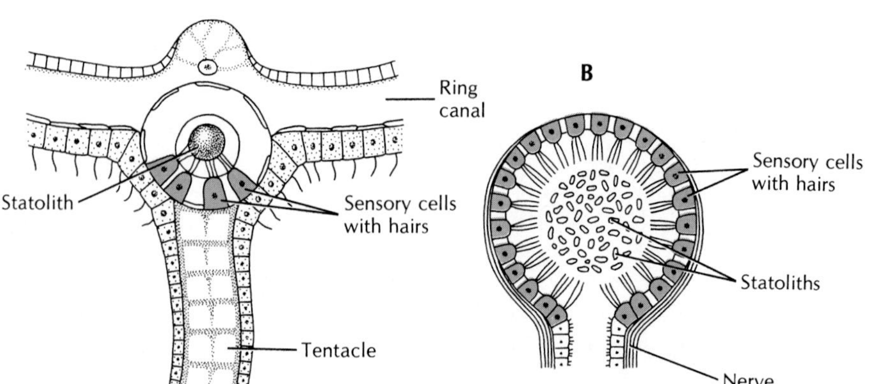

FIGURE 10-24

Types of statocysts, static balance organs of invertebrates. **A,** Statocyst of the medusa of the hydrozoan *Obelia.* **B,** Statocyst of the bivalve mollusc *Pecten.*

number of hair cells stimulated, whereas the **timbre,** or quality, of a tone is produced by the pattern of the hair cells stimulated by sympathetic vibration. This latter characteristic of tone enables us to distinguish between different human voices and different musical instruments, even though the notes in each case may be of the same pitch and loudness.

Sense of equilibrium

In the invertebrates, specialized sense organs for monitoring gravity and low-frequency vibrations often appear as **statocysts.** Each is a simple sac lined with hair cells and containing a heavy calcareous structure, the **statolith** (Figure 10-24). The delicate, hair-like filaments of the sensory cells are activated by the shifting position of the statolith when the animal changes position. Statocysts are found in many different invertebrate phyla from the radiates to the arthropods; all are built on similar principles.

In vertebrates, the organ of equilibrium is the **labyrinth.** It consists of two small chambers (saccule and utricle) and three **semicircular canals** (Figure 10-21, *B*). The utricle and saccule are static balance organs that, like invertebrate statocysts, give information about the position of the head or body with respect to the force of gravity. As the head is tilted one direction or another, stony accretions press on different groups of hair cells; these send nerve impulses to the brain, which interprets this information with reference to head position.

The semicircular canals of vertebrates are designed to respond to **rotational acceleration** and are relatively insensitive to linear acceleration. The three semicircular canals are at right angles to each other, one for each axis of rotation. They are filled with fluid (endolymph), and within each canal is a bulblike enlargement, the **ampulla,** which contains hair cells. The hair cells are embedded in a gelatinous membrane, the **cupula,** which projects into the fluid. When the head is rotated, the fluid in the canal at first tends not to move because of inertia (Figure 10-25). Since the cupula is attached, its free end is pulled in the direction opposite to the direction of rotation. Bending of the cupula distorts and excites the hair cells embedded in it, and this stimulation increases the discharge rate over the afferent nerve fibers leading from the ampulla to the brain. This produces the sensation of rotation. Since the three canals of each ear are in different planes, similar acceleration in any direction stimulates at least one ampulla.

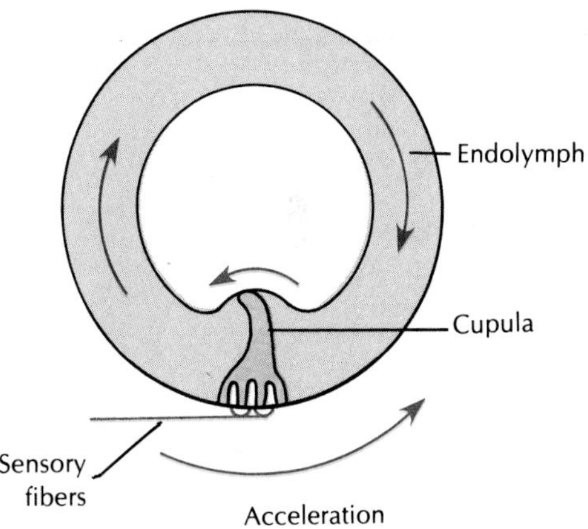

FIGURE 10-25

Semicircular canal, showing bending of cupula on ampulla during angular acceleration.

PHOTORECEPTION: VISION

Light-sensitive receptors are called **photoreceptors.** These receptors range all the way from simple light-sensitive cells scattered randomly on the body surface of the lowest invertebrates (**dermal light sense**) to the exquisitely developed vertebrate eye. Although dermal light receptors contain little photochemical substance and are far less sensitive than optic receptors, they are important in locomotory orientation, pigment distribution in chromatophores, photoperiodic adjustment of reproductive cycles, and other behavior changes in many lower invertebrates.

More highly organized eyes, many capable of excellent image formation, are based on one or the other of two different principles: a single-lens, camera-type eye such as those of molluscs and vertebrates; or a multifaceted (compound) eye as in arthropods. The arthropod **compound** eyes are composed of many independent visual units called **ommatidia** (Figure 22-28, p. 485). The eye of a bee contains about 15,000 of these units, each of which views a separate narrow sector of the visual field. Such eyes form a mosaic of images of varying brightness from the separate units. Resolution (that is, the ability to see objects sharply) is probably good for close objects but falls off rapidly with increasing distance. However, the compound eye is especially well suited to picking up motion, as anyone knows who has tried to swat a fly.

One of several marvels of the vertebrate eye is its capacity to compress the enormous range of light intensities presented to it into a narrow range that can be handled by the optic nerve fibers. Light intensity between a sunny noon and starlight differs more than 10 billion to 1. Rods quickly saturate with high light intensity, but the cones do not; they shift their operating range with changing ambient light intensity so that a high-contrast image is perceived over a broad range of light conditions. This is made possible by complex interactions among the network of nerve cells that lie between the cones and the ganglion cells that generate the retinal output to the brain.

The eyes of certain annelids, molluscs, and all vertebrates are built like a camera — or rather we should say that a camera is modeled somewhat after the vertebrate eye. It contains a light-tight chamber with a lens system in front, which focuses an image of the visual field on a light-sensitive surface (the retina) in back (Figure 10-26). Because eyes and cameras are based on the same laws of optics, we can wear eyeglasses to correct optical defects.

The spherical eyeball is built of three layers: (1) a tough outer white **sclera** that provides support and protection, (2) the middle **choroid** coat, containing blood vessels for nourishment, and (3) the light-sensitive **retina** (Figure 10-26). The **cornea** is a transparent anterior modification of the sclera. A circular, pigmented curtain, the **iris,** regulates the size of the light opening, the **pupil.** Just behind the iris is the **lens,** a transparent, elastic oval disc that, with the aid of **ciliary muscles,** can alter the curvature of the lens and bend the rays to focus them on the retina. In land vertebrates the cornea actually does most of the bending of light rays, whereas the lens adjusts the focus for near and far objects. Between the cornea and the lens is the outer chamber filled with watery **aqueous humor;** between the lens and the retina is the much larger inner chamber filled with viscous **vitreous humor.**

The **retina** is composed of photoreceptors, the **rods** (Figure 10-27, *A*) and **cones.** Approximately 125 million rods and 7 million cones are present in each human eye. Cones are primarily concerned with color vision in ample light; rods, with colorless vision in dim light. The retina is actually made up of three sets of neurons in series with each other: (1) photoreceptors (rods and cones), (2) intermediate neurons, and (3) ganglionic neurons whose axons form the optic nerve (Figure 10-27, *B*).

The **fovea centralis,** the region of keenest vision, is located in the center of the retina, in direct line with the center of the lens and cornea. It contains only cones. The acuity of an animal's eyes depends on the density of cones in the fovea. The human fovea and that of a lion contain approximately 150,000 cones per square millimeter. But many water and field birds have up to 1 million cones per square millimeter. Their eyes are as good as human eyes would be if aided by eight-power binoculars.

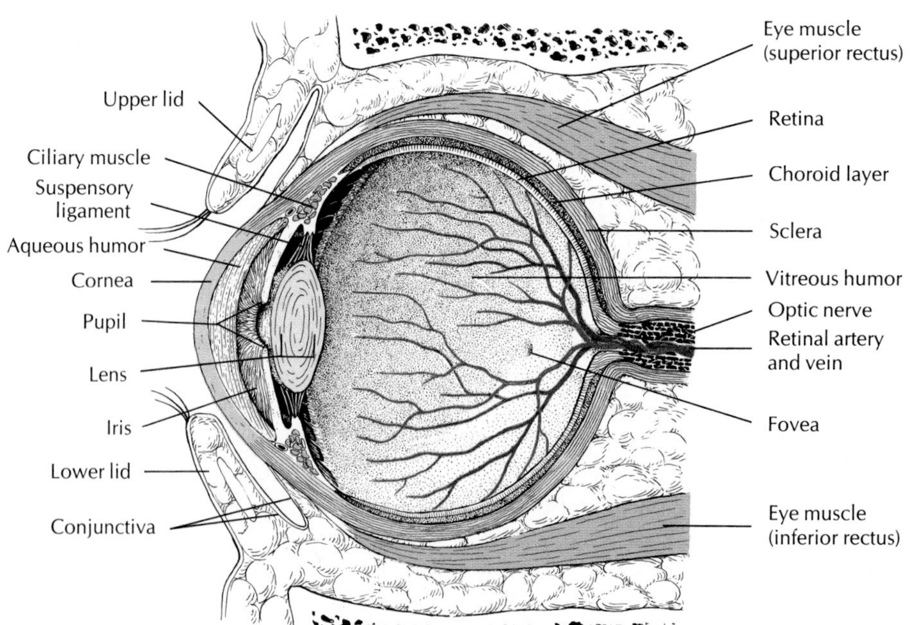

FIGURE 10-26

Section through human eye.

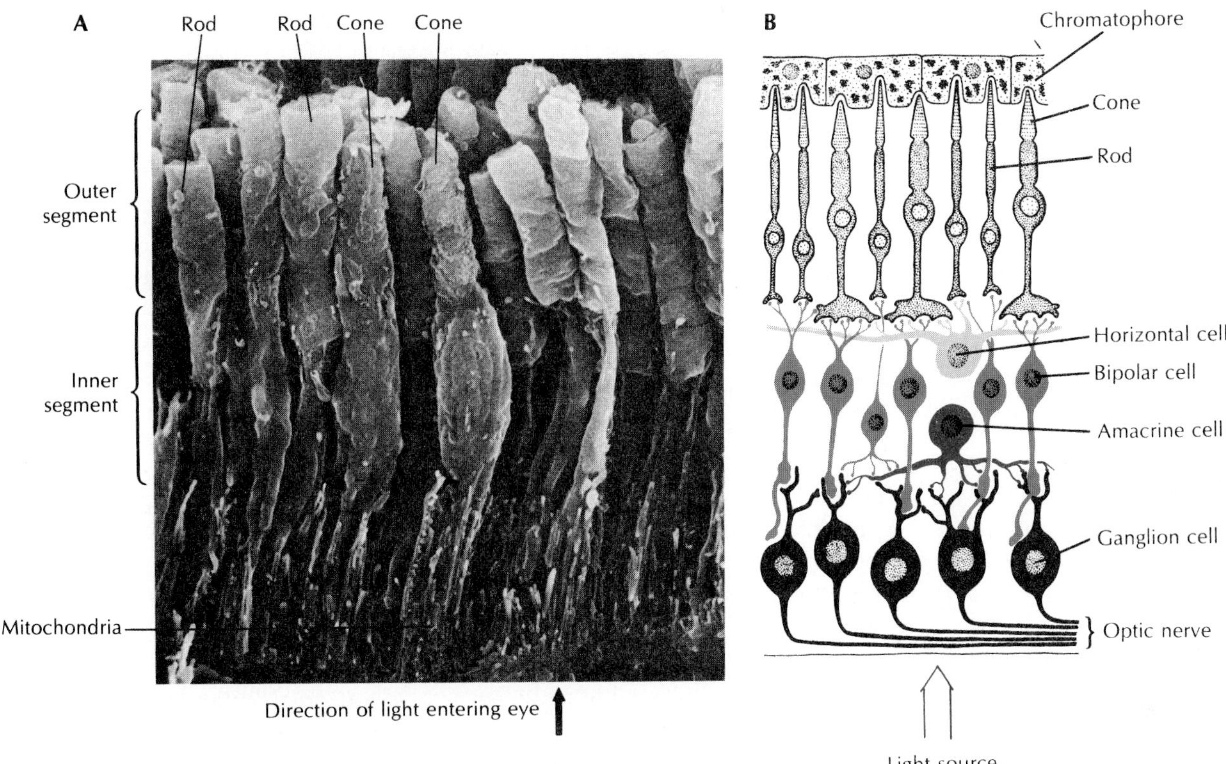

FIGURE 10-27

A, Scanning electron micrograph of photoreceptor cells of the guinea pig's retina. (×3500.) Rods can be distinguished from cones by the thin inner segment of the former and the swollen inner segments (resulting from more numerous mitochondria) of the latter. The photosensitive pigments are located in the outer segments of both types of cells. **B,** Structure of the primate retina, showing the organization of intermediate neurons that connect the photoreceptor cells to the ganglion cells of the optic nerve.

A, Courtesy of A. Spira and M. Patten.

At the peripheral parts of the retina only rods are found. This is why we can see better at night by looking out of the corners of our eyes; the rods, adapted for high sensitivity with dim light, are brought into use.

Chemistry of vision

Each rod contains a light-sensitive pigment known as **rhodopsin.** Each rhodopsin molecule consists of a large protein, **opsin,** which behaves as an enzyme, and a small carotenoid molecule, **retinal,** a derivative of vitamin A. When a quantum of light strikes a rod and is absorbed by the rhodopsin molecule, retinal is isomerized, that is, the shape of the molecule is changed. This triggers the enzymatic activity of opsin, which sets in motion a biochemical sequence of several steps. This complex sequence, the details of which have only recently been worked out (Stryer, 1987), behaves as an excitatory cascade that vastly amplifies the energy of a single photon to generate a nerve impulse in the rod.

The amount of intact rhodopsin in the retina depends on the intensity of light reaching the eye. The dark-adapted eye contains much rhodopsin and is very sensitive to weak light. Conversely most of the rhodopsin is broken down in the light-adapted eye. It takes approximately half an hour for the light-adapted eye to accommodate to darkness, while the rhodopsin level is gradually built up.

The composition of the cone pigments has not yet been completely clarified but they are believed to be similar to rhodopsin, containing retinal combined with a special protein, **cone opsin.** Cones function to perceive color and require 50 to 100 times more light for stimulation than do rods. Consequently, night vision is almost totally rod vision; this is why the landscape illuminated by moonlight appears in shades of black and white only. Unlike humans, who have both day and night vision, some vertebrates specialize for one or the other. Strictly nocturnal animals, such as bats and owls, have pure rod retinas. Purely diurnal forms, such as the common gray squirrel and some birds, have only cones. They are of course, virtually blind at night.

Color vision

How does the eye see colors? According to the trichromatic hypothesis of color vision, there are three different types of cones, each reacting most strongly to red, green, or violet light. Colors are perceived by comparing the levels of excitation of the three different kinds of cones. This comparison is made both in nerve circuits in the retina and in the visual cortex of the brain. Color vision is present in some members of all vertebrate groups with the possible exception of the amphibians. Bony fishes and birds have particularly good color vision. Surprisingly, most mammals are color blind; exceptions are primates and a few other species such as squirrels.

Summary

The nervous system is a rapid communications system that interacts continuously with the endocrine system in the control and coordination of body function. The basic unit of nervous integration in all animals is the neuron, a highly specialized cell designed to conduct self-propagating impulses to other cells. The nerve impulse is transmitted from one nerve to another across synapses. The thin gap between nerves at these junction points is bridged by a chemical transmitter, which is released from the synaptic knob.

The most primitive organization of neurons into a system is the nerve net of cnidarians, basically a plexus of nerve cells that, with additions, is the basis of the nervous systems of several primitive invertebrate phyla. With the appearance of ganglia (nerve centers) in the flatworms, nervous systems became differentiated into central and peripheral divisions. In vertebrates, the central nervous system consists of the brain and spinal cord. In its remarkable evolutionary history, the vertebrate brain has developed from a primitive three-part, linear brain to a complex multicomponent structure. In mammals, the cerebral cortex has become vastly enlarged and has assumed the most important integrative activities of the nervous system. It completely overshadows the primitive brain, which is consigned to the role of relay center and to serving numerous unconscious but nonetheless vital functions such as breathing, blood pressure, and heart rate.

In humans the left cerebral hemisphere is usually specialized for language and mathematical skills while the right hemisphere is specialized for visual-spatial and musical skills.

The peripheral nervous system connects the central nervous system to receptors and effector organs. It is divided broadly into an afferent system, which conducts sensory signals to the central nervous system, and an efferent system, which conveys motor impulses to effector organs. The autonomic nervous system is a motor system with its own separate set of fibers. It is subdivided into anatomically distinct sympathetic and parasympathetic systems, each of which sends fibers to most body organs. Generally the sympathetic system governs excitatory activities and the parasympathetic system governs maintenance and restoration of body resources.

Sensory organs are receptors designed especially to respond to internal or environmental change. The most primitive and ubiquitous sense in the animal kingdom is chemoreception. Chemoreceptors may be contact receptors such as the vertebrate sense of taste, or distance receptors such as smell, which detects airborne molecules. Most hypotheses of olfaction postulate some kind of molecular integration between odorant and receptor, but the mechanism of odor reception remains unknown.

The receptors for touch, pain, equilibrium, and hearing are all mechanical force receptors. Touch and pain receptors are characteristically simple structures, but hearing and equilibrium are highly specialized senses based on special hair cells that respond to mechanical deformation. Sound waves received by the ear are mechanically amplified and transmitted to the inner ear where different areas of the cochlea respond to different sound frequencies. Equilibrium receptors, also located in the inner ear, consist of two saclike static balance organs and three semicircular canals that detect movement.

Vision receptors (photoreceptors) are associated with special pigment molecules that photochemically decompose in the presence of light and, in doing so, trigger nerve impulses in optic fibers. The advanced compound eye of arthropods is especially well suited to detecting motion in the visual field. Vertebrates have a camera eye with focusing optics. The photoreceptor cells of the retina are of two kinds: rods, designed for high sensitivity with dim light, and cones, primarily concerned with color vision. Cones predominate in the fovea centralis of the human eye, the area of keenest vision. Rods are more abundant in the peripheral areas of the retina.

REVIEW QUESTIONS

1. Define the following terms: neuron, axon, dendrite, myelin sheath, afferent neuron, efferent neuron, association neuron.
2. What are glial cells in the nervous system and what functions do they perform?
3. Explain how the permeability properties of a nerve fiber give rise to the resting (equilibrium) potential of the fiber. What is the importance of the sodium pump in maintaining the resting potential?
4. What ionic and electrical changes occur during the passage of an action potential along a nerve fiber?
5. Describe the microstructure of a chemical synapse. Summarize what happens when an action potential arrives at a synapse.
6. Describe the cnidarian (radiate) nervous system. How is the tendency toward centralization of the nervous system manifested in flatworms, annelids, molluscs, and arthropods?
7. How does the vertebrate spinal cord differ morphologically from the nerve cord of invertebrates?
8. Name the components of a typical reflex arc. What is the difference between a reflex *arc* and a reflex *act?*
9. Name the major functions known to be associated with the following brain structures: medulla, cerebellum, tectum, thalamus, hypothalamus, cerebrum, limbic system.
10. What functional activities are known to be associated with the left and the right hemispheres of the cerebral cortex?
11. What is the autonomic nervous system and what activities does it carry out that distinguish it from the central nervous system? Why can the autonomic nervous system be described as a "two-neuron" system?
12. Chemoreception in vertebrates and insects is mediated through the clearly distinguishable senses of taste and smell. Contrast these two senses in humans in terms of anatomical location and nature of the receptors and sensitivity to chemical molecules.
13. Explain how the ultrasonic detectors of certain nocturnal moths are adapted to help them escape an approaching bat.
14. Outline the place theory of pitch discrimination as an explanation of the human ear's ability to distinguish between sounds of different frequencies.
15. Explain how the semicircular canals of the ear are designed to detect rotation of the head in any directional plane.
16. Explain what happens when light strikes a dark-adapted rod that leads to the generation of a nerve impulse. What is the difference between rods and cones in their sensitivity to light?

SELECTED REFERENCES

Bullock, T.H., R. Orkand, and A. Grinnell. 1977. Introduction to nervous systems. San Francisco, W.H. Freeman & Co. *Excellent comparative treatment.*

Dunant, Y., and M. Israel. 1985. The release of acetylcholine. Sci. Am. **252**:58-66 (April). *Recent studies have altered prevailing views of the events at a synapse during impulse transmission.*

Hudspeth, A.J. 1983. The hair cells of the inner ear. Sci. Am. **248**:54-64 (Jan.). *How these biological transducers work.*

Milne, L., and M. Milne. 1972. The senses of animals and men. New York, Atheneum Publishers. *Well-written account of animal senses. Undergraduate level.*

Nathan, P. 1982. The nervous system, ed. 2. Oxford, Eng., Oxford University Press. *One of the best of several semipopular accounts of the nervous system.*

Stebbins, W.C. 1983. The acoustic sense of animals. Cambridge, Mass., Harvard University Press. *Broadly comparative introduction to the physics, physiology, natural history, and evolution of hearing.*

Stevens, C.F. 1979. The neuron. Sci. Am. **241**:54-65 (Sept.). *Neuron function is detailed.*

Stryer, L. 1987. The molecules of visual excitation. Sci. Am. **257**:42-50 (July). *Describes the cascade of molecular events following light absorption by a rod cell that leads to a nerve signal.*

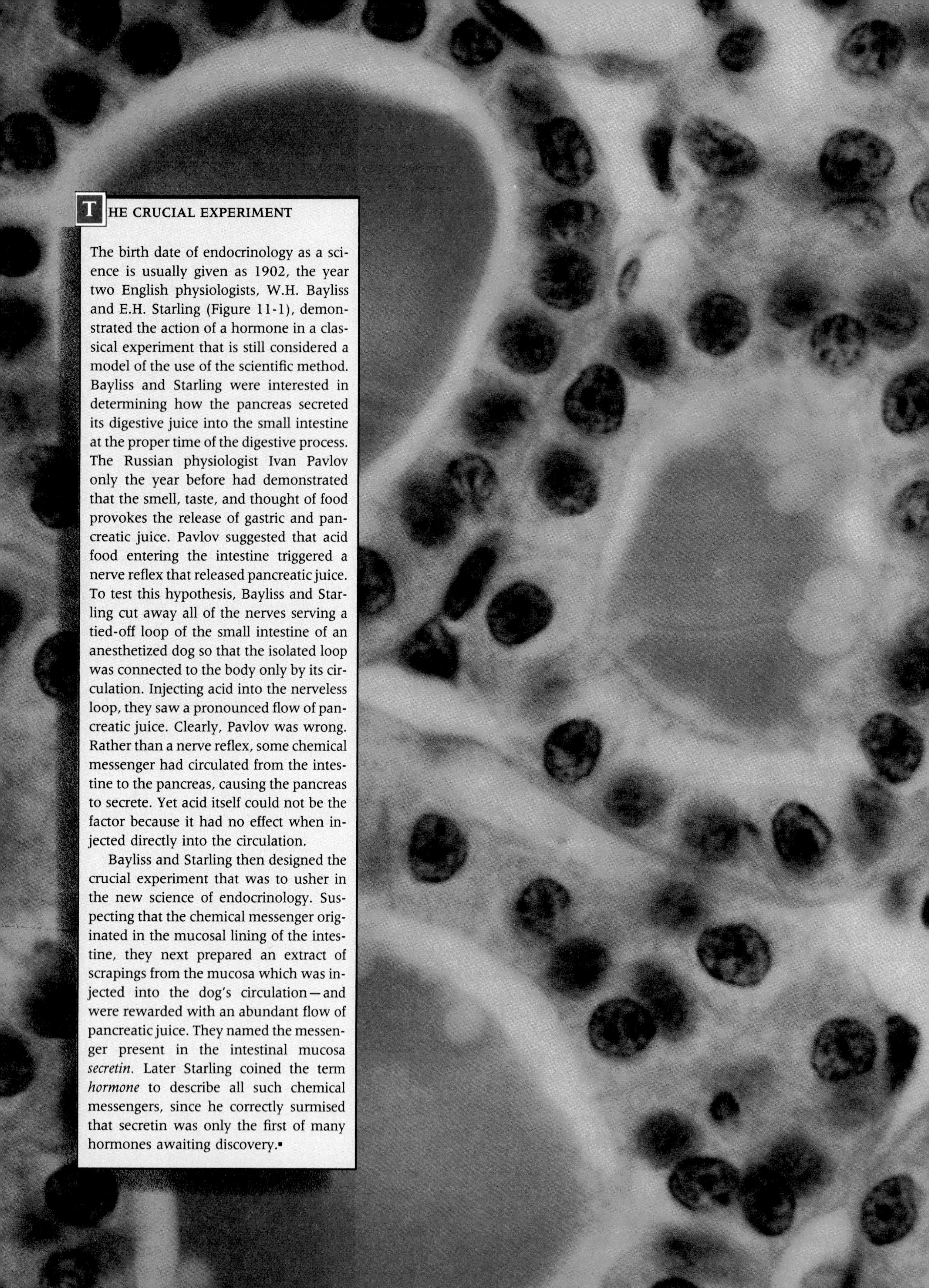

THE CRUCIAL EXPERIMENT

The birth date of endocrinology as a science is usually given as 1902, the year two English physiologists, W.H. Bayliss and E.H. Starling (Figure 11-1), demonstrated the action of a hormone in a classical experiment that is still considered a model of the use of the scientific method. Bayliss and Starling were interested in determining how the pancreas secreted its digestive juice into the small intestine at the proper time of the digestive process. The Russian physiologist Ivan Pavlov only the year before had demonstrated that the smell, taste, and thought of food provokes the release of gastric and pancreatic juice. Pavlov suggested that acid food entering the intestine triggered a nerve reflex that released pancreatic juice. To test this hypothesis, Bayliss and Starling cut away all of the nerves serving a tied-off loop of the small intestine of an anesthetized dog so that the isolated loop was connected to the body only by its circulation. Injecting acid into the nerveless loop, they saw a pronounced flow of pancreatic juice. Clearly, Pavlov was wrong. Rather than a nerve reflex, some chemical messenger had circulated from the intestine to the pancreas, causing the pancreas to secrete. Yet acid itself could not be the factor because it had no effect when injected directly into the circulation.

Bayliss and Starling then designed the crucial experiment that was to usher in the new science of endocrinology. Suspecting that the chemical messenger originated in the mucosal lining of the intestine, they next prepared an extract of scrapings from the mucosa which was injected into the dog's circulation—and were rewarded with an abundant flow of pancreatic juice. They named the messenger present in the intestinal mucosa *secretin*. Later Starling coined the term *hormone* to describe all such chemical messengers, since he correctly surmised that secretin was only the first of many hormones awaiting discovery.▪

CHAPTER 11

Chemical Coordination

Endocrine System

A

B

FIGURE 11-1

Founders of endocrinology.
A, Sir William H. Bayliss (1860-1924).
B, Ernest H. Starling (1866-1927).

From Fulton, J.F., and L.G. Wilson. 1966. Selected readings in the history of physiology. Springfield, Ill., Charles C Thomas, Publisher.

HORMONAL INTEGRATION

The endocrine system is the second great integrative system controlling the body's activities. Endocrine glands secrete **hormones** (Gr. *hormon,* to excite), chemical messengers that are transported by the blood to some part of the body where they initiate definite physiological responses.

Endocrine glands are small, well-vascularized ductless glands composed of groups of cells arranged in cords or plates. Since the endocrine glands have no ducts, their only connection with the rest of the body is by the bloodstream; they must capture their raw materials from the blood and secrete their finished hormonal products into it. Consequently, it is not surprising that the endocrine glands receive enormous blood flows. The thyroid gland is said to have the highest blood flow per unit of tissue weight of any organ in the body. **Exocrine glands,** in contrast, are provided with ducts for discharging their secretions onto a free surface. Examples of exocrine glands are sweat glands and sebaceous glands of skin, salivary glands, and the various enzyme-secreting glands lining the wall of the stomach and intestine.

The classical definition of a hormone given above, like so many other generalizations in biology, may have to be altered as new information appears. For one thing, some hormones, such as certain neurosecretions, may never enter the general circulation at all. Furthermore, there is good evidence that many of the traditional hormones, such as insulin, are synthesized in minute amounts in a variety of nonendocrine tissues (nerve cells, for example) where they may function as local **tissue factors:** substances that stimulate cell growth or some biochemical process. Most hormones, however, are blood borne and therefore diffuse into every tissue space in the body. This is quite unlike the discrete action of the nervous system with its network of cablelike nerve fibers that selectively send messages to specific points.

Compared to the nervous system, the endocrine system is slow-acting because of the time required for a hormone to reach the appropriate tissue, cross the capillary endothelium, and diffuse through tissue fluid to, and sometimes into, cells. Thus the minimum response time is a matter of seconds and

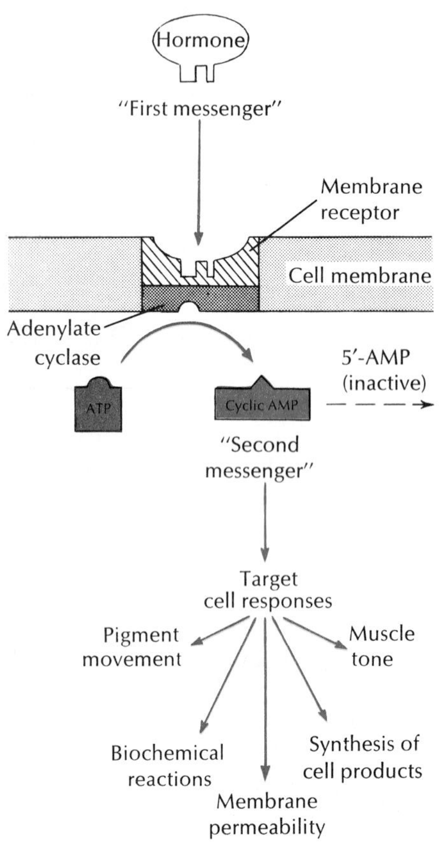

FIGURE 11-2

Second-messenger concept of hormone action. Many hormones act through cyclic AMP. The hormone is carried by the bloodstream from the endocrine gland to the target cell where it selectively combines with an immobile receptor in the plasma membrane. This interaction stimulates the enzyme adenylate cyclase to catalyze the formation of cyclic AMP (second messenger). Cyclic AMP acts intracellularly to initiate any of several changes, depending on the kind of cell.

may be much longer. Furthermore, hormone responses in general are much longer lasting than those under nervous control. Where a sustained effect is required, as in many metabolic and growth processes, or where some concentration or secretion rate must be maintained at a particular level, we expect to find endocrine control.

However, the nervous and endocrine systems really function as a single, united system. There is no sharp separation between the two. As we shall see, the nervous system is itself an endocrine organ that controls most endocrine function. Conversely, several hormones act on the nervous system and may significantly affect many kinds of animal behavior.

MECHANISMS OF HORMONE ACTION

The widespread distribution of hormones in the body makes it possible for certain hormones, such as the growth hormone of the pituitary gland, to affect most, if not all, cells during specific stages of cellular differentiation. Other hormones, however, produce highly specific responses only in certain target cells and at certain times. But how is this specificity accomplished? Since hormones are distributed everywhere by the circulation, there must be some mechanism that limits hormone action to specific target cells. This is provided by **receptor molecules** located on or in the responding cells. A hormone will engage only those cells that display the receptor, which, by virtue of its specific molecular shape, will bind with the hormone molecule. Other cells are insensitive to the hormone's presence because they lack the specific receptors.

Cell surface receptors and the second messenger concept

In many hormone actions, when the hormone arrives at its target cell, it binds to a receptor site on the membrane. The hormone-receptor complex activates an enzyme, **adenylate cyclase,** which is coupled to the membrane receptor (Figure 11-2). Adenylate cyclase converts ATP in the cytoplasm to **cyclic AMP.** This compound is formed when two of three phosphate groups are split from ATP and the remaining phosphate combines with a carbon atom of the adjacent ribose unit of AMP to form a ring, making it "cyclic" (Figure 11-3). The cyclic AMP thus generated acts as a "second messenger" that relays the hormone's message to the cell's biochemical machinery, where it alters (usually stimulates) some cellular process (Figure 11-2). Since many molecules of cyclic AMP may be manufactured after a single hormone molecule has been bound, the message is amplified, perhaps many thousands of times.

Cyclic AMP mediates the actions of many hormones, including glucagon, epinephrine, adrenocorticotropic hormone (ACTH), thyroid-stimulating hormone (TSH), melanophore-stimulating hormone (MSH), and vasopressin. With the exception of epinephrine, these are all peptides — small proteins but much too large to penetrate the cell membrane. All act *indirectly* through an immobile receptor on the cell surface.

FIGURE 11-3

Formation of cyclic AMP.

ATP → Adenylate cyclase → CYCLIC AMP

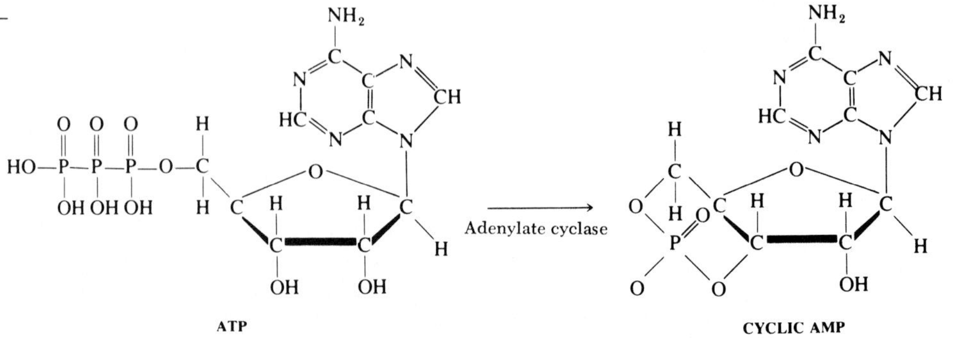

Cytoplasmic receptors

All of the steroid hormones (for example, estrogen, testosterone, aldosterone, and the arthropod hormone ecdysone) diffuse into cells where they bind selectively to **cytoplasmic receptor** molecules found only in the target cells. The hormone-receptor complex can then diffuse into the nucleus where it binds directly to certain proteins (nonhistones) in the chromosomes. As a result, gene transcription is increased, and messenger RNA molecules are synthesized on specific sequences of DNA. Moving from the nucleus into the cytoplasm, the newly formed messenger RNA initiates the formation of new proteins, thus setting in motion the hormone's observed effect (Figure 11-4). Thyroxine, a modified peptide is also believed to act through this mechanism.

As compared to hormones acting through the cyclic AMP mechanism, steroids have a more *direct* effect on protein synthesis because they combine with a mobile receptor and move with it into the nucleus to couple with chromosomal proteins.

All hormones are low-level signals. Even when an endocrine gland is secreting maximally, the hormone is so greatly diluted by the large volume of blood it enters that its plasma concentration seldom exceeds 10^{-9} M (or one billionth of a 1 M concentration). Since hormones have far-reaching and often powerful influences on cells, it is evident that their effects are vastly amplified at the cellular level.

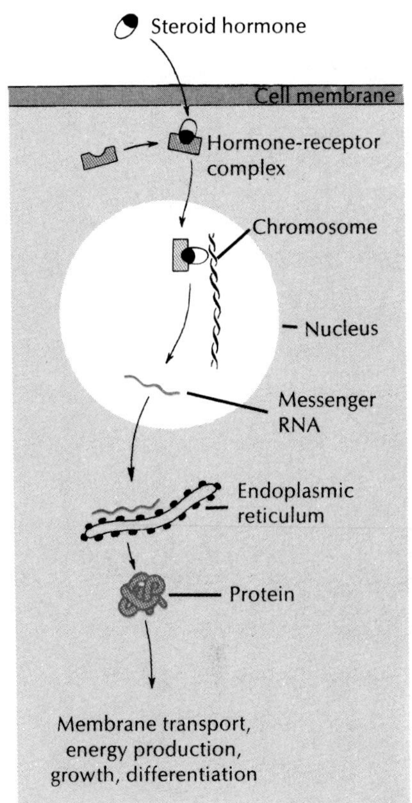

FIGURE 11-4

The concept of gene regulation by steroid hormones and thyroxine. The hormone penetrates the cell membrane to combine with a mobile cytoplasmic receptor. The complex enters the nucleus where it stimulates the transcription of messenger RNA. This is translated to specific proteins in the cytoplasm.

HOW HORMONE SECRETION RATES ARE CONTROLLED

Hormones influence cell functions by altering the rates of a large range of biochemical processes. Some affect enzyme activity and thus alter cellular metabolism, some change membrane permeability, some regulate the synthesis of cellular proteins, and some stimulate the release of hormones from other endocrine glands. Since these are all dynamic processes that must adapt to changing metabolic demands, they must be regulated, not merely activated, by the appropriate hormones. This is achieved by varying hormone output from the gland. However, the concentration of hormone in the plasma depends on two factors: its rate of secretion and the rate at which it is inactivated and removed from the circulation. Consequently, an endocrine gland requires information about the level of its own hormone(s) in the plasma to control its secretion.

Many hormones, especially those of the pituitary gland, are controlled by **negative feedback systems** that operate between the glands secreting the hormones and the target cells. A feedback pattern is one in which the output is constantly compared with a set point. For example, ACTH, secreted by the pituitary, stimulates the adrenal gland (the target cells) to secrete cortisol. As the cortisol level in the plasma rises, it acts on the pituitary gland to inhibit the release of ACTH. Thus any deviation from the set point (a specific plasma level of cortisol) leads to corrective action in the opposite direction. Such a system is highly effective in preventing extreme oscillations in hormone output. However, hormonal feedback systems are not identical to a rigid "closed loop" system like the thermostat that controls the central heating system in a house, because they may be altered by input from the nervous system or by metabolites or other hormones.

Invertebrate Hormones

During the last half century physiologists have shown that many invertebrates have endocrine integrative systems that approach the complexity of the vertebrate endocrine system. Not surprisingly, however, there are few, if any, homologies between invertebrate and vertebrate hormones. The invertebrate phyla have different functional systems, growth patterns, and reproductive processes from vertebrates and have been separated from them phylogenetically for a vast span of time.

In many invertebrate phyla, the principal source of hormones is **neurosecretory cells,** specialized nerve cells capable of synthesizing and secreting hormones. Their products, called neurosecretions, or neurosecretory hormones, are discharged directly into the circulation. Neurosecretion is in fact important in vertebrates as well. It is an ancient physiological activity, and because it serves as a crucial link between the nervous and endocrine systems, we believe that hormones first evolved as nerve-cell secretions. Later, nonnervous endocrine glands appeared, especially among the vertebrates, but remained chemically linked to the nervous system by the neurosecretory hormones.

Neurosecretory hormones have been found in all the larger invertebrate phyla including the cnidarians, flatworms, nematodes, molluscs, annelids, arthropods, and echinoderms. Most extensively studied, however, have been the insects, and we will limit our discussion to that group.

In insects, as in other arthropods, growth is a series of steps in which the rigid, nonexpansible exoskeleton is periodically discarded and replaced with a new, larger one. Almost all insects undergo a process of metamorphosis, in which there is a series of juvenile stages, each requiring the formation of a new exoskeleton, and each ending with a molt. In some orders the change to the adult form is gradual. In others the adult is separated from the larval stages by a quiescent form, the pupa, and the change to the adult is abrupt. Hormonal control of both types is the same.

Insect physiologists have discovered that molting and metamorphosis are controlled by the interaction of two hormones, one favoring growth and the differentiation of adult structures and the other favoring the retention of juvenile structures. These two hormones are the **molting hormone** (also referred to as **ecdysone** [ek'duh-sone]), produced by the prothoracic gland, and the **juvenile hormone,** produced by the corpora allata (Figure 11-5). The structure of both

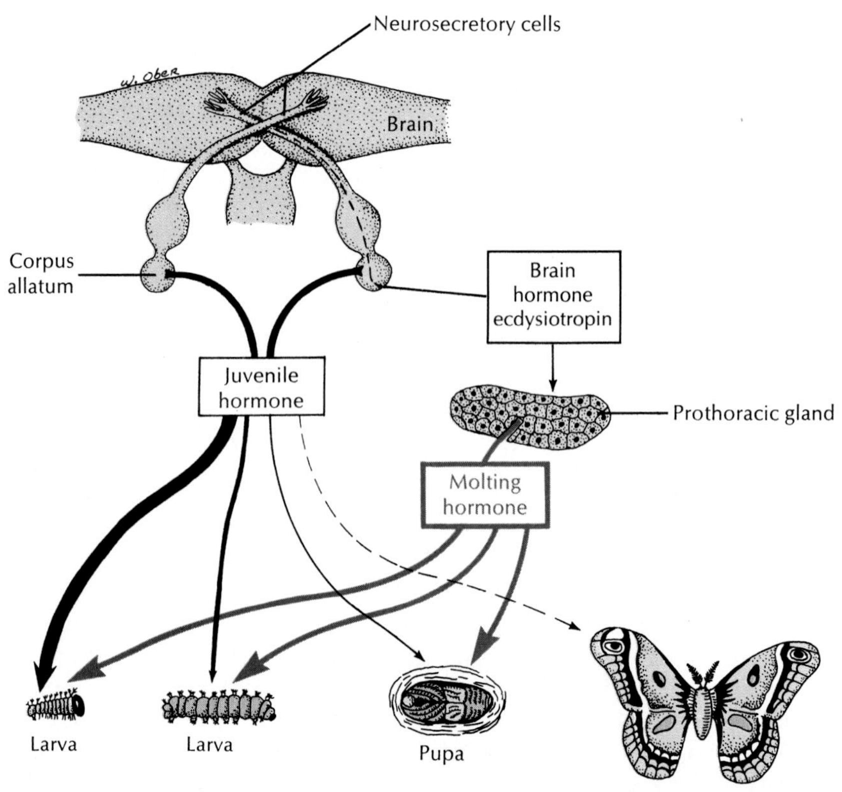

FIGURE 11-5

Endocrine control of molting in a moth. Many moths mate in the spring or summer, and eggs soon hatch into the first of several larval stages (called instars). After the final larval molt, the last and largest larva (caterpillar) spins a cocoon in which it pupates. The pupa overwinters, and an adult emerges in the spring to start a new generation. Two hormones interact to control molting and pupation. The molting hormone, produced by the prothoracic gland and stimulated by a separate brain hormone (ecdysiotropin), favors molting and the formation of adult structures. These effects are inhibited, however, by the juvenile hormone, produced by the corpora allata. Juvenile hormone output declines with successive molts, and the larva undergoes adult differentiation.

Juvenile hormone of silkworm

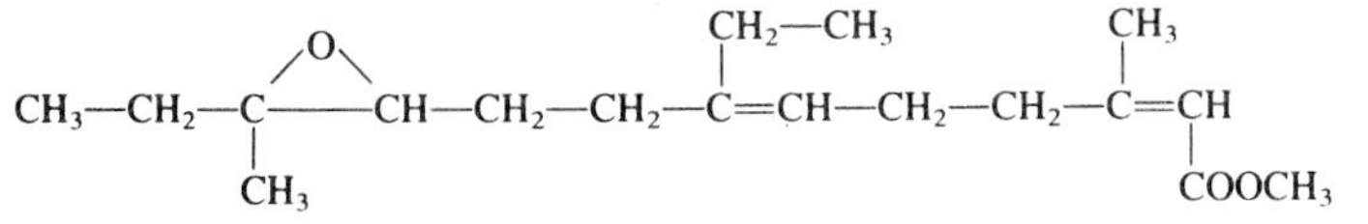

$$CH_3{-}CH_2{-}\overset{\overset{\displaystyle O}{\frown}}{\underset{\underset{\displaystyle CH_3}{|}}{C}{-}CH}{-}CH_2{-}CH_2{-}\underset{}{\overset{\overset{\displaystyle CH_2-CH_3}{|}}{C}}{=}CH{-}CH_2{-}CH_2{-}\overset{\overset{\displaystyle CH_3}{|}}{C}{=}\underset{\underset{\displaystyle COOCH_3}{|}}{CH}$$

Molting hormone (α-ecdysone) of silkworm

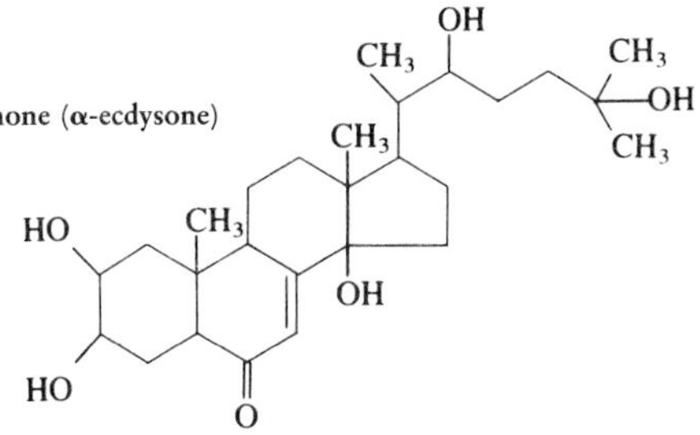

hormones has been determined. It required extraction from 1000 kg (about 1 ton) of silkworm pupae to show that the molting hormone is a steroid. The juvenile hormone has an entirely different structure. (See top of page.)

The molting hormone is under the control of ecdysiotropin. This hormone is a polypeptide that is produced by neurosecretory cells of the brain, transported down axons, and stored in the corpora allata, from where it is released into the blood and stimulates the release of molting hormone. Molting hormone appears to act directly on the chromosomes to set in motion the changes resulting in a molt. The molting hormone favors the development of adult structures. It is held in check, however, by the juvenile hormone, which favors the development of juvenile characteristics. During juvenile life the juvenile hormone predominates and each molt yields another larger juvenile. Finally the output of juvenile hormone decreases and the final juvenile molt occurs.

The precise location of ecdysiotropin in the brain of pupal tobacco hornworms was revealed by N. Agui by delicate microdissection. Using a human eyebrow hair, he was able to isolate the single cell in each brain hemisphere that contained ecdysiotropin activity. Thus only two cells, each about 20 μm in diameter, produce this insect's total supply of ecdysiotropin. In an age when sophisticated instrumentation has removed much of the tedium (and some of the creativity) from research, it is refreshing to learn that certain biological mysteries succumb only to skillful use of the human hand.

■ Vertebrate Endocrine Glands and Hormones

In the remainder of this chapter we describe some of the best understood and most important of the vertebrate hormones. The hormones of reproduction are discussed in the next chapter. Space does not permit us to deal with all the hormones and hormonelike substances that have been discovered. The mammalian hormonal mechanisms are the best understood, since laboratory mammals and humans have always been the objects of the most intensive research. Research with the lower vertebrates has shown that all vertebrates share similar endocrine organs. All vertebrates have a pituitary gland, for example, and all have thyroid glands, adrenal glands (or the special cells of which they are composed), and gonads. Nevertheless, there are some important differences in the functional roles that the hormones of these glands have among the different vertebrates.

HORMONES OF THE PITUITARY GLAND AND HYPOTHALAMUS

The pituitary gland, or **hypophysis,** is a small gland (0.5 g in humans) lying in a well-protected position between the roof of the mouth and the floor of the brain (Figure 11-6). It is a two-part gland having a double embryological origin. The **anterior pituitary gland** (adenohypophysis) consists of the **anterior lobe** and the **intermediate lobe;** both are derived embryologically from the roof of the mouth. The **posterior pituitary gland** (neurohypophysis) arises from a ventral portion of the brain, the **hypothalamus,** and is connected to it by a stalk, the **infundibulum.** Although the anterior pituitary lacks any *anatomical* connection to the brain, it is nonetheless *functionally* connected to it by a special portal circulatory system. A portal circulation is one that delivers blood from one capillary bed to another (Figure 11-6).

Anterior pituitary gland

The anterior lobe, despite its minute dimensions, produces at least six protein hormones, all but one of which are **tropic hormones.** Tropic (pronounced trōpic,

FIGURE 11-6

Human pituitary gland. The posterior lobe is connected directly to the hypothalamus by neurosecretory fibers. The anterior lobe is indirectly connected to the hypothalamus by a portal circulation beginning in the base of the hypothalamus and ending in the anterior pituitary.

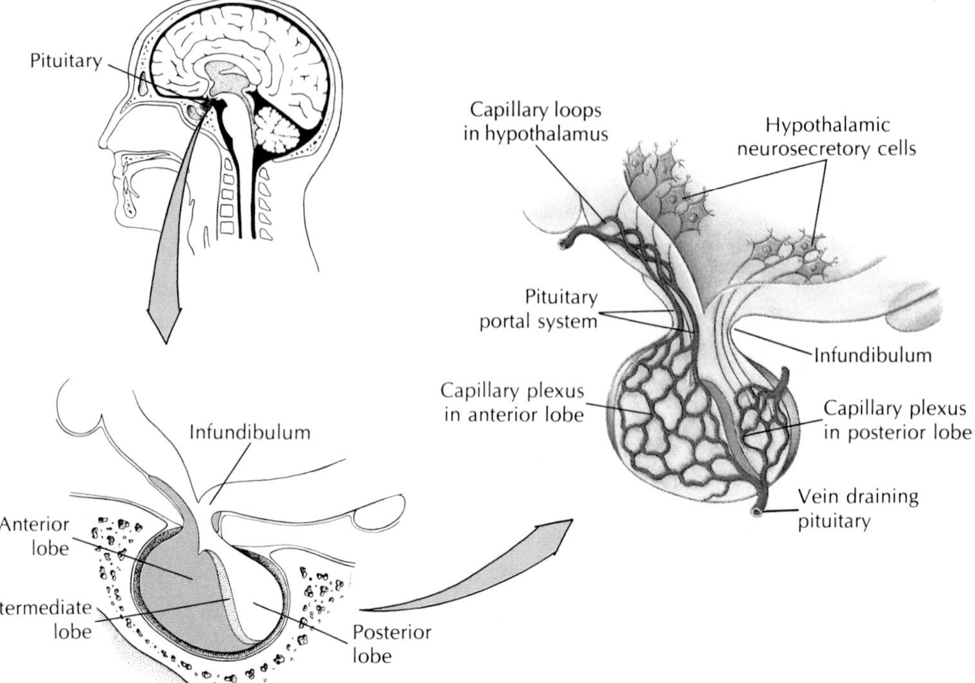

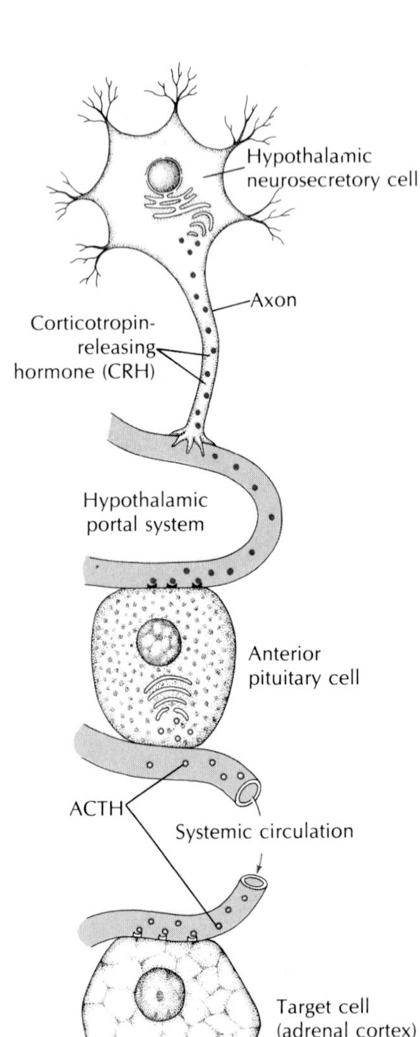

FIGURE 11-7

Relationship of hypothalamic, pituitary, and target-gland hormones. The hormone sequence controlling the release of cortisol from the adrenal cortex is used as an example.

from the Greek *tropē,* to turn) hormones are those that regulate the release of hormones from other endocrine glands (Table 11-1).

The **thyroid-stimulating hormone (TSH),** or **thyrotropin,** stimulates the production of thyroid hormones by the thyroid gland. The **adrenocorticotropic hormone** (ACTH) stimulates the adrenal cortex. Two of the tropic hormones are commonly called **gonadotropins** because they act on the gonads (ovary of the female, testis of the male). These are the **follicle-stimulating hormone** (FSH) and the **luteinizing hormone** (LH). (In the male, the luteinizing hormone goes by a different name, interstitial cell–stimulating hormone [ICSH], but it is the same hormone chemically.) A fifth tropic hormone is **prolactin,** which stimulates milk production by the female mammary glands and has a variety of other effects in the lower vertebrates. The functions of the two gonadotropins and prolactin are discussed in Chapter 13 in connection with the hormonal control of reproduction.

The sixth hormone of the anterior lobe is the **growth hormone** (also called somatotropic hormone). This hormone performs a vital role in governing body growth through its stimulatory effect on cellular mitosis and protein synthesis, especially in new tissue of young animals. If produced in excess, the growth hormone causes giantism. A deficiency of this hormone in the human child results in a pituitary midget.

Hypothalamus and neurosecretion

Because of the strategic importance of the pituitary gland in influencing most of the hormonal activities in the body, the pituitary gland was once called the body's "master gland." This description is not appropriate, however, since the anterior lobe hormones are regulated by a higher council, the neurosecretory centers of the hypothalamus. The hypothalamus is itself under the ultimate control of the brain. The hypothalamus contains groups of neurosecretory cells, which are specialized giant nerve cells (Figure 11-7). These cells manufacture polypeptide hormones, called **releasing hormones** (or "factors"), which then travel within the nerve fibers to their endings in the lower portion of the hypothalamus. Here they enter a capillary network to complete their journey to the

Table 11-1 Hormones of the vertebrate pituitary and hypothalamus—chemical nature and actions

	Hormone	Chemical nature	Principal action
Adenohypophysis Anterior lobe	Thyroid-stimulating hormone (TSH)	Glycoprotein	Stimulates thyroid to secrete thyroid hormones
	Adrenocorticotropic hormone (ACTH)	Polypeptide	Stimulates adrenal cortex to secrete steroid hormones
	Gonadotropins 1. Follicle-stimulating hormone (FSH)	Glycoprotein	Stimulates gamete production and secretion of sex hormones
	2. Luteinizing hormone (LH, ICSH)	Glycoprotein	Stimulates ovulation and secretion of sex hormones
	Prolactin (LTH)	Protein	Stimulates mammary gland growth and secretion in mammals; various reproductive and nonreproductive functions in lower vertebrates
	Growth hormone (GH)	Protein	Stimulates growth
Intermediate lobe	Melanophore-stimulating hormone (MSH)	Polypeptide	Pigment dispersion in melanophores of ectotherms; function unclear in endotherms
Neurohypophysis Posterior lobe (hormones secreted by hypothalamus, stored in and released from posterior lobe)	Oxytocin	Octapeptide	Stimulates milk ejection and uterine contraction
	Vasopressin (ADH)	Octapeptide	Antidiuretic effect on kidney
	Vasotocin	Octapeptide	Antidiuretic activity
	Mesotocin and others in lower vertebrates	Octapeptide	Functions uncertain
Hypothalamus	Thyrotropin-releasing hormone (TRH) Corticotropin-releasing hormone (CRH) Follicle-stimulating hormone–releasing hormone (FSH-RH) Luteinizing hormone–releasing hormone (LH-RH) Prolactin release–inhibiting factor (PIF) Prolactin releasing factor (PRF) Melanophore-stimulating hormone–releasing factor (MRF) Melanophore-stimulating hormone–release-inhibiting factor (MIF) Growth hormone–releasing factor (GH-RF) Growth hormone release–inhibiting hormone (GH-RIF)	All polypeptides	Control release of anterior and intermediate lobe hormones

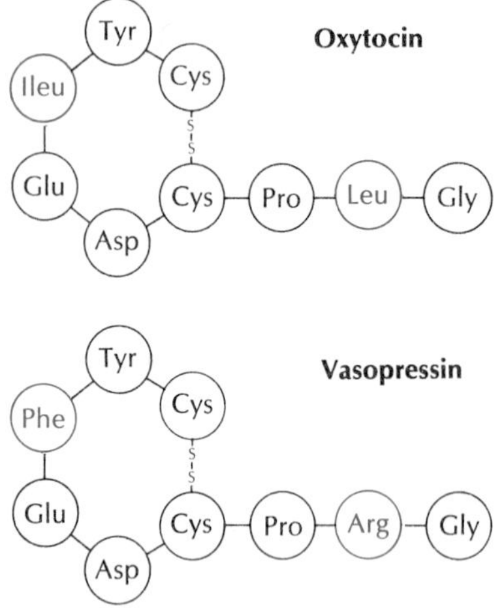

FIGURE 11-8

Posterior lobe hormones of humans. Both oxytocin and vasopressin consist of eight amino acids (the two sulfur-linked cysteine molecules are considered to be a single amino acid, cystine). Oxytocin and vasopressin are identical except for amino acid substitutions in the blue positions. The abbreviations represent amino acids.

The radioimmunoassay technique developed by Solomon Berson and Rosalyn Yalow about 1960 has revolutionized endocrinology and neurochemistry. First, antibodies to the hormone of interest (insulin, for example) are prepared by injecting guinea pigs or rabbits with the hormone. Then, a fixed amount of radioactively labeled insulin and unlabeled insulin atibodies is mixed with the sample of blood plasma to be measured. The native insulin in the blood plasma and the radioactive insulin compete for antibodies. The more insulin there is in the sample, the less radioactive insulin will bind to the antibodies. Bound and unbound insulin are then separated, and their radioactivities are measured together with those of appropriate standards to determine the amount of insulin present in the blood sample. The method is so incredibly sensitive that it can measure the equivalent of a cube of sugar dissolved in one of the Great Lakes.

anterior pituitary by way of a short pituitary portal system. The hypothalamic hormones then stimulate or inhibit the release of the various anterior pituitary hormones. Ten hypothalamic releasing hormones in all have been discovered since the demonstration in 1955 of a corticotropin-releasing hormone. There appears to be one or more releasing hormones regulating each of the six anterior pituitary hormones (Table 11-1). Several of the releasing hormones have now been isolated in pure state and characterized chemically. All are peptides.

Posterior pituitary

The hypothalamus is also the source of two hormones released by the posterior lobe of the pituitary. They are formed within neurosecretory cells of the hypothalamus and then transported down the infundibular stalk and into the posterior lobe, ending in proximity to blood capillaries, which the hormones enter when released (Figure 11-6). In a sense the posterior lobe is not a true endocrine gland, but a storage and release center for hormones manufactured entirely in the hypothalamus. The two posterior lobe hormones of mammals, oxytocin and vasopressin, are chemically very much alike; both are polypeptides consisting of eight amino acids and are referred to as octapeptides (Figure 11-8). These hormones are among the fastest-acting hormones in the body, since they are capable of producing a response within seconds of their release from the posterior lobe.

Oxytocin has two important specialized reproductive functions in adult female mammals. It stimulates contraction of the uterine smooth muscles during parturition (birth of the young). In clinical practice, oxytocin may be used to induce labor, to facilitate delivery, and to prevent uterine hemorrhage after birth. The second action of oxytocin is that of milk ejection by the mammary glands in response to suckling. Although present, oxytocin has no known function in the male.

Vasopressin, the second posterior lobe hormone, acts on the kidney to restrict urine flow, as already described on p. 212. It is therefore often referred to as the **antidiuretic hormone** (ADH). Vasopressin has a second, weaker effect of increasing the blood pressure through its generalized constrictor effect on the smooth muscles of the arterioles. Although the name "vasopressin" unfortunately suggests that the vasoconstrictor action is the hormone's major effect, it is probably of little physiological importance, except perhaps to help sustain the blood pressure during a severe hemorrhage.

All the lower vertebrates, except the most primitive fishes, also secrete two posterior lobe octapeptides. However, there is some variation in chemical structure; the two posterior lobe hormones secreted by fishes, for example, are not identical to oxytocin and vasopressin of mammals. Although a total of 10 different posterior lobe hormones have been identified from the various vertebrate groups, never more than two are found in any one animal.

Of all the posterior lobe hormones, **vasotocin** has the widest phylogenetic distribution and is believed to be the parent hormone from which the other octapeptides have evolved. It is found in all vertebrate classes except mammals. It is a water-balance hormone in amphibians, especially toads, in which it acts to conserve water by (1) increasing permeability of the skin (to promote water absorption from the environment), (2) stimulating water reabsorption from the urinary bladder, and (3) decreasing urine flow. The action of vasotocin is best understood in amphibians, but it appears to play some water-conserving role in birds and reptiles as well.

Brain neuropeptides

The blurred distinction between the endocrine and nervous systems is nowhere more evident than in the brain, where numerous hormonelike neuropeptides recently have been discovered. More than a dozen neuropeptides (short chains of amino acids) have been identfiied, and many are known to lead double lives. They are capable of behaving both as hormones, carrying signals from gland cells to their targets, and as neurotransmitters, relaying signals between nerve cells. For example, both oxytocin and vasopressin have been discovered at widespread sites in the brain by radioimmunochemical methods. Apparently related to this is the fascinating observation that people and experimental animals injected with minute quantities of vasopressin experience enhanced learning and improved memory. As far as we can tell, this effect of vasopressin in brain tissue has nothing to do with its well-known antidiuretic function in the kidney (p. 212).

Just as amazing was the discovery in the cerebral cortex and hippocampus of several hormones, such as gastrin and cholecystokinin (p. 277), which long had been supposed to function only in the gut. We have a good idea of what these hormones do in the gastrointestinal tract, but what functional roles do they play in the brain?

Among the dramatic developments in this field was the discovery in 1975 of the endorphins and enkephalins, substances that bind with opiate receptors and are important in pain reception (see marginal note on p. 253). The endorphins and enkephalins are also found in brain circuits that modulate several other functions unrelated to pain, such as control of blood pressure, body temperature, and body movement. Even more intriguing, the endorphins are derived from the same chemical precursor that gives rise to the anterior pituitary hormone ACTH. It is clear that we have discovered in the brain a complex family of compounds whose functions and interrelationships are not yet clear. This is currently an active area of biomedical research, and we seem to be on the threshold of some exciting biological discoveries.

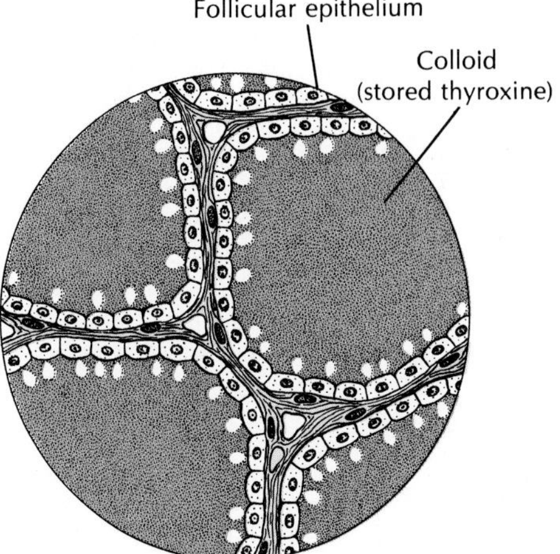

Inactive follicles

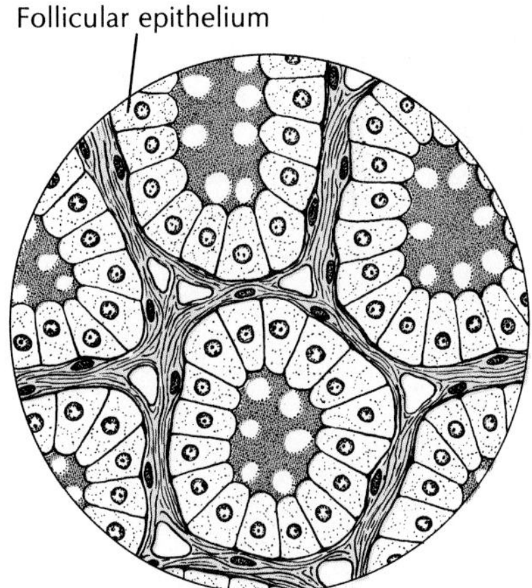

Active follicles

FIGURE 11-9

Appearance of thyroid gland follicles viewed through the microscope. (Approximately ×350.) When inactive, the follicles are distended with colloid, the storage form of thyroxine, and the epithelial cells are flattened. When active, the colloid disappears as thyroxine is secreted into the circulation, and the epithelial cells become greatly enlarged.

HORMONES OF METABOLISM

Another important group of hormones adjusts the delicate balance of metabolic activities in the body. The rates of chemical reactions within cells are often regulated by long sequences of enzymes. Although such sequences are complex, each step in a pathway is mostly self-regulating, as long as the equilibrium between substrate, enzyme, and product remains stable. However, hormones may alter the activity of crucial enzymes in a metabolic process, thus accelerating or inhibiting the entire process. It should be emphasized that hormones never initiate enzymic processes. They simply alter their rate, speeding them up or slowing them down. The most important hormones of metabolism are those of the thyroid, parathyroid, adrenal glands, and pancreas.

Thyroid hormones

The two thyroid hormones **thyroxine** and **triiodothyronine** are secreted by the thyroid gland. This large endocrine gland is located in the neck region of all vertebrates. The thyroid is made up of thousands of tiny spherelike units, called follicles, where thyroid hormone is synthesized, stored, and released into the bloodstream as needed. The size of the follicles, and the amount of stored thyroxine they contain, depends on the activity of the gland (Figure 11-9).

One of the unique characteristics of the thyroid is its high concentration of **iodine;** in most animals this single gland contains well over half the body store of iodine. The epithelial cells of the thyroid follicles actively trap iodine from the blood and combine it with the amino acid tyrosine, creating the two thyroid hormones. Each molecule of thyroxine contains four atoms of iodine. Triiodothyronine is identical to thyroxine, except that it has three instead of four iodine atoms. Thyroxine is formed in much greater amounts than triidothyronine, but both hormones have two important actions. One is to promote normal growth and development of the nervous system of growing animals. The other is to stimulate the metabolic rate.

Thyroxine

HO—(benzene ring, I, I)—O—(benzene ring, I, I)—CH_2—CH(NH_2)—COOH

Undersecretion of thyroid hormone dramatically impairs growth, especially of the nervous system. The human **cretin,** a mentally retarded dwarf, is the tragic product of thyroid malfunction from a very early age. Conversely, the oversecretion of thyroid hormones causes precocious development, particularly in lower vertebrates. Frogs and toads undergo a dramatic metamorphosis from aquatic tadpole without lungs or legs to semiterrestrial or terrestrial adult with lungs, four legs, and a completely remodeled alimentary canal. This transformation occurs when the thyroid gland becomes active at the end of larval premetamorphosis. Stimulated by a rise in the blood thyroxine level, metamorphosis and climax occur (Figure 11-10).

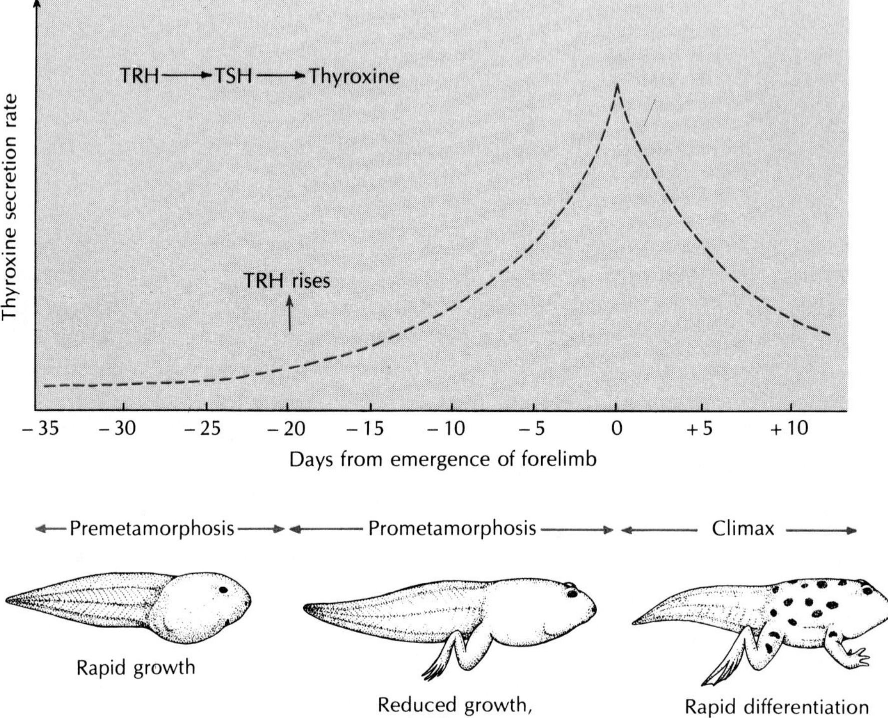

FIGURE 11-10

Effect of thyroxine on frog growth and metamorphosis. The release of TRH from the hypothalamus at the end of premetamorphosis sets in motion the hormonal changes (increased TSH and thyroxine) leading to metamorphosis. Thyroxine levels are maximum at the time the forelimbs emerge.

Modified from Bentley, P.J. 1976. Comparative vertebrate endocrinology. Cambridge, Eng., Cambridge University Press.

The control of oxygen consumption and heat production in birds and mammals is the best-known action of the thyroid hormones. The thyroid maintains metabolic activity of homeotherms (birds and mammals) at a normal level. Too much thyroid hormone will speed up body processes as much as 50%, resulting in irritability, nervousness, fast heart rate, intolerance of warm environments, and loss of body weight despite increased appetite. Too little thyroid hormone slows metabolic activities, which can result in loss of mental alertness, slowing of the heart rate, muscular weakness, increased sensitivity to cold, and weight gain. One important function of the thyroid gland is to help animals adapt to cold environments by increasing their heat production. Thyroxine in some way causes cells to produce more heat and store less chemical energy (ATP); in other words, thyroxine *reduces* the efficiency of the cellular oxidative phosphorylation system (p. 52). This is why many cold-adapted mammals have heartier appetites and eat more food in winter than in summer even though their activity is about the same in both seasons. In winter, a larger portion of the food is being converted directly into body-warming heat.

The synthesis and release of thyroxine and triidothyronine are governed by thyroid-stimulating hormone (TSH) from the anterior pituitary gland (Table 11-1). TSH controls the thyroid through a beautiful example of negative feedback. If the thyroxine level in the blood decreases, more TSH is released, which returns the thyroxine level to normal. Should the thyroxine level rise too high, it acts on the anterior pituitary to inhibit TSH release. With declining TSH output, the thyroid is less stimulated and the blood thyroxine level returns to normal. Such a system is obviously very effective in damping out oscillations in hormone output by the target gland. It can be overridden, however, by neural stimuli, such as exposure to cold, which can directly stimulate increased release of TSH.

The control of thyroid activity involves another component, the thyrotropin-releasing hormone (TRH) of the hypothalamus. As noted earlier, TRH is part of a higher regulatory council that controls the tropic hormones of the anterior pituitary. But, if the releasing hormone controls the anterior pituitary, what controls release of the releasing hormone? At present, there is no general agreement on the answer, although there is evidence that the thyroid hormones have a negative feedback effect on the hypothalamus as well as on the anterior pituitary.

Some years ago, a condition called **goiter** was common among people living in the Great Lakes region of the United States and Canada, as well as in other parts of the earth such as the Swiss Alps. This type of goiter is an enlargement of the thyroid gland caused by a deficiency of iodine in the food and water. In striving to produce thyroid hormone with not enough iodine available, the gland hypertophies, sometimes so much that the entire neck region becomes swollen (Figure 11-11). Iodine-deficiency goiter is seldom seen in North America because of the widespread use of iodized salt. However, it is estimated that even today 200 million people suffer from goiter worldwide, mostly in high mountain areas of Latin America, Europe, and Asia.

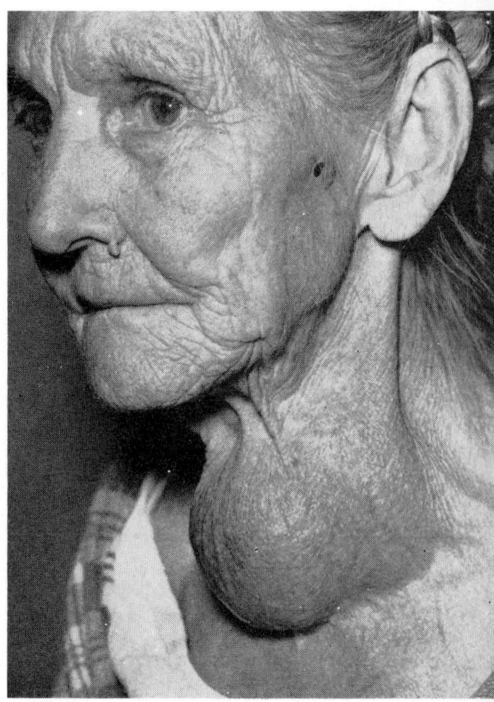

FIGURE 11-11

A large goiter caused by iodine deficiency. By enlarging enormously, the thyroid gland can extract enough iodine from the blood to synthesize the body's thyroid hormone requirement.

From Prior, J.A., et al. 1981. Physical diagnosis. St. Louis, The C.V. Mosby Co.

Hormonal regulation of calcium metabolism

Closely associated with the thyroid gland and often buried within it are the parathyroid glands. These tiny glands occur as two pairs in humans but vary in number and position in other vertebrates. They were discovered at the end of the nineteenth century when the fatal effects of "thyroidectomy" were traced to the unknowing removal of the parathyroid glands as well as the thyroid gland.

In many animals, including humans, removal of the parathyroid glands causes the blood calcium to decrease rapidly. This results in a serious increase in nervous system excitability, severe muscular spasms and tetany, and finally death.

Actually, three hormones are involved in the stabilization of both calcium and phosphorus in the blood. They are **parathyroid hormone** (PTH), produced by the parathyroid gland; **calcitonin,** produced by specialized cells (C cells) in the thyroid gland; and a hormonal metabolite of vitamin D called **1,25-dihydroxyvitamin D** (1,25-$[OH]_2D$). Before we consider how these factors interact, it will be helpful to summarize mineral metabolism in bone, a densely packed storehouse of both calcium and phosphorus.

Bone contains approximately 98% of the body calcium and 80% of the phosphorus. Although bone is second only to teeth as the most durable material in the body, as evidenced by the survival of fossil bones for millions of years, it is in a state of constant turnover in the living body. Bone-building cells (**osteoblasts**) withdraw calcium and phosphorus (as phosphate) from the blood and deposit them in a complex crystalline form around previously formed organic fibers. Bone-resorbing cells (**osteoclasts**), present in the same bone, tear down bone by engulfing it and releasing the calcium and phosphate into the blood. These opposing activities allow bone to remodel itself constantly, especially in the growing animal, for structural improvements to counter new mechanical stresses on the body. They additionally provide a vast and accessible reservoir of minerals that can be withdrawn as the body needs them for its general cellular requirements.

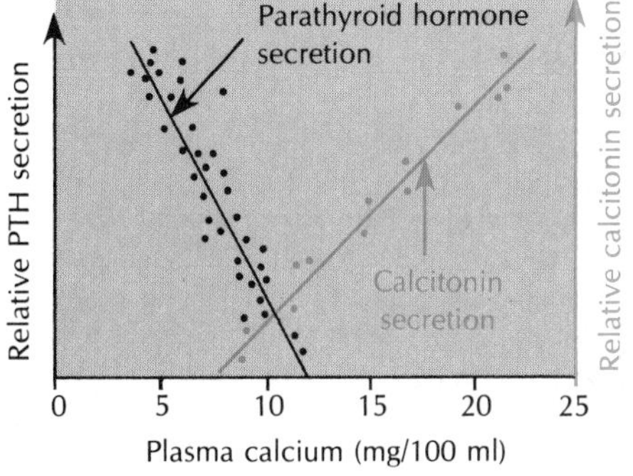

FIGURE 11-12

How parathyroid hormone (PTH) and calcitonin secretion rates respond to changes in blood calcium level in a mammal.

After Copp, D.H. 1969. J. Endocrinol. **43**:137-161.

If the blood calcium should decrease slightly, the parathyroid gland increases its output of parathyroid hormone. This stimulates the osteoclasts to destroy bone adjacent to these cells, thus releasing calcium and phosphate into the bloodstream and returning the blood calcium level to normal. Should the calcium in

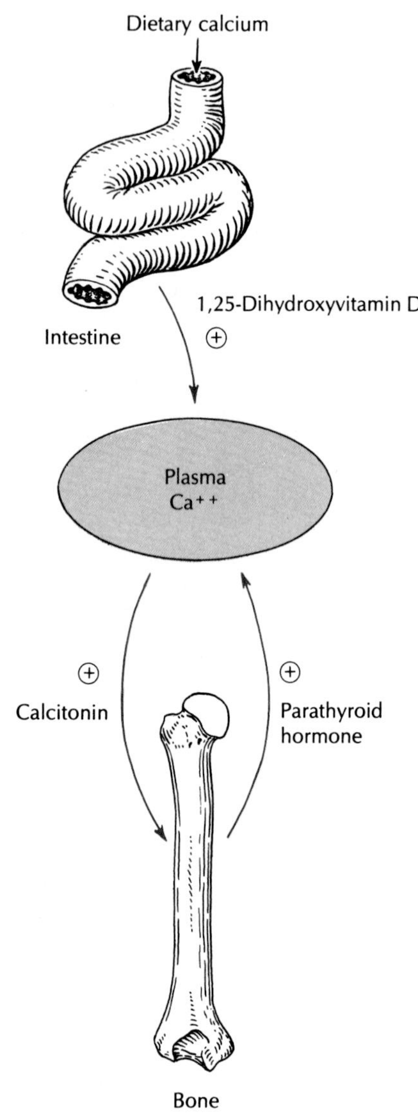

FIGURE 11-13

Regulation of blood calcium in birds and mammals.

the blood rise above normal, the parathyroid gland decreases its output of parathyroid hormone. The parathyroid hormone level varies inversely with blood calcium level, as shown in Figure 11-12.

The second calcium-regulating hormone, calcitonin, is secreted when the blood level of calcium begins to rise too high (Figure 11-12). In mammals, calcitonin lowers the blood calcium level by inhibiting bone resorption by the osteoclasts. It thus protects the body against a dangerous increase in the blood calcium level, just as parathyroid hormone protects it from a dangerous decrease in blood calcium. The two act together to smooth out oscillations in blood calcium (Figure 11-13). In lower vertebrates, the physiological role of calcitonin is uncertain.

The third factor involved in calcium metabolism, 1,25-dihydroxyvitamin D, is an active hormonal form of vitamin D. Vitamin D, like all vitamins, is a dietary requirement. But unlike other vitamins, vitamin D may also be synthesized in the skin from a precursor by irradiation with ultraviolet light from the sun. Vitamin D is then converted in a two-step oxidation to 1,25-dihydroxyvitamin D. This steroid hormone is essential for active calcium absorption by the gut (Figure 11-13). It also promotes the synthesis of a protein that transports calcium in the blood.

A deficiency of vitamin D causes rickets, a disease characterized by low blood calcium and weak, poorly calcified bones that tend to bend under postural and gravitational stresses. Rickets has been called a disease of northern winters, when sunlight is minimal. It was once common in the smoke-darkened cities of England and Europe.

Hormones of the adrenal cortex

The vertebrate adrenal gland is a double gland consisting of two very different kinds of tissue: **interrenal** tissue, called **cortex** in mammals, and **chromaffin** tissue, called **medulla** in mammals (Figure 11-14). The mammalian terminology of cortex (meaning "bark") and medulla (meaning "core") arose because in this group of vertebrates the interrenal tissue completely surrounds the chromaffin tissue like a cover. Although in the lower vertebrates the interrenal and chromaffin tissue are usually separated, the mammalian terms "cortex" and "medulla" are so firmly fixed in our vocabulary that we commonly use them for all vertebrates instead of the more correct terms "interrenal" and "chromaffin."

Biochemists have found that the adrenal cortex contains at least 30 different compounds, all of them closely related lipoid compounds known as steroids. Only a few of these compounds, however, are true steroid *hormones;* most are various intermediates in the synthesis of steroid hormones from **cholesterol** (Figure 11-15). The corticosteroid hormones are commonly classified into three groups, according to their function:

1. **Glucocorticoids,** such as **cortisol** (Figure 11-15) and **corticosterone,** are concerned with food metabolism, inflammation, and stress. They cause the conversion of nonglucose compounds, particularly amino acids and fats, into glucose. This process, called **gluconeogenesis,** is extremely important, since most of

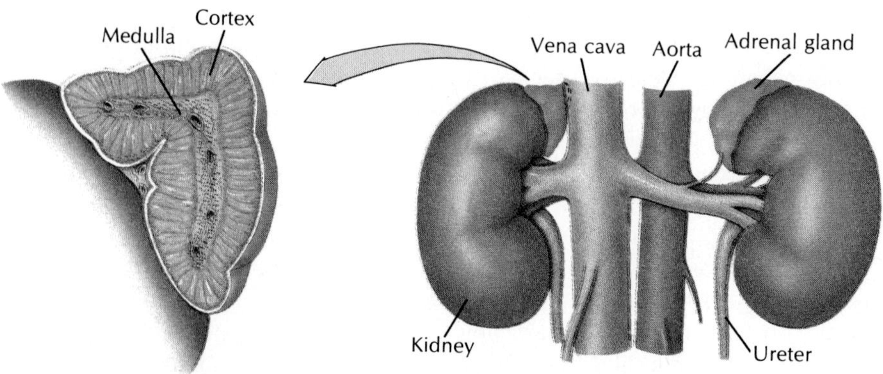

FIGURE 11-14

Paired adrenal glands of human, showing gross structure and position on the upper poles of the kidneys. Steroid hormones are produced by the outer cortex. The sympathetic hormones epinephrine and norepinephrine are produced by the inner medulla.

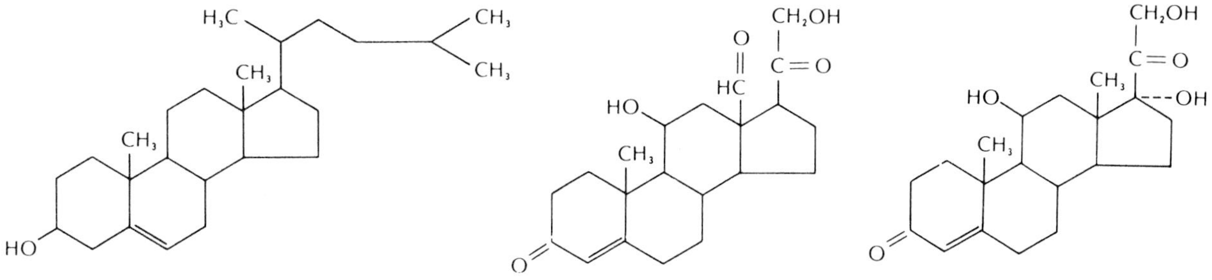

FIGURE 11-15

Hormones of the adrenal cortex. Cortisol (a glucocorticoid) and aldosterone (a mineralocorticoid) are two of several steroid hormones synthesized from cholesterol in the adrenal cortex.

the body's stored energy reserves are in the form of fats and proteins. Cortisol, cortisone, and corticosterone are also **antiinflammatory.** Because several diseases of humans are inflammatory diseases (for example, allergies, hypersensitivity, and arthritis, p. 180), these corticosteroids have important medical applications. They must be used with great care, however, since, if administered in excess, they may suppress the body's normal repair processes and lower resistance to infectious agents.

2. **Mineralocorticoids,** the second group of corticosteroids, are those that regulate salt balance. **Aldosterone** (Figure 11-15) and **deoxycorticosterone** are the most important steroids of this group. They promote the tubular reabsorption of sodium and chloride and the tubular excretion of potassium by the kidney. Since sodium often is in short supply in the animal diet and potassium in excess, it is obvious that the mineralocorticoids play vital roles in preserving the correct balance of blood electrolytes. We may also note that the mineralocorticoids *oppose* the antiinflammatory effect of cortisone. In other words, they promote the *inflammatory* defense of the body to various noxious stimuli. Although these opposing actions of the corticosteroids seem self-defeating, they actually are not. They are necessary to maintain readiness of the body's defenses for any stress or disease threat, yet prevent these defenses from becoming so powerful that they turn against the body's own tissues.

3. **Sex hormones,** such as testosterone, estrogen, and progesterone, are produced primarily by the ovaries and testes (p. 319). The adrenal cortex is also a minor source of certain steroids that mimic the action of testosterone. These sex hormone–like secretions are of little physiological significance, except in certain disease states of humans.

The synthesis and secretion of the corticosteroids are controlled principally by ACTH of the anterior pituitary (Figure 11-7). As with pituitary control of the thyroid, a negative feedback relationship exists between ACTH and the adrenal cortex: an increase in the level of corticosteroids suppresses the output of ACTH; a decrease in the blood steroid level increases ACTH output. ACTH is also controlled by the corticotropin-releasing hormone (CRH) of the hypothalamus.

The adrenal steroid hormones, especially the glucocorticoids, are remarkably effective in relieving the *symptoms* of rheumatoid arthritis, allergies, and various connective tissue, skin, and blood disorders. Following the report in 1948 that cortisone dramatically relieved the pain and crippling effects of advanced arthritis, the steroid hormones were hailed by the media as "wonder drugs." Optimism was soon dimmed, however, when it became apparent that severe side effects always attended long-term administration of the antiinflammatory steroids. Steroid therapy lulls the adrenal cortex into inactivity and may permanently impair the body's capacity to produce its own steroids. Today steroid therapy is applied with caution, since it is realized that the inflammatory response is a necessary part of the body's defenses.

Adrenal medulla hormones

The adrenal medulla secretes two structurally similar hormones, **epinephrine** (adrenaline) and **norepinephrine** (noradrenaline). Norepinephrine is also released at the endings of sympathetic nerve fibers throughout the body, where it serves as a neurotransmitter to carry neural signals across the gap that separates the fiber and the organ it innervates. The adrenal medulla has the same embryological origin as sympathetic nerves; in many respects the adrenal medulla is nothing more than an overgrown sympathetic nerve ending.

It is not surprising then that the adrenal medulla hormones have the same general effects on the body that the sympathetic nervous system has. These effects center on responses to emergencies, such as fear and other strong emotional states, flight from danger, fighting, and exposure to cold, pain, lack of oxygen,

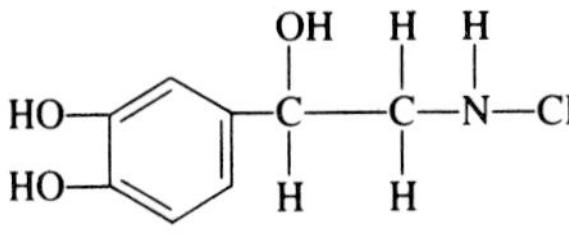

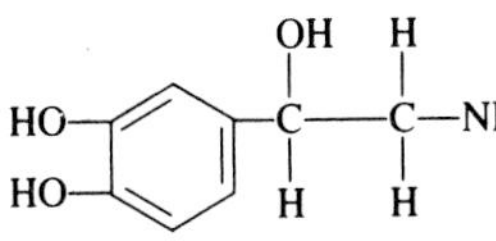

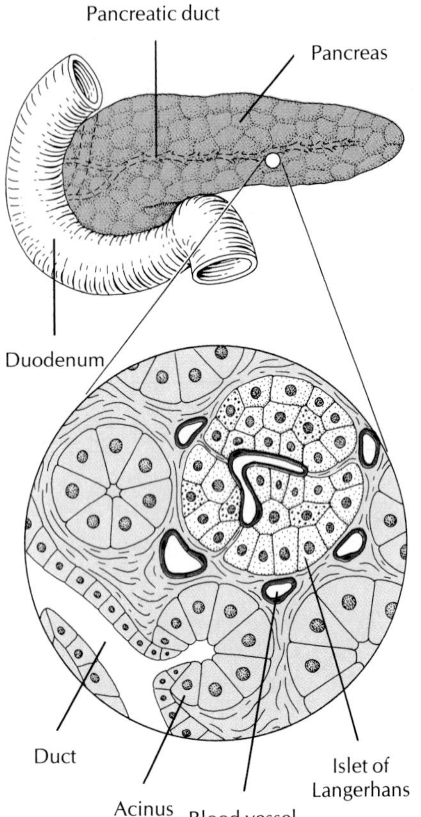

FIGURE 11-16

The pancreas is composed of two kinds of glandular tissue: exocrine acinar cells, which secrete digestive juices that enter the intestine through the pancreatic duct, and the endocrine islets of Langerhans. The islets of Langerhans secrete the hormones insulin and glucagon directly into the blood circulation.

In 1982, insulin became the first hormone product of genetic engineering (recombinant DNA technology) to be marketed for human use. The new recombinant insulin has the exact structure of human insulin and therefore will not produce immunogenic reactions, which has often been a problem for diabetics receiving insulin purified from pig or cow pancreas.

and blood loss. Walter B. Cannon, of homeostasis fame (p. 198), termed these "fight or flight" responses that are appropriate for survival. To study the sympathetic nervous system in cats, Cannon would expose the cats suddenly to barking dogs. In such situations the cats exhibited several physiological responses. Their bronchioles expanded and respiration increased, delivering more oxygen to the blood; heart rate and blood pressure increased, and skeletal muscle capillaries dilated, bringing more blood with oxygen and nutrients into active muscles; blood flow to nonessential organs (intestines, for example) decreased; liver glycogen was mobilized, releasing glucose for quick energy; and mental alertness was enhanced. Humans behave similarly when excited. These effects are attributable both to the rapid release into the blood of epinephrine from the adrenal medulla and to increased activity of the sympathetic nervous system.

Insulin from the islet cells of the pancreas

The pancreas is both an exocrine and an endocrine organ. The *exocrine* portion produces pancreatic juice, a mixture of digestive enzymes that is conveyed by ducts to the digestive tract. Scattered within the extensive exocrine portion of the pancreas are numerous small islets of tissue, called **islets of Langerhans** (Figure 11-16). This is the *endocrine* portion of the gland. The islets are without ducts and secrete their hormones directly into blood vessels that extend throughout the pancreas.

Two polypeptide hormones are secreted by different cell types within the islets: **insulin,** produced by the **beta cells,** and **glucagon,** produced by the **alpha cells.** Insulin and glucagon have antagonistic actions of great importance in the metabolism of carbohydrates and fats. Insulin is essential for the use of blood glucose by cells, especially skeletal muscle cells. It lowers the blood glucose level by promoting glucose transport into body cells. Without insulin, the cells cannot use glucose, even if there is an abnormally high blood glucose level (hyperglycemia), and sugar appears in the urine. Lack of insulin also inhibits the uptake of amino acids by skeletal muscle, and fats and muscle are broken down to provide energy. The body cells actually starve while the urine abounds in the very substance the body craves. The disease, called diabetes mellitus ("sweet running through") afflicts nearly 5% of the human population in varying degrees of severity. If left untreated, it leads inexorably to emaciation, coma, and death.

The first extraction of insulin in 1921 by two Canadians, Frederick Banting and Charles Best, was one of the most dramatic and important events in the history of medicine. Many years earlier two German scientists, J. von Mering and O. Minkowski, discovered that surgical removal of the pancreas of dogs invariably caused severe symptoms of diabetes, resulting in the animal's death within a few weeks. Many attempts were made to isolate the diabetes preventive factor, but all failed because powerful protein-splitting digestive enzymes in the exocrine portion of the pancreas destroyed the hormone during extraction procedures. Following a hunch, Banting, in collaboration with Best and his physiology professor J.J.R. Macleod, tied off the pancreatic ducts of several dogs. This caused the exocrine portion of the gland with its hormone-destroying enzyme to degenerate, but left the islets' tissue healthy, since they were independently served by their own blood supply. Banting and Best then successfully extracted insulin from the glands. Injected into another dog, the insulin immediately lowered the blood sugar level (Figure 11-17). Their experiment paved the way for the commercial extraction of insulin from slaughterhouse animals. It meant that millions of persons with diabetes, previously doomed to invalidism or death, could look forward to more normal lives.

Glucagon, the second hormone of the pancreas, has several effects on carbohydrates and fat metabolism that are opposite to the effects of insulin. For example, glucagon raises the blood glucose level, whereas insulin lowers it. Glucagon and insulin do not have the same effects in all vertebrates and in some glucagon is lacking altogether. Glucagon is an example of a hormone that operates through the cyclic AMP second-messenger system.

FIGURE 11-17

Charles H. Best and Sir Frederick Banting in 1921 with the first dog to be kept alive by insulin.

From Fulton, J.F., and L.G. Wilson. 1966. Selected readings in the history of physiology. Springfield, Ill., Charles C Thomas, Publisher.

HORMONES OF DIGESTION

Gastrointestional function is coordinated by a family of hormones produced by endocrine cells scattered through the gut. Although together they constitute the largest endocrine organ in the body, they have long been neglected by endocrinologists because it has been impossible to study them by applying the classical method of surgical removal of a gland, followed by examination of the effect of its absence. However, by the mid-1970s seven gut hormones had been chemically purified or defined. Recently several of these hormones were discovered in the nervous system where they may serve as neurotransmitters. If this dual role of the gut hormones is confirmed by current research, it would be another example of nature's conservative capacity to put the same cellular products to completely different uses in unrelated systems.

The three best-understood gut hormones are gastrin, cholecystokinin (CCK), and secretin (Figure 11-18). **Gastrin** is a small polypeptide hormone produced in the mucosa of the stomach. When food enters the stomach, gastrin stimulates powerful contractions by the stomach musculature and the secretion of hydrochloric acid by the stomach wall. Gastrin is an unusual hormone in that it exerts its action on the same organ from which it is secreted. **CCK** is also a polypeptide hormone having a striking structural resemblance to gastrin, suggesting that the two arose by duplication of ancestral genes. CCK has two distinct functions. It stimulates gallbladder contraction and thus increases the flow of bile salts into the intestine; it also stimulates an enzyme-rich secretion from the pancreas. The third gut hormone is **secretin,** the first hormone to be discovered (p. 262). Like CCK, it stimulates pancreatic secretion, but instead of being rich in enzymes, the secretion contains bicarbonate, which rapidly neutralizes stomach acid.

All of the gastrointestinal hormones are peptides that bind to surface receptors on target tissues and act through the second messenger, cyclic AMP.

Recently the gastrointestinal hormone CCK has been associated with memory enhancement in mammals. Animals fed immediately after leaving a task remember the task better than other animals not fed after the task, an effect mimicked by injecting CCK into the abdomen. The effect is adaptive, it is suggested, because it would help animals remember how they successfully found food in the past.

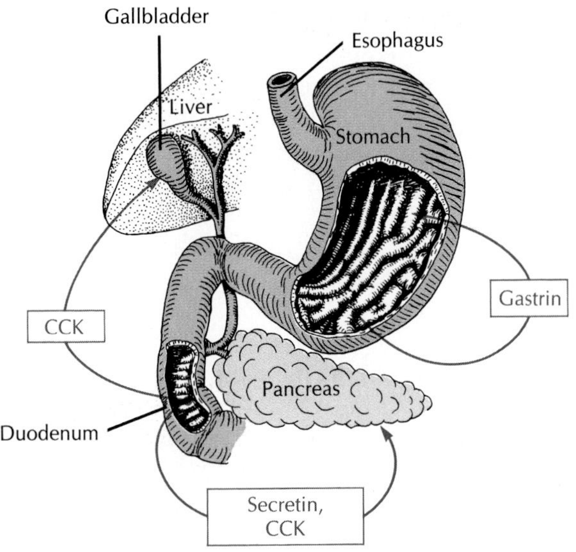

FIGURE 11-18

Three hormones of digestion. Arrows show source and target of three gastrointestinal hormones.

Summary

Hormones are chemical messengers synthesized by special endocrine glands and transported by the bloodstream to target cells where they affect cell function by altering specific biochemical processes. Specificity of response is ensured by the presence of protein receptors on or in the target cells that bind only selected hormones. Hormone effects are vastly amplified in the target cells by acting through one or the other of two basic mechanisms. Many hormones, including epinephrine, glucagon, vasopressin, and some anterior pituitary hormones, cause production of cyclic AMP, a "second messenger" that relays the hormones' message from a surface receptor to the cell's biochemical machinery. The alternative mechanism relays the action of steroid hormones by operating through cytoplasmic receptors. A hormone-receptor complex is formed that moves into the cell's nucleus to induce protein synthesis by setting gene transcription in motion.

Most invertebrate hormones are products of neurosecretory cells. The best-understood invertebrate endocrine system is that controlling molting and metamorphosis in insects. An insect juvenile grows by passing through a series of molts under the control of two hormones, one (the molting hormone) favoring molting to an adult and the other (the juvenile hormone) favoring retention of juvenile characteristics. The molting hormone is under the control of a brain neurosecretory hormone called ecdysiotropin.

The vertebrate endocrine system is orchestrated by the pituitary gland. The anterior lobe of the pituitary produces at least six hormones. Five of these are tropic hormones that regulate subservient endocrine glands: adrenocorticotropic hormone (ACTH), which stimulates the adrenal cortex; thyroid-stimulating hormone (TSH); follicle-stimulating hormone (FSH) and luteinizing hormone (LH), which act on the ovaries and testes; and prolactin, which plays several diverse roles, including the stimulation of milk production. A sixth anterior pituitary hormone is the growth hormone that governs body growth. The release of all of the anterior lobe hormones is regulated in part by hypothalamic neurosecretory products called releasing hormones. The hypothalamus also produces two neurosecretory hormones, oxytocin and vasopressin, which are stored and released from the posterior lobe of the pituitary.

The recent application of ultrasensitive radioimmunochemical techniques has revealed many neuropeptides in the brain, several of which behave as neurotransmitters in the brain but as hormones elsewhere in the body.

Several hormones play important roles in regulating cellular metabolic activities. The two thyroid hormones, thyroxine and triiodothyronine, promote normal growth and nervous system development, and they control the rate of cellular metabolism. Calcium metabolism is regulated principally by two antagonistic hormones: parathyroid hormone from the parathyroid glands and calcitonin from the thyroid gland. A hormonal derivative of vitamin D, 1,25-dihydroxyvitamin D, is essential for calcium absorption from the gut.

The steroid hormones of the adrenal cortex are glucocorticoids, which stimulate glucose formation from nonglucose sources; mineralocorticoids, which regulate blood electrolyte balance; and certain of the sex hormones. The adrenal medulla is the source of epinephrine and norepinephrine, which have many effects, including assisting the sympathetic nervous system in emergency responses.

Sugar metabolism is regulated by the antagonistic action of two pancreatic hormones. Insulin is needed for cellular use of blood glucose and the uptake of amino acids by muscle. Glucagon opposes the action of insulin.

Several gut hormones coordinate gastrointestinal function. They include gastrin, which stimulates acid secretion by the stomach, and CCK and secretin, which regulate gallbladder contraction and pancreatic secretion.

REVIEW QUESTIONS

1. Provide definitions for the following: hormone, endocrine gland, exocrine gland, hormone receptor molecule.
2. Outline the famous experiment of Bayliss and Starling that marks the birth of endocrinology. What might their *hypothesis* have been?
3. Hormone receptor molecules are the key to understanding the specificity of hormone action on target cells. Describe and distinguish between receptors located on the cell surface and those located in the cytoplasm of target cells. Name two hormones whose action is mediated through each receptor type.
4. What is the importance of feedback systems in the control of hormone output? Offer an example of a hormonal feedback pattern.
5. Explain how the three hormones involved in insect growth—molting hormone, juvenile hormone, and ecdysiotropin—interact in molting and metamorphosis.
6. Name six hormones produced by the anterior lobe of the pituitary gland. Explain how the secretion of these hormones is controlled by neurosecretory cells in the hypothalamus.
7. Describe the chemical nature and function of two posterior lobe hormones: oxytocin and vasopressin. What is distinctive about the way these neurosecretory hormones are secreted as compared with the neurosecretory release hormones that control the anterior pituitary hormones?
8. What are endorphins and enkephalins?
9. What are the two most important functions of the thyroid hormones?
10. Explain how you would interpret the graph in Figure 11-12 to show that PTH and calcitonin act in a complementary way to control the blood calcium level.
11. Describe the principal functions of the two major groups of adrenal corticosteroids: the glucocorticoids and the mineralocorticoids. To what extent do these names provide clues to their function?
12. Where are the hormones epinephrine and norepinephrine produced, and what is their relationship to the sympathetic nervous system and its response to emergencies?
13. Explain the actions of the hormones of the islets of Langerhans on the blood glucose level. What is the consequence of insulin insufficiency as in the disease diabetes mellitus?
14. Name three hormones of the gastrointestinal tract, and explain how they assist in the coordination of gastrointestinal function.

SELECTED REFERENCES

Bloom, F.E. 1981. Neuropeptides. Sci. Am. **245**:148-168 (Oct.). *Recent research on the brain peptides is described.*

Goldsworthy, G.J., J. Robinson, and W. Mordue. 1981. Endocrinology. New York, John Wiley & Sons, Inc. *Concise comparative approach.*

Gorbman, A., W.W. Dickhoff, S.R. Vigna, N.B. Clark, and C. Ralph. 1983. Comparative endocrinology. New York, John Wiley & Sons. *Authoritative and up-to-date text although the comparative treatment is limited to vertebrates.*

Hadley, M.E. 1984. Endocrinology: Englewood Cliffs, N.J., Prentice-Hall, Inc. *Undergraduate level textbook in vertebrate endocrinology.*

Highnam, K.C., and L. Hill. 1977. The comparative endocrinology of the invertebrates, ed. 2. New York, American Elsevier Publishers, Co., Inc. *Clearly written, well-illustrated comparative treatment.*

Snyder, S.H. 1985. The molecular basis of communication between cells, Sci. Am. **253**:132-141 (Oct.). *Review of vertebrate hormonal communication.*

THE LENGTHENING SHADOW OF ONE MAN

As long as people have walked the earth, their lives have been touched by, indeed interwoven with, the lives of other animals. They hunted and fished them, domesticated them, ate and were eaten by them, made pets of them, revered them, hated and feared them, immortalized them in art, song, and verse, fought them, and loved them. The very survival of ancient people depended on knowledge of wild animals. To stalk them, humans had to know the ways of the quarry. As the hunting societies of primitive people gave way to agricultural civilizations, people remained aware of their interrelationship with other animals.

This is still evident today. Zoos attract more visitors than ever before; wildlife television shows are increasingly popular; game-watching safaris to Africa constitute a thriving enterprise; and millions of pet animals share our homes—more than half a million pet dogs in New York City alone. Although people have always been interested in the behavior of animals, the science of animal behavior—ethology—is a newcomer to biology. Charles Darwin, with the uncanny insight of genius, prepared for the reception of ethology by showing how natural selection would favor specialized behavioral patterns for survival. Darwin's pioneering book, *The Expression of the Emotions of Man and Animals,* published in 1872, mapped out a strategy for behavioral research still in use today. But science in 1872 was unprepared for Darwin's central insight that behavioral patterns, no less than bodily structures, are selected for and have evolutionary histories. Another 60 years would pass before such concepts would begin to flourish within the science of ethology.

It was Ralph Waldo Emerson who said that an institution is the lengthening shadow of one man. For Charles Darwin the shadow is long indeed, for he brought into being entire fields of knowledge, such as genetics, psychology, ecology, and, after a long gestation, ethology. Above all, he altered the way we think about ourselves, the earth we inhabit, and the animals that share it with us.▪

CHAPTER 12

Animal Behavior

In 1973, the Nobel Prize in Physiology or Medicine was awarded to three pioneering zoologists, Karl von Frisch, Konrad Lorenz, and Niko Tinbergen (Figure 12-1). The citation stated that these three were the principal architects of the new science of ethology. It was the first time any contributor to the behavioral sciences was so honored, and it meant that the discipline of animal behavior, which really takes its roots from the work of Charles Darwin, had arrived.

Ethology, meaning literally "character study," was first used in the late eighteenth century to signify the interpretation of character through the study of gesture. It was an appropriate term for the new field of animal behavior, which had as its objective the study of motor patterns, that is, actions of animals, with the anticipation that such study would reveal the true characters of animals just as the interpretation of human gestures might reveal the true characters of people. This decidedly restricted interpretation of ethology as a purely descriptive study of the "habits" of animals was considerably modified during ethology's epoch-making period, 1935 to 1950. Today we may define ethology as a discipline that involves the study of the total repertoire of behavior, both simple and complex, that animals employ in their natural environment as they resolve the problems of survival and reproduction.

The aim of ethologists has been to describe the behavior of an animal in its *natural habitat.* Most ethologists have been naturalists. Their laboratory has been the out-of-doors, and early ethologists gathered their data by field observation. They also conducted experiments, often with nature providing the variables, but increasingly ethologists have manipulated the variables for their own purposes by using animal models, playing recordings of animal vocalizations, altering the habitat, and so on. Modern ethologists also conduct many experiments in the laboratory where they can test their predictions under closely controlled conditions. However, ethologists usually take pains to compare laboratory observations with observations of free-ranging animals in undisturbed natural environments. They recognize that it makes no more sense to try to study the natural behavior of an animal divorced from its natural surroundings than it does to try to interpret a structural adaptation of an animal apart from the function it serves.

A

B

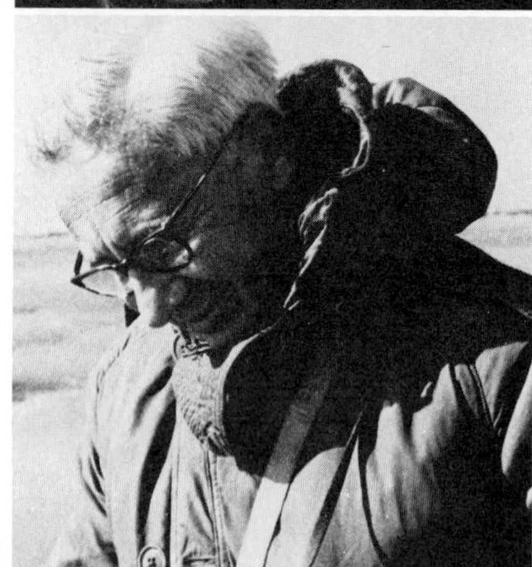
C

FIGURE 12-1

A, Konrad Lorenz. **B,** Karl von Frisch. **C,** Niko Tinbergen.

A, Photograph by Thomas McAvoy, Life Magazine © 1955 Time Inc.; **B,** photograph courtesy W.S. Hoar; **C,** photograph by Lary Shaffer.

With infinite patience von Frisch, Lorenz, Tinbergen, and their colleagues watched and cataloged the activities and vocalizations of animals during feeding, courtship, and nest building, as well as seemingly insignificant behavioral movements such as head scratching, stretching postures, and turning and shaking movements. Thus these studies concentrated largely on innate motor patterns used for communication within a species.

One of the great contributions of von Frisch, Lorenz, and Tinbergen was to demonstrate that behavioral traits are measurable entities like anatomical or physiological traits. This was to become the central theme of ethology: behavioral traits can be isolated and measured and they have evolutionary histories. They showed that behavior is not the wavering, transient, unpredictable phenomenon often depicted by earlier writers. In short, behavior is genetically mediated. It is apparent that, if behavior is determined by genes in the same way that genes determine morphological and physiological characters (ethologists marshaled abundant evidence showing that it is), then behavior evolves and is adaptive. Thus modern behavioral study is founded on the recognition that the Darwinian view of evolution holds for behavioral traits, as well as for anatomical and functional characters.

Principles of Ethology

The ethologists, through step-by-step analysis of the behavior of animals in nature, focused on the relatively invariant components of behavior. From such studies emerged several concepts that were first popularized in Tinbergen's influential book, *The Study of Instinct* (1951).

The basic concepts of ethology can be approached by considering the egg-rolling response of the greylag goose (Figure 12-2), described by Lorenz and Tinbergen in a famous paper published in 1938. If Lorenz and Tinbergen presented a female greylag goose with an egg a short distance from her nest, she would rise, extend her neck until the bill was just over the egg and then contract her neck, pulling the egg carefully into the nest.

Although this behavior appeared to be intelligent, Tinbergen and Lorenz noticed that if they removed the egg once the goose had begun her retrieval, or if the egg being retrieved slipped away and rolled down the outer slope of the nest, the goose would continue the retrieval movement without the egg until she was again settled comfortably on her nest. Then, seeing that the egg had not been retrieved, she would begin the egg-rolling pattern all over again.

Thus the bird performed the egg-rolling behavior as if it were a program which, once initiated, had to run to completion. Lorenz and Tinbergen termed this stereotyped behavior a **fixed action pattern** (FAP): a motor pattern that is mostly invariable in its performance. The goose did not have to learn the movement; it was a "prewired," or innate, skill.

Further experiments by Tinbergen disclosed that the greylag goose was not terribly discriminating about what she retrieved. Almost any smooth and rounded object placed outside the nest would trigger the egg-rolling behavior; even a small toy dog and a large yellow balloon were dutifully retrieved. But once the goose settled down on such objects, they obviously did not feel right and she discarded them.

Lorenz and Tinbergen realized that the presence of the egg outside the nest must act as a stimulus, or trigger, that released the fixed action pattern. Lorenz termed the triggering stimulus a **releaser:** a simple feature in the environment that would trigger a certain innate behavior. Or, because the animal usually responded to some specific aspect of the releaser (sound, shape, or color, for example) the effective stimulus was called a **sign stimulus.** Ethologists have described hundreds of examples of sign stimuli. In every case the response is highly predictable. For example, the alarm call of adult herring gulls always releases a crouching freeze response in the chicks. Or, to cite an example given in an earlier chapter (p. 254), certain nocturnal moths take evasive maneuvers or drop to the ground when they hear the ultrasonic cries of bats that feed on them; most other sounds do not release this response.

Sometimes exaggerated sign stimuli release an exaggerated response. Tinbergen found that an oyster catcher, offered a choice between its own egg and a giant egg four times as large, would attempt to incubate the giant egg while ignoring its own (Figure 12-3). Similarly, a greylag goose, allowed to choose between a goose egg and a volleyball near its nest, would invariably recover the volleyball. These examples of exceptionally effective signals are called **supernormal stimuli.** Obviously a goose has no need for a volleyball in its nest, but these experiments do give us some insight into the nature of sign stimuli: they are usually simple cues, a fact that reduces the chance that the signal might be misunderstood. Natural selection will favor those combinations of genes that boost the signal value (in this case, large size of the object), even though this may occasionally lead to inappropriate behavior.

FIGURE 12-2

Egg-rolling movement of the greylag goose (*Anser anser*).

From Lorenz, K., and N. Tinbergen. 1938. Zeit. Tierpsychol. **2**:1-29.

Lorenz and Tinbergen proposed that the nervous system of animals must have special filtering units, or "centers," capable of releasing a fixed action pattern when stimulated by the appropriate sign stimulus. They called this filter-trigger complex an **innate releasing mechanism** (IRM). The most important characteristic of the IRM is that it behaves like a programmed motor message that, once begun, produces a coordinated muscle performance without requiring any further sensory input. And the response is completely functional the first time it is performed (the animal "knows" how with no learning), once the animal is the right age and in the proper motivational state (only an adult *female* greylag goose *in a nest* will respond to an egg near the nest).

The IRM dramatically illustrates the stereotyped, predictable, and programmed nature of much animal behavior. This is even more evident when an IRM is released inappropriately. In the spring the male three-spined stickleback selects a territory that it defends vigorously against other males. The underside of the male becomes bright red, and the approach of another red-bellied male will release a threat posture or even an aggressive attack. Tinbergen's suspicion that the red belly of the male served as a releaser for aggression was reinforced when a passing red postal truck evoked attacks from the males in his aquarium. Tinbergen then carried out experiments using a series of models, which he presented to the males. He found that they vigorously attacked any model bearing a red stripe, even a plump lump of wax with a red underside. Yet a carefully made model that closely resembled a male stickleback but lacked the red belly was ignored (Figure 12-4). Tinbergen discovered other examples of IRMs released by simple sign stimuli. Male English robins furiously attacked a bundle of red feathers placed in their territory but ignored a stuffed juvenile robin without the red feathers (Figure 12-5).

We have seen in the examples above that there are costs to the releaser-IRM system because it may lead to improper responses. Fortunately for red-bellied sticklebacks and red-breasted English robins, their aggressive response toward red works appropriately most of the time because red objects are uncommon in the worlds of these animals. But why don't these and other animals simply *reason* out the correct response rather than relying on stereotyped responses? Under conditions that are relatively consistent and predictable, automatic preprogrammed responses may be most efficient. Even if they can or could, thinking about or learning the correct response may take too much time. Releasers have the advantage of focusing the animal's attention on the relevant signal, and IRMs enable an animal to respond rapidly when speed may be essential for survival.

FIGURE 12-3

An oyster catcher (*Haematopus ostralegus*) attempts to roll a giant egg model into its nest while ignoring its own egg.

From Tinbergen, N. 1951. The study of instinct. Oxford, Eng., Oxford University Press.

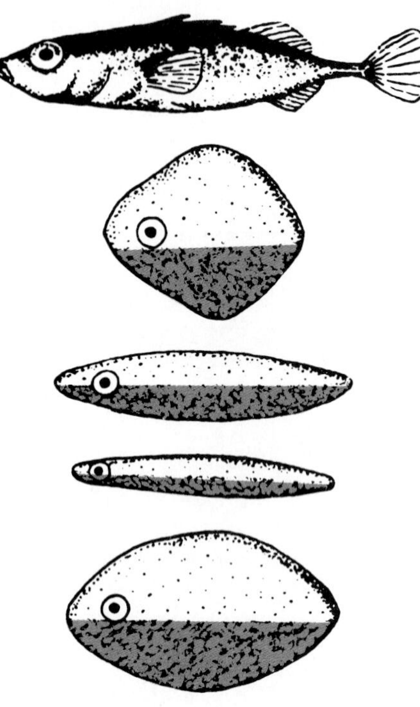

FIGURE 12-4

Stickleback models used to study stereotyped behavior. The carefully made model of the stickleback (*top*) without a red belly, is attacked much less frequently by a territorial male stickleback than the four simple red-bellied models.

From Tinbergen, N. 1951. The study of instinct. Oxford, Eng., Oxford University Press.

The concepts of releasers, IRMs, and FAPs, however, have been vigorously criticized by some neurophysiologists who were reluctant to believe that the IRM was a specific center of nerve cells (as Lorenz believed) designed to perform just one motor program (the FAP) in response to a specific sign stimulus. Instead, physiologists argued that the central nervous system operates as an integrated whole and that other neural processes also play important roles. But while the IRM concept has been modified somewhat in recent years, there is increasing evidence that IRMs do exist in a sense and that they generate FAPs. For example, in the large nudibranch *Tritonia* (a sluglike mollusc), a giant nerve cell has been discovered that, when stimulated, plays out a complete stereotyped escape-swimming response that is normally evoked when the animal comes in contact with its enemy, a sea star. This cell looks very much like the IRM envisioned by early ethologists. In vertebrates, executive neurons or brain "centers" for IRMs *may* not exist, but the brain works as though they do.

INSTINCT AND LEARNING

From the beginning, the mostly invariable and predictable (that is, stereotyped) nature of IRMs suggested to ethologists that they were dealing with inherited, or **innate,** behavior. Many kinds of stereotyped behavior appear suddenly in animals and are indistinguishable from similar behavior performed by older, experienced individuals. Orb-weaving spiders "know" how to build their webs without practice, and male crickets "know" how to court females without lessons from more experienced crickets or by learning from trial and error. To such behaviors the term **instinct,** meaning "driven from within," is applied. It is easy to understand why instinctive behavior is an important adaptation for survival, especially in lower forms that never know their parents. They must be equipped to respond to the world immediately and correctly as soon as they emerge into it. It is also evident that more advanced animals with longer lives and with parental care or other opportunities for social interactions may improve or change their behavior by learning.

Learning and the diversity of behavior

Learning can be defined as the modification of behavior through experience. We should add that, like innate behavior, behavioral modifications by learning are, as a rule, adaptive. But unlike innate patterns, which emerge completely functional the first time the animal performs them, learning requires a change in

FIGURE 12-5

Two models of the English robin. The bundle of red feathers is attacked by male robins, whereas the stuffed juvenile bird (*right*) without a red breast is ignored.

From Tinbergen, N. 1951. The study of instinct. Oxford, Eng., Oxford University Press; modified from Lack, D. 1943. The life of the robin. Cambridge, Eng., Cambridge University Press.

previously existing activity. For example, newly hatched gull chicks crouch down in response to moving objects overhead, an innate and clearly adaptive response to the danger from overflying predators. As the chick grows, it becomes more discriminating and begins to lose its general fear of overflying birds. Its loss of sensitivity to this particular stimulus is a simple kind of learning called **habituation.** For gull chicks, habituation to overflying objects is appropriate, since running and hiding from every passing shadow would consume time and energy better applied to more productive activities.

However, gull chicks do not come to ignore *all* birds flying overhead. Absolute habituation in this instance would be less appropriate than no habituation at all, since birds of prey are genuine predators of gull chicks. Chicks that fail to hide from an overflying hawk are less likely to survive than chicks that have retained a healthy fear of this predator.

Lorenz and Tinbergen discovered that chicks distinguish predatory birds from harmless songbirds and ducks on the basis of shape, especially the length of the head and neck. When chicks were presented with a silhouette having a hawklike shape and a short neck such that the head protruded only slightly in front of the wings, they crouched in alarm. Long-necked silhouettes were ignored or aroused only mild interest (Figure 12-6). In another experiment, a model was built having symmetrical wings, with the head and tail shaped so that either the front or the rear of the dummy could be regarded as the head or as the tail (Figure 12-7). When sailed in the direction of the long-necked head, the model resembled a goose and caused no alarm; when sailed in the direction of the short-necked model, however, it resembled a hawk and did cause alarm. Obviously both the shape of the bird and the direction of its motion are important in recognition.

Sometimes learning is demonstrated in surprising and amusing ways. Tinbergen describes one such incident*:

> In order to sail our models, which crossed a meadow where the birds were feeding or resting, at a height of about 10 yards along a wire, running from one tree to another 50 yards away, either Lorenz or I had to climb a tree and mount the dummy we wanted to test out. One family of geese (which also reacted to some of our dummies) very soon associated tree-climbing humans with something dreadful to come, and promptly called the alarm and walked off when one of us went up.

These observations with models might suggest incorrectly that gull chicks (as well as pheasant and turkey chicks that react similarly) have an innate ability to distinguish short-necked predators from harmless birds having longer necks. But, in fact, subsequent experiments demonstrated that newly hatched chicks, which are at first alarmed by anything that passes overhead—even a falling leaf—gradually become habituated to familiar objects. The chick learns that songbirds and shorebirds are common and harmless features of its world. But they never become accustomed to short-necked predators because these are seldom seen. It is the unfamiliar that arouses fear.

Learning depends on experience encountered by the organism as it interacts with its environment. It also depends on internal programming because the things an organism learns to do best are determined by the animal's genotype. The nervous system must be designed to facilitate the acquisition of learning at specific stages in the organism's development. In other words, through its genetically determined development, the brain possesses properties that prepare for its eventual use in the modification of behavior. Learned behavior, like instinctive behavior, contains both genetic and environmental components.

*Tinbergen, N. 1961. The herring gull's world: a study of the social behavior of birds. New York, Basic Books, Inc., Publishers.

FIGURE 12-6

Models used by Lorenz and Tinbergen for the study of predator reactions in young fowl. Young gull chicks crouch in alarm when hawk silhouettes (*red*) pass overhead but ignore the shapes of harmless birds (*black*).

From Tinbergen, N. 1951. The study of instinct. Oxford, Eng., Oxford University Press.

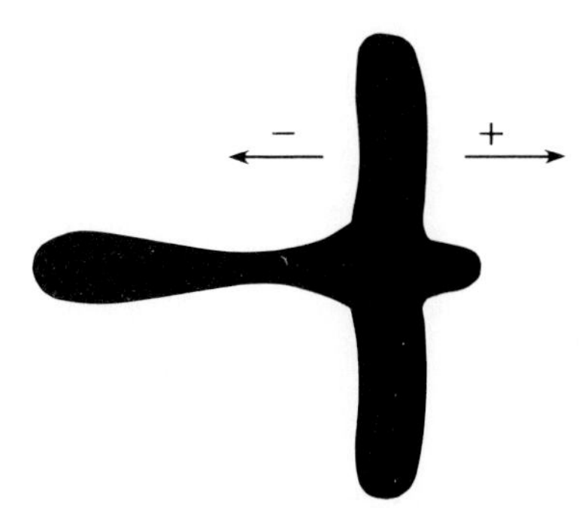

FIGURE 12-7

Model that drew positive responses from gull chicks when sailed to the right (in the direction of the short-necked model), but none when sailed to the left (in the direction of the long-necked model).

One kind of learned behavior that clearly illustrates the interaction of heredity and environment is **imprinting.** As soon as a newly hatched gosling or duckling is strong enough to walk, it follows its mother away from the nest. After it has followed the mother for some time, it follows no other animal (Figure 12-8). But, if the eggs are hatched in an incubator or if the mother is separated from the eggs as they hatch, the goslings follow the first large object they see. As they grow, the young geese prefer the artificial "mother" to anything else, including their true mother. The goslings are said to be imprinted on the artificial mother.

FIGURE 12-8

Canada goose, *Branta canadensis,* with her imprinted young.

Photograph by L.L. Rue III.

Imprinting was observed at least as early as the first century AD when the Roman naturalist Pliny the Elder wrote of "a goose which followed Lacydes as faithfully as a dog." Konrad Lorenz was the first to study the imprinting phenomenon objectively and systematically. When Lorenz hand-reared goslings they formed an immediate and permanent attachment to him and waddled after him wherever he went (Figure 12-1, *A*). They could no longer be induced to follow their own mother or another human being. Lorenz found that the imprinting period is confined to a brief *sensitive* period in the individual's early life and that once established the imprinted bond is mostly retained for life.

What imprinting shows is that the goose or duck brain (or the brain of numerous other birds and mammals that show imprinting-like behavior) is designed to accommodate the imprinting experience. The animal's genotype is provided with an internal plasticity that permits the animal to recognize its mother soon after hatching. Natural selection favors the evolution of animals having a brain structure that imprints in this way because following the mother and obeying her commands are important for survival. The fact that a gosling can be made to imprint to a mechanical toy duck or a human being under artificial conditions is a cost to the system that can be tolerated; the disadvantages of the system's simplicity are outweighed by the advantages of its reliability.

Let us cite one final example to complete our consideration of instinct and learning. The males of many species of birds have characteristic territorial songs that identify the singers to the other birds and announce territorial rights to other males of that species. Like many other songbirds, the male white-crowned sparrow must learn the song of its species by hearing the song of its father. If the sparrow is hand-reared in acoustic isolation in the laboratory, it develops an abnormal song (Figure 12-9). But if the isolated bird is allowed to hear recordings of normal white-crowned sparrow songs during a critical period of 10 to 50 days

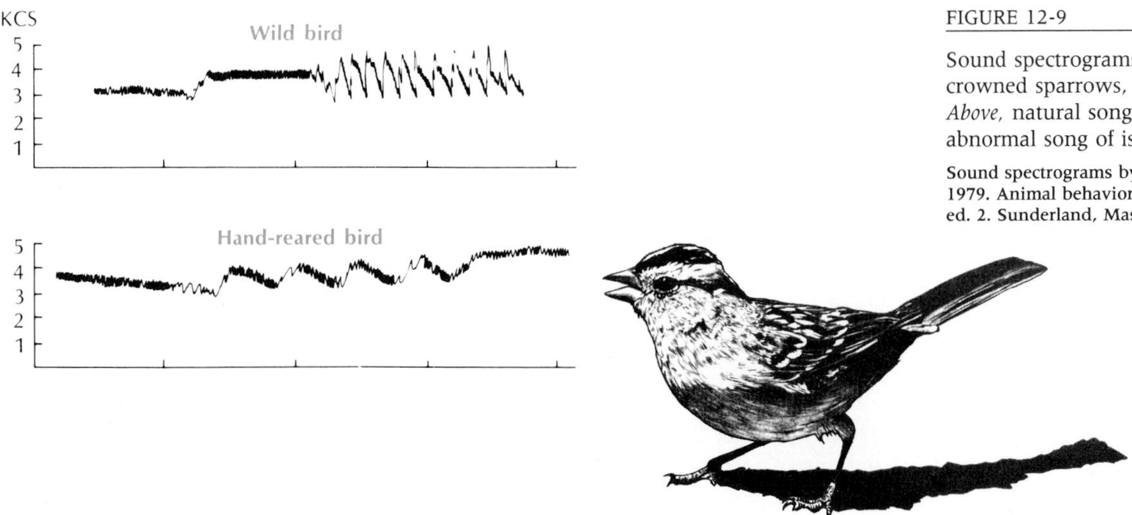

FIGURE 12-9

Sound spectrograms of songs of white-crowned sparrows, *Zonotrichia leucophrys. Above,* natural songs of wild bird; *below,* abnormal song of isolated bird.

Sound spectrograms by M. Konishi from Alcock, J. 1979. Animal behavior: an evolutionary approach, ed. 2. Sunderland, Mass., Sinauer Associates, Inc.

after hatching, it learns to sing normally. It even imitates the local dialect it hears.

It might appear from this that song characteristic is determined by learning alone. However, if during the critical learning period, the isolated male white-crowned sparrow is played a recording of another sparrow species, even a closely related one, it does not learn the song. It learns only the song appropriate to its own species. Thus, although the song must be learned, the brain has been programmed in its development to recognize and learn vocalizations produced by males of its species alone. The sparrow *learns* by example but its attention has been *innately* narrowed to focus on the appropriate example. Learning the wrong song would result in behavioral chaos, and natural selection quickly eliminates those genotypes that permit such mistakes to occur.

Social Behavior

When we think of "social" animals we are likely to think of highly structured honeybee colonies, herds of antelope grazing on the African plains (Figure 12-10), schools of herring, or flocks of starlings. But social behavior of animals *of the same species* living together is by no means limited to such obvious examples in which individuals influence one another.

In the broad sense, any kind of interaction resulting from the response of one animal to another of the same species represents social behavior. Even a pair of rival males squaring off for a fight over the possession of a female is a social interaction, although our perceptual bias as people might encourage us to label it antisocial. Thus social aggregations are only one kind of social behavior, and indeed not all animal aggregations are social.

Clouds of moths attracted to a light at night, barnacles attracted to a common float, or trout gathering in the coolest pool of a stream are animal groupings responding to *environmental* signals. Social aggregations, on the other hand, depend on *animal* signals. They remain together and do things together by influencing one another.

Of course, not all animals showing sociality are social to the same degree. All sexually reproducing species must at least cooperate enough to achieve fertilization; but among some animals, breeding is about the only adult sociality to occur. Alternatively, swans, geese, albatrosses, and beavers, to name just a few, form strong monogamous bonds that last a lifetime. Whether or not adult sociality is strongly or weakly developed, the most persistent social bonds usually form

FIGURE 12-10

Mixed herd of topi and common zebra grazing on the savannah of tropical Africa.

Photograph by C.P. Hickman, Jr.

between mothers and their young, and these bonds for birds and mammals usually terminate at fledging or weaning.

ADVANTAGES OF SOCIALITY

Living together may be beneficial in many ways. Each species profits in its own particular way; what confers adaptive value to one species may not for another. One obvious benefit for social aggregations is defense, both passive and active, from predators. Musk-oxen that form a passive defensive circle when threatened by a wolf pack are much less vulnerable than an individual facing the wolves alone.

FIGURE 12-11

An infant yellow baboon (*Papio cyanocephalus*) "jockey rides" its mother. Later, as the infant is weaned, the mother-infant bond weakens and the infant will be refused rides.

Photograph by C.P. Hickman, Jr.

As an example of active defense, a breeding colony of gulls, alerted by the alarm calls of a few, attack predators *en masse;* this is certain to discourage a predator more effectively than individual attacks. The members of a prairie dog town, although divided into social units called coteries, cooperate by warning each other with a special bark when danger threatens. Thus every individual in a social organization benefits from the eyes, ears, and noses of all other members of the group.

Sociality offers several benefits to animal reproduction. It facilitates encounters between males and females which, for solitary animals, may consume much time and energy. Sociality also helps synchronize reproductive behavior through the mutual stimulation that individuals have on one another. Among colonial birds the sounds and displays of courting individuals set in motion prereproductive endocrine changes in other individuals. Because there is more social stimulation, large colonies of gulls produce more young per nest than do small colonies. Furthermore, the parental care that social animals provide their offspring increases survival of the brood (Figure 12-11). Social living provides opportunities for individuals to give aid and to share food with young other than their own. Such interactions within a social network have resulted in some intricate cooperative behavior among parents, their young, and their kin.

Of the many other advantages of social organization noted by ethologists, we will mention only a few in this brief treatment: cooperation in hunting for food; huddling for mutual protection from severe weather; opportunities for division of labor, which is especially well developed in the social insects with their caste systems; and the potential for learning and transmitting useful information through the society. This last advantage of social organization is discussed in the following example.

On the Galápagos Islands, hunting by humans in the last century had so greatly thinned the giant tortoise population on one island that the few surviving males and females seldom, if ever, met. Lichens grew on the females' backs because there were no males to scrub them off during mating! Research personnel saved the tortoise from inevitable extinction by collecting them together in a pen, where they began to reproduce.

Observers of a seminatural colony of macaque monkeys in Japan recount an interesting example of acquiring and passing tradition in a society. The macaques were provisioned with sweet potatoes and wheat at a feeding station on the beach of an island colony. One day a young female named Imo was observed washing the sand off a sweet potato in seawater. The behavior was quickly imitated by Imo's playmates and later by Imo's mother. Still later when the young members of the troop became mothers they waded into the sea to wash their potatoes; their offspring imitated them without hestiation. The tradition was firmly established in the troop (Figure 12-12).

Some years later, Imo, an adult, discovered that she could separate wheat from sand by tossing a handful of sandy wheat in the water; allowing the sand to sink, she could scoop up the floating wheat to eat. Again, within a few years, wheat-sifting became a tradition in the troop.

Imo's peers and social inferiors copied her innovations most readily. The adult males, her superiors in the social hierarchy, would not adopt the practice but continued laboriously to pick wet sand grains off their sweet potatoes and scout the beach for single grains of wheat.

If social living offers so many benefits, why haven't all animals become social through natural selection? The answer is that a solitary existence offers its own set of advantages. In the diverse array of ecological situations in nature, species extract their own optimal ways of life. Species that survive by camouflage from potential predators profit by being well spaced out. Large predators benefit from a solitary existence for a different reason, their requirement for a large supply of prey. Thus there is no overriding adaptive advantage to sociality that inevitably selects against the solitary way of life. It depends on the ecological situation.

FIGURE 12-12

Japanese macaque washing sweet potatoes. The tradition began when a young female named Imo began washing sand from the potatoes before eating them. Younger members of the troop quickly imitated the behavior.

AGGRESSION AND DOMINANCE

Many animal species are social because of the numerous benefits that sociality offers. This requires cooperation. At the same time animals, like governments, tend to look out for their own interests. In short, they are in competition with one another because of limitations in the common resources that all require for life. Animals may compete for food, water, sexual mates, or shelter, when such requirements are limited in quantity and are therefore worth fighting over.

Much of what animals do to resolve competition is called **aggression,** which we may define as an offensive physical action, or threat, to force others to abandon something they own or might attain. Many ethologists consider aggression to be part of a somewhat more inclusive interaction called **agonistic** (Gr., contest) **behavior,** referring to any activity related to fighting, whether it be aggression, defense, submission, or retreat.

Contrary to the widely held notion that aggressive behavior aims at the destruction or at least defeat of an opponent, most aggressive encounters are ritualized duels that lack the atmosphere of violence that we usually associate with fighting. Many species possess specialized weapons such as teeth, beaks, claws, or horns that are used for protection from, or predation on, other species. Although potentially dangerous, such weapons are seldom used in any effective way against members *of their own species.*

Animal aggression within the species seldom results in injury or death because animals have evolved many symbolic aggressive displays that carry mutually understood meanings. Fights over mates, food, or territory become ritualized jousts rather than bloody, no-holds-barred battles. When fiddler crabs spar for territory, their large claws usually are only slightly opened. Even in the most intense fighting when the claws are used, the crabs grasp each other in a way that prevents reciprocal injury. Rival male poisonous snakes engage in stylized bouts by winding themselves together; each attempts to butt the other's head with its own until one becomes so fatigued that it retreats. The rivals never bite each other. Many species of fish contest territorial boundaries with lateral display threats, the males puffing themselves up to look as threatening as possible. The encounter is usually settled when either animal perceives itself to be obviously inferior, folds up its fins, and swims off. Rival giraffes engage in largely symbolic "necking" matches in which two males standing side by side wrap and unwrap their necks around each other. (Figure 12-13). Neither uses its potentially lethal hooves on the other, and neither is injured.

FIGURE 12-13

Male Masai giraffes, *Giraffa camelopardalis,* fight for social dominance. Such fights are largely symbolic, seldom resulting in injury. **Photograph by L.L. Rue III.**

Thus animals fight as though programmed by rules that prevent serious injury. Fights between rival bighorn rams are spectacular to watch, and the sound of clashing horns may be heard for hundreds of meters (Figure 12-14). But the skull is so well protected by the massive horns that injury occurs only by accident. Nevertheless, despite these constraints, aggressive encounters can on occasion be true fights to the death. If African male elephants are unable to resolve dominance conflicts painlessly with ritual postures, they may resort to incredibly violent battles, with each trying to plunge its tusks into the most vulnerable parts of the opponent's body.

More commonly, however, the loser of a ritualized encounter may simply run away, or signal defeat by a specialized subordination ritual. If it becomes evident to him that he is going to lose anyway, he is better off communicating his sub-

FIGURE 12-14

Male bighorn sheep, *Ovis canadensis,* fight for social dominance during the breeding season.

Photograph by L.L. Rue III.

mission as quickly as possible and avoiding the cost of a real thrashing. Such submissive displays that signal the end of a fight may be almost the opposite of threat displays (Figure 12-15). In his book *Expression of the Emotions in Man and Animals* (1872), Charles Darwin described the seemingly opposite nature of threat and appeasement displays as the "principle of antithesis." The principle remains accepted by ethologists today.

Why doesn't the victor of an aggressive contest kill its opponent? A defeated wolf or dog presents its vulnerable neck to the victor as a sign of complete submission. Although the dominant wolf could easily kill the defeated foe and thus remove a competitor, it never does so. The display of submission has effectively inhibited further aggression by the winner. The best explanation for aggressive restraint is that the winner has little to gain by continuing the fight. His superiority is already assured. By continuing the aggression he merely endangers himself, since a defeated opponent fighting for his life might inflict a wound. It is not difficult to see how natural selection would favor genes that induce aggressive restraint. Aggression that is inappropriate runs counter to the maximization of individual fitness. It is maladaptive and consequently is selected against.

The winner of an aggressive competition is dominant to the loser, the subordinant. For the victor, dominance means enhanced access to all the contested resources that contribute to reproductive success: food, mates, territory, and so on. In a social species, dominance interactions often take the form of a dominance hierarchy. One animal at the top wins enounters with all other members in the social group; the second in rank wins all but those with the top-ranking individual.

Such a simple, linear hierarchy was first observed in chicken societies by Schjelderup-Ebbe, who called the hierarchy a "peck-order." Once social ranking is established, actual pecking diminishes and is replaced by threats, bluffs, and bows. Top hens and cocks get unquestioned access to feed and water, dusting areas, and the roost. The system works because it reduces the social tensions that would constantly surface if animals had to fight all the time over social position.

Still, life for the subordinates in any social order is likely to be hard. They are the expendables of the social group. They almost never get a chance to reproduce, and when times get difficult they are the first to die. During times of food scarcity, the death of the weaker members helps to protect the resource for the stronger members. Rather than sharing food, the population excess is sacrificed. This is not viewed by contemporary behaviorists as resulting from some direct, purposeful "good for the species" process, however; rather it results as a *consequence* of the individual advantage that the stronger, dominant individuals possess during such circumstances. It is still a matter of survival of the fittest.

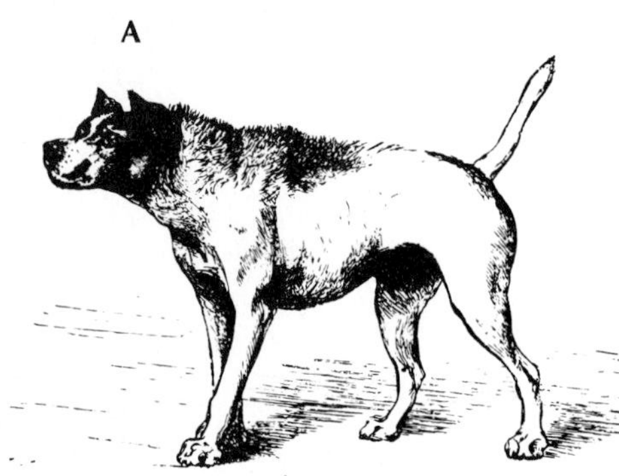

FIGURE 12-15

Darwin's principle of antithesis as exemplified by the postures of dogs. **A,** A dog approaches another dog with hostile, aggressive intentions. **B,** The same dog is in a humble and conciliatory state of mind. The signals of aggressive display have been reversed.

From Darwin, C. 1872. Expression of the emotions in man and animals. New York, D. Appleton & Co.

TERRITORIALITY

Territorial ownership is another facet of sociality in animal populations. A **territory** is a fixed area from which intruders of the same species are excluded. This involves defending the area from intruders and spending long periods of time on the site being conspicuous. Territorial defense has been observed in numerous animals: insects, crustaceans, other invertebrates, fish, amphibians, lizards, birds, and mammals, including humans.

Territoriality is generally an alternative to dominance behavior, although both systems may be observed operating in the same species. A territorial system may work well when the population is low, but break down with increasing population density to be replaced with dominance hierarchies with all animals occupying the same space.

Like every other competitive endeavor, territoriality carries both costs and advantages. It is beneficial when it ensures access to limited resources, *unless* the territory boundaries cannot be maintained with little effort. The presumed benefits of a territory are, in fact, numerous: uncontested access to a foraging area; enhanced attractiveness to females, thus reducing the problems of pair-bonding, mating, and rearing the young; reduced disease transmission; reduced vulnerability to predators. But the advantages of holding a territory begin to wane if the individual must spend most of the time in boundary disputes with neighbors.

Sometimes the space defended moves with the individual. This individual distance, as it is called, can be observed as the spacing between swallows or pigeons on a wire, in gulls lined up on the beach, or in people lined up for a bus.

Most of the time and energy required for territoriality are expended when the territory is first established. Once the boundaries are located they tend to be respected, and aggressive behavior diminishes as territorial neighbors come to recognize each other. Indeed, neighbors may look so peaceful that an observer who was not present when the territories were established may conclude (incorrectly) that the animals are not territorial. A "beachmaster" sea lion (that is, a dominant male with a harem) seldom quarrels with his neighbors who have their own territories to defend. However, he must be on constant vigilance against bachelor bulls who challenge the beachmaster for harem privileges.

Of all vertebrate classes, birds are the most conspicuously territorial. Most male songbirds establish territories in the early spring and defend these vigorously against all males of the same species during spring and summer when mating and nesting are at their height. A male song sparrow, for example, has a territory of approximately three fourths of an acre. In any given area, the number of song sparrows remains approximately the same year after year. The population remains stable because the young occupy territories of adults that die or are killed. Any surplus in the song sparrow population is excluded from territories and thus not able to mate or nest.

Sea birds such as gulls, gannets, boobies, and albatrosses occupy colonies that are divided into very small territories just large enough for nesting (Figure 12-16). These birds' territories cannot include their fishing grounds, since they all forage in the sea where the food is always shifting in location and shared by all.

Territorial behavior is not as prominent with mammals as it is with birds. Mammals are less mobile than birds, and this makes it more difficult for them to patrol a territory for trespassers. Instead, many mammals have **home ranges.** A home range is the total area an individual traverses in its activities. It is not an exclusive defended preserve but overlaps with the home ranges of other individuals of the same species.

For example, the home ranges of baboon troops overlap extensively, although a small part of each range becomes the recognized territory of each troop for its exclusive use. Home ranges may shift considerably with the seasons. A baboon troop may have to shift to a new range during the dry season to obtain water and better grass. Elephants, before their movements were restricted by humans, made long seasonal migrations across the African savannah to new feeding ranges. However, the home ranges established for each season are remarkably consistent in size.

FIGURE 12-16

Gannett nesting colony. Note precise spacing of nests with each occupant just beyond pecking distance of its neighbors.

Photograph by C.P. Hickman, Jr.

ANIMAL COMMUNICATION

Social animals, including people, must be able to communicate with each other. Only through communication can one animal influence the behavior of another. Compared to the enormous communicative potential of human speech, however, non-human communication is severely restricted. Whereas human communication is based mainly, although by no means exclusively, on sounds, animals may communicate by sounds, scents, touch, and movement. Indeed any sensory channel may be used, and in this sense animal communication has richness and variety.

Unlike our language, which is composed of words with definite meanings that may be rearranged to generate an almost infinite array of new meanings and images, communication of other animals consists of a limited repertoire of signals. Typically, each signal conveys one and only one message. These messages cannot be divided or rearranged to construct *new kinds* of information. A single message from the sender may, however, contain several bits of relevant information for the receiver.

The song of a cricket announces to an unfertilized female the species of the sender (males of different species have different songs), his sex (only males sing), his location (source of the song), and social status (only a male able to defend the area around his burrow sings from one location). This is all crucial information to the female and accomplishes a biological function. But there is no way for the male to alter his song to provide additional information concerning food, predators, or habitat, which might improve his mate's chances of survival and thus enhance his own fitness.

The limitations of communication are especially evident in the invertebrates and lower vertebrates. Signals are characteristically stereotyped, and the responses highly predictable and constant throughout the species. This does not mean that such communication is always lacking in intensity and versatility, however. Of the two contrasting examples that follow, mate attraction in silkworm moths illustrates an extreme case of stereotyped, single-message communication that has evolved to serve a single biological function: mating. Yet, in the same group, the insects, we find one of the most sophisticated and complex of all nonhuman communication systems, the symbolic language of bees.

Chemical sex attraction in moths

Virgin female silkworm moths have special glands that produce a chemical sex attractant to which the males are sensitive. Adult males smell with their large bushy antennae, covered with thousands of sensory hairs that function as receptors. Most of these receptors are sensitive to the chemical attractant (a complex alcohol called bombykol, from the name of the silkworm *Bombyx mori*) and to nothing else.

To attract the male, the female merely sits quietly and emits a minute amount of bombykol, which is carried downwind. When a few molecules reach the male's antennae, he is stimulated to fly upwind in search of the female. His search is at first random, but, when by chance he approaches within a few hundred yards of the female, he encounters a concentration gradient of the attractant. Guided by the gradient, he flies toward the female, finds her, and copulates with her.

In this example of chemical communication the attractant, referred to as a pheromone, bombykol, serves as a signal to bring the sexes together. Its effectiveness is assured because natural selection favors the evolution of males with antennal receptors sensitive enough to detect the attractant at great distances (several miles). Males with a genotype that produces a less sensitive sensory system fail to locate a female and thus are reproductively eliminated from the population.

The significance of the bee dances was discovered in the 1920s by the German zoologist Karl von Frisch, one of the recipients of the 1973 Nobel Prize. Despite detailed and extensive experiments by von Frisch that supported his original interpretations of the bee dances, the experiments have been criticized, especially by the American biologist A. Wenner, who suggested that the correlation between dance symbolism and food location is accidental. He argued that foraging bees bring back odors characteristic of the food source, and that recruits are stimulated by dance to go search for flowers bearing those odors. Few biologists were prepared to accept this interpretation, but Wenner's beneficial skepticism stimulated more rigorously controlled experiments by J.L. Gould, which established more conclusively than ever before that the bee dances communicated both distance and direction information. The dances of bees are among the true wonders of the natural world.

Language of the bees

Honeybees are able to communicate the location of food resources when these sources are too distant to be located easily by individual bees. Communication is done by dances, which are mainly of two forms. The form having the most communicative richness is the **waggle dance** (Figure 12-17). Bees most commonly execute these dances when a forager has returned from the rich source, carrying either nectar in her stomach or pollen grains packed in basketlike spaces formed by hairs on her legs. The waggle dance is roughly in the pattern of a figure-of-eight made against the vertical surface on the comb inside the hive. One cycle of the dance consists of three components: (1) a circle with a diameter about three times the length of the bee, (2) a straight run while waggling the abdomen from side to side, and (3) another circle, turning in the opposite direction from the first. This dance is repeated many times with the circling alternating clockwise and counterclockwise.

The straight, waggle run is the important information component of the dance. Waggle dances are performed almost always in clear weather, and the direction of the straight run is related to the position of the sun. If the forager has located food directly toward the sun, she will make her waggle run straight upward over the vertical surface of the comb. If food was located 60 degrees to the right of the sun, her waggle run is 60 degrees to the right of vertical. We see then that the waggle run points at the same angle relative to the vertical as the food is located relative to the sun.

Distance information about the food source is also coded into bee dances. If the food is close to the hive (less than 50 m), the forager employs a simpler dance called the **round dance.** The forager simply turns a complete clockwise circle, then turns, and completes a counterclockwise circle, a performance that is repeated many times. Other workers cluster around the scout and become stimulated by the dance as well as by the odor of nectar and pollen grains from flowers she has visited. The recruits then fly out and search in all directions but do not stray far. The round dance carries the message that food is to be found in the vicinity of the hive.

If the food source is farther away, the round dances become waggle dances, which provide both distance and direction information. The tempo of the waggle dance is inversely related to the food distance. If the food is about 100 m away, each figure-of-eight cycle lasts about 1.25 seconds; if 1000 m away, it lasts about 3 seconds; and if about 8 km (5 miles) away, it lasts 8 seconds. When food is plentiful, the bees may not dance at all. But when food is scarce, the dancing becomes intense, and the other workers cluster about the returning scouts and follow them through the dance patterns.

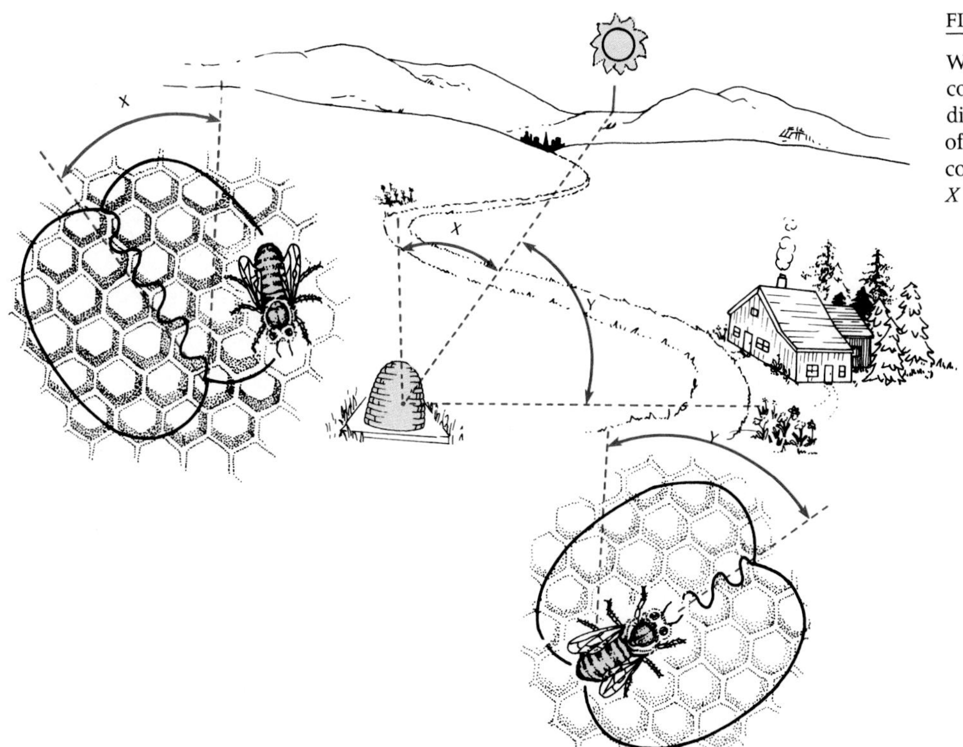

FIGURE 12-17

Waggle dance of the honeybee used to communicate both the direction and distance of a food source. The straight run of the waggle dance indicates direction according to the position of the sun (angles *X* and *Y*).

Communication by displays

A display is a kind of behavior or series of behaviors that serves a communicative purpose. The release of sex attractant by the female moth and the dances of bees just described are examples of displays; so are the alarm calls of herring gulls, song of the white-crowned sparrow, courtship dance of the sage grouse, and "eyespots" on the hind wings of certain moths that are quickly exposed to startle potential predators. Of course, just about anything an animal does communicates *something* to other animals that see, hear, or smell it. A true display, on the other hand, is a behavior pattern that has been modified through evolution to make it increasingly effective in serving a communicative function. This process is called **ritualization.** Through ritualization, simple movements or traits become more intensive, conspicuous, or precise, and their original undifferentiated function acquires signal value. The result of such intensification is to reduce the possibility of misunderstanding.

The elaborate pair-bonding behavior of the blue-footed boobies (Figure 12-18) exemplifies this point. These displays are performed with maximum intensity when the birds come together after a period of separation. The male at right in the illustration is sky pointing: the head and tail are pointed skyward and the wings are swiveled forward in a seemingly impossible position to display their glossy upper surfaces to the female. This is accompanied by a high, piping whistle. The female at left, for her part, is parading. She goose steps with exaggerated slow deliberation, lifting each brilliant blue foot in turn, as if holding it aloft momentarily for the male to admire. Such highly personalized displays, performed with droll solemnity, appear comical, even inane to the observer. Indeed the boobies, whose name is derived from the Spanish word "bobo" meaning clown, presumably were so designated for their amusing antics.

Needless to say, for the birds, amusement plays no part in the ceremonies. The exaggerated nature of the displays ensures that the message is not missed or misunderstood. Such displays are essential to establish and maintain a strong pair bond between male and female. This requirement also explains the repetitious nature of the displays that follow one another throughout courtship and until egg laying. Redundancy of displays maintains a state of mutual stimulation between male and female, ensuring the degree of cooperation necessary for copulation and subsequent incubation and care of the young. A sexually aroused male has little success with an indifferent female.

FIGURE 12-18

A pair of Galápagos blue-footed boobies, *Sula nebouxii*, displays to each other. The male (*right*) is sky pointing; the female (*left*) is parading. Such vivid, stereotyped, communicative displays serve to maintain reciprocal stimulation and cooperative behavior during courtship, mating, nesting, and care of the young.

Communication between humans and other animals

One uncertainty in studies of animal communication is understanding what sensory channel an animal is using. The signals may be visual displays, odors, vocalizations, tactile vibrations, or electrical currents (as, for example, among certain fishes). Even more difficult is establishing two-way communication between animals and humans, since the investigator must translate meanings into symbols that the animal can understand. Furthermore, people are poor social partners for most other animals. However, in the 1970s a female chimpanzee named Washoe was taught to employ *and combine* gestures, using words from the American Sign Language for the deaf, much as people use spoken words. The discovery that manual gestures and expressive motions were much more appropriate than vocalizations in communicating with apes was considered a major breakthrough in behavioral research. Since Washoe, sign-language studies have been extended to other chimpanzees and to gorillas; several have acquired "vocabularies" of several hundred reliable signs, some invented by the apes themselves.

Whether apes are able to combine signals (that is, "words") to produce sentences—an ability that is crucial to true language—cannot yet be answered conclusively. Furthermore, although apes use signals for requests, commands, and pleas, they do not appear able to make statements aimed at changing the behavior or beliefs of the other party. Critics of the first ape-language studies charged that most of the signed utterances of apes were prompted by the teacher. They argued that the investigators had fallen victim to the "Clever Hans" error, after a horse prodigy of the turn of the century that appeared to do sums in his head and tap out answers with his hoof, but was in fact responding to subtle and inadvertent cues from the trainer. More recent studies appear to have dealt satisfactorily with this criticism by isolating the ape from the teacher during the training sessions.

The animal behaviorist Irven De Vore has reported how choosing the proper channel for dialogue can have more than academic interest*:

> One day on the savanna I was away from my truck watching a baboon troop when a young juvenile came and picked up my binoculars. I knew if the glasses disappeared into the troop they'd be lost, so I grabbed them back. The juvenile screamed. Immediately every adult male in the troop rushed at me—I realized what a cornered leopard must feel like. The truck was 30 or 40 feet away. I had to face the males. I started smacking my lips very loudly, a gesture that says as strongly as a baboon can, "I mean you no harm." The males came charging up, growling, snarling, showing their teeth. Right in front of me they halted, cocked their heads to one side—and started lip-smacking back to me. They lip-smacked. I lip-smacked, "I mean you no harm." "I mean *you* no harm." It was, in retrospect, a marvelous conversation. But while my lips talked baboon, my feet edged toward the truck until I could leap inside and close the door.

We alluded earlier in this chapter to the stereotyped, mechanical nature of instinctive behavior. Yet in watching the interactions between mates during courtship, nest building, and care of the young, it is difficult for the observer to avoid anthropomorphic interpretations of behavior in terms of human behavior by using words such as "love," "deceit," "happiness," and "gentleness." This is not necessarily false, especially for the higher primates, but the ethologist must always take care to interpret every animal response by the simplest mechanism that is known to work.

Animals are capable of highly organized behavior in the absence of any intelligent appreciation of its purpose. Sometimes instinctive behavior misfires, and such incidents often emphasize its stereotyped nature. The following excerpt contains a perfect example of the automatic release of inappropriate behavior in a gannet colony†:

> A male of an old pair flew into his nest. Normally, he would bite his mate on the head with some violence and then go through a long and complicated meeting ceremony, an ecstatic display confined to members of a pair. Unfortunately, the female had caught her lower mandible in a loop of fish netting that was firmly anchored in the structure of the nest. Every time she tried to raise her head to perform the meeting ceremony with her mate she merely succeeded in opening her upper mandible whilst the lower remained fixed in the netting. So she apparently

*Irven De Vore in The marvels of animal behavior. 1972. Washington, D.C., National Geographic Society.

†Nelson, B. 1968. Galápagos: islands of birds. London, Longmans, Green & Co., Ltd. Reprinted by permission of William Morrow & Company.

> threatened the male with widely gaping beak and he immediately responded by attacking her. With each attack she lowered and turned away her bill (the way in which a female gannet appeases an aggressive male). At once the male stopped biting her and she again turned to greet him but simply repeated the beak-opening and drew another attack. And so it went on despite the fact that these two birds had been mated for years and that the netting, the cause of all the trouble, was clearly visible.

Despite such limitations to the adaptiveness of instinctive behavior, it obviously functions beautifully most of the time. In recognizing that reasoning and insight are not required for effective, highly organized behavior, we should not conclude that lower animals are, as Descartes proclaimed in the seventeenth century, nothing more than machines. Although the gannet in this example lacked the "intelligence" to free his mate by purposefully disentangling her beak from the netting, he was capable of appropriately analyzing the thousands of strategic choices he must make during his lifetime: how to find and hold a mate, where to build a nest and how to defend it, how to locate and catch evasive marine food, and what to do when the environment changes. All this and more requires endless behavioral adjustments to new situations. Conceivably this might be accomplished by a machine, but only by one of staggering complexity.

Summary

Ethology is the study of the behavior, both innate and learned, of animals in their natural habitat. Ethologists have shown that behavioral traits are genetically determined and thus are adaptive and evolve by natural selection.

Much innate (instinctive or inherited) behavior of animals is stereotyped; that is, it is highly predictable and invariable in performance. Ethologists have observed and cataloged numerous stereotyped motor acts, called fixed action patterns, which are triggered, or "released," by specific, and usually simple, environmental stimuli, called sign stimuli. Stereotyped behavior is triggered and coordinated through a "prewired" neural circuit classically called an innate releasing mechanism.

Innate behavior may be modified by learning through experience. A simple kind of learning behavior is habituation, which is the reduction or elimination of a behavioral response in the absence of any reward or punishment. The modification of the alarm response of herring gull chicks is described as an example of habituation. Another form of learning is imprinting, the lasting recognition bond that forms early in life between the young of many social animals and their mothers.

Two male hippopotamuses, *Hippopotamus amphibius,* in territorial battle.
Photograph by C.P. Hickman, Jr.

Social behavior is the behavior of a species when the members interact with one another. In social organizations, animals tend to remain together, communicate with each other, and usually resist intrusions by "outsiders." The advantages of sociality include cooperative defense from predators, improved reproductive performance and parental care of the young, cooperative searching for food, and transmission of useful information through the society. Because social animals compete with one another for common resources (such as food, sexual mates, and shelter), conflicts are often resolved by a form of overt hostility called aggression. Most aggressive encounters between conspecifics are stylized bouts involving more bluff than intent to injure or kill. Dominance hierarchies, in which a priority of access to common resources is established by aggression, is common in social organizations. Territoriality is a related alternative to dominance. A territory is a defended area from which intruders of the same species are excluded.

Communication, often considered the essence of social organization, is the means by which animals influence the behavior of other animals, using sounds, scents, visual displays, touch, or other sensory signals. As compared to the richness of human language, animals communicate with a very limited repertoire of signals. One of the most famous examples of animal communication is that of the symbolic dances of honeybees. Birds communicate by calls and songs and, especially, by visual displays. By ritualization, simple movements have evolved into conspicuous signals having definite meanings.

REVIEW QUESTIONS

1. Define the term ethology as it is used today, and comment on the aims of, and methods employed by, ethologists.
2. The egg-rolling behavior of greylag geese is an excellent example of stereotyped behavior. Interpret this behavior within the framework of classical ethology, using these terms: releaser, sign stimulus, innate releasing mechanism, and fixed action pattern. Interpret the territorial defense behavior of male three-spined sticklebacks in the same context.
3. Using silhouettes of birds, Lorenz and Tinbergen found that the perception of "hawkness" by gull chicks was mostly a matter of relative neck and tail length. Short-necked silhouettes evoked alarm, whereas long-necked silhouettes were ignored. These findings could be interpreted to mean that gull chicks enter the world with an innate fear of hawks. What is the evidence against this interpretation?
4. Two kinds of simple learning are habituation and imprinting. Distinguish between these two types of learning, and offer an example of each from the living world.
5. The idea that behavior must be *either* innate or learned has been called the "nature versus nurture" controversy. Cite evidence that such a strict dichotomy does not exist. Comment on the role of an animal's genetic instructions in determining innate and learned behavior.
6. Discuss the advantages of sociality for animals. If social living has so many advantages, who do many animals live alone successfully?
7. Suggest why aggression, which might seem to be a counterproductive form of behavior, exists among social animals. Do you think aggression might be more characteristic of K-selected or r-selected species (refer to Chapter 5)?
8. What is the selective advantage, to the probable winner as well as the loser, of ritualized aggression over unrestrained fighting?
9. Of what use is a territory to an animal, and how is a territory established and kept? What is the difference between territory and home range?
10. Comment on the limitations of animal communication as compared to that of human communication.
11. The dance language used by returning forager honeybees to specify the location of food is a remarkable example of complex communication among "simple" animals. How is direction and distance information coded into the waggle dance of the bees?
12. What is meant by "ritualization" in display communication? What is the adaptive significance of ritualization?
13. Early efforts by humans to communicate vocally with chimpanzees were almost total failures. Recently, however, researchers have learned how to communicate successfully with apes. How was this done?

SELECTED REFERENCES

Alcock, J. 1989. Animal behavior: an evolutionary approach, ed. 4. Sunderland, Mass., Sinauer Associates, Inc. *Clearly written and well-illustrated discussion of the genetics, physiology, ecology, and history of behavior in an evolutionary perspective.*

Drickamer, L.C., and S.H. Vessey. 1986. Animal behavior: concepts, processes, and methods, ed. 2. Belmont, Calif., Wadsworth Publishing Company. *Comprehensive, with helpful discussions on the methods and experimental approaches used to answer behavioral questions.*

Eaton, G.G. 1976. The social order of Japanese macaques. Sci. Am. **235**:96-106 (Oct.). *Long-term observations of a troop of macaques.*

Gould, J.L. 1982. Ethology: the mechanisms and evolution of behavior. New York, W.W. Norton & Co. *This is an excellent introduction to classical and modern ethology.*

Grier, J.W. 1984. Biology of animal behavior. St. Louis, Times Mirror/Mosby. *Lucid broad perspective of animal behavior with strong emphasis on ecological and evolutionary aspects.*

Griffin, D.R. 1984. Animal thinking. Cambridge, Mass., Harvard University Press. *An important and provocative book about a controversial question.*

Lorenz, K.Z. 1952. King Solomon's ring. New York, Thomas Y. Crowell Co., Inc. *One of the most delightful books ever written about the behavior of animals.*

Savage-Rumbaugh, E.S. 1986. Ape language: from conditioned response to symbol. New York, Columbia University Press. *Details the author's studies as well as the general area of ape language.*

OMNE VIVUM EX OVO

If we admit that all living things are mortal, that every organism is endowed with a life span that must end, we must also acknowledge the indispensability of reproduction. Like Samuel Butler, who concluded that a chicken is just an egg's way of making another egg, many biologists consider reproduction to be the ultimate objective of all life processes.

In 1651, late in a long life, William Harvey, the English physiologist who earlier had ushered in experimental physiology by explaining the circuit of the blood, published a treatise on reproduction. He asserted that all life evolved from the egg—*omne vivum ex ovo.* This was curiously insightful, since Harvey had no means for visualizing the eggs of many animals, in particular the microscopic mammalian egg, which is no larger than a speck of dust to the unaided eye. Further, argued Harvey, the egg is launched into its developmental course by some influence from the semen, a conclusion that was either remarkably perceptive or a lucky guess, since sperm were invisible to Harvey. Such ideas differed sharply from existing notions of biogenesis, which saw life springing from many sources of which eggs were but one. Harvey was describing characteristics of sexual reproduction in which two parents, male and female, must come together physically to ensure fusion of gametes from each parent.

Despite the importance of Harvey's aphorism that all life arises from eggs, it was too sweeping to be wholly correct. Life springs from the reproduction of preexisting life, and reproduction may not be bound up in eggs and sperm. Nonsexual reproduction, the creation of new, genetically identical individuals by budding or fragmentation or fission from a single parent, is common, indeed characteristic, among the lower forms of life. Most animals, however, have found sex to be the winning strategy, probably because sexual reproduction provides the diversity needed for survival in a world of constant change.▪

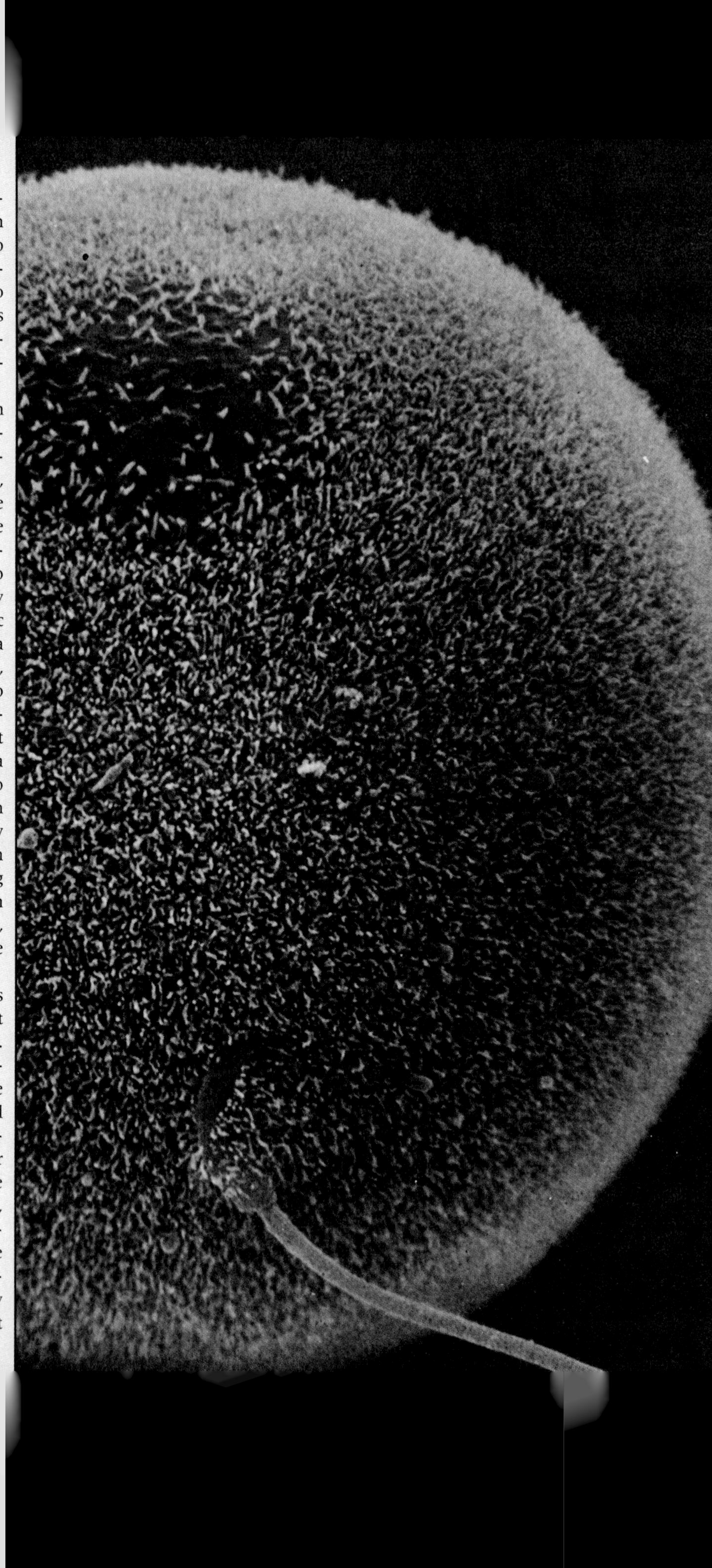

CHAPTER 13

Reproduction and Development

Reproduction is one of the unique yet ubiquitous miracles of life. Evolution is inextricably linked to reproduction because the ceaseless replacement of aging predecessors with new life enabled animals to respond and evolve in a changing environment as the Earth itself changed over the ages. Most animals reproduce sexually, which, for multicellular animals at least, appears to offer important advantages over asexual reproduction, as we will explain in the first part of this chapter.

Following sexual reproduction, the fertilized egg is guided by the genetic material from its parents to develop into a functioning and reproducing adult. The second part of this chapter deals with development: the organization of eggs and early embryos, the genetic control of development, and, finally, the embryology of amniotes (reptiles, birds, and mammals), especially humans.

Nature of the Reproductive Process

The two fundamental modes of reproduction are asexual and sexual. In **asexual** reproduction there is only one parent and there are no special reproductive organs or cells. Each organism is capable of producing genetically identical copies of itself as soon as it becomes an adult. The production of copies is marvelously simple and direct and typically rapid. **Sexual** reproduction (Figure 13-1) involves two parents as a rule, each of which contributes special **sex cells** (also called germ cells, or **gametes**) that in union develop into a new individual. The **zygote** formed from this union receives genetic material from *both* parents and accordingly is different from both. The combination of genes produces a genetically unique individual, still bearing the characteristics of the species but also bearing traits that make it different from its parents.

Sexual reproduction, by recombining the parental characters, tends to multiply variations and makes possible a richer and more diversified evolution. Mechanisms for interchange of genes between individuals are much more limited in organisms with only asexual reproduction. This would seem to explain why asexual reproduction is restricted mostly to unicellular forms, which can multiply rapidly enough to offset the disadvantages of repeated replication of identical products.

A

B

FIGURE 13-1

Sexual reproduction involves two parents. Fertilization—union of egg and sperm—may be external as in amphibians, **A,** or internal as in mammals, **B.**

A, Photograph by C.P. Hickman, Jr.; B, photograph by L.L. Rue III.

Of course, in those asexual organisms such as fungi and bacteria, which are haploid (bear only one set of genes, p. 40), mutations are immediately expressed and evolution can proceed quickly. In sexual animals, on the other hand, a gene mutation is often not expressed immediately, since it may be masked by its normal partner on the homologous chromosome. There is only a remote chance that both members of a gene pair will mutate in the same way at the same moment.

ASEXUAL REPRODUCTION

Asexual reproduction is found only among the simpler forms of life, such as bacteria, protozoa, cnidarians, platyhelminths, ectoprocts, and a few others. Even in those animal phyla where it occurs, most members employ sexual reproduction as well. In these groups, asexual reproduction ensures rapid increase in numbers when the differentiation of the organism has not advanced to the point of forming highly specialized gametes. Asexual reproduction is absent among the vertebrates and higher invertebrates.

It would be a mistake to conclude that asexual reproduction is in any way a "defective" form of reproduction relegated to the minute forms of life that have not yet discovered the joys of sex. Given the facts of their abundance, that they have persisted on earth for 3.5 billion years, and that they form the roots of the food chain on which all higher forms depend, the single-celled asexual organisms are both resoundingly successful and supremely important. For these forms, the advantages of asexual reproduction are its rapidity (many bacteria divide every half-hour) and simplicity (no sex cells to produce and no time and energy expended in finding a mate).

The forms of asexual reproduction in invertebrates are binary fission, budding (both internal and external), fragmentation, and multiple fission. **Binary fission** is common among bacteria and protozoa (Figure 15-6, p. 353) and to a limited extent among metazoa. In this method the body of the parent is divided into two approximately equal parts, each of which grows into an individual similar to the parent. **Budding** is an unequal division of the organism. The new individual arises as an outgrowth (bud) from the parent. This bud develops organs like those of the parent and then usually detaches itself. If the bud is formed on the surface of the parent, it is an external bud; but in some cases internal buds, or **gemmules,** are produced. Gemmules are collections of many cells surrounded by a dense covering in the body wall. When the body of the parent disintegrates, each gemmule gives rise to a new individual. External budding is common in the cnidarians and internal budding in the freshwater sponges. Ectoprocta also have a form of internal bud called a statoblast. **Fragmentation** is a method in which an organism breaks into two or more parts, each capable of becoming a complete animal. This method is found among the flatworms, ribbon worms, and echinoderms. In **multiple fission** the nucleus divides repeatedly before division of the cytoplasm, thereby giving rise to many daughter cells almost simultaneously. Multiple fission occurs in a number of protozoan forms.

SEXUAL REPRODUCTION

The essential feature of sexual reproduction is the involvement of *two genetically different parents that combine their genetic material to produce a cell having a new genotype.* The individuals sharing parenthood are characteristically of different **sexes,** male and female. The distinction between male and female is based, not on any differences in parental size or appearance, but on the size and mobility of the gametes they produce. The **ovum** (egg) is produced by the female. Ova are large (because of stored yolk to sustain early development), nonmotile, and produced in relatively small numbers. The **spermatozoon** (sperm) is produced by the male. Sperm are small, motile, and produced in enormous numbers. Each is a stripped-down package of highly condensed genetic material designed for the single mission of finding and fertilizing the egg.

There is another crucial event that distinguishes sexual from asexual reproduction: **meiosis,** a distinctive type of gamete-producing nuclear division. As described earlier (p. 64), meiosis differs from ordinary cell division (mitosis) in being a double division. The chromosomes split once, but the cell divides *twice,* producing four cells, each with half the original number of chromosomes (the haploid number). Meiosis is followed by fertilization in which two haploid gametes are combined to restore the normal (diploid) chromosomal number of the species.

The new cell (zygote), which now begins to divide by mitosis, has equal numbers of chromosomes from each parent and accordingly is different from each. It is a unique individual bearing a random assortment of parental characters. This is the great strength of sexual reproduction, the "master adaptation" that keeps feeding new varieties into the population. Neither meiosis nor fertilization ever occurs in asexual reproduction.

Many protozoans reproduce by both sexual and asexual modes of reproduction. When sexual reproduction does occur, it may or may not involve male and female gametes. Sometimes two mature sexual parents merely join together to exchange nuclear material or merge cytoplasm. It is not possible in these cases to distinguish sexes.

The male-female distinction is more clearly evident in the metazoa. Organs that produce the gametes are known as **gonads.** The gonad that produces the sperm is called the **testis** (Figure 13-2) and that which forms the egg, the **ovary** (Figure 13-3). The gonads represent the **primary sex organs,** the only sex organs found in certain groups of animals. Most metazoa, however, have various **accessory sex organs** (such as penis, vagina, oviduct, and uterus) that transfer and receive gametes. In the primary sex organs the gametes undergo many complicated changes during their development, the details of which are described in a later discussion. In our present discussion we will distinguish biparental reproduction from two alternatives: parthenogenesis and hermaphroditism.

Biparental reproduction

Biparental reproduction is the common and familiar method of sexual reproduction involving separate and distinct male and female individuals. Each has its own reproductive system and produces only one kind of sex cell, spermatozoon or ovum, but never both. Nearly all vertebrates and many invertebrates have separate sexes, and such a condition is called **dioecious.**

Parthenogenesis

Parthenogenesis ("virgin origin") is the development of an embryo from an unfertilized egg. Spontaneous, or natural, parthenogenesis is known to occur in rotifers; some nematodes, crustaceans, and insects; and several species of fish, amphibians, and desert lizards. Often several generations of parthenogenetic reproduction alternate with biparental reproduction in which the egg is fertilized.

From time to time claims arise that spontaneous parthenogenetic development to term has occurred in humans. A British investigation of about 100 cases in which the mother denied having had intercourse revealed that in nearly every case the child possessed characters not present in the mother, and consequently must have had a father. Nevertheless, mammalian eggs very rarely will spontaneously start developing into embryos without fertilization. The most remarkable instance of parthenogenetic development among the higher vertebrates has been found in turkeys in which certain strains, selected for their ability to develop without sperm, grow to reproducing adults.

In many species of freshwater rotifers, for example, diploid females produce meiotically reduced eggs at some time of the year (p. 414). If not fertilized, these eggs will develop into haploid males. If fertilized by a male, the egg develops into a dormant embryo that hatches, usually the following season, into a diploid female.

The queen bee is fertilized only once by a male (drone) or sometimes by more than one drone. She stores the sperm in her seminal receptacles, and as she lays her eggs she can either fertilize the eggs or allow them to pass unfertilized. The fertilized eggs become females (queens or workers), and the unfertilized eggs become males (drones) (p. 499).

In some cases parthenogenesis appears to be the only form of reproduction. Certain populations of whiptail lizards in the American Southwest are clones consisting solely of females. The females are diploid because, during meiosis, the chromosomes duplicate themselves to become tetraploid before undergoing the meiotic reductional division to form eggs. The eggs, now diploid, begin development without fertilization.

Hermaphroditism

Animals that have both male and female organs in the same individual are called hermaphrodites, and the condition is called hermaphroditism (from a combination of the names of the Greek god Hermes and goddess Aphrodite). In contrast to the dioecious state of separate sexes, hermaphrodites are **monoecious,** meaning that both male and female organs are in the same organism. Many invertebrate animals (most flatworms, some hydroids, annelids, and crustaceans) and a few vertebrates (some fishes) are hermaphroditic. Most avoid self-fertilization by exchanging sex cells with each other. For example, although the earthworm bears both male and female organs, its eggs are fertilized by the copulating mate and vice versa. Another way of preventing self-fertilization is by developing the eggs and sperm at different times.

Variety may make sexual reproduction a winning strategy for the unstable environment, but some biologists believe that for higher animals sexual reproduction is unnecessary and may even be maladaptive. In animals (humans, for example) in which most of the young survive to reproductive age, there is no demand for novel recombinations to cope with changing habitats. One offspring appears as successful as the next in each habitat. Significantly, parthenogenesis has evolved in several species of fish and in a few amphibians and reptiles. Such species are exclusively parthenogenetic, suggesting that where it has been possible to overcome the numerous constraints to making the transition, bisexuality loses out. Would all vertebrates do better to live sexlessly, avoiding the costs of sex? Possibly, but despite the costs, it is unlikely that higher animals if given the choice would opt for sexless lives—especially the males, which would necessarily become extinct. Woody Allen has pointed out that bisexuality, if nothing else, doubles the chance of getting a date on Saturday night.

WHAT GOOD IS SEX?

The question "What good is sex?" appears to have an easy answer: it serves the purpose of reproduction. But if we rephrase the question to ask, "Why do so many animals reproduce sexually rather than asexually?" the answer is not so apparent. Sexual reproduction is so nearly universal among animals that one can infer that it must be highly advantageous. Yet it is easier to list disadvantages to sex than advantages. Sexual reproduction is complicated, requires more time, and uses much more energy than asexual reproduction. Mating partners must come together, or at least come within proximity of each other, and coordinate their activities to produce young. Many biologists believe that an even more troublesome problem is the "cost of meiosis." A female that reproduces nonsexually (by parthenogenesis) passes all of her genes to her offspring. But when she reproduces sexually the genome is divided during meiosis and only half her genes flow to the next generation. Another cost is wastage in the production of males, many of whom fail to reproduce and thus consume resources that could be applied to the reproduction of asexual forms.

Clearly, the costs of sexual reproduction are substantial. How are they offset? Biologists have disputed this question for years without producing an answer that satisfies everyone. Many biologists believe that sexual reproduction, with its breakup and recombination of genomes, keeps producing novel genotypes that *in times of environmental change* may survive and reproduce while most others die. Variability, advocates of this viewpoint argue, is sexual reproduction's trump card.

But is variability worth the biological costs of sexual reproduction? The underlying problem keeps coming back: asexual organisms, because they can have more offspring in a given time, appear to be more fit in Darwinian terms. And yet most metazoan animals are determinedly committed to sexuality. There is considerable evidence that asexual reproduction is most successful in colonizing new environments. When habitats are empty what matters most is rapid reproduction; variability matters little. But as habitats become more crowded, competition among species for resources increases. Selection becomes more intense and genetic variability—new genotypes produced by recombination in sexual reproduction—furnishes the diversity on which natural selection can act. There are many invertebrates that use both sexual and asexual reproduction, thus enjoying the advantages each has to offer. Whatever the explanation for sex, it seems here to stay.

Plan of Reproductive Systems

The basic plan of the reproductive systems is similar in all animals, although differences in reproductive habits, methods of fertilization, etc. have produced many variations. In vertebrate animals the reproductive and excretory systems are often referred to as the **urinogenital system** because of their close anatomical connection. This association is very striking during embryonic development.

The reproductive and excretory systems of the male are usually more intimately connected than they are in the female. For example, in male fishes and amphibians the duct that drains the kidney (**wolffian duct**) also serves as the sperm duct. In male reptiles, birds, and mammals in which the kidney develops its own independent duct (**ureter**) to carry away waste, the old wolffian duct becomes exclusively a sperm duct (**vas deferens**). In all these forms, with the exception of mammals, the ducts open into a **cloaca,** a common chamber into which the intestinal, reproductive, and excretory canals empty (the word derived, appropriately, from the Latin word meaning "sewer"). The **oviduct** of the female is an independent reproductive duct that, like the vas deferens, opens into the cloaca. In higher mammals there is no cloaca; instead the urinogenital system has its own opening separate from the anal opening.

The plan of the reproductive system in vertebrates includes (1) **gonads,** which produce the sperm and eggs, (2) **ducts** to transport the gametes, (3) **accessory organs** for transferring and receiving gametes, (4) **accessory glands** (exocrine and endocrine) to provide secretions necessary to facilitate and synchronize the reproductive process, and (5) **organs** for storage before and after fertilization. This plan is modified among the various vertebrates, and some of the items may be lacking altogether.

Most aquatic vertebrates have no need for a penis, since sperm and eggs are liberated into the water in close proximity to each other. However, in terrestrial (and some aquatic) vertebrates that bear their young alive or enclose the egg within a shell, sperm must be transferred to the female. In most birds, this is a rather haphazard process of simply presenting cloaca to cloaca. Only the reptiles and mammals have a true penis. In mammals, the normally flaccid organ is erected when engorged with blood. Many mammals, although not humans, possess a bone in the penis (baculum), which presumably helps with rigidity. The baculum has a highly variable shape in different species and is a favored object for bewildering comparative anatomy students during practical examinations.

HUMAN REPRODUCTIVE SYSTEM

Male reproductive system

The paired **testes** are the locus of sperm production. Each testis is made up of numerous **seminiferous tubules,** in which the sperm develop (Figure 13-2), and the **interstitial** tissue lying along the tubules, which produces the male sex hormone (testosterone). The two testes are housed in the scrotal sac (scrotum), which in many mammals hangs down as an appendage of the body. This strange and seemingly insecure arrangement provides an environment of slightly lower temperature, since in humans, as in many other mammals, sperm apparently do not form at temperatures maintained within the body.

The sperm are conveyed from the seminiferous tubules to the **vasa efferentia,** small tubes passing to a coiled **vas epididymis** (one for each testis). The epididymis is connected by a **vas deferens** to the **urethra.** From this point the urethra serves to carry both sperm and urinary products through the **penis** or external intromittent organ.

Three pairs of glands open into the reproductive channels—**seminal vesicles,**

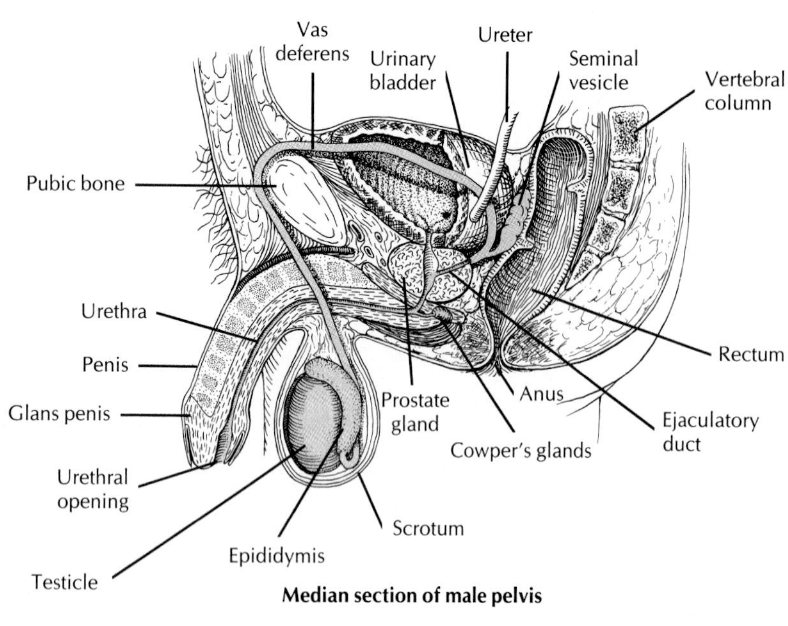

Median section of male pelvis

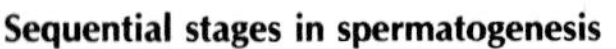

Sequential stages in spermatogenesis

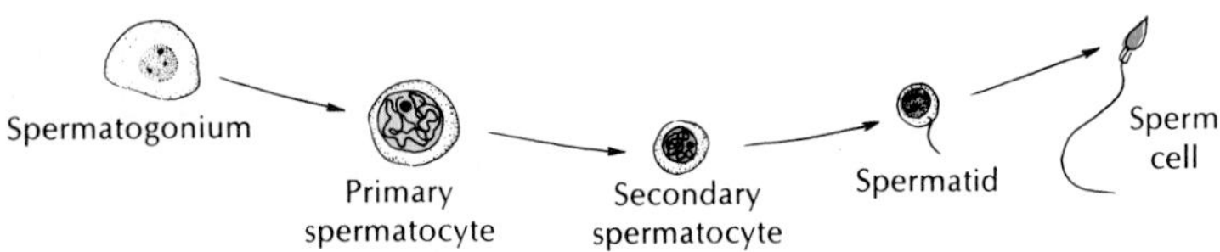

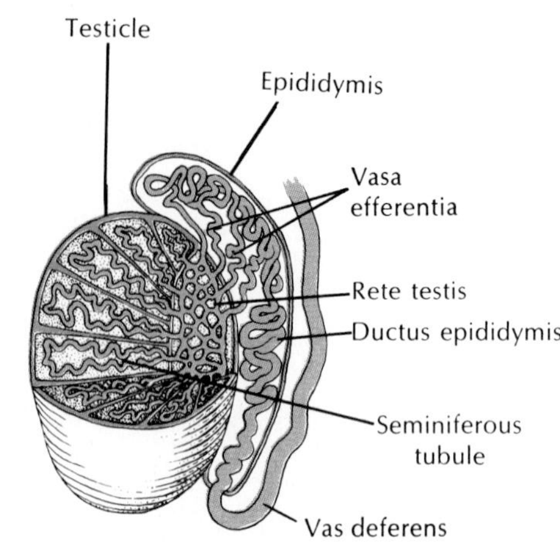

Section of left testis

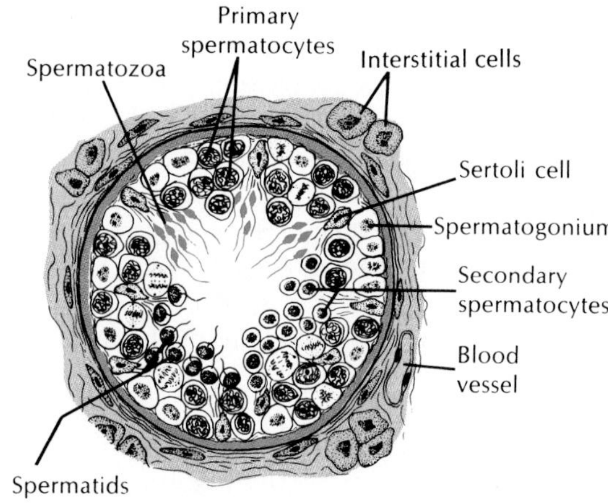

Cross section of one seminiferous tubule, showing different stages in spermatogenesis

FIGURE 13-2

Human male reproduction system.

prostate glands, and **Cowper's glands.** Fluid secreted by these glands furnishes food to the sperm, lubricates the passageways for the sperm, and counteracts the acidity of the urine so that the sperm are not harmed.

Female reproductive system

The paired ovaries, slightly smaller than the male testes, contain many thousands of eggs (ova). Each egg develops within a **graafian follicle** that enlarges and finally ruptures to release the mature egg (Figure 13-3). During the fertile period of the woman approximately 13 eggs mature each year, and usually the ovaries alternate in releasing an egg. Since the female is fertile for only some 30 years, only approximately 400 eggs have a chance to reach maturity; the others degenerate and are absorbed.

The **oviducts,** or fallopian tubes, are egg-carrying tubes with funnel-shaped openings for receiving the eggs when they emerge from the ovary. The oviduct is lined with cilia for propelling the egg in its course. The two ducts open into the upper corners of the **uterus,** or womb, which is specialized for housing the embryo during the 9 months of its intrauterine existence. It is provided with thick muscular walls, many blood vessels, and a specialized lining—the **endometrium.** The uterus varies with different mammals. It was originally paired but tends to fuse in higher forms.

The **vagina** is a muscular tube adapted for receiving the male's penis and for serving as the birth canal during expulsion of the fetus from the uterus. Where

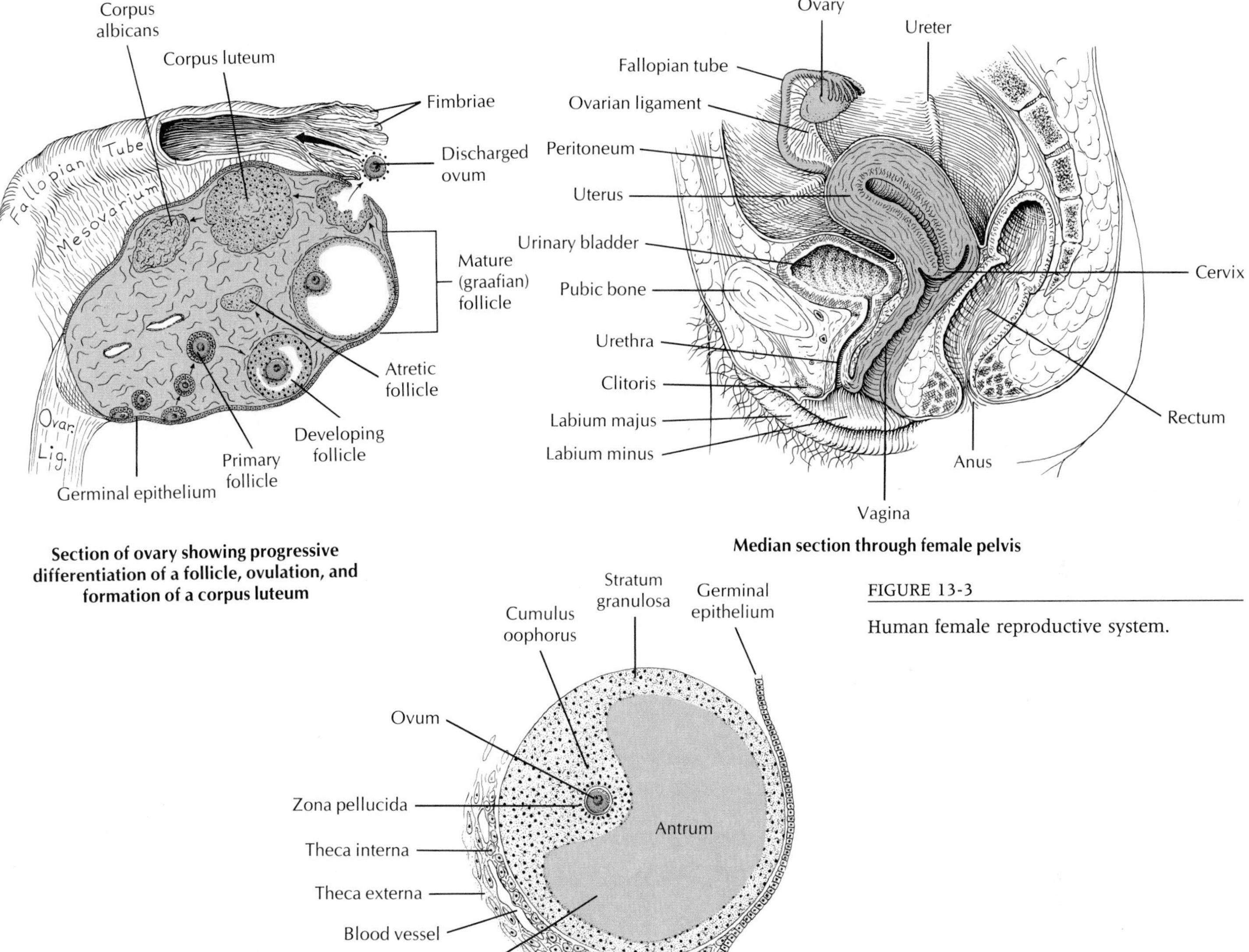

Section of ovary showing progressive differentiation of a follicle, ovulation, and formation of a corpus luteum

Median section through female pelvis

Section of mature follicle

FIGURE 13-3

Human female reproductive system.

the vagina and the uterus meet, the uterus projects down into the vagina to form the **cervix.**

The external genitalia of the female, or vulva, include folds of skin, the **labia majora** and **labia minora,** and a small erectile organ, the **clitoris.** The opening into the vagina is normally reduced in size in the virgin state by a membrane, the **hymen.**

Origin of Reproductive Cells

The animal body has two basically different types of cells: the somatic (body) cells, which are differentiated for specialized functions and die with the individual; and the germinal cells, some of which may contribute to the formation of a zygote and thereby to a new generation. The germinal cells are set aside at the beginning of embryonic development, usually in the endoderm, and migrate to the gonads. The germinal cells, or primordial germ (sex) cells, develop into eggs

FIGURE 13-4

Section of a seminiferous tubule containing male germ cells. More than 200 long, highly coiled seminiferous tubules are packed in each human testis. This scanning electron micrograph reveals numerous tails of mature spermatozoa in the tubule's central cavity that have differentiated from germ cells in the periphery of the tubule.

From Tissues and organs: a text-atlas of scanning electron microscopy, by Richard G. Kessel and Randy H. Kardon. W.H. Freeman & Co. Copyright 1979.

and sperm—nothing else. The other cells of the gonads are somatic cells. They cannot form eggs or sperm, but they are necessary aids in the development of the sex cells (gametogenesis).

GAMETOGENESIS

The series of transformations that results in the formation of mature gametes (sex, or germ, cells) is called gametogenesis.

Although the same essential processes are involved in the maturation of both sperm and eggs, there are some important differences. Gametogenesis in the testis is called **spermatogenesis** and in the ovary it is called **oogenesis** (o'uh-gen'uh-sis).

Spermatogenesis

The walls of the seminiferous tubules contain the differentiating sex cells arranged in a stratified layer five to eight cells deep. The outermost layers contain **spermatogonia** (Figure 13-2), which have increased in number by ordinary mitosis. Each spermatogonium increases in size and becomes a **primary spermatocyte.** Each primary spermatocyte then undergoes the first meiotic division, as described before (p. 63), to become two **secondary spermatocytes.**

Each secondary spermatocyte enters the second meiotic division, without the intervention of a resting period. The resulting cells are called **spermatids,** and each contains the haploid number (23 in humans) of chromosomes. A spermatid may have all maternal, all paternal, or both maternal and paternal chromosomes in varying proportions. Without further divisions the spermatids are transformed into mature sperm by losing a great deal of cytoplasm, condensing the nucleus into a head, and forming a whiplike, flagellar tail (Figures 13-4 and 13-5).

From following the divisions of spermatogenesis one can see that each primary spermatocyte gives rise to four functional sperm, each with the haploid number of chromosomes (Figure 13-6).

Oogenesis

The early germ cells in the ovary are called **oogonia,** which increase in number by ordinary mitosis (Figure 13-6). Each oogonium contains the diploid number of chromosomes. In females after puberty, one of these oogonia typically

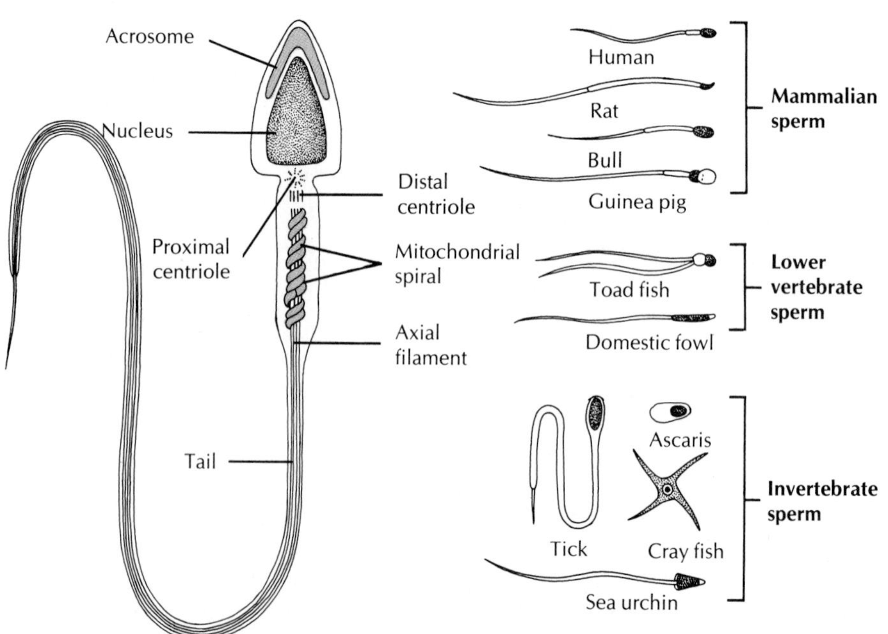

FIGURE 13-5

Types of sperm. *Left,* Structure of a typical mammalian spermatozoon.

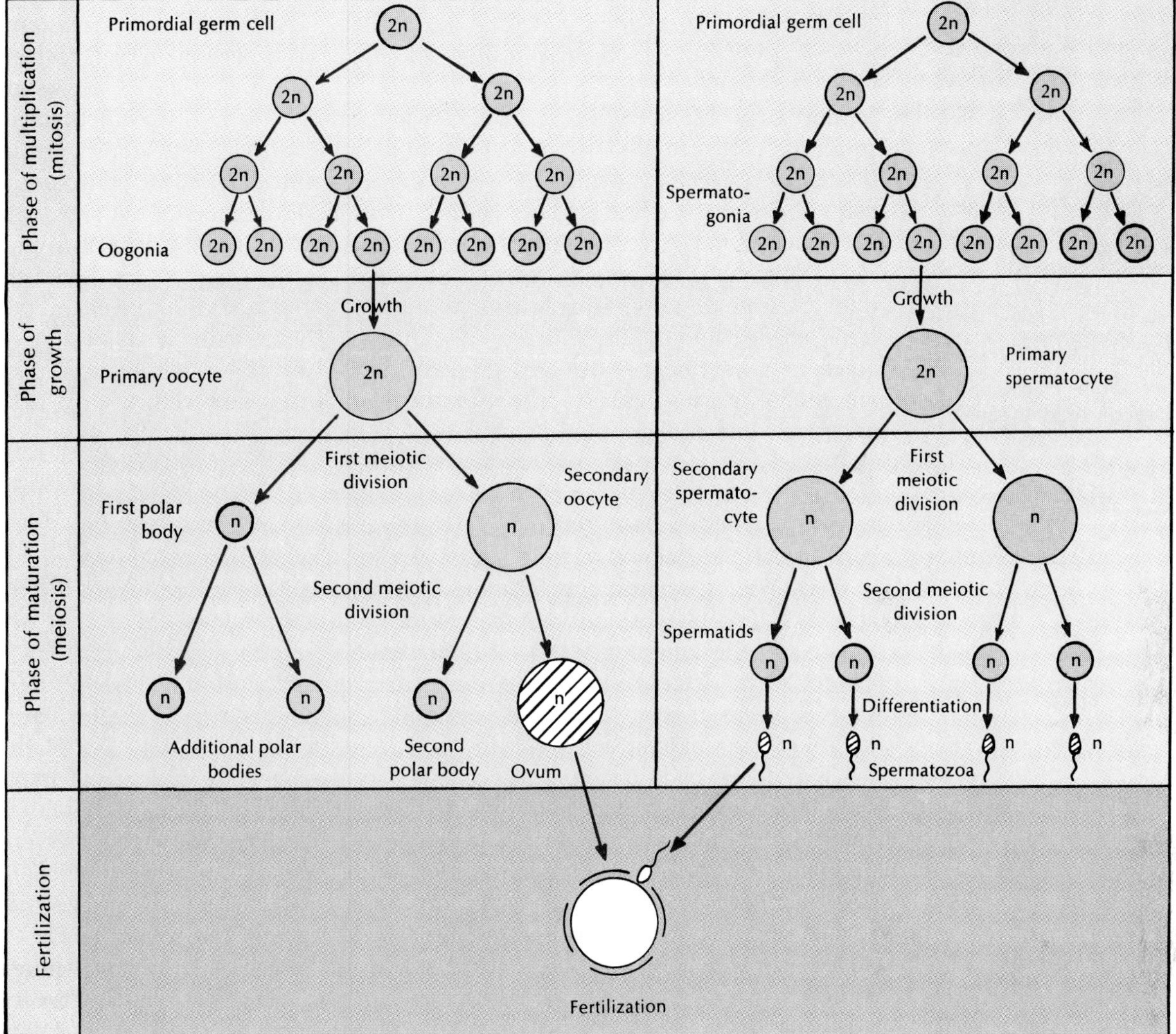

FIGURE 13-6

Gametogenesis compared in eggs and sperm. *n*, Haploid chromosome number.

develops each menstrual month into a functional egg. After the oogonia cease to increase in number, they grow in size and become **primary oocytes.** Before the first meiotic division, the chromosomes in each primary oocyte meet in pairs, paternal and maternal homologues, just as in spermatogenesis. When the first meiotic division occurs, the cytoplasm is divided unequally. One of the two daughter cells, the **secondary oocyte,** is large and receives most of the cytoplasm; the other is very small and is called the **first polar body** (Figure 13-6). Each of these daughter cells, however, has received half the nuclear material or chromosomes.

In the second meiotic division, the secondary oocyte divides into a large **ootid** and a small polar body. If the first polar body also divides, which sometimes happens, there are three polar bodies and one ootid. The ootid grows into a functional **ovum;** the polar bodies are nonfunctional and disintegrate. The formation of the nonfunctional polar bodies is necessary to enable the egg to rid itself of excess chromosomes, and the unequal cytoplasmic division makes possible a large cell with sufficient yolk for the development of the young. Thus the mature ovum has the haploid number of chromosomes, the same as the sperm. However, each primary oocyte gives rise to only *one* functional gamete instead of four as in spermatogenesis.

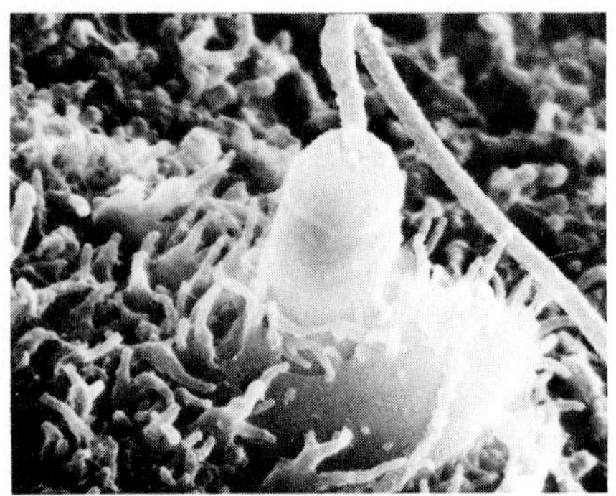

FIGURE 13-7

The dramatic moment of fertilization of a sea urchin egg. This scanning electron micrograph shows the sperm head attached to the egg surface. A fertilization cone (shown lifting from the surface) and the rapidly elongating microvilli soon engulf the sperm. (×29,000.)

Courtesy Gerald Schatten, Florida State University.

The Developmental Process

The phenomenon of development is a remarkable and in many ways awesome process. How is it possible that a tiny, spherical, fertilized human egg, scarcely visible to the naked eye, can unfold into a fully formed, unique person, consisting of thousands of billions of cells, each cell performing a predestined functional or structural role? How is this marvelous unfolding controlled? Obviously all the information needed is contained within the egg, principally in the genes of the egg's nucleus. The fabric of genes is DNA. Thus all development originates from the structure of the nuclear DNA molecules and in the egg cytoplasm surrounding the nucleus. But knowing where the blueprint for development resides is very different from understanding how this control system guides the conversion of a fertilized egg into a fully differentiated animal. This remains a major—many consider it *the* major—unsolved problem of biology. It has stimulated a vast amount of research on the processes and phenomena involved; from it have emerged some early and, in many cases, tentative answers.

OOCYTE MATURATION

During oogenesis the egg becomes a highly specialized, very large cell containing condensed food reserves for subsequent growth. The nucleus also grows rapidly in size during egg maturation, although not as much as the cell as a whole. Large amounts of both DNA and RNA accumulate during oogenesis.

Most of this enormous food accumulation and nucleic acid synthesis occurs before the meiotic, or maturation, divisions begin. When the maturation divisions do occur, they are, as already described, highly unequal: the single mature ovum retains a haploid set of chromosomes and the vast bulk of the cytoplasm, whereas the other three sets of haploid chromosomes are cast off as small, cytoplasm-starved polar bodies. Because all three polar bodies lack stored nutrients, none are capable of further development. This obviously undemocratic hoarding of all accumulated food reserves by just one of four otherwise genetically equal cells is, of course, a device for avoiding a decrease in ovum size once all the nutrients have been packaged inside at the end of the growth phase.

In most vertebrates, the egg does not actually complete all the meiotic divisions before fertilization occurs. The general rule is that the egg completes the first meiotic division and proceeds to the metaphase stage of the second meiotic division, at which point further progress stops. The second meiotic division is completed and the second polar body extruded only if the egg is activated by fertilization.

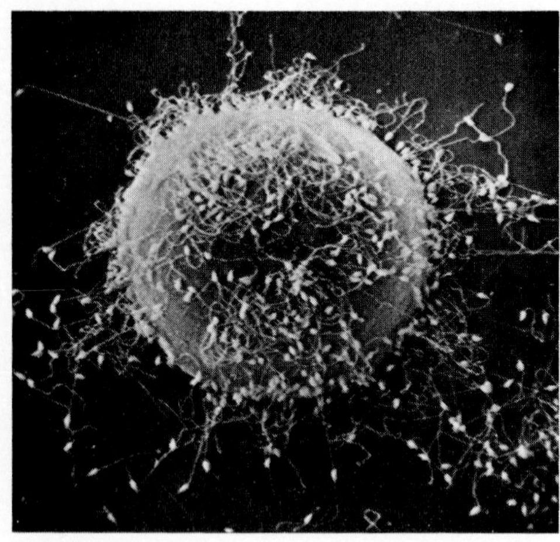

FIGURE 13-8

Binding of sperm to the surface of a sea urchin ovum. Only one sperm penetrates the egg surface, the others being blocked from entrance by the rapid formation of a fertilization membrane. Unsuccessful sperm soon fall away from the egg. (×1425.)

Courtesy Gerald Schatten, Florida State University.

FERTILIZATION AND ACTIVATION

Fertilization is the union of male and female gametes to form a **zygote.** This process accomplishes two things: it activates the process of development and provides for the recombination of paternal and maternal genes. Thus it restores the original diploid number of chromosomes characteristic of the species.

For a species to survive, it must ensure that fertilization occurs and that enough progeny survive to maintain a healthy population. Many marine fishes simply set their eggs and sperm adrift in the ocean and rely on the random swimming movements of sperm to make chance encounters with eggs. Even though an egg is a large target for a sperm, the enormous dispersing effect of the ocean, the short life span of the gametes (usually just a few minutes for fish gametes), and the limited range of the tiny sperm all conspire against an egg and a sperm coming together. Accordingly each male releases millions of sperm at spawning. The odds against fertilization are further reduced by coordinating the time and place of spawning of both parents.

FIGURE 13-9

Fertilization of an egg. **A,** Many sperm swim to the egg. **B,** The first sperm to penetrate the protective jelly envelope and contact the egg membrane causes the fertilization cone to rise and engulf the sperm head. **C,** The fertilization membrane begins to form at the site of penetration and spreads around the entire egg preventing the entrance of additional sperm. **D,** Male and female pronuclei approach one another, lose their nuclear membranes, swell, and fuse. **E,** The mitotic spindle forms, signaling creation of a zygote and heralding the first cleavage of a new embryo.

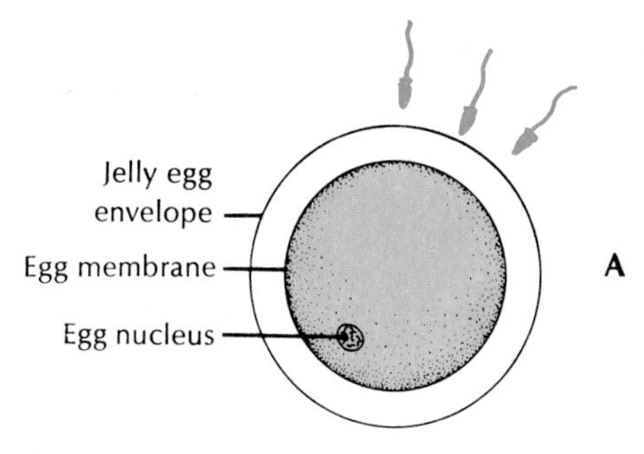

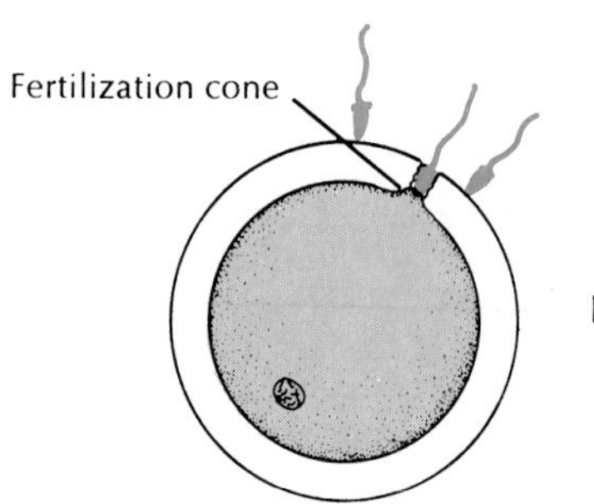

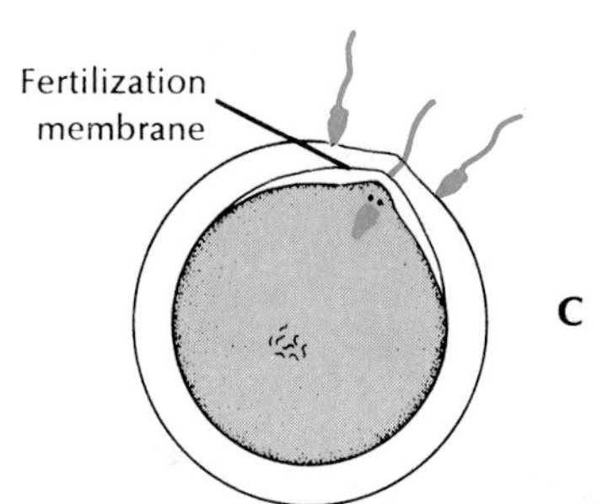

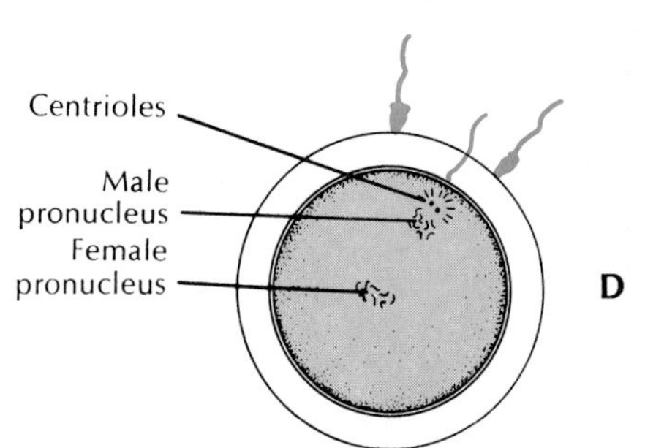

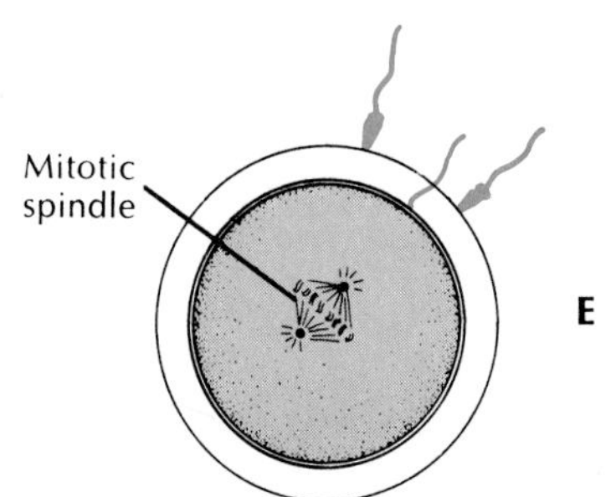

Ensuring that some eggs are fertilized, however, is not enough. The ocean is a perilous environment for a developing animal, and most never make it to maturity. Thus, the females produce huge numbers of eggs. The codfish of the North American East Coast regularly spawns 4 to 6 million eggs, of which only two or three, on the average, will reach maturity. A single oyster may lay 25 to 60 million eggs at a time. Animals that provide more protection to their young produce fewer eggs. Internal fertilization, characteristic of flatworms, gastropod molluscs, and some other invertebrates, as well as sharks, reptiles, birds, and mammals, avoids dispersion of the gametes and protects them. However, even with internal fertilization vast numbers of sperm must be released by the male into the female tract. Furthermore, the events of ovulation and insemination must be closely synchronized and the gametes must remain viable for several hours to accomplish fertilization. Sperm may have to travel a considerable distance to reach the egg in the female genital tract, many parts of which may be hostile to sperm. Experiments with rabbits have shown that of the approximately 10 million sperm released into the female vagina, only about 100 reach the site of fertilization.

Activation—restoration of metabolic activity in the quiescent eggs—is a dramatic event. In sea urchin eggs, the contact of the spermatozoon with the egg surface sets off almost instantaneous changes in the outer layer of the egg, the egg cortex. At the point of contact, a **fertilization cone** appears into which the sperm head is later drawn (Figure 13-7). As soon as the first sperm enters the egg, important changes occur in the egg membrane that block the entrance of additional sperm, which, in marine eggs especially, may quickly surround the egg and bind to its surface (Figure 13-8). The entrance of more than one sperm (**polyspermy**) must be prevented, since the union of more than two haploid nuclei would be ruinous for normal development. In the sea urchin egg, entrance of the first sperm is instantly followed by an electrical potential change in the egg membrane that prevents additional sperm from fusing with the membrane. This is followed by the elevation of a **fertilization membrane** (Figure 13-9), a permanent barrier that completes the egg's insurance against lethal polyspermy. In mammals, the events leading to fertilization are different from those described for the sea urchin but the results are the same: the first sperm to penetrate the outer coat of the mammalian egg triggers a reaction that chemically alters the outer coat and plasma membrane, blocking the entrance to more than one sperm.

Sperm entrance is a crucial event in development. Within seconds it produces a complete molecular reorganization of the egg cortex. Inhibitors that have kept the egg in its quiescent, suspended-animation state are removed and normal activity resumes. Stored nucleic acids begin producing protein, the male and female pronuclei fuse, and the egg, now a zygote, enters into cleavage.

CLEAVAGE AND EARLY DEVELOPMENT

During cleavage, the zygote divides repeatedly to convert the large, unwieldy cytoplasmic mass into a large number of small, maneuverable cells (called **blas-**

tomeres) clustered together like a mass of soap bubbles (see Figures 13-13 and 13-14). There is no growth during this period, only subdivision of mass, which continues until normal somatic cell size is attained. At the end of cleavage the zygote has been divided into many hundreds or thousands of cells (approximately 1000 in polychaete worms, 9000 in amphioxuses, and 700,000 in frogs). There is a rapid increase in DNA content during cleavage because the number of nuclei and the amount of DNA are doubled with each division. Apart from this, there is little change in chemical composition or displacement of constituent parts of the egg cytoplasm during cleavage. **Polarity,** that is, a polar axis, is present in the egg, and this establishes the direction of cleavage and subsequent differentiation of the embryo.

"Behold Driesch grinding the eggs of Loeb's favorite sea urchin up between plates of glass, pounding and breaking and deforming them in every way. And when he ceased from thus abusing them, they proceeded with their orderly and normal development. Is any machine conceivable, Driesch asks, which could thus be torn down,...have its parts all disarranged and transposed, and still have them act normally? One cannot imagine it. But of the living egg, fertilized or not, we can say that there lie latent within it all the potentialities presumed by Aristotle, and all of the sculptor's dream of form, yes, and the very power of the sculptor's arm."
From Peattie, D.C. 1935. *An Almanac for Moderns*. New York, G.P. Putnams Sons.

Significance of the cortex

Early cleavage proceeds independent of nuclear genetic information, guided instead by information deposited in the egg during its maturation. It was once believed that the visible particulate material in the cytoplasm had determinative properties. However, it was soon discovered that if the egg's cytoplasm is thoroughly displaced by centrifugation, the embryo still develops normally. The only elements not affected by strong centrifugation of the egg are the plasma membrane and the **cortex,** a gel-like layer just beneath the plasma membrane. The embryo will also develop perfectly when up to 50% of the egg's cytoplasm is removed with a micropipette.

These experiments show conclusively that the cortical layer of the egg, and not the cytoplasm, contains an invisible but dynamic organization that determines the pattern of cleavage. Cortical organization is at first unstable but soon becomes regionally fixed and irreversible. Thus, as cleavage progresses, the cortex becomes segregated into territories having specific determinative properties. This explains why different blastomeres bear different cytodifferentiation properties.

Radial and spiral cleavage

The eggs of the great majority of metazoan animal groups cleave by one of two patterns: radial or spiral. In **radial cleavage,** the cleavage planes are symmetrical to the polar axis and produce tiers, or layers, of cells on top of each other (Figure 13-10, *A*). Radial cleavage is also said to be **regulative** (or **indeterminate**) because each blastomere of the early embryo, if separated from the others, can adjust or "regulate" its development into a complete and well-proportioned embryo (Figure 13-11, *A*). This happens because the blastomeres are equipotential in the sense that there is no definite relation between the position of any of the *early* blastomeres and the specific tissue it will form in the developing embryo. Radial cleavage with regulative development is characteristic of vertebrates and echinoderms.

Spiral cleavage, found in several invertebrate groups, is different from radial. The cleavage planes are oblique to the polar axis and typically produce quartets of unequal cells that come to lie not on top of each other but in the furrows between the cells (Figure 13-10, *B*). Spirally cleaving embryos also differ from radial embryos in having a **mosaic** (or **determinate**) form of development. This means that the organ-forming regions of the egg cytoplasm become strictly localized in the egg, even before the first cleavage division. The result is that if the early blastomeres are separated, each will continue to develop for a time as if it were still part of the whole. Each forms a defective, partial embryo (Figure 13-11, *B*). With such mosaic eggs it is often possible to map out the fates of specific areas on the uncleaved egg surface that are known to be presumptive for specific structures (Figure 13-12). Localization happens much later in regulative cleavage. Once localization is established in either type of egg, blastomeres cannot be dam-

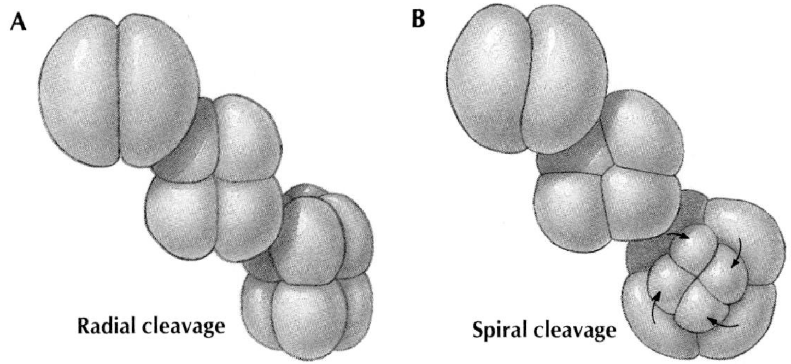

FIGURE 13-10

Radial and spiral cleavage patterns shown at 2-, 4-, and 8-cell stages. **A,** Radial cleavage, typical of echinoderms and chordates. **B,** Spiral cleavage, typical of molluscs, annelids, and other protostomes. Arrows indicate clockwise movement of small cells (micromeres) following division of large cells (macromeres).

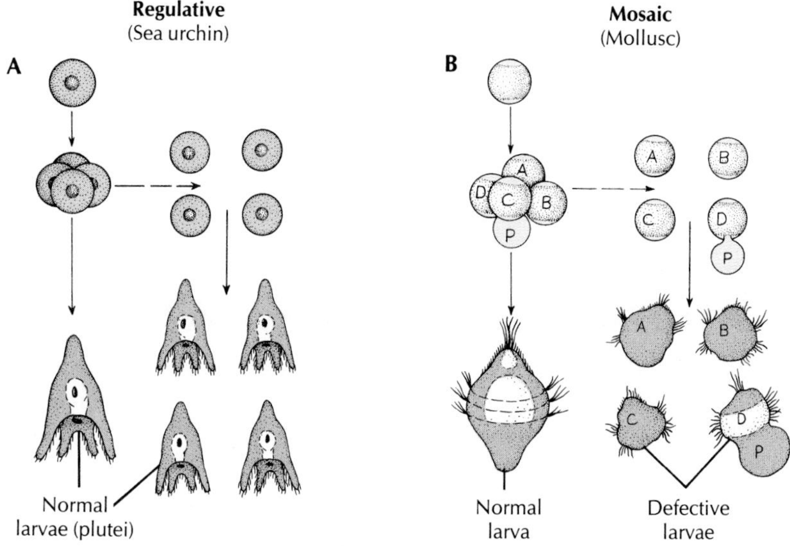

FIGURE 13-11

Regulative and mosaic development. **A,** regulative (indeterminate) development. If the early blastomeres of a sea urchin embryo are separated, each will develop into a complete larva. **B,** Mosaic (determinate) development. When the blastomeres of a mollusc embryo are separated, each gives rise to a partial, defective larva.

aged or removed from the embryo without corresponding defects in development.

The importance of these two cleavage patterns extends well beyond the differences we have described. They signal a fundamental evolutionary dichotomy, the divergence of bilateral metazoan animals into two separate lineages. Spiral cleavage is found in segmented worms, molluscs, and several other invertebrate phyla, all of which are included in the **Protostomia** ("primary mouth") division of the Animal Kingdom (see the illustration inside the front cover of this book). Radial cleavage (and a modified pattern called biradial) is characteristic of the **Deuterostomia** ("secondary mouth") division of the Animal Kingdom, a grouping that includes the echinoderms (sea stars and their kin), the chordates, and two minor phyla. Other distinguishing hallmarks of these two divisions will be summarized in Chapter 14 (see especially pp. 336-337).

BLASTULA FORMATION

Cleavage, however modified by different cleavage patterns, results in a cluster of cells. In most animals, with continued divisions the cells arrange themselves into a hollow ball called a **blastula** (Figure 13-13). The central fluid-filled cavity of the blastula is called a **blastocoel.** At this point the embryo consists of a few hundred to several thousand cells poised for further development.

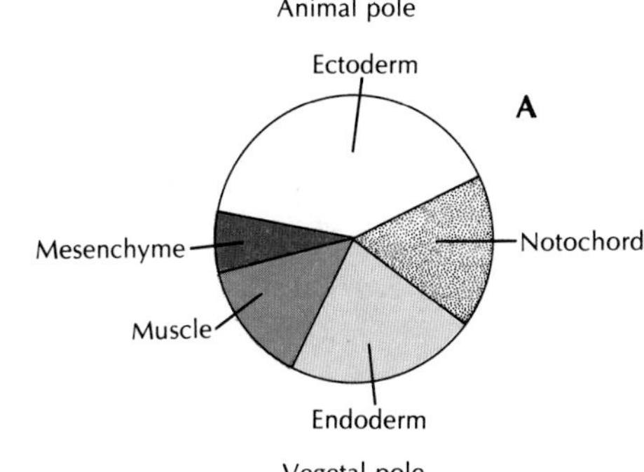

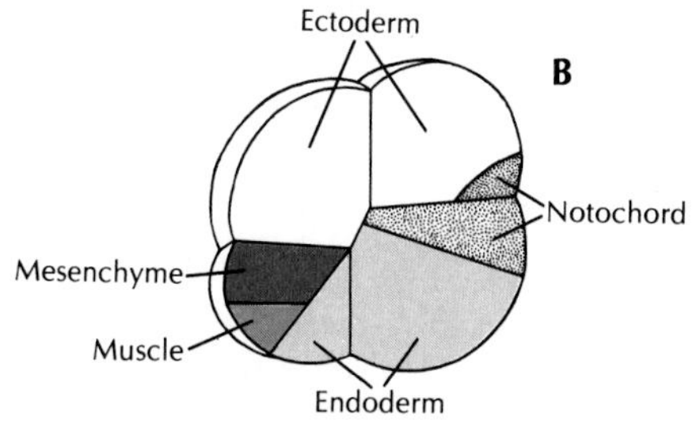

FIGURE 13-12

Segregation of organ-forming regions in **A,** an ascidian (tunicate) egg and, **B,** an 8-cell embryo. Although the tunicates are deuterostomes, they have mosaic eggs like the protostomes. Mapping of prospective regions in this way would be impossible in a highly regulative mammalian egg.

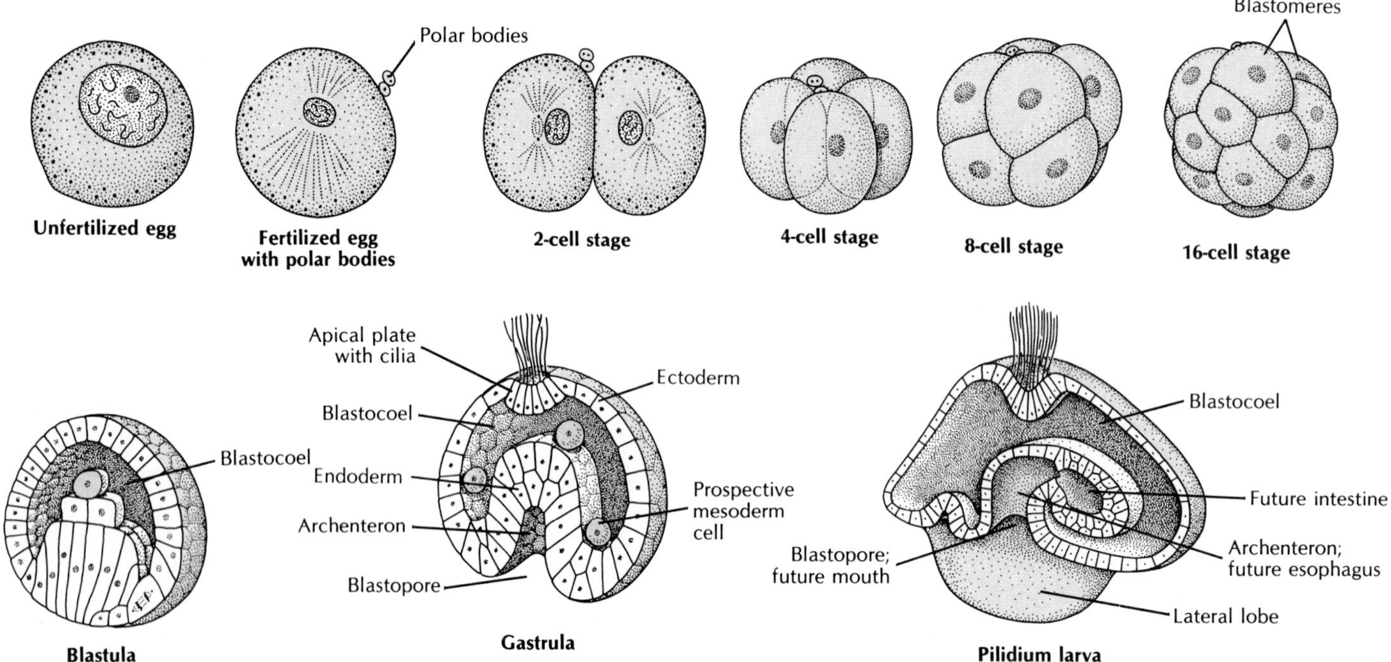

FIGURE 13-13

Early embryology of a nemertean worm, a protostome with spiral cleavage.

GASTRULATION AND THE FORMATION OF GERM LAYERS

Gastrulation is a regrouping process in which cells of the blastula become rearranged into new, orderly associations by morphogenetic movements.

The early cell movements in gastrulation are remarkably similar in most protostomes and deuterostomes. In the nemertean worm, a protostome with spiral cleavage (Figure 13-13), cells at the vegetal pole indent and migrate inward, much like pushing in the side of a soft tennis ball. This inward migration creates an internal cavity called the **archenteron.** The archenteron is the primitive gut, and its opening to the outside, the **blastopore,** will become the future mouth of the worm. The gastrula is now an embryo of two **germ layers.** The outer layer is the **ectoderm;** it will give rise to the epithelium of the body surface, and to the nervous system. The inner layer that forms the archenteron is the **endoderm;** it will give rise to the epithelial lining of the digestive tube. The cavity between these two layers is the old blastocoel. A third germ layer, the **mesoderm,** arises from cells in the blastocoel that pinch off from either side of the blastopore. The mesoderm will later differentiate into muscles, blood vessels, and the reproductive organs of the future adult body.

The assignment of early embryonic layers to specific "germ layers" (not to be confused with "germ cells," which are the eggs and sperm) is for the convenience of embryologists and of no concern to the embryo. Whereas the three germ layers normally differentiate into the tissue and organs described here, it is not the germ layer itself that determines differentiation but rather the precise position of an embryonic cell with relation to other cells that controls the real fate of that cell. If for example a piece of ectoderm is grafted into another region of the embryo, it may form organs that normally come from a different germ layer.

In the frog, a deuterostome with radial cleavage (Figure 13-14), the morphogenetic movements of gastrulation are greatly influenced by the mass of inert yolk in the vegetal half of the embryo. Cleavage divisions are slowed in this half so that the resulting blastula consists of many small cells in the animal half and fewer large cells in the vegetal half (Figure 13-14). Cells on the surface begin to sink inward (invaginate) at the blastopore and continue to move inward as a sheet to form a flattened cavity, the archenteron. Although the cell movements are different, the result of gastrulation is the same in the frog and the nemertean worm: an embryo of two germ layers (endoderm and ectoderm) is formed. As invagination proceeds, the third germ layer, the mesoderm, rolls over the lateral and ventral lip of the blastopore and, once inside, penetrates between the endoderm on the inside and the ectoderm on the outside. The three germ layers now formed are the primary structural layers that play crucial roles in the further differentiation of the embryo.

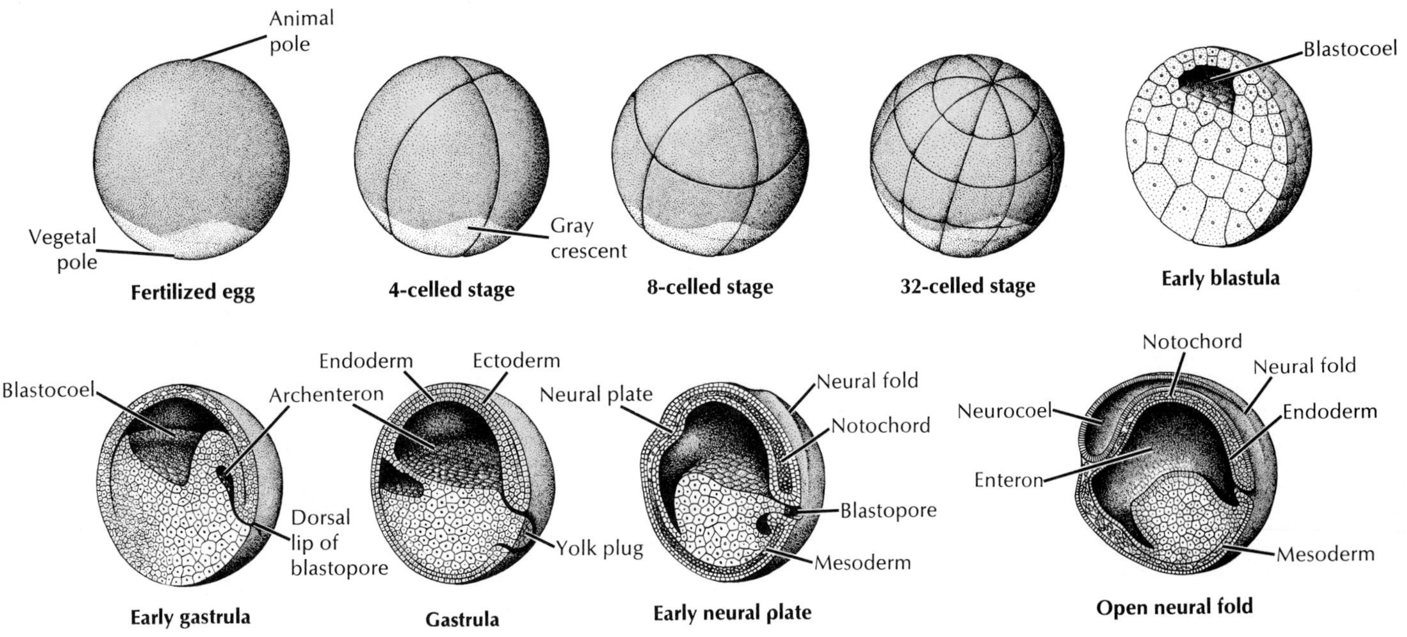

FIGURE 13-14

Early embryology of a frog, a deuterostome with radial cleavage.

In the most primitive of the true metazoa, the Cnidaria and the Ctenophora, only two germ layers are formed, the endoderm and the ectoderm. These animals are **diploblastic.** In all more advanced metazoa, the mesoderm also appears, either from pouches of the archenteron or from other cells associated with endoderm formation. This three-layered condition is called **triploblastic.**

DIFFERENTIATION AND GENE EXPRESSION

We saw that the egg cortex contains the primary guiding force for early development and synthetic activities during this period are supported by nucleic acids and reserves laid down in the egg cytoplasm *before* fertilization. However, continued differentiation past cleavage or the blastula stage requires the action of genetic information in the nuclear chromosomes. There is considerable evidence to show that this is so. If the maternal nucleus is removed from an activated but unfertilized frog egg, the anucleate egg will cleave more or less normally but arrest at the outset of gastrulation, when nuclear information is required.

Hybridization experiments have also been carried out in which the maternal nucleus is removed from the egg by microsurgery, just after the egg has been fertilized, but before the pronuclei have time to fuse. Such a zygote contains *maternal* cytoplasm and a *paternal* (sperm) nucleus. The egg will cleave to the bastula stage normally; whether it will gastrulate depends on the compatibility of the sperm with the maternal cytoplasm. If the egg was fertilized with a sperm of the same species, it develops into a haploid (but otherwise normal) tadpole. If it was fertilized with a sperm of a closely related species, it may develop to the neurula stage before development stops. If fertilized with a distantly related species, its development will be arrested in the blastula or early gastrula stage. These differences are explained as varying degrees of immunological incompatibilities that appear when the paternal (sperm) genes begin coding for proteins in the embryo. The more foreign the sperm, the greater the incompatibility and the earlier the stage of arrest.

Finally, the most direct evidence that nuclear genes become active at or shortly before gastrulation is that the production of messenger RNA increases

Two Oxford scientists, J.B. Gurdon and R. Laskey, were able to grow a normal, reproductive adult frog from an unfertilized egg containing the nucleus of a differentiated intestinal cell of a frog tadpole. The technique they used had been developed by R. Briggs and T.J. King at Indiana University in 1952. Using minute glass tools, they were able to pluck out the nucleus from an egg and replace it with a nucleus from a fully differentiated cell. These experiments are of great significance because they demonstrate that a cell, or actually the nucleus from a cell, can be forced backward from its specialized state and once again make available all of its genetic information.

sharply at this time, followed by the production of transfer and ribosomal RNA. This has been demonstrated with the use of radioactively labeled RNA precursors.

Control of differentiation

As cells differentiate, they obviously use only a part of the instructions their nuclei contain. Cells that are differentiating into a thyroid gland are not concerned with that part of the genome that codes for striated muscle, for example. The unneeded genes are in some way switched off. They are not destroyed, however, because a nucleus from a cell at an advanced stage of development transplanted into an activated egg whose nucleus had been removed can support development of a complete organism (see marginal note).

The basic problem is **gene expression.** What determines whether a particular blastomere of, say, a 100-cell embryo will differentiate into muscle or skin or thyroid gland? If genes are the same in all nuclei of an early embryo, the only way differences can develop is through interaction between nuclei and the surrounding cytoplasm. We have already seen that the basic polarity of the egg and the organizing qualities of the egg cortex provide an early opportunity for such interaction.

Let us briefly review what is now known of the transmission of genetic information (see p. 77-83). Genetic information is coded in the sequence of nucleotides in DNA molecules. DNA serves as a template for the synthesis of messenger RNA in the nucleus. Messenger RNA then migrates out through nuclear pores into the cytoplasm, where it attaches to a ribosome. Here the messenger RNA serves as a template for the synthesis of specific proteins. In this way cytoplasmic proteins that may be specific for the cell are formed.

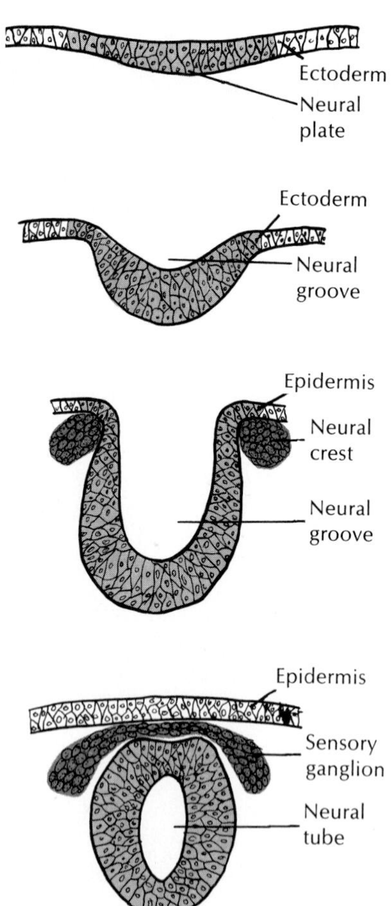

FIGURE 13-15

Development of the neural tube and neural crest from neural plate ectoderm (cross section).

Evidence to date suggests that at the beginning of development most nuclear genes are inactive. Only small amounts of messenger RNA are being produced on the DNA templates. As development proceeds, new cytoplasmic proteins appear, indicating that more genic DNA is producing messenger RNA. Evidently, different genes are activated (or "derepressed") in different parts of the young embryo, and this differential gene activity is responsible for embryonic differentiation. The effect a gene has when it is "turned on" depends on the status of the embryo at that moment. A gene turned on in the region of the thyroid gland in, say, the second week of development will not have the same effect as that same gene would have if activated in the same region during the third week. Thus, what a gene does is not just a property of the gene itself but also relates to the chemical conditions of the cell at a particular time in the developmental sequence.

What is the mechanism by which genes are turned on and off at the right moment and place? We do not know at present, although several possible genetic control mechanisms have been identified (p. 83-85). Whatever the mechanism is, it seems certain that the kinds of cytoplasm present in different cells determine when genes come into action. While there is much yet to learn about the developmental process, this much seems clear: development is a highly integrated process in which the formation of each structure depends not on a linear sequence of steps but on complex interactions between many such sequences.

DEVELOPMENT OF SYSTEMS AND ORGANS

Gastrulation is a critical time of orderly and integrated cell movements. By folding and splitting processes, the layers of cells that compose the rapidly growing embryo form the ectoderm, mesoderm, and endoderm. These rapidly differentiate into rudimentary, and later functional, tissues and organs. During this process, cells become increasingly committed to specific fates. Cells that previously had the potential to develop into a variety of structures lose this potential and assume commitments to become, for example, kidney cells, intestinal cells, or blood cells.

Derivatives of ectoderm: nervous system and nerve growth

In vertebrates, the brain, spinal cord, and nearly all the outer epithelial structures of the body develop from the primitive ectoderm. They are among the earliest organs to appear. Just above the notochord, the ectoderm thickens to form a **neural plate** (Figure 13-15). The edges of this plate rise up, fold, and join together at the top to create an elongated, hollow **neural tube.** The neural tube gives rise to most of the nervous system: anteriorly it enlarges and differentiates into the brain, cranial nerves, and eyes; posteriorly it forms the spinal cord and spinal motor nerves. Sensory nerves arise from special **neural crest** cells pinched off from the neural tube before it closes.

How are the billions of nerve axons in the body formed? What directs their growth? Since a single nerve axon may be several feet in length (for example, motor nerves running from the spinal cord to the toes), at one time it seemed impossible that a single cell could spin out so far. It was suggested that a nerve fiber grew from a series of preformed protoplasmic bridges along its route. The answer had to await the development of one of the most powerful tools available to biologists, the cell culture technique.

The tissue culture technique developed by Ross G. Harrison is now used extensively by scientists in all fields of active biomedical research, not just by embryologists. The great impact of the technique has been felt only in recent years. Harrison was twice considered for the Nobel Prize (1917 and 1933), but he failed ever to receive the award because, ironically, the tissue culture method was then believed to be "of rather limited value."

In 1907 embryologist Ross G. Harrison discovered that he could culture living neuroblasts (embryonic nerve cells) for weeks outside the body by placing them in a drop of frog lymph hung from the underside of a cover slip. Watching nerves grow for periods of days, he saw that each nerve fiber was the outgrowth of a single cell. As the fibers extended outward, materials for growth flowed down the axon center to the growing tip where they were incorporated into new protoplasm.

The second question—what directs nerve growth—has taken longer to unravel. An idea held well into the 1940s was that nerve growth is a random, diffuse process. It was believed that the nervous system developed as an equipotential network, or blank slate, that later would be shaped by usage into a functional system. The nervous system just seemed too incredibly complex for us to imagine how nerve fibers could find their way selectively to so many predetermined destinations. Yet it appears that this is exactly what they do! Recent work with invertebrate nervous systems indicates that each of the billions of nerve cell axons acquires a distinct identity that in some way directs it along a specific pathway to its destination. Many years ago Harrison observed that a growing nerve axon terminated in a "growth cone," from which extend numerous tiny thread-like pseudopodial processes (filopodia) (Figure 13-16). These are constantly reaching out and searching the environment in all directions for the pathway having the molecular labels that only that axon's growth cone recognizes. This chemical guidepost system, which must, of course, be genetically directed, is just one example of the amazing precision that characterizes the entire process of differentiation.

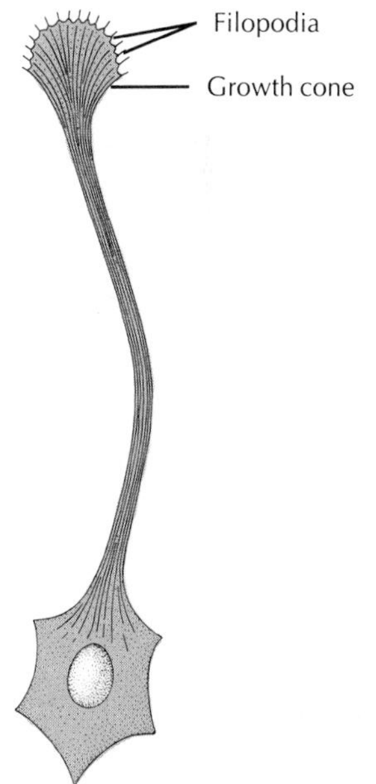

FIGURE 13-16

Growth cone at the growing tip of a nerve axon. Materials for growth flow down the axon from the cell body to the growth cone from which numerous thread-like filopodia extend. These appear to serve as a pioneering guidance system for the developing nerve fiber.

Derivatives of endoderm: digestive tube and survival of gill arches

In the frog embryo the primitive gut makes its appearance during gastrulation with the formation of an internal cavity, the **archenteron** (Figure 13-14). From this simple endodermal cavity develop the lining of the digestive tract, the lining of the pharynx and lungs, most of the liver and pancreas, the thyroid and parathyroid glands, and the thymus.

The **alimentary canal** develops from the primitive gut and is folded off from the yolk sac by the growth and folding of the body wall (Figure 13-17). The ends of the tube open to the exterior and are lined with ectoderm, whereas the rest of the tube is lined with endoderm. The **lungs, liver,** and **pancreas** arise from the foregut.

Among the most intriguing derivatives of the digestive tract are the pharyngeal (gill) arches and pouches, which make their appearance in the early embry-

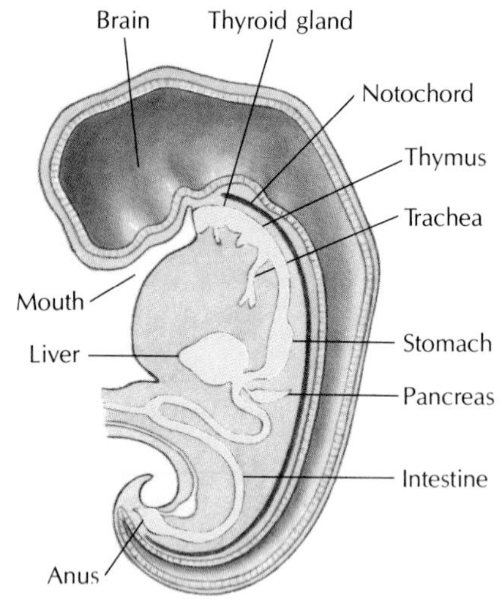

FIGURE 13-17

Derivatives of the alimentary canal of a human embryo.

onic stages of all vertebrates (Figure 4-12, p. 101). In fishes, the gill arches develop into gills and supportive structures and serve as respiratory organs. When the early vertebrates moved onto land, gills were unsuitable for aerial respiration and were replaced by lungs.

Why then do gill arches persist in the embryos of terrestrial vertebrates? Certainly not for the convenience of biologists who use these and other embryonic structures to reconstruct lines of vertebrate descent. Even though the gill arches serve no respiratory function in either the embryos or adults of terrestrial vertebrates, they remain as necessary primordia for a great variety of other structures. For example, the first arch and its endoderm-lined pouch (the space between adjacent arches) form the upper and lower jaws and inner ear of higher vertebrates. The second, third, and fourth gill pouches contribute to the tonsils, parathyroid gland, and thymus. We can understand then why gill arches and other fishlike structures appear in early mammalian embryos. Their original function has been abandoned, but the structures are retained for new purposes. It is the great conservatism of early embryonic development that has so conveniently provided us with a telescoped evolutionary history.

Derivatives of mesoderm: support, movement, and beating heart

The intermediate germ layer, the mesoderm, forms the vertebrate skeletal, muscular, and circulatory structures and the kidneys. As vertebrates have increased in size and complexity, the mesodermally derived supportive, movement, and transport structures make up an even greater proportion of the body bulk.

Most **muscles** arise from the mesoderm along each side of the spinal cord (Figure 13-18). The mesoderm divides into a linear series of somites (38 in humans), which by splitting, fusion, and migration become the muscles of the body and axial parts of the skeleton.

Although the primitive mesoderm appears after the ectoderm and endoderm, it gives rise to the first functional organ, the embryonic heart. Guided by the underlying endoderm, clusters of precardiac mesodermal cells move ameba-like into a central position between the underlying primitive gut and the overlying neural tube. Here the heart is established, first as a single, thin tube.

Even while the cells group together, the first twitchings are evident. In the chick embryo, a favorite and nearly ideal animal for experimental embryology studies, the primitive heart begins to beat on the second day of the 21-day incubation period; it begins beating before any true blood vessels have formed and before there is any blood to pump. As the ventricle primordium develops, the spontaneous cellular twitchings become coordinated into a feeble but rhythmical beat. Then, as the atrium develops behind the ventricle, followed by the development of the sinus venosus behind the atrium, the heart rate quickens. Each new heart chamber has an intrinsic beat that is faster than its predecessor.

Finally a specialized area of heart muscle called the **sinus** node develops in the sinus venosus and takes command of the entire heartbeat. This becomes the heart's **pacemaker.** As the heart builds up a strong and efficient beat, vascular channels open within the embryo and across the yolk. Within the vessels are the first primitive blood cells suspended in plasma.

The early development of the heart and circulation is crucial to continued embryonic development because without a circulation the embryo could not obtain materials for growth. Food is absorbed from the yolk and carried to the embryonic body; oxygen is delivered to all the tissues, and carbon dioxide and other wastes are carried away. The embryo is totally dependent on these extraembryonic support systems, and the circulation is the vital link between them.

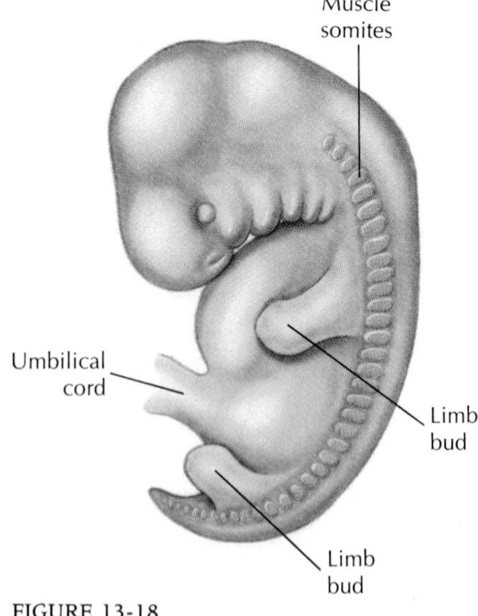

FIGURE 13-18

Human embryo showing muscle somites, which differentiate into muscles and axial skeleton.

Amniotes and the Amniotic Egg

Reptiles, birds, and mammals form a natural grouping of vertebrates called **amniotes,** meaning that they develop an amnion, one of the extraembryonic membranes that make the development of these forms unique among animals.

As rapidly growing living organisms, embryos have the same basic animal requirements as adults—food, oxygen, and disposal of wastes. For the embryos of the many invertebrates that develop in contact with the external environment, gas exchange is a simple matter of direct diffusion. The egg may contain only enough yolk to sustain the embryo through hatching, when it must begin to feed itself. Beyond hatching, the embryo (called a free-swimming **larva**) is on its own.

Yolk enclosed in a membranous **yolk sac** is a conspicuous feature of all fish embryos (Figure 13-19). The yolk is gradually used up as the embryo grows; the yolk sac shrinks and finally is enclosed within the body of the embryo. Bird and reptile eggs are also provided with large amounts of yolk to support early development. In birds the yolk reaches relatively massive proportions, since it must nourish a baby bird to a much more advanced stage of development at hatching than a larval fish.

In abandoning an aquatic life for a land existence the first terrestrial animals had to evolve a sophisticated egg containing a complete set of life-support systems. Thus appeared the egg of amniotes, equipped to protect and support the growth of embryos on dry land. In addition to the yolk sac containing the nourishing yolk are three other membranous sacs—amnion, chorion, and allantois. All are referred to as extraembryonic membranes because they are accessory structures that develop beyond the embryonic body and are discarded when the embryo hatches.

The **amnion** is a fluid-filled bag that encloses the embryo and provides a private aquarium for development (Figure 13-20). Floating freely in this aquatic environment, the embryo is fully protected from shocks and adhesions. The evolution of this structure was crucial to successful habitation on land.

The **allantois,** another component in the support system for embryos of land animals, is a bag that grows out of the hindgut of the embryo (Figure 13-20). It

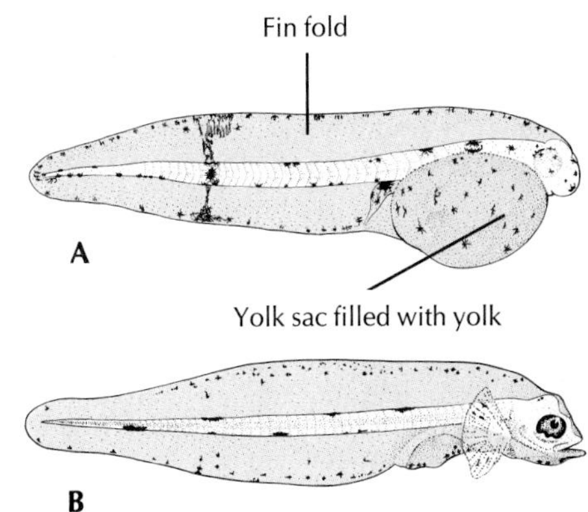

FIGURE 13-19

Fish embryos showing yolk sac. **A,** The 1-day-old larva of a marine flounder has a large yolk sac. **B,** By the time the 10-day-old larva has developed a mouth and primitive digestive tract, its yolk supply has been exhausted. It must catch food to survive and continue growing.

From Hickman, C.P., Jr. 1959. The larval development of the sand sole, *Paralichthys melanostictus,* Washington State Fisheries Research Papers 2:38-47.

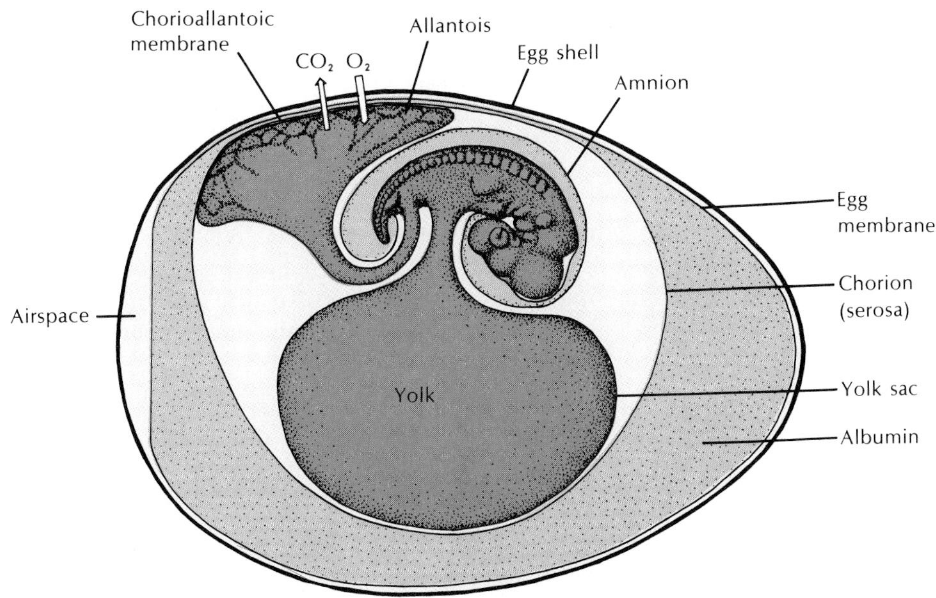

FIGURE 13-20

The egg of an amniote (chick) at an early stage of development, showing the extraembryonic membranes (amnion, yolk sac, allantois, and chorion). Additional egg membranes and the egg shell are secreted as the egg proceeds down the hen's oviduct. The porous shell allows gaseous exchange of oxygen and carbon dioxide. Circulatory channels from the embryo's body to the allantois and yolk sac are not shown.

collects the wastes of metabolism (mostly uric acid). At hatching, the young animal breaks its connection with the allantois and leaves it and its refuse behind in the shell.

The **chorion** (also called **serosa**) is an outermost extraembryonic membrane that completely encloses the rest of the embryonic system. It lies just beneath the shell (Figure 13-20).

An important consequence of the evolution of the shelled egg is that the egg must be fertilized *before* the eggshell is wrapped around it, if it is to develop. Thus the male must introduce the sperm into the female tract so that the sperm can reach the egg before it passes to that part of the oviduct where the shell is secreted. Hence, as one biologist put it, it is the eggshell and not the devil that deserves blame for the happy event we know as sex!

As the embryo grows and its need for oxygen increases, the allantois and chorion fuse together to form a **chorioallantoic membrane.** This double membrane is provided with a rich vascular network, connected to the embryonic circulation. Lying just beneath the porous shell, the vascular chorioallantoic membrane serves as a kind of lung across which oxygen and carbon dioxide can freely exchange. Although nature did not intend it, the chorioallantoic membrane of the chicken egg has been used extensively by generations of experimental embryologists as a place to culture chick and mammalian tissues.

The great importance of the amniotic egg to the establishment of a land existence cannot be overemphasized. Amphibians must return to water to lay their eggs. But the reptiles, even before they took to land, developed the amniotic egg with its self-contained aquatic environment enclosed by a tough outer shell. Protected from drying out and provided with yolk for nourishment, such eggs could be laid on dry land, far from water. Reptiles were thus freed from aquatic life and could become the first true terrestrial tetrapods.

The amniotic egg, for all its virtues, has one basic vulnerability: placed neatly in a nest, it makes fine food for other animals. The mammals evolved an alternative for early development: allow the embryo to grow within the protective confines of the mother's body. This has resulted in important modifications in development of mammals as compared with other vertebrates. The earliest mammals, descended from early reptiles, were egg layers. Even today the most primitive mammals, the monotremes (for example, duckbill platypus and spiny anteater, Figure 13-21, *A*), lay large yolky eggs that closely resemble bird eggs. In the marsupials (pouched mammals such as the opossum and kangaroo, Figure 13-21, *B*), the embryos develop for a time within the mother's uterus. But the embryo does not "take root" in the uterine wall, as do the embryos of the most advanced **placental mammals,** and consequently it receives little nourishment from the mother. The young of marsupials are therefore born in a very immature state after less than 2 weeks of development and are sheltered and nourished in a pouch of the abdominal wall.

All other mammals, the placentals, nourish their young in the uterus by means of a **placenta.**

A

B

FIGURE 13-21

Mammals without placentas. **A,** The echidna, or spiny anteater, is a monotreme that deposits her eggs in a pouch. **B,** The gray kangaroo carries her joey in a pouch.

Photographs by C.P. Hickman, Jr.

Reproductive Patterns

The diversity of reproduction in animals is extraordinary, probably exceeding that of any other body system. The most imaginative, and certainly the most complex, evolutionary adaptations seem to have occurred in the area of sexuality. Diversity is expressed in numerous ways, including structure and function of reproductive systems, breeding patterns, methods of care and nourishment of the offspring, and behavior.

MATERNAL SUPPORT OF THE EMBRYO

The great majority of invertebrates, as well as many vertebrates, lay their eggs in the environment for development; these animals are called **oviparous** ("egg-birth"). Fertilization may be either external or internal. While many oviparous animals simply abandon their eggs rather indiscriminately, others display extreme care in finding places that will provide immediate and suitable sources of food for the young when they hatch. Some animals retain their eggs in the body (usually the oviduct) while they develop, with the embryo deriving all of its nourishment from yolk stored within the egg. These animals are called **ovoviviparous** ("egg-live-birth"). Ovoviviparity occurs in several invertebrate groups (e.g., various annelids, brachiopods, insects, and gastropod molluscs) and is common among certain fishes and reptiles. In the third pattern, **viviparous** ("live-birth"), the egg develops in the oviduct or uterus with the embryo deriving its nourishment directly from the mother. Usually some kind of intimate anatomical relationship is established between the developing embryo and the mother. In both ovoviviparity and viviparity, fertilization must be internal, and the mother gives birth to young in an advanced stage of development. Viviparity is confined mostly to the mammals and elasmobranch fishes, although viviparous invertebrates (some scorpions, for example), amphibians, and reptiles are known. Development of the embryos within the mother's body, whether ovoviviparous or viviparous, obviously affords more protection to the offspring than egg-laying.

BREEDING CYCLES

The reproductive activities of almost all animals are cyclic. Usually breeding cycles are seasonal and coordinated so that the young are born or hatched at a time of year when conditions are most favorable for growth and survival. In mammals we find two different kinds of reproductive cycles: the **estrous cycle,** which is characteristic of most mammals, and the **menstrual cycle,** which is characteristic of primates only. These two cyclic patterns differ in two ways. First, in the estrous cycle, but not in the menstrual cycle, the female is receptive to the male (that is, she is "in heat") only at restricted periods of the year. Second, the menstrual cycle, but not the estrous cycle, ends with the collapse and sloughing of the uterine lining (endometrium), leading to the menstrual discharge. In the estrous animal each cycle ends with the uterine lining simply reverting to its original state, without the bleeding characteristic of the menstrual cycle.

Human Development

HOW HORMONES COORDINATE DEVELOPMENT

Sexual reproduction is a complex process requiring intricate cuing, especially in the female. The cues are hormones produced by the brain (hypothalamus), pituitary, and gonads. Study of how these hormones interact has made possible the development of the most effective and reliable contraception yet found, the "pill."

The ovaries produce two kinds of steroid sex hormones—**estrogens** and **progesterone** (Figure 13-22). Estrogens are responsible for the development of the female accessory sex structures (uterus, oviducts, and vagina) and secondary sex characters, such as breast development and characteristic bone growth, fat deposition, and hair distribution. Progesterone is responsible for preparing the uterus to receive the developing embryo. These hormones are controlled by the **pituitary gonadotropins,** follicle-stimulating hormone (FSH), and luteinizing hormone (LH) (Figure 13-23). The gonadotropins are in turn governed by neurosecretory centers in the hypothalamus of the brain. These cells produce neurohormones (polypeptide-releasing hormones, p. 268) that control synthesis and release of FSH and LH. Through this control system environmental factors such as light, nutrition, and stress may influence reproductive cycles. In many mammals, two separate releasing hormones have been identified: follicle-stimulating hormone–releasing hormone (FSH-RH) and luteinizing hormone–releasing hormone (LH-RH). In humans, however, it appears that both FSH and LH are controlled by a single releasing hormone, LH-RH.

The menstrual cycle (L. *mensis,* month) consists of three distinct phases: men-

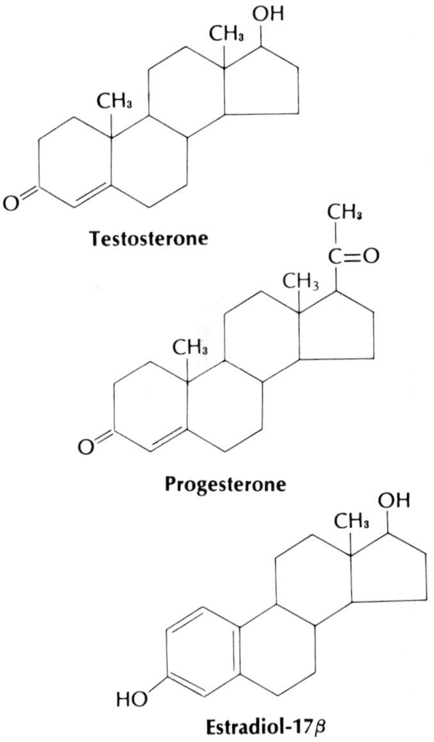

OH
CH_3
CH_3
O
Testosterone

CH_3
C=O
CH_3
CH_3
O
Progesterone

OH
CH_3
HO
Estradiol-17β

FIGURE 13-22

These three sex hormones show the basic four-ring steroid structure. The female sex hormone estradiol-17β (an estrogen) is a C_{18} (18-carbon) steroid with an aromatic A ring (first ring to left). The male sex hormone testosterone is a C_{19} steroid with a carbonyl group (C=O) on the A ring. The female pregnancy hormone progesterone is a C_{21} steroid, also bearing a carbonyl group on the A ring.

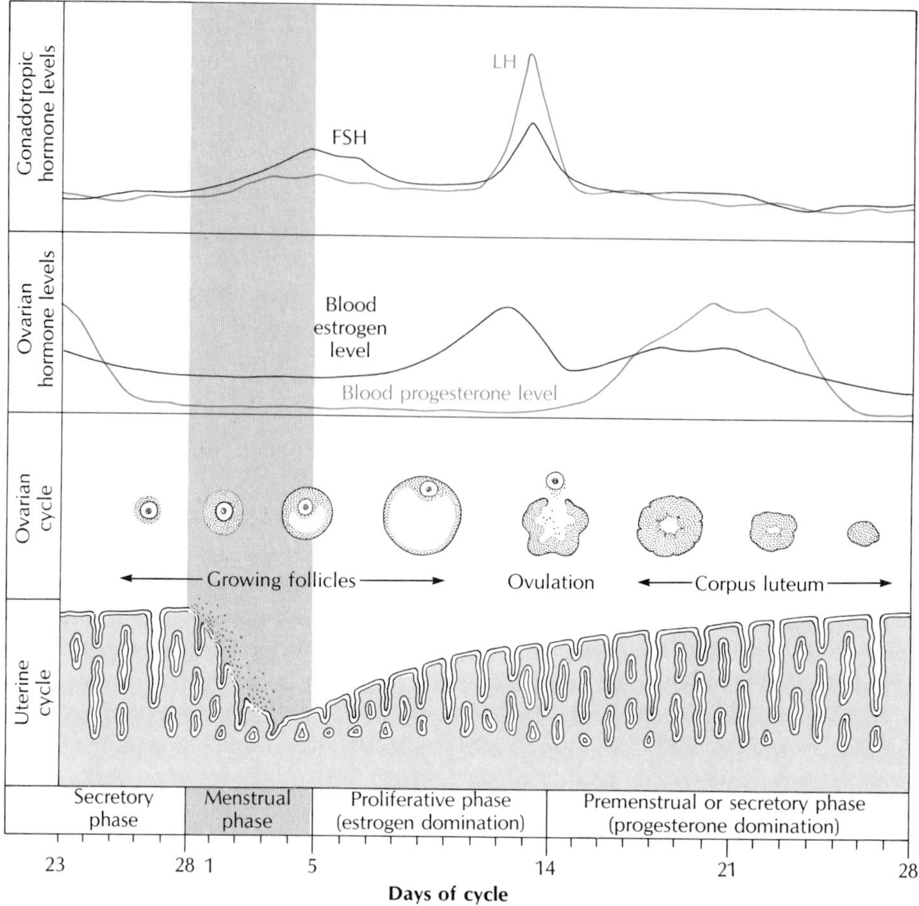

FIGURE 13-23

Human menstrual cycle, showing changes in blood hormone levels and uterine endometrium during the 28-day cycle. FSH promotes maturation of the ovarian egg follicles, which secrete estrogen. Estrogen prepares the uterine endometrium and causes a surge in LH, which in turn stimulates the corpus luteum to secrete progesterone. Progesterone production will persist only if the egg is fertilized; without pregnancy progesterone declines and menstruation follows.

strual phase, follicular phase, and **luteal phase.** Menstruation (the "period") signals the menstrual phase, when part of the lining of the uterus (endometrium) degenerates and sloughs off, producing the menstrual discharge. By day 3 of the cycle the blood levels of FSH and LH begin to rise slowly, prompting some of the ovarian follicles to begin growing. As the follicles grow, they begin to secrete estrogen. As estrogen levels in the blood increase, the uterine endometrium begins to thicken and uterine glands within the endometrium enlarge. By day 10 most of the ovarian follicles that began to develop at day 3 now degenerate, leaving only one (sometimes two or three) to continue ripening until it appears like a blister on the surface of the ovary.

At day 13 or 14 in the cycle, a surge of LH from the pituitary induces the largest follicle to rupture (**ovulation**), releasing the egg onto the ovarian surface. Now follows a critical period, for unless the mature egg is fertilized within a few hours it will die. During the luteal phase, a **corpus luteum** ("yellow body") forms from the wall of the follicle that ovulated (Figure 13-23). The corpus luteum, responding to the continued stimulation of LH, secretes progesterone in addition to estrogen. Progesterone ("before carrying [gestation]"), as its name implies, stimulates the uterus to undergo the final maturational changes that prepare it for gestation.

The uterus is thus fully ready to house and nourish the embryo by the time the latter settles out onto the uterine surface, usually about 7 days after ovulation. If fertilization has *not* occurred, the corpus luteum disappears, and its hormones are no longer secreted. Since the uterine lining (endometrium) depends on progesterone and estrogen for its maintenance, their disappearance causes the uterine lining to deteriorate, leading to the menstrual discharge. However, if the egg has been fertilized and has implanted, the corpus luteum continues to supply the essential sex hormones needed to maintain the mature uterine lining. During the first few weeks of pregnancy the developing placenta itself begins to produce the sex hormones progesterone and estrogen and soon replaces the corpus luteum in this function.

As pregnancy advances, progesterone and estrogen stimulate the breasts to prepare for milk production. The actual secretion and release of milk after birth (lactation) is the result of two other hormones, **prolactin** and **oxytocin** (Table 11-1, p. 269). Milk is not secreted during pregnancy because the placental sex hormones inhibit the release of prolactin by the pituitary. The placenta, like the corpus luteum that preceded it, thus becomes a special endocrine gland of pregnancy. After delivery, many mammals eat the placenta (afterbirth), a behavior that serves to remove telltale evidence of a birth from potential predators.

The male sex hormone **testosterone** (Figure 13-22) is manufactured by the **interstitial cells** of the testes. Testosterone is necessary for the growth and development of the male accessory sex structures (penis, sperm ducts, and glands), development of secondary male sex characters (hair distribution, voice quality, and bone and muscle growth), and male sexual behavior. The same pituitary hormones that regulate the female reproductive cycle, FSH and LH, are also produced in the male, where they guide the growth of the testes and testosterone secretion.

EARLY DEVELOPMENT OF HUMAN EMBRYO

The eggs of all placental mammals, although relatively enormous on a cellular scale, are small by egg standards. The human egg is approximately 0.1 mm in diameter and barely visible to the unaided eye. It contains very little yolk. After fertilization in the oviduct, the dividing egg begins a 5-day journey down the oviduct toward the uterus, propelled by a combination of ciliary action and muscular peristalsis. Cleavage is very slow: 24 hours for the first cleavage and 10 to 12 hours for each subsequent cleavage. By comparison, frog eggs cleave once every hour. Cleavage produces a small solid ball of 20 to 30 cells (called the **morula**). Later, a fluid-filled cavity appears, the **segmentation cavity** (Figure 13-24). This is comparable to the blastocoel of a frog's egg. The embryo is now called a **blastocyst.**

At this point, development of the mammalian embryo departs radically from that of lower vertebrates. A mass of cells, called the **inner cell mass**, develops on one side of the peripheral cell, or **trophoblast**, layer (Figure 13-24). The inner cell mass forms the embryo, whereas the surrounding trophoblast forms the placenta. When the blastocyst is approximately 6 days old and composed of approximately 100 cells, it contacts and implants into the uterine endometrium (Figure 13-24). Very little is known of the forces involved in implantation, or why, incidentally, the intrauterine birth control devices (coil and loop) are so effective in preventing successful implantation. On contact, the trophoblast cells proliferate rapidly and produce enzymes that break down the epithelium of the uterine endometrium. This allows the blastocyst to sink into the endometrium. By the eleventh or twelfth day the blastocyst, totally buried, has eroded through the walls of capillaries and small arterioles; this releases a pool of blood that bathes the embryo. At first the minute embryo derives what nourishment it requires by direct diffusion from the surrounding blood. But very soon a remarkable fetal-maternal structure, the placenta, develops to assume these exchange tasks.

The widely heralded first "test-tube baby," born in England in July 1978, was a medical and technological triumph, the result of years of collaborative research by R.G. Edwards and P. Steptoe. More properly termed external human fertilization (EHF), the procedure is used to overcome infertility caused usually by a malfunction or malformation of the oviduct that prevents fertilization. First, eggs must be collected from the ovary using a special instrument that is introduced through a small incision in the abdominal wall. The eggs are placed in a specially compounded solution in a laboratory dish and fertilized with the husband's sperm. The eggs are allowed to develop to the 8-cell stage (about 2 days) until the uterus has reached a receptive condition. Two or three embryos are then introduced into the uterus via a tube passed through the cervix and are allowed to implant. The results of this technically simple step are unpredictable, and only about 30% of the embryos result in a pregnancy. Nevertheless, thousands (an estimated 20,000 by 1987) of "test tube babies" have been born to delighted parents.

FIGURE 13-24

Early development of the human embryo.

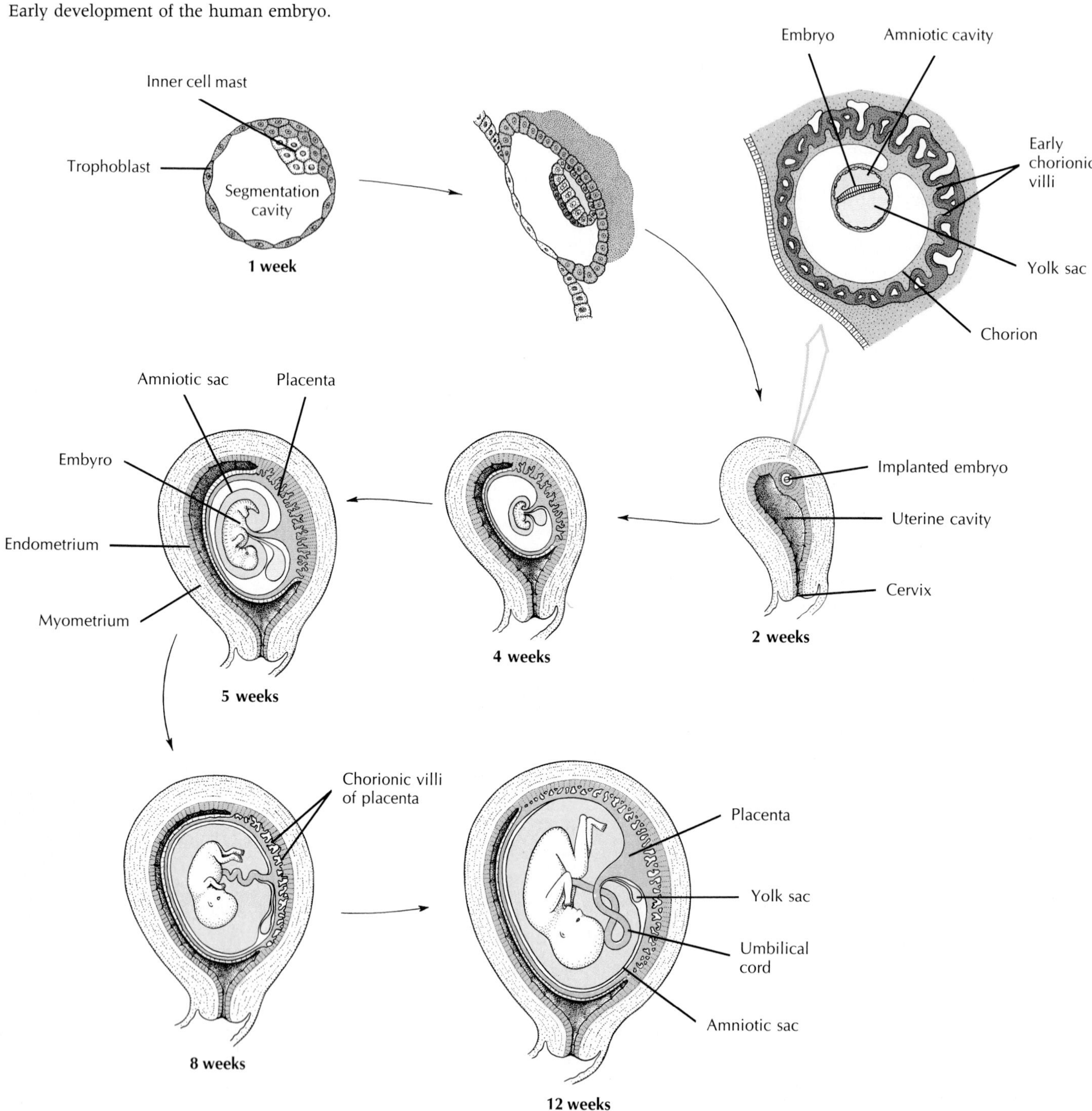

PLACENTA

The placenta is a marvel of biological engineering. Serving as a provisional lung, intestine, and kidney for the embryo, it performs elaborate selective activities without allowing the maternal and fetal blood to intermix (very small amounts of fetal blood may regularly escape into the maternal system without causing harm). The placenta permits the entry of foodstuffs, hormones, vitamins, and oxygen and the exit of carbon dioxide and metabolic wastes. Its action is highly selective because it allows some materials to enter that are chemically quite similar to others that are rejected.

The two circulations are physically separated at the placenta by an exceedingly thin membrane (in places only 2 μm thick), across which materials are transferred by diffusive interchange. The transfer occurs across the thousands of tiny fingerlike projections, called **chorionic villi**, which develop from the original chorion membrane (Figure 13-24). These projections sink like roots into the uterine endometrium after the embryo implants. As development proceeds and embryonic demands for food and gas exchange increase, the great proliferation of villi in the placenta vastly increases its total surface area. Although the human placenta at term measures only 18 cm (7 inches) across, its total absorbing surface is approximately 13 square meters — 50 times the surface area of the skin of the newborn infant.

Since the mammalian embryo is protected and nourished by the placenta, what becomes of the various embryonic membranes of the amniotic egg whose functions are no longer required? Surprisingly perhaps, all of these special membranes are still present, although they may be serving a new function. Two extraembryonic membranes, the allantois and chorion, have been totally redesigned. The allantois is no longer needed as a urinary bladder. Instead it becomes the stalk, or **umbilical cord**, that links the embryo physically and functionally with the placenta. The chorion, the outermost membrane, forms most of the placenta itself. The yolk sac is incorporated into the umbilical cord, and during early development it is the source of lymphocytes that later migrate to the spleen and liver (later still they migrate again to the bone marrow) to originate the body's immune system. The amnion remains unchanged, a protective water jacket in which the embryo floats.

One of the most intriguing questions the placenta presents is why it is not rejected by the mother's tissues. The placenta is a uniquely successful foreign transplant, or **allograft.** It (as well as the embryo) is genetically foreign in the sense that it consists of tissue derived not only from the mother but also from the father. We should expect it to be rejected by the uterine tissues, just as a piece of a child's skin is rejected by the child's mother when a surgeon attempts a grafting transplant (p. 177). The placenta in some way circumvents the normal rejection phenomenon, a matter of the greatest interest to immunologists seeking ways to transplant tissues and organs successfully.

Human twins may come from one zygote (identical twins) or two zygotes (nonidentical, or fraternal, twins). Triplets, quadruplets, and quintuplets may include a pair of identical twins. The other babies in such multiple births usually come from separate zygotes. Fraternal twins do not resemble each other more than other children born separately in the same family, but identical twins are, of course, strikingly alike and always of the same sex. Embryologically, each member of fraternal twins has its own placenta, chorion, and amnion. Usually (but not always) identical twins share the same chorion and the same placenta, but each has its own amnion. Sometimes identical twins fail to separate completely and form Siamese twins, in which the organs of one may be a mirror image of the organs of the other. The frequency of twin births in comparison to single births is approximately 1 in 86, that of triplets approximately 1 in 86^2, and that of quadruplets approximately 1 in 86^3.

PREGNANCY AND BIRTH

Pregnancy may be divided into four phases. The first phase of 6 or 7 days is the period of cleavage and blastocyst formation; it ends when the blastocyst implants in the uterus. The second phase of 2 weeks is the period of gastrulation and formation of the neural plate. The third phase, called the **embryonic period**, is a crucial and sensitive period of primary organ system differentiation. This phase ends at approximately the eighth week of pregnancy when the primordia of all the chief organs have been formed. The last phase, known as the **fetal period**, is characterized by rapid growth, proportional changes in body parts, and final preparation for birth.

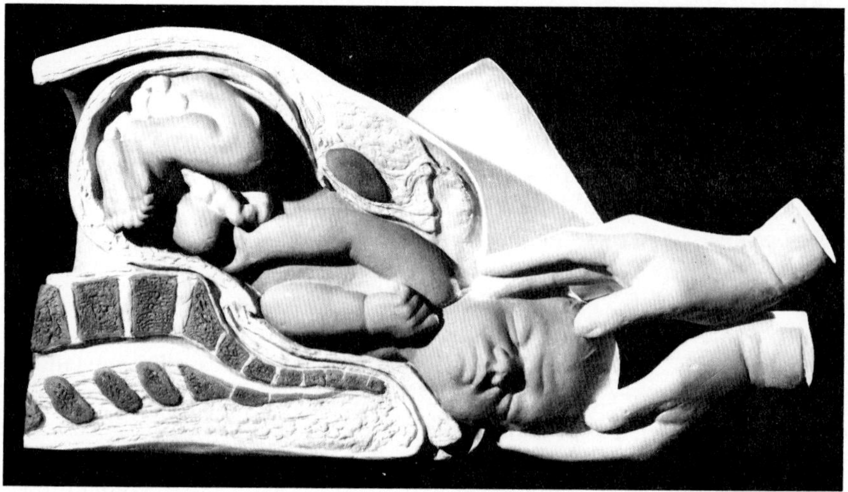

FIGURE 13-25

Birth of a baby.

Courtesy American Museum of Natural History.

Oral contraceptives (the "pill") are usually combined preparations of estrogen and progesterone that act to decrease the output of the pituitary gonadotropins FSH and LH. This prevents the ovarian follicles from ripening and ovulation from occurring. Birth control is highly effective, with a failure rate of less than 1%, if the treatment procedure is adhered to properly.

During pregnancy the placenta gradually takes over most of the functions of regulating growth and development of the uterus and the fetus. As an endocrine gland it secretes estrogen and progesterone, hormones that are secreted by the ovaries and corpus luteum in the early periods of pregnancy. The placenta also produces **chorionic gonadotropin,** a hormone that assumes the role of the LH and FSH pituitary hormones, which cease their secretions about the second month of pregnancy. Chorionic gonadotropin maintains the corpus luteum so that it may continue secreting the progesterone and estrogen that are necessary for the integrity of the placenta. In the later stages of pregnancy the placenta becomes a totally independent endocrine organ, requiring support from neither the corpus luteum nor the pituitary.

What stimulates birth? Why does pregnancy not continue indefinitely? What factors produce the onset of labor (the rhythmical contractions of the uterus)? Thus far, we have only tentative answers to these questions. It has long been known that estrogens stimulate uterine contractions, whereas progesterone, also secreted by the placenta, blocks uterine activity. Just before birth there is a sharp increase in the plasma level of estrogens and a decrease in progesterone. This appears to remove the "progesterone block" that keeps the uterus quiescent throughout pregnancy. Thus labor contractions can proceed.

The first major signals that birth is imminent are the so-called labor pains, caused by the rhythmical contractions of the uterine musculature. These are usually slight at first and occur at intervals of 15 to 30 minutes. They gradually become more intense, longer in duration, and more frequent. They may last anywhere from 6 to 24 hours, usually longer with the first child.

Childbirth occurs in three stages. In the first stage the neck (cervix), or opening of the uterus into the vagina, is enlarged by the pressure of the baby in its bag of amniotic fluid, which may be ruptured at this time. In the second stage the baby is forced out of the uterus and through the vagina to the outside (Figure 13-25). In the third stage the placenta, or afterbirth, is expelled from the mother's body, usually within 10 minutes after the baby is born.

Summary

Reproduction is a universal property of all life. Asexual reproduction, characteristic of simple life forms, is a rapid and direct process by which a single organism, by budding, fission, or sporulation, produces genetically identical copies of itself. Sexual reproduction, however, is far more widespread among animals and has been firmly maintained throughout metazoan evolution. In sexual reproduction, sex cells (gametes) produced by two parents are combined to form a zygote that develops into a new genotype that is different from either parent. By combining parental characters in this way, sexual reproduction reshuffles and amplifies genetic diversity; this is important for evolution. This appears to be the principal advantage of sex, but, because sexual reproduction carries with it certain "costs," there is no general agreement among biologists concerning the mechanisms responsible for the origin and maintenance of sex.

In sexual reproduction, germ cells mature in the gonads by mitosis and meiosis to produce mature, nonmotile eggs and motile sperm. When these are combined at fertilization, metabolic activity in the quiescent egg is suddenly renewed. The egg divides repeatedly to produce numerous small cells (blastomeres) that have specific developmental fates. As development proceeds, certain parts of each cell's genome are expressed while the remainder are switched off in some way. This differential gene expression allows individual cells, or groups of cells, to become committed to specific directions of differentiation in an orderly and integrated manner.

Cleavage is followed by gastrulation, a process of cellular reorganization, when three germ layers (ectoderm, mesoderm, endoderm) are formed. Each germ layer differentiates into tissues and organs: the ectoderm gives rise to the skin and nervous system; the endoderm becomes the alimentary canal, pharynx, lungs, and certain glands; and the mesoderm forms the muscular, skeletal, and excretory systems.

In the invertebrates and lower vertebrates (fish and amphibians), embryological development terminates with the emergence of a larva, nymph, or juvenile. Larvae, especially, are important as a means of dispersal for the species. A new departure appeared with the evolution of reptiles: the evolution of an egg having a set of special extraembryonic membranes that provide prolonged support for the growth of the embryo within the egg. With this advancement, reptiles, and the birds that evolved from them, could lay their eggs on land and become completely freed of the aquatic environment for reproduction. In all but the most primitive mammals, the egg develops in the oviduct or uterus with the embryo deriving its nourishment directly from the mother by way of a complex fetal-maternal structure, the placenta.

The management of sexual reproduction is complex because two kinds of sex cells are housed in separate animals. This, together with the seasonal or cyclic nature of reproduction, has required the evolution of precise hormonal mechanisms that control the production of sex cells, signal readiness for mating, and prepare ducts and glands for successful fertilization of eggs. In vertebrates, the pituitary gland, through gonadotropic hormones, links the neurosecretory centers of the brain with the endocrine tissues of the gonads. Placental mammals have extended endocrine regulation to optimize conditions for the prolonged maternal support of the young, both during pregnancy and subsequent postnatal care.

REVIEW QUESTIONS

1. Define asexual reproduction, and describe four forms of asexual reproduction in invertebrates.
2. Define sexual reproduction, and explain why meiosis contributes to one of its great strengths.
3. Explain why gene mutations in asexual organisms lead to much more rapid evolutionary change than do gene mutations in sexual forms.
4. Define the two alternatives to typical biparental reproduction—parthenogenesis and hermaphroditism—and offer a specific example of each from the animal kingdom.
5. Define the following: dioecious, monoecious.
6. Name the general location, and give the function of the following reproductive structures: seminiferous tubules, vas deferens, urethra, seminal vesicles, graafian follicle, oviducts, endometrium.
7. Draw or trace Figure 13-6, putting in only the circles and arrows. Then, completely label the events in oogenesis and spermatogenesis without reference to Figure 13-6. How do oogenesis and spermatogenesis differ?
8. Describe the events that follow contact of a spermatozoon with an egg and explain what is meant by the term "activation."
9. Compare and explain the results of removing one blastomere from a 4-cell stage embryo showing regulative development and from a 4-cell stage embryo showing mosaic development.
10. Distinguish between radial and spiral cleavage. Why are these two patterns of cleavage said to signal an important evolutionary dichotomy in the animal kingdom?
11. Give the evidence for this statement: "Nucleocytoplasmic interactions form the basis of the organized differentiation of tissues."
12. Explain what the "growth cone" that Ross Harrison observed at the ends of growing nerve fibers has to do with the direction of nerve growth.
13. Name two organ system derivatives of each of the three primary germ layers (ectoderm, mesoderm, and endoderm).
14. What are the four extraembryonic membranes of the amniotic egg of a bird or reptile, and what is the function of each membrane?
15. Explain how the female hormones FSH, LH, and estrogen interact during the menstrual cycle to bring about ovulation and, subsequently, formation of the corpus luteum.
16. Explain the function of the corpus luteum in the menstrual cycle. Compare what happens to the corpus luteum when the ovulated egg is fertilized and when it is not fertilized.
17. What is the male sex hormone, and what is its function?
18. Define the following: blastocyst, inner cell mass, trophoblast, placenta, chorionic villi.
19. What is chorionic gonadotropin, and what is its functional role during pregnancy?

SELECTED REFERENCES

Austin, C.R., and R.V. Short (eds.), 1986. Reproduction in mammals, Book 5: Manipulating reproduction, ed. 2. New York, Cambridge University Press. *Includes chapters on contraception and alleviating infertility in humans.*

Bell, G. 1982. The masterpiece of nature: the evolution and genetics of sexuality. Berkeley, University of California Press. *Scholarly synthesis of ideas on the meaning of sex. Advanced treatment.*

Blüm, V. 1986. Vertebrate reproduction: a textbook. New York, Springer-Verlag (translated from the German by A.C. Whittle). *Broad, readable treatment of structure and function of vertebrate reproductive systems.*

Forsyth, A. 1986. A natural history of sex: the ecology and evolution of sexual behavior. New York, Charles Scribner's Sons. *Engagingly written, factually accurate account of the sex lives of animals from protozoa to humans, abounding in imagery and analogy. Highly recommended.*

Gilbert, S.F. 1986. Developmental biology, ed. 2. Sunderland, Mass., Sinauer Associates. *Combines descriptive and mechanistic aspects; good selection of examples from many animal groups.*

Goodman, C.S., and M.J. Bastiani. 1984. How embryonic nerve cells recognize one another. Sci. Am. **251**:58-66 (Dec.). *Research with insect larvae shows that developing neurons follow pathways having specific molecular labels.*

Halliday, T. 1982. Sexual strategy. Survival in the wild. Chicago, University of Chicago Press. *Semipopular treatment of sexual strategies, especially vertebrate mating systems, presented in a framework of natural selection. Well-chosen illustrations.*

Slack, J.M.W. 1984. From egg to embryo: determinative events in early development. New York, Cambridge University Press. *Emphasis on pattern formation; comparative approach.*

Wassarman, P.M. 1988. Fertilization in mammals. Sci. Am. **259**:78-84 (Dec.). *Describes the sequence of events preceding and following sperm and egg fusion that, among other things, prevents polyspermy.*

PART THREE

The Invertebrate Animals

NEW DESIGNS FOR LIVING

Thirty-two phyla of multicellular animals are recognized by zoologists today, each phylum distinguished by a distinctive body plan and array of biological properties that set it apart from all other phyla. Nearly all are the survivors of perhaps 100 phyla that were generated 600 million years ago during the Cambrian explosion, which is the most important evolutionary event in the history of animal life. Within the space of a few million years virtually all of the major body plans that we see today, together with many other novel plans that we know only from the fossil record, were established. Entering a world sparse in species and mostly free of competition, these new life forms began widespread experimentation, producing new themes in animal architecture. There has been nothing since like the Cambrian explosion. Later bursts of speciation that followed major extinction events produced only variations on established themes.

Once forged, a major body plan becomes a limiting determinant of body form for descendants of that ancestral line. Molluscs beget only molluscs and birds beget birds, nothing else. Despite the appearance of structural and functional adaptations for distinctive ways of life, the evolution of new forms always develops within the architectural constraints of a phylum's ancestral pattern. This is why we shall never see molluscs that fly or birds confined within a protective shell.▪

CHAPTER 14

Classification, Phylogeny, and Body Plans of Animals

ANIMAL CLASSIFICATION

Because most biologists have worked in the temperate zones, the majority of named species are from those areas of the world. However, tropical habitats are more diverse (have larger numbers of species), especially tropical rain forests. Today tropical rain forests are being destroyed at an unprecedented rate. Tragically, because of habitat loss, many species are becoming extinct before they are ever known to humanity.

Zoologists have named over 1.5 million species of animals, and thousands more are added to the list each year. We estimate that up to 4 million living species remain to be named, and some investigators even put the number at 30 million.

To communicate with each other about the diversity of life, biologists have found it a practical necessity not only to name living organisms but to classify them. It is not just that the desire to put things into some kind of order is a fundamental activity of the human mind. A system of classification is a storage, retrieval, and communication system for biological information. This is the science of **taxonomy,** concerned with naming each kind of organism by a uniformly adopted system of rules and guidelines that best expresses the degree of similarity of organisms. The somewhat broader science of **systematics** embraces both taxonomy, which is concerned with classification, and evolutionary biology, which attempts to determine the evolutionary relationships of organisms.

LINNAEUS AND THE DEVELOPMENT OF CLASSIFICATION

Although the history of human efforts to distinguish and name plants and animals must have been rooted in the beginnings of language, the great Greek philosopher and biologist Aristotle was the first to attempt seriously the classification of organisms on the basis of structural similarities. Following the Dark Ages in Europe, the English naturalist John Ray (1627-1705) introduced a more comprehensive classification system and a modern concept of species. The flowering of systematics in the eighteenth century culminated in the work of Carolus Linnaeus (1707-1778) (Figure 14-1), who gave us our modern scheme of classification.

Linnaeus was a Swedish botanist at the University of Uppsala. He had a great talent for collecting and classifying objects, especially flowers. Linnaeus

FIGURE 14-1

Carolus Linnaeus (1707-1778). This portrait was made of Linnaeus at age 68, 3 years before his death.

Courtesy British Museum (Natural History) London.

worked out a fairly extensive system of classification of both plants and animals. His scheme of classification, published in his great work *Systema Naturae,* emphasized morphological characters as a basis for arranging specimens in collections. Actually his classification was largely arbitrary and artificial, and he believed strongly in the constancy of species.

Linnaeus' scheme of arranging organisms into an ascending series of groups of ever-increasing inclusiveness is the **hierarchical system** of classification. Species were grouped into genera, genera into orders, and orders into classes. This taxonomic hierarchy has been considerably expanded since Linnaeus' time. The major categories, or **taxa** (sing., **taxon**), now used are as follows, in descending series: **kingdom, phylum** (pl., **phyla**), **class, order, family, genus** (pl., **genera**), and **species**. This hierarchy of seven ranks can be subdivided into finer categories, such as superclass, subclass, infraclass, superorder, suborder, and so on. In all, more than 30 taxa are recognized. For very large and complex groups, such as the fishes and insects, these additional ranks are required to express recognized degrees of evolutionary divergence. Unfortunately they also contribute complexity to the system.

According to Linnaeus' system, each species was given a distinctive name. He recognized four classes of vertebrates and two classes of invertebrates. Since his knowledge of animals was limited, his lower groups (such as the genera) were very broad and included animals that we now realize are only distantly related. As a result, much of his classification has been drastically altered, yet the basic principle of this scheme is still followed.

Linnaeus' system for naming species is known as the **binomial nomenclature** system. Each species has a Latinized name composed of two words (hence binomial). The first word is the genus, written with a capital initial letter; the second word is the specific name that is peculiar to the species and is written with a small initial letter (Table 14-1). The generic name is always a noun, and the specific name is usually an adjective that must agree in gender with the genus. For instance, the scientific name of the common robin is *Turdus migratorius* (L. *turdus,* thrush; *migratorius,* of the migratory habit).

There are times when a species is divided into subspecies, in which case a **trinomial nomenclature** is employed (see katydid example, Table 14-1). Thus to distinguish the southern form of the robin from the eastern robin, the scientific term *Turdus migratorius achrustera* (duller color) is employed for the southern type. The trinomial nomenclature is really an addition to the Linnaean system, which is basically binomial. The generic, specific, and subspecific names are printed in italics (underlined if handwritten or typed).

It is important to recognize that *only* the species is binomial. All ranks above the species are uninomial nouns, written with a capital initial letter. We must also note that the second word of a species is an epithet that has no meaning by itself. The scientific name of the white-breasted nuthatch is *Sitta carolinensis.* The "carolinensis" may be and is used in combination with other genera to mean "of Carolina," as for instance *Parus carolinensis* (Carolina chickadee) and *Anolis carolinensis* (green anole, a lizard). The genus name, on the other hand, may stand alone to designate a taxon that may include several species. No two species of organisms, living or extinct, may share the same binomial species name.

Linnaeus wrote, "The Author of Nature, when He created species, imposed on His Creations an eternal law of reproduction and multiplication within the limits of their proper kinds. He did indeed in many instances allow them the power of sporting in their outward appearances, *but never that of passing from one species to another*" (emphasis ours). It is interesting too that, whereas Linnaeus was working from a creationist model, the same system is used today to show phylogenetic relationships. This is an example of facts supporting different interpretations, rather than speaking for themselves. As new evidence accumulates, inadequate interpretations must be discarded.

SPECIES

Despite the central importance of the species concept in biology, biologists do not agree on a single rigid definition that applies to all cases. Before Darwin's time, the species was considered a primeval pattern, or archetype, divinely created. With gradual acceptance of the concept of organic evolution, scientists realized that species were not fixed, immutable units but have evolved one from another. Sometimes the gaps between species are so subtle that they can be distinguished

Table 14-1 Examples of taxonomic categories to which representative animals belong

	Human	Gorilla	Southern leopard frog	Katydid
Phylum	Chordata	Chordata	Chordata	Arthropoda
Subphylum	Vertebrata	Vertebrata	Vertebrata	Uniramia
Class	Mammalia	Mammalia	Amphibia	Insecta
Subclass	Eutheria	Eutheria		
Order	Primates	Primates	Salientia	Orthoptera
Suborder	Anthropoidea	Anthropoidea		
Family	Hominidae	Pongidae	Ranidae	Tettigoniidae
Subfamily			Raninae	
Genus	*Homo*	*Gorilla*	*Rana*	*Scudderia*
Species	*Homo sapiens*	*Gorilla gorilla*	*Rana pipiens*	*Scudderia furcata*
Subspecies			*Rana pipiens sphenocephala*	*Scudderia furcata furcata*

only by the most careful examination. In other instances, a species is so distinctive in every way that it is clearly unique and only remotely related to other species. Consequently, the criteria of taxonomy have undergone gradual changes.

At first each species was supposed to have been represented by a **type** that was used as a fixed standard. The type specimen was duly labeled and deposited in some prestigious center such as a museum (Figure 14-2). Anyone classifying a particular group would always take the pains to compare specimens with the available type specimens. Since variations from the type specimen nearly always occurred, these differences were supposed to be attributable to imperfections during embryonic development and were considered of minor significance.

The **typological** (or **morphological**) **species concept** of classifying organisms based on the fixity or constancy of a species persisted for a long period (and still does to some extent). During this time, though, the idea that species represent lineages in evolutionary descent was becoming more firmly established, resulting in the development of the **biological species concept**. Taxonomists began to think of species as *groups of interbreeding natural populations that are reproductively isolated from other such groups.* This recognizes that a species is considered a population of unique individuals that may change to a greater or lesser extent when placed in a different environment. It is still important, nevertheless, for a person who describes a species to deposit a representational specimen (type) in a museum so that later taxonomists can examine an actual specimen that the original author believed was a member of that species.

The modern concept of species is, of course, the antithesis of that of the typologist to whom variations from the type specimen are illusions caused by small mistakes during embryonic development. In population studies, the type is considered an abstract average of *real* variations that occur within the interbreeding population. Thus the species must be regarded as an **interbreeding population** made up of individuals of common descent and sharing intergrading characteristics.

The biological species concept is not without difficulties. For one thing, it does

FIGURE 14-2

Type specimens of crustaceans in the Smithsonian Institution's Natural History Museum in Washington, D.C.

Photograph by C.P. Hickman, Jr.

To avoid problems with the biological species concept, the evolutionary species concept has been proposed. *An evolutionary species is a single lineage of ancestor-descendant populations that maintains its identity from other such lineages and that has its own evolutionary tendencies and historical fate* (Wiley, 1981). In this definition, "identity" means any quality a species uses to keep its line separated from other species. Each species "knows" its identity and has recognition systems in appearance and behavior and ecological roles to maintain reproductive isolation. The evolutionary species concept has the advantage of embracing both asexual and sexual organisms.

not apply to organisms that reproduce asexually because there is no way to test the interbreeding criterion in uniparental species. For another, it is a "static" concept that does not flex easily with the slow, usually unmeasurable, changes that occur in a species through time. Furthermore, even in species that reproduce sexually, reproductive isolation may be only partial. Not infrequently populations are found that are in an intermediate stage of differentiation between races, which can interbreed, and species, which cannot interbreed. They cannot be classed as definite races or as definite species, so are sometimes called semispecies.

Despite these problems, the biological species concept works satisfactorily most of the time. Reproductive barriers do exist. Sometimes different species can be made to hybridize, but they produce sterile offspring. A well-known example is the sterile mule, produced by a cross between a horse and an ass. Even when fertile hybrids are possible, they are normally prevented by various external barriers, such as anatomical and, especially, behavioral incompatibilities that create obstacles to successful mating.

BASIS FOR FORMATION OF TAXA

Classification emphasizes the natural relationships of animals. Descent from a common ancestor explains similarity in character; the more recent the descent, the more closely the animals are grouped in taxonomic units. The genera of a particular family show less diversity than do the families of an order. This is because the common ancestor of families within an order is more remote than the common ancestor of different genera within a family. The same applies to higher categories. The common ancestor of the various vertebrate classes, for example, must be much older than the common ancestor of orders of mammals within the class Mammalia.

It is apparent that this criterion of evolutionary ancestry is not a very definite one for use in setting up taxa. There is no way to define a class that does not apply equally well to an order or a family. The genera of certain ancient groups, such as the molluscs, may actually be much older than orders and classes of other groups. The taxonomist must arrange the classifications so that all members of a taxon, as far as can be determined, are related to one another more than they are to the members of any other taxon of the same rank. Obviously the assignment of taxonomic rank depends on the opinion of the taxonomist making the study, and this is one reason classification has been called as much an art as a science. As more is learned about animals and their relationships, changes in classification are required. This brings instability to biological nomenclature and reduces its efficiency as a reference system.

Authorities on various groups sometimes differ in their opinions as to what constitutes the proper higher classification of their group, that is, how the subphyla, superclasses, classes, subclasses, and so on within the phylum should be arranged. Authors of textbooks (such as this one) must choose between two or more arrangements and thus even current texts often differ to some degree in classification.

The **law of priority** also brings about frequent changes. The first name proposed for a taxonomic unit that is published and meets other proper specifications has priority over all subsequent names proposed. The rejected duplicate names are called **synonyms.** It is disturbing sometimes to find that a species which has been well established for years must undergo a change in terminology when some industrious systematist discovers that on the basis of priority or for some other reason the species is, according to this "law," misnamed.

Despite such difficulties, the hierarchical system of classification is both accepted and highly functional. To reduce confusion in the field of taxonomy and to lay down a uniform code of rules for the classification of animals, the International Commission of Zoological Nomenclature was established in 1898. This commission meets from time to time to formulate rules and to make decisions in connection with taxonomic work.

TRADITIONAL AND MODERN APPROACHES TO SYSTEMATICS

The science of systematics is charged with two tasks. On the one hand it is required to name organisms and place them in some kind of order; this is taxonomy. On the other hand, systematics is supposed to help unravel the course of evolution. It is an unsteady mix of the practical and the theoretical. Most biologists firmly believe that systematics should legitimately concern itself with both these functions, but the difficulty of doing so has placed stresses on established taxonomic procedures. As a result, three different methods have been developed, none perfect, in an effort to develop a meaningful classification system.

Evolutionary taxonomy

Since Darwin's time, **evolutionary taxonomy** has been considered the traditional approach to classification. It bases taxonomy on evolutionary theory, and its goal is to reconstruct evolutionary history as closely as possible. Ancestor-descendant relationships, established from the fossil record and from morphological and biochemical homologies (see marginal note) are often used to construct phylogenetic trees showing the pattern of evolutionary descent. The lines in the phylogenetic tree portray evolutionary species and the branches represent speciation events, that is, the appearance of new species. A geological time scale is usually included, and sometimes the number of species in a group at any time is suggested by the widening and thinning of the branch lines (see Figure 28-1, pp. 592-593).

Homologous structures, processes, or molecules are derived from a common ancestral form (p. 101). They may or may not have similar functions, such as the forelimbs of dogs and humans, for example, as contrasted with those of birds and whales. _Analogous_ structures have a common function, such as the wings of an insect and those of a bird, but were not derived from a common ancestor.

Ideally, the taxon depicted in a phylogenetic tree is **monophyletic** ("one tribe"); that is, all the organisms in the taxon have descended from a common ancestor. For example, all birds are believed to have descended from a common thecodont (reptile) ancestor; birds are therefore a monophyletic group (Figure 29-3, pp. 610-611). Sometimes, however, a taxon is suspected to have arisen from more than one common ancestor. Such a group is called **polyphyletic** ("many tribes"). For example, mammals were once thought to have arisen independently as many as nine different times. If this were true (and this view is no longer in favor), the mammals would be an artificial taxon composed of animals of more than one origin that share characters (such as hair and mammary glands) that were independently acquired.

The traditional evolutionary approach to taxonomy has been criticized for the sometimes arbitrary and subjective way in which classification schemes had been applied. Evolutionary taxonomy is also vulnerable to circular reasoning: hypotheses about taxonomic relationships are used to establish phylogeny, and phylogeny is then used to determine taxonomy. Out of the search for objective remedies to these problems grew two new methodologies: **numerical taxonomy** and **cladistics.**

Numerical taxonomy

In all taxonomic work it is first necessary to recognize **characters** that can be used to describe an organism adequately and place it in the proper taxon. A character is any feature or attribute that can be described, measured, weighed, pictured, counted, scored, or otherwise defined about an organism. In numerical taxonomy as many arbitrarily chosen, equally weighted characters as possible (well over 100) are selected, coded, and fed into a computer, and an analysis is made of the calculations. Species or groups are then clustered by similarity. This approach aims simply at producing meaningful groupings of organisms and makes no attempt to reconstruct the evolutionary history of groups.

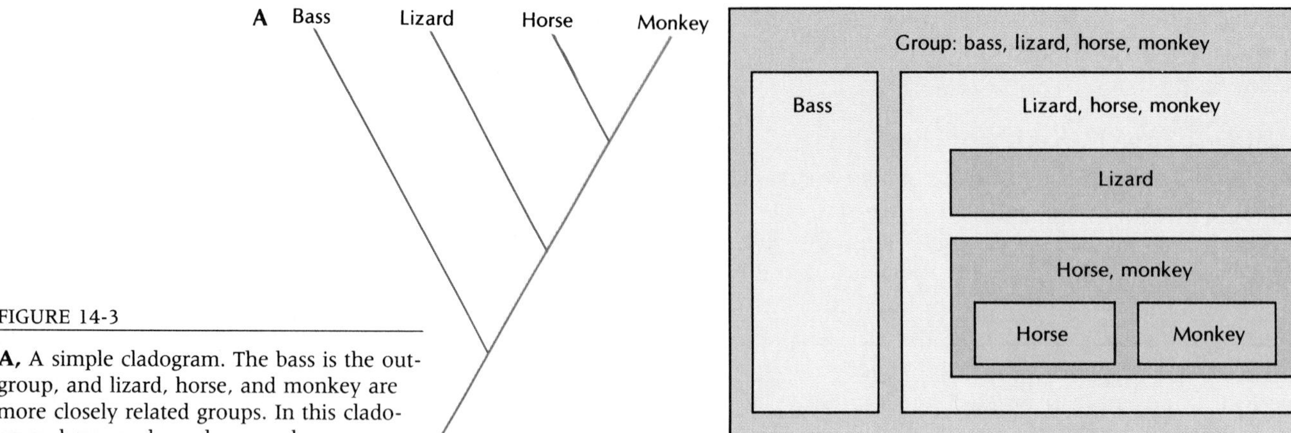

FIGURE 14-3

A, A simple cladogram. The bass is the outgroup, and lizard, horse, and monkey are more closely related groups. In this cladogram, horse and monkey are shown as most closely related and would have the largest number of shared derived characteristics. The lizard would have a smaller number of shared derived characteristics with the horse and monkey, and we could estimate that it had branched off from the common ancestral line of all three at an earlier stage. All characteristics shared by the bass, lizard, horse, and monkey would be considered primitive and thus would have been possessed by the common ancestor of all four. **B,** The "nested sets" of the cladistic hierarchy can be constructed from the cladogram.

Cladistics

An alternative to numerical taxonomy is the method of **cladistics** (Gr. *cladus,* branch). Unlike numerical taxonomy this approach bases classification exclusively on phylogeny. With it one can set up a kind of branching pattern called a **cladogram,** which resembles a phylogenetic tree but is not one (Figure 14-3). The cladogram is a series of branches, each branch representing the splitting of a parental species into two daughter species. A cladogram is meant to show only relative degrees of relationship and not actual historical events. Therefore, unlike traditional "phylogenetic trees," cladograms carry no time scale (compare Figures 28-1 and 29-3).

Although the theory of cladistics may seem complicated, its rationale is fairly simple. In practice the investigator is interested in studying the evolutionary relationships of certain groups of animals, for example, species within a genus or genera within a family. Another genus or species is selected, which is believed to be only distantly related to the groups of interest; this distantly related taxon is called the **outgroup.**

In the simplified example of four vertebrate animals shown in Figure 14-3, the bass would be considered the outgroup because we would suspect that it is only distantly related to the other three vertebrates. Then a variety of characteristics are chosen that will be used to compare among the more closely related animals and the outgroup. Any characteristic that the bass shares with the lizard, horse, and monkey—vertebrae or jaws, for example—would be considered **primitive** (or ancestral) because it is a characteristic possessed by the common ancestor of all four vertebrates. However, characteristics shared by the related group, but not possessed by the outgroup, are **derived.** In the example, the horse and monkey are most closely related because they have the greatest number of *shared* derived characteristics. The lizard would have a smaller number of derived characteristics shared with the horse and monkey, and so we could conclude that it branched off from the common ancestral line of all three at an earlier stage. For example, all three have tetrapod limbs and amniotic eggs, but only the horse and monkey have hair and mammary glands.

From this sort of analysis a branching cladogram is constructed that shows successive splits of phyletic lines. Unlike a traditional phylogenetic tree, a cladogram does not attempt to show actual historical events, and the branch points do not identify particular ancestors.

Kingdoms of Life

Since Aristotle's time, it has been traditional to assign every living organism to one of two kingdoms—plant or animal. However, the two-kingdom system has outlived its usefulness. Although it is easy to place rooted, photosynthetic organisms such as trees, flowers, mosses, and ferns among the plants and to place food-ingesting, motile forms such as worms, fishes, and mammals among the animals, unicellular organisms present difficulties. Some forms are claimed both for the plant kingdom by botanists and for the animal kingdom by zoologists. An example is *Euglena* (p. 356) and its phytoflagellate kin, which are motile, like animals, but have chlorophyll and photosynthesize, like plants. Other groups such as the bacteria were rather arbitrarily assigned to the plant kingdom.

It was inevitable that biologists would try to resolve these problems by separating problem groups into new kingdoms. This was first done in 1866 by Ernst Haeckel who proposed the new kingdom Protista to include all single-celled organisms. At first the bacteria and cyanobacteria, forms that lack nuclei bounded by a membrane, were included with nucleated unicellular organisms. Finally the important differences between the anucleate bacteria and cyanobacteria (prokaryotes) and all other organisms that have cells with membrane-bound nuclei (eukaryotes) were recognized. The prokaryote-eukaryote distinction is actually much more profound than the plant-animal distinction of the traditional system. The many differences between prokaryotes and eukaryotes are summarized on p. 30.

In 1969 R.H. Whittaker proposed a five-kingdom system that incorporated the basic prokaryote-eukaryote distinction (Figure 14-4). The kingdom Monera contains the prokaryotes; the eukaryotes are divided among the remaining four kingdoms. The kingdom Protista contains the unicellular eukaryotic organisms (protozoa and eukaryotic algae). The multicellular organisms are split into three kingdoms on the basis of mode of nutrition and other fundamental differences in organization. The kingdom Plantae includes multicellular photosynthesizing organisms, higher plants, and multicellular algae. Kingdom Fungi contains the molds, yeasts, and fungi that obtain their food by absorption. The invertebrates (except the protozoa) and the vertebrates comprise the kingdom Animalia. Most of these forms ingest their food and digest it internally, although some parasitic

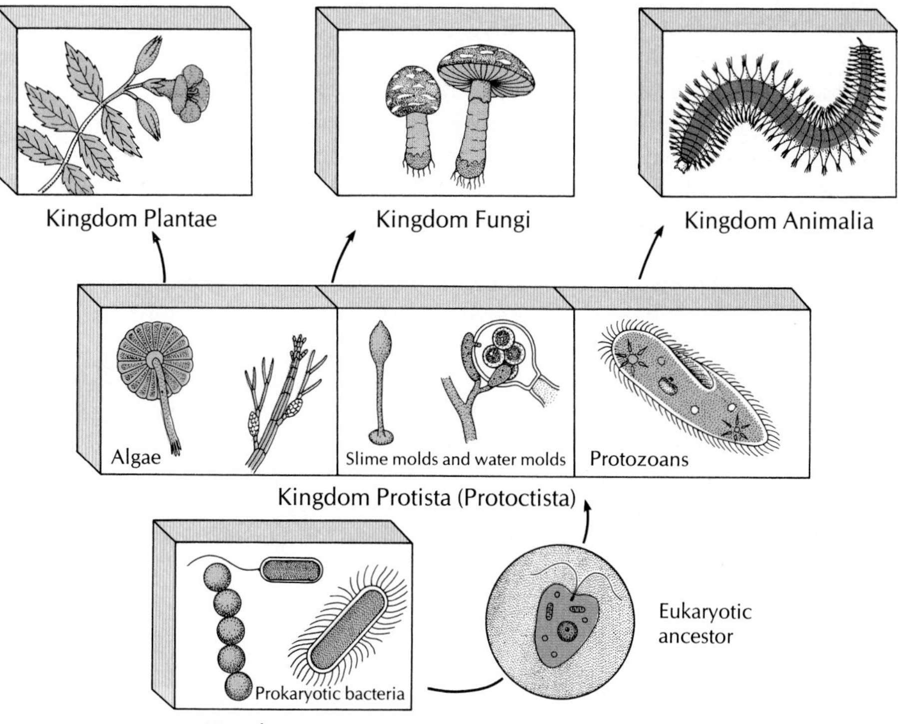

FIGURE 14-4

Five-kingdom system of classification, showing postulated evolutionary relationships.

forms are absorptive. The supposed evolutionary relationships of the five kingdoms are shown in Figure 14-4. The Protista are believed to have given rise to all three multicellular kingdoms, which therefore have evolved independently.

Exclusion of the unicellular protozoa from the animal kingdom, in which they have been traditionally included, presents a didactic problem for the zoologist. In the five-kingdom system, the phylum Protozoa has been divided into seven groups that are each elevated to phylum status. This change emphasizes the fluid nature of the hierarchical system of classification pointed out earlier in this chapter. In fact, the latest classification adopted by the Society of Protozoologists organizes the protozoa into seven other phyla on the basis of fundamental differences between groups (Chapter 15). On the other hand, the protozoa share many animal-like characteristics. Most ingest food; many have specialized organelles and advanced locomotory systems, portending tissue differentiation in multicellular forms; many reproduce sexually; and some flagellate forms are colonial with division of labor among cell types, again suggestive of a metazoan pattern. Because of their animal-like characteristics, we include the protozoa in this book.

Whether or not the protozoa should be classified as a single animal phylum, as their name suggests, or split into several phyla is a matter of opinion. At present many zoologists retain the protozoa as a single phylum of animals.

LARGER DIVISIONS OF THE ANIMAL KINGDOM

Although the phylum is often considered to be the largest, most distinctive taxonomic unit, zoologists often find it convenient to combine phyla under a few large groups because of certain common embryological and anatomical features. Such large divisions may have a logical basis, because the members of some of these arbitrary groups are not only united by common traits, but evidence also indicates some relationship in phylogenetic descent.

- **Subkingdom Protozoa** (unicellular) — protozoan phyla
- **Subkingdom Metazoa** (multicellular) — all other phyla
  - **Branch Mesozoa** — phylum Mesozoa, the mesozoans
  - **Branch Parazoa** — phylum Porifera, the sponges, and phylum Placozoa
  - **Branch Eumetazoa** — all other phyla
    - **Grade Radiata** — phyla Cnidaria, Ctenophora
    - **Grade Bilateria** — all other phyla
      - **Division Protostomia** — characteristics in Table 14-2
        - **Acoelomates** — phyla Platyhelminthes, Nemertea
        - **Pseudocoelomates** — phyla Rotifera, Gastrotricha, Kinorhyncha, Gnathostomulida, Nematoda, Nematomorpha, Acanthocephala, Entoprocta, Loricifera.
        - **Eucoelomates** — phyla Mollusca, Annelida, Arthropoda, Priapulida, Echiurida, Sipuncula, Tardigrada, Pentastomida, Onychophora, Pogonophora, Phoronida, Ectoprocta, Brachiopoda
      - **Division Deuterostomia** — characteristics in Table 14-2
        - Phyla Echinodermata, Chaetognatha, Hemichordata, Chordata

As in the outline, the bilateral animals are customarily divided into protostomes and deuterostomes on the basis of their embryological development (Table 14-2). However, some of the phyla are difficult to place into one of these two categories because they possess some of the characteristics of each group (Chapter 23).

Animal Phylogeny

Phylogenetics ("tribe origin") is the science of genealogical relationships among lineages of life forms. The reconstruction of lines of descent leading to living

Table 14-2 Basis for distinction between divisions of bilateral animals*

Protostomes	Deuterostomes
Mouth from, at, or near blastopore	New mouth from stomodeum
Anus new formation	Anus from or near blastopore
Cleavage mostly spiral	Cleavage mostly radial
Embryology mostly determinate (mosaic)	Embryology usually indeterminate (regulative)
In coelomate protostomes coelom forms as split in mesodermal bands; schizocoelous	All coelomate, coelom from fusion of enterocoelous pouches (except chordates, which are schizocoelous)
Endomesoderm usually from a particular blastomere designated 4d	Endomesoderm from enterocoelous pouching (except chordates)
Includes phyla Platyhelminthes, Nemertea, Annelida, Mollusca, Arthropoda, minor phyla	Includes phyla Echinodermata, Hemichordata, Chaetognatha, and Chordata

*Embryos of many of the advanced members of each group may have some or all of these characteristics obscured or lost.

organisms is not unlike attempts to ferret out the ancestral history (genealogy) of one's own family. Neither plant nor animal species nor members of one's own family arose spontaneously, but rather they represent branching of ancestral forms. The phylogeny of an animal group, then, represents our concept of the path its evolution has taken. It is the evolutionary history of the group.

Unfortunately, the evolutionary origin of most animal phyla is shrouded in the obscurity of Precambrian times. Because fossil records are fragmentary, our reconstructions of patterns of evolutionary relationships must rely to a large extent on evidence from comparative morphology and embryology. Relationships are naturally more easily established within the smaller taxonomic units (such as species, genera, orders) than in the classes and phyla.

Other techniques are now in use in systematics that offer great promise in solving problems of phylogenetic relationships and evolution. Research is being carried out in animal behavior, comparative biochemistry, serology, cytology, genetic homology, and comparative physiology with this aim in view.

For example, it has been shown that there are certain homologies among polynucleotide sequences in the DNA molecules of such different forms as fish and humans. These sequences appear to be genes that have been retained with little change throughout vertebrate evolution. Possible phenotypical expressions of these homologous sequences are bilateral symmetry, notochord, and hemoglobin, as well as others. By using a single strand of DNA from one species and short radioactive pieces of a DNA strand from another species and mixing the strands together, it was found that some of the smaller strands paired with similar regions on the large strand, indicating that the paired parts had common genes.

Another recent technique involves the recognition of mRNA by transfer RNA of another species. These new biochemical methods of classification are used to complement, rather than replace, the older, more established methods. In general, molecular evidences have not agreed very well with the fossil evidence. Nevertheless, the new molecular approach provides a potentially powerful tool for the systematist.

Although the representation of ancestral relationships of animals in the form of a "family tree" seems obvious today, it was not apparent to biologists before classification became founded on evolutionary theory. Even Darwin never attempted a pictorial diagram of animal relationships. Yet, despite all the shortcomings that any phylogenetic tree must possess, especially the danger of depicting and accepting highly tentative relationships as dogmatic certainty, such a scheme is a valuable tool. A family tree ties the taxa together in an evolutionary blueprint. Constructing a family tree is a creative activity based on judgment and experience and as such is always subject to modification as new information becomes available.

In each of the following chapters on animal phyla, we include a summary of that group's origins and the relationships within the group. You should make frequent reference to the geological time scale on the inside back cover of this book and to the phylogenetic tree of multicellular animals on the inside front cover. We have based conclusions about group histories on recent informed opinion. Although we recognize that family trees can give false impressions, we have nonetheless used them in the absence of a better alternative. They may be viewed as educated speculations and as such, with a certain measure of skepticism; at the same time they are not science fiction. They are derived from careful morphological reasoning and a thorough understanding of general biological principles by scientists who have devoted their lives to this form of detective work. Evolution, with its idea of life transforming itself through the ages, is supported by a vast wealth of fossil and living evidence. It is after all the framework of biology.

Animal Body Plans

As pointed out in the prologue to this chapter, while the diversity of animal body form is enormous, the possibilities for designs for living are constrained by ancestral history. Nevertheless, animals are shaped by their particular habitat and way of life. A worm that adopts a parasitic life in a vertebrate's intestine will look and function very differently from a free-living member of the same group. Yet both will share the distinguishing hallmarks of the phylum. Here we will consider the limited number of body plans that underlie the diversity of animal form and examine some of the common architectural themes that animals share.

The major advances in body architecture are, in sequence, multicellularity, bilateral symmetry, the "tube-within-a-tube" plan, and, finally, the eucoelomate (true coelom) body plan. These advancements with their principal alternatives can be arranged in a branching pattern as shown in Figure 14-5.

ANIMAL SYMMETRY

Symmetry refers to balanced proportions, or the correspondence in size and shape of parts on opposite sides of a median plane. **Spherical symmetry** means that any plane passing through the center divides the body into equivalent, or mirrored, halves. This type of symmetry is found chiefly among some of the protozoa and is rare in other groups of animals. Spherical forms are best suited for floating and rolling.

Radial symmetry (Figure 14-6) applies to forms that can be divided into similar halves by more than two planes passing through the longitudinal axis. These are the tubular, vase, or bowl shapes found in some sponges and in the hydras, jellyfish, sea urchins, and the like, in which one end of the longitudinal axis is usually the mouth. A variant form is **biradial symmetry** in which, because of some part that is single or paired rather than radial, only two planes passing through the longitudinal axis produce mirrored halves. Sea walnuts, which are more or less globular in form but have a pair of tentacles, are an example. Radial

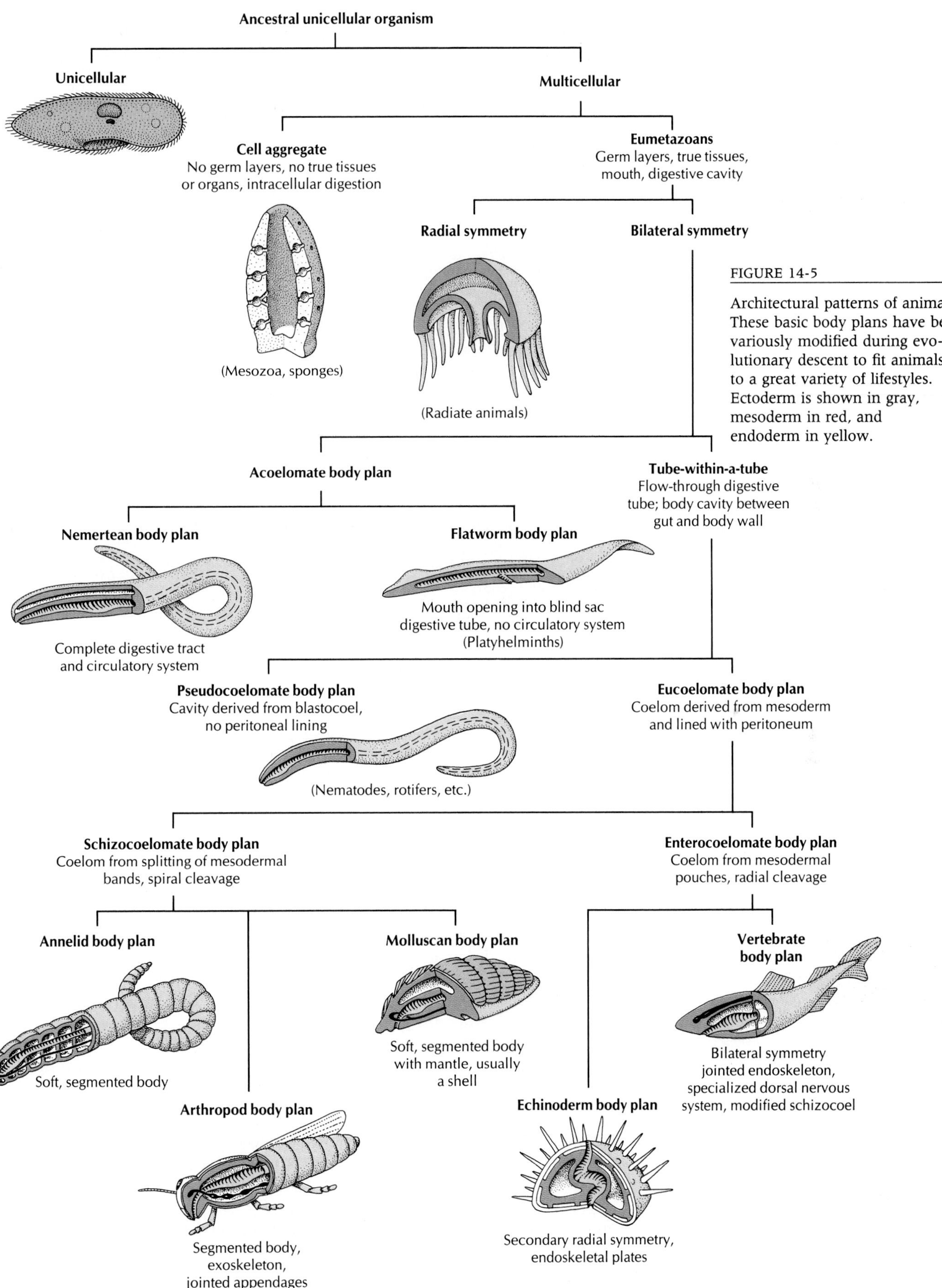

FIGURE 14-5

Architectural patterns of animals. These basic body plans have been variously modified during evolutionary descent to fit animals to a great variety of lifestyles. Ectoderm is shown in gray, mesoderm in red, and endoderm in yellow.

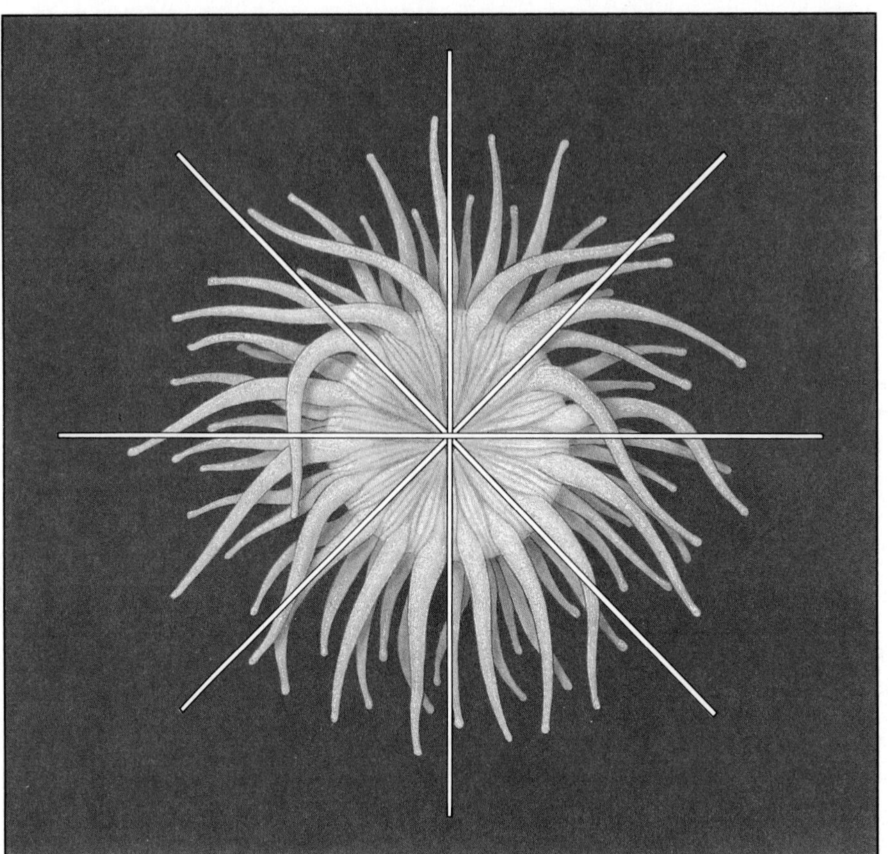

FIGURE 14-6

The planes of symmetry as illustrated by a radically symmetrical animal.

and biradial animals are usually sessile, freely floating, or weakly swimming. The two phyla that are primarily radial, Cnidaria and Ctenophora, are called the **Radiata.** The echinoderms (sea stars and their kin) are primarily bilateral animals (their larvae are bilateral) that have become secondarily radial as adults and so are excluded from the true Radiata.

The appearance of **bilateral symmetry** in animal evolution was a major advancement because bilateral animals are much better fitted for directional (forward) movement than are radially symmetrical animals. In bilateral symmetry only a sagittal plane can divide the animal into two mirrored portions — right and left halves (Figure 14-7). Bilateral animals make up all of the higher phyla and are collectively called the **Bilateria.** Bilateral symmetry is strongly associated with **cephalization,** discussed below.

Let us review some of the convenient terms used for locating regions of animal bodies (Figure 14-7). **Anterior** is used to designate the head end; **posterior,** the opposite or tail end; **dorsal,** the back side; and **ventral,** the front or belly side. **Medial** refers to the midline of the body, **lateral** to the sides. **Distal** parts are farther from the middle of the body than some point of reference; **proximal** parts are nearer. **Pectoral** refers to the chest region or the area supported by the forelegs, and **pelvic** refers to the hip region or the area supported by the hind legs. A **frontal plane** (also sometimes called coronal plane) divides a bilateral body into dorsal and ventral halves by running through the anteroposterior axis and the right-left axis at right angles to the **sagittal plane,** the plane dividing an animal into right and left halves. A **transverse plane** would cut through a dorsoventral and a right-left axis at right angles to both the sagittal and frontal planes and would result in anterior and posterior portions.

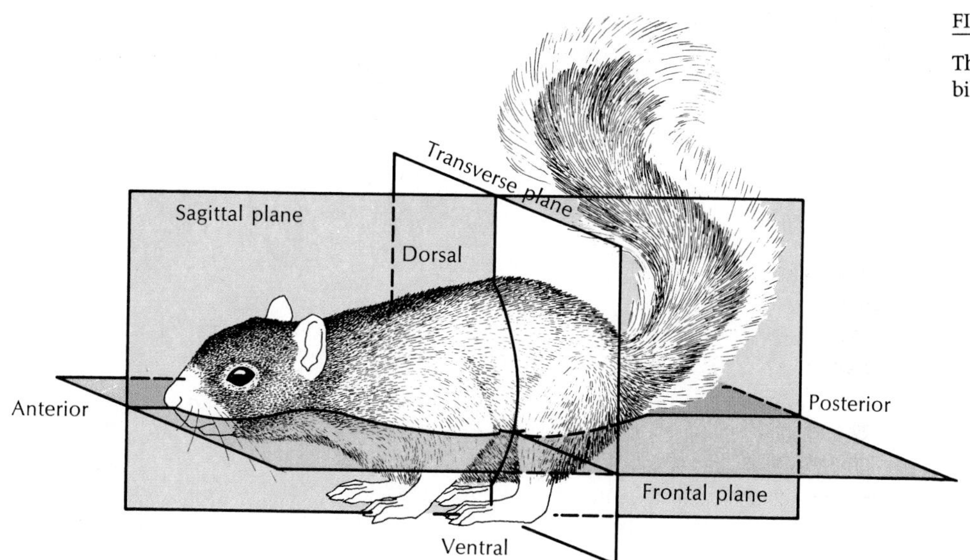

FIGURE 14-7

The planes of symmetry as illustrated by a bilateral animal.

BODY CAVITIES

The bilateral animals can be grouped according to their body-cavity type or lack of body cavity (Figure 14-5). In higher animals the main body cavity is the **coelom,** a fluid-filled space that surrounds the gut. The evolution of the coelom, like bilateral symmetry, was a major advancement in animal architecture because it provides a "tube-within-a-tube" arrangement (Figure 14-5) that allows much greater body flexibility than is possible in animals lacking an internal body cavity. The coelom also provides space for visceral organs and permits greater size and complexity by exposing more cells to surface exchange. The fluid-filled coelom additionally serves as a hydrostatic skeleton in some forms, especially many worms, aiding in such activities as movement and burrowing.

As shown in Figure 14-5, the presence or absence of a coelom is a key determinant in the evolutionary advancement of the Bilateria.

Acoelomate Bilateria

The more primitive bilateral animals do not have a true coelom. In fact, the flatworms and a few others have *no body cavity* surrounding the gut (Figure 14-8, *top*). The region between the ectodermal epidermis and the endodermal digestive tract is completely filled with mesoderm in the form of a spongy mass of space-filling cells called **parenchyma.** The parenchyma is derived from an inwandering of ectodermal cells from the general surface of the early embryo. In at least some acoelomates, the parenchymal cells are cell bodies of muscle cells (see p. 400).

Pseudocoelomate Bilateria

Nematodes and several other phyla have a cavity surrounding the gut, but it is not lined with mesodermal peritoneum. It is derived from the blastocoel of the embryo and represents a persistent blastocoel. This type of body cavity is called a **pseudocoel,** and its possessors also have a "tube-within-a-tube" arrangement (Figure 14-8, *center*).

Eucoelomate Bilateria

The remainder of the bilateral animals possess a **true coelom** lined with mesodermal peritoneum (Figure 14-8, *bottom*). The true coelom may be formed by one

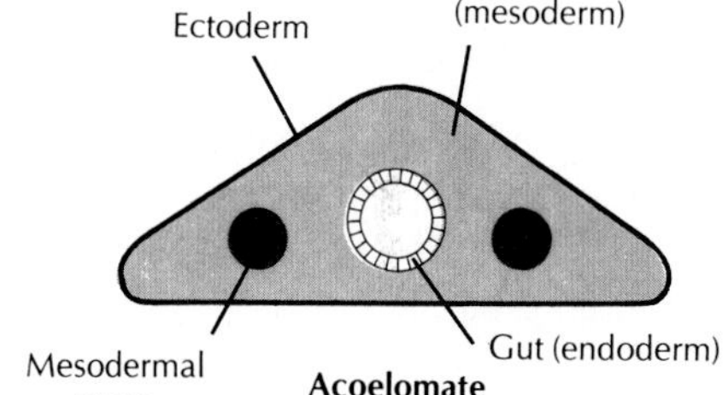

Mesodermal organ
Mesoderm (muscle)
Ectoderm
Pseudocoel (from blastocoel)
Gut (endoderm)
Pseudocoelomate

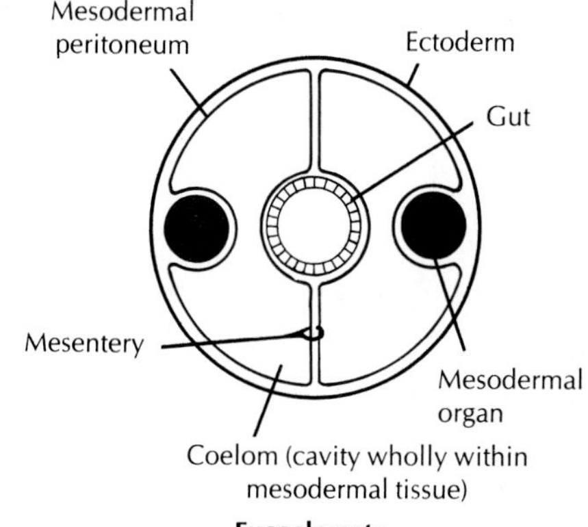

FIGURE 14-8

Acoelomate, pseudocoelomate, and eucoelomate body plans.

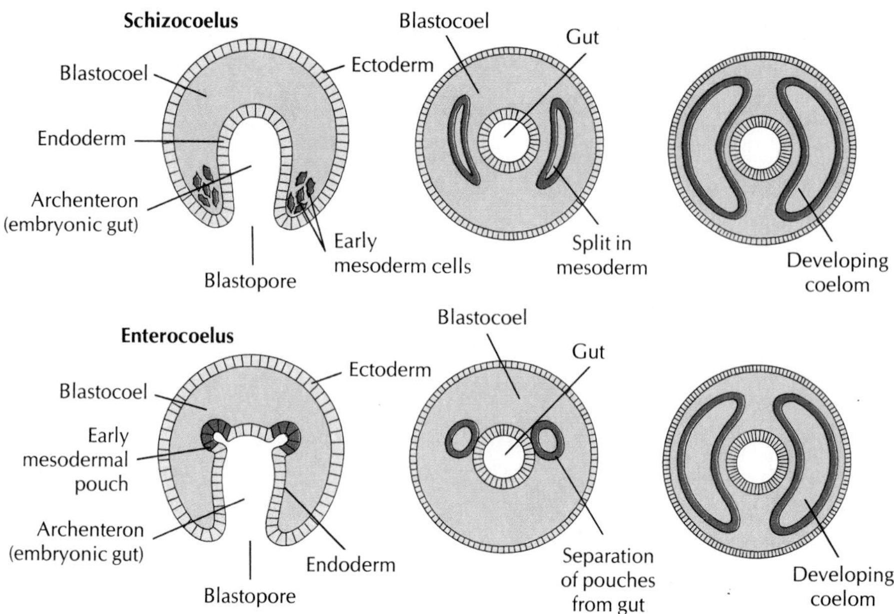

FIGURE 14-9

Types of mesoderm and coelom formation. In schizocoelous formation, mesoderm originates from wall of archenteron near blastopore and proliferates into band of tissue which splits to form coelom. In enterocoelous formation, most mesoderm originates as series of pouches from archenteron; these pinch off and enlarge to form coelom. In both formations, coeloms expand to obliterate blastocoel.

of two methods—**schizocoelous** or **enterocoelous** (Figure 14-9), or by modifying these methods. The two terms are descriptive, for *schizo* comes from the Greek *schizein,* to split; *entero* is derived from the Greek *enteron,* meaning gut; and *coelous* comes from the Greek *koilos,* meaning hollow or cavity. In schizocoelous formation the coelom arises, as the word implies, from the splitting of mesodermal bands that originate from cells in the blastopore region. (Mesoderm is one of the three primary germ layers that appear very early in the development of all bilateral animals, lying between the innermost endoderm and outermost ectoderm, p. 312.) In enterocoelous formation the coelom comes from pouches of the archenteron, or primitive gut.

Once development is complete, the schizocoelous and enterocoelous formations are indistinguishable. Both give rise to a true coelom lined with a mesodermal peritoneum (from the Greek *peritonaios,* meaning stretched around) and having mesenteries in which the visceral organs are suspended.

METAMERISM (SEGMENTATION)

Metamerism is the serial repetition of similar body segments along the longitudinal axis of the body. Each segment is called a metamere, or **somite.** In forms such as the earthworm and other annelids, in which metamerism is most clearly represented, the segmental arrangement includes both external and internal structures of several systems. There is repetition of muscles, blood vessels, nerves, and the setae of locomotion. Some other organs, such as those of sex, may be repeated in only a few somites. In higher animals much of the segmental arrangement has become obscure.

True metamerism is found in only three phyla: Annelida, Arthropoda, and Chordata, although superficial segmentation of the ectoderm and the body wall may be found among many diverse groups of animals.

CEPHALIZATION

The differentiation of a head end is called **cephalization** and is found chiefly in bilaterally symmetrical animals. The concentration of nervous tissue and sense organs in the head bestows obvious advantages to an animal moving through its environment head first. This is the most efficient positioning of instruments for sensing the environment and responding to it. Usually the mouth of the animal is located on the head as well, since so much of an animal's activity is concerned with procuring food. Cephalization is always accompanied by differentiation along an anteroposterior axis (**polarity**). Polarity usually involves gradients of activities between limits, such as between the anterior and the posterior ends.

Some Helpful Definitions

In the discussions on evolutionary relationships in ensuing chapters, terms such as lower, higher, primitive, advanced, specialized, generalized, adaptation, and fitness will be used frequently, and an understanding of their meaning may be critical to an understanding of the discussion to follow.

The terms **lower** and **higher** usually refer to a group's relative position on a phylogenetic tree, that is, the level at which it is believed to have branched off a main stem of evolution. For example, sponges and jellyfish are usually considered to be members of "lower" phyla because they are believed to have been among the earliest metazoa. In other words, they originated near the base of the family tree of the animal kingdom.

The terms **primitive** and **advanced** are often used when discussing relationships within a particular group. A primitive species is one that possesses a great many of the same characteristics believed to have belonged to the ancestral stock from which it evolved. An advanced species is one that has undergone considerable change from the primitive condition, usually because of adaptation to a changed environment or to a different mode of living. Among the molluscs, for example, the chitons are considered to be more primitive (that is, more like the hypothetical mollusc ancestor) than the snails, clams, or octopuses. A primitive species is not "less perfect" than an advanced species, since it may be as well adapted to its own type of environment as the advanced one is to the environment for which it has become especially adapted. Also, a species or group may be primitive in some respects and advanced in others.

Specialized might refer either to an organism or to one or more of its parts that has become adapted to a particular ecological niche or to a particular function. A more **generalized** species or structure may share the characteristics of two or more distinct groups or structures. For example, many of the small aquatic crustaceans have a number of similar feathery trunk appendages, all serving several functions, such as swimming, respiration, filter feeding, or egg bearing. Such multipurpose appendages would be considered generalized in comparison with the highly specialized defensive chelipeds or sensory antennae of the crayfish, the pollen-collecting legs of the honeybee, or the digging forelegs of the mole cricket. Each of these specialized appendages is adapted for one primary function.

An **adaptation** is any characteristic of an organism taken in the context of how the characteristic helps the organism survive and reproduce. Many adaptations are possessed in common by a wide variety of animals, but the word is most often used in reference to special adaptations for a particular habitat or environment. Indeed, the more *special adaptations* the animal has, the more *specialized* the animal is. Any characteristic or any change in a characteristic is said to have **adaptive value**, if it tends toward greater fitness of the organism for a particular niche or habitat.

An organism is fitted to its environment when it is adjusted or adapted to it. Its **fitness** for any particular place or mode of living is the degree of its adjustment, suitability, or adaptation to that particular environment or niche.

Summary

Systematics is concerned with the speciation, classification, and phylogeny of animals. It includes a system of uniform naming of kinds of animals and the erection of a classification scheme that reflects evolutionary relationships. The scheme is hierarchical, and the most widely used groupings (taxa) are as follows, in order of increasing inclusiveness: species, genus, family, order, class, phylum, and kingdom. Subdivisions of all of these are commonly used as necessary. The basic binomial system of nomenclature, giving each kind of organism a generic and a specific name, originated with Carolus Linnaeus and is still in use today. The biological species concept has replaced the older typological concept of a species. A biological species is an interbreeding (or potentially interbreeding) population that is reproductively isolated from other such groups. It is not immutable through time but changes during the course of evolution. If a species is given more than one name, the earliest name published with an appropriate description of the organism is considered the correct name, and other, later names become synonyms.

There are today three major schools of taxonomy. The traditional approach, called evolutionary taxonomy, attempts to reconstruct actual ancestor-descendant relationships, which may be depicted in the form of a phylogenetic tree. A second approach, called numerical taxonomy, strives to group taxa according to a statistical analysis of their total similarities and dissimilarities. Numerical taxonomy does not infer evolutionary relationships. A third approach is cladistics, whose proponents attempt to achieve phylogenetically significant groupings. Cladistics discriminates between primitive and derived characteristics of organisms. Although morphology, embryology, and the fossil record are still used as criteria for assessing evolutionary relationships, other characteristics such as behavioral, serological, biochemical, cytological, and physiological are increasingly used in modern systematics. A highly useful device for visualizing evolutionary relationships and derivations from ancestral groups is the phylogenetic tree.

The broadest, most inclusive categories of living organisms have been traditionally considered the plant and animal kingdoms. However, most biologists now prefer the five-kingdom scheme: Monera (prokaryotes), Protista (protozoa and eukaryotic algae), Fungi, Plantae, and Animalia. In this book we will consider the kingdom Animalia to contain the subkingdoms Protozoa (unicellular phyla) and Metazoa (multicellular phyla). Among the Metazoa, important groupings are the radial and the bilateral animals, and among the bilateral, the Protostomia and the Deuterostomia.

Every organism has an inherited body plan that may be described in terms of broadly inclusive characteristics, such as symmetry, presence or absence of body cavities, partitioning of body fluids, presence or absence of segmentation, and degree of cephalization.

The following terms have specific meanings in evolutionary biology, each of which should be carefully distinguished and understood: higher, lower, primitive, advanced, specialized, generalized, adaptation, and fitness.

REVIEW QUESTIONS

1. Distinguish between systematics and taxonomy.
2. Arrange the following in the proper hierarchical arrangement, from the most inclusive taxon to the least inclusive: genus, kingdom, order, species, class, phylum, family.
3. Define "species" in accord with the biological species concept.
4. Contrast biological species and evolutionary species.
5. What is the law of priority?
6. Distinguish between monophyletic and polyphyletic.
7. How does cladistics differ from numerical taxonomy?
8. Name six sources of evidence that might be used to support a proposed phylogeny of an animal group.
9. Name the five kingdoms of organisms, and tell what each contains.
10. Match the animal group with its body plan:

______ Unicellular	a. Cnidaria
______ Cell aggregate	b. Echinoderm (e.g., sea stars)
______ Radiate	c. Protozoa
______ Blind sac digestive tube, acoelomate	d. Annelid
______ Tube-within-a-tube, pseudocoelomate	e. Sponge
______ Eucoelomate, secondary radial symmetry	f. Flatworm
______ Schizocoelous eucoelomate	g. Nematode

11. Distinguish between spherical, radial, biradial, and bilateral symmetry.
12. Use the following terms to identify regions on your body and on the body of a frog: anterior, posterior, dorsal, ventral, lateral, distal, and proximal.
13. How would frontal, sagittal, and transverse planes divide your upright body?
14. What is meant by metamerism? Name three phyla showing metamerism.
15. Define and contrast the following terms in an evolutionary context: higher, lower; primitive, advanced; specialized, generalized; adaptation, adaptive value; fitness.

SELECTED REFERENCES

Brooks, D.R., J.N. Caira, T.R. Platt, and M.H. Pritchard. 1984. Principles and methods of phylogenetic systematics: a cladistics workbook. Lawrence, Kan., University of Kansas Museum of Natural History Special Pub. No. 12. *A practical introduction designed to help students learn to use cladistics.*

McMenamin, M.A.S. 1987. The emergence of animals. Sci. Am. **256**:94-102 (April). *Describes the burst of new body plans at the beginning of the Cambrian period.*

Ross, H.H. 1974. Biological systematics. Reading, Mass., Addison-Wesley Publishing Co., Inc. *Theory and practice of systematics are presented and exemplified from animals, plants, and microorganisms. Very comprehensive and useful.*

Wiley, E.O. 1981. Phylogenetics: the theory and practice of phylogenetic systematics. New York, John Wiley & Sons, Inc. *Excellent, thorough presentation of cladistic theory.*

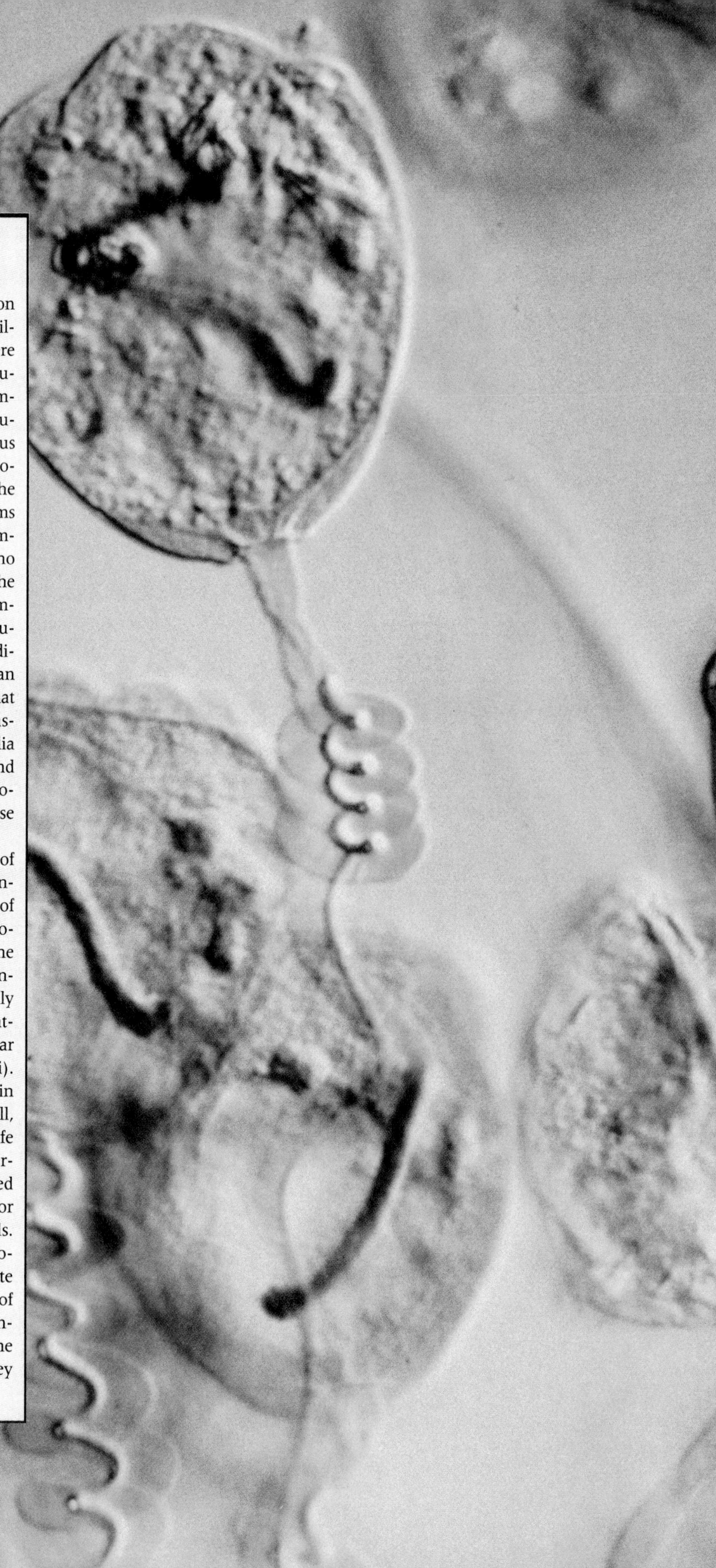

EMERGENCE OF EUKARYOTES AND A NEW LIFE PATTERN

The first reasonable evidence for life on earth dates from approximately 3.5 billion years ago. These first cells were prokaryotic (meaning "before nucleus") — bacteria-like organisms of simple construction that propagated asexually by binary fission. After an enormous time span of slow, almost stagnant, evolution, the prokaryotes gave rise to the eukaryotes of which the earliest forms were protozoa. Protozoa, though the simplest of eukaryotes, were (and are) in no sense themselves simple organisms. The eukaryotic cell is profoundly more complex than its prokaryotic counterpart. Eukaryotic cells are larger and have, in addition to a membrane-bound nucleus, an extensive array of advancements that prokaryotes lack; these include cytoplasmic organelles, mitochondria, true cilia and flagella, cell division by mitosis, and (usually) sexual reproduction with meiosis and production of gametes. All of these are present in the protozoan phyla.

Protozoa are a diverse assemblage of unicellular organisms with puzzling affinities that relate to the obscure origin of eukaryotic cells; the transition from prokaryotic to eukaryotic cells is one of the great unexplained mysteries of evolutionary history. Today, protozoa are formally assigned to the kingdom Protista, separating them from the three multicellular kingdoms (animals, plants, and fungi). Yet, protozoa are distinctly animal-like in several respects: they lack a cell wall, have at least one motile stage in the life cycle, and most ingest their food. Furthermore, multicellular animals have evolved from protozoa — in itself an argument for considering protozoa to be animals. Because of their ancient beginnings, protozoa have had ample time to radiate adaptively into a bewildering array of morphological forms within the constraints of a single cell, unaware of the troublesome classification tangle they were to bestow on today's zoologists.▪

CHAPTER 15

Unicellular Animals

Protozoa

Some protozoa ("first animals") are among the most primitive forms of animal life known today. Although the protozoa are unicellular, they are by no means simple. They have many complicated structures, and physiologically they are quite complex. Within the cytoplasm of a protozoan there is specialization and division of labor. Specialized structures within the cytoplasm are called **organelles,** and each is fitted for a specific function, just as specialized organs or groups of cells perform special functions in the metazoa. Particular organelles may perform as skeletons, sensory systems, conduction mechanisms, contractile systems, organs of locomotion, and defense mechanisms. Because of the lack of specialized cells and organs, however, the protozoa are considered to be at the **protoplasmic level** of organization (p. 150).

The problem of where to place the fence-straddling autotrophic flagellates is conveniently eliminated in the five-kingdom system, which assigns all protozoa, together with the unicellular algae and fungi, to the kingdom Protista (p. 335).

Among the protozoa, the dividing line between animals and plants becomes quite vague and arbitrary. Many of the flagellates are autotrophic (**holophytic**), that is, they contain chlorophyll and, like plants, can manufacture their own organic molecules by photosynthesis. They may be considered plants by botanists and animals by zoologists, each to suit the convenience of their taxonomic systems. Perhaps in the early evolution of animals some of these autotrophic flagellates may have lost their green chloroplasts to become colorless heterotrophs that must either absorb preformed organic nutrients from their environment (**saprozoic**) or feed on other plant or animal matter (**holozoic**).

Most protozoa live singly, but many, particularly among the flagellates, live in distinct colonies of from several to hundreds of individual zooids. The distinctions between such protozoan colonies and the simplest metazoa are also vague and arbitrary. In a protozoan colony, some cells may be specialized for reproduction, but the rest can perform all other body functions independent of each other. In the simplest metazoa, there may be only slightly more interdependence, but in the higher and more complex metazoa, most cells become highly specialized and depend strongly on the integrity of the organism for survival.

The Protozoan Phyla

Most protozoa occupy a hidden world; they are microscopic, usually from 3 to 300 μm long. The largest are among an order in the subphylum Sarcodina (Foraminiferida), some of which have shells 100 to 125 mm in diameter, and some extinct protozoa reached a size of several centimeters.

Over 35,000 living species of protozoa have been described, of which about 10,000 are parasitic. These figures may represent only a fraction of the total number of species that exist. Some protozoologists believe that there may be more protozoan species than all other species together because each species of the higher phyla may have several species of parasitic protozoa, and many protozoa bear parasites themselves. In reality, there may be more parasitic species than free-living ones.

CLASSIFICATION

Traditionally, four main groups of protozoa have been recognized: the flagellates (Mastigophora), amebas (Sarcodina), spore formers (Sporozoa), and ciliates (Ciliophora). The following classification is more accurate phylogenetically.

Phylum Sarcomastigophora (sar'ko-mas-ti-gof'o-ra) (Gr. *sarkos,* flesh, + *mastix,* whip, + *phora,* bearing). Flagella, pseudopodia, or both types of locomotory organelles; usually with only one type of nucleus; typically no spore formation; sexuality, when present, essentially syngamy.

Subphylum Mastigophora (mas-ti-gof'o-ra) (Gr. *mastix,* whip, + *phora,* bearing). One or more flagella typically present in adult stages; autotrophic or heterotrophic or both; reproduction usually asexual by fission.

Class Phytomastigophorea (fi'to-mas-ti-go-for'e-a) (Gr. *phyton,* plant, + *mastix,* whip, + *phora,* bearing). Plantlike flagellates, usually bearing chromoplasts, which contain chlorophyll. Examples: *Chilomonas, Euglena, Volvox, Ceratium, Peranema.*

Class Zoomastigophorea (zo'o-mas-ti-go-for'e-a) (Gr. *zōon,* animal, + *mastix,* whip, + *phora,* bearing). Flagellates without chromoplasts; one to many flagella; ameboid forms with or without flagella in some groups; species predominantly symbiotic. Examples: *Trichomonas, Trichonympha, Trypanosoma, Leishmania, Dientamoeba.*

Subphylum Opalinata (o'pa-lin-a'ta) (NF *opaline,* like opal in appearance, + *-ata,* group suffix). Body covered with longitudinal rows of cilium-like organelles; parasitic; cytostome (cell mouth) lacking; two to many nuclei of one type. Examples: *Opalina, Protoopalina.*

Subphylum Sarcodina (sar-ko-di'na) (Gr. *sarkos,* flesh, + *ina,* belonging to). Pseudopodia typically present; flagella present in developmental stages of some; cortical zone of cytoplasm relatively undifferentiated compared with other major taxa; body naked or with external or internal skeleton; free living or parasitic.

Superclass Actinopoda (ak'ti-nop'o-da) (Gr. *aktis, aktinos,* ray, + *pous, podos,* foot). Often spherical; usually planktonic; pseudopodia in form of axopodia, with microtubular supporting structure. Examples: *Actinosphaerium, Actinophrys, Thalassicolla.*

Superclass Rhizopoda (ri-zop'o-da) (Gr. *rhiza,* root, + *pous, podos,* foot). Locomotion by one of several pseudopodial types or by protoplasmic flow without production of discrete pseudopodia. Examples: *Amoeba, Entamoeba, Arcella.*

Phylum Labyrinthomorpha (la'bi-rinth-o-morf'a) (Gr. *labyrinth,* maze, labyrinth, + *morph,* form + *a,* suffix). Small group living on algae; mostly marine or estuarine.

Phylum Apicomplexa (a′pi-com-plex′a) (L. *apex,* tip or summit, + *complex,* twisted around + *a,* suffix). Characteristic set of organelles (apical complex) associated with anterior end present in some developmental stages; cilia and flagella absent except for flagellated microgametes in some groups; cysts often present; all species parasitic.

Class Perkinsea (per-kin′se-a). Small group parasitic in oysters.

Class Sporozoea (spor′o-zo′e-a) (Gr. *sporos,* seed, + *zōon,* animal). Spores or oocysts typically present that contain infective sporozoites; locomotion of mature organisms by body flexion, gliding, or undulation of longitudinal ridges; flagella present only in microgametes of some groups; pseudopodia ordinarily absent, if present they are used for feeding, not locomotion; one or two host life cycles. Examples: *Monocystis, Gregarina, Eimeria, Plasmodium, Toxoplasma, Babesia.*

Phylum Myxozoa (mix-o-zo′a) (Gr. *myxa,* slime, mucus, + *zōon,* animal). Parasites of lower vertebrates, especially fishes, and invertebrates.

Phylum Microspora (mi-cros′por-a)(Gr. *micro,* small, + *sporos,* seed). Parasites of invertebrates, especially arthropods, and lower vertebrates.

Phylum Ascetospora (as-e-tos′por-a) (Gr. *asketos,* curiously wrought, + *sporos,* seed). Small group that is parasitic in invertebrates and a few vertebrates.

Phylum Ciliophora (sil-i-of′or-a) (L. *cilium,* eyelash, + Gr. *phora,* bearing). Cilia or ciliary organelles in at least one stage of life cycle; two types of nuclei, with rare exception; binary fission across rows of cilia; budding and multiple fission also occur; sexuality involving conjugation, autogamy, and cytogamy; nutrition heterotrophic; contractile vacuole typically present; most species free living, but many commensal, some parasitic. (This is a very large group, now divided by the Society of Protozoologists into three classes and numerous orders and suborders. The classes are separated on technical characteristics of the ciliary patterns, expecially around the cytostome, the development of the cytostome, and other characteristics.) Examples: *Paramecium, Colpoda, Tetrahymena, Balantidium, Stentor, Blepharisma, Epidinium, Euplotes, Vorticella, Carchesium, Trichodina, Podophrya, Ephelota.*

■ General Features

ECOLOGICAL RELATIONSHIPS

Protozoa seem to be found wherever life exists. They need moist habitats and are found in seawater, fresh water, soil and decaying organic matter and plants, and as parasites in every kind of animal, including other protozoa. They may be free living or symbiotic, solitary or colonial, and sessile or free swimming.

All forms of symbiosis (p. 139, Chapter 5) occur among protozoa. An example of mutualism is a green alga harbored by a species of *Paramecium.* The alga photosynthesizes carbohydrates for its host and receives shelter in return. Other examples are the protozoa that live in the gut of termites and digest the wood that the insects eat. Many species of protozoa live as commensals, and many others are parasites either inside or on the surface of other animals. One large protozoan phylum, Apicomplexa, and three smaller phyla are entirely parasitic. A number of parasitic protozoa are extremely important disease agents in humans and domestic animals.

Some protozoa contribute to the contamination of water, affecting its taste and odor, and some disease-producing protozoa are transmitted in water. There are also certain dinoflagellates that, when occurring in unusually large numbers, cause "red tides," producing toxins that sometimes cause widespread destruction of fish and other sea life. The toxins can accumulate in shellfish and become poisonous to humans and other mammals that eat the shellfish.

Although the name "red tide" was originally applied to situations in which the organisms reproduced in such profusion (producing a "bloom") that the water turned red from their color, any instance of a dinoflagellate or algal bloom producing detectable levels of toxic substances is now called a red tide. The water may be red, brown, yellow, or not remarkably colored at all, and the phenomenon has nothing to do with tides! Red tides have resulted in considerable economic losses to the shellfish industry. Another dinoflagellate produces a toxin that is concentrated in the food chain (Chapter 5). Humans can be severely poisoned and even die after eating fish containing the toxin. The illness, known as ciguatera, is especially associated with consumption of large fishes caught near coral reefs.

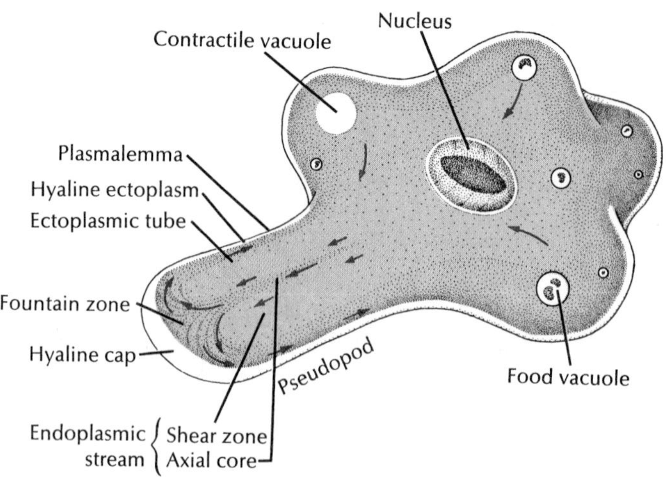

FIGURE 15-1

Ameba in active locomotion. Arrows indicate direction of streaming protoplasm. First sign of new pseudopodium is thickening of ectoplasm to form clear hyaline cap, into which the fluid endoplasm flows. As endoplasm reaches forward tip, it fountains out and is converted into ectoplasm, forming a stiff outer tube that lengthens as the forward flow continues. Posteriorly the ectoplasm is converted into fluid endoplasm, replenishing the flow. Substratum is necessary for ameboid movement.

LOCOMOTION

Pseudopodia

Pseudopodia are the chief means of locomotion of members of the subphylum Sarcodina, and they can be formed by some flagellates. The characteristic movement with pseudopodia is called **ameboid movement.**

In many amebas, a peripheral, nongranular layer of cytoplasm, the **ectoplasm,** can be distinguished, which encloses the central, granular **endoplasm.** The ectoplasm is in the gel state (jellylike semisolid), and the endoplasm is in the sol state (fluid). When a pseudopodium is beginning to form, an ectoplasmic projection called the **hyaline cap** appears, and a stream of endoplasm begins to flow into it (Figure 15-1). At the advancing end of the forming pseudopodium, the endoplsmic stream fountains out toward the periphery. Here it converts to the gel state, thus forming a stiff, and continually lengthening, ectoplasmic tube around the inner core of fluid endoplasm. At some point the tube becomes anchored through its plasmalemma (the flexible plasma membrane of the sarcodines) to the substratum, and the animal is drawn forward. At the temporary "tail end" of the pseudopodium the ectoplasmic tube is being converted back to streaming protoplasm, thus replenishing the forward flow. In essence, then, the ameba creates a tube, anchors it, and flows through it as it moves forward. At any time the direction of flow can be reversed, its length shortened, and new pseudopodia started (Figure 15-2).

FIGURE 15-2

Ameba "changes its mind." This series of a single ameba shows the change of direction in protoplasmic flow as tentative pseudopodia are advanced and retracted. Note the pseudopodium advancing toward a retreating rotifer.

Photograph by F.M. Hickman.

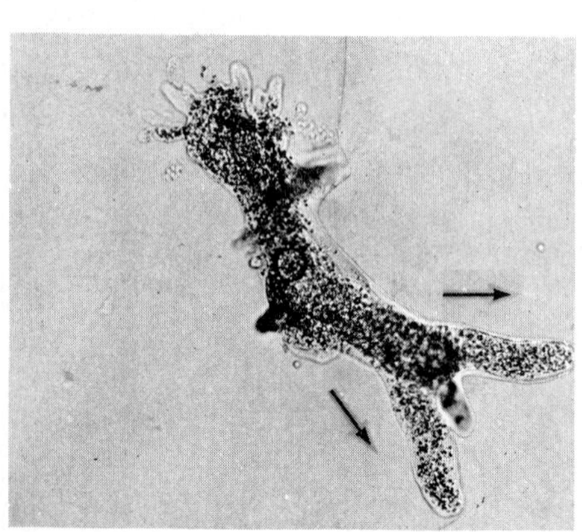

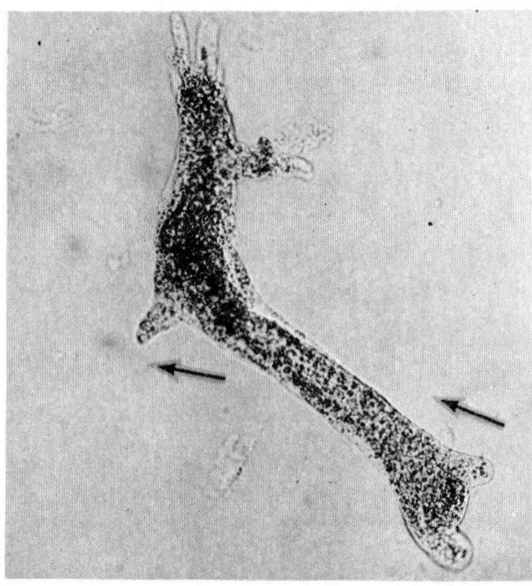

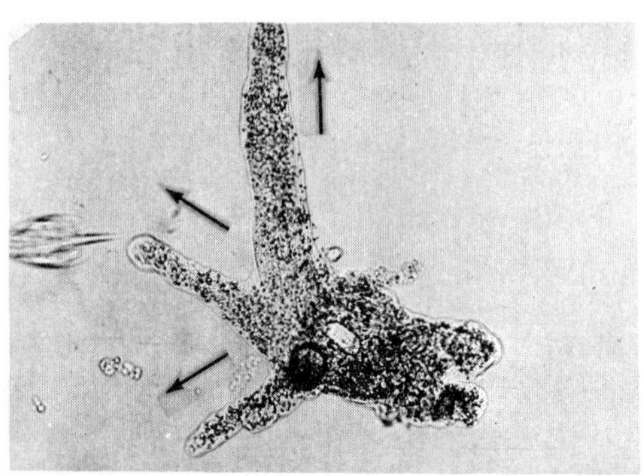

Ameboid movement, like that of muscle cells, may be effected by tiny filaments of actin and myosin moving past each other. Presumably, filaments in the gel phase would pull filaments in the solated endoplasm toward the direction of pseudopod extension. Support for this hypothesis comes from experiments in which extracts of amebas are treated with ATP and show gelation, contractility, streaming of particles, and extrusion of water. Oriented filaments have been shown in the gel phase by electron microscopy, and myosin filaments have been identified in *Amoeba proteus* (Gr. *amoibe,* change).

Although well exemplified and most easily studied in the ameba, ameboid movement is of very wide occurrence and of extreme importance in multicellular animals. Much of the defense against disease in the human body depends on ameboid white blood cells, and ameboid cells in many other animals, vertebrate and invertebrate, play similar roles (p. 177).

Cilia and Flagella

Cilia and **flagella** comprise the other major locomotory mechanisms of protozoa, and they are no less important than pseudopodia for multicellular animals. Not only do many small metazoa use cilia for locomotion, the cilia of many create water currents for their feeding and respiration. Ciliary movement is vital to many species in such functions as handling food, reproduction, excretion, osmoregulation, and so on.

As pointed out in Chapter 2, cilia and flagella are structurally the same. Each contains nine pairs of longitudinal microtubules arranged in a circle around a central pair (Figure 15-3), and this is true for all flagella and cilia in the animal kingdom, with a few notable exceptions. This "9 + 2" tube of microtubules in the flagellum or cilium is its **axoneme;** the axoneme is covered by a membrane continuous with the cell membrane covering the rest of the organism. At about the point where the axoneme enters the cell proper, the central pair of fibrils ends at a small plate within the circle of nine pairs (Figure 15-3). Also at about that point, another fibril joins each of the nine pairs, so that these form a short tube extending from the base of the flagellum into the cell and consisting of nine *triplets* of fibrils. The short tube of nine triplets is the **kinetosome** and is exactly the same in structure as the **centriole** (p. 35). The centrioles of some flagellates may give rise to the kinetosomes, or the kinetosomes may function as centrioles. All typical flagella and cilia have a kinetosome at their base, regardless of whether they are borne by a protozoan or metazoan cell. The kinetosomes of protozoa have older, traditional names (blepharoplast, basal body, basal granule) that are still in common usage.

The current theory to account for ciliary and flagellar movement is the **sliding-microtubule hypothesis.** The movement is powered by the release of chemical bond energy in ATP (p. 47). Two little arms are visible in electron micrographs on each of the pairs of peripheral tubules in the axoneme (Figure 15-3), and these bear the enzyme adenosine triphosphatase (ATPase) to cleave the ATP. When the bond energy in ATP is released, the arms "walk along" one of the microtubules in the adjacent pair, causing it to slide relative to the other tubule in the pair. Shear resistance, causing the axoneme to bend when the microtubules slide past each other, is provided by "spokes" from one of the tubules in each doublet projecting toward the central pair of microtubules. These are also visible in electron micrographs.

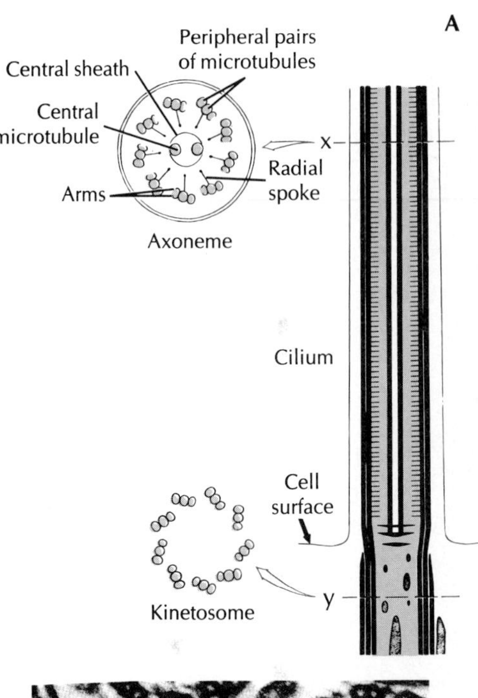

FIGURE 15-3

A, The axoneme is composed of nine pairs of microtubules plus a central pair, and it is enclosed within the cell membrane. The central pair ends at about the level of the cell surface in a basal plate (axosome). The peripheral microtubules continue inward for a short distance to comprise two of each of the triplets in the kinetosome (at level *y* in **A**). **B,** Electron micrograph of section through several flagella, corresponding to section at *x* in **A.** (×133,000.)

Electron micrograph courtesy I.R. Gibbons, Harvard University.

NUTRITION AND DIGESTION

Protozoa exhibit a wide variety of food-getting habits. The phytoflagellates, such as *Euglena* and *Volvox,* are chiefly producers (p. 130). (*Euglena* is derived from the Greek *eu,* true, good, + *glēnē,* eyeball, and refers to the stigma, or eyespot. *Volvox* is derived from the Latin *volvere,* to roll, and refers to their characteristic move-

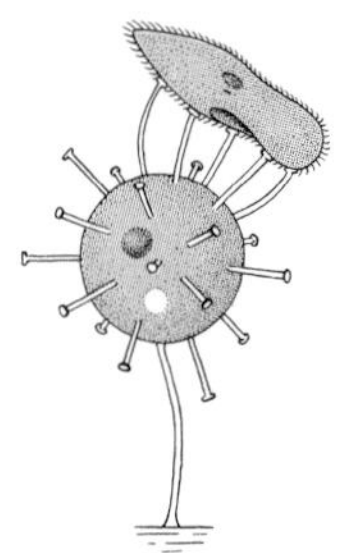

Podophrya

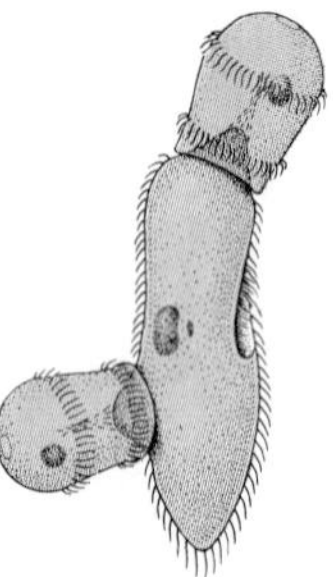

Didinium

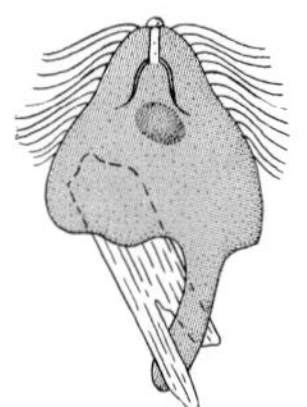

Leidyopsis

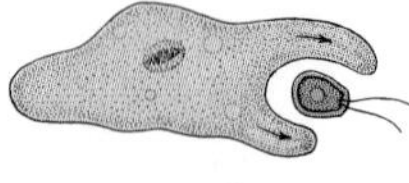

Amoeba

FIGURE 15-4

Some feeding methods among protozoa. *Amoeba* surrounds a small flagellate with pseudopodia. *Leidyopsis,* a flagellate living in the intestine of termites, forms pseudopodia and ingests wood chips. *Didinium,* a ciliate, feeds only on *Paramecium,* which it swallows through a temporary cytostome in its anterior end. Sometimes more than one *Didinium* feed on the same *Paramecium. Podophrya* is a suctorian ciliophoran. Its tentacles attach to its prey and suck prey cytoplasm into the body of the *Podophrya,* where it is pinched off to form food vacuoles. Technically, all of these methods are types of phagocytosis.

ment.) Endowed with the necessary chlorophyll, they can produce their own food by photosynthesis in the same manner as green plants (autotrophic, holophytic nutrition). Some of them are also "beggars" with the ability, when deprived of light, to absorb nutrients from the water in which they live (saprozoic nutrition). Some protozoa are parasites, or thieves, that absorb nutrients from their hosts.

The majority of protozoa, however, ingest solid or liquid food through a temporary or permanent oral opening, the **cytostome,** or "cell mouth" (holozoic nutrition). These are the browsers, hunters, and trappers. All have some adaption for ingestion (phagocytosis), whether by the cytostome, as seen in the ciliates, or by the use of pseudopodia that surround and engulf the food, as seen in the amebas and other sarcodines (Figure 15-4).

The hunters may move by flagella, cilia, or pseudopodia. A slow-moving but surprisingly adept ameba captures with its pseudopodia much more swiftly moving ciliates or rotifers. *Didinium* (Gr. *di,* two, + *din,* whirling, + *-ium,* small), a ciliate, attaches immediately when it encounters its prey, *Paramecium* (Gr. *paramekes,* oblong), in accidental collisions. Trapping is done by means of water currents created by cilia or flagella that sweep food particles toward the mouth (filter-feeding). Filter-feeding is used by both motile and sessile forms, but some of the sessile forms are especially well adapted for it. Other trappers are found among the shelled sarcodines, which have especially adapted pseudopodia that branch and rejoin to form protoplasmic nets for trapping their prey.

Food is digested within **food vacuoles,** which generally also contain some of the environmental water. Lysosomes, which are produced by Golgi bodies and contain digestive enzymes, attach to the food vacuoles and release enzymes therein. The digested products are absorbed into the surrounding cytoplasm, and indigestible wastes are expelled to the outside. Many protozoa have a permanent site for waste expulsion, the **cytoproct.**

EXCRETION AND OSMOREGULATION

Water balance, or osmoregulation, is a function of the one or more **contractile vacuoles** (see Figures 15-1, 15-10, and 15-17) possessed by most protozoa, particularly freshwater forms, which live in a hypoosmotic environment. These vacuoles are often absent in marine or parasitic protozoa, which live in a nearly isosmotic medium.

The contractile vacuoles are usually located in the ectoplasm and act as pumps to remove excess water from the cytoplasm. They are filled by droplets fed by a system of collecting canals in some species and from smaller vesicles in others (Figure 15-5). When full they empty through a canal to the outside. The rate of pulsation varies; in *Paramecium* the posterior vacuole may pulsate faster than the anterior one because of water delivered there along with ingested food. Vacuoles in marine forms pulsate more slowly than in similar freshwater forms.

Nitrogenous wastes from metabolism are apparently eliminated by diffusion through the cell membrane, but it is likely that some may also be emptied by way of the contractile vacuoles.

RESPIRATION

There are no respiratory organelles in protozoa. Exchange of oxygen and carbon dioxide between the organism and its environment occurs by diffusion through the cell membrane.

REPRODUCTION

Reproduction in most protozoa is primarily by asexual cell division. It is comparable in many respects to cell division in the multicellular animals. The protozoa, however, often have certain structural specializations (organelles) — such as flagella, cilia, contractile vacuole, and gullet — that may be divided equally or unequally between the two daughter cells, so that a certain amount of differentiation or regeneration may be necessary to make the new animal complete. Some of these organelles are self-reproducing, but others are lost by resorption (dedifferentiation), then differentiated anew in each of the daughter organisms.

Three general modes of cell division are recognized: binary fission, budding, and multiple fission. All of these are basically asexual processes. In many species, however, asexual reproduction is often followed at certain periods by some form of sexual reproduction that may or may not be necessary for the continued existence of the organisms.

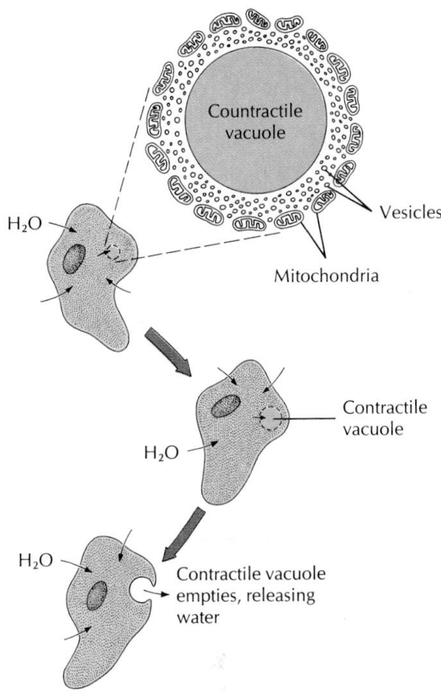

FIGURE 15-5

How an ameba pumps out water. Water enters the ameba's body by osmosis and is removed by the rhythmic filling and emptying of the contractile vacuole. The contractile vacuole of *Amoeba proteus* is surrounded by tiny vesicles that fill with fluid, which is then emptied into the vacuole. Note the numerous mitochondria that are believed to provide energy needed to adjust the salt content of the tiny vesicles.

Asexual reproduction

Binary fission. Binary fission, the most common type of reproduction among protozoa, involves the division of the organism — both its nucleus and cytoplasm — into two essentially equal daughter organisms (Figures 15-6 and 15-7). Binary fission may be transverse (across ciliary rows, as in most ciliates) or between flagellar groups or rows (as in Mastigophora). The nucleus divides by mitosis, but the nuclear membrane may persist through the division in some forms. In many forms the chromosomes are similar in structure and behavior to metazoan chromosomes; in others the chromosomes are granular and highly atypical.

Budding. Budding is quite similar to binary fission except that the division products are unequal. The daughter cell is usually much smaller than the parent.

Multiple fission (schizogony, sporogony). In multiple fission, the nucleus divides a number of times, followed by the division of the cytoplasm (cytokinesis) to form as many cells as there are nuclei. It occurs in some species of flagellates and several Sarcodina and is characteristic of the Sporozoea. **Sporogony** is dis-

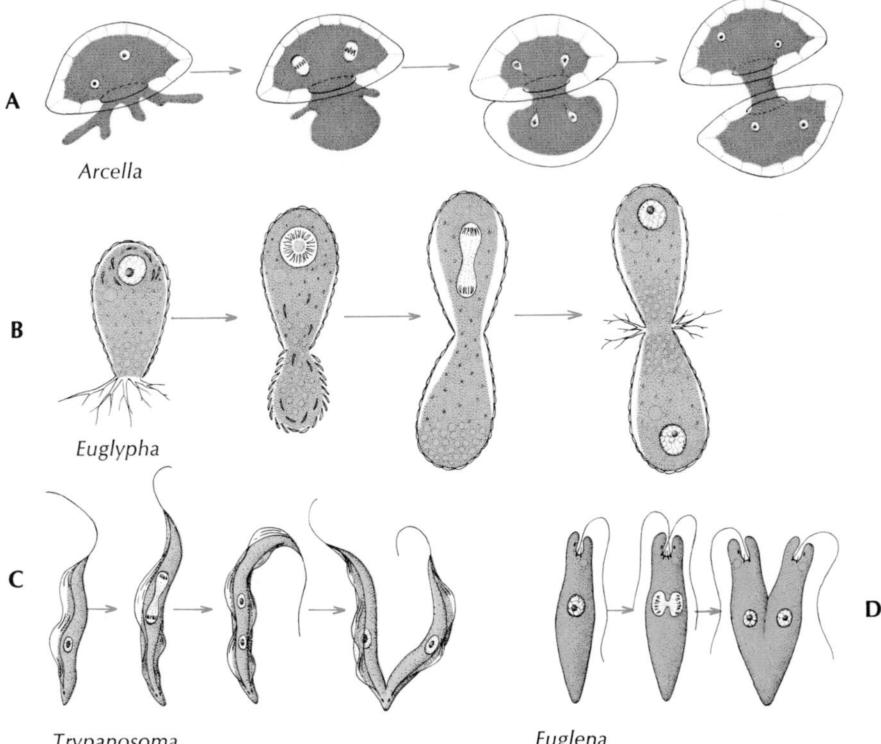

FIGURE 15-6

Binary fission in some sarcodines and flagellates. **A,** The two nuclei of *Arcella* divide as some of its cytoplasm is extruded and begins to secrete a new shell for the daughter cell. **B,** The shell of another sarcodine, *Euglypha,* is constructed of secreted platelets. Secretion of the platelets for the daughter cell is begun before the cytoplasm begins to move out the aperture. As these are used to construct the shell of the daughter cell, the nucleus divides. **C,** *Trypanosoma* has a kinetoplast (part of the mitochondrion) near the kinetosome of its flagellum close to its posterior end in the stage shown. All of these parts must be replicated before the cell divides. **D,** Division of *Euglena.*

FIGURE 15-7

Binary fission in a ciliophoran *(Paramecium).* Division is transverse, across rows of cilia.

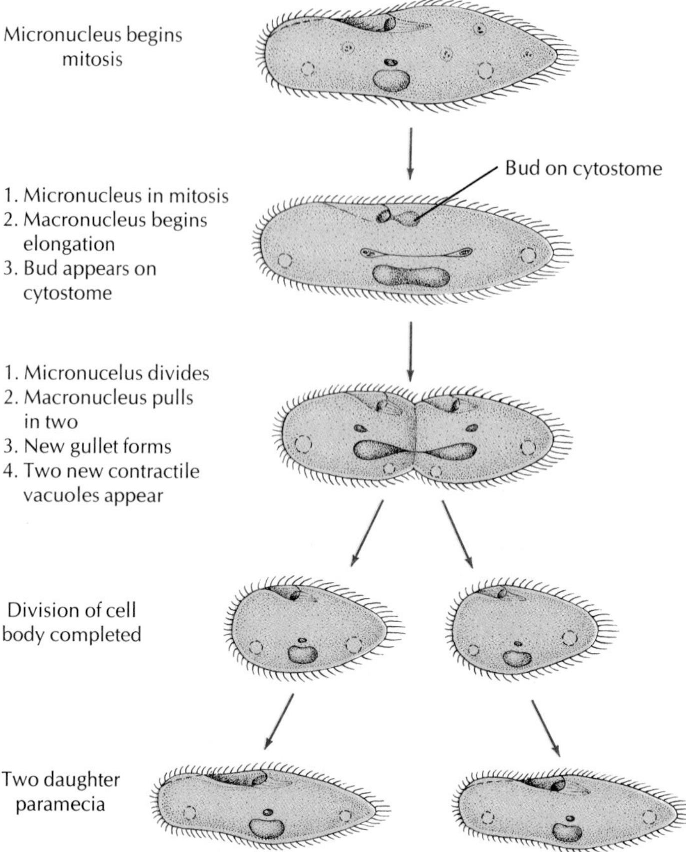

FIGURE 15-8

Life cycle of *Volvox.* Asexual reproduction occurs in spring and summer when specialized diploid reproductive cells divide to form young colonies that remain in the mother colony until large enough to escape. Sexual reproduction occurs largely in autumn when haploid sex cells develop. The fertilized ova may encyst and so survive the winter, developing into a mature asexual colony in the spring. In some species the colonies have separate sexes; in others both eggs and sperm are produced in the same colony.

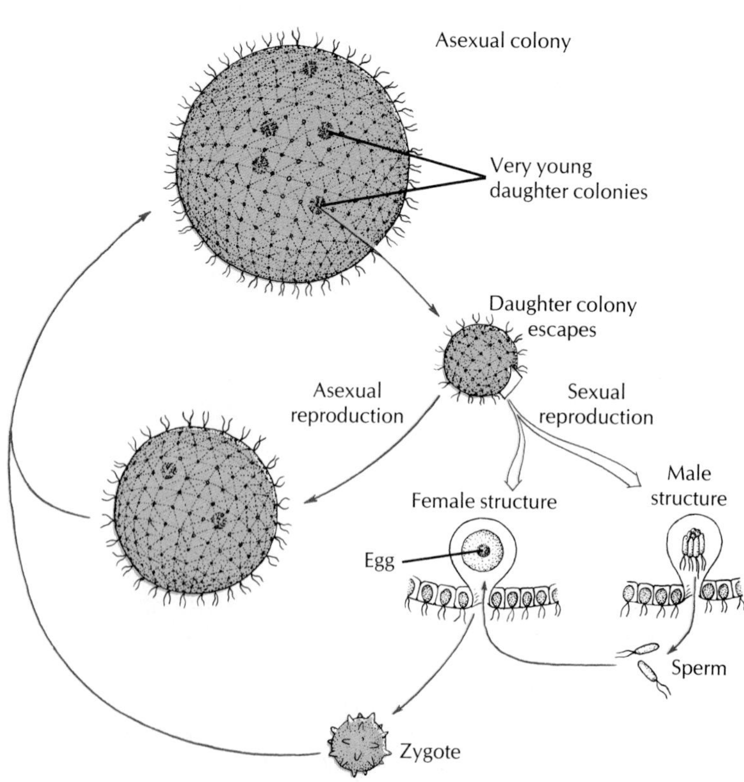

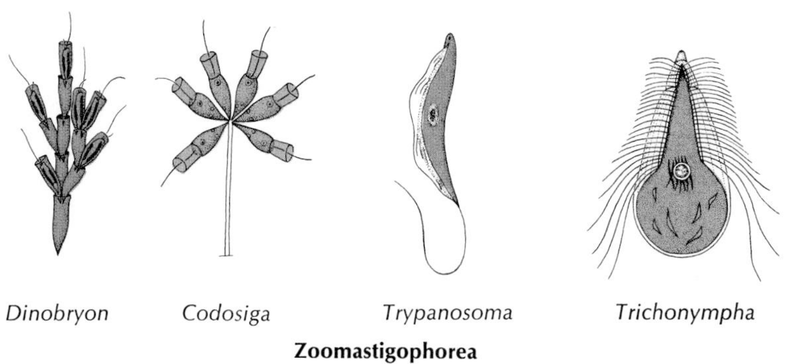

tinguished from schizogony in that sporogony takes place at some time following fertilization of gametes (zygote formation) and often involves spore formation (p. 358).

Protozoan colonies. Protozoan colonies are formed when the daughter zooids derived asexually remain associated instead of moving apart and living a separate existence (Figures 15-8 and 15-9). Protozoan colonies vary from individuals embedded together in a gelatinous substance to those which have protoplasmic connections among them. The arrangement of the individuals results in a variety of colony types, each characteristic of the protozoan that forms it. Usually the individuals of a colony are structurally and physiologically the same, although there may be some division of labor, such as differentiation of reproductive and somatic zooids. Division of labor, however, may be carried so far that it is difficult to distinguish between a protozoan colony and a metazoan individual.

Sexual reproduction

Sexual reproduction involves that special kind of nuclear division called **meiosis** (p. 63), in which the number of chromosomes is reduced by half (diploid number reduced to haploid number). The union of gamete nuclei (zygote formation) restores the diploid condition. In multicellular animals, reduction division occurs during the formation of gametes. This kind of meiosis is found in protozoa among some sarcodines and flagellates and in the Opalinata and the ciliates. If meiosis occurs in the first division after zygote formation, all intervening stages from that point to gamete formation are haploid. This kind of meiosis is found in some flagellates and all Sporozoea. The gametes may be similar in appearance or unlike.

The fertilization of one individual gamete by another is called **syngamy,** but some sexual phenomena in protozoa do not involve that process. Examples are **autogamy,** in which gametic nuclei arise by meiosis and fuse to form a zygote within the same organisms that produced them; **parthenogenesis,** the development of an organism from a gamete without fertilization; and **conjugation,** in which there is an exchange of gametic nuclei between paired organisms (conjugants). The process of conjugation is detailed in the discussion of the paramecium.

LIFE CYCLES

Many protozoa have very complex life cycles; others have simple ones. A simple life cycle may consist of an active phase and a **cyst.** A cyst is a resistant, quiescent stage within a cyst wall. In some cases the cyst may be lacking. *Amoeba* has a relatively simple life cycle. The more complex life cycles include two or more stages in the active phase and a reproductive phase that may include sexual as well as asexual phenomena. For example, some protozoa have both a ciliated

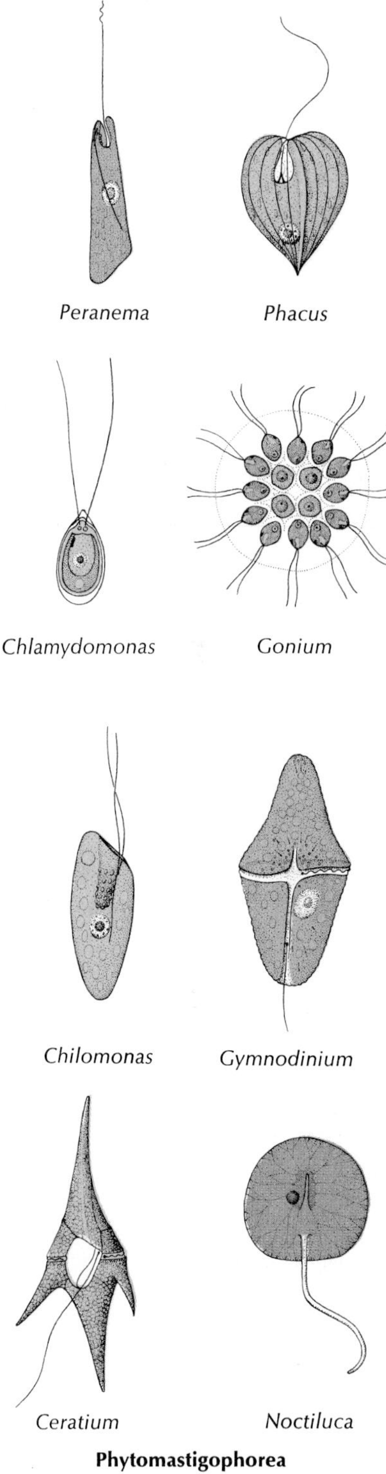

FIGURE 15-9

Some flagellate protozoa. *Gonium, Dinobryon,* and *Codosiga* are colonial.

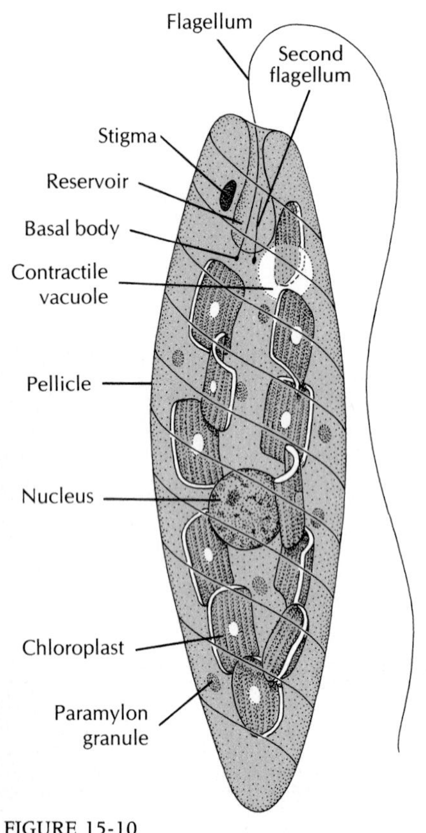

FIGURE 15-10

Euglena. Features shown are a combination of those visible in living and stained preparations.

Two subspecies of *Trypanosoma brucei (T. b. rhodesiense and T. b. gambiense)* cause clinically distinct forms of African trypanosomiasis (sleeping sickness) in humans. Another subspecies, *T. b. brucei,* along with several other trypanosomes, causes a similar disease in domestic animals. This makes agriculture very difficult in large areas of Africa. *Trypanosoma cruzi* causes American trypanosomiasis, or Chagas' disease, a very serious disease in South and Central America.

and a nonciliated stage; others have ameboid and flagellate stages; and still others, free-swimming and sessile stages. The most complex protozoan life cycles are found in the parasitic class Sporozoea—a good example of which is *Plasmodium* (Gr. *plasma,* something molded) (see Figure 15-14), the malarial parasite.

Encystment is common among protozoa, helping them withstand unfavorable conditions. There is usually a complex series of events that occurs when the organism encysts. It becomes quiescent, and many organelles, such as cilia, flagella, and the contractile vacuole, may dedifferentiate and disappear. A cyst wall is secreted over the surface so that the animal can withstand desiccation, temperature changes, and other harsh conditions. Reproductive cycles, such as budding, fission, and syngamy, may also occur in the encysted condition. The cysts of some species may be viable for many years.

■ Phylum Sarcomastigophora

SUBPHYLUM MASTIGOPHORA

The flagellates are protozoa that move by means of flagella. The name of the group, "Mastigophora," means "whip bearing." These organisms are common in both fresh and marine waters. The group is subdivided into the **phytoflagellates** (Phytomastigophorea), most of which contain chlorophyll and are thus plantlike, and the **zooflagellates** (Zoomastigophorea),which, lacking chlorophyll, are holozoic or saprozoic and are thus animal-like (Figure 15-9).

The phytoflagellates are commonly called the "green flagellates," although they may appear green, yellow, brown, or even colorless, according to the pigments present. They are producers, manufacturing most of what they need themselves, and they make up much of the base of the food chains of pelagic communities (Chapter 5). In the process of **photosynthesis,** they use light as the energy source, and usually carbon dioxide and nitrate or ammonium ions are the carbon and nitrogen sources, respectively, from which they synthesize their carbon compounds. As in *Euglena* (Figure 15-10), these forms are seen to have one or more characteristic **chloroplasts,** colored bodies that contain the chlorophyll and sometimes other pigments. Most of the green flagellates have a light-sensitive **stigma,** or eyespot, a shallow pigment cup that allows light from only one direction to strike a light-sensitive receptor.

Although phytoflagellates are primarily autotrophic, as described above, some are also saprozoic, some holozoic, and some use a combination of methods.

The zooflagellates, colorless because they lack chloroplasts, are holozoic or saprozoic. Many are parasitic, such as the various species of the leaf-shaped *Trypanosoma* (Gr. *trypanon,* auger, + *sōma,* body) (Figure 15-6, *C*), a blood parasite that causes African sleeping sickness. Some have pseudopodia as well as flagella.

The movement of the flagellum is varied, making it a versatile propulsive organ. The flagellar beat is undulating, proceeding from one end to the other as succeeding waveforms or as a corkscrew or spiral form. In various groups the undulation may begin at the tip or the base of the flagellum, and the protozoan may be pushed or pulled through the water by this beat. Many species have tiny, hairlike projections from the flagella to increase the efficiency of the beat. There may be more than one flagellum; *Trichomonas* (Gr. *thrix,* hair, + *monas,* single), for example, has four of them extending forward and one trailing backward, which apparently helps anchor and steer the animal. Flagella may also form rows somewhat like ciliary rows.

Most flagellates reproduce asexually by longitudinal binary fission (Figure 15-6, *C* and *D*). However, multiple fission is seen in some stages of trypanosomes and others. The sexual process is rare except in some of the colonial forms such as

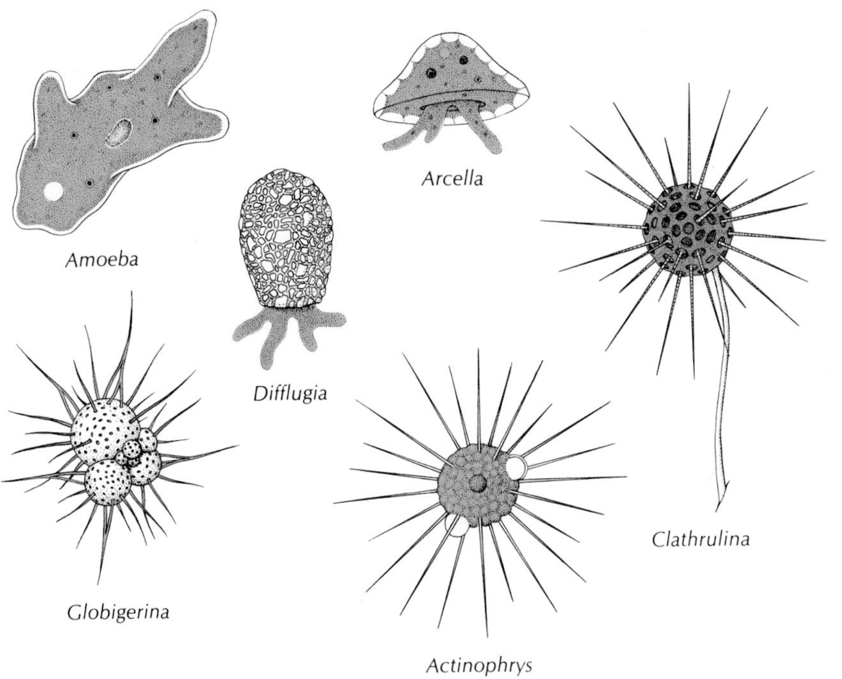

FIGURE 15-11

A group of sarcodines.

A

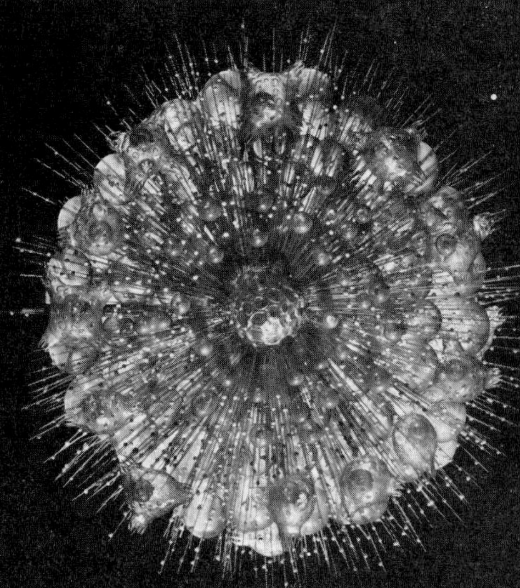
B

FIGURE 15-12

A, *Globigerina bulloides,* a pelagic foraminiferan that builds a many-chambered test, largely of calcium carbonate. **B,** *Trypanosphaera regina,* a pelagic radiolarian, has an internal siliceous skeleton.

Glass models courtesy American Museum of Natural History.

Volvox (Figure 15-8). In *Volvox* certain reproductive zooids develop into eggs or bundles of sperm. Each fertilized egg (zygote) undergoes repeated divisions to produce a small colony, which is released in the spring. A number of asexual generations may follow before sexual reproduction occurs again.

SUBPHYLUM SARCODINA

The sarcodines characteristically move and feed by means of pseudopodia (Figures 15-1 and 15-11). Classification of sarcodines is based to a great extent on the characteristics of their pseudopodia and whether they are naked (covered only by a cell membrane) or possess protective **tests** or internal skeletons. Sarcodines are found in both fresh and salt water and in moist soils. Some are planktonic; some prefer a substratum. A few are parasitic.

Nutrition in amebas and other sarcodines is holozoic; that is, they ingest and digest liquid or solid foods. Most amebas are omnivorous, living on algae, bacteria, protozoa, rotifers, and other microscopic organisms. An ameba may take in food at any part of its body surface merely by putting out a pseudopodium to enclose the food **(phagocytosis).** The enclosed food particle, along with some of the environmental water, becomes a food vacuole, which is carried about by the streaming movements of endoplasm. As digestion occurs within the vacuole by enzymatic action, water and digested materials pass into the cytoplasm. Undigested particles are eliminated through the cell membrane.

Most sarcodines reproduce by binary fission. Sporulation and budding occur in some of the sarcodines.

Not all sarcodines are naked. Some are covered with a protective test, or shell, with openings for the pseudopodia. The tests may be secreted siliceous material reinforced with sand grains (Figure 15-11). The **foraminiferans** (Foraminiferida [L. *foramen,* hole, + *fero,* bear]) are mostly marine forms that secrete complex many-chambered tests of calcium carbonate (Figure 15-12, *A*). They usually add sand grains to the secreted material, using great selectivity in choosing colors.

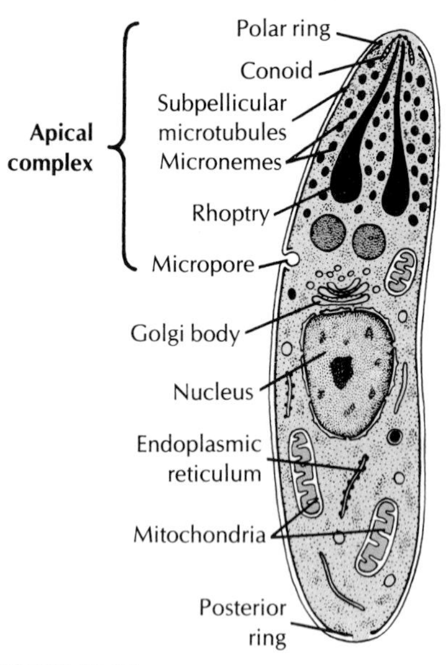

FIGURE 15-13

Diagram of an apicomplexan sporozoite or merozoite at the electron microscope level, illustrating the apical complex. The polar ring, conoid, micronemes, rhoptries, subpellicular microtubules, and micropore (cytostome) are all considered components of the apical complex.

Slender pseudopodia extend through openings in the test and then run together to form a protoplasmic net, in which they ensnare their prey. They are beautiful little creatures with many slender pseudopodia radiating out from a central test. The **radiolarians** are marine forms, mostly living in plankton, that have intricate and beautiful siliceous skeletons (Figure 15-12, *B*). (Radiolaria [L. *radiolus,* small ray], used here as a common name, are now separated into two classes by the Society of Protozoologists.) A central capsule that separates the inner and outer cytoplasm is perforated to allow cytoplasmic continuity.

The radiolarians are the oldest known group of animals, and they and the foraminiferans have left excellent fossil records. For millions of years the tests of dead protozoa have been dropping to the sea floor, forming deep-sea sediments that are estimated to be from 600 to 3600 m deep (approximately 2000 to 12,000 feet) and containing as many as 50,000 foraminiferans per gram of sediment. Many limestone and chalk deposits on land were laid down by these small creatures when the land was covered by the sea.

Some of the amebas are endoparasitic, mostly in the intestine of humans or other animals. Most are harmless, but *Entamoeba histolytica* (Gr. *entos,* within, + *amoibē,* change; *histos,* tissue, + *lysis,* a loosing), the most common amebic parasite harmful to humans, causes amebic dysentery. Occasionally, certain free-living amebas cause an almost always fatal brain disease. They apparently gain access through the nose while a person is swimming in a pond or lake.

■ Phylum Apicomplexa

CLASS SPOROZOEA

All apicomplexans are endoparasites, and their hosts are found in many animal phyla. The presence of a certain combination of organelles, the **apical complex,** distinguishes this phylum (Figure 15-13). The apical complex is usually present only in certain developmental stages of the organisms; for example, **merozoites** and **sporozoites** (Figure 15-14). Some of the structures, especially the **rhoptries** and **micronemes,** apparently are of aid in penetration of the host's cells or tissues.

Locomotor organelles are not as obvious in this group as they are in other protozoa. Pseudopodia occur in some intracellular stages, and gametes of some species are flagellated. Tiny contractile fibrils can form waves of contraction across the body surfaces to propel the organism through a liquid medium.

The life cycle usually includes both asexual and sexual reproduction, and there is sometimes an invertebrate intermediate host. At some point in the life cycle, most of them develop a **spore (oocyst)**, which is infective for the next host and is often protected by a resistant coat.

Plasmodium is the sporozoan parasite that causes **malaria.** The vectors (carriers) of the parasites are female *Anopheles* mosquitoes, and the life cycle is depicted in Figure 15-14. Malaria is one of the most important diseases in the world, with 100 million cases occurring each year. It causes 1 million deaths per year in the world. Though this represents a great decline from the situation 25 years ago, recent years have seen a resurgence caused by increasing resistance of mosquitoes to insecticides, increasing numbers of *Plasmodium* strains that are resistant to drugs, and socioeconomic conditions in tropical countries that interfere with malaria control efforts.

Toxoplasma (Gr. *toxo,* a bow, + *plasma,* molded) is a common parasite in the intestinal tissues of cats, but this parasite can produce extraintestinal stages, as well. The extraintestinal stages can develop in a wide variety of animals other than cats—for example, rodents, cattle, and humans. Gametes and oocysts are not produced by the extraintestinal forms, but they can initiate the intestinal cycle in a cat if the cat eats infected prey. In humans *Toxoplasma* causes few or no

Although malaria has been recognized as a disease and a scourge of humanity since antiquity, the discovery that it was caused by a protozoan was made only 100 years ago by Charles Louis Alphonse Laveran, a French army physician. At that time, the mode of transmission was still mysterious, and "bad air" (hence the name malaria) was a popular candidate. Ronald Ross, an English physician in the Indian Medical Service, determined some years later that the malarial organism was carried by *Anopheles* mosquitoes. Because Ross knew nothing about mosquitoes or their normal parasites, his efforts were long frustrated by trying to use the wrong kinds of mosquitoes and confusion caused by the other parasites he found in them. That he persisted is a tribute to his determination. The discovery was nonetheless momentous, and it earned Ross the Nobel Prize in 1902 and knighthood in 1911.

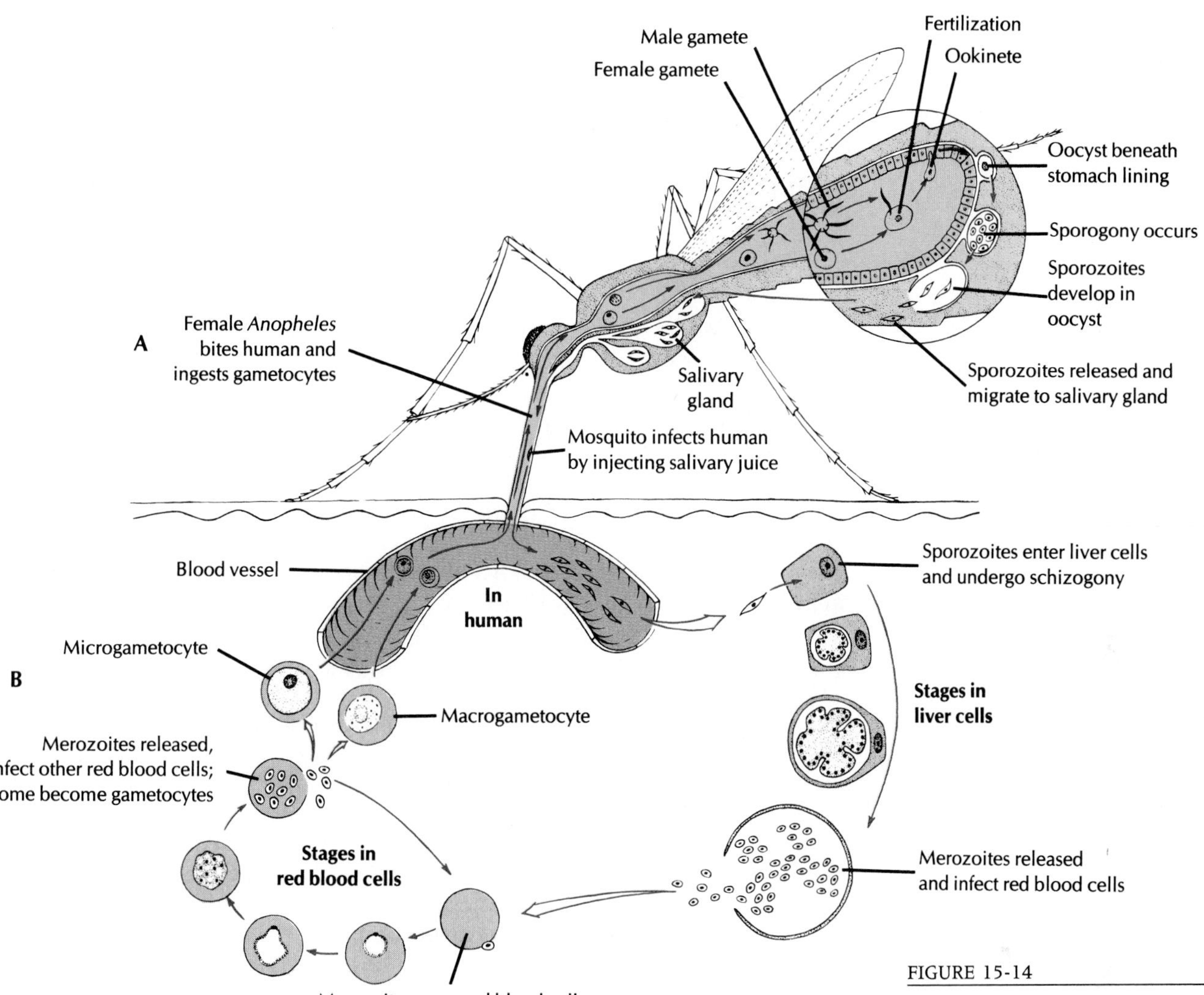

FIGURE 15-14

Life cycle of *Plasmodium vivax,* one of the protozoa (class Sporozoea) that causes malaria in humans. **A,** Sexual cycle produces sporozoites in body of mosquito. **B,** Sporozoites infect humans and reproduce asexually, first in liver cells then in red blood cells. Malaria is spread by the *Anopheles* mosquito, which sucks up gametocytes along with human blood, then, when biting another victim, leaves sporozoites in the new wound.

ill effects except in a woman infected during pregnancy, particularly in the first trimester. Such infection greatly increases the chances of a birth defect in the baby; it is now believed that 2% of all mental retardation in the United States is a result of congenital toxoplasmosis. Humans can become infected from eating insufficiently cooked beef, pork, or lamb or from accidentally ingesting oocysts from the feces of cats. Pregnant women should not eat raw meat or empty cats' litterboxes.

Other common sporozoan parasites are the **coccidians** (Coccidia [Gr. *kokkos,* kernel or berry]), which infect epithelial tissues in both vertebrates and invertebrates, and the **gregarines,** which live mainly in the digestive tract and body cavity of certain invertebrates.

■ Phylum Ciliophora

The ciliates are a large and interesting group, with a great variety of forms living in all types of freshwater and marine habitats. They are the most structurally complex and diversely specialized of all the protozoa. The majority are free living, although some are commensal (p. 139) or parasitic. They are usually solitary and motile, but some are colonial and some are sessile. There is a great diversity of

FIGURE 15-15

Some representative ciliates. *Euplotes* have stiff cirri used for crawling about. Contractile myonemes in ectoplasm of *Stentor* and in stalks of *Vorticella* allow great expansion and contraction. Note the macronuclei, long and curved in *Euplotes* and *Vorticella*, shaped like a string of beads in *Stentor*.

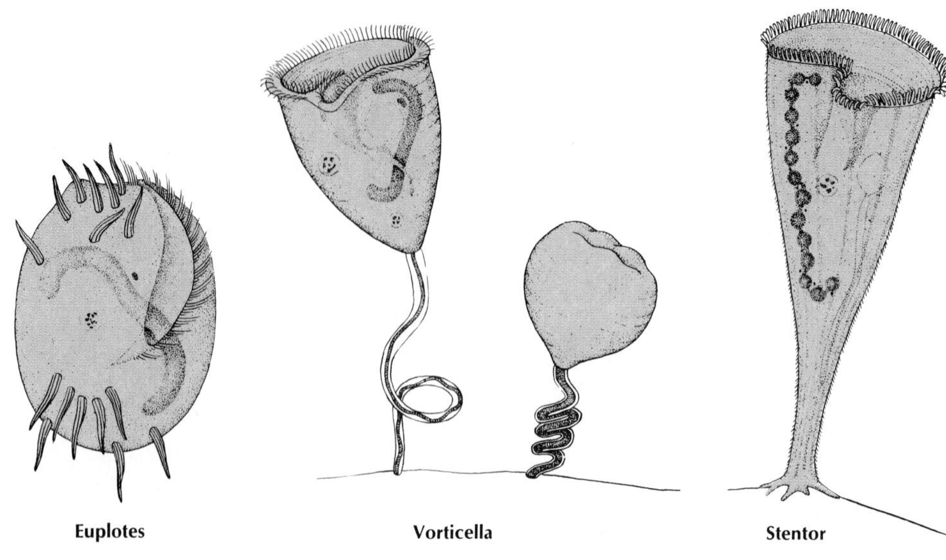

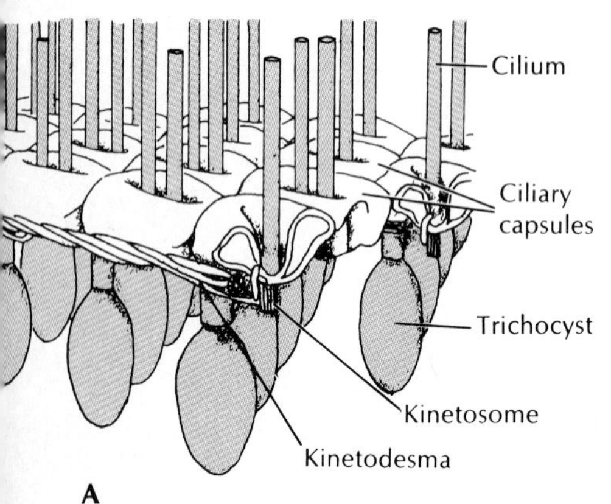

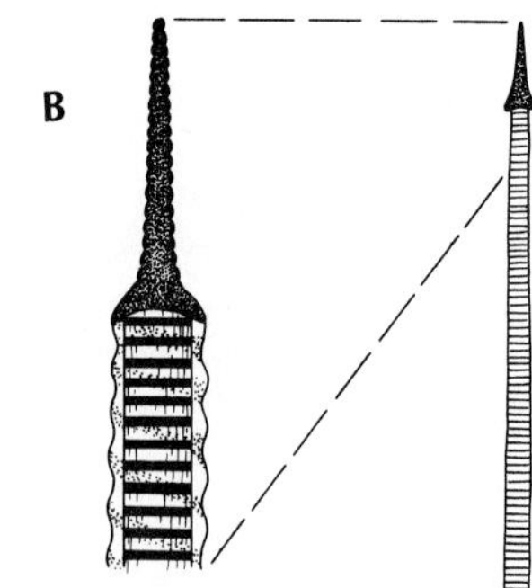

FIGURE 15-16

Infraciliature and associated structures in ciliates. **A,** Structure of the pellicle and its relation to the infraciliature system. **B,** Expelled trichocyst.

shape and size. In general they are larger than most other protozoa, but they range from very small (10 to 12 μm) up to 3 mm long. All have cilia that beat in a coordinated rhythmical manner, although the arrangement of the cilia may vary.

Ciliates are always multinucleate, possessing at least one **macronucleus** and one **micronucleus,** but varying from one to many of either type. The macronuclei are apparently responsible for metabolic, synthetic, and developmental functions. Macronuclei are varied in shape among the different species (Figures 15-15 and 15-17). The micronuclei participate in sexual reproduction and give rise to macronuclei after exchange of micronuclear material between individuals. The micronuclei divide mitotically, and the macronuclei divide amitotically.

Ciliates are covered by a secreted envelope or **pellicle,** which may be very thin or, in some species, may form a thickened armor. The cilia are short and usually arranged in longitudinal or diagonal rows (Figure 15-16, *A*). Their structure is exactly comparable to flagella, with axoneme and kinetosome, except that cilia are shorter. Cilia may cover the surface of the animal or may be restricted to the oral region or to certain bands.

In some forms the cilia are fused into a sheet called an **undulating membrane** or into smaller **membranelles,** both used to propel food into the **cytopharynx** (gullet). In other forms there may be fused cilia forming stiffened tufts called **cirri,** often used in locomotion by the creeping ciliates (Figure 15-15).

The kinetosomes and a structural system of fibrils just beneath the pellicle make up the **infraciliature** (Figure 15-16, *A*). The infraciliature apparently does not coordinate the ciliary beat, as formerly believed. Coordination of the ciliary movement seems to be by waves of depolarization of the cell membrane moving down the animal, similar to the conduction of a nerve impulse (p. 239).

Most ciliates are holozoic. Most of them possess a cytostome (mouth) that in some forms is a simple opening and in others is connected to a gullet or ciliated groove. The mouth in some is strengthened with stiff, rodlike structures for swallowing larger prey; in others ciliary water currents carry microscopic food particles toward the mouth, as in the paramecia. *Didinium* has an oral cone, by which it attaches to its prey, then it forms a temporary cytostome for engulfing the paramecia on which it feeds (Figure 15-4). Suctorians paralyze their prey and then ingest their contents through tubelike tentacles by a complex feeding mechanism that apparently combines phagocytosis with a sliding filament action of microtubules in the tentacles (Figure 15-4). In any case the food is digested within food vacuoles.

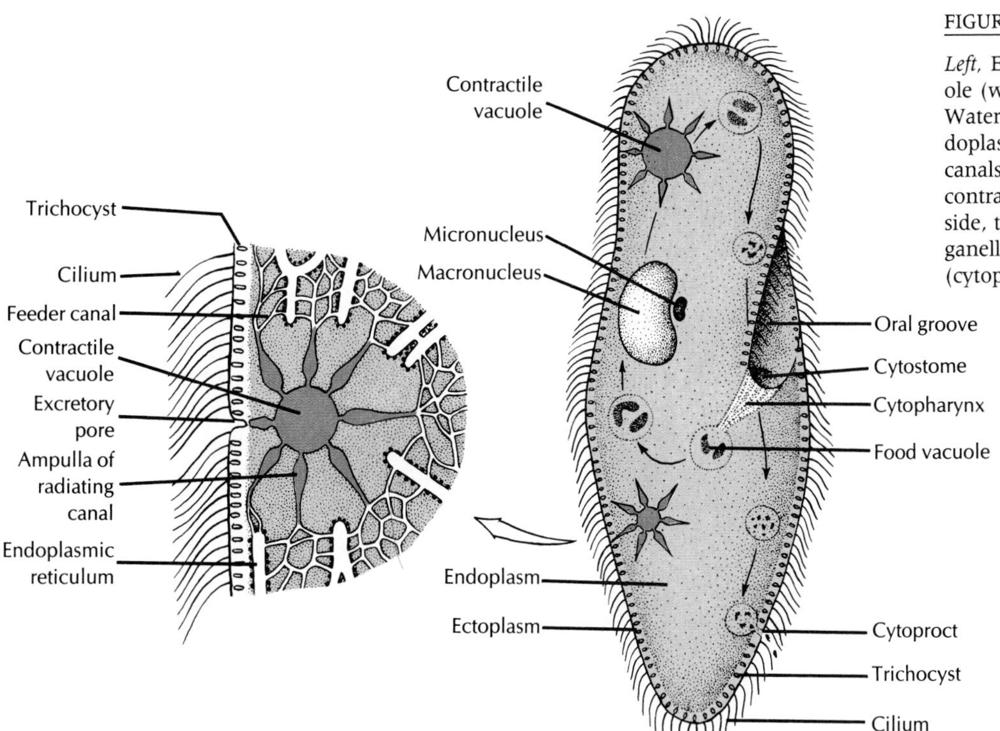

FIGURE 15-17

Left, Enlarged section of a contractile vacuole (water-expulsion vesicle) of *Paramecium.* Water is believed to be collected by the endoplasmic reticulum, emptied into feeder canals and then into the vesicle. The vesicle contracts to empty its contents to the outside, thus serving as an osmoregulatory organelle. *Right, Paramecium,* showing gullet (cytopharynx), food vacuoles, and nuclei.

In some of the ciliates small bodies called **trichocysts** are located in the ectoplasm (Figures 15-16 and 15-17). When discharged, they form long threadlike filaments that pass through the pellicle and harden, except for the tips, which are sticky for attachment (Figure 15-16). In some ciliates the discharged trichocysts seem to help anchor the animal while feeding; in others they are apparently used to paralyze other small organisms for defense or capture of prey.

Contractile vacuoles are found in all freshwater and some marine ciliates; the number varies from one to many among the different species. In most ciliates the vacuole is fed by one or more collecting canals (Figure 15-17). The vacuoles occupy a fixed position, and each discharges through a more or less permanent pore.

REPRODUCTION AND LIFE CYCLES

The life cycles of ciliates usually involve both **asexual binary fission** and a type of sexual reproduction called **conjugation** (Figure 15-18). Conjugation is the temporary union of two individuals for the purpose of exchanging chromosomal material. During union the micronucleus of each individual undergoes meiosis, giving rise to four haploid micronuclei, three of which degenerate. The remaining micronucleus then divides into two haploid pronuclei, one of which is exchanged for a pronucleus of the conjugant partner. When the exchanged pronucleus unites with the pronucleus of the partner, the diploid number of chromosomes is restored. The two partners, each with fused pronuclei (now comparable to a zygote), separate, and each divides twice by mitosis, each one thereby giving rise to four daughter paramecia.

Conjugation always involves two individuals of different **mating types.** This prevents inbreeding. Most ciliate species are divided into several varieties, each variety made up of two mating types. Mating occurs only between two mating types of each variety.

The advantage of sexual reproduction is that it permits gene recombinations, thus promoting genetic diversity in a population. Seasonal change or a deteriorating environment usually stimulates sexual reproduction.

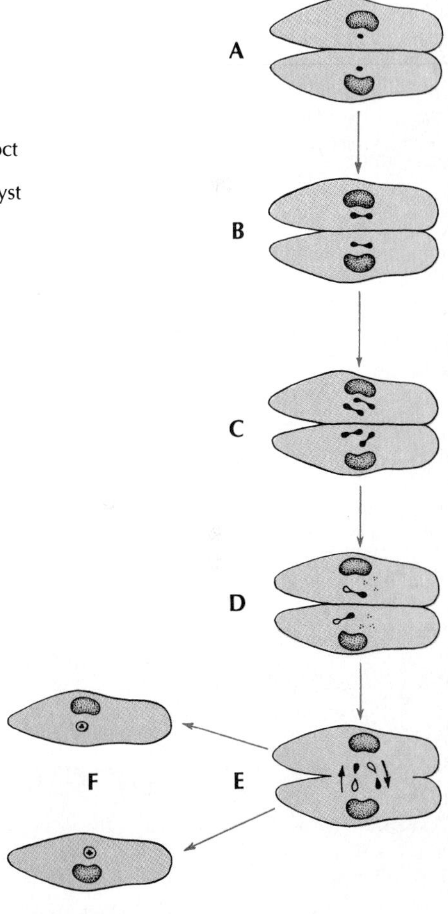

FIGURE 15-18

Scheme of conjugation in *Paramecium.* **A,** Two individuals come in contact on oral surface. **B** and **C,** Micronuclei divide twice (meiosis), resulting in four haploid nuclei in each partner. **D,** Three micronuclei degenerate; remaining one divides to form "male" and "female" pronuclei. **E,** Male pronuclei are exchanged between conjugants. **F,** Male and female pronuclei fuse, and individuals separate. Subsequently, old macronuclei are absorbed and replaced by new macronuclei.

Phylogeny and Adaptive Radiation

PHYLOGENY

Protozoa are usually placed at the base of the phylogenetic tree of eukaryotes. Doubtless, multicellular animals were derived from a protozoan or protozoan-like ancestor, perhaps more than once. Colonial protozoa, particularly among flagellates (Figure 15-8), show various degrees of cell aggregation and some differentiation that suggest the body plans of early metazoa.

With the exception of certain shell-bearing Sarcodina, such as foraminiferans and radiolarians, protozoa have left no fossil records; thus our conjectures about protozoan evolution must be based on living forms. Mastigophorans may be the oldest of all protozoa, perhaps having arisen from a combination of bacteria and spirochaetes, but the group is probably polyphyletic. Most of the phytoflagellates are more closely related to the multicellular algae than they are to zooflagellates. Some colorless phytoflagellates have chlorophyll-bearing relatives, and some autotrophic forms are facultatively saprophytic in darkness. Hence, the common origin of animals and plants may lie in the Phytomastigophorea. That the amebas were probably derived from the flagellates is indicated by the ameboid stages in some flagellates and flagellated stages of some amebas. However, the different orders of Sarcodina may have arisen independently from different kinds of flagellates. Apicomplexa, which are all specialized parasites, probably come from flagellated ancestors: they often have ameboid feeding stages and flagellated gametes. The origin of the ciliates is somewhat obscure, but the basic structural similarity of the flagellum to the cilium is undeniable.

ADAPTIVE RADIATION

Some of the wide range of adaptations of protozoa have been illustrated in the previous pages. The Sarcodina range from bottom-dwelling, naked species to planktonic forms such as the foraminiferans and radiolarians with beautiful, intricate tests. There are many symbiotic species of amebas. Flagellates likewise show adaptations for a similarly wide range of habitats, with the added variation of photosynthetic ability in many species of Phytomastigophorea. The fine line between plants and animals at this level is shown by our ability to turn a "plant" into an "animal" by experimentally destroying the chloroplasts of *Euglena.*

Within a single-cell body plan, the division of labor and specialization of organelles are carried furthest by the ciliates, the most complex of all protozoa. Specializations for intracellular parasitism have been adopted by the Sporozoea.

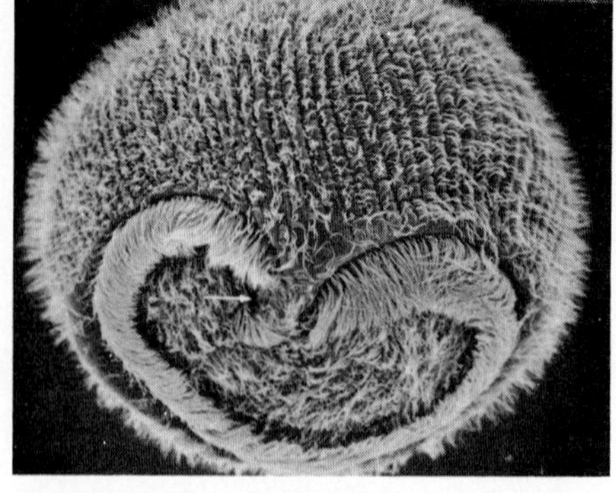

Scanning electron micrograph of the anterior end of the ciliate, *Stentor coeruleus.* In life, the prominent oral ciliature directs food toward the cytostome *(arrow).* (×500.)

Photograph by J.J. Paulin, University of Georgia/BPS.

Summary

Protozoa are unicellular animals at the protoplasmic level of organization. Traditionally considered a single, heterogeneous phylum, the protozoa have been recently assigned to seven phyla, the largest and most important of which are the Sarcomastigophora (flagellates and amebas), the Apicomplexa (coccidians, malaria organisms, and others), and the Ciliophora (ciliates). Ecologically, protozoa are especially important as segments of aquatic food chains, for example in plankton, and as parasitic, disease-causing organisms.

Pseudopodial or ameboid movement is a locomotory and food-gathering mechanism in protozoa and plays a vital role as a defense mechanism in metazoa. It is accomplished by an incompletely known process of microfilaments moving past each other, and it requires expenditure of energy from ATP. Ciliary movement is likewise important in both protozoa and metazoa. The most widely accepted mechanism to account for ciliary movement is the sliding microtubule hypothesis.

Various protozoa feed by holophytic, holozoic, or saprozoic means. They expel excess water that enters their bodies by contractile vacuoles. Respiration and waste elimination are through the body surface. Protozoa reproduce asexually by binary fission, multiple fission, and budding; sexual processes are common.

Most phytoflagellates are photosynthetic, and many zooflagellates are important parasites. They move by beating one or more flagella. Sarcodines move by pseudopodia; many are important members of planktonic communities, and some are parasites. Many have a test, or shell. All apicomplexans are parasitic, and they include *Plasmodium*, which causes malaria. The Ciliophora move by means of cilia or ciliary organelles. They are a large and diverse group, and many are complex in structure.

REVIEW QUESTIONS

1. Explain why a protozoan may be very complex, even though it is composed of only one cell.
2. Distinguish the following protozoan phyla: Sarcomastigophora, Apicomplexa, Ciliophora.
3. Explain the transitions of endoplasm and ectoplasm in ameboid movement.
4. Contrast the structure of an axoneme to that of a kinetosome.
5. What is the sliding microtubule hypothesis?
6. Explain how protozoa eat, digest their food, osmoregulate, and respire.
7. Distinguish the following: sexual and asexual reproduction; binary fission, budding, and multiple fission.
8. What is the survival value of encystment?
9. Contrast and give an example of phytoflagellates and zooflagellates.
10. Name three kinds of sarcodines, and tell where they are found (their habitats).
11. Outline the general life cycle of malaria organisms.
12. What is the public health importance of *Toxoplasma?*
13. Define the following with reference to ciliates: macronucleus, micronucleus, pellicle, undulating membrane, cirri, infraciliature, trichocysts, conjugation.
14. What are indications that the Sarcodina, Apicomplexa, and Ciliophora may have been derived from ancestral Phytomastigophorea?

SELECTED REFERENCES

Barnes, R.D. 1987. Invertebrate zoology, ed. 5. Philadelphia, Saunders College/Holt, Rinehart & Winston. *Chapter 2 has good coverage of protozoa.*

Bonner, J.T. 1983. Chemical signals of social amoebae. Sci. Am. **248:**114-120 (April). *Social amebas of two different species secrete different chemical compounds that act as aggregation signals. The evolution of the aggregation signals in these protozoa may provide clues to the origin of the diverse chemical signals (neurotransmitters and hormones) in more complex organisms.*

Harrison, G. 1978. Mosquitoes, malaria and man: a history of the hostilities since 1880. New York, E.P. Dutton. *A fascinating story, well told.*

Lee, J.J., S.H. Hutner, and E.C. Bovee (eds.). 1985. An illustrated guide to the protozoa. Lawrence, Kan., Society of Protozoologists. *A comprehensive guide and essential reference for students of the protozoa.*

Pennak, R.W. 1989. Fresh-water invertebrates of the United States, ed. 3. New York, John Wiley & Sons, Inc. *A reference work with considerable attention devoted to the freshwater protozoa that one might collect from a pond.*

Schmidt, G.D., and L.S. Roberts. 1989. Foundations of parasitology, ed. 4. St. Louis, The C.V. Mosby Co. *A parasitology text containing a good account of protozoan parasites.*

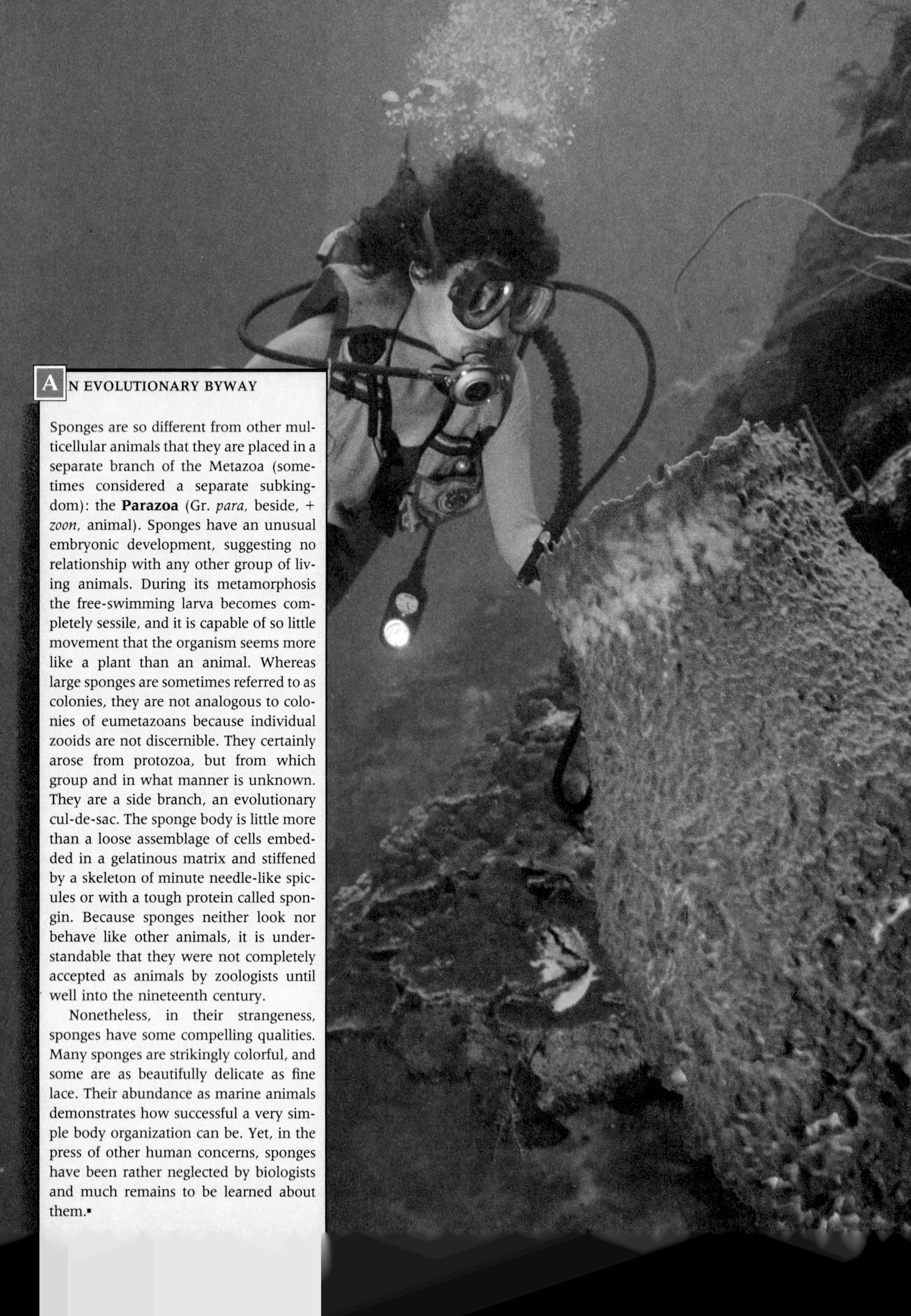

AN EVOLUTIONARY BYWAY

Sponges are so different from other multicellular animals that they are placed in a separate branch of the Metazoa (sometimes considered a separate subkingdom): the **Parazoa** (Gr. *para,* beside, + *zoon,* animal). Sponges have an unusual embryonic development, suggesting no relationship with any other group of living animals. During its metamorphosis the free-swimming larva becomes completely sessile, and it is capable of so little movement that the organism seems more like a plant than an animal. Whereas large sponges are sometimes referred to as colonies, they are not analogous to colonies of eumetazoans because individual zooids are not discernible. They certainly arose from protozoa, but from which group and in what manner is unknown. They are a side branch, an evolutionary cul-de-sac. The sponge body is little more than a loose assemblage of cells embedded in a gelatinous matrix and stiffened by a skeleton of minute needle-like spicules or with a tough protein called spongin. Because sponges neither look nor behave like other animals, it is understandable that they were not completely accepted as animals by zoologists until well into the nineteenth century.

Nonetheless, in their strangeness, sponges have some compelling qualities. Many sponges are strikingly colorful, and some are as beautifully delicate as fine lace. Their abundance as marine animals demonstrates how successful a very simple body organization can be. Yet, in the press of other human concerns, sponges have been rather neglected by biologists and much remains to be learned about them.▪

CHAPTER 16

Sponges

Sponges, in contrast to protozoa, are many-celled animals, or metazoa. However, they have the simplest type of metazoan organization — a **cellular level** of organization. Since the sponge lacks true tissues or organs, the division of labor is confined to a few types of cells that have become specialized for certain functions.

Phylum Porifera — Sponges

Sponges belong to the phylum Porifera (po-rif'er-a) (L. *porus,* pore, + *fera,* bearing). They are sessile and their bodies bear myriads of tiny pores and canals that comprise a filter-feeding system adequate for their inactive lifestyle. They depend on the water currents carried through their unique canal systems to bring them food and oxygen and to carry away their body wastes. Their bodies are little more than masses of cells embedded in a gelatinous matrix and stiffened by a skeleton of minute **spicules** of calcium or silica or by "spongy" fibers of a collagenous substance called **spongin.** They have no organs or true tissues, and even their cells show a certain degree of independence. As sessile animals with only negligible body movement, they have not evolved a nervous system or sense organs and have only the simplest of contractile elements.

Most sponges are colonial, and they vary in size from a few millimeters to the great loggerhead sponges, which may reach 2 meters or more across. Many sponge species are brightly colored because of pigments in the surface cells. Red, yellow, orange, green, and purple sponges are not uncommon. However, the color fades quickly when the sponges are removed from the water. Some sponges, including the simplest and most primitive, are radially symmetrical, but many are quite irregular in shape. Some stand erect, some are branched or lobed, and others are low, even encrusting, in form. Some bore holes into shells or rocks.

The sponges are an ancient group, with an abundant fossil record extending back to the Cambrian period and even, according to some claims, the Precambrian.

CHARACTERISTICS

1. Multicellular; body a loose aggregation of cells of mesenchymal origin
2. Body with pores (ostia), canals, and chambers that serve for passage of water
3. All aquatic; mostly marine
4. Radial symmetry or none
5. Epidermis of flat pinacocytes; most interior surfaces lined with flagellated collar cells (choanocytes) that create water currents; a gelatinous protein matrix called mesoglea contains amebocytes, collencytes, and skeletal elements
6. Skeletal structure of fibrillar collagen (a protein) and calcareous or siliceous crystalline spicules, often combined with variously modified collagen (spongin)
7. No organs or true tissues; digestion intracellular; excretion and respiration by diffusion
8. Reactions to stimuli apparently local and independent; nervous system probably absent
9. All adults sessile and attached to substratum
10. Asexual reproduction by buds or gemmules and sexual reproduction by eggs and sperm; free-swimming ciliated larvae

FIGURE 16-1

This orange demosponge, *Mycale laevis*, often grows beneath platelike colonies of the stony coral, *Montastrea annularis*. The large oscula of the sponge are seen at the edges of the plates. Unlike some other sponges, *Mycale* does not burrow into the coral skeleton and may actually protect the coral from invasion by more destructive species. Pinkish radioles of a Christmas tree worm, *Spirobranchus giganteus* (phylum Annelida, class Polychaeta), also project from the coral colony. An unidentified reddish sponge can be seen to the right of the Christmas tree worm.

Photograph by L.S. Roberts.

CLASSIFICATION

Class Calcarea (cal-ca're-a) (L. *calcis,* lime, + Gr. *spongos,* sponge) **(Calcispongiae).** Have spicules of calcium carbonate that often form a fringe around the osculum (main water outlet); spicules needle-shaped or three- or four-rayed; all three types of canal systems (asconoid, syconoid, leuconoid) represented; all marine. Examples: *Sycon, Leucosolenia.*
Class Hexactinellida (hex-ak-tin-el'i-da) (Gr. *hex,* six, + *aktis,* ray) **(Hyalospongiae).** Have six-rayed, siliceous spicules extending at right angles from a central point; spicules often united to form network; body often cylindrical or funnel shaped. Flagellated chambers in simple syconoid or leuconoid arrangement. Habitat mostly deep water; all marine. Examples: Venus' flower basket *(Euplectella), Hyalonema.*
Class Demospongiae (de-mo-spun'je-e) (Gr. *demos,* people, + *spongos,* sponge). Have siliceous spicules that are not six rayed, or spongin, or both. Leuconoid-type canal systems. One family found in fresh water; all others marine. Examples: *Thenea, Cliona, Spongilla, Myenia,* and all bath sponges.
Class Sclerospongiae (skler'o-spun'je-e) (Gr. *skleros,* hard, + *spongos,* sponge). Secrete massive basal skeleton of calcium carbonate, with living tissue extending into skeleton from 1 mm to 3 cm or more, extending above skeleton less than 1 mm; have siliceous spicules similar to Demospongiae (sometimes absent), and spongin fibers; leuconoid organization; inhabit caves, crevices, tunnels, and deep water on coral reefs. Examples: *Astrosclera, Calcifibrospongia.*

ECOLOGICAL RELATIONSHIPS

Most of the 5000 or more sponge species are marine, although some 150 species of the Demospongiae live in fresh water. Marine sponges are much more abundant than most people realize; they are present in all seas and at all depths, and a few even exist in brackish water. Although the larvae are free swimming, the adults are always attached, usually to rocks, shells, corals, or other submerged objects (Figure 16-1). Some benthic forms even grow on sand and mud bottoms.

Many animals (crabs, nudibranchs, mites, bryozoans) live as commensals or parasites in or on sponges. The larger sponges particularly tend to harbor a large variety of invertebrate commensals. On the other hand, sponges grow on many other living animals, such as molluscs, barnacles, brachiopods, corals, or hydroids. Some crabs attach pieces of sponge to their carapaces for camouflage and for protection, since most predators seem to find sponges distasteful. Some reef fishes, however, are known to graze on shallow-water sponges.

Certainly one reason for the success of sponges as a group is that they have few enemies. Because of a sponge's elaborate skeletal framework and often noxious odor or taste, most potential predators find sampling a sponge about as pleasant as eating a mouthful of glass splinters embedded in fibrous gelatin.

FORM AND FUNCTION

The only body openings of these unusual animals are pores, usually many tiny ones called **ostia** for incoming water, and one to a few large ones called **oscula** (sing., **osculum**) for water outlet. These openings are connected by a system of canals, some of which are lined with peculiar flagellated collar cells called **choanocytes,** whose flagella maintain a current of environmental water through the canals. Water enters the canals through a multitude of tiny incurrent pores **(dermal ostia)** and leaves by way of one or more large oscula. The choanocytes (described in more detail below) not only keep the water moving but also trap and phagocytose food particles that are carried in the water. The cells lining the passageways are very loosely organized. Collapse of the canals is prevented by the skeleton, which, depending on the species, may be made up of needlelike calcareous or siliceous spicules, a meshwork of organic spongin fibers, or a combination of the two.

Sessile animals make few movements and therefore need little in the way of nervous, sensory, or locomotor parts. Sponges apparently have lived as sessile animals from their earliest appearance and have never acquired specialized nervous or sensory structures, and they have only the very simplest of contractile systems.

Types of canal systems

Most sponges fall into one of three types, based on their systems of canals—asconoid, syconoid, or leuconoid (Figure 16-2).

Asconoids—flagellated spongocoels. The asconoid sponges have the simplest type of organization. They are small and tube shaped. Water moves through ostia in donut-shaped cells known as **porocytes** and enters a large cavity called the **spongocoel,** which is lined with choanocytes. The choanocyte flagella pull the water through the ostia, and the choanocytes filter out food particles (p.

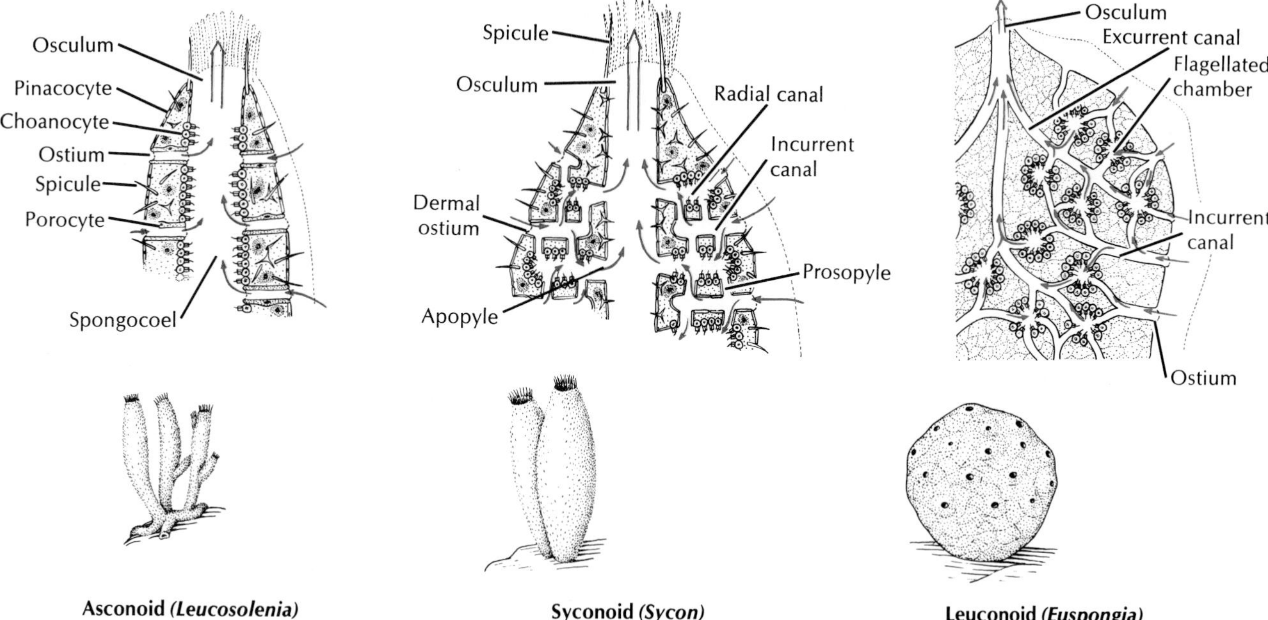

FIGURE 16-2

Three types of sponge structure. The degree of complexity from simple asconoid to complex leuconoid type has involved mainly the water-canal and skeletal systems, accompanied by outfolding and branching of the collar cell layer. The leuconoid type is considered the major plan for sponges, because it permits greater size and more efficient water circulation.

FIGURE 16-3

Leucosolenia canariensis (class Calcarea) is common on Caribbean reefs in caves and under ledges.

Photograph by L.S. Roberts.

369). The water is then expelled through the single osculum (Figure 16-2). *Leucosolenia* (Gr. *leukos*, white, + *solen*, pipe) is an asconoid type of sponge (Figure 16-3). Its slender, tubular individuals grow in groups attached by a common **stolon**, or stem, to objects in shallow seawater. Asconoids are found only in the Calcarea.

Syconoids—flagellated canals. Syconoid sponges look somewhat like larger editions of asconoids, from which they were derived. They have the tubular body and single osculum, but the body wall, which is thicker and more complex than that of asconoids, receives water through **incurrent canals** that deliver it to the choanocyte-lined **radial canals,** which empty into the spongocoel (Figures 16-2 and 16-4). The spongocoel in syconoids is lined with epithelial-type cells, rather than the choanocytes as found in asconoids. Syconoids are found in Calcarea and Hexactinellida. *Sycon* (Gr. *sykon*, a fig) is a commonly studied example of the syconoid type of sponge (Figure 16-4).

Leuconoids—flagellated chambers. Leuconoid organization is the most complex of the sponge types and the best adapted for increase in sponge size. Most leuconoids form large colonial masses, each member of the mass having its own osculum, but individual members are poorly defined and often impossible to distinguish (Figure 16-1). Clusters of flagellated chambers are filled from **incurrent canals** and discharge water into **excurrent canals** that eventually lead to the osculum (Figure 16-2). Most sponges are of the leuconoid type, which occurs in the Calcarea, the Demospongiae, and the Sclerospongiae.

These three types of canal systems—asconoid, syconoid, and leuconoid—are correlated with the evolution of sponges, from simple to complex forms. Evolutionary changes involved increasing the flagellated surfaces in proportion to the volume, thus providing more collar cells to meet the food demands. This was achieved by the out-pushing of the spongocoel of a simple sponge such as the asconoid type to form the radial canals (lined with flagellated cells) of the syconoid type. Further folding of the body wall produced the complex canals and chambers of the leuconoid type.

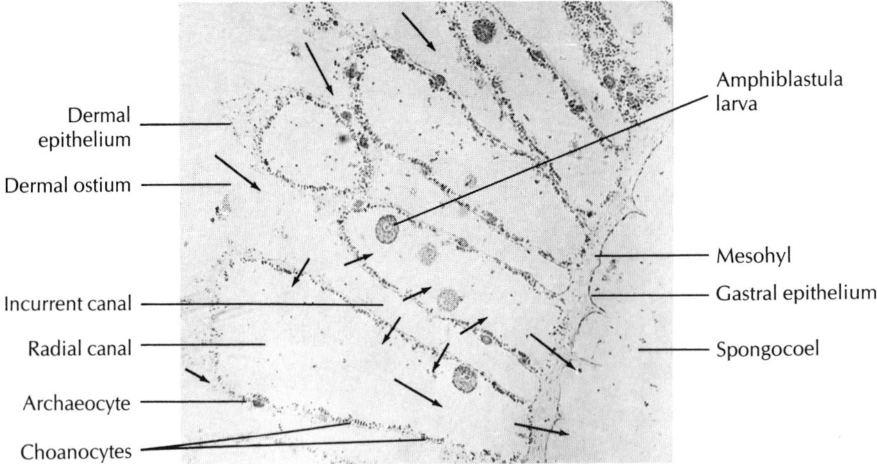

FIGURE 16-4

Cross section through wall of sponge *Sycon*, showing canal system. Photomicrograph of stained slide.

Types of cells

Sponge cells are loosely arranged in a gelatinous matrix called **mesohyl** (also called mesoglea, or mesenchyme) (Figure 16-5). The mesohyl is the "connective tissue" of the sponges; in it are found various ameboid cells, fibrils, and skeletal elements. There are several types of cells in sponges.

Pinacocytes. The nearest approach to a true tissue in sponges is found in the arrangement of the **pinacocyte** (Figure 16-5) cells of the external epithelium. These are thin, flat, epithelial-type cells that cover the exterior surface and some interior surfaces. Some are T-shaped, with their cell bodies extending into the mesohyl. Pinacocytes are somewhat contractile and help regulate the surface area of the sponge. Some of the pinacocytes are modified as contractile **myocytes,** which are usually arranged in circular bands around the oscula or pores, where they help regulate the rate of water flow.

Porocytes. The tubular cells that pierce the wall of asconoid sponges, through which water flows, are called **porocytes** (Figure 16-2).

Choanocytes. The **choanocytes,** which line the flagellated canals and chambers, are ovoid cells with one end embedded in the mesohyl and the other exposed. The exposed end bears a flagellum surrounded by a collar (Figures 16-5 and 16-6). The electron microscope shows the collar to be made up of adjacent microvilli, connected to each other by delicate microfibrils, so that the collar forms a fine filtering device for straining food particles from the water (Figure 16-6, *B* and *C*). The beat of the flagellum pulls water through the sievelike collar and forces it out through the open top of the collar. Particles too large to enter the collar become trapped in secreted mucus and slide down the collar to the base where they are phagocytosed by the cell body. Larger particles have already been screened out by the small size of the dermal pores and prosopyles. The food engulfed by the cells is passed on to a neighboring archaeocyte for digestion.

Archaeocytes. Archaeocytes are ameboid cells that move about in the mesohyl (Figure 16-5) and carry out a number of functions. They can phagocytose particles at the external epithelium and receive particles for digestion from the choanocytes. Archaeocytes apparently can differentiate into any of the other types of more specialized cells in the sponge. Some, called **sclerocytes,** secrete spicules. Others, called **spongocytes,** secrete the spongin fibers of the skeleton, and **collencytes** secrete fibrillar collagen.

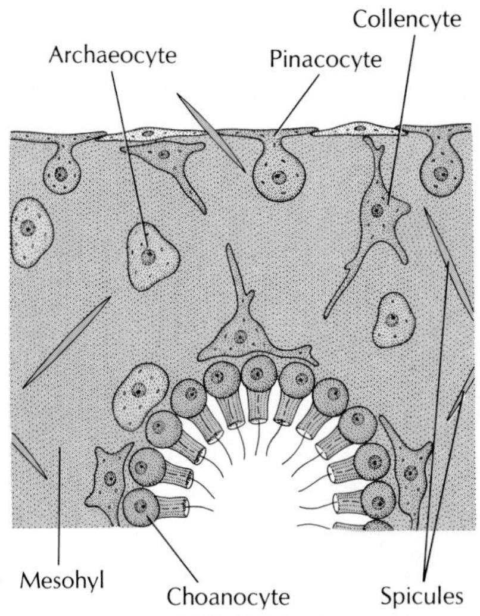

FIGURE 16-5

Small section through sponge wall, showing four types of sponge cells. Pinacocytes are protective and contractile; choanocytes create water currents and engulf food particles; archaeocyte have a variety of functions; collencytes appear to have a contractile function.

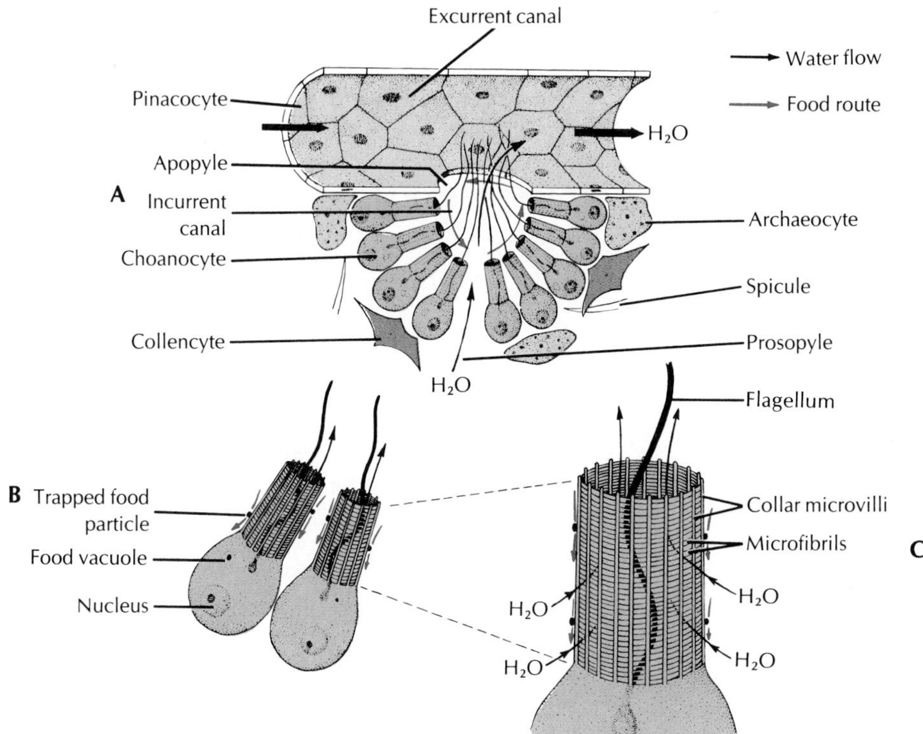

FIGURE 16-6

Food trapping by sponge cells. **A,** Cutaway section of canals showing cellular structure and direction of water flow. **B,** Two choanocytes. **C,** Structure of the collar. Small red arrows indicate movement of food particles.

Types of skeletons

The skeleton supports the sponge, preventing collapse of the canals and chambers. The major structural protein in the animal kingdom is collagen, and fibrils of collagen are found throughout the intercellular matrix of all sponges. In addition, various Demospongiae secrete a form of collagen traditionally known as spongin. Demospongiae also secrete siliceous spicules, as do Sclerospongiae. Calcareous sponges secrete spicules composed mostly of crystalline calcium carbonate that have one, three, or four rays (Fig 16-7). Glass sponges have siliceous spicules with six rays arranged in three planes at right angles to each other. There are many variations in the shape of spicules, and these structural variations are important taxonomically.

Sponge physiology

Sponges feed primarily on particles suspended in the water pumped through their canal systems. Detritus particles, planktonic organisms, and bacteria are consumed nonselectively in the size range from 50 μm (average diameter of ostia) to 0.1 μm (width of spaces between the microvilli of the choanocyte collar). Pinacocytes may phagocytose particles at the surface, but most of the larger particles are consumed in the canals by archaeocytes that move close to the lining of the canals. The smallest particles, accounting for about 80% of the particulate organic carbon, are phagocytosed by the choanocytes. Sponges can also absorb dissolved nutrients from the water passing through the system. Digestion is entirely **intracellular** (occurs within cells), a chore performed by the archaeocytes.

There are no respiratory or excretory organs; both functions are apparently carried out by diffusion in individual cells. Contractile vacuoles have been found in archaeocytes and choanocytes of freshwater sponges.

All the life activities of the sponge depend on the current of water flowing through the body. A sponge pumps a remarkable amount of water. Some large sponges have been found to filter 1500 liters of water a day.

Reproduction and development

All sponges are capable of both sexual and asexual reproduction. In sexual reproduction ova are fertilized by motile sperm in the mesohyl; there the fertilized ova develop into flagellated larvae, which break loose and are carried away by water currents. The larvae swim about for some time, then settle, become attached, and grow into adults. Some sponges are **monoecious** (having both male and female sexes in one individual), and some are **dioecious** (having separate sexes).

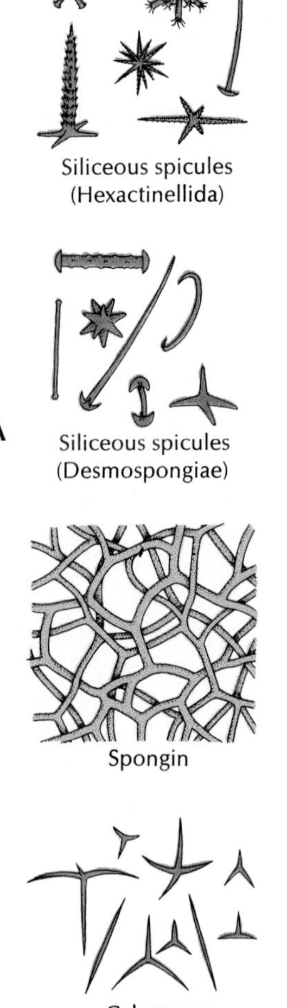

B

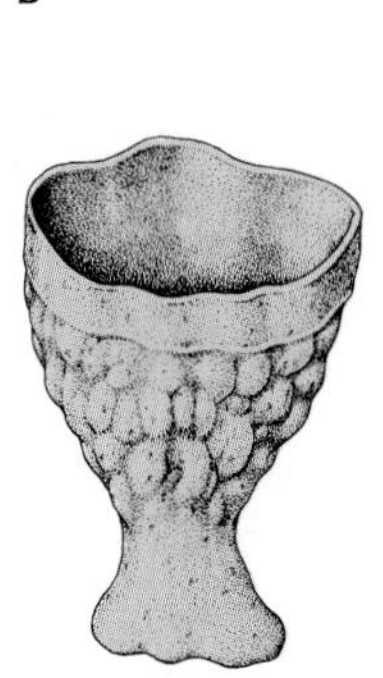

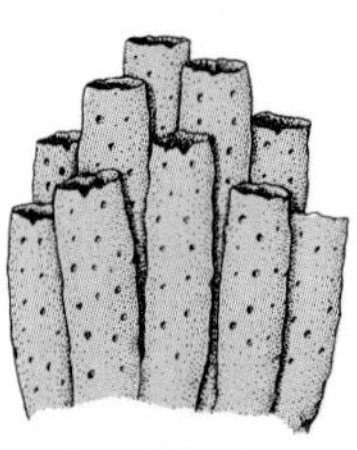

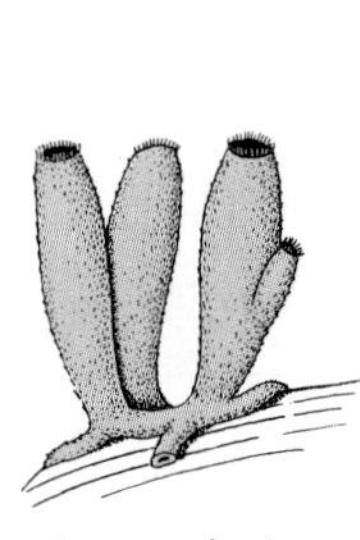

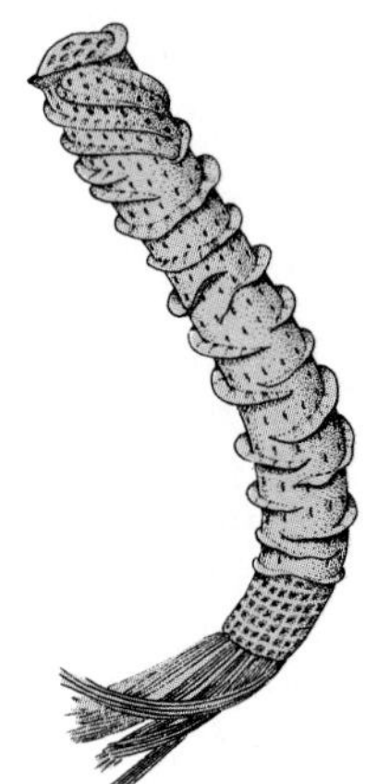

FIGURE 16-7

A, Types of spicules found in sponges. There is amazing diversity, complexity, and beauty of form among the many types of spicules. **B,** Some sponge body forms.

The loose organization of sponges is ideally suited for the regeneration of injured and lost parts, and for asexual reproduction. Sponges reproduce asexually by forming external buds that become detached, or they may remain to form colonies. In addition to external buds, which all sponges can form, freshwater sponges and some marine sponges reproduce asexually by the regular formation of internal buds, called **gemmules.** These dormant masses of encapsulated cells are produced during unfavorable conditions. They can survive periods of drought and freezing; later, with the return of favorable conditions for growth, the cells in the gemmules escape and develop into new sponges.

CLASS CALCAREA (CALCISPONGIAE)

Calcarea are the calcareous sponges, so called because their spicules are composed of calcium carbonate. The spicules are straight monaxons or three or four rayed (Figure 16-7, *A*). The sponges tend to be small— 10 cm or less in height— and tubular or vase shaped. They may be asconoid, syconoid, or leuconoid in structure. Although many are drab, some are bright yellow, red, green, or lavender. *Leucosolenia* (Figure 16-3) and *Sycon* are common examples.

CLASS HEXACTINELLIDA (HYALOSPONGIAE)

The glass sponges are nearly all deep-sea forms. Most of them are radially symmetrical and range from 7 to 10 cm to more than 1 meter in length. Their distinguishing feature, reflected in the alternate class name, is the skeleton of six-rayed siliceous spicules that are bound together in an exquisite glasslike latticework (Figure 16-7). The living tissue is a network produced by the fusion of the pseudopodia of many types of archaeocytes, forming chambers lined with choanocytes. The chambers open into the spongocoel, and the osculum is unusually large. Glass sponges include both syconoid and leuconoid types. Their structure is adapted to the slow, constant current of the sea bottom, because the channels and pores are relatively large and permit an easy flow of water. The beautiful *Euplectella* (NL from Gr. *euplektos,* well-plaited), or the Venus' flower basket, is a classic example of this class.

CLASS DEMOSPONGIAE

Demospongiae contain approximately 80% of all sponge species, including most of the larger sponges. Their skeleton may be of siliceous spicules, spongin fibers, or both. All members of the class are leuconoid, and all are marine except one family, the freshwater Spongillidae. Freshwater sponges are widely distributed in well-oxygenated ponds and streams, where they are found encrusting plant stems and old pieces of submerged wood. They resemble a bit of wrinkled scum, pitted with pores, and are brownish or greenish in color. Freshwater sponges die and disintegrate in late autumn, leaving gemmules to survive the winter.

The marine Demospongiae are varied in both color and shape. Some are encrusting; some are tall and fingerlike; and some are shaped like fans, vases, cushions, or balls (Figure 16-8). Some are boring sponges that bore into molluscan shells and encrust. Loggerhead sponges may grow several meters in diameter. The so-called bath sponges belong to the group called horny sponges, which have only spongin skeletons. They can be cultured by cutting out pieces of the individual sponges, fastening them to a weight, and dropping them into the proper water conditions. It takes many years for them to grow to market size. Most of the commercial "sponges" now on the market are synthetic.

CLASS SCLEROSPONGIAE

The Sclerospongiae are a small group of sponges that secrete a massive skeleton and are thus often called coralline sponges. Living tissue may extend as far as 3 cm into the skeleton but only 1 mm above it. With leuconoid organization and

A

B

C

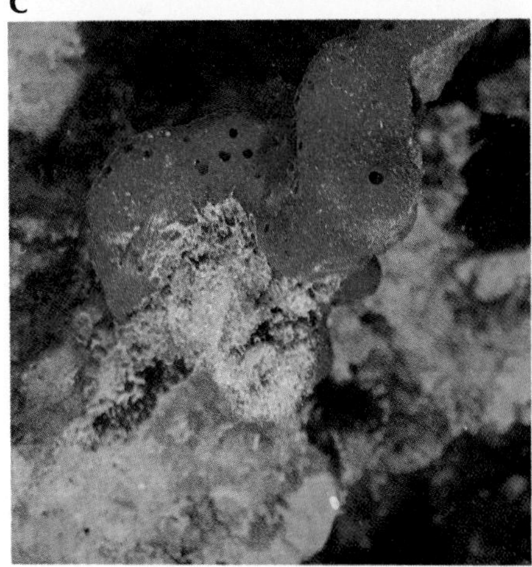

FIGURE 16-8

Marine Demospongiae on Caribbean coral reefs. **A,** *Aplysina longissima* is a colorful sponge growing at moderate depths. **B,** *Hemectyon ferox* is irregular in shape and its oscula form small, volcano-like cones. It is toxic and may cause skin irritation if touched. **C,** *Haliclona rubens* forms irregular, fingerlike growths with randomly placed oscula.

Photographs by L.S. Roberts.

siliceous spicules and spongin in most, sclerosponges live in caves, crevices, and other such cryptic habitats on coral reefs or in deep water. They are thought to be relict representatives of ancient groups with a geological history extending from the Cambrian period.

■ Phylogeny and Adaptive Radiation

PHYLOGENY

The origin of sponges dates back to the Cambrian period, and the Devonian period saw the rapid development of many glass sponges. That sponges are related to flagellate protozoa is shown by their phagocytic method of nutrition and the resemblance between their flagellated larvae and colonial protozoa. One hypothesis, that sponges were derived from choanoflagellates (protozoa that bear collars and flagella), has now been largely discarded. Choanocytes of sponges do not acquire collars until late in embryonic development, and collar cells are not unique to sponges, being found in certain corals and echinoderms.

Another hypothesis is that sponges derived from a hollow, free-swimming colonial flagellate, such as may have given rise to the ancestral stocks of other metazoa. Certainly the sponge larvae resemble such flagellate colonies. Sponges undergo a very curious process called inversion during their embryonic development, and the same process occurs in the colonial phytoflagellate *Volvox.* The development of the water canals and the movement of the flagellated cells to the interior to become choanocytes may have occurred as the sponges began to assume a sessile existence. Whatever the origin, it is obvious that the sponges diverged early from the main line leading to other metazoa. That they are remote, phylogenetically, from other metazoa is shown by their low level of organization, the independent nature of their cells, the absence of organs, and their body structure built around a system of water canals. They became a "dead-end" phylum.

ADAPTIVE RADIATION

The Porifera have been a highly successful group that has branched out into several thousand species and a variety of marine and freshwater habitats. Their diversification centers largely around their unique water-current system and its various degrees of complexity. The asconoid body plan is not efficient enough to move water through a spongocoel with a large cross-sectional area; therefore, asconoid sponges are all small. Development of the syconoid plan, with its flagellated canals, was an improvement, but the complex canals and flagellated chambers of the leuconoid sponges provided a more efficient means of moving large amounts of water through the colony. Their increased capacity for feeding and gaseous exchange have made larger size and adaptive radiation possible among the leuconoid sponges.

■ Summary

Porifera are an abundant marine group with some freshwater representatives. They have various specialized cells, but these are not organized into tissues or organs. They depend on the flagellar beat of their choanocytes to circulate water through their bodies for food gathering and respiratory gas exchange. They are supported by secreted skeletons of fibrillar collagen, collagen in the form of large fibers or filaments (spongin), calcareous or siliceous spicules, or a combination of spicules and spongin in most species.

Sponges are an ancient group, remote phylogenetically from other metazoa. They have been successful, and their adaptive radiation is centered on elaboration of the water circulation and filter feeding system. Most sponges are leuconoid, and the efficiency of this type has enabled some sponges to become very large.

REVIEW QUESTIONS

1. Give six characteristics of sponges.
2. Briefly describe asconoid, syconoid, and leuconoid body types in sponges.
3. What sponge body type is most efficient and makes possible the largest body size?
4. Define the following: ostia, osculum, spongocel, mesohyl, pinacoderm.
5. Define the following: pinacocytes, choanocytes, archaeocytes, sclerocytes, collencytes.
6. What material is found in the skeleton of all sponges?
7. Describe the skeletons of each of the classes of sponges.
8. Describe how sponges feed, respire, and excrete.
9. What is a gemmule?
10. What is the largest class of sponges, and what is its body type?
11. What are possible ancestors to sponges? Justify your answer.

SELECTED REFERENCES

Barnes, R.D. 1987. Invertebrate zoology, ed. 5. Philadelphia, Saunders College/Holt Rinehart & Winston. *Chapter 4 covers sponges.*

Bergquist, P.R. 1978. Sponges. Berkeley, Calif., University of California Press. *Excellent monograph on sponge structure, classification, evolution, and general biology.*

Hartman, W.D. 1982. Porifera. In S.P. Parker (ed.). Synopsis and classification of living organisms, vol. 1. New York, McGraw-Hill Book Co. *Review of sponge classification.*

Long, M.E., and D. Doubelet. 1977. Consider the sponge. Natl. Geogr. **151**(3):392-407. *Beautiful color photographs of sponges.*

Pennak, R.W. 1989. Fresh-water invertebrates of the United States, ed. 3. New York, John Wiley & Sons, Inc. *A reference work with considerable attention devoted to Porifera.*

A FEARSOME TINY WEAPON

Although the radiate animals of the phylum Cnidaria are more highly organized than sponges, they are still quite primitive animals. Most radiates are sessile; those that are unattached, such as jellyfish, can swim only feebly. None can chase their prey. Indeed, we might easily get the false impression that the cnidarians, burdened as they appear to be with such limitations, were placed on earth to provide easy meals for other animals. The truth is, however, many cnidarians are very effective predators that are able to kill and eat prey that are much more highly organized, swift, and intelligent. They manage these feats because they possess tentacles that bristle with tiny, remarkably sophisticated weapons called nematocysts.

A nematocyst is secreted within the cell that contains it. As it is secreted, it is endowed with the potential energy to power its discharge. It is as though a factory manufactured a gun, cocked and ready with a bullet in its chamber, as it is rolled off the assembly line. Like the cocked gun, the completed nematocyst requires only a small stimulus to make it fire. Rather than a bullet, a tiny thread is discharged from the nematocyst. Achieving a velocity of 2 meter/sec and an acceleration of 40,000 $\times$ gravity, it instantly penetrates its prey and injects a paralyzing toxin. A small animal unlucky enough to brush against one of the tentacles is suddenly pierced with hundreds or even thousands of nematocysts and quickly immobilized. The cnidarian then leisurely swallows and digests its victim. Some nematocyst threads can penetrate human skin, resulting in sensations ranging from minor irritation to great pain, even death, depending on the species. A fearsome, but wondrous, tiny weapon.■

CHAPTER 17

Radiate Animals

Cnidarians and Ctenophores

The two phyla Cnidaria and Ctenophora make up the radiate animals, which are characterized by **primary radial** or **biradial symmetry,** and represent the most primitive of the eumetazoans. Radial symmetry, in which the body parts are arranged concentrically around the oral-aboral axis, is particularly suitable for **sessile** or sedentary animals. Biradial symmetry is basically a type of radial symmetry in which only two planes through the oral-aboral axis divide the animal into mirror images because of the presence of some part that is single or paired. All other eumetazoa have the primary bilateral symmetry, that is, they are bilateral or were derived from an ancestor that was bilateral. For sessile or free-floating animals there is a selective advantage in having a radiate form, since it is equally receptive, responsive, and protected on all sides.

Neither phylum has advanced generally beyond the **tissue level of organization,** although a few organs occur. In general the ctenophores have a more complex structural grade than that of the cnidarians.

Phylum Cnidaria

Because the word "Cnidaria" alludes to that outstanding characteristic of these animals, the nematocyst, it is preferred to Coelenterata, which describes ctenophores just as well as cnidarians.

The phylum Cnidaria (ny-dar'e-a) (Gr. *knide,* nettle, + L. *-aria* [pl. suffix] like or connected with) is a large and interesting group of more than 9000 species. It takes its name from the cells called **cnidocytes,** which contain the stinging organelles **(nematocysts)** that are so characteristic of the phylum. Nematocysts are *formed* and *used* only by cnidarians and by one species of ctenophore. Another name for the phylum, Coelenterata (se-len'te-ra'ta) (Gr. *koilos,* hollow, + *enteron,* gut, + L. *-ata* [pl. suffix] characterized by), is used less frequently than formerly.

The cnidarians are generally regarded as being close to the basic stock of the metazoan line. Although their organization has a structural and functional simplicity not found in other metazoa, they are a rather successful phylum, forming a significant proportion of the biomass in some locations. They are widespread in marine habitats, and there are a few in fresh water. As noted in the prologue, they are quite efficient predators, even though they are sessile or slow moving. The phylum includes some of nature's strangest and loveliest creatures—the branching, plantlike hydroids; the flowerlike sea anemones; the jellyfishes; and those architects of the ocean floor, sea whips, sea fans, sea pansies, and all the hard corals whose thousands of years of calcareous house building have produced great reefs and coral islands.

The name "medusa" was suggested by a fancied resemblance to the Gorgon Medusa, a mythological lass with snaky tresses that turned to stone any who gazed upon them.

CHARACTERISTICS

1. Entirely aquatic, some in fresh water but mostly marine
2. **Radial symmetry** or biradial symmetry around a longitudinal axis with **oral** and **aboral** ends; no definite head
3. Two basic types of individuals: **polyps** and **medusae**
4. Exoskeleton or endoskeleton of chitinous, calcareous, or protein components in some
5. Body with two layers, epidermis and gastrodermis, with mesoglea **(diploblastic)**; mesoglea with cells and connective tissue (ectomesoderm) in some **(triploblastic)**
6. **Gastrovascular cavity** (often branched or divided with septa) with a single opening that serves as both mouth and anus; extensible tentacles usually encircling the mouth or oral region
7. Special stinging cell organelles called **nematocysts** in either or both epidermis and gastrodermis; nematocysts abundant on tentacles, where they may form batteries or rings
8. **Nerve net** with symmetrical and asymmetrical synapses; with some sensory organs; diffuse conduction
9. Muscular system (epitheliomuscular type) of an outer layer of longitudinal fibers at base of epidermis and an inner one of circular fibers at base of gastrodermis; modifications of this plan in higher cnidarians, such as separate bundles of independent fibers in the mesoglea
10. Reproduction by asexual budding (in polyps) or sexual reproduction by gametes (in all medusae and some polyps); sexual forms monoecious or dioecious; **planula larva;** holoblastic cleavage
11. No excretory or respiratory systems
12. No coelomic cavity

CLASSIFICATION

Class Hydrozoa (hy-dro-zo'a) (Gr. *hydra,* water serpent, + *zōon,* animal). Solitary or colonial; asexual polyps and sexual medusae, although one or the other of the types may be suppressed; hydranths with no mesenteries; medusae (when present) with a velum; both freshwater and marine. Examples: *Hydra, Obelia, Physalia, Tubularia.*

Class Scyphozoa (sy-fo-zo'a) (Gr. *skyphos,* cup, + *zōon,* animal). Solitary; polyp stage reduced or absent; bell-shaped medusae without velum; gelatinous mesoglea much enlarged; margin of bell or umbrella typically with eight notches that are provided with sense organs; all marine. Examples: *Aurelia, Cassiopeia, Rhizostoma.*

Class Cubozoa (ku'bo-zo'a) (Gr. *kybos,* a cube, + *zōon,* animal). Solitary; polyp stage reduced; bell-shaped medusae square in cross section, with tentacle or group of tentacles hanging from a bladelike pedalium at each corner of the umbrella; margin of umbrella entire; all marine. Examples: *Tripedalia, Carybdea, Chironex, Chiropsalmus.*

Class Anthozoa (an-tho-zo'a) (Gr. *anthos,* flower, + *zōon,* animal). All polyps, no medusae; solitary or colonial; enteron subdivided by at least eight mesenteries or septa bearing nematocysts; gonads endodermal; all marine.

Subclass Zoantharia (zo'an-tha'ri-a) (NL). With simple unbranched tentacles. Sea anemones and hard corals. Examples: *Metridium, Adamsia, Astrangia, Acropora.*

Subclass Ceriantipatharia (se-ri-ant′i-pa-tha′ri-a) (NL, combination of Ceriantharia and Antipatharia). With simple unbranched tentacles; mesenteries unpaired; tube anemones and black or thorny corals. Examples: *Cerianthus, Antipathes, Stichopathes.*

Subclass Alcyonaria (al′cy-o-na′ri-a) (NL). With eight pinnate tentacles. Soft and horny corals. Examples: *Tubipora, Alcyonium, Gorgonia, Renilla.*

ECOLOGICAL RELATIONSHIPS

Cnidarians are found most abundantly in shallow marine habitats, especially in warm temperatures and tropical regions. There are no terrestrial species. Colonial hydroids are found usually in shallow coastal water attached to mollusc shells, rocks, wharves, and other animals, but some species are found at great depths. Floating and free-swimming medusae are found in open seas and lakes, often a long distance from the shore. Colonies such as the Portuguese man-of-war have floats by which the wind and currents carry them.

Some molluscs and flatworms eat hydroids bearing nematocysts and use these stinging structures for their own defense. Some other animals feed on cnidarians, although cnidarians rarely serve as food for humans.

Cnidarians sometimes live symbiotically with other animals, often as commensals on the shell or other surface of their host (Figure 17-1). Algae may live as mutuals in the tissues of cnidarians, notably in some freshwater hydras and in many anthozoans. The presence of the algae in reef-building corals limits the occurrence of coral reefs to relatively shallow, clear water where there is sufficient light for the photosynthetic requirements of the algae. These kinds of corals form an essential component of coral reefs, and reefs are extremely important habitats in tropical waters. Additional comments on coral reefs are given later in the chapter.

FIGURE 17-1

A hermit crab with its cnidarian commensals. The shell is blanketed with polyps of the hydrozoan *Hydractinia milleri.*

Photograph by R.M. Harbo.

ECONOMIC IMPORTANCE

Although many cnidarians have little economic importance, reef-building corals are an important exception. Fish and other animals associated with reefs provide substantial amounts of food for humans, and reefs are of economic value as tourist attractions. Coral rock is sometimes used as a building material, and quite lovely jewelry can be made from some corals. Jewelry manufacture has led to serious overharvesting of black corals (class Anthozoa, order Ceriantipatharia) in many areas.

FORM AND FUNCTION

Polymorphism in cnidarians

One of the most interesting—and sometimes puzzling—aspects of this phylum is the dimorphism and often polymorphism displayed among many of its members. In general, cnidarian forms fit into one of two morphological types—the **polyp,** or hydroid form, which is adapted for a sedentary or sessile life, and the **medusa,** or jellyfish form, which is adapted for a floating or free-swimming existence (Figure 17-2).

Most polyps have tubular bodies with a mouth at one end surrounded by tentacles. The **aboral** end (the end opposite the mouth, or **oral** end) is usually attached to a substratum by a pedal disc or other device (Figure 17-2). Polyps may live singly, or they may live in colonies. In colonies of some species, there are more than one kind of individual, each specialized for a certain function, such as feeding, reproduction, or defense.

Medusae are free swimming and have bell-shaped or umbrella-shaped bodies and tetramerous symmetry (body parts arranged in fours). The mouth is usually centered on the concave side, and tentacles extend from the rim of the umbrella.

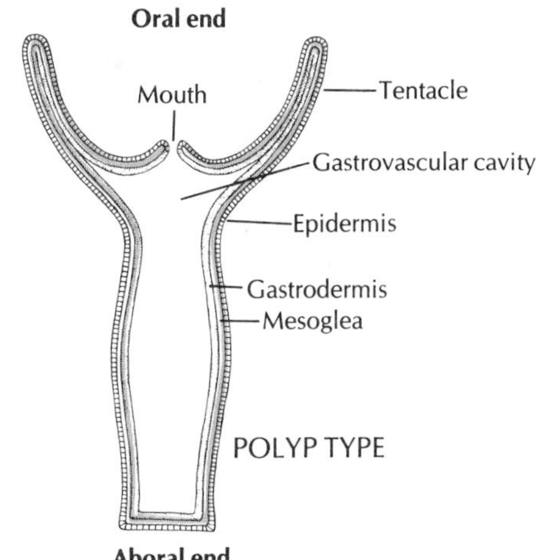

FIGURE 17-2

Comparison of the polyp and medusa types of individuals.

The sea anemones and corals (class Anthoza) are all polyps, and the true jellyfishes (class Scyphozoa) are all medusae but may have a polyp larval stage. The colonial hydroids of class Hydrozoa, however, sometimes have life histories that feature both the polyp stage and the free-swimming medusa stage—similar to a Jeckyll-and-Hyde existence. A species that has both the attached polyp and the floating medusa within its life history can take advantage of the feeding and distribution possibilities of both the pelagic (open-water) and the benthic (bottom) types of environment.

Superficially the polyp and medusa seem very different. But actually each has retained the saclike body plan that is basic to the phylum (Figure 17-2). The medusa is essentially an unattached, upside-down polyp with the tubular portion widened and flattened into the bell shape; the mouth, in this inverted condition, faces downward, and the circle of tentacles fringing the rim also hangs downward.

Both the polyp and the medusa possess the three body wall layers typical of the cnidarians, but the jellylike layer of mesoglea is much thicker in the medusa, constituting the bulk of the animal and making it more buoyant. It is because of this mass of mesogleal "jelly" that the medusae are commonly called jellyfishes.

Nematocysts—the stinging organelles

One of the most characteristic structures in the entire cnidarian group is the stinging organelle called the **nematocyst** (Figure 17-3). Over 20 different types of nematocysts have been described in the cnidarians so far; they are important in taxonomic determinations. The nematocyst is a tiny capsule composed of material similar to chitin and containing a coiled tubular "thread" or filament, which is a continuation of the narrowed end of the capsule. This end of the capsule is covered by a little lid, or **operculum.** The inside of the undischarged thread may bear tiny barbs, or spines.

The nematocyst is enclosed in the cell that has secreted it, the **cnidocyte** (during its development, the cnidocyte is properly called the **cnidoblast**). Most cnidocytes are provided with a triggerlike **cnidocil,** which is a modified flagellum with a kinetosome at its base. Contact of the cnidocil with an object such as prey provides tactile stimulation for the nematocyst to discharge. Cnidocytes with their nematocysts are borne in invaginations of ectodermal cells and, in some forms, in gastrodermal cells, and they are especially abundant on the tentacles.

FIGURE 17-3

Left, Cnidocyte, bearing nematocyst. *Right,* Portion of the body wall of a hydra. Cnidocytes arise in the epidermis from interstitial cells.

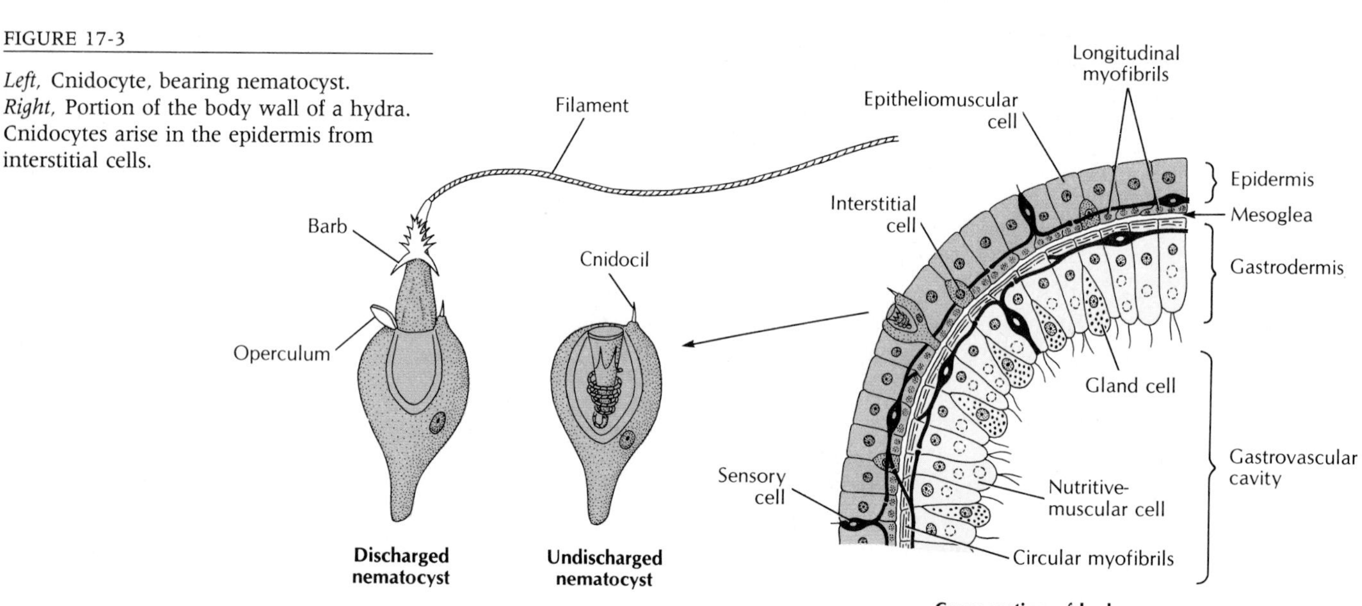

When a nematocyst is discharged, its cnidocyte is absorbed and a new one replaces it. Nematocysts used in defense and food capture require chemical stimulation (presence of organic compounds from other animals) to discharge, as well as tactile stimulation. Adhesive nematocysts usually do not discharge in food capture.

The mechanism of nematocyst discharge is remarkable. Inside the capsule, there is an osmotic pressure of 140 atmospheres. When stimulated to discharge, the nematocyst membrane becomes permeable to water, and the high internal osmotic pressure causes water to rush into the capsule. The operculum opens, the increase in *hydrostatic pressure* within the capsule pushes the thread out with great force, and the thread turns inside out as it goes. At the everting end of the thread, the barbs flick to the outside like tiny switchblades. This minute but awesome weapon then injects poison when it penetrates the prey.

The nematocysts of most cnidarians are not harmful to humans, but the stings of the Portuguese man-of-war (see Figure 17-13) and certain jellyfish (see Figure 17-17) are quite painful and in some cases may be dangerous.

Note again the distinction between osmotic and hydrostatic pressure (p. 36). The nematocyst is never required actually to contain 140 atmospheres of hydrostatic pressure within itself; such a hydrostatic pressure would doubtless cause it to explode. As the water rushes in during discharge, the osmotic pressure falls rapidly, while the hydrostatic pressure rapidly increases.

Not all nematocysts have barbs or inject poison. Some, for example, do not penetrate the prey but rapidly recoil like a spring after discharge, serving to grasp and hold any part of the prey caught in the coil. Others are sticky (adhesive types).

Nerve net

The nerve net of the cnidarians is one of the best examples in the animal kingdom of a diffuse nervous system. This network of interlacing nerve cells is found both at the base of the epidermis and at the base of the gastrodermis, forming two interconnected nerve nets. Nerve processes (axons) end on other nerve cells at synapses or at junctions with sensory cells or effector organs (nematocysts or epitheliomuscular cells). Nerve impulses are transmitted from one cell to another by release of a neurotransmitter from small vesicles on one side of the synapse or junction (p. 240). One-way transmission between nerve cells in higher animals is assured because the vesicles are located only on one side of the synapse; the synapses are asymmetrical. However, cnidarian nerve nets are peculiar in that many of the synapses are symmetrical; that is, clusters of neurotransmitter vesicles lie inside both synaptic membranes, allowing transmission across the synapse in either direction (Figure 17-4). Another peculiarity of cnidarian nerves is the absence of any sheathing material (myelin) on the axons.

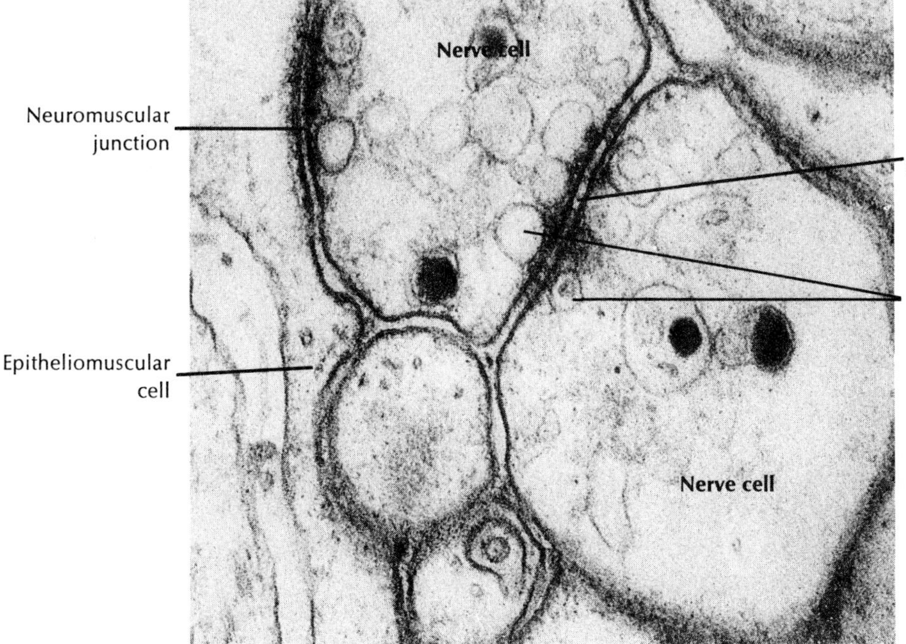

FIGURE 17-4

Transmission electron micrograph of neuronal synapse and neuromuscular junction in the scyphistoma of the jellyfish *Aurelia*. Contrast the symmetrical two-way interneuronal synapse (vesicles on both sides) to the asymmetrical junction of the nerve with the epitheliomuscular cell. (×67,300.)

From Westfall, J.A. 1973. Ultrastructural evidence for neuromuscular systems in coelenterates. Am. Zool. 13:237.

There is no concentrated grouping of nerve cells to suggest a "central nervous system." Nerves are grouped, however, in the "ring nerves" of hydrozoan medusae and in the marginal sense organs of scyphozoan medusae. In some cnidarians the nerve nets form two or more systems: in Scyphozoa there is a fast conducting system to coordinate swimming movements and a slower one to coordinate movements of tentacles.

The nerve cells of the net have synapses with slender sensory cells that receive external stimuli, and the nerve cells have junctions with epitheliomuscular and nutritive-muscular cells and nematocysts. Together with the contractile fibers of the epitheliomuscular and nutritive-muscular cells, the sensory-nerve cell net combination is often referred to as a **neuromuscular system,** an important landmark in the evolution of the nervous system. The nerve net is never completely lost from higher forms. Annelids have it in their digestive systems. In the human digestive system it is represented by nerve plexuses in the musculature. The rhythmical peristaltic movements of the stomach and intestine are coordinated by this counterpart of the cnidarian nerve net.

Body structure

The mouth opens into the **gastrovascular cavity** (coelenteron), which communicates with the cavities in the tentacles. The mouth may be surrounded by an elevated **manubrium** or by elongated **oral lobes.**

Body wall. The body wall surrounding the gastrovascular cavity consists of an outer **epidermis** (ectodermal) and an inner **gastrodermis** (endodermal) with **mesoglea** between them (Figure 17-3).

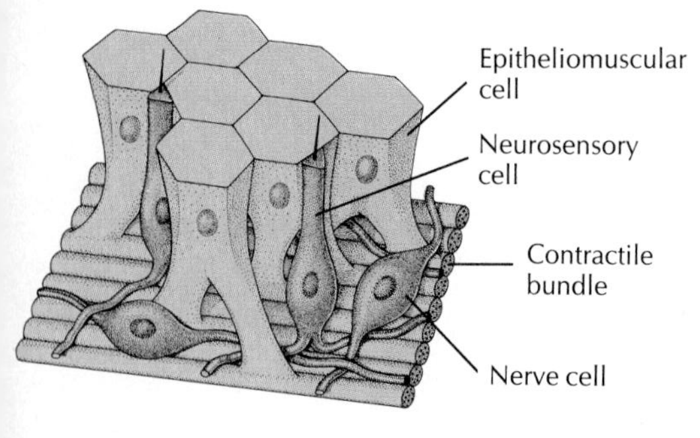

FIGURE 17-5

Epitheliomuscular and nerve cells in hydra.

EPIDERMIS. The epidermis is made up of small cubical cells and may be covered with a delicate cuticle or a secreted, chitinous sheath. This layer contains several types of cells—epitheliomuscular, interstitial, gland, cnidocyte, and sensory and nerve cells.

Epitheliomuscular cells (Figure 17-5) make up most of the epidermis and serve both for covering and for muscular contraction. The bases of most of these cells contain myofibrils, whose contractions shorten the body or tentacles.

Interstitial cells are the undifferentiated stem cells found among the bases of the epitheliomuscular cells. Differentiation of the interstitial cells gives rise to cnidoblasts, sex cells, buds, nerve cells, and others, but generally not to epitheliomuscular cells (which reproduce themselves).

Gland cells are tall cells around the mouth and in the pedal disc of hydra. They secrete mucus or adhesive material.

Cnidocytes containing nematocysts are found throughout the epidermis. They may be between the epitheliomuscular cells or housed in invaginations of these cells, and they are most abundant on the tentacles. There are three functional types of nematocysts in the hydra: penetrants, which penetrate the prey and inject poison (Figure 17-3); volvents, which recoil and entangle the prey; and glutinants, which secrete an adhesive substance used in locomotion and attachment (glutinants).

Sensory cells are scattered among the other epidermal cells, especially around the mouth and the tentacles. The free end of each sensory cell bears a flagellum, which is the sensory receptor for chemical and tactile stimuli. The other end branches into fine processes, which synapse with the nerve cells.

Nerve cells of the epidermis are often multipolar (have many processes), and in more highly organized cnidarians the cells may be bipolar (with two processes). Their processes (axons) form synapses with sensory cells and other nerve cells and junctions with epitheliomuscular cells and cnidocytes. Both one-way and two-way synapses with other nerve cells are present (Figure 17-4).

GASTRODERMIS. The gastrodermis, a layer of cells lining the gastrovascular cavity, is made up chiefly of large, flagellated, columnar epithelial cells with irregular flat bases. The cells of the gastrodermis include nutritive-muscular, interstitial, gland cells, and, in classes other than Hydrozoa, cnidocytes.

Nutritive-muscular cells are usually tall columnar cells and have laterally extended bases containing myofibrils. In hydrozoans the myofibrils run at right angles to the body or tentacle axis and so form a circular muscle layer. However, this muscle layer is very weak, and longitudinal extension of the body and tentacles is brought about mostly by increasing the volume of water in the gastrovascular cavity. The water is brought in through the mouth by the beating of the flagella on the nutritive-muscular cells in hydrozoans or by ciliated cells in the pharynx of anthozoans. Thus, the water in the gastrovascular cavity serves as a **hydrostatic skeleton.** The flagella on the free end of each cell also serve to circulate food and fluids in the digestive cavity. The cells often contain large numbers of food vacuoles. Gastrodermal cells of many cnidarians contain algal cells. In the green hydra (*Chlorohydra* [Gr. *chloros*, green, + *hydra*, a mythical nine-headed monster slain by Hercules]), these are green algae called **zoochlorellae,** but in marine cnidarians they are a type of dinoflagellate (p. 349) called **zooxanthellae.** Both are cases of mutualism, with the algae furnishing organic compounds they have synthesized to their cnidarian hosts.

Interstitial cells scattered among the bases of the nutritive cells can transform into other cell types. **Gland cells** secrete digestive enzymes; feeding and digestion are discussed later.

MESOGLEA. The mesoglea lies between the epidermis and gastrodermis and is attached to both layers. It is gelatinous, or jellylike, and has no fibers or cellular elements in hydrozoan polyps. It is thicker in medusae and has elastic fibers, and in scyphozoan medusae it has ameboid cells. The mesoglea of anthozoans is a mesenchyme containing ameboid cells.

Locomotion

Colonial polyps are permanently attached, but the hydra can move about freely by gliding on its basal disc, aided by mucus secretions. Sea anemones can move similarly on their basal discs. Hydras can also use a "measuring worm" movement, looping along by bending over and attaching their tentacles to the substratum. They may even turn handsprings or detach and, by forming a gas bubble on the basal disc, float to the surface.

Most medusae can move freely, and they swim by contracting the bell, expelling water from the concave, oral side. The muscular contractions are antagonized by the compressed mesoglea and the elastic fibers within it. Usually, they contract several times and move generally upward, then sink slowly. Cubozoan medusae, however, can swim strongly.

Feeding and digestion

Cnidarians prey on a variety of organisms of appropriate size; larger species are usually capable of killing and eating larger prey. Normally, the prey is drawn into the gastrovascular cavity where gland cells discharge enzymes onto the food. Digestion is started in the gastrovascular cavity **(extracellular digestion),** but many food particles are drawn into the nutritive-muscular cells of the gastrodermis, where **intracellular digestion** occurs. Ameboid cells may carry undigested particles to the gastrovascular cavity, where they are eventually expelled with other indigestible matter.

The effectiveness of cnidarians in snaring prey was amply demonstrated recently in a fjord in the western Baltic Sea where a heavy bloom of jellyfish reduced the larval herring population to less than half its former density in just a few weeks. One jellyfish only 4.2 cm in diameter contained 68 larvae!

Reproduction

Most cnidarians are dioecious, and many shed their gametes directly into the water. Ova may be retained by the female and brooded for some period after fertilization. Gonads are epidermal in hydrozoans and gastrodermal in the other

FIGURE 17-6

A colony of *Tubularia crocea.* Medusa buds do not detach from *Tubularia* polyps, but instead shed their gametes in place.

Photograph by R. Harbo.

Over 230 years ago, Abraham Trembley was astonished to discover that isolated sections of the stalk of hydra could regenerate and each become a complete animal. Since then, over 2000 investigations of hydra have been published, and the organism has become a classic model for the study of morphological differentiation. The mechanisms governing morphogenesis have great practical importance, and the simplicity of hydra lends itself to these investigations. Substances controlling development (morphogens), such as those determining which end of a cut stalk will develop a mouth and tentacles, have been discovered, and they may be present in the cells in extremely low concentrations (10^{-10}M).

groups. The embryo characteristically develops into a free-swimming **planula** larva (Figure 17-9).

Cnidarians are capable of asexual reproduction, usually by budding, but sea anemones commonly practice a peculiar form of fission known as pedal laceration (p. 390).

CLASS HYDROZOA

The majority of hydrozoans are marine and colonial in form, and the typical life cycle includes both the asexual polyp and the sexual medusa stages. Some, however, such as the freshwater hydra, have no medusa stage. Some marine hydroids do not have free medusae (Figure 17-6), whereas some hydrozoans only occur as medusae and have no polyp.

The hydra, although not a typical hydrozoan, has become a favorite as an introduction to Cnidaria because of its size and ready availability. Combining its study with that of a representative colonial marine hydroid such as *Obelia* (Gr. *obelias,* round cake) gives an excellent idea of the class Hydrozoa.

Hydra, a freshwater hydrozoan

The common freshwater hydra (Figure 17-7) is a solitary polyp and one of the few cnidarians found in fresh water. Its normal habitat is the underside of aquatic leaves and lily pads in cool, clean fresh water of pools and streams. The hydra family is found throughout the world, with 10 species occurring in the United States.

The body of the hydra can extend to a length of 25 to 30 mm or can contract to a tiny, jellylike mass. It is a cylindrical tube with the lower (aboral) end drawn out into a slender stalk, ending in a basal or **pedal disc** for attachment. This pedal disc is provided with gland cells to enable the hydra to adhere to a substratum and also to secrete a gas bubble for floating. In the center of the disc there may be an excretory pore. The **mouth,** located on a conical elevation called the **hypostome,** is encircled by six to ten hollow tentacles that, like the body, can be greatly extended when the animal is hungry.

The mouth opens into the **gastrovascular cavity,** which communicates with the cavities in the tentacles. In some individuals **buds** may project from the sides, each with a mouth and tentacles like the parent. Testes or ovaries, when present, appear as rounded projections on the surface of the body (Figure 17-7).

Hydras feed on a variety of small crustaceans, insect larvae, and annelid worms. The hydra awaits its prey with tentacles extended (Figure 17-8). The food organism that brushes against its tentacles may find itself harpooned by scores of nematocysts that render it helpless, even though it may be larger than the hydra. The tentacles move toward the mouth, which slowly widens. Well-moistened with mucus secretions, the mouth glides over and around the prey, totally engulfing it.

The activator that actually causes the mouth to open is the reduced form of **glutathione,** which is found to some extent in all living cells. Glutathione is released from the prey through the wounds made by the nematocysts, but only those animals releasing enough of the chemical to activate the feeding response are eaten by the hydra. This explains how a hydra distinguishes between *Daphnia,* which it relishes, and some other forms that it refuses. When glutathione is added to water containing hydras, each hydra will go through the motions of feeding, even though no prey is present.

In asexual reproduction, buds appear as outpocketings of the body wall and develop into young hydras that eventually detach from the parent. In sexual reproduction, temporary gonads (Figure 17-7) usually appear in the autumn, stimulated by the lower temperatures and perhaps also by the reduced aeration of stagnant waters. Eggs in the ovary usually mature one at a time and are fertilized by sperm shed into the water. A cyst forms around the embryo before it breaks loose from the parent, enabling it to survive the winter. Young hydras hatch out in spring when the weather is favorable.

FIGURE 17-7

Hydra with developing bud and ovary.

Courtesy Carolina Biological Supply Co.

Hydroid colonies

Far more representative of class Hydrozoa than the hydras are those hydroids which have a medusa stage in their life cycle. *Obelia* is often used in laboratory exercises for beginning students to illustrate the hydroid type (Figure 17-9).

A typical hydroid has a base, a stalk, and one or more terminal polyps (zooids). The base by which the colonial hydroids are attached to the substratum is a rootlike stolon, or **hydrorhiza,** which gives rise to one or more stalks called **hydrocauli.** The living cellular part of the hydrocaulus is the tubular **coenosarc,** composed of the three typical cnidarian layers surrounding the coelenteron (gastrovascular cavity). The protective covering of the hydrocaulus is a nonliving chitinous sheath, or **perisarc.** Attached to the hydrocaulus are the individual zooids. Most of the zooids are feeding polyps called **hydranths,** or **gastrozooids.** They may be tubular, bottle shaped, or vaselike, but all have a terminal mouth and a circlet of tentacles. In some forms, such as *Obelia,* the perisarc continues as a protective cup around the polyp into which it can withdraw for protection (Figure 17-9). In others the polyp is naked.

The hydranths, much like miniature hydras, capture and ingest prey, such as tiny crustaceans, worms, and larvae, thus providing nutrition for the entire colony. After partial digestion in the hydranth, the digestive broth passes into the common gastrovascular cavity where intracellular digestion occurs.

Circulation within the gastrovascular cavity is a function of the flagellated gastrodermis but is also aided by rhythmical contractions and pulsations of the body, which occur in many hydroids.

In contrast to hydras, new individuals that bud do not detach from the parent; thus the size of the colony is increased. The new polyps may be hydranths or reproductive polyps known as **gonangia.** Medusae are produced by budding within the gonangia. The young medusae leave the colony as free-swimming individuals that mature and produce gametes (eggs and sperm) (Figure 17-9). In some species the medusae remain attached to the colony and shed their gametes there. In other species the medusae never develop, the gametes being shed by male and female gonophores. Embryonation of the zygote results in a ciliated planula larva that swims about for a time. Then it settles down to a substratum to

FIGURE 17-8

A, Hungry hydra catches an unwary water flea with the stinging cells of its tentacles. Mouth, *arrow.* **B,** Water flea is swallowed whole. **C,** Hydra is full, but not too full to capture a protozoan for dessert.

Photographs by F.M. Hickman.

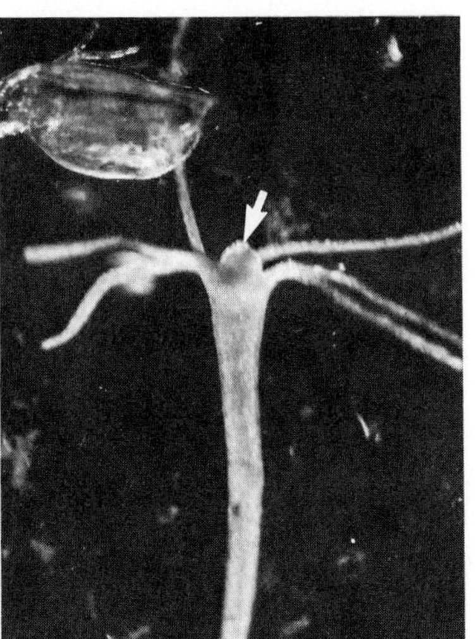

A

B

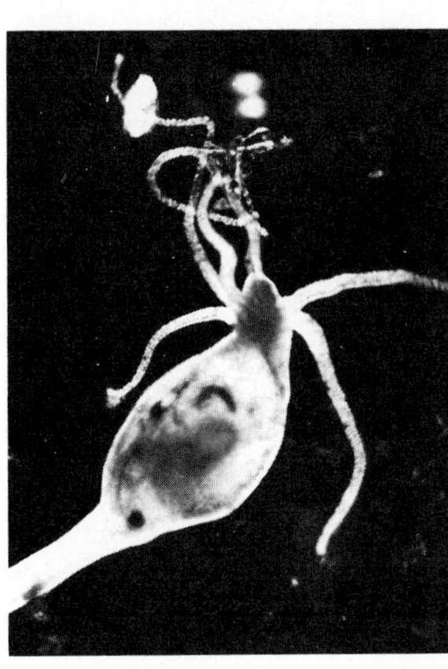

C

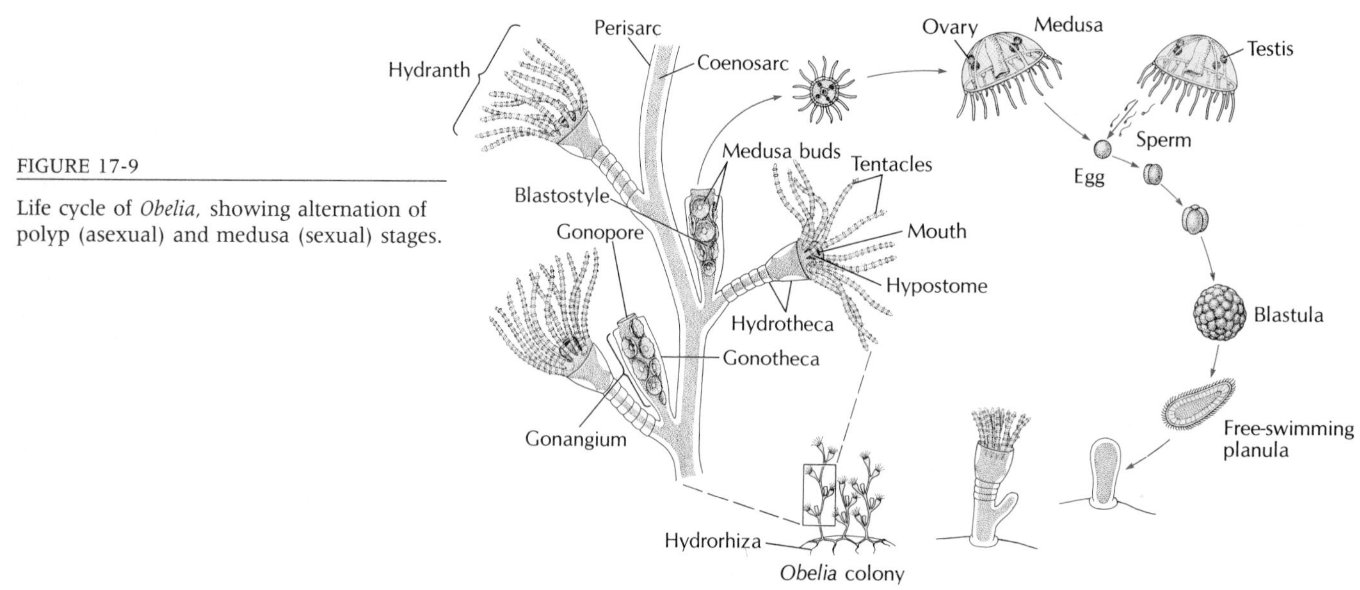

FIGURE 17-9

Life cycle of *Obelia,* showing alternation of polyp (asexual) and medusa (sexual) stages.

develop into a minute polyp that gives rise, by asexual budding, to the hydroid colony, thus completing the life cycle.

Hydroid medusae are usually smaller than scyphozoan medusae, ranging from 2 or 3 mm to several centimeters in diameter (Figure 17-10). The margin of the bell projects inward as a shelflike **velum,** which partly closes the open side of the bell and is used in swimming (Figure 17-11). Muscular pulsations that alternately fill and empty the bell propel the animal forward, aboral side first, with a sort of "jet propulsion." The tentacles attached to the bell margin are richly supplied with nematocysts.

The mouth opening at the end of a suspended **manubrium** leads to a stomach and four radial canals that connect with a ring canal around the margin. This in turn connects with the hollow tentacles. Thus, the coelenteron is continuous from mouth to tentacles, and the entire system is lined with gastrodermis. Nutrition is similar to that of the hydranths.

The nerve net is usually concentrated into two nerve rings at the base of the velum. The bell margin is liberally supplied with sensory cells. It usually also bears two kinds of specialized sense organs—**statocysts,** which are small organs of equilibrium (Figure 17-11, *B*), and **ocelli,** which are light-sensitive organs.

FIGURE 17-10

Bell medusa, *Polyorchis penicillatus,* medusa stage of an unknown attached polyp.

Photograph by D. Gotshall.

Other hydrozoans

Some hydrozoans form floating colonies, such as *Physalia* (Gr. *physallis,* bladder), or Portuguese man-of-war (Figure 17-12). These colonies include several types of modified medusae and polyps. *Physalia* has a rainbow-hued float, said to be a modified polyp, which carries it along at the mercy of the winds and currents. It contains an air sac filled with secreted gas and acts as a carrier for the generations of individuals that bud from it and hang suspended in the water. There are several types of individuals, including the feeding polyps, reproductive polyps, long stinging tentacles, and the so-called jelly polyps. Many swimmers have experienced the uncomfortable sting that these colonial floaters can inflict. The pain, along with the panic of the swimmer, can increase the danger of drowning.

Other hydrozoans secrete massive calcareous skeletons that resemble true corals (Figure 17-13). They are sometimes called **hydrocorals.**

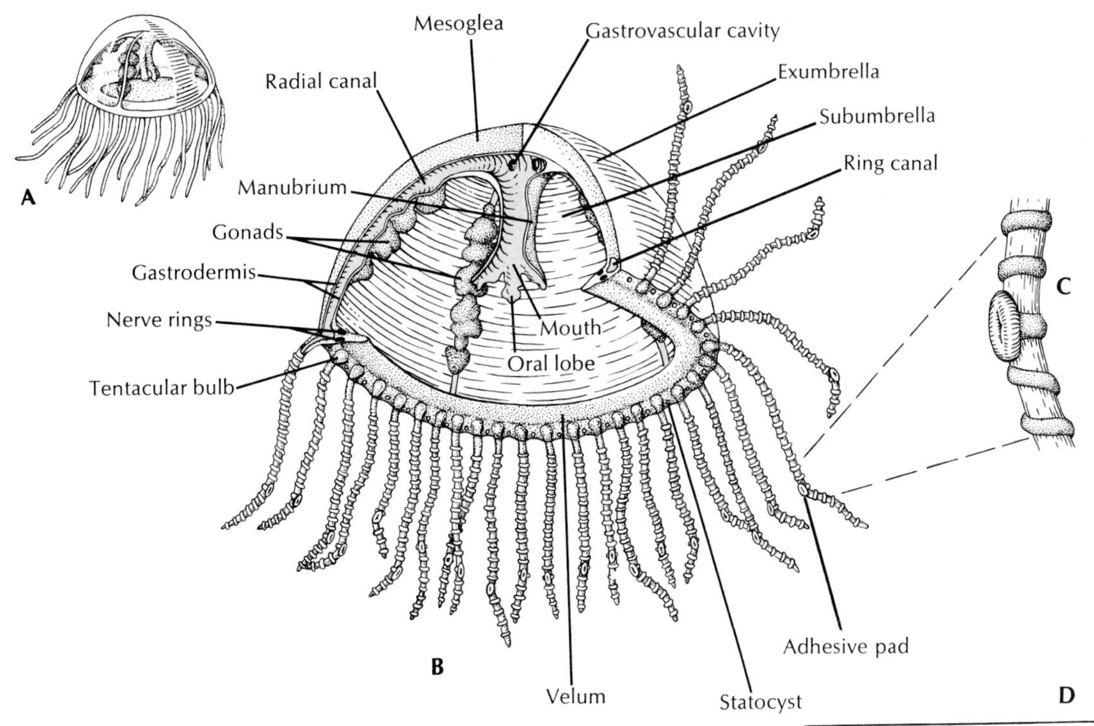

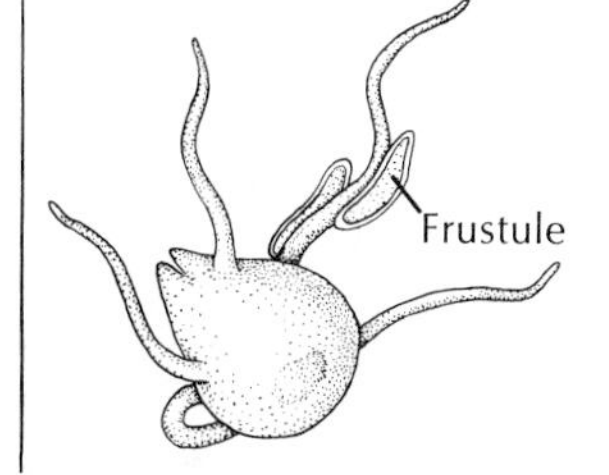

FIGURE 17-11

Structure of *Gonionemus.* **A,** Medusa with typical tetramerous arrangement. **B,** Cutaway view showing morphology. **C,** Portion of a tentacle with its adhesive pad and ridges of nematocysts. **D,** Tiny polyp, or hydroid stage, that develops from the planula larva. It can bud off more polyps (frustules) or produce medusa buds.

FIGURE 17-12

Three Portuguese man-of-war colonies, *Physalia physalis* (order Siphonophora, class Hydrozoa), lie stranded in shallow water. Colonies often drift onto southern ocean beaches, where they are a hazard to bathers. Each colony of medusa and polyp types is integrated to act as one individual. As many as a thousand zooids may be found in one colony. The nematocysts secrete a powerful neurotoxin.

Photograph by C. Lane.

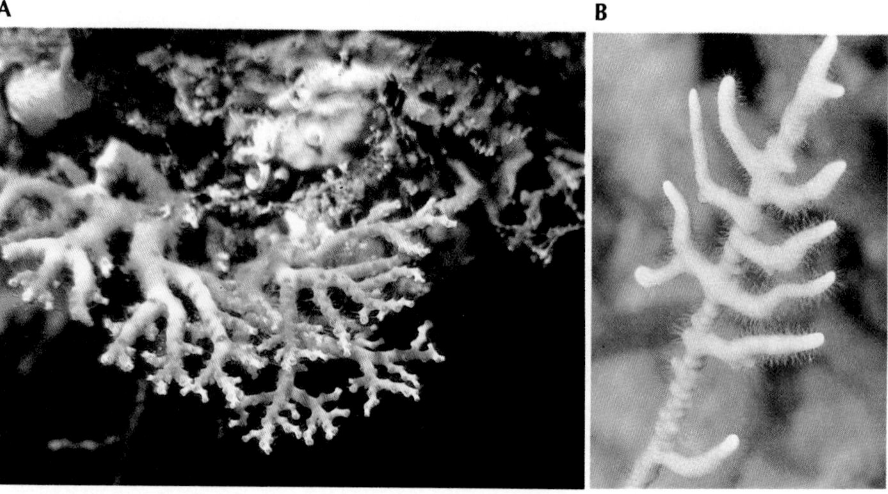

FIGURE 17-13

These hydrozoans form calcareous skeletons that resemble true coral. **A,** *Stylaster roseus* (order Stylasterina) occurs commonly in caves and crevices in coral reefs. These fragile colonies may be white, pink, purple, red, or red with white tips. **B,** Species of *Millepora* (order Milleporina) form branching or platelike colonies and often grow over the horny skeleton of some alcyonarians (see p. 391), as is shown here. They are generously supplied with powerful nematocysts that produce a burning sensation on human skin, justly earning the common name fire coral.

Photographs by L.S. Roberts.

CLASS SCYPHOZOA

Class Scyphozoa (si-fo-zo'a) (Gr. *skyphos,* cup) includes most of the larger jellyfishes, or "cup animals." A few, such as *Cyanea* (Gr. *kyanos,* dark blue substance), may attain a bell diameter exceeding 2 m and tentacles 60 to 70 m long (Figure 17-14). Most scyphozoans, however, range from 2 to 40 cm in diameter. Most are found floating in the open sea, some even at depths of 3000 m, but one unusual order is sessile and attaches by a stalk to seaweeds and other objects on the sea bottom. Their coloring may range from colorless to striking orange and pink hues.

Scyphomedusae, unlike the hydromedusae, have no velum. The bell ranges from shallow to a deep helmet shape, and in many the margin is scalloped, each notch bearing a sense organ called a **rhopalium** and a pair of lobelike projections called lappets. *Aurelia* (L. *aurum,* gold) has eight such notches (Figure 17-15); others may have four or sixteen. Each rhopalium bears a statocyst for balance, two sensory pits containing concentrations of sensory cells, and sometimes an ocellus (simple eye) for photoreception. The mesoglea is thick and contains cells as well as fibers. The stomach is usually divided into pouches containing small tentacles with nematocysts.

The mouth is centered on the subumbrellar side. The manubrium is usually drawn out into four frilly **oral lobes** that are used in food capture and ingestion. The marginal tentacles may be many or few and may be short, as in *Aurelia,* or long, as in *Cyanea.* The tentacles, manubrium, and often the entire body surface of scyphozoans are well supplied with nematocysts. Scyphozoans feed on all sorts of small organisms, from protozoa to fishes. Capture of prey involves stinging and manipulation with tentacles and oral arms, but the methods vary. *Aurelia* feeds on small planktonic animals. These are caught in the mucus of the umbrella surface, carried to "food pockets" on the umbrella margin by cilia, and picked up from the pockets by the oral lobes whose cilia carry the food to the gastrovascular cavity. Flagella in the gastrodermis layer keep a current of water moving to bring food and oxygen into the stomach and carry out wastes.

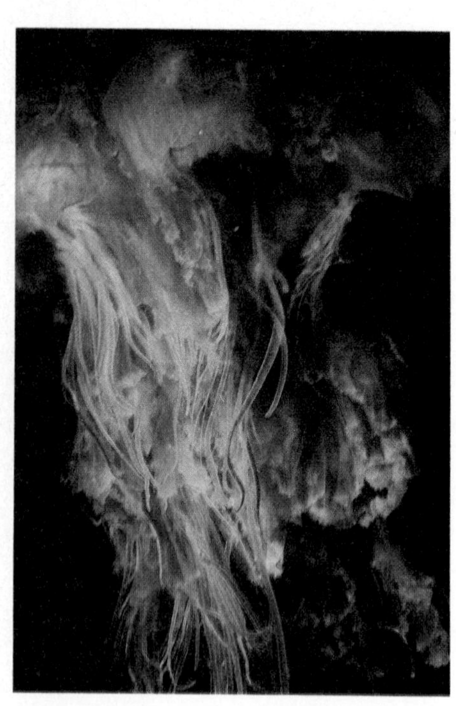

FIGURE 17-14

Giant jellyfish, *Cyanea capillata.* A North Atlantic species of *Cyanea* reaches a bell diameter exceeding 2 m. It is known as the "sea blubber" by fishermen.

Photograph by R. Harbo.

Internally four **gastric pouches** containing nematocysts connect with the stomach in scyphozoans, and a complex system of **radial canals** that branch from the pouches to the **ring canal** (Figure 17-15) completes the gastrovascular cavity, through which nutrients circulate.

The sexes are separate, and fertilization and early development occur in the gastric pouches or on the frilled oral arms. In most scyphozoans, a ciliated planula larva develops into a little polyp larva called a **scyphistoma** (Figure 17-16),

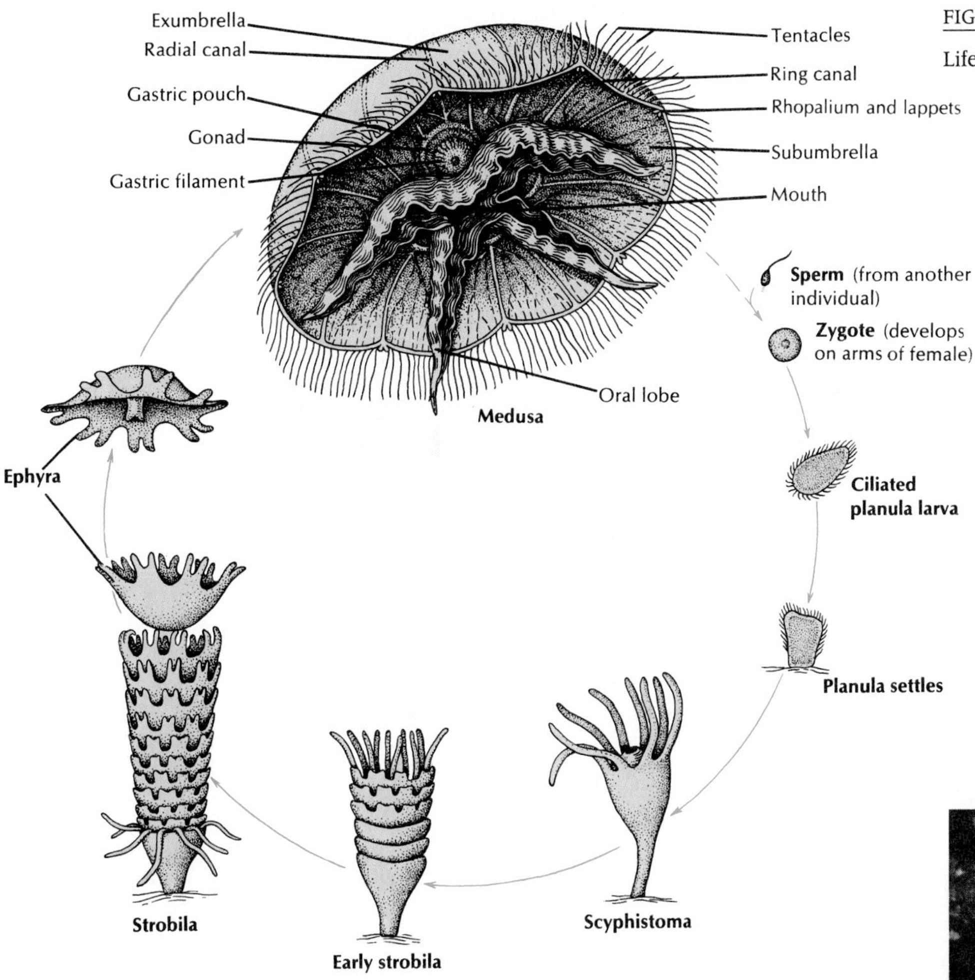

FIGURE 17-15

Life cycle of *Aurelia,* a marine scyphozoan.

which looks somewhat like a hydra. During the summer the scyphistoma produces more scyphistomas by budding. In winter and spring the scyphistoma buds off a series of juvenile, saucer-shaped medusae, and the series of buds is known as a **strobila.** The juvenile medusae, called **ephyrae,** break loose, swim away, and grow into mature, sexual medusae (Figure 17-16).

CLASS CUBOZOA

The class Cubozoa was until recently considered an order (Cubomedusae) of Scyphozoa. In transverse section the bells are almost square (Figure 17-17). A tentacle or group of tentacles is found at each corner of the square at the umbrella margin, and the base of each tentacle is differentiated into a flattened, tough blade called a **pedalium** (Figure 17-17). Cubomedusae are strong swimmers and voracious predators, feeding mostly on fish. *Chironex fleckeri* (Gr. *cheir,* hand, + *nexis,* swimming) is known as the sea wasp. Its stings are quite dangerous and sometimes fatal, with death occurring rapidly (usually within a few minutes). Most of the fatalities resulting from stings have been reported from tropical Australian waters.

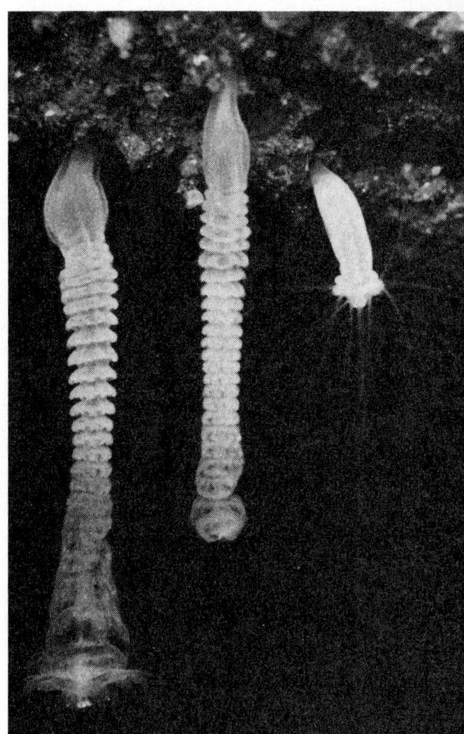

FIGURE 17-16

Scyphistoma *(right)* and strobilas *(left)* of *Aurelia aurita* hanging down from a rock. The strobila at left has an ephyra nearly ready to separate.

Photograph by D.P. Wilson.

CLASS ANTHOZOA

The anthozoans, or "flower animals," are polyps with a flowerlike appearance (Figure 17-18). There is no medusa stage. Anthozoans are all marine and are found all over the world in both deep and shallow water. They vary greatly in size

FIGURE 17-17

Carybdea, a cubozoan medusa.

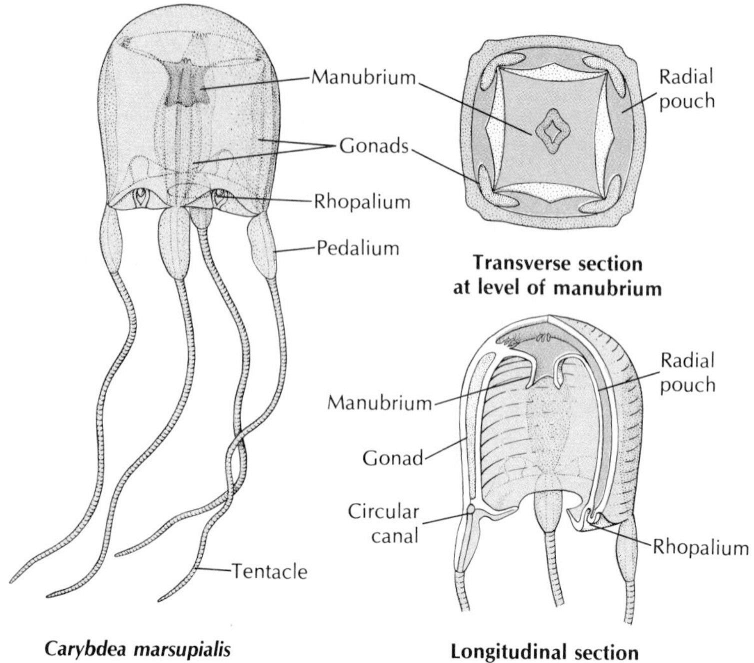

The subclass Ceriantipatharia has been created from the Ceriantharia and Antipatharia, formerly considered orders of Zoantharia. Ceriantharians are tube anemones and live in soft bottom sediments, buried to the level of the oral disc. Antipatharians are the thorny or black corals. They are colonial and have a skeleton of a horny material. Both of these groups are small in numbers of species and are limited to the warmer waters of the sea.

and may be solitary or colonial. Many of the forms are supported by skeletons.

The class has three subclasses—**Zoantharia,** made up of the sea anemones, stony corals, and others, **Alcyonaria,** which includes the sea fans, sea pens (Figure 17-19), sea pansies, and other soft corals, and the **Ceriantipatharia,** which includes the tube anemones and the black or thorny corals. The zoantharians have a **hexamerous** plan (of six or multiples of six) or polymerous symmetry and have simple tubular tentacles arranged in one or more circlets on the oral disc. The alcyonarians are **octomerous** (built on a plan of eight) and always have eight pinnate (featherlike) tentacles arranged around the margin of the oral disc.

The gastrovascular cavity is large and partitioned by mesenteries, or **septa,** that are inward extensions of the body wall. The walls and septa contain both circular and longitudinal muscle fibers.

FIGURE 17-18

Sea anemones are the familiar and colorful "flower animals" of tide pools, rocks, and pilings of the intertidal zone. Most, however, are subtidal, their beauty seldom revealed to human eyes. These are rose anemones, *Tealia piscivora.*

Photograph by D.W. Gotshall.

FIGURE 17-19

Orange sea pen *Ptilosarcus gurneyi* (order Pennatulacea, class Anthozoa). Sea pens are colonial forms that inhabit soft bottoms. **A,** Entire colony. The base of the fleshy body of the primary polyp is buried in the bottom. It gives rise to numerous secondary, branching polyps, shown enlarged in **B.** The pinnate tentacles characteristic of the subclass Alcyonaria are apparent.

A, Photograph by J.L. Rotman; **B,** photograph by R. Harbo.

The mesoglea is a mesenchyme containing ameboid cells. There is a general tendency toward biradial symmetry in the septal arrangement and in the shape of the mouth and pharynx. There are no special organs for respiration or excretion.

Sea anemones

Sea anemone polyps are larger and heavier than hydrozoan polyps (Figures 17-18 and 17-20). Most of them range from 5 mm or less to 100 mm in diameter, and from 5 mm to 200 mm long, but some grow much larger. Some of them are quite colorful. Anemones are found in coastal areas all over the world, especially in the warmer waters, and they attach by means of their pedal discs to shells, rocks, timber, or whatever submerged substrata they can find. Some burrow in the bottom mud or sand.

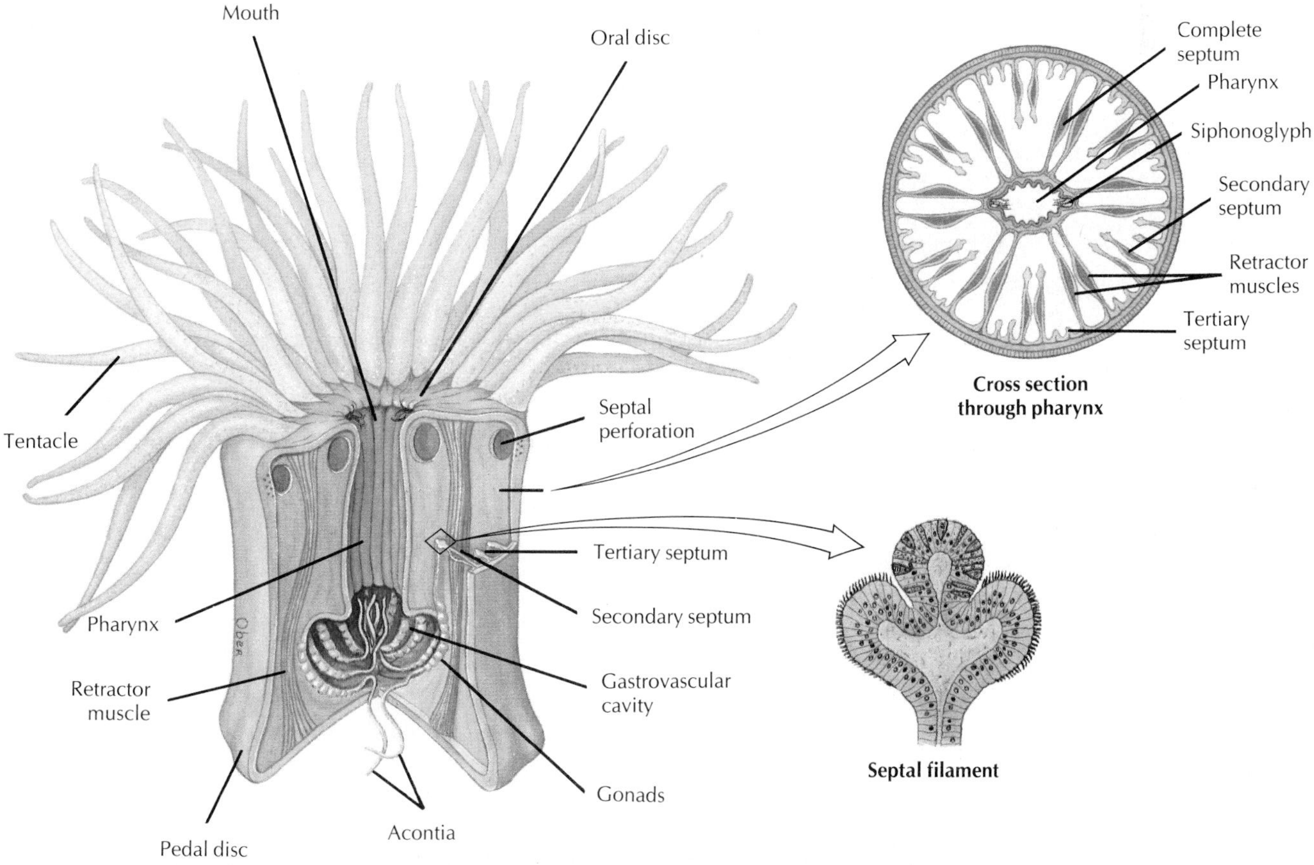

FIGURE 17-20

Structure of a sea anemone. The free edges of the septa and the acontia threads are equipped with nematocysts to complete the paralyzation of prey begun by the tentacles.

Anemones form some interesting mutualistic relationships with other organisms. Many anemones house unicellular algae in their tissues (as do reef-building corals), from which they undoubtedly derive some nutrients. Some hermit crabs place anemones on the snail shells in which the crabs live, gaining some protection by the presence of the anemone, while the anemone dines on particles of food dropped by the crab. The clownfishes of the tropical Indo-Pacific form associations with large anemones. Somehow, these fishes do not trigger discharge of the anemone's nematocysts, but if some other fish is so unfortunate as to brush the anemone's tentacles, it is likely to become a meal.

FIGURE 17-21

A predator becomes prey. A rose anemone *Tealia lofotensis* consumes a leather star *Dermasterias imbricata.*

Photograph by D. Gotshall.

Sea anemones are cylindrical in form with a crown of tentacles arranged in one or more circles around the mouth on the flat **oral disc** (Figure 17-20). The slit-shaped mouth leads into a **pharynx.** At one or both ends of the mouth is a ciliated groove called a **siphonoglyph,** which extends into the pharynx. The siphonoglyphs create water currents directed into the pharynx. The cilia elsewhere on the pharynx direct water outward. The currents thus created carry in oxygen and remove wastes. They also help maintain an internal fluid pressure or a hydrostatic skeleton that serves as a support for opposing muscles.

The pharynx leads into a large **gastrovascular cavity** that is divided into radial chambers by means of pairs of septa that extend vertically from the body wall toward the pharynx (Figure 17-20). These chambers communicate with each other and are open below the pharynx. In many anemones the lower ends of the septal edges are prolonged into **acontia threads,** also provided with nematocysts and gland cells, that can be protruded through the mouth or through pores in the body wall to help overcome prey or provide defense. The pores also aid in the rapid discharge of water from the body when the animal is endangered and contracts to a small size.

Sea anemones are carnivorous, feeding on fish or almost any live animals of suitable size (Figure 17-21). Some species live on minute forms caught by ciliary currents.

The sexes are separate in some sea anemones, and some are hermaphroditic. The gonads are arranged on the margins of the septa. Fertilization is external in some species, whereas in others the sperm enter the gastrovascular cavity to fertilize the eggs. The zygote develops into a ciliated larva. Asexual reproduction commonly occurs by **pedal laceration.** Small pieces of the pedal disc break off as the animal moves, and these each regenerate a small anemone.

FIGURE 17-22

A, Cup coral *Tubastrea aurea.* The polyps form clumps resembling groups of sea anemones. Although often found on coral reefs, *Tubastrea* is not a reef-building coral and has no symbiotic zooxanthellae in its tissues. **B,** The polyps of *Montastrea cavernosa* are tightly withdrawn in the daytime but open to feed at night, as in **C.**

Photographs by L.S. Roberts.

A

B

C

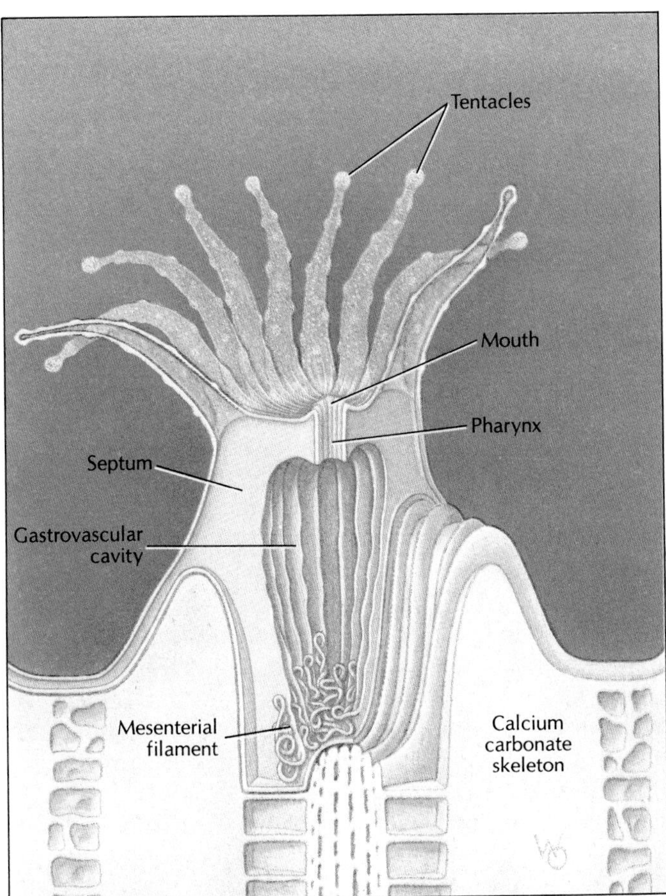

FIGURE 17-23

Polyp of a zoantharian coral (order Scleractinia) showing calcareous cup (exoskeleton), gastrovascular cavity, sclerosepta, mesenterial septa, and mesenterial filaments.

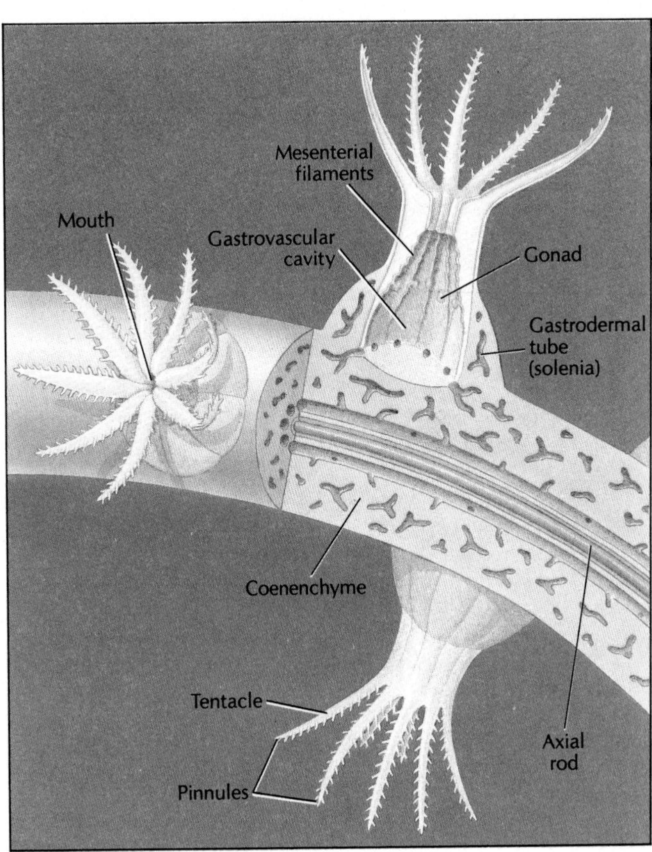

FIGURE 17-24

Polyps of an alcyonarian coral (octocoral). Note the eight pinnate tentacles and solenia. They have an endoskeleton of limy spicules often with a horny protein, which may be in the form of an axial rod.

Zoantharian corals

The zoantharian corals include the true stony corals (order Scleractinia) (Figures 17-22 and 17-23). The stony corals might be described as miniature sea anemones that live in calcareous cups they have secreted, but they differ from anemones in that scleractinians have no siphonoglyph. Since the skeleton is secreted below the living tissue, rather than within it, the calcareous material is an exoskeleton. In the colonial corals, the skeleton may become very large and massive, building up over many years, with the living coral forming a sheet of tissue over the surface. The gastrovascular cavities of the polyps are all connected through this sheet of tissue.

Alcyonarian corals

Alcyonarians are often referred to as octocorals because of their strict octomerous symmetry, with eight pinnate tentacles and eight unpaired, complete septa (Figure 17-24). They are all colonial, and the gastrovascular cavities of the polyps communicate through a system of gastrodermal tubes. The tubes run through an extensive mesoglea in most alcyonarians, and the surface of the colony is covered by epidermis. The skeleton is secreted in the mesoglea and consists of limey spicules, fused spicules, or a horny protein, often in combination. Thus the skeletal support of most alcyonarians is an endoskeleton. The variation in pattern among the species of alcyonarians lends great variety to the form of the colonies.

FIGURE 17-25

Red gorgonian, *Lophogorgia chilensis.* The colonial gorgonian, or horny, corals are conspicuous components of reef faunas, especially of the West Indies.
Photograph by D. Gotshall.

The graceful beauty of the alcyonarians—in hues of yellow, red, orange, and purple—helps create the "submarine gardens" of the coral reefs (Figure 17-25).

Coral reefs

Coral reefs are among the most productive of any ecosystem, and they have a diversity of life forms rivaled only by the tropical rain forest. They are large formations of calcium carbonate (limestone) in shallow tropical seas laid down by living organisms over thousands of years; living plants and animals are confined to the top layer of reefs where they add more calcium carbonate to that deposited by their predecessors. The most important organisms that take dissolved calcium carbonate from seawater and precipitate it to form reefs are the **reef-building corals** and **coralline algae.** Reef-building corals have mutualistic algae **(zooxanthellae)** living in their tissues. Coralline algae are several types of red algae, and they may be encrusting or form upright, branching growths. Not only do they contribute to the total mass of calcium carbonate, but their deposits help to hold the reef together. Some alcyonarians and hydrozoans (especially *Millepora* [L. *mille,* thousands, + *porus,* pore] spp., the "fire coral" [Figure 17-13]) contribute in some measure to the calcareous material, and an enormous variety of other organisms contributes small amounts. However, reef-building corals seem essential to the formation of large reefs, since such reefs do not occur where these corals cannot live.

Reef-building corals require warmth, light, and the salinity of undiluted seawater. This limits coral reefs to shallow waters between 30 degrees north and 30 degrees south latitude and excludes them from areas with upwelling of cold water or areas near major river outflows with attendant low salinity and high turbidity.

The microscopic zooxanthellae are very important to the corals: their photosynthesis and fixation of carbon dioxide furnish food molecules for their hosts, they recycle phosphorus and nitrogenous waste compounds that otherwise would be lost, and they enhance the ability of the coral to deposit calcium carbonate. Can you understand now why reef-building corals cannot live in deep water and are killed by turbid water and siltation? Interestingly, some deposits of coral reef limestone, particularly around Pacific islands and atolls, reach great thickness—even thousands of feet. Clearly, the corals and other organisms could not have grown from the bottom in the abyssal blackness of the deep sea and reached shallow water where light could penetrate. Charles Darwin was the first to realize that such reefs began their growth in *shallow* water around volcanic islands; then, as the islands slowly sank beneath the sea, the growth of the reefs kept up with the rate of sinking, thus accounting for the depth of the deposits.

Phylum Ctenophora

GENERAL RELATIONSHIPS

Ctenophora (te-nof'o-ra) (Gr. *kteis, ktenos,* comb, + *phora,* pl. of bearing) comprise a small group of fewer than 100 species. All are marine forms. They take their name from the eight rows of comblike plates they bear for locomotion. Common names for ctenophores are "sea walnuts" and "comb jellies." Ctenophores, along with cnidarians, represent the only two phyla having primary radial symmetry, in contrast to other metazoa, which have primary bilateral symmetry.

Ctenophores do not have nematocysts, except in one species. In common with the cnidarians, ctenophores have not advanced beyond the tissue grade of organization. There are no definite organ systems in the strict meaning of the term.

The ctenophores are strictly marine animals and are all free swimming except for a few creeping forms. They occur in all seas, but especially in warm waters.

Although feeble swimmers are more common in surface waters, ctenophores are sometimes found at considerable depths. They use their ciliated comb plates to propel themselves mouth-end forward. Some highly modified forms such as *Cestum* use sinuous body movements as well as their comb plates in locomotion (Figure 17-26).

The fragile, transparent bodies of ctenophores (Figure 17-27) are easily seen at night when they emit light (luminesce).

FORM AND FUNCTION

Pleurobrachia (Gr. *pleuron,* side + L. *brachia,* arms), a pretty little sea walnut and a member of class Tentaculata, is often used as a representative example of the

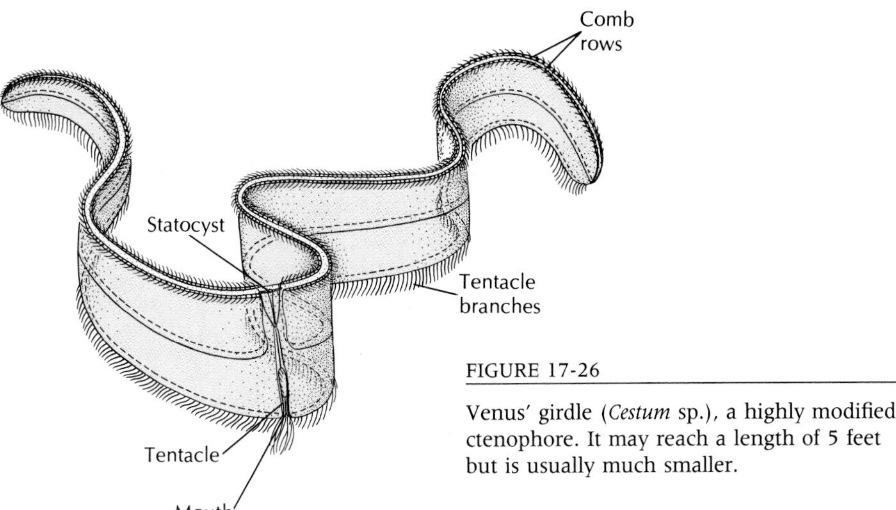

FIGURE 17-26

Venus' girdle (*Cestum* sp.), a highly modified ctenophore. It may reach a length of 5 feet but is usually much smaller.

ctenophores (Figure 17-28). Its surface bears eight longitudinal rows of transverse plates bearing long fused cilia and called **comb plates.** The beating of the cilia in each row starts at the aboral end and proceeds along the rows to the oral end, thus propelling the animal forward. All rows beat in unison. A reversal of the wave direction drives the animal backward. Ctenophores may be the largest animals that swim exclusively by cilia.

Two long tentacles are carried in a pair of tentacle sheaths (Figure 17-28) from which they can stretch to a length of perhaps 15 cm. The surface of the tentacles bears specialized glue cells called **colloblasts,** which secrete a sticky substance that facilitates the catching of small prey organisms. When covered with food, the tentacles contract and the food is wiped off on the mouth. The gastrovascular cavity consists of a pharynx, stomach, and a system of gastrovascular canals. Rapid digestion occurs in the pharynx, then partly digested food is circulated through the rest of the system where digestion is completed intracellularly. Residues are regurgitated or expelled through small pores in the aboral end.

A nerve net system similar to that of the cnidarians includes a subepidermal plexus that is concentrated under each comb plate.

The sense organ at the aboral pole is a **statocyst,** or organ of equilibrium, and is also concerned with the beating of the comb rows but does not trigger their beat. Other sensory cells are abundant in the epidermis.

All ctenophores are monoecious, bearing both an ovary and a testis. Gametes are shed into the water, except in a few species that brood their eggs, and there is a free-swimming larva.

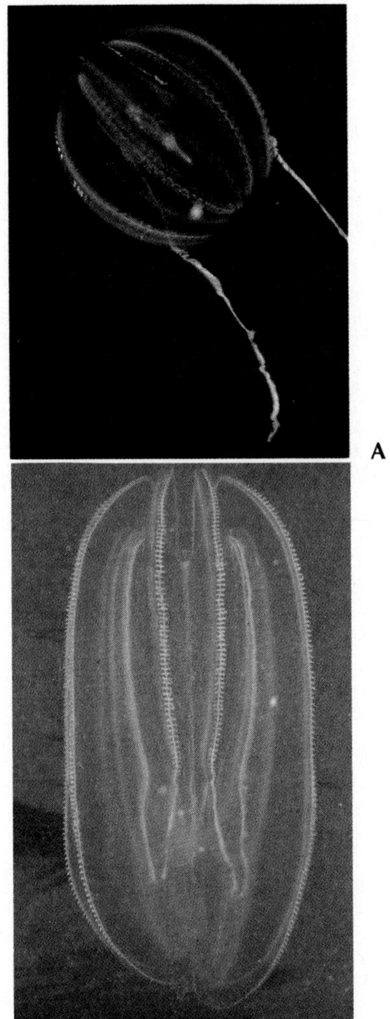

FIGURE 17-27

A, Comb jelly *Pleurobrachia* sp. (order Cydippida, class Tentaculata). Its fragile beauty is especially evident at night when it luminesces from its comb rows. **B,** *Mnemiopsis* sp. (order Lobata, class Tentaculata).

A, Photograph by J.L. Rotman; **B,** photograph by K. Sandved.

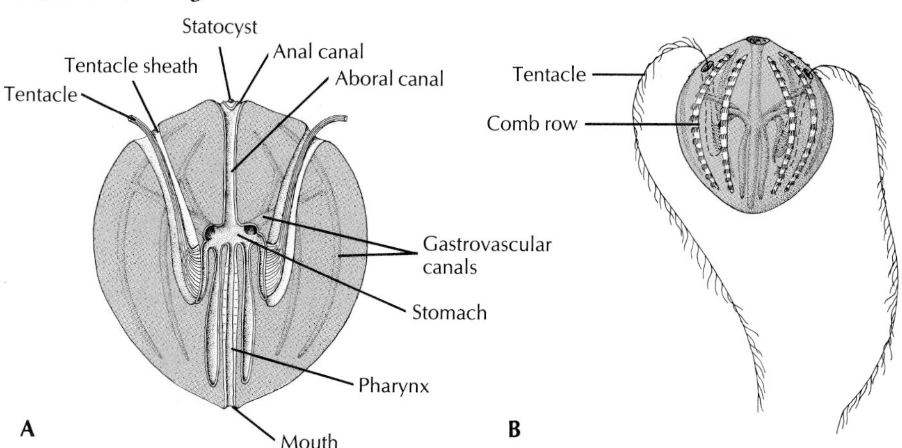

FIGURE 17-28

The comb jelly *Pleurobrachia,* a ctenophore. **A,** Hemisection. **B,** External view.

Phylogeny and Adaptive Radiation

PHYLOGENY

The origin of the cnidarians and ctenophores is obscure, although the most widely supported hypothesis today is that the radiate phyla arose from a radially symmetrical, planula-like ancestor. It may well have been that such an ancestor could have been common to the radiates and to the higher metazoa, the latter having been derived from a branch whose members habitually crept about on the sea bottom. Such a habit would select for bilateral symmetry. Others became sessile or free floating, conditions for which radial symmetry is a selective advantage. A planula larva in which an invagination formed to become the gastrovascular cavity would correspond roughly to a cnidarian with an ectoderm and an endoderm.

Members of a hydrozoan order having only a medusa stage (order Trachylina) are sometimes considered the most primitive of modern cnidarians because of their direct development from the larva to the medusa. Such a group could have given rise to the three classes of cnidarians, with further development of the polyp and subsequent loss of the medusa occurring in the anthozoan line.

In light of their many resemblances to the cnidarians, the ctenophores may have originated from the same ancestral stock, such as the trachyline medusae, but conclusive evidence is not available.

ADAPTIVE RADIATION

In their evolution neither phylum has deviated very far from its basic plan of structure. In the Cnidaria, both the polyp and medusa are constructed on the same scheme. Likewise, the ctenophores have adhered to the arrangement of the comb plates and their biradial symmetry.

Nonetheless, the cnidarians are a successful phylum in terms of numbers of individuals and species, demonstrating a surprising degree of diversity considering the simplicity of their basic body plan. They are efficient predators, many feeding on prey quite large compared to themselves. Some are adapted for feeding on small particles. The colonial form of life is well explored, with some colonies growing to great size among the corals, and others, such as the siphonophores, showing astonishing polymorphism and specialization of individuals within the colony.

Summary

The phyla Cnidaria and Ctenophora have a primary radial symmetry; radial symmetry is an advantage for sessile or free-floating organisms. The Cnidaria are surprisingly efficient predators because they possess stinging organelles called nematocysts. Both phyla are essentially diploblastic, with a body wall composed of epidermis and gastrodermis and a mesoglea between. The digestive-respiratory (gastrovascular) cavity has a mouth and no anus. They are at the tissue level of organization. Cnidarians have two basic body types (polyp and medusa), and in many hydrozoans and scyphozoans the life cycle involves both the asexually reproducing polyp and the sexually reproducing medusa.

The unique organelle, the nematocyst, is contained coiled within a capsule. On discharge, some types of nematocysts penetrate the prey and inject poison. There is no concentrated central nervous system in cnidarians; the nerves are spread through the body in a netlike arrangement. Some synapses can transmit impulses in either direction.

Most hydrozoans (class Hydrozoa) are colonial and marine, but the freshwater hydras are commonly demonstrated in class laboratories and have the typical polyp structure. The body is approximately cylindrical in shape, with the mouth surrounded by nematocyst-bearing tentacles at one end. Hydras reproduce asexually by budding and sexually after formation of gonads in the cylinder wall; they have no medusa stage. The more typical form, found in most marine hydrozoans, is that of a branching colony containing many polyps (hydranths), and a free-swimming medusa stage. An example of a floating, colonial hydrozoan with numerous body forms (both polyp and medusa) in the colony is the Portuguese man-of-war, *Physalia.*

The scyphozoans (class Scyphozoa) are the typical jellyfishes, in which the medusa is the dominant body form. Many scyphozoans have an inconspicuous polyp stage that reproduces asexually (scyphistoma). Scyphozoan medusae differ from hydrozoan medusae in that scyphozoans have complex sensory structures called rhopalia and do not have a velum.

The anthozoans (class Anthozoa) are all marine and are polypoid; there is no medusa stage. The subclass Zoantharia, with hexamerous or polymerous symmetry, contains the sea anemones, which are mostly solitary and do not have a secreted skeleton, and the stony corals, which are mostly colonial and secrete a calcareous exoskeleton. The subclass Alcyonaria, with octomerous symmetry, contains the soft and horny corals, many of which are important and beautiful components of coral reefs.

The Ctenophora are biradial organisms, which swim by means of eight comb rows, formed of modified cilia. These weakly swimming animals capture small prey with the aid of adhesive cells called colloblasts.

Cnidaria and Ctenophora are probably derived from an ancestor that resembled the planula larva of the cnidarians, and the most primitive cnidarians are among the Hydrozoa. In spite of their relatively simple level or organization, the cnidarians are a successful and important phylum.

A delicately colored anemone, *Condylactis gigantea,* extends its tentacles from beneath a stony coral, *Siderastraea.*

Photograph by L.S. Roberts.

REVIEW QUESTIONS

1. Explain the selective advantage of radial symmetry for sessile animals.
2. Give eight characteristics of the phylum Cnidaria.
3. Name and distinguish the classes in the phylum Cnidaria.
4. Distinguish between the polyp and medusa forms.
5. Explain the mechanism of nematocyst discharge.
6. What is an unusual feature of the nervous system of cnidarians?
7. Diagram a hydra, labeling the main body parts.
8. Name and give the functions of the main cell types in the epidermis and in the gastrodermis of cnidarians.
9. What stimulates feeding behavior in hydras?
10. Define the following with regard to hydroids: hydrorhiza, hydrocaulus, coenosarc, perisarc, hydranth, gonangium, manubrium, statocyst, ocellus.
11. Distinguish the following from each other: statocyst and rhopalium; scyphomedusae and hydromedusae; scyphistoma and ephyrae; velum and pedalium; Zoantharia and Alcyonaria.
12. Define the following with regard to sea anemones: siphonoglyph; septa; acoritia threads; pedal laceration.
13. Specifically, what kinds of organisms are most important in deposition of calcium carbonate on coral reefs?
14. Why do reef-building corals grow only in relatively shallow water?
15. How do ctenophores swim, and how do they obtain food?

SELECTED REFERENCES

Barnes, R.D. 1987. Invertebrate zoology, ed. 5. Philadelphia, Saunders College/Holt, Rinehart & Winston. *Has a chapter on cnidarians and ctenophores plus a chapter on coral reefs.*

Colin, P.I. 1978. Caribbean reef invertebrates and plants. Neptune City, N.J., T.R.H. Publications, Inc. *Many color photographs, very good for sight recognition of Caribbean reef corals and other invertebrates.*

Goreau, T.F., N.I. Goreau, and T.J. Goreau. 1979. Corals and coral reefs. Sci. Am. **241**:124-135 (Aug.). *A good summary of the biology, ecology, and physiology of reef corals.*

Gosner, K.L. 1979. A field guide to the Atlantic seashore: invertebrates and seaweeds of the Atlantic coast from the Bay of Fundy to Cape Hatteras. The Peterson Field Guide Series. Boston, Houghton Mifflin Co. *A helpful aid for students of the invertebrates found along the northeastern coast of the United States.*

Kaplan, E.H. 1982. A field guide to coral reefs of the Caribbean and Florida. The Peterson Field Guide Series. Boston, Houghton Mifflin Co. *More than just a field guide, this little book has much information on biology of coral reefs.*

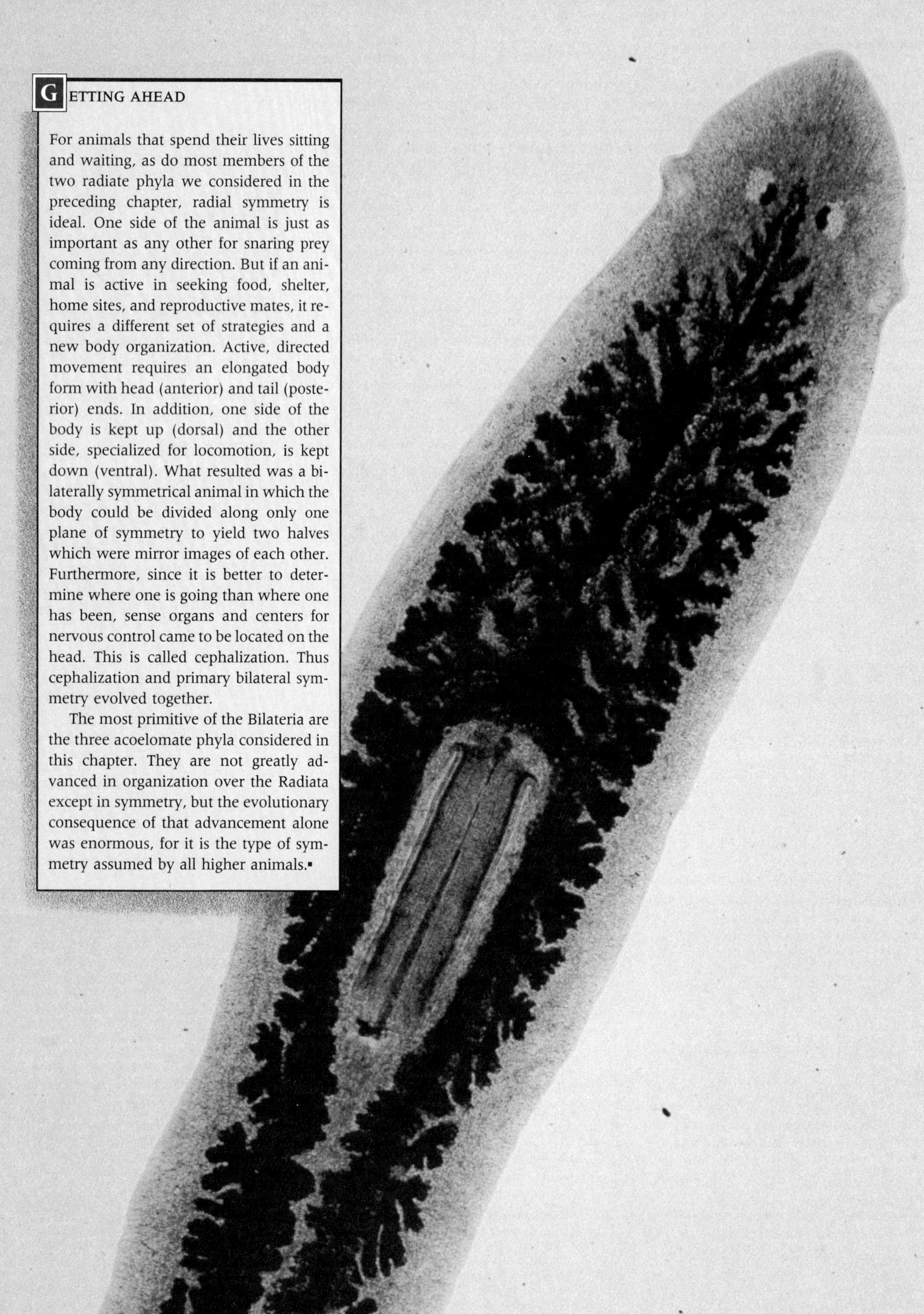

GETTING AHEAD

For animals that spend their lives sitting and waiting, as do most members of the two radiate phyla we considered in the preceding chapter, radial symmetry is ideal. One side of the animal is just as important as any other for snaring prey coming from any direction. But if an animal is active in seeking food, shelter, home sites, and reproductive mates, it requires a different set of strategies and a new body organization. Active, directed movement requires an elongated body form with head (anterior) and tail (posterior) ends. In addition, one side of the body is kept up (dorsal) and the other side, specialized for locomotion, is kept down (ventral). What resulted was a bilaterally symmetrical animal in which the body could be divided along only one plane of symmetry to yield two halves which were mirror images of each other. Furthermore, since it is better to determine where one is going than where one has been, sense organs and centers for nervous control came to be located on the head. This is called cephalization. Thus cephalization and primary bilateral symmetry evolved together.

The most primitive of the Bilateria are the three acoelomate phyla considered in this chapter. They are not greatly advanced in organization over the Radiata except in symmetry, but the evolutionary consequence of that advancement alone was enormous, for it is the type of symmetry assumed by all higher animals.▪

CHAPTER 18

Acoelomate Animals

Flatworms, Ribbon Worms, and Jaw Worms

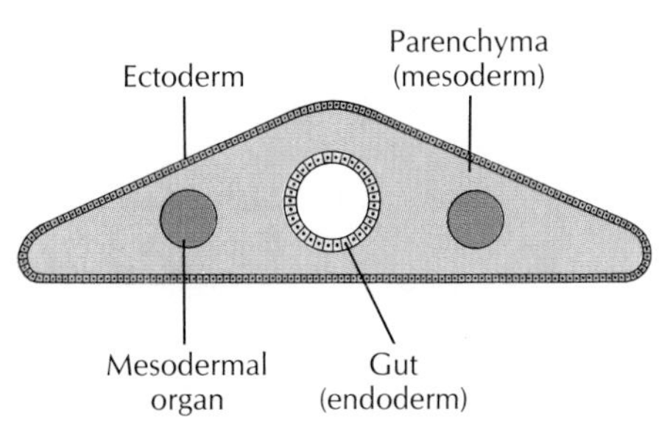

FIGURE 18-1

Acoelomate body plan.

The Platyhelminthes (Gr. *platys,* flat, + *helmins,* worm), or flatworms, the Nemertea (Gr. *Nemertes,* one of the nereids, unerring one), or ribbon worms, and the Gnathostomulida (Gr. *gnathos,* jaw, + *stoma,* mouth, + L. *ulus,* dim.), or jaw worms, have only one internal space. This is the digestive cavity, with the region between the ectoderm and endoderm filled with mesoderm in the form of muscle fibers and their cell bodies (mesenchyme or parenchyma) (Figure 18-1). Since they lack a coelom or a pseudocoel, they are termed **acoelomate** animals, and because they have three well-defined germ layers, they are **triploblastic.** Acoelomates show more specialization and division of labor among their organs than do the radiate animals because the mesoderm makes more elaborate organs possible; thus the acoelomates are said to have reached the organ-system level of organization.

These phyla belong to the protostome division of the Bilateria and have spiral and determinate cleavage. They have some centralization of the nervous system, with a concentration of nerves anteriorly and a ladder-type arrangement of trunks and connectives down the body. They have an excretory (or osmoregulatory) system, and the nemerteans are the most primitive phylum with a circulatory system. Nemerteans also have a one-way digestive system, with an anus as well as a mouth.

■ Phylum Platyhelminthes

The term "worm" has been loosely applied to elongated, bilateral invertebrate animals without appendages. At one time zoologists considered worms (Vermes) to be a group in their own right. Such a group included a highly diverse assortment of forms. This unnatural assemblage has been reclassified into various phyla. By tradition, however, zoologists still refer to the various groups of these animals as flatworms, ribbon worms, roundworms, segmented worms, and the like.

The Platyhelminthes were derived from an ancestor that probably had many cnidarian-like characteristics, perhaps an ancestor shared by the cnidarians. Nonetheless, replacement of the jellylike mesoglea of cnidarians with a cellular, mesodermal parenchyma laid the basis for a more complex organization. Parenchyma is a form of tissue containing more cells and fibers than the mesoglea of the cnidarians.

Flatworms range in size from a millimeter or less to some of the tapeworms that are many meters in length. Their flattened bodies may be slender, broadly leaflike, or long and ribbonlike.

CHARACTERISTICS

1. Three germ layers (**triploblastic**)
2. **Bilateral symmetry;** definite polarity of anterior and posterior ends
3. **Body flattened dorsoventrally;** oral and genital apertures mostly on ventral surface
4. Body with multiple reproductive units in one class (Cestoda)
5. Epidermis may be cellular or syncytial (ciliated in some); **rhabdites** in epidermis of most Turbellaria; epidermis a syncytial **tegument** in Monogenea, Trematoda, and Cestoda
6. Muscular system of mesodermal origin, in the form of a sheath of circular, longitudinal, and oblique layers beneath the epidermis or tegument
7. No internal body space (acoelomate) other than digestive tube; spaces between organs filled with parenchyma
8. Digestive system incomplete (gastrovascular type); absent in some
9. Nervous system consisting of a **pair of anterior ganglia** with **longitudinal nerve cords** connected by transverse nerves and located in the parenchyma in most forms; similar to cnidarians in primitive forms
10. Simple sense organs; eyespots in some
11. Excretory system of two lateral canals with branches bearing **flame cells (protonephridia);** lacking in some primitive forms
12. Respiratory, circulatory, and skeletal systems lacking; lymph channels with free cells in some trematodes
13. Most forms monoecious; reproductive system complex, usually with well-developed gonads, ducts, and accessory organs; internal fertilization; development direct in free-swimming forms and those with a single host in the life cycle; usually indirect in internal parasites in which there may be a complicated life cycle often involving several hosts
14. Class Turbellaria mostly free living; classes Monogenea, Trematoda, and Cestoda entirely parasitic.

CLASSIFICATION

Class Turbellaria (tur'bel-lar'e-a) (L. *turbellae* [pl.], stir, bustle, + *-aria,* like or connected with). The turbellarians. Usually free-living forms with soft flattened bodies; covered with ciliated epidermis containing secreting cells and rodlike bodies (rhabdites); mouth usually on ventral surface sometimes near center of body; no body cavity except intercellular spaces in parenchyma; mostly hermaphroditic, but some have asexual fission. Examples: *Dugesia* (planaria), *Microstomum, Planocera.*

Class Monogenea (mon'o-gen'e-a) (Gr. *mono,* single, + *gene,* origin, birth). The monogenetic flukes. Body covered with a syncytial tegument without cilia; body usually leaflike to cylindrical in shape; posterior attachment organ with hooks, suckers, or clamps, usually in combination; monoecious; development direct, with single host and usually with free-swimming, ciliated larva; all parasitic, mostly on skin or gills of fish. Examples: *Dactylogyrus, Polystoma, Gyrodactylus.*

Class Trematoda (trem'a-to'da) (Gr. *trematodes,* with holes, + *eidos,* form). The digenetic flukes. Body covered with a syncytial tegument without cilia;

leaflike or cylindrical in shape; usually with oral and ventral suckers, no hooks; alimentary canal usually with two main branches; mostly monoecious; development indirect, with first host a mollusc, final host usually a vertebrate; parasitic in all classes of vertebrates. Examples: *Fasciola, Clonorchis, Schistosoma.*

Class Cestoda (ses-to'da) (Gr. *kestos,* girdle, + *eidos,* form). The tapeworms. Body covered with nonciliated, syncytial tegument; general form of body tapelike; scolex with suckers or hooks, sometimes both, for attachment; body usually divided into series of proglottids; no digestive organs; usually monoecious; parasitic in digestive tract of all classes of vertebrates; development indirect with two or more hosts; first host may be vertebrate or invertebrate. Examples: *Diphyllobothrium, Hymenolepis, Taenia.*

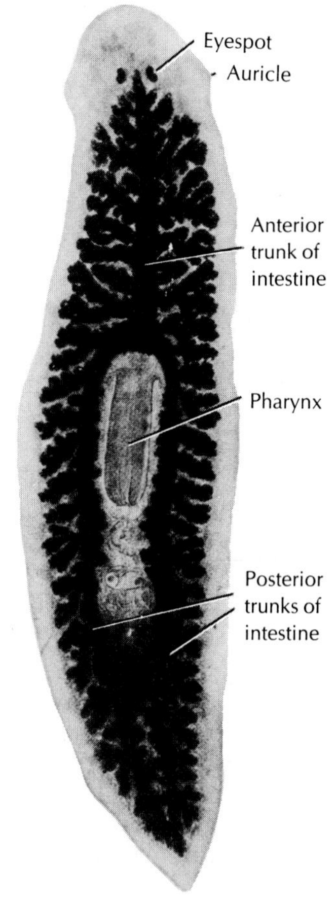

FIGURE 18-2

Stained planarian.
Courtesy Carolina Biological Supply Co.

ECOLOGICAL RELATIONSHIPS

The flatworms include both free-living and parasitic forms, but the free-living members are found exclusively in the class Turbellaria. A few turbellarians are symbiotic (commensals or parasites), but the majority are adapted as bottom dwellers in marine or fresh water or live in moist places on land. Many, especially of the larger species, are found on the underside of stones and other hard objects in freshwater streams or in the littoral zones of the ocean.

Relatively few turbellarians live in fresh water. Planarians (Figure 18-2) and some others frequent streams and spring pools; others prefer flowing water of mountain streams. Some species occur in fairly hot springs. Terrestrial turbellarians are found in fairly moist places under stones and logs.

All members of the classes Monogenea and Trematoda (the flukes) and the class Cestoda (the tapeworms) are parasitic. Most of the Monogenea are ectoparasites, but all the trematodes and cestodes are endoparasitic. Many species have indirect life cycles with more than one host; the first host is often an invertebrate, and the final host is usually a vertebrate. Humans serve as hosts for a number of species. Certain larval stages may be free living.

FORM AND FUNCTION

Body form

The body of turbellarians is covered by a ciliated epidermis resting on a basement membrane. It contains rod-shaped **rhabdites** (Figure 18-3) that, when discharged into water, swell and form a protective mucous sheath around the body. Single-cell mucous glands open on the surface of the epidermis. The body covering in all the other classes is a **tegument** (Figure 18-4), which does not bear cilia in the adult. The cell bodies are sunk beneath the outer layer and superficial muscle layers and communicate with the outer layer (**distal cytoplasm**) by processes extending between the muscles. Because the distal cytoplasm is continu-

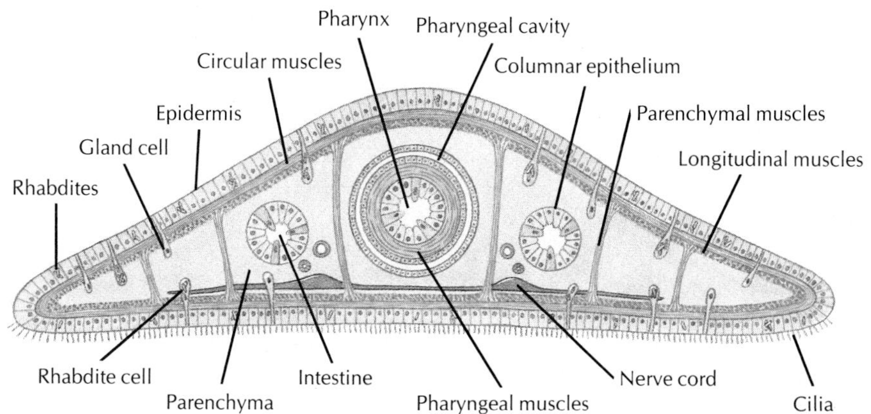

FIGURE 18-3

Cross section of planarian through pharyngeal region, showing relationships of body structures.

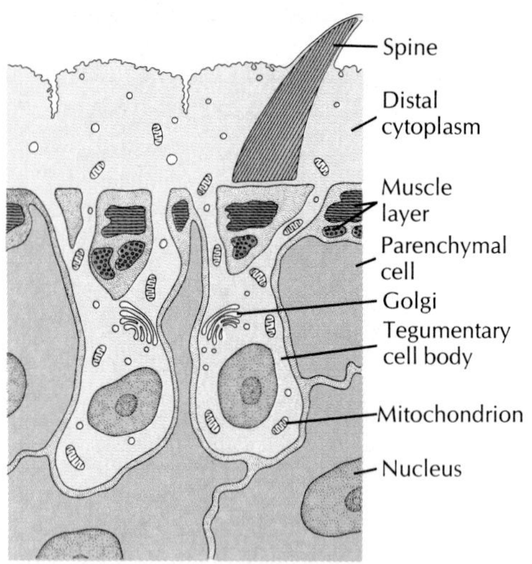

FIGURE 18-4

Diagrammatic drawing of the structure of the tegument of a trematode *Fasciola hepatica*.

Based on a drawing by L.T. Threadgold, from Schmidt, G.D., and L.S. Roberts. 1989. Foundations of parasitology, ed. 4. St. Louis, The C.V. Mosby Co.

ous, with no intervening cell membranes, the tegument is **syncytial.** This peculiar epidermal arrangement is probably related to adaptations for parasitism in ways that are still unclear.

In the body wall beneath the basement membrane are layers of **muscle fibers** that run circularly, longitudinally, and diagonally. Other muscle fibers may cross through the body from one side of the outer muscle sheath to the other (Figure 18-3). A meshwork of **parenchyma** cells, developed from mesoderm, fills the spaces between the muscles and visceral organs. Parenchyma cells in some, perhaps all, flatworms are not a separate cell type, but are the noncontractile portions (cell bodies) of muscle cells.

Nutrition and digestion

The digestive system includes a mouth, pharynx, and intestine. The muscular **pharynx** opens posteriorly just inside the mouth, through which it can extend, in the turbellarians (Figure 18-5). The mouth is usually at the anterior end in flukes, and the pharynx is not protrusible. The intestine may be simple or branched.

Intestinal secretions contain proteolytic enzymes for some **extracellular digestion.** Food is sucked into the intestine, where cells of the gastrodermis often phagocytose it and complete the digestion (**intracellular**). Undigested food is egested through the pharynx. The entire digestive system is lacking in the tapeworms. They must absorb all of their nutrients as small molecules (predigested by the host) directly through their tegument.

Excretion and osmoregulation

Except in the primitive turbellarians, the osmoregulatory system consists of canals with tubules that end in **flame cells** (**protonephridia**) (Figure 18-5, *A*). The flame cell surrounds a small space into which a tuft of cilia projects. The space is continuous with the tubule complex, and the beat of the cilia (resembling a flickering flame) provides the force to drive fluids through the system. The tubules lead into collecting ducts that finally open to the outside by pores. Among the various flatworms, there may be a single protonephridium or from one to four pairs. They open to the outside through one to many **nephridiopores.** This sys-

FIGURE 18-5

Structure of a planarian. **A,** Reproductive and excretory systems, shown in part. **B,** Digestive tract and ladder-type nervous system. Pharnyx shown in resting position. **C,** Pharynx extended through ventral mouth.

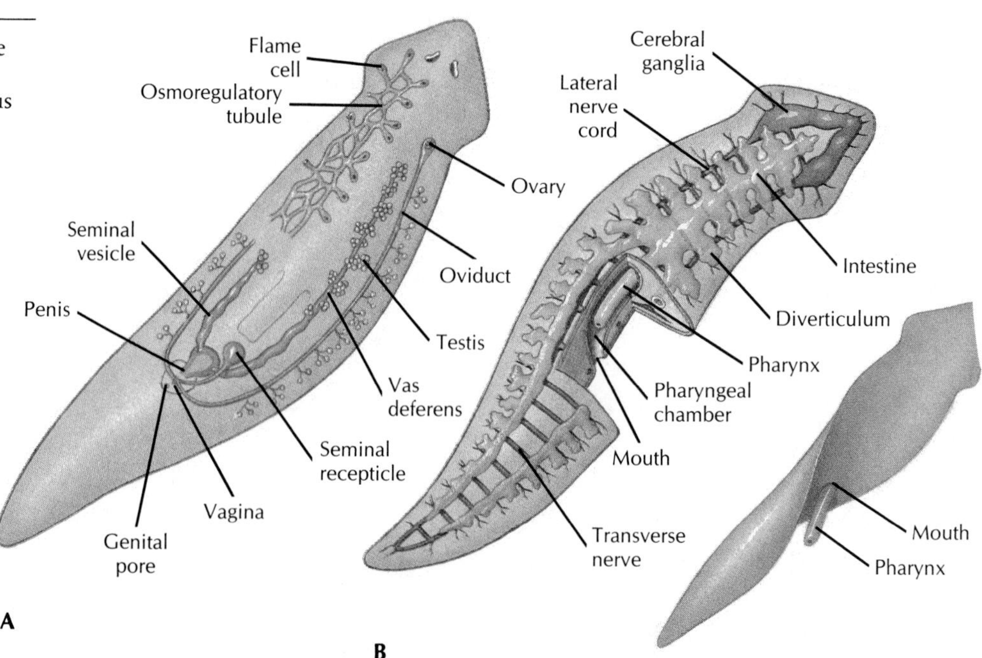

tem is probably osmoregulatory in most forms; it is reduced or absent in marine turbellarians, which do not have to expel excess water.

Metabolic wastes are largely removed by diffusion through the body wall.

Nervous system

The most primitive flatworm nervous system, found in some turbellarians, is a **subepidermal nerve plexus** resembling the nerve net of the cnidarians. Other flatworms have, in addition to a nerve plexus, one to five pairs of **longitudinal nerve cords** lying under the muscle layer (Figure 18-5, *B*). The more advanced flatworms tend to have the lesser number of nerve cords. Connecting nerves form a "ladder-type" pattern. The brain is a mass of ganglion cells arising anteriorly from the nerve cords. Except in primitive turbellarians, which have a diffuse system, the neurons are organized into sensory, motor, and association types—an important advance in the evolution of the nervous system.

Sense organs

Active locomotion in flatworms has favored not only cephalization in the nervous system but also advancements in the development of sense organs. Light-sensitive eyespots, (ocelli), are found in the turbellarians (Figure 18-2) and some flukes.

Tactile cells and chemoreceptive cells are abundant over the body, and in planarians they form definitive organs on the **auricles** (the earlike lobes on the sides of the head). Some also have **statocysts** for equilibrium and **rheoreceptors** for sensing water current direction.

Reproduction

Many flatworms reproduce both asexually and sexually. Many freshwater turbellarians can reproduce by fission, merely constricting behind the pharynx and separating into two animals, each of which regenerates the missing parts. In some forms such as *Stenostomum* and *Microstomum,* the individuals do not separate at once but remain attached, forming chains of zooids (Figure 18-6, *B* and *C*). Flukes reproduce asexually in the snail intermediate host (described further below), and some tapeworms, such as *Echinococcus,* can bud off thousands of juveniles in the intermediate host.

Most flatworms are monoecious (hermaphroditic) but practice cross-fertilization. In the more advanced turbellarians and in the other flatworm classes, the fertilized ovum and some yolk cells (produced by the **yolk glands**), become enclosed in a shell (capsule). In some freshwater planarians the capsules are attached by little stalks to the underside of stones or plants, and embryos emerge as juveniles that resemble mature adults. In some marine turbellarians the egg develops into a ciliated free-swimming larva. In the other flatworms, the embryo becomes a larva, which may be ciliated or not, according to the group.

CLASS TURBELLARIA

Turbellarians are mostly free-living worms that range in length from 5 mm or less to 50 cm. Usually covered with ciliated epidermis, these are mostly creeping worms that combine muscular with ciliary movements to achieve locomotion. The mouth is on the ventral side. Unlike the flukes (trematodes) and tapeworms (cestodes), they have simple life cycles.

The order Acoela (Gr. *a,* without, + *koilos,* hollow) is often regarded as having changed least from the ancestral form. Its members are small and have a mouth but no gastrovascular cavity or excretory system. Food is merely passed through the mouth or pharynx into temporary spaces that are surrounded by a syncytial mesenchyme where gastrodermal phagocytic cells digest the food intracellularly. The order has a syncytial epidermis and a diffuse nervous system.

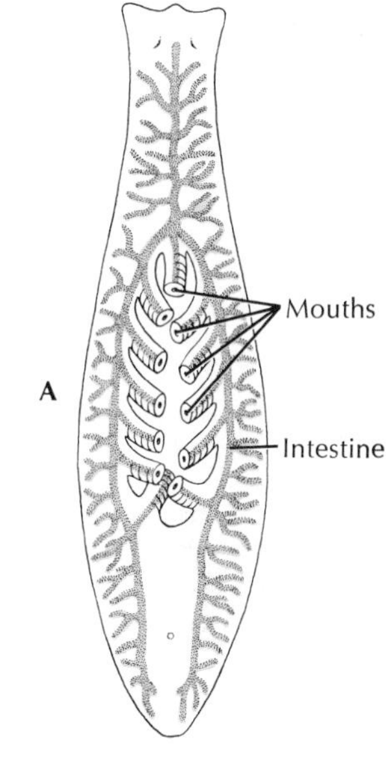

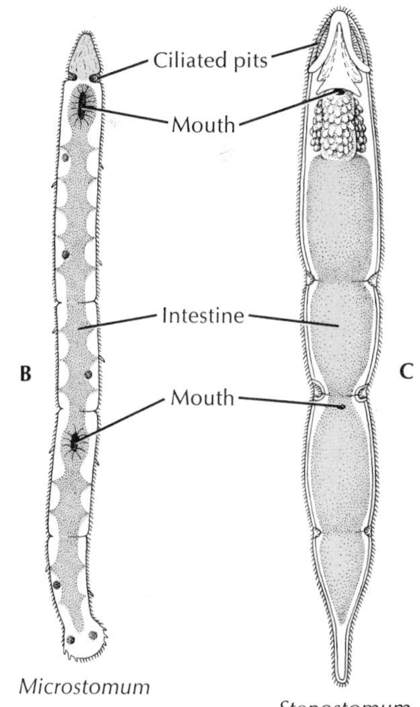

FIGURE 18-6

Some small freshwater turbellarians. **A,** *Phagocata* has numerous mouths. **B** and **C,** Incomplete fission results for a time in a series of attached zooids.

The freshwater planarians, such as *Dugesia* (formerly called *Euplanaria* but changed by priority to *Dugesia* after Duges, who first described the form in 1830), belong to one of the more advanced orders and are used extensively in introductory laboratories.

Planarians move by gliding, head slightly raised, over a slime track secreted by the marginal adhesive glands. The beating of the epidermal cilia in the slime track drives the animal along. Rhythmical muscular waves can also be seen passing backward from the head as it glides.

Planarians are mainly carnivorous, feeding largely on small crustaceans, nematodes, rotifers, and insects. They can detect food from some distance by means of chemoreceptors. They entangle their prey in mucous secretions from the mucous glands and rhabdites. Gripping its prey with its anterior end, the planarian wraps its body around the prey, extends its proboscis, and sucks up minute food particles.

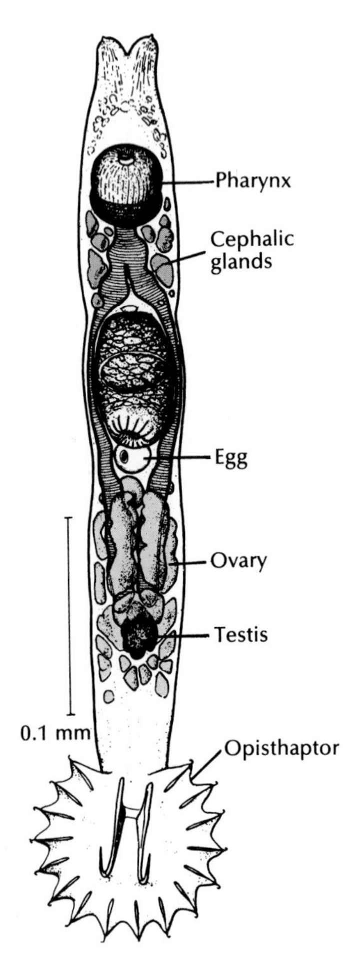

FIGURE 18-7

A monogenetic fluke, *Gyrodactylus cylindriformis,* ventral view.

From Mueller, J.F., and H.J. Van Cleave. 1932. Roosevelt Wildlife Annals.

CLASS MONOGENEA

The monogenetic flukes traditionally have been placed as an order of the Trematoda, but they are sufficiently different to deserve a separate class. They are all parasites, primarily of the gills and external surfaces of fish (Figure 18-7). A few are found in the urinary bladders of frogs and turtles, and one has been reported from the eye of a hippopotamus. Although widespread and common, monogeneans seem to cause little damage to their hosts under natural conditions. However, like numerous other fish pathogens, they become a serious threat when their hosts are crowded together, as in fish farming.

The life cycles of monogeneans are direct with a single host, as suggested by the name of the group, which means "single birth." The egg hatches a ciliated larva that attaches to the host or swims around awhile before attachment.

CLASS TREMATODA

Trematodes are all parasitic flukes, and as adults they are almost all found as endoparasites of vertebrates. They are chiefly leaflike in form and are structurally similar in many respects to the more advanced Turbellaria. A major difference is found in the tegument (described earlier, Figure 18-4).

Other structural adaptations for parasitism are apparent: various penetration glands or glands to produce cyst material; organs for adhesion such as suckers and hooks; and increased reproductive capacity. Otherwise, trematodes retain several turbellarian characteristics, such as a well-developed alimentary canal (but with the mouth at the anterior, or cephalic, end) and similar reproductive, excretory, and nervous systems, as well as a musculature and parenchyma that are only slightly modified from those of the Turbellaria. Sense organs are poorly developed.

Of the three subclasses of Trematoda, two are small and poorly known groups, but Digenea (Gr. *dis,* double, + *genos,* race) is a large group with many species of medical and economic importance.

Digenea

With rare exceptions, digenetic trematodes have an indirect life cycle, the first (**intermediate**) host being a mollusc and the final (**definitive**) host being a vertebrate. In some species a second, and sometimes even a third, intermediate host intervenes. The group has been very successful and they inhabit, according to species, a wide variety of sites in their hosts: all parts of the digestive tract, respiratory tract, circulatory system, urinary tract, and reproductive tract.

One of the world's most amazing biological phenomena is the digenean life cycle. Although the cycles of different species vary widely in detail, a typical example would include the adult, egg, miracidium, sporocyst, redia, cercaria, and

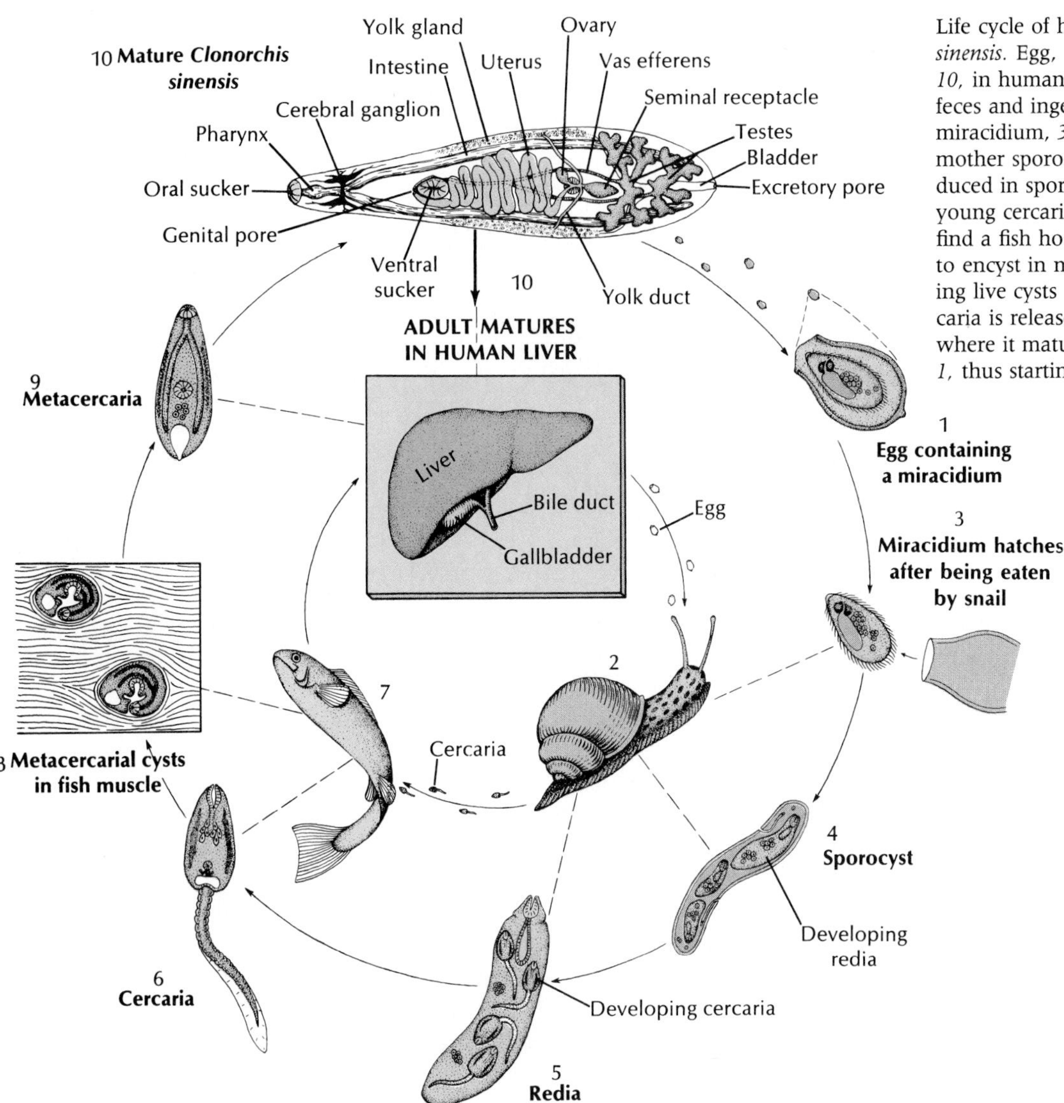

FIGURE 18-8

Life cycle of human liver fluke, *Clonorchis sinensis.* Egg, *1,* shed from adult trematode, *10,* in human bile ducts, is carried out in feces and ingested by snail, *2,* in which miracidium, *3,* hatches and becomes mother sporocyst. *4,* Young rediae are produced in sporocyst, grow, *5,* and produce young cercariae. Cercariae leave snail, *6,* find a fish host, *7,* and burrow under scales to encyst in muscle, *8.* When fish containing live cysts is eaten by humans, metacercaria is released, *9,* and enters bile duct, where it matures, *10,* to shed eggs in feces, *1,* thus starting another cycle.

metacercaria stage (Figure 18-8). The egg usually passes from the definitive host in the excreta and must reach water to develop further. There, it hatches to a free-swimming, ciliated larva, the **miracidium.** The miracidium penetrates the tissues of a snail, where it transforms into a **sporocyst.** The sporocyst reproduces asexually to yield either more sporocysts or a number of **rediae.** The rediae, in turn, reproduce asexually to produce more rediae or to produce **cercariae.** In this way a single egg can give rise to an enormous number of progeny. The cercariae emerge from the snail and penetrate a second intermediate host or encyst on vegetation or other objects to become **metacercariae,** which are juvenile flukes. The adult grows from the metacercaria when that stage is eaten by the definitive host.

Some of the most serious parasites of humans and domestic animals belong to the Digenea, of which only two will be discussed.

Clonorchis (Gr. *clon,* branch, + *orchis,* testis) (Figure 18-8) is the most important liver fluke of humans and is common in many regions of the Orient, especially in China, southern Asia, and Japan. Cats, dogs, and pigs are also often infected. The adult lives in the bile passages, and eggs containing the miracidia

A

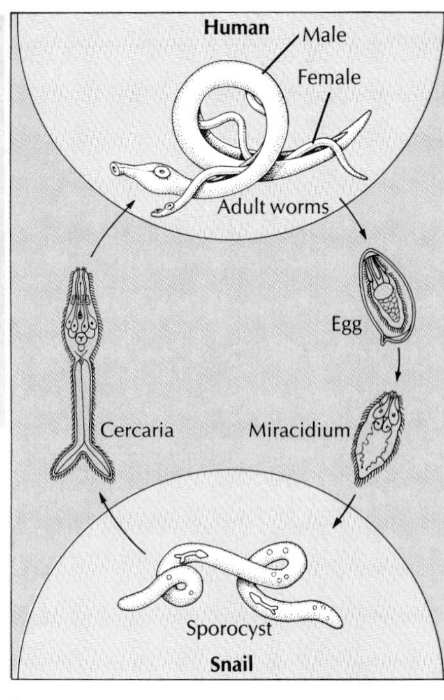

B

FIGURE 18-9

A, Adult male and female *Schistosoma mansoni* in copulation. The blood flukes differ from most other flukes in being dioecious. The male is broader and heavier and has a large, ventral *gynecophoric canal,* posterior to the ventral sucker. The gynecophoric canal embraces the long, slender female (darkly stained individual) during insemination and oviposition. **B,** Life cycle of *Schistosoma mansoni.*

A, AFIP No. 56-3334.

are shed in the feces. If ingested by certain freshwater snails, the sporocyst and redia stages are passed, and free-swimming cercariae emerge. Cercariae that manage to find a suitable fish encyst in the skin or muscles as metacercariae. When the fish is eaten raw, the juveniles migrate up the bile duct to mature and may survive there for 15 to 30 years. The effect of the flukes on humans depends mainly on the extent of the infection. A heavy infection may cause a pronounced cirrhosis of the liver and may result in death.

Schistosomiasis, an infection with blood flukes of the genus *Schistosoma* (Gr. *schistos,* divided, + *soma,* body) (Figure 18-9), ranks as one of the major infec-

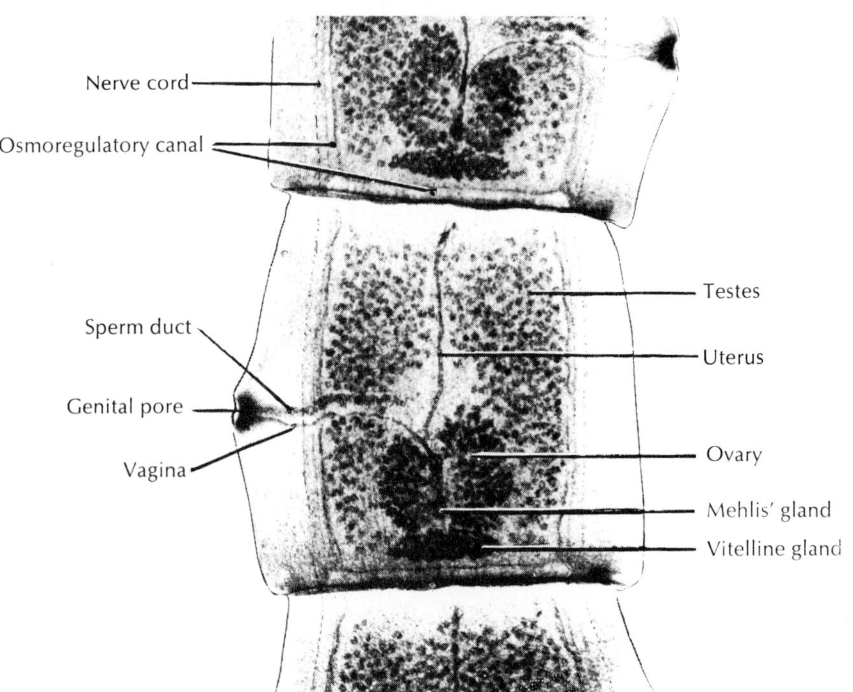

FIGURE 18-10

Photomicrograph of mature proglottids of *Taenia pisiformis,* dog tapeworm.

Courtesy General Biological Supply House, Inc.

tious diseases in the world, with 200 million people infected. The disease is widely prevalent over much of Africa and parts of South America, the West Indies, the Middle East, and the Far East. It is spread when eggs shed in human feces and urine get into water containing host snails (Figure 18-9, *B*). Cercariae that contact human skin penetrate through the skin to enter blood vessels, which they follow to certain favorite regions depending on the type of fluke. The inch-long adults may live for years in the human host, causing such disturbances as severe dysentery, anemia, liver enlargement, bladder inflammation, and brain damage.

An avian schistosome common in the northern lakes of the United States often enters the skin of human bathers in its search for a suitable bird host, causing a skin irritation known as "swimmer's itch." In this case the human is a dead end in the fluke's life cycle because the fluke cannot develop further in the human.

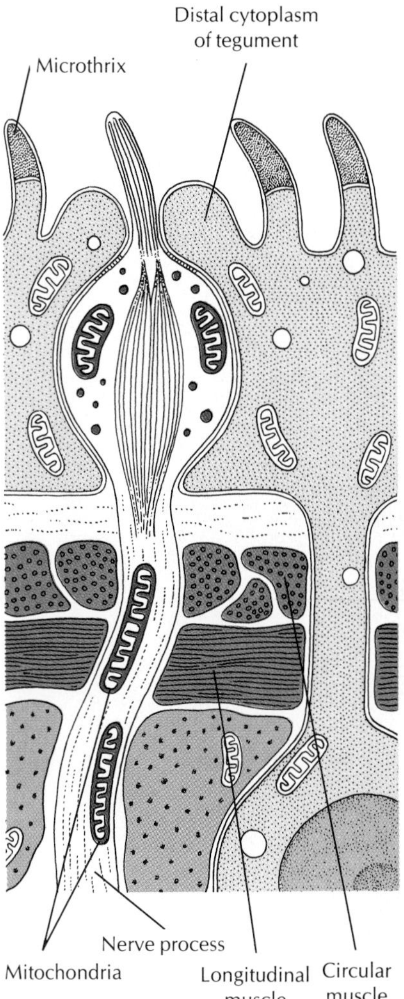

FIGURE 18-11

Schematic drawing of a longitudinal section through a sensory ending in the tegument of *Echinococcus granulosus*.

CLASS CESTODA

Cestoda, or tapeworms, differ in many respects from the preceding classes: they usually have long flat bodies made up of many reproductive units, or **proglottids** (Figure 18-10), and there is a complete lack of a digestive system. As in Monogenea and Trematoda, there are no external, motile cilia in the adult, and the tegument is of a distal cytoplasm with sunken cell bodies beneath the superficial muscle layer (Figure 18-4). In contrast to the monogeneans and trematodes, however, their entire surface is covered with minute projections called **microtriches** (sing., **microthrix**) (Figure 18-11). The microtriches greatly amplify the surface area of the tegument, which is a vital adaptation of the tapeworm, since it must absorb all its nutrients across the tegument.

Tapeworms are nearly all monoecious. They have well-developed muscles, and their excretory system and nervous system are somewhat similar to those of other flatworms. They have no special sense organs but do have sensory endings in the tegument that are modified cilia (Figure 18-11). One of their most specialized structures is the **scolex,** or holdfast, which is the organ of attachment. It is usually provided with suckers or suckerlike organs and often with hooks or spiny tentacles (Figure 18-12).

With rare exceptions, all cestodes require at least two hosts, and the adult is a parasite in the digestive tract of vertebrates. Often one of the intermediate hosts is an invertebrate.

The main body of the worms, the chain of proglottids, is called the **strobila** (Figure 18-12). Typically, there is a **germinative zone** just behind the scolex where new proglottids are formed. As new proglottids are differentiated in front of it, each individual segment moves posteriorly in the strobila, and its gonads mature. The proglottid is usually fertilized by another proglottid in the same or a different strobila. The shelled embryos form in the uterus of the proglottid, and they are either expelled through a uterine pore, or the entire proglottid is shed from the worm as it reaches the posterior end of the strobila.

More than 1000 species of tapeworms are known to parasitologists. Almost all vertebrate species are infected. Normally, adult tapeworms do little harm to their hosts. The most common tapeworms found in humans are given in Table 18-1.

In the beef tapeworm, *Taeniarhynchus saginatus* (Gr. *tainia,* band, ribbon, + *rhynchos,* snout, beak), eggs shed from the human host are ingested by cattle (Figure 18-13). The six-hooked larvae (oncospheres) hatch, burrow into blood or lymph vessels, and migrate to skeletal muscle where they encyst to become "bladder worms" (cysticerci). Each of these juveniles develops an invaginated scolex and remains quiescent until the uncooked muscle is eaten by humans. In the new host the scolex evaginates, attaches to the intestine, and matures in 2 or

GUTLESS WONDER

Though lacking skeletal strengths
Which we associate with most
Large forms, tapeworms go to
great lengths
To take the measure of a host.

Monotonous body sections
In a limp mass-production line
Have nervous and excretory connections
And the means to sexually combine

And to coddle countless progeny
But no longer have the guts
To digest for themselves or live free
Or know a meal from soup to nuts.

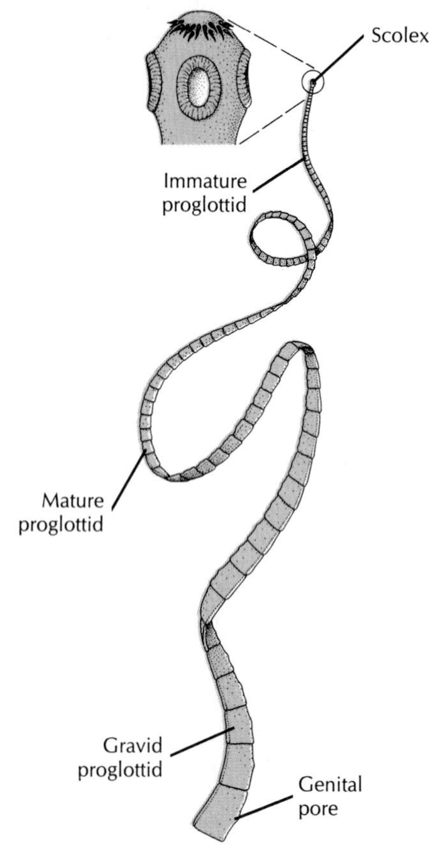

FIGURE 18-12

A tapeworm, showing strobila and scolex. The scolex is the organ of attachment.

Table 18-1 Common cestodes of humans

Common and scientific name	Means of infection; prevalence in humans
Beef tapeworm (*Taeniarhynchus saginatus*)	Eating rare beef; most common of all tapeworms in humans
Pork tapeworm (*Taenia solium*)	Eating rare pork
Fish tapeworm (*Diphyllobothrium latum*)	Eating rare or poorly cooked fish; fairly common in Great Lakes region of United States, and other areas of world where raw fish is eaten
Dog tapeworm (*Dipylidium caninum*)	Unhygienic habits of children (larvae in flea and louse); moderate frequency
Dwarf tapeworm (*Vampirolepis nana*)	Juveniles in flour beetles; common
Unilocular hydatid (*Echinococcus granulosus*)	Cysts of juveniles in humans; infection by contact with dogs; common wherever humans are in close relationship with both dogs and ruminants
Multilocular hydatid (*Echinococcus multilocularis*)	Cysts of juveniles in humans; less common than unilocular hydatid

3 weeks; then ripe proglottids may be expelled daily for many years. Humans become infected by eating raw or rare infested ("measly") beef. The adult worm may attain a length of 7 m or more, folded back and forth in the host intestine.

The pork tapeworm, *Taenia solium,* uses humans as definitive hosts and pigs as intermediate hosts. Humans can also serve as intermediate hosts by ingesting the eggs from contaminated hands or food or, in persons harboring an adult worm, by regurgitating segments into the stomach. The juveniles may encyst in the central nervous system, where great damage may result.

■ Phylum Nemertea (Rhynchocoela)

Nemerteans are often called the ribbon worms. Their name refers to the unerring aim of the proboscis, a long muscular tube that can be thrust out swiftly to grasp the prey. The phylum is also called Rhynchocoela (Gr. *rhynchos,* beak, + *koilos,* hollow), which also refers to the proboscis. They are thread-shaped or ribbon-shaped worms; nearly all of them are marine. Some live in secreted gelatinous tubes. There are about 650 species in the group.

Nemertean worms are usually less than 20 cm long, although a few are several meters in length. Some are brightly colored, although most are dull or pallid.

A few of the nemerteans are found in moist soil and fresh water, but by far the larger number are marine. At low tide they are often coiled up under stones. It seems probable that they are active at high tide and quiescent at low tide.

With few exceptions, the general body plan of the nemerteans is similar to that of Turbellaria. Like the latter, their epidermis is ciliated and has many gland cells. Another striking similarity is the presence of flame cells in the excretory system. Recently rhabdites have been found in several nemerteans. In the marine forms there is a ciliated larva that has some resemblance to the trochophore larva found in annelids and molluscs. Other flatworm characteristics are the presence

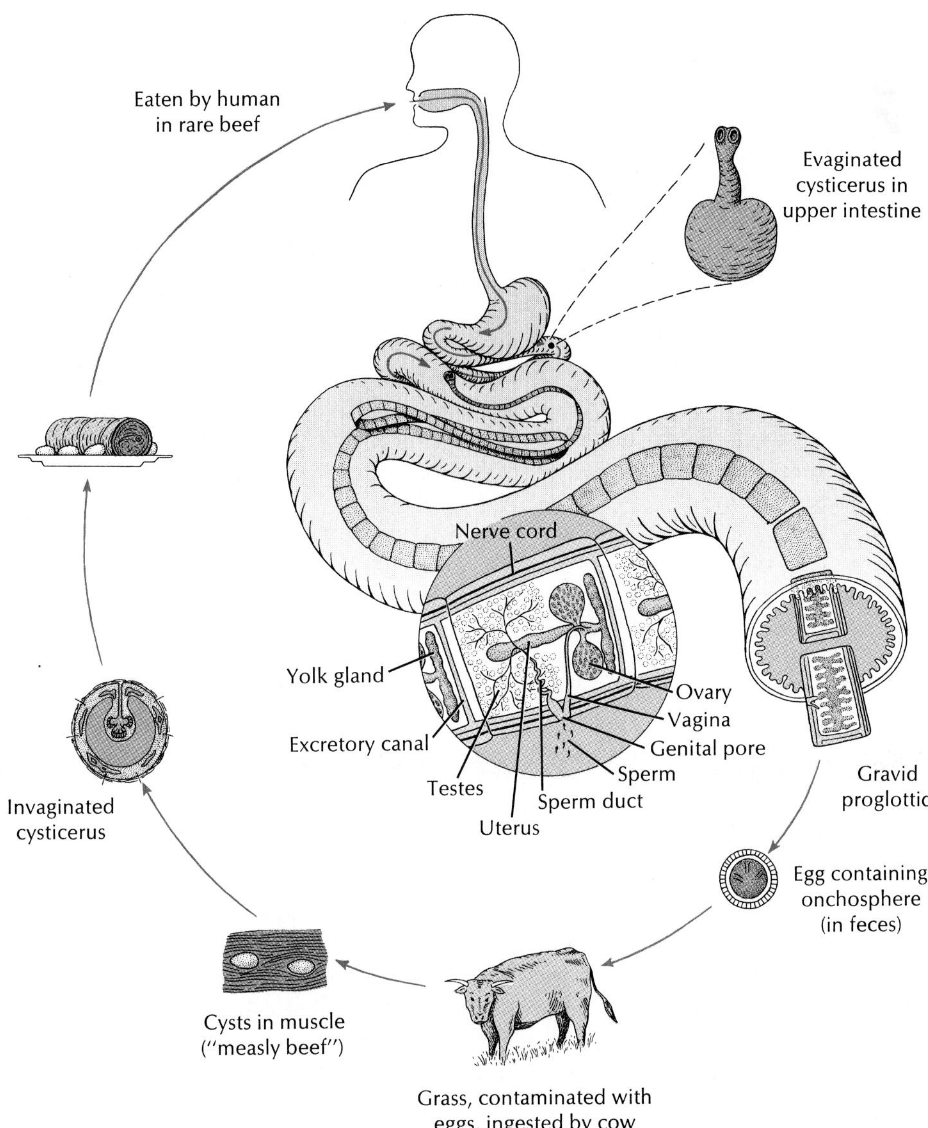

FIGURE 18-13

Life cycle of beef tapeworm, *Taeniarhynchus*. Ripe proglottids break off in human intestine, pass out in feces, crawl out of feces onto grass, and are ingested by cattle. Eggs hatch in cow's intestine, freeing oncospheres, which penetrate into muscles and encyst, developing into "bladder worms." Human eats infected rare beef, and cysticercus is freed in intestine where it attaches to the intestine wall, forms a strobila, and matures.

of bilateral symmetry and a mesoderm and the lack of coelom. All in all, the present evidence seems to indicate that the nemerteans come from an ancestral form closely related to Platyhelminthes. Nevertheless, they differ from flatworms in several important aspects.

FORM AND FUNCTION

Many nemerteans are difficult to examine because they are so long and fragile (Figure 18-14). *Amphiporus* (Gr. *amphi,* on both sides, + *poros,* pore), one of the smaller genera that ranges from 2 to 10 cm in length, is fairly typical of the nemertean structure (Figure 18-15). Its body wall consists of ciliated epidermis and layers of circular and longitudinal muscles. Locomotion consists largely of gliding over a slime track, although larger species move by muscular contractions.

The mouth is anterior and ventral, and the **digestive tract** is **complete,** extending the length of the body and ending at the anus. The development of an anus marks a significant advancement over the gastrovascular systems of the lower phyla. Regurgitation of wastes is no longer necessary; ingestion and eges-

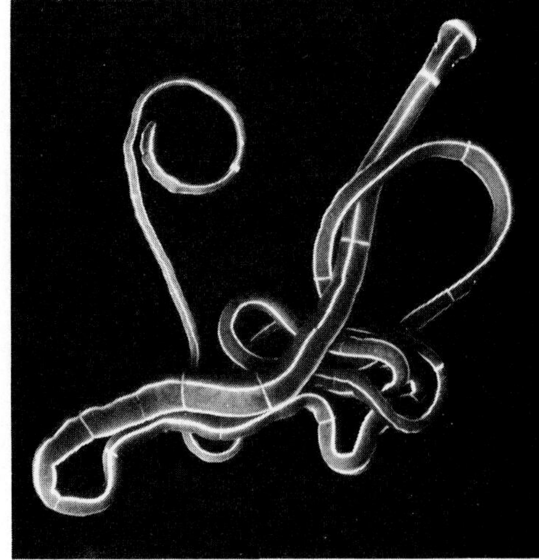

FIGURE 18-14

The ribbon worm *Tubulanus annulatus* (phylum Nemertea) may be several feet long. The anterior end has the enlarged, rounded cephalic lobe.

Courtesy Encyclopedia Britannica Films, Inc.

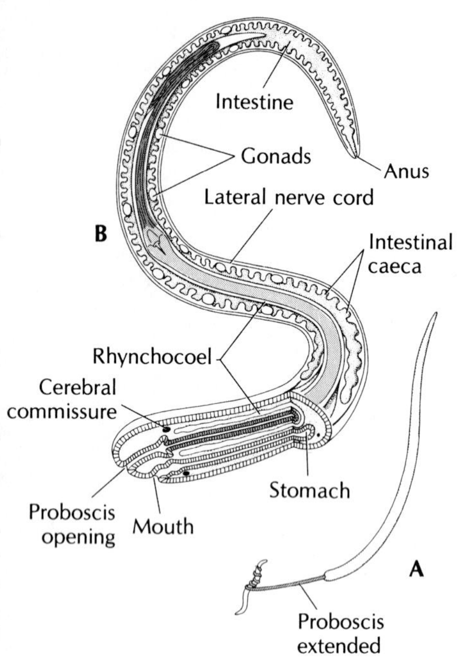

FIGURE 18-15

Ribbon worm *Amphiporus* (phylum Nemertea). **A,** Ribbon worm with proboscis extended to catch prey. **B,** Internal structure of female ribbon worm.

tion can occur simultaneously. Food is moved through the intestine by cilia. Digestion is largely extracellular.

Nemerteans are carnivorous, feeding primarily on annelids and other small invertebrates. They seize their prey with a **proboscis** that lies in an interior cavity of its own, the **rhynchocoel,** above the digestive tract (but not connected with it). The proboscis itself is a long, blind muscular tube that opens at the anterior end at a proboscis pore above the mouth. (In a few nemerteans the esophagus opens through the proboscis pore rather than through a separate mouth.) Muscular pressure on the fluid in the rhynchocoel causes the long tubular proboscis to be everted through the proboscis pore. Eversion of the proboscis exposes a sharp barb, called a stylet (absent in some nemerteans). The sticky, slime-covered proboscis coils around the prey and stabs it repeatedly with the stylet, which pours a toxic secretion on the prey. As the proboscis is withdrawn by a retractor muscle, the prey is brought close to the mouth so that it can be engulfed and swallowed. The proboscis withdraws to its own cavity, ready to be shot out again in defense or in search of food.

Unlike other acoelomates, the nemerteans have a true circulatory system, and the irregular flow is maintained by the contractile walls of the vessels. Many flame-bulb protonephridia are closely associated with the circulatory system, so that their function appears to be truly excretory (dispose of metabolic wastes), in contrast to their apparently osmoregulatory role in Platyhelminthes.

The nervous system is composed of a pair of ganglia, and one or more pairs of longitudinal nerve cords are connected by transverse nerves.

Some species reproduce asexually by fragmentation and regeneration. In contrast to flatworms, most nemerteans are dioecious.

Phylum Gnathostomulida

The first species of the Gnathostomulida (nath'o-sto-myu'lid-a) (Gr. *gnatho,* jaw, + *stoma,* mouth, + L. *-ulus,* diminutive) was observed in 1928 in the Baltic, but its description was not published until 1956. Since then these animals have been found in many parts of the world, including the Atlantic coast of the United States, and over 80 species in 18 genera have been described.

Gnathostomulids, often called jaw worms, are delicate wormlike animals and are 0.5 to 1 mm long (Figure 18-16). They live in the interstitial spaces of very fine sandy coastal sediments and silt and can endure conditions of very low oxygen. They are often found in large numbers and frequently in association with gastrotrichs, nematodes, ciliates, tardigrades, and other small forms.

Lacking a pseudocoel, a circulatory system, and an anus, the gnathostomulids show some similarities to the turbellarians, but they have some characteristics reminiscent of rotifers and share other features with sponges and cnidarians. The phyletic relationships of the Gnathostomulida are still obscure.

Phylogeny and Adaptive Radiation

PHYLOGENY

Platyhelminthes and Nemertea are apparently closely related, with the flatworms being the more primitive. There can be little doubt that the bilaterally symmetrical flatworms were derived from a radial ancestor, perhaps one very similar to the planula larva of the cnidarians. Some investigators believe that this **planuloid ancestor** may have given rise to one branch of descendants that were sessile or free floating and radial, which became the Cnidaria, and another that acquired a creeping habit and bilateral symmetry, which became the Platyhelminthes. Bilateral symmetry is a selective advantage for creeping or swimming animals because sensory structures are concentrated on the anterior end, the end which first encounters environmental stimuli (cephalization).

The transformation from a planuloid ancestor to an early platyhelminth involved a number of body modifications, such as an oral-aboral flattening, with the oral end becoming the ventral surface and the ventral surface adapting for locomotion with the aid of cilia and muscles. The small flatworms of the order Acoela seem to meet many of the requirements of an early ancestor of Platyhelminthes. They have several characteristics in common with the planula larva of the cnidarians, such as no epidermal basement membrane, no digestive system, a nerve plexus under the epidermis, and no distinct gonads. The acoeloid ancestor gave rise to the other orders of Turbellaria and the other classes in the phylum. It may be that the ancestral cestodes never had a digestive tract and therefore did not lose it in their adaptation to parasitism. The Nemertea probably arose from flatworm stock.

The Gnathostomulida occupy a position intermediate between the Platyhelminthes and the pseudocoelomate phyla (Chapter 19). Their protonephridial system is similar to that of gastrotrichs, and their jaws show similarities to the rotifer mastax. However, their complete ciliation, arrangement of body wall muscles, cleavage pattern, and reproductive system strongly suggest a relationship to the primitive Turbellaria.

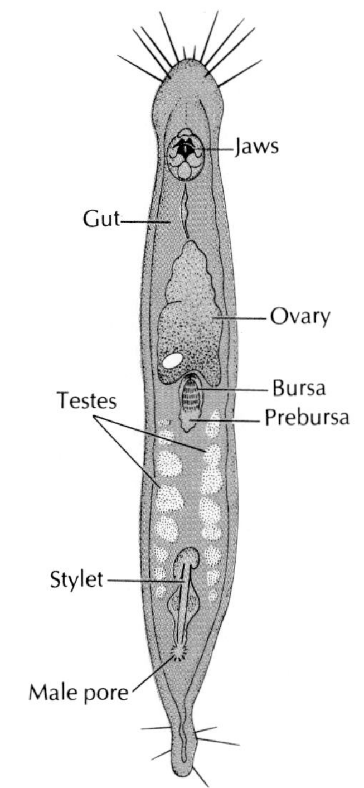

FIGURE 18-16

Gnathostomula jenneri (phylum Gnathostomulida) is a tiny member of the interstitial fauna between grains of sand or mud. Species in this family are among the most commonly encountered jaw worms, found in shallow water and down to depths of several hundred meters.

From Sterrer, W.E. 1972. Systematics and evolution within the Gnathostomulida. Syst. Zool. **21**:151.

ADAPTIVE RADIATION

Because of their body shape and metabolic requirements, early flatworms must have been well preadapted for parasitism and gave rise to symbiotic lines on numerous occasions. These lines produced descendants that were extremely successful as parasites, and many flatworms became very highly specialized for that mode of existence.

Although the ribbon worms have advanced beyond the flatworms in their complexity of organization, they have been dramatically less successful as a group. Perhaps the proboscis was so efficient as a predator tool that there was little selective pressure to explore parasitism, or perhaps some critical preadaptations were simply not present.

Likewise, the jaw worms have not radiated nor been nearly as successful as the flatworms. However, they have exploited the marine interstitial environment, particularly zones of very low oxygen concentration.

Summary

The Platyhelminthes (flatworms), the Nemertea (ribbon worms), and the Gnathostomulida (jaw worms) are the most primitive phyla that are bilaterally symmetrical, a condition of adaptive value for actively crawling or swimming animals. They have neither a coelom nor a pseudocoel and are thus acoelomate. They are triploblastic and at the organ-system level of organization.

The body surface of turbellarian flatworms is a cellular epithelium, at least in part ciliated, containing mucous cells and rod-shaped rhabdites which function together in locomotion. Members of all the other classes of flatworms are covered by a nonciliated, syncytial tegument with a vesicular distal cytoplasm and cell bodies beneath superficial muscle layers. Digestion is extracellular and intracellular in most; cestodes must absorb predigested nutrients across their tegument because they have no digestive tract. Osmoregulation is by flame-cell protonephridia, and removal of metabolic wastes and respiration occur across the body wall. Except for the primitive turbellarians, flatworms have a ladder-type nervous system with motor, sensory, and association neurons. Various types of sensory receptors are found in flatworms, including photosensitive ocelli in some. Most flatworms are hermaphroditic, and asexual reproduction occurs in some groups.

Members of the class Turbellaria are mostly free living and carnivorous. The Monogenea are important ectoparasites of fishes and have a direct life cycle (without intermediate hosts). These contrast with the digenetic trematodes, which have a mollusc intermediate host and almost always a vertebrate definitive host. The great amount of asexual reproduction that occurs in the intermediate host helps to increase the chances that some of the offspring will reach a definitive host. Aside from the tegument, digeneans share many basic structural characteristics with the Turbellaria. The Digenea includes a number of important parasites of humans and domestic animals.

Cestodes (tapeworms) generally have a scolex at their anterior end, followed by a long chain of proglottids, each of which contains a complete set of reproductive organs of both sexes. Cestodes live as adults in the digestive tract of vertebrates. They have microtriches on their tegument, which increase its surface area for absorption. The shelled larvae are passed in the feces, and the juveniles develop in a vertebrate or invertebrate intermediate host.

Members of the Nemertea have a complete digestive system with an anus and a true circulatory system. They are free living, mostly marine, and they capture their prey by ensnaring it with their long, eversible proboscis.

The Gnathostomulida are a curious phylum of little wormlike marine animals living among sand grains and silt. They have no anus, and they share certain characteristics with such widely diverse groups as turbellarians, rotifers, sponges, and cnidarians.

The flatworms and the cnidarians both probably evolved from a common ancestor (planuloid), some of whose descendants became sessile or free floating and radial, while others became creeping and bilateral. The more advanced flatworms were probably derived from an ancestor resembling the turbellarian order Acoela, whereas the nemerteans arose from platyhelminth stock.

REVIEW QUESTIONS

1. Why is bilateral symmetry of adaptive value for actively motile animals?
2. Match the terms in the right column with the classes in the left column:

______Turbellaria	a. Endoparasitic
______Monogenea	b. Free-living and commensal
______Trematoda	c. Ectoparasitic
______Cestoda	

3. Why is the Acoela considered the most primitive of the Turbellaria?
4. Briefly describe the body plan of flatworms.
5. How do flatworms digest their food?
6. Briefly describe the osmoregulatory system and the nervous system and sense organs of Platyhelminthes.
7. Contrast asexual reproduction in Turbellaria, Trematoda, and Cestoda.
8. Contrast the typical life cycle of Monogenea with that of a digenetic trematode.
9. Describe and contrast the tegument of trematodes and cestodes.
10. Answer the following questions with respect to both *Clonorchis* and *Schistosoma.* How do humans become infected? What is the general geographical distribution? What are the main disease conditions produced?
11. Define each of the following: scolex, microtriches, proglottids, strobila.
12. Give three ways in which nemerteans differ from platyhelminths.
13. Explain how a planuloid ancestor could have given rise to both the Cnidaria and the Platyhelminthes.
14. How would the acoeloid ancestor have differed from the planuloid ancestor?
15. Where do gnathostomulids live?

SELECTED REFERENCES

Arme, C., and P.W. Pappas (eds.). 1983. Biology of the Eucestoda, 2 vols. New York, Academic Press, Inc. *The most up-to-date reference available on the biology of tapeworms. Advanced; not an identification manual.*

Barnes, R.D. 1987. Invertebrate zoology, ed. 5. Philadelphia, Saunders College/Holt, Rinehart & Winston. *Accounts of the flatworms and nemerteans compose Chapters 7 and 8, and the gnathostomids are covered in Chapter 9.*

Desowitz, R.S. 1981. New Guinea tapeworms and Jewish grandmothers. New York, W.W. Norton & Co. *Accounts of parasites and parasitic diseases of humans. Entertaining and instructive. Recommended for all students.*

Schmidt, G.D., and L.S. Roberts. 1989. Foundations of parasitology, ed. 4. St. Louis, The C.V. Mosby Co. *Good accounts of the form and function of trematodes and cestodes; emphasizes species of medical and veterinary importance.*

Smyth, J.D. and D.W. Halton. 1983. The physiology of trematodes, ed. 2. Cambridge, England, Cambridge University Press. *Thorough treatment of trematode physiology.*

Strickland, G.T. 1984. Hunter's tropical medicine, ed. 6. Philadelphia. W.B. Saunders Co. *A valuable source of information on parasites of medical importance.*

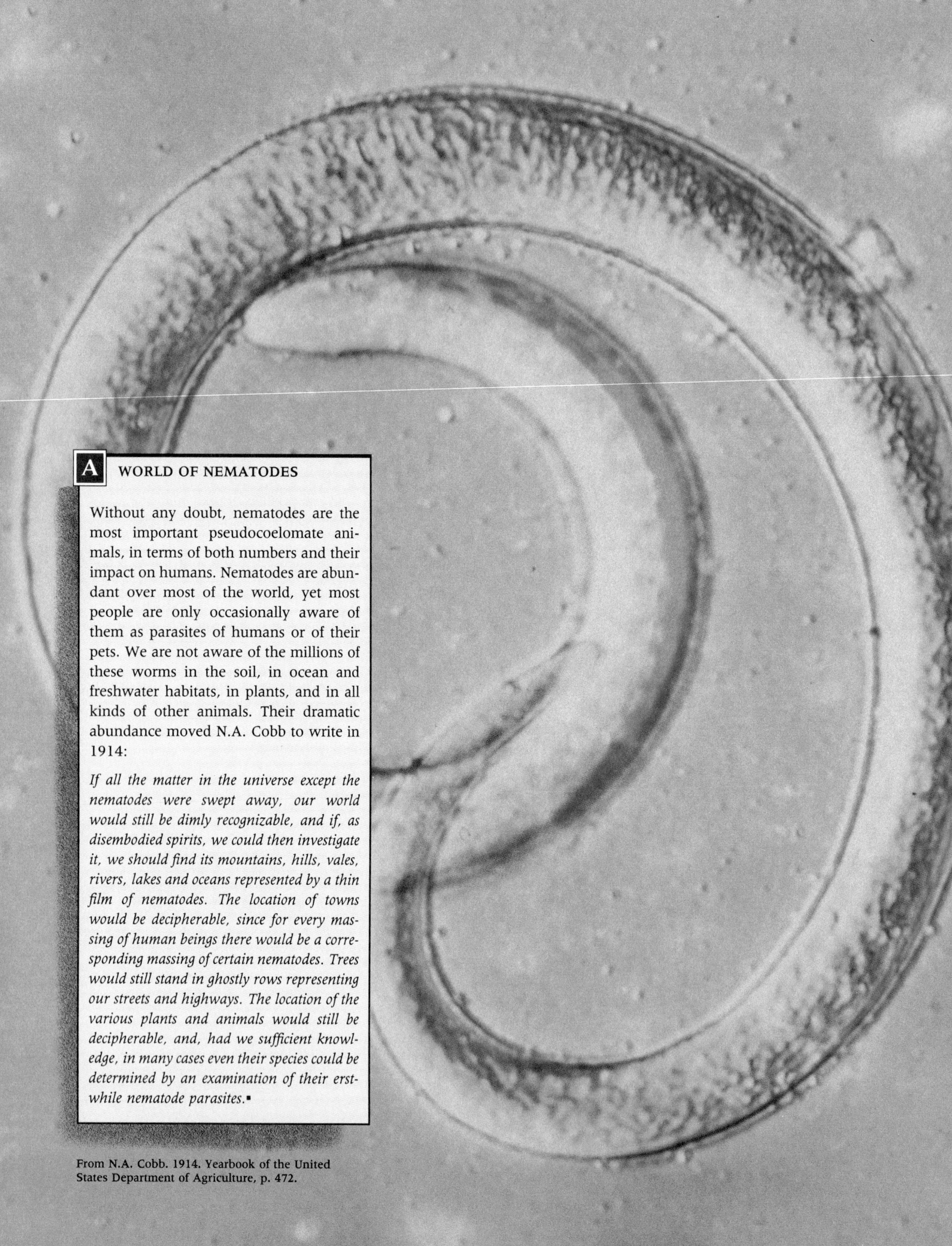

A WORLD OF NEMATODES

Without any doubt, nematodes are the most important pseudocoelomate animals, in terms of both numbers and their impact on humans. Nematodes are abundant over most of the world, yet most people are only occasionally aware of them as parasites of humans or of their pets. We are not aware of the millions of these worms in the soil, in ocean and freshwater habitats, in plants, and in all kinds of other animals. Their dramatic abundance moved N.A. Cobb to write in 1914:

If all the matter in the universe except the nematodes were swept away, our world would still be dimly recognizable, and if, as disembodied spirits, we could then investigate it, we should find its mountains, hills, vales, rivers, lakes and oceans represented by a thin film of nematodes. The location of towns would be decipherable, since for every massing of human beings there would be a corresponding massing of certain nematodes. Trees would still stand in ghostly rows representing our streets and highways. The location of the various plants and animals would still be decipherable, and, had we sufficient knowledge, in many cases even their species could be determined by an examination of their erstwhile nematode parasites.▪

From N.A. Cobb. 1914. Yearbook of the United States Department of Agriculture, p. 472.

CHAPTER 19

Pseudocoelomate Animals

Vertebrates and higher invertebrates have a true **coelom,** or peritoneal cavity, which is formed within the mesoderm during embryonic development and is therefore lined with a layer of mesodermal epithelium, the **peritoneum** (Figure 19-1). The pseudocoelomate phyla have a **pseudocoel** rather than a true coelom. It is derived from the embryonic blastocoel (p. 311) rather than from a secondary cavity within the mesoderm. It is a space not lined with peritoneum, between the gut and the mesodermal and ectodermal components of the body wall. A fluid-filled space surrounding the internal organs may serve several functions, such as distribution of nutrients, removal of wastes, and space to store gametes during their maturation. A very important function of such a space in some animals is to serve as a hydrostatic skeleton (p. 159).

Eight distinct phyla of animals belong to the pseudocoelomate category. These are Rotifera, Gastrotricha, Kinorhyncha, Loricifera, Nematoda, Nematomorpha, Acanthocephala, and Entoprocta. They are a heterogeneous assemblage of animals. Most of them are small; some are microscopic; some are fairly large. Some, such as the nematodes, are found in freshwater, marine, terrestrial, and parasitic habitats; others, such as the Acanthocephala, are strictly parasitic. Some have unique characteristics such as the lacunar system of the acanthocephalans and the ciliary corona of the rotifers.

Even in such a diversified grouping, however, there are a few characteristics in common. In all there is a body wall of epidermis (often syncytial) and muscles surrounding the pseudocoel. The digestive tract is complete (except in Acanthocephala), and it, along with the gonads and excretory organs, is within the pseudocoel and bathed in perivisceral fluid. The epidermis in many secretes a nonliving cuticle with some specializations such as bristles, spines, and the like.

A constant number of cells or nuclei in the individuals of a species, a condition known as **eutely,** is common in several of the groups, and in most of them there is an emphasis on the longitudinal muscle layer.

FIGURE 19-1

Acoelomate, pseudocoelomate, and eucoelomate body plans.

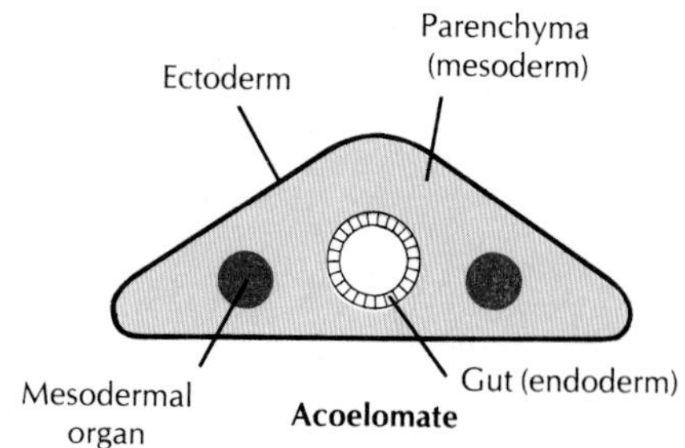

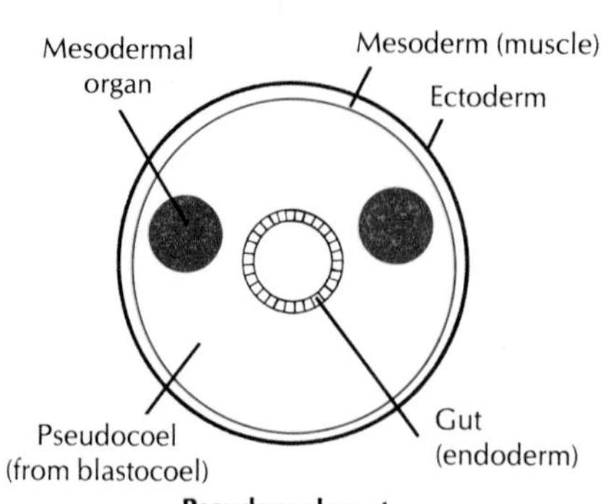

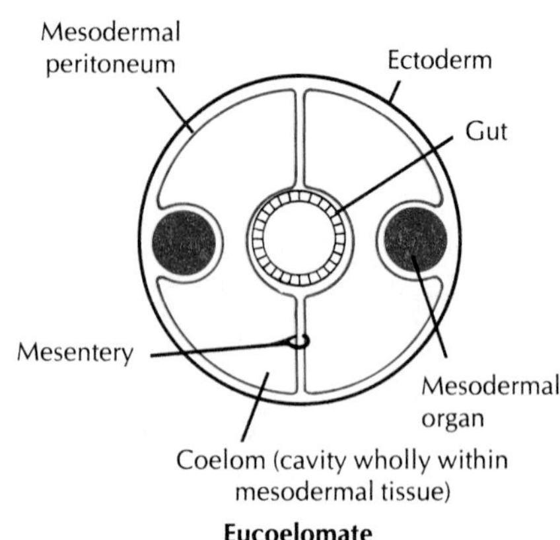

Phylum Rotifera

Rotifera (ro-tif'e-ra) (L. *rota,* wheel, + *-fera,* those which bear) derive their name from the characteristic ciliated crown, or **corona,** that, when beating, often gives the impression of rotating wheels. Rotifers range in size from 40 μm to 3 mm in length, but most are between 100 and 500 μm long. Some have beautiful colors, although most are transparent, and some have odd and bizarre shapes. Their shapes are often correlated with their mode of life. The floaters are usually globular and saclike; the creepers and swimmers are somewhat elongated and wormlike; and the sessile types are commonly vaselike, with a cuticular envelope. Some are colonial.

Rotifers are a cosmopolitan group of about 2000 species, some of which are found throughout the world. Most of the species are freshwater inhabitants, but a few are marine; some are terrestrial, and some are epizoic (live on the surface of other animals) or parasitic. Most often they are benthic, occurring on the bottom or in vegetation of ponds or along the shores of large freshwater lakes where they swim or creep about on the vegetation.

The body is usually made up of a **head, trunk,** and **foot.** The head region bears the corona. The corona may form a ciliated funnel with its upper edges folded into lobes bearing bristles, or the corona may be made up of a pair of ciliated discs (Figure 19-2). The cilia create currents of water toward the mouth that draw in small planktonic forms for food. The corona may be retractile. The mouth, surrounded by some part of the corona, opens into a modified muscular pharynx called a **mastax,** which is a unique characteristic of the rotifers. The mastax is equipped with a set of intricate jaws used for grasping and chewing. The trunk contains the visceral organs, and the terminal foot, when present, is segmented and, in some, is ringed to joints that can telescope to shorten. The one to four toes secrete a sticky substance from the **pedal glands** for attachment.

Rotifers have a pair of protonephridial tubules with flame bulbs, which eventually drain into a **cloacal bladder** that collects excretory and digestive wastes. They have a bilobed "brain" and sense organs that include eyespots, sensory pits, and papillae.

Rotifers are dioecious. The males, however, are few in number and with few exceptions are degenerate, having neither mouth nor digestive organs.

Members of one family reproduce parthenogenetically; that is, females produce only diploid eggs that have not undergone reduction division, cannot be fertilized, and develop only into females. Such eggs are called **amictic eggs.** Most other rotifers can produce two kinds of eggs—amictic eggs, which develop parthenogenetically into females, and **mictic eggs,** which have undergone meiosis and are haploid. Mictic eggs, if unfertilized, develop quickly and parthenogenetically into males; if fertilized, they secrete a thick shell and become dormant for

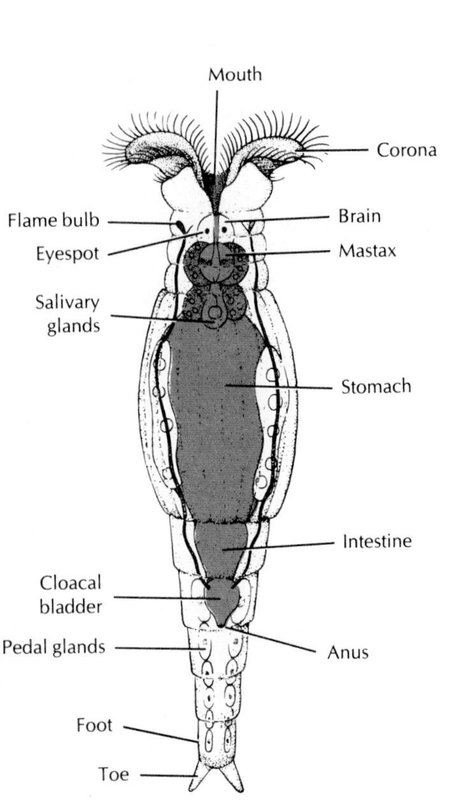

FIGURE 19-2

Structure of *Philodina,* a common rotifer.

several months before developing into females. Such dormant eggs can withstand desiccation and other adverse conditions and permit rotifers to live in temporary ponds that dry up during certain seasons.

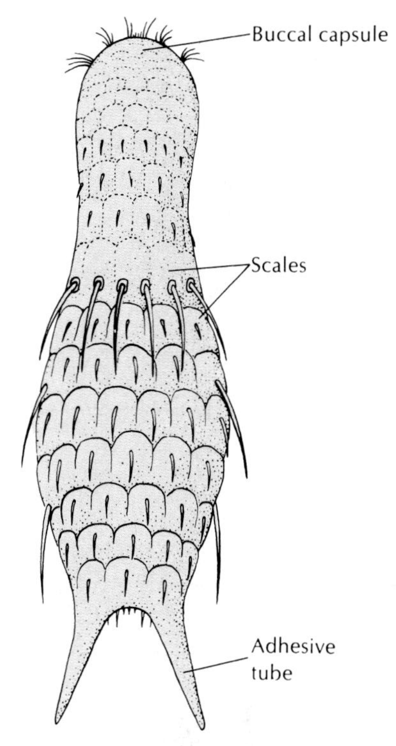

FIGURE 19-3

Chaetonotus, a common gastrotrich.

Phylum Gastrotricha

The phylum Gastrotricha (gas-trot're-ka) (Gr. *gaster,* belly, + *thrix,* hair) is a small group (about 400 species) of microscopic animals, approximately 65 to 500 μm long (Figure 19-3). They are usually bristly or scaly in appearance, flattened on the ventral side, and many move by gliding on ventral cilia. Others move in a leechlike fashion by briefly attaching the posterior end by means of adhesive glands. There are both marine and freshwater species, and they are common in lakes, ponds, and seashore sands. They feed on bacteria, diatoms, and small protozoa.

Phylum Kinorhyncha

There are only about 100 species of kinorhynchs (Kinorhyncha [Gr.*kineo,* to move, + *rhynchos,* beak or snout]). They are tiny marine worms, usually less than 1 mm long, that prefer mud bottoms. They have no external cilia, but their cuticle is divided into 13 segments (Figure 19-4). Kinorhynchs burrow into the mud by extending the head, anchoring it by its recurved spines(**scalids**), and drawing the body forward until the head is retracted. They feed on organic sediment in the mud, and some feed on diatoms.

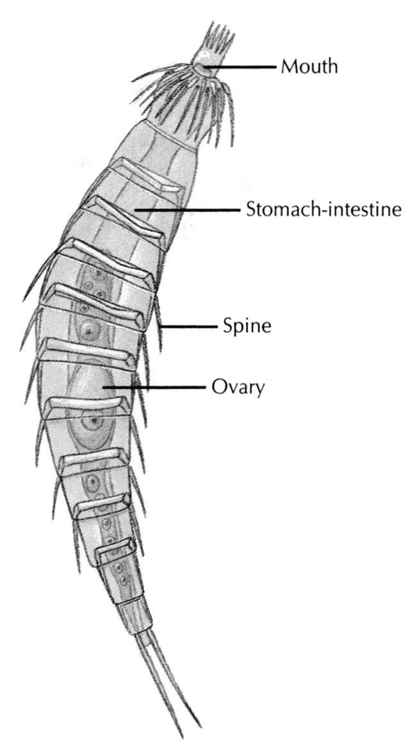

FIGURE 19-4

Echinoderes, a kinorhynch, is a minute marine worm. Segmentation is superficial. Head with its circle of spines is retractile.

Phylum Loricifera

The Loricifera (L. *lorica,* corselet, + Gr., *phora,* bearing) are the most recently described phylum of animals (1983). They live in the spaces between the grains of marine gravel, to which they cling tightly. Though they were described from specimens collected off the coast of France, they are apparently widely distributed in the world. They have oral styles and scalids rather similar to those of the kinorhynchs, and the entire forepart of the body can be retracted into the circular **lorica** (Figure 19-5).

Phylum Nematoda — Roundworms

Nematodes, (Gr., *nēmatos,* thread) are present in nearly every conceivable kind of ecological niche. Approximately 15,000 species have been named, but the American zoologist Libbie Hyman estimated that, if all species were known, the number would be nearer 500,000. They live in the sea, in fresh water, and in soil, from polar regions to the tropics, and from mountaintops to the depths of the sea. Good topsoil may contain billions of nematodes per acre. Nematodes also parasitize virtually every type of animal and many plants. The effects of nematode infestation on crops, domestic animals, and humans make this phylum one of the most important of all parasitic animal groups.

CHARACTERISTICS

1. Body bilaterally symmetrical, cylindrical in shape
2. Body covered with a secreted, flexible, nonliving **cuticle**
3. Motile **cilia and flagella completely lacking;** some sensory endings derived from cilia present
4. **Muscles in body wall running in longitudinal direction** only
5. Excretory system of either one or more **gland cells** opening by an excre-

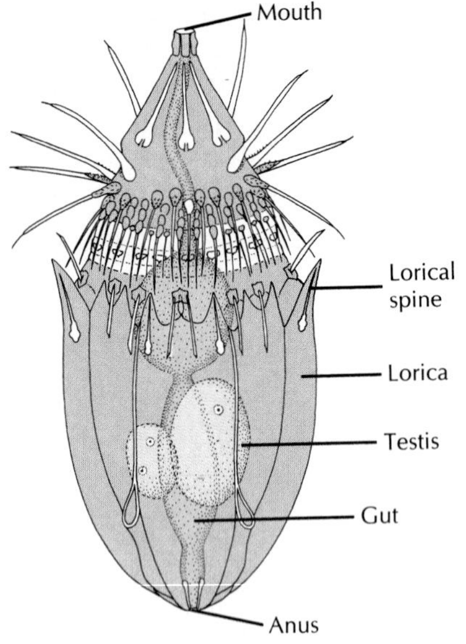

FIGURE 19-5

Dorsal view of adult loriciferan, *Nanaloricus mysticus*.

Redrawn from Kristensen, R.M. 1983. Loricifera, a new phylum with Aschelminthes characters from the meiobenthos. Zeitsch. zool. Syst. Evol. 21:163.

tory pore, a **canal system** without gland cells, or both gland cells and canals together; **flame cell protonephridia lacking**

6. **Pharynx** usually muscular and **triradiate** in cross section
7. Male reproductive tract opening into rectum to form a cloaca; female reproductive tract opening a separate gonopore
8. Fluid in pseudocoel enclosed by cuticle forming a **hydrostatic skeleton**

FORM AND FUNCTION

Most nematodes are under 5 cm long, and many are microscopic, but some parasitic nematodes are over a meter in length.

Use of the pseudocoel as a hydrostatic organ is highly developed, and much of the functional morphology of the nematodes can be best understood in the context of the high **hydrostatic pressure** in the pseudocoel.

The outer body covering is a relatively thick, noncellular **cuticle,** secreted by the underlying epidermis. The cuticle is of great functional importance to the worm, serving to contain the high hydrostatic pressure exerted by the fluid in the pseudocoel. The several layers of the cuticle are primarily of **collagen,** a structural protein also abundant in vertebrate connective tissue. Three of the layers are comprised of criss-crossing fibers, which confer some longitudinal elasticity on the worm but severely limit its capacity for lateral expansion.

Beneath the epidermis is a layer of **longitudinal muscles.** There are no circular muscles in the body wall. The muscles are arranged in four bands, or quadrants, marked off by four epidermal cords that project inward to the pseudocoel.

The fluid-filled pseudocoel, in which the internal organs lie, constitutes a hydrostatic skeleton. Hydrostatic skeletons, found in many invertebrates, lend support by transmitting the force of muscle contraction to the enclosed, noncompressible fluid. Normally, muscles are arranged antagonistically, so that movement is effected by contraction of one group of muscles and relaxation of the other. However, nematodes do not have circular body wall muscles to antagonize the longitudinal muscles; therefore, the cuticle must serve that function. Compression of the cuticle on the side of the muscular contraction and stretching on the opposite side are the forces that return the body to resting position when the muscles relax; this produces the characteristic thrashing motion seen in nematode movement. An increase in efficiency of this system can only be achievedby an increase in hydrostatic pressure. Consequently, the hydrostatic pressure in the nematode pseudocoel is much higher than is usually found in other kinds of

FIGURE 19-6

Structure of a nematode as illustrated by *Ascaris* female. *Ascaris* has two ovaries and uteri, which open to the outside by a common genital pore.

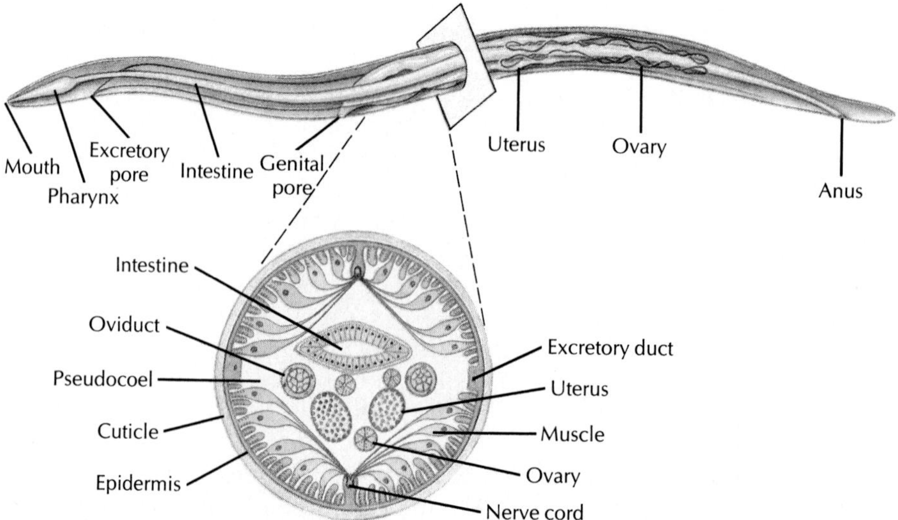

Table 19-1 Common parasitic nematodes of humans in North America

Common and scientific names	Mode of infection; prevalence
Hookworm *(Ancylostoma duodenale* and *Necator americanus)*	Contact with juveniles in soil that burrow into skin; common in southern states
Pinworm *(Enterobius vermicularis)*	Inhalation of dust containing ova and by contamination with fingers; most common parasite in United States
Intestinal roundworm *(Ascaris lumbricoides)*	Ingestion of embryonated ova in contaminated food; common in rural areas of Appalachia and southeastern states
Trichina worm *(Trichinella spiralis)*	Ingestion of infected pork muscle; occasional in humans throughout North America
Whipworm *(Trichuris trichiura)*	Ingestion of contaminated food or by unhygienic habits; usually common wherever *Ascaris* is found

animals with hydrostatic skeletons but that also have antagonistic muscle-groups.

The alimentary canal of the nematode consists of a **mouth** (Figure 19-6), a muscular **pharynx,** a long nonmuscular **intestine,** a short **rectum,** and a terminal **anus.** Food material is sucked into the pharynx when the muscles in in its anterior portion contract rapidly and open the lumen. Relaxation of the muscles anterior to the food mass closes the lumen of the pharynx, forcing the food posteriorly toward the intestine. The intestine is one cell layer thick. Food matter is moved posteriorly by body movements and by additional food being passed into the intestine from the pharynx. Defecation is accomplished by muscles that simply pull the anus open, and the expulsive force is provided by pseudocoelomic pressure.

The adults of many parasitic nematodes have an anaerobic energy metabolism; thus, a citric acid cycle and cytochrome system characteristic of aerobic metabolism are absent. Energy is derived through glycolysis and probably through some incompletely known electron transport sequences. Interestingly, some free-living nematodes and free-living stages of parasitic nematodes are obligate aerobes (require oxygen) and have a citric acid cycle and cytochrome system.

A **ring of nerve tissue** and ganglia around the pharynx gives rise to small nerves to the anterior end and to two **nerve cords,** one dorsal and one ventral. Some sensory organs are elaborately developed around the lips and around the posterior end.

Most nematodes are dioecious. The male is smaller than the female, and its posterior end usually bears a pair of copulatory spicules. Fertilization is internal, and eggs are usually stored in the uterus until deposition. After embryonation a juvenile worm hatches from the egg. There are four juvenile stages, each separated by a molt, or shedding, of the cuticle. Many parasitic nematodes have free-living juvenile stages, and others require an intermediate host to complete their life cycles.

NEMATODE PARASITES OF HUMANS

Nearly all vertebrates and many invertebrates are parasitized by nematodes. A number of these are very important pathogens of humans and domestic animals. A few nematodes are common in humans in North America (Table 19-1), but

Other ascarids are common in wild and domestic animals. Species of *Toxocara*, for example, are found in dogs and cats. Their life cycle is generally similar to that of *Ascaris*, but the juveniles often do not complete their tissue migration in adult dogs, remaining in the host's body in a stage of arrested development. Pregnancy in the female, however, stimulates the juveniles to wander, and they infect the embryos in the uterus. The puppies are then born with worms. These ascarids also survive in humans but do not complete their development, leading to an occasionally serious condition in children known as *visceral larva migrans*. This is a good argument for pet owners to practice hygienic disposal of canine wastes!

they and many others usually abound in tropical countries. Only a few will be mentioned in this discussion.

Ascaris lumbricoides—the large roundworm of humans

The large human roundworm (Gr. *askaris*, intestinal worm) is one of the most common worm parasites of humans (Figures 19-6 and 19-7). Recent surveys have shown a prevalence of up to 64% in some areas of the southeastern United States, and it is estimated that 1.26 billion people are infected worldwide. *Ascaris suum* (Figure 19-8), found in pigs, is morphologically similar and was long considered the same species. Females of both are up to 30 cm in length and can produce 200,000 eggs a day. The adults live in their host's small intestine, and the eggs pass out in the feces. They are extremely resistant to adverse conditions other than direct sunlight and high temperatures and can survive for months or years in the soil.

When embryonated eggs are eaten with uncooked vegetables or when children put soiled fingers or toys in their mouths, the tiny juveniles hatch in the intestine. They penetrate the intestinal wall and are carried through the heart in the blood to the lungs, where they break out into the alveoli. They may cause a serious pneumonia at this stage. From the alveoli, the juveniles make their way up the bronchi, trachea, and pharynx, to be swallowed and finally reach the intestine again and grow to maturity. In the intestine the worms cause abdominal symptoms and allergic reactions, and in large numbers they may cause intestinal blockage.

Hookworms

Hookworms are so named because the anterior end curves dorsally, suggesting a hook. The most common species is *Necator americanus* (L. *necator*, killer), whose females are up to 11 mm long. The males can reach 9 mm in length. Large plates in their mouths (Figure 19-9) cut into the intestinal mucosa of the host where they suck blood and pump it through their intestines, partially digesting it

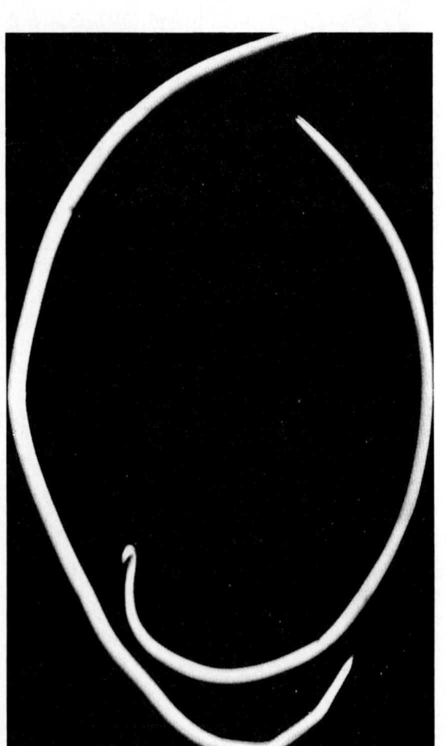

FIGURE 19-7

Intestinal roundworm *Ascaris lumbricoides*, male and female. Male *(right)* is smaller and has characteristic sharp kink in the end of the tail. The females of this large nematode may be over 30 cm long.

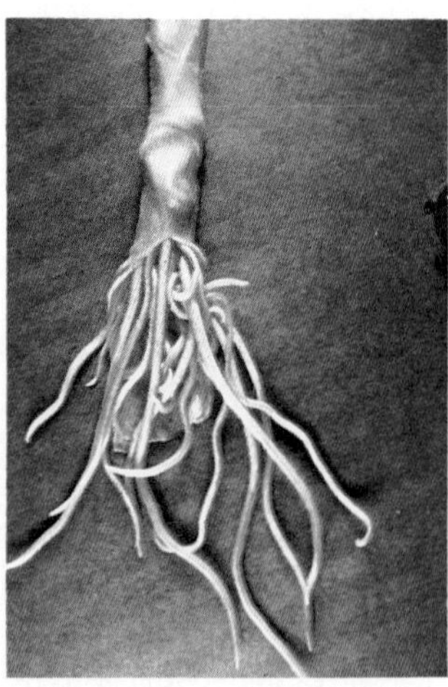

FIGURE 19-8

Intestine of a pig, nearly completely blocked by *Ascaris suum*. Such heavy infections are also fairly common with *A. lumbricoides* in humans.

Original photograph courtesy G.W. Kelley, Jr. From H. Zaiman (ed.). A pictorial presentation of parasites.

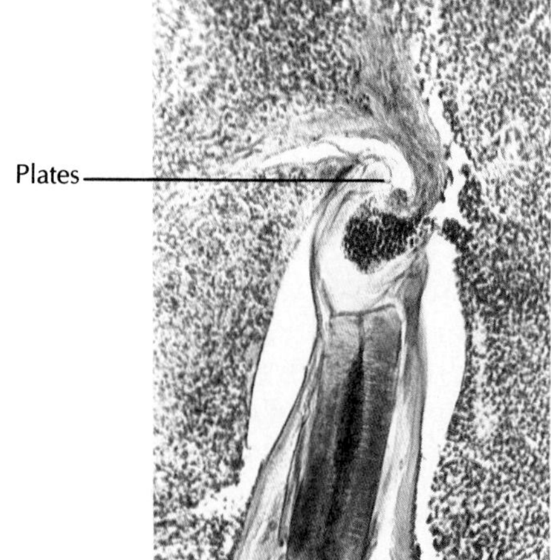

FIGURE 19-9

Section through anterior end of hookworm attached to dog intestine. Note cutting plates of mouth pinching off bit of mucosa from which the thick muscular pharynx sucks blood. Esophageal glands secrete an anticoagulant to prevent blood clotting.

From H. Zaiman (ed.). A pictorial presentation of parasites.

and absorbing the nutrients. They suck much more blood than they need for food, and heavy infections cause anemia in the patient. Hookworm disease in children may result in retarded mental and physical growth and general loss of energy.

Eggs are passed in the feces, and the juveniles hatch in the soil, where they live on bacteria. When human skin comes in contact with infested soil, the juveniles burrow through the skin to the blood, and reach the lungs and finally the intestine in a manner similar to that described for *Ascaris.*

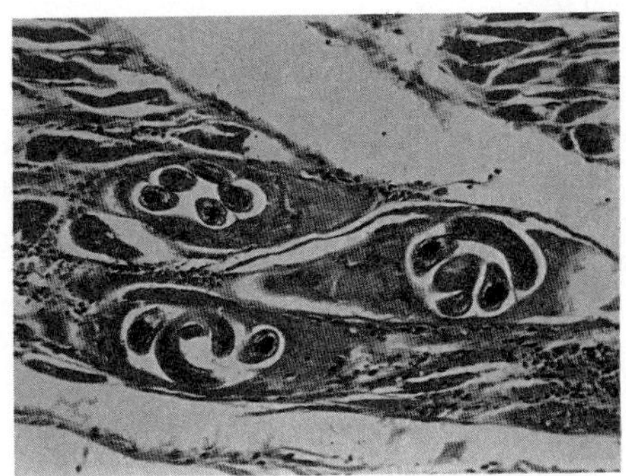

FIGURE 19-10

Muscle infected with trichina worm *Trichinella spiralis,* human case. Note degeneration of muscle fibers and inflammatory reaction. Juveniles may live 10 to 20 years in these cysts. If eaten in poorly cooked meat, the juveniles are liberated in the intestine. They quickly mature and release many young into the blood of the host.

Original photograph courtesy H. Zaiman. From H. Zaiman (ed.). A pictorial presentation of parasites.

Trichina worm

Trichinella spiralis (Gr. *trichinos,* of hair, + *-ella,* diminutive) is the tiny nematode responsible for the potentially lethal disease **trichinosis.** Adult worms burrow in the mucosa of the small intestine where the female produces living young. The juveniles penetrate into blood vessels and are carried to the skeletal muscles where they coil up within a cyst that eventually becomes calcified (Figure 19-10). When meat containing cysts with live worms is swallowed, the juveniles are liberated into the intestine where they mature.

In addition to humans, *Trichinella spiralis* can infect many other mammals, including hogs, rats, cats, and dogs. Humans most often acquire the parasite by eating improperly cooked pork. Hogs become infected by eating garbage containing pork scraps with cysts or by eating infected rats.

Heavy infections may cause death, but lighter infections are much more common—about 2.4% of the population of the United States is infected.

Pinworms

The pinworm, *Enterobius vermicularis* (Gr. *enteron,* intestine, + *bios,* life), causes relatively little disease but is the most common helminth parasite in the United States, estimated at 30% in children and 16% in adults. The adults (Figure 19-11, *A*) live in the large intestine and cecum. The females, up to about 12 mm in length, migrate to the anal region at night to lay their eggs (Figure 19-11, *B*). Scratching the resultant itch effectively contaminates the hands and bedclothes. Eggs develop rapidly and become infective within 6 hours at body temperature. When they are swallowed, they hatch in the duodenum, and the worms mature in the large intestine.

Diagnosis of most intestinal roundworms is usually made by examination of a small bit of feces under the microscope and finding characteristic eggs. However, pinworm eggs are not often found in the feces because the female deposits them on the skin around the anus. The "Scotch tape method" is more effective. The sticky side of cellulose tape is applied around the anus to collect the eggs, then the tape is placed on a glass slide and examined under the microscope. Several drugs are effective against this parasite, but all members of a family should be treated at the same time, since the worm easily spreads through a household.

Filarial worms

At least eight species of filarial nematodes infect humans, and some of these are major causes of diseases. Some 250 million people in tropical countries are infected with *Wuchereria bancrofti* (named for Otto Wucherer) or *Brugia malayi* (named for S.L. Brug), which places these species among the scourges of humanity. The worms live in the lymphatic system, and the females are as long as 100 mm. The disease symptoms are associated with inflammation and obstruction of the lymphatic system. The females release live young, the tiny microfilariae, into

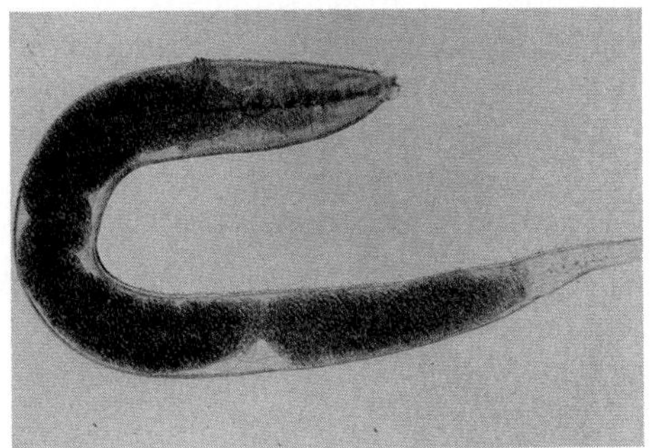

A

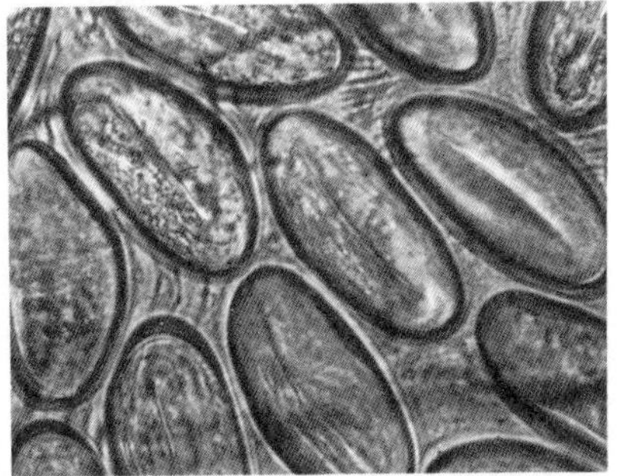

B

FIGURE 19-11

Pinworms, *Enterobius vermicularis.* **A,** Female worm from human large intestine (slightly flattened in preparation), magnified about 20 times. **B,** Group of pinworm eggs, which are usually discharged at night around the anus of the host, who, by scratching during sleep, may get fingernails and clothing contaminated. This may be the most common and widespread of all human helminth parasites.

A, Photograph by R. Calentine/Visuals Unlimited; B, courtesy H. Zaiman. From H. Zaiman (ed.). A pictorial presentation of parasites.

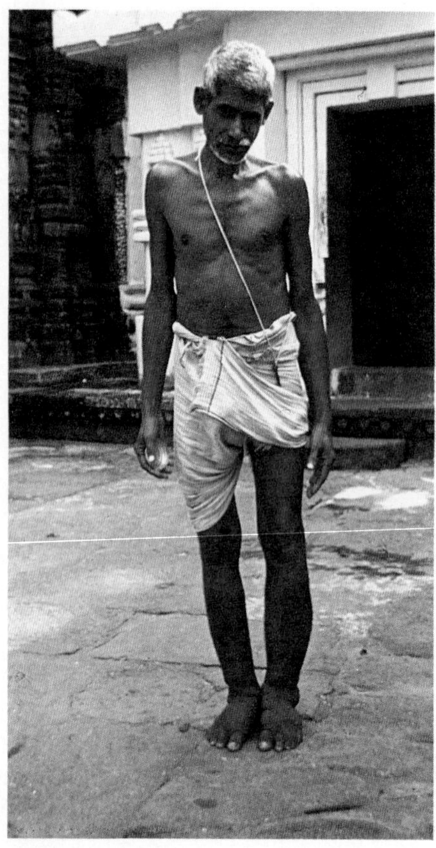

FIGURE 19-12

Elephantiasis of leg caused by adult filarial worms of *Wuchereria bancrofti,* which live in lymph passages and block the flow of lymph. Tiny juveniles, called microfilariae, are picked up with blood meal of mosquitoes, where they develop to infective stage and are transmitted to a new host.

Original photograph courtesy M.G. Schultz. From H. Zaiman (ed.). A pictorial presentation of parasites.

the blood and lymph. The microfilariae are picked up by mosquitoes as the insects feed, and they develop in the mosquitoes to the infective stage. They escape from the mosquito when it is feeding again on a human and penetrate the wound made by the mosquito bite.

The dramatic manifestations of elephantiasis are occasionally produced after long and repeated exposure to the worms. The condition is marked by an excessive growth of connective tissue and enormous swelling of affected parts, such as the scrotum, legs, arms, and more rarely, the vulva and breasts (Figure 19-12).

Another filarial worm causes river blindness and is carried by the blackfly. It infects more than 30 million people in parts of Africa, Arabia, Central America, and South America.

■ Phylum Nematomorpha

The Nematomorpha (nem'a-to-mor'fa) (Gr. *nema,* thread, + *morphe,* form) (about 100 species) are sometimes called "horsehair worms," based on an old superstition that the worms arise from horsehairs that happen to fall into the water. Certainly, their appearance suggests a horsehair because they are quite long and slender (5 to 100 cm long, by only 1 to 3 mm diameter). They were long included with the nematodes, with which they share several characteristics, but they differ morphologically from nematodes in some important respects.

Worldwide in distribution, horsehair worms are free living as adults and parasitic in arthropods as juveniles. Adults have a vestigial digestive tract and do not feed, but they will live almost anywhere in wet or moist surroundings if oxygen is adequate. Juveniles do not emerge from the arthropod host unless there is water nearby, and adults are often seen wriggling slowly about in ponds or streams. Juveniles of freshwater forms use various terrestrial insects as hosts, while marine forms use certain crabs.

■ Phylum Acanthocephala

The members of the phylum Acanthocephala (a-kan'tho-sef'a-la) (Gr. *akantha,* spine or thorn, + *kephale,* head) are commonly known as "spiny-headed worms." The phylum derives its name from one of its most distinctive features, a cylindrical invaginable **proboscis** (Figure 19-13) bearing rows of recurved spines, by which the worms attach themselves to the intestine of their hosts. All acanthocephalans are endoparasitic, living as adults in the intestines of vertebrates.

Over 500 species are known, most of which parasitize fish, birds, and mammals, and the phylum is worldwide in distribution. However, no species is normally a parasite of humans, although rarely humans are infected with species that usually occur in other hosts.

Various species range in size from less than 2 mm to over 1 m in length, with the females of a species usually being larger than the males. In life, the body is usually bilaterally flattened, with numerous transverse wrinkles. The worms are usually cream color but may be yellowish or brown from absorption of pigments from the intestinal contents.

The body wall is syncytial, and its surface is punctured by minute crypts 4 to 6 μm deep, which greatly increase the surface area of the tegument. About 80% of the thickness of the tegument is the radial fiber zone, which contains a **lacunar system** of ramifying fluid-filled canals (Figure 19-13). The function of the lacunar system is unclear, but it may serve in distribution of nutrients to the peculiar, tube-like muscles in the body wall of these organisms.

Excretion is across the body wall in most species. In one family there is a pair

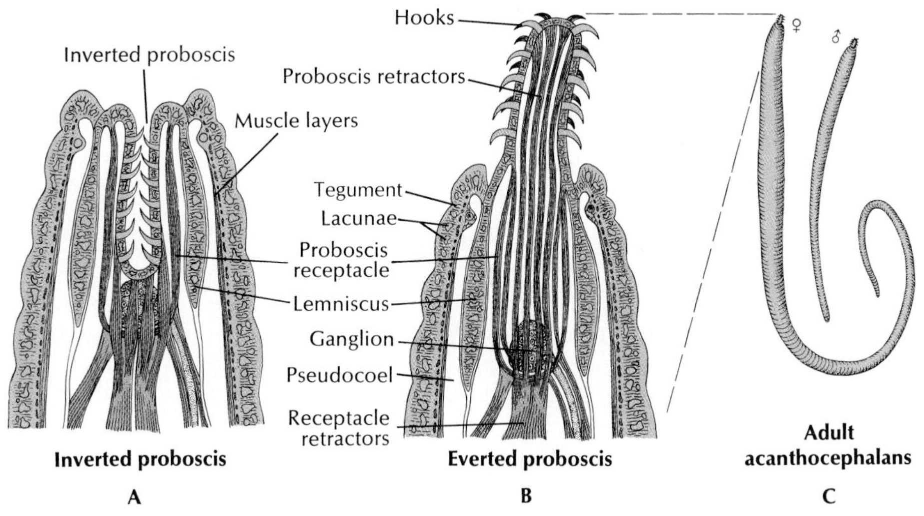

FIGURE 19-13

Structure of a spiny-headed worm (phylum Acanthocephala). **A** and **B,** Eversible spiny proboscis by which the parasite attaches to the intestine of the host, often doing great damage. Since they lack a digestive tract, food is absorbed through the tegument. **C,** Male is typically smaller than female.

of **protonephridia** with flame cells that unite to form a common tube that opens into the sperm duct or uterus.

Since acanthocephalans have no digestive tract, they must absorb all nutrients through their tegument. They can absorb various molecules by specific membrane transport mechanisms, and their tegument can carry out pinocytosis.

Sexes are separate. The male has a protrusible penis, and at copulation the sperm travel up the genital duct and escape into the pseudocoel of the female. The zygotes develop into shelled acanthor larvae. The shelled larvae escape from the vertebrate host in the feces, and, if eaten by a suitable arthropod, they hatch and work their way into the hemocoel where they grow to juvenile acanthocephalans. Either development ceases until the arthropod is eaten by a suitable host, or they may pass through several transport hosts in which they encyst until eaten.

Phylum Entoprocta

Entoprocta (en'to-prok'ta) (Gr. *entos,* within, + *proktos,* anus) is a small phylum of less than 100 species of tiny, sessile animals that, superficially, look much like hydroid cnidarians, but their tentacles are ciliated and tend to roll inward (Figure 19-14). Most entoprocts are microscopic, and none is more than 5 mm long. They are all stalked and sessile forms; some are colonial, and some are solitary. All are ciliary feeders.

With the exception of one genus all entoprocts are marine forms that have a wide distribution from the polar regions to the tropics. Most marine species are restricted to coastal and brackish waters and often grow on shells and algae. Some are commensals on marine annelid worms. Freshwater entoprocts (Figure 19-14) occur on the underside of rocks in running water.

The body of the entoproct is cup shaped, bears a crown, or circle, of ciliated tentacles, and is attached to a substratum by a stalk in solitary species. In colonial species, the stalks of the individuals are connected to the stolon of the colony. The tentacles and stalk are continuations of the body wall.

The gut is U shaped and ciliated, and both the mouth and anus open within the circle of tentacles. Feeding is accomplished by capturing food particles in the current created by the tentacular cilia, then passing the particles along the tentacles to the mouth. Entoprocts have a pair of protonephridia but no circulatory or respiratory organs.

Some species are monoecious, some dioecious, and some appear to be pro-

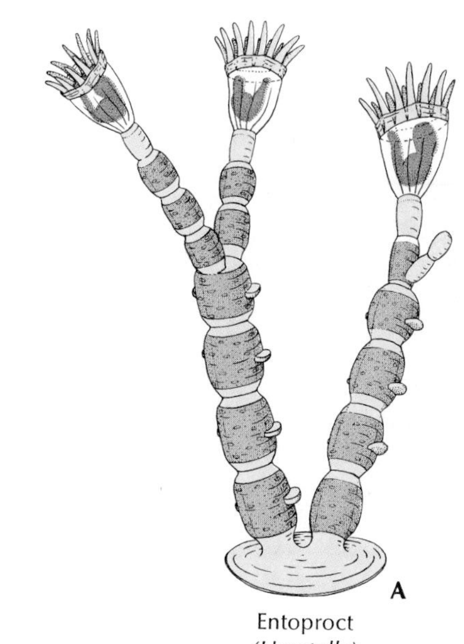

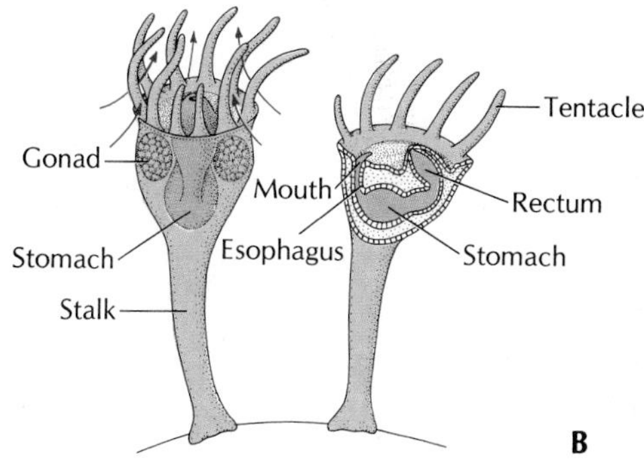

FIGURE 19-14

A, *Urnatella,* a freshwater entoproct, forms small colonies of two or three stalks from a basal plate. **B,** *Loxosomella,* a solitary entoproct. Both solitary and colonial enteroprocts can reproduce asexually by budding, as well as sexually.

A, Modified from Con, C. 1936. Kamptozoa. In H.G. Bronn (ed.). Klassen und Ordnungen des Tier-Reichs, vol. 4, part 2. Leipzig, Akademische Verlagsgesselschaft.

tandric; that is, the gonad at first produces sperm and then eggs. Cleavage is modified spiral and determinate (mosaic), and a trochophore-like larva is produced, similar to that of some molluscs and annelids.

Phylogeny and Adaptive Radiation

PHYLOGENY

The phylogenetic relationships of the Acanthocephala and the Entoprocta are problematic, and they are probably distant from the other phyla covered in this chapter. Except for the Loricifera (which were then undiscovered), the remaining pseudocoelomate phyla were grouped into a single phylum by Hyman. All of these phyla share a certain combination of characteristics, including the fact that they are usually wormlike, have a cuticle secreted by an epidermis with underlying muscles that are not arranged in regular circular and longitudinal layers, have a brain that is a nerve ring around the digestive tract, have mosaic cleavage and eutely, and lack a muscle layer in the intestine.

Hyman contended that the evidences of relationships were so concrete and specific that they could not be disregarded. Nevertheless, most authors now consider that differences between the groups are sufficient to merit phylum status for each. These phyla may well have derived originally from the protostome line by way of an acoelomate common ancestor resembling some of the primitive flatworms. The closest relatives of the loriciferans seem to be the kinorhynchs.

ADAPTIVE RADIATION

Certainly, the most impressive adaptive radiation in this group of phyla is shown in the nematodes. They are by far the most numerous in terms of both individuals and species, and they have been able to adapt to almost every habitat available to animal life. Their basic pseudocoelomate body plan, with the cuticle, hydrostatic skeleton, and longitudinal muscles, has proved generalized and plastic enough to adapt to an enormous variety of physical conditions. Free-living lines gave rise to parasitic forms on at least several occasions, and virtually all potential hosts have been exploited. All types of life cycle occur—from the simple and direct to the complex, with intermediate hosts; from normal dioecious reproduction to parthenogenesis, hermaphroditism, and alternation of free-living and parasitic generations. A major factor contributing to the evolutionary opportunism of the nematodes has been their extraordinary capacity to survive conditions suboptimal for viability.

Summary

The phyla covered in this chapter possess a body cavity called a pseudocoel, which is derived from the embryonic blastocoel, rather than a secondary cavity in the mesoderm (coelom). Several of the groups exhibit eutely, a constant number of cells or nuclei in adult individuals of a given species.

The phylum Rotifera is comprised of small, mostly freshwater organisms with a ciliated corona, which creates currents of water to draw planktonic food toward the mouth. The mouth opens into a muscular pharynx, or mastax, that is equipped with jaws.

The Gastrotricha, Kinorhyncha, and Loricifera are small phyla of tiny, aquatic pseudocoelomates. Gastrotrichs move by cilia or adhesive glands, and kinorhynchs anchor and then pull themselves by the spines on their head. Loriciferans can withdraw their bodies into the lorica.

By far the largest and most important of this group of phyla are the nematodes, of which there may be as many as 500,000 species in the world. They are more or less cylindrical, tapering at the ends, and covered with a tough, secreted cuticle. The body-wall muscles are longitudinal only, and to function well in

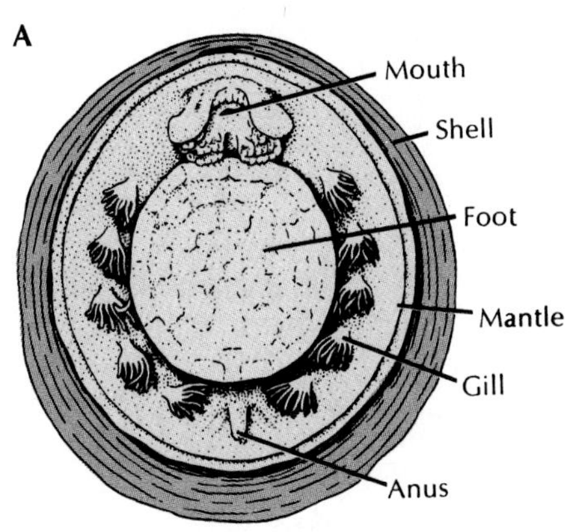

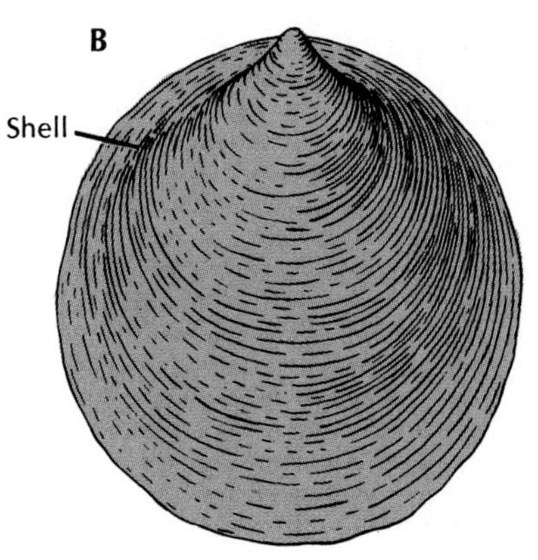

FIGURE 20-6

Neopilina, class Monoplacophora. Living specimens range from 3 mm to about 3 cm in length. **A,** Ventral view. **B,** Dorsal view.

Class Polyplacophora — Chitons

The chitons are somewhat flattened and have a convex dorsal surface that bears eight articulating limy **plates,** or **valves,** which give them their name (Figures 20-7 and 20-8). The term Polyplacophora means "bearing many plates" in contrast to the Monoplacophora, which bear one shell (*mono,* single). The plates overlap posteriorly and are usually dull colored to match the rocks to which the chitons cling.

Most chitons are small (2 to 5 cm); the largest rarely exceeds 30 cm. They are commonly found on rocky surfaces in intertidal regions, although some live at great depths. Chitons are stay-at-home organisms, straying only very short distances for feeding. In feeding, a sensory subradular organ protrudes from the mouth to explore for algae. When some are found, the radula is then projected to scrape them off. The chiton clings tenaciously to its rock with the broad flat foot. If detached, it can roll up like an armadillo for protection.

The mantle forms a **girdle** around the margin of the plates, and in some species mantle folds cover part or all of the plates. On each side of the broad ventral foot and lying between the foot and the mantle is a row of gills suspended from the roof of the mantle cavity. With the foot and the mantle margin adhering tightly to the substrate, these grooves become closed chambers, open only at the ends. Water enters the grooves anteriorly, flows across the gills, and leaves posteriorly, thus bringing to the gills a continuous supply of oxygen.

Blood pumped by the three-chambered heart reaches the gills by way of an aorta and sinuses. Two kidneys (metanephridia) carry waste from the pericardial cavity to the exterior. Two pairs of longitudinal nerve cords are connected in the buccal region. Sense organs include shell eyes on the surface of the shell (in some) and a pair of **osphradia** (chemosensory organs for sampling water).

Sexes are separate in chitons. Sperm shed by males in the excurrent water enter the gill grooves of the females by incurrent openings. Eggs are shed into the sea singly or in strings or masses of jelly.

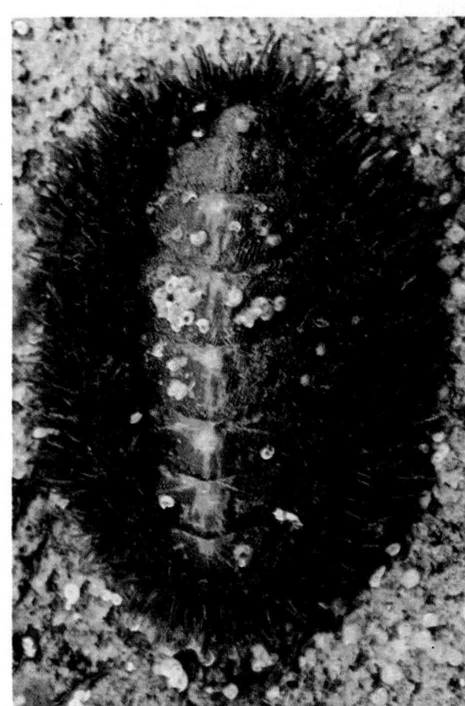

FIGURE 20-7

Mossy chiton, *Mopalia muscosa.* The upper surface of the mantle, or "girdle," is covered with hairs and bristles, an adaptation for defense.

Photograph by R. Harbo.

Class Scaphopoda

The Scaphopoda (ska-fop'o-da), commonly called the tusk shells or tooth shells, are sedentary marine molluscs that have a slender body covered with a mantle and a tubular shell open at both ends. Here the molluscan body plan has taken a new direction, with the mantle wrapped around the viscera and fused to form a tube. Most scaphopods are 2.5 to 5 cm long, although they range from 4 mm to 25 cm long.

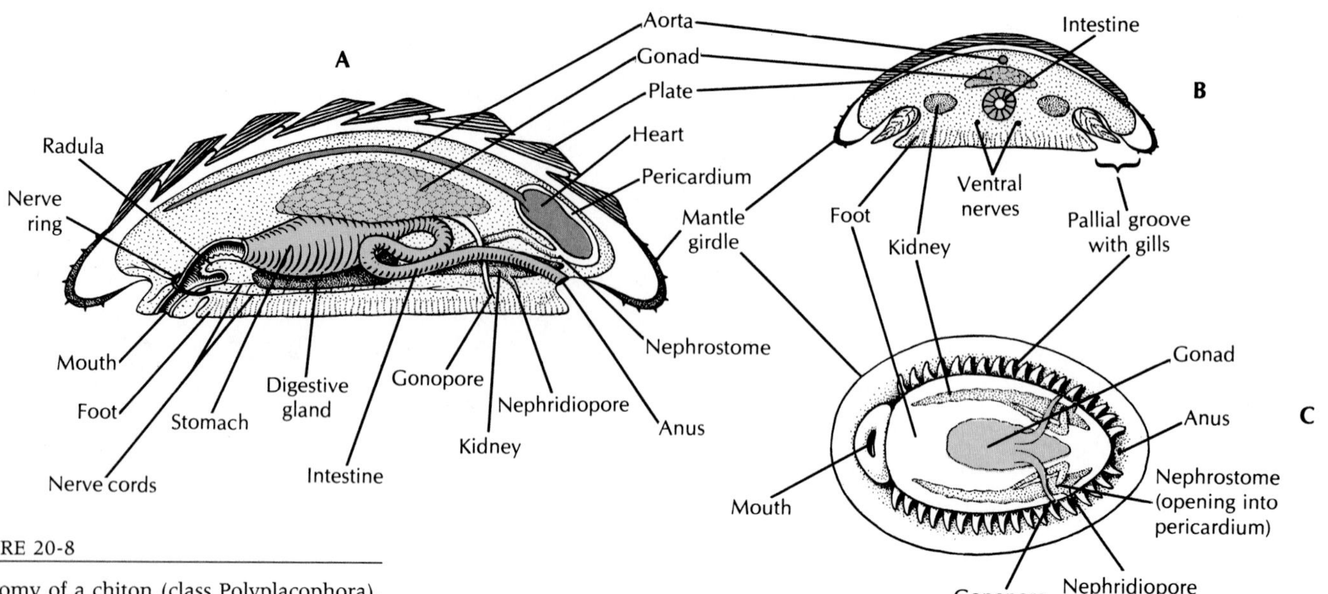

FIGURE 20-8

Anatomy of a chiton (class Polyplacophora). **A,** Longitudinal section. **B,** Transverse section. **C,** External ventral view.

The foot, which protrudes through the larger end of the shell, is used to burrow into mud or sand, always leaving the small end of the shell exposed to the water above (Figure 20-9). Respiratory water is circulated through the mantle cavity both by movements of the foot and by ciliary action. Gaseous exchange occurs in the mantle. Most of the food is detritus and protozoa from the substrate. It is caught on the cilia of the foot or on the mucus-covered, ciliated knobs of long tentacles.

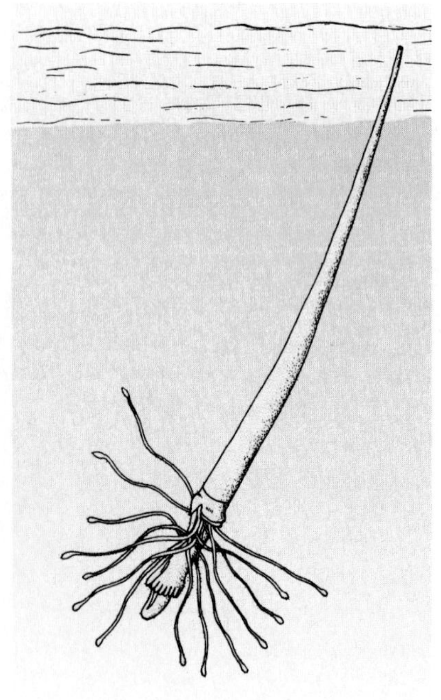

FIGURE 20-9

The tusk shell, *Dentalium,* a scaphopod. It burrows into soft mud or sand and feeds by means of its prehensile tentacles. Respiratory currents of water are drawn in by ciliary action through the small open end of the shell, then expelled through the same opening by muscular action.

Class Gastropoda

Among the molluscs the class Gastropoda is by far the largest and most successful, containing about 35,000 living and 15,000 fossil species. It is made up of members of such diversity that there is no single general term in our language that can apply to them as a group. They include snails, limpets, slugs, whelks, conchs, periwinkles, sea slugs, sea hares, sea butterflies, and others. They range from some of the most primitive of marine molluscs to the highly evolved terrestrial air-breathing snails and slugs.

Gastropods are usually sluggish, sedentary animals because most of them have heavy shells and slow locomotor organs. The shell, when present, is always of one piece (univalve) and may be coiled or uncoiled. Some snails have an **operculum,** a horny plate that covers the shell aperture when the body is withdrawn into the shell. It protects the body and prevents water loss. These animals are basically bilaterally symmetrical, but because of **torsion,** a twisting process that occurs in the early larval stage, the visceral mass has become asymmetrical.

FORM AND FUNCTION

Torsion

Of all the molluscs, only gastropods undergo torsion. Torsion is a peculiar phenomenon that moves the mantle cavity to the front of the body and then twists the visceral organs and mantle cavity in a 180-degree rotation to produce the typical gastropod asymmetry.

Torsion occurs very early in development, usually in the veliger stage, and in some species the entire action may take only a few moments. Before torsion

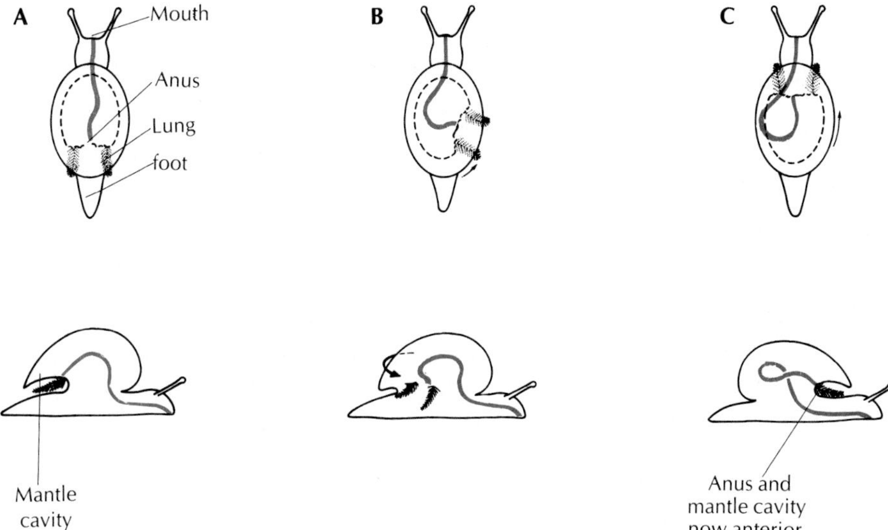

FIGURE 20-10

Torsion in gastropods. **A,** Ancestral condition before torsion. **B,** Intermediate condition. **C,** Early gastropod, torsion complete; direction of crawling now tends to carry waste products back into mantle cavity, resulting in fouling.

occurs, the embryo is bilaterally symmetrical with an anterior mouth and a posterior anus and mantle cavity (Figure 20-10, *A*). Torsion is brought about by an uneven growth of the right and left muscles that attach the shell to the head-foot.

During the first stage of torsion there is a ventral flexure, bringing the anal region downward and then forward (Figure 20-10, *B*), so that the anus opens anteriorly below (ventral to) the head and mouth. In the second phase the mantle cavity and associated viscera rotate 180 degrees, so that the ventral structures shift up the right side to a dorsal position and the dorsal structures shift down the left side to a ventral position (Figure 20-10, *C*). This also shifts the originally left gill, kidney, and other organs to the right side and the originally right organs to the left side, bends the digestive tract, and twists the nerve cords into a figure eight. The mantle cavity with its anal opening now faces forward and lies above the head and mouth. Such a torsion allows the sensitive head end of the animal to be drawn into the protection of the mantle cavity, with the tougher foot forming a barrier to the outside. This may have been the evolutionary pressure that selected for such a strange realignment of body parts, but we are not sure. Other possible explanations have been suggested, but it is clear that whatever the selective advantage torsion conferred on gastropods, a great selective disadvantage accompanied it: **fouling.** After torsion, the anus and excretory pore are placed in a position to drop wastes on the head, in front of the gills and osphradium. How these animals have evolved mechanisms to avoid fouling is an interesting story in adaptive radiation among gastropods, to which we will refer again.

Coiling

The coiling, or spiral winding, of the shell and visceral hump is not the same as torsion. Coiling may occur in the larval stage at the same time as torsion, but the fossil record shows that coiling was a separate evolutionary event and originated in gastropods earlier than torsion did. Nevertheless, all living gastropods have descended from coiled, torted ancestors, whether or not they now show these characteristics.

Early gastropods had a bilaterally symmetrical shell with all the whorls lying in a single plane (Figure 20-11, *A*). Such a shell was not very compact, since each whorl had to lie completely outside the preceding one. Curiously, a few modern species have secondarily returned to that form. The compactness problem of the planospiral shell was solved by a shape in which each succeeding whorl was at

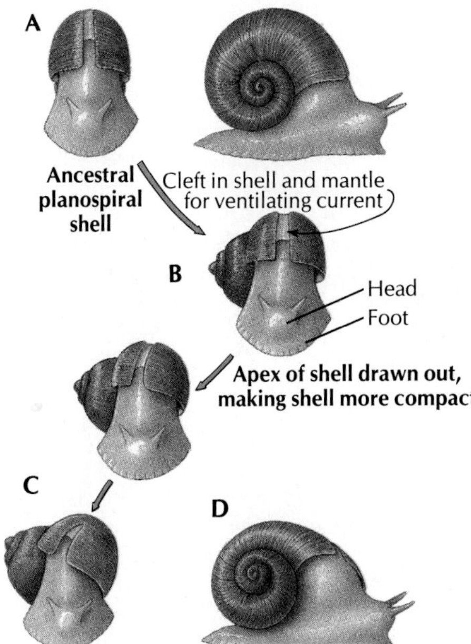

FIGURE 20-11

Evolution of shell in gastropods. **A,** Earliest coiled shells were planospiral, each whorl lying completely outside the preceding whorl. **B,** Better compactness was achieved by snails in which each whorl lay partially to the side of the preceding whorl. **C** and **D,** Better weight distribution resulted when shell was moved upward and posteriorly.

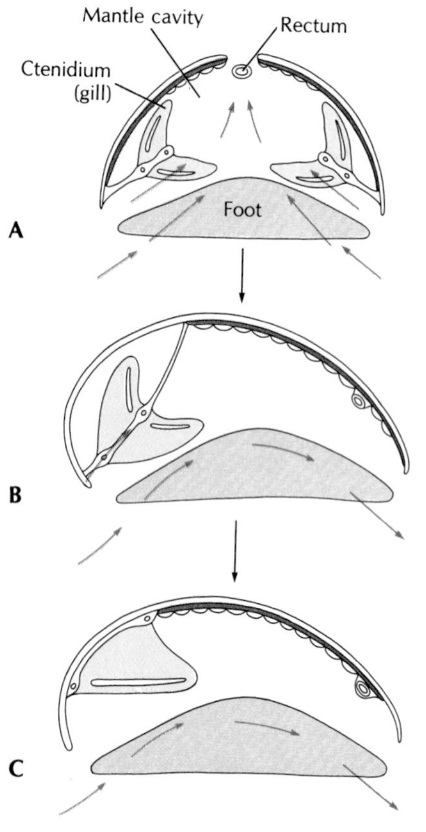

FIGURE 20-12

Evolution of the ctenidia in gastropods. **A,** Primitive prosobranchs with two ctenidia and excurrent water leaving the mantle cavity by a dorsal slit or hole. **B,** Condition after one ctendium had been lost. **C,** Condition in advanced prosobranchs, in which filaments on one side of remaining gill are lost, and axis is attached to mantle wall.

the side of the preceding one (Figure 20-11, *B*). However, this shape was clearly unbalanced, hanging as it was with much weight over to one side. Better weight distribution was achieved by shifting the shell upward and posteriorly, with the shell axis oblique to the longitudinal axis of the foot (Figure 20-11, *C*). The weight and bulk of the main body whorl, the largest whorl of the shell, pressed on the right side of the mantle cavity, however, and apparently interfered with the organs on that side. Accordingly, the gill, auricle, and kidney of the right side have been lost in all except primitive living gastropods, leading to a condition of **bilateral asymmetry.**

Adaptations to avoid fouling

Although the loss of the right gill was probably an adaptation to the mechanics of carrying the coiled shell, that condition made possible a way to avoid fouling, which is displayed in most modern prosobranchs (the largest subclass of marine snails). Water is brought into the left side of the mantle cavity and out the right side, carrying with it the wastes from the anus and nephridiopore, which lie near the right side (Figure 20-12). Some of the more primitive prosobranchs (those with two gills, such as abalone) (Figure 20-13, *A*) avoid fouling by venting the excurrent water through a dorsal slit or hole in the shell above the anus (Figure 20-12). The opisthobranchs (nudibranchs and others) have evolved an even more curious "twist"; after undergoing torsion as larvae, they develop various degrees of *detorsion* as adults. The pulmonates (most freshwater and terrestrial snails) have lost the gill altogether, and the vascularized mantle wall has become a lung. The anus and nephridiopore open near the opening of the lung to the outside (pneumostome), and waste is expelled forcibly with air or water from the lung.

Feeding habits

Feeding habits of gastropods are as varied as their shapes and habitats, but all include the use of some adaptation of the radula. The majority of gastropods are herbivorous, rasping off particles of algae. Some herbivores are grazers, some are browsers, some are planktonic feeders. The abalone (Figure 20-13) holds seaweed with the foot and breaks off pieces with the radula. Some snails are scavengers, living on dead and decayed flesh; others are carnivorous, tearing their prey apart with their radular teeth. Some, such as the oyster borer and the moon snail (Figure 20-13, *B*), have an extensible proboscis for drilling holes in the

A

B

FIGURE 20-13

A, Red abalone, *Haliotus rufescens.* This huge, limpetlike snail is prized as food and extensively marketed. Abalones are strict vegetarians, feeding especially on sea lettuce and kelp. **B,** Moon snail, *Polinices lewisii.* A common inhabitant of West Coast sand flats, the moon snail is a predator of clams and mussels. It uses its radula to drill neat holes through its victim's shell, through which the proboscis is then extended to eat the bivalve's fleshy body.

Photographs by D. Gotshall.

shells of the bivalves whose soft parts they find delectable. Some even have a spine for opening the shells. Most of the pulmonates (air-breathing snails) (Figure 20-19) are herbivorous, but some live on earthworms and other snails.

Some of the sessile gastropods, such as the slipper shells, are ciliary feeders that use the gill cilia to draw in particulate matter, which is rolled into a mucous ball and carried to the mouth. Some of the sea butterflies secrete a mucous net to catch small planktonic forms and then draw the web into the mouth.

After maceration by the radula or by some grinding device, such as the so-called gizzard in the sea hare (Figure 20-15) and in others, digestion is usually extracellular in the lumen of the stomach or digestive glands. In ciliary feeders the stomachs are sorting regions and most of the digestion is intracellular in the digestive glands.

FIGURE 20-14

Cone shell, *Conus californicus.* Note the extended siphon through which water is drawn into the mantle cavity.

Photograph by D.W. Behrens.

Internal form and function

Respiration in most gastropods is carried out by a gill (two gills in primitive prosobranchs), although some aquatic forms lack gills and depend on the skin. The pulmonates have a lung. Freshwater pulmonates must surface to expel a bubble of gas from the lung and curl the edge of the mantle around the **pneumostome** (pulmonary opening in the mantle cavity) to form a siphon for taking in air.

Most gastropods have a single nephridium (kidney). The circulatory and nervous systems are well developed (Figure 20-16). The nervous system includes three pairs of ganglia connected by nerves. Sense organs include eyes, statocysts, tactile organs, and chemoreceptors.

Not all snails are slow-moving herbivores. The poisonous cone shells (Figure 20-14) are predators that immobilize their prey by stabbing them with a single radular tooth through which a highly toxic venom is injected. The sting is very painful, even lethal, to humans.

There are both dioecious and hermaphroditic gastropods. During copulation in hermaphroditic species there is an exchange of **spermatophores** (bundles of sperm), so that self-fertilization is avoided. Many forms perform courtship ceremonies. Most land snails lay their eggs in holes in the ground or under logs. Some aquatic gastropods lay their eggs in gelatinous masses; others enclose them in gelatinous capsules or in parchment egg cases. Marine gastropods go through a free-swimming veliger larval stage during which torsion and coiling occur.

MAJOR GROUPS OF GASTROPODS

Much the largest subclass of gastropods is the Prosobranchia, and almost all of these are marine. The few freshwater species can usually be distinguished from pulmonates because the prosobranchs have an operculum. Familiar examples of

A

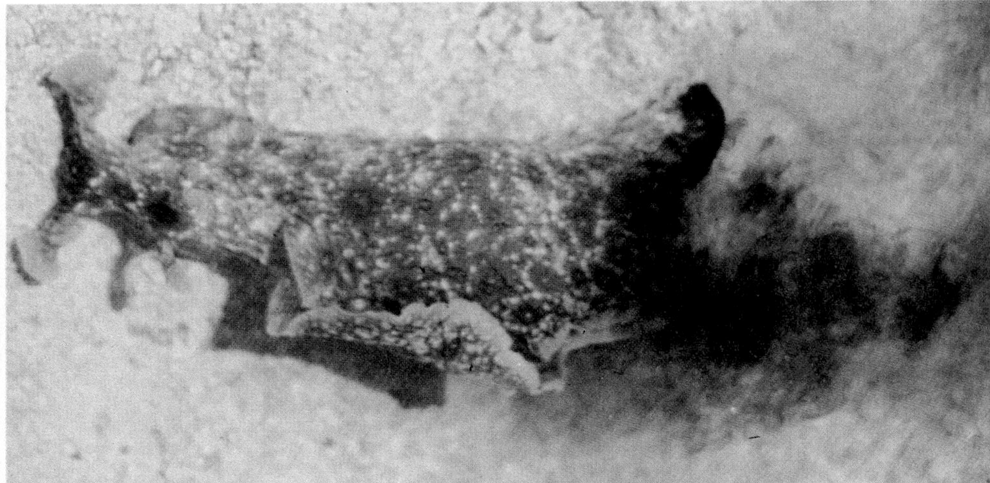

B

FIGURE 20-15

A, The sea hare, *Aplysia dactylomela,* crawls and swims across a coral reef, assisted by large, winglike parapodia, here curled above the body. **B,** When attacked, the sea hare squirts a copious protective secretion from its "purple gland" in the mantle cavity.

Photographs by C.P. Hickman, Jr.

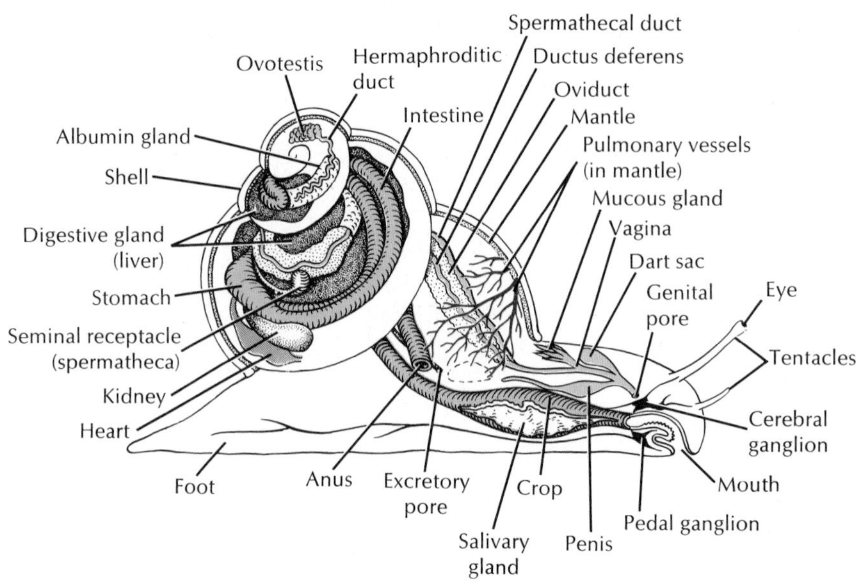

FIGURE 20-16

Anatomy of a pulmonate snail.

A

B

FIGURE 20-17

A, Keyhole limpet, *Diodora aspera,* a prosobranch gastropod with a hole in the apex through which the water leaves the shell. **B,** The flamingo tongue, *Cyphoma gibbosum,* is a showy inhabitant of Caribbean coral reefs, where it is associated with gorgonians. This snail has a smooth creamy orange to pink shell that is normally covered by the brightly marked mantle.

A, Photograph by R. Harbo; **B,** photograph by L.S. Roberts.

marine prosobranchs are the periwinkles, limpets (Figure 20-17, *A*), whelks, conchs, abalones (Figure 20-13, *A*), slipper shells, oyster borers, rock shells, and cowries.

The subclass Opisthobranchia is an odd assemblage of molluscs that include sea slugs, sea hares, sea butterflies, canoe shells, and others. The opisthobranchs are all marine. At present 8 to 12 orders of opisthobranchs are recognized, but for convenience they can be divided into two classical groups: **tectibranchs,** with gill and shell usually present, and **nudibranchs,** in which there is no shell or true gill, but in which secondary gills are present along the dorsal side or around the anus. Among the tectibranchs is the large sea hare *Aplysia,* (see Figure 20-15), which has large earlike anterior tentacles and a vestigial shell. Nudibranchs rank among the most beautiful and colorful of the molluscs (Figure 20-18). Having lost the gill, the body surface is often increased for gaseous exchange by small projections (**cerata**), secondary gills around the anus, or a ruffling of the mantle edge.

The third subclass (Pulmonata) contains most land and freshwater snails and slugs. Lacking gills, their mantle cavity has become a lung, which fills with air by contraction of the mantle floor. The aquatic species have one pair of nonretractile tentacles, at the base of which are the eyes; land forms have two pairs of tentacles, with the posterior pair bearing the eyes (Figures 20-16 and 20-19).

■ Class Bivalvia (Pelecypoda)

The Bivalvia (bi-val've-a) are also known as Pelecypoda (pel-e-sip'o-da) (Gr. *pelekus,* hatchet, + *pous, podos,* foot). They are the bivalved (two-shelled) molluscs that include the mussels, clams, scallops, oysters, and shipworms and range in size from tiny seed shells 1 to 2 mm in length to the giant, South Pacific clams *Tridacna,* mentioned before (Figure 20-28). Most bivalves are sedentary **filter feeders** that depend on ciliary currents produced by the gills to bring in food materials. Unlike the gastropods, they have no head, no radula, and very little cephalization (Figure 20-22).

Most bivalves are marine, but many live in brackish water and in streams, ponds, and lakes.

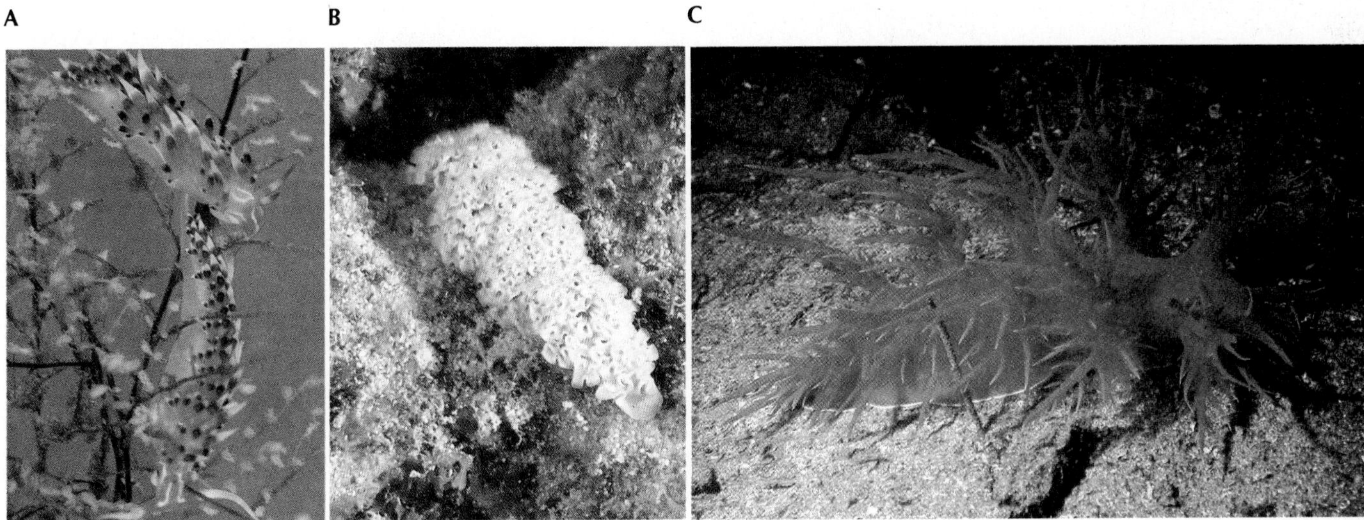

FIGURE 20-18

"Nudibranchs." **A,** Aeolid nudibranchs crawl across a hydroid colony. Their long, dorsal cerata contain nematocysts, which the animals obtain from their cnidarian diet. **B,** *Tridachia crispata* feeds on algae but does not destroy the chloroplasts in the algal cells it eats. The chloroplasts continue to function and contribute to the nutrition of the gastropod. **C,** The giant nudibranch, *Dendronotus irus,* may reach 30 cm in length.

A, Photograph by K. Sandved; **B,** photograph by L.S. Roberts; **C,** photograph by R. Harbo.

FORM AND FUNCTION

Shell

Bivalves are laterally compressed, and their two shells (**valves**) are held together dorsally by a **hinge ligament** that causes the valves to gape ventrally. The valves are drawn together by **adductor muscles** that work in opposition to the hinge ligament (Figure 20-20, *C* and *D*). Projecting above the hinge ligament on each valve is the **umbo,** which is the oldest part of the shell. The valves function largely for protection, but those of the shipworms (Figure 20-23) have microscopic teeth for rasping wood, and the rock borers use spiny valves for boring into rock. A few bivalves such as scallops use their shells for locomotion by clapping the valves together so that they move in spurts.

Mantle

The mantle hangs down from each side of the visceral mass, over the gills and sides of the foot, and adheres to the valves. The posterior edges of the mantle folds are modified to form dorsal **excurrent** and ventral **incurrent siphons** for regulating water flow (Figures 20-20, *A,* and 20-21). In some marine bivalves the mantle is drawn out into long muscular siphons to allow the clam to burrow into the mud or sand and extend the siphons to the water above. Cilia on the gills and inner surface of the mantle direct the flow of water over the gills.

FIGURE 20-19

Pulmonate land snail. Note two pairs of tentacles; the second larger pair bear the eyes.

Photograph by C.P. Hickman, Jr.

Locomotion

Most bivalves move by extending the slender muscular foot between the valves (Figure 20-20, *A*). Blood swells the end of the foot to anchor it in mud or sand, and then longitudinal muscles contract to shorten the foot and pull the animal forward. In most bivalves the foot is used for burrowing, but a few creep. Some bivalves are sessile; oysters attach their shells to a surface by secreting cement, and mussels attach themselves by secreting a number of slender **byssal threads** (Figure 20-25).

FIGURE 20-20

Tagelus plebius, the stubby razor clam (class Bivalvia). **A,** External view of right valve. **B,** Inside of left shell showing scars where muscles were attached. The mantle was attached to the pallial line. **C,** and **D,** Sections showing function of adductor muscles and hinge ligament. In **C** the adductor muscle is contracted, pulling the valves together. In **D** the adductor muscle is relaxed, allowing the hinge ligament to pull the valves apart.

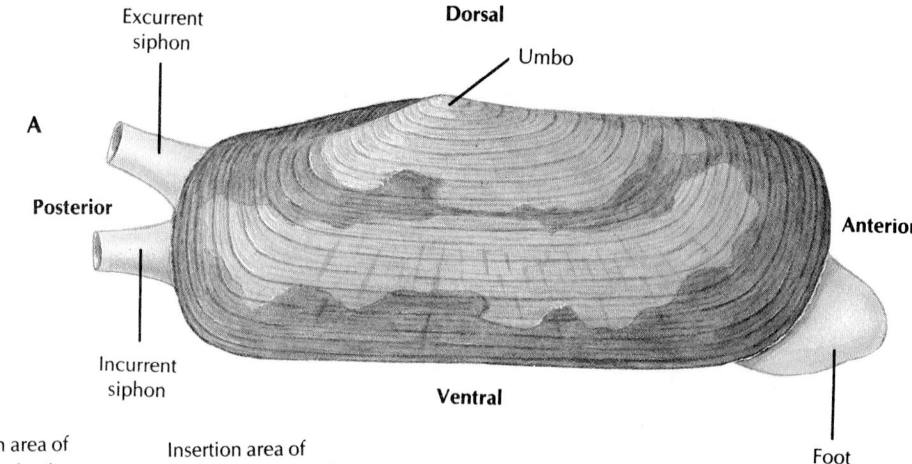

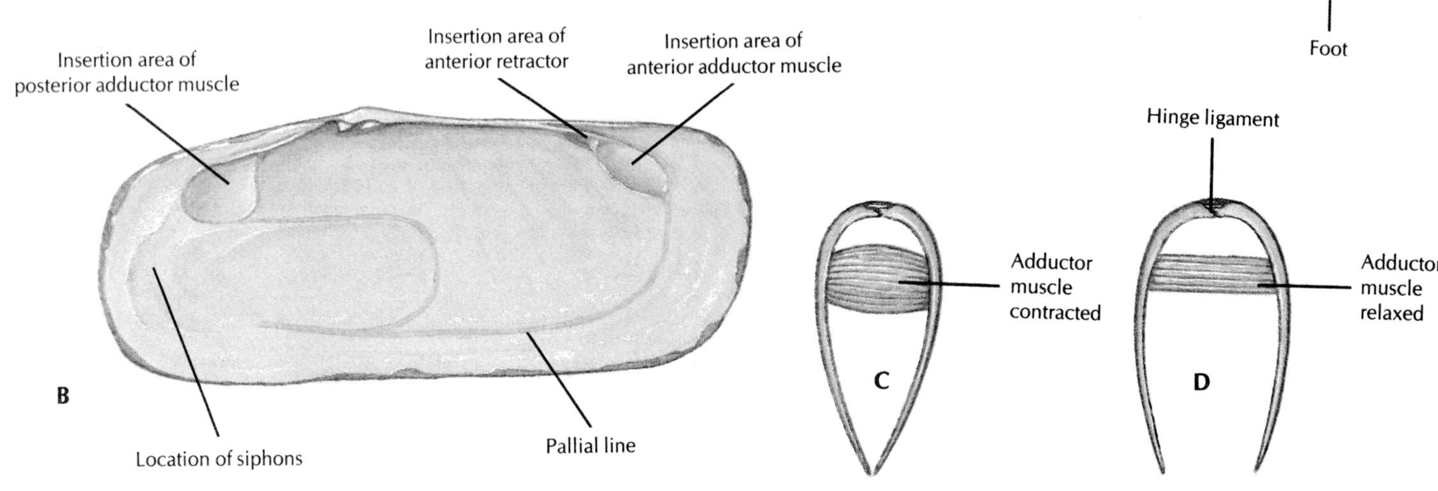

FIGURE 20-21

In the northwest ugly clam, *Entodesma saxicola,* the incurrent and excurrent siphons are clearly visible.

Photograph by R. Harbo.

Feeding and digestion

Most bivalves are filter feeders that secrete mucus to trap food particles brought in by the gill currents (Figure 20-22). In the stomach the mucus and food particles are kept whirling by a rotating gelatinous rod, called a **crystalline style.** As layers of the rotating style dissolve, certain digestive enzymes are freed for extracellular digestion. Food particles detached from the spinning mass are sorted in the ciliated ridges of the stomach from which suitable particles are directed to the **digestive gland** for intracellular digestion.

Internal features and reproduction

Bivalves have a three-chambered heart that pumps blood through the gills and mantle for oxygenation and to the kidneys for waste elimination (Figure 20-26). The three pairs of ganglia are widely separated, and sense organs are poorly developed. A few bivalves have ocelli. The steely blue eyes of the scallops (Figure 20-24), located around the mantle edge, are remarkably advanced, equipped with cornea, lens, and retina.

Sexes are separate, and fertilization is usually external. Marine embryos go through three free-swimming larval stages—**trochophore, veliger larva,** and young **spat**—before reaching adulthood. In freshwater clams fertilization is internal, and some of the gill tubes become temporary brood chambers. There the zygotes develop into tiny bivalved **glochidium larvae,** which are discharged with the excurrent flow (Figure 20-27). If the larvae come in contact with a passing fish, they hitchhike a ride as parasites in the fish's gills for the next 20 to 70 days before sinking to the bottom to become sedentary adults.

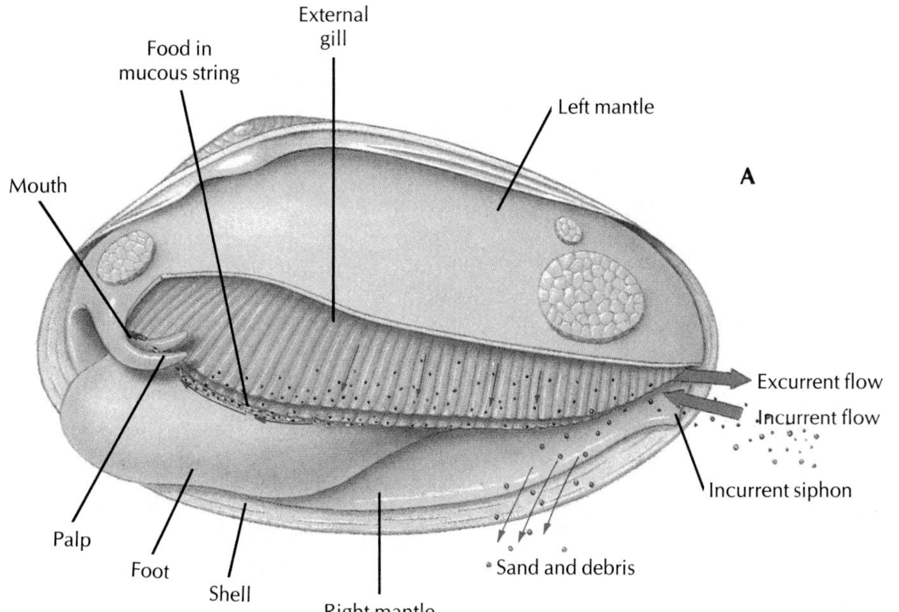

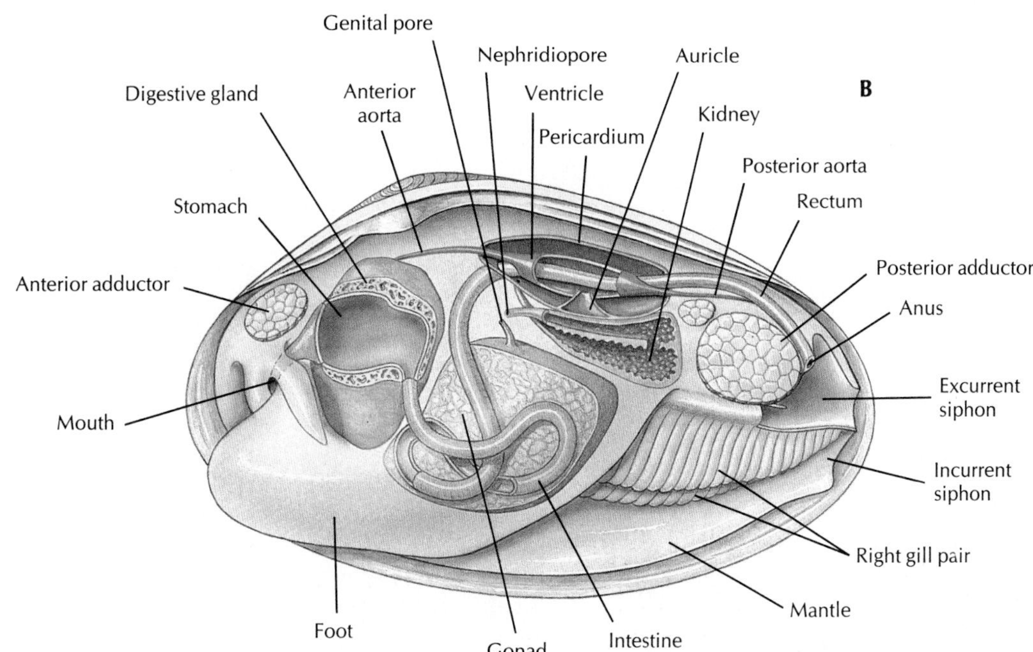

FIGURE 20-22

A, Feeding mechanism of freshwater clam. Left valve and mantle are removed. Water enters the mantle cavity posteriorly and is drawn forward by ciliary action to the gills and palps. As water enters the tiny openings of the gills, food particles are sieved out and caught up in strings of mucus that are carried by cilia to the palps and directed to the mouth. Sand and debris drop into the mantle cavity and are removed by cilia. **B,** Clam anatomy.

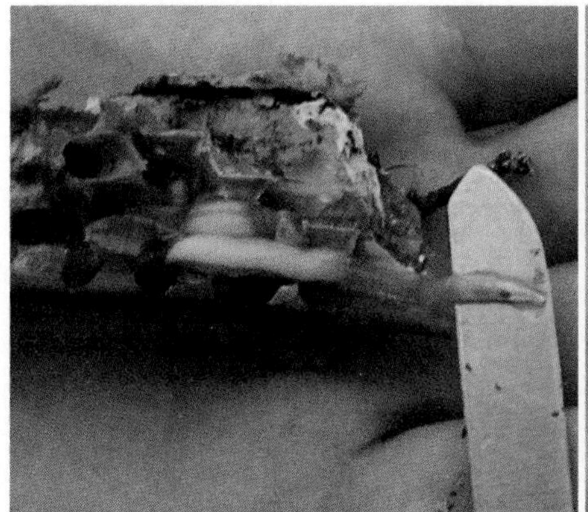

A

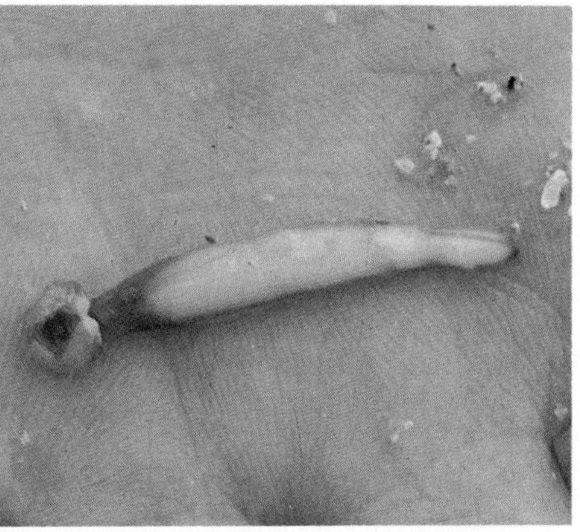

B

FIGURE 20-23

A, Shipworms are bivalves that burrow in wood, causing great damage to unprotected wooden hulls and piers. **B,** The two small, anterior valves, seen at left, are used as rasping organs to extend the burrow.

Photographs by L.S. Roberts

FIGURE 20-24

Representing a group that has evolved from burrowing ancestors, the surface-dwelling bivalve *Pecten* sp. has developed sensory tentacles and a series of blue eyes along its mantle edges.

Photograph by L.S. Roberts

FIGURE 20-25

Mussels, *Mytilus edulis,* occur in northern oceans around the world; they form dense beds in the intertidal zone. A host of marine creatures live protected beneath attached mussels.

Photograph by R. Harbo.

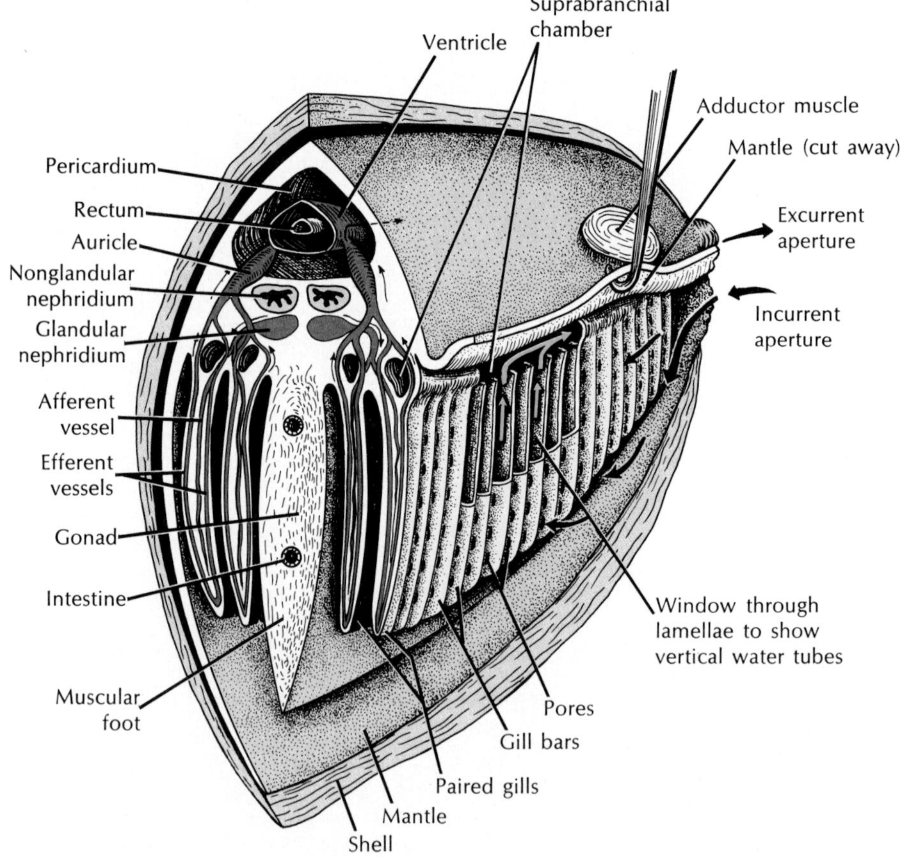

FIGURE 20-26

Section through heart region of clam showing relation of circulatory and respiratory systems. *Blood circulation:* Heart ventricle pumps blood to sinuses of foot, viscera, and mantle; on its return to the heart, blood passes through kidney or gills. *Water currents:* Water drawn in by cilia enters gill pores, passes up gill tubes, and out excurrent aperture. Blood in gills exchanges carbon dioxide for oxygen.

■ Class Cephalopoda

The Cephalopoda (sef-a-lop'o-da) are the most advanced of the molluscs—in fact, in some respects they are the most advanced of all the invertebrates. They include the squids, octopuses, nautiluses, and cuttlefishes. All are marine, and all are active predators.

Cephalopods are the "head-footed" molluscs in which the modified foot is concentrated in the head region. It takes the form of a funnel for expelling water from the mantle cavity. The anterior margin of the head is drawn out into a circle or crown of arms or tentacles.

Cephalopods range in size from 2 to 3 cm up to the giant squid, *Architeuthis*, which is the largest invertebrate known. The squid *Loligo* (L., cuttlefish) is about 30 cm long (Figure 20-30, *A*).

Fossil records of cephalopods go back to Cambrian times. The earliest shells were straight cones; others were curved or coiled, culminating in the coiled shell similar to that of the modern *Nautilus* (Gr. *nautilos*, sailor)—the only remaining member of the once flourishing nautiloids (Figure 20-29). Cephalopods without shells or with internal shells (such as octopuses and squids) are believed to have evolved from some early straight-shelled nautiloid.

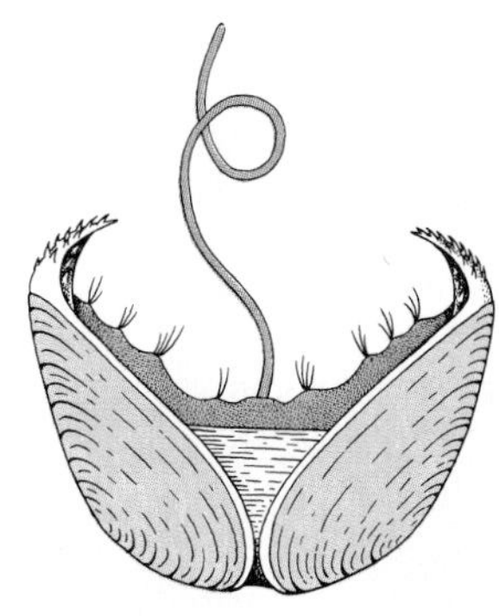

FIGURE 20-27

Glochidium, or larval form, of freshwater clam. When the larva is released from brood pouch of mother, it may become attached to a fish's gill by clamping its valves closed. It remains as a parasite on the fish for several weeks. Its size is approximately 0.3 mm.

FORM AND FUNCTION

Shell

Although early nautiloid shells were heavy, they were made buoyant by a series of **gas chambers,** as is that of the *Nautilus* (Figure 20-29, *B*), enabling the animal to swim while carrying its shell. The shell of *Nautilus*, although coiled, is quite different from that of a gastropod. The shell is divided by transverse septa into internal chambers (Figure 20-29, *B*). The living animal inhabits only the last chamber. As it grows, it moves forward, secreting behind it a new septum. The chambers are connected by a cord of living tissue called the **siphuncle,** which extends

The enormous size of the giant squid, *Architeuthis*, has already been mentioned. These animals are very poorly known because no one has ever been able to study a living specimen. The anatomy has been studied from stranded animals, from those captured in the nets of fishermen, and from specimens found in the stomach of sperm whales. The mantle length is 5 to 6 m, and the head is up to 1 m long. They have the largest eyes in the animal kingdom: up to 25 cm (10 inches) in diameter. They apparently eat fish and other squids, and they are an important food item for sperm whales. They are thought to live on or near the bottom at a depth of 1000 m, but some have been observed swimming at the surface.

FIGURE 20-28

Giant clam, *Tridacna* sp., lies buried in coral rock with only the richly colored fluted siphonal edge visible. These bivalves bear enormous numbers of symbiotic single-cell plants (zooxanthellae) that provide much of the clam's nutriment.

Photography by L. S. Roberts.

FIGURE 20-29

Nautilus, a cephalopod. **A,** Live *Nautilus,* feeding on a fish. **B,** Longitudinal section, showing gas-filled chambers of shell, and diagram of body structure.

A, Courtesy M. Butschler, Vancouver Public Aquarium.

A

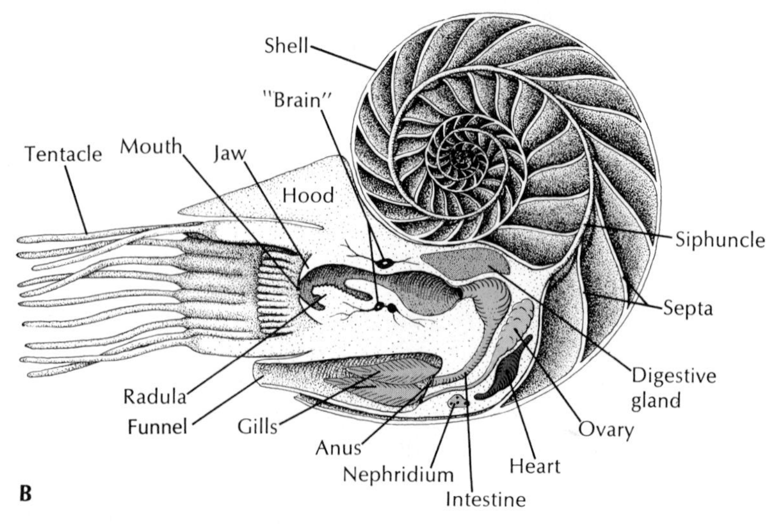

FIGURE 20-30

A, Squid *Loligo opalescens.* **B,** Lateral view of squid anatomy, with the left half of the mantle removed.

A, Photograph by D. Gotshall.

A

B

Anterior aorta
Digestive gland
Arm
Brain
Salivary gland and duct
Esophagus
Stomach
Mantle artery
Kidney
Cecum
Jaw
Mantle
Pen
Gonad
Radula
Statocyst
Funnel with valve
Anus
Gill
Ink sac
Intestine
Gill heart
Gonoduct
Systemic heart
Posterior vena cava
Tentacle

FIGURE 20-31

The cuttlefish, *Sepia* sp., has an internal shell familiar to keepers of caged birds as "cuttlebone."

Photograph by L.S. Roberts.

from the visceral mass. Cuttlefishes also have a small coiled or curved shell, but it is entirely enclosed by the mantle. In the squids most of the shell has disappeared, leaving only a thin, horny strip called a **pen,** which is enclosed by the mantle. In *Octopus* (Gr. *oktos,* eight + *pous, podos,* foot) the shell has disappeared entirely.

Locomotion

Most cephalopods swim by forcefully expelling water from the mantle cavity through a ventral **funnel**—a sort of jet propulsion method. The funnel is mobile and can be pointed forward or backward to control direction; speed is controlled by the force with which water is expelled.

Squids and cuttlefishes are excellent swimmers. The squid body is streamlined and built for speed (Figure 20-30). Cuttlefishes swim more slowly (Figure 20-31). Both squids and cuttlefishes have lateral fins that can serve as stabilizers, but they are held close to the body for rapid swimming.

Nautilus is active at night; its gas-filled chambers keep the shell upright. Although not as fast as the squid, it moves surprisingly well.

Octopus has a rather globular body and no fins (Figure 20-1, *E*). The octopus can swim backwards by spurting jets of water from its funnel, but it is better adapted to crawling about over the rocks and coral, using the suction discs on its arms to pull or to anchor itself. Some deep-water octopods have the arms webbed like an umbrella and swim in a sort of medusa fashion.

External features

During the embryonic development of the cephalopod, the head and foot become indistinguishable. The ring around the mouth, which bears the arms and tentacles, is considered to be derived from the anterior margin of the head.

In *Nautilus* the head with its 60 to 90 or more tentacles can be extruded from the opening of the body compartment of the shell (Figure 20-29). Its tentacles have no suckers but are made adhesive by secretions. They are used in searching for, sensing, and grasping food. Beneath the head is the funnel. The mantle, mantle cavity, and visceral mass are sheltered by the shell. Two pairs of gills are located in the mantle cavity.

Cephalopods other than nautiloids have only one pair of gills. Octopods have 8 arms with suckers; squids and cuttlefishes (decapods) have 10 arms: 8 arms with suckers and a pair of long retractile tentacles. The thick mantle covering the trunk fits loosely at the neck region allowing intake of water into the mantle cavity. When the mantle edges contract closely about the neck, water is expelled through the funnel. The water current thus created provides oxygenation for the gills in the mantle cavity, jet power for locomotion, and a means of carrying wastes and sexual products away from the body.

Color changes

There are special pigment cells called **chromatophores** in the skin of most cephalopods, which by expanding and contracting produce color changes. They are controlled by the nervous system and perhaps by hormones. Some color changes are protective to agree with background hues; most are behavioral and are associated with alarm or with courtship. Many deep-sea squids are also bioluminescent.

Ink production

All cephalopods, except *Nautilus,* have an ink sac that empties into the rectum. The sac contains an ink gland that secretes into the sac **sepia,** a dark fluid containing the pigment melanin. When the animal is alarmed, it releases a cloud of ink through the anus to form a "smokescreen" to confuse the enemy or perhaps to dull its senses.

It was formerly believed that the gas in the chambers of *Nautilus* was the product of secretion by the siphuncle, but recent investigation has shown that the function of the siphuncle is *not* to secrete gas but to *bail fluid out* of the unoccupied chambers. The shell and septa are composed of calcium carbonate and protein, and as the animal grows, the mantle covering the visceral mass secretes more septa and creates additional chambers. Each new chamber is initially filled with a fluid similar in ionic composition to that of the *Nautilus'* blood (and of seawater). The mechanism of fluid removal appears to involve the active secretion of ions into tiny intercellular spaces in the siphuncular epithelium, so that a very high local osmotic pressure is produced, and the water is drawn out of the chamber by osmosis. The gas in the chamber is only the respiratory gas from the siphuncle tissue that diffuses into the chamber as the fluid is removed. Thus the gas pressure in the chamber is 1 atmosphere or less because it is in equilibrium with the gases dissolved in the seawater surrounding the *Nautilus,* which are in turn in equilibrium with air at the surface of the sea, despite the fact that the *Nautilus* may be swimming at 400 m beneath the surface. That the shell can withstand implosion by the surrounding 41 atmospheres (about 600 pounds per square inch), and that the siphuncle can remove water against this pressure are marvelous feats of natural engineering!

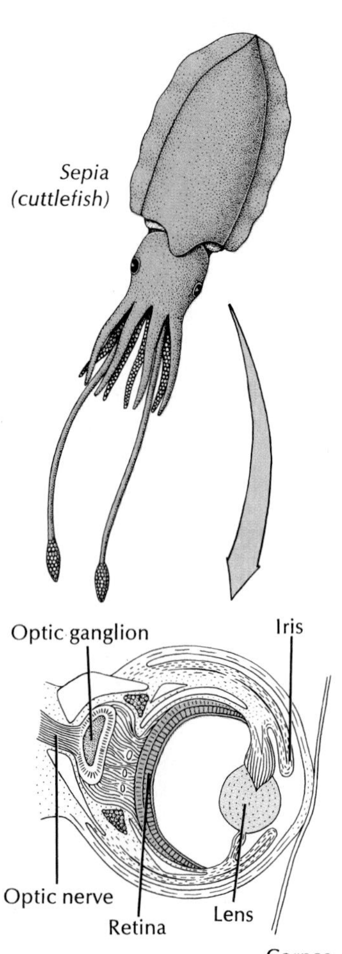

FIGURE 20-32

Eye of a cuttlefish (*Sepia*). The structure of cephalopod eyes shows a high degree of convergent evolution with the eyes of vertebrates.

Feeding and nutrition

Cephalopods are predaceous, feeding chiefly on small fishes, molluscs, crustaceans, and worms. Their arms, which are used in food capture and handling, have a complex musculature and are capable of delicately controlled movements. They are highly mobile and used for swiftly seizing the prey and bringing it to the mouth. Strong, beaklike **jaws** can bite or tear off pieces of flesh, which are then pulled into the mouth by the tonguelike action of the **radula** (Figure 20-30, *B*). Octopods and cuttlefishes have salivary glands that secrete a poison for immobilizing prey. Digestion is extracellular and occurs in the stomach and cecum.

Internal features and reproduction

The circulatory and nervous systems of cephalopods are more advanced than in other molluscs. They have the most complex brain among the invertebrates (Figure 20-30, *B*). Their best-developed sense organs are the eyes, which, except in *Nautilus,* which has relatively simple eyes, are able to form images. Cephalopod eyes are remarkably like vertebrate eyes, with cornea, lens, chambers, and retina (Figure 20-32).

Sexes are separate in cephalopods. In the male seminal vesicle the spermatozoa are encased in spermatophores and stored in a sac that opens into the mantle cavity. During copulation one arm of the adult male plucks a spermatophore from his own mantle cavity and inserts it into the mantle cavity of the female near the oviduct opening (Figure 20-33). Before copulation males often undergo color displays, apparently directed against rival males. Eggs are fertilized as they leave the oviduct and are usually attached to stones or other objects to develop (Figure 20-34). Some octopods tend their eggs.

Table 20-1 Summary comparison of certain features in living molluscs

	Class		
Feature	**Caudofoveata**	**Solenogastres**	**Monoplacophora**
Shell	Spicules	Spicules	One piece, cap shaped
Radula	Present	Absent	Present
Foot	Pedal Shield	pedal Groove	Flat, broad
Mantle cavity and gills	Posterior	Absent	Along sides of foot
Circulatory system	Open	Open	Open
Head	Reduced	Reduced	Reduced

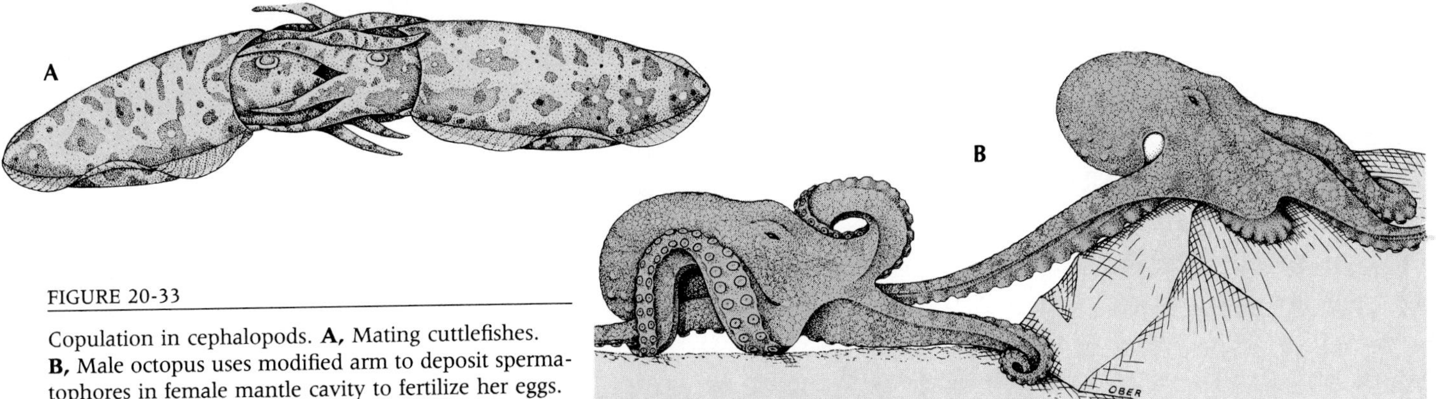

FIGURE 20-33

Copulation in cephalopods. **A,** Mating cuttlefishes. **B,** Male octopus uses modified arm to deposit spermatophores in female mantle cavity to fertilize her eggs. Octopuses often tend their eggs during development.

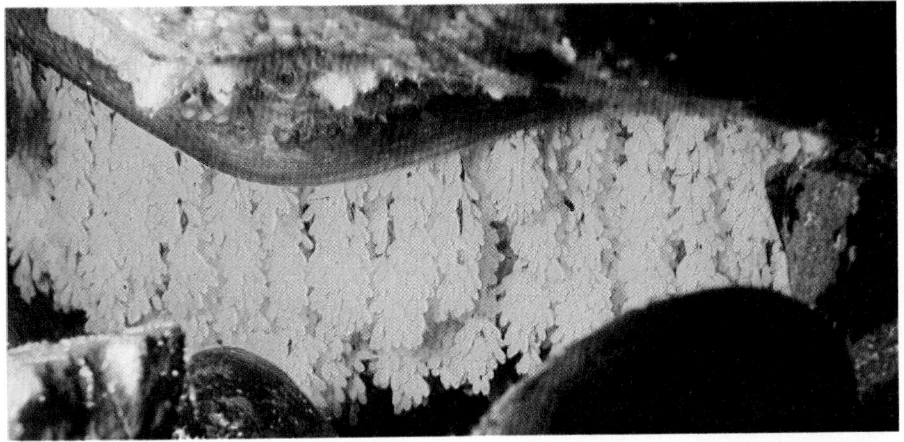

FIGURE 20-34

Octopus den with eggs. The female guards the eggs during incubation.

Photograph by R. Harbo.

Class				
Polyplacophora	**Scaphopoda**	**Gastropoda**	**Bivalvia**	**Cephalopoda**
8 plates	Tubular	One piece, spiral or cap shaped; reduced or absent in some	2 valves, hinged dorsally	Well developed with septa, reduced, or absent
Present	Present	Present	Absent	Present
Flat, broad	Cone shaped	Flat, broad	Hatchet shaped	Funnel
Along sides of foot	Tubular, gills absent	Torted; detorted in some; one or both lost in some	Large, adapted for filter feeding	Large, adapted for swimming
Open	Open	Open	Open	Closed; branchial hearts present
Reduced	Reduced	Well developed; with sense organs	Reduced	Highly developed with sense organs; arms

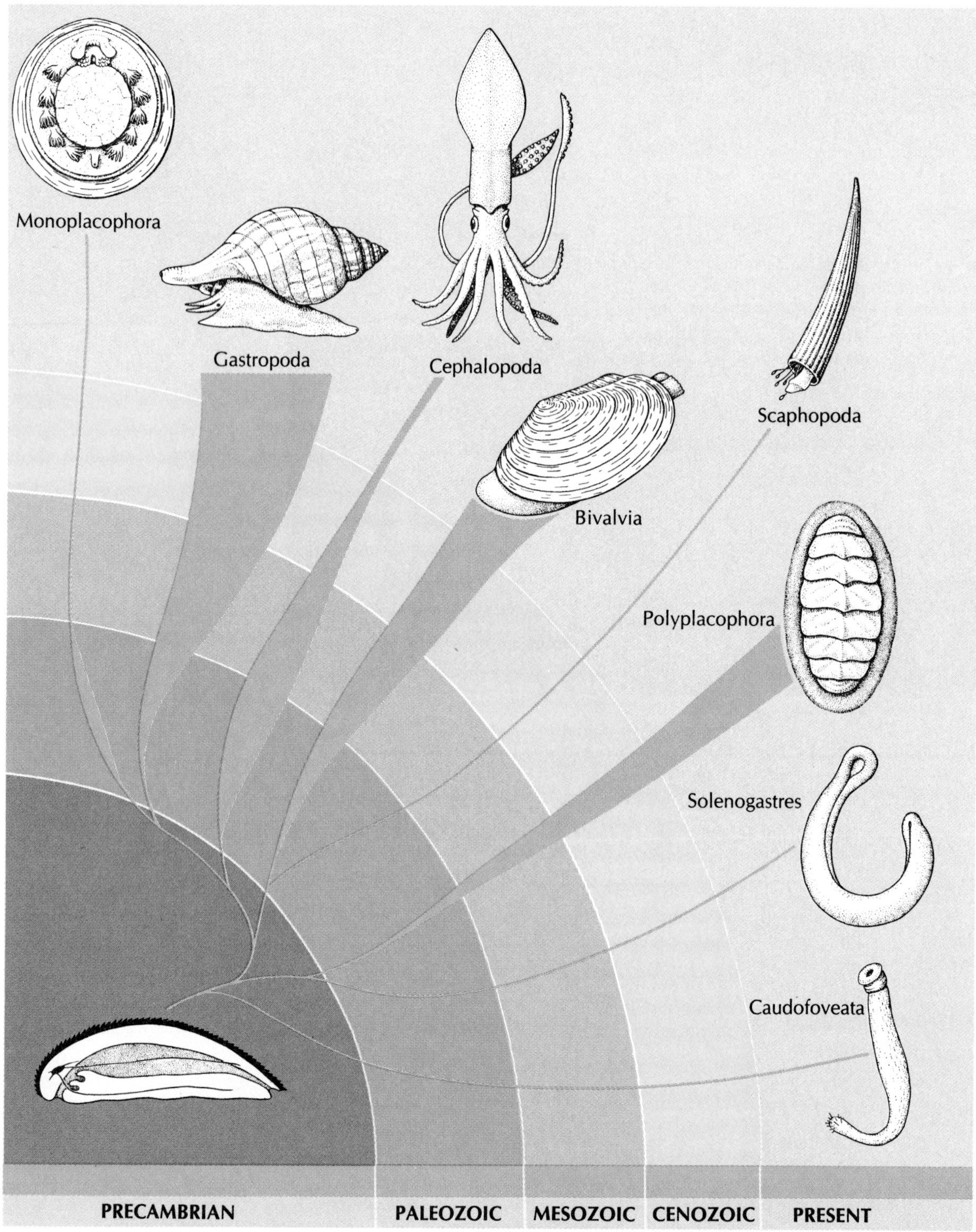

FIGURE 20-35

The classes of Mollusca, showing their derivations and relative abundance.
Derivations after K.J. Boss, personal communication.

Phylogeny and Adaptive Radiation

The same type of embryonic cleavage and the trochophore larva are found in both the annelids and the molluscs and are strong evidence that the two phyla are related. The ladderlike nervous system of some molluscs and the calcareous integumentary scales of the Solenogastres and Caudofoveata resemble those of some turbellarians. It is probable that a flatworm type of ancestor gave rise to the two main protostome groups: the nonsegmented molluscs and the segmentally arranged annelid-arthropod stem. Because there is no characteristic suggesting metamerism in the development of any known molluscan larva, most zoologists agree that the molluscs were not derived from a metameric ancestor but split off from the annelid-arthropod line before the appearance of metamerism.

The primitive ancestral mollusc was a more or less wormlike, dorsoventrally flattened organism with a ventral gliding surface and a dorsal mantle with a chitinous cuticle and calcareous scales (Figure 20-35). It had a posterior mantle cavity with two gills; a complete, straight gut; a radula; a ladderlike nervous system; and an open circulatory system with a heart. Among living molluscs the primitive condition is most nearly approached by the Caudofoveata. The Solenogastres have lost the radula and gills (Table 20-1). They and the Caudofoveata both branched off from the ancestral stem before the advent of the solid shell and the distinct head with sensory organs. Primitive monoplacophorans apparently gave rise to the higher molluscan classes, as well as more advanced monoplacophorans, such as *Neopilina*. Torsion in gastropods may have arisen from an ability to shift the mantle cavity to the side by muscular action, providing a space into which the head could be withdrawn to escape predators or desiccation. The Gastropoda may be polyphyletic, being composed of several groups independently derived from monoplacophorans.

The Scaphopoda and Bivalvia are groups that have exploited the burrowing habitat, rather than crawling on the bottom. The mantle edges of the bivalves dropped curtainlike over the sides of the body, while the shell became bivalved. The lateral flattening and the foot modifications were adaptations for burrowing, and bivalves have achieved great success as filter feeders. Some of the bivalves, such as oysters and scallops, have returned to the surface of the sea bottom and have acquired adaptations for that existence.

In the cephalopods the mantle cavity was brought ventrally toward the head, with a concomitant increase in length of the visceral mass. The foot became highly modified as the funnel. The development of the chambered shell was very important in contributing to their freedom from the substratum and ability to swim. Their highly developed nervous system and sense organs are correlated with their predatory and swimming habits.

Summary

The Mollusca is one of the largest and most diverse phyla, its members ranging in size from very small organisms to the largest of invertebrates. Their basic body divisions are the visceral hump, usually covered by a shell, the head, and the foot. The majority are marine, but some are freshwater, and a few are terrestrial. They occupy a variety of niches; a number are economically important, and a few are medically important as hosts of parasites.

The molluscs are coelomate (have a coelom), although their coelom is limited to the area around the heart. The evolutionary development of a coelom was important because it enabled better organization of visceral organs and, in many of the animals that have it, an efficient hydrostatic skeleton. All of the most advanced invertebrates and vertebrates are coelomates.

Marine clam, *Donax denticulatus,* with siphons and foot extended, class Bivalvia.
Photograph by K. Sandved.

The mantle and mantle cavity are important characteristics of molluscs. The mantle secretes the shell and overlies a part of the visceral hump to form a cavity housing the gills. The mantle cavity has been modified into a lung in some molluscs. The foot is usually a ventral, solelike, locomotory organ, but it may be variously modified, as in the cephalopods, where it has become a funnel. The radula is found in all molluscs except bivalves and solenogasters and is a protrusible, tonguelike organ with teeth used in feeding. Except in the cephalopods, which have a closed circulatory system, the circulatory system of molluscs is open, with a heart and blood sinuses. Molluscs usually have a pair of nephridia connecting with the coelom and a complex nervous system with a variety of sense organs. The primitive larva of molluscs is the trochophore, and most marine molluscs have a more advanced larva, the veliger.

The classes Caudofoveata and Solenogastres are small groups of wormlike molluscs with no shell. The Scaphopoda is a slightly larger class with a tubular shell, open at both ends, and the mantle wrapped around the body.

The class Monoplacophora is a tiny, univalve marine group showing pseudometamerism. The Polyplacophora are more common, marine organisms with shells in the form of a series of eight plates. They are rather sedentary animals with a row of gills along each side of their foot.

The Gastropoda are the most successful and largest class of molluscs. Their interesting evolutionary history includes torsion, or the twisting of the posterior end to the anterior, so that the anus and head are at the same end, and coiling, an elongation and spiraling of the visceral hump. Torsion has led to the survival problem of fouling, which is the release of excreta over the head and in front of the gills, and this has been solved in various ways among different gastropods. Among the solutions to fouling are bringing water into one side of the mantle cavity and out the other (many prosobranchs), some degree of detorsion (opisthobranchs), and conversion of the mantle cavity into a lung (pulmonates).

The class Bivalvia are marine and freshwater, and they have their shell divided into two valves joined by a dorsal ligament and held together by an adductor muscle. Most of them are filter feeders, drawing water through their gills by ciliary action.

The members of the class Cephalopoda are the most advanced molluscs; they are all predators and many can swim rapidly. Their tentacles capture prey by adhesive secretions or by suckers. They swim by forcefully expelling water from their mantle cavity through a funnel, which was derived from the foot.

There is strong embryological evidence that the molluscs are related to the annelids, although the molluscs are not metameric. The enormous diversity of molluscs can be derived from a hypothetical ancestral mollusc that showed the basic body plan.

REVIEW QUESTIONS

1. How does the coelom develop embryologically? Why was the evolutionary development of the coelom important?
2. Give several characteristics of molluscs that distinguish the phylum.
3. Name five ways in which molluscs are important to humans.
4. Distinguish among the following classes of molluscs: Polyplacophora, Gastropoda, Bivalvia, Cephalopoda.
5. Define the following: odontophore, periostracum, prismatic layer, nacreous layer, trochophore, veliger, glochidium.
6. Briefly describe the habitat and habits of a typical chiton.
7. Define the following with respect to gastropods: operculum, torsion, fouling, bilateral asymmetry.
8. Describe three ways to avoid fouling that have evolved in gastropods.
9. Distinguish among prosobranchs, opisthobranchs, and pulmonates.
10. Briefly describe how a typical bivalve feeds and how it burrows.
11. What is the function of the siphuncle of cephalopods?
12. Describe how cephalopods swim and how they eat.
13. What cephalopod characteristics are particularly valuable for actively swimming, predaceous animals?
14. To what other major invertebrate groups are molluscs related, and what is the nature of the evidence for the relationship?
15. Briefly describe the characteristics of the primitive ancestral mollusc, and tell how each class of molluscs differs from the primitive condition with respect to each of the following: shell, radula, foot, mantle cavity and gills, circulatory system, head.

SELECTED REFERENCES

Abbott, R.T. 1974. American seashells, ed. 2. New York, Van Nostrand Reinhold Co., Inc. *Identification guide to 1,500 Atlantic and Pacific species.*

Barnes, R.D. 1987. Invertebrate zoology, ed. 5. Philadelphia, Saunders College/Holt, Rinehart & Winston. *Chapter 11 covers molluscs.*

Gosline, J.M., and M.D. DeMont. 1985. Jet-propelled swimming in squids. Sci. Am. **252:**96-103 (Jan.). *Mechanics of swimming in squids are analyzed; elasticity of collagen in mantle increases efficiency.*

Morris, P.A. (W.J. Clench [ed.]) 1973. A field guide to shells of the Atlantic and Gulf coasts and the West Indies, ed. 3. Boston, Houghton Mifflin Co. *An excellent revision of a popular handbook.*

Roper, C.R.E., and K.J. Boss. 1982. The giant squid. Sci. Am. **246:**96-105 (April). *Many mysteries remain about the deep-sea squid,* Architeuthis, *because it has never been studied alive. It can reach a weight of 1000 pounds and a length of 18 m, and its eyes are as large as automobile headlights.*

Ward, P., L. Greenwald, and O.E. Greenwald. 1980. The buoyancy of the chambered nautilus. Sci. Am. **243:**190-203 (Oct.). *Reviews recent discoveries on how the nautilus removes the water from a chamber after secreting a new septum.*

Yonge, C.M. 1975. Giant clams. Sci. Am. **232:**96-105 (Apr.). *Details the fascinating morphological adaptations of this bivalve for life with its mutualistic zooxanthellae.*

Yonge, C.M., and T.E. Thompson. 1976. Living marine molluscs. London, William Collins Sons & Co., Ltd. *Interesting book about both the more common molluscs and those with particularly interesting adaptive features.*

DIVIDING UP THE BODY

Although a spacious, fluid-filled coelom provided an efficient hydrostatic skeleton for burrowing, precise control of body movements was not possible in the earliest coelomates. The force of muscle contraction in one area was carried throughout the body by the fluid in the undivided coelom. This defect was remedied when a series of partitions (septa) evolved in the ancestral annelids. When the septa divided the coelom into a series of compartments, components of most other body systems, such as circulatory, nervous, and excretory, were repeated in each segment. This body plan is known as *metamerism*.

The evolutionary advent of metamerism was highly significant because it made possible development of much greater complexity in structure and function. Metamerism not only increased the efficiency of burrowing, it made possible independent and separate movements by the separate segments. The need for fine control of movements led, in turn, to the evolution of a more sophisticated nervous system. Moreover, repetition of body parts gave the organisms a built-in redundancy, as in some human-made systems. This provided a safety factor: if one segment should fail, the others could still function. Thus an injury to one part would not necessarily be fatal.

The evolutionary potential of the metameric body plan is amply demonstrated by the fact that the most successful of all invertebrates, members of the phylum Arthropoda, are descended from a metameric, annelid-like ancestor. Metamerism also arose independently in the deuterostome line, and its possessors gave rise to the most successful of the deuterostomes, the vertebrates.▪

CHAPTER 21

Segmented Worms

Annelids

Annelida (an-nel'i-da) (L. *annellus,* little ring, + *-ida,* suffix) consists of the segmented worms. It is a large phylum, numbering approximately 9000 species, the most familiar of which are the earthworms and freshwater worms (oligochaetes) and the leeches (hirudineans). However, approximately two thirds of the phylum is composed of the marine worms (polychaetes), which are less familiar to most people. Among the latter are many curious members; some are strange, even grotesque, whereas others are graceful and beautiful. They include the clam worms, plumed worms, parchment worms, scaleworms, lugworms, and many others. The annelids are true coelomates and belong to the protostome branch, with spiral and mosaic cleavage. They are a highly developed group in which the nervous system is more centralized and the circulatory system more complex than those of the phyla we have studied thus far.

Annelids are sometimes called "bristle worms" because, with the exception of the leeches, most annelids bear tiny chitinous bristles called **setae** (L. *seta,* hair or bristle). Short needlelike setae help anchor the somites during locomotion to prevent backward slipping; long, hairlike setae aid aquatic forms in swimming. Since many annelids either are burrowers or live in secreted tubes, the stiff setae also aid in preventing the worm from being pulled out or washed out of its home. Robins know from experience how effective the earthworms' setae are.

CHARACTERISTICS

1. Body **metamerically** segmented; symmetry bilateral
2. Body wall with outer circular and inner longitudinal muscle layers; outer transparent moist cuticle secreted by epithelium
3. **Chitinous setae,** often present on fleshy appendages called **parapodia;** setae absent in leeches
4. Coelom (schizocoel) well developed and divided by septa, except in leeches; coelomic fluid supplies turgidity and functions as hydrostatic skeleton
5. **Blood system closed** and segmentally arranged; respiratory pigments (hemoglobin, hemerythrin, or chlorocruorin) often present; amebocytes in blood plasma
6. Digestive system complete and not metamerically arranged
7. Respiratory gas exchange through skin, **gills,** or **parapodia**
8. Excretory system typically a **pair of nephridia for each metamere**
9. Nervous system with a double ventral nerve cord and a pair of ganglia with lateral nerves in each metamere; brain, a pair of dorsal cerebral ganglia with connectives to cord
10. Sensory system of tactile organs, taste buds, statocysts (in some), photoreceptor cells, and eyes with lenses (in some)
11. Hermaphroditic or separate sexes; larvae, if present, are trochophore type; asexual reproduction by budding in some; spiral and mosaic cleavage

CLASSIFICATION

Class Polychaeta (pol'e-ke'ta) (Gr. *polys,* many, + *chaitē,* long hair). Mostly marine; head distinct and bearing eyes and tentacles; most segments with parapodia (lateral appendages) bearing tufts of many setae; clitellum absent; sexes usually separate; gonads transitory; asexual budding in some; trochophore larva usually; mostly marine. Examples: *Nereis* (Figure 21-4), *Aphrodita, Glycera, Arenicola* (Figure 21-9), *Chaetopterus* (Figure 21-8), *Amphitrite* (Figure 21-5).

Class Oligochaeta (ol'i-go-ke'ta) (Gr. *oligos,* few, + *chaitē,* long hair). Body with conspicuous segmentation; number of segments variable; setae few per metamere; no parapodia; head absent; coelom spacious and usually divided by intersegmental septa; hermaphroditic; development direct, no larva; chiefly terrestrial and freshwater. Examples: *Lumbricus* (Figure 21-10), *Stylaria* (Figure 21-14, *A,*), *Aeolosoma* (Figure 21-14, *B*), *Tubifex* (Figure 21-14, *C*).

Class Hirudinea (hir'u-din'e-a) (L. *hirudo,* leech, + *-ea,* characterized by) — **leeches.** Body with fixed number of segments (34) with many annuli; body with anterior and posterior suckers usually; clitellum present; no parapodia; setae absent (except *Acanthobdella*); coelom closely packed with connective tissue and muscle; development direct; hermaphroditic; terrestrial, freshwater, and marine. Examples: *Hirudo* (Figure 21-17), *Placobdella* (Figure 21-2, *D*), *Macrobdella.*

ECOLOGICAL RELATIONSHIPS

Annelids are worldwide in distribution, occurring in the sea, fresh water, and terrestrial soil. Some marine annelids live quietly in tubes or burrow into bottom mud or sand. Some of these feed on organic matter in the mud through which they burrow; others are filter feeders with elaborate ciliary or mucous devices for trapping food. Many are predators, either pelagic or hiding in crevices of coral or rock except when hunting. Freshwater annelids burrow in mud or sand, live among vegetation, or swim about freely. The most familiar annelids are the terrestrial earthworms, which move about through the soil. Some leeches are bloodsuckers, and others are carnivores; most of them live in fresh water.

ECONOMIC IMPORTANCE

Much of the economic importance of annelids is indirect, deriving from their ecological roles. Many are members of grazing food chains or detritus food chains (p. 131), serving as prey for other organisms of more direct interest to humans, such as fishes. Consequently, a lively market in some polychaetes and oligochaetes as fish bait thrives. The burrows of earthworms increase drainage and aeration of soils, and the migrations of worms help mix the soil and distribute organic matter to deeper layers (p. 458). Some marine annelids that burrow serve an analogous role in the sea; the lugworm (*Arenicola*, Figure 21-9) is sometimes called the "earthworm of the sea." The new medical uses for leeches (p. 463) have led to a revival of the market in blood-sucking leeches and establishment of "leech farms" where these organisms are raised in captivity.

BODY PLAN

The annelid body typically has a head or prostomium, segmented body, and terminal portion bearing the anus. New segments form just in front of the terminal portion; thus, the oldest segments are at the anterior end and the youngest segments are at the posterior end. Neither the prostomium nor the terminal portion are considered metameres.

The body wall is made up of strong circular and longitudinal muscles adapted for swimming, crawling, and burrowing and is covered with epidermis and a thin, outer layer of nonchitinous cuticle (Figure 21-1).

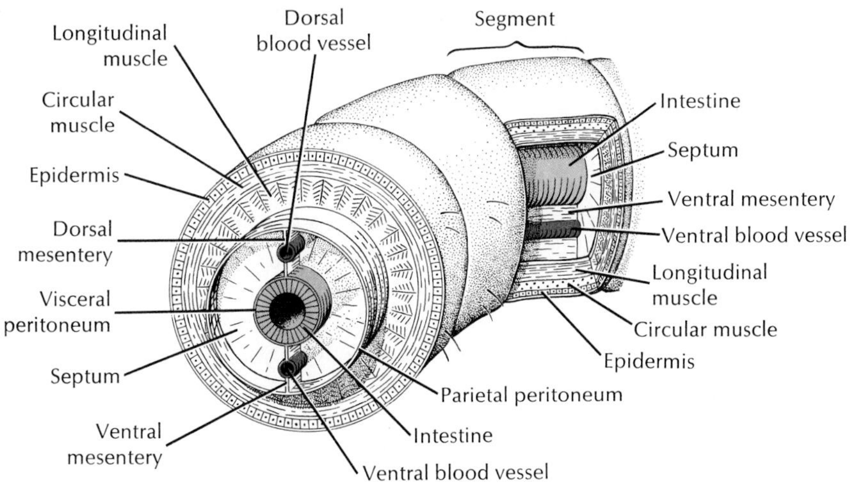

FIGURE 21-1

Annelid body plan.

In most annelids the coelom develops embryonically as a split in the mesoderm on each side of the gut **(schizocoel),** forming a pair of coelomic compartments in each segment. Each compartment is surrounded with **peritoneum** (a layer of mesodermal epithelium), which lines the body wall, forms dorsal and ventral **mesenteries** (double-membrane partitions that support the gut), and covers all the organs (Figure 21-1). Where the peritonea of adjacent segments meet, the **septa** are formed. These are perforated by the gut and longitudinal blood vessels. Not only is the coelom metamerically arranged, but practically every body system is affected in some way by this segmental arrangement.

Except in the leeches, the coelom is filled with fluid and serves as a **hydrostatic skeleton.** Because the volume of the fluid is essentially constant, contraction of the longitudinal body wall muscles causes the body to shorten and become larger in diameter, whereas contraction of the circular muscles causes it to lengthen and become thinner. Separation of the hydrostatic skeleton into a

FIGURE 21-2

Living segmented worms. **A,** Marine polychaete tubeworms, *Serpula vermicularis,* with red and white tentacular crowns, live in coiled calcareous tubes. **B,** The sedentary polychaete, *Amphitrite,* shown here removed from the tube it builds in sand. Note the long tentacles that it extends over the surface to pick up food. **C,** The fire worm, *Hermodice carunculata,* has fine, brittle setae that can break off in human skin and cause irritation. It feeds on corals and other sedentary invertebrates. **D,** A freshwater leech, *Placobdella,* parasitizing a snapping turtle.

A, photograph by R. Harbo; **B,** photograph by C.P. Hickman; **C,** photograph by L.S. Roberts; **D,** photograph by J.H. Gerard.

metameric series of coelomic cavities by the septa increases its efficiency greatly because the force of local muscle contraction is not transferred throughout the length of the worm. Widening and elongation can occur in restricted areas. Crawling motions are effected by alternating waves of contraction by longitudinal and circular muscles (peristaltic contraction) passing down the body. Segments in which longitudinal muscles are contracted widen and anchor themselves against burrow walls or other substrata while other segments, in which circular muscles are contracted, elongate and stretch forward. Forces powerful enough for burrowing as well as locomotion can thus be generated. Swimming forms use undulatory rather than peristaltic movements in locomotion.

FIGURE 21-3

This feather-duster worm has a tough, noncalcareous tube (family Sabellidae).

Photograph by K. Lucas/Biological Photo Service.

Class Polychaeta

The polychaetes are the largest and most primitive class of annelids, with more than 5300 described species, most of them marine. Although the majority of them are from 5 to 10 cm long, some are less than a millimeter, and others may be as long as 3 m. Some are brightly colored in reds and greens; others are dull or iridescent. Some are picturesque, such as the "featherduster" worms (Figure 21-3).

Polychaetes live under rocks, in coral crevices, or in abandoned shells, or they burrow into mud or sand; some build their own tubes on submerged objects or in bottom material; some adopt the tubes or homes of other animals; some are pelagic, making up a part of the planktonic population. They are extremely abundant in some areas; for example, a square meter of mud flat may contain thou-

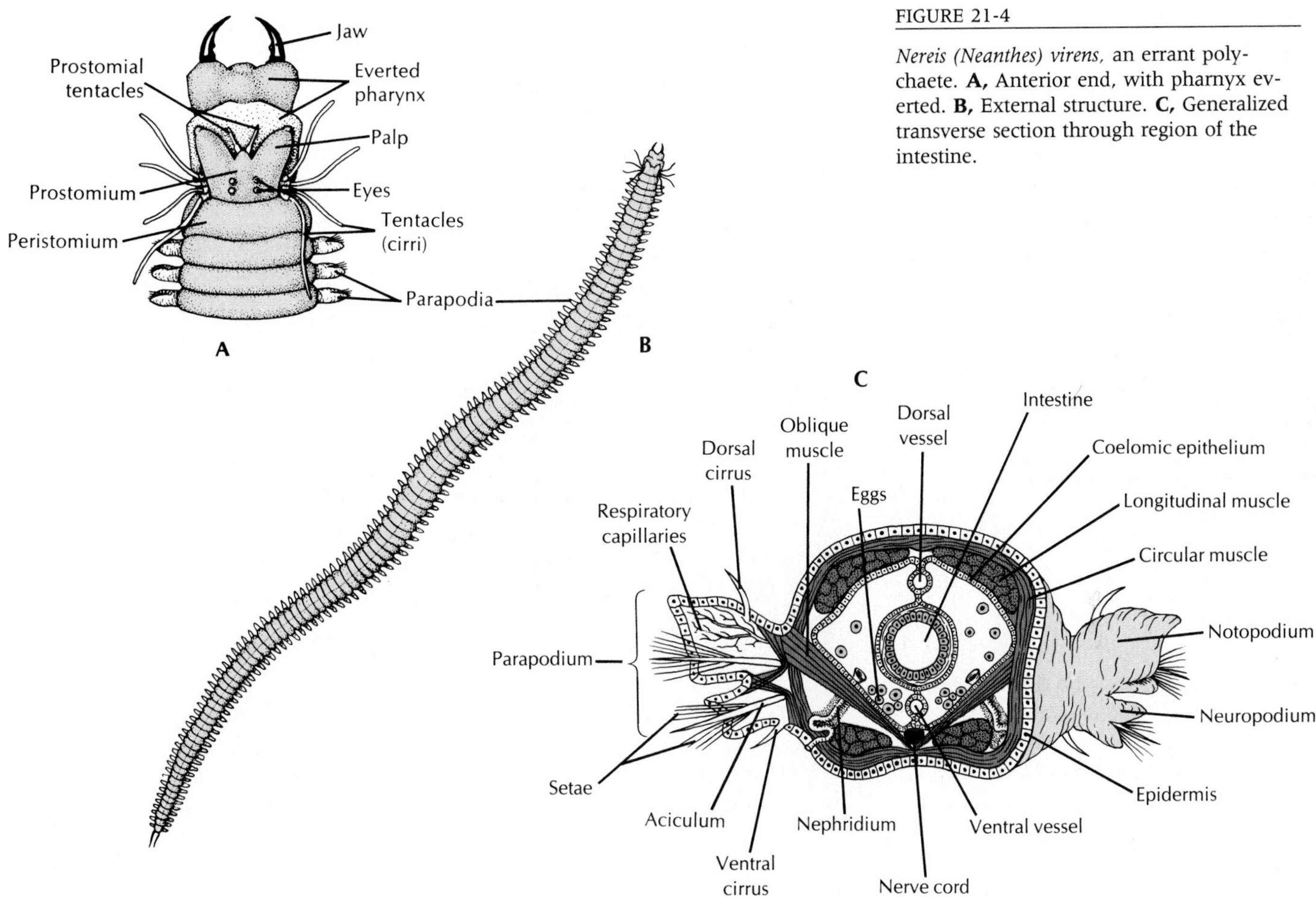

FIGURE 21-4

Nereis (Neanthes) virens, an errant polychaete. **A,** Anterior end, with pharnyx everted. **B,** External structure. **C,** Generalized transverse section through region of the intestine.

sands of polychaetes. They play a significant part in marine food chains, since they are eaten by fish, crustaceans, hydroids, and many others.

Polychaetes differ from other annelids in having a well-differentiated head with specialized sense organs; paired, paddlelike appendages, called **parapodia,** on most segments; and no clitellum (Figure 21-4). As their name implies, they have many setae, usually arranged in bundles on the parapodia. They show a pronounced differentiation of some body somites and a specialization of sensory organs practically unknown among clitellates (oligochaetes and leeches).

The polychaete typically has a head, or **prostomium,** which may or may not be retractile and which often bears eyes, antennae, and sensory palps (Figure 21-4).

The first segment **(peristomium)** surrounds the mouth and may bear setae, palps, or, in predatory forms, chitinous jaws. Ciliary feeders may bear a tentacular crown that may be opened like a fan or withdrawn into the tube.

The trunk is segmented and most segments bear parapodia, which may have lobes, cirri, setae, and other parts on them (Figure 21-4, *C*). Parapodia are composed of two main parts—a dorsal **notopodium** and ventral **neuropodium**—either of which may be prominent or reduced in a given species. The parapodia are used in crawling, swimming, or anchoring in tubes. They usually serve as the chief respiratory organs, although some polychaetes may also have gills. *Amphitrite,* for example, has three pairs of branched gills and long extensible tentacles (Figure 21-5). *Arenicola,* the lugworm (Figure 21-9), which burrows through the sand leaving characteristic castings at the entrance to its burrow, has paired gills on certain somites.

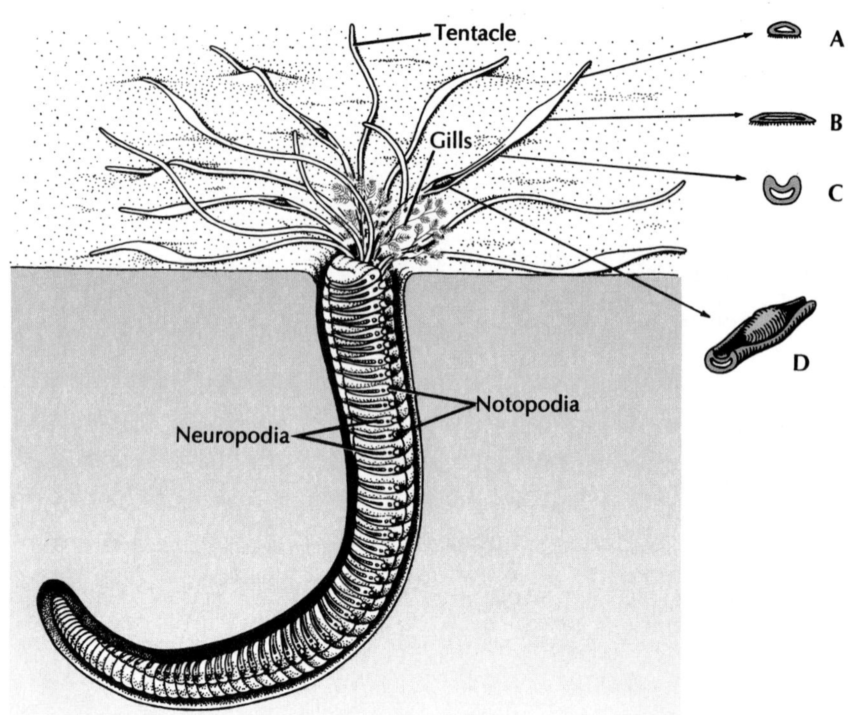

FIGURE 21-5

Amphitrite, which builds its tubes in mud or sand, extends long grooved tentacles out over the mud to pick up bits of organic matter. The smallest particles are moved along food grooves by cilia, larger particles by peristaltic movement. Its plumelike gills are blood red. **A,** Section through exploratory end of tentacle. **B,** Section through tentacle in area adhering to substratum. **C,** Section showing ciliary groove. **D,** Particle being carried toward mouth.

Some polychaetes live most of the year as sexually unripe animals called atokes, but during the breeding season a portion of the body develops into a sexually ripe worm called an epitoke, which is swollen with gametes (Figure 21-6). One example is the palolo worm, which lives in burrows among the coral reefs of the South Seas. During the reproductive cycle, the posterior somites become swollen with gametes. During the swarming period, which occurs at the beginning of the last quarter of the October-November moon, these epitokes break off and swim to the surface. Just before sunrise, the sea is literally covered with them, and at sunrise they burst, freeing the eggs and sperm for fertilization. The anterior portions of the worms regenerate new posterior sections. A related form swarms in the Atlantic in the third quarter of the June-July moon. Swarming is of great adaptive value because the synchronous maturation of all the epitokes ensures the maximum number of fertilized eggs. However, it is very hazardous; many types of predators have a feast. In the meantime, the atoke remains safe in its burrow to produce another epitoke at the next cycle!

Sense organs are more highly developed in polychaetes than in oligochaetes and include eyes and statocysts. Eyes, when present, may range from simple eyespots to well-developed organs. Usually the eyes are retinal cups, with rodlike photoreceptor cells lining the cup wall and directed toward the lumen of the cup.

In contrast to clitellates, polychaetes have no permanent sex organs, possess no permanent ducts for their sex cells, and usually have separate sexes. Gonads appear as temporary swellings of the peritoneum and shed their gametes into the coelom. Gametes are carried outside through gonoducts, through nephridia, or by rupture of the body wall. Fertilization is external, and development is indirect with a trochophore larva.

Some polychaetes are free-moving pelagic forms, some are active burrowers and crawlers, and some are sedentary, living in tubes or burrows that they rarely (or never) leave. An example of an active, predatory worm is *Nereis* (Greek mythology, a Nereid, or daughter of Nereus, ancient sea god), the clam worm (Figure 21-4). *Nereis* has a muscular, eversible pharynx equipped with jaws that can be thrust out with surprising speed and dexterity for capturing prey.

Most sedentary tube and burrow dwellers are particle feeders, using ciliary or mucoid methods of obtaining food. The principal food source is plankton and detritus. Some, like *Amphitrite* (Greek mythology, sea goddess) (Figures 21-2, *B* and 21-5), with head peeping out of the mud, send out long extensible tentacles over the surface. Cilia and mucus on the tentacles entrap particles found on the sea bottom and move them toward the mouth.

The fanworms, or "featherduster" worms, are beautiful tubeworms, fascinating to watch as they emerge from their secreted tubes and unfurl their lovely tentacular crowns to feed. A slight disturbance, sometimes even a passing shadow, causes them to duck quickly into the safety of the homes they have built. Food attracted to the feathery arms, or **radioles,** by ciliary action is trapped in mucus and is carried down ciliated food grooves to the mouth (Figure 21-7). Particles too large for the food grooves are carried along the margins and dropped

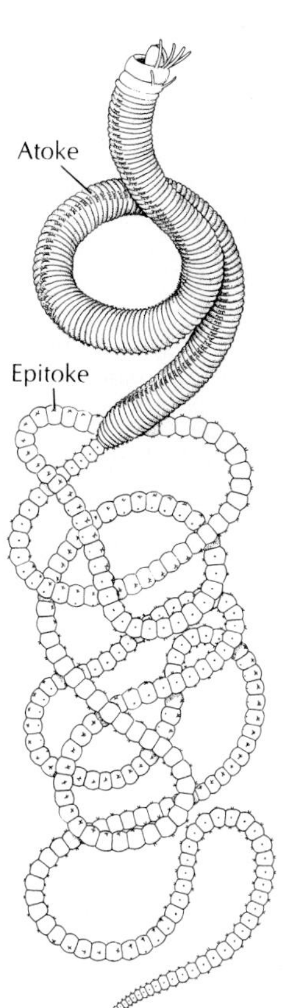

FIGURE 21-6

Eunice viridis, the Samoan palolo worm. The posterior segments make up the epitokal region, consisting of segments packed with gametes. Each segment has one eyespot on the ventral side. Once a year the worms swarm, and the epitokes detach, rise to the surface, and discharge their ripe gametes, leaving the water milky. By the next breeding season, the epitokes are regenerated.

After W.M. Woodworth, 1907; from Fauvel, P. 1959. Annélides polychètes. Reproduction. In P.P. Grasse (ed.). Traite de Zoologie, vol 5, part l. Paris, Masson et Cie.

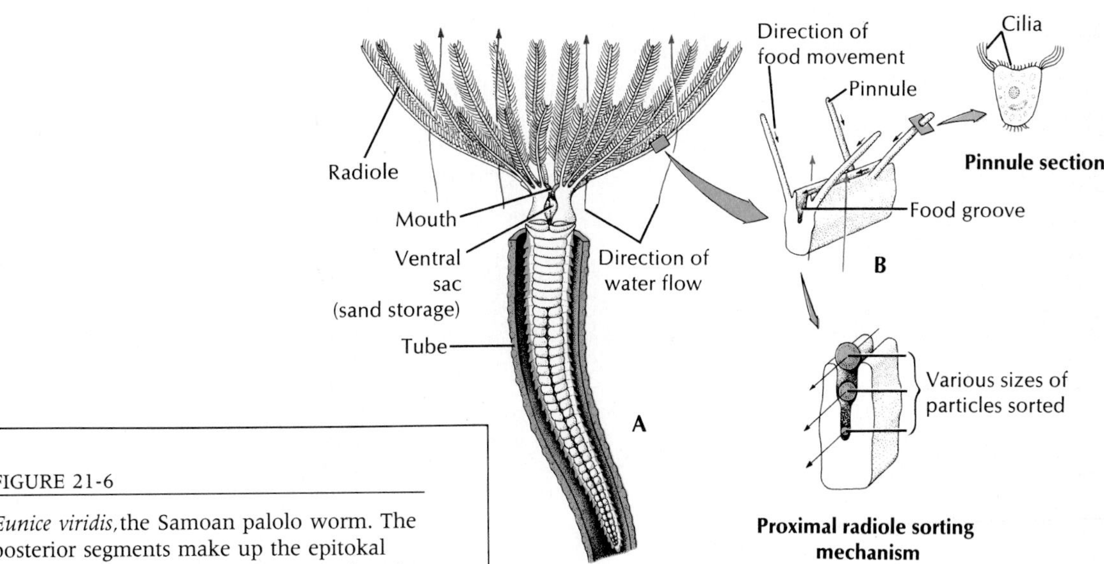

FIGURE 21-7

Sabella, a polychaete ciliary feeder. **A,** Anterior view of the crown. Cilia direct small food particles along grooved radioles to mouth and discard larger particles. Sand grains are directed to storage sacs and later are used in tube building. **B,** Distal portion of radiole showing ciliary tracts of pinnules and food grooves.

FIGURE 21-8

Chaetopterus, a sedentary polychaete (in U-tube), and *Phascolosoma,* a sipunculan worm (in center). *Chaetopterus* lives in a parchment tube through which it pumps water with its three pistonlike fans. The fans beat 60 times per minute to keep water currents moving. The winglike notopodia of the twelfth segment continuously secrete a mucous net that strains out food particles. As the net fills with food, the food cup rolls it into a ball and, when the ball is large enough (about 3 mm), the food cup bends forward and deposits the ball in a ciliated groove to be carried by cilia to the mouth and swallowed.

Courtesy American Museum of Natural History.

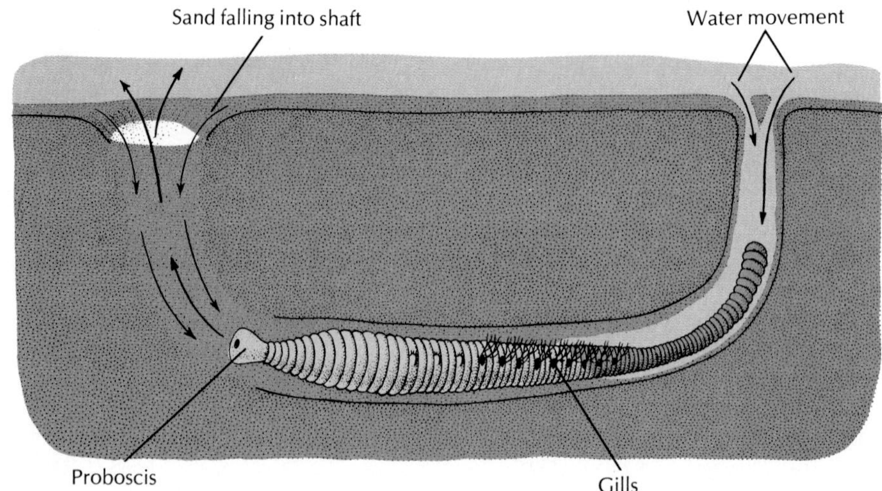

FIGURE 21-9

Arenicola, the lugworm, lives in an L-shaped burrow in intertidal mud flats. It burrows by successive eversions and retractions of its proboscis. By peristaltic movements it keeps water coming in the open end of the tube and filtering through the sand at the head end. The worm then ingests the food-laden sand.

off. Further sorting may occur near the mouth where only the small particles of food enter the mouth, and sand grains are stored in a sac to be used later in enlarging the tube.

Some worms, such as *Chaetopterus* (Gr. *chaitē*, hair or mane, + *pteron*, wing), secrete mucous filters through which they pump water to collect edible particles (Figure 21-8). The lugworm *Arenicola* (L. *arena*, sand, + *colere*, to inhabit) lives in an L-shaped burrow in which, by peristaltic movements, it keeps water filtering down through the sand and out the open end of the burrow. It ingests the food-laden sand brought by the water current (Figure 21-9).

Tube dwellers secrete many types of tubes. Some are parchmentlike; some are firm, calcareous tubes attached to rocks or other surfaces (Figure 21-2, *A*); and some are simply grains of sand or bits of shell or seaweed cemented together with mucous secretions. Many burrowers in sand and mud flats simply line their burrows with mucus (Figure 21-9).

Aristotle called earthworms the "intestines of the soil." Some 22 centuries later Charles Darwin published his observations in his classic *The Formation of Vegetable Mould Through the Action of Worms.* He showed how worms enrich the soil by bringing subsoil to the surface and mixing it with the topsoil. An earthworm can ingest its own weight in soil every 24 hours, and Darwin estimated that from 10 to 18 tons of dry earth per acre pass through their intestines annually, thus bringing up potassium and phosphorus from the subsoil and also adding to the soil nitrogenous products from their own metabolism. They expose the mold to the air and sift it into small particles. They also drag leaves, twigs, and organic substances into their burrows closer to the roots of plants. Their activities are important in aerating the soil. Darwin's views were at odds with his contemporaries, who thought earthworms were harmful to plants. But recent research has amply confirmed Darwin's findings, and earthworm management is now practiced in many countries.

■ Class Oligochaeta

The more than 3000 species of oligochaetes are found in a great variety of sizes and habitats. They include the familiar earthworms and many species that live in fresh water. Most are terrestrial or freshwater forms, but some are parasitic, and a few live in marine or brackish water.

Oligochaetes, with few exceptions, bear setae, which may be long or short, straight or curved, blunt or needlelike, or arranged singly or in bundles. Whatever the type, they are less numerous in oligochaetes than in polychaetes, as is implied by the class name, which means "few setae." Aquatic forms usually have longer setae than do earthworms.

EARTHWORMS

The most familiar of the oligochaetes are the earthworms ("night crawlers"), which burrow in moist, rich soil, emerging at night to feed on surface detritus and vegetation and to breed. In damp, rainy weather they stay near the surface, often with mouth or anus protruding from the burrow. In very dry weather they may burrow several feet underground, coil up in a slime chamber, and become dormant. *Lumbricus terrestris* (L. *lumbricum*, earthworm), the form commonly studied in school laboratories, is approximately 12 to 30 cm long (Figure 21-10). Giant tropical earthworms may have from 150 to 250 or more segments and may grow to as much as 3 m in length. They usually live in branched and interconnected tunnels.

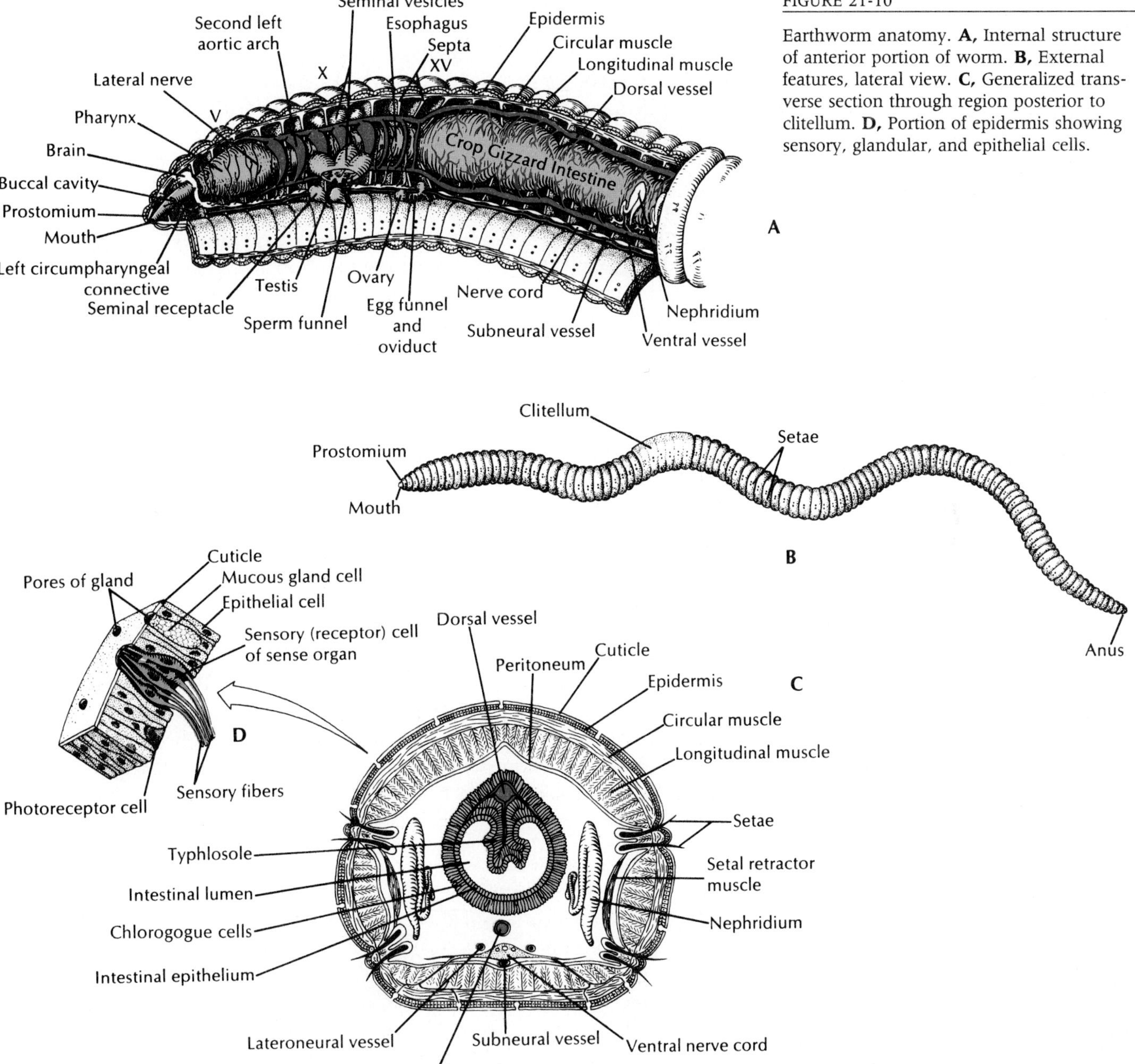

FIGURE 21-10

Earthworm anatomy. **A,** Internal structure of anterior portion of worm. **B,** External features, lateral view. **C,** Generalized transverse section through region posterior to clitellum. **D,** Portion of epidermis showing sensory, glandular, and epithelial cells.

Form and function

In earthworms the mouth is overhung by a fleshy prostomium at the anterior end, and the anus is on the terminal end (Figure 21-10, *B*). In most earthworms, each segment bears four pairs of chitinous setae (Figure 21-10, *C*), although in some oligochaetes each segment may have up to 100 or more. Each seta is a bristlelike rod set in a sac within the body wall and moved by tiny muscles (Figure 21-11). The setae project through small pores in the cuticle to the outside. In locomotion and burrowing, setae anchor parts of the body to prevent slipping. Earthworms move by peristaltic movement. Contraction of circular muscles in the anterior end lengthens the body, thus pushing the anterior end forward where it is anchored by setae; contractions of longitudinal muscles then shorten

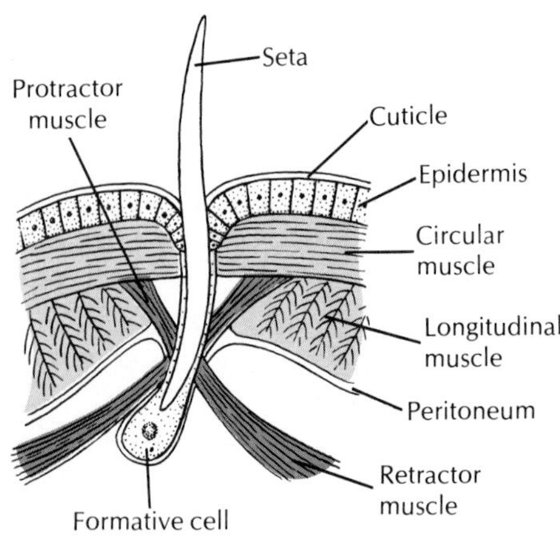

FIGURE 21-11

Seta with its muscle attachments showing relation to adjacent structures. Setae lost by wear and tear are replaced by new ones, which develop from formative cells.

Modified from Stephenson, J. 1972. The Oligochaeta. New York, Wheldon & Wesley Ltd., Stechert-Hafner Service Agency, Inc; and others.

the body, thus pulling the posterior end forward. As these waves of contraction pass along the entire body, it is gradually moved forward.

As in other annelids, the digestive tract is unsegmented and extends the length of the worm. The food of the earthworm is mainly decayed organic matter and bits of vegetation drawn in by the muscular **pharynx** (Figure 21-10, *A*). **Calciferous glands** along the esophagus secrete calcium ions into the gut lumen, reducing the excess calcium in the blood from the soil ingested by the earthworm. The food passes from the esophagus to a thin-walled storage organ, the **crop,** then into the muscular **gizzard**, which grinds the food into small pieces. The absorptive area of the intestine is increased by the **typhlosole,** an infolding of its dorsal side (Figure 21-10, *C*). **Chloragogue tissue** is found in the typhlosole and around the intestine. Chloragogue cells synthesize glycogen and fat and can break free to distribute these nutrients through the coelom. They also serve an excretory function.

Annelids have a double transport system — the coelomic fluid and circulatory system. Food, wastes, and respiratory gases are carried by both in varying degrees. The blood is carried in a closed system of blood vessels, including capillary systems in the tissues. There are five main longitudinal trunks, of which the **dorsal blood vessel** is the main pumping organ (Figure 21-10, *A* and *C*). The blood contains colorless ameboid cells and a dissolved respiratory pigment, **hemoglobin.** The blood of other annelids may have respiratory pigments other than hemoglobin.

The organs of excretion are the **nephridia,** a pair of which is found in each somite except the first three and the last one. Each one occupies parts of two successive somites (Figure 21-12). A ciliated funnel, known as the **nephrostome,** lies just anterior to an intersegmental septum and leads by a small ciliated tubule through the septum into the somite behind, where it connects with the main part of the nephridium. This part of the nephridium is made up of several loops of increasing size, which finally terminate in a bladderlike structure leading to an aperture, the **nephridiopore;** this opens to the outside near the ventral row of setae. By means of cilia, wastes from the coelom are drawn into the nephrostome

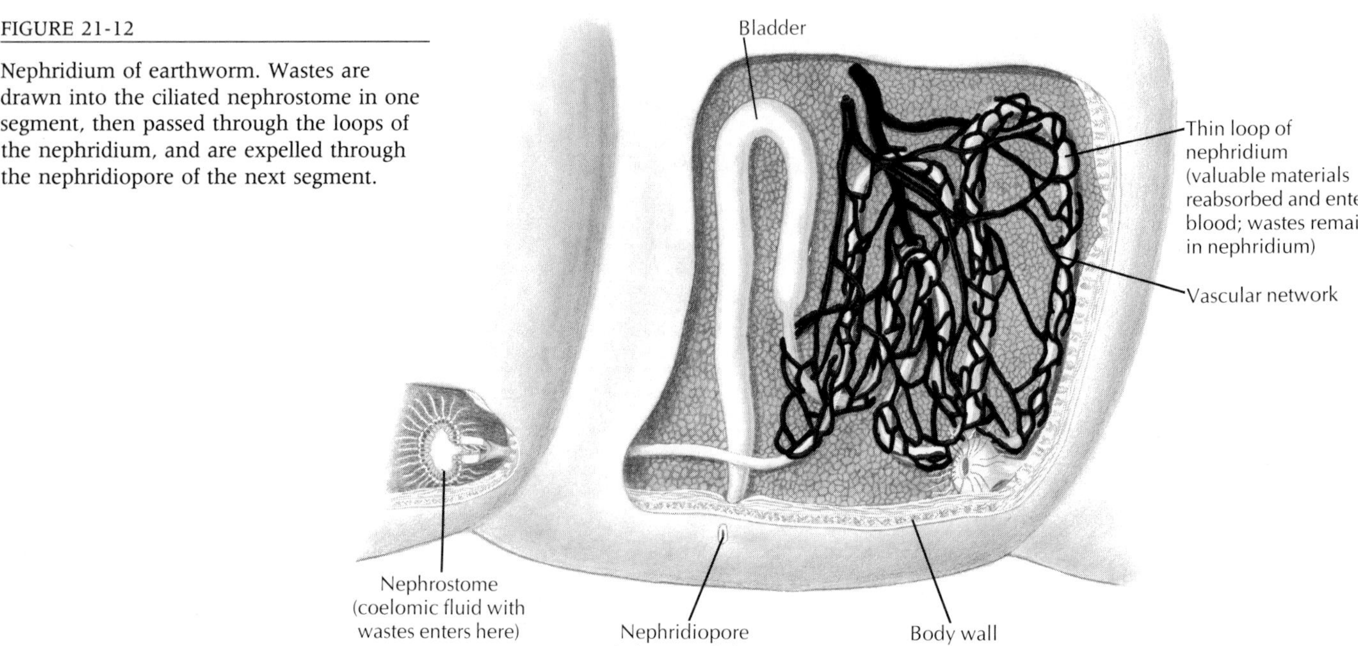

FIGURE 21-12

Nephridium of earthworm. Wastes are drawn into the ciliated nephrostome in one segment, then passed through the loops of the nephridium, and are expelled through the nephridiopore of the next segment.

and tubule. In the tubule water and salts are resorbed, forming a dilute urine that is discharged to the outside through the nephridiopore.

The plan of the nervous system in oligochaetes is typical of all annelids. There is a pair of **cerebral ganglia** (brain) above the pharynx joined to the ventral nerve cord by a pair of connectives around the pharynx (Figure 21-10, *A*). The double nerve cord has a pair of ganglia in each somite, giving off segmental nerves containing both sensory and motor fibers. For rapid escape movements most annelids are provided with from one to several very large axons commonly called **giant axons,** or giant fibers, located in the ventral nerve cord. The speed of conduction in these giant nerve fibers is tremendously increased over that of small axons.

Earthworms are hermaphroditic and exchange sperm during copulation, which usually occurs at night. When mating, the worms extend their anterior ends from their burrows and bring their ventral surfaces together (Figure 21-13). They are held together by mucus secreted by the **clitellum** and by special ventral setae, which penetrate each other's bodies in the regions of contact. Sperm are discharged and travel to the seminal receptacles of the other worm in its seminal grooves. After copulation each worm secretes around its clitellum, first a mucous tube and then a tough, chitinlike band that forms a **cocoon** (Figure 21-13). As the cocoon passes forward, eggs from the oviducts, albumin from the skin glands, and sperm from the mate (stored in the seminal receptacles) are poured into it. Fertilization of the eggs now takes place within the cocoon. When the cocoon

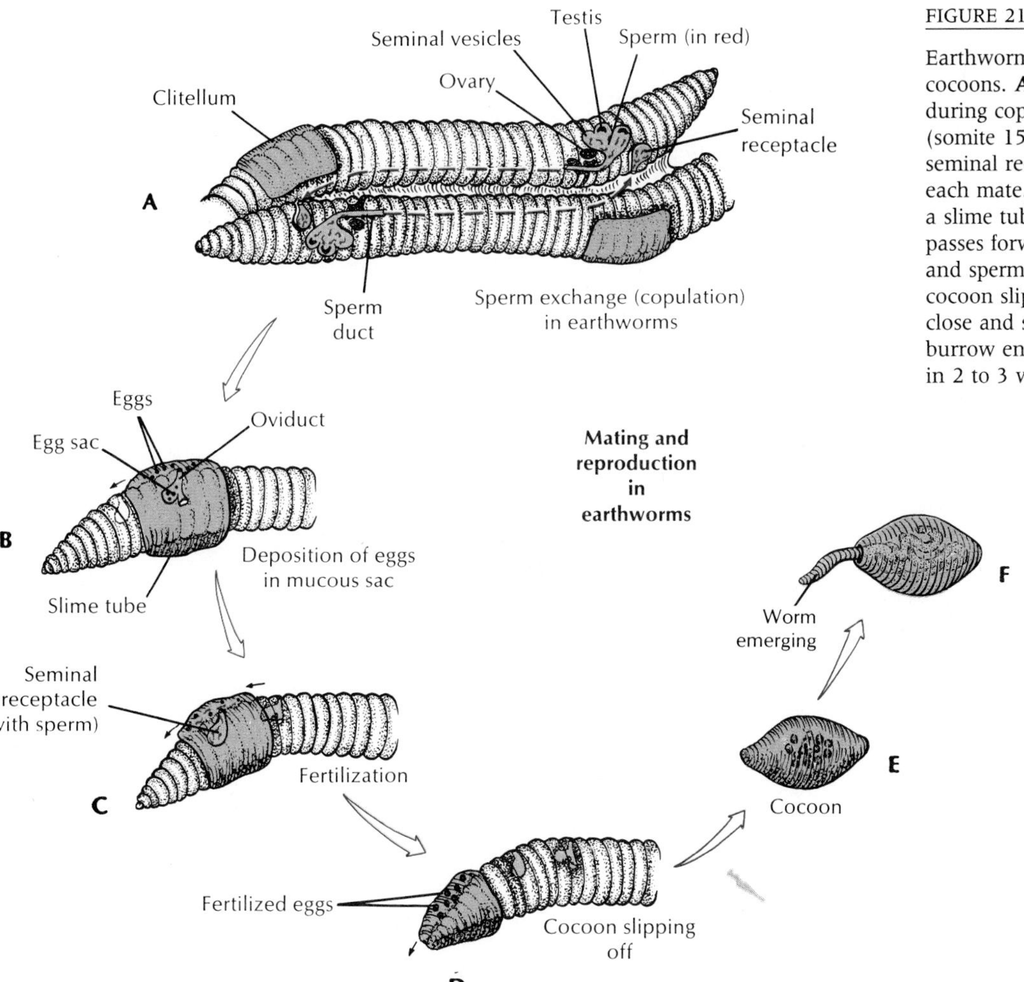

FIGURE 21-13

Earthworm copulation and formation of egg cocoons. **A,** Mutual insemination occurs during copulation; sperm from genital pore (somite 15) pass along seminal grooves to seminal receptacles (somites 9 and 10) of each mate. **B** and **C,** After worms separate, a slime tube formed over the clitellum passes forward to receive eggs from oviducts and sperm from seminal receptacles. **D,** As cocoon slips off over anterior end, its ends close and seal. **E,** Cocoon is deposited near burrow entrance. **F,** Young worms emerge in 2 to 3 weeks.

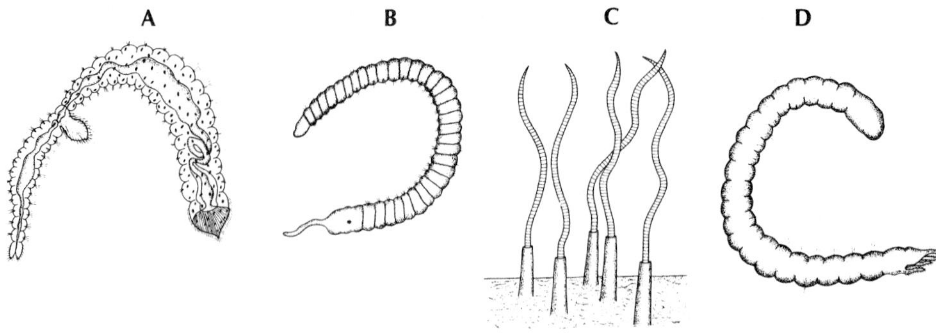

FIGURE 21-14

Some freshwater oligochaetes. **A,** *Stylaria* has the prostomium drawn out into a long snout. **B,** *Aeolosoma* uses cilia around the mouth to sweep in food particles, and it buds off new individuals asexually. **C,** *Tubifex* lives head down in long tubes. **D,** *Aulophorus* is provided with ciliated anal gills.

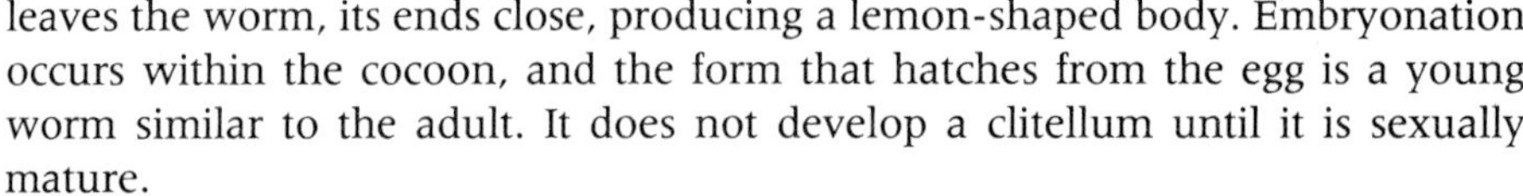
leaves the worm, its ends close, producing a lemon-shaped body. Embryonation occurs within the cocoon, and the form that hatches from the egg is a young worm similar to the adult. It does not develop a clitellum until it is sexually mature.

FRESHWATER OLIGOCHAETES

Freshwater oligochaetes usually are smaller and have more conspicuous setae than do the earthworms. They are more mobile than earthworms and tend to have better-developed sense organs. They are generally benthic forms that creep about on the bottom or burrow into the soft mud. Aquatic oligochaetes provide an important food source for fishes. A few are ectoparasitic.

Some aquatic forms have **gills.** In some the gills are long, slender projections from the body surface. Others have ciliated posterior gills (Figure 21-14, *D*), which they extend from their tubes and use to keep the water moving. Most forms respire through the skin as do the earthworms.

The chief foods are algae and detritus, which they may pick up by extending a mucus-coated pharynx. Burrowers swallow mud and digest the organic material. Some ciliary feeders that use currents produced by cilia at the anterior end of the body to sweep food particles into the mouth (Figure 21-14, *B*).

FIGURE 21-15

The world's largest leech, *Haementeria ghilianii,* on the arm of Dr. Roy K. Sawyer, who found it in French Guiana, South America.

Photograph by T. Branning.

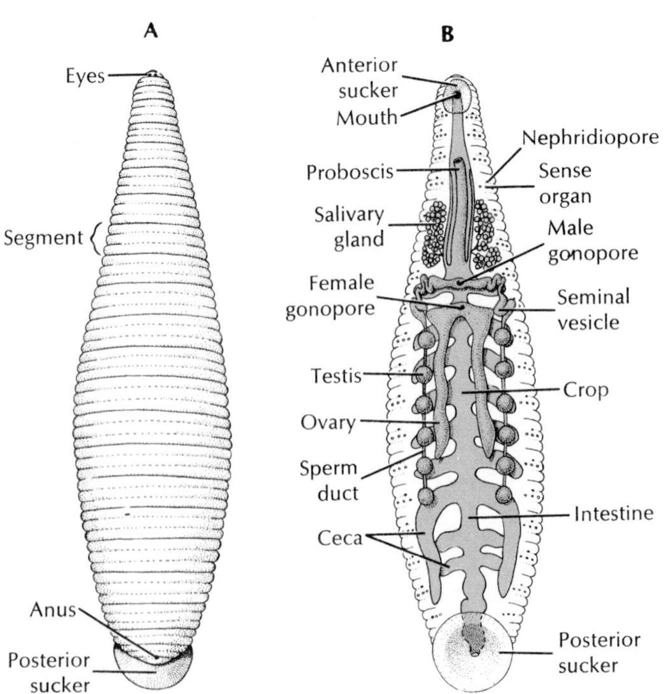

FIGURE 21-16

Structure of a leech, *Placobdella.* **A,** External appearance, dorsal view. (See also Figure 21-2, *D.*) **B,** Internal structure, ventral view.

Class Hirudinea—The Leeches

Leeches, numbering about 500 species, are found predominantly in freshwater habitats, but a few are marine, and some have even adapted to terrestrial life in moist, warm areas. Most leeches are between 2 and 6 cm in length, but some are smaller, and some reach 20 cm or more (Figure 21-15). They are found in a variety of patterns and colors—black, brown, red, or olive green. They are usually flattened dorsoventrally (Figure 21-16).

Leeches have a fixed number of segments, usually 34, and typically have both an anterior and a posterior sucker. They have no parapodia and, except in one primitive genus (*Acanthobdella*), they have no setae. *Acanthobdella* has five anterior coelomic compartments separated by septa, but septa have disappeared in all the other leeches. The coelom has become filled with connective tissue and muscle, substantially reducing its effectiveness as a hydrostatic skeleton.

Many leeches live as carnivores on small invertebrates; some are temporary parasites, sucking blood from vertebrates; and some are permanent parasites, never leaving their host. Most leeches have a muscular, protrusible proboscis or a muscular pharynx with three jaws armed with teeth. They feed on the body juices of their prey, penetrating its surface with their proboscis or jaws and sucking the fluids with their powerful, muscular pharynx. Blood sucking leeches (Figure 21-17) secrete an anticoagulant in their saliva. Predatory leeches feed frequently, but those that feed on blood of vertebrates consume large meals (up to several times their body weight) and digest the food slowly. The slow digestion of their meals may be due to the absence of secretion by their gut of amylases, lipase, or endopeptidases. In fact, they apparently depend mostly on bacteria in their gut for digestion of the blood meal.

For centuries the "medicinal leech" (*Hirudo medicinalis*) was used for blood letting because of the mistaken idea that bodily disorders and fevers were caused by an excess of blood. Leech collecting and leech culture in ponds were practiced in Europe on a commercial scale during the nineteenth century. Wordsworth's poem "The Leech-Gatherer" was based on this use of the leech. The use of leeches has an advantage over the lancing of veins; the operation is painless because of anesthetic components in the leech's saliva. A 10- to 12-cm long leech can extend to a much greater length when distended with blood, and the amount of blood it can suck is considerable.

Leeches are once again being used medically. When fingers or toes are severed, microsurgeons can reconnect arteries but not the more delicate veins. Leeches are used to relieve congestion until the veins can grow back into the healing digit.

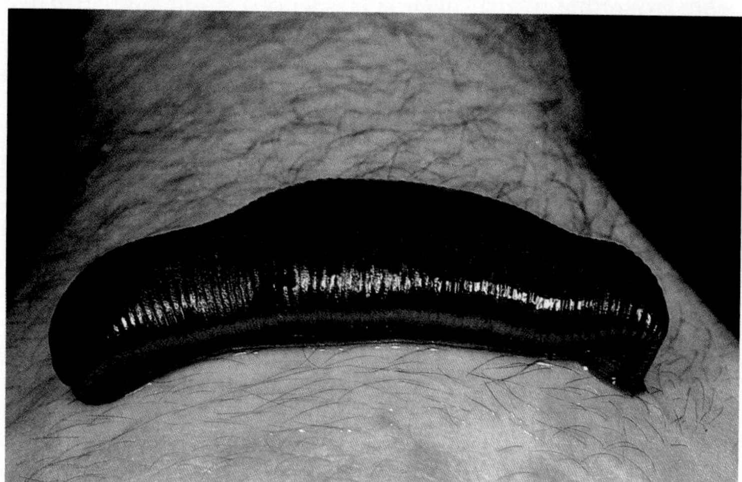

FIGURE 21-17

Hirudo medicinalis feeding on blood from human arm.

Photograph by C.P. Hickman, Jr.

Leeches are hermaphroditic but practice cross-fertilization during copulation. Sperm are transferred by a penis or by hypodermic impregnation. Leeches have a clitellum, but it is evident only during the breeding season. After copulation, the clitellum secretes a cocoon that receives the eggs and sperm. Cocoons are buried in bottom mud, attached to submerged objects or, in terrestrial species, placed in damp soil. Development is similar to that of oligochaetes.

Leeches are highly sensitive to stimuli associated with the presence of a prey or host. They are attracted by and will attempt to attach to an object smeared with appropriate host substances, such as fish scales, oil secretions, or sweat. Those that feed on the blood of mammals are attracted by warmth, and the terrestrial haemadipsid leeches of the tropics will converge on a person standing in one place.

Phylogeny and Adaptive Radiation

PHYLOGENY

There are so many similarities in the early development of the molluscs, annelids, and primitive arthropods that there seems little doubt about their close relationship. It is believed that the common ancestor of the three phyla was some type of flatworm. Many marine annelids and molluscs have an early embryogenesis typical of protostomes, in common with some marine flatworms, suggesting a real, if remote, relationship. Annelids share with the arthropods an outer secreted cuticle and a similar nervous system, and there is a similarity between the lateral appendages (parapodia) of many marine annelids and the appendages of certain primitive arthropods. The most important resemblance, however, probably lies in the segmented plan of the annelid and the arthropod body structure.

Although the polychaetes are more primitive in some respects, such as in their reproductive system, some authorities have argued that the ancestral annelids were more similar to the oligochaetes in overall body plan and that these gave rise to the polychaetes and leeches. Others maintain that the oligochaetes developed from polychaete stock. The leeches are closely related to the oligochaetes and probably evolved from them in connection with a swimming existence and the abandonment of a burrowing mode of life.

ADAPTIVE RADIATION

Annelids are an ancient group that has undergone extensive adaptive radiation. The basic body structure, particularly of the polychaetes, lends itself to great modification. As marine worms, polychaetes have a wide range of habitats in an environment that is not physically or physiologically demanding. Unlike the earthworms, whose environment imposes strict physical and physiological selective pressure, the polychaetes have been free to experiment and thus have achieved a wide range of adaptive features.

A basic adaptive feature in the evolution of annelids is their septal arrangement, resulting in fluid-filled coelomic compartments. Fluid pressure in these compartments is used as a hydrostatic skeleton in precise movements such as burrowing and swimming. Powerful circular and longitudinal muscles have been adapted for flexing, shortening, and lengthening the body.

There is a wide variation in feeding adaptations, from the sucking pharynx of the oligochaetes and the chitinous jaws of carnivorous polychaetes to the specialized tentacles and cirri of the ciliary feeders.

In polychaetes the parapodia have been adapted in many ways and for many functions, chiefly locomotion and respiration.

In leeches many of their adaptations, such as suckers, cutting jaws, pumping pharynx, distensible gut, and others, are related to their predatory and bloodsucking habits.

■ Summary

The phylum Annelida is a large, cosmopolitan group containing the marine polychaetes, the earthworms and freshwater oligochaetes, and the leeches. Certainly the most important structural advancement of this group is metamerism, the division of the body into a series of similar segments, each of which contains a repeated arrangement of many organs and systems. The coelom is also highly developed in the annelids, and this, together with the septal arrangement of fluid-filled compartments and a well-developed body-wall musculature, has provided the annelids with an effective hydrostatic skeleton for precise burrowing and swimming movements. The primitive metamerism of the annelids lays the groundwork for the much more specialized metamerism of the arthropods to be considered in the next chapter.

Polychaetes are the largest class of annelids and are mostly marine. They have many setae on each somite, which are borne on paired parapodia. Parapodia show a wide variety of adaptations among polychaetes, including specialization for swimming, respiration, crawling, maintaining position in a burrow, pumping water through a burrow, and as accessory feeding organs. Some polychaetes are mostly predaceous and have an eversible pharynx with jaws. Other polychaetes rarely leave the burrows or tubes in which they live. Several styles of deposit and filter feeding are shown among the members of this group. Polychaetes are dioecious, have a primitive reproductive system, no clitellum, practice external fertilization, and their larva is a trochophore.

The class Oligochaeta contains the earthworms and many freshwater forms; they have a small number of setae per segment (compared to the Polychaeta) and no parapodia. The circulatory system is closed, and the dorsal blood vessel is the main pumping organ. There is a pair of nephridia in most somites. Earthworms contain the typical annelid nervous system: dorsal cerebral ganglia connected to a double, ventral nerve cord with segmental ganglia, running the length of the worm. Oligochaetes are hermaphroditic and practice cross-fertilization. The clitellum plays an important role in reproduction, including secretion of mucus to surround the worms during copulation and secretion of a cocoon to receive the eggs and sperm and in which embryonation occurs. A small, juvenile worm hatches from the cocoon.

The leeches (class Hirudinea) are mostly freshwater, although a few are marine and a few are terrestrial. They feed mostly on fluids; many are predators, some are temporary parasites, and a few are permanent parasites. The hermaphroditic leeches reproduce in a fashion similar to oligochaetes, with cross-fertilization and cocoon formation by the clitellum.

Embryological evidence supports a phylogenetic relationship of the annelids with the molluscs and arthropods.

A beautiful Christmas-tree worm, *Spirobranchus giganteus,* extends its feathery plumes from the mouth of its tube to trap minute food particles on the delicate radioles.

Photograph by K. Sandved.

REVIEW QUESTIONS

1. Name eight of the most important characteristics of the phylum Annelida.
2. Distinguish among the classes of the phylum Annelida.
3. Describe the annelid body plan, including the body wall, segments, coelom and its compartments, and coelomic lining.
4. Explain how the hydrostatic skeleton of the annelids helps them to burrow. How is the efficiency for burrowing increased by metamerism?
5. Apart from their utility as a hydrostatic skeleton, what was the evolutionary significance of metamerism and the coelom to their earliest possessors?
6. Define each of the following: prostomium, peristomium, radioles, parapodium, neuropodium, notopodium.
7. Describe at least three ways that various polychaetes obtain food.
8. Explain the function of each of the following in earthworms: pharynx, calciferous glands, crop, gizzard, typhlosole, chloragogue tissue.
9. Describe the main features of each of the following in earthworms: circulatory system, nervous system, excretory system.
10. Describe the function of the clitellum and the cocoon.
11. Describe the ways in which leeches obtain food.
12. What are the main differences in reproduction and development among the three classes of annelids?
13. What are the phylogenetic relationships among the molluscs, annelids, and arthropods? What is the evidence for these relationships?

SELECTED REFERENCES

Barnes, R.D. 1987. Invertebrate zoology, ed. 5. Philadelphia, Saunders College/Holt, Rinehart & Winston. *Chapter 10 covers annelids.*

Dales, R.P. 1967. Annelids, ed. 2. London, The Hutchinson Publishing Group, Ltd. *A concise account of the annelids.*

Gosner, K.L. 1979. A field guide to the Atlantic seashore: invertebrates and seaweeds of the Atlantic coast from the Bay of Fundy to Cape Hatteras. The Peterson Field Guide Series. Boston, Houghton Mifflin Co. *Good for identification of polychaetes from the eastern coast of the United States.*

Nicholls, J.C., and D. Van Essen. 1974. The nervous system of the leech. Sci. Am. **230**:38-48 (Jan.). *Because the leech has large nerve cells and only a few neurons perform a given function, its nervous system is particularly appropriate for experimental studies.*

Pennak, R.W. 1989. Freshwater invertebrates of the United States, ed. 3. New York, John Wiley & Sons, Inc. *A brief account with keys to families and genera of freshwater annelids.*

A WINNING COMBINATION

*Tunis, Algeria — Treating it as an invading army, Tunisia, Algeria, and Morocco have mobilized to fight the most serious infestation of locusts in over 30 years. Billions of the insects have already caused extensive damage to crops and are threatening to inflict great harm to the delicate economies of North Africa.**

The staggering losses occasionally inflicted by the billions of locusts in Africa serve as only one reminder of our ceaseless struggle with the dominant group of animals on earth today: the insects. With nearly 1 million species recorded, and probably as many yet remaining to be classified, insects far outnumber all the other species of animals in the world combined. Numbers of individuals are equally enormous. Some scientists have estimated that there are 200 million insects for every single human alive today! Insects have an unmatched ability to adapt to all land environments and to virtually all climates. Having originally evolved as land animals, insects developed wings and invaded the air 150 million years before flying reptiles, birds, or mammals. Many have exploited freshwater habitats, where they are now widely prevalent; only in the sea are the numbers of insects more limited.

How can we account for the enormous success of these creatures? In common with other arthropods, insects have a combination of valuable structural and physiological adaptations, including a versatile exoskeleton, segmentation, an efficient respiratory system, and highly developed sensory organs. In addition, insects have a waterproofed cuticle, and many have extraordinary abilities to survive adverse environmental conditions. These adaptations and others will be described in this chapter.▪

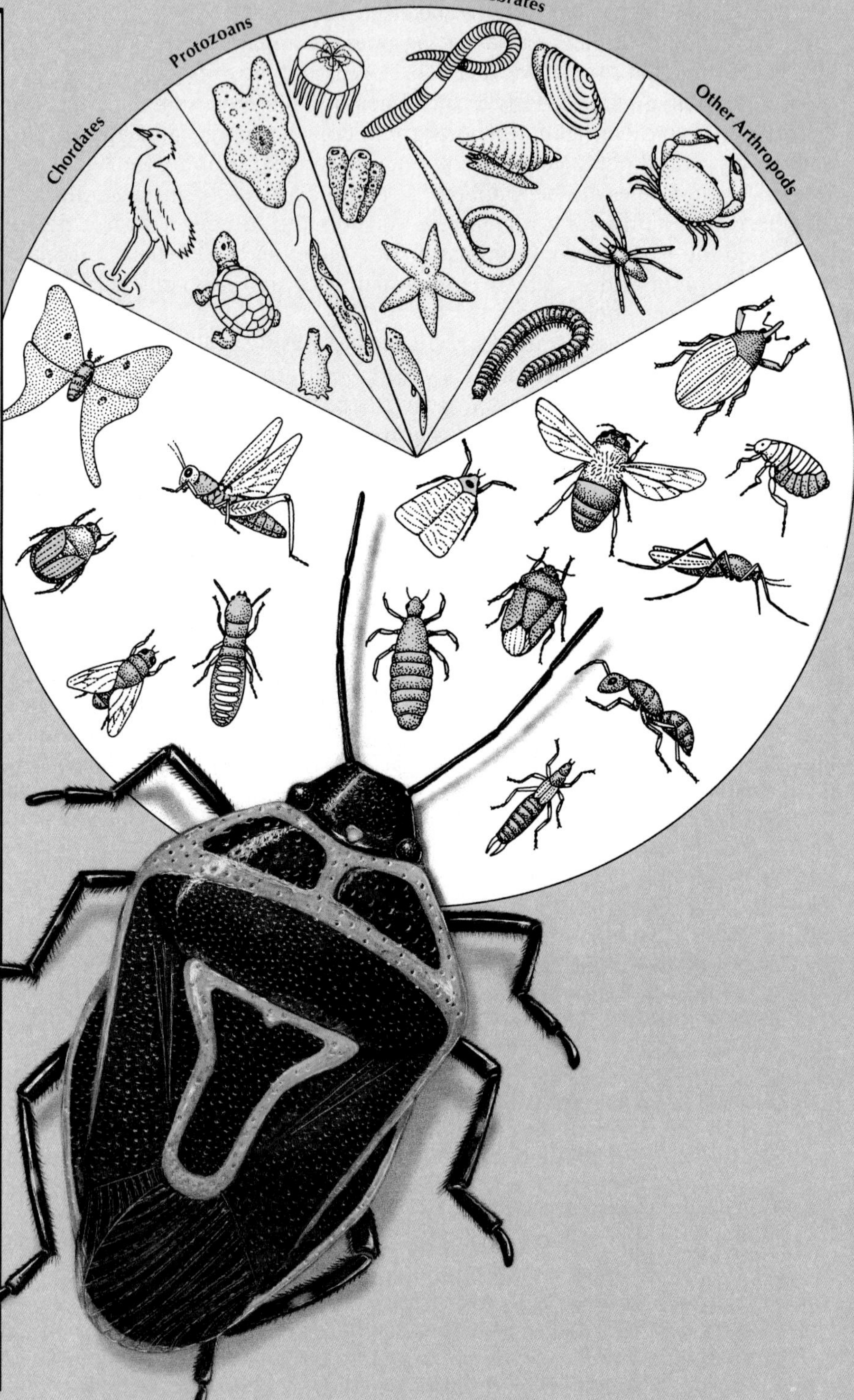

*From the *New York Times*, 20 April 1988.

CHAPTER 22

Arthropods

The phylum Arthropoda (ar-throp'o-da) (Gr. *arthron,* join,+ *pous, podos,* foot) embraces the largest assemblage of living animals on earth. It includes the spiders, scorpions, ticks, mites, crustaceans, millipedes, centipedes, insects, and some small groups. In addition there is a rich fossil record extending back to the very late Precambrian period (Figure 22-1).

Arthropods are eucoelomate protostomes with well-developed organ systems, and their cuticular exoskeleton containing chitin is a prominent characteristic. Like the annelids, they are conspicuously segmented; their primitive body pattern is a linear series of similar somites, each with a pair of jointed appendages. However, unlike the annelids, the arthropods have embellished the segmentation theme: there is much variation in the pattern of somites and appendages in the phylum. Often the somites are combined or fused into functional groups, called **tagmata,** for specialized purposes. Appendages, too, are frequently differentiated and specialized for walking, swimming, flying, eating, and so on.

Few arthropods exceed 60 cm in length, and most are far smaller. The largest is a Japanese crab (*Macrocheira kaempferi*), which has approximately a 3.7 m span; the smallest is a parasitic mite, which is less than 0.1 mm long.

Arthropods are usually active, energetic animals. However we judge them, whether by their great diversity or their wide ecological distribution or their vast numbers of species, the answer is the same: they are the most successful of all animals.

Although arthropods compete with us for food supplies and spread serious diseases, they are essential in pollination of many food plants, and they also serve as food, yield drugs and dyes, and create products such as silk, honey, and beeswax.

A

B

FIGURE 22-1

Fossils of early arthropods. **A,** Trilobites. These animals were abundant in the mid-Cambrian period. **B,** Eurypterid fossil. Eurypterids flourished in Europe and North America from Ordovician to Permian periods.

Photographs by A.J. Copley/Visuals Unlimited.

CHARACTERISTICS

1. Bilateral symmetry; **metameric body,** tagmata of head and trunk; head, thorax, and abdomen; or cephalothorax and abdomen
2. **Jointed appendages;** primitively, one pair to each somite (metamere), but number often reduced; appendages often modified for specialized functions
3. **Exoskeleton of cuticle** containing protein, lipid, chitin, and often calcium carbonate secreted by underlying epidermis and shed (molted) at intervals
4. **Complex muscular system,** with exoskeleton for attachment; **striated muscles** for rapid action; smooth muscles for visceral organs; **no cilia**
5. **Reduced coelom;** most of body cavity consisting of hemocoel (sinuses, or spaces, in the tissues) filled with blood
6. **Complete digestive system;** mouthparts modified from appendages and adapted for different methods of feeding
7. Open circulatory system, with dorsal **contractile heart,** arteries, and hemocoel
8. Respiration by **body surface, gills, tracheae** (air tubes), or **book lungs**
9. Paired excretory glands called **coxal, antennal,** or **maxillary glands** present in some, homologous to metameric nephridial system of annelids; some with other excretory organs, called **malphigian tubules**
10. Nervous system of annelid plan, with dorsal brain connected by a ring around the gullet to a double nerve chain of ventral ganglia; fusion of ganglia in some species; well-developed sensory organs
11. Sexes usually separate, with paired reproductive organs and ducts; usually internal fertilization; **oviparous** or **ovoviviparous;** often with **metamorphosis;** parthenogenesis in a few forms

CLASSIFICATION

Subphylum Trilobita (tri'lo-bi'ta) (Gr. *tri-*, three, + *lobos,* lobe)—**trilobites.** All extinct forms; Cambrian to Carboniferous; body divided by two longitudinal furrows into three lobes; distinct head, thorax, and abdomen; biramous (two-branched) appendages.

Subphylum Chelicerata (ke-lis'e-ra'ta) (Gr. *chēle,* claw, + *keras,* horn, + *ata,* group suffix)—**eurypterids, horseshoe crabs, spiders, ticks.** First pair of appendages modified to form chelicerae; pair of pedipalps and four pairs of legs; no antennae, no mandibles; cephalothorax and abdomen often with segments fused.

Class Merostomata (mer'o-sto'ma-ta) (Gr. *meros,* thigh, + *stoma,* mouth, + *ata,* group suffix)—**aquatic chelicerates.** Cephalothorax and abdomen; compound lateral eyes; appendages with gills; sharp telson; **subclasses Eurypterida** (all extinct) and **Xiphosurida,** the horseshoe crabs.

Class Pycnogonida (pik'no-gon'i-da) (Gr. *pyknos,* compact, + *gony,* knee, angle)—**sea spiders.** Small (3 to 4 mm), but some reach 500 mm; body chiefly cephalothorax; tiny abdomen; usually four pairs of long walking legs (some with five or six pairs); one pair of subsidiary legs (ovigers) for egg bearing; mouth on long proboscis; four simple eyes; no respiratory or excretory system. Example: *Pycnogonum.*

Class Arachnida (ar-ack'ni-da) (Gr. *arachnē,* spider)—**scorpions, spiders, mites, ticks, harvestmen.** Four pairs of legs; segmented or unsegmented abdomen with or without appendages and generally distinct from cephalothorax; respiration by gills, tracheae, or book lungs; excretion by malpighian tubules or coxal glands; dorsal bilobed brain connected to ventral ganglionic mass with nerves; simple eyes; sexes separate; chiefly oviparous; no true metamorphosis. Examples: *Argiope, Centruroides.*

Subphylum Crustacea (crus-ta'she-a) (L. *crusta,* shell, + *acea,* group suffix)—**crustaceans.** Mostly aquatic, with gills; cephalothorax usually with dorsal carapace; biramous appendages, modified for various functions. Head appendages consisting of two pairs of antennae, one pair of mandibles, and two pairs of maxillae. Sexes usually separate. Development primitively with nauplius stage (see résumé of classes, pp. 477-480).

Subphylum Uniramia (yu-ni-ra'me-a) (L. *unus,* one, + *ramus,* a branch)—**insects and myriapods.** All appendages uniramous; head appendages consisting of one pair of antennae, one pair of mandibles, and one or two pairs of maxillae.

Class Diplopoda (di-plop'o-da) (Gr. *diploos,* double, + *pous, podos,* foot)—**millipedes.** Subcylindrical body; head with short antennae and simple eyes; body with variable number of somites; short legs, usually two pairs of legs to a somite; separate sexes. Examples: *Julus, Spirobolus.*

Class Chilopoda (ki-lop'-o-da) (Gr. *cheilos,* lip, + *pous, podos,* foot)—**centipedes.** Dorsoventrally flattened body; variable number of somites, each with one pair of legs; one pair of long antennae; separate sexes. Examples: *Cermatia, Lithobius, Geophilus.*

Class Pauropoda (pau-rop'o-da) (Gr. *pauros,* small, + *pous, podos,* foot)—**pauropods.** Minute (1 to 1.5 mm), cylindrical body consisting of double segments and bearing nine or ten pairs of legs; no eyes. Example: *Pauropus.*

Class Symphyla (sim'fi-la) (Gr. *syn,* together, + *phylon,* tribe)—**garden centipedes.** Slender (1 to 8 mm) with long, threadlike antennae; body consisting of 15 to 22 segments with 10 to 12 pairs of legs; no eyes. Example: *Scutigerella.*

Class Insecta (in-sek'ta) (L. *insectus,* cut into)—**insects.** Body with distinct head, thorax, and abdomen; pair of antennae; mouthparts modified for different food habits; head of six fused somites; thorax of three somites; abdomen with variable number, usually 11 somites; thorax with two pairs of wings (sometimes one pair or none) and three pairs of jointed legs; separate sexes; usually oviparous; gradual or abrupt metamorphosis. (A brief description of insect orders is given on pp. 502-505).

ECOLOGICAL RELATIONSHIPS

The arthropods are found in all types of environment from low ocean depths to very high altitudes and from the tropics far into both north and south polar regions. Some species are adapted for life in the air; others for life on land or in fresh, brackish, and marine waters; others live in or on the bodies of plants and other animals. Some live in places where no other form could survive.

Although all types—carnivorous, omnivorous, and symbiotic—occur in this vast group, the majority are herbivorous. Most aquatic arthropods depend on algae for their nourishment, and the majority of land forms live chiefly on plants. In diversity of ecological distribution the arthropods have no rivals.

WHY HAVE ARTHROPODS BEEN SO SUCCESSFUL?

The success of the arthropods is demonstrated by their diversity, number of species, wide distribution, variety of habitats and feeding habits, and power of adaptation to changing conditions. Some of the structural and physiological patterns that have been helpful to them follow, but of these, the most important is the exoskeleton.

1. *A versatile exoskeleton.* The arthropods possess an exoskeleton that is highly protective without sacrificing mobility. The skeleton is the **cuticle,** an outer covering secreted by the underlying epidermis. This versatile structure may be adapted for biting mouthparts, sensory organs, copulatory organs, wings, and ornamental purposes.

The cuticle is made up of an inner and usually thicker **endocuticle** (or **procuticle**) and an outer, relatively thin **epicuticle.** The endocuticle contains **chitin** bound with protein. Chitin is a tough, resistant, nitrogenous polysaccharide that is insoluble in water, alkalis, and weak acids. Thus the endocuticle is not only flexible and lightweight but also affords protection, particularly against dehydration. In most crustaceans, the endocuticle in some areas is also impregnated with **calcium salts,** which reduce its flexibility. In the hard shells of lobsters and crabs, for instance, this calcification is extreme. The outer epicuticle is composed of protein and often lipids. The protein is stabilized and hardened by a chemical process called tanning, adding further protection. Both the endocuticle and epicuticle are laminated, that is, composed of several layers each (Figure 6-10, p. 157).

The cuticle may be soft and permeable or may form a veritable coat of armor. Between body segments and between the segments of appendages it is thin and flexible, permitting free movement of the joints. In crustaceans and insects the cuticle forms ingrowths for muscle attachment. It may also line the foregut and hindgut, line and support the trachea, and be adapted for a variety of purposes.

The nonexpansible cuticular exoskeleton does, however, impose important conditions on growth. To grow, an arthropod must shed its outer covering at intervals and grow a larger one—a process called **ecdysis,** or **molting.** Arthropods molt from four to seven times before reaching adulthood, and some continue to molt after that. An exoskeleton is also relatively heavy and becomes proportionately heavier with increasing size. This tends to limit the ultimate body size.

2. *Segmentation and appendages for more efficiency and better locomotion.* Typically each somite has a pair of jointed appendages, but this arrangement is often modified, with both segments and appendages adapted for special functions. The limb segments are essentially hollow levers that are moved by rapid-action, internal, striated muscles. The jointed appendages are equipped with sensory hairs and have been modified for sensory functions, food handling, swift and efficient walking legs, and swimming appendages. This affords a wider capacity for adjustment to varied habitats.

3. *Air piped directly to cells.* Most terrestrial arthropods have a highly efficient tracheal system of air tubes, which delivers oxygen directly to the tissues and cells and makes a high metabolic rate possible. Aquatic arthropods breathe mainly by some form of gill that is quite efficient.

4. *Highly developed sensory organs.* Sensory organs are found in great variety, from the compound (mosaic) eye to those simple senses that have to do with touch, smell, hearing, balancing, chemical reception, and so on. Arthropods are keenly alert to what goes on in their environment.

5. *Complex behavior patterns.* Arthropods exceed most other invertebrates in the complexity and organization of their activities. Innate (unlearned) behavior unquestionably controls much of what they do, but learning also plays an important part in the lives of some species.

6. *Reduced competition through metamorphosis.* Many arthropods have indirect development, including a larval form quite different from the adult in structure. The larval form is often adapted for eating a different kind of food from that of the adult, resulting in less competition within a species.

7. *Power of flight.* Many insects have wings that develop from the cuticle. This exploitation of the aerial environment has allowed the possessors of wings to invade many niches unavailable to earthbound animals and has made possible wide geographic distribution.

Subphylum Trilobita

The trilobites (Figure 22-1) probably had their beginnings a million or more years before the Cambrian period in which they flourished. They have been extinct some 200 million years, but were abundant during the Cambrian and Ordovician periods. Their name refers to the trilobed shape of the body, caused by a pair of longitudinal grooves. They were bottom dwellers, probably scavengers. Most of them could roll up like pill bugs.

Subphylum Chelicerata

The chelicerate arthropods make up a very ancient group that includes the eurypterids (extinct), horseshoe crabs, spiders, ticks, and mites, scorpions, sea spiders, and others. They are characterized by having six pairs of appendages that include a pair of **chelicerae,** a pair of **pedipalps,** and **four pairs of walking legs** (a pair of chelicerae and five pairs of walking legs in horseshoe crabs). They have **no mandibles** and **no antennae.** Most chelicerates suck up liquid food from their prey.

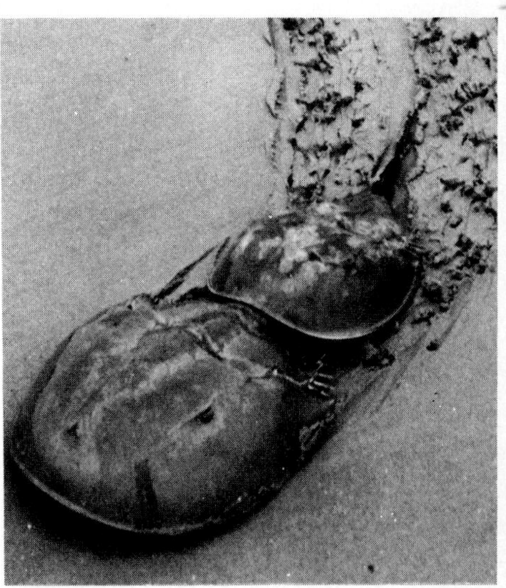

FIGURE 22-2

Mating of horseshoe crabs, *Limulus.* At high tide the larger female digs a depression in the sand for her eggs, which the smaller male fertilizes externally before the hole fills up with sand. The male *(right)* is clasping the female *(left)* as she selects the spot for her eggs. Sometimes she is followed by several males.

Photograph by L.L. Rue III

CLASS MEROSTOMATA

Eurypterids (subclass Eurypterida)

The eurypterids, or giant water scorpions (Figure 22-1), lived 200 to 500 million years ago and some of them were perhaps the largest of all arthropods, reaching a length of 3 m. They had some resemblances to the marine horseshoe crabs (Figure 22-2) and to the scorpions. Some fossil evidence suggests that this group may have given rise to the Arachnida.

Horseshoe crabs (subclass Xiphosurida)

The xiphosurids are an ancient marine group that dates from the Cambrian period. There are only three genera (five species) living today. *Limulus* (Figure 22-2), which lives in shallow water along the North American Atlantic coast, goes back practically unchanged to the Triassic period. Horseshoe crabs have an unsegmented, horseshoe-shaped **carapace** (hard dorsal shield) and a broad abdomen, which has a long spinelike **telson,** or tailpiece. On some of the abdominal appendages **book gills** (flat leaflike gills) are exposed. Horseshoe crabs can swim awkwardly by means of the abdominal plates and can walk on their walking legs. They feed at night on worms and small molluscs and are harmless to humans.

CLASS PYCNOGONIDA—SEA SPIDERS

Pycnogonids are curious little marine animals that are much more common than most of us realize. They stalk about on their four pairs of long, thin walking legs, sucking juices from hydroids and soft-bodied animals with their large suctorial proboscis (Figure 22-3). Their odd appearance is enhanced by the much reduced abdomen attached to the elongated cephalothorax. Most are only a few millimeters long, although some are much larger. They are common in all oceans.

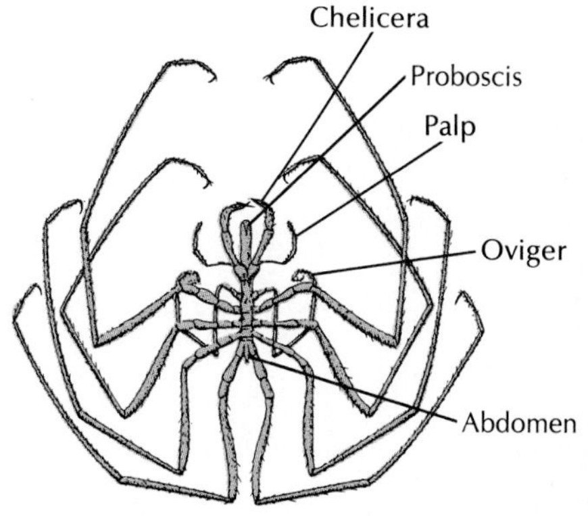

FIGURE 22-3

Pycnogonid, *Nymphon* sp. In this genus all the anterior appendages (chelicerae, palps, and ovigers) are present in both sexes, although ovigers are often not present in females of other genera.

From Hedgpeth, J.W. 1982. In Parker, S.P. (ed.). Synopsis and classification of living organisms. New York, McGraw-Hill Book Co.

CLASS ARACHNIDA

Arachnids have been an extremely successful group, with over 50,000 species described so far. They include the spiders, scorpions, pseudoscorpions, whip scorpions, ticks, mites, daddy longlegs (harvestmen), and others. The arachnid tagmata are a cephalothorax and an abdomen.

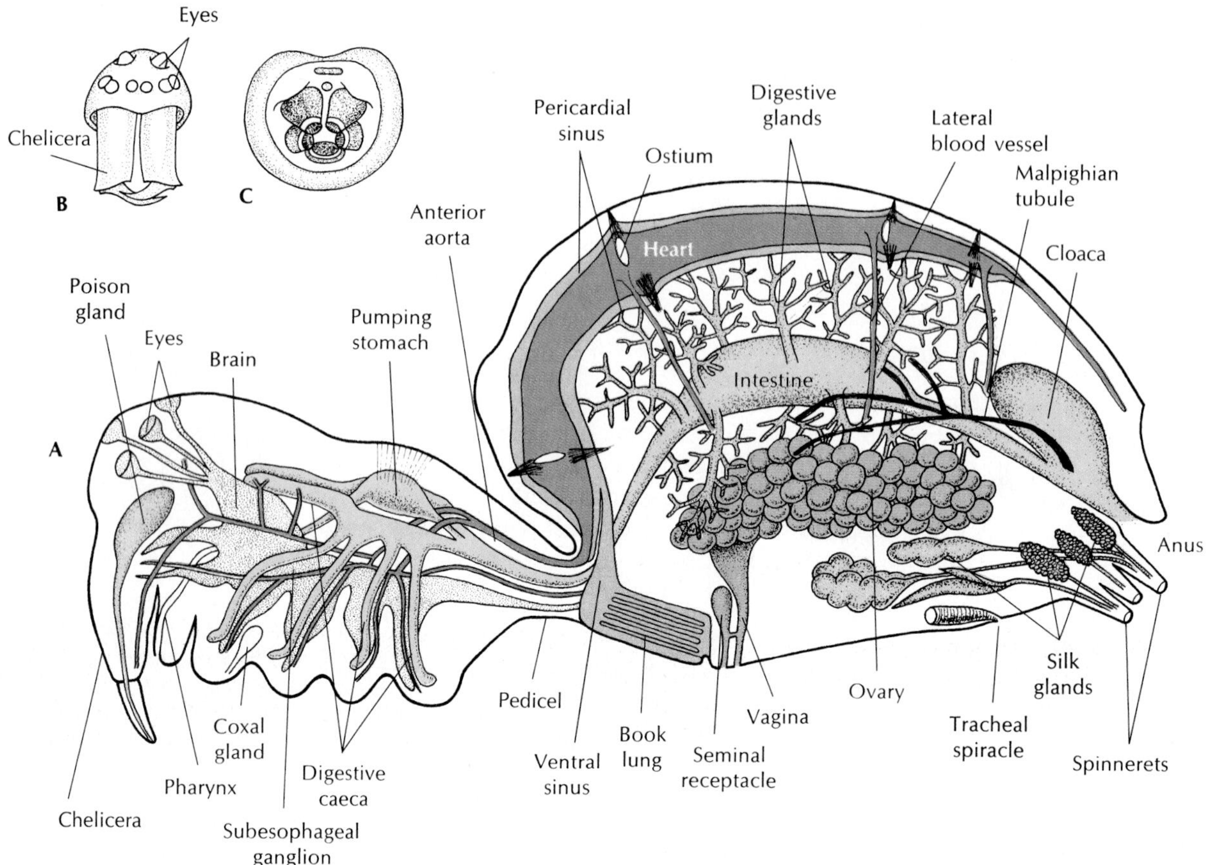

FIGURE 22-4

A, Internal structure of the spider. **B,** Anterior view of head, showing eyes and chelicerae with fangs. **C,** Ventral view of spinnerets. The type of chelicerae and the number, arrangement, and/or locations of eyes, lung slits, tracheal spiracles, and spinnerets are all identifying characteristics used in classifying spiders.

Spiders (order Araneae)

The spiders are a large group of 35,000 species, distributed all over the world. The cephalothorax and abdomen show no external segmentation, and the tagimata are joined by a narrow, waistlike **pedicel** (Figure 22-4).

All spiders are predaceous and feed largely on insects (Figure 22-5). Their chelicerae function as fangs and are provided with ducts from poison glands, with which they effectively dispatch their prey. Some spiders chase their prey, others ambush them, and many trap them in a net of silk. After the spider seizes its prey

FIGURE 22-5

A, A crab spider, *Misumenoides aleatorius,* on a zinnia awaits its insect prey. Its coloration matches the petals among which it lies, thus deceiving insects that visit the flowers in search of pollen or nectar. **B,** A jumping spider, *Eris aurantius.* This species has excellent vision and stalks an insect until it is close enough to leap with unerring precision, fixing its chelicerae into its prey.

Photographs by J.H. Gerard.

with its chelicerae and injects venom, it then liquefies the tissues with a digestive fluid and sucks up the resulting broth into the stomach. Spiders with teeth at the bases of the chelicerae crush or chew up the prey, aiding digestion by enzymes from the mouth.

Spiders breathe by means of **book lungs** or **tracheae** or both. Book lungs, which are unique in spiders, consist of many parallel air pockets extending into a blood-filled chamber (Figure 22-4). Air enters the chamber by a slit in the body wall. The tracheae make up a system of air tubes that carry air directly to the tissues from openings called **spiracles.** The tracheae are similar to those in insects (p. 492), but are much less extensive.

Spiders and insects have a unique **excretory system of malpighian tubules** (Figure 22-4), which work in conjunction with specialized rectal glands. Potassium and other solutes and waste materials are secreted into the tubules, which drain the fluid, or "urine," into the intestine. The rectal glands reabsorb most of the potassium and water, leaving behind such wastes as uric acid. By this cycling of water and potassium, species living in dry environments may conserve body fluids, producing a nearly dry mixture of urine and feces. Many spiders also have **coxal glands,** which are modified nephridia, that open at the coxa, or base, of the first and third walking legs.

FIGURE 22-6

A wolf spider, *Lycosa aspera.* Several of its eight eyes are visible as shiny black bulbs on its head.

Photograph by J.H. Gerard.

Spiders usually have eight **simple eyes,** each provided with a lens, optic rods, and a retina (Figure 22-4, *B,* and 22-6). They are used chiefly for perception of moving objects, but some, such as those of the hunting and jumping spiders, may form images. Since vision is usually poor, a spider's awareness of its environment depends a great deal on its hairlike **sensory setae.** Every seta on its surface, whether or not it is actually connected to receptor cells, is useful in communicating some information about the surroundings, air currents, or changing tensions in the spider's web. By sensing the vibrations of its web, the spider can judge the size and activity of its entangled prey or can receive the message tapped out on a silk thread by a prospective mate.

Web-spinning habits. The ability to spin silk is an important factor in the lives of spiders, as it is in some other arachnids. Two or three pairs of spinnerets containing hundreds of microscopic tubes run to special abdominal **silk glands** (Figure 22-4, *A* and *C*). A protein secretion emitted as a liquid hardens on contact with air to form the silk thread. Spiders' silk threads are stronger than steel threads of the same diameter and are said to be second in strength only to fused quartz fibers. The threads will stretch one fifth of their length before breaking.

The spider web used for trapping insects is familiar to most people. The kind of net varies with the species. Some are primitive and consist merely of a few strands of silk radiating out from a spider's burrow or place of retreat. Others spin beautiful, geometric orb webs. However, spiders use silk threads for many purposes besides web making. They use them to line their nests; form sperm webs or egg sacs; build draglines; make bridge lines, warning threads, molting threads, attachment discs, or nursery webs; or to wrap up their prey securely (Figure 22-7). Not all spiders spin webs for traps. Some, such as the wolf spiders (Figure 22-6), jumping spiders (Figure 22-5, *B*), and fisher spiders (Figure 22-8), simply chase and catch their prey.

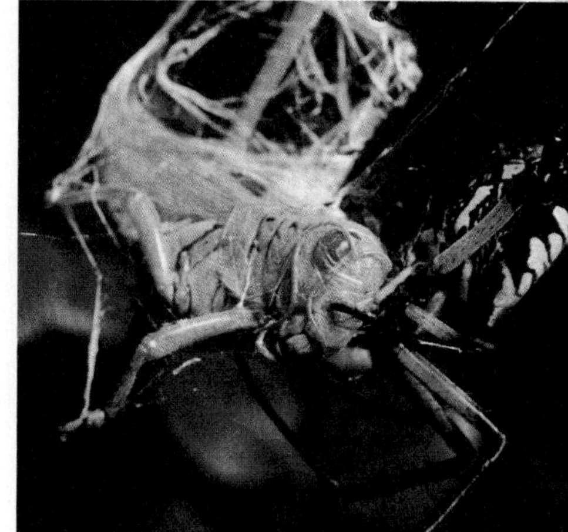

FIGURE 22-7

A grasshopper, snared and helpless in the web of a golden garden spider *(Argiope aurantia),* is wrapped in silk while still alive. If the spider is not hungry, the prize will be saved for a later meal.

Photograph by J.H. Gerard.

Reproduction. Before mating, the male spins a small web, deposits a drop of sperm on it, and then picks the sperm up and stores it in the special cavities of his pedipalps. When he mates, he inserts the pedipalps into the female genital opening to store the sperm in his mate's seminal receptacles. Before mating, there is usually a courtship ritual. The female lays her fertilized eggs in a silken cocoon, which she may carry about or may attach to a web or plant. A cocoon may contain hundreds of eggs, which hatch in approximately 2 weeks. The young usually remain in the egg sac for a few weeks and molt once before leaving it. Several molts occur before adulthood.

FIGURE 22-8

A fisher spider, *Dolomedes triton,* feeds on a minnow. This handsome spider feeds mostly on aquatic and terrestrial insects but occasionally captures small fishes and tadpoles. It pulls its paralyzed victim from the water, pumps in digestive enzymes, then sucks out the predigested contents.

Photograph by J.H. Gerard.

Are spiders really dangerous? It is truly amazing that such small and helpless creatures as the spiders have generated so much unreasoned fear in the human heart. Spiders are timid creatures, which, rather than being dangerous enemies to humans, are actually allies in the continuing battles with insects. The venom produced to kill the prey is usually harmless to humans. Even the most poisonous spiders bite only when threatened or when defending their eggs or young. The American tarantulas (Figure 22-9), despite their fearsome appearance, are *not* dangerous. They rarely bite, and their bite is not considered serious.

There are, however, two genera in the United States that can give severe or even fatal bites: *Latrodectus* (L. *latro,* robber + *dektes,* biter), and *Loxosceles* (Gr. *loxos,* crooked, + *skelos,* leg). The most important species are *Latrodectus mactans,* the **black widow,** and *Loxosceles reclusa,* the **brown recluse.** The black widow is moderate to small in size and shiny black, with a bright orange or red "hourglass" on the underside of the abdomen (Figure 22-10, *A*). The venom is neurotoxic; that is, it acts on the nervous system. About four or five out of each 1000 bites reported have proved fatal.

The brown recluse, which is smaller than the black widow, is brown, and bears a violin-shaped dorsal stripe on its cephalothorax (Figure 22-10, *B*). Its venom is hemolytic rather than neurotoxic, producing death of the tissues and skin surrounding the bite. Its bite can be mild to serious and occasionally fatal.

FIGURE 22-9

A tarantula, *Dugesiella hentzi.*

Photograph by J.H. Gerard.

Scorpions (order Scorpionida)

Although scorpions are more common in tropical and subtropical regions, some also occur in temperate zones. Scorpions are generally selective, hiding in burrows or under objects by day and feeding at night. They feed largely on insects and spiders, which they seize with the clawlike pedipalps and tear up with the jawlike chelicerae.

A

B

FIGURE 22-10

A, A black widow spider, *Latrodectus mactans,* suspended on her web. Note the orange "hourglass" on the ventral side of her abdomen. **B,** The brown recluse spider, *Loxosceles reclusa,* is a small venomous spider. Note the small violin-shaped marking on its cephalothorax. The venom is hemolytic and dangerous.

Photographs by J.H. Gerard.

The scorpion body consists of a rather short cephalothorax, which bears the appendages and from one to six pairs of eyes and a clearly segmented abdomen. The abdomen is divided into a broader **preabdomen** and tail-like **postabdomen,** which ends in a stinging apparatus used to inject venom (Figure 22-11, *A*). The venom of most species is not harmful to humans, although that of certain species of *Androctonus* in Africa and *Centruroides* in Mexico, Arizona, and New Mexico can be fatal unless antivenom is available.

Scorpions bring forth living young, which are carried on the back of the mother until after the first molt.

Harvestmen (order Opiliones)

Harvestmen, often known as "daddy longlegs," are common in the United States and other parts of the world (Figure 22-11, *B*). These curious creatures are easily distinguished from spiders by the fact that their abdomen and cephalothorax are broadly joined, without the constriction of the pedicel, and their abdomen shows external segmentation. They have four pairs of long, spindly legs, and without apparent ill effect, they can cast off one or more of these if they are grasped by a predator (or human hand). The ends of their chelicerae are pincer-like, and they feed much more as scavengers than do most arachnids.

A

B

FIGURE 22-11

A, The striped scorpion, *Centruroides vittatus* (order Scorpionida). This species is not considered dangerous to humans. **B,** A harvestman, *Leiobunum* sp. (order Opiliones). Harvestmen run rapidly on their stiltlike legs. They are especially noticeable during the harvesting season, hence the common name.

A, Photograph by J.H. Gerard; **B,** photograph by C.P. Hickman, Jr.

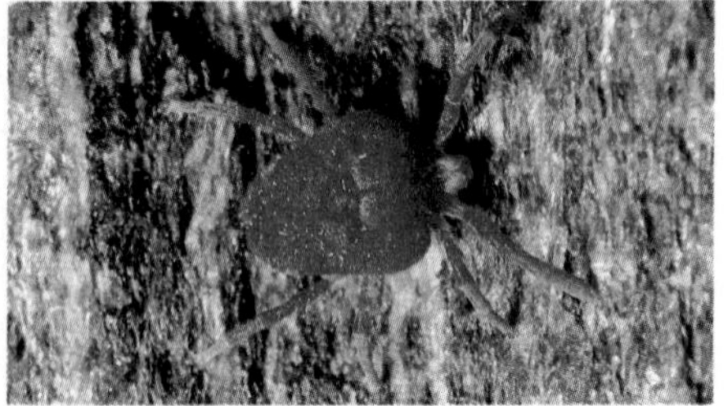

FIGURE 22-12

A, A wood tick, *Dermacentor variabilis* (order Acarina). **B,** A red velvet (harvest) mite, *Trombidium* sp.

Photographs by J.H. Gerard.

Ticks and mites (order Acarina)

Acarines differ from all other arachnids in having their cephalothorax and abdomen completely fused, with no sign of external division or segmentation (Fig. 22-12). Their mouthparts are carried on a little anterior projection, the **capitulum.** They are found almost everywhere—in both fresh and salt water, on vegetation, on the ground, and parasitic in vertebrates and invertebrates. Over 25,000 species have been described, many of which are of importance to humans, but this is probably only a fraction of the species that exist.

Ticks are usually larger than mites. They pierce the skin of vertebrates and suck up the blood until enormously distended; then they drop off and digest the meal. After molting, they are ready for another meal. Some ticks are important disease carriers. Texas cattle fever is caused by a protozoan parasite transmitted by the tick *Boophilus annulatus.* The wood tick *Dermacentor* (Fig. 22-12, *A*) carries Rocky Mountain spotted fever, caused by a rickettsia (a bacteria-like organism). Tularemia (rabbit fever, caused by a bacterium) is also carried by a species of *Dermacentor.* Lyme disease also results from a bacterium and is transmitted by *Ixodes dammini.*

Most mites are less than 2 mm long, and they are very diverse in their habits and habitats. Spider mites, or red spiders, are destructive to plants, piercing and sucking juices from the cells of leaves. The hair follicle mite, *Demodex,* causes a skin disease in dogs (mange), but another species in humans, whose prevalence approaches 100%, is usually symptomless. The parasitic larvae of chiggers (Trombiculidae) cause great skin irritation. Many other parasitic mites are known to infest humans and other animals, and some carry diseases from host to host.

Lyme disease was first reported in 1975 from the town of Lyme, Connecticut and has since spread rapidly. It is now the most frequently diagnosed tick-borne disease in the United States and has been recorded in Australia, the Soviet Union, China, Japan, and Africa. Early symptoms usually include an expanding skin rash, and if left untreated, the disease can become serious, with symptoms resembling rheumatoid arthritis. Fortunately, it can be cured in all stages by treatment with antibiotics.

■ Subphylum Crustacea

The crustaceans have traditionally been included as a class in the subphylum Mandibulata, along with the insects and myriapods. The members of all of these groups have, at least, a pair of antennae, mandibles, and maxillae on the head. Whether the Mandibulata constitutes a natural (phylogenetically related) grouping has been debated, but we will not explore that debate here.

The 30,000 or more species of Crustacea (L. *crusta,* shell) include lobsters, crayfish, shrimp, crabs, water fleas, copepods, barnacles, and some others. It is the only arthropod subphylum that is primarily aquatic; they are mainly marine, but there are many freshwater and a few terrestrial species. The majority are free living, but many are sessile, commensal, or parasitic. Crustaceans are often very important components of aquatic ecosystems, and several have considerable economic importance.

Crustaceans are the only arthropods with two pairs of antennae (Figure 22-13). In addition to the antennae and mandibles, they have two pairs of maxillae on the head, followed by a pair of appendages on each body segment (although appendages on some somites are absent in some groups). All appendages, except perhaps the first antennae (antennules), are primitively **biramous** (two main branches), and at least some of the appendages of present-day adults show that condition. Organs specialized for respiration, if present, are in the form of gills. Crustaceans lack malpighian tubules.

Most crustaceans have between 16 and 20 somites, but some primitive forms have 60 segments or more. The more advanced crustaceans tend to have fewer segments and increased tagmatization. The major tagmata are the **head, thorax,** and **abdomen,** but these are not homologous throughout the subphylum (or even within some classes) because of varying degrees of fusion of somites, for example, as in the cephalothorax.

In many crustaceans, the dorsal cuticle of the head may extend posteriorly and around the sides of the animal to cover or be fused with some or all of the thoracic and abdominal somites. This covering is called the **carapace.** In some

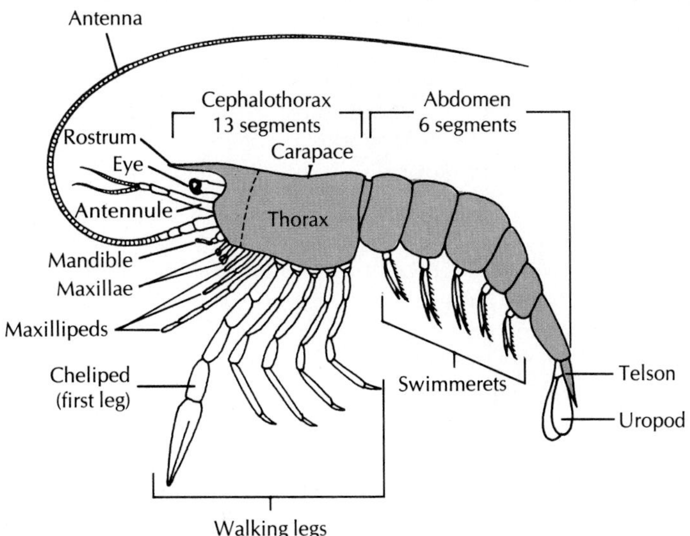

FIGURE 22-13

Archetypical plan of the Malacostraca. Note that the maxillae and maxillipeds have been separated diagrammatically to illustrate general plan. Typically in the living animal only the third maxilliped is visible externally. In the order Decapoda the cephalothorax is covered by the carapace, as shown here.

groups the carapace forms clamshell-like valves that cover most or all of the body. In the decapods (including lobsters, shrimp, crabs, and others) the carapace covers the entire cephalothorax but not the abdomen.

BRIEF RÉSUMÉ OF MAJOR CRUSTACEAN CLASSES

Class Branchiopoda

The class Branchiopoda (bran'kee-op'o-da) (Gr. *branchia*, gills, + *pous, podos*, foot) also represents a primitive crustacean type. Four orders are recognized: **Anostraca** (fairy shrimp and brine shrimp, Figure 22-14), which lack a carapace;

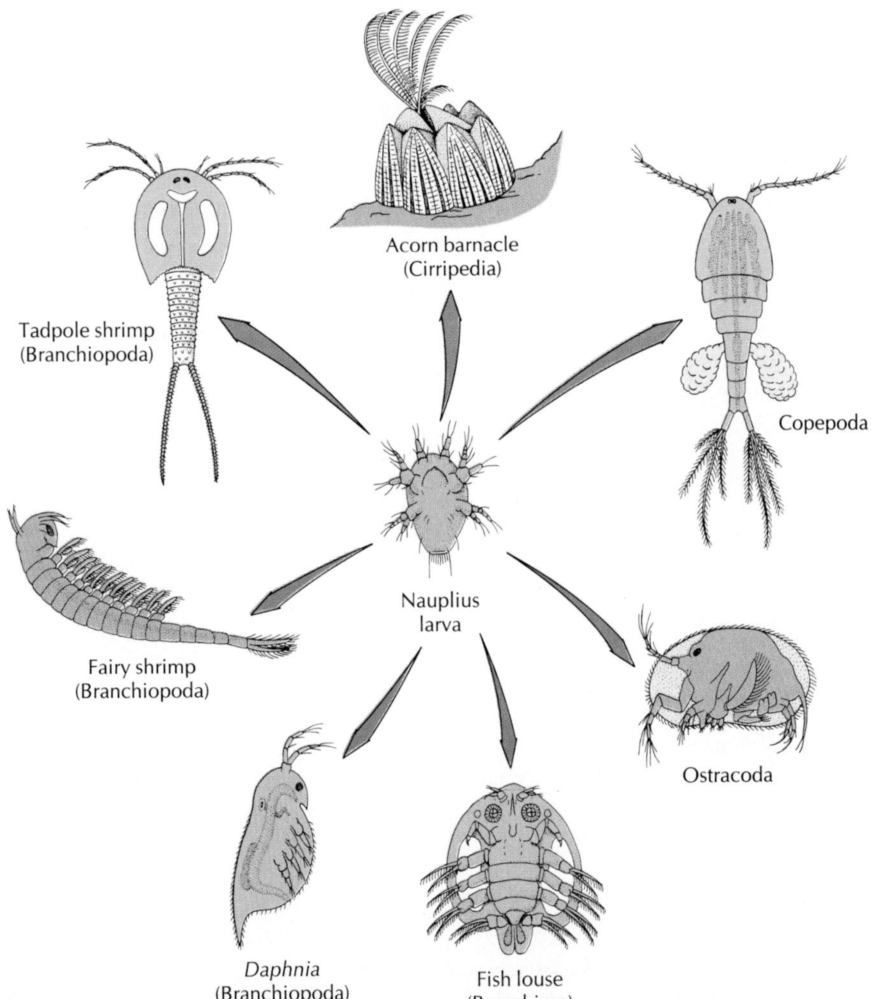

FIGURE 22-14

Group of smaller crustaceans. A nauplius hatches from the egg in most cases, except in the Branchiura, in which the nauplius is passed embryonically, and a postlarva hatches. Not drawn to same scale.

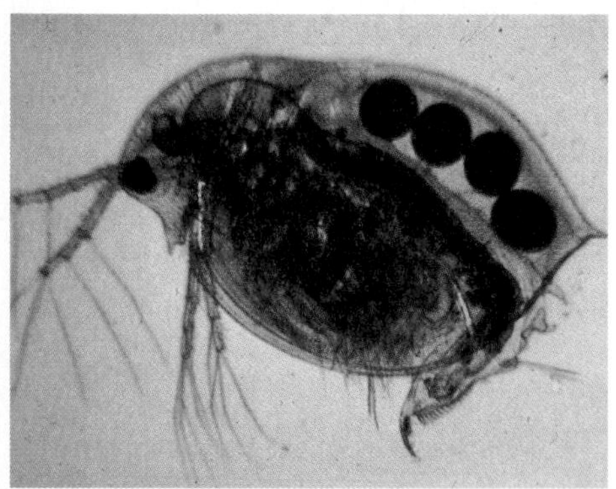

FIGURE 22-15

A water flea, *Daphnia* (order Cladocera), photographed with polarized light. These microscopic forms occur in great numbers in northern lakes and are an important component of the food chain leading to fishes.

Courtesy Carolina Biological Supply Co.

A

B

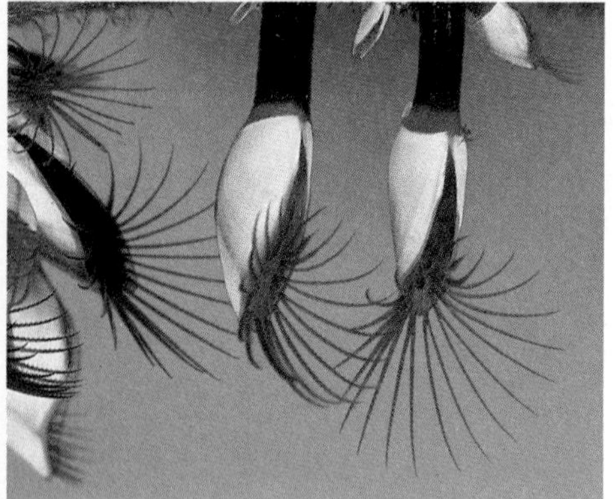

FIGURE 22-16

A, Acorn barnacles, *Semibalanus cariosus,* are found on rocks along the Pacific coast of North America. **B,** Pelagic goose barnacles, *Lepas fascicularis,* attach to floating timbers and other objects by long stalks that may reach 25 cm when fully extended.

A, Photograph by R.M. Harbo; **B,** photograph by K. Sandved.

Notostraca (tadpole shrimp such as *Triops,* Figure 22-14), whose carapace forms a large dorsal shield covering most of the trunk somites; **Conchostraca** (clam shrimp such as *Lynceus*), whose carapace is bivalve and usually encloses the entire body; and **Cladocera** (water fleas such as *Daphnia,* Figures 22-14 and 22-15), with a carapace typically covering the entire body but not the head. Branchiopods have reduced first antennae and second maxillae. Their legs are flattened and leaflike (**phyllopodia**) and are the chief respiratory organs (hence, the name branchiopods). The legs also are used in filter feeding in most branchiopods, and in groups other than the cladocerans, they are used for locomotion as well. The most important and successful order is the Cladocera, which often forms a large segment of the freshwater zooplankton.

Class Ostracoda

Members of the class Ostracoda (os-trak'o-da) (Gr. *ostrakodes,* testaceous, that is, having a shell) are, like the conchostracans, enclosed in a bivalve carapace and resemble tiny clams, 0.25 to 8 mm long (Figure 22-14). Ostracods show considerable fusion of trunk somites, and numbers of thoracic appendages are reduced to two or none.

Class Copepoda

The class Copepoda (ko-pep'o-da) (Gr. *kōpē,* oar, + *pous, podos,* foot) is an important group of Crustacea, second only to the Malacostraca in number of species. The copepods are small (usually a few millimeters or less in length), rather elongate, tapering toward the posterior end, lacking a carapace, and retaining the simple, median, nauplius eye in the adult (Figure 22-14). They have four pairs of rather flattened, biramous, thoracic swimming appendages, and a fifth, reduced pair. The abdomen bears no legs. The Copepoda have large numbers of symbiotic, as well as free-living, species. Many of the parasites are highly modified, and the adults may be so highly modified (and may depart so far from the description just given) that they can hardly be recognized as arthropods. Ecologically, the free-living copepods are of extreme importance, often dominating the primary consumer level (herbivore, p. 131) in aquatic communities.

Class Branchiura

The class Branchiura (brank-i-yu'ra) (Gr. *branchia,* gills, + *ura,* tail) is a small group of primarily fish parasites, which, despite its name, has no gills (Figure 22-14). Members are usually 5 to 10 mm long and may be found on marine or freshwater fish. They typically have a broad, shieldlike carapace, compound eyes, four biramous thoracic appendages for swimming, and a short, unsegmented abdomen. The second maxillae have become modified as suction cups.

Class Cirripedia

The class Cirripedia (sir-ri-ped'i-a) (L. *cirrus,* curl of hair, + *pes, pedis,* foot) includes the barnacles, which are usually enclosed in a shell of calcareous plates, as well as three smaller orders of burrowing or parasitic forms. Barnacles are sessile as adults and may be attached to the substrate by a stalk (goose barnacles) (Figure 22-16, *B*) or directly (acorn barnacles) (Figures 22-14 and 22-16, *A*). Typically, the carapace (mantle) surrounds the body and secretes a shell of calcareous plates. The head is reduced, the abdomen absent, and the thoracic legs are long, many-jointed cirri with hairlike setae. The cirri are extended through an opening between the calcareous plates to filter from the water the small particles on which the animal feeds (Figure 22-16, *B*).

Class Malacostraca

The class Malacostraca (mal-a-kos'tra-ka) (Gr. *malakos,* soft, + *ostrakon,* shell) is the largest class of Crustacea and shows great diversity. We will mention

A B

FIGURE 22-17

A, Four pill bugs, *Armadillidium vulgare,* (order Isopoda), common terrestrial forms. **B,** A freshwater sow bug, *Asellus* sp., an aquatic isopod.

A, Photograph by C.P. Hickman, Jr.; B, photograph by J.H. Gerard.

only 4 of its 12 to 13 orders. The trunk of malacostracans usually has eight thoracic and six abdominal somites, each with a pair of appendages. There are many marine and freshwater species.

The **Isopoda** (i-sop'o-da) (Gr. *isos,* equal, + *pous, podos,* foot) are commonly dorsoventrally flattened, lack a carapace, and have sessile compound eyes. Their abdominal appendages bear gills. Common land forms are the sow bugs or pill bugs (*Porcellio* and *Armadillidium,* Figure 22-17, *A*), which live under stones and in damp places. *Asellus* (Figure 22-17, *B*) is common in fresh water, and *Ligia* is abundant on sea beaches and rocky shores.

The **Amphipoda** (am-fip'o-da) (Gr. *amphis,* on both sides, + *pous, podos,* foot) resemble isopods in that the members have no carapace and have sessile compound eyes. However, they are usually compressed laterally, and their gills are in the thoracic position, as in other malacostracans. There are many marine amphipods (Figure 22-18), such as the beach flea, *Orchestia,* and numerous freshwater species.

FIGURE 22-18

Marine amphipods. **A,** Free-swimming amphipod, *Anisogammarus* sp. **B,** Skeleton shrimp, *Caprella* sp., shown on a bryozoan colony, resemble praying mantises. **C,** *Phronima,* a marine pelagic amphipod, takes over the tunic of a salp (subphylum Urochordata, Chapter 25). Swimming by means of its abdominal swimmerets, which protude from the opening of the barrel-shaped tunic, the amphipod maneuvers to catch its prey.

A, Photograph by R. Harbo; B and C, photographs by K. Sandved.

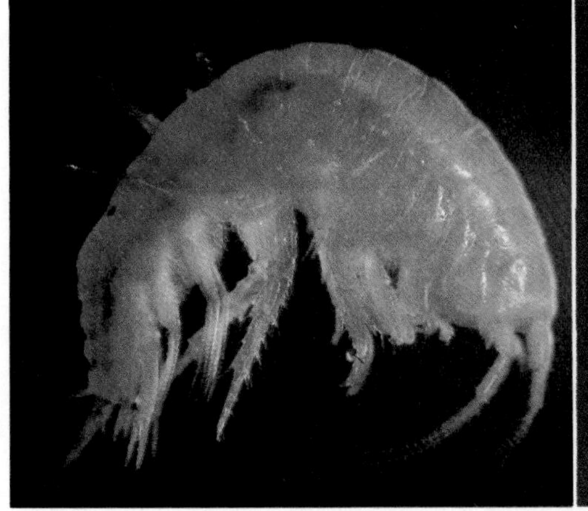

A B C

The **Euphausiacea** (yu-faws-i-a'se-a) (Gr. *eu,* well, + *phausi,* shining bright, + *acea,* L. suffix, pertaining to) is a group of only about 90 species, but they are important as the oceanic plankton known as "krill." They are about 3 to 6 cm long (Figure 22-19) and commonly occur in great oceanic swarms, where they are eaten by baleen whales and many fishes.

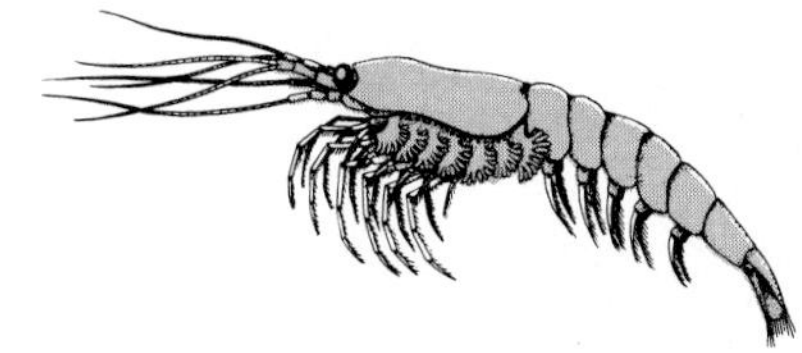

FIGURE 22-19

Meganyctiphanes, class Malacostraca, order Euphausiacea, the northern krill.

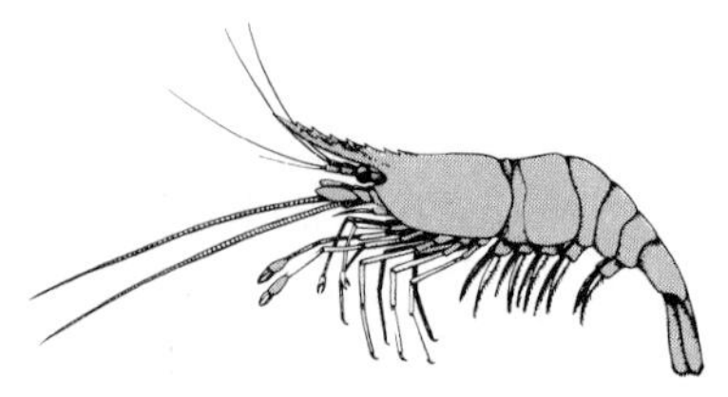

FIGURE 22-20

Palaemonetes, class Malacostraca, order Decapoda, a shrimp.

The **Decapoda** (de-cap'o-da) (Gr. *deka,* ten, + *pous, podos,* foot) have five pairs of walking legs of which the first is often modified to form pincers (**chelae**) (Figures 22-13 and 22-21). These are the lobsters, crayfishes (see Figure 22-26), shrimps (Figure 22-20), and crabs, the largest of the crustaceans (Figure 22-21). True crabs differ from the others in having a broader carapace and a much-reduced abdomen (Figure 22-21, *A, C* and *D*). Familiar examples are the fiddler crabs *Uca,* which burrow in sand just below the high-tide level (Figure 22-21, *C*), decorator crabs, which cover their carapaces with sponges and sea anemones for camouflage, and spider crabs, such as *Libinia.* Hermit crabs (Figure 22-21, *B*) have become adapted to live in snail shells; their abdomens, which lack a hard exoskeleton, are protected by the snail shell.

The terminology applied by various workers to crustacean appendages has not been blessed with uniformity. At least two systems are in wide use. Alternative terms to those we have used, for example, are protopodite, exopodite, endopodite, basipodite, coxopodite, and epipodite. The first and second pairs of antennae may be referred to as the antennules and antennae, and the first and second maxillae are often called maxillules and maxillae. A rose by any other name. . . .

FIGURE 22-21

Decapod crustaceans. **A,** The bright orange tropical rock crab, *Grapsus grapsus,* is a conspicuous exception to the rule that most crabs bear cryptic coloration. **B,** The hermit crab, *Dardanus venosus,* which has a soft abdominal exoskeleton, lives in a snail shell that it carries about and into which it can withdraw for protection. **C,** The male fiddler crab, *Uca* sp., uses its enlarged cheliped to wave territorial displays and in threat and combat. **D,** The massive chelipeds of box crabs, *Calappa ocellata,* are well adapted for crushing the shells of their snail prey. Here the crab is "exhaling" water from its branchial chamber. **E,** Northern lobster, *Hoarus americanus.*

A and **C,** Photographs by C.P. Hickman, Jr.; **B** and **D,** photographs by K. Sandved; **E,** photograph by J.H. Gerard.

FORM AND FUNCTION

Appendages

Some of the modifications of crustacean appendages may be illustrated by those of crayfishes and lobsters. The **swimmerets,** or abdominal appendages, retain the primitive biramous condition. Such an appendage consists of inner and outer branches, called the **endopod** and **exopod,** which are attached to one or more basal segments collectively called the **protopod** (Figure 22-22).

There are many modifications of this plan. In the most primitive crustaceans, all of the trunk appendages tend to be similar in structure and adapted for swimming. The evolutionary trend, shown in the crayfishes, has been toward reduction in number of appendages and toward a variety of modifications that fit them for many functions. Some are foliaceous (flat and leaflike), as are the maxillae; some are biramous, as are the swimmerets, maxillipeds, uropods, and antennae; some have lost one branch and are **uniramous,** as are the walking legs.

In the crayfish we find the first three thoracic appendages, called **maxillipeds,** serving along with the two pairs of maxillae as food handlers; the other five pairs of appendages are lengthened and strengthened for walking and defense (Figure 22-23). The first pair of walking legs, called **chelipeds,** are enlarged with a strong claw, or chela, for defense. The abdominal swimmerets serve not only for locomotion, but in the male the first pair is modified for copulation, and in the female they all serve as a nursery for attached eggs and young. The last pair of appendages, called **uropods,** are wide and are used as rudders for swift backward movements, and, with the telson, they form a protective device for eggs or young on the swimmerets.

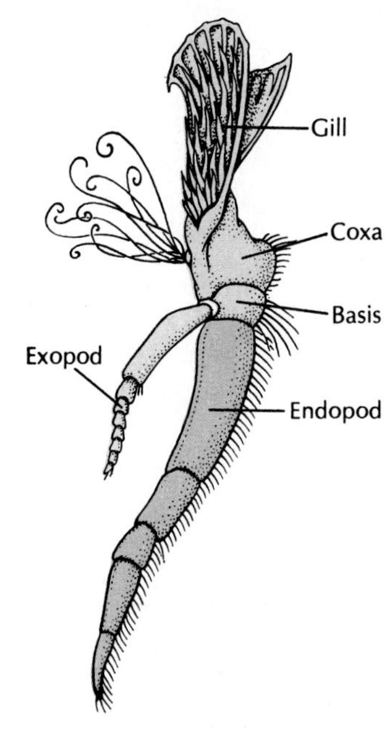

FIGURE 22-22

Parts of a biramus crustacean appendage (third maxilliped of a crayfish).

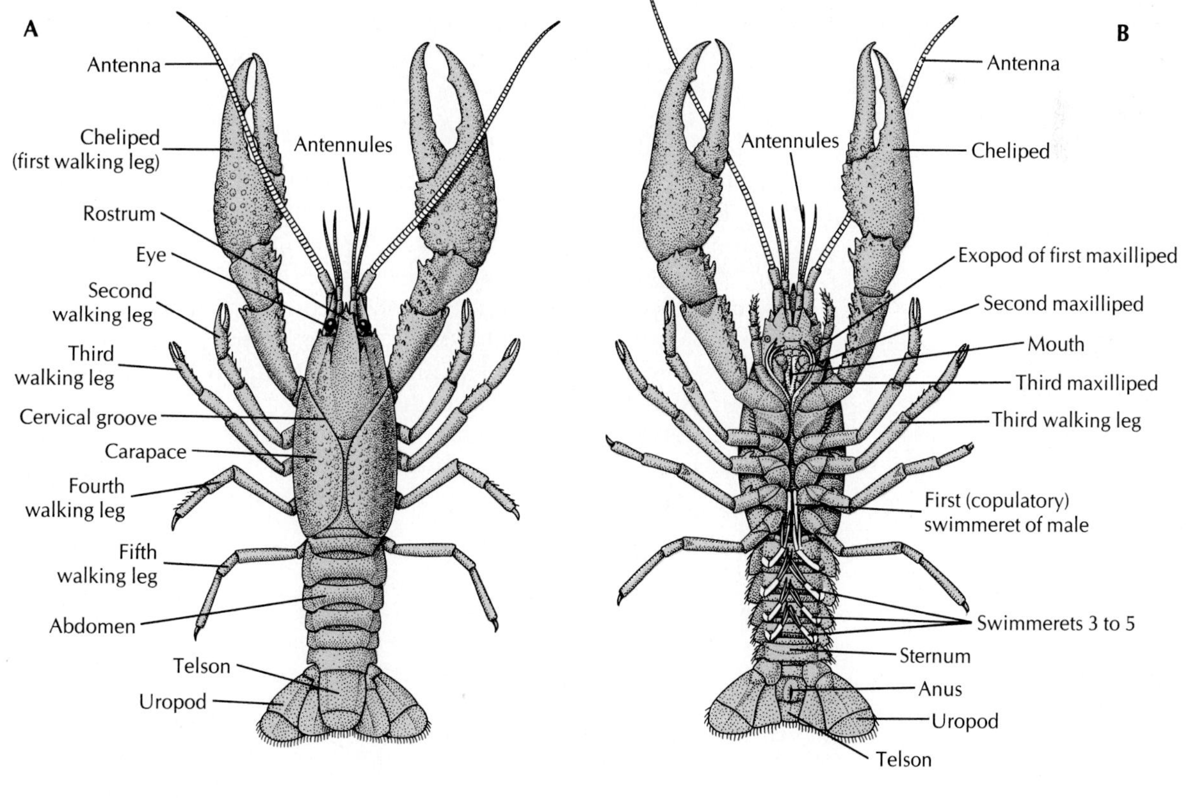

FIGURE 22-23

External structure of the crayfish. **A,** Dorsal view. **B,** Ventral view.

Ecdysis

The problem of growth despite a restrictive exoskeleton is solved in crustaceans, as in other arthropods, by ecdysis (Gr. *ekdyein,* to strip off), the periodic shedding of the old cuticle and the formation of a larger new one. Molting occurs most frequently during larval stages and less often as the animal reaches adulthood. Although the actual shedding of the cuticle is periodic, the molting process and the preparations for it, involving the storage of reserves and changes in the integument, are a continuous process going on during most of the animal's life.

During each **premolt** period the old cuticle becomes thinner as inorganic salts are withdrawn from it and stored in the tissues. Other reserves, both organic and inorganic, are also accumulated and stored. The underlying epidermis begins to grow by cell division; it secretes first a new inner layer of epicuticle and then enzymes that digest away the inner layers of old endocuticle (Figure 22-24). Gradually a new cuticle is formed inside the degenerating old one. Finally the actual ecdysis occurs as the old cuticle ruptures, usually along the middorsal line, and the animal backs out of it (Figure 22-25). By taking in air or water the animal swells to stretch the new larger cuticle to its full size. During the **postmolt** period the cuticle is thickened, the outer layer is hardened by tanning, and the inner layer is strengthened as salvaged inorganic salts and other constituents are redeposited. Usually the animal is very secretive during the postmolt period when its defenseless condition makes it particularly vulnerable to predation.

That ecdysis is under hormonal control has been demonstrated in both crustaceans and insects, but the process is often initiated by a stimulus perceived by the central nervous system. The action of the stimulus in decapods is to decrease the production of a **molt-inhibiting hormone** from neurosecretory cells in the **X-organ** of the eye stalk. The hormone is released from the sinus gland, also in the eye stalk. When the level of molt-inhibiting hormone drops, the **Y-organs** near the mandibles are stimulated to produce **molting hormone.** This initiates the processes leading to premolt. The Y-organs are homologous to the prothoracic glands of insects, which produce ecdysone (p. 266).

FIGURE 22-24

Cuticle secretion and resorption in preecdysis. **A,** Interecdysis condition. **B,** Old endocuticle separates from epidermis, which secretes new epicuticle. **C,** As new endocuticle is secreted, molting fluid dissolves old endocuticle, and the solution products are resorbed. **D,** At ecdysis, little more than the old epicuticle is left to discard. In postecdysis, new cuticle is stretched and unfolded, and more endocuticle is secreted.

Modified from Schmidt, G.D., and L.S. Roberts, 1989. Foundations of parasitology, ed. 4, St. Louis, The C.V. Mosby Co.

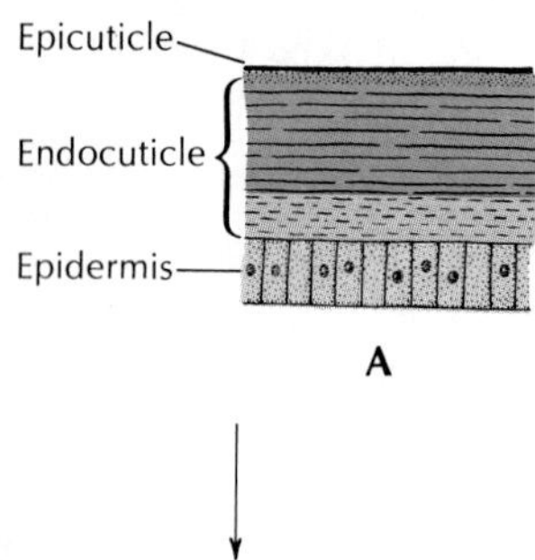

Other endocrine functions

Body color of crustaceans is largely a result of pigments in special branched cells (**chromatophores**) in the epidermis. Color change is achieved by concentration of the pigment granules in the center of the cells, which causes a lightening effect, or by dispersal of the pigment throughout the cells, which causes a darkening effect. The pigment behavior is controlled by hormones from neurosecretory cells in the eye stalk. Neurosecretory hormones also control pigment in the eyes for light and dark adaptation, and other neurosecretory hormones control the rate and amplitude of the heartbeat.

Androgenic glands, which are not neurosecretory, are found in male malacostracans, and their secretion stimulates the expression of male sexual characteristics. If androgenic glands are artificially implanted in a female, her ovaries

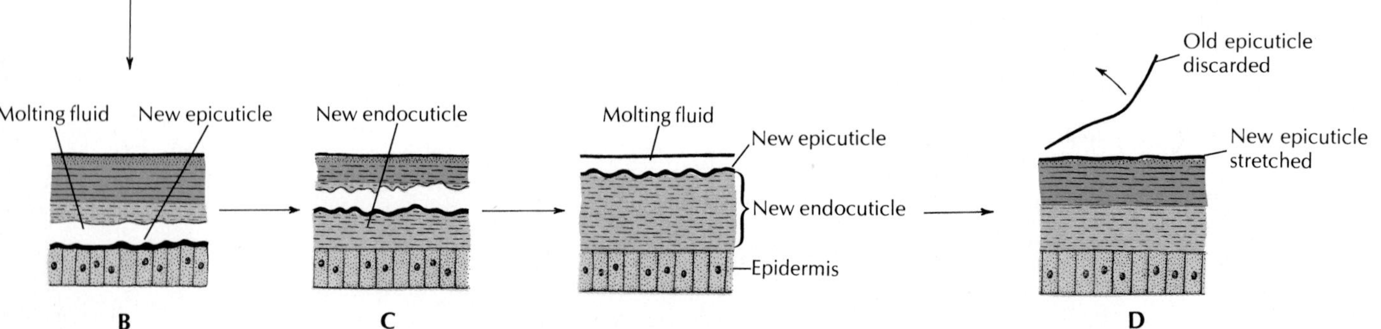

transform to testes and begin to produce sperm, and her appendages begin to take on male characteristics at the next molt.

Feeding habits

Feeding habits and adaptations for feeding vary greatly among crustaceans. Many forms can shift from one type of feeding to another depending on environment and food availability, but fundamentally the same set of mouthparts is used by all. The mandibles and maxillae are involved in the actual ingestion; maxillipeds hold and crush food. In predators the walking legs, particularly the chelipeds, serve in food capture.

Many crustaceans, both large and small, are predatory, and some have interesting adaptations for killing their prey. One shrimplike form, *Lygiosquilla,* has on one of its walking legs a specialized digit that can be drawn into a groove and released suddenly to pierce a passing prey. The pistol shrimp *Alpheus* has one enormously enlarged chela that can be cocked like the hammer of a gun and snapped with a force that stuns its prey.

The food of crustaceans ranges from plankton, detritus, and bacteria, used by the **filter feeders,** to larvae, worms, crustaceans, snails, and fishes, used by **pred-**

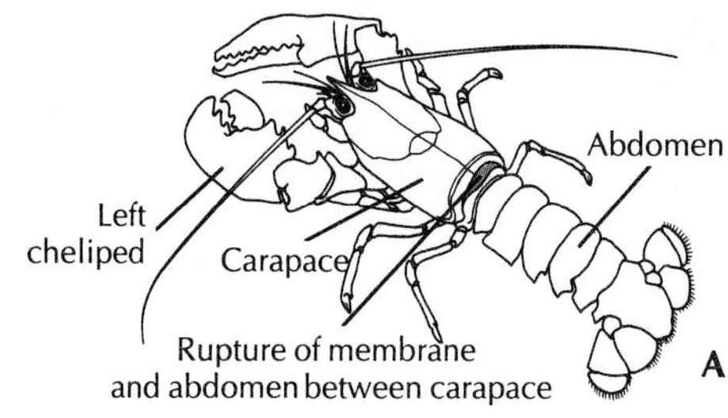

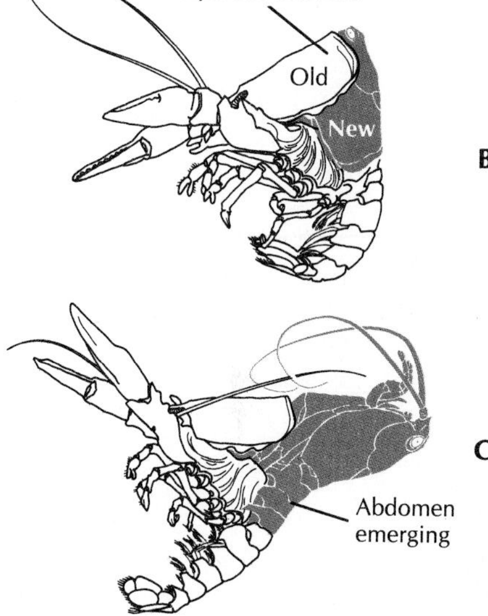

FIGURE 22-25

Molting sequence in the northern lobster, *Homarus americanus.* **A,** Membrane between carapace and abdomen ruptures, and carapace begins slow elevation. This step may take up to 2 hours. **B** and **C,** Head, thorax, and finally abdomen are withdrawn. This process usually takes no more than 15 minuts. Immediately after ecdysis, chelipeds are desiccated and body is very soft. Lobster now begins rapid absorption of water so that within 12 hours body increases about 20% in length and 50% in weight. Water will be replaced by living tissue in succeeding weeks.

Drawn from photographs by D.E. Aiken.

A

B

FIGURE 22-26

A, River crayfish, *Orconectes virilis.*
B, Painted shrimp, *Hymenocera picta. Hymenocera* is known to feed on sea stars (phylum Echinodermata, p. 523)

A, Photograph by J.H. Gerard; B, photograph by K. Sandved.

ators, and dead animal and plant matter, used by **scavengers.** Filter feeders, such as the fairy shrimps, water fleas, and barnacles, use their legs, which bear a thick fringe of setae, to create water currents that sweep food particles through the setae. The mud shrimp *Upogebia* uses long setae on its first two pairs of thoracic appendages to strain food material from water circulated through its burrow by movements of its swimmerets.

Crayfishes have a two-part stomach. The first contains a **gastric mill** in which food, already torn up by the mandibles, can be further ground up by three calcareous teeth into particles fine enough to pass through a filter of setae in the second part of the stomach; the food particles then pass into the intestine for chemical digestion.

Respiration, excretion, and circulation

The **gills** of crustaceans vary in shape — treelike, leaflike, or filamentous — all provided with blood vessels or sinuses. They are usually attached to the appendages and kept ventilated by the movement of the appendages in the water. The **branchial chambers** are usually protected by the overlapping carapace. Some smaller crustaceans breathe through the general body surface.

Excretory and **osmoregulatory** organs in crustaceans are paired glands located in the head, with excretory pores opening at the base of either the antennae or the maxillae, and they are called **antennal glands** or **maxillary glands** (Figure 22-27). The antennal glands of decapods are also called **green glands.** They resemble the coxal glands of the chelicerates. The waste product is mostly ammonia with some urea and uric acid. Some wastes diffuse through the gills as well as through the excretory glands.

Circulation, as in other arthropods, is an **open system** consisting of a heart, either compact or tubular, and arteries, which transport blood to different areas of the hemocoel. Some smaller crustaceans lack a heart. An open circulatory system is less dependent on heartbeats because the movement of organs and limbs circulates the blood more effectively in open sinuses than in capillaries. The blood may contain as respiratory pigments either hemocyanin or hemoglobin (hemocyanin in decapods), and the blood has the property of clotting to prevent loss of blood in minor injuries.

Nervous and sensory systems

A cerebral ganglion above the esophagus sends nerves to the anterior sense organs and is connected to a subesophageal ganglion by a pair of connectives around the esophagus. A double ventral nerve cord has a ganglion in each segment that sends nerves to the viscera, appendages, and muscles (Figure 22-27). Giant fiber systems are common among the crustaceans.

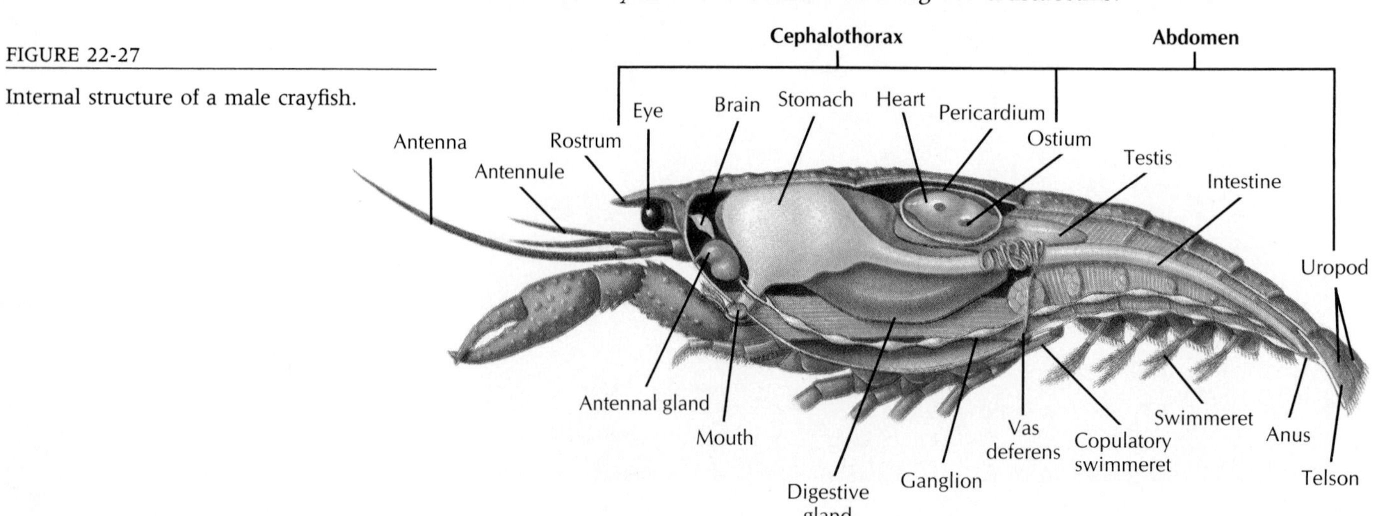

FIGURE 22-27

Internal structure of a male crayfish.

The sensory organs are well developed. There are two types of eyes—the median, or nauplius, eye and the compound eye. The **median eye** is found in the nauplius larvae and in some adult forms and may be the only adult eye, as in the copepods. It is usually a group of three pigment cups containing retinal cells; they may or may not have a lens.

Most crustaceans have **compound eyes** similar to insect eyes. In crabs and crayfishes they are located on the ends of movable eyestalks (Figure 22-21). Compound eyes are precise instruments, different from vertebrate eyes, yet especially adept at detecting motion; they are able to analyze polarized light. The convex corneal surface gives a wide visual field, particularly in the stalked eyes where the surface may cover an arc of 200 degrees or more.

The compound eye is composed of many tapering units called **ommatidia** set close together (Figure 22-28). The facets, or corneal surfaces, of the ommatidia give the surface of the eye the appearance of a fine mosaic. Most crustacean eyes are adapted either to bright or to dim light, depending on their diurnal or nocturnal habits, but some are able, by means of screening pigments, to adapt, to some extent at least, to both bright and dim light. The number of ommatidia varies from a dozen or two in some small crustaceans to 15,000 or more in a large lobster. Some insects have approximately 30,000.

Other sensory organs include statocysts, tactile setae on the cuticle of most of the body, and chemosensitive setae, especially on the antennae, antennules, and mouthparts.

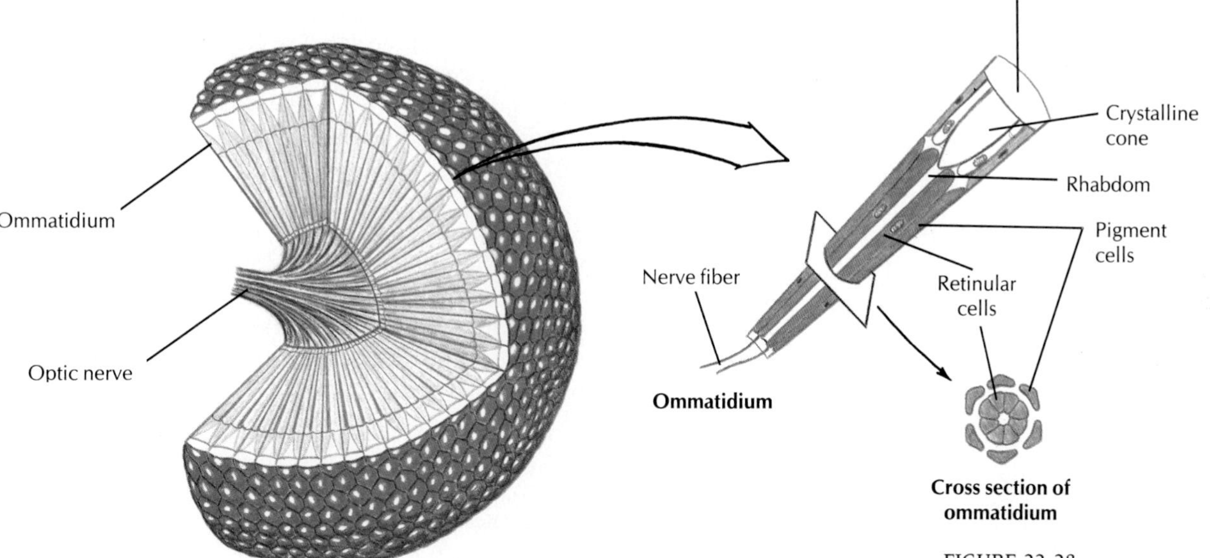

FIGURE 22-28

Compound eye of an insect. A single ommatidium is shown enlarged to the right.

Reproduction and life cycles

Most crustaceans have separate sexes, and there are a variety of specializations for copulation among the different groups. The barnacles are monoecious but generally practice cross-fertilization. In some of the ostracods males are scarce, and reproduction is usually parthenogenetic. Most crustaceans brood their eggs in some manner—branchiopods and barnacles have special brood chambers, the copepods have egg sacs attached to the sides of the abdomen (Figure 22-14), and the malacostracans usually carry eggs and young attached to their appendages.

The form that hatches from the egg of a crayfish is a tiny juvenile with the same form as the adult and a complete set of appendages and somites. However, the egg of most crustaceans produces a larva that must go through a series of changes, either gradual or abrupt over the course of the series of molts, to assume the adult form. The primitive larva of the crustaceans is the **nauplius** (Figure 22-14). It has an unsegmented body, a frontal eye, and three pairs of appendages, representing the two pairs of antennae and the mandibles. The form of the developmental stages and postlarvae of different groups of Crustacea are varied and have special names.

Subphylum Uniramia

The appendages of members of the Uniramia are unbranched, as the name implies. The subphylum includes the insects and the myriapods. The term **myriapod** (Gr. *myrias,* a myriad, + *podos,* foot) means many footed and refers to several classes that have a head and trunk and paired appendages on all trunk segments except the last. They include the Chilopoda (centipedes), Diplopoda (millipedes), Pauropoda (pauropods), and Symphyla (symphylans).

Myriapods and insects are both primarily terrestrial, and they have had a remarkable success story in an environment that has, as a rule, proved most inhospitable to invertebrates. They have evolved along two main lines. One is characterized by a body arrangement of **two tagmata,** the head and trunk, with paired appendages on all but the last trunk segments—the **myriapods.** The other evolutionary branch features a body of **three tagmata,** head, thorax, and abdomen, with the abdominal appendages missing or greatly reduced—the **insects.**

The head of the myriapods and insects resembles the crustacean head but has only **one pair of antennae,** instead of two. It also has the **mandibles** and two pairs of **maxillae** (one pair of maxillae in millipedes). The legs are all **uniramous.**

Respiratory exchange is by body surface and tracheal systems, although juveniles, if aquatic, may have gills.

CENTIPEDES—CLASS CHILOPODA

The centipedes are active predators with a preference for moist places such as under logs or stones, where they feed on earthworms, insects, etc. Their bodies are somewhat flattened dorsoventrally, and they may contain from a few to 177 somites (Figure 22-29). Each somite, except the one behind the head and the last

A

FIGURE 22-29

A, Centipede, *Scolopendra* (class Chilopoda). Most segments have one pair of appendages each. First segment bears a pair of poison claws, which in some species can inflict serious wounds. Centipedes are carnivorous. **B.** Head of centipede.

A, Photograph by Adrian Wenner/Visuals Unlimited.

B

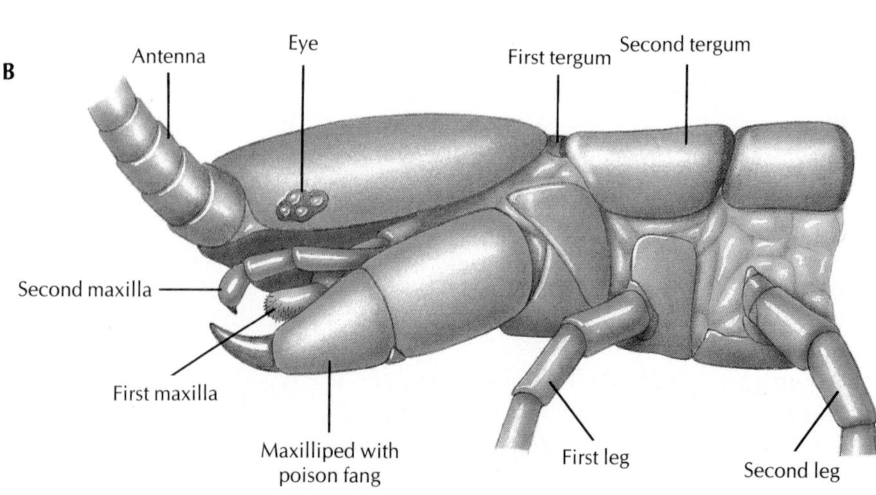

two, bears one pair of appendages. Those of the first body segment are modified to form poison claws, which they use to kill their prey. Most species are harmless to humans.

The head bears a pair of eyes, each consisting of a group of ocelli. Respiration is by tracheal tubes with a pair of spiracles in each somite. Sexes are separate. Some are oviparous (females release eggs from which young later hatch); some are viviparous (living young, instead of eggs, are released). The young are similar to the adult. The common house centipede *Scutigera*, with 15 pairs of legs, and *Scolopendra* (Figure 22-29), with 21 pairs of legs, are familiar genera.

MILLIPEDES—CLASS DIPLOPODA

Diplopods, or "double-footed" arthropods, are commonly called millipedes, which literally means "thousand feet" (Figure 22-30). Although they do not have that many legs, they do have a great many. Their cylindrical bodies are made up of 25 to 100 segments. The four thoracic segments bear only one pair of legs each, but the abdominal segments each have two pairs, a condition that may have evolved from fusion of somites. There are two pairs of spiracles on each abdominal somite, each opening into an air chamber that gives off tracheal tubes.

A

B

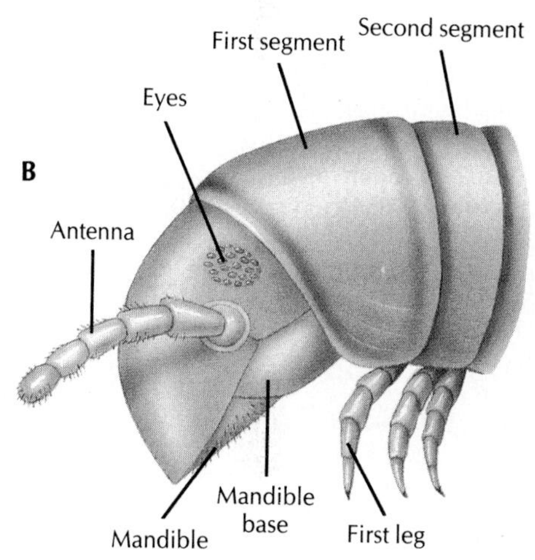

FIGURE 22-30

A, Seven-inch long millipedes from Ecuador, mating. Note the typical doubling of appendages on most segments. **B,** Head of millipede.

Photograph by K. Sandved.

Millipedes are less active than centipedes and are generally herbivorous, living on decayed plant and animal matter and sometimes living plants. They prefer dark moist places under stones and logs. Their eggs are laid in a nest and carefully guarded by the female. The larval forms have only one pair of legs to a somite.

INSECTS—CLASS INSECTA

The insects are the most successful biologically of all the groups of arthropods. There are more species of insects than species in all the other classes of animals combined. The number of insect species named has been estimated at close to 1 million, with thousands, perhaps millions, of other species yet to be discovered and classified.

It is difficult to appreciate fully the significance of this extensive group and its role in the biological pattern of animal life. The study of insects (**entomology**) occupies the time and resources of skilled men and women all over the world. The struggle between humans and their insect competitors seems to be an endless one, yet paradoxically, insects are so interwoven into the economy of nature in so many useful roles that we would have a difficult time without them.

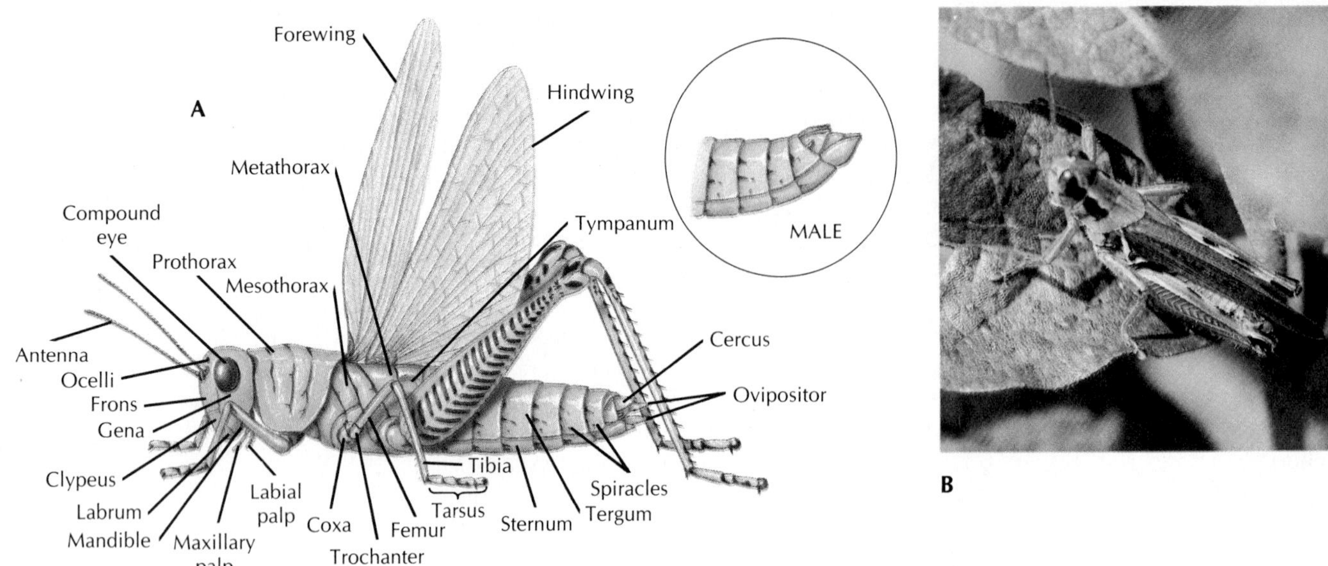

FIGURE 22-31

A, External features of a female grasshopper. The terminal segment of a male with external genitalia is shown in inset. **B,** A spur-throated grasshopper, *Melanoplus* sp. (order Orthoptera). This large genus contains many agricultural pest species.

B, Photograph by C.P. Hickman, Jr.

Insects differ from other arthropods in having **three pairs of legs** and usually **two pairs of wings** on the thoracic region of the body (Figure 22-31), although some have one pair of wings or none. In size insects range from less than 1 mm to 20 cm in length—the majority being less than 2.5 cm long.

Distribution and adaptability

Insects have spread into practically all habitats that will support life, but only a relatively few are marine. They are common in brackish water, in salt marshes, and on sandy beaches. They are abundant in fresh water, soils, forests, and plants, and they are found even in deserts and wastelands, on mountaintops, and as parasites in and on the bodies of plants and animals, including other insects.

Their wide distribution is made possible by their powers of flight and their highly adaptable nature. In many cases they can easily surmount barriers that are impassable to many other animals. Their small size and well-protected eggs allow them to be carried great distances by wind, water, and other animals.

The amazing adaptability of insects is evidenced by their wide distribution and enormous diversity of species. Such diversity enables this vigorous group to take advantage of all available resources of food and shelter.

A great deal of the success of insects can be attributed to the adaptive qualities of the cuticular exoskeleton, as is the case in the Crustacea. However, the great exploitation of the terrestrial environment by insects has been made possible by the array of adaptations they possess to withstand its rigors. For example, their epicuticle has a waxy and a varnish layer and they can close their spiracles; both characteristics minimize evaporative water loss. They extract the utmost in fluid from food and fecal material, and many can retain the water produced in oxidative energy metabolism. Many can enter a resting stage (**diapause**) and lie dormant during inhospitable conditions.

External features

The insect tagmata are the **head, thorax,** and **abdomen**. The cuticle of each body segment is typically composed of four plates (**sclerites**), a dorsal notum (**tergum**), a ventral **sternum**, and a pair of lateral **pleura.** The pleurae of abdominal segments are membranous rather than sclerotized.

The head usually bears a pair of relatively large compound eyes, a pair of antennae, and usually three ocelli. Mouthparts typically consist of a **labrum**, a pair each of **mandibles** and **maxillae**, a **labium**, and a tonguelike **hypopharynx.** The type of mouthparts an insect possesses determines how it feeds. Some of these modifications will be discussed later.

The thorax is composed of the **prothorax, mesothorax,** and **metathorax**, each bearing a pair of legs (Figure 22-31). In most insects the mesothorax and metathorax each bear a pair of wings. The wings are cuticular extensions formed by the epidermis. They consist of a double membrane that contains veins of thicker cuticle which serve to strengthen the wing. Although these veins vary in their patterns among the different species, they are constant within a species and serve as one means of classification and identification.

Legs of insects are often modified for special purposes. Terrestrial forms have walking legs with terminal pads and claws as in beetles. These pads may be sticky for walking upside down, as in houseflies. The hind legs of the grasshoppers and crickets are adapted for jumping (Figure 22-31, *B*). The mole cricket has the first pair of legs modified for burrowing in the ground. Water bugs and many beetles have paddle-shaped appendages for swimming. For grasping its prey, the forelegs of the praying mantis are long and strong (Figure 22-32).

A

B

FIGURE 22-32

A, Praying mantis (order Orthoptera) feeding on an insect. **B,** Praying mantis laying eggs.

Photographs by R. West.

Wings and the flight mechanism

Insects share the power of flight with birds and flying mammals. However, their wings have evolved in a different manner from that of the limb buds of birds and mammals and are not homologous to them. Insect wings are formed by outgrowth from the body wall of the mesothoracic and metathoracic segments and are composed of cuticle.

Most insects have two pairs of wings, but the Diptera (true flies) have only one pair, the hind wings being represented by a pair of small **halteres** (balancers) that vibrate and are responsible for equilibrium during flight. The males of the scale insects also have one pair of wings but no halteres. Some insects are wingless. Ants and termites, for example, have wings only on males, and on females during certain periods; workers are always wingless. Lice and fleas are always wingless.

Wings may be thin and membranous, as in flies and many others (Figure 22-46); thick and horny, as in the front wings of beetles (Figure 22-43); parchmentlike, as in the front wings of grasshoppers; covered with fine scales, as in butterflies and moths; or with hairs, as in caddis flies.

Wing movements are controlled by a complex of muscles in the thorax. **Direct flight muscles** are attached to a part of the wing itself. **Indirect flight muscles** are not attached to the wing; instead they cause wing movement by altering the shape of the thorax. The wing is hinged at the thoracic tergum and also slightly laterally on a pleural process, which acts as a fulcrum (Figure 22-33). In all insects, the upstroke of the wing is effected by contracting indirect muscles that pull the tergum down toward the sternum (Figure 22-33, *A*). In dragonflies and cockroaches the downstroke is accomplished by contracting direct muscles attached to the wings lateral to the pleural fulcrum. In Hymenoptera and Diptera all the flight muscles are indirect. The downstroke occurs when the sternotergal muscles relax and longitudinal muscles of the thorax arch the tergum (Figure 22-33, *B*), pulling up the tergal articulations relative to the pleura. The downstroke in beetles and grasshoppers involves both direct and indirect muscles.

Flight muscle contraction has two basic types of neural control: **synchronous** and **asynchronous.** Synchronous muscles are found in the larger insects such as dragonflies and butterflies, in which a single volley of nerve impulses stimulates a muscle contraction and thus one wing stroke. Asynchronous muscles are found

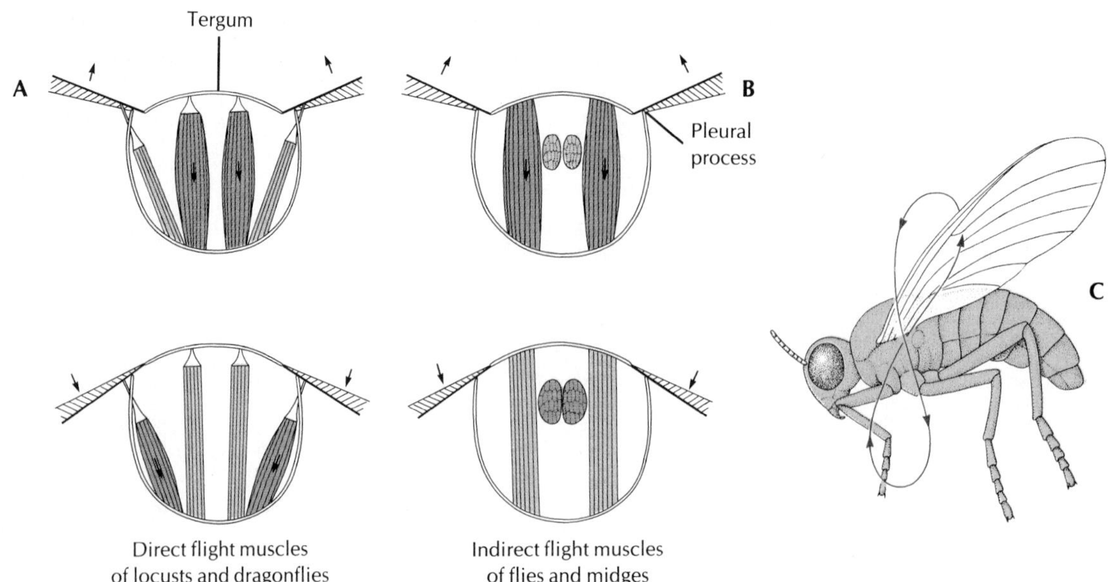

FIGURE 22-33

A, Flight muscles of insects such as cockroaches, in which upstroke is by indirect muscles and downstroke is by direct muscles. **B,** In insects such as flies and bees, both upstroke and downstroke are by indirect muscles. **C,** The figure-eight path followed by the wing of a flying insect during the upstroke and downstroke.

in the more specialized insects. Their mechanism of action is complex and depends on the storage of potential energy in resilient parts of the thoracic cuticle. As one set of muscles contracts (moving the wing in one direction), they stretch the antagonistic set of muscles, causing them to contract (and move the wing in the other direction). Because the muscle contractions are not phase-related to nervous stimulation, only occasional nerve impulses are necessary to keep the muscles responsive to alternating stretch activation. Thus extremely rapid wing beats are possible. For example, butterflies (with synchronous muscles) may beat as few as four times per second. Insects with asynchronous muscles, such as flies and bees, may vibrate at 100 beats per second or more.

Obviously flying entails more than a simple flapping of the wings; a forward thrust is necessary. As the indirect flight muscles alternate rhythmically to raise and lower the wings, the direct flight muscles alter the angle of the wings so that they act as lifting airfoils during both the upstroke and the downstroke, twisting the leading edge of the wings downward during the downstroke and upward during the upstroke. This produces a figure-eight movement (Figure 22-33, *C*) that aids in spilling air from the trailing edges of the wings. The quality of the forward thrust depends, of course, on several factors, such as variations in wing venation, how much the wings are tilted, how they are feathered, and so on.

Flight speeds vary. The fastest flyers usually have narrow, fast-moving wings with a strong tilt and a strong figure-eight component. Sphinx moths and horseflies are said to achieve approximately 48 km (30 miles) per hour and dragonflies approximately 40 km (25 miles) per hour. Some insects are capable of long continuous flights. The migrating monarch butterfly *Danaus plexippus* (Gr. after Danaus, mythical king of Arabia) travels south for hundreds of miles in the fall, flying at a speed of approximately 10 km (6 miles) per hour.

Internal form and function

Nutrition. The digestive system (Figure 22-34) consists of a **foregut** (mouth with salivary glands, **esophagus, crop** for storage, and **gizzard** for grinding), **midgut** (stomach and gastric ceca), and **hindgut** (intestine, rectum, and anus). The foregut and hindgut are lined with cuticle, so absorption of food is confined largely to the midgut, although some absorption may take place in all sections. The majority of insects feed on plant juices and plant tissues. Such a food habit is called **phytophagous.** Some insects feed on specific plants; others, such as grasshoppers, will eat almost any plant. The caterpillars of many moths and butterflies

Proventriculus
Ostia
Ovary
Colon
Brain
Aorta
Crop
Heart
Anus
Esophagus
Ovipositor
Salivary gland
Vagina
Labrum
Labium
Gastric ceca
Malpighian tubules
Oviduct
Rectum
Mouth
Subesophageal ganglion
Thoracic ganglion
Midgut
Abdominal ganglion
Seminal receptacle

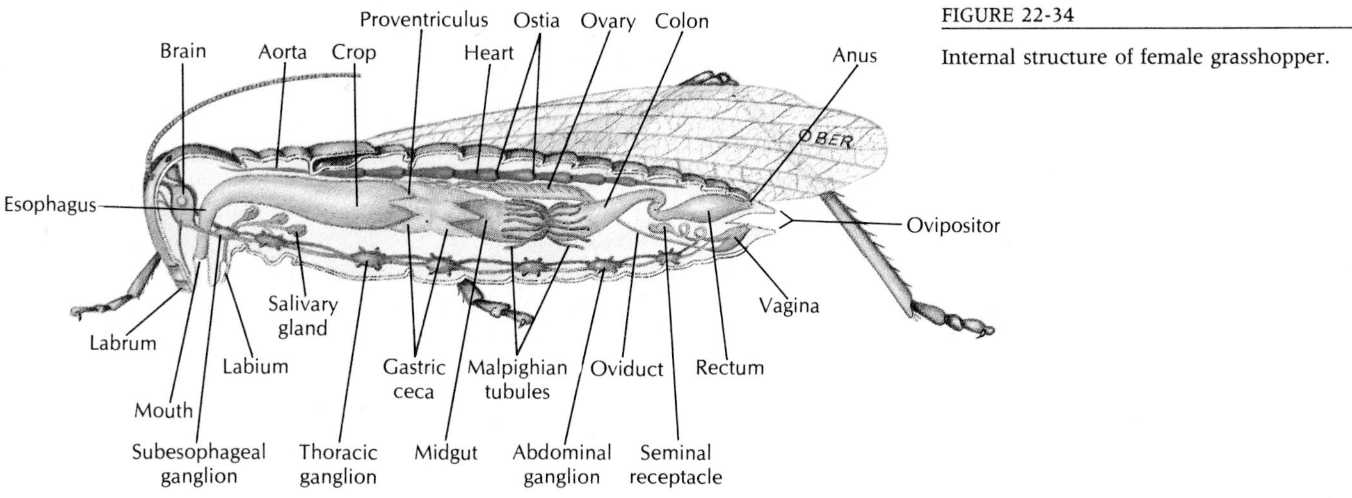

FIGURE 22-34

Internal structure of female grasshopper.

eat the foliage of only certain plants. Certain species of ants and termites cultivate fungus gardens as a source of food.

Many beetles and the larvae of many insects live on dead animals (**saprophagous**). A number of insects are **predaceous**, catching and eating other insects as well as other types of animals.

Many insects, adults as well as larvae, are **parasitic.** Adult fleas, for instance, live on the blood of birds and mammals, and the larvae of many varieties of wasps live on spiders and caterpillars. In turn, many insects are parasitized by other insects. Some of the latter are beneficial to humans by controlling the numbers of injurious insects. When parasitic insects are themselves parasitized by other insects, the condition is known as **hyperparasitism.**

The feeding habits of insects are determined to some extent by their mouthparts, which are highly specialized for each type of feeding.

Biting and chewing mouthparts, such as those of the grasshopper and many

FIGURE 22-35

Four types of insect mouthparts.

Illustrations by George Venable.

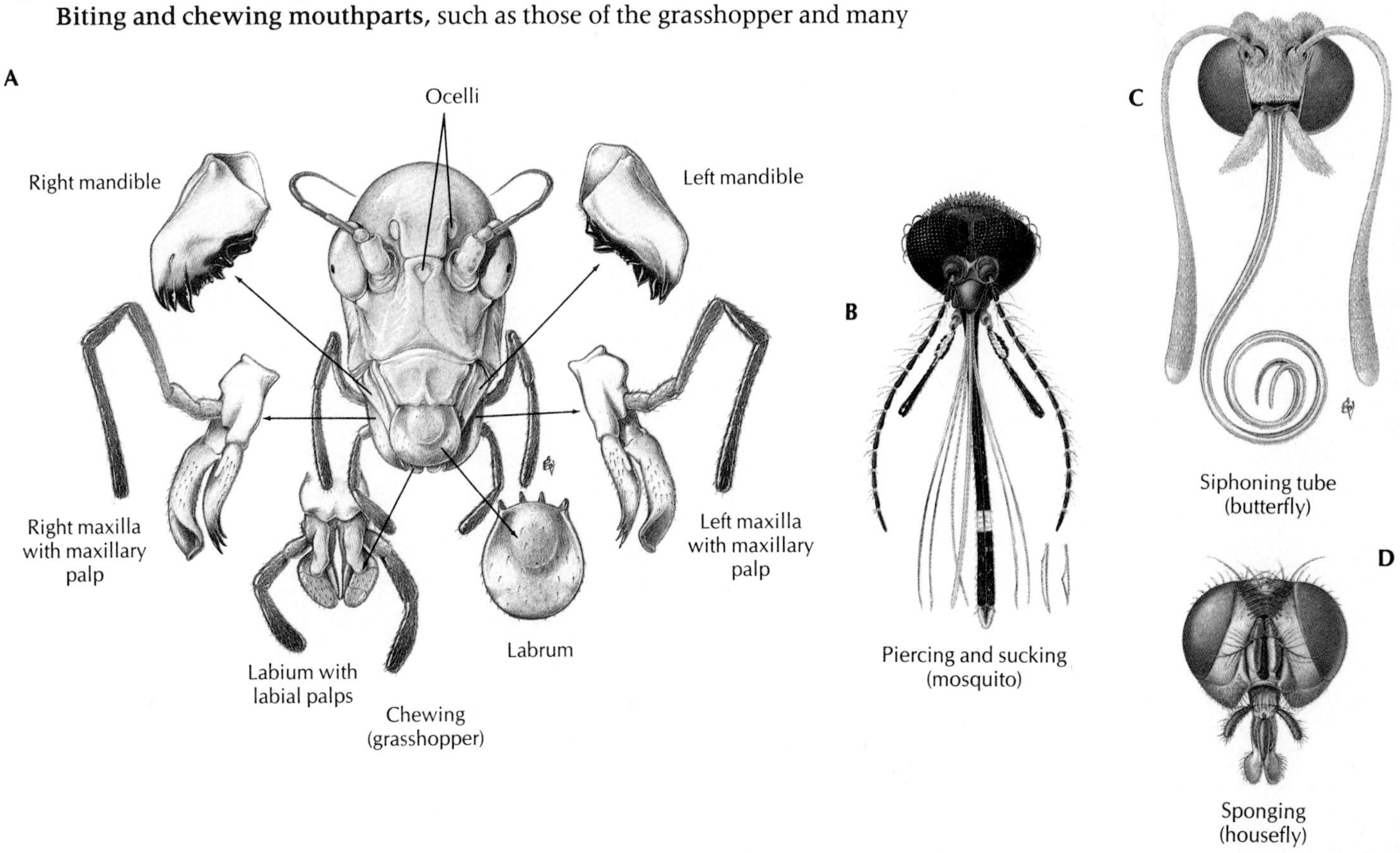

A
Ocelli
Right mandible
Left mandible
Right maxilla with maxillary palp
Left maxilla with maxillary palp
Labium with labial palps
Labrum
Chewing (grasshopper)
B
Piercing and sucking (mosquito)
C
Siphoning tube (butterfly)
D
Sponging (housefly)

herbivorous insects, are adapted for seizing and crushing food (Figure 22-35, *A*). The mandibles of chewing insects are strong, toothed plates whose edges can bite or tear while the maxillae hold the food and pass it toward the mouth. Enzymes secreted by the salivary glands add chemical action to the chewing process.

Sucking mouthparts are greatly varied. Houseflies and fruit flies have no mandibles; the labium is modified into two soft lobes containing many small tubules that sponge up liquids with a capillary action much as the holes of a commercial sponge do (Figure 22-35, *D*). Horseflies, however, are fitted not only to sponge up surface liquids but to bite into the skin with slender, tapering mandibles and then sponge up blood. Mosquitoes combine **piercing** by means of needlelike stylets and sucking through a food channel (Figure 22-35, *B*). In honeybees the labium forms a flexible and contractile "tongue" covered with many hairs. When the bee plunges its proboscis into nectar, the tip of the tongue bends upward and moves back and forth rapidly. Liquid enters the tube by capillarity and is drawn up continuously by a pumping pharynx. In butterflies and moths mandibles are usually absent, and the maxillae are modified into a long sucking proboscis (Figure 22-35, *C*) for drawing nectar from flowers. At rest the proboscis is coiled up into a flat spiral. In feeding it is extended, and fluid is pumped up by pharyngeal muscles.

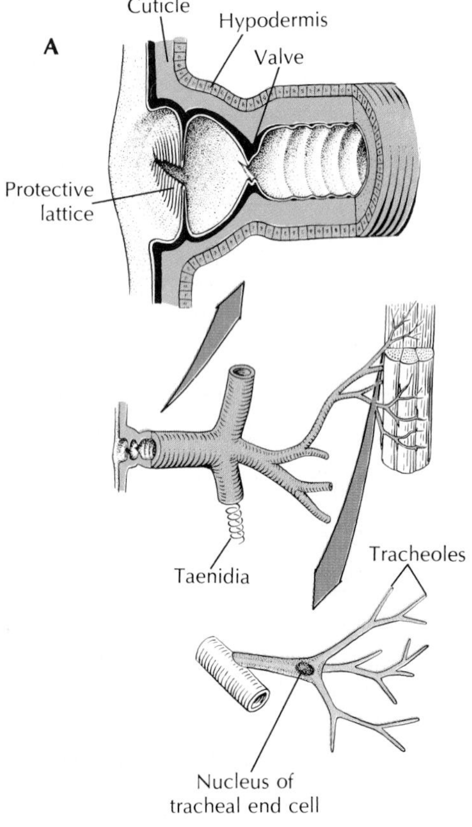

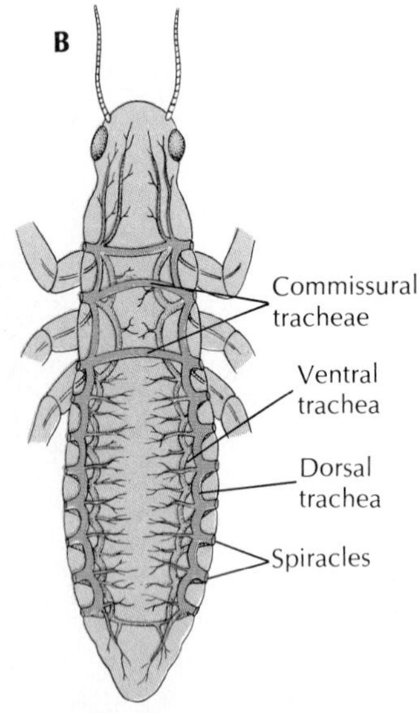

FIGURE 22-36

A, Relationship of spiracle, tracheae, taenidia (chitinous bands that strengthen the tracheae), and tracheoles (diagrammatic). **B,** Generalized arrangement of insect tracheal system (diagrammatic). Air sacs and tracheoles not shown.

Gas exchange. Terrestrial animals require efficient respiratory systems that permit rapid oxygen–carbon dioxide exchange but at the same time restrict water loss. In insects this is the function of the **tracheal system,** an extensive network of thin-walled tubes that branch into every part of the body (Figure 22-36). The tracheal trunks open to the outside by paired **spiracles,** usually two on the thorax and seven or eight on the abdomen. A spiracle may be merely a hole in the integument, as in primary wingless insects, but it is usually provided with a valve or some sort of closing mechanism that cuts down water loss. The evolution of such a device must have been very important in enabling insects to move into drier habitats. The **tracheae** are composed of a single layer of cells and are lined with cuticle that is shed, along with the outer cuticle, during the molt. They are supported by spiral thickenings of the cuticle, the **taenidia,** that prevent their collapse. The tracheae branch out into smaller tubes, ending in very fine, fluid-filled tubules called **tracheoles** (not lined with cuticle), which branch into a fine network over the cells. Scarcely any living cell is located more than a few micrometers away from a tracheole. In fact, the ends of some tracheoles actually indent the membranes of the cells they supply, so that they terminate close to the mitochondria. The tracheal system affords an efficient system of transport without the use of oxygen-carrying pigments in the blood.

In some very small insects gas transport occurs entirely by diffusion along a concentration gradient. As oxygen is used, a partial vacuum develops in the tracheae, and air is sucked in through the spiracles. Larger or more active insects employ some ventilation device for moving air in and out of the tubes. Usually muscular move-ments in the abdomen perform the pumping action that draws air in or expels it.

The tracheal system is primarily adapted for air breathing, but many insects (nymphs, larvae, and adults) live in water. In small, soft-bodied aquatic nymphs, the gaseous exchange may occur by diffusion through the body wall, usually into and out of a tracheal network just under the integument. The aquatic nymphs of stoneflies and mayflies are equipped with **tracheal gills,** which are thin extensions of the body wall containing a rich tracheal supply. The gills of dragonfly nymphs are ridges in the rectum (rectal gills) where gas exchange occurs as water moves in and out.

Excretion and water balance. Malpighian tubules (Figure 22-34) are typical of most insects. As in the spiders (p. 473), malpighian tubes are very efficient, both as excretory organs and as a means of conserving body fluids—an important factor in the success of terrestrial animals.

Since water requirements vary among different types of insects, this ability to cycle water and salts is very important. Insects living in dry environments may resorb nearly all water from the rectum, producing a nearly dry mixture of urine and feces. Leaf-feeding insects take in and excrete quantities of fluid. Freshwater larvae need to excrete water and conserve salts. Insects that feed on dry grains need to conserve water and excrete salt.

Nervous system. The nervous system in general resembles that of the larger crustaceans, with a similar tendency toward fusion of ganglia (Figure 22-34). A giant fiber system has been demonstrated in a number of insects. There is also a stomodeal nervous system that corresponds in function with the autonomic nervous system of vertebrates. Neurosecretory cells located in various parts of the brain have an endocrine function, but, except for their role in molting and metamorphosis, little is known of their activity.

Sense organs. The sensory perceptions of insects are usually keen. Organs receptive to mechanical, auditory, chemical, visual, and other stimuli are well developed. They are scattered over the body but are especially numerous on the appendages.

Photoreceptors include both ocelli and compound eyes. The compound eyes are large and constructed of ommatidia, like those of crustaceans (p. 485). Apparently, visual acuity in insect eyes is much lower than that of human eyes, but most flying insects rate much higher than humans in flicker-fusion tests. Flickers of light become fused in the human eye at a frequency of 45 to 55 per second, but in bees and blowflies they do not fuse until 200 to 300 per second. This should be an advantage in analyzing a fast-changing landscape.

Most insects have three ocelli on the head, and they also have dermal light receptors on the body surface, but not much is known about them.

Sounds may be detected by sensitive hairlike **sensilla** or by tympanic organs sensitive to sonic or ultrasonic sound. Sensilla are modifications in the cuticular surface for the reception of sensory stimuli other than light and are supplied with one or more neurons. Tympanic organs, found in grasshoppers (Figure 22-31, *A*), crickets, cicadas, butterflies, moths, etc. involve a number of sensory cells extending to a thin tympanic membrane that encloses an air space in which vibrations can be detected.

Chemoreceptive sensilla, which are peglike setae, are especially abundant on the antennae, mouthparts, or legs. Mechanical stimuli, such as contact pressure, vibrations, and tension changes in the cuticle, are picked up by sensilla or by sensory cells in the epidermis. Insects also sense temperature, humidity, body position (proprioception), gravity, etc.

Reproduction. Sexes are separate in insects, and fertilization is usually internal. Insects have various means of attracting mates. The female moth gives off a chemical (pheromone, p. 252) that can be detected for a great distance by the male (the amazing sensitivity of this system is described on p. 292). Fireflies use flashes of light; some insects find each other by means of sounds or color signals and by various kinds of courtship behavior.

Sperm are usually deposited in the vagina of the female at the time of copulation (Figure 22-37). In some orders the sperm are encased in spermatophores that may be transferred at copulation or deposited on the substratum to be picked up by the female. The silverfish deposits a spermatophore on the ground, then spins signal threads to guide the female to it. During evolutionary transition from aquatic to terrestrial life, spermatophores were widely used, with copulation evolving much later.

Usually the sperm are stored in the seminal receptacle of the female in numbers sufficient to fertilize more than one batch of eggs. Many insects mate only once during their lifetime, and none mates more than a few times.

Insects usually lay a great many eggs. The queen honeybee, for example, may lay more than 1 million eggs during her lifetime. On the other hand, some flies

Although the diving beetle *Dytiscus* (Gr. *dytikos,* able to swim) can fly, it spends most of its life in the water as an excellent swimmer. It uses an "artificial gill" in the form of a bubble of air held under its wing covers. The bubble is kept stable by a layer of hairs on top of the abdomen and is in contact with the spiracles on the abdomen. Oxygen from the bubble diffuses into the tracheae and is replaced by diffusion of oxygen from the water. Thus the bubble can last for several hours before the beetle must surface to replace it. Mosquito larvae are not good swimmers but live just below the surface, putting out short breathing tubes like snorkels to the surface for air.

A

B

FIGURE 22-37

Copulation in insects. **A,** Bluet damselflies *Enallagma* sp. (order Odonata) are common throughout North America. **B,** *Omura congrua* (order Orthoptera) are a kind of grasshopper found in Brazil.

Photographs by K. Sandved.

A

B

C

D

FIGURE 22-38

Complete metamorphosis. **A,** A monarch butterfly, *Danaus plexippus* (order Lepidoptera), lays eggs on a milkweed plant. **B,** The hatched larvae feed on milkweed leaves. **C,** When mature, the larva hangs on the milkweed plant and transforms into a chrysalis, or pupa, an inactive stage that does not feed and is covered by a cocoon or protective covering. **D,** An adult has emerged. The wings quickly expand and harden, pigmentation develops, and the butterfly goes on its way.

A, Photograph by C.P. Hickman Jr.; **B** to **D,** photographs by J.H. Gerard.

are ovoviviparous and bring forth only a single offspring at a time. Forms that make no provision for the care of the young may lay many more eggs than those which provide for the young or those that have a very short life cycle.

Most species normally lay their eggs in a particular type of place to which they are guided by visual, chemical, or other clues. Butterflies and moths lay their eggs on the specific kind of plant on which the caterpillar must feed. The tiger moth may look for a pigweed and the sphinx moth for a tomato or tobacco plant and the monarch butterfly for a milkweed plant (Figure 22-38). Insects whose immature stages are aquatic lay their eggs in water. A tiny braconid wasp lays her eggs on the caterpillar of the sphinx moth where they will feed and pupate in tiny white cocoons (Figure 22-39). The ichneumon wasp, with unerring accuracy, seeks out a certain kind of larva in which her young will live as internal parasites. Her long ovipositors may have to penetrate 1 to 2 cm of wood to find and deposit her eggs in the larva of a wood wasp or a wood-boring beetle (Figure 22-40).

Metamorphosis and growth

Although metamorphosis is not limited to insects, they illustrate it more dramatically than any other group. The transformation of a caterpillar into a beautiful moth or butterfly is indeed an astonishing change.

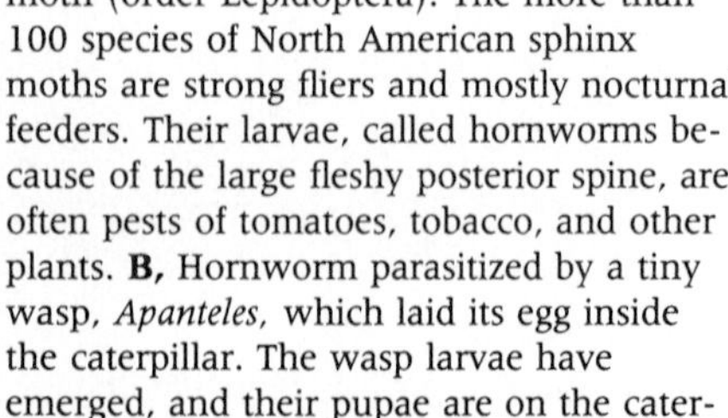

A

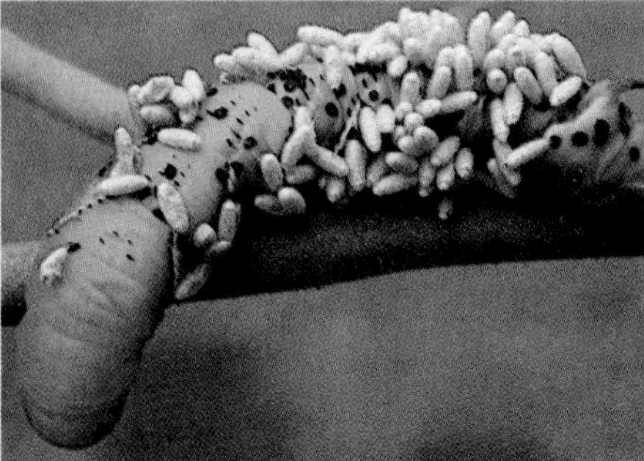

B

FIGURE 22-39

A, Hornworm, larval stage of a sphinx moth (order Lepidoptera). The more than 100 species of North American sphinx moths are strong fliers and mostly nocturnal feeders. Their larvae, called hornworms because of the large fleshy posterior spine, are often pests of tomatoes, tobacco, and other plants. **B,** Hornworm parasitized by a tiny wasp, *Apanteles,* which laid its egg inside the caterpillar. The wasp larvae have emerged, and their pupae are on the caterpillar's skin. Young wasps emerge in 5 to 10 days, but the caterpillar usually dies.

A, Photograph by C.P. Hickman, Jr.; **B,** photograph by J.H. Gerard.

Early development occurs within the egg, and the hatching young escape from the egg in various ways. During the postembryonic development most insects change in form; that is, they undergo **metamorphosis**. To grow during this period they must undergo a number of molts, and each stage of the insect between molts is called an **instar**.

Approximately 88% of insects undergo a **complete metamorphosis**, which separates the physiological processes of growth (**larva**) from those of differentiation (**pupa**) and reproduction (**adult**). Each stage functions efficiently without competition with the other stages, because the larvae often live in entirely different surroundings and eat different foods from the adults. The wormlike larvae, which usually have chewing mouthparts, are known as caterpillars, maggots, bagworms, fuzzy worms, grubs, etc. After a series of instars during which the wings are developing internally, the larva forms a case or cocoon about itself and becomes a pupa, or chrysalis, a nonfeeding stage in which many insects pass the winter. When the final molt occurs the full-grown adult emerges (Figure 22-38), pale and with wings wrinkled. In a short time the wings expand and harden, and the insect is on its way. The stages, then, are **egg, larva** (several instars), **pupa,** and **adult.** The adult undergoes no further molting. Insects that undergo complete metamorphosis are said to be **holometabolous** (Gr. *holo,* complete + *metabolē,* change).

Some insects undergo a type of **gradual**, or **incomplete, metamorphosis.** These include the grasshoppers, cicadas, and mantids, which have terrestrial young, and mayflies, stoneflies (Figure 22-41, *A*), and dragonflies (Figure 22-41, *B*), which lay their eggs in water. The young are called **nymphs** (or **naiads**, if aquatic, Figure 22-41, *C*), and their wings develop externally as budlike outgrowths in the early instars and increase in size as the animal grows by successive molts and becomes a winged adult (Figure 22-42). The aquatic naiads have tracheal gills or other modifications for aquatic life. The stages are **egg, nymph** (several instars), and **adult.** Insects that undergo gradual metamorphosis are called **hemimetabolous** (Gr. *hemi,* half, + *metabolē,* change).

A few insects, such as silverfish and springtails, undergo direct development. The young, or juveniles, are similar to the adults except in size and sexual maturation. The stages are egg, juvenile, and adult. Such insects include the wingless insects (apterygote orders).

Metamorphosis in insects is controlled and regulated by hormones. There are three major endocrine organs involved in development through the juvenile instars and eventually to the emergence of the adult. These organs are the **brain, ecdysial (prothoracic) glands,** and **corpora allata.** The action of the hormones produced is described on p. 266. The hormonal control of molting and metamorphosis is the same in both holometabolous and hemimetabolous insects.

Diapause. Many animals can enter a state of dormancy during adverse conditions, and there are periods in the life cycle of many insects when a particular stage can remain dormant for a long time because external conditions of climate, moisture, and the like are too harsh for normal activity. Most insects enter such a

FIGURE 22-40

An ichneumon wasp with the end of the abdomen raised to thrust her long ovipositor into wood to find a tunnel made by the larva of a wood wasp or wood-boring beetle. She can bore 13 mm or more into the wood to lay her eggs in the larva of the wood-boring beetle, which will become host for the ichneumon larvae. Other ichneumon species attack spiders, moths, flies, crickets, caterpillars, and other insects.

Photograph by C.P. Hickman, Jr.

FIGURE 22-41

A, A stonefly, *Perla* sp. (order Plecoptera). **B,** A ten-spot dragonfly, *Libellula pulchella* (order Odonata). **C,** A dragonfly naiad. Both stoneflies and dragonflies have aquatic larvae, called naiads, which undergo gradual metamorphosis.

A, Photograph by C.P. Hickman, Jr.; **B,** photograph by J.H. Gerard; **C,** courtesy Caroline Biological Supply Co.

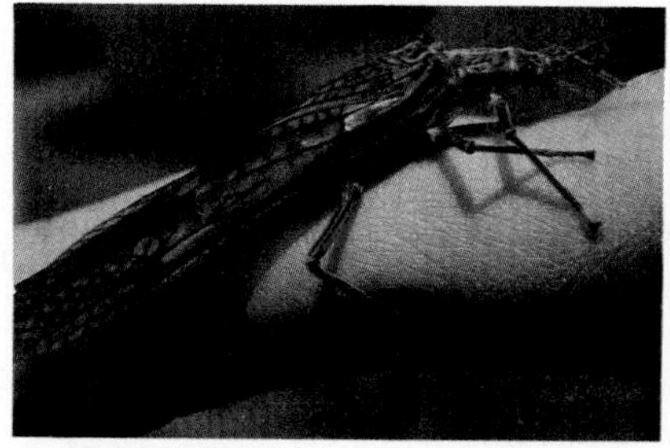

A

B

C

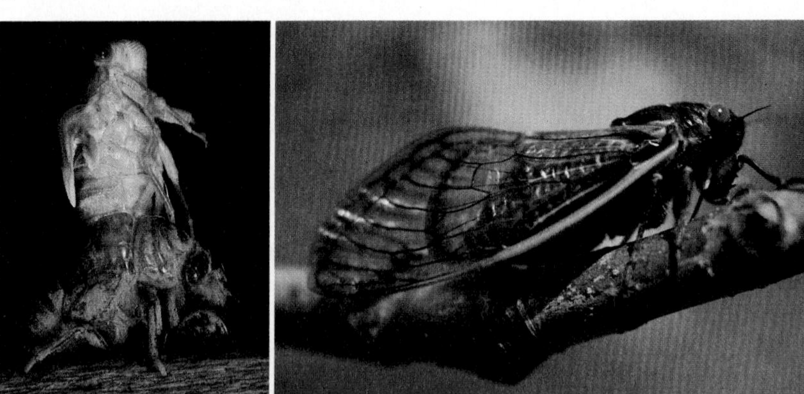

FIGURE 22-42

A, Ecdysis in the dog-day cicada, *Tibicen pruinosa* (order Homoptera). The old cuticle splits along a dorsal midline as a result of increased blood pressure and of air forced into the thorax by muscle contraction. The emerging insect is pale, and its new cuticle is soft. The wings will be expanded by blood pumped into veins, and the insect enlarges by taking in air. **B,** The seventeen-year cicada, *Magicicada septendecim.*

A, Photograph by J.H. Gerard; B, photograph by L.L. Rue III.

stage when some factor of the environment, such as temperature, becomes unfavorable, and the state continues until conditions again become favorable.

However, some species have a prolonged arrest of growth that occurs irrespective of the environment, that is, whether or not favorable conditions prevail. This type of dormancy is called **diapause** (di'a-poz) (Gr. *dia,* through, dividing into two parts, + *pausis,* a stopping), and it is an important adaptation to survive adverse environmental conditions. Diapause is genetically determined in each species, but it is usually set off by some external signal, such as shortening day length. Diapause always occurs at the end of an active growth stage or the molting cycle so that, when the diapause period is over, the insect is ready for another molt.

Behavior and communication

The keen sensory perceptions of insects make them extremely responsive to many stimuli. The stimuli may be internal (physiological) or external (environmental), and the responses are governed by both the physiological state of the animal and the pattern of nerve pathways traveled by the impulses. Many of the responses are simple, such as orientation toward or away from the stimulus, for example, attraction of a moth to light, avoidance of light by a cockroach, or attraction of carrion flies to the odor of dead flesh.

Much of the behavior of insects, however, is not a simple matter of orientation but involves a complex series of responses. A pair of tumble bugs, or dung beetles, chew off a bit of dung, roll it into a ball, and roll the ball laboriously to where they intend to bury it after laying their eggs in it (Figure 22-43). The cicada slits the bark of a twig and then lays an egg in each of the slits. The female potter wasp *Eumenes* scoops up clay into pellets, carries them one by one to her building site, and fashions them into dainty little narrow-necked clay pots, into each of which she lays an egg. Then she hunts and paralyzes a number of caterpillars, pokes them into the opening of a pot, and closes up the opening with clay. Each egg, in its own protective pot, hatches to find a well-stocked larder of food awaiting it.

FIGURE 22-43

Tumble bugs, or dung beetles, *Canthon pilularis* (order Coleoptera), chew off a bit of dung, roll it into a ball, and then roll it to where they will bury it in soil. One beetle pushes while the other pulls. Eggs are laid in the ball, and the larvae feed on the dung. Tumble bugs are black, an inch or less in length, and common in pasture fields.

Photograph by J.H. Gerard.

Much of such behavior is "innate," that is, entire sequences of actions apparently have been genetically programmed. However, a great deal more learning is involved than was once believed. The potter wasp, for example, must learn where she has left her pots if she is to return to fill them with caterpillars one at a time. Social insects, which have been studied extensively, have been found capable of most of the basic forms of learning used by mammals. The exception is insight learning. Apparently insects cannot, when faced with a new problem, reorganize their memories to construct a new response.

Insects communicate with other members of their species by means of chemical, visual, auditory, and tactile signals. **Chemical signals** take the form of **pheromones,** which are substances secreted by one individual that affect the behavior or physiological processes of another individual. Pheromones include sex attractants, releasers of certain behavior patterns, trail markers, alarm signals, territorial markers, and the like. Like hormones, pheromones are effective in minute quantities. Social insects, such as bees, ants, wasps, and termites, can recognize a nestmate—or an alien in the nest–by means of identification pheromones. An intruder from another species is violently attacked at once. If the intruder is from the same species but another colony there may be a variety of responses. It may be attacked and killed; it may be investigated but finally accepted; or it may be accepted but given less food than the others until it has had time to acquire the colony odor. Caste determination in termites, and to some extent in ants and bees, is determined by pheromones. In fact, pheromones are probably a primary integrating force in populations of social insects. Many insect pheromones have been extracted and chemically identified.

Sound production and reception (phonoproduction and phonoreception) in insects have been studied extensively, and it is evident that, although a sense of hearing is not present in all insects, this means of communication is meaningful to insects that use it. Sounds serve as warning devices, advertisement of territorial claims, or courtship songs. The sounds of crickets and grasshoppers seem to be concerned with courtship and aggression. Grasshoppers rub the femur of the third pair of legs over the ridges of the forewings. Male crickets scrape the rough edges of the forewings together to produce their characteristic chirping. The hum of mosquitoes is caused by the rapid vibration of their wings. The long, drawn-out sound of the male cicada, a recruitment call, is produced by the vibrating membranes in a pair of organs located on the ventral side of the basal abdominal segment.

One of the types of communication by means of **visual signals** is in the form of **bioluminescence** (production of light by living organisms), known in certain kinds of flies, springtails, and beetles. The best known of the luminescent beetles are the fireflies, or lightning bugs (which are neither flies nor bugs, but beetles), in which the flash of light is a means of locating a prospective mate. Each species has its own characteristic flashing rhythm produced on the ventral side of the last abdominal segments. The male flashes his species-specific pattern while flying. If a female flashes the proper answer after the appropriate interval, he flies toward her, giving his signal again. The dialogue usually culminates in copulation. An

Some insects can memorize and perform in sequence tasks involving multiple signals in various sensory areas. Worker honeybees have been trained to walk through mazes that involved five turns in sequence, using such clues as the color of a marker, the distance between two spots, or the angle of a turn. The same is true of ants. Workers of one species of *Formica* learned a six-point maze at a rate only two or three times slower than that of laboratory rats. The foraging trips of ants and bees often wind and loop about in a circuitous route, but once the forager has found food, the return trip is relatively direct. One investigator suggests that the continuous series of calculations necessary to figure the angles, directions, distance, and speed of the trip and to convert it into a direct return could involve a stopwatch, a compass, and integral vector calculus. How the insect does it is unknown.

A B

FIGURE 22-44

Mimicry in butterflies. **A,** The monarch butterfly is distasteful to, and avoided by, birds because as a caterpillar it fed on the acrid milkweed. **B,** It is mimicked by the smaller viceroy butterfly, *Limenitis archippus,* which feeds on willows and is presumably tasteful to birds, but is not eaten because it so closely resembles the monarch in color and markings. This kind of mimicry is called "Batesian" mimicry.

Photographs by J.H. Gerard.

FIGURE 22-45

Firefly *femme fatale, Photuris versicolor,* eating a male *Photinus tanytoxus,* which she has attracted with false mating signals.

Photograph by J.E. Lloyd, copyright 1975, American Association for the Advancement of Science.

interesting instance of aggressive mimicry has been observed in females of several species of *Photuris,* which prey on male fireflies of other species by mimicking the female mating signals of the prey species and then by capturing and devouring the luckless males that court them (Figure 22-45).

There are many forms of **tactile communication,** such as tapping, stroking, grasping, and antennae touching, which evoke responses varying from recognition to recruitment and alarm. The food-begging behavior of the larvae of the ant *Formica* is to tap the mandibles of the worker with their own mouthparts, thus triggering the regurgitation of liquid food by the worker. When an adult *Formica* worker taps another worker with its antennae, the signal is to stop moving about.

Social behavior. Insects rank very high in the animal kingdom in their organization of social groups, and cooperation within the more complex groups depends heavily on chemical and tactile communication. Social communities are not all complex, however. Some community groups are temporary and uncoordinated, as are the hibernating associations of carpenter bees or the feeding gatherings of aphids. Some are coordinated for only brief periods, such as the tent caterpillars *Malacosoma,* that join in building a home web and a feeding net. However, all these are still open communities with social behavior.

In the true societies of the higher orders, such as honeybees, ants, and termites, a complex social life is necessary for the perpetuation of the species. Such societies are closed. In them all stages of the life cycle are involved, the communities are usually permanent, all activities are collective, and there is reciprocal communication. There is a high degree of efficiency in the division of labor. Such a society is essentially a family group in which the mother or perhaps both parents remain with the young, sharing the duties of the group in a cooperative manner. The society is usually characterized by polymorphism, or **caste** differentiation, along with differences in behavior that are associated with the division of labor.

The honeybees have one of the most complex organizations in the insect world. Instead of lasting one season, their organization continues for a more or less indefinite period. As many as 60,000 to 70,000 honeybees may be found in a single hive. Of these, there are three castes—a single sexually mature female, or **queen,** a few hundred **drones,** which are sexually mature males, and the **workers,** which are sexually inactive genetic females (Figure 22-46).

The workers take care of the young, secrete wax with which they build the six-sided cells of the honeycomb, gather the nectar from flowers, manufacture honey, collect pollen, and ventilate and guard the hive. Each worker appears to be responsible for a specific task, depending on its age, but during its lifetime of a few weeks it performs all of the various tasks.

FIGURE 22-46

Queen bee surrounded by her court. The queen is the only egg layer in the colony. The attendants, attracted by her pheromones, constantly lick her body. As food is transferred from these bees to others, the queen's presence is communicated throughout the colony.

Photograph by K. Lorenzen © 1979 Educational Images, Lyons Falls, N.Y.

One drone, sometimes more, fertilizes the queen during the mating flight, at which time enough sperm is stored in her seminal receptacle to last her a lifetime. Drones have no stings and are usually driven out or killed by the workers at the end of the summer. A queen may live as long as five seasons, laying thousands of eggs in that time. She is responsible for keeping the hive going through the winter, and only one reigning queen is tolerated in a hive at one time.

Castes are determined partly by fertilization and partly by what is fed to the larvae. Drones develop from unfertilized eggs (and consequently are haploid); queens and workers develop from fertilized eggs (and thus are diploid). Female larvae that are destined to become queens are fed **royal jelly**, a secretion from the salivary glands of the nurse workers. Royal jelly differs from the "worker jelly" fed to ordinary larvae, but the components in it that are essential for queen determination have not yet been identified. Honey and pollen are added to the worker diet about the third day of larval life. Female workers are prevented from maturing sexually by pheromones in the "queen substance," which is produced by the queen's mandibular glands. Royal jelly is produced by the workers only when the level of "queen substance" pheromone in the colony drops. This occurs when the queen becomes too old, dies, or is removed. Then the workers start enlarging a larval cell and feeding the larva the royal jelly that produces a new queen.

Honeybees have evolved an efficient system of communication by which, through certain body movements, their scouts inform the workers of the location and quantity of food sources (see pp. 292-293).

Termite colonies contain several castes, consisting of fertile individuals, both males and females, and sterile individuals (Figure 22-47). Some of the fertile individuals may have wings and may leave the colony, mate, lose their wings, and as **king** and **queen** start a new colony. Wingless fertile individuals may under certain conditions substitute for the king or queen. Sterile members are wingless and become **workers** and **soldiers**. Soldiers have large heads and mandibles and serve for the defense of the colony. As in bees and ants, caste differentiation is caused by extrinsic factors. Reproductive individuals and soldiers secrete inhibiting pheromones that are passed throughout the colony to the nymphs through a mutual feeding process, called **trophallaxis**, so that they become sterile workers. Workers also produce pheromones, and if the level of

A

B

FIGURE 22-47

A, Termite workers, *Reticulitermes flavipes* (order Isoptera), eating yellow pine. Workers are wingless sterile adults that tend the nest, care for the young, and so forth. **B,** Termite queen (*Macrotermes bellicosus* from Ghana) becomes a distended egg-laying machine. The queen and several workers and soldiers are shown here.

A, Photograph by J.H. Gerard; **B,** photograph by K. Sandved.

FIGURE 22-48

A weaver ant nest in Australia.
Photograph by L.S. Roberts.

"worker substance" or "soldier substance" falls, as might happen after an attack by marauding predators, for example, compensating proportions of the appropriate caste are produced in the next generation.

The phenomenon of trophallaxis, or exchange of nutrients, appears to be common among all social insects because it integrates the colony by passage of pheromones. The process involves feeding of the young by the queen and workers, which in turn may receive a drop of saliva from the young. It may also involve mutual licking, grooming, and the like.

Ants also have highly organized societies Superficially, they resemble termites, but they are quite different (belong to a different order) and can be distinguished easily. In contrast to termites, ants are usually dark in color, are hard bodied, and have a constriction between the thorax and abdomen.

In ant colonies the males die soon after mating and the queen either starts her own new colony or joins some established colony and does the egg laying. The sterile females are wingless workers and soldiers that do the work of the colony—gather food, care for the young, and protect the colony. In many larger colonies there may be two or three types of individuals within each caste.

Ants have a varied diet. In some the larvae are the real food digesters for the colony, as they can digest solid food and the adults feed on liquid foods. Nutrients are distributed among the members by means of trophallaxis.

Ants have evolved some striking patterns of "economic" behavior, such as making slaves, farming fungi, herding "ant cows" (aphids), sewing their nests together with silk (Figure 22-48), and using tools.

Insects and human welfare

Beneficial insects. Although most of us think of insects primarily as pests, humanity would have great difficulty in surviving if all insects were suddenly to disappear. Some of them produce useful materials: honey and beeswax from bees, silk from silkworms, and shellac from a wax secreted by the lac insects. More important, however, insects are necessary for the cross-fertilization of many fruits and other crops.

Insects and higher plants very early in their evolution formed a relationship of mutual adaptations that have been to each other's advantage. Insects exploit flowers for food, and flowers exploit insects for pollination. Each floral develop-

FIGURE 22-49

Fly maggots feeding on a deer carcass (order Diptera).
Photograph by L.L. Rue III.

ment of petal and sepal arrangement is correlated with the sensory adjustment of certain pollinating insects. Among these mutual adaptations are amazing devices of allurements, traps, specialized structure, precise timing, and so on.

Many predaceous insects, such as tiger beetles, aphid lions, ant lions, praying mantises, and ladybird beetles, destroy harmful insects. Some insects control harmful ones by parasitizing them or by laying their eggs where their young, when hatched, may devour the host. Dead animals are quickly taken care of by maggots hatched from eggs laid on carcasses (Figure 22-49).

Insects and their larvae serve as an important source of food for many birds, fish, and other animals.

Harmful insects. Harmful insects include those that eat and destroy plants and fruits, such as grasshoppers, chinch bugs, corn borers, boll weevils, grain weevils, San Jose scale, and scores of others (Figure 22-50). Practically every cultivated crop has some insect pest. Lice, bloodsucking flies, warble flies, botflies, and many others attack humans or domestic animals or both. Malaria, carried by the *Anopheles* mosquito, is still one of the world's killers; yellow fever and filariasis are also transmitted by mosquitoes. Fleas carry plague, which at many times in history has almost wiped out whole populations. The housefly is the vector for typhoid and the louse for typhus fever; the tsetse fly carries African sleeping sickness; and certain bloodsucking bugs are carriers of Chagas' disease. In addition there is a tremendous destruction of food, clothing, and property by weevils, cockroaches, ants, clothes moths, termites, and carpet beetles. Not the least of the insect pests is the bedbug, *Cimex,* a bloodsucking hemipterous insect that humans contracted, probably early in their evolution, from bats that shared their caves.

Control of insects. Because all insects are an integral part of the ecological communities to which they belong, their total destruction would probably do more harm than good. Food chains would be disturbed, some of our most loved birds would disappear, and the biological cycles by which dead animal and plant matter disintegrates and returns to enrich the soil would be seriously impeded. The beneficial role of insects in our environment has often been overlooked, and in our zeal to control the pests we have indiscriminately sprayed the landscape with extremely effective "broad-spectrum" insecticides that eradicate the beneficial, as well as the harmful, insects. We have also found, to our chagrin, that many of the chemicals we have used persist in the environment and accumulate as residues in the bodies of animals higher up in the food chains. Also, many strains of insects have developed a resistance to the insecticides in common use.

In recent years an effort has been made to be more selective in the choice and use of pesticides that are specific in their targets. In addition to chemical control, other methods of control have been under intense investigation and experimentation.

FIGURE 22-50

Insect pests. **A,** Japanese beetles, *Popillia japonica* (order Coleoptera) are serious pests of fruit trees and ornamental shrubs. They were introduced into the United States from Japan in 1917. **B,** Walnut caterpillars, *Datana ministra* (order Lepidoptera), defoliating a hickory tree. **C,** Corn ear worms, *Heliothis zea* (order Lepidoptera). An even more serious pest of corn is the infamous corn borer, an import from Europe in 1908 or 1909.

A, Photograph by L.L. Rue III; B and C, photographs by J.H. Gerard.

A

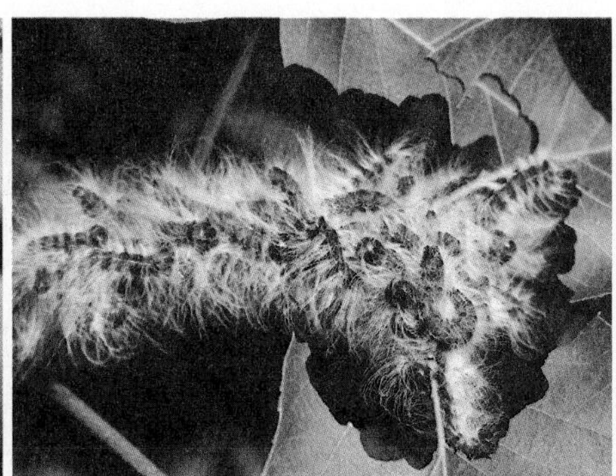

B

C

The sterile male approach has been used effectively in eradicating screwworm flies, a livestock pest. Large numbers of male insects, sterilized by irradiation, are introduced into the natural population; females that mate with the sterile flies lay infertile eggs.

The development of **insect-resistant crops** is one area of investigation. So many factors, such as yield and quality, are involved in developing resistant crops that the teamwork of specialists from many related fields is required. Some progress, however, has been made.

Several types of biological controls have been developed and are under investigation. All of these areas present problems but also show great possibilities. One is the use of bacterial and viral pathogens. A bacterium, *Bacillus thuringiensis,* has been used to control several lepidopteran pests (cabbage looper, imported cabbage worm, tomato worm), and several more show some potential. Many viruses have been isolated that seem to have potential as insecticides. However, specific viruses are difficult to rear and could be expensive to put into commercial production.

Introduction of natural predators or parasites of the insect pests has met with some success. In the United States the vedalia beetle was brought from Australia to counteract the work of the cottony-cushion scale on citrus plants, and numerous instances of control by use of insect parasites have been recorded.

Another approach to biological control is to interfere with the reproduction or behavior of insect pests with sterile males or with naturally occurring organic compounds that act as hormones or pheromones. Such research, although very promising, is slow because of our limited understanding of insect behavior and the problems of isolating and identifying complex compounds that are produced in such minute amounts. Nevertheless, pheromones will probably play an important role in biological pest control in the future.

A systems approach referred to as **integrated pest management** is receiving increased attention. This involves integrated utilization of all possible, practical techniques to contain pest infestations at a tolerable level, for example, cultural techniques (resistant plant varieties, crop rotation, tillage techniques, timing of sowing, planting, or harvesting, and others), use of biological controls, and sparing use of insecticides.

BRIEF REVIEW OF MAJOR INSECT ORDERS

Insects are divided into orders on the basis of wing structure, mouthparts, metamorphosis, and so on. Entomologists do not all agree on the names of the orders or on the limits of each order. Some tend to combine and others to divide the groups. However, the following synopsis of the major orders is one that is rather widely accepted.

Subclass Apterygota (ap-ter-i-go'ta) (Gr. *a,* not, + *pterygōtos,* winged) (**Ametabola**). Primitive **wingless** insects, without metamorphosis or with a primitive type of metamorphosis. Examples are springtails and silverfish (Figure 22-51).

Subclass Pterygota (ter-i-go'ta) (Gr. *pterygōtos,* winged) (**Metabola**). **Winged insects** (some secondarily wingless) **with metamorphosis;** includes 97% of all insects.

FIGURE 22-51

A, Silverfish *Lepisma* (order Thysanura) is often found in homes. **B** to **D,** Springtails (order Collembola). **B,** *Anurida.* **C** and **D,** *Orchesetta* in resting and leaping positions.

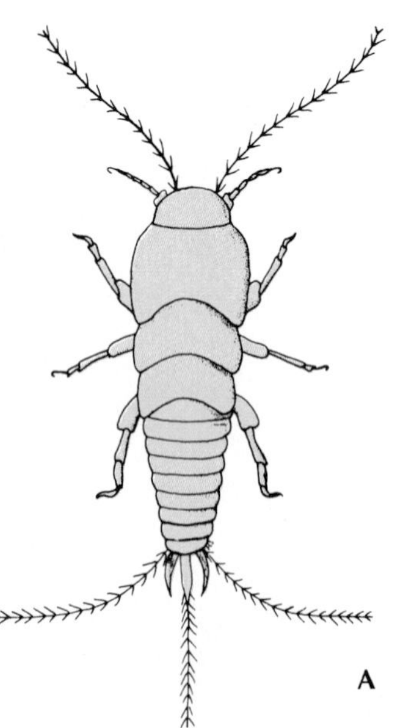

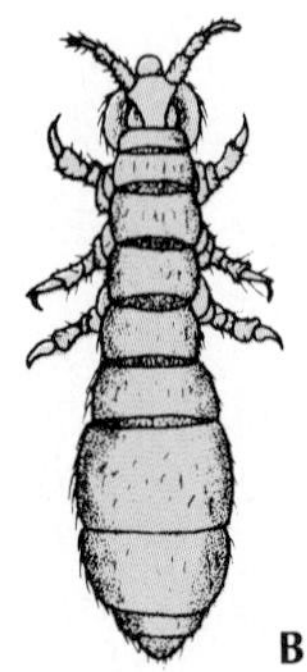

Superorder Exopterygota (ek-sop-ter-i-go'ta) (Gr. *exo,* outside, + *pterygōtos,* winged) (**Hemimetabola**). **Metamorphosis gradual; wings develop externally** on nymphs; compound eyes present on nymphs.

Order Ephemeroptera (e-fem-er-op'ter-a) (Gr. *ephēmeros,* lasting but a day, + *pteron,* wing) — **mayflies** (Figure 22-52). Wings membranous; forewings larger than hind wings; adult mouthparts vestigial; nymphs aquatic, with lateral tracheal gills.

Order Odonata (o-do-na'ta) (Gr. *odontos,* tooth, + *ata,* characterized by) — **dragonflies** (Figure 22-41, *B*), **damselflies** (Figure 22-37, *A*). Large; membranous wings are long, narrow, net veined, and similar in size; long and slender body; aquatic nymphs with aquatic gills and prehensile labium for capture of prey.

Order Orthoptera (or-thop'ter-a) (Gr. *orthos,* straight, + *pteron,* wing) — **grasshoppers, locusts, crickets, cockroaches, walkingsticks, praying mantises** (Figures 22-31 and 22-32). Wings when present, with forewings thickened and hind wings folded like a fan under forewings; chewing mouthparts.

Order Isoptera (i-sop'ter-a) (Gr. *isos,* equal, + *pteron,* wing) — **termites** (Figure 22-47). Small; membranous, narrow wings similar in size with few veins; wings shed at maturity; erroneously called "white ants"; distinguishable from true ants by broad union of thorax and abdomen; complex social organization.

Order Mallophaga (mal-lof'a-ga) (Gr. *mallos,* wool, + *phagein,* to eat) — **biting lice** (Figure 22-53). As large as 6 mm; wingless; chewing mouthparts; legs adapted for clinging to host; live on birds and mammals.

Order Anoplura (an-o-plu'ra) (Gr. *anoplos,* unarmed, + *oura,* tail) — **sucking lice** (Figure 22-54). Depressed body; as large as 6 mm; wingless; mouthparts for piercing and sucking; adapted for clinging to warm-blooded host; includes the head louse, body louse, crab louse, others.

Order Hemiptera (he-mip'ter-a) (Gr. *hemi,* half + *pteron,* wing) (**Heteroptera**) — **true bugs** (Figure 22-55). Size 2 to 100 mm; wings present or absent; forewings with basal portion leathery, apical portion membranous; hind wings membranous; at rest, wings held flat over abdomen; piercing-sucking mouthparts; many with odorous scent glands; include water scorpions, water striders, bedbugs, squash bugs, assassin bugs, chinch bugs, stinkbugs, plant bugs, lace bugs, others.

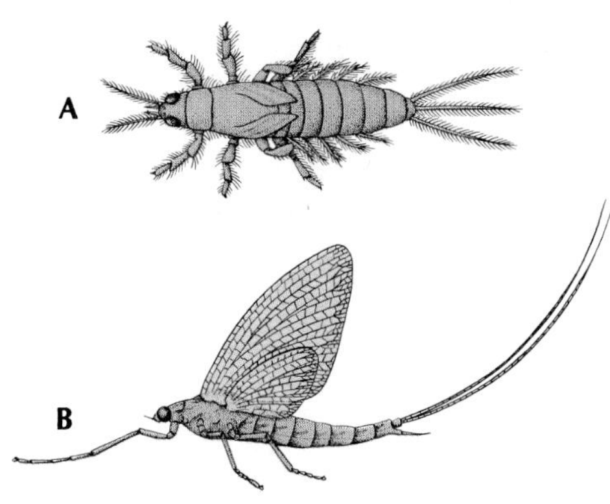

FIGURE 22-52

Mayfly (order Ephemeroptera) naiad, **A,** and adult, **B.**

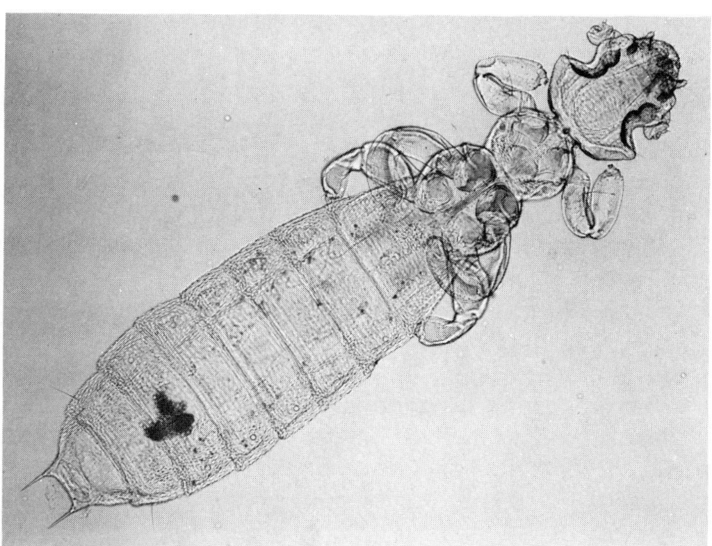

FIGURE 22-53

Gliricola porcelli (order Mallophaga), a chewing louse of guinea pigs. Antennae are normally held in the deep grooves on the sides of the head.

Photograph by Jay Georgi, from Schmidt, G.D., and L.S. Roberts. 1989. Foundations of parasitology, ed. 4. St. Louis, The C.V. Mosby Co.

FIGURE 22-54

Pediculus humanus (order Anoplura), the head and body louse of humans.

Photograph by Warren Buss, from Schmidt, G.D., and L.S. Roberts. 1989. Foundations of parasitology, ed. 4. St. Louis, The C.V. Mosby Co.

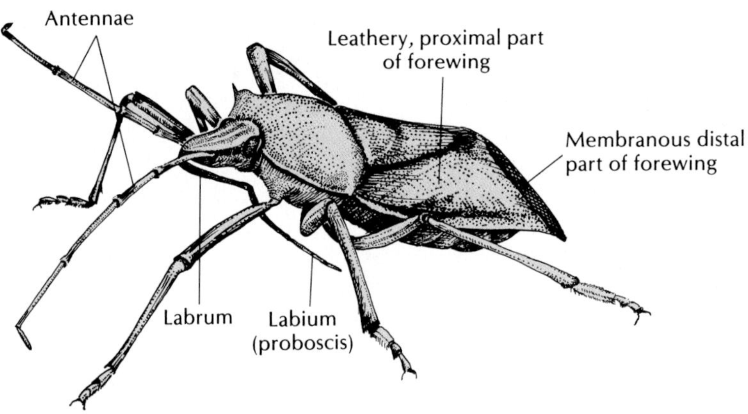

FIGURE 22-55

Diagram of a typical bug (order Hemiptera). The labium forms a tube within which lie the maxillae and mandibles.

Drawing by Ian Grant, from Schmidt, G.D., and L.S. Roberts. 1989. Foundations of parasitology, ed. 4. St. Louis, The C.V. Mosby Co.

Order Homoptera (ho-mop'ter-a) (Gr. *homos,* same, + *pteron,* wing)—**cicadas** (Figure 22-42), **aphids, scale insects, leafhoppers.** (Often included as suborder under Hemiptera.) If winged, either membranous or thickened front wings and membranous hind wings; wings held rooflike over body; piercing-sucking mouthparts; all plant eaters; some destructive; a few serving as source of shellac, dyes, etc.; some with complex life histories.

Superorder Endopterygota (en-dop-ter-i-go'ta) (Gr. *endon,* inside, + *pterygōtos,* winged) **(Holometabola). Metamorphosis complete; wings develop internally;** larvae without compound eyes.

Order Neuroptera (neu-rop'ter-a) (Gr. *neuron,* nerve, + *pteron,* wing)—**dobsonflies, ant lions, lacewings** (Figure 22-56). Medium to large size; similar, membranous wings with many cross veins; chewing mouthparts; dobsonflies with greatly enlarged mandibles in males, and with aquatic larvae; ant lion larvae (doodlebugs) make craters in sand to trap ants.

Order Coleoptera (ko-le-op'ter-a) (Gr. *koleos,* sheath, + *pteron,* wing)—**beetles** (Figure 22-50, *A*), **fireflies** (Figure 22-45), **weevils.** The largest order of animals in the world; front wings (elytra) thick, hard, opaque; membranous hind wings folded under front wings at rest; mouthparts for biting and chewing; includes ground beetles, carrion beetles, whirligig beetles, darkling beetles, stag beetles, dung beetles (Figure 22-43), diving beetles, boll weevils, others.

FIGURE 22-56

Green lacewing fly, *Chrysopa oculata* (order Neuroptera), adult. The larvae feed extensively aphids and other small arthropods.

Photograph by Peter J. Bryant, Biological Photo Service.

Order Lepidoptera (lep-i-dop'ter-a) (Gr. *lepidos,* scale, + *pteron,* wing) — **butterflies and moths** (Figures 22-38 and 22-50, *B* and *C*). Membranous wings covered with overlapping scales, wings coupled at base; mouthparts a sucking tube, coiled when not in use; larvae (caterpillars) with chewing mandibles for plant eating, stubby prolegs on the abdomen, and silk glands for spinning cocoons; antennae knobbed in butterflies and usually plumed in moths.

Order Diptera (dip'ter-a) (Gr. *dis,* two, + *pteron,* wing) — **true flies.** Single pair of wings, membranous and narrow; hind wings reduced to inconspicuous balancers (halteres); sucking mouthparts or adapted for sponging or lapping or piercing; legless larvae called maggots (Figure 22-49) or, when aquatic, wigglers; include crane flies, mosquitoes, moth flies, midges, fruit flies, flesh flies, houseflies, horseflies, botflies, blowflies, and many others.

Order Trichoptera (tri-kop'ter-a) (Gr. *trichos,* hair, + *pteron,* wing) — **caddis flies** (Figure 22-57). Small, soft bodied; wings, well-veined and hairy, folded rooflike over hairy body; chewing mouthparts; aquatic larvae construct cases of leaves, sand, gravel, bits of shell, or plant matter, bound together with secreted silk or cement; some make silk feeding nets attached to rocks in stream.

Order Siphonaptera (si-fon-ap'ter-a) (Gr. *siphon,* a siphon, + *apteros,* wingless) — **fleas** (Figure 22-58). Small; wingless; bodies laterally compressed; legs adapted for leaping; no eyes; ectoparasitic on birds and mammals; larvae legless and scavengers.

Order Hymenoptera (hi-men-op'ter-a) (Gr. *hymen,* membrane, + *pteron,* wing) — **ants, bees, wasps** (Figure 22-46). Very small to large; membranous, narrow wings coupled distally; subordinate hind wings; mouthparts for biting and lapping up liquids; ovipositor sometimes modified into stinger, piercer, or saw; both social and solitary species; most larvae legless, blind, and maggotlike.

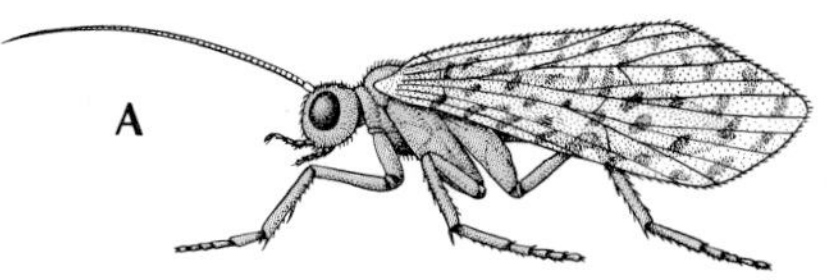

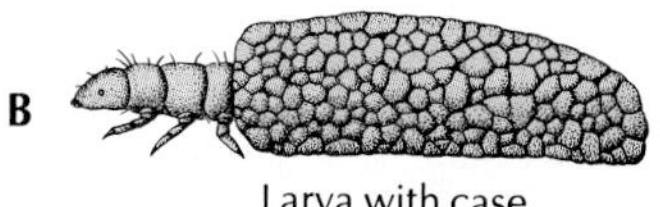

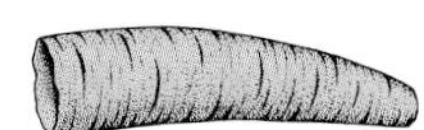

FIGURE 22-57

A, Caddis fly (order Trichoptera). **B,** Several types of larval cases built by aquatic caddis fly larvae, often on the underside of stones.

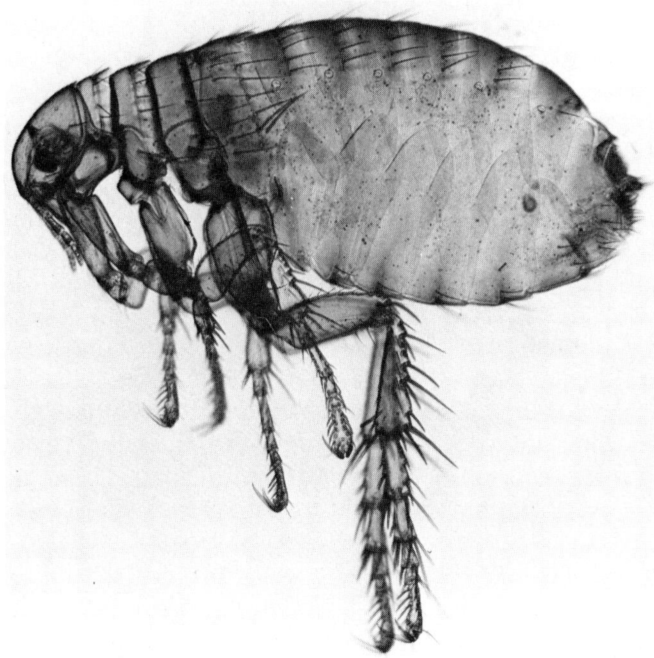

FIGURE 22-58

Female *Pulex irritans* (order Siphonaptera).

Photograph by Jay Georgi, from Schmidt, G.D., and L.S. Roberts. 1989. Foundations of parasitology, ed. 4. St. Louis, The C.V. Mosby Co.

Phylogeny and Adaptive Radiation

PHYLOGENY

The similarities between the annelids and the arthropods give strong support to the hypothesis that both phyla originated from a line of coelomate segmented protosomes, which in time diverged to form a protoannelid line with laterally located parapodia and a protoarthropod line with more ventrally located parapodia. The protoannelid line gave rise eventually to the polychaetes and other annelids. The protoarthropod line apparently diverged further into three or four branches: (1) trilobites, (2) chelicerates, (3) crustaceans, and (4) insects and myriapods. The phylum Onychophora (Chapter 23) may have also come from part of the protoarthropod line, and some workers consider them part of the Uniramia. Onychophorans show similarities to both annelids and arthropods.

Insect fossils, while not abundant, have been found in numbers sufficient to give a general idea of their evolutionary history. Although a variety of marine arthropods, such as trilobites, crustaceans, and xiphosurans, was present in the Cambrian period, the first terrestrial arthropods — the scorpions and millipedes — did not appear until the Silurian period. The first insects, which were wingless, date from the Devonian period. By the Carboniferous period, some 340 million years ago, several orders of winged insects, most of which are now extinct, had appeared.

It is believed that the insects arose from a myriapod (centipedes and millipedes) ancestor that had paired leglike appendages. Both myriapods and insects have clearly defined heads provided with antennae and mandibles. However, the evolution of insects involved specialization of the next three segments to become the locomotor segments (thorax) and a loss or reduction of appendages on the rest of the body (abdomen). The primitively wingless apterygotes are undoubtedly the most primitive of the insects, and traits similar to those of the myriapods are found in them. Probably some ancestral apterygote form gave rise to two major lines of winged insects, which differed in their ability to flex their wings.

W.S. Bristowe estimated that at certain seasons a Sussex field that had been undisturbed for several years had a population of 2 million spiders to the acre. He concluded that so many could not successfully compete except for the many specialized adaptations they had evolved, including adaptations to cold and heat, wet and dry conditions, and light and darkness. Some capture large insects, some only small ones; web builders snare mostly flying insects, whereas hunters seek those that live on the ground. Some lay eggs in the spring, others in the late summer. Some feed by day, others by night, and some have developed flavors that are distasteful to birds or to certain predatory insects. As it is with the spiders, so has it been with other arthropods; their adaptations are many and diverse and contribute in no small way to their long success.

ADAPTIVE RADIATION

Annelids show little specialization or fusion of somites and relatively little differentiation of appendages. However, in arthropods the adaptive trend has been toward tagmatization of the body by differentiation or fusion of somites, giving rise in more advanced groups to such tagmata as head and trunk; head, thorax, and abdomen; or cephalothorax (fused head and thorax) and abdomen. Primitive arthropods tend to have similar appendages, whereas the more advanced forms have appendages specialized for specific functions, or some appendages may be lost entirely.

Much of the amazing diversity in arthropods seems to have developed because of modification and specialization of their cuticular exoskeleton and their jointed appendages, thus resulting in a wide variety of locomotor and feeding adaptations.

Summary

The Arthropoda is the largest, most successful phylum in the world. They are metameric, coelomate protostomes with well-developed organ systems. Most show marked tagmatization. They are extremely diverse and occur in all habitats capable of supporting life. Perhaps more than any other single factor, the success of the arthropods is accounted for by the adaptations made possible by their cuticular exoskeleton. Other important elements in their success are jointed appendages, tracheal respiration, efficient sensory organs, complex behavior, metamorphosis, and the ability to fly.

Members of the subphylum Chelicerata have no antennae, and their main feeding appendages are chelicerae. In addition, they have a pair of pedipalps (which may be similar to walking legs) and four pairs of walking legs. The great majority of living chelicerates are in the class Arachnida: spiders (order Araneae), scorpions (order Scorpionida), harvestmen (order Opiliones), ticks and mites (order Acarina), and others.

The tagmata of spiders (cephalothorax and abdomen) show no external segmentation and are joined by a waistlike pedicel. Chelicerae of spiders are provided with poison glands for paralyzing or killing their prey. Spiders can spin silk, which they use for a variety of purposes.

The cephalothorax and abdomen of ticks and mites are completely fused, and the mouthparts are borne on the anterior capitulum. They are the most numerous of any arachnids; some are important disease carriers, and others are serious plant pests.

The Crustacea is a large, primarily aquatic subphylum of arthropods. Crustaceans bear two pairs of antennae, mandibles, and two pairs of maxillae on the head. Their appendages are primitively biramous, and the major tagmata are the head, thorax, and abdomen. Many have a carapace and respire by means of gills.

Members of the class Copepoda lack a carapace and abdominal appendages. They are abundant and are among the most important of the primary consumers in many freshwater and marine ecosystems. The Malacostraca are the largest crustacean class, and the most important orders are the Isopoda, Amphipoda, Euphausiacea, and Decapoda. All have both abdominal and thoracic appendages. Decapods include crabs, shrimp, lobster, crayfish, and others; they have five pairs of walking legs (including the chelipeds) on their thorax.

All arthropods must periodically cast off their cuticle (ecdysis) and grow larger before the newly secreted cuticle hardens. Premolt and postmolt periods are hormonally controlled, as are several other structures and functions.

There are many predators, scavengers, filter feeders, and parasites among the Crustacea. Respiration is through the body surface or by gills, and excretory organs take the form of maxillary or antennal glands. Circulation, as in other arthropods, is through an open system of sinuses (hemocoel), and a dorsal, tubular heart is the chief pumping organ. Most crustaceans have compound eyes composed of units called ommatidia.

Members of the subphylum Uniramia have uniramous appendages and bear one pair of antennae, a pair of mandibles, and two pairs of maxillae (one pair of maxillae in millipedes) on the head. The tagmata are the head and trunk in the myriapods and head, thorax, and abdomen in the insects.

The Insecta is the largest class of the world's largest phylum. Insects are easily recognized by the combination of their tagmata and the possession of three pairs of thoracic legs.

The evolutionary success of insects is largely explained by several features allowing them to exploit terrestrial habitats, such as waterproofing their cuticle and other mechanisms to minimize water loss and the ability to become dormant during adverse conditions.

Feeding habits vary greatly among insects, and there is an enormous variety of specialization of mouthparts reflecting the particular feeding habits of a given insect. They breathe by means of a tracheal system, which is a system of tubes that opens by spiracles on the thorax and abdomen. Excretory organs are malpighian tubules.

Sexes are separate in insects, and fertilization is usually internal. Almost all insects undergo metamorphosis during development. In hemimetabolous (gradual) metamorphosis, the juvenile instars are called nymphs, and the adult emerges at the last nymphal molt. In holometabolous (complete) metamorphosis, the juvenile instars are larvae, and the last larval molt gives rise to a nonfeeding stage (pupa). A winged adult emerges at the final, pupal, molt. Both types of metamorphosis are hormonally controlled.

Insects are important to human welfare, particularly because they pollinate food crop plants, control populations of other, harmful insects by predation and parasitism, and serve as food for other animals. Many insects are harmful to human interests because they feed on crop plants, and many are carriers of important diseases affecting humans and domestic animals.

Arthropods evolved from one or more annelid-like ancestors. Modern insects and myriapods show certain basic similarities, and insects are probably descended from one or more myriapod-like ancestors.

Adaptive radiation and the evolutionary success of the arthropods have been enormous.

REVIEW QUESTIONS

1. Give 10 characteristics of arthropods.
2. Name the subphyla of arthropods, and give a few examples of each.
3. Briefly discuss the contribution of the cuticle to the success of arthropods, and name some other factors that have contributed to their success.
4. What is a trilobite?
5. What appendages are characteristic of chelicerates?
6. Briefly describe the appearance of each of the following: eurypterids, horseshoe crabs, pycnogonids.
7. Tell the mechanism of each of the following with respect to spiders: feeding, excretion, sensory reception, web-spinning, reproduction.
8. Distinguish each of the following orders from each other: Araneae, Scorpionida, Opiliones, Acarina.
9. What are some ways Acarina affect humans?
10. What are the tagmata and the appendages on the head of crustaceans? What are some other important characteristics of Crustacea?
11. Of the classes of Crustacea, the Branchiopoda, Ostracoda, Copepoda, Cirripedia, and Malacostraca are the most important. Distinguish them from each other.
12. Distinguish among the Isopoda, Amphipoda, and Decapoda.
13. Define each of the following: swimmeret, endopod, exopod, maxilliped, cheliped, uropod, nauplius.
14. Describe the molting process in Crustacea, including the action of the hormones.
15. Tell the mechanism of each of the following with respect to crustaceans: feeding, respiration, excretion, circulation, sensory reception, reproduction.
16. Distinguish the following from each other: Diplopoda, Chilopoda, Insecta.
17. Define each of the following with respect to insects: sclerite, notum, tergum, sternum, pleura, labrum, labium, hypopharynx, haltere, instar, diapause.
18. Explain why wings powered by indirect flight muscles can beat much more rapidly than those powered by direct flight muscles.
19. What different modes of feeding are found in insects, and how are these reflected in their mouthparts?
20. Describe each of the following with respect to insects: respiration, excretion and water balance, sensory reception, reproduction.
21. Explain the difference between holometabolous and hemimetabolous metamorphosis in insects, including the stages in each.
22. Describe and give an example of each of the four ways insects communicate with each other.
23. What are the castes found in honeybees and in termites, and what is the function of each? What is trophallaxis?
24. Name several ways in which insects are beneficial to humans and several ways they are detrimental.
25. What are ways in which detrimental insects can be controlled? What is integrated pest management?
26. Whether the arthropods constitute one or several phyla and whether the concept of the Mandibulata (including crustaceans, insects, and myriapods) is valid are the subjects of debate. What is the nature of the arguments for each position?

SELECTED REFERENCES

Askew, R.R. 1971. Parasitic insects. New York, American Elsevier Publishers, Inc. *Very interesting book, good coverage of insect parasites of insects.*

Barnes, R.D. 1987. Invertebrate zoology, ed. 5. Philadelphia, Saunders College/Holt, Rinehart & Winston. *Chapters 12 through 16 cover arthropods.*

Blum, M.S. (ed.) 1985. Fundamentals of insect physiology. New York, John Wiley & Sons. *Good, multi-authored text on insect physiology. Recommended.*

Chapman, R.F. 1982. The insects: structure and function. Cambridge, Mass., Harvard University Press. *Comprehensive text on morphology and physiology of insects.*

Foelix, R.F. 1982. Biology of spiders. Cambridge, Mass., Harvard University Press. *Attractive, comprehensive book with extensive references; of interest to both amateurs and professionals.*

Hadley, N.F. 1986. The arthropod cuticle. Sci. Am. **234:**100-107 (Mar.). *Modern studies on the chemistry and structure of arthropod cuticle help to explain its remarkable properties.*

Harwood, R.F., and M.T. James. 1979. Entomology in human and animal health, ed. 7. New York, Macmillan, Inc. *One of the best texts available in medical entomology.*

Holldobler, B.K., and E.O. Wilson. 1977. Weaver ants. Sci. Am. **237:**146-154 (Dec.). *Colonies of weaver ants are highly organized, and individuals communicate with complex pheromonal mechanisms. They use their own larvae as shuttles to weave leaves into large nests.*

Kingsolver, J.G. 1985. Butterfly engineering. Sci. Am. **253:**106-113 (Aug.). *Butterfly feeding, regulation of internal temperature, and flying were studied according to engineering principles.*

Pennak, R.W. 1989. Freshwater invertebrates of the United States, ed. 3. New York, John Wiley & Sons, Inc. *Unusually clear taxonomic keys and illustrative drawings. This edition does not cover insects.*

Rosenthal, G.A. 1986. The chemical defenses of higher plants. Sci. Am. **254:**94-99 (Jan.). *Reviews the fascinating relationships of the secondary metabolites produced by plants to the insects that feed on them. Some insects are repelled, while others can use the chemicals to deter their own predators.*

Shear, W.A. (ed.). 1986. Spiders: webs, behavior, and evolution. Stanford, California, Stanford University Press. *Web building in all its many forms and patterns is detailed.*

SOME EVOLUTIONARY EXPERIMENTS

The early Cambrian Period, about 570 million years ago, was a most fertile time in evolutionary history. For 3 billion years before this, evolution had forged little more than bacteria and blue-green algae, and a vast number of potential niches awaited occupants. Then, within the space of a few million years, all of the major phyla, and probably all of the smaller phyla, became established. This was the Cambrian explosion, the greatest evolutionary "bang" the world has known. In fact, the fossil record suggests that more phyla existed in the Paleozoic Era than exist now, but some disappeared during major extinction events that punctuated the evolution of life on earth. The greatest of these disruptions was the Permian extinction about 230 million years ago. Thus evolution has led to many "experimental models." Some of these models failed because they were unable to survive in changing conditions. Others gave rise to abundant and dominant species and individuals that inhabit the world today. Still others met with limited success, and small numbers of species persist or were successful but are now in decline. This chapter deals with this last group of organisms, the most distal twigs of our hypothetical evolutionary tree.

The great evolutionary flow that began with the appearance of the coelom and led to the three huge phyla of molluscs, annelids, and arthropods produced other lines as well. Those that have survived are small and lack great economic and ecological importance; they are usually grouped together as "lesser protostomes." They have probably diverged at different times from the annelid-arthropod line, all following different adaptive currents and none enjoying much evolutionary success.

Three other phyla—Phoronida, Ectoprocta, and Brachiopoda—are included in this chapter mainly for convenience. They are apparently related to each other since they all possess a crown of cilated tentacles, called a lophophore, used in food capture and respiration. The brachiopods were very successful in the Paleozoic but began to decline thereafter. The exception to the common theme of this chapter is the phylum Ectoprocta. They arose in the Cambrian and became widespread in the Paleozoic. They remain an abundant and successful group today.▪

CHAPTER 23

Lesser Protostomes and Lophophorates

This chapter includes a brief discussion of 10 coelomate phyla whose position in the phylogenetic lines of the animal kingdom is somewhat problematic. Other than having derived from the main protostome line, we can say little of the relationships of seven of the phyla. The Sipuncula, Echiura, and Pogonophora show some relationship to the annelids. The Priapulida were considered pseudocoelomate until the presence of a peritoneum was reported in 1961. The Pentastomida, Onychophora, and Tardigrada show some arthropod characteristics. They have often been grouped together and called Pararthropoda because they have unjointed limbs with claws (at some stage) and a cuticle that undergoes molting. The relationships of the lophophorate phyla are also obscure, since none of the known protostome stocks would seem to be likely ancestors. It is possible that they separated from the protostome line around the time the deuterostome ancestors diverged; some of their characteristics suggest deuterostome affinities.

The Lesser Protostomes

Four of the lesser protostome phyla are benthic (bottom-dwelling) marine worms. The Pentastomida is an entirely parasitic group, and the onychophorans are terrestrial (but limited to damp areas). Tardigrades are found in marine, freshwater, and terrestrial habitats.

PHYLUM SIPUNCULA

The phylum Sipuncula (sy-pun'kyu-la) (L. *sipunculus*, little siphon) consists of about 330 species of benthic marine worms, predominantly littoral or sublittoral. Sometimes called "peanut worms," they live sedentary lives in burrows in mud or sand (Figure 23-1), occupy borrowed snail shells, or live in coral crevices or among vegetation. Some species construct their own rock burrows by chemical and perhaps mechanical means. More than half the species are restricted to tropical zones. Some are tiny, slender worms, but the majority range from 15 to 30 cm in length.

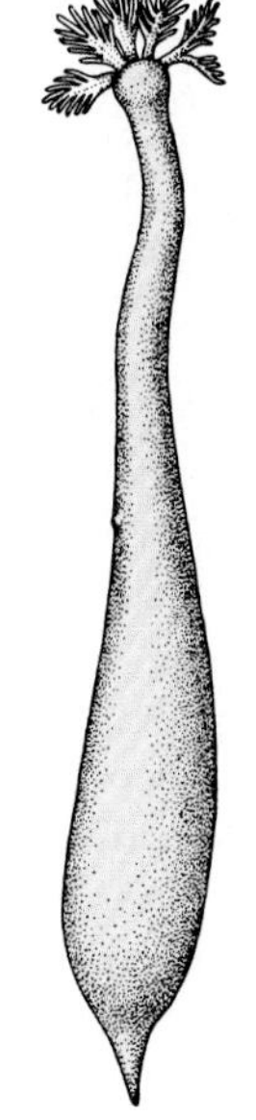

FIGURE 23-1

Dendrostomum, a sipunculan.

Sipunculans are not metameric, nor do they possess setae. Their head is in the form of an **introvert,** which is crowned by ciliated tentacles surrounding the mouth (Figure 23-1). They are largely deposit feeders, extending the introvert and tentacles from their burrow to explore and feed. They have a cerebral ganglion, nerve cord, and pair of nephridia; the coelomic fluid contains red blood cells bearing a respiratory pigment known as hemerythrin.

The sipunculan larvae are trochophores, and their early embryological development indicates affinities to the Annelida and Echiura. They appear to have diverged from a common ancestor of the three phyla before metamerism evolved.

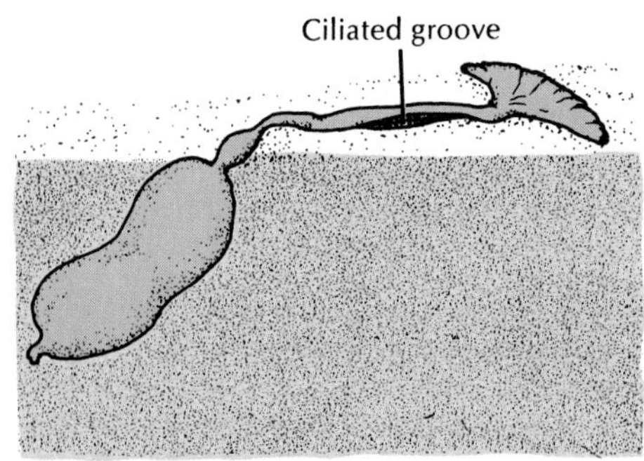

FIGURE 23-2

Tatjanellia (phylum Echiurida) is a detritus feeder. Lying buried in the sand, it explores the surface with its long proboscis, which picks up organic particles and carries them along a ciliated groove to the mouth.

After Zenkevitch; modified from Dawydoff, C. 1959. Classe des Echiuriens. In Grassé, P. (ed.). Traité de Zoologie, vol. 5. Paris, Masson et Cie.

PHYLUM ECHIURA

The phylum Echiura (ek-ee-yur'a) (Gr. *echis,* viper, + *oura,* tail) consists of marine worms that burrow into mud or sand or live in empty snail shells or sand dollar tests, rocky crevices, and so on. They are found in all oceans—most commonly in littoral zones of warm waters. They vary in length from a few millimeters to 40 to 50 cm.

Although there are only about 100 species, echiurans are more diverse than sipunculans. Their bodies are cylindrical. Anterior to the mouth is a flattened, extensible proboscis, which, unlike the introvert of sipunculans, cannot be retracted into the trunk. Echiurans are often called "spoonworms" because of the shape of the contracted proboscis in some of them. The proboscis has a ciliated groove leading to the mouth. While the animal lies buried, the proboscis can extend out over the mud for exploration and deposit feeding (Figure 23-2). *Urechis* (Gr. *oura,* tail, + *echis,* viper), however, secretes a mucous net in a U-shaped burrow through which it pumps water and strains out food particles. *Urechis* is sometimes called the "fat innkeeper" because it has characteristic species of commensals living with it in its burrow, including a crab, fish, and polychaete annelid.

Echiurans, with the exception of *Urechis,* have a **closed circulatory system** with a contractile vessel; most have one to three pairs of nephridia (some have many pairs), and all have a nerve ring and ventral nerve cord. A pair of anal sacs arises from the rectum and opens into the coelom; they are believed to be respiratory in function and possibly accessory nephridial organs.

Among the most amazing animals found in the deep-water, Pacific rift communities (Chapter 5, p. 126) are the giant pogonophorans, *Riftia pachyptila.* They measure up to 3 m in length and 2 to 3 cm in diameter. Although each animal has approximately 2.28 × 10^5 tentacles, it is unlikely that sufficient dissolved nutrients would be present in seawater to support such a large animal. It has been discovered that the bodies of the worms contain large numbers of symbiotic, chemoautotrophic bacteria in a highly vascularized organ in the trunk called the trophosome. The bacteria apparently oxidize the sulfide from the vent water and make use of the ATP and reducing power generated by sulfur oxidation to reduce and fix carbon dioxide. Thus *Riftia* fixes enough carbon in the trophosome to nourish the rest of the worm.

PHYLUM POGONOPHORA

The phylum Pogonophorora (po-go-nof'e-ra) (Gr. *pōgōn,* beard, + *phora,* bearing), or beardworms, was entirely unknown before the twentieth century. The first specimens to be described were collected from deep-sea dredgings in 1900 off the coast of Indonesia. They have since been discovered in several seas, including the western Atlantic off the U.S. eastern coast. Some 80 species have been described so far.

Most pogonophores live in the bottom ooze on the ocean floor, usually at depths of more than 200 m. Their length varies from 5 to 85 cm, with a usual diameter of a fraction of a millimeter. They are sessile and secrete very long chitinous tubes in which they live, probably extending the anterior end only for feeding.

The body is divided into a short forepart, a long, very slender trunk, and a small, segmented opisthosoma (Figure 23-3). They are covered with a cuticle similar in structure to that of annelids and sipunculans, and they bear setae on the trunk and opisthosoma similar to those of annelids. The body is divided into a series of coelomic compartments. The forepart bears from one to many tentacles.

The most remarkable feature of the pogonophores, unparalleled among any other free-living metazoans, is the complete absence of a mouth or digestive tract. How the animals feed is quite unclear, but it appears that digestion (if any) must be external, with absorption through the tentacles. There is a well-developed, closed, blood vascular system. Sexes are separate.

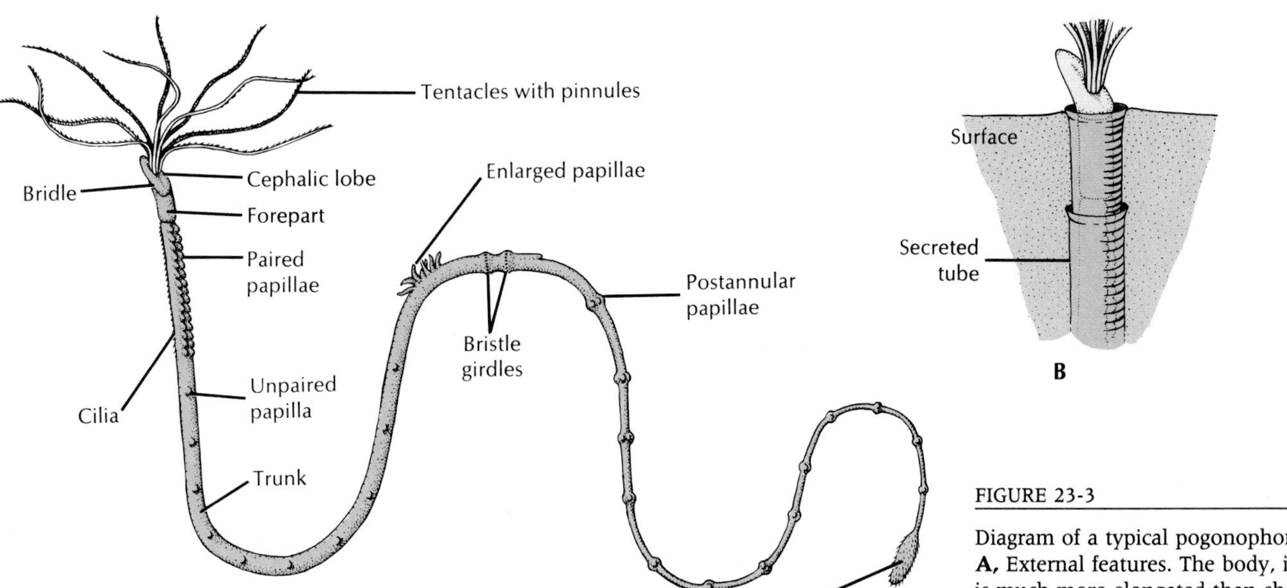

FIGURE 23-3

Diagram of a typical pogonophoran. **A,** External features. The body, in life, is much more elongated than shown in this diagram. **B,** Position in tube.

Pogonophores have photoreceptor cells very similar to those of annelids (oligochaetes and leeches), and the structure of the cuticle, the makeup of the setae, and the segmentation of the opisthosoma all point strongly toward a close relationship with the annelids.

PHYLUM PRIAPULIDA

The Priapulida (pri'a-pyu'li-da) (Gr. *priapos,* phallus, + *-ida,* pl. suffix) are a small group (only nine species) of marine worms found chiefly in the colder water of both hemispheres. Their cylindrical bodies are rarely more than 12 to 15 cm long. Most of them are burrowing predaceous animals that usually orient themselves upright in the mud with the mouth at the surface.

Long thought to be pseudocoelomate, they were judged coelomate when nuclei were found in the membranes lining the body cavity, the membranes thus representing a peritoneum. However, a later report maintains that the nuclei originate from amebocytes and that the membranes are acellular. Therefore the status of the priapulids is still unsettled.

The body includes a proboscis, trunk, and usually one or two caudal appendages (Figure 23-4). The eversible proboscis usually ends with rows of curved spines around the mouth; it is used in sampling the surroundings, as well as for the capture of small soft-bodied prey. Priapulids are not metameric.

There is no circulatory system, but coelomocytes in the body fluids contain the respiratory pigment hemerythrin.

Sexes are separate, and the embryogenesis is only poorly known. Their relationship to other coelomates is obscure. Some authorities believe that they are remnants of groups that were once more successful and widely distributed.

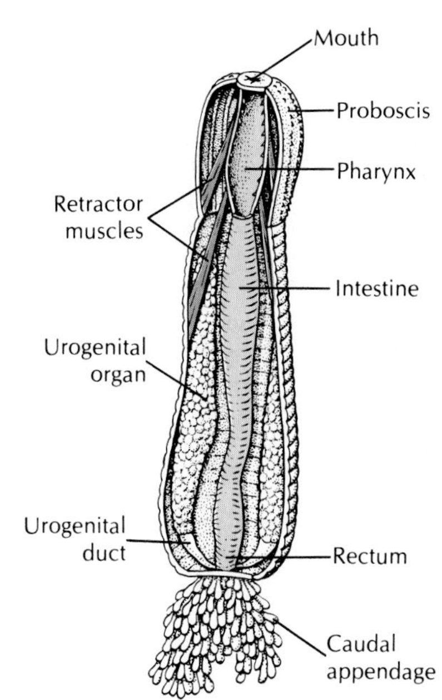

FIGURE 23-4

Major internal structures of *Priapulus.*

PHYLUM PENTASTOMIDA

The wormlike Pentastomida (pen-ta-stom'i-da) (Gr. *pente,* five, + *stoma,* mouth) are parasites, 2.5 to 12 cm long, that are found in the lungs and nasal passages of carnivorous vertebrates—most commonly in reptiles (Figure 23-5). Some human infections have been found in Africa and Europe. The intermediate host is usually a vertebrate that is eaten by the final host.

They have four clawlike appendages at their anterior end, and their body is covered by a chitinous cuticle, which is periodically molted during juvenile stages. Pentastomids show arthropod affinities, but there is little agreement as to

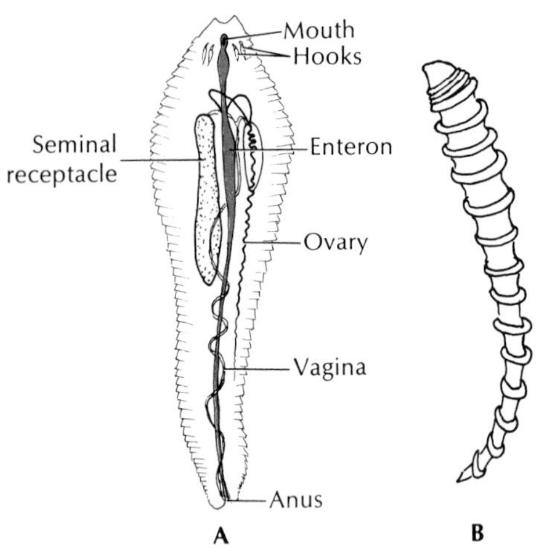

FIGURE 23-5

Two pentastomids. **A,** *Pentastomum,* found in lungs of snakes and other vertebrates. Female is shown with some internal structures. **B,** Female *Armillifer,* a pentastomid with pronounced body rings. In parts of Africa and Asia, humans are parasitized by immature stages; adults (10 cm long or more) live in lungs of snakes. Human infection may occur from eating snakes or from contaminated food and water.

where they fit in that phylum. On the basis of the structure of their spermatozoa, it has been suggested that pentastomids are closest to the crustacean class Branchiura. Some authorities consider Pentastomida a subphylum of Arthropoda.

PHYLUM ONYCHOPHORA

Members of the phylum Onychophora (on-i-kof'o-ra) (Gr. *onyx,* claw, + *phorein,* to bear) are called "velvet worms" or "walking worms." They are about 70 species of caterpillar-like animals, 1.4 to 15 cm long, that live in rain forests and other tropical and semitropical leafy habitats.

The fossil record of the onychophorans shows that they have changed little in their 500 million-year history. They were originally marine animals and were probably far more common than they are now. Today they are all terrestrial and are extremely retiring, coming out only at night or when the air is nearly saturated with moisture.

Onychophorans are covered by a soft cuticle, which contains chitin and protein. Their wormlike bodies are carried on 14 to 43 pairs of stumpy, unjointed legs, each ending with a flexible pad and two claws (Figure 23-6). The head bears a pair of flexible antennae with annelid-like eyes at the base.

They are air breathers, using a **tracheal system** that connects with pores scattered over the body. The tracheal system, although similar to that of the arthropods, has probably evolved independently. Other arthropod characteristics are the open circulatory system with a tubular heart, a hemocoel for a body cavity, and the large brain. Annelid-like characteristics are segmentally arranged nephridia, a muscular body wall, and pigment-cup ocelli.

Onychophorans are dioecious. In some species there is a placental attachment between mother and young, and the young are born "alive" (viviparous); in others the young are also not encased in a shell when released from the mother, but they develop in the uterus without attachment (ovoviviparous).

Onychophorans have been of unusual interest to zoologists because they share so many characteristics with both the annelids and the arthropods. They have been called, a bit too hopefully perhaps, the "missing link" between these two groups. However, because of their similarities to both phyla, it seems likely that they were derived from or close to an annelid-like ancestor of the arthropods.

PHYLUM TARDIGRADA

Tardigrada (tar-di-gray'da) (L. *tardus,* slow, + *gradus,* step), or "water bears," are minute forms usually less than a millimeter in length. Most of the 300 to 400 species are terrestrial forms that live in the water film that surrounds mosses and lichens. Some live in fresh water, and a few are marine. They share many characteristics with the arthropods.

The body bears eight short, **unjointed legs,** each with claws (Figure 23-7). Unable to swim, they creep about awkwardly, clinging to the substrate with their claws. A pair of sharp stylets and a sucking pharynx adapt them for piercing and sucking plant cells or small prey such as nematodes and rotifers.

There is a body covering of nonchitinous **cuticle** that is molted several times during the life cycle. As in the arthropods, muscle fibers are attached to the cuticular exoskeleton, and the body cavity is a hemocoel.

The annelid-type nervous system is surprisingly complex, and in some species there is a pair of eyespots. Circulatory and respiratory organs are lacking.

Females may deposit their eggs in the old cuticle as they molt or attach them to a substrate. Embryonic formation of the coelom is enterocoelous, a deutero-

FIGURE 23-6

Peripatus, a caterpillar-like onychophoran that has both annelid and arthropod characteristics.

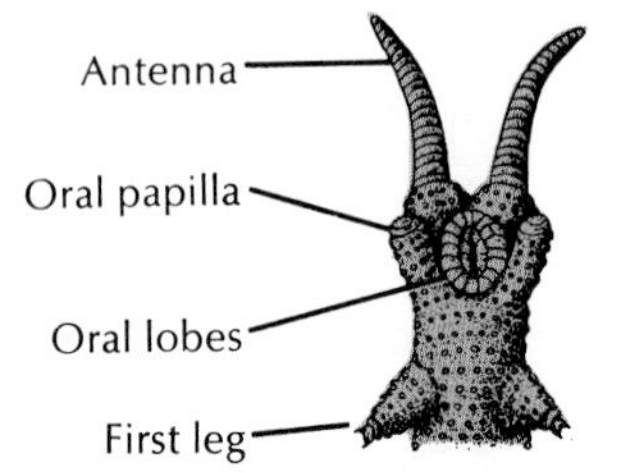

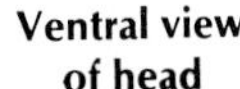

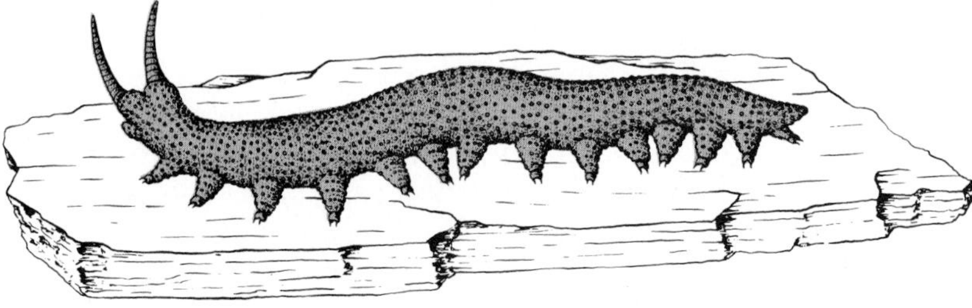

stome characteristic. Nevertheless, their numerous arthropod-like characteristics lead most authors to ally them with the Arthropoda.

One of the most intriguing features of terrestrial tardigrades is their capacity to enter a state of suspended animation, called **cryptobiosis,** during which metabolism is barely perceptible. Under gradual drying conditions the water content of the body is reduced from 85% to only 3%, movement ceases, and the body becomes barrel shaped. In a cryptobiotic state tardigrades can withstand harsh environmental conditions: temperature extremes, ionizing radiations, oxygen deficiency, etc., and many survive for years. Activity resumes when moisture is again available.

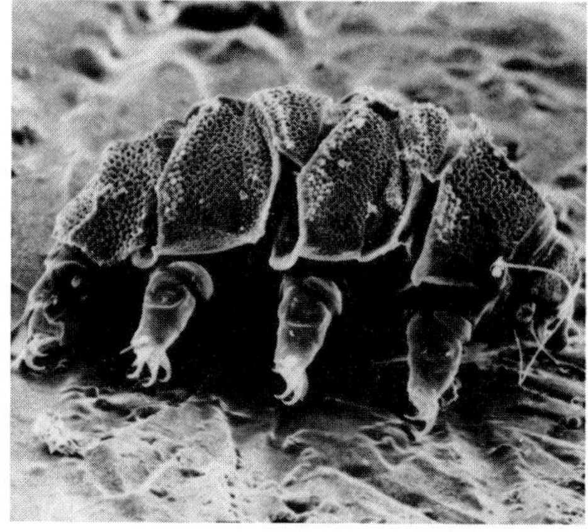

FIGURE 23-7

Scanning electron micrograph of *Echiniscus maucci,* phylum Tardigrada. This species is 300 to 500 μm long. Unable to swim, it clings to moss or water plants with its claws, and if the environment dries up, it goes into a state of suspended animation and "sleeps away" the drought.

Photograph by Diane R. Nelson and Robert O. Schuster.

The Lophophorates

The three lophophorate phyla might appear to have little in common. The **phoronids** (phylum Phoronida) are wormlike marine forms that live in secreted tubes in sand or mud or attached to rocks or shells. The **ectoprocts** (phylum Ectoprocta) are minute forms, mostly colonial, whose protective cases often form encrusting masses on rocks, shells, or plants. The **brachiopods** (phylum Brachiopoda) are bottom-dwelling marine forms that superficially resemble molluscs because of their bivalved shells. All have a free-swimming larval stage but are sessile as adults.

One might wonder why these three apparently widely different types of animals may be considered together. They are all eucoelomates; all have some protostome characteristics; all are sessile; and none has a distinct head. But these characteristics are also shared by other phyla. What really sets them apart from other phyla is the common possession of a **ciliary feeding device** called a **lophophore** (Gr. *lophas,* crest or tuft, + *phorein,* to bear).

A lophophore is a unique arrangement of ciliated tentacles borne on a ridge (a fold of the body wall), which surrounds the mouth but not the anus. The lophophore with its crown of tentacles contains within it an extension of the coelom, and the thin, ciliated walls of the tentacles not only comprise an efficient feeding device but also serve as a respiratory surface for exchange of gases between the environmental water and the coelomic fluid. The lophophore can usually be extended for feeding or withdrawn for protection.

PHYLUM PHORONIDA

The phylum Phoronida (fo-ron'i-da) (Gr. *phoros,* bearing, + L. *nidus,* nest) consists of approximately 10 species of small wormlike animals that live on the bottom of shallow coastal waters, especially in temperate seas. The phylum name refers to the tentacled lophophore. Phoronids range from a few millimeters to 30 cm in length. Each worm secretes a leathery or chitinous tube in which it lies free, but which it never leaves (Figure 23-8). The tubes may be anchored singly or in a tangled mass on rocks, shells, or pilings or buried in the sand. The tentacles on the lophophore are thrust out for feeding, but if the animal is disturbed it can withdraw completely into its tube.

The lophophore is made up of two parallel ridges curved in a horseshoe shape, the bend located ventrally and the mouth lying between the two ridges. The cilia on the tentacles direct a water current toward a groove between the two ridges, which leads toward the mouth. Plankton and detritus caught in this current become entangled in mucus and are carried by the cilia to the mouth.

The coelomic cavity is subdivided by mesenteric partitions into compartments similar to those of the deuterostomes. The phoronids have a closed system of contractile blood vessels but no heart; the red blood contains hemoglobin.

Cleavage seems to be related to both the spiral and the radial types. Clearly, in terms of numbers of individuals and of species, the phoronids are the least successful of the lophophorate phyla.

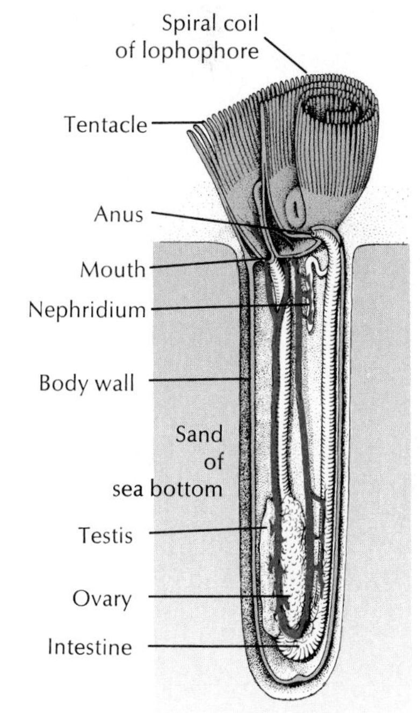

FIGURE 23-8

Internal structure of *Phoronis* (phylum Phoronida).

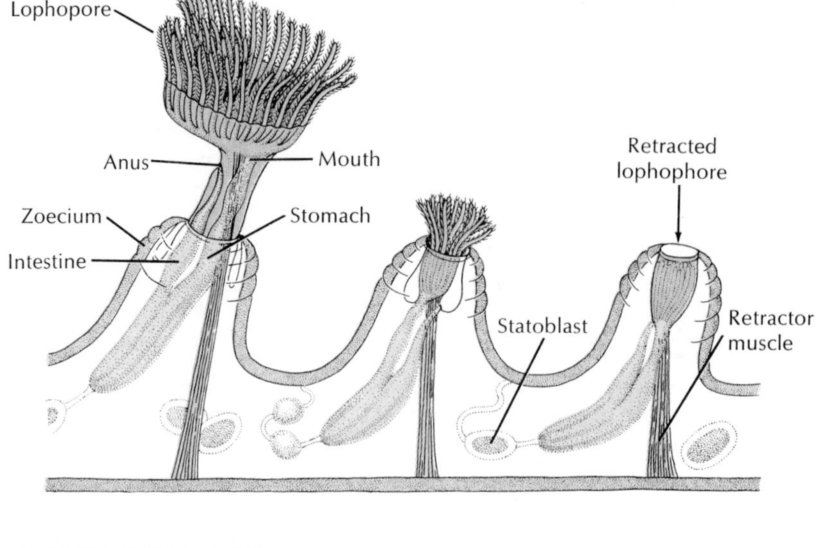

FIGURE 23-9

Small portion of freshwater colony of *Plumatella* (phylum Ectoprocta), which grows on the underside of rocks. These tiny individuals disappear into their chitinous zoecia when disturbed.

PHYLUM ECTOPROCTA

The Ectoprocta (ek'to-prok'ta) (Gr. *ektos,* outside, + *proktos,* anus) have long been called bryozoans (Gr. *bryon,* moss, + *zōon,* animal), or moss animals, a term that originally included the Entoprocta also.

Of the 4000 or so species of ectoprocts, few are more than 0.5 mm long; all are aquatic, both freshwater and marine, but are largely found in shallow waters; and most, with very few exceptions, are colony builders. Ectoprocts, unlike the other phyla considered in this chapter, have been very successful. They have left a rich fossil record since the Ordovician era. Marine forms today exploit all kinds of firm surfaces, such as shells, rock, large brown algae, mangrove roots, and ship bottoms. They are one of the most important groups of fouling organisms on boat hulls; they decrease efficiency of the hull passing through the water and make periodic scraping of the hull necessary.

Each member of a colony **(zooid)** lives in a tiny chamber, called a **zoecium,** which is secreted by its epidermis (Figure 23-9). The zoecium may, according to the species, be gelatinous, chitinous, or stiffened with calcium and possibly also impregnated with sand.

Some colonies form limy encrustations on seaweed, shells, and rocks (Figure 23-10); others form fuzzy or shrubby growths or erect, branching colonies that look like seaweed. Some ectoprocts might easily be mistaken for hydroids but can be distinguished under a microscope by the fact that their tentacles are ciliated (Figure 23-11). In some freshwater forms the individuals are borne on finely branching stolons that form delicate tracings on the underside of rocks or plants. Other freshwater ectoprocts are embedded in large masses of gelatinous material. Although the zooids are minute, the colonies may be several centimeters in diameter; some encrusting colonies may be a meter or more in width, and erect forms may reach 30 cm or more in height.

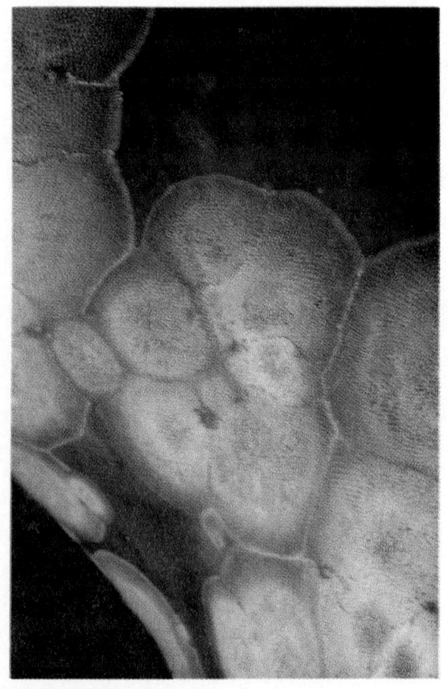

FIGURE 23-10

Skeletal remains of a colony of *Membranipora,* a marine encrusting form of Ectoprocta. Each little oblong zoecium is the calcareous former home of a tiny ectoproct.
Photograph by R. Harbo.

The ectoproct lives in a type of jack-in-the-box existence, popping up to feed and then quickly withdrawing into its little chamber, which often has a tiny trapdoor (operculum) that shuts to conceal its inhabitant. To extend the tentacular crown, certain muscles contract, which increases the hydraulic pressure within the body cavity and pushes the lophophore out. Other muscles can contract to withdraw the crown to safety with great speed.

The lophophore ridge tends to be circular in marine ectoprocts and U-shaped in freshwater species. When feeding, the animal extends the lophophore, and the tentacles spread out to a funnel shape (Figure 23-11). Cilia on the tentacles draw water into the funnel and out between the tentacles. Food particles trapped in mucus in the funnel are drawn into the mouth, both by the pumping action of the muscular pharynx and by the action of cilia in the pharynx.

Respiratory, vascular, and excretory organs are absent. Gaseous exchange occurs through the body surface, and, since the ectoprocts are small, the coelomic fluid is adequate for internal transport. The coelom is divided by a septum into an anterior portion in the lophophore and a larger posterior portion. Pores in the walls between adjoining zooids permit exchange of materials by way of the coelomic fluid.

Most colonies are made up of feeding individuals, but polymorphism also occurs. One type of modified zooid resembles a bird beak that snaps at small invading organisms that might foul a colony. Another type has a long bristle that sweeps away foreign particles.

Most ectoprocts are hermaphroditic. Some species shed eggs into the seawater, but most brood their eggs, some within the coelom and some externally in a special zoecium in which the embryo develops. Marine species have radial cleavage but a highly modified trochophore larva.

Brooding is often accompanied by degeneration of the lophophore and gut of the adults, the remains of which contract into minute dark balls, or **brown bodies.** Later, new internal organs may be regenerated in the old chambers. The brown bodies may remain passive or may be taken up and eliminated by the new digestive tract—an unusual kind of storage excretion.

Freshwater species reproduce both sexually and asexually. Asexual reproduction is by budding or by means of **statoblasts,** which are hard, resistant capsules containing a mass of germinative cells that are formed during the summer and fall (Figure 23-9). When the colony dies in late autumn, the statoblasts are released, and in spring they can give rise to new polypides and eventually to new colonies.

FIGURE 23-11

Ciliated lophophore of *Flustrella,* a marine ectoproct.

Photograph by J.A. Cooke.

PHYLUM BRACHIOPODA

The Brachiopoda (brak-i-op'o-da) (Gr. *brachiōn,* arm, + *pous, podos,* foot), or lamp shells, are an ancient group. Compared with the fewer than 300 species now living, some 30,000 fossil species, which flourished in the Paleozoic and Mesozoic seas, have been described. The brachiopods were once a very successful phylum, but they are now apparently in decline. Some modern forms have changed little from the early ones. The genus *Lingula* (L., little tongue) (Figure 23-12, *A*) has existed virtually unchanged for over 400 million years. Most modern brachiopod shells range between 5 and 80 mm, but some fossil forms reached 30 cm in length.

Brachiopods are all attached, bottom-dwelling, marine forms that mostly prefer shallow water. Their name, which means "arm-footed," refers to the arms of the **lophophore.** Externally brachiopods resemble the bivalved molluscs in having two calcareous shell valves secreted by the mantle. They were, in fact, classed with the molluscs until the middle of the nineteenth century. Brachiopods, however, have **dorsal** and **ventral valves** instead of right and left lateral valves as do the bivalve molluscs and, unlike the bivalves, most of them are attached to a substrate either directly or by means of a fleshy stalk called a **pedicel** (or pedicle).

In most brachiopods the ventral (pedicel) valve is slightly larger than the dorsal (brachial) valve, and one end projects in the form of a short pointed beak that is perforated where the fleshy stalk passes through (Figure 23-12, *B*). In many the shape of the pedicel valve is that of the classical oil lamp of Greek and Roman times, so that the brachiopods came to be known as the "lamp shells."

There are two classes of brachiopods based on shell structure. The shell valves of Articulata are connected by a hinge with an interlocking tooth-and-socket arrangement (articular process); those of Inarticulata lack the hinge and are held together by muscles only.

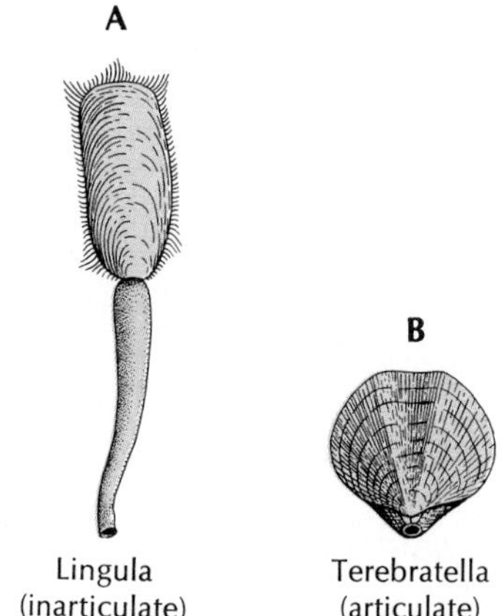

FIGURE 23-12

Brachiopods. **A,** *Lingula,* an inarticulate that normally occupies a burrow. The contractile pedicel can withdraw the body into the burrow. **B,** An articulate brachiopod, *Terebratella.* The valves have a tooth and socket articulation and a short pedicel projects through one valve to attach to the substratum.

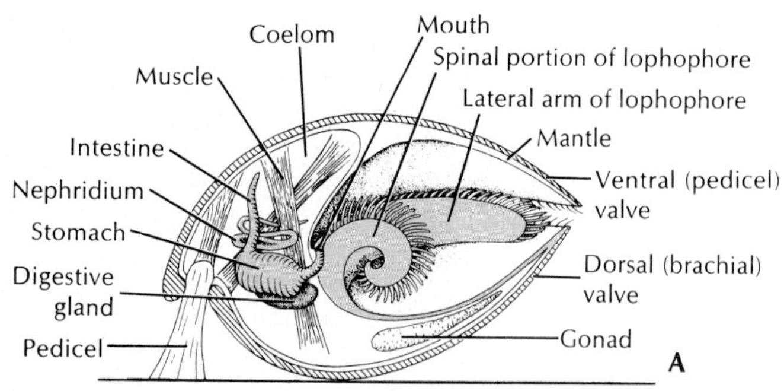

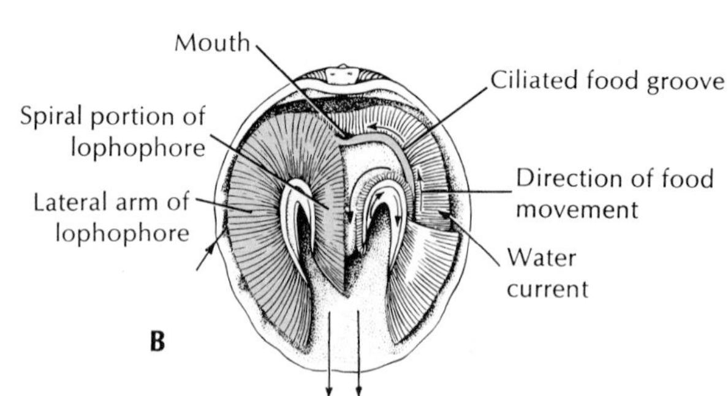

FIGURE 23-13

Phylum Brachiopoda. **A,** An articulate brachiopod (longitudinal section). **B,** Feeding and respiratory currents. Large arrows show water flow over lophophore; small arrows indicate food movement toward mouth in ciliated food groove.

B, Modified from Russell-Hunter, W.D. 1969. A biology of higher invertebrates, New York, Macmillan, Inc.

The body occupies only the posterior part of the space between the valves (Figure 23-13, *A*), and extensions of the body wall form mantle lobes that line and secrete the shell. The large horseshoe-shaped lophophore in the anterior mantle cavity bears long ciliated tentacles used in respiration and feeding. Ciliary water currents carry food particles between the gaping valves and over the lophophore. Food is caught in mucus on the tentacles and carried in a ciliated food groove along the arm of the lophophore to the mouth (Figure 23-13).

Sexes are separate and paired gonads discharge gametes through the nephridia. The development of brachiopods is similar in some ways to that of the deuterostomes, with radial, mostly equal, holoblastic cleavage and the coelom forming enterocoelically in the articulates. The free-swimming larva of the articulates resembles the trochophore.

PHYLOGENY AND ADAPTIVE RADIATION OF THE LOPHOPHORATES

The possession of a lophophore by all three phyla is evidence of their relationship if the structure is homologous in all three. As a group they seem to occupy a phylogenetic position somewhere between the protostomes and the deuterostomes. The coelom of some deuterostomes (echinoderms and hemichordates) is divided into three compartments. The anterior compartment is presumably repressed in the lophophorates, since there is no head. The lack of a head may be correlated with their ciliary method of feeding. In their embryogenesis they display both protostome and deuterostome characteristics, with radial cleavage and, in the brachiopods, enterocoelic development of the coelom. All have a trochophore type of larva. The ancestral deuterostome may have been similar to a lophophorate.

All lophophorates are **filter feeders** and most of their evolutionary diversification has been guided by this function. The tubes of phoronids vary according to their habitats. Various ectoprocts tend to build their protective exoskeletons of chitin or gelatin, which may or may not be impregnated with calcium and sand. Brachiopod variations occur largely in their shells and lophophores.

■ Summary

The 10 small, coelomate phyla covered by this chapter are grouped together here for convenience. The Sipuncula, Echiura, and Pogonophora are related to the annelids; the Pentastomida, Onychophora, and Tardigrada seem closer to the arthropods; and the lophophorates (Phoronida, Ectoprocta, and Brachiopoda) have some characteristics of both protostomes and deuterostomes. The relationship of the Priapulida to other phyla is obscure, and it is yet unclear whether priapulids have a true coelom.

Sipunculans, echiurans, and priapulids are burrowing marine worms; none show metamerism. Pogonophorans live in tubes on the deep ocean floor and have no digestive tract. They are metameric.

The Pentastomida are wormlike parasites that are clearly related to the arthropods, possibly closest to the branchiuran Crustacea. Onychophora are caterpillar-like animals that are metameric and show both annelid and arthropod characteristics. Tardigrades are minute, mostly terrestrial animals that have a hemocoel, as do arthropods.

The Phoronida, Ectoprocta, and Brachiopoda all bear a lophophore, which is a crown of ciliated tentacles surrounding the mouth but not the anus and containing an extension of the mesocoel. They are also sessile as adults, have a U-shaped digestive tract, and have free-swimming larvae. The lophophore functions both as a respiratory and a feeding structure.

Phoronida are the least successful of the lophophorates, living in tubes mostly in shallow coastal waters.

Ectoprocts are abundant in marine habitats, living on a variety of submerged substrata, and a number of species are common in fresh water. They are colonial.

Brachiopods were a very successful phylum in the Paleozoic era but have been declining since the early Mesozoic era. They have a dorsal and a ventral shell and are usually attached to the substrate directly or by means of a pedicel.

The lophophorates have coelomic compartments that apparently correspond to two of the three compartments found in some deuterostomes. Their embryogenesis shows both protostome and deuterostome characteristics.

***Peripatus*, a caterpillar-like onychophoran with both annelid and arthropod characteristics.**
Courtesy Ward's Natural Science Establishment, Inc.

REVIEW QUESTIONS

1. Give three distinctive characteristics of the following, and tell where each lives: Sipuncula, Echiura, Pogonophora, Priapulida, Pentastomida, Onychophora, Tardigrada.
2. What do the members of each of the aforementioned groups feed upon?
3. What is the evidence that the Sipuncula and Echiura diverged from the protostome line before the origin of the annelids? Why are these phyla considered closely related?
4. What is the largest pogonophoran known? Where is it found, and how is it nourished?
5. Name three characteristics of onychophorans that are annelid-like and three that are arthropod-like.
6. What is the survival value of cryptobiosis in tardigrades?
7. What characteristics do the three lophophorate phyla have in common? What characteristics distinguish them from each other?
8. Define each of the following: lophophore, zoecium, zooid, brown bodies, statoblasts.
9. What are the coelomic compartments found in the lophophorates?
10. What is the difference in orientation of the valves of brachiopods compared to bivalve molluscs?
11. What is the evidence for placing the lophophorates in a phylogenetic position between the protostomes and the deuterostomes?

SELECTED REFERENCES

Barnes, R.D. 1987. Invertebrate zoology, ed. 5. Philadelphia, Saunders College/Holt, Rinehart & Winston. *Chapters 17 and 18 cover lesser protostomes and lophophorates.*

Hackman, R.H., and M. Goldberg. 1975. Cryptobiosis. Sci. Am. **225**:30-36 (Dec.). *Cryptobiotic nematodes, rotifers, and tardigrades can withstand adverse conditions of astonishing rigor, yet perceptible metabolism continues in their state of suspended animation.*

Hyman, L.H. 1959. The invertebrates, vol. 5. Smaller coelomate groups. New York, McGraw-Hill Book Co. *Phoronida, Ectoprocta, Brachiopoda, and Sipuncula are covered in this volume. An excellent reference despite its age.*

Jones, M.L. 1981. *Riftia pachyptila* Jones: observations on the vestimentiferan worm from the Galapagos Rift. Science **213**:333-336. *The article concerns primarily the structure of a giant pogonophoran from a deep-sea thermal vent community. Following it in the same issue of* Science *are five more papers on the worm's nutrition, blood, and bacterial symbionts.*

Nielson, C. 1977. The relationships of the Entoprocta, Ectoprocta and Phoronida. Am. Zool. **17**:149-150. *On the basis of their development, the author considers phoronids deuterostomes and reunites Entoprocta with Ectoprocta.*

Pennak, R.W. 1989. Freshwater invertebrates of the United States, ed. 3. New York, John Wiley & Sons, Inc. *Accounts of the freshwater ectoprocts and the tardigrades, with keys to the common species.*

Richardson, J.R. 1986. Brachiopods. Sci. Am. **255**:100-106 (Sept.). *Reviews brachiopod biology and adaptations and contends that in the next few million years there may be an increase in the number of species, rather than a further decline.*

A DESIGN TO PUZZLE THE ZOOLOGIST

The distinguished American zoologist Libbie Hyman once described the echinoderms as a "noble group especially designed to puzzle the zoologist." With a combination of characteristics that should delight the most avid reader of science fiction, the echinoderms would seem to confirm Lord Byron's observation that

Tis strange — but true;
for truth is always strange;
Stranger than fiction.

Despite the adaptive value of bilaterality for free moving animals, and the merits of radial symmetry for sessile animals, echinoderms confounded the rules by becoming free moving but radial. That they evolved from a bilateral ancestor there can be no doubt, for their larvae are bilateral. They undergo a bizarre metamorphosis to a radial adult in which there is a 90° reorientation in body axis, with a new mouth arising on the left side, and a new anus appearing on the right side.

A compartment of the coelom has been transformed in echinoderms into a unique water-vascular system that uses hydraulic pressure to power a multitude of tiny tube feet used in food gathering and locomotion. An endoskeleton of dermal ossicles may fuse together to invest the echinoderm in armor, or it may be reduced in some to microscopic bodies. Many echinoderms have miniature jaw-like pincers (pedicellariae) scattered on their body surface, often stalked and some equipped with poison glands, that keep their surface clean by snapping at animals that would settle there.

Such a constellation of characteristics is wholly unique in the animal kingdom. It has both defined and limited the evolutionary potential of the echinoderms. Despite the vast amount of research that has been devoted to them, we are still far from understanding many aspects of echinoderm biology.▪

CHAPTER 24

Echinoderms and Lesser Deuterostomes

DEUTEROSTOMES

The Echinodermata, along with the chordates and two smaller phyla, Hemichordata (acorn worms and pterobranchs) and Chaetognatha (arrowworms), belong to the Deuterostomia branch of the animal kingdom (see inside front cover). They are all coelomates. Primitively, the deuterostomes share the following embryological features: anus developing from or near the blastopore, and mouth developing elsewhere; coelom budded off from the archenteron (enterocoel); radial and regulative cleavage; and endomesoderm (mesoderm derived from or with the endoderm) from enterocoelic pouches. Although these features have been obscured in the embryogenesis of the more advanced vertebrates, their presence in some of the lower vertebrates clearly indicates that vertebrates are deuterostomes. Thus, although they are only distantly related, the Echinodermata is the only *major* invertebrate group showing affinities with the vertebrates. Nonetheless, their evolutionary history has taken the echinoderms to the point where they are very much unlike any other animal group.

Phylum Echinodermata

The Echinodermata (e-ki'no-der'ma-ta) (Gr. *echinos,* sea urchin, hedgehog, + *derma,* skin) are marine forms and include the sea stars, brittle stars, sea urchins, sea cucumbers, and sea lilies. Echinoderms have a unique constellation of characteristics that are found in no other phylum: (1) an endoskeleton of plates or ossicles, usually spiny, (2) the water-vascular system, (3) the pedicellariae, (4) the dermal branchiae, and (5) radial or biradial symmetry. The water-vascular system and the dermal ossicles have been particularly important in determining the evolutionary potential and limitations of this phylum. Their larvae are bilateral and undergo a metamorphosis to a radial adult.

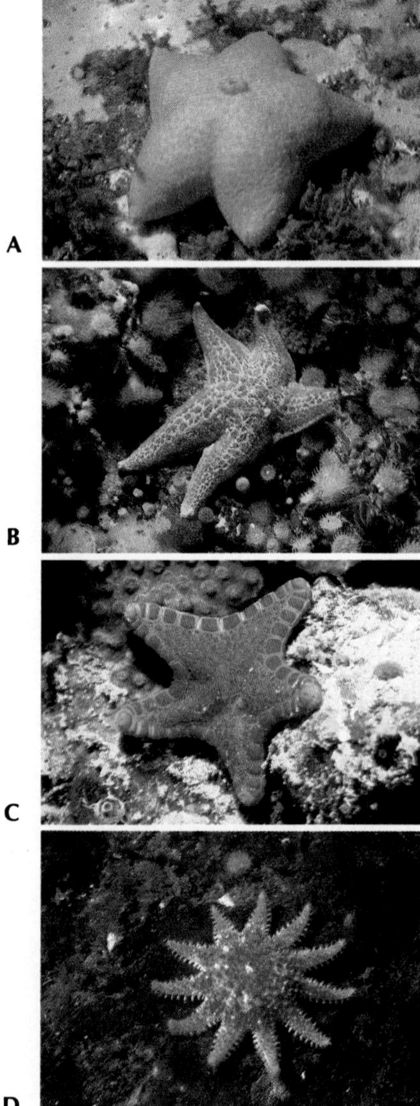

FIGURE 24-1

Some sea stars (subclass Asteroidea) from the Pacific. **A,** Cushion star *Pteraster tesselatus* can secrete incredible quantities of mucus as a defense reaction. **B,** Leather star *Dermasterias imbricata* lacks spines and feeds on sea anemones. **C,** *Pentagonaster duebeni* from the Great Barrier Reef is brilliant red and orange. **D,** *Crossaster papposus,* one of the sun stars, feeds on other sea stars.

A, B, and D, Photographs by R. Harbo; C, photograph by L.S. Roberts.

CHARACTERISTICS

1. Body not metameric with **radial, pentamerous symmetry** characterized by five or more radiating areas
2. **No head or brain;** few specialized sensory organs
3. Nervous system with circumoral ring and radial nerves
4. **Endoskeleton** of **dermal calcareous ossicles** with **spines** or of calcareous **spicules** in dermis; covered by an epidermis (ciliated in most); **pedicellariae** (in some)
5. A unique **water-vascular system** of coelomic origin that extends from the body surface as a series of tentacle-like projections **(podia, or tube feet)**
6. Locomotion by tube feet, which project from the ambulacral areas, or by movement of spines, or by movement of arms, which project from central disc of body
7. Digestive system usually complete; axial or coiled; anus absent in ophiuroids
8. Coelom extensive, forming the perivisceral cavity and the cavity of the water-vascular system; coelom of enterocoelous type
9. So-called **hemal system** present, of uncertain function but playing little, if any, role in circulation, and surrounded by extensions of coelom **(perihemal** sinuses)
10. Respiration by **dermal branchiae, tube feet, respiratory tree** (holothuroids), and **bursae** (ophiuroids)
11. Excretory organs absent
12. Sexes separate (except a few hermaphroditic); fertilization usually external
13. Development through free-swimming, bilateral, larval stages (some with direct development); metamorphosis to radial adult or subadult form

CLASSIFICATION

The following classification is abbreviated and omits groups represented by fossil forms only.

Subphylum Crinozoa (krin'o-zo'a) (Gr. *krinon,* lily, + *zōon,* animal). Radially symmetrical with rounded or cup-shaped theca and brachioles or arms. Attached during part or all of life by stem. Oral surface directed upward.

Class Crinoidea (krin-oi'de-a) (Gr. *krinon,* lily, + *eidos,* form, + *-ea,* characterized by)—**sea lilies and feather stars.** Aboral attachment stalk of dermal ossicles; mouth and anus on oral surface; spines, madreporite, and pedicellariae absent.

Subphylum Asterozoa (as'ter-o-zo'a) (Gr. *aster,* star, + *zōon,* animal). Radially symmetrical, star-shaped echinoderms, unattached as adults.

Class Stelleroidea (stel'ler-oi'de-a) (L. *stella,* star, + Gr. *eidos,* form).

Subclass Somasteroidea (som'ast-er-oi'de-a) (Gr. *soma,* body, + *aster,* star, + *eidos,* form). Mostly extinct sea stars with single living species.

Subclass Asteroidea (as'ter-oi'de-a) (Gr. *aster,* star, + *eidos,* form, + *-ea,* characterized by)—**sea stars.** Arms not sharply marked off from the central disc; ambulacral grooves open, with tube feet on oral side; tube feet often with suckers; anus and madreporite aboral; pedicellariae present.

Subclass Ophiuroidea (o'fe-u-roi'de-a) (Gr. *ophis,* snake, + *oura,* tail, + *eidos,* form)—**brittle stars and basket stars.** Arms sharply marked off from the central disc; ambulacral grooves closed, covered by ossicles; tube feet without suckers and not used for locomotion; pedicellariae absent.

Subphylum Echinozoa (ek'in-o-zo'a) (Gr. *echinos,* sea urchin, hedgehog, + *zōon,* animal). Unattached, globoid, discoid, or cucumber-shaped echinoderms without arms.

Class Echinoidea (ek′i-noi′de-a) (Gr. *echinos,* sea urchin, hedgehog, + *eidos,* form) — **sea urchins, sea biscuits, and sand dollars.** More or less globular or disc-shaped echinoderms with no arms; compact skeleton or test with closely fitting plates; movable spines; ambulacral grooves covered by ossicles, closed; tube feet often with suckers; pedicellariae present.

Class Holothuroidea (hol′o-thu-roi′de-a) (Gr. *holothourion,* sea cucumber, + *eidos,* form) — **sea cucumbers.** Cucumber-shaped echinoderms with no arms; spines absent; microscopic ossicles embedded in thick, muscular, body wall; anus present; ambulacral grooves closed; tube feet often with suckers; circumoral tentacles (modified tube feet); pedicellariae absent; madreporite plate internal.

ECOLOGICAL RELATIONSHIPS

Echinoderms are all marine; they have no ability to osmoregulate and are thus rarely found in waters that are brackish. Virtually all benthic, they are found in all oceans of the world and at all depths, from the intertidal to the abyssal regions.

Some sea stars (Figure 24-1) are particle feeders, but many are predators, feeding particularly on sedentary or sessile prey. Brittle stars (Figure 24-6) are the most active echinoderms, moving by their arms, and they may be scavengers, browsers, or deposit or filter feeders. Some brittle stars are commensals in large sponges. Compared to other echinoderms, sea cucumbers (Figure 24-11) are greatly extended in the oral-aboral axis and are oriented with that axis more or less parallel to the substrate and lying on one side. Most are suspension or deposit feeders. "Regular" sea urchins (Figure 24-8), which are radially symmetrical, prefer hard bottoms and feed chiefly on algae or detritus. "Irregular" urchins (sand dollars and heart urchins) (Figure 24-9), which have become secondarily bilateral, are usually found on sand and feed on small particles. Sea lilies and feather stars (Figures 24-13 and 24-14) stretch their arms out and up like a flower's petals and feed on plankton and suspended particles.

CLASS STELLEROIDEA

Sea stars — subclass Asteroidea

Although sea stars are not considered the most primitive living echinoderms, they demonstrate the basic features of echinoderm structure and function very well, and they are easily obtainable. Thus we will consider them first, then comment on the major differences shown by the other groups.

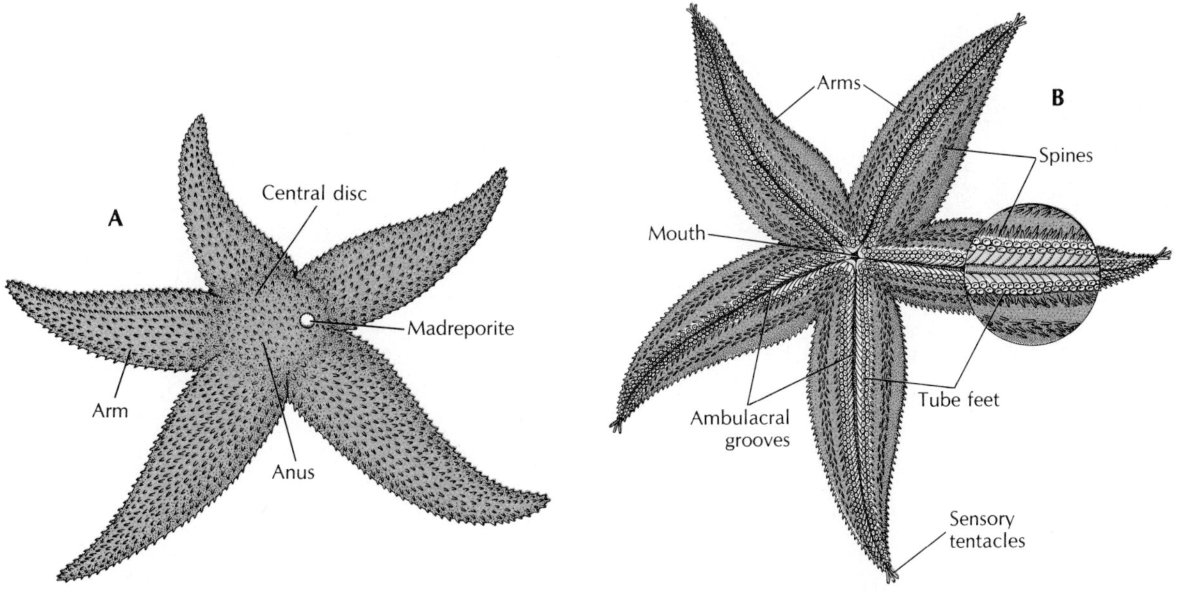

FIGURE 24-2

External anatomy of asteroid. **A,** Aboral view. **B,** Oral view.

Sea stars are familiar along the shorelines, where sometimes large numbers of them may aggregate on the rocks. They also live on muddy or sandy bottoms and among coral reefs. They are often brightly colored and range in size from a centimeter in greatest diameter to about a meter across from arm tip to opposite arm tip.

Form and function. Reflecting their radial pentamerous symmetry, sea stars typically have five arms (rays), but there may be more (Figure 24-1, *D*). The arms merge gradually with the central disc (Figure 24-2, *A*). Beneath the epidermis of the sea star is a mesodermal endoskeleton of small calcareous plates, or **ossicles,** bound together with connective tissue. From these ossicles project the spines and tubercles that are responsible for the spiny surface. The **ambulacral** (am-bu-la'kral) **grooves** radiate out along the arms from the centrally located mouth on the under, or oral, side of the animal. The **tube feet (podia)** project from the grooves, which are bordered by movable spines. Viewed from the oral side, each ambulacral groove (Figure 24-2, *B*) can be seen between the rows of tube feet.

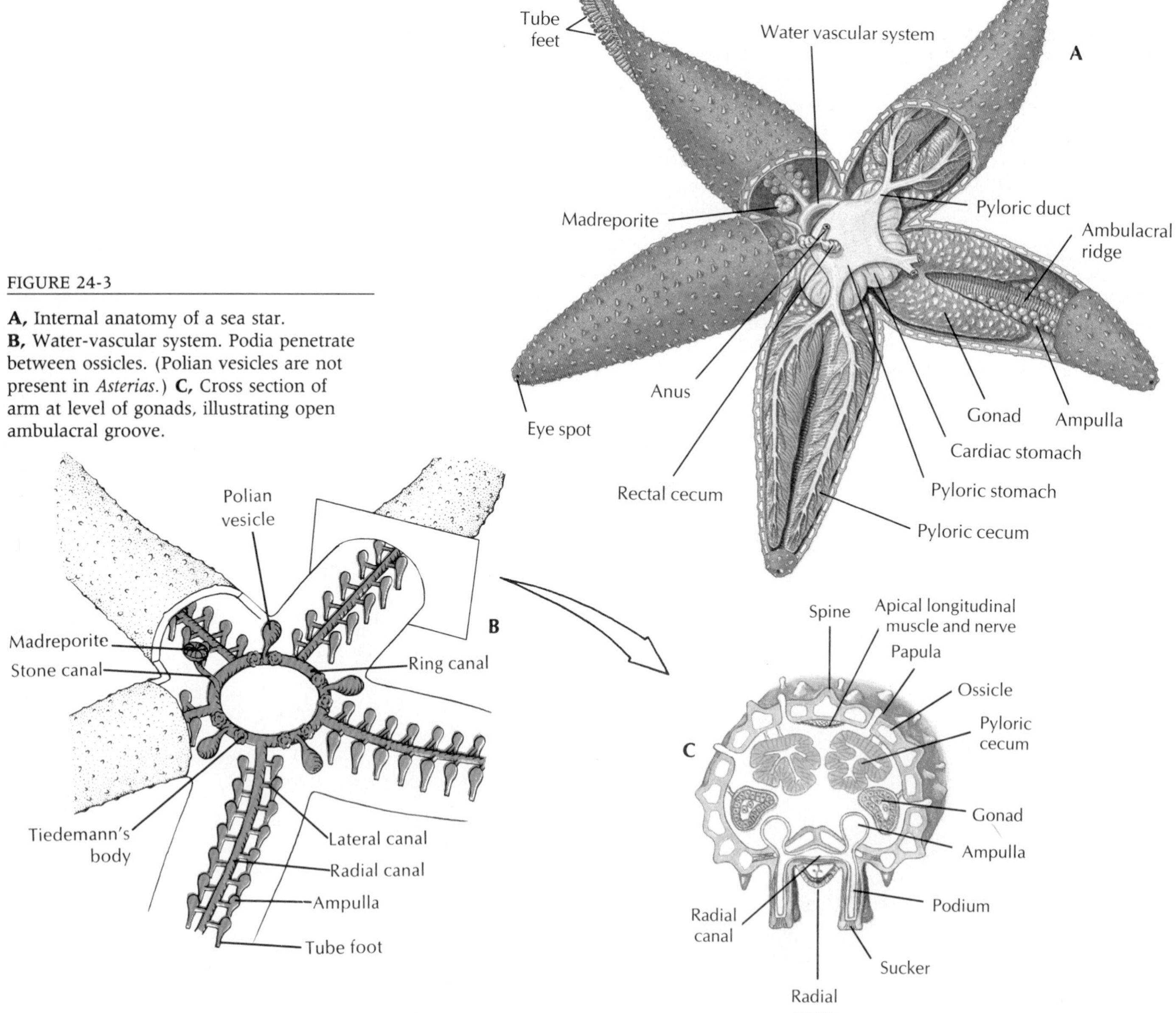

FIGURE 24-3

A, Internal anatomy of a sea star. **B,** Water-vascular system. Podia penetrate between ossicles. (Polian vesicles are not present in *Asterias.*) **C,** Cross section of arm at level of gonads, illustrating open ambulacral groove.

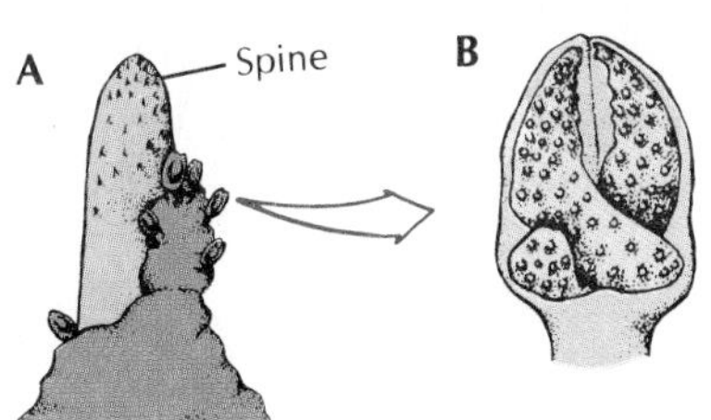

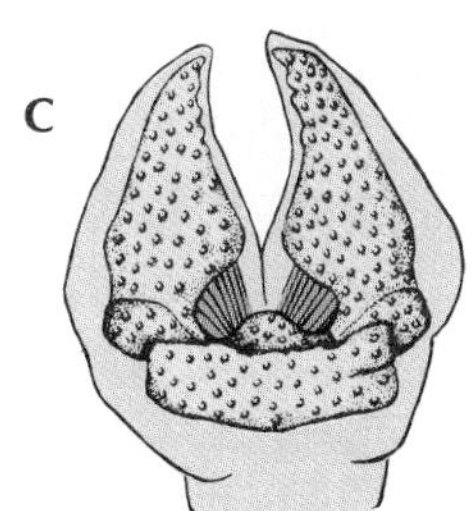

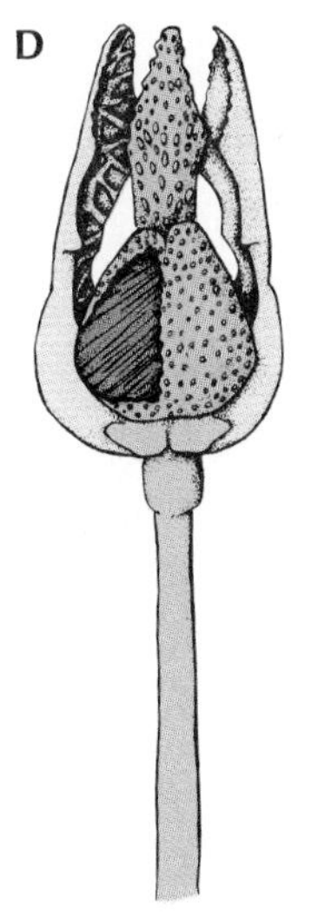

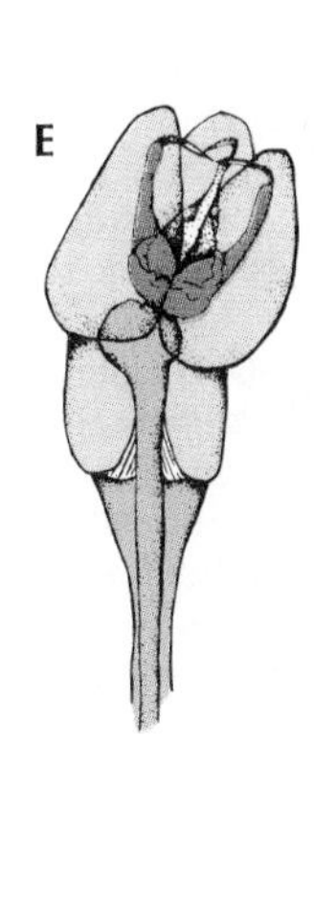

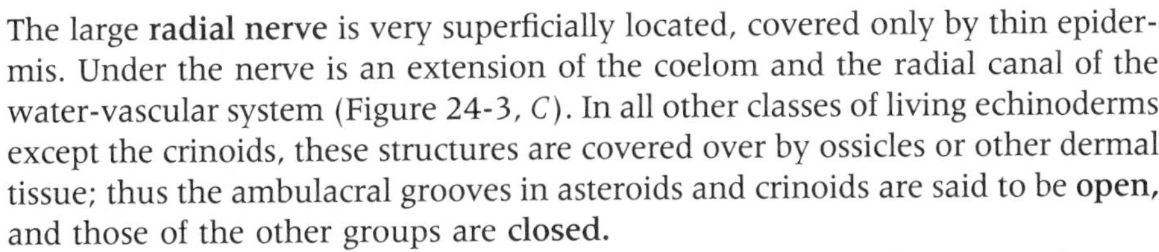

FIGURE 24-4

Pedicellariae of sea stars and sea urchins. **A** and **B**, Scissors-type pedicellariae of *Asterias;* size relative to spine is shown in **A.** **C,** Forceps-type pedicellaria of *Asterias.* **D,** Tridactyl pedicellaria of *Strongylocentrotus,* cutaway showing muscle. **E,** Globiferous pedicellaria of *Strongylocentrotus.*

Drawings by T. Doyle

The large **radial nerve** is very superficially located, covered only by thin epidermis. Under the nerve is an extension of the coelom and the radial canal of the water-vascular system (Figure 24-3, *C*). In all other classes of living echinoderms except the crinoids, these structures are covered over by ossicles or other dermal tissue; thus the ambulacral grooves in asteroids and crinoids are said to be **open,** and those of the other groups are **closed.**

The aboral surface is usually rough and spiny, although the spines of many species are flattened, so that the surface appears smooth. Around the bases of the spines in many sea stars are groups of minute pincerlike **pedicellariae** (ped-e-cell-ar'e-ee), bearing tiny jaws manipulated by muscles (Figure 24-4). These help keep the body surface free of debris, protect the papulae, and sometimes aid in food capture. The **papulae** (pap'u-lee) **(dermal branchiae** or **skin gills)** are soft, delicate projections of the coelomic cavity, covered only with epidermis and lined internally with peritoneum; they extend out through spaces between the ossicles and are concerned with respiration (Figure 24-3, *C*). Also on the aboral side are the inconspicuous **anus** and the circular **madreporite** (Figure 24-3, *A*), a calcareous sieve leading to the water-vascular system.

The coelomic compartments of the larval echinoderm give rise to several structures in the adult, one of which is a spacious body **coelom** filled with fluid. The coelomic fluid is circulated around the body cavity and into the papulae by the ciliated peritoneal lining. Exchange of respiratory gases and excretion of nitrogenous waste, principally ammonia, take place by diffusion through the thin walls of the papulae and tube feet.

Water-vascular system. The water-vascular system is another coelomic compartment and is unique to the echinoderms. It is a system of canals and specialized tube feet that shows exploitation of hydraulic mechanisms to a greater degree than in any other animal group. In sea stars the primary functions of the water-vascular system are locomotion and food gathering, as well as those of respiration and excretion.

Structurally, the water-vascular system opens to the outside through small pores in the madreporite. The madreporite of asteroids is on the aboral surface (Figure 24-2, *A*), and leads into the **stone canal,** which descends toward the **ring canal** around the mouth (Figure 24-3, *B*). **Radial canals** diverge from the ring canal, one into the ambulacral groove of each ray. **Polian vesicles** are also attached to the ring canal of most asteroids (but not *Asterias*) and apparently serve as fluid reservoirs for the water-vascular system.

The function of the madreporite is still obscure. One suggestion is that it allows rapid adjustment of hydrostatic pressure within the water-vascular system in response to changes in external hydrostatic pressure resulting from depth changes, as in tidal fluctuations.

A series of small **lateral canals,** each with a one-way valve, connects the radial canal to the cylindrical **podia** or **tube feet,** along the sides of the ambulacral groove in each ray. Each podium is a hollow, muscular tube, the inner end of which is a muscular sac, the **ampulla,** that lies within the body coelom (Figure 24-3), and the outer end of which usually bears a sucker. Some species lack the suckers. The podia pass to the outside between the ossicles in the ambulacral groove.

Feeding and digestive system. The mouth on the oral side leads into a two-part stomach located in the central disc (Figure 24-3). In some species, the large,

Locomotion by means of tube feet illustrates the interesting exploitation of hydraulic mechanisms by echinoderms. The valves in the lateral canals prevent backflow of fluid into the radial canals. The tube foot has in its walls connective tissue that maintains the cylinder at a relatively constant diameter. On contraction of muscles in the ampulla, fluid is forced into the podium, extending it. Conversely, contraction of the longitudinal muscles in the tube foot retracts the podium, forcing fluid back into the ampulla. Contraction of muscles in one side of the podium bends the organ toward that side. Small muscles at the end of the tube foot can raise the middle of the disclike end, thus creating a suction-cup effect when the end is applied to the substrate. It has been estimated that by combining mucous adhesion with suction, a single podium can exert a pull equal to 25 to 30 g. Coordinated action of all or many of the tube feet is sufficient to draw the animal up a vertical surface or over rocks.

On a soft surface, such as muck or sand, the suckers are ineffective (and numerous sand-dwelling species have no suckers), so the tube feet are employed as legs.

The mesocoel and metacoel compartments of the coelom, as found in the lophophorates, are present in echinoderms. Echinoderms also have a more anterior compartment, the protocoel, although the protocoel often is not completely separated from the mesocoel. In echinoderms the metacoel, mesocoel, and protocoel are referred to as somatocoel, hydrocoel, and axocoel. During metamorphosis of sea stars, the paired somatocoels become the oral and aboral coelomic cavities, the right axocoel and hydrocoel are lost, and the left axocoel and hydrocoel become the water-vascular system and perihemal channels.

lower **cardiac stomach** can be everted. The smaller upper **pyloric stomach** connects with **digestive ceca** located in the arms. Digestion is largely extracellular, occurring in the digestive ceca. A short **intestine** leads from the stomach to the inconspicuous **anus** on the aboral side. Some species lack an intestine and anus.

Many sea stars are carnivorous and feed on molluscs, crustaceans, polychaetes, echinoderms, other invertebrates, and sometimes small fish, but many show particular preferences. Some feed on brittle stars, sea urchins, or sand dollars, swallowing them whole and later regurgitating undigestible ossicles and spines. Some attack other sea stars, and if they are small compared to their prey, they may attack and begin eating at the end of one arm.

Many asteroids feed heavily on molluscs, and *Asterias* is a significant predator on commercially important clams and oysters. When feeding on a bivalve, a sea star will hump over its prey, attaching its podia to the valves, and then exert a steady pull, using its feet in relays. A force of some 1300 g can thus be exerted. In half an hour or so the adductor muscles of the bivalve fatigue and relax. With a very small gap available, the star inserts its soft everted stomach into the space between the valves, wraps it around the soft parts of the shellfish, and secretes digestive juices to start digesting the tissues of the bivalve. After feeding, the sea star draws in its stomach by contraction of the stomach muscles and relaxation of body wall muscles.

Some sea stars feed on small particles, either entirely or in addition to carnivorous feeding. Plankton or other organic particles coming in contact with the animal's oral or aboral surface are carried by the epidermal cilia to the ambulacral grooves and then to the mouth.

Hemal system. Although the hemal system is characteristic of echinoderms, its function remains unclear. It has little or nothing to do with circulation of body fluids, despite its name, which means "blood." It is a system of tissue strands enclosing unlined channels and is itself enclosed in another coelomic compartment, the perihemal channels or sinuses. The main channel of the hemal system connects aboral, gastric, and oral rings that give rise to branches to the gonads, stomach ceca, and arms, respectively.

Nervous and sensory system. The nervous system in echinoderms comprises three subsystems, each made up of a nerve ring and radial nerves placed at different levels in the disc and arms. An epidermal nerve plexus, or nerve net, connects the systems. Sense organs include ocelli at the arm tips and sensory cells scattered all over the epidermis.

Reproductive system and regeneration and autonomy. Most sea stars have separate sexes. A pair of gonads lies in each interradial space (Figure 24-3), and fertilization is external.

Echinoderms can regenerate lost parts. Sea star arms can regenerate readily, even if all are lost. Stars also have the power of **autonomy** (the ability to break off part of their own bodies) and can cast off an injured arm near the base. It may take months to regenerate a new arm.

If an arm is broken off or removed, and it contains a part of the central disc (about one fifth), the arm can regenerate a complete new sea star! In former times fishermen used to dispatch sea stars they collected from their oyster beds by chopping them in half with a hatchet—a worse-than-futile activity. Some sea stars reproduce asexually under normal conditions by cleaving the central disc, each part regenerating the rest of the disc and missing arms.

Development. In some species the liberated eggs are brooded, either under the oral side of the animal or in specialized aboral structures, and development is direct; but in most species the embryonating eggs are free in the water and hatch to free-swimming larvae.

Early embryogenesis shows the typical primitive deuterostome pattern. Gastrulation is by invagination, and the anterior end of the archenteron pinches off to become the coelomic cavity, which expands in a U shape to fill the blastocoel. Each of the legs of the U, at the posterior end, constricts to become a separate vesicle, and these eventually give rise to the main coelomic compartments of the body. The anterior portions of the U give rise to the water-vascular system and the perihemal channels. The free-swimming larva has cilia arranged in bands, and these tracts become extended into larval arms as development continues. The larva grows three adhesive arms and a sucker at its anterior end and attaches to the substratum. While it is thus attached by this temporary stalk, it undergoes metamorphosis.

Metamorphosis involves a dramatic reorganization of a bilateral larva into a radial juvenile. The anteroposterior axis of the larva is lost. *What was the left side becomes the oral surface, and the larval right side becomes the aboral surface* (Figure 24-5). Correspondingly, the larval mouth and anus disappear, and a new mouth and anus form on what were originally the left and right sides, respectively. As internal reorganization proceeds, short, stubby arms and the first podia appear. The animal then detaches from its stalk and begins life as a young sea star.

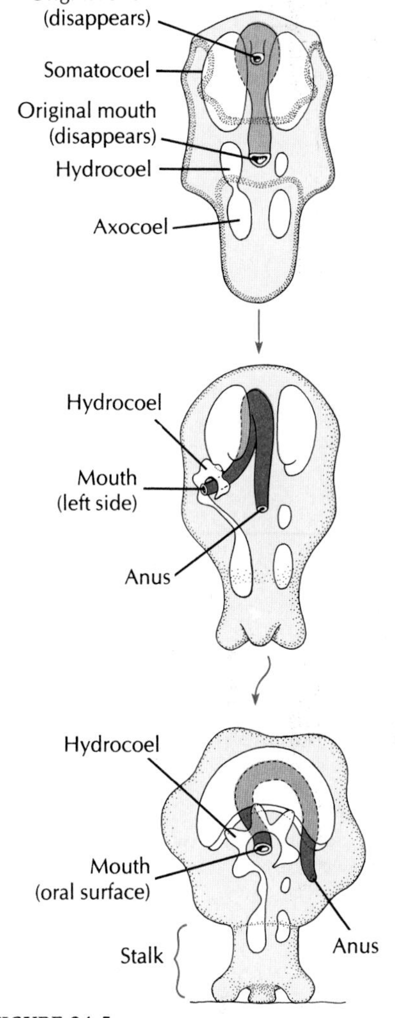

FIGURE 24-5

Sea star metamorphosis.

Brittle stars—subclass Ophiuroidea

The brittle stars are the largest of the major groups of echinoderms in numbers of species, and they are probably the most abundant also. They abound in all types of benthic marine habitats, even carpeting the abyssal sea bottom in many areas.

Apart from the typical possession of five arms, the brittle stars are surprisingly different from the asteroids. The arms of brittle stars are slender and sharply set off from the central disc (Figure 24-6). They have no pedicellariae or papulae, and their ambulacral grooves are closed, covered with arm ossicles. The tube feet are without suckers; they aid in feeding but are of limited use in locomotion. In contrast to that in the asteroids, the madreporite of the ophiuroids is located on the oral surface, on one of the oral shield ossicles (Figure 24-7).

Each of the jointed arms consists of a column of articulated ossicles connected by muscles and covered by plates. Locomotion is by arm movement.

FIGURE 24-6

A, This brittle star, *Ophiopholis aculeata,* has its bursae swollen with eggs, which it is ready to expel. The arms have been broken and are regenerating. **B,** Oral view of a basket star, *Gorgonocephalus eucnemis,* showing pentaradial symmetry.

Photographs by R. Harbo.

A

B

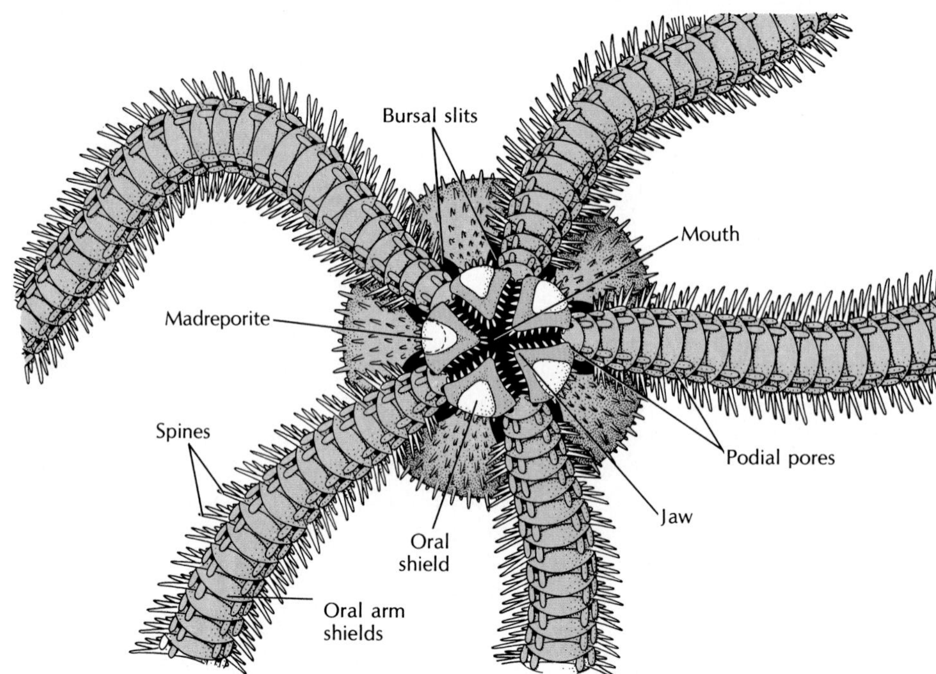

FIGURE 24-7

Oral view of spiny brittle star *Ophiothrix.*

The mouth is surrounded by five movable plates that serve as **jaws** (Figure 24-7). There is no anus. The skin is leathery, with dermal plates and spines arranged in characteristic patterns. Surface cilia are mostly lacking.

The visceral organs are confined to the central disc, since the rays are too slender to contain them. The **stomach** is saclike and there is no intestine. Indigestible material is cast out of the mouth.

Five pairs of **bursae** (peculiar to ophiuroids) open toward the oral surface by **genital slits** at the bases of the arms. Water circulates in and out of these sacs for exchange of gases. On the coelomic wall of each bursa are small **gonads** that discharge into the bursa their ripe sex cells, which pass through the genital slits into the water for fertilization. Sexes are usually separate; a few ophiuroids are hermaphroditic. The ciliated bands of the larva extend onto delicate, beautiful larval arms, like those of larval echinoids. During the metamorphosis to the juvenile, there is no temporarily attached phase, as in the asteroids.

Water-vascular, nervous, and hemal systems are similar to those of the sea stars.

Brittle stars tend to be secretive, living on hard bottoms where no light penetrates. They are generally negatively phototropic and work themselves into small crevices between rocks, becoming more active at night. They are commonly fully exposed on the bottom in the permanent darkness of the deep sea. Ophiuroids feed on a variety of small particles, either browsing food from the bottom or filter feeding. Podia are important in transferring food to the mouth. Some brittle stars extend arms into the water and catch suspended particles in mucous strands between the arm spines.

Regeneration and autonomy are even more pronounced in brittle stars than in sea stars. Many seem very fragile, releasing an arm or even part of the disc at the slightest provocation. Some can reproduce asexually by cleaving the disc; each progeny then regenerates the missing parts.

SEA URCHINS, SAND DOLLARS, AND HEART URCHINS— CLASS ECHINOIDEA

The echinoids have a compact body enclosed in an endoskeletal **test,** or shell. The dermal ossicles, which have become closely fitting plates, make up the test.

FIGURE 24-8

Purple sea urchin *Strongylocentrotus purpuratus* is common along the Pacific coast of North America where there is heavy wave action.

Photograph by R. Harbo.

Echinoids lack arms, but their tests reflect the typical five-part plan of the echinoderms in their five ambulacral areas. Rather than extending from the oral surface to the tips of the arms, as in asteroids, the ambulacral areas follow the contours of the test from the mouth around to the aboral side, ending at the area around the anus **(periproct).** The majority of living species of sea urchins are referred to as "regular"; they are hemispherical in shape, radially symmetrical, and have medium to long spines (Figure 24-8). Sand dollars and heart urchins (Figure 24-9) are "irregular" because the orders to which they belong have become secondarily bilateral; their spines are usually very short. Regular urchins move by means of their tube feet, with some assistance from their spines, and irregular urchins move chiefly by their spines. Some echinoids are quite colorful.

Echinoids have wide distribution in all seas, from the intertidal regions to the deep oceans. Regular urchins often prefer rocky or hard bottoms, whereas sand dollars and heart urchins like to burrow into a sandy substrate.

The echinoid test is a compact skeleton of 10 double rows of plates that bear movable, stiff spines (Figure 24-10). The five pairs of ambulacral rows have pores (Figure 24-10) through which the long tube feet extend. The spines are moved by small muscles around the bases.

There are several kinds of **pedicellariae,** the most common of which are three jawed and are mounted on long stalks.

The mouth of regular urchins is surrounded by five converging teeth. In some sea urchins branched **gills** (modified podia) encircle the peristome, although these are of little importance in respiratory gas exchange. The **anus, genital openings,** and **madreporite** are located aborally in the periproct region (Figure 24-10). The sand dollars also have teeth, and the mouth is located at about the center of the oral side, but the anus has shifted to the posterior margin or even the oral side of the disc, so that an anteroposterior axis and bilateral symmetry can be recognized. Bilateral symmetry is even more accentuated in the heart urchins, with the anus near the posterior end on the oral side and the mouth moved away from the oral pole toward the anterior end (Figure 24-9).

Inside the test (Figure 24-10) is the coiled digestive system and a complex chewing mechanism (in the regular urchins and in sand dollars), called **Aris-**

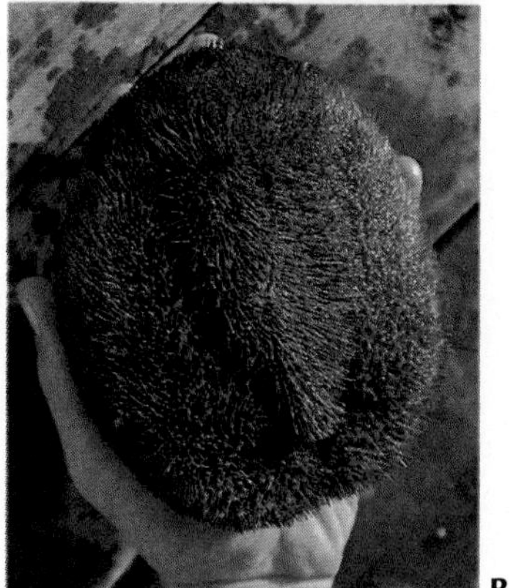

FIGURE 24-9

An irregular echinoid *Meoma,* one of the largest heart urchins (test up to 18 cm). *Meoma* occurs in the West Indies and from the Gulf of California to the Galápagos Islands. **A,** Aboral view. **B,** Oral view. Note curved mouth at anterior end and periproct at posterior end.

Photographs by L.S. Roberts.

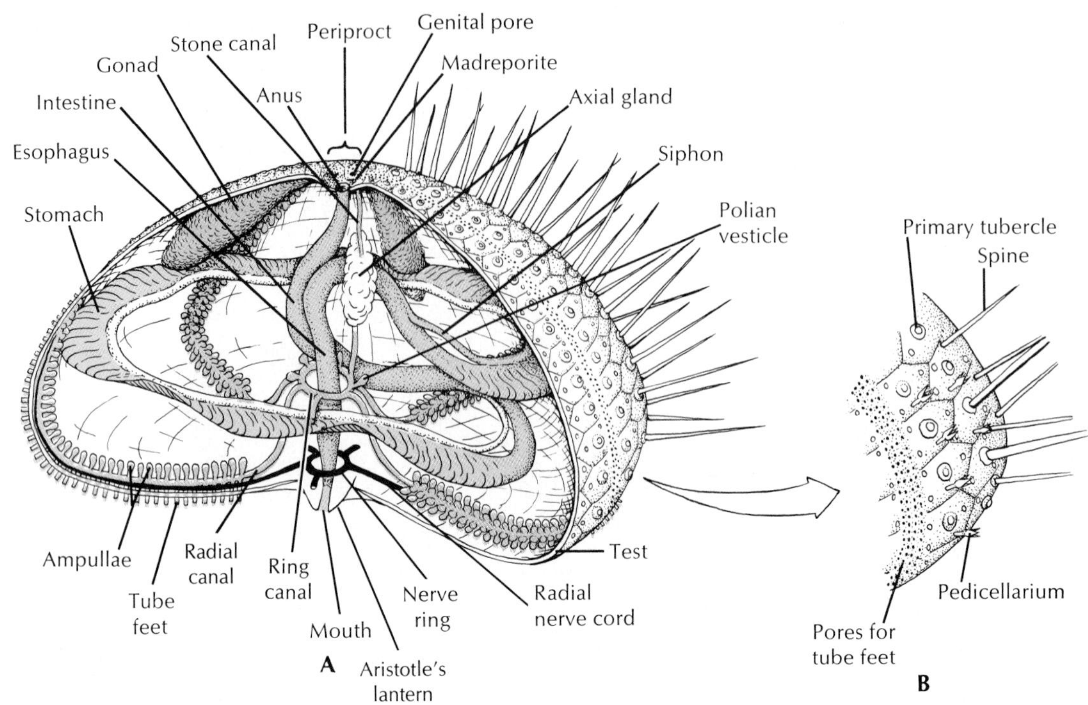

FIGURE 24-10

A, Internal structure of the sea urchin; water-vascular system in red. **B,** Detail of portion of test.

totle's lantern, to which the teeth are attached. A ciliated siphon connects the esophagus to the intestine and enables the water to bypass the stomach to concentrate the food for digestion in the intestine. Sea urchins eat algae and other organic material, and sand dollars collect fine particles on ciliated tracts.

The hemal and nervous systems are basically similar to those of the asteroids. The ambulacral grooves are closed, and the radial canals of the water-vascular system run just beneath the test, one in each of the ambulacral radii (Figure 24-10).

Sexes are separate, and both eggs and sperm are shed into the sea for external fertilization. The larvae may live a planktonic existence for several months and then metamorphose quickly into young urchins.

SEA CUCUMBERS—CLASS HOLOTHUROIDEA

In a phylum characterized by odd animals, class Holothuroidea contains members that both structurally and physiologically are among the strangest of all. These animals have a remarkable resemblance to the vegetable after which they are named (Figure 24-11). Compared to the other echinoderms, the holothurians are greatly elongated in the oral-aboral axis, and the ossicles are much reduced in most, so that the animals are soft bodied. Some species characteristically crawl on the surface of the sea bottom, others are found beneath rocks, and some are burrowers.

The body wall is usually leathery, with the tiny ossicles embedded in it, although a few species have large ossicles forming a dermal armor. Because of the elongate body form of the sea cucumbers, they characteristically lie on one side. In most species the tube feet are well developed only in the ambulacra normally applied to the substratum. Thus a secondary bilaterality is present, albeit of quite different origin from that of the irregular urchins.

The **oral tentacles** are 10 to 30 retractile, modified tube feet around the mouth. The body wall contains circular and longitudinal muscles along the ambulacra.

The **coelomic cavity** is spacious and fluid filled. The digestive system empties posteriorly into a muscular **cloaca** (Figure 24-12). A **respiratory tree** composed

FIGURE 24-11

Sea cucumber (class Holothuroidea). Common along the Pacific coast of North America, *Parastichopus californicus* grows up to 50 cm in length. Its tube feet on the dorsal side are reduced to papillae and warts.

Photograph by R. Harbo.

of two long, many-branched tubes also empties into the cloaca, which pumps seawater into it. The respiratory tree serves both for respiration and excretion and is not found in any other group of living echinoderms. Gas exchange also occurs through the skin and tube feet.

The hemal system is more well developed in holothurians than in other echinoderms. The water-vascular system is peculiar in that the madreporite lies free in the coelom.

The sexes are separate, but some holothurians are hermaphroditic. Among the echinoderms, only the sea cucumbers have a single gonad, and this is considered a primitive character. Fertilization is external.

Sea cucumbers are sluggish, moving partly by means of their ventral tube feet and partly by waves of contraction in the muscular body wall. The more sedentary species trap suspended food particles in the mucus of their outstretched oral tentacles or pick up particles from the surrounding bottom. They then stuff the tentacles into their pharynx, one by one, sucking off the food material. Others crawl along, grazing the bottom with their tentacles.

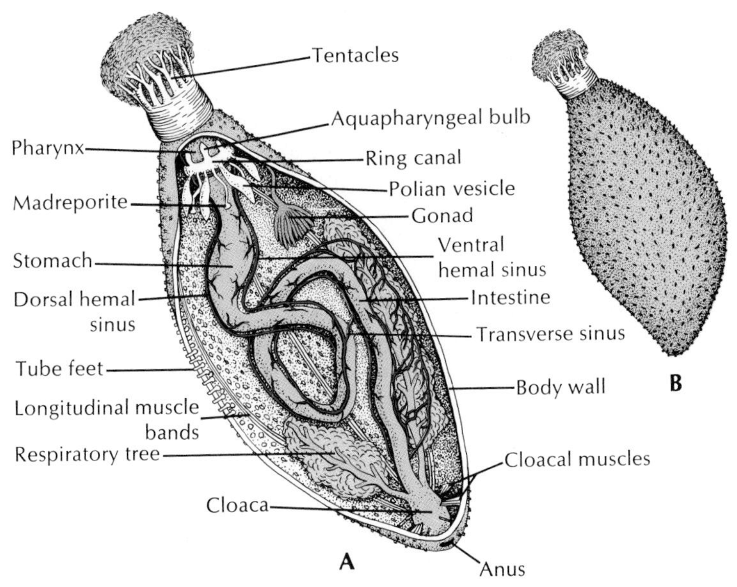

FIGURE 24-12

Anatomy of the sea cucumber *Thyone.*
A, Internal view; hemal system in red.
B, External view.

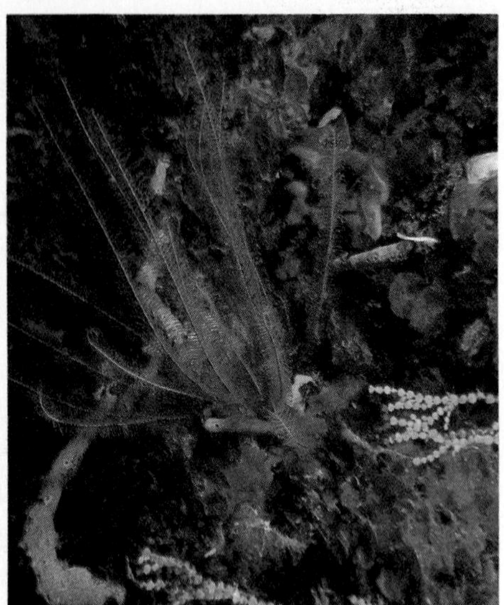

FIGURE 24-13

Nemaster spp. are crinoids found on Caribbean coral reefs. They extend their arms into the water to catch food particles while maintaining their body in a crevice.
Photograph by D.W. Gotshall.

SEA LILIES AND FEATHER STARS—CLASS CRINOIDEA

The crinoids are the most primitive of the living echinoderms. As fossil records reveal, crinoids were once far more numerous than now. They differ from other echinoderms by being attached during a substantial part of their lives. Many crinoids are deep-water forms, but feather stars may inhabit shallow waters, especially in the Indo-Pacific and the West Indian—Caribbean regions, where the largest numbers of species are found.

The body disc is covered with a leathery skin containing calcareous plates. The epidermis is poorly developed. Five flexible arms branch to form many more arms, each with many lateral **pinnules** arranged like barbs on a feather (Figure 24-13). Sessile forms have a long, jointed **stalk** attached to the aboral side of the body **(calyx)** (Figure 24-14). This stalk is made up of plates, appears jointed, and may bear **cirri.** Madreporite, spines, and pedicellariae are absent.

The upper (oral) surface bears the mouth and the anus. With the aid of tube feet and mucous nets, crinoids feed on small organisms that are caught in the ambulacral grooves. The **ambulacral grooves** are open and ciliated and serve to carry food to the mouth. Tube feet in the form of tentacles are also found in the grooves. The **water-vascular system** has the echinoderm plan. Sense organs are scanty and primitive.

The sexes are separate, and the gonads are primitive. The larvae are free swimming for a time before they become attached and metamorphose. Most living crinoids are from 15 to 30 cm long, but some fossil species had stalks 25 m in length.

PHYLOGENY AND ADAPTIVE RADIATION

Phylogeny

Based on the embryological evidence of the bilateral larvae of echinoderms, there can be little doubt that their ancestors were bilateral and that their radial symmetry is secondary. Some recent investigators believe that the radial symmetry arose in a free-moving echinoderm ancestor and that sessile groups were derived several times independently from the free-moving ancestors. However, this view does not account for the adaptive significance of radial symmetry for the sessile existence. The more traditional view is that the first echinoderms were sessile, became radial as an adaptation to that existence, and then gave rise to the free-moving groups. Certainly, the most primitive living echinoderms are the crinoids, which are both sessile and radial as adults, and the existence of a transitory stalked phase in the asteroids, which also have some primitive characteristics such as open ambulacra, supports the traditional sequence. It is believed also that the endoskeleton was an adaptation for a sessile existence, and that the original function of the water-vascular system was in feeding.

The nature of the hypothetical preechinoderm has also been subject to debate. Hyman and others have suggested that it was like a lophophorate animal, with tentacles around the mouth. According to this idea, the water-vascular system was derived from the tentacles around one side, which were five in number and contained extensions of the mesocoel, as in the lophophorates.

The flower-shaped body of the sea lily is attached to the substratum by a stalk. During metamorphosis feather stars also become sessile and attached, but after several months they detach and become free-moving. Although they may remain attached in the same location for long periods, they are capable of crawling and swimming short distances. They swim by alternate sweeping of their long, feathery arms.

Adaptive radiation

The radiation of the echinoderms has been determined by the limitations and potentials of their most important characteristics: radial symmetry, the water-vascular system, and their dermal endoskeleton. If their ancestors had a brain and specialized sense organs, these were lost in the adoption of radial symmetry. Thus, it is not surprising that there are large numbers of creeping, benthic forms with filter feeding, deposit feeding, scavenging, and herbivorous habits, but comparatively few predators. The relative success of the asteroids as predators is impressive and probably attributable to the extent to which they have exploited the hydraulic mechanism of the tube feet.

Lesser Deuterostomes

The term lesser (or minor) deuterostomes is usually used in reference to the hemichordates and chaetognaths, but they are "lesser" only in the sense that they have a relatively small number of species. They are actually widespread and contain some commonly found invertebrate forms. Often smaller groups deserve much more attention than we usually give them, because they too can contribute much to our understanding of evolutionary diversity and relationships.

The hemichordates were formerly included as a subphylum of the Chordata, but they are probably more closely related to the Echinodermata. The Chaetognatha apparently are not closely related to any other group.

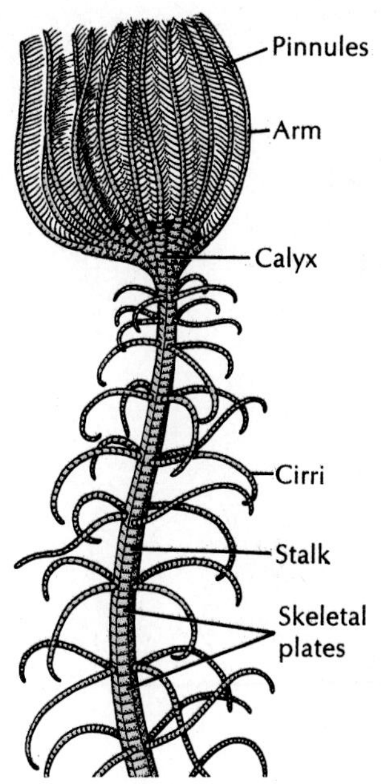

FIGURE 24-14

A stalked crinoid with portion of stalk. Modern crinoid stalks rarely exceed 60 cm, but fossil forms were as much as 20 m long.

Phylum Chaetognatha

A common name for the Chaetognatha (ke-tog'na-tha) (Gr. *chaitē,* long flowing hair, + *gnathos,* jaw) is arrowworms. This small group (65 species) of marine animals is considered by some to be related to the nematodes and by others to be related to the annelids. However, arrowworms actually seem to be aberrant and show no distinct relations to any other group. Only their embryology indicates their position as deuterostomes.

Their small, straight bodies resemble miniature torpedoes, or darts, ranging from 2.5 to 10 cm in length.

Most arrowworms are adapted for a planktonic existence. They usually swim to the surface at night and descend during the day. Much of the time they drift passively, but they can dart forward in swift spurts, using the caudal fin and longitudinal muscles—a fact that no doubt contributes to their success as planktonic predators. Horizontal fins bordering the trunk are used in flotation rather than in active swimming.

The body of the arrowworm is unsegmented and is made up of the head, trunk, and postanal tail (Figure 24-15). Teeth and chitinous spines are found on the head around the mouth. When the animal captures prey, the teeth and raptorial spines spread apart and then snap shut with startling speed. Arrowworms are voracious feeders, living on other planktonic forms, especially copepods, and even small fish.

The body is covered with a thin cuticle. Arrowworms are fairly advanced worms in that they have a complete digestive system, well-developed coelom, and nervous system with a nerve ring containing several ganglia. Vascular, respiratory, and excretory systems, however, are entirely lacking.

Arrowworms are hermaphroditic with either cross-fertilization or self-fertilization. The juveniles develop directly without metamorphosis. There is no true peritoneum lining the coelom. Cleavage is radial, complete, and equal.

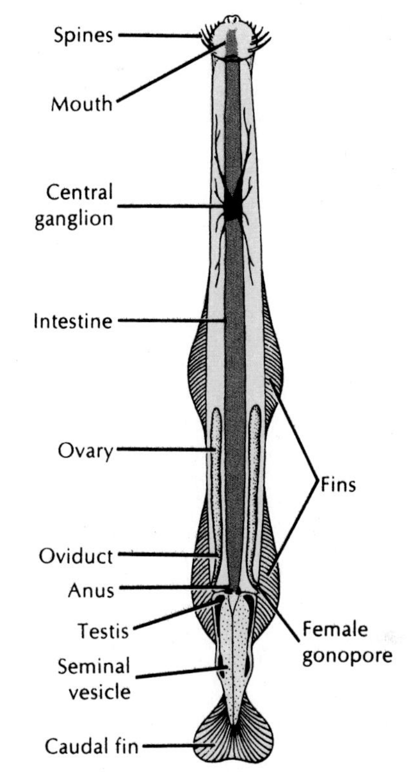

FIGURE 24-15

Arrowworm *Sagitta.*

Phylum Hemichordata—the Acorn Worms

The Hemichordata (hem'i-kor-da'ta) (Gr. *hemi,* half, + *chorda,* string, cord) are marine animals that were formerly considered a subphylum of the chordates, based on their possession of gill slits, a rudimentary notochord, and a dorsal nerve cord. However, it is now generally agreed that the so-called hemichordate "notochord" is really an evagination of the mouth cavity and not homologous with the chordate notochord, so the hemichordates are given the rank of a separate phylum.

Hemichordates are wormlike bottom dwellers, living usually in shallow waters. Some are colonial and live in secreted tubes. Many are sedentary or sessile. They are widely distributed, but their secretive habits and fragile bodies make collecting them difficult.

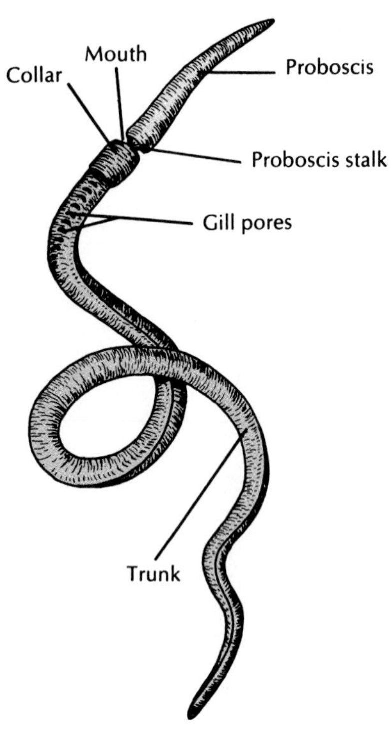

FIGURE 24-16

External lateral view of the acorn worm, *Saccoglossus* (phylum Hemichordata).

Members of class Enteropneusta (acorn worms) range from 20 mm to 2.5 m in length and 3 to 200 mm in breadth. Members of class Pterobranchia (pterobranchs) are smaller, usually from 5 to 14 mm, not including the stalk. About 70 species of enteropneusts and three small genera of pterobranchs are recognized.

Hemichordates have the typical tricoelomate structure of deuterostomes.

CLASS ENTEROPNEUSTA

The enteropneusts, or acorn worms (Figure 24-16), are sluggish wormlike animals that live in burrows or under stones, usually in mud or sand flats of intertidal zones.

The mucus-covered body is divided into a tonguelike **proboscis,** a short **collar,** and a long **trunk** (protosome, mesosome, and metasome.)

In the posterior end of the proboscis is a small coelomic sac (protocoel) into which extends the buccal diverticulum, a slender, blindly ending pouch of the gut that reaches forward into the buccal region and was formerly believed to be a notochord. A row of gill pores is located dorsolaterally on each side of the trunk just behind the collar (Figure 24-16). These open from a series of gill chambers that in turn connect with a series of gill slits in the sides of the pharynx. The primary function of these structures is not respiration, but food-gathering (Figure 24-17). Food particles caught in mucus and brought to the mouth by ciliary action on the proboscis and collar are strained out of the branchial water that leaves through the gill slits. The particles are then directed along the ventral part of the pharynx and esophagus to the intestine.

A middorsal vessel carries blood forward to a heart above the buccal diverticulum. The blood is then driven into a network of sinuses that may have an excretory function, then posteriorly through a ventral blood vessel and through a network of sinuses in the gut and body wall. Acorn worms have dorsal and ventral nerve cords, and the dorsal cord is hollow in some. Sexes are separate. In some enteropneusts there is a free-swimming larva that closely resembles the larva of sea stars.

CLASS PTEROBRANCHIA

The basic plan of the class Pterobranchia is similar to that of the Enteropneusta, but certain differences are correlated with the sedentary mode of life of pterobranchs. Only two genera are known in any detail. In both genera there are arms with tentacles containing an extension of the coelomic compartment of the mesosome, as in a lophophore (Figure 24-18). One genus has a single pair of gill slits, and the other has none. Both live in tubes from which they project their proboscis and tentacles to feed by mucociliary mechanisms. Some species are dioecious, others are monoecious, and asexual reproduction by budding occurs.

PHYLOGENY

Hemichordates share characteristics with both the echinoderms and the chordates. They share with the chordates the gill slits, which are used primarily for

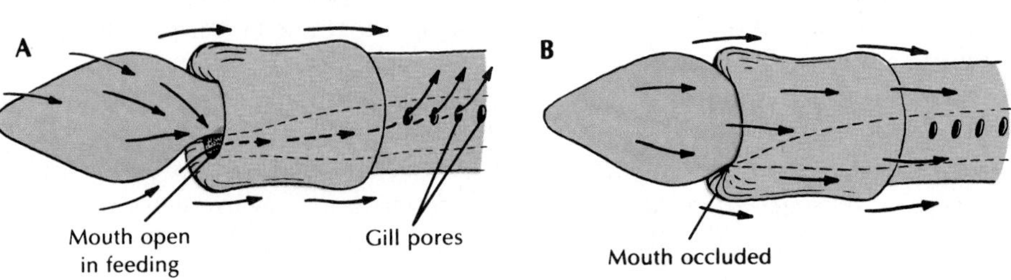

FIGURE 24-17

Food currents of enteropneust hemichordate. **A,** Side view of acorn worm with mouth open, showing direction of currents created by cilia on proboscis and collar. Food particles are directed toward mouth and digestive tract. Rejected particles move toward outside of collar. Water leaves through gill pores. **B,** When mouth is occluded, all particles are rejected and passed onto the collar. Nonburrowing and some burrowing hemichordates use this feeding method.

Modified from Russell-Hunter, W.D. 1969. A biology of the higher invertebrates. New York, Macmillan, Inc.

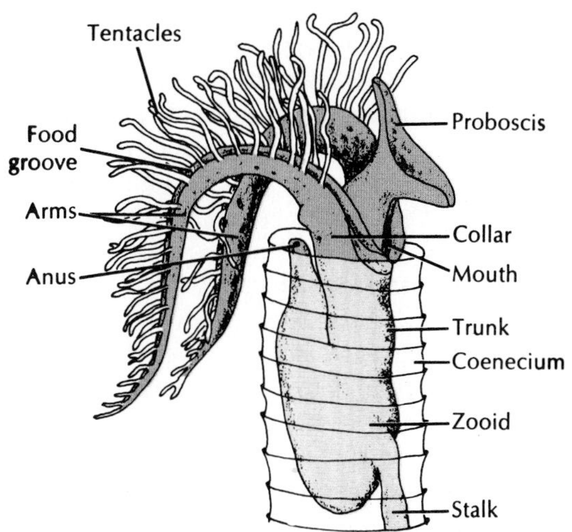

FIGURE 24-18

Rhabdopleura, a pterobranch hemichordate in its tube. Individuals live in branching tubes connected to each other by stolons. They protrude the tentacled arms for ciliary-mucus feeding.

filter feeding and only secondarily for breathing, as in some of the protochordates. The short dorsal, sometimes hollow, nerve cord in the collar zone foreshadows the nerve cord of the chordates. The early embryogenesis is remarkably like that of echinoderms, and the early larva is almost identical with the larva of asteroids.

Within the phylum, the class Pterobranchia is considered more primitive than the class Enteropneusta and shows affinities with the Ectoprocta, Brachiopoda, and Phoronida because of its lophophore-like tentacles and sessile habits. Some believe that the pterobranchs may be similar to the common ancestors of both the hemichordates and the echinoderms.

Summary

The phyla Echinodermata, Chordata, Hemichordata, and Chaetognatha show the characteristics of the Deuterostomia division of the animal kingdom. The echinoderms are an important marine group sharply distinguished from other phyla of animals. They have a radial symmetry but were derived from bilateral ancestors. They fill a variety of benthic niches, including particle feeders, browsers, scavengers, and predators.

The sea stars (class Stelleroidea, subclass Asteroidea) can be used to illustrate the echinoderms. They usually have five arms, and the arms merge gradually with a central disc. They have no head and few specialized sensory organs. The mouth is central on the under (oral) side of the body. They have an endoskeleton of dermal ossicles, open ambulacral areas, pedicellariae, and papulae. The water-vascular system is an elaborate hydraulic system derived from one of the coelomic cavities. The madreporite, opening to the outside, is connected to the ring canal around the esophagus by way of the stone canal, and radial canals extend from the ring canal along each ambulacral area. Branches from the radial canals lead to the many tube feet, structures that are important in locomotion, food-gathering, respiration, and excretion. Many sea stars are predators, while others feed on small particles. Sexes are separate, and the reproductive systems are very simple. The bilateral, free-swimming larva becomes attached, transforms to a radial juvenile, then detaches and becomes a motile sea star.

Brittle stars (subclass Ophiuroidea) have slender arms that are sharply set off from the central disc; they have no pedicellariae or ampullae, and their ambulacral grooves are closed. Their madreporite is on the oral side. They crawl by means of their arms, and they can move around more rapidly than other echinoderms.

Brittle stars (subclass Ophiuroidea) are usually dull gray or brown, but some are quite showy, like these specimens from Puerto Rico.

Photograph by K. Sandved.

In sea urchins (class Echinoidea), the dermal ossicles have become closely fitting plates, and there are no arms. Some urchins (sand dollars and heart urchins) are evolving toward a return to bilateral symmetry.

Sea cucumbers (class Holothuroidea) have very small dermal ossicles and a soft body wall. They are greatly elongated in the oral-aboral axis and lie on their side. Because three of their ambulacral areas characteristically lie against the substratum, sea cucumbers have also undergone some return to bilateral symmetry.

Sea lilies and feather stars (class Crinoidea) are the only group of living echinoderms, other than asteroids, with open ambulacral areas. They are mucociliary particle feeders and lie with their oral side up. They become attached to the substratum by a stem during metamorphosis, although the feather stars later detach from the stem.

The arrowworms (phylum Chaetognatha) are a small group but are important as a component of marine plankton.

Members of the phylum Hemichordata are marine worms that were formerly considered chordates because their buccal diverticulum was believed to be a notochord. In common with the chordates some of them do have gill slits and a hollow, dorsal nerve cord. The hemichordates are important phylogenetically because they show affinities with the chordates, echinoderms, and lophophorates.

REVIEW QUESTIONS

1. What is the constellation of characteristics possessed by echinoderms that is found in no other phylum?
2. How do we know that echinoderms were derived from an ancestor with bilateral symmetry?
3. Distinguish the following groups of echinoderms from each other: Crinoidea, Asteroidea, Ophiuroidea, Echinoidea, Holothuroidea.
4. What is an ambulacral groove, and what is the difference between open and closed ambulacral grooves?
5. Trace or make a rough copy of Figure 24-3, *A*, without the labels, then from memory label the parts of the water-vascular system of the sea star.
6. Name the structures involved in the following functions in sea stars, and briefly describe the action of each: respiration, feeding and digestion, excretion, reproduction.
7. Compare the structures and functions in question no. 6 as they are found in brittle stars, sea urchins, sea cucumbers, and crinoids.
8. Match the groups in the left column with *all* correct answers in the right column.

___ Crinoidea	a. Closed ambulacral grooves
___ Asteroidea	b. Oral surface generally upward
___ Ophiuroidea	c. With arms
___ Echinoidea	d. Approximately globular or disc-shaped
___ Holothuroidea	e. Elongated in oral-aboral axis
	f. With pedicellariae
	g. Madreporite internal
	h. Without arms
	i. Madreporite on oral plate

9. Define the following: pedicellariae, madreporite, respiratory tree, Aristotle's lantern.
10. What is some evidence that the ancestral echinoderm was sessile?
11. What are four morphological characteristics of Chaetognatha?
12. What is the ecological importance of arrowworms?
13. Distinguish the Enteropneusta from the Pterobranchia.
14. What is evidence that the Hemichordata are related both to the echinoderms and the lophophorate phyla?

SELECTED REFERENCES

Barnes, R.D. 1987. Invertebrate zoology, ed. 5. Philadelphia, Saunders College/Holt, Rinehart & Winston. *Chapters 19 and 20 cover the echinoderms and lesser deuterostomes.*

Barrington, E.J.W. 1979. Invertebrate structure and function, ed. 2. New York, John Wiley & Sons, Inc. *Excellent account of function in echinoderms and hemichordates.*

Davidson, E.H., B.R. Hough-Evans, and R.J. Britten. 1982. Molecular biology of the sea urchin embryo. Science **217**:17-26. *Many fundamental insights into the process of embryogenesis have been revealed through studies of sea urchins.*

Fell, H.B. 1982. Echinodermata. In Parker, S.P. (ed.). Synopsis and classification of living organisms, vol. 2. New York, McGraw-Hill Book Co. *Good, brief outline of echinoderms.*

Lawrence, J. 1987. A functional biology of echinoderms. Baltimore, The Johns Hopkins University Press. *Treats many aspects of the fascinating biology of these organisms.*

Russell-Hunter, W. D. 1979. A life of invertebrates. New York, Macmillan, Inc. *Very good comparative summary of the structure and function of the echinoderm groups. Also includes good discussion of function in hemichordates.*

PART FOUR

The Vertebrate Animals

IT'S A LONG WAY FROM AMPHIOXUS

Along the more southern coasts of North America, half-buried in sand on the sea floor, lives a small, fishlike, translucent animal quietly filtering organic particles from seawater. Inconspicuous, of little commercial value and largely unknown, this creature is nonetheless one of the famous animals of classical zoology. It is amphioxus, an animal that wonderfully exhibits the four distinctive hallmarks of the phylum Chordata—dorsal, tubular nerve cord overlying a supportive notochord, gill slits for filter feeding, and a postanal tail for propulsion—all wrapped up in one primitive form with textbook simplicity. Amphioxus is an animal that might have been designed by a zoologist for the classroom. During the nineteenth century, with interest in vertebrate ancestry running high, amphioxus was considered by many to be the direct ancestor of the vertebrates. Its exalted position was later acknowledged by Philip Pope in a poem sung to the tune of "Tipperary." It ends with the refrain:

It's a long way from Amphioxus
It's a long way from us.
It's a long way from Amphioxus
To the meanest human cuss.
It's good-bye fins and gill slits,
Welcome skin and hair.
It's a long way from Amphioxus
But we came from there.

But amphioxus' place in the sun was not to endure. For one thing, amphioxus lacks one of the most important of vertebrate characteristics, a distinct head with special sense organs and the equipment for shifting to an active predatory mode of life. This, together with several specialized features, suggests to zoologists today that amphioxus lies in an offshoot, rather than in the main line, of chordate descent. It seems that we are a very long way indeed from amphioxus. Nevertheless, while amphioxus is denied the ultimate ancestral award, we believe that it more closely resembles the earliest prevertebrate than any other animal we know.▪

CHAPTER 25

Chordates

Vertebrate Beginnings

Animals most familiar to people belong to the great phylum Chordata (korda'ta) (L. *chorda,* cord). Humans are members and share the characteristic from which the phylum derives its name—the **notochord** (Gr. *nōton,* back, + L. *chorda,* cord) (Figure 25-1). This structure is possessed by all members of the phylum, either in the larval or embryonic stages or throughout life. The notochord is a rodlike, semirigid body of cells enclosed by a fibrous sheath, which extends, in most cases, the length of the body between the gut tract and the central nervous system. Its primary purpose is to support and to stiffen the body.

The structural plan of chordates retains many of the features of higher invertebrates, such as bilateral symmetry, anteroposterior axis, coelom tube-within-a-tube arrangement, metamerism, and cephalization. Yet, whereas the kinship of chordates and nonchordates is clear, it has not been possible to establish the exact relationship with certainty.

Two possible lines of descent have been proposed. Earlier speculations that focused on the arthropod-annelid-mollusc group (Protostomia branch) of the invertebrates have fallen from favor. It is now believed that only the echinoderm group (Deuterostomia branch) deserves serious consideration as a chordate ancestor. The echinoderms and chordates share several important characteristics: **radial cleavage** (p. 310), **anus derivation from the first embryonic opening (blastopore) with mouth derivation from an opening of secondary origin, and coelom formed by fusion of enterocoelous pouches** (except in vertebrates in which the coelom is basically schizocoelous). These common characteristics indicate a natural unity among the Deuterostomia.

As a whole, there is more fundamental unity of plan throughout all the organs and systems of the phylum Chordata than there is in many of the invertebrate phyla. Ecologically the chordates are among the most adaptable of organic forms and are able to occupy most kinds of habitats. From a purely biological viewpoint, chordates are of primary interest because they illustrate so well the broad biological principles of evolution, development, and relationship. They represent as a group the background of the human species.

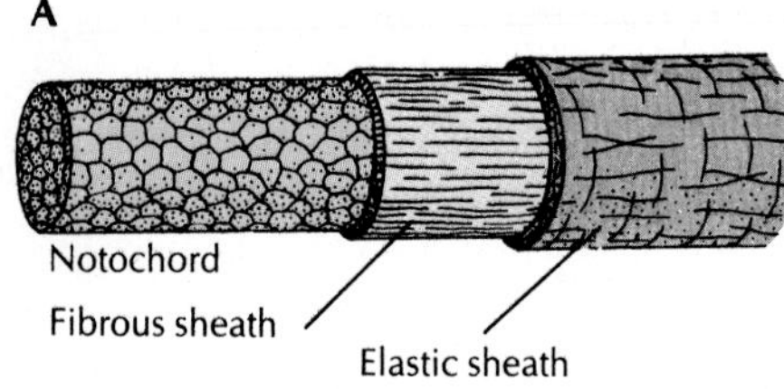

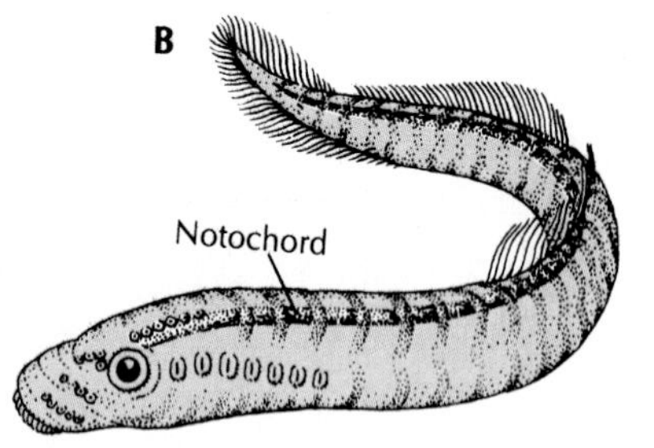

FIGURE 25-1

A, Structure of notochord and its surrounding sheaths. Cells of notochord proper are thick walled, pressed together closely, and filled with semifluid. Stiffness is caused mainly by turgidity of fluid-filled cells and surrounding connective tissue sheaths. This primitive type of endoskeleton is charactertistic of all chordates at some stage of the life cycle. The notochord provides longitudinal stiffening of the main body axis, a base for trunk muscles, and an axis around which the vertebral column develops.
B, In hagfishes and lampreys it persists throughout life, but in higher vertebrates it is largely replaced by the vertebrae. In humans slight remnants are found in nuclei pulposi of intervertebral discs. Its method of formation is different in the various groups of animals. In amphioxus it originates from the endoderm; in birds and mammals it arises as an anterior outgrowth of the embryonic primitive streak.

CHARACTERISTICS

1. Bilateral symmetry; segmented body; three germ layers; well-developed coelom
2. **Notochord** (a skeletal rod) present at some stage in life cycle
3. **Single, dorsal, tubular nerve cord;** anterior end of cord usually enlarged to form brain
4. **Pharyngeal gill slits** present at some stage in life cycle and may or may not be functional
5. **Postanal tail,** usually projecting beyond the anus at some stage and may or may not persist
6. **Segmented muscles** in an unsegmented trunk
7. **Ventral heart,** with dorsal and ventral blood vessels; closed blood system
8. Complete digestive system
9. A cartilage or bony **endoskeleton** present in the majority of members (vertebrates)

CLASSIFICATION

There are three subphyla under the phylum Chordata. Two of these subphyla are small, lack a vertebral column, and are of interest primarily as borderline or first chordates (protochordates). Since the subphyla lack a cranium, they are also referred to as Acrania. The third subphylum, Vertebrata, is subdivided into eight classes.

Phylum Chordata

Group Protochordata (Acrania)

Subphylum Urochordata (u'ro-kor-da'ta) (Gr. *oura,* tail, + L. *chorda,* cord, + *ata,* characterized by) **(Tunicata)—tunicates.** Notochord and nerve cord in free-swimming larva only; ascidian adults sessile, encased in tunic.

Subphylum Cephalochordata (sef'a-lo-kor-da'ta) (Gr. *kephalē,* head, + L. *chorda,* cord)—**lancelets** (amphioxus). Notochord and nerve cord found along entire length of body and persist throughout life; fishlike in form.

Group Craniata

Subphylum Vertebrata (ver'te-bra'ta) (L. *vertebratus,* backboned). Bony or cartilaginous vertebrae surrounding spinal cord; notochord in all embryonic stages, persisting in some of the fish; may also be divided into two groups (superclasses) according to presence of jaws.

Superclass Agnatha (ag'na-tha) (Gr. *a,* without, + *gnathos,* jaw) **(Cyclostomata)—hagfishes, lampreys.** Without true jaw or appendages.

Class Myxini (mik-sy'ny) (Gr. *myxa,* slime)—**hagfishes.** Terminal mouth with four pairs of tentacles; nasal sac with duct to pharynx; five to 15 pairs of gill pouches; partially hermaphroditic.

Class Cephalaspidomorphi (sef-a-lass'pe-do-morf'y) (Gr. *kephalē,* head, + *aspidos,* shield, + *morphē,* form)—**lampreys.** Sucking mouth with horny teeth; nasal sac not connected to mouth; seven pairs of gill pouches.

Superclass Gnathostomata (na'tho-sto'ma-ta) (Gr. *gnathos,* jaw, + *stoma,* mouth). With jaws and (usually) paired appendages—**jawed fishes, all tetrapods.**

Class Chondrichthyes (kon-drik'thee-eez) (Gr. *chondros,* cartilage, + *ichthys,* a fish)—**sharks, skates, rays, chimaeras.** Streamlined body with heterocercal tail; cartilaginous skeleton; five to seven gills with separate openings; no operculum; no swim bladder.

Class Osteichthyes (os'te-ik'thee-eez) (Gr. *osteon,* bone + *ichthys,* a fish)

Table 25-1 Divisions of the phylum Chordata

<table>
<tr><th>Urochordata (tunicates)</th><th>Cephalochordata (lancelets)</th><th>Myxini (hagfishes)</th><th>Cephalaspidomorphi (lampreys)</th><th>Chondrichthyes (sharks)</th><th>Osteichthyes (bony fishes)</th><th>Amphibia (amphibians)</th><th>Reptilia (reptiles)</th><th>Aves (birds)</th><th>Mammalia (mammals)</th></tr>
<tr><td colspan="10">Chordata</td></tr>
<tr><td colspan="2">Protochordata</td><td colspan="8">Vertebrata</td></tr>
<tr><td colspan="2">Acraniata</td><td colspan="8">Craniata</td></tr>
<tr><td colspan="2"></td><td colspan="2">Agnatha</td><td colspan="6">Gnathostomata</td></tr>
<tr><td colspan="4"></td><td colspan="2">Pisces</td><td colspan="4">Tetrapoda</td></tr>
<tr><td colspan="2"></td><td colspan="5">Anamniota</td><td colspan="3">Amniota</td></tr>
</table>

(Teleostomi) — **bony fishes.** Primitively streamlined body but variously modified; mostly ossified skeleton; single gill opening on each side covered with operculum; usually swim bladder or lung.

Class Amphibia (am-fib'e-a) (Gr. *amphi,* both or double, + *bios,* life) — **amphibians.** Ectothermic tetrapods; respiration by lungs, gills, or skin; development through larval stage; skin moist, containing mucous glands, and lacking scales.

Class Reptilia (rep-til'e-a) (L. *repere,* to creep) — **reptiles.** Ectothermic tetrapods possessing lungs; embryo develops within shelled egg; no larval stage; skin dry, lacking mucous glands, and covered by epidermal scales.

Class Aves (ay'veez) (L. pl. of *avis,* bird) — **birds.** Endothermic vertebrates with front limbs modified for flight; body covered with feathers; scales on feet.

Class Mammalia (ma-may'lee-a) (L. *mamma,* breast) — **mammals.** Endothermic vertebrates possessing mammary glands; body with hair; well-developed brain.

The basic taxonomic division of the phylum Chordata is only one of several systems used to characterize different groups within the phylum. These are summarized in Table 25-1. A fundamental separation is the Protochordata from the Vertebrata. Since the former lack a well-developed head, they are also called Acraniata. All vertebrates have a well-developed skull case enclosing the brain and are called Craniata.

Two basic branches of the vertebrates are the Agnatha, forms lacking jaws (hagfishes and lampreys), and Gnathostomata, forms having jaws (all other vertebrates). These two groups represent early separations within the Vertebrata. The Gnathostomata in turn can be subdivided into Pisces, jawed vertebrates with limbs (if any) in the shape of fins; and the Tetrapoda (Gr. *tetras,* four, + *podos,* foot), jawed vertebrates with two pairs of limbs. Still another fundamental separation among the vertebrates is based on embryological patterns. The embryos of reptiles, birds, and mammals develop within a special fluid-filled membranous sac, the amnion. These are called the Amniota, and their shelled eggs can be laid on land. Fishes and amphibians, the Anamniota, lack this important adaptation.

FOUR CHORDATE HALLMARKS

Four distinctive characteristics that set chordates apart from all other phyla are the **notochord; single, dorsal, tubular nerve cord; pharyngeal gill slits;** and

postanal propulsive tail. These characteristics are always found at some embryonic stage, although they may be altered or may disappear altogether in later stages of the life cycle.

Notochord

The notochord is a flexible, rodlike structure, extending the length of the body; it is the first part of the endoskeleton to appear in the embryo. The notochord is an axis for muscle attachment, and because it can bend without shortening, it permits undulatory movements of the body. In most of the protochordates and in primitive vertebrates, the notochord persists throughout life (Figure 25-1). In all jawed vertebrates a series of cartilaginous or bony vertebrae is formed from the connective tissue sheath around the notochord and replaces the notochord as the chief mechanical axis of the body.

Dorsal, tubular nerve cord

In most invertebrate phyla that have a nerve cord, it is ventral to the alimentary canal and is solid, but in the chordates the single cord is dorsal to the alimentary canal and notochord and is a tube (although the hollow center may be nearly obliterated during growth). In the vertebrates the anterior end becomes enlarged to form the brain. The hollow cord is produced in the embryo by the infolding of the ectodermal cells on the dorsal side of the body above the notochord. The nerve cord passes through the protective neural arches of the vertebrae, and the anterior brain is surrounded by a bony or cartilaginous cranium.

Pharyngeal gill slits

Pharyngeal gill slits are perforated slitlike openings that lead from the pharyngeal cavity to the outside. They are formed by the invagination of the outside ectoderm and the evagination of the endodermal lining of the pharynx. The two pockets break through when they meet to form the slit. In higher vertebrates these pockets may not break through, and only grooves are formed instead of slits; these are visible even in the human fetus. All traces of them usually disappear.

The perforated pharynx evolved as a filter-feeding apparatus and is used as such in the protochordates. Water with suspended food particles is drawn by ciliary action through the mouth and flows out through the gill slits where food is trapped in mucus. Later, in the vertebrates, ciliary action was replaced by a muscular pump that drives the water through the pharynx by expanding and contracting the pharyngeal cavity. Also modified were the blood vessels that carry blood through the gill bars. In the protochordates these are simple vessels surrounded by connective tissue. In the early fishes a capillary network was added with only thin, gas-permeable walls separating water outside from blood inside. These adaptations led to the development of **internal gills,** completing the conversion of the pharynx from a filter-feeding apparatus in protochordates to a respiratory organ in aquatic vertebrates.

Postanal tail

The postanal tail, together with somatic musculature and the stiffening notochord, provides the motility that larval tunicates and *Amphioxus* need for their free-swimming existence. As a structure added to the body behind the anus, it clearly has evolved specifically for propulsion in water. Its efficiency is later increased in fishes with the addition of fins. The tail is evident in humans only as a vestige (the coccyx, a series of small vertebrae at the end of the spinal column) but most other mammals have a waggable tail as adults.

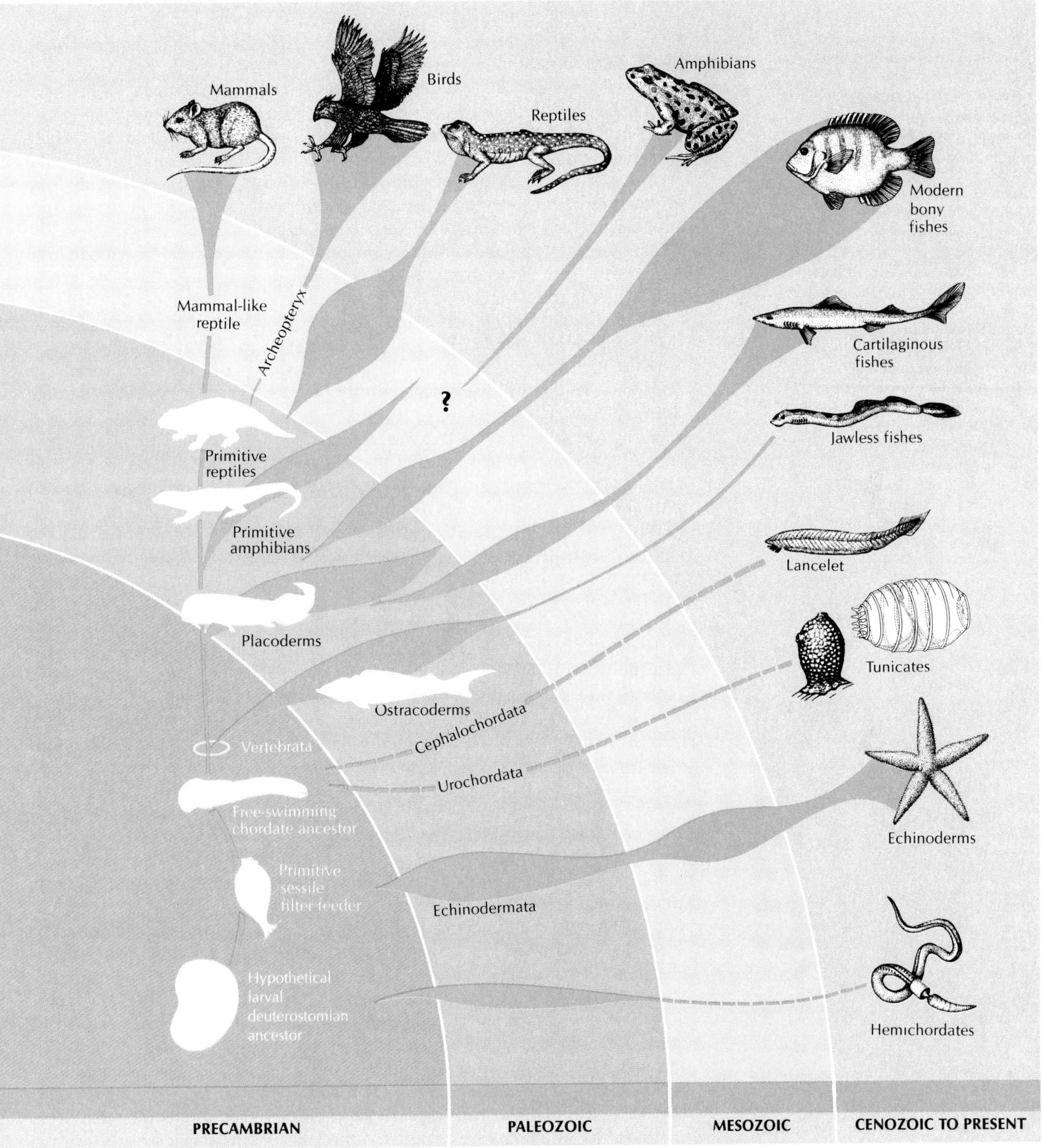

FIGURE 25-2

Family tree of the chordates, suggesting probable origin and relationships. Other schemes have been suggested and are possible. *White,* Extinct groups. *Black,* Living groups. *Tan,* Lines of descent. The relative success in numbers of species of each group through geological time, as indicated by the fossil record, is suggested by the bulging and thinning of that group's line of descent. Dashed lines indicate a poor or nonexistent fossil record.

■ Ancestry and Evolution of the Chordates

Most of the early efforts to pin together invertebrate and chordate kinship are now recognized as based on similarities due to analogy rather than homology. Analogous structures are those which perform similar functions but have altogether different origins (such as wings of birds and butterflies). Homologous structures on the other hand share a common origin but may look quite different (at least superficially) and perform quite different functions. For example, all vertebrate forelimbs are homologous because they are derived from a pentadactyl limb, even though they may be modified as differently as the human arm and a bird's wing. Homologous structures share a genetic heritage; analogous structures do not. Obviously, only homologous similarities have any bearing in ancestral connections.

Since the middle of the nineteenth century when the theory of organic evolution became the focal point for ferreting out relationships between groups of living organisms, zoologists have debated the question of chordate origins. It has been very difficult to reconstruct lines of descent because the earliest protochordates were in all probability soft-bodied creatures that stood little chance of being preserved as fossils even under the most ideal conditions. Consequently, such reconstructions largely come from the study of living organisms, especially from an analysis of early developmental stages that tend to be more insulated from evolutionary changes than the differentiated adult forms that they become.

The earliest speculations understandably focused on the most successful and in many respects most advanced of invertebrate groups, the Arthropoda. It was quickly recognized that, if an arthropod (a cockroach, for example) with its segmented body, ventral nerve cord, and dorsal heart were turned on its back, the basic plan of a vertebrate would result. Later, the ancestral award was transferred to the Annelida, because this group shares a basic body plan with the arthropods and in addition has an excretory system that strikingly resembles that of primitive vertebrates.

Although the annelid-vertebrate hypothesis continued to receive support as late as 1922, it contained unresolvable difficulties. An inverted annelid has its brain and mouth in the wrong relative positions. The annelid's nerve cord is ventral but connects to a dorsal brain by way of circumpharyngeal connectives through which the digestive tube passes (Figure 21-10, *A*, p. 459). When an annelid is inverted, as was required for the annelid-vertebrate hypothesis, the mouth ends up on top of the head and the brain below. Despite efforts to explain this discrepancy, the annelid hypothesis was eventually discarded as was the arthropod idea before it.

Early in this century when further theorizing became rooted in developmental patterns of animals, it soon became apparent that the chordates must have evolved from within the deuterostome branch of the animal kingdom. As explained earlier (p. 337), the Deuterostomia, a grouping that includes the echinoderms, hemichordates, chaetognaths, and chordates, share several important embryological features that clearly separate them from the Protostomia. Thus the deuterostomes are almost certainly a natural grouping of interrelated animals which have their common origin in ancient precambrian seas. There are several lines of evidence suggesting that somewhat later, at the base of the Cambrian period some 570 million years ago, the first distinctive chordates appeared, evolved from an early echinoderm or early hemichordate, or a common ancestor of both (Figure 25-2).

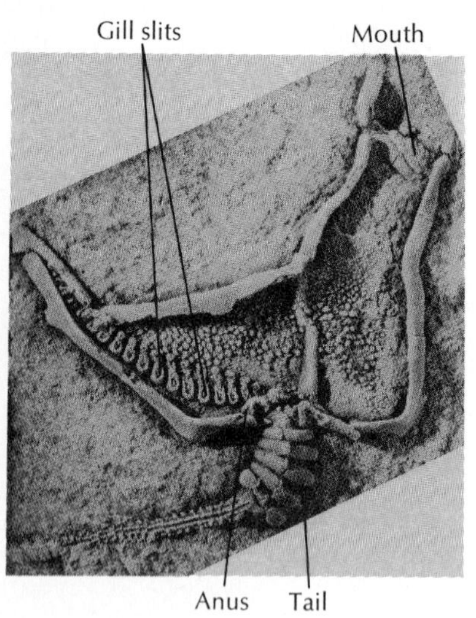

FIGURE 25-3

Fossil of a primitive echinoderm, *Cothurnocystis*, that lived during the Ordovician period (450 million years BP). It shows affinities with both echinoderms and chordates and may belong to a group that was ancestral to the chordates.

Courtesy of R.P.S. Jeffries, British Museum.

While modern echinoderms look nothing at all like modern chordates, there is one curious group of fossil echinoderms, the Stylophora (Gr. *stylos*, pillar, + *phoros*, to bear), which have gill slits and possibly other chordate attributes. These small nonsymmetrical forms have a head resembling a long-toed medieval boot, a series of branchial slits covered with flaps much like the gill openings of sharks, a postanal tail, and structures that might be interpreted as a notochord and muscle blocks (Figure 25-3). These creatures apparently used their gill slits for filter feeding, as do the primitive chordates today. Are these the long-sought chordate ancestors? Most zoologists remain unconvinced, contending that these fossil adult echinoderms are already too specialized to have given rise directly to the chordate line.

Although we may never know the identity of the chordate ancestor, we do know a great deal about two living protochordate groups that descended from it. These we will now consider.

Subphylum Urochordata (Tunicata)

The urochordates ("tail-chordates"), more commonly called tunicates, number some 1500 species. They are found in all seas from near the shoreline to great depths. Most of them are sessile as adults, although some are free living. The name "tunicate" is suggested by the usually tough, nonliving **tunic,** or test, that surrounds the animal and contains cellulose. As adults, tunicates are highly specialized chordates, for in most species only the larval form, which resembles a microscopic tadpole, bears all the chordate hallmarks. During adult metamorphosis, the notochord (which, in the larva, is restricted to the tail, hence the group name Urochordata) and the tail disappear altogether, while the dorsal nerve cord becomes reduced to a single ganglion.

Urochordata is divided into three classes—**Ascidiacea** (Gr. *askiolion,* little bag, + *acea,* suffix), **Larvacea** (L. *larva,* ghost, + *acea,* suffix), and **Thaliacea** (Gr. *thalia,* luxuriance, + *acea,* suffix). Of these, the members of **Ascidiacea,** commonly known as the ascidians, or sea squirts, are by far the most common and are the best known. Ascidians may be solitary, colonial, or compound. Each of the solitary and colonial forms has its own test, but among the compound forms many individuals may share the same test. In some of these compound ascidians each member has its own incurrent siphon, but the excurrent opening is common to the group.

FIGURE 25-4

An adult "sea squirt," the sea vase *Ciona intestinalis.* Ascidians are popularly called sea squirts because prodding or squeezing them provokes spurts of water from the siphons. The remarkably transparent tunic of this species clearly reveals vertical muscle bands in the mantle and the branchial sac inside.
Photograph by C.P. Hickman, Jr.

The typical solitary ascidian (Figure 25-4) is globose in form and is attached by its base to hard substrates such as rocks, pilings, or the bottoms of ships. Lining the tunic is an inner membrane, the **mantle.** On the outside are two projections: the **incurrent** and **excurrent siphons** (Figure 25-5). Water enters the incurrent siphon and passes into the branchial sac (pharynx) through the mouth. On the midventral side of the branchial sac is a groove, the endostyle, which is ciliated and secretes mucus. As the mucous sheet is carried by cilia across the inner surface of the pharynx to the dorsal side, it sieves out small food particles from the water passing through the slits in the wall of the branchial sac. Then the mucus with its entrapped food is collected and passed posteriorly to the esophagus. The water, now largely cleared of food particles, is driven by cilia into the atrial cavity and finally out the excurrent siphon. The intestine leads to the anus near the excurrent siphon.

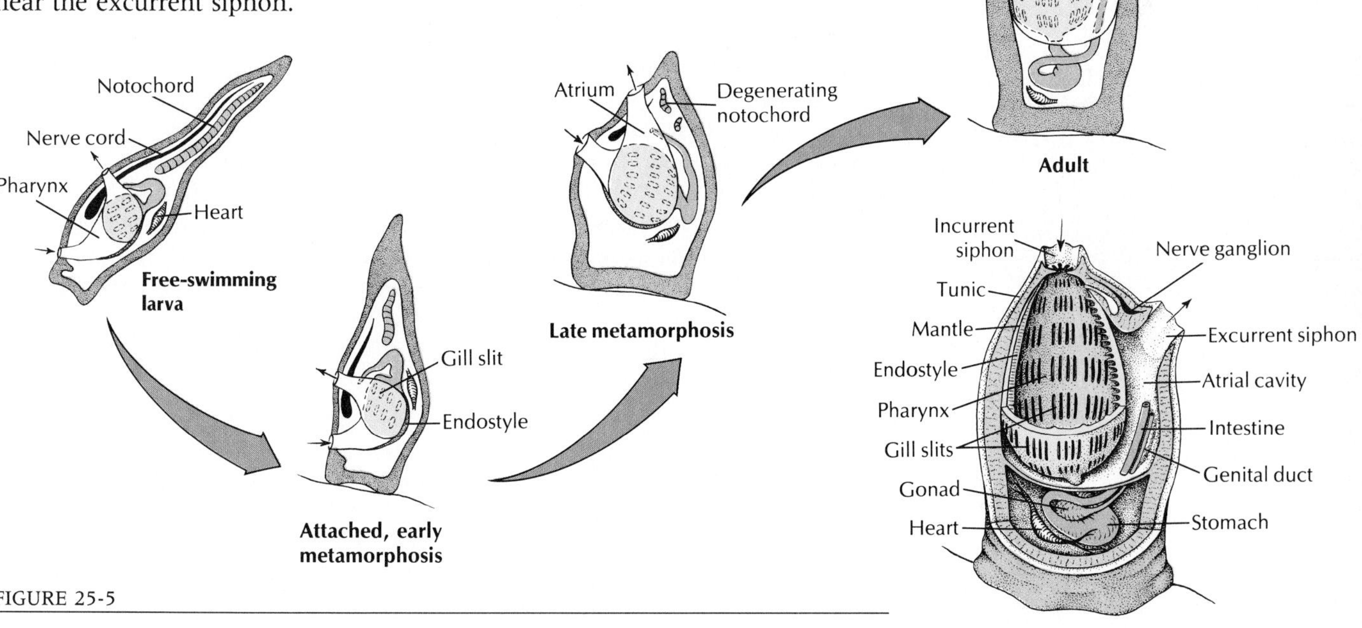

FIGURE 25-5

Metamorphosis of a solitary ascidian from a free-swimming tadpole larva stage.

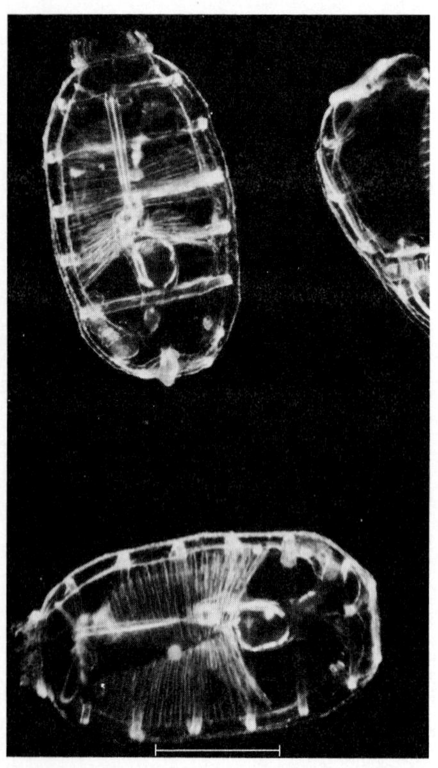

FIGURE 25-6

A thaliacean *Doliolum nationalis*—a member of the class Thaliacea.

Photograph by D.P. Wilson.

The circulatory system contains a ventral **heart** near the stomach and two large vessels, one on either side of the heart. The action of the heart is peculiar in that it drives the blood first in one direction for a period of time, then reverses and drives the blood in the opposite direction. This reversal of blood flow is found in no other chordate. The excretory system is a type of nephridium near the intestine. The nervous system is restricted to a nerve ganglion and a few nerves that lie on the dorsal side of the pharynx. A notochord is lacking. The animals are hermaphroditic. The germ cells are carried out the excurrent siphon into the surrounding water, where cross-fertilization occurs.

Of the four chief characteristics of chordates, adult sea squirts have only one, the pharyngeal gill slits. However, the larval form gives away the secret of their true relationship. The tadpole larva (Figure 25-5) is an elongate, transparent form with a head and all four chordate characteristics: a notochord, hollow dorsal nerve cord, propulsive postanal tail, and a large pharynx with endostyle and gill slits. The larva does not feed but swims about for some hours before fastening itself vertically by its adhesive papillae to some solid object. It then metamorphoses to become the sessile adult.

The remaining two classes of the Urochordata—**Larvacea** and **Thaliacea**—are mostly small, transparent animals of the open sea (Figure 25-6). Some are small tadpolelike forms resembling the larval stage of ascidians. Others are spindle shaped or cylindrical forms surrounded by delicate muscle bands. They are mostly carried along by the ocean currents and as such form a part of the plankton. Many are provided with luminous organs and emit a beautiful light at night.

Subphylum Cephalochordata

The cephalochordates are the marine lancelets: slender, laterally compressed, translucent animals about 5 to 7 cm in length (Figure 25-7) that inhabit the sandy bottoms of coastal waters around the world. Lancelets originally bore the generic name *Amphioxus* (Gr. *amphi,* both ends, + *oxys,* sharp) but later surrendered by priority to *Branchiostoma* (Gr. *branchia,* gills, + *stoma,* mouth). This left amphioxus as a convenient colloquial name for all of the approximately 25 species in the diminutive subphylum. Four species of amphioxus are found in North American coastal waters.

Amphioxus is especially interesting because it has the four distinctive characteristics of chordates in simple form, and in other ways it may be considered an early blueprint of the phylum. Water enters the mouth, driven by cilia in the buccal cavity, then passes through numerous gill slits in the pharynx where food is trapped in mucus, which is then moved by cilia into the intestine. Here the smallest food particles are separated from the mucus and passed into the midgut caecum (diverticulum) where they are phagocytized and digested intracellularly. As in the tunicates, the filtered water passes first into an atrium, then leaves the body by an atriopore (the excurrent siphon of tunicates).

The closed circulatory system is complex for so simple a chordate. The flow pattern is remarkably similar to that of the primitive fishes, although there is no heart. Blood is pumped forward in the ventral aorta by peristaltic-like contractions of the vessel wall, then passes upward through the branchial arteries (aortic arches) in the gill bars to the dorsal aorta. From here the blood is distributed to the body tissues by capillary plexi and then is collected in veins, which return it to the ventral aorta. The blood is colorless, lacking both erythrocytes and hemoglobin.

The nervous system is centered around a hollow nerve cord lying above the notochord. Pairs of spinal nerve roots emerge at each trunk myomeric (muscle) segment. Sense organs are simple, unpaired bipolar receptors located in various

A

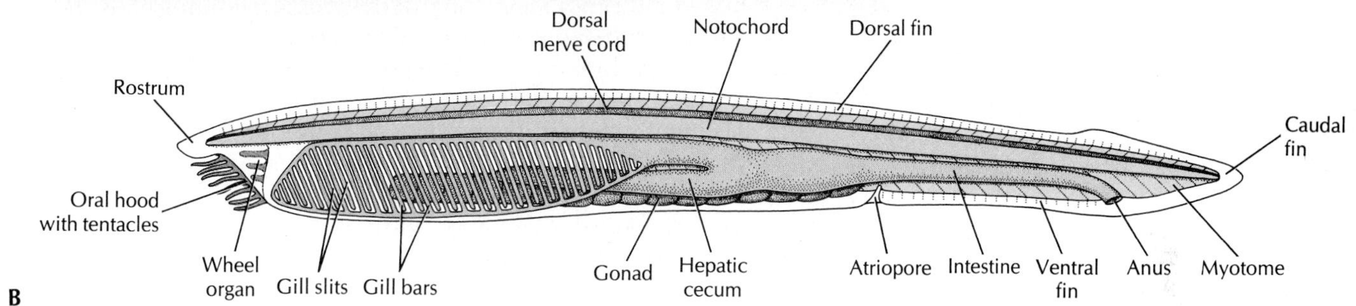

B

FIGURE 25-7

Amphioxus. This interesting bottom-dwelling cephalochordate possesses the four distinctive chordate characteristics (notochord, dorsal nerve cord, pharyngeal gill slits, and postanal tail). The vertebrate ancestor is thought to have had a similar body plan. **A,** Living specimen. **B,** External structure.

parts of the body. The "brain" is a simple vesicle at the anterior end of the nerve cord.

Sexes are separate in amphioxus. The sex cells are set free in the atrium, then pass out the atriopore to the outside where fertilization occurs. The larvae hatch soon after the eggs are fertilized and gradually assume the shape and size of adults.

Just where amphioxus belongs in the phylogeny of chordates is a much disputed point. As pointed out in the prologue to this chapter (p. 538), amphioxus is no longer considered to lie in the main line of chordate descent. Nevertheless, no other chordate shows the basic diagnostic characteristics of the chordates so well. In addition to the four chordate anatomical hallmarks, amphioxus possesses several structural features that foreshadow the vertebrate plan. Among these are the midgut diverticulum that secretes digestive enzymes, segmented trunk musculature, separation of dorsal and ventral spinal roots in the vertebrate style, and the basic circulatory pattern of more advanced chordates. The earliest vertebrate ancestor may well have emerged from a larval form that closely resembled that of amphioxus.

■ Subphylum Vertebrata

The third subphylum of the chordates is the large and eminently successful Vertebrata, the subject of the next five chapters of this book. The subphylum Vertebrata shares the basic chordate characteristics with the other two subphyla, but in addition it has a number of features that the others do not share. The characteristics that give the members of this group the name "Vertebrata" or "Craniata" are a spinal column of vertebrae, which forms the chief skeletal axis of the body, a braincase, or cranium.

CHARACTERISTICS

1. Chief diagnostic features of chordates—**notochord, dorsal nerve cord, pharyngeal gill slits,** and **postanal tail**—all present at some stage of the life cycle

2. **Integument** basically of two divisions, an outer **epidermis** of stratified epithelium from the ectoderm and an inner **dermis** of connective tissue derived from the mesoderm; many modifications of skin among the various classes, such as glands, scales, feathers, claws, horns, and hair
3. Notochord more or less replaced in jawed vertebrates by the spinal column of vertebrae composed of cartilage or bone or both; distinctive **endoskeleton** consisting of vertebral column with the cranium, visceral arches, limb girdles, and two pairs of jointed appendages
4. **Many muscles** attached to the skeleton to provide for movement
5. Complete digestive system ventral to the spinal column and provided with large digestive glands, liver, and pancreas
6. Circulatory system consisting of the **ventral heart** of two to four chambers; a closed blood vessel system of arteries, veins, and capillaries; blood fluid containing **red blood corpuscles** with hemoglobin and white corpuscles; paired aortic arches connecting the ventral and dorsal aortas and giving off branches to the gills among the aquatic vertebrates; in the terrestrial types modification of the aortic arch plan into pulmonary and systemic systems
7. Well-developed **coelom** largely filled with the visceral systems
8. Excretory system consisting of **paired kidneys** (mesonephric or metanephric types in adults) provided with ducts to drain the waste to cloaca or anal region
9. Brain typically divided into five vesicles
10. Ten or twelve pairs of **cranial nerves** with both motor and sensory functions usually; a pair of spinal nerves for each primitive myotome (segmental block of striated muscle); an **autonomic nervous system** in control of involuntary functions of internal organs
11. **Endocrine system** of ductless glands scattered through the body
12. Nearly always separate sexes; each sex containing paired gonads with ducts that discharge their products either into the cloaca or into special openings near the anus
13. **Body plan** consisting typically of **head, trunk,** and **postanal tail; neck** present in some, especially terrestrial forms; two pairs of appendages usually, although entirely absent in some; coelom divided into a pericardial space and a general body cavity; mammals with a thoracic cavity

ADAPTATIONS THAT HAVE GUIDED VERTEBRATE EVOLUTION

From the earliest fishes to the most advanced mammals, the evolution of the vertebrates has been guided by the specialized basic adaptations of the living endoskeleton, efficient respiration, advanced nervous system, and paired limbs.

Living endoskeleton

The endoskeleton of vertebrates, as in the echinoderms, is an internal supportive structure and framework for the body. This is a departure in animal architecture, since invertebrate skeletons generally enfold the body. Exoskeletons and endoskeletons have their own particular set of advantages and limitations which are related to size (see marginal note on p. 159). For vertebrates the living endoskeleton possesses an overriding advantage over the secreted, nonliving exoskeleton of arthropods: growing with the body as it does, the endoskeleton permits almost unlimited body size with much greater economy of building materials. Some vertebrates have become the most massive animals on earth. The endoskeleton forms an excellent jointed scaffolding for muscles, and the muscles in turn protect the skeleton and cushion it from potentially damaging impact.

We should note that the vertebrates have not wholly lost the protective function of a firm external covering. The skull and the thoracic rib cage enclose and protect vulnerable organs. Most vertebrates are further protected with a tough integument, often bearing nonliving structures such as scales, hair, and feathers that may provide insulation as well as physical security.

Pharynx and efficient respiration

The replacement of the ciliated filter-feeding apparatus of the protochordates with a muscular buccal pump and internal gills provided the vertebrates with an exceedingly effective respiratory device. Some fish gills can extract as much as 85% of the oxygen from the water in its single pass across the gill surface, an efficiency that far surpasses even the best invertebrate gill. This improved respiratory efficiency was essential to support the elevated metabolism of the active, predatory vertebrates.

Advanced nervous system

No single system in the body is more strongly associated with functional and structural advancement than is the nervous system. The protochordate nervous system consisted of a brainless nerve cord and rudimentary sense organs, which were for the most part chemosensory in function. When the protovertebrates switched to a predatory life style, new sensory, motor, and integrative controls were now essential for the location and capture of larger prey items. In short, the protovertebrates developed a new head, complete with a brain and external paired sense organs especially designed for distance reception. These included paired eyes with lenses and inverted retinas; pressure receptors, such as paired ears designed for equilibrium and later redesigned to include sound reception; and chemical receptors, including taste receptors and exquisitely sensitive olfactory organs.

Paired limbs

Pectoral and pelvic appendages are present in most vertebrates in the form of paired fins or legs. They originated as swimming stabilizers and later became prominently developed into legs for travel on land. Jointed limbs are especially suited for life on land because they permit finely graded levering motions against a substrate.

Ancestry and Evolution of the Vertebrates

CANDIDATES FOR THE VERTEBRATE ANCESTRAL STOCK

As we have seen, many developmental similarities between the protochordates and echinoderms seem to place the echinoderms firmly in the position of the ancestral stock to the chordates. But what is the origin of the vertebrates, the most spectacularly successful group in the animal kingdom?

Position of amphioxus

There is agreement among students of evolution that the vertebrates arose gradually from a ciliated filter-feeding animal resembling protochordates. Without question, the most important development in the transition to the mobile vertebrate predatory life-style was the emergence of a distinct head. This new head appears as an addition to the existing protochordate body. With it came three structural advancements, all crucial to the subsequent success of the vertebrates: (1) an improved nervous system with complex paired sense organs for the detection of prey and enhanced environmental awareness, (2) a well-developed cranial skeleton and, later, jaws for the capture of prey, and (3) a muscular buccal pump and effective gill circulation for efficient gas exchange.

Although none of these structures is present in any protochordate, amphioxus was long considered to be the most suitable structural ancestor of the vertebrates for the reasons mentioned earlier. But amphioxus fell from favor when it was recognized that it is too specialized to provide a suitable framework for the vertebrate body plan. Still, many zoologists today believe that the vertebrates arose gradually from a filter-feeding chordate that *resembled* present day cephalochordates.

Garstang's hypothesis of chordate larval evolution

Neoteny is well known among living amphibians. For example, the larvae of the American axolotl (a salamander) may become sexually mature and reproduce without ever undergoing metamorphosis. In some lakes, generation after generation reproduce in the neotenous larval state, and true adults never occur. In other lakes having different environmental conditions, the axolotls metamorphose to reproducing adults. Other examples of neoteny are known; indeed, many believe the human species is neotenous! Certainly, neoteny *could* have occurred among the early chordates, and if it did, the sessile adult stage would be dropped, leaving a new free-swimming chordate.

After amphioxus, attention in the early part of this century focused on the alternative protochordate group, the tunicates. At first glance, more unlikely candidates for vertebrate ancestors could hardly be imagined. As adults, most tunicates superficially resemble sponges far more than they resemble any vertebrate. They even lack most of the anatomical hallmarks that would distinguish them as chordates. Curiously, however, their tadpole larvae have all the right qualifications for membership into the prevertebrate club: notochord, hollow dorsal nerve cord (vesicle), gill slits, postanal tail, brain, and sense organs (balance organ and a single eye complete with lens). These larvae, being motile for a few hours, are able to seek out new habitats before settling down to metamorphose into sessile adults, thus helping to spread the species far and wide.

In 1928 W. Garstang in England suggested that at some point in the ancestry of the tunicates, some tadpole larvae "forgot" that they were supposed to metamorphose into adults and became sexually mature before they reached the adult stage. This phenomenon—accelerated development of the reproductive system—is called **neoteny.** Then, with continued larval evolution, they began a new life mode as completely planktonic organisms. Such an animal, Garstang contended, would be the ancestral vertebrate (Figure 25-8).

Garstang departed from previous thinking by suggesting that evolution may occur in the larval stages of animals. Zoologists were slow to accept this idea because they knew that early developmental stages of animals are well insulated against evolutionary change. Nevertheless, neoteny is known to occur in several groups of animals (see marginal note), including some groups of the modern protochordates.

Thus many zoologists believe that the transition between the protochordates and the earliest vertebrates occurred by neoteny through a primitive protochordate larva resembling that of an ascidian or amphioxus. Unable to narrow the search further, we turn to the earliest known chordate fossils and to the vertebrates themselves.

THE EARLIEST VERTEBRATES—JAWLESS OSTRACODERMS

The earliest vertebrate fossils are fragments of bony armor discovered in Ordovician rock in Russia and in the United States. They were small, jawless creatures

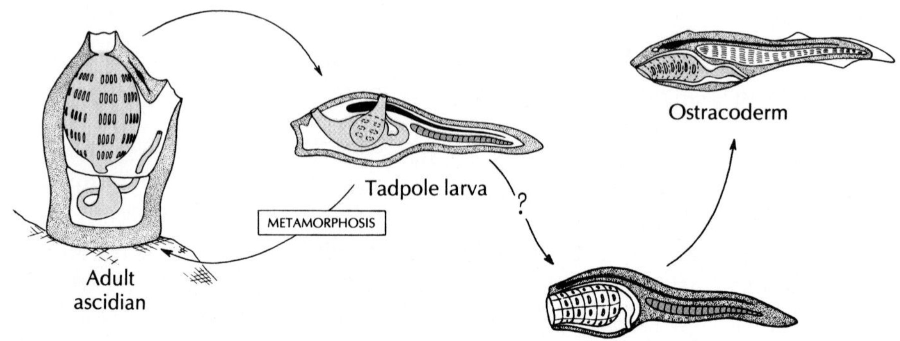

FIGURE 25-8

Garstang's hypothesis of vertebrate origins. Adult ascidians live on the sea floor but reproduce through a free-swimming tadpole larva. More than 500 million years ago, some larvae began to reproduce in the swimming stage. These evolved into the ostracoderms, the first known vertebrates.

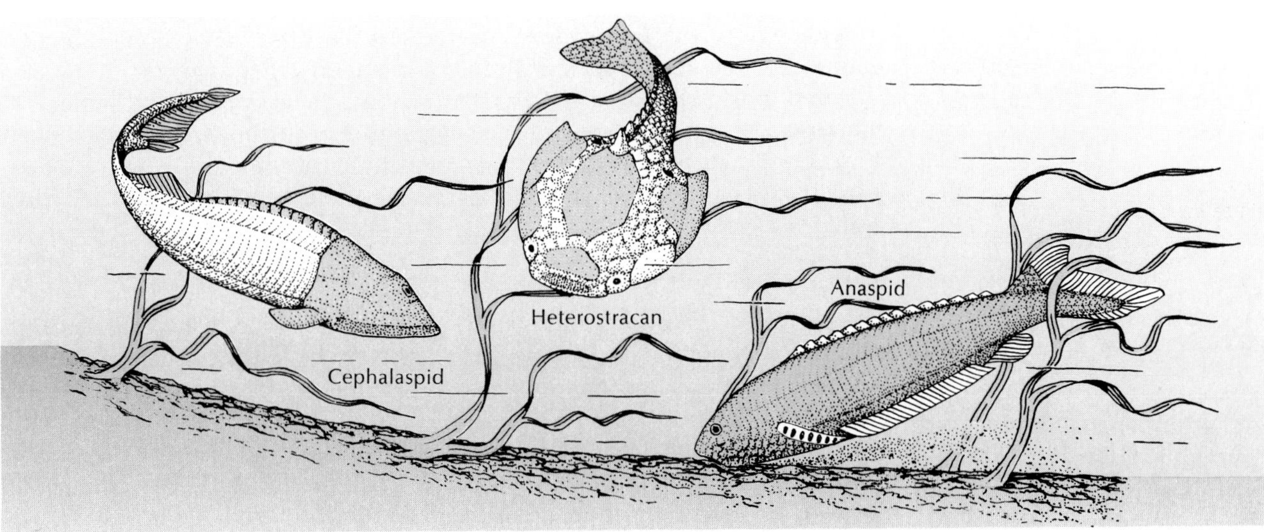

FIGURE 25-9

Three ostracoderms, jawless fishes of Silurian and Devonian times. Representatives of three of the best known ostracoderm groups are illustrated as they might have appeared while feeding on the floor of a Devonian sea. They are depicted here as filter feeders, drawing water and organic debris in the mouth, straining out the organic matter, and expelling the water through the gill openings and between the ventral head plates. However, some aspects of their anatomy suggest they might have been predators of soft-bodied animals.

collectively called ostracoderms (os-trak'o-derm) (Gr. *ostrakon,* shell, + *derma,* skin), which belongs to the Agnatha division of the vertebrates. These earliest ostracoderms, called **heterostracans** (Figure 25-9), lacked paired lateral fins that subsequent fishes found so important for stability. Their swimming movements must have been clumsy, although sufficient to propel them along the ocean bottom where they searched for food. They were probably filter-feeders, although their highly differentiated brain and relatively advanced sense organs suggest to some authorities that they may have been mobile predators that fed on soft-bodied animals. During the middle of the Paleozoic era, the heterostracans underwent a major radiation, but were eventually eclipsed by another ostracoderm group, the **cephalaspids.**

The cephalaspids improved the efficiency of a benthic life by evolving paired fins. These fins, located just behind the head (Figure 25-9), provided control over pitch and yaw that ensured well-directed forward movement. A typical cephalaspid was a small fish covered with armor: the head by a solid shield and the body by bony plates. However, it lacked an axial skeleton and vertebrae, still depending on the notochord for axial support. It was also jawless and toothless.

A third group of ostracoderms were the anaspids (Figure 25-9). Even smaller than the cephalaspids, they were more streamlined and more closely resembled modern-day jawless fishes (the lamprey, for example) than any other ostracoderm.

FIGURE 25-10

Early jawed fishes of the Devonian period, 400 million years ago. The placoderm (*left*) and a related acanthodian (*right*) were highly mobile and voracious forms. Although more successful than the less maneuverable ostracoderms, they eventually failed in competition with their successors, the bony and cartilaginous fishes.

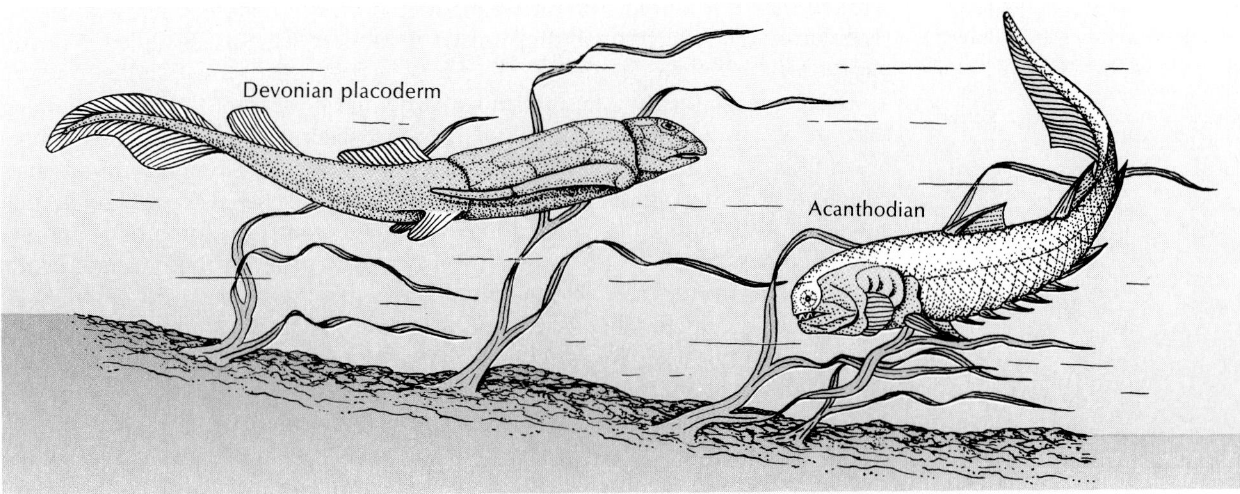

As a group, the ostracoderms were basically fitted for a simple, bottom-feeding life. Yet, despite their anatomical limitations, they enjoyed a respectable radiation in the Silurian and Devonian periods. Their overall contribution was enormous, because they provided a blueprint for subsequent vertebrate evolution. But they could not survive the competition of the more advanced jawed fishes that began to dominate the Devonian period, and in the end they disappeared.

EARLY JAWED VERTEBRATES

All jawed vertebrates, whether extinct or living, are collectively called gnathostomes ("jaw mouth") in contrast to jawless vertebrates, the agnathans ("without jaw"). The latter are also often referred to as cyclostomes ("circle mouth").

The first jawed vertebrates to appear in the fossil record were the **placoderms** (plak'o-derm) (Gr. *plax,* plate, + *derma,* skin) (Figure 25-10). The advantages of jaws are obvious; they allow predation on large and active forms of food. Possessors of jaws would enjoy a great advantage over jawless vertebrates, which were restricted to a wormlike existence of sifting out organic debris and small organisms in the bottom mud.

Jaws arose through modifications of the first two of the serially repeated cartilaginous gill arches. The beginnings of this trend can, in fact, be seen in some of the jawless ostracoderms where the mouth became bordered by strong dermal plates that could be manipulated somewhat like jaws with the gill arch musculature. The more anterior arches were gradually modified to permit more efficient seizing, and the skin surrounding the mouth was modified into teeth. Eventually the anterior gill arches became bent into the characteristic position of vertebrate jaws, as seen in the placoderms (Figure 25-11).

Placoderms and their relatives evolved into a great variety of forms, some large and grotesque in appearance. They were armored fish covered with diamond-shaped scales or with large plates of bone. All became extinct by the end of the Paleozoic era.

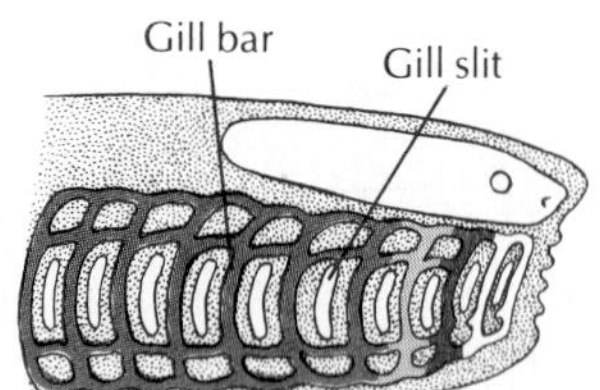

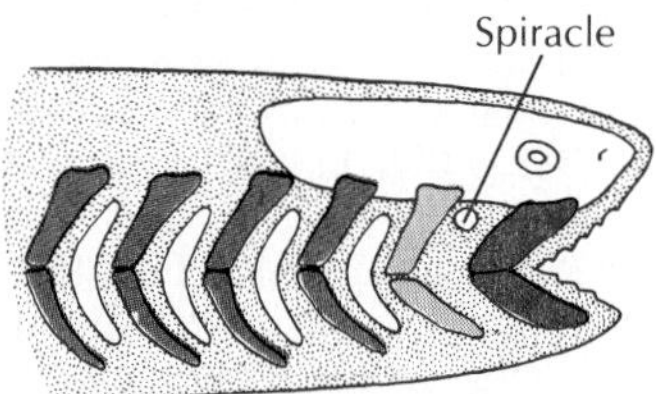

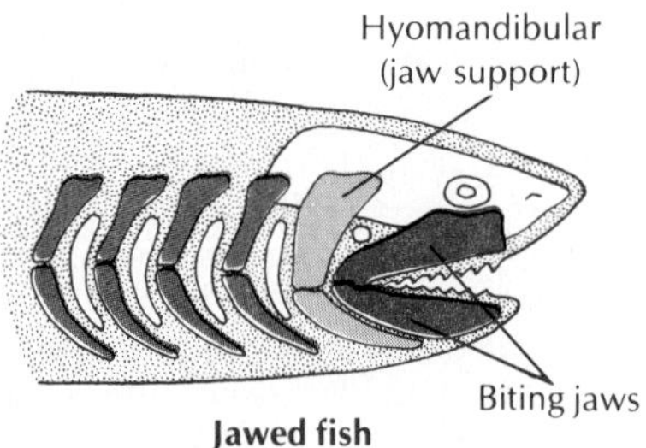

FIGURE 25-11

How the vertebrates got their jaw. The mud-loving Silurian fish converted the first gill opening into a spiracle that opened on the upper head, allowing water to be drawn in without fouling the gills. The first gill bars, no longer required for gill support, became enlarged and armed with teeth. Relics of this transformation are seen during the development of sharks.

EVOLUTION OF MODERN FISHES AND TETRAPODS

Reconstruction of the origins of the vast and varied assemblage of modern living vertebrates is, as we have seen, based largely on fossil evidence. Unfortunately the fossil evidence for the earliest vertebrates is often incomplete and tells us much less than we would like to know about subsequent trends in evolution. Affinities become much easier to establish as the fossil record improves. For instance, the descent of birds and mammals from reptilian ancestors has been worked out in a highly convincing manner from the relatively abundant fossil record available. By contrast, the ancestry of modern fishes is shrouded in uncertainty.

Despite the difficulty of establishing early lines of descent for the vertebrates, they are clearly a natural, monophyletic group, distinguished by a great number of common characteristics. They have almost certainly descended from a common ancestor, the nature of which we have already discussed. Very early in their evolution, the vertebrates divided into two great stems, the agnathans and the gnathostomes. These two groups differ from each other in many fundamental ways, in addition to the obvious lack of jaws in the former group and their presence in the latter. Thus both groups are very old and of approximately the same age. On this basis we cannot say that agnathans are more "primitive" than gnathostomes, even though the latter have continued on a marvelous evolutionary advance that produced most of the modern fishes, all of the tetrapods, and the reader of this book. Although the agnathans are represented today only by the hagfishes and the lampreys, these creatures too are successful in their own way.

Summary

The phylum Chordata is named for the rodlike notochord that forms a stiffening body axis at some stage in the life cycle of every chordate. All chordates share four distinctive hallmarks that set them apart from all other phyla: notochord, dorsal tubular nerve cord, pharyngeal gill slits, and postanal tail. Two of the three chordate subphyla are invertebrates and lack a well-developed head. They are the Urochordata (tunicates), most of which are sessile as adults, but all of which have a free-swimming larval stage; and the Cephalochordata (lancelets), fishlike forms that include the famous amphioxus.

The chordates are believed to have descended from echinoderm-like ancestors probably in the Precambrian period, but the true origin of the chordates is not yet, and may never be, known with certainty. Taken as a whole, the chordates have a greater fundamental unity of organ systems and body plan than have many of the invertebrate phyla.

The subphylum Vertebrata, backboned members of the animal kingdom, are characterized as a group by having a well-developed head, and by their comparatively large size, high degree of motility, and distinctive body plan, which embodies several distinguishing features that have led to the eminent evolutionary success of the group. Most important of these are the living endoskeleton that allows continuous growth and provides a sturdy framework for efficient muscle attachment and action; a pharynx perforated with gill slits (lost or greatly modified in higher vertebrates) with vastly increased respiratory efficiency; advanced nervous system with clear separation of the brain and spinal cord; and paired limbs. The vertebrates are believed to have evolved by neoteny from a protochordate larval ancestor such as the ascidian tadpole larva or the larval amphioxus.

A compound sea squirt, *Botryllus schlosseri*, common in shallow coastal waters and rock tide pools. Each of the star-shaped patterns represents a colony of individuals united centrally where they share a common test. (Approximately ×2.)
Photograph by D.P. Wilson.

REVIEW QUESTIONS

1. What is the evidence that the four deuterostome phyla — Echinodermata, Chaetognatha, Hemichordata, and Chordata — should be considered a natural grouping?
2. Name four hallmarks shared by all chordates, and explain the function of each.
3. Name the two invertebrate chordate subphyla, and give three distinguishing characteristics of each.
4. Both sea squirts (urochordates) and lancelets (cephalochordates) are filter feeders. Describe the feeding apparatus of a sea squirt and explain in what ways its mode of feeding is similar to, and different from, that of amphioxus.
5. Explain why it is necessary to know the life history of a sea squirt to understand why sea squirts are chordates.
6. Explain why the body plan of amphioxus has been considered to represent a blueprint of the early vertebrate body plan.
7. Distinguish among the terms Acraniata, Craniata, Agnatha, and Gnathostomata.
8. List four adaptations that guided vertebrate evolution, and explain how each has contributed to the success of vertebrates.
9. Discuss Garstang's hypothesis of chordate larval evolution.
10. Distinguish between ostracoderms and placoderms.
11. Describe the evolutionary origin of the vertebrate jaw.

SELECTED REFERENCES

Alldredge, A. 1976. Appendicularians. Sci. Am. **235**:94-102 (July). *Describes the biology of larvaceans which build gossamer houses for trapping food.*

Bone, Q. 1979. The origin of chordates. Oxford Biology Readers, No. 18. New York, Oxford University Press. *Excellent synthesis of hypotheses and range of disagreements bearing on an unsolved riddle.*

Jeffries, R.P.S. 1986. The ancestry of the vertebrates. Cambridge, Cambridge University Press. *Jeffries argues that the stylophorans (which he calls Calcichordata) are the direct ancestors of the vertebrates, a view that most zoologists are not yet willing to accept. Still, this book is an excellent summary of the various deuterostome groups and of the competing hypotheses on vertebrate ancestry.*

Northcutt, R.G., and C. Gans. 1983. The genesis of neural crest and epidermal placodes: a reinterpretation of vertebrate origins. Quart. Rev. Biol. **58**:1-28. *Describes the protochordate-vertebrate transition and the origin of the vertebrate head.*

WHAT IS A FISH?

In common (and especially older) usage, the term "fish" has often been used to describe a member of a mixed assortment of water-dwelling animals. We speak of jellyfish, cuttlefish, starfish, crayfish, and shellfish, knowing full well that when we use *fish* in such combinations we are not referring to true fish. In earlier times, even biologists did not make such a distinction. Sixteenth-century natural historians classified seals, whales, amphibians, crocodiles, even hippopotamuses, as well as a host of aquatic invertebrates, as fishes. Later biologists were more discriminating, eliminating first the invertebrates and then the higher vertebrates from the concept. Today we recognize fishes as gill-breathing, ectothermic, aquatic vertebrates that possess fins, and have skin that is usually covered with scales.

Fishes are the undisputed masters of the aquatic environment. Because fishes live in a habitat that is basically alien to humans, people have rarely appreciated the remarkable success of these vertebrates. Plato considered fishes "senseless beings . . . which have received the most remote habitations as a punishment for their extreme ignorance." The average North American today is apt to be unconscious of and uninformed about fishes, a perspective further nurtured by wildlife television shows and popular books on natural history that devote most attention to birds and mammals. Nevertheless, whether appreciated by humans or not, the world's fishes have enjoyed an effusive proliferation that has produced nearly 23,000 species—more than all other species of vertebrates put together—with adaptations that have fitted them to almost every conceivable aquatic niche. No other animal group threatens their domination of the seas.▪

CHAPTER 26

Fishes

The life of a fish is bound to its body form. Their mastery of river, lake, and ocean is revealed in the many ways that fishes have harmonized their life design to the physical properties of their aquatic surroundings. Suspended in a medium that is eight hundred times more dense than air, a trout or pike can hang motionless without using energy, varying its neutral buoyancy by adding or removing air from the swim bladder. Or it may dart forward or at angles, using its fins as brakes and tilting rudders. With excellent organs for salt and water exchange, bony fishes can steady and fine-tune their body fluid composition in their chosen freshwater or seawater environment. Fish gills are the most effective respiratory devices in the animal kingdom for extracting oxygen from a medium that contains less than 1/20 as much oxygen as air. Fishes have excellent olfactory and visual senses and a unique lateral line system, which with its exquisite sensitivity to water currents and vibrations provides a "distance touch" in water. Thus, in mastering the physical problems of their element, fishes evolved a basic body plan and set of physiological strategies that both shaped and constrained the evolution of the vertebrate groups that descended from them.

Ancestry and Relationships of Major Groups of Fishes

The fishes are of ancient ancestry, having descended from an unknown common ancestor that may have arisen from a free-swimming larval tunicate (hypotheses of chordate and vertebrate origins were described in Chapter 25). Whatever their origin, during the Cambrian period, or perhaps even in the Precambrian period more than 600 million years ago, the earliest fishlike vertebrates branched into the jawless agnathans and the jawed gnathostomes (Figure 26-1). All vertebrates have descended from one or the other of these two ancestral stems.

The agnathans include the extinct ostracoderms (p. 550) and living **hagfishes** and **lampreys.** The ancestry of hagfishes and lampreys is uncertain. Although they look superficially similar, they are so different from each other in form and function that they are placed in separate classes by most ichthyologists.

All the remaining fishes have jaws (gnathostomes). They have all descended from one or more early jawed ancestors, possibly from the placo-

FIGURE 26-1

Family tree of the fishes, showing the evolution through geological time. Widened areas in the lines of descent indicate periods of adaptive radiation and the relative success of each group. The lungfishes, for example, flourished in the Devonian period, but declined and are today represented by only three surviving genera. The sharks and rays radiated during the Carboniferous period and at that time were rulers of the sea (their successful contemporaries, the early chondrosteans, were freshwater fishes). The sharks came dangerously close to extinction during the Permian period but staged a recovery in the Mesozoic era and are a secure group today. Johnny-come-latelies in fish evolution are the spectacularly successful modern fishes, or teleosts, which make up most living fishes.

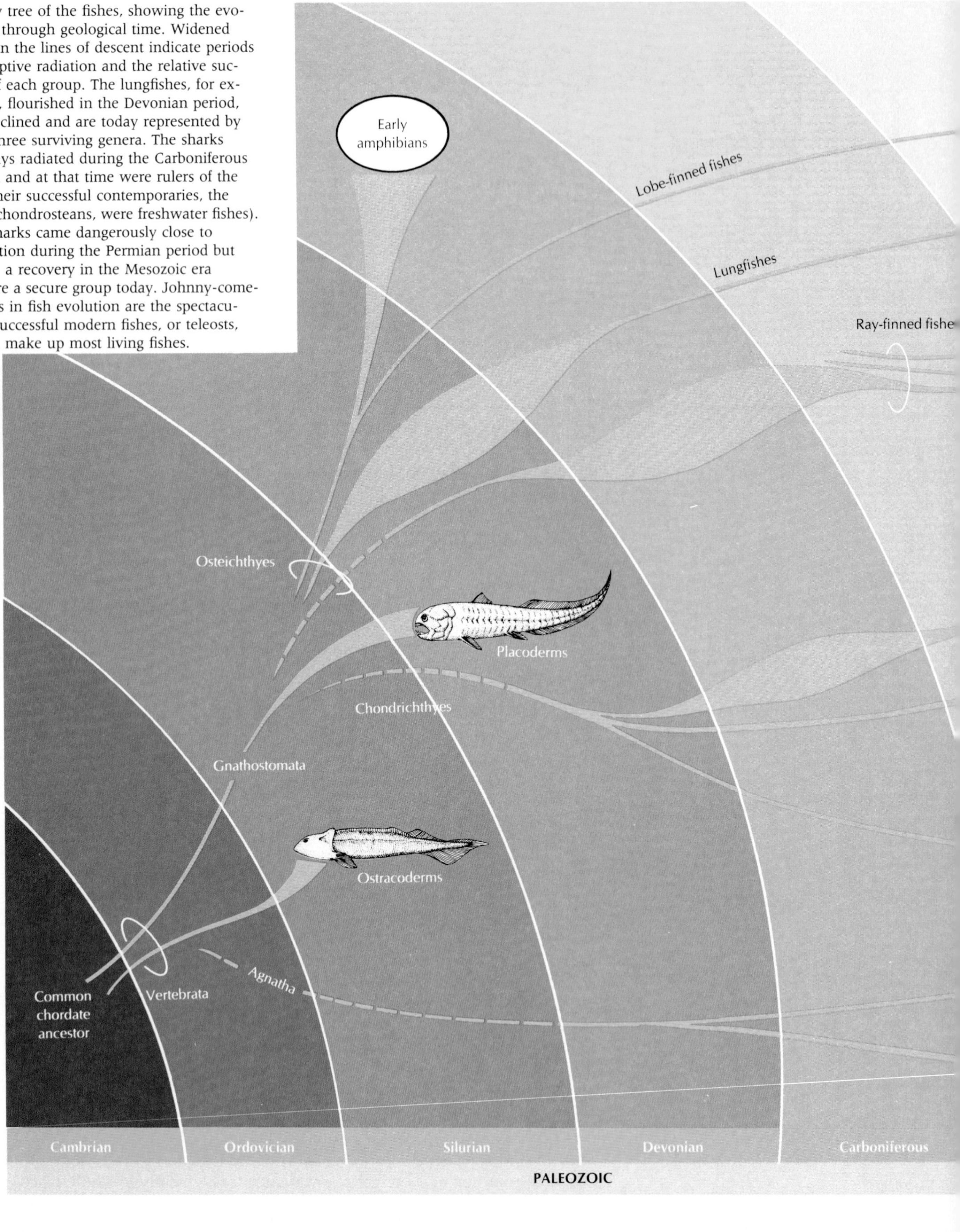

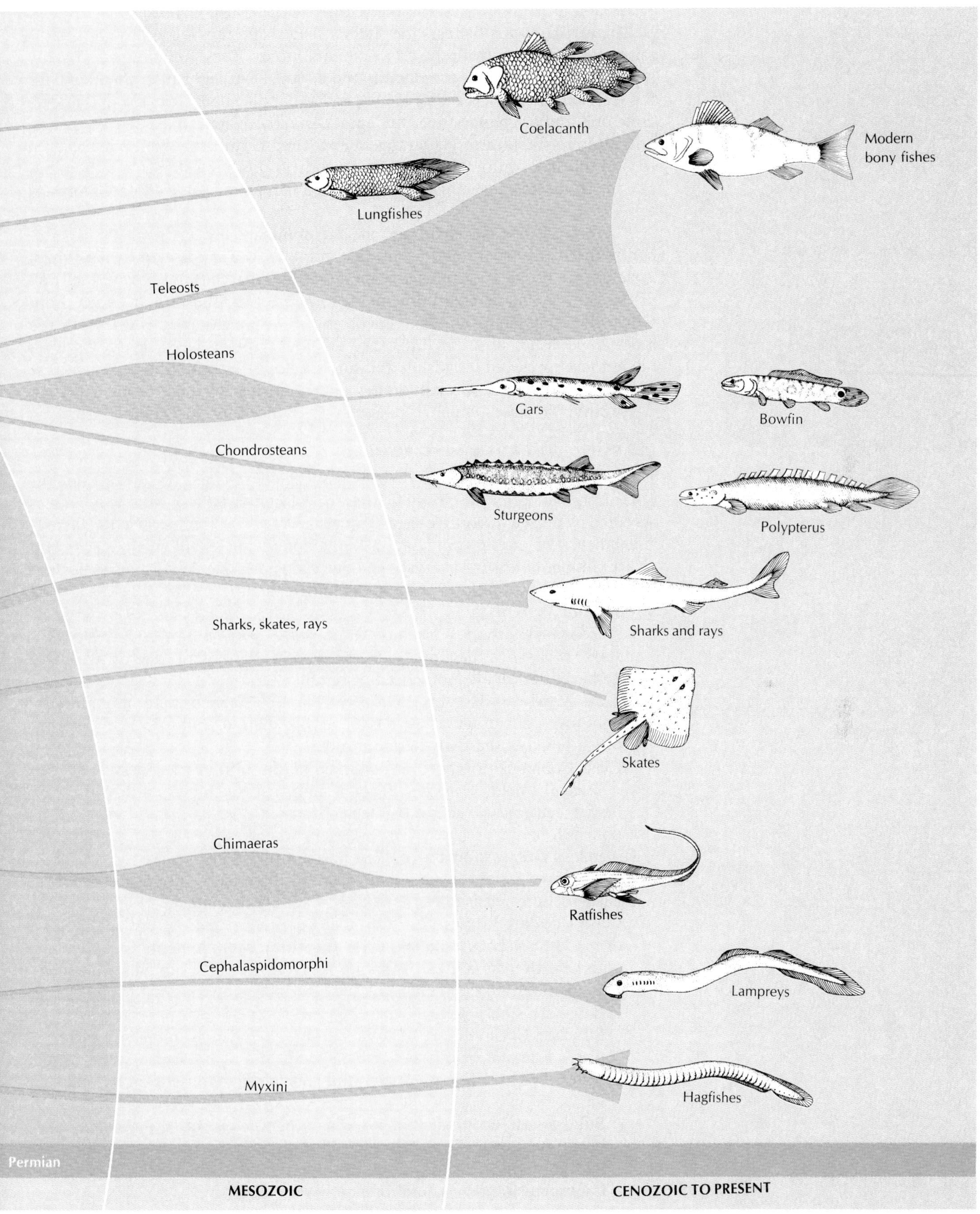
Coelacanth
Modern bony fishes
Lungfishes
Teleosts
Holosteans
Gars
Bowfin
Chondrosteans
Sturgeons
Polypterus
Sharks, skates, rays
Sharks and rays
Skates
Chimaeras
Ratfishes
Cephalaspidomorphi
Lampreys
Myxini
Hagfishes
Permian
MESOZOIC
CENOZOIC TO PRESENT

derms (p. 552), and are divided into two major groups, the cartilaginous fishes (class Chondrichthyes) and the bony fishes (class Osteichthyes).

The **cartilaginous fishes** — the sharks, skates, and rays — compose one natural, compact subclass with the class Chondrichthyes. These animals have lost the heavy armor of the early placoderms and adopted cartilage rather than bone for the skeleton, an active predatory habit, and a sharklike body form that has undergone only minor changes over the ages. Distantly related to the sharks, skates, and rays are the bizarre yet strangely appealing **chimaeras,** which make up a second subclass within the class Chondrichthyes. Both the sharks and their kin, and the chimaeras, first appeared in the Devonian period some 400 million years ago.

The **bony fishes** (Osteichthyes) are the dominant fishes today. We can recognize three great stems of descent: first, there are the **ray-finned fishes,** which radiated into the vast assemblage of modern fishes that today occupy virtually all of the earth's aquatic environments; second are the **lobe-finned fishes,** from which the amphibians are descended; and third are the **lungfishes.** Both the lobe-finned fishes and the lungfishes are relic groups. Only seven survivors remain (six species of lungfishes and one species of lobe-fin, the coelacanth) — meager evidence of important stocks that flourished in the Devonian period. These survivors have remained relatively unchanged for 400 million years.

CLASSIFICATION OF LIVING FISHES

The following broad classification is a composite of schemes by several contemporary ichthyologists but mostly follows that of Nelson (1984). No one scheme is accepted by even a majority of ichthyologists. When we contemplate the incredible difficulty of ferreting out relationships among some 23,000 living species and a vast number of fossils of varying age, we can appreciate why fish classification has been and will continue to be undergoing change.

Subphylum Vertebrata

Superclass Agnatha (ag'na-tha) (Gr. *a,* not, + *gnathos,* jaw) **(Cyclostomata).** No jaws; cartilaginous skeleton; ventral fins absent; one or two pairs of semicircular canals; notochord persistent.

Class Myxini (mik-sy'ny) (Gr. *myxa,* slime) — **hagfishes.** Terminal mouth with four pairs of tentacles; nasal sac with duct to pharynx; 5 to 15 pairs of gill pouches; partially hermaphroditic.

Class Cephalaspidomorphi (sef-a-lass'pe-do-morf'e) (Gr. *kephalē,* head, + *aspidos,* shield, + *morphē,* form) — **lampreys.** Sucking mouth with horny teeth; nasal sac not connected to mouth; seven pairs of gill pouches.

Superclass Gnathostomata (na'tho-sto'ma-ta) (Gr. *gnathos,* jaw, + *stoma,* mouth). Jaws present; usually paired limbs; three pairs of semicircular canals; notochord persistent or replaced by vertebral centra.

Class Chondrichthyes (kon-drik'thee-eez) (Gr. *chondros,* cartilage, + *ichthys,* a fish) — **sharks, skates, rays,** and **chimaeras.** Streamlined body with heterocercal tail; cartilaginous skeleton; five to seven gills with separate openings, no operculum, no swim bladder, intestine with spiral valve.

Subclass Elasmobranchii (e-laz'mo-brank'ee-i) (Gr. *elasmos,* a metal plate, + *branchia,* gills) — **sharks, skates,** and **rays.** Placoid scales or no scales; five to seven gill arches and gills in separate clefts along pharynx.

Subclass Holocephali (hol'-o-sef'a-li) (Gr. *holos,* entire, + *kephalē,* head) — **chimaeras** or **ratfishes.** Gill slits covered with operculum; jaws with tooth plates; single nasal opening; without scales; accessory clasping organs in male; lateral line an open groove.

Class Osteichthyes (os'te-ik'thee-eez) (Gr. *osteon,* bone, + *ichthys,* a fish) **(Teleostomi) — bony fishes.** Primitively fusiform body but variously modified; skeleton mostly ossified; single gill opening on each side covered with operculum; usually swim bladder or lung.

Subclass Crossopterygii (cros'sop-ter-ij'ee-i) (Gr. *krossi,* fringe or tassels, + *pteryx,* fin, wing) **— lobed-finned fishes.** Paired fins lobed with internal skeleton of basic tetrapod type; three-lobed diphycercal tail; skeleton with much cartilage; vestigial air bladder; intestine with spiral valve; spiracle present.

Subclass Dipneusti (dip-nyu'sti) (Gr. *di-,* two, + *pneustikos,* of breathing) **— lungfishes.** All median fins fused to form diphycercal tail; fins lobed or of filaments; teeth of grinding plates; air bladder of single or paired lobes and specialized for breathing; intestine with spiral valve; spiracle absent.

Subclass Actinopterygii (ak'ti-nop-te-rij'ee-i) (Gr. *aktis,* ray, + *pteryx,* fin, wing) **— ray-finned fishes.** Paired fins supported by dermal rays and without basal lobed portions; nasal sacs open only to outside.

Jawless Fishes — Superclass Agnatha

The living members of the Agnatha are represented by some 70 species almost equally divided between two classes: Myxini (hagfishes) and Cephalaspidomorphi (lampreys) (Figure 26-2). Members of both groups lack jaws, internal ossification, scales, and paired fins, and both share porelike gill openings and an eel-like body form. At the same time there are so many important differences, some of which are indicated in the following list, that they have been assigned to separate vertebrate classes.

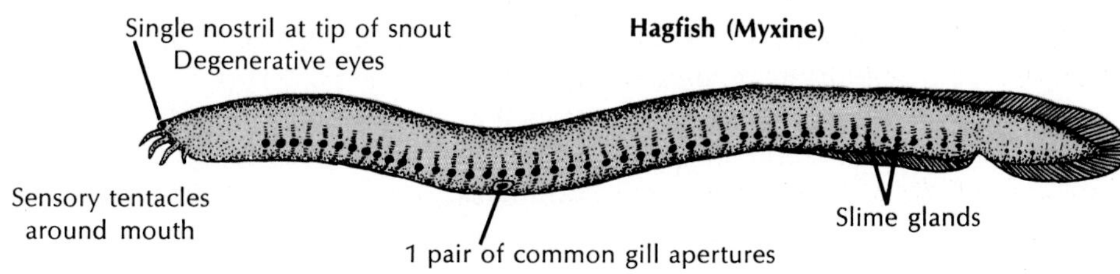

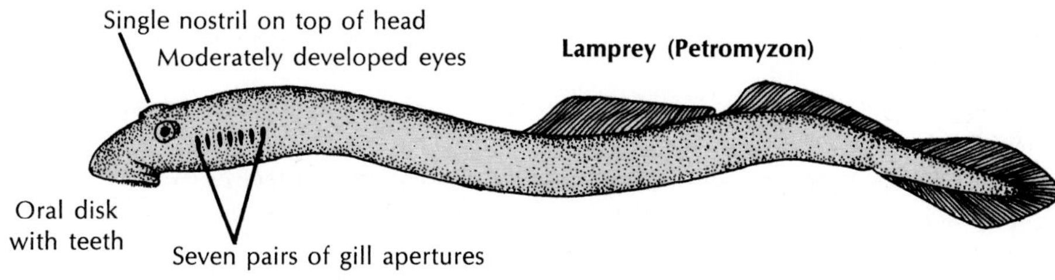

FIGURE 26-2

Comparison of hagfish (class Myxini) and lamprey (class Cephalaspidomorphi) representatives of the superclass Agnatha.

CHARACTERISTICS OF THE AGNATHA

1. Slender, **eel-like** body
2. Median fins but **no** paired appendages.
3. **Fibrous** and **cartilaginous skeleton;** notochord persistent
4. Biting mouth with two rows of eversible teeth in hagfish; suckerlike oral disc with well-developed teeth in lampreys
5. Heart with one atrium and one ventricle; hagfish with three accessory hearts; aortic arches in gill region
6. Five to 16 pairs of gills and a single pair of gill apertures in hagfish; 7 pairs of gills in lampreys

7. **Pronephric kidney** anteriorly and **mesonephric kidney** posteriorly in hagfish; mesonephric kidney only in lampreys
8. Dorsal nerve cord with differentiated brain; 8 to 10 pairs of cranial nerves
9. Digestive systems **without stomach;** intestine with spiral valve and cilia in lampreys; both lacking in intestine of hagfish
10. Sense organs of taste, smell, hearing; eyes poorly developed in hagfish but moderately well developed in lampreys; one pair of **semicircular canals** (hagfish) or two pairs (lampreys)
11. External fertilization; both ovaries and testes present in an individual but gonads of only one sex functional and no larval stage in hagfish; separate sexes and long larval stage with radical metamorphosis in lampreys.

HAGFISHES—CLASS MYXINI

The hagfishes are an entirely marine group that feed on dead or dying fishes, annelids, molluscs, and crustaceans. Thus they are neither parasitic, like lampreys, nor predaceous; rather they are scavengers. There are only 32 species of hagfishes, of which the best known in North America are the Atlantic hagfish *Myxine glutinosa* (Gr. *myxa,* slime) (Figure 26-2) and the Pacific hagfish *Eptatretus stouti* (NL, *ept*<Gr. *hepta,* seven, + *tretos,* perforated).

Hagfishes have long been of interest to physiologists. They are the only vertebrates that are isosmotic to seawater like marine invertebrates. They are also the only vertebrates having both pronephric and mesonephric kidneys in the adult (p. 206), and, to boost blood flow, there are no fewer than four sets of hearts positioned at strategic points in the low-pressure circulatory system.

While the unique anatomical and physiological features of the strange hagfishes are of interest to biologists, hagfishes have not endeared themselves to either sports or commercial fishermen. In the earlier days of commercial fishing mainly by gill nets and set lines, hagfish often bit into the bodies of captured fish and ate out the contents, leaving behind a useless sack of skin and bones. But as large and efficient otter trawls came into use, hagfishes ceased to be an important pest.

Hagfishes are renowned for their capacity to generate enormous quantities of slimy mucus from the special slime glands positioned along the body. A single hagfish may produce half a liter of slime within minutes of capture. Slime production is important to the animal for feeding and in defense, but is a nuisance to the sportsman unfortunate enough to hook one.

The reproductive biology of hagfishes remains largely a mystery, despite a prize, still unclaimed, offered more than 100 years ago by the Copenhagen Academy of Science for information on the animal's breeding habits. It is known that while both male and female gonads are found in the same animal, only one gonad becomes functional. The females produce small numbers of surprisingly large, yolky eggs up to 3 cm in diameter. There is no larval stage, and growth is direct.

LAMPREYS—CLASS CEPHALASPIDOMORPHI

Of the approximately 40 species of lampreys distributed around the world, by far the best known to North Americans is the destructive marine lamprey, *Petromyzon marinus,* of the Great Lakes. The name *Petromyzon* (Gr. *petros,* stone, + *myzon,* sucking) refers to the lamprey's habit of grasping a stone with its mouth to hold position in a current. There are 20 species of lampreys in North America of which about half are parasitic; the rest are species that spawn and die soon after becoming adults without ever feeding.

In North America all lampreys, marine as well as freshwater forms, spawn in the spring in shallow gravel and sand in freshwater streams. The males begin nest building and are joined later by females. Using their oral discs to lift stones and pebbles and using vigorous body vibrations to sweep away light debris, they form an oval depression. At spawning, with the female attached to a rock to maintain position over the nest, the male attaches to the dorsal side of her head. As the eggs are shed into the nest, they are fertilized by the male. The sticky eggs adhere to pebbles in the nest and soon become covered with sand. The adults die soon after spawning.

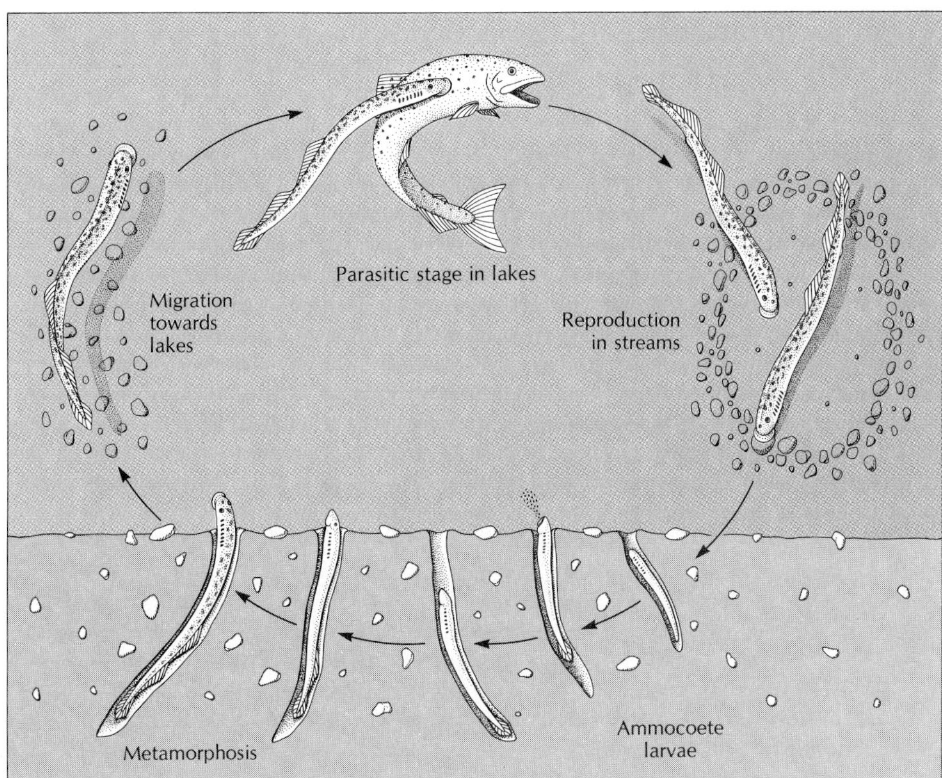

FIGURE 26-3

Life cycle of the sea lamprey, *Petromyzon marinus.*

The eggs hatch in approximately 2 weeks into small larvae **(ammocoetes)** (Figure 26-3), which stay in the nest until they are approximately 1 cm long; they then burrow into the mud and sand and emerge at night to feed. The larval period lasts from 3 to 7 years before the larva rapidly metamorphoses into an adult.

Parasitic lampreys either migrate to the sea, if marine, or else remain in fresh water, where they attach themselves by their suckerlike mouth to fish and with their sharp horny teeth rasp away flesh and suck blood (Figure 26-4). To promote the flow of blood, the lamprey injects an anticoagulant into the wound. When gorged, the lamprey releases its hold but leaves the fish with a wound that may prove fatal. The parasitic freshwater adults live a year or more before spawning and then die; the marine forms may live longer.

The nonparasitic lampreys do not feed after emerging as adults, since their alimentary canal degenerates to a nonfunctional strand of tissue. Within a few months and after spawning, they die.

The invasion of the Great Lakes above Lake Ontario by the landlocked sea lamprey *Petromyzon marinus* in this century had a devastating effect on the fisheries. Lampreys first entered the Great Lakes after the Welland Canal around Niagara Falls, a barrier to further western migration, was deepened between 1913 and 1918. Moving first through Lake Erie to Lakes Huron, Michigan, and Superior, sea lampreys, accompanied by overfishing, caused the total collapse of a multimillion dollar lake trout fishery in the early 1950s. Other less valuable fish species were attacked and destroyed in turn. After reaching a peak abundance in 1951 in Lakes Huron and Michigan and in 1961 in Lake Superior, the sea lampreys began to decline, due in part to depletion of their food and in part to the effectiveness of control measures (mainly chemical larvicides placed in selected spawning streams). Lake trout are now recovering, but wounding rates are still high in some lakes.

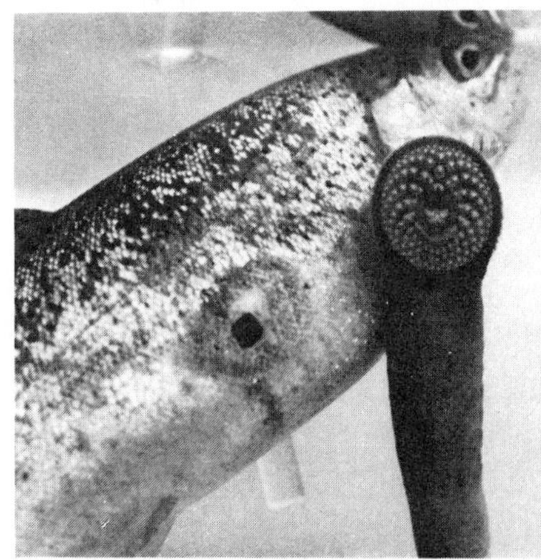

FIGURE 26-4

Sea lamprey detached from rainbow trout to show feeding wound that had penetrated the body cavity and perforated gut. The trout died from the wound. Note the chitinous teeth on the underside of the lamprey head.

Photograph by L.L. Marking, courtesy United States Bureau of Sports Fisheries and Wildlife, Fish Control Laboratory, La Crosse, Wis.

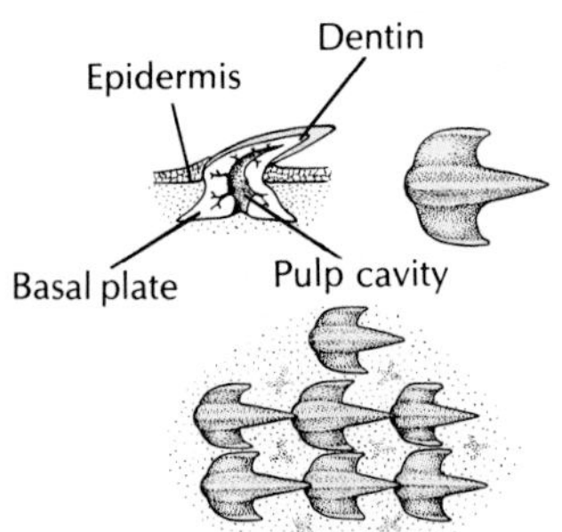

Placoid scales
(cartilaginous fishes)

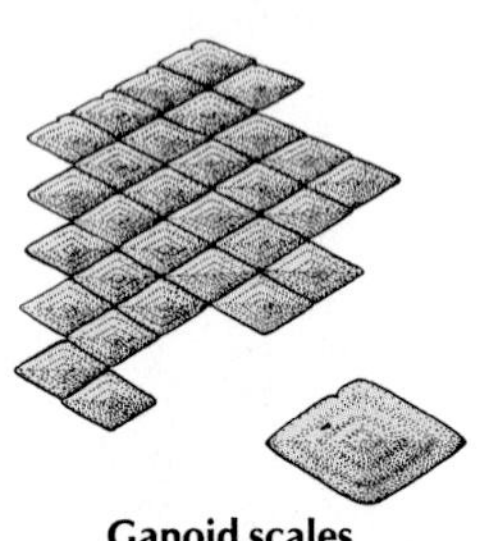

Ganoid scales
(primitive bony fishes)

Cycloid scales
(teleost fishes)

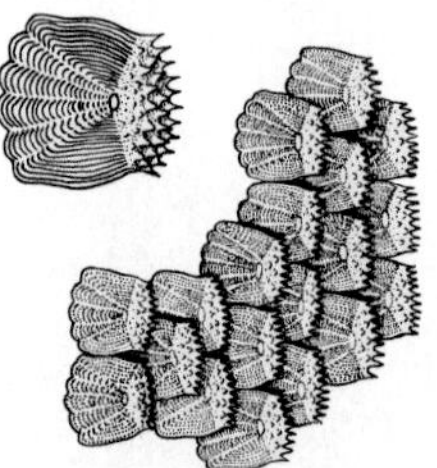

Ctenoid scales
(teleost fishes)

FIGURE 26-5

Types of fish scales. Placoid scales are small, conical toothlike structures characteristic of Chondrichthyes. Diamond-shaped ganoid scales, present in primitive bony fishes such as the gar, are composed of layers of silvery enamel (ganoin) on the upper surface and bone on the lower. Advanced bony fishes have either cycloid or ctenoid scales. These are thin and flexible and are arranged in overlapping rows.

Cartilaginous Fishes — Class Chondrichthyes

There are nearly 800 living species in the class Chondrichthyes, an ancient, compact, and highly developed group. Although a much smaller and less diverse assemblage than the bony fishes, their impressive combination of well-developed sense organs, powerful jaws and swimming musculature, and predaceous habits ensures them a secure and lasting niche in the aquatic community. One of their distinctive features is their cartilaginous skeleton. Although there is some calcification here and there, bone is entirely absent throughout the class — a curious evolutionary feature, since the Chondrichthyes are derived from ancestors having well-developed bone.

SHARKS, SKATES, AND RAYS — SUBCLASS ELASMOBRANCHII

With the exception of whales, sharks and their kin are the largest living vertebrates. The whale shark may reach 15 m in length. The dogfish sharks so widely used in zoological laboratories rarely exceed 1 m. There are approximately 760 living species of elasmobranchs.

CHARACTERISTICS OF THE SHARKS, SKATES, AND RAYS

1. **Body fusiform** (except skates and rays) with a **heterocercal** caudal fin
2. **Ventral mouth; two olfactory sacs that do not connect to the mouth cavity;** jaws present
3. Skin with **placoid scales** (Figure 26-5) and **mucous glands;** teeth of modified placoid scales
4. **Entirely cartilaginous endoskeleton**
5. Digestive system with a J-shaped stomach and intestine with a spiral valve (Figure 26-6)
6. Circulatory system of several pairs of aortic arches; two-chambered heart
7. Respiration by means of five to seven pairs of gills with **separate and exposed gill slits,** no operculum
8. No swim bladder or lung.
9. Brain of two olfactory lobes, two cerebral hemispheres, two optic lobes, a cerebellum, and a medulla oblongata; 10 pairs of cranial nerves
10. Separate sexes; oviparous, ovoviviparous, or viviparous; direct development; **internal fertilization**

Although sharks to most people have a sinister appearance and a fearsome reputation, they are at the same time among the most gracefully streamlined of all fishes (Figure 26-6). Sharks are heavier than water and must always keep swimming forward to avoid sinking. The asymmetrical **heterocercal tail,** in which the vertebral column turns upward and extends into the dorsal lobe of the tail (Figure 26-16), provides the necessary lift as it sweeps to and fro in the water, and the broad head and flat pectoral fins act as planes to provide head lift.

Sharks are well equipped for their predatory life. The tough leathery skin is covered with numerous knifelike **placoid scales** (Figure 26-5) that are modified anteriorly to form replaceable rows of teeth in both jaws. Placoid scales in fact consist of dentine enclosed by an enamel-like substance, and they very much resemble the teeth of other vertebrates. Sharks have a keen sense of smell used to guide them to food. Vision is less acute than in most bony fishes, but a well-developed **lateral line system** serves as a "distance touch" in water for detecting and locating objects and moving animals (predators, prey, and social partners). It is composed of a canal system extending along the side of the body and over the head. The canal opens at intervals to the surface. Inside are special receptor organs **(neuromasts)** that are extremely sensitive to vibrations and currents in the water. Sharks can also detect and accurately attack prey by sensing the bioelectric fields that surround all animals.

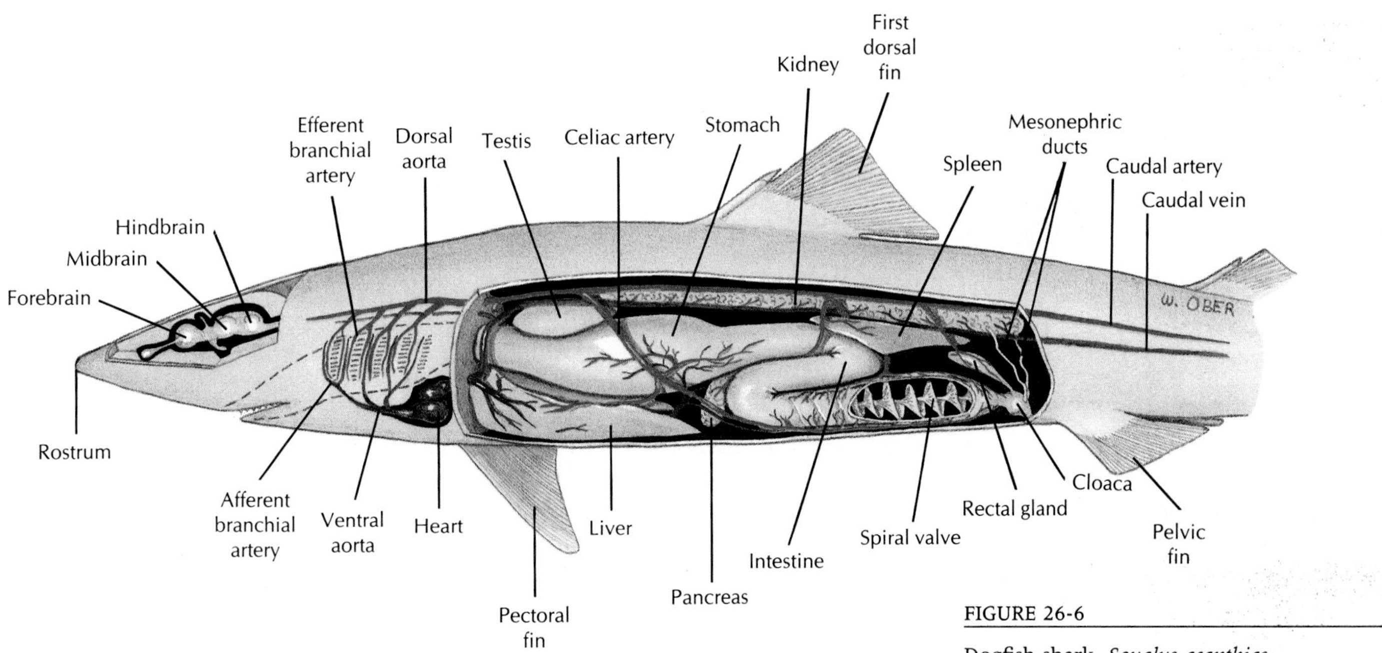

FIGURE 26-6

Dogfish shark, *Squalus acanthias.*

Skates and rays belong to a separate order from the sharks. Skates and rays are distinguished by their dorsoventrally flattened bodies and the much-enlarged pectoral fins that behave as wings in swimming. The gill openings are on the underside of the head, and the **spiracles** (on top of the head) are unusually large. Respiratory water is taken in through these spiracles to prevent clogging the gills, because the mouth is often buried in sand. The teeth are adapted for crushing the prey—mainly molluscs, crustaceans, and an occasional small fish.

In the stingrays and eagle rays (Figure 26-7), the caudal and dorsal fins have disappeared, and the tail is slender and whiplike. The stingray tail is armed with one or more saw-toothed spines that can inflict dangerous wounds. Electric rays have certain dorsal muscles modified into powerful electric organs, which can give severe shocks and stun their prey.

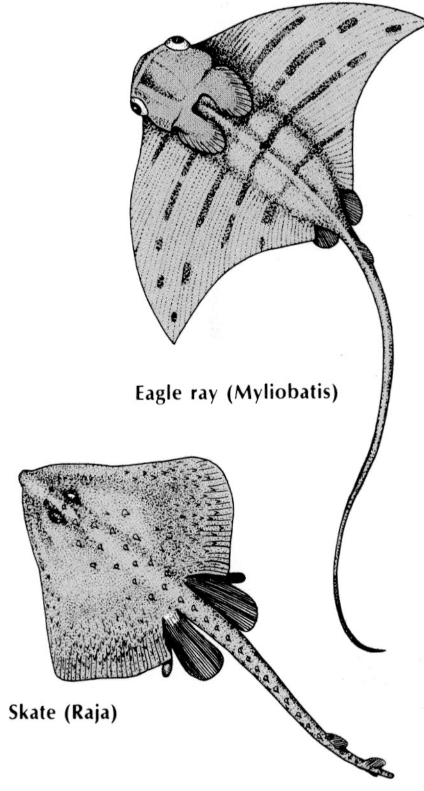

FIGURE 26-7

Skates and rays are specialized for life on the sea floor. They are flattened dorsoventrally and move by undulations of greatly expanded winglike pectoral fins. Eagle rays are among the largest of fishes, and are feared by tropical pearl divers who reportedly have been attacked and eaten by them.

CHIMAERAS—SUBCLASS HOLOCEPHALI

The members of the small group of chimaeras (L., monster), approximately 25 species in all, are distinguished by such suggestive names as ratfish (Figure 26-8), rabbitfish, spookfish, and ghostfish. They are remnants of an aberrant line that diverged from the placoderms at least 300 million years ago (Carboniferous or Devonian periods). Fossil chimaeras (ky-meer'-uz) were first found in the Jurassic, reached their zenith in the Cretaceous and early Tertiary periods (120 million to 50 million years ago), and have declined ever since. Anatomically they present an odd mixture of sharklike and bony fish-like features. Their food is a mixed diet of seaweed, molluscs, echinoderms, crustaceans, and fishes. Chimaeras are not commercial species and are seldom caught. Despite their grotesque shape, they are beautifully colored with a pearly iridescence and have vivid emerald-green eyes.

■ Bony Fishes—Class Osteichthyes

In few other major animal groups do we see better examples of adaptive radiation than among the bony fishes. Their adaptations have fitted them for every aquatic habitat except the most completely inhospitable. Body form alone is indicative of

FIGURE 26-8

Chimaera, or ratfish, of North American west coast. This species is one of the most handsome of chimaeras, which tend toward bizarre appearances.

this diversity. Some have fusiform (streamlined) bodies and other adaptations for reducing friction. Predaceous, pelagic fish have trim, elongate bodies and powerful tail fins and other mechanical advantages for swift pursuit. Sluggish bottom-feeding forms have flattened bodies for movement and concealment on the ocean floor. The elongate body of the eel is an adaptation for wriggling through mud and reeds and into holes and crevices. Some, such as pipefishes, are so slender that they are easily mistaken for filaments of marine algae waving in the current. Many other remarkable body forms are obviously cryptic or mimetic adaptations for concealment from predators or as predators. Such few examples cannot begin to express the amazing array of physiological and anatomical specializations for defense and offense, food gathering, navigation, and reproduction in the diverse aquatic habitats to which bony fishes have adapted themselves.

The Osteichthyes are divided into three clearly distinct groups: lobe-finned fishes, lungfishes, and ray-finned fishes.

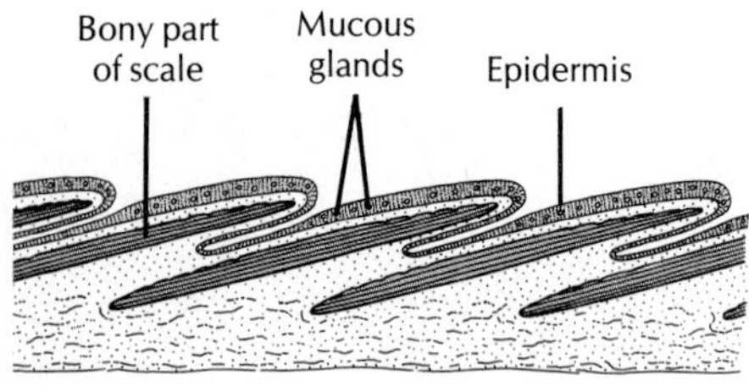

FIGURE 26-9

Section through the skin of a bony fish, showing the overlapping scales *(red)*. The scales lie in the dermis and are covered by epidermis.

CHARACTERISTICS

1. **More or less bony skeleton,** numerous vertebrae; usually **homocercal tail**
2. Skin with mucous glands and embedded **dermal scales** (Figure 26-9) of three types: **ganoid, cycloid, or ctenoid;** some without scales, no placoid scales
3. Both median and paired fins with **fin rays of cartilage or bone**
4. **Terminal mouth** with many homodont teeth (some toothless); jaws present; olfactory sacs paired and may or may not open into mouth
5. Respiration by gills supported by bony gill arches and covered by a **common operculum**
6. **Swim bladder** often present with or without duct connected to pharynx
7. Circulation consisting of a two-chambered heart, arterial and venous systems, and four pairs of aortic arches
8. Nervous system of a brain with small olfactory lobes and cerebrum and large optic lobes and cerebellum; 10 pairs of cranial nerves
9. Separate sexes; paired gonads; usually external fertilization; larval forms may differ greatly from adults

Bony fishes vary greatly in size. Some of the minnows are less than 2 cm long as adults; other fishes may exceed 3 m in length. The swordfish is one of the largest and may attain a length of 4 m. Most fishes, however, fall between 2 and 30 cm in length.

CLASSIFICATION

Class Osteichthyes (os'-te-ik'thee-eez)—**bony fishes.** Three subclasses, 42 living orders.

Subclass Crossopterygii (cros'sop-te-rij'ee-i)—**lobe-finned fishes.** Four extinct orders, one living order (Coelacanthimorpha) containing one species, *Latimeria chalumnae.*

Subclass Dipneusti (dip-nyu'sti)—**lungfishes.** Six extinct orders; two living orders, containing three genera: *Neoceratodus, Lepidosiren,* and *Protopterus.*

Subclass Actinopterygii (ak'ti-nop-te-rij'ee-i)—**ray-finned fishes.** Three superorders and 38 living orders.

Superorder Chondrostei (kon-dros'tee-i) (Gr. *chondros,* cartilage, + *osteon,* bone)—**primitive ray-finned fishes.** Ten extinct orders; two living orders containing the bichir *(Polypterus),* sturgeons, and paddlefishes.

Superorder Holostei (ho-los'tee-i) (Gr. *holos,* entire, + *osteon,* bone)—**intermediate ray-finned fishes.** Four extinct orders; two living orders containing the bowfin *(Amia)* and the gars *(Lepidosteus).*

Superorder Teleostei (tel′e-os′tee-i) (Gr. *teleos,* perfect, + *osteon,* bone)—**climax bony fishes.** Body covered with thin scales without bony layer (cycloid or ctenoid) or scaleless; dermal and chondral parts of skull closely united; caudal fin mostly homocercal; terminal mouth; notochord a mere vestige; swim bladder mainly a hydrostatic organ and usually not opened to the esophagus; endoskeleton mostly bony. According to Nelson (1984), there are 50 living orders, 435 families, and approximately 21,800 living species, representing 96% of all living fishes.

FIGURE 26-10

Coelacanth, *Latimeria chalumnae.* This surviving marine relic of a group that flourished some 350 million years ago has fleshy-based ("lobed") fins with which its ancestors used to pull themselves across land from pond to pond.

Courtesy Jerry Smith, artist, Vancouver Public Aquarium, British Columbia.

LOBE-FINNED FISHES—SUBCLASS CROSSOPTERYGII

The lobe-finned fishes are represented today by a single known species, the famous coelacanth *Latimeria chalumnae* (Figure 26-10). Since the last coelacanths (seal′a-canths) were believed to have become extinct 70 million years ago, the astonishment of the scientific world can be imagined when the remains of a coelacanth were found on a trawler off the coast of South Africa in 1938. An intensive search was begun in the Comoro Islands area near Madagascar where, it was learned, native Comoran fishermen occasionally caught them with hand lines at great depths. Numerous specimens have now been caught, many in excellent condition, although none has been kept alive beyond a few hours after capture. The "modern" marine coelacanth is a descendant of the Devonian freshwater stock that reached its evolutionary peak in the Mesozoic era and then disappeared—or so it was believed until 1938.

The lobe-finned fishes occupy an important position in vertebrate evolution because the amphibians—indeed all tetrapod vertebrates—arose from one or more of their ancient members. The lobe-finned fishes had lungs as well as gills, which would have been a decided advantage to survival during the Devonian period, a capricious time of alternating droughts and floods. These fishes used their strong lobed fins as four legs to scuttle from one disappearing swamp to another that offered more promise for a continuing aquatic existence.

LUNGFISHES—SUBCLASS DIPNEUSTI

The lungfishes are another relic group of fishes represented today by only three surviving genera (Figure 26-11). Like the lobe-finned fishes to which they are related, the lungfishes were a distinct group during the swampy Devonian period when their lungs would have been a distinct asset for survival.

Of the three surviving lungfishes, the least specialized is *Neoceratodus* (Gr. *neos,* new, + *keratos,* horn, + *ōdēs,* form), the living Australian lungfish, which may attain a length of 1.5 m. This lungfish is able to survive in stagnant, oxygen-poor water by coming to the surface and gulping air into its single lung, but it cannot live out of water.

The South American lungfish, *Lepidosiren* (L. *lepidus,* pretty, + *siren,* Siren, mythical mermaid), and the African lungfish, *Protopterus* (Gr. *prōtos,* first, + *pteron,* wing), can live out of water for long periods of time. *Protopterus* lives in African streams and rivers that run completely dry during the dry season, with their mud beds baked hard by the hot tropical sun. The fish burrows down at the approach of the dry season and secretes a copious slime that is mixed with mud to form a hard cocoon in which it remains dormant until the rains return (Figure 26-12).

Neoceratodus (Australia)
Direct descendant of ancient lungfish
Cannot withstand complete drying up of water

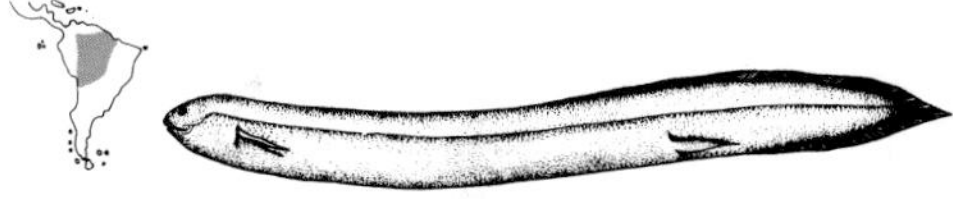

Lepidosiren (South America)
Side branch of lungfish evoluton
Can burrow in mud when water dries up

Protopterus (Africa)
Side branch of lungfish evolution
Can burrow in mud when water dries up

FIGURE 26-11

The three surviving lungfishes. The approximate range of each genus is shown on map inserts.

RAY-FINNED FISHES—SUBCLASS ACTINOPTERYGII

Ray-finned fishes are an enormous assemblage containing all of our familiar bony fishes. The ray-finned fishes had their beginnings in the Devonian freshwater lakes and streams. The ancestral forms were small, bony, heavily armored fishes with functional lungs as well as gills. In their evolution, the ray-finned fishes have passed through three stages.

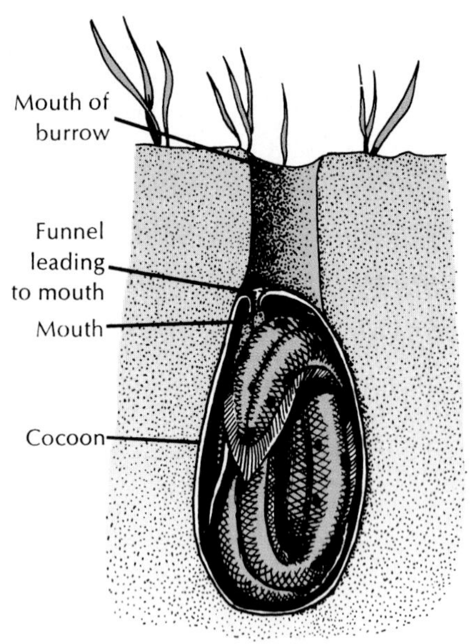

FIGURE 26-12

Dormant (aestivating) African lungfish, *Protopterus.* As water disappears during dry season, *Protopterus* digs a burrow, coils tightly, and secretes mucus from the skin and lips. Combined with mud, the mucus dries into a stiff, papery cocoon. A hollow tube of mucus extruded from the mouth forms a direct opening for air. Aestivation may last several months, during which nitrogenous wastes gradually accumulate in body tissues. When rains return, rising water level softens the cocoon and *Protopterus* breaks out quickly, croaks several times with evident pleasure, and swims off.

The most primitive of ray-finned fishes are the chondrosteans, represented today by the freshwater and marine sturgeons and freshwater paddlefishes (Figure 26-13) and the bichir, or reedfish, *Polypterus* (Gr. *poly,* many, + *pteros,* winged), of African rivers (Figure 26-1). *Polypterus* is an interesting relic with a lunglike swim bladder and many other primitive characteristics; it resembles an ancestral ray-finned fish more than does any other living descendant. There is no satisfactory explanation for the survival to the present of certain fish such as this and the coelacanth *Latimeria* when all of their kin perished millions of years ago.

The holosteans are a second, less primitive group of ray-finned fishes. There were several lines of descent within this group, which flourished during the Triassic and Jurassic periods. They declined toward the end of the Mesozoic era as their successors, the teleosts, crowded them out. But they left two surviving lines, the bowfin, *Amia* (Gr., tunalike fish) (Figure 26-1), of shallow, weedy waters of the Great Lakes and southeastern United States, and the gars of eastern North and Central America (Figure 26-14).

The teleosts are the third group of modern bony fishes (Figure 26-15). Diversity appeared early in teleost evolution, foreshadowing the truly incredible variety of body forms among teleosts today. The skeleton of primitive fish was largely ossified but returned to a partly cartilaginous condition among many of the chondrosteans and holosteans. Teleosts, however, have an internal skeleton almost completely ossified like the primitive bony fish.

There are a number of other distinctive characteristics of the teleosts. The heavylike armorlike scales of more primitive fish have been replaced by light, thin, and flexible **cycloid** and **ctenoid** scales (Figure 26-5). These look much alike, except that ctenoid scales have comblike ridges on the exposed edge that may be an adaptation to improve swimming efficiency by reducing frictional resistance to movement through the water. Some teleosts, such as catfishes and sculpins, lack scales altogether. Nearly all teleosts have a **homocercal** tail, with the upper and lower lobes of about equal size (Figure 26-16). The lungs of primitive forms have been transformed in the teleosts to a swim bladder with a buoyancy function. Teleosts have highly maneuverable fins for control of body movement. In small teleosts the fins are often provided with stout, sharp spines, thus making themselves prickly mouthfuls for would-be predators. With these adaptations (and many others), teleosts have evolved into the most successful of fishes.

FIGURE 26-13

Paddlefish, *Polyodon,* of the Mississippi River may reach a weight of 90 kg. This plankton feeder is commercially valuable, but river pollution has greatly reduced catches.

Photograph by C.P. Hickman, Jr.

FIGURE 26-14

Longnose gar, *Lepisosteus osseus,* inhabitant of the Great Lakes region and the Mississippi basin.

Photograph by Patrice/Visuals Unlimited.

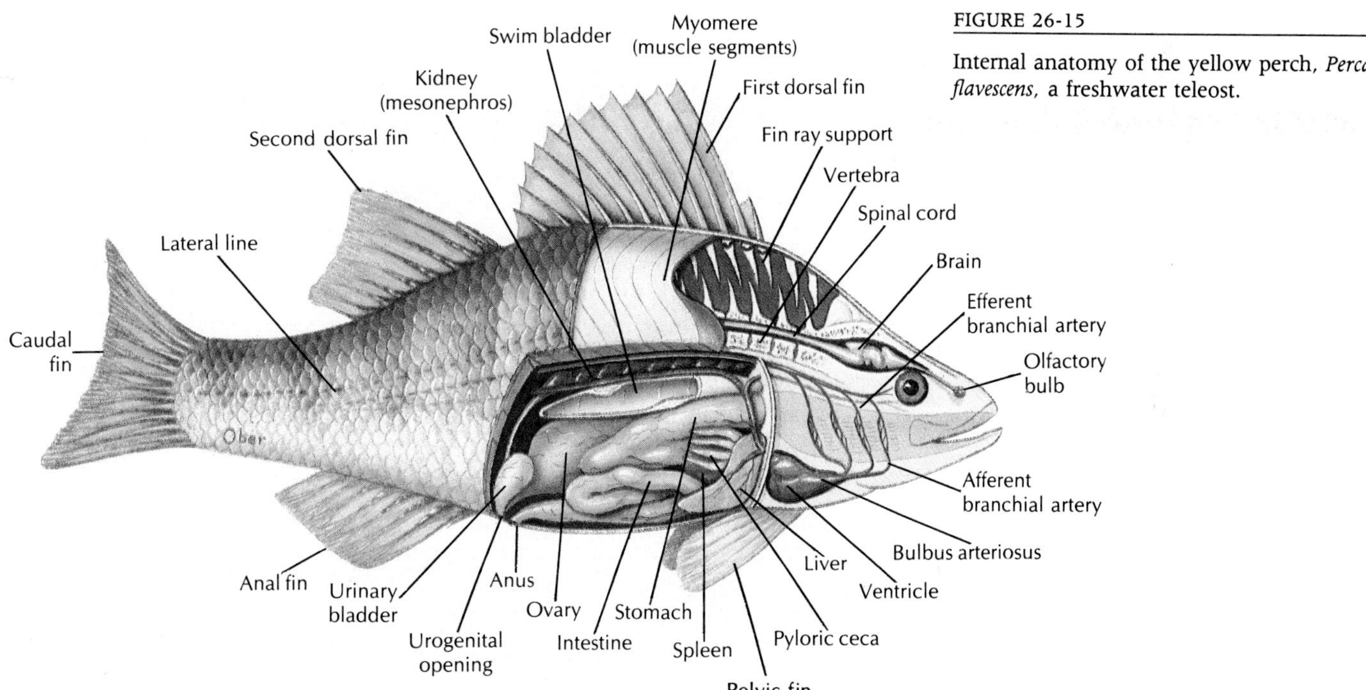

FIGURE 26-15

Internal anatomy of the yellow perch, *Perca flavescens,* a freshwater teleost.

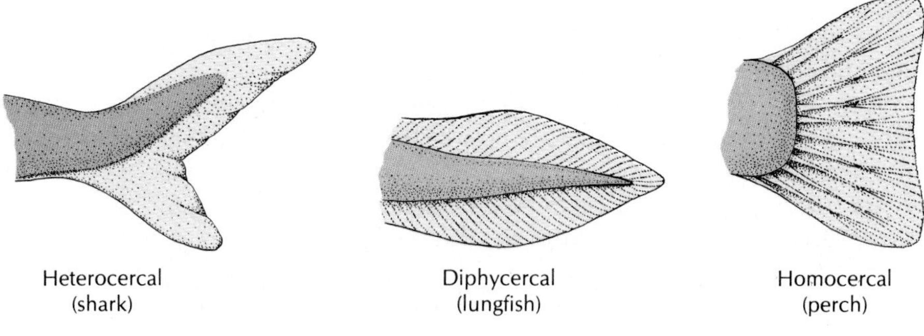

FIGURE 26-16

Types of caudal fins among fishes.

Measuring fish cruising speeds accurately is best done in a "fish wheel," a large ring-shaped channel filled with water that is turned at a speed equal and opposite to that of the fish. Much more difficult to measure are the sudden (though brief) bursts of speed that most fish can make to capture prey or to avoid being captured. A hooked bluefin tuna was once "clocked" at 66 km per hour (41 mph); swordfish and marlin are thought to be capable of incredible bursts of speed approaching, or even exceeding, 110 km per hour (68 mph). Such high speeds can be sustained for no more than one to five seconds.

Structural and Functional Adaptations of Fishes

LOCOMOTION IN WATER

To the human eye, some fishes appear capable of swimming at extremely high speeds. But our judgment is unconsciously tempered by our own experience that water is a highly resistant medium to move through. Most fishes, such as a trout or a minnow, can swim maximally about 10 body lengths per second, obviously an impressive performance by human standards. Yet when these speeds are translated into kilometers per hour it means that a 30 cm (1 foot) trout can swim only about 10.4 km (6.5 miles) per hour. As a general rule, the larger the fish, the faster it can swim.

The propulsive mechanism of a fish is its trunk and tail musculature. The axial, locomotory musculature is composed of zigzag muscle bands (myotomes) that on the surface take the shape of a W lying on its side (Figure 26-17). Internally the muscle bands are deflected forward and backward in a complex fashion that apparently promotes efficiency of movement. The muscles are bound to broad sheets of tough connective tissue, which in turn tie to the highly flexible vertebral column.

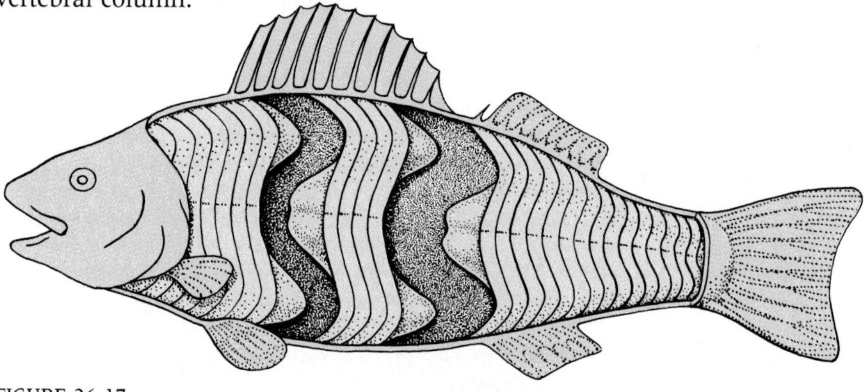

FIGURE 26-17

Trunk musculature of a teleost fish. Segmented myotomes are arranged as bands, each shaped like a W lying on its side. The musculature has been dissected away in two places to show internal anterior and posterior deflection of myotomes.

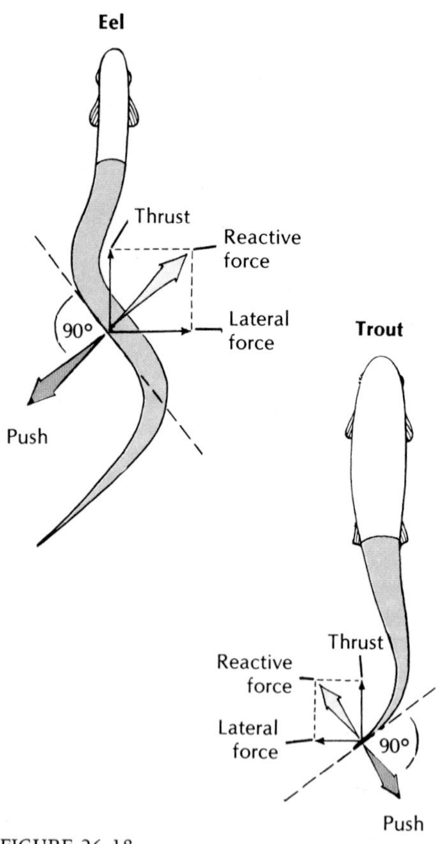

FIGURE 26-18

Movements of swimming fishes, showing the forces developed by an eel-shaped and spindle-shaped fish.

From Pough, F.H., J.B. Heiser, and W.N. McFarland. 1989. Vertebrate life, ed. 3. New York, Macmillan Publishing Company.

Understanding how fishes swim can be approached by studying the motion of a very flexible fish such as an eel (Figure 26-18). The movement is serpentine, not unlike that of a snake, with waves of contraction moving backward along the body by alternate contraction of the myotomes on either side. The anterior end of the body bends less than the posterior end, so that each undulation increases in amplitude as it travels along the body. While undulations move backward, the bending of the body pushes laterally against the water, producing a **reactive force** that is directed forward, but at an angle. It can be analyzed as having two components: **thrust,** which is used to overcome drag and propels the fish forward, and **lateral force,** which tends to make the fish's head "yaw," or deviate from the course in the same direction as the tail. This side-to-side head movement is very obvious in a swimming eel or shark, but many fishes have a large, rigid head with enough surface resistance to minimize yaw.

The movement of an eel is reasonably efficient at low speed, but its body shape generates too much frictional drag for rapid swimming. Fishes that swim rapidly, such as trout, are less flexible and limit the body undulations mostly to the caudal region (Figure 26-18). Muscle force generated in the large anterior muscle mass is transferred through tendons to the relatively nonmuscular caudal peduncle and tail where thrust is generated. This form of swimming reaches its highest development in the tunas, whose bodies do not flex at all. Virtually all the thrust is derived from powerful beats of the tail fin (Figure 26-19). Many fast oceanic fishes such as marlin, swordfish, amberjacks, and wahoo have swept-back tail fins shaped much like a sickle. Such fins are the aquatic counterpart of the high–aspect ratio wings of the swiftest birds (p. 618).

Swimming is the most economical form of animal locomotion, largely because aquatic animals are almost perfectly supported by their medium and need expend little energy to overcome the force of gravity. If we compare the energy cost per kilogram of body weight of traveling 1 km by different forms of locomotion, we find swimming to cost only 0.39 kcal (salmon) as compared to 1.45 kcal for flying (gull) and 5.43 kcal for walking (ground squirrel). However, part of the unfinished business of biology is understanding how fish and aquatic mammals are able to move through the water while creating almost no turbulence. The secret lies in the way aquatic animals bend their bodies and fins (or flukes) to swim and in the friction-reducing properties of the body surface.

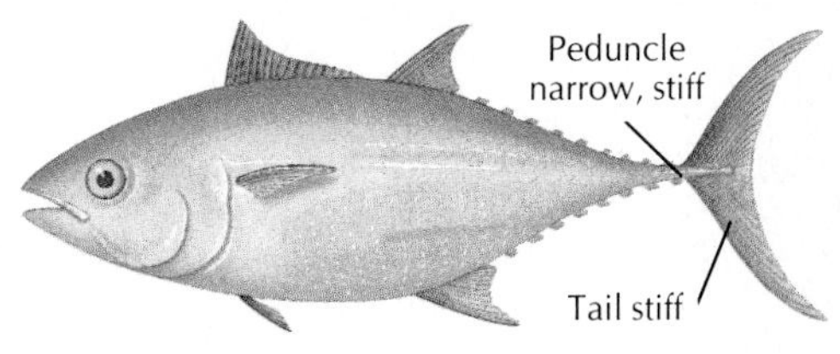

FIGURE 26-19

Bluefin tuna, showing adaptations for fast swimming. Powerful trunk muscles pull on the slender tail stalk. Since the body does not bend, all of the thrust comes from beats of the stiff, sickle-shaped tail.

NEUTRAL BUOYANCY AND THE SWIM BLADDER

All fishes are slightly heavier than water because their skeletons and other tissues contain heavy elements that are present only in trace amounts in natural waters. To keep from sinking, sharks must always keep moving forward in the water. The asymmetrical (heterocercal) tail of a shark provides the necessary tail lift as it sweeps to and fro in the water, and the broad head and flat pectoral fins (Figure 26-6) act as angled planes to provide head lift. Sharks are also aided in buoyancy by having very large livers containing a special fatty hydrocarbon called **squalene** that has a density of only 0.86. The liver thus acts like a large sack of buoyant oil that helps compensate for the shark's heavy body.

By far the most efficient flotation device is a gas-filled space. The **swim bladder** (or gas bladder as it is often called) serves this purpose in the bony fishes. It arose from the paired lungs of the primitive Devonian bony fishes (Figure 26-20). Lungs were probably a ubiquitous feature of the Devonian freshwater bony fishes when, as we have seen, the alternating wet and dry climate probably made such an accessory respiratory structure essential for life. Swim bladders are present in most oceanic (pelagic) bony fishes but absent in most bottom dwellers such as flounders and sculpins.

By adjusting the volume of gas in the swim bladder, a fish can achieve neutral buoyancy and remain suspended indefinitely at any depth with no muscular effort. There are severe technical problems, however. If the fish descends to a greater depth, the swim bladder gas is compressed, so that the fish becomes heavier and tends to sink. Gas must be added to the bladder to establish a new equilibrium buoyancy. If the fish swims up, the gas in the bladder expands, making the fish lighter. Unless the fish can release the gas, it would rise with ever-increasing speed until, having lost control, the helpless fish would pop out of the water.

Fishes adjust gas volume in the swim bladder in two ways. The less specialized fishes (trout, for example) have a **pneumatic duct** that connects the swim bladder to the esophagus (Figure 26-20); these forms must come to the surface and gulp air to charge the bladder and obviously are restricted to relatively shallow depths. More specialized teleosts have lost the pneumatic duct. In these fishes the gas must originate in the blood and be secreted into the swim bladder. Gas exchange depends on two highly specialized areas: a **gas gland** that secretes gas into the bladder and a **resorptive area,** or "oval," that can remove gas from the bladder. The gas gland is supplied by a remarkable network of blood capillaries, called the **rete mirabile** ("marvelous net") that functions as a countercurrent exchange system to trap gases, especially oxygen, and prevent their loss to the circulation.

The amazing effectiveness of this device is exemplified by a fish living at a

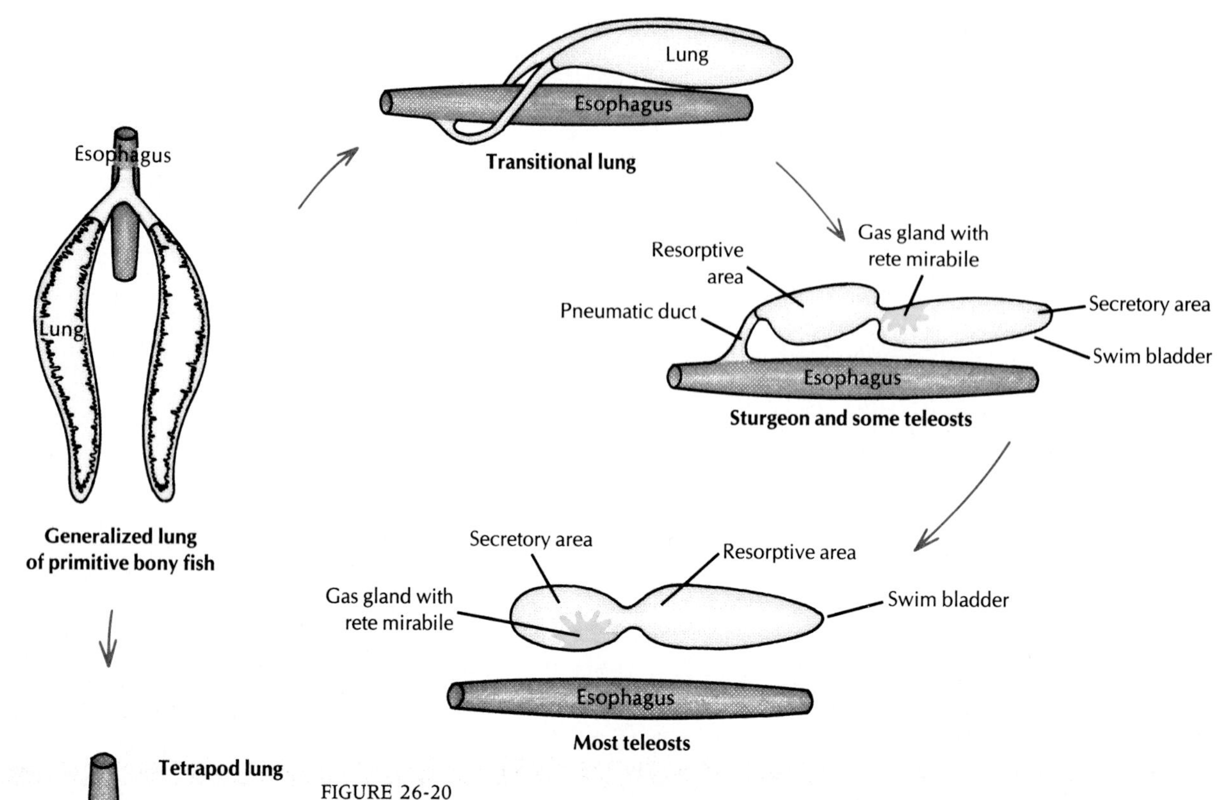

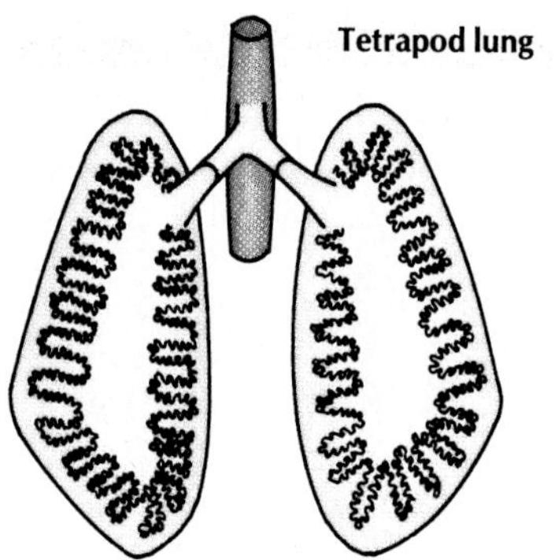

FIGURE 26-20

Evolution of lung and swim bladder. Most primitive fishes were provided with lungs, adaptations for the oxygen-depleted environments that existed during the evolution of the Osteichthyes. The lung originated as an outpocketing of the foregut. From this early, generalized lung two lines of evolution appeared. One led to the swim bladder of the modern teleost fishes. Various transitional stages show that the swim bladder shifted to a dorsal position above the esophagus, becoming a buoyancy organ. The pneumatic duct has been lost in most teleosts, and the swim bladder, with specialized gas secretion and reabsorption areas, is served by an independent blood supply. The second line of evolution has led to the tetrapod lung found in land forms. There has been extensive internal folding, but no radical change in lung position.

depth of 2400 m (8000 ft). To keep the bladder inflated at that depth, the gas inside (mostly oxygen, but also variable amounts of nitrogen, carbon dioxide, argon, and even a little carbon monoxide) must have a pressure exceeding 240 atmospheres, much greater than the pressure in a fully charged steel gas cylinder. Yet the oxygen pressure in the fish's blood cannot exceed 0.2 atmosphere — equal to the oxygen pressure at the sea surface.

Physiologists, who were at first baffled by the secretion mechanism, now understand how it operates. In brief, the gas gland secretes lactic acid, which enters the blood, causing a localized high acidity in the rete mirabile that forces hemoglobin to release its load of oxygen. The capillaries in the rete are arranged so that the released oxygen accumulates in the rete, eventually reaching such a high pressure that the oxygen diffuses into the swim bladder. The final gas pressure attained in the swim bladder depends on the length of the rete capillaries; they are relatively short in fishes living near the surface but extremely long in deep-sea fishes.

RESPIRATION

Fish gills are composed of thin, branched **gill filaments** covered with thin epidermal membrane that is folded repeatedly into platelike lamellae (Figure 26-

21). These are richly supplied with blood vessels. The gills are located inside the pharyngeal cavity and covered on the outside with a movable flap, the **operculum.** This arrangement protects the delicate gill filaments, streamlines the body, and provides a pumping system for moving water through the mouth, across the gills, and out from beneath the raised opercular flap. Instead of opercular flaps as in bony fishes, the elasmobranchs have a series of **gill slits** out of which the water flows. In both elasmobranchs and bony fishes the branchial mechanism is arranged to pump water continuously and smoothly over the gills, even though to an observer it appears that fish breathing is pulsatile.

The flow of water is opposite to the direction of blood flow (**countercurrent flow**), the best arrangement for extracting the greatest possible amount of oxygen from the water. Some bony fishes can remove as much as 85% of the oxygen from the water passing over their gills. Very active fishes, such as herring and mackerel, can obtain sufficient water for their high oxygen demands only by continuously swimming forward to force water into the open mouth and across the gills. This is called **ram ventilation.** Such fishes are asphyxiated if placed in an aquarium that restricts free swimming movements, even though the water is saturated with oxygen.

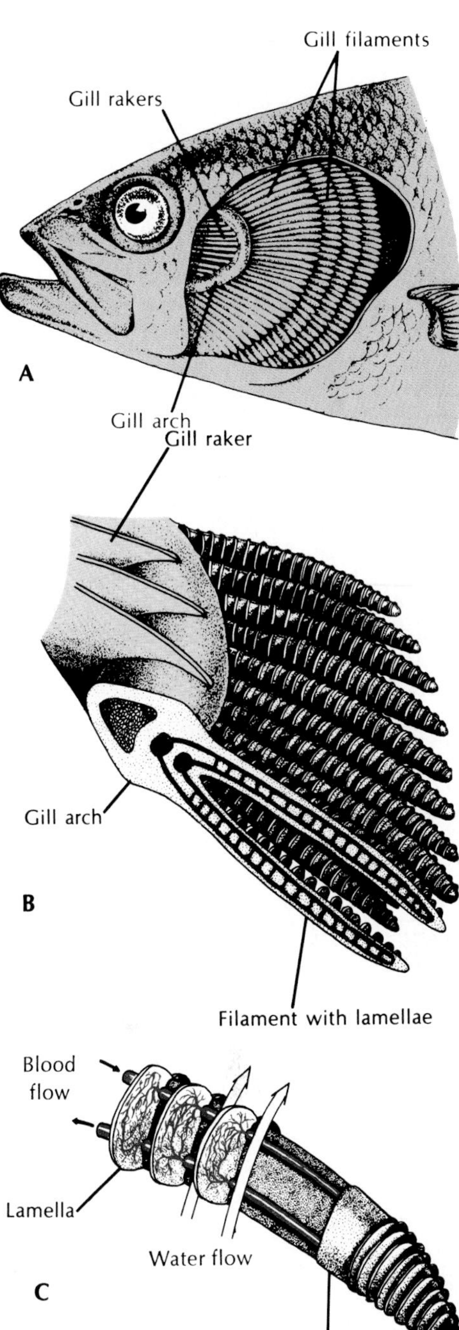

FIGURE 26-21

Gills of fish. Bony, protective flap covering the gills (operculum) has been removed, **A,** to reveal branchial chamber containing the gills. There are four gill arches on each side, each bearing numerous filaments. A portion of gill arch, **B,** shows gill rakers that project forward to strain out food and debris, and gill filaments that project to the rear. A single gill filament, **C,** is dissected to show the blood capillaries within the platelike lamellae. Direction of water flow *(large arrows)* is opposite the direction of blood flow.

MIGRATION

Eel

For centuries naturalists had been puzzled about the life history of the freshwater eel, *Anguilla* (ang-wil'la) (L., eel), a common and commercially important species in coastal streams of the North Atlantic. Eels are **catadromous** (Gr. *kata,* down, + *dromos,* running): they spend most of their lives in fresh water but spawn in salt water. Each fall, large numbers of eels were seen swimming down the rivers toward the sea, but no adults ever returned. Each spring countless numbers of young eels, called "elvers," each approximately the size of a wooden matchstick, appeared in the coastal rivers and began swimming upstream. Beyond the assumption that eels must spawn somewhere at sea, the location of their breeding grounds was totally unknown.

The first clue was provided by two Italian scientists, Grassi and Calandruccio, who in 1896 reported that elvers were not, in fact, larval eels; rather they were relatively advanced juveniles. The true larval eels, the Italians discovered, were tiny, leaf-shaped, completely transparent creatures that bore absolutely no resemblance to an eel. They had been called **leptocephali** by early naturalists who never suspected their true identity. In 1905 Johann Schmidt, supported by the Danish government, began a systematic study of eel biology, which he continued until his death in 1933. Through the cooperation of captains of commercial vessels plying the Atlantic, thousands of the leptocephali were caught in different areas of the Atlantic with the plankton nets that Schmidt supplied. By noting in what areas of the ocean larvae in different stages of development were captured, Schmidt and his colleagues eventually reconstructed the spawning migrations.

When the adult eels leave the coastal rivers of Europe and North America, they swim steadily and apparently at great depth for 1 to 2 months until they reach the Sargasso Sea, a vast area of warm oceanic water southeast of Bermuda (Figure 26-22). Here, at depths of 1000 feet or more, the eels spawn and die. The minute larvae then begin an incredible journey back to the coastal rivers of Europe. Drifting with the Gulf Stream and preyed on constantly by numerous predators, they reach the middle of the Atlantic after 2 years. By the end of the third year they reach the coastal waters of Europe where the leptocephali metamorphose into elvers, with an unmistakable eel-like body form (Figure 26-22). The males and females part company; the males remain in the brackish waters of coastal rivers and estuaries while the females continue up the rivers, often pen-

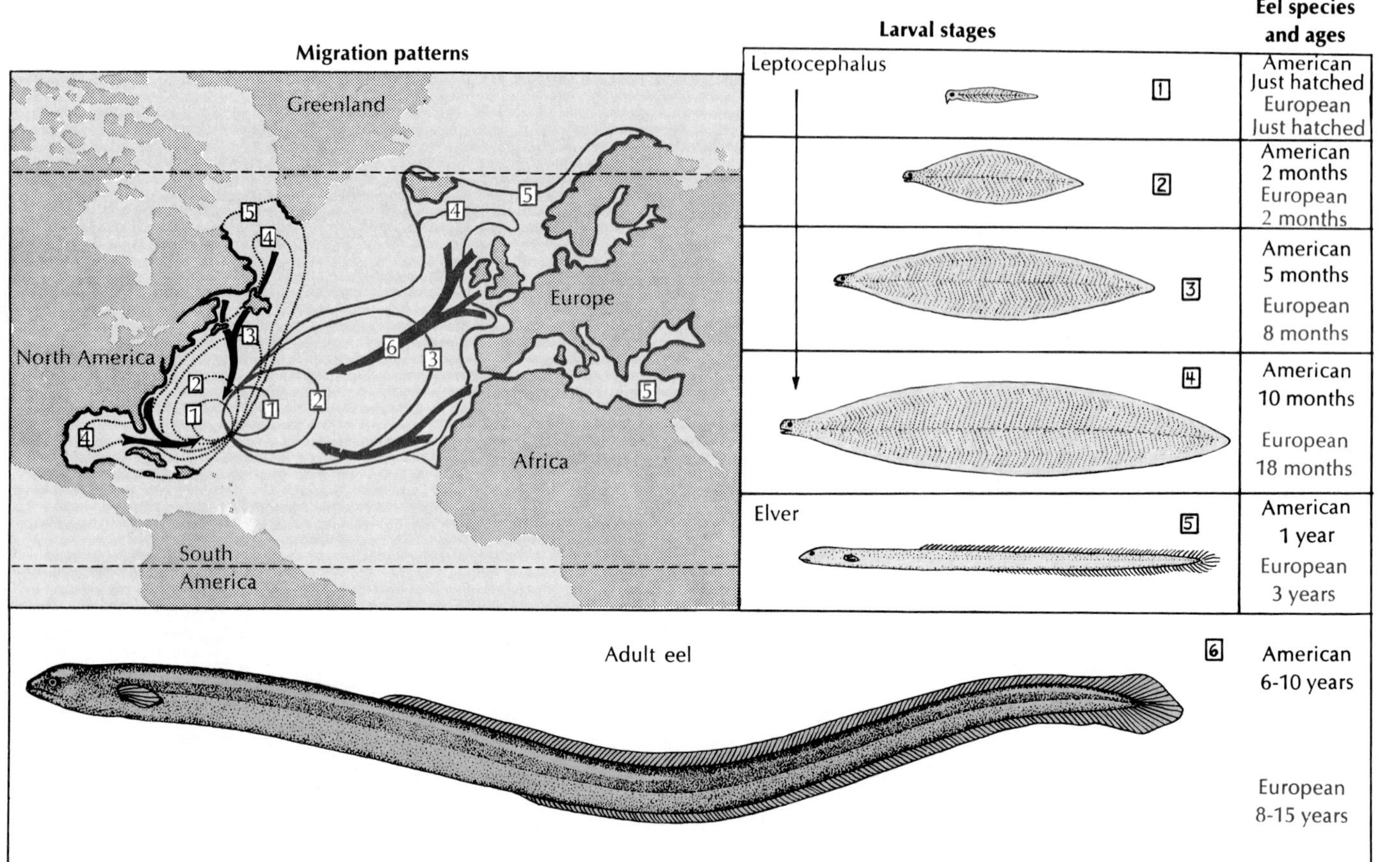

FIGURE 26-22

Life histories of the European eel, *Anguilla anguilla*, and American eel, *Anguilla rostrata*. *Red*, Migration patterns of European species. *Black*, Migration patterns of American species. Boxed numbers refer to stages of development. Note that the American eel completes its larval metamorphosis and sea journey in 1 year. It requires nearly 3 years for the European eel to complete its much longer journey.

etrating hundreds of miles upstream. After 8 to 15 years of growth, the females, now 1 m or more in length, return to the sea to join the smaller males; both return to the ancestral breeding grounds thousands of miles away to complete the life cycle. Since the Sargasso Sea is much closer to the North American coastline than it is to Europe, it requires only approximately 8 months for the American eel to make the journey.

Homing salmon

The life history of salmon is nearly as remarkable and fully as intriguing as that of the eel. Salmon are **anadromous** (Gr. *ana*, up, + *dromos*, running); that is, they spend their adult lives at sea but return to fresh water to spawn. The Atlantic salmon *(Salmo salar)* and the Pacific salmon (six species of the genus *Oncorhynchus* [on-ko-rink'us] [Gr. *onkos*, hooked, + *rhynchos*, snout]) have this practice, but there are important differences among the six species. The Atlantic salmon (as well as the closely related steelhead trout) make repeated upstream spawning runs over a period of several years. The six Pacific salmon species (king, sockeye, silver, humpback, chum, and Japanese masu) each make a single spawning run, after which they die.

Schmidt found the American eel *(Anguilla rostrata)* could be distinguished from the European eel *(Anguilla anguilla)* because it had fewer vertebrae. Recent enzyme electrophoresis analysis of eel larvae collected from the Sargasso Sea has confirmed not only the existence of separate species, but also Schmidt's belief that the European and American eels spawn in partially overlapping areas of the Sargasso Sea.

The virtually infallible homing instinct of the Pacific salmon is legend; after migrating downstream as a smolt, a sockeye salmon ranges many hundreds of miles over the Pacific for nearly 4 years, grows to 2 to 5 kg in weight, and then returns almost unerringly to spawn in the headquarters of its parent stream. Some straying does occur and is an important means of increasing gene flow and populating new streams.

Experiments by A.D. Hasler and others have shown that homing Pacific salmon are guided upstream by the characteristic odor of their parent stream.

Finally reaching the spawning beds of their parents where they themselves were hatched, they spawn and die. After hatching the following spring, the fry transform into smolts (a more advanced juvenile stage) before and during the downstream migration. At this time they are imprinted (p. 286) with the distinctive odor of the stream, which is apparently a mosaic of compounds released by the characteristic vegetation and soil in the watershed of the parent stream. They also seem to imprint on the odors of other streams they pass while migrating downriver and use these odors in reverse sequence as a map during their upriver migration as returning adults.

How do salmon find their way to the mouth of the coastal river from the trackless miles of the open ocean? Salmon move hundreds of miles away from the coast, much too far to be able to detect their parent stream odor. There are experiments suggesting that some migrating fish, like birds, can navigate by orienting to the position of the sun. However, migrant salmon have been observed to navigate on cloudy days and at night, indicating that sun navigation, if used at all, cannot be the salmon's only navigational cue. Fish also (again, like birds) appear able to detect and navigate according to the earth's magnetic field. Finally, fishery biologists concede that salmon may not require precise navigational abilities at all, but instead may use ocean currents, temperature gradients, and food availability to reach the general coastal area where "their" river is located. From this point, they would navigate by their imprinted odor map.

FIGURE 26-23

Spawning of Pacific salmon and development of the egg and young.

Courtesy Joey Morgan/Hoot Productions, Ltd. From Childerhose, R.J., and M. Trim. 1979. Pacific salmon. Seattle, University of Washington Press. Reproduction by permission of the Minister of Supply and Services Canada.

Female chum digging her nest.

Courting chum salmon. Male is nosing female's vent, possibly checking for spawning readiness.

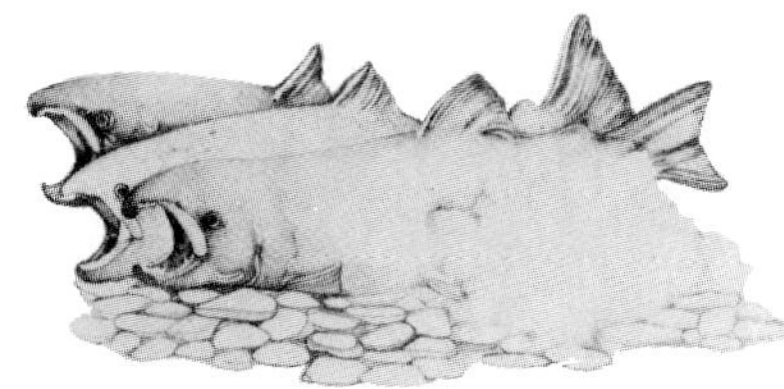

Two coho salmon fertilizing eggs spawned from female between them.

REPRODUCTION AND GROWTH

In a group as diverse as the fishes, it is no surprise to find extraordinary variations on the basic theme of sexual reproduction. Most fishes favor a simple theme: they are **dioecious,** with **external fertilization** and **external development** of the eggs and embryos. This mode of reproduction is called **oviparous** (meaning "egg-producing"). However, as tropical fish enthusiasts are well aware, the ever-popular guppies and mollies of home aquaria bear their young alive after development in the ovarian cavity of the mother. These forms are said to be **ovoviviparous,** meaning "live egg–producing." There are even some sharks that develop some kind of placental attachment through which the young are nourished during gestation. These forms, like placental mammals, are **viviparous** ("alive-producing").

Let us return to the much more common oviparous mode of reproduction. Many marine fishes are extraordinarily profligate egg producers. Males and females come together in great schools and, without mating, release vast numbers of germ cells into the water to drift with the current. Large female cod may release 4 to 6 million eggs at a single spawning. Less than one in a million will survive the numerous perils of the ocean to reach productive maturity.

Unlike the minute, buoyant, transparent eggs of pelagic marine teleosts, those of near-shore species are larger, typically yolky, nonbuoyant, and adhesive. On the whole, fishes living in coastal waters where wave action and along-shore currents are prevalent dispose of their eggs in a more conservative manner. Some

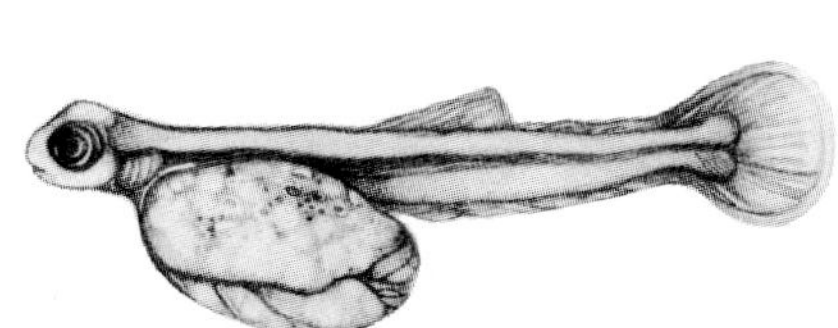

Developing alvin. When yolk is absorbed, salmon emerges from gravel bed as fry and begins feeding.

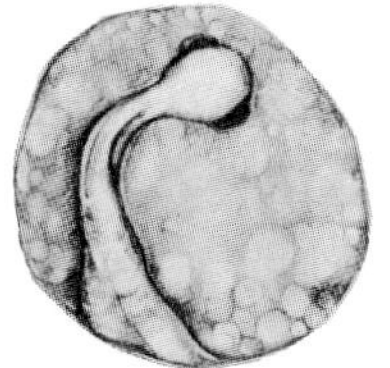

Developing embryo overlying large yolk sac.

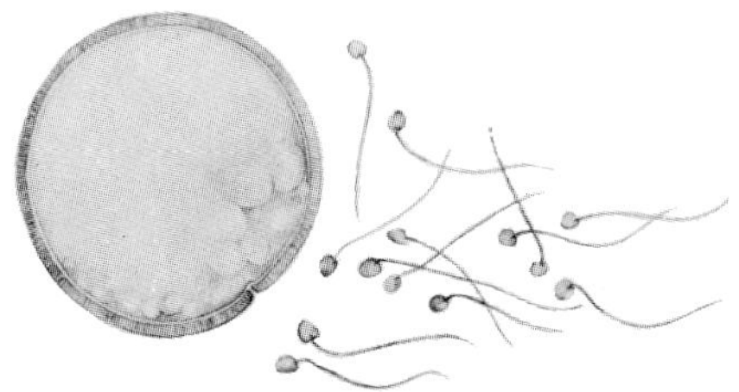

At fertilization sperm are guided to opening in egg (micropyle) by chemical attractant.

bury their eggs, many attach them to vegetation, some deposit them in nests, and some even incubate them in their mouths. Many coastal species guard their eggs. Intruders expecting an easy meal of eggs may be met with a vivid and often belligerent display by the guard, which is almost always the male.

Freshwater fishes almost invariably produce nonbuoyant eggs. Those, such as perch, that provide no parental care simple scatter their myriads of eggs among weeds or along the bottom. Freshwater fishes that do provide some form of egg care produce fewer, larger eggs that enjoy a better chance for survival.

Elaborate preliminaries to mating are the rule for freshwater fishes. The female Pacific salmon, for example, performs a ritualized mating "dance" with her breeding partner after arriving at the spawning bed in a fast-flowing, gravel-bottomed stream (Figure 26-23). She then turns on her side and scoops out a nest with her tail. As the eggs are laid by the female, they are fertilized by the male. After the female covers the eggs with gravel, the exhausted fish dies and drifts downstream.

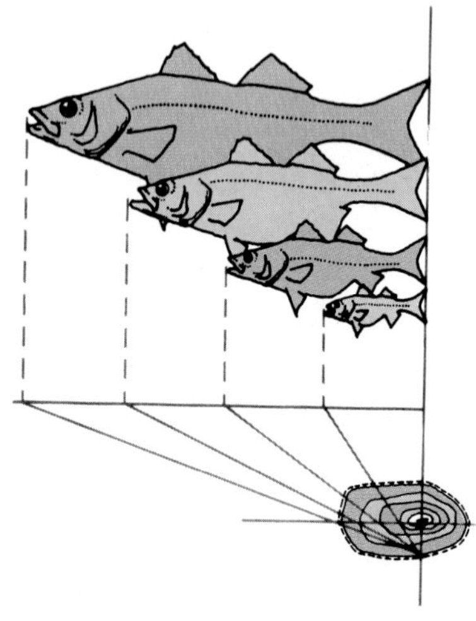

FIGURE 26-24

Scale growth. Fish scales disclose seasonal changes in growth rate. Growth is interrupted during winter, producing year marks (annuli). Each year's increment in scale growth is a ratio to the annual increase in body length. Otoliths (ear stones) and certain bones can also be used in some species to determine age and growth rate.

Soon after the eggs are fertilized and laid, they take up water and the outer surface of the eggs hardens. Following embryonic development, the fish hatches carrying a semitransparent sac of yolk, so large that active movement is nearly impossible. Not until the yolk is absorbed and the mouth and digestive tract are formed does the larva begin searching for its own food. The larva then undergoes a **metamorphosis,** especially dramatic in many marine species such as the freshwater eel described previously (Figure 26-22). Body shape and color are refashioned, and the animal, now a juvenile, bears the unmistakable body form of the adult.

Growth is temperature dependent. Consequently, fish living in temperate regions (such as the United States) grow rapidly in summer when temperatures are high and food is abundant but nearly stop growing in winter. Seasonal growth is reflected as annual rings in the scales (Figure 26-24), a distinctive record of convenience to fishery biologists wishing to determine a fish's age.

Summary

Fishes are ectothermic, gill-breathing aquatic vertebrates with fins. They are the oldest vertebrates, having originated from an unknown chordate ancestor in the Cambrian period or possibly earlier. Four classes of fishes are recognized. Most primitive are the jawless hagfishes (class Myxini) and lampreys (class Cephalaspidomorphi), remnant groups having an eel-like body form without paired fins; a cartilaginous skeleton (although their ancestors, the ostracoderms, had bony skeletons); a notochord that persists throughout life; and a disclike mouth adapted for sucking or biting. All other vertebrates have jaws, a major development in vertebrate evolution. Members of the class Chondrichthyes (sharks, rays, skates, and chimaeras) are a successful group having a cartilaginous skeleton (a degenerative feature), paired fins, excellent sensory equipment, and an active, characteristically predaceous habit. The fourth class of fishes are the bony fishes (class Osteichthyes), which may be subdivided into three stems of descent. Two stems are relic groups, the lobe-finned fishes (represented by one living species, the coelacanth) and the lungfishes. The third stem is the ray-finned fishes, a huge and diverse modern assemblage containing nearly all of the familiar freshwater and marine fishes.

The modern bony fishes (teleost fishes) have radiated into nearly 21,000 species that reveal an enormous diversity of adaptations, body form, behavior, and habitat preference. Fishes swim by undulatory contractions of the body muscles, which generate thrust (propulsive force) and lift (lateral force). Flexible fishes oscillate the whole body, but in more rapid swimmers the undulations are limited to the caudal region or tail fin alone.

Most pelagic bony fishes achieve neutral buoyancy in water using a gas-filled swim bladder, the most effective gas-secreting device known in the animal kingdom. The gills of fishes, having efficient countercurrent flow between water and blood, facilitate high rates of oxygen exchange.

Most fishes are migratory to some extent, and some, such as freshwater eels and anadromous salmon, make remarkable migrations of great length and precision. Fishes reveal an extraordinary range of sexual reproductive strategies. Most fishes are oviparous, but ovoviviparous and viviparous fishes are not uncommon. The reproductive investment may be in large numbers of germ cells with low survival (many marine fishes) or in fewer germ cells with greater parental care for better survival (freshwater fishes).

REVIEW QUESTIONS

1. Sketch a family tree for the fishes, placing the following groups in their proper relationship to each other: hagfishes; lampreys; gnathostomes; sharks, skates, and rays; lobe-finned fishes; lungfishes; ray-finned fishes. Refer to Figure 26-1.
2. Compare the morphology of hagfishes and lampreys, pointing out similarities and differences in the following features and systems: body shape, skeleton, notochord, mouth, circulatory system, kidneys, digestive system, eyes, reproduction.
3. Describe the life cycle of the sea lamprey, *Petromyzon marinus*, and the history of its invasion of the Great Lakes.
4. In what ways are sharks well equipped for their predatory lifestyle?
5. Give the common name(s) of the group or groups of fishes included within each of the following taxa: Agnatha, Osteichthyes, Cephalaspidomorphi, Chondrichthyes, Myxini, Gnathostomata.
6. Explain how the bony fishes differ from the sharks, skates, and rays in the following systems or features: skeleton, tail shape, scales, buoyancy, respiration, position of mouth, reproduction.
7. Describe the discovery of a living lobe-finned fish, the coelacanth. What is the evolutionary significance of this group?
8. Where on earth are the three surviving genera of lungfishes found?
9. Match the ray-finned fishes in the right column with the group to which each belongs in the left column:

___ chondrosteans	a. Perch
___ holosteans	b. Sturgeon
___ teleosts	c. Gar
	d. Salmon
	e. Paddlefish
	f. Bowfin

10. Give three characteristics of modern teleost fishes that distinguish them from the more primitive ray-finned fishes (chondrosteans and holosteans).
11. Compare the swimming movement of the eel with that of the trout, and explain why the latter is more efficient for rapid locomotion.
12. Explain the purpose and functioning of the swim bladder in teleost fishes. How is gas volume adjusted in the swim bladder?
13. What is meant by "countercurrent flow" as it applies to fish gills?
14. Describe the life cycle of the European eel. How does the life cycle of the American eel differ from the European?
15. How do adult Pacific salmon find their way back to their parent stream to spawn?
16. Give the mode of reproduction in fishes described by each of the following terms: oviparous, ovoviviparous, viviparous.
17. Reproduction in marine pelagic fishes and in freshwater fishes are distinctively different from each other. How and why do they differ?

SELECTED REFERENCES

Bone, Q., and N.B. Marshall 1982. Biology of fishes. New York, Chapman & Hall. *Concise, well-written, and well-illustrated primer on the functional processes of fishes.*

Horn, M.H., and R.N. Gibson. 1988. Intertidal fishes. Sci. Am. **258**:64-70 (Jan.). *Describes the special adaptations of fishes living in a demanding environment.*

Moyle, P.B., and J.J. Cech, Jr. 1982. Fishes: an introduction to ichthyology. Englewood Cliffs, N.J., Prentice-Hall, Inc. *Textbook written in a lively style and stressing function and ecology rather than morphology; good treatment of the fish groups.*

Nelson, J.S. 1984. Fishes of the world, ed. 2. New York, John Wiley & Sons, Inc. *A modern and authoritative classification of all major groups of fishes.*

THE DRAMA OF METAMORPHOSIS

The earliest chorus of frogs beside a pond on a spring evening heralds one of nature's dramatic events. Masses of frog eggs soon hatch into limbless, gill-breathing, herbivorous, fishlike tadpole larvae. Warmed by the late spring sun, they feed and grow. Then, almost imperceptibly, a remarkable transformation unfolds. Hindlegs appear and gradually lengthen. The tail shortens. The long, coiled intestine diminishes to perhaps 15% of its original length as the diet shifts from vegetarian to carnivorous. Larval teeth are lost, and the gills are replaced by lungs. The eyes develop lids. Forelegs emerge. In a matter of weeks the aquatic tadpole will, should it escape the many perils of the pond, complete its metamorphosis into an adult frog.

Metamorphosis, meaning literally "change of form," is a profound postembryonic reorganization of body structure and physiology which prepares an animal for life in a new habitat. In many invertebrates, the highly motile or planktonic larval forms transform into adults that have restricted distributions, often for a specialized habitat or food requirement. For amphibians, metamorphosis permits escape from temporary aquatic habitats that are essential for reproduction, but which may be quite unsuitable for the carnivorous and wide-ranging adults. Metamorphosis is common in protochordates and fishes, but of all vertebrates, the transformation of amphibian from tadpole larva to adult frog is the most radical and dramatic.

Springtime metamorphosis marks the end of winter and the irrepressible annual rebirth of life on earth. Perhaps, too, we glimpse in these transformations a reenactment of a Paleozoic drama when, over countless generations, the first amphibians equipped themselves for life on land.▪

CHAPTER 27

Amphibians

The class Amphibia (Gr. *amphi,* both or double, + *bios,* life) consists of some 3900 species, all that remain of a group of vertebrates that flourished and diversified during the warm Carboniferous period, some 300 million years ago. In the warm, tropical conditions of the mid-Paleozoic period, with abundant insect food, and unburdened by competition from better adapted vertebrates yet to appear, the early amphibians gradually evolved the adaptations that fitted them for life on land.

Even now, after millions of years of evolution, the amphibians are not completely land adapted; they are quasiterrestrial, hovering between aquatic and land environments. This double life is expressed in their name. Structurally they are intermediate between the fully aquatic fishes and the fully terrestrial reptiles. Although adapted for a terrestrial existence, few can stray far from moist conditions. Many, however, have developed strategies for keeping their eggs out of open water where the larvae would be exposed to predators.

The amphibians are grouped into three living orders: the tailed salamanders and newts (order Caudata), least specialized and most aquatic of all amphibians; the tailless frogs and toads (order Anura), largest and most successful group of amphibians and closest to the stock from which the higher quadrupeds descended; and the highly specialized, secretive, earthwormlike tropical caecilians (order Gymnophiona).

■ Movement onto Land

The movement from water to land is perhaps the most dramatic event in animal evolution, since it involves the invasion of a habitat that in many respects is more hazardous for life. Life was conceived in water, animals are mostly water in composition, and all cellular activities occur in water. Nevertheless, organisms eventually invaded the land, carrying their watery composition with them. To survive and maintain this fluid matrix, various structural, functional, and behavioral changes had to evolve. Considering that almost every system in the body required some modification, it is remarkable that all vertebrates are basically alike in fundamental structural and functional pattern: whether aquatic or terrestrial, vertebrates are obviously descendants of the same evolutionary branch.

Amphibians were not the first to move onto land. Insects made the transition earlier and plants much earlier still. The pulmonate snails were experimenting with land as a suitable place to live about the same time the early amphibians were. Yet of all these, the amphibian story is of particular interest because their descendants became the most successful and advanced animals on earth.

PHYSICAL CONTRAST BETWEEN AQUATIC AND LAND HABITATS

Beyond the obvious difference in water content of aquatic and terrestrial habitats there are several sharp differences between the two environments that are significant to animals attempting to move from water to land.

Greater oxygen content of air

Air contains at least 20 times more oxygen than water. Air has approximately 210 ml of oxygen per liter; water contains 3 to 9 ml per liter, depending on temperature, presence of other solutes, and degree of saturation. Furthermore, because of the great density and viscosity of water, and low rates of gaseous diffusion in water as compared to air, aquatic animals must work hard for their oxygen. They expend far more energy extracting oxygen from water than land animals expend removing oxygen from air.

Greater density of water

Although water is a resistant medium to swim through, its high density, only a little less than that of animal protoplasm, buoys up the body. One of the major problems encountered by land animals was the need to develop strong limbs and remodel the skeleton to support their bodies in air.

Constancy of temperature in water

Natural bodies of water, containing a medium with high thermal capacity, experience less fluctuation in temperature than air. The temperature of the oceans remains almost constant day after day. In contrast, both the range and fluctuation in temperature are acute on land. Its often harsh cycles of freezing, thawing, and warm temperatures, compounded by the unpredictable availability of water, presented severe thermal challenges to terrestrial animals, which had to evolve behavioral and physiological strategies to protect themselves from temperature extremes. The most successful protective strategy, homeothermy (regulated, constant body temperature), appeared in the birds and mammals.

Variety of land habitats

The variety of cover and shelter on land provided many unoccupied niches. The rich offerings of terrestrial habitats include coniferous, temperate, and tropical forests, grasslands, deserts, mountains, oceanic islands, and polar regions. Even so, the earth's hydrosphere (oceans, seas, lakes, rivers, and ice sheets), although offering a less diverse range of habitats, contains the greatest diversity of living things on earth.

Opportunities for breeding on land

The provision of safe shelter for the protection of vulnerable eggs and young is much more readily accomplished on land than in water habitats.

Origin and Relationships

The movement onto land required structural modifications of the fish body plan. Unlike a fish, which is supported and wetted by its medium and supplied with dissolved oxygen, a terrestrial animal must support its own weight, resist drying and rapid temperature change, and extract oxygen from air.

APPEARANCE OF LUNGS

The Devonian period, beginning some 400 million years ago, was a time of mild temperatures and alternating droughts and floods. During dry periods, pools and streams progressively dried up, water became foul, and the dissolved oxygen disappeared. Only those fishes able to use the abundance of atmospheric oxygen could survive such conditions. Gills were unsuitable because in air the filaments collapsed into clumps that soon dried out.

Virtually all survivors of this period, including the lobe-finned fishes and the lungfishes, had developed a kind of lung as an outgrowth of the pharynx. It was relatively simple to enhance the efficiency of the air-filled cavity by improving its vascularity with a rich capillary network and by supplying it with arterial blood from the last (sixth) pair of aortic arches. Oxygenated blood was returned directly to the heart by a pulmonary vein to form a complete pulmonary circuit. Thus the double circulation characteristic of all tetrapods originated—a systemic circulation, serving the body, and a pulmonary circulation, supplying the lungs.

DEVELOPMENT OF LIMBS FOR TRAVEL ON LAND

The evolution of limbs was also a product of difficult times during the Devonian period. When pools dried up, fishes were forced to move to another pool that still contained water. Only the lobe-finned fishes (crossopterygians) were preadapted for the task. They had strong lobed fins (used originally as swimming stabilizers) that could be adapted as paddles to lever their way across land in search of water. The pectoral fins were especially well developed, containing a series of skeletal elements in the fins and pectoral girdle that clearly foreshadowed the pentadactyl limb of tetrapods. We should note that the development first of strong fins and later of limbs did not happen so that fish could colonize land but rather as a necessary adaptation so they could find water and continue living like fish. Land travel was simply and paradoxically a means for survival in water. But lungs and limbs were fortunate developments and essential specializations that preadapted vertebrates for life on land.

EARLIEST AMPHIBIANS

All evidence points to the lobe-finned fishes, the crossopterygians, as ancestors of the modern amphibians. The lobe-finned fishes, abundant and successful in the Devonian period, possessed lungs and strong, mobile fins. Their skull and tooth structure was similar to that of the earliest known amphibians, the Labyrinthodontia, a distinct salamander-like group of the late Devonian period.

A representative of this group was a 350 million–year–old fossil called *Ichthyostega* (Gr. *ichthyos*, fish, + *stegos*, covering) (Figures 27-1 and 27-2). *Ichthyostega* possessed several new adaptations that equipped it for life on land. It had jointed, pentadactyl limbs for crawling on land, a more advanced ear structure for picking up airborne sounds, a shortening and strengthening of the skull, and a lengthening of the snout that announced improved olfactory powers for detecting

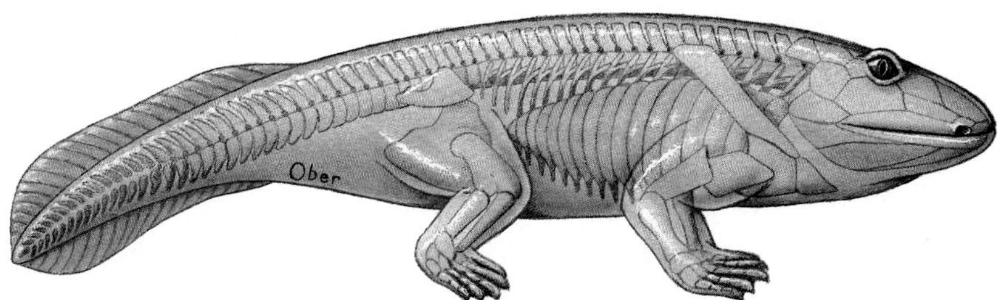

FIGURE 27-1

Reconstruction of the skeleton and body of the very early amphibian, *Ichthyostega*.

Modified from Jarvik, E. 1955. Sci. Monthly **80:**141-154.

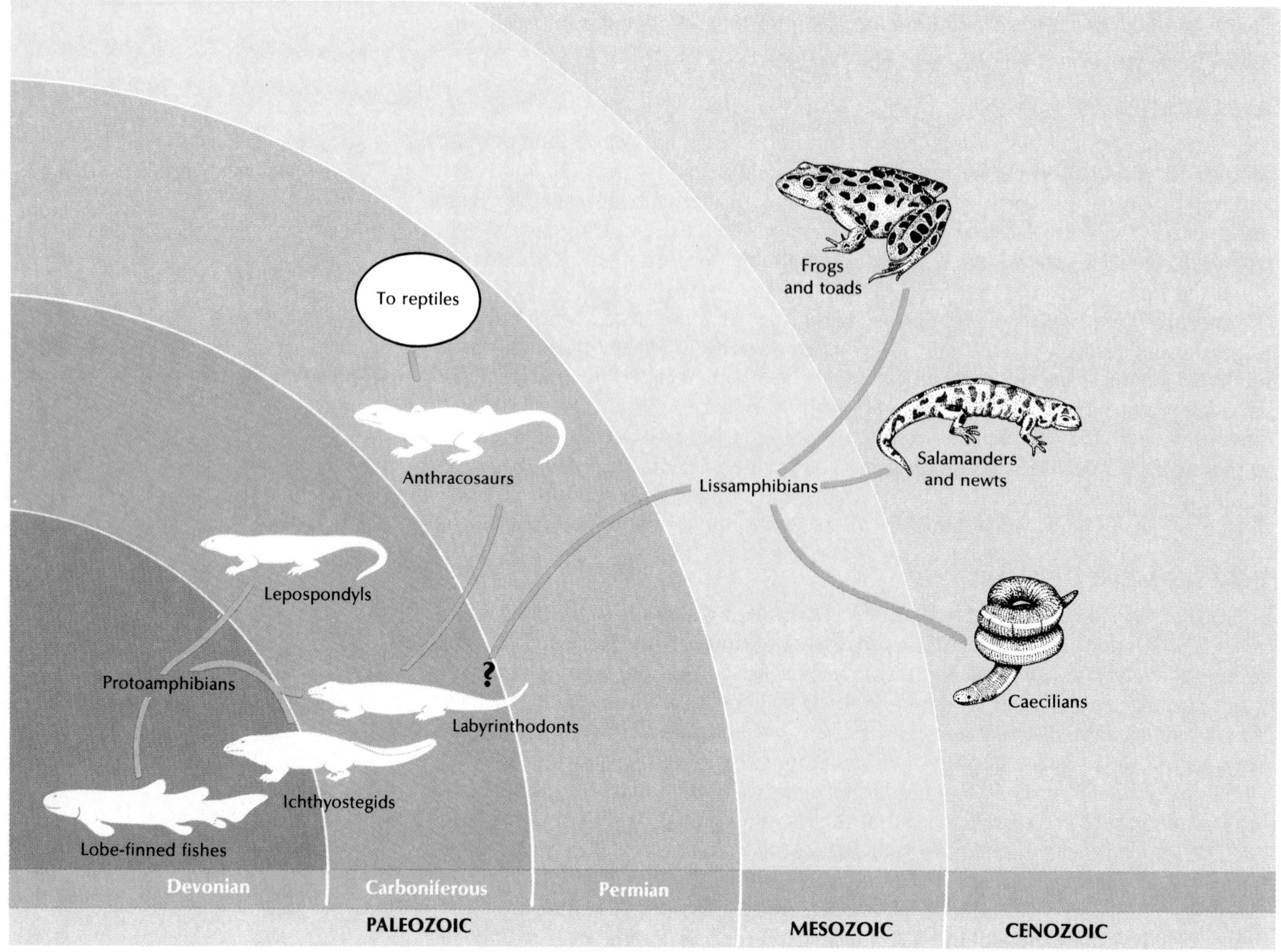

FIGURE 27-2

Evolution of the amphibians. The early amphibians arose from the lobe-finned fishes and radiated successfully during the Carboniferous period (the Age of the Amphibians). Modern amphibians have radiated from the lissamphibians of the Mesozoic era, a group of uncertain connection to the Paleozoic amphibians.

dilute airborne odors. Yet *Ichthyostega* was still fishlike in retaining a fish tail complete with fin rays and in having an operculum covering functional gills.

The capricious Devonian period was followed by the Carboniferous period, characterized by a warm, wet climate during which mosses and large ferns grew in profusion on a swampy landscape. Conditions were ideal for the amphibians. They radiated quickly into a great variety of species, feeding on the abundance of insects, insect larvae, and aquatic invertebrates available: this was the age of amphibians.

With water everywhere, however, there was little selective pressure to encourage movement onto land, and many amphibians actually improved their adaptations for living in water. Their bodies became flatter for moving about in shallow water. Many of the salamanders developed weak limbs. The tail became better developed as a swimming organ. Even the anurans (frogs and toads), which are the most terrestrial of all amphibians, developed specialized hind limbs with webbed feet better suited for swimming than for movement on land. All groups of amphibians use their thin skin as an accessory breathing organ. This specialization was encouraged by the swampy surroundings of the Carboniferous period but presented serious desiccation problems for life on land.

AMPHIBIANS' CONTRIBUTION TO VERTEBRATE EVOLUTION

Amphibians have met the problems of independent life on land only halfway. To be sure, they made several important contributions to the transition that required the evolution of their own descendants, the reptiles, to complete. Nevertheless, the amphibians are basically aquatic animals. They are ectothermic; that is, their body temperature is determined by and varies with the environmental temperature. The skin of most amphibians is thin, moist, and unprotected from desiccation in air. An intact frog loses water nearly as rapidly as a skinless frog. Most important, the amphibians remain chained to the aquatic environment by their mode of reproduction. Eggs are shed directly into the water or laid in moist surroundings and are externally fertilized (with very few exceptions). The larvae that hatch typically pass through an aquatic tadpole stage.

Many amphibians have developed ingenious devices for laying their eggs elsewhere to give their young protection and a better chance for life. They may lay eggs under logs or rocks, in the moist forest floor, in flooded treeholes, in pockets on the mother's back (Figure 27-3), or in folds of the body wall. One species of Australian frog even broods its young in its stomach.

However, it remained for the reptiles to complete the conquest of land with the development of a shelled (amniotic) egg (p. 317), which finally freed the vertebrates from a reproductive attachment to the aquatic environment. With the appearance of reptiles at the end of the Paleozoic era, the halcyon era for the amphibians began to fade. The reptiles captured rule of both water and land and removed most amphibians from both environments. From the survivors have descended the three modern orders of amphibians.

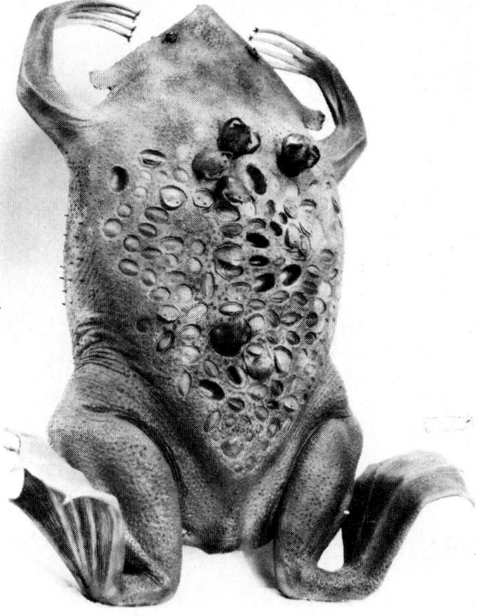

FIGURE 27-3

The female Surinam frog, carrying young on her back. As eggs are laid, the male assists in positioning them on the rough back skin of the female. The skin swells, enclosing the eggs. The approximately 60 young pass through the tadpole stage beneath the skin and emerge as froglets. The Surinam frog is found mainly in Amazon and Orinoco river systems of equatorial South America.

Courtesy American Museum of Natural History.

CHARACTERISTICS

1. Skeleton mostly bony; ribs present in some, absent in most living forms
2. Body forms vary greatly from an elongated trunk with distinct head, neck, and tail to a compact, depressed body with fused head and trunk and no intervening neck
3. **Usually four limbs (tetrapod);** webbed feet often present
4. **Smooth and moist skin with many glands,** some of which may be poison glands; pigment cells (chromatophores) common; **no scales,** except concealed dermal ones in some
5. Mouth usually large with small teeth in upper or both jaws; **two nostrils open into anterior part of mouth cavity**
6. Respiration by lungs (absent in some salamanders), skin, and gills in some, either separately or in combination
7. **Circulation with three-chambered heart,** two atria and one ventricle, and a double circulation (systemic and pulmonary) through the heart
8. Ectothermic; many frogs behaviorally thermoregulate
9. Separate sexes; metamorphosis usually present; predominantly oviparous; rarely ovoviviparous; **eggs with jellylike membrane coverings**

CLASSIFICATION

Order Gymnophiona (jim'no-fy'o-na) (Gr. *gymnos,* naked, + *ophineos,* of a snake) **(Apoda)—caecilians.** Body wormlike; limbs and limb girdles absent; dermal scales present in skin of some; tail short or absent; 95-285 vertebrae; pantropical; 6 families, 34 genera, approximately 160 species.

Order Caudata (caw-dod'uh) (L., *caudatus,* having a tail) **(Urodela)—salamanders.** Body with head, trunk, and tail; no scales; usually two pairs of equal limbs; 10-60 vertebrae; predominantly holarctic; 9 living families, 62 genera, approximately 350 species.

Order Anura (uh-nur'uh) (Gr., *an,* without, + *oura,* tail) **(Salientia)—frogs, toads.** Head and trunk fused; no tail; no scales; two pairs of limbs; large mouth; lungs; 6-10 vertebrae including urostyle (coccyx); cosmopolitan, predominantly tropical; 21 living families; 301 genera; approximately 3400 species.

Structure and Natural History

CAECILIANS—ORDER GYMNOPHIONA (APODA)

The little-known order Gymnophiona contains approximately 160 species of burrowing, wormlike creatures commonly called caecilians (Figure 27-4). They are distributed in tropical forests of South America (their principal home), Africa, and Southeast Asia. Blind or nearly blind, slender-bodied, and limbless, they feed on worms and small invertebrates underground. In their native tropical habitats they are seldom seen by humans.

FIGURE 27-4

A caecilian (order Gymnophiona). These legless and wormlike amphibians may reach lengths of 55 cm and diameters of 2 cm. Body folds give the superficial appearance of segmented worms.

SALAMANDERS AND NEWTS—ORDER CAUDATA (URODELA)

The name Caudata (L. *cauda*, tail) and the alternative order name Urodela (Gr. *oura*, tail, + *dēlos*, visible) both define those amphibians that retain their tail throughout life, as opposed to the frogs and toads that lose their tail at metamorphosis.

The salamanders and newts are the least specialized of all the amphibians. Although salamanders are found in almost all temperate and tropical regions of the world, most species occur in North America. Salamanders are typically small; most of the common North American salamanders are less than 15 cm long. Some aquatic forms are considerably longer, and the carnivorous Japanese giant salamander may exceed 1.5 m in length and weigh more than 35 kg (about 80 lb).

Caudates have primitive limbs set at right angles to the body with forelimbs and hindlimbs of approximately equal size. In some the limbs are rudimentary. One group, the sirens, with minute forelimbs and no hindlimbs at all, is so different from other urodeles that some authorities place them in a completely separate order.

Salamanders and newts prey on worms, small arthropods (especially insects), and small molluscs. Most eat only things that are moving.

Breeding behavior

Some caudates are wholly aquatic throughout their life cycle, but most are terrestrial, living in moist places under stones and rotten logs, usually not far from

FIGURE 27-5

Courtship and sperm transfer in the pygmy salamander *Desmognathus wrighti*. After judging the female's receptivity by the presence of her chin on his tail base, the male deposits a spermatophore on the ground, then moves forward a few paces. **A,** The white mass of the sperm atop a gelatinous base is visible at the level of the female's forelimb. The male moves ahead, the female following until the spermatophore is at the level of her vent. **B,** The female has recovered the sperm mass in her vent, while the male arches his tail, tilting the female upward and presumably facilitating recovery of the sperm mass.

Photographs by Lynne Houck.

water. They do not show as great a diversity of breeding habits as do frogs and toads. The eggs of most salamanders are fertilized internally, usually after the female picks up a packet of sperm **(spermatophore)** that previously has been deposited by the male on a leaf or stick (Figure 27-5). Aquatic species lay their eggs in clusters or stringy masses in the water. Terrestrial species deposit eggs in small grapelike clusters under logs or in excavations in soft earth (Figure 27-6). Unlike frogs and toads that hatch into fishlike tadpole larvae, the embryos of caudates resemble their parents when they hatch from their eggs. The larvae undergo metamorphosis in the course of development, but it is not nearly so revolutionary a change as is the metamorphosis of frog and toad tadpoles to the adult body form. American newts often have a terrestrial stage interposed between the aquatic larvae and the aquatic, breeding adults (Figure 27-7).

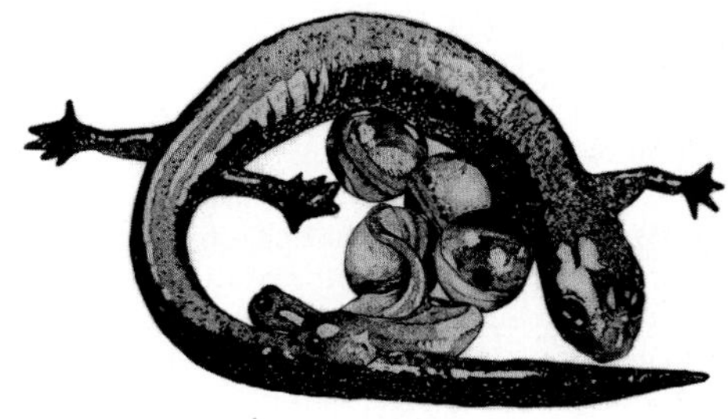

FIGURE 27-6

Red-backed salamander, *Plethodon cinereus,* encircles her hatching brood in their nest of forest humus. Unlike most amphibians, which show no further interest in their eggs after laying them, females of the large North American family Plethodontidae remain with their eggs during incubation and may even return to the same nest site year after year.

Drawn from a photograph by P.A. Zahl.

Respiration

All salamanders hatch with gills, but during development these are lost in all except the aquatic forms or in those that fail to undergo a complete metamorphosis. Gills would be useless for terrestrial salamanders, since the filaments would collapse and dry out, becoming functionless. Lungs, the characteristic respiratory organ of terrestrial vertebrates, replace the larval gills in most adult amphibians.

Yet some salamanders have dispensed with lungs altogether and thus bear the distinction of being the only vertebrates to have neither lungs nor gills. Members of the large family Plethodontidae (woodland salamanders), a group containing most of the familiar North American salamanders (Figure 27-6), are completely lungless, and some members of other salamander families exhibit reductions in lung development. In all amphibians the skin contains extensive vascular nets that serve in varying degrees for the respiratory exchange of oxygen and carbon dioxide. In lungless salamanders the efficiency of **cutaneous respiration** is increased by the penetration of a capillary network into the epidermis or by the thinning of the epidermis over superficial dermal capillaries. Cutaneous respiration is supplemented by the pumping of air in and out of the mouth where the respiratory gases exchange across the vascularized membranes of the buccal (mouth) cavity **(buccopharyngeal breathing).** Lungless plethodontid salamanders are believed to have originated in swift streams of the Appalachian mountains, where lungs would have been a disadvantage by providing too much buoy-

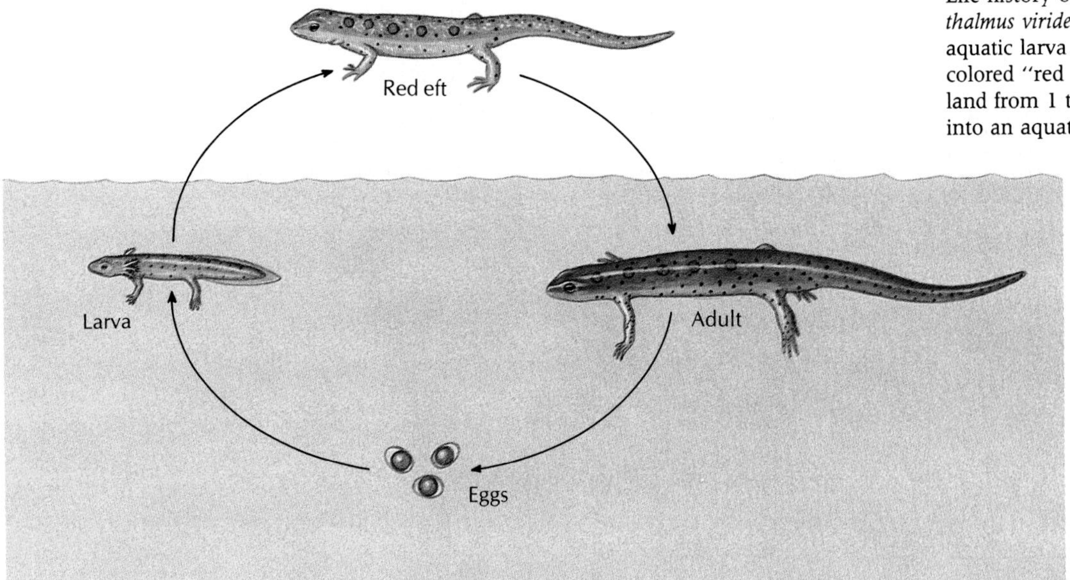

FIGURE 27-7

Life history of the red-spotted newt, *Notophthalmus viridescens.* In many habitats the aquatic larva metamorphoses into a brightly colored "red eft" stage, which remains on land from 1 to 3 years before transforming into an aquatic adult.

Although the British zoologist Walter Garstang is best remembered for his theories on the origin of chordates and vertebrates, he also wrote numerous clever, unorthodox poems about larval forms and their development. Some of his verses were so complex as to require reference books to unravel, but the following charming poem about "The Axolotl" can be enjoyed by all.

Ambystoma's a giant newt who
rears in swampy waters,
As other newts are wont to do,
a lot of fishy daughters:
These axolotls, having gills,
pursue a life aquatic,
But, when they should transform
to newts, are naughty and
erratic.

They change upon compulsion,
if the water grows too foul,
For then they have to use their
lungs, and go ashore to prowl;
But when a lake's attractive,
nicely aired, and full of food,
They cling to youth perpetual,
and rear a tadpole brood.

ancy, and the water is so cool and well oxygenated that cutaneous respiration alone was sufficient for life.

Neoteny

Whereas most salamanders complete their development to the definitive adult body form by metamorphosis, there are some species that become sexually mature while retaining their gills and other larval characteristics. This condition is called **neoteny** (Gr. *neos,* young, + *teinen,* to extend). Some are permanent larvae, a genetically fixed condition in which the developing tissues fail to respond to the thyroid hormone that, in other amphibians, stimulates metamorphosis. This condition is called **obligatory neoteny.**

Examples of permanent larvae are mud puppies of the genus *Necturus* (Gr. *nekton,* swimming, + *oura,* tail) (Figure 27-8), which live on bottoms of ponds and lakes and keep their external gills throughout life, and the amphiuma (Gr. *amphi,* double, + *pneuma,* breath) of the southeastern United States, which with its nearly useless, rudimentary legs superficially resembles an eel more than an amphibian.

There are other species of salamanders that become sexually mature and breed in the larval state, but unlike the permanent larvae, they may metamorphose to adults if environmental conditions change. Examples are species of the genus *Ambystoma* (Gr. *ambyx,* cup, + *stoma,* mouth) and of the genus *Triturus* (N.L. *Triton,* a sea god, + *oura,* tail). The American axolotl, *Ambystoma tigrinum,* widely distributed over Mexico and the southwestern United States, remains in the aquatic, gill-breathing, and fully reproductive larval form unless the water begins to dry up; then it metamorphoses to an adult, loses its gills, develops lungs, and assumes the appearance of an ordinary salamander.

FROGS AND TOADS — ORDER ANURA (SALIENTIA)

The more than 3400 species of frogs and toads that make up the order Anura are the most familiar and most successful of amphibians. The name of the order, Anura, refers to an obvious group characteristic, the absence of tails in adults (although all pass through a tailed larval stage during development). Frogs and toads are specialized for a jumping mode of locomotion, as suggested by the alternative order name, Salientia, meaning leaping.

The Anura are further distinguished from the Caudata by their larvae and a dramatic metamorphosis during development. The eggs of most frogs hatch into a tadpole ("polliwog") stage, with a long, finned tail, both internal and external gills, no legs, specialized mouthparts for herbivorous feeding (salamander larvae, in distinction, are carnivorous), and a highly specialized internal anatomy. Tadpoles look and act altogether differently from adult frogs. The metamorphosis of the frog tadpole to the adult frog is thus a striking transformation. Neoteny is never exhibited in frogs and toads as it is among salamanders.

The Anura are an old group — fossil frogs are known from the Jurassic period, 150 million years ago — and today they are a secure and successful group. Frogs and toads occupy a great variety of habitats despite their aquatic mode of reproduction and water-permeable skin, which prevent them from wandering too far afield from sources of water, and their ectothermy, which bars them from polar and subarctic habitats.

Notwithstanding their success as a distinct group, the frogs and toads are really a specialized side branch of amphibian evolution, and despite their popularity for education purposes — approximately 20 million are used each year in the United States alone — they are not good representatives of the vertebrate body plan. They lack a visible neck, the caudal vertebrae are fused into a urostyle, ribs are absent in most species, and the hind legs are much enlarged for leaping locomotion. The primitive and unspecialized salamander would be a much superior choice for the zoology laboratory were not frogs so readily available.

FIGURE 27-8

The mud puppy, *Necturus maculosus,* is an example of a permanent neotenic larva. Gills are retained throughout life.

Photograph by K. Lucas/ Biological Photo Service.

A

B

FIGURE 27-9

Two common North American frogs. **A,** Bullfrog, *Rana catesbeiana,* largest American frog and mainstay of the frog leg epicurean market (family Ranidae). **B,** Green tree frog *Hyla cinerea,* a common inhabitant of swamps of the southeastern United States (family Hylidae). Note adhesive pads on the feet.

A, Photograph by Joe McDonald/ Visuals Unlimited; B, Photograph by C.P. Hickman, Jr.

FIGURE 27-10

American toad *Bufo americanus* (family Bufonidae). This principally nocturnal yet familiar amphibian feeds on large numbers of insect pests as well as on snails and earthworms. The warty skin contains numerous poison glands that produce a surprisingly poisonous milky fluid, providing the toad excellent protection from a variety of potential predators.

Photograph by C.P. Hickman, Jr.

Frogs and toads are divided into 12 families. The best-known frog families in North America are Ranidae, containing most of our familiar frogs (Figure 27-9, *A*), and Hylidae, the tree frogs (Figure 27-9, *B*). True toads, belonging to the family Bufonidae (L. *bufo,* toad, + *idae,* suffix), have short legs, stout bodies, and thick skins usually with prominent warts (Figure 27-10). However, the term "toad" is used rather loosely to refer to more or less terrestrial members of several other families.

The largest anuran is the West African *Gigantorana (=Conraua) goliath* (Gr., *gigantos,* giant, + *rana,* frog), which is more than 30 cm long from tip of nose to anus (Figure 27-11). This giant eats animals as big as rats and ducks. The smallest frog recorded is *Phyllobates limbatus* (Gr. *phyllon,* leaf, + *batēs,* climber), which is only approximately 1 cm long. This tiny frog, which is more than covered by a dime, is found in Cuba. Our largest American frog is the bullfrog, *Rana catesbeiana* (Figure 27-9, *A*), which reaches a head and body length of 20 cm.

In addition to their importance in biomedical research and education, frogs have long served the epicurean frog leg market. The mainstay of this market is the bullfrog which is in such heavy demand in Europe and the United States—the worldwide harvest is an estimated 200 million bullfrogs annually—that its populations have fallen drastically as the result of excessive exploitation, and the draining and pollution of wetlands. Most are Asian bullfrogs imported from India and Bangladesh, some 80 million collected each year from rice fields in Bangladesh alone. With so many insect-eating frogs removed from the ecosystem, rice production is threatened from uncontrolled, flourishing insect populations. In the United States, attempts to raise bullfrogs in farms have not been successful, mainly because bullfrogs are voracious eating machines that normally will only accept living prey, such as insects, crayfish, and other frogs.

Distribution and behavior

Perhaps the most abundant and successful of frogs are the some 260 species of the genus *Rana* (Gr., frog), found all over the temperate and tropical regions of the world except in New Zealand, the oceanic islands, and southern South America. The leopard frog *Rana pipiens* has a wide variety of habitats and, with all its subspecies and forms, is the most widespread of all the North American frogs. This is the species most commonly used in biology laboratories and for classical electrophysiological research. It has been found in some form in nearly every state, although it is sparingly represented along the extreme western part of the Pacific coast. It also extends far into northern Canada and as far south as Panama.

Within the range of a species, frog populations are often restricted to certain habitats (for instance, to certain streams or pools) and may be absent or scarce in similar habitats of the range. The pickerel frog *(Rana palustris)* is especially noteworthy because it is known to be abundant only in certain localized regions.

Most of our larger frogs are solitary in their habitats except during the breed-

FIGURE 27-11

Gigantorana goliath (family Ranidae) of West Africa, the world's largest frog. This specimen weighed 3.3 kg (approximately 7½ pounds).

Courtesy American Museum of Natural History.

ing season. During the breeding period most of them, especially the males, are very noisy. Each male usually takes possession of a particular perch, where he may remain for hours or even days trying to attract a female to that spot by his calls. (Both males and females have a vocal apparatus, but is is better developed in males, and only males of some species possess vocal sacs which serve as resonating chambers to amplify the sound [Figure 27-12].) At times frogs are mainly silent, and their presence is not detected until they are disturbed. When they enter the water, they dart about swiftly and reach the bottom of the pool, where they kick up a cloud of muddy water. In swimming, they hold the forelimbs near the body and kick backward with the webbed hind limbs, which propel them forward. When they come to the surface to breathe, only the head and foreparts are exposed, and, since they usually take advantage of any protective vegetation, they are difficult to see.

During the winter months most frogs hibernate in the soft mud of the bottom of pools and streams. The wood frog hibernates under stones, logs, and stumps in the forest area. Naturally their life processes are at a very low ebb during the hibernation period, and such energy as they need is derived from the glycogen and fat stored in their bodies during the spring and summer months.

Adult frogs have numerous enemies, such as snakes, aquatic birds, turtles, raccoons, humans, and many others; only a few tadpoles survive to maturity. Although usually defenseless, many frogs and toads in the tropics and subtropics are aggressive, jumping and biting at predators. Some defend themselves by feigning death. Most anurans can blow up their lungs so that they are difficult to swallow. When disturbed along the margin of a pond or brook, a frog often remains quite still; when it thinks it is detected, it jumps, not always into the water where enemies may be lurking but rather into grassy cover on the bank. When held in the hand a frog may cease its struggles for an instant to put its captor off guard and then leap violently, at the same time voiding its urine. The best protection is their ability to leap and their use of poison glands. Bullfrogs in captivity do not hesitate to snap at tormenters and are capable of inflicting painful bites.

While native American amphibians continue to disappear as wetlands are drained, an exotic frog introduced into southern California has found the climate quite to its liking. The African clawed frog *Xenopus laevis* (Figure 27-13) is a voracious, aggressive, primarily aquatic frog that is rapidly displacing native frogs and fish from several waterways and is spreading rapidly. The species was introduced into North America in the 1940s when they were used extensively in human pregnancy tests. When more efficient tests appeared in the 1960s, some hospitals simply dumped surplus frogs into nearby streams, where the prolific breeders have become almost indestructible pests.

FIGURE 27-12

Inflated vocal sac of a singing spring peeper *Hyla crucifer*. The sac, which functions primarily as a resonator, is connected to the mouth cavity by slitlike valves.

Photograph by C.P. Hickman, Jr.

FIGURE 27-13

African clawed frog, *Xenopus laevis.* The claws, an unusual feature in frogs, are on the hind feet. This is an alien species in California, where it is considered a serious pest.

Photograph by C.P. Hickman, Jr.

Reproduction

Because frogs and toads are ectothermic, they breed, feed, and grow only during the warmer seasons of the year. One of the first drives after the dormant period is breeding. In the spring males croak and call vociferously to attract females. When their eggs are mature, the females are clasped by the males, a mating embrace called **amplexus** (L. embrace) (Figure 27-14). As the female lays the eggs, the male discharges seminal fluid containing sperm over the eggs to fertilize them. The eggs are laid in large masses, usually anchored to vegetation.

Development of the fertilized egg (zygote) begins almost immediately (Figure 27-15). By repeated division (cleavage) the egg is converted into a hollow ball of cells (blastula). This undergoes continued differentiation to form an embryo with a tail bud. At 6 to 9 days, depending on the temperature, a tadpole hatches from the protective jelly coats that had surrounded the original fertilized egg.

FIGURE 27-14

A male green frog *Hyla cinerea* clasps a larger female during the breeding season in a South Carolina swamp. Clasping (amplexus) is maintained until the female deposits her eggs. Like most tree frogs, these are capable of rapid and marked color changes; the male here, normally green, has darkened during amplexus.

Photograph by C.P. Hickman, Jr.

At the time of hatching, the tadpole has a distinct head and body with a compressed tail. The mouth is located on the ventral side of the head and is provided with horny jaws for scraping off vegetation from objects for food. Behind the mouth is a ventral adhesive disc for clinging to objects. In front of the mouth are two deep pits, which later develop into the nostrils. Swellings are found on each side of the head, and these later become external gills. There are finally three pairs of external gills, which are later transformed into internal gills that become covered with a flap of skin (the operculum) on each side. On the right side the operculum completely fuses with the body wall, but on the left side a small opening, the **spiracle** (L. *spiraculum,* air hole) remains, through which water flows after entering the mouth and passing the internal gills. The hindlegs appear first, while the forelimbs are hidden for a time by the folds of the operculum. During metamorphosis the tail is resorbed, the intestine becomes much shorter, the mouth undergoes a transformation into the adult condition, lungs develop, and the gills are resorbed. The leopard frog usually completes its metamorphosis within 3 months; the bullfrog takes 2 or 3 years to complete the process.

Migration of frogs and toads is correlated with their breeding habits. Males usually return to a pond or stream in advance of the females, which they then attract by their calls. Some salamanders are also known to have a strong homing instinct; guided by olfactory cues, they return year after year to the same pool for reproduction. The initial stimulus for migration in many cases is attributable to a

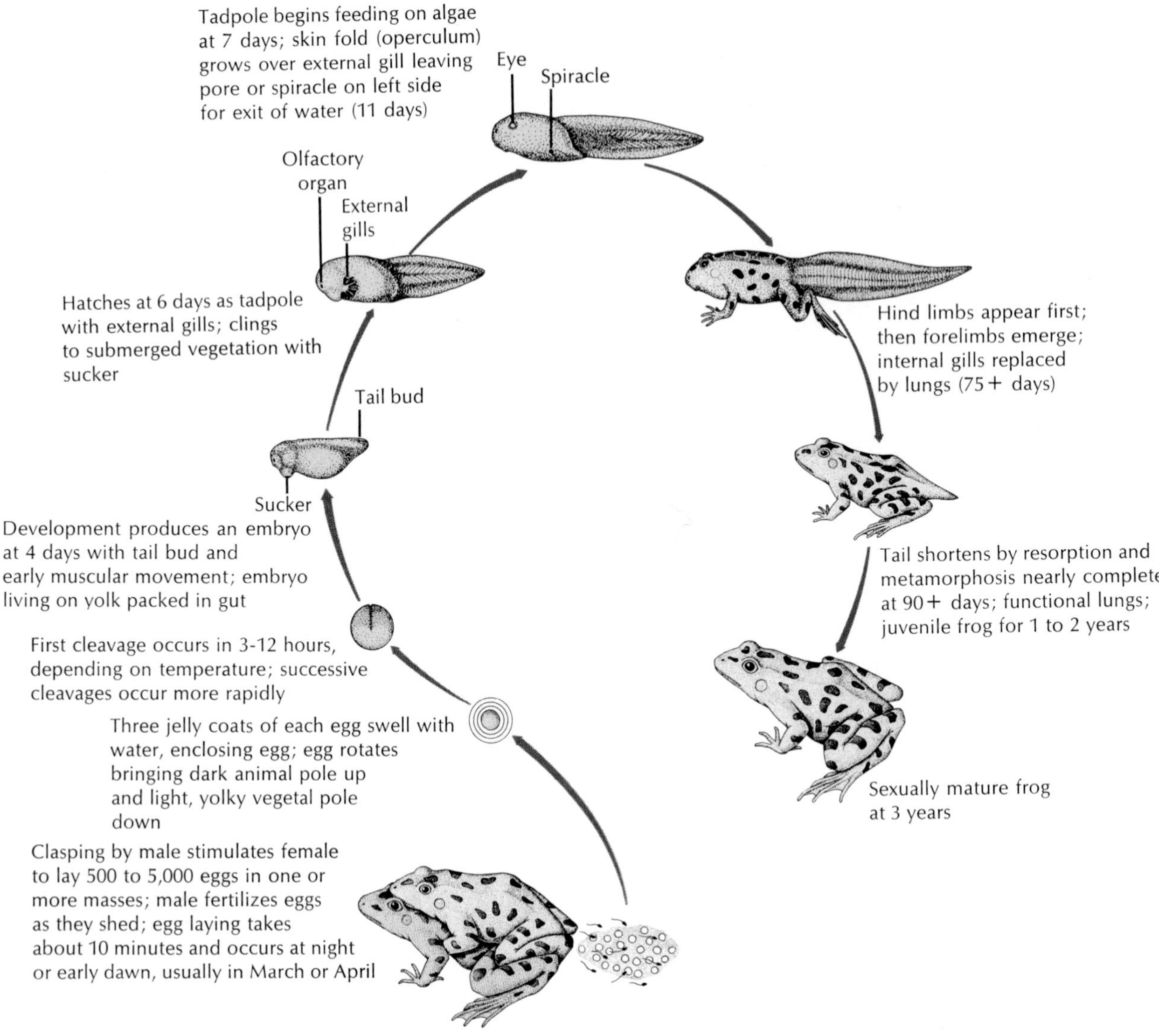

FIGURE 27-15

Life cycle of a leopard frog.

seasonal cycle in the gonads plus hormonal changes that increase the frogs' sensitivity to temperature and humidity changes.

Summary

The amphibians are ectothermic quadrupedal vertebrates with glandular skin, and they breathe by lungs, gills, or skin. They are a transitional group, neither fully aquatic nor fully terrestrial. Although the group enjoyed a lengthy period of prominence in the late Paleozoic era, their aquatic mode of reproduction prevented a successful conquest of land. Most amphibians must return to water to lay their eggs, and their larvae are clearly fishlike. Direct development that omits the aquatic larval stage is not uncommon, however. Despite their insignificance as a group today, the amphibians made several important contributions to vertebrate evolution, including a shift from gill to lung breathing, separation of pulmonary and systemic circulations, development of tetrapod limbs, restructuring of the skeleton to support the body in air, and redesigning of the senses of vision, smell, and hearing for use on land.

The modern amphibians consist of three evolutionary lines. The caecilians (order Gymnophiona) are a small tropical group of highly specialized, wormlike forms. The salamanders (order Caudata) are tailed amphibians that have retained the generalized four-legged body plan of their Paleozoic ancestors. Fertilization is internal following spermatophore transfer from male to female. Some salamanders are neotenic, becoming sexually mature while retaining gills and other larval characteristics.

The most successful of modern amphibians are the frogs and toads (order Anura). All are specialized for a jumping mode of locomotion. Most are adapted for wet habitats, but many toads and the tree frogs tolerate arid environments. Vocal cords are highly developed in anurans, especially in males which vocalize to attract females. Fertilization is external. The eggs hatch into tadpole larvae that undergo a dramatic metamorphosis to the adult body form.

Longtailed salamander, *Eurycea longicauda*, a common plethodontid salamander.

Photograph by C.P. Hickman, Jr.

REVIEW QUESTIONS

1. Compared to the aquatic habitat, the terrestrial habitat offers both advantages and problems for an animal making the transition from water to land. Summarize how these differences might have influenced the early evolution of amphibians.
2. Describe the evolution of lungs and strong limbs in amphibians from fish ancestors and comment on the selective pressures responsible.
3. What evidence can you offer to support the following statement: "While amphibians made an important start toward movement onto land, they are still aquatic animals in many respects"?
4. Give the literal meaning of the name Gymnophiona. What animals are included in this amphibian order, what do they look like, and where do they live?
5. What is the literal meaning of the order names Caudata and Anura? What features distinguish the members of these two orders from each other?
6. Describe the breeding behavior of a typical woodland salamander.
7. Explain the meaning of the term "neoteny" as it applies to the salamanders. What is the difference in the kind of neoteny that occurs in the mudpuppy *Necturus* and the kind that occurs in the American axolotl *Ambystoma?*
8. Briefly describe the reproductive behavior of frogs. In what important ways do frog and salamander reproductions differ?

SELECTED REFERENCES

Duellman, W.E., and L. Trueb. 1986. Biology of amphibians. New York, McGraw-Hill Book Company. *Important comprehensive sourcebook of information on amphibians, extensively referenced and illustrated.*

Gibbons, W. 1983. Their blood runs cold: adventures with reptiles and amphibians. University, Ala., University of Alabama Press. *Delightful account of the personal experience of a herpetologist, filled with engaging stories and interesting facts.*

Goin, C.J., O.B. Goin, and G.R. Zug. 1978. Introduction to herpetology, ed. 3. San Francisco, W.H. Freeman & Co. *Basic introductory text for the study of amphibians and reptiles.*

Halliday, T.R., and K. Adler (eds.). 1986. The encyclopedia of reptiles and amphibians. New York, Facts on File, Inc. *Excellent authoritative reference work with high-quality illustrations.*

THE LAND EGG

The amphibians, with lungs and well-developed legs, pointed the path to the complete conquest of land. But, with shell-less eggs and gill-breathing larvae, their development remained hazardously aquatic. The reptiles developed an egg that could be laid on land. This shelled egg, more than any other adaptation, unshackled the early reptiles from the aquatic environment by freeing the developmental process from water-dwelling stages. In fact, these water-dwelling stages were not eliminated but simply enclosed within a fluid-filled membranous cavity, the amnion, in which the developing embryo floats. Its food supply is provided from a yolk sac. Another membranous sac, the allantois, serves both as a surface for gas exchange and as a chamber for the storage of nitrogenous wastes that accumulate during development. Finally, surrounding and protecting everything is a firm or leathery, but porous, shell.

The early reptiles that developed this land egg (or "amniotic" egg as it is often called) must certainly have enjoyed an immediate advantage over the amphibians. No longer requiring water for reproduction, reptiles could disperse over the land and bury their eggs in nests that were both free from predation by water-inhabiting species and hidden from land-based predators. With the evolution of this ultimate adaptation, the vertebrates' conquest of land was assured.▪

CHAPTER 28

Reptiles

The class Reptilia (rep-til'e-a) (L. *repto,* to creep) are the first truly terrestrial vertebrates. With some 7000 species (approximately 300 species in the United States and Canada) occupying a great variety of aquatic and terrestrial habitats, they are clearly a successful group. Nevertheless, reptiles are perhaps remembered best for what they once were, rather than for what they are now. The age of reptiles, which lasted 160 million years, encompassing the Jurassic and Cretaceous periods of the Mesozoic era, saw the appearance of a great radiation of reptiles, many of huge stature and awesome appearance, that completely dominated life on land. However, they declined relatively quickly toward the end of the Cretaceous period.

Out of the dozen or so principal reptile groups that evolved, four remain today. The most successful of these are the lizards and snakes of the order Squamata. A second group is the crocodilians (order Crocodilia) that have survived almost unchanged for 200 million years. To a third group belong the turtles of the order Testudines, also an ancient group that has undergone little change from its early reptile ancestors. The last group, order Rhynchocephalia, is a relict stock represented today by a sole survivor, the tuatara of New Zealand.

Origin and Adaptive Radiation

Biologists generally agree that reptiles arose from the labyrinthodont amphibians sometime before the Permian period, which began approximately 280 million years ago. The oldest "stem reptiles" were the captorhinomorphs (Gr. *kaptō,* to gulp down, + *rhinos,* nose, + *morphē,* form) (Figure 28-1). These were small, lizardlike animals, which probably fed mainly on the insects that were undergoing an adaptive radiation at the same time and were already numerous. Early in their evolution, the captorhinomorphs rather quickly (in a geological time sense) diverged into several specialized lineages and entered a long and spectacular evolutionary history that carried them to the present. Their descendants included the dinosaurs, marine reptiles, flying reptiles (pterosaurs [Gr. *pteron,* wing, + *sauros,* lizard]), and mammal-like reptiles

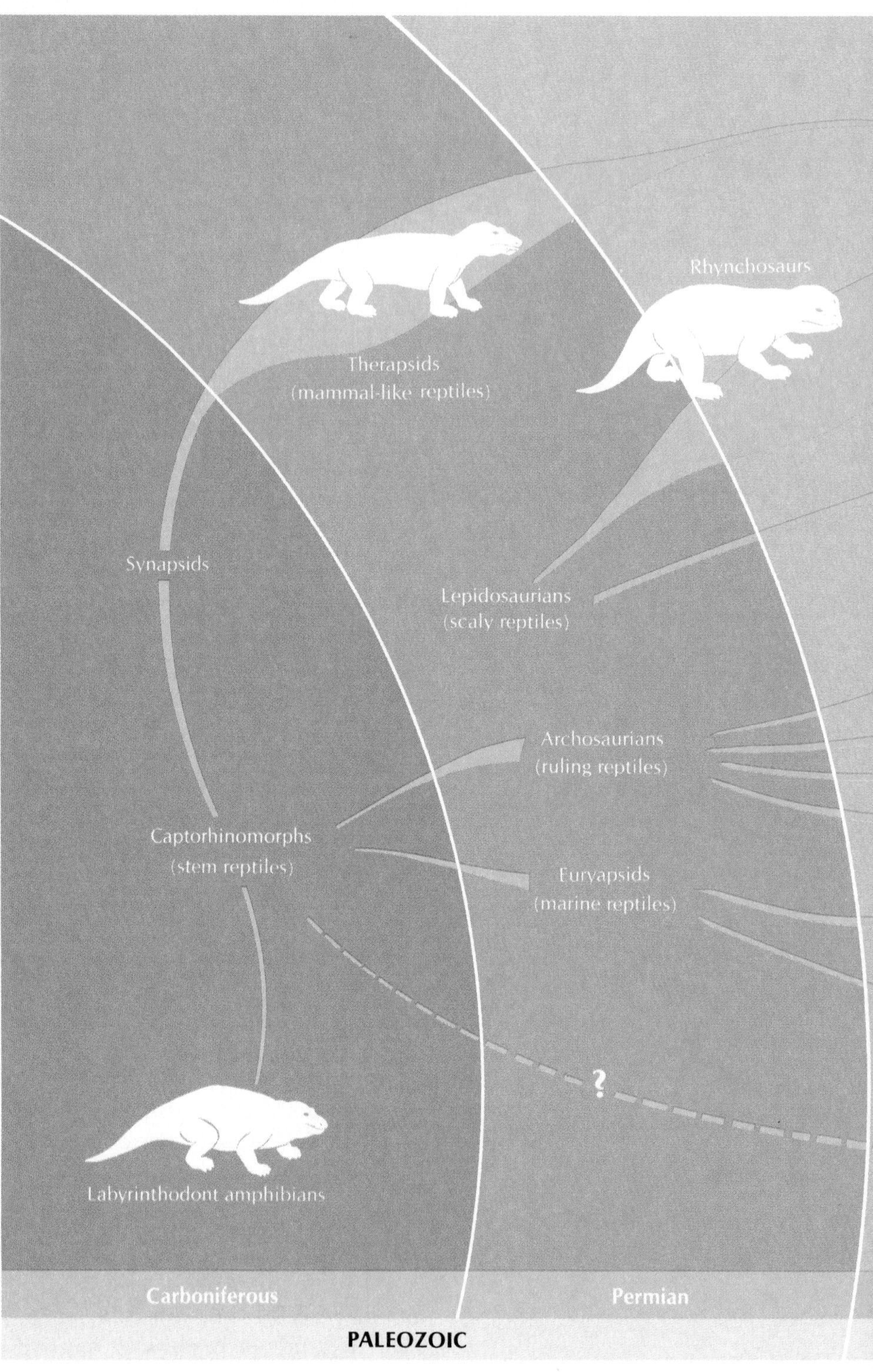

FIGURE 28-1

Evolution of the reptiles. The transition from certain labyrinthodont amphibians to reptiles occurred in the late Paleozoic to Mesozoic eras. The transition was effected by developing an amniotic egg that made land existence possible, although this egg may well have developed before the earliest reptiles had ventured far on land. Explosive radiation by the reptiles may have been due partly to the increased variety of ecological habitats into which they could move. The fossil record shows that lines arising from stem reptiles led to marine reptiles, dinosaurs, pterosaurs, crocodilians, turtles, lizards and snakes, and rhynchosaurs. Another radiation led to the mammals. Of this great assemblage, the only reptiles now in existence belong to four orders: Testudines, Crocodilia, Squamata, and Rhynchocephalia. How are the mighty fallen!

(therapsids [Gr. *theraps*, attendant, + *-ida*, suffix]), which produced the ancestral stock of the mammals. All were highly successful groups.

The adaptive radiation of reptiles, especially pronounced in the Triassic period (which followed the Permian period), corresponded with the appearance of new ecological habitats. These were provided by the climatic and geological changes that were taking place at that time, such as a variable climate from hot to cold, mountain building and terrain transformations, and a varied assortment of plant life.

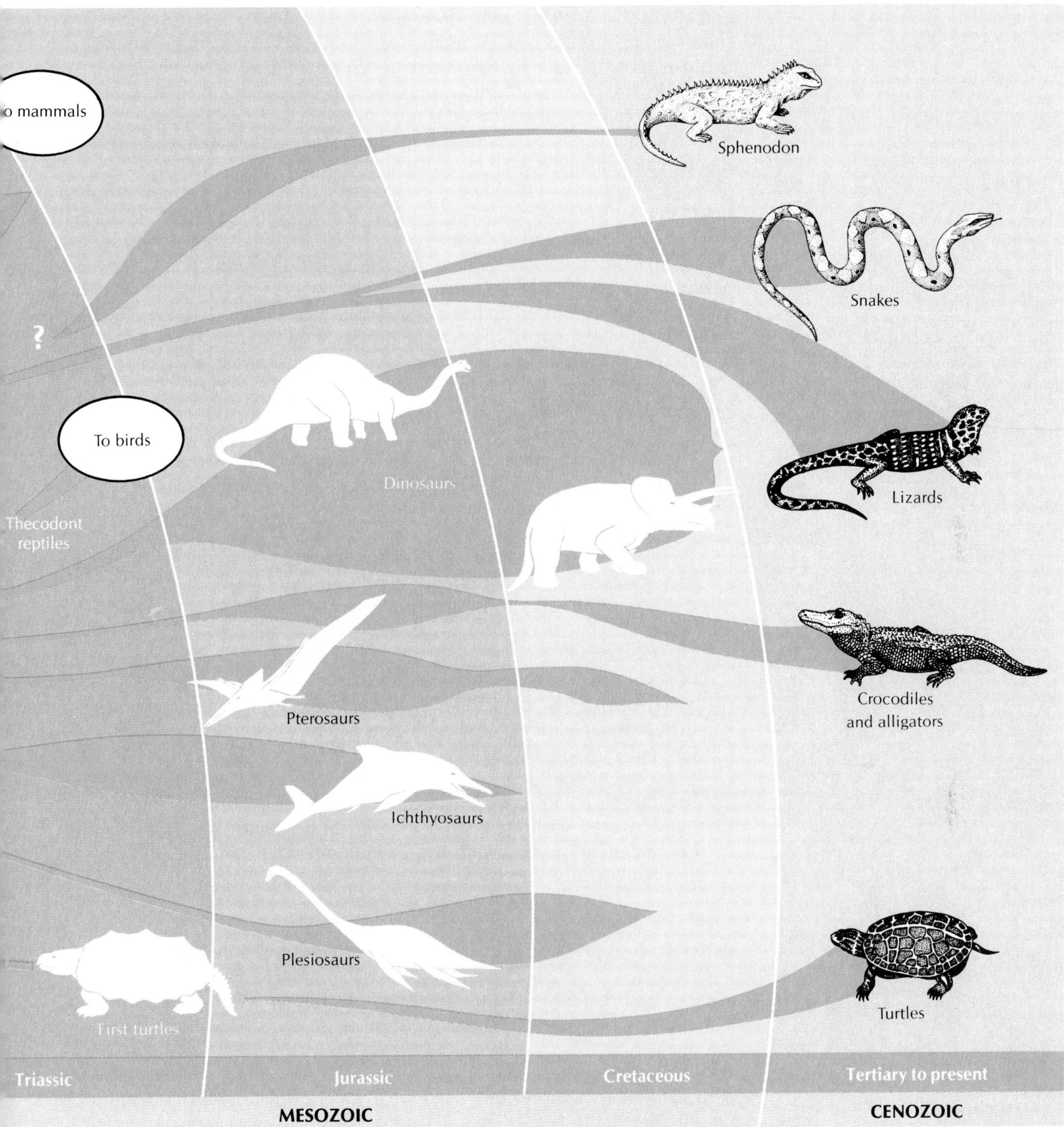

The Mesozoic era was the age of the great ruling reptiles. Then suddenly they disappeared near the close of the Cretaceous period approximately 65 million years ago. What caused their demise? The most exotic explanation is the recent hypothesis of L.W. Alvarez that the earth was struck with a huge asteroid several miles in diameter. The impact would have injected an enormous quantity of rock dust into the stratosphere where it would circle the globe for months, turning day into night. The absence of sunlight would have shut off photosynthesis, attacking food chains at their base. Both marine algae and land plants would have died,

Classical views of the dinosaurs and other Mesozoic reptiles as witless, cold-blooded beasts with inferior adaptability have recently been undergoing revision. Many biologists believe that the dinosaurs were alert creatures showing complex behavior equal to that of today's lizards and crocodilians. Some certainly traveled in organized family groups. Speculations drawn from ecological and paleontological data that the dinosaurs were warm blooded (endothermic) have generally been discounted, although because of their large size the dinosaurs probably enjoyed fairly stable body temperatures.

and all herbivorous and carnivorous animals would have become extinct except those feeding on insects or decaying vegetation. This hypothesis is supported by the fossil record, which reveals that many groups of marine microorganisms that depended directly or indirectly on photosynthesis, as well as all animals larger than about 25 kg, became extinct at this time. But the Alvarez hypothesis, despite this and other supportive evidence, remains controversial and debate continues among paleontologists.

Other changes were also taking place at this time. The ruling reptiles may not have been sufficiently adaptable to survive the combined effects of changing climate, the rapid spread of modern plants, and the appearance of aggressive and intelligent mammals. Competition from the aggressive mammals must have been particularly severe. Yet some reptiles survived. Turtles had their protective shells, snakes and lizards evolved in habitats of dense forests and rocks where they could meet the competition of any tetrapod, and crocodiles, because of their size, stealth, and aggressiveness, had few enemies in their aquatic habitats.

CHARACTERISTICS

1. Body varied in shape, compact in some, elongated in others; body covered with an exoskeleton of **horny epidermal scales** with the addition sometimes of bony dermal plates; **integument with few glands**
2. **Paired limbs, usually with five toes,** adapted for climbing, running, or paddling; absent in snakes and some lizards
3. Well-ossified skeleton; ribs with sternum forming a complete thoracic basket; **skull with one occipital condyle**
4. Respiration by lungs; **no gills;** cloaca secondarily used for respiration by some
5. **Three-chambered heart; crocodiles with four-chambered heart;** usually one pair of aortic arches
6. **Metanephric (paired) kidney; uric acid as main nitrogenous waste**
7. **Ectothermic;** some snakes and lizards behaviorally thermoregulate
8. Nervous system with the optic lobes on the dorsal side of brain; **12 pairs of cranial nerves**
9. Separate sexes; **internal fertilization**
10. **Eggs covered with leathery or calcium-impregnated shells; extraembryonic membranes (amnion, chorion, yolk sac,** and **allantois)** present during embryonic life

CLASSIFICATION

Order Testudines (tes-tu'din-eez) (L. *testudo,* tortoise) **(Chelonia).** Body in a bony case of dermal plates with dorsal carapace and ventral plastron; jaws without teeth but with horny sheaths; immovable quadrate (articular surface of lower jaw); vertebrae and ribs fused to shell; anus consisting of a longitudinal slit. Turtles (330 species).

Order Squamata (squa-ma'ta) (L. *squamatus,* scaly, + *-ata,* characterized by). Skin with outer layer of horny epidermal scales or plates which is shed periodically; teeth attached to jaws; quadrate freely movable; vertebrae usually concave in front; anus consisting of a transverse slit. Snakes (2700 species), lizards (3100 species).

Order Crocodilia (crok'o-dil'e-a) (L. *crocodilus,* crocodile, + *-ia,* pl. suffix). Four-chambered heart; vertebrae usually concave in front; forelimbs usually with five digits, hindlimbs with four digits; immovable quadrate; anus consisting of longitudinal slit. Crocodiles, gavials, and alligators (25 species).

Order Rhynchocephalia (rin'ko-se-fay'le-a) (Gr. *rhynchos,* snout, + *kephalē,* head). Biconcave vertebrae, immovable quadrate; parietal eye fairly well developed and easily seen; anus consisting of transverse slit. *Sphenodon punctatum*—only species existing.

HOW REPTILES SHOW ADVANCEMENTS OVER AMPHIBIANS

Reptiles have tough, dry, scaly skin, offering protection against desiccation and physical injury. The skin consists of a thin **epidermis,** which is shed periodically, and a much thicker, well-developed **dermis** (Figure 28-2). The dermis is provided with **chromatophores,** the color-bearing cells that give many lizards and snakes their colorful hues. It is also the layer that, unfortunately for their bearers, is converted into alligator and snakeskin leather, so esteemed for expensive pocketbooks and shoes. The characteristic **scales** of reptiles are mostly derived from the epidermis and thus are not homologous to fish scales, which are bony, dermal structures. In some reptiles, such as alligators, the scales remain throughout life, growing gradually to replace wear. In others, such as snakes and lizards, new scales grow beneath the old, which are then shed at intervals. In snakes the old skin (epidermis and scales) is turned inside out when discarded; lizards split out of the old skin, leaving it mostly intact and right side out, or it may slough off in pieces.

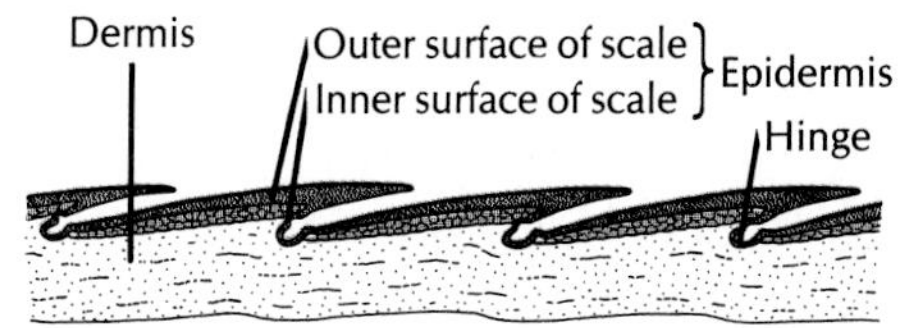

FIGURE 28-2

Section of the skin of a reptile showing the overlapping epidermal scales.

The shelled egg of reptiles contains food and protective membranes for supporting embryonic development on dry land. Reptiles lay their eggs in sheltered locations on land. The young hatch as lung-breathing juveniles rather than as aquatic larvae. The great significance of this adaptation was described in the prologue to this chapter. The appearance of the shelled (amniotic) egg widened the division between evolving amphibians and reptiles and, probably more than any other adaptation, contributed to the decline of amphibians and the ascendance of reptiles. The shelled egg is illustrated and described in some detail on pp. 317 and 318.

Reptilian jaws are efficiently designed for applying crushing or gripping force to prey. The jaws of fish and amphibians were designed for quick jaw closure, but once the prey was seized, little static force could be applied. In reptiles jaw muscles became larger, longer, and arranged for much better mechanical advantage.

Reptiles have some form of copulatory organ, permitting internal fertilization. Internal fertilization is obviously a requirement for a shelled egg, since the sperm must reach the egg before the egg is enclosed. Sperm from the paired testes are carried by the vas deferens to the copulatory organ (penis), which is an evagination of the cloacal wall. The female system consists of paired ovaries and oviducts. The glandular walls of the oviducts secrete albumin (a water reserve) and shells for the large eggs.

Reptiles have a more efficient circulatory system and higher blood pressures than amphibians. In all reptiles the right atrium, which receives unoxygenated blood from the body, is completely partitioned from the left atrium, which receives oxygenated blood from the lungs. In the crocodilians there are two completely separated ventricles as well (Figure 28-3); in other reptiles the ventricle is incompletely separated. The crocodilians are thus the first vertebrates with a four-chambered heart. Even in those reptiles with incomplete separation of the ventricles, flow patterns within the heart prevent admixture of pulmonary (oxygenated) and systemic (unoxygenated) blood; all reptiles therefore have two functionally separate circulations.

Reptile lungs are better developed than those of amphibians. Reptiles depend almost exclusively on the lungs for gas exchange; this is supplemented by **pharyngeal respiration** in some of the aquatic turtles. The lungs have a larger respiratory surface in reptiles than amphibians, and breathing is by **negative pressure;** that is, the air is *sucked* into the lungs, as in higher vertebrates, rather than *forced* in by mouth muscles, as in the amphibians. **Cutaneous respiration** (gaseous exchange through the skin), so important to most amphibians, was largely abandoned by the reptiles.

The reptilian kidneys are of the advanced metanephros type with their own passageways (ureters) to the exterior. The kidneys are very efficient in

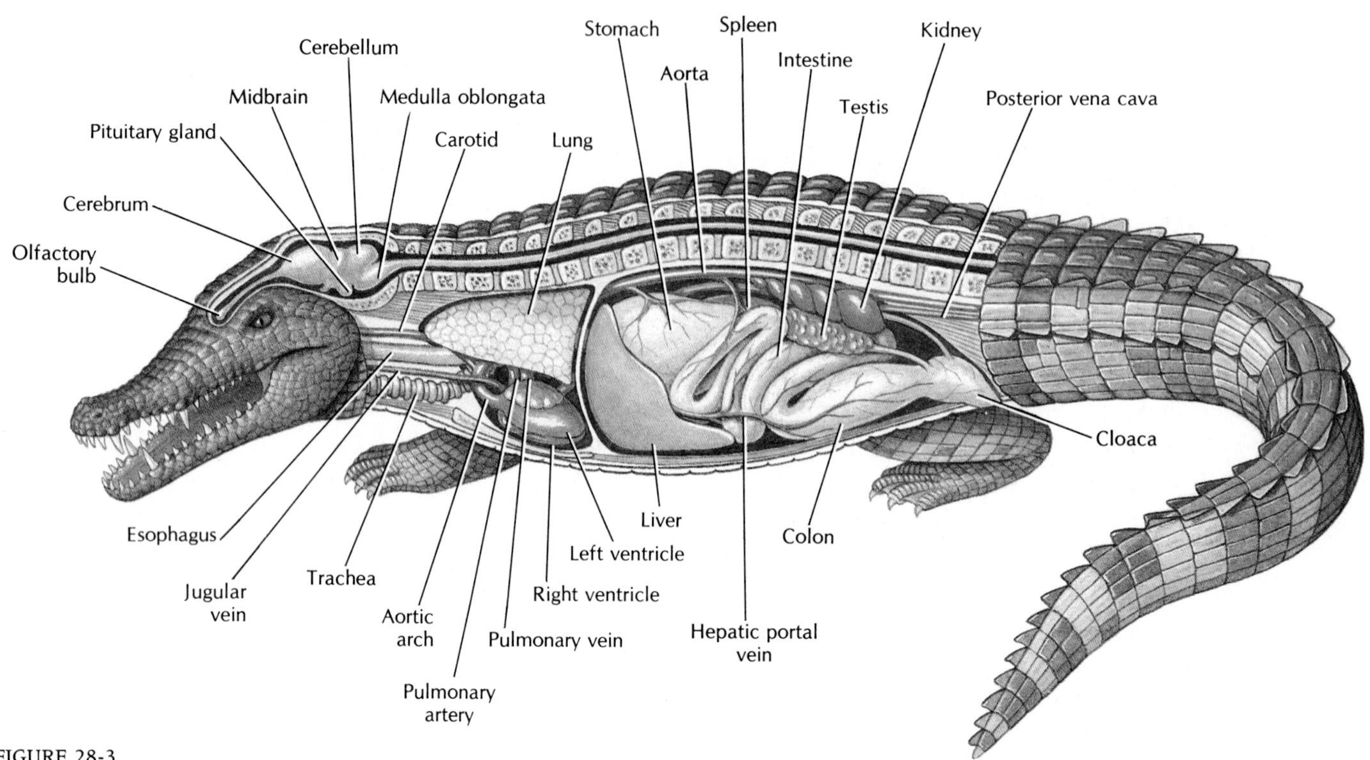

FIGURE 28-3

Internal structure of the male crocodile.

producing small volumes of urine, thus conserving precious water. Nitrogenous wastes are excreted as **uric acid,** rather than urea or ammonia. Uric acid has a low solubility and precipitates out of solution readily; as a result the urine of many reptiles is a semisolid paste.

All reptiles, except the limbless members, have better body support than the amphibians and more efficiently designed limbs for travel on land. Many of the dinosaurs walked on powerful hindlimbs alone.

The reptilian nervous system is considerably more advanced than that of the amphibian. Although the reptile's brain is small, the **cerebrum** is increased in size relative to the rest of the brain. The crocodilians have the first true cerebral cortex **(neopallium).** Central nervous system connections are more advanced, permitting complex kinds of behavior unknown in the amphibians.

■ Characteristics and Natural History of Reptilian Orders

TURTLES—ORDER TESTUDINES

The terms "turtle," "tortoise," and "terrapin" are applied variously to different members of the turtle order. In North American usage, they are all correctly called turtles. The term "tortoise" is frequently given to land turtles, especially the large forms. British usage of the terms, for example, is different: "tortoise" is the inclusive term, whereas "turtle" is applied only to the aquatic members.

The turtles are an ancient group that have plodded on from the Triassic period to the present with very little change in their early basic morphology. They are enclosed in shells consisting of a dorsal **carapace** (Sp. *carapacho,* covering) and a ventral **plastron** (Fr., breastplate). Clumsy and unlikely as they appear to be within their protective shells, they are nonetheless a varied and successful group that seems able to accommodate to the presence of humans. The shell is so much a part of the animal that it is built in with the thoracic vertebrae and ribs (Figure 28-4). Like a medieval coat of armor, the shell offers protection for the head and appendages, which, in most turtles, can be retracted into it. But since the ribs are fused to the shell, the turtle cannot expand its chest to breathe but must force air in and out of the lungs by alternately contracting the flank and shoulder muscles.

Lacking teeth, the turtle jaw is provided with tough, horny plates for gripping food (Figure 28-5). Sound perception is poor in turtles, and most turtles are mute

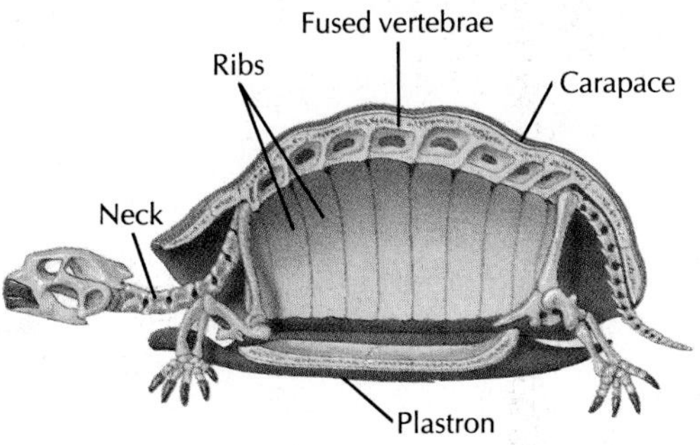

FIGURE 28-4

Skeleton and shell of a turtle, showing fusion of vertebrae and ribs with the carapace. The long and flexible neck allows the turtle to withdraw its head into its shell for protection.

(the biblical "voice of the turtle" refers to the turtledove). Compensating for poor hearing is a good sense of smell and color vision. Turtles are oviparous and fertilization is internal by means of a penis on the ventral wall of the cloaca.

The great marine turtles, buoyed by their aquatic environment, may reach 2 m in length and 725 kg in weight. One is the leatherback. The green turtle, so named because of its greenish body fat, may exceed 360 kg, although most individuals of this economically valuable and heavily exploited species seldom live long enough to reach anything approaching this size. Some land tortoises may weigh several hundred kilograms, such as the giant tortoises of the Galápagos Islands (Figure 28-6) that so intrigued Darwin during his visit there in 1835. Most tortoises are rather slow moving; 1 hour of determined trudging carries a large Galápagos tortoise approximately 300 m. A low metabolism probably explains in part the longevity of turtles, the Methuselahs of vertebrates. At least five species are known to live a century or more.

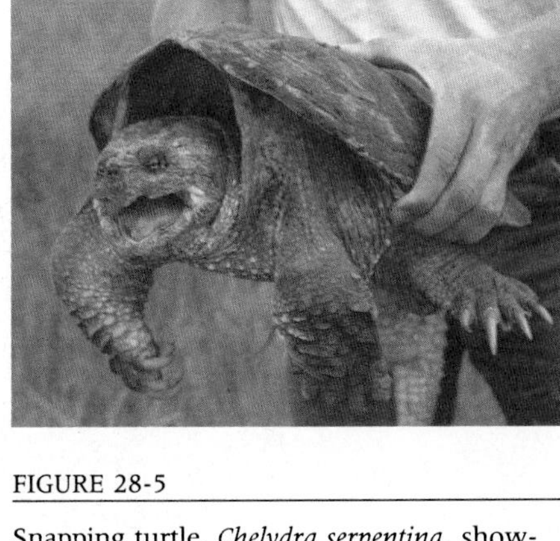

FIGURE 28-5

Snapping turtle, *Chelydra serpentina*, showing the absence of teeth. Instead, the jaw edges are covered with a horny plate.

Photograph by C.P. Hickman, Jr.

LIZARDS AND SNAKES—ORDER SQUAMATA

The lizards and snakes are the most recent products of reptile evolution and by all odds the most successful, comprising approximately 95% of all known living reptiles. The modern lizards began their adaptive radiation during the Cretaceous period when the great dinosaurs were at the climax of their dominance of land.

FIGURE 28-6

Mating Galápagos tortoises. The male has a concave plastron that fits over the highly convex carapace of the female, helping to provide stability during mating.

Photograph by C.P. Hickman, Jr.

The immediate success of lizards was probably caused by a versatile and highly mobile jaw apparatus, which allowed them to capture large, struggling prey. The snakes appeared during the late Cretaceous period probably from burrowing lizards, although the early fossil record of this group is poor. Two adaptations in particular characterize the snakes. One is the extreme elongation of the body. The other is the highly mobile jaw apparatus that allows a snake to swallow prey much larger than the snake's own diameter.

The order Squamata is divided into suborders Sauria, which includes the lizards, and Serpentes, which includes the snakes. In addition to the difference in jaw articulation and structure, lizards have movable eyelids, external ear openings, and legs (usually); snakes lack these characteristics.

Lizards—suborder Sauria

FIGURE 28-7

Tokay, *Gekko gecko,* of Southeast Asia has a true voice and is named after the strident repeated *to-kay, to-kay* call.

Photograph by J.H. Gerard.

The lizards are an extremely diversified group, including terrestrial, burrowing, aquatic, arboreal, and aerial members. Among the more familiar groups in this varied suborder are the **geckos** (Figure 28-7), small, agile, mostly nocturnal forms with adhesive toe pads that enable them to walk upside down and on vertical surfaces; the **iguanas,** New World lizards, often brightly colored with ornamental crests, frills, and throat fans (this group includes the remarkable marine iguana of the Galápagos Islands shown in Figure 28-8); the **skinks,** with elongate bodies and reduced limbs; and the **chameleons,** a group of arboreal lizards, mostly of Africa and Madagascar (Figure 28-9). The chameleons are entertaining creatures that catch insects with their sticky-tipped tongues, which can be flicked accurately and rapidly to a distance greater than their own body length.

Unlike the turtles, snakes, and crocodilians, which have distinctive body forms and ways of life, the lizards have radiated extensively into a variety of habitats and reveal great diversity of body form. Some are even limbless like snakes (Figure 28-10). Nevertheless, lizards can be distinguished from snakes in several ways. One difference is the lower jaw. Even in limbless lizards, the two halves are firmly united at the mandibular symphysis. This makes it impossible for lizards to swallow oversize meals as snakes are able to do with their "floating" jaw elements. Most lizards have movable eyelids, whereas a snake's eyes are permanently covered with a transparent cap. Lizards have keen vision for daylight, although one group, the nocturnal geckos, has pure rod retinas for night vision. Most lizards have an external ear that snakes lack. However, as with other

FIGURE 28-8

Marine iguanas, *Amblyrhynchus cristatus,* of the Galápagos Islands. This is the only marine lizard in the world. It has special salt-removing glands in the eye orbits and long claws that enable it to cling to the bottom while feeding on seaweed, its exclusive diet. It may dive to depths exceeding 10 m (33 feet) and remain submerged more than 30 minutes.

Photograph by C.P. Hickman, Jr.

reptiles, hearing does not play an important role in the lives of most lizards. Geckos are exceptions because the males are strongly vocal (to announce territory and discourage the approach of other males), and they must, of course, hear their own vocalizations.

Many lizards live in the world's hot and arid regions, aided by several adaptations for desert life. Since their skin lacks glands, water loss by this avenue is much reduced. They produce a semisolid urine with a high content of crystalline uric acid. This is an excellent adaptation for conserving water and is found in other groups living successfully in arid habitats (birds, insects, and pulmonate snails). Some, such as the Gila monster of the southwestern United States deserts, store fat in their tails, which they draw on during drought to provide both energy and metabolic water (Figure 28-11). The way many lizards keep their body temperature relatively constant by behavioral thermoregulation was described in Chapter 8.

FIGURE 28-9

A chameleon snares a dragonfly. After cautiously edging close to its target, the chameleon suddenly lunges forward, anchoring its tail and feet to the branch. A split second later, it launches its sticky-tipped, foot-long tongue to trap the prey. The eyes of this common European chameleon *(Chamaeleo chamaeleon)* are swiveled forward to provide binocular vision and excellent depth perception.

Photograph by J. Andrada.

FIGURE 28-10

A glass lizard, *Ophisaurus* sp., of the southeastern United States. This legless lizard feels stiff and brittle to the touch and has an extremely long, fragile tail that readily fractures when the animal is struck or seized. Most specimens, such as this one, have only a partly regenerated tip to replace a much longer tail previously lost. Glass lizards can be readily distinguished from snakes by the deep, flexible groove running along each side of the body. They feed on worms, insects, spiders, birds' eggs, and small reptiles.

Photograph by L.L. Rue III.

FIGURE 28-11

Gila monster, *Heloderma suspectum,* of southwestern United States desert regions and the related Mexican bearded lizard are the only venomous lizards known. These brightly colored, clumsy-looking lizards feed principally on birds' eggs, nesting birds and mammals, and insects. Unlike poisonous snakes, the Gila monster secretes venom from glands in its lower jaw. The chewing bite is painful to humans but seldom fatal.

Photograph by L.L. Rue III.

Snakes do not bite pieces from their prey; they swallow the prey whole (Figure 28-12). The two halves of the lower jaw (mandibles) are joined only by muscles and skin, allowing them to spread widely apart. Furthermore, many of the skull bones are so loosely articulated that the entire skull can flex asymmetrically to accommodate oversize prey. Since the snake must keep breathing during the slow process of swallowing, the tracheal opening is located forward between the two mandibles. And to protect the brain from protesting struggles of the prey, it is entirely enclosed within a bony case. Altogether, the feeding equipment of a snake is perhaps its most remarkable specialization.

Snakes—suborder Serpentes

Snakes are entirely limbless and lack both the pectoral and pelvic girdles (the latter persists as vestiges in pythons and boas). The numerous vertebrae of snakes, shorter and wider than those of tetrapods, permit quick lateral undulations through grass and over rough terrain. The ribs increase rigidity of the vertebral column, providing more resistance to lateral stresses. The elevation of the neural spine gives the numerous muscles more leverage. In addition to the specialized skull that enables them to swallow prey several times their own diameter, snakes also differ from lizards in having no movable eyelids (snakes' eyes are permanently covered with upper and lower transparent eyelids fused together) and no external ears. Snakes are totally deaf, although they are sensitive to low-frequency vibrations conducted through the ground.

Snakes employ a unique set of special senses to hunt down their prey. In addition to being deaf, most snakes have relatively poor vision, with the tree-living snakes of the tropical forest being a conspicuous exception (Figure 28-13). In fact, some arboreal snakes possess excellent binocular vision that helps them track prey through the branches where scent trails would be difficult to follow. But most snakes live on the ground and rely on chemical senses to hunt food. In addition to the usual olfactory areas in the nose, which are not well developed, there are the **Jacobson's organs,** a pair of pitlike organs in the roof of the mouth. These are lined with an olfactory epithelium and are richly innervated. The **forked tongue,** flicking through the air, picks up scent particles and conveys them to the mouth (Figure 28-14); the tongue is then drawn past Jacobson's organs or the tips of the forked tongue are inserted directly into the organs. Information is then transmitted to the brain where scents are identified.

Snakes of the subfamily Crotalinae within the family Viperidae are called **pit vipers** because of special heat-sensitive **pit organs** on their heads, between the nostrils and the eyes (Figures 28-14 and 28-15). All of the best-known North American poisonous snakes are pit vipers, such as the several species of rattlesnakes, the water moccasin, and the copperhead. The pits are supplied with a dense packing of free nerve endings from the fifth cranial nerve. They are exceedingly sensitive to radiant energy (long-wave infrared) and can distinguish temperature differences smaller than 0.003° C from a radiating surface. Pit vipers use the pits to track warm-blooded prey and to aim strikes, which they can make as effectively in total darkness as in daylight.

All vipers have a pair of teeth, modified as fangs, on the maxillary bones. These lie in a membrane sheath when the mouth is closed. When the viper

FIGURE 28-12

Black rat snake, *Elaphe obsoleta obsoleta,* swallowing a chipmunk.

Photograph by L.L. Rue III.

FIGURE 28-13

Parrot snake, *Leptophis ahaetulla.* The slender body of this Central American tree snake is an adaptation for sliding along branches without weighing them down.

Photograph by C.P. Hickman, Jr.

FIGURE 28-14

A timber rattlesnake, *Crotalus horridus,* flicks its tongue to smell its surroundings. Scent particles trapped on the tongue's surface are transferred to Jacobson's organs, olfactory organs in the roof of the mouth. Note the heat-sensitive pit organ between the nostril and eye.

Photograph by L.L. Rue III.

strikes, a special muscle and bone lever system erects the fangs when the mouth opens (Figure 28-16). The fangs are driven into the prey by the thrust, and venom is injected into the wound through a canal in the fangs.

A viper immediately releases its prey after the bite and follows it until it is paralyzed or dies. Then the snake swallows it whole. Approximately 8000 bites and 12 deaths from pit vipers are reported each year in the United States.

The tropical and subtropical countries are the homes of most species of snakes, both the venomous and nonvenomous varieties. Even there, less than one third of the snakes are venomous; the nonvenomous snakes kill their prey by constriction (Figure 28-17) or by grasping and swallowing. Their diet tends to be restricted; many feed principally on rodents, whereas others feed on other reptiles, fishes, frogs, and insects. Some African, Indian, and Neotropical snakes have become specialized as egg eaters.

Poisonous snakes are usually divided into four groups based on the type of fangs. The vipers (family Viperidae) have tubular fangs at the front of the mouth; this group includes the American pit vipers previously mentioned and the Old World true vipers, which lack facial heat-sensing pits. Among the latter are the common European adder and the African puff adder. A second family of poisonous snakes (family Elapidae [Gr. *elaps,* a fish, + *idae,* suffix]) has short, permanently erect fangs so that the venom must be injected by chewing. In this group are the cobras (Figure 28-18), mambas, coral snakes, and kraits. The highly poisonous sea snakes are usually placed in a third family (Hydrophiidae). The

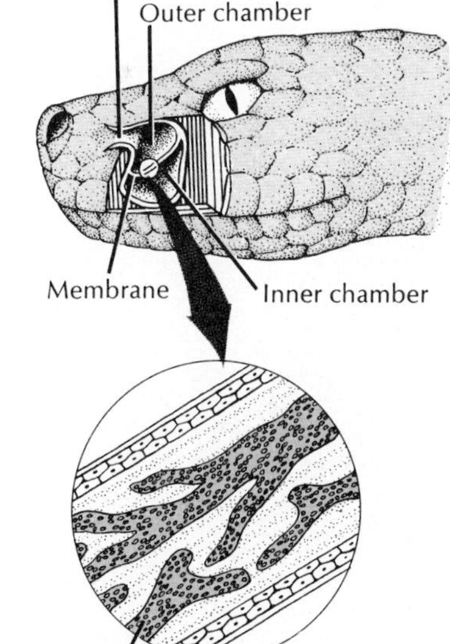

FIGURE 28-15

Pit organ of rattlesnake, a pit viper. Cutaway shows location of a deep membrane that divides the pit into inner and outer chambers. Heat-sensitive nerve endings are concentrated in the membrane.

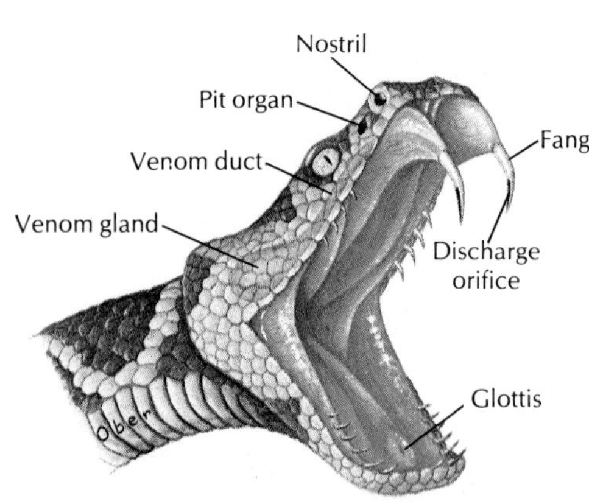

FIGURE 28-16

Head of rattlesnake showing the venom apparatus. The venom gland, a modified salivary gland, is connected by a duct to the hollow fang.

FIGURE 28-17

Nonvenomous African house snake, *Boaedon fuluginosus,* constricting a mouse before swallowing it.

Photograph by C.P. Hickman, Jr.

very large family Colubridae, which contains most of the familiar nonvenomous snakes, does include at least two poisonous (and very dangerous) snakes—the African boomslang and the African vine snake—that have been responsible for human fatalities. Both are rear-fanged snakes that normally use their venom to quiet struggling prey.

The saliva of all harmless snakes possesses limited toxic qualities, and it is logical that evolution should have stressed this toxic tendency. There are two types of snake venom. The **neurotoxic** type acts mainly on the nervous systems, affecting the optic nerves (causing blindness) or the phrenic nerve of the diaphragm (causing paralysis of respiration). The **hemolytic** type breaks down the red blood corpuscles and blood vessels and produces extensive hemorrhaging of blood into the tissue spaces. Many venoms have both neurotoxic and hemolytic properties. The toxicity of a venom is determined by the median lethal dose on laboratory animals (LD_{50}). By this standard the venoms of the Australian tiger snake and some of the sea snakes appear to be the most deadly of poisons drop for drop. However, several larger snakes are more dangerous. The aggressive king cobra, which may exceed 5.5 m in length, is the largest and probably the most dangerous of all poisonous snakes. In India, where snakes come in constant contact with people, some 200,000 snakebites cause more than 9000 deaths each year.

FIGURE 28-18

Yellow cobra, *Naja flava,* of Africa. Cobras erect the front of the body and flatten the neck as a threat display and before attacking. Although the cobra's strike range is limited, all cobras are dangerous because of the extreme toxicity of the venom.

Photograph by L.L. Rue III.

Most snakes are **oviparous** (L. *ovum,* egg, + *parere,* to bring forth) species that lay their shelled, elliptical eggs beneath rotten logs, under rocks, or in holes dug in the ground (Figure 28-19). Most of the remainder, including all the American pit vipers except the tropical bushmaster, are **ovoviviparous** (L. *ovum,* egg, + *vivus,* living, + *parere,* to bring forth), giving birth to well-formed young. A very few snakes are **viviparous** (L. *vivus,* living, + *parere,* to bring forth); in these a placenta-like structure forms, permitting the exchange of materials between the embryonic and maternal bloodstreams. Snakes are able to store sperm and can lay several clutches of fertile eggs at long intervals after one mating.

CROCODILES AND ALLIGATORS—ORDER CROCODILIA

The modern crocodiles are the largest living reptiles. They are what remains of a group that was once abundant in the Jurassic and Cretaceous periods.

Most crocodiles have relatively long slender snouts; alligators (Sp. *el lagarto,*

FIGURE 28-19

Milk snake, *Lampropeltis doliata,* with eggs. The eggs, normally hidden in a nest, have been exposed by the photographer. This distinctively banded snake feeds largely on rodents and is one of our most beneficial snakes.

Photograph by L.L. Rue III.

FIGURE 28-20

Nile crocodile, *Crocodylus niloticus,* basking. Note the slender snout and the lower jaw tooth that fits *outside* the upper jaw; alligators lack this feature.

Photograph by C.P. Hickman, Jr.

lizard) have shorter and broader snouts. With their powerful jaws and sharp teeth, they are formidable predators. The "man-eating" members of the group are found mainly in Africa and Asia. The Nile crocodile *(Crocodylus niloticus)* (Figure 28-20) grows to a great size (to 1000 kg in weight and 5 m in length) and is very much feared. It is swift and aggressive and eats any bird or mammal it can drag from the shore to water, where the prey is violently torn into pieces that can be swallowed. Crocodiles are known to attack animals as large as cattle, deer, and people.

Alligators are usually less aggressive than crocodiles. They are unusual among reptiles in being able to make definite vocalizations. The male alligator can give loud bellows in the mating season. Vocal sacs are found on each side of the throat and are inflated when he calls. In the United States, *Alligator mississipiensis* (Figure 28-21) is the only species of alligator; *Crocodylus acutus,* restricted to extreme southern Florida, is the only species of crocodile.

Alligators and crocodiles are oviparous. Usually from 20 to 50 eggs are laid in

How doth the little crocodile
Improve his shining tail,
And pour the waters of the Nile
On every golden scale!
How cheerfully he seems to grin,
How neatly spread his claws,
And welcomes little fishes in,
With gently smiling jaws!

Lewis Carroll

FIGURE 28-21

American alligator, *Alligator mississipiensis,* an increasingly noticeable resident of the rivers, bayous, and swamps of the southeastern United States.

Photograph by C.P. Hickman, Jr.

a mass of dead vegetation. The eggs are approximately 8 cm long. The penis of the male is an outgrowth of the ventral wall of the cloaca.

THE TUATARA—ORDER RHYNCHOCEPHALIA

The order Rhynchocephalia is represented by a single living species, the tuatara (*Sphenodon punctatum* [Gr. *sphēnos,* wedge, + *odontos,* tooth]) of New Zealand (Figure 28-22). This animal is the sole survivor of a group of primitive reptiles that otherwise became extinct 100 million years ago. The tuatara was once widespread on the North Island of New Zealand but is now restricted to islets of Cook Strait and off the northern coast of North Island, where, under strict protection from the New Zealand government, it should survive.

Tuataras have a lizardlike form, 66 cm long or less, and live in burrows often shared with petrels. They are slow-growing animals with a long life; one is recorded to have lived 77 years.

The tuatara has captured the interest of biologists because of its numerous primitive features that are almost identical to those of Mesozoic fossils 200 million years old. These include a primitive skull structure found in early Permian reptiles that were ancestors to the modern lizards. It also bears a well-developed parietal eye, complete with evidences of a retina (Figure 28-22), and a complete palate. It lacks a copulatory organ. A specialized feature is the teeth, which are fused wedgelike to the edge of the jaws rather than being set in sockets. *Sphenodon* represents one of the slowest rates of evolution known among the vertebrates.

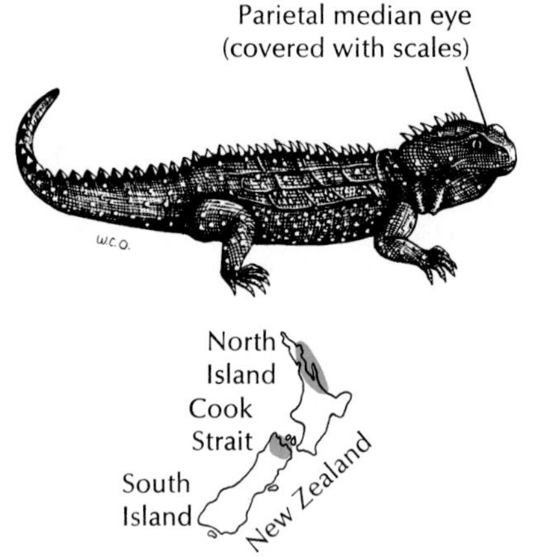

FIGURE 28-22

Tuatara, *Sphenodon punctatum,* the only living representative of the order Rhynchocephalia. This "living fossil" reptile has, on top of the head, a well-developed parietal "eye" with retina, lens, and nervous connections to the brain. Although covered with scales, this third eye is sensitive to light. The parietal eye may have been an important sense organ in early reptiles. The tuatara is found today only on certain islands off the coastline of New Zealand.

Summary

The reptiles evolved from labyrinthodont amphibians during the late Paleozoic era, some 300 million years ago. Their success as the first truly terrestrial vertebrates is attributed to the evolution of the shelled, amniotic egg, whose internal membranes provide the developing embryo with its own independent water and food supplies.

Reptiles are also distinguished from amphibians by their dry, scaly skin that prevents water loss; more powerful jaws; internal fertilization; and more advanced circulatory, respiratory, excretory, and nervous systems. Like amphibians, reptiles are ectotherms, but most exercise considerable behavioral control over their body temperature.

The modern reptiles are descendants of survivors of the Mesozoic age of reptiles when a burst of reptile evolution produced a worldwide fauna of great diversity. The turtles (order Testudines) with their distinctive shells have changed little in design since the Triassic period. Turtles are a small group of long-lived terrestrial, semiaquatic, aquatic, and marine species. They lack teeth. All are oviparous and all, including the marine forms, bury their eggs.

The lizards and snakes (order Squamata) comprise 95% of all living reptiles. Lizards are a diversified and successful group adapted for walking, running, climbing, swimming, and burrowing. They are distinguished from snakes by typically having two pairs of legs (some species are legless), united halves of the lower jaw, movable eyelids, external ears, and absence of fangs. Snakes, in addition to being entirely limbless, are characterized by their elongate bodies and an elastic connection that permits the two halves of the lower jaw to spread widely during swallowing. Most snakes rely on the chemical senses, especially Jacobson's organs, to hunt prey, rather than on the weakly developed visual and auditory senses. Pit vipers have unique infrared-sensing organs for tracking warm-bodied prey. Many snakes are venomous.

The crocodiles and alligators (order Crocodilia) are the largest living reptiles and have the most complex social behavior. Most are endangered species today.

The tuatara of New Zealand (order Rhynchocephalia) is a relict species and sole survivor of a group that otherwise disappeared 100 million years ago. It bears several primitive features that are almost identical to those of Mesozoic fossil reptiles.

REVIEW QUESTIONS

1. Name the major groups of reptiles, both extinct and living, that descended from the captorhinomorphs (stem reptiles) (refer to both Figure 28-1 and to the text).
2. At the end of the Mesozoic the ruling reptiles disappeared along with several invertebrate groups and several lineages of plants. What may have been responsible for these extinctions?
3. Describe at least six ways in which the reptiles are more advanced functionally or structurally than the amphibians.
4. Many zoologists believe that the evolution of the shelled (amniotic) egg was the single most important adaptation contributing to the great success of the reptiles. Describe this egg and explain its importance.
5. Describe the principal characteristics of turtles (order Testudines) that would distinguish them from any other reptile order.
6. In what ways are lizards and snakes (order Squamata) better adapted to hot and arid desert conditions than amphibians?
7. Name three anatomical characteristics of snakes that distinguish them from any lizard (remember that some lizards are legless!).
8. What is the function of Jacobson's organ of snakes?
9. What is the function of the "pit" of pit vipers?
10. What is the difference in the structure or location of the fangs of a rattlesnake, a cobra, and an African boomslang?
11. What is meant by the abbreviation LD_{50}?
12. Most snakes are oviparous, but some are ovoviviparous or viviparous. What do these terms mean and what would you have to know to be able to assign a particular snake to one of these reproductive modes?
13. How do crocodiles and alligators differ from each other?
14. Why is the tuatara (order Rhynchocephalia) of special interest to biologists? Where would you have to go to see one in its natural habitat?

SELECTED REFERENCES

Gibbons, W. 1983. Their blood runs cold: adventures with reptiles and amphibians. University, Ala., University of Alabama Press. *Lots of interesting reptile lore in this engaging book.*

Goin, C.J., O.B. Goin, and G.R. Zug. 1978. Introduction to herpetology, ed. 3. San Francisco, W.H. Freeman & Co., Publishers. *Basic college-level textbook.*

Halliday, T.R., and K. Adler (eds.). 1986. The encyclopedia of reptiles and amphibians. New York, Facts on File, Inc. *Comprehensive and beautifully illustrated treatment of the reptilian groups with helpful introductory sections on origins and characteristics.*

Lillywhite, H.B. 1988. Snakes, blood circulation and gravity. Sci. Am. **259**:92-98 (Dec.). *Even long snakes are able to maintain blood circulation when the body is extended vertically (head up posture) through special circulatory reflexes that control blood pressure.*

Russell, D.A. 1982. The mass extinctions of the late Mesozoic. Sci. Am. **246**:58-65 (Jan.). *Details the evidence that the catastrophic disruption of the biosphere and the extinction of the dinosaurs at the close of the Mesozoic were caused by the impact on earth of a large asteroid.*

Seigel, R.A., J.T. Collins, and S.S. Novak (eds.). 1987. Snakes: ecology and evolutionary biology. New York, Macmillan Publishing Company. *Contributed chapters on snake biology including techniques for capture, maintenance, and study.*

LONG TRIP TO A SUMMER HOME

Perhaps it was ordained that birds, having mastered flight, would use this power to make the long and arduous seasonal migrations that have captured human wonder and curiosity. For the advantages of migration are many. Moving between southern wintering areas and northern, summer breeding regions enables birds to live in an optimal climate all the time, with abundant and unfailing sources of food available to sustain their intense metabolism. In the far North long summer days and large multitudes of insects combine to provide parents with ample food for rearing their young. There are few predators of birds in the far North, and the brief once-a-year appearance of vulnerable young birds does not encourage the buildup of predator populations. Migration also vastly increases the amount of space available for breeding and reduces aggressive territorial behavior. Finally, migration favors homeostasis—the balancing of physiological processes that maintains internal stability—by allowing birds to avoid climatic extremes.

Still, the wonder of the migratory pageant remains, and there is much yet to learn about its mechanisms. What triggers migration, and how do birds know to store sufficient fuel for the journey? How did the sometimes difficult migratory routes originate, and what cues do birds use in navigation? What was the origin of this instinctive force to follow the retreat of winter northward? For it is instinct that drives the migratory waves in spring and fall, instinctive blind obedience that carries most birds successfully to their northern nests, while countless others fail and die, winnowed by nature and natural selection.▪

CHAPTER 29

Birds

Of the vertebrates, birds (class Aves [ay'veez] [L. pl. of *avis,* bird]) are the most studied, the most melodious, and many think the most beautiful. With 8600 species distributed over nearly the entire earth, birds far outnumber all other vertebrates except the fishes. Birds are found in forests and deserts, in mountains and prairies, and on all the oceans. Four species are known to have visited the North Pole, and one, a skua, was seen at the South Pole. Some birds live in total blackness in caves, finding their way about by echolocation, and others dive to depths greater than 45 m to prey on aquatic life.

The single unique feature that distinguishes birds from other animals is their feathers. If an animal has feathers, it is a bird; if it lacks feathers, it is not a bird. No other vertebrate group bears such an easily recognizable and foolproof identification tag.

There is great uniformity of structure among birds. Despite approximately 130 million years of evolution, during which they proliferated and adapted themselves to specialized ways of life, we have no difficulty recognizing a bird as a bird. In addition to feathers, all birds have forelimbs modified into wings (although they may not be used for flight); all have hind limbs adapted for walking, swimming, or perching; all have horny beaks; and all lay eggs. Probably the reason for this great structural and functional uniformity is that birds evolved into flying machines. This fact greatly restricts diversity, so much more evident in other vertebrate classes. For example, birds do not begin to approach the diversity seen in their warm-blooded evolutionary peers, the mammals, a group that includes forms as unlike as a whale, porcupine, bat, and giraffe.

Birds share with mammals the most complex development of organ systems in the animal kingdom. But a bird's entire anatomy is designed around flight and its perfection. An airborne life for a large vertebrate is a highly demanding evolutionary challenge. A bird must, of course, have wings for support and propulsion. But wings alone are not enough. Bones must be light and hollow, yet they must serve as a rigid airframe. To meet the intense metabolic demands of flight, the bird must have a highly efficient respiratory system for rapid gas exchange and removal of excess body heat. It must move its blood rapidly in a high-pressure circulatory system. It must eat an energy-rich diet and digest it quickly and efficiently. Above all, birds must have a finely tuned nervous system and acute senses, especially superb vision, to meet the demands of hurtling headfirst and at high speed across the landscape.

Origin and Relationships

Approximately 150 million years ago, a flying animal drowned and settled to the bottom of a tropical lake in what is now Bavaria, Germany. It was rapidly covered with a fine silt and eventually fossilized. There it remained until discovered in 1861 by a workman splitting slate in a limestone quarry. The fossil was approximately the size of a crow, with a skull not unlike that of modern birds except that the beaklike jaws bore small bony teeth set in sockets like those of reptiles (Figure 29-1). The skeleton was decidedly reptilian with a long bony tail, clawed fingers, and abdominal ribs. It might have been classified as a reptile except that it carried the unmistakable imprint of **feathers,** those marvels of biological engineering that only birds possess. The finding was dramatic because it showed beyond reasonable doubt that birds had evolved from reptiles.

Archaeopteryx (ar-kee-op'ter-ix), meaning "ancient wing," as the fossil was named, was an especially fortunate discovery because the fossil record of birds is disappointingly meager. The lightweight and fragile bones of birds quickly disintegrate, so that only under the most favorable conditions do they fossilize. Nevertheless, by 1952 more than 780 different fossil species had been recorded. Although most of these are relatively recent fossils, enough intermediate forms are known to provide a reasonable picture of bird evolution from the Jurassic period, when *Archaeopteryx* lived, to recent times. By the close of the Cretaceous period, approximately 63 million years ago, the characteristics of modern birds

FIGURE 29-1

Archaeopteryx, the 150-million-year-old ancestor of modern birds. **A,** Cast of the second and most nearly perfect fossil of *Archaeopteryx,* which was discovered in a Bavarian stone quarry. **B,** Reconstruction of *Archaeopteryx.*

A, Courtesy American Museum of Natural History.

had been thoroughly molded (see Figure 29-3). There remained only the emergence and proliferation of the modern orders of birds. Hundreds of thousands of bird species have appeared and nearly as many have disappeared, following *Archaeopteryx* to extinction. Only a minute fraction of these nameless species have been discovered as fossils.

Most paleontologists agree that the ancestors of both birds and dinosaurs were derived from a stem group of archosaurian reptiles called thecodonts. (See Figure 28-1, p. 599.) Birds probably have a monophyletic origin (that is, evolved from a single ancestor). However, existing birds are broadly divided into two groups: (1) **ratite** (rat'ite) (L. *ratitus*, marked like a raft, from *ratis*, raft), the flightless ostrichlike birds that have a flat sternum with poorly developed pectoral muscles, and (2) **carinate** (L. *carina*, keel), the flying birds that have a keeled sternum on which the powerful flight muscles insert. This division originated from the view that the flightless birds (ostrich, emu, kiwi, rhea, etc.) represented a separate line of descent that never attained flight. This idea is now completely rejected. The ostrichlike ratites clearly have descended from flying ancestors. Furthermore, not all carinate, or keeled, birds can fly and many of them even lack keels (Figure 29-2). Flightlessness has appeared independently among many groups of birds; the fossil record reveals flightless wrens, pigeons, parrots, cranes, ducks, auks, and even a flightless owl. Penguins are flightless although they use their wings to "fly" through the water. Flightlessness has almost always evolved on islands where few terrestrial predators are found. The flightless birds living on continents today are the large ratites (ostrich, rhea, cassowary, emu), which can run fast enough to escape predators. The ostrich can run 70 km (42 miles) per hour, and claims of speeds of 96 km (60 miles) per hour have been made.

The bodies of flightless birds are dramatically redesigned. All of the restrictions of flight are removed. The keel of the sternum is lost, and the heavy flight muscles (as much as 17% of the body weight of flying birds), as well as other specialized flight apparatus, have disappeared. Since body weight is no longer a restriction, flightless birds tend to become large. Several extinct flightless birds were enormous: the giant moas of New Zealand weighed more than 225 kg (500 pounds), and the elephantbird of Madagascar, the largest bird that ever lived, probably weighed nearly 450 kg (about 1000 pounds) and stood nearly 2 m tall.

It may seem paradoxical that birds with their agile, warm-blooded, colorful, and melodious way of life should have descended from lethargic, cold-blooded, and silent reptiles. Yet the numerous anatomical affinities of the two groups are abundant evidence of close kinship and led the great English zoologist Thomas Henry Huxley to call birds merely "glorified reptiles." This unflattering description has never pleased bird lovers, who answer, "But how wondrously glorified."

FIGURE 29-2

One of the strangest birds in a strange land, the flightless cormorant, *Nannopterum harrisi*, of the Galápagos Islands dries its wings after a fishing forage. It is a superb swimmer, propelling itself through the water with its feet to catch fish and octopuses. The flightless cormorant is an example of a carinate bird (having a keeled sternum) that has lost the keel and the ability to fly.

Photograph by C.P. Hickman, Jr.

CHARACTERISTICS

1. Usually spindle-shaped body with four divisions: head, neck, trunk, and tail; **neck disproportionately long** for balancing and food gathering.
2. Paired limbs, with the **forelimbs usually adapted for flying**; posterior pair variously adapted for perching, walking, and swimming; foot with four toes (chiefly)
3. Thin integument of epidermis and dermis; no sweat glands; skin naked or **covered with feathers** of epidermal origin; legs usually covered with scales.
4. **Fully ossified skeleton with air cavities**; each jaw covered with horny sheath, forming a **beak; no teeth;** ribs with strengthening processes; sternum well developed with keel or reduced with no keel.
5. Well-developed nervous system with brain and 12 pairs of cranial nerves
6. Circulatory system of **four-chambered heart**, with the **right aortic arch persisting**; nucleated red blood cells
7. Respiration by slightly expansible lungs, ventilated by thin **air sacs** among the visceral organs and skeleton; **syrinx (voice box)** near junction of trachea and bronchi
8. Excretory system of **metanephric kidney**; ureters open into cloaca; **no bladder**; semisolid urine; uric acid main nitrogenous waste
9. Separate sexes; **females of most birds with left ovary and oviduct only**

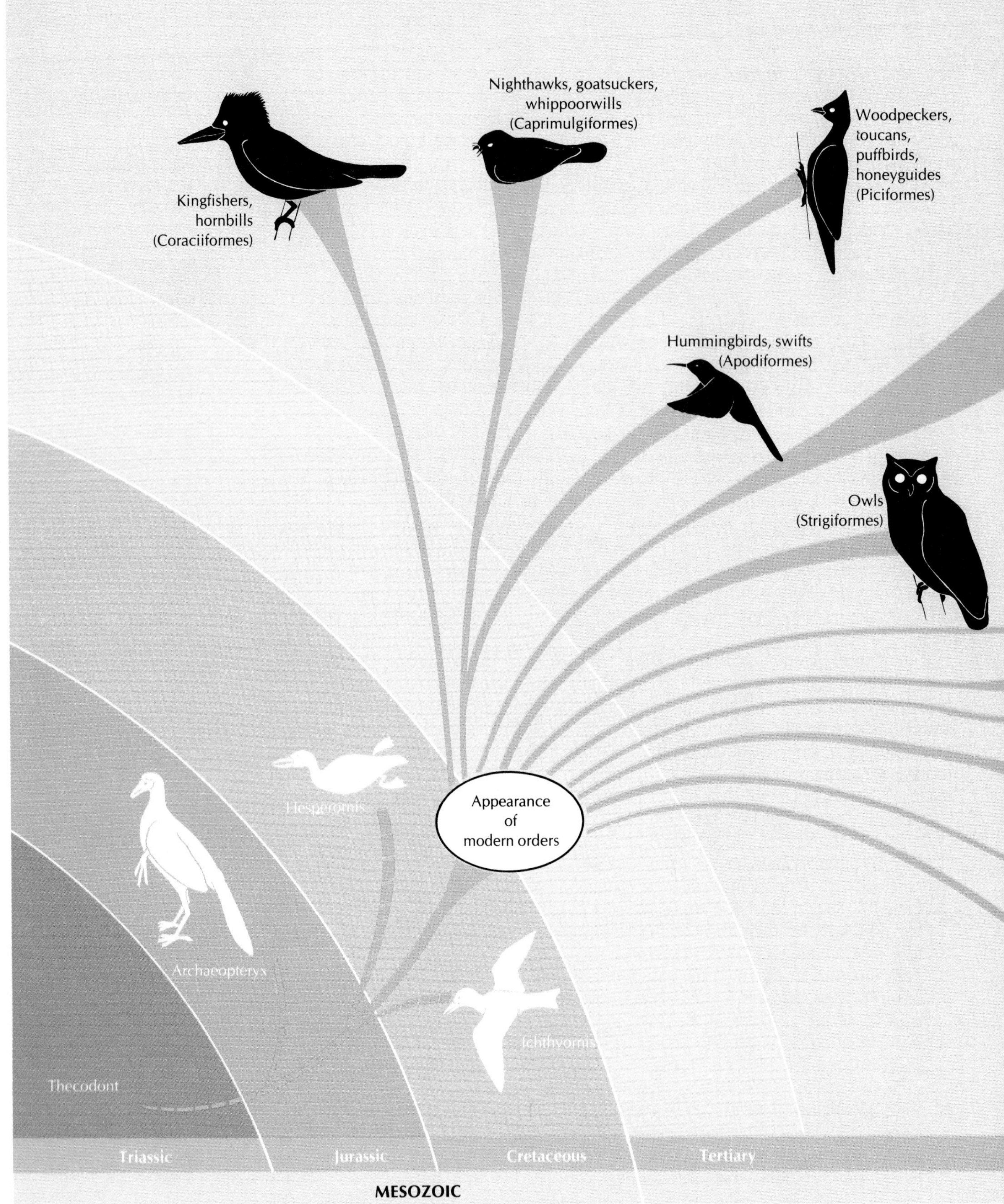

FIGURE 29-3

Evolution of modern birds. Birds are believed to have descended from thecodont reptiles, a group having affinities with certain of the dinosaurs. The earliest known bird is *Archaeopteryx,* which bears many reptilian features. Fossil birds of the Cretaceous period (for exam-

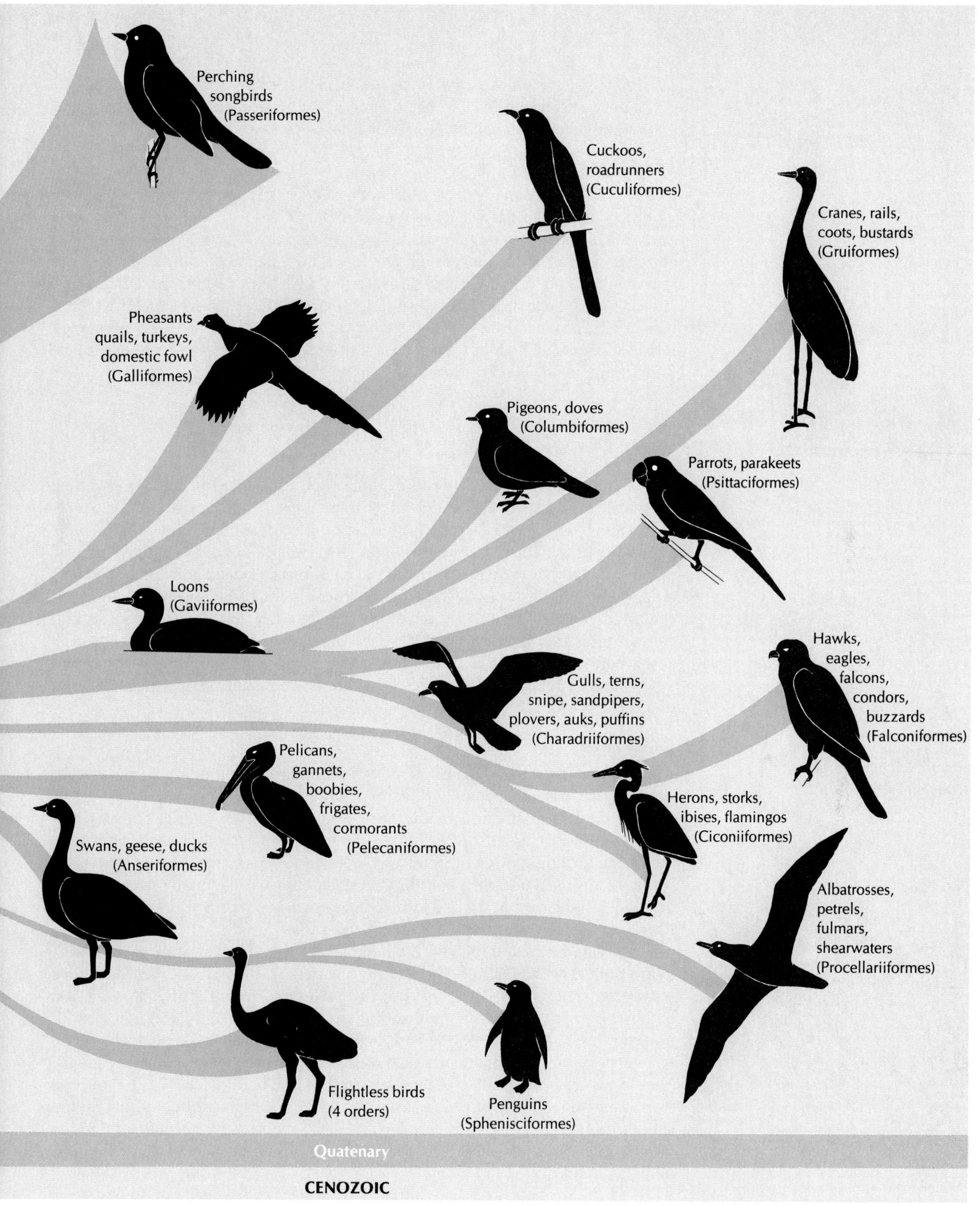

ple, *Hesperornis,* a diving bird, and *Ichthyornis,* a gull-like bird) were much like modern birds. Evolution of modern bird orders occurred rapidly during the Cretaceous and early Tertiary periods.

10. Internal fertilization; **eggs with much yolk and hard calcareous shells;** embryonic membranes in egg during development; **external incubation;** young active at hatching (**precocial**) or helpless and naked (**altricial**)

CLASSIFICATION

Class Aves (birds) is made up of more than 8600 species distributed among some 27 orders of living birds and a few fossil orders. Of the 27 living bird orders, four are **ratite**, or flightless, birds (ostriches, rheas, cassowaries and emus, and kiwis), although as was noted above, flightlessness is not restricted to these groups. The remaining 23 orders are **carinate** birds (birds with a keeled sternum). Of these, by far the largest is the order Passeriformes containing the perching songbirds; these make up 60% of all bird species. Most perching songbirds have a well-developed syrinx and feet adapted for perching on thin stems and twigs. Nineteen of the best known orders of birds are shown in Figure 29-3.

■ Adaptations of Bird Structure and Function for Flight

Just as an airplane must be designed and built according to rigid aerodynamic specifications if it is to fly, so too must birds meet stringent structural requirements if they are to stay airborne. All the special adaptations found in flying birds contribute to two things: more power and less weight. Flight by humans became possible when they developed an internal combustion engine and learned how to reduce the weight-to-power ratio to a critical point. Birds did this millions of years ago. But birds must do much more than fly. They must feed themselves and convert food into high-energy fuel; they must escape predators; they must be able to repair their own injuries; they must be able to air-condition themselves when overheated and heat themselves when too cool; and, most important of all, they must reproduce.

FEATHERS

A feather is very lightweight, yet possesses remarkable toughness and tensile strength. The most typical of bird feathers are the **contour feathers**, vaned feathers that cover and streamline the bird's body. A contour feather consists of a hollow **quill**, or calamus, emerging from a skin follicle, and a **shaft**, or rachis, which is a continuation of the quill and bears numerous **barbs** (Figure 29-4). The barbs are arranged in closely parallel fashion and spread diagonally outward from both sides of the central shaft to form a flat, expansive, webbed surface, the **vane**. There may be several hundred barbs in the vane.

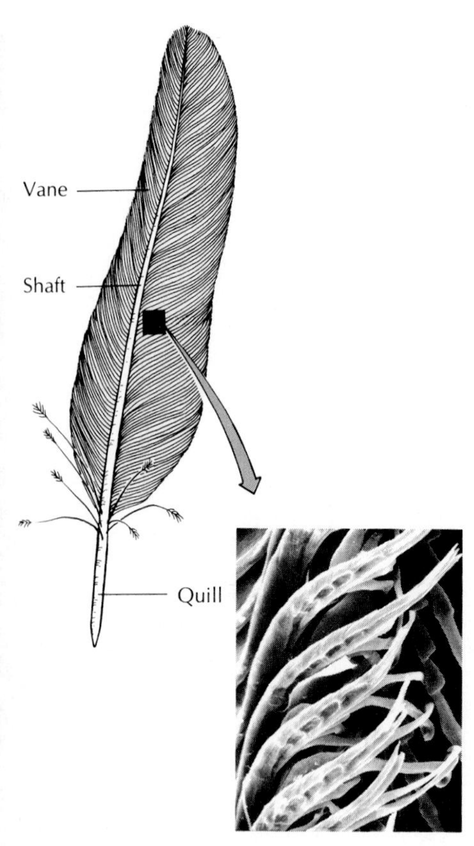

FIGURE 29-4

Contour feather. Scanning electron micrograph of vane shows interlocking mechanism between adjacent barbs. Minute hooks on barbules hold opposing rows of barbules loosely together to form continuous surface of vane.

Photograph courtesy of E.H. Burtt, Jr.

If the feather is examined with a microscope, each barb appears to be a miniature replica of the feather, with numerous parallel filaments, called **barbules**, set in each side of the barb and spreading laterally from it. There may be 600 barbules on each side of the barb, adding up to more than 1 million barbules for the feather. The barbules of one barb overlap the barbules of a neighboring barb in a herringbone pattern and are held together with great tenacity by tiny hooks. Should two adjoining barbs become separated—and considerable force is needed to pull the vane apart—they are instantly zipped together again by drawing the feather through the fingertips. The bird, of course, does it with its bill, and much of a bird's time is occupied with preening to keep its feathers in perfect condition.

Feathers are epidermal structures that evolved from the reptilian scale; a developing feather closely resembles a reptile scale when growth is just beginning. We can imagine that in its evolution the scale elongated and its edges frayed outward until it became the complex feather of birds. Strangely enough, although modern birds possess both scales (especially on their feet) and feathers, no intermediate stage between the two has been discovered on either fossil or living forms.

When fully grown, a feather is a dead structure. Tough as it is, it eventually frays and may even break from wear. All birds renew their feathers at regular intervals, usually in late summer after the nesting season. Many birds also undergo a second partial molt just before the mating season to equip them with their breeding finery, so important for courtship display. The shedding of feathers is a highly ordered process and is scheduled to least disturb the demands for self-protection and food gathering. Except in penguins, which molt all their short feathers at once and lose 50% of the body weight while waiting solemnly in one spot for new feathers to grow, feathers are discarded gradually, thus avoiding the appearance of bare spots. Flight and tail feathers are lost in exact pairs, one from each side, so that balance is maintained (Figure 29-15). Replacements emerge before the next pair is lost, and most birds can continue flying unimpaired during the molting period. However, many water birds (ducks, geese, loons, and others)

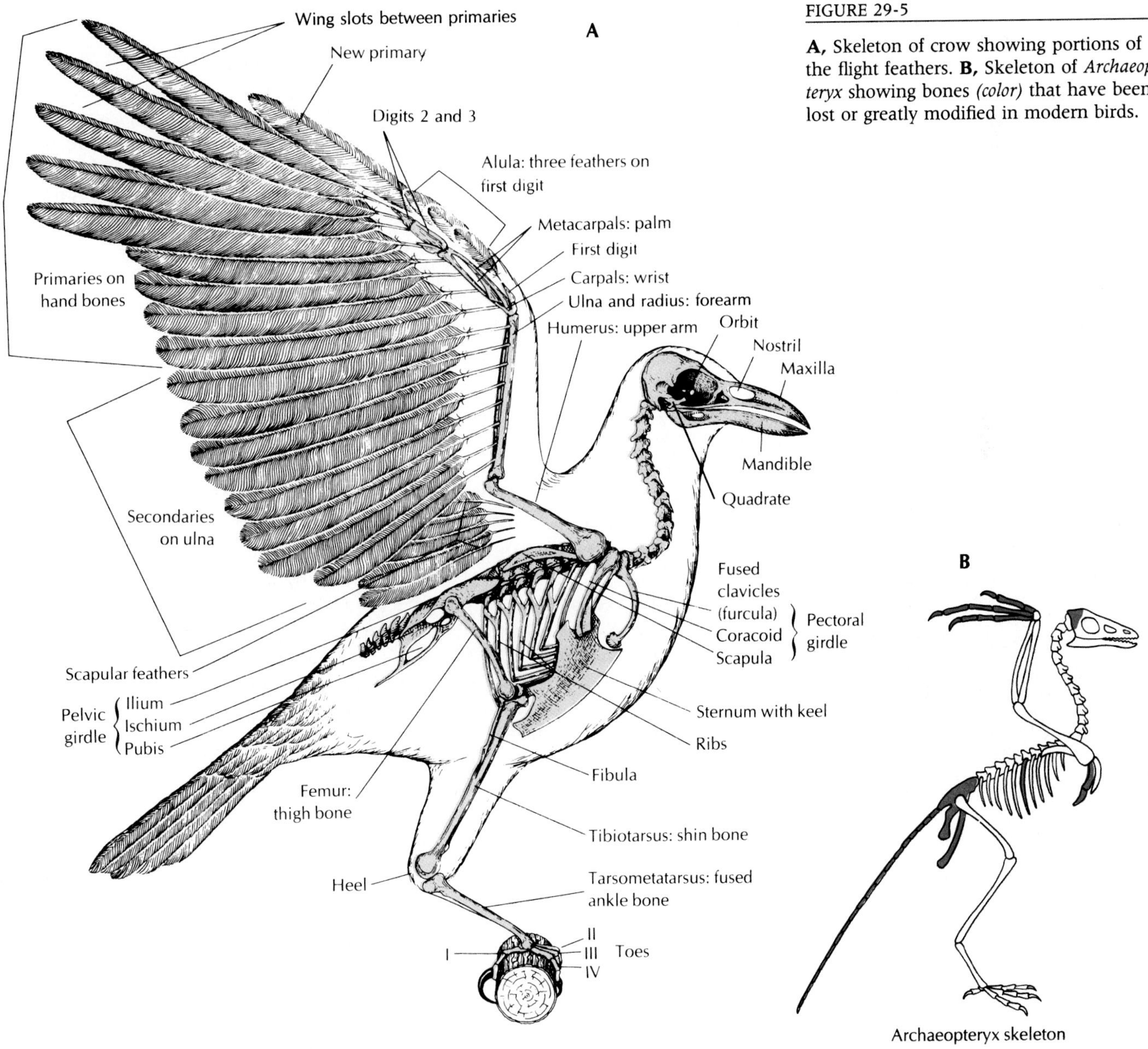

FIGURE 29-5

A, Skeleton of crow showing portions of the flight feathers. **B,** Skeleton of *Archaeopteryx* showing bones *(color)* that have been lost or greatly modified in modern birds.

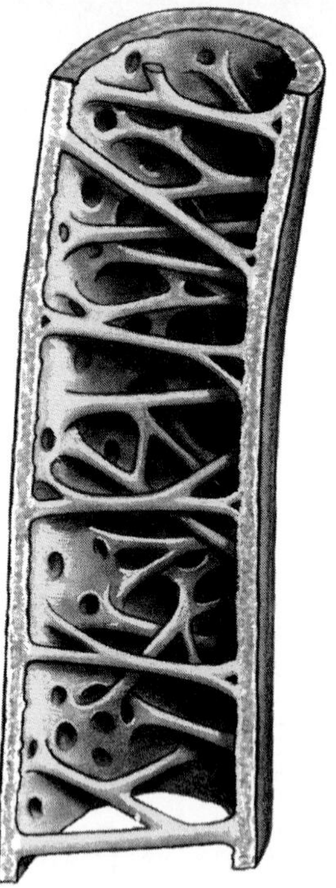

FIGURE 29-6

Hollow wing bone of a songbird showing the stiffening struts and air spaces that replace bone marrow. Such "pneumatized" bones are remarkably light and strong.

FIGURE 29-7

The yellow-shafted flicker, *Colaptes auratus*, in common with other woodpeckers, has a heavy skull, chisel-edged beak, strong feet with two opposing toes (instead of the more common single toe), and stiff tail feathers that serve as a third, bracing leg.

Photograph by L.L. Rue III.

lose all their primary feathers at once and are grounded during the molt. Many prepare for this by moving before molting to isolated bodies of water where they can find food and more easily escape enemies.

SKELETON

One of the major adaptations that allows a bird to fly is its light skeleton (Figure 29-5). Bones are phenomenally light, delicate, and laced with air cavities. Such **pneumatized** bones are nevertheless strong (Figure 29-6). The skeleton of a frigate bird with a 7-foot wingspan weighs only 114 g (4 ounces), less than the weight of all its feathers. A pigeon skull weighs only 0.21% of its body weight; the skull of a rat by comparison weighs 1.25% of its body weight.

The bird skull is mostly fused into one piece. The braincase and orbits are large to accommodate a bulging cranium and the large eyes needed for quick motor coordination and superior vision. The anterior skull bones are elongated to form a beak. The lower **mandible** is a complex of several bones that hinge on two small movable bones, the **quadrates**. This provides a double-jointed action that permits the mouth to open widely. The upper jaw is usually fused to the forehead, but in some birds—parrots, for instance—the upper jaw is hinged also. This adaptation allows greater flexibility of the beak in food manipulation and provides insect-catching species with a wider gap for successful feeding on the wing.

The beaks of birds are strongly adapted to specialized food habits—from generalized types, such as the strong, pointed beaks of crows, to grotesque, highly specialized ones in flamingos, hornbills, and toucans. The beak of a woodpecker is a straight, hard, chisel-like device. Anchored to a tree trunk with its tail serving as a brace, the woodpecker delivers powerful, rapid blows to build nests or expose the burrows of wood-boring insects (Figure 29-7). It then uses its long, flexible, barbed tongue to seek out insects in their galleries. The woodpecker's skull is especially thick to absorb shock.

The vertebral column of birds is highly specialized for flight. Its most distinctive feature is its rigidity. Most of the vertebrae except the **cervicals** (neck vertebrae) are fused together and with the pelvic girdle to form a stiff but light framework to support the legs and provide rigidity for flight. To assist in this rigidity, the ribs are mostly fused with the vertebrae, pectoral girdle, and sternum. Except for the flightless birds, the sternum bears a large, thin keel that provides for the attachment of the powerful flight muscles. Of the body box, only the 8 to 24 (according to the species) vertebrae of the neck remain fully flexible.

The bones of the forelimbs have become highly modified for flight. They are not only hollow, but also reduced in number, and several are fused together. Despite these alterations, the bird wing is clearly a rearrangement of the basic vertebrate tetrapod limb from which it arose, and all the elements—upper arm, forearm, wrist, and fingers—are represented in modified form (Figure 29-5). The birds' legs have undergone less pronounced modification than the wings, since they are still designed principally for walking, as well as for perching, scratching, food gathering, and occasionally for swimming, as were those of their reptilian ancestors.

MUSCLE SYSTEM

The locomotor muscles of the wings are relatively massive to meet the demands of flight. The largest of these is the **pectoralis**, which depresses the wings in flight (Figure 29-8). Its antagonist is the **supracoracoideus** muscle, which raises the wing. Surprisingly perhaps, this latter muscle is not located on the backbone (anyone who has been served the back of a chicken knows it offers little meat) but is positioned under the pectoralis on the breast. It is attached by a tendon to the upper side of the humerus of the wing so that it pulls from below by an ingenious "rope-and-pulley" arrangement. Both of these muscles are anchored to the keel. Thus, with the main muscle mass low in the body, aerodynamic stability is improved.

The main leg muscle mass is located in the thigh, surrounding the femur, and a smaller mass lies over the tibiotarsus (shank, or "drumstick"). Strong but thin tendons extend downward through sleevelike sheaths to the toes. Consequently the feet are nearly devoid of muscles, thus explaining the thin, delicate appearance of the bird leg. This arrangement places the main muscle mass near the bird's center of gravity and at the same time allows great agility to the slender, lightweight feet. Since the feet are made up mostly of bone, tendon, and tough, scaly skin, they are highly resistant to damage from freezing. When a bird perches on a branch, an ingenious toe-locking mechanism (Figure 29-9) is activated, which prevents a bird from falling off its perch when asleep. The same mechanism causes the talons of a hawk or owl to automatically sink deeply into its victim as the legs bend under the impact of the strike. The powerful grip of a bird of prey was described by L. Brown*:

> When an eagle grips in earnest, one's hand becomes numb, and it is quite impossible to tear it free, or to loosen the grip of the eagle's toes with the other hand. One just has to wait until the bird relents, and while waiting one has ample time to realize that an animal such as a rabbit would be quickly paralyzed, unable to draw breath, and perhaps pierced through and through by the talons in such a clutch.

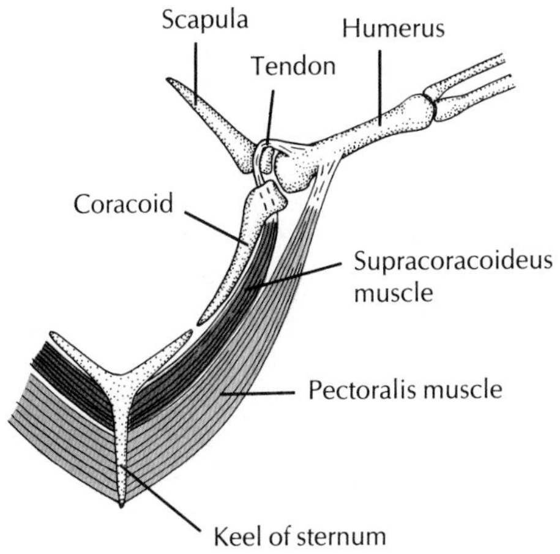

Scapula
Humerus
Tendon
Coracoid
Supracoracoideus muscle
Pectoralis muscle
Keel of sternum

FIGURE 29-8

Flight muscles of a bird are arranged to keep the center of gravity low in the body. Both major flight muscles are anchored on the sternum keel. Contraction of the pectoralis muscle pulls the wing downward. Then, as the pectoralis relaxes, the supracoracoideus muscle contracts and, acting as a pulley system, pulls the wing upward.

DIGESTIVE SYSTEM

Birds eat an energy-rich diet in large amounts and process it rapidly and thoroughly with efficient digestive equipment. A shrike can digest a mouse in 3 hours, and berries pass completely through the digestive tract of a thrush in just 30 minutes. There are no teeth in the mouth, and the poorly developed salivary glands mainly secrete mucus for lubricating the food and the slender, horn-covered **tongue**. There are few taste buds, although all birds can taste to some extent. From the short **pharynx** a relatively long, muscular, elastic **esophagus** extends to the **stomach**. In many birds there is an enlargement (**crop**) at the lower end of the esophagus, which serves as a storage chamber.

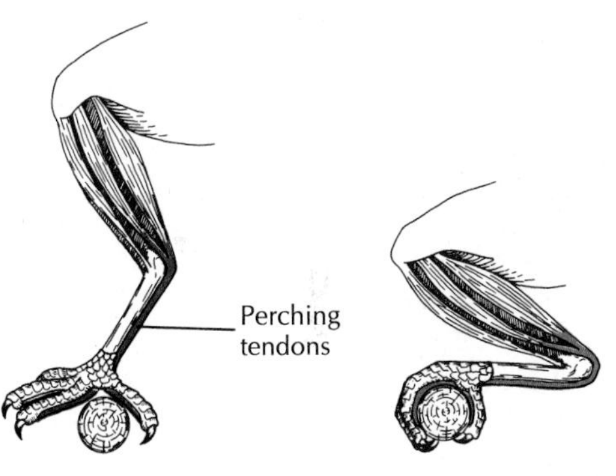

FIGURE 29-9

Perching mechanism of a bird. When a bird settles on a branch, tendons automatically tighten, closing the toes around the perch.

In pigeons, doves, and some parrots the crop not only stores food but produces milk by the breakdown of epithelial cells of the lining. This "bird milk" is regurgitated by both the male and female into the mouth of the young squabs. The milk has a much higher fat content than cow's milk.

The stomach proper consists of a **proventriculus**, which secretes gastric juice, and the muscular **gizzard**, which is lined with horny plates that serve as millstones for grinding the food. To assist in the grinding process, birds swallow coarse, gritty objects or pebbles, which lodge in the gizzard. Certain birds of prey, such as owls, form pellets of indigestible materials, mainly bones and fur, in the proventriculus and eject them through the mouth. At the junction of the intestine with the rectum there are paired **ceca**, which may be well developed in herbivorous birds. The terminal part of the digestive system is the **cloaca**, which also receives the genital ducts and ureters.

CIRCULATORY SYSTEM

The general plan of circulation in birds is not greatly different from that of mammals. The four-chambered heart is large with strong ventricular walls; thus birds share with mammals a complete separation of the respiratory and systemic circulations. The heartbeat is extremely fast, and as in mammals there is an inverse relationship between heart rate and body weight. For example, a turkey has a heart rate at rest of approximately 93 beats per minute, a chicken has 250 beats

*Brown, L. 1970. Eagles. New York, Arco Publishing Co., Inc.

per minute, and a black-capped chickadee has a rate of 500 beats per minute when asleep, which may increase to a phenomenal 1000 beats per minute during exercise. Blood pressure in birds is roughly equivalent to that in mammals of similar size. Birds' blood contains **nucleated, biconvex erythrocytes.** (Mammals, the only other endotherm vertebrates, have biconcave erythrocytes without nuclei that are somewhat smaller than those of birds.) The **phagocytes**, or mobile ameboid cells of the blood, are very active and efficient in birds in the repair of wounds and in destroying microbes.

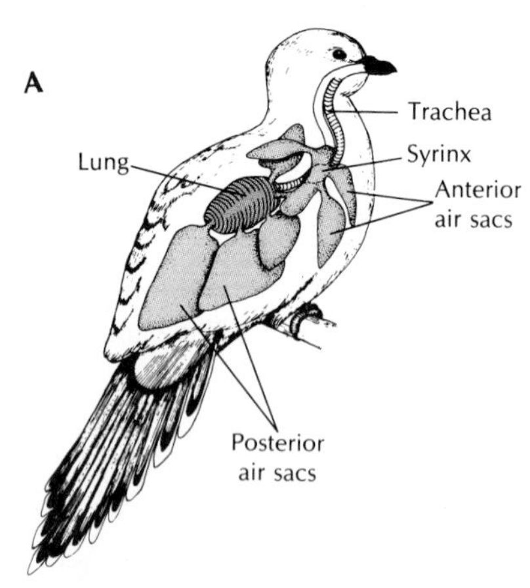

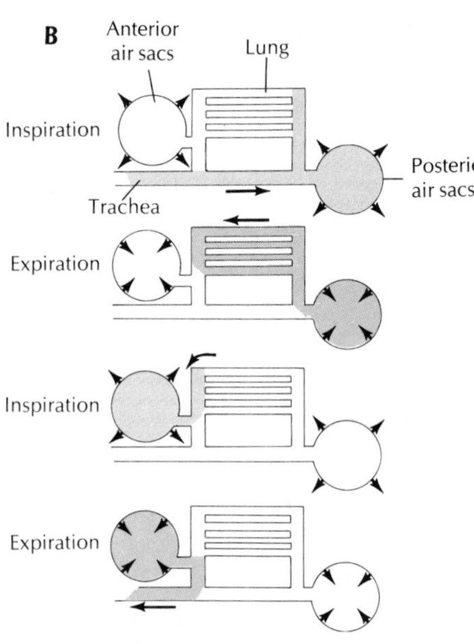

FIGURE 29-10

Respiratory system of a bird. **A,** Lungs and air sacs. One side of the bilateral air sac system is shown. **B,** Movement of a single volume of air through the bird's respiratory system. Two full respiratory cycles are required to move the air through the system.

B, from Schmidt-Nielsen, K. 1979. Animal physiology. New York, Cambridge University Press.

RESPIRATORY SYSTEM

The respiratory system of birds differs radically from the lungs of reptiles and mammals and is marvelously adapted for meeting the high metabolic demands of flight. In birds the finest branches of the bronchi, rather than ending in saclike alveoli as in mammals, are developed as tube-like **parabronchi** through which the air flows continuously. Also unique is the extensive system of nine interconnecting **air sacs** that are located in pairs in the thorax and abdomen and even extend by tiny tubes into the centers of the long bones (Figure 29-10, *A*). The air sacs are connected to the lungs in such a way that perhaps 75% of the inspired air bypasses the lungs and flows directly into the air sacs, which serve as reservoirs for fresh air. On expiration, some of this fully oxygenated air is shunted through the lung, while the used air passes directly out (Figure 10-10, *B*). The advantage of such a system is obvious: the lungs receive fresh air during both inspiration and expiration. An almost continuous stream of oxygenated air passes through a system of richly vascularized parabronchi. Although many details of the bird's respiratory system are not fully understood, it is clearly the most efficient of all the vertebrates.

EXCRETORY SYSTEM

The relatively large paired metanephric kidneys are composed of many thousands of **nephrons**, each consisting of a renal corpuscle and a nephric tubule. As in other vertebrates, the urine is formed by glomerular filtration followed by the selective modification of the filtrate in the tubule (the details of this sequence are found on pp. 209-212).

The birds, like the reptiles from which they evolved, excrete their nitrogenous wastes as uric acid rather than urea, an adaptation that originated with the evolution of the shelled egg. In the shelled egg, all excretory products must remain in the eggshell with the growing embryo. If urea were produced, it would quickly accumulate in solution to toxic levels. Uric acid, however, crystallizes *out* of solution and can be stored harmlessly within the egg shell. Thus, from an embryonic necessity was born an adult virtue. Because of uric acid's low solubility, a bird can excrete 1 g of uric acid in only 1.5 to 3 ml of water, whereas a mammal may require 60 ml of water to excrete 1 g of urea. Uric acid is combined with fecal material in the **cloaca**. Excess water is reabsorbed in the cloaca, resulting in the formation of a white paste. Thus, despite having kidneys that are much less effective in true concentrative ability than mammalian kidneys, birds can excrete uric acid nearly 3000 times more concentrated than that in the blood. Even the most effective mammalian kidneys, those of certain desert rodents, can excrete urea only about 25 times the plasma concentration.

Marine birds (also marine turtles) have evolved a unique method for excreting the large loads of salt eaten with their food and in the seawater they drink. Seawater contains approximately 3% salt and is three times saltier than a bird's body fluids. Yet the bird kidney cannot concentrate salt in urine above approximately 0.3%. The problem is solved by special **salt glands**, one located above each eye (Figure 29-11). These glands are capable of excreting a highly concentrated solution of sodium chloride—as much as twice the concentration of seawater. The salt solution runs out the internal or external nostrils, giving gulls, petrels, and other sea birds a perpetual runny nose.

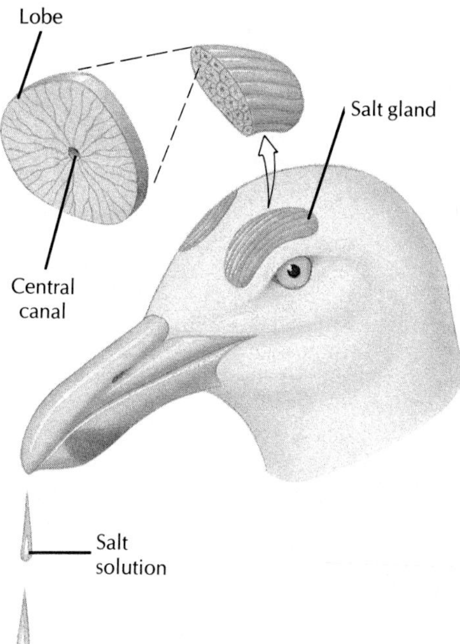

FIGURE 29-11

Salt glands of a marine bird (gull). One salt gland is located above each eye. Each gland consists of several lobes arranged in parallel. One lobe is shown in cross section, much enlarged. Salt is secreted into many radially arranged tubules, then flows into a central canal that leads into the nose.

NERVOUS AND SENSORY SYSTEM

A bird's nervous and sensory system accurately reflects the complex problems of flight and a highly visible existence, in which it must gather food, mate, defend territory, incubate and rear young, and correctly distinguish friend from foe. The brain of a bird has well-developed **cerebral hemispheres, cerebellum,** and **midbrain tectum** (optic lobes). The **cerebral cortex**—the portion in mammals that becomes the chief coordinating center—is thin, unfissured, and poorly developed in birds. But the core of the cerebrum, the **corpus striatum,** has enlarged into the principal integrative center of the brain, controlling such activities as eating, singing, flying, and all the complex instinctive reproductive activities. Relatively intelligent birds, such as crows and parrots, have larger cerebral hemispheres than do less intelligent birds such as chickens and pigeons. The **cerebellum** is a crucial coordinating center where muscle-position sense, equilibrium sense, and visual cues are all assembled and used to coordinate movement and balance. The **optic lobes,** laterally bulging structures of the midbrain, form a visual association apparatus comparable to the visual cortex of mammals.

Except in flightless birds, ducks, and vultures, the senses of smell and taste are poorly developed in birds. This deficiency, however, is more than compensated for by good hearing and superb vision, the keenest in the animal kingdom. The organ of hearing, the **cochlea,** is much shorter than the coiled mammalian cochlea, yet birds can hear roughly the same range of sound frequencies as humans. Actually the bird ear far surpasses our capacity to distinguish differences in intensities and to respond to rapid fluctuations in pitch.

The bird eye resembles that of other vertebrates in gross structure but is relatively larger, less spherical, and almost immobile; instead of turning their eyes, birds turn their heads with their long flexible necks to scan the visual field. The light-sensitive **retina** (Figure 29-12) is generously equipped with rods (for dim light vision) and cones (for color vision). Cones predominate in day birds, and rods are more numerous in nocturnal birds. A distinctive feature of the bird eye is the **pecten,** a highly vascularized organ attached to the retina and jutting out into the vitreous humor (Figure 29-12). The pecten is thought to provide nutrients to the eye. It may do more, but its function remains largely a mystery.

The **fovea,** or region of keenest vision on the retina, is placed (in birds of prey and some others) in a deep pit, which makes it necessary for the bird to focus exactly on the subject. Many birds moreover have two sensitive spots (foveae) on the retina (Figure 29-12)—the central one for sharp monocular views and the posterior one for binocular vision. The visual acuity of a hawk is believed to be eight times that of humans (enabling it to see clearly a crouching rabbit 2 km away), and an owl's ability to see in dim light is more than 10 times that of the human eye.

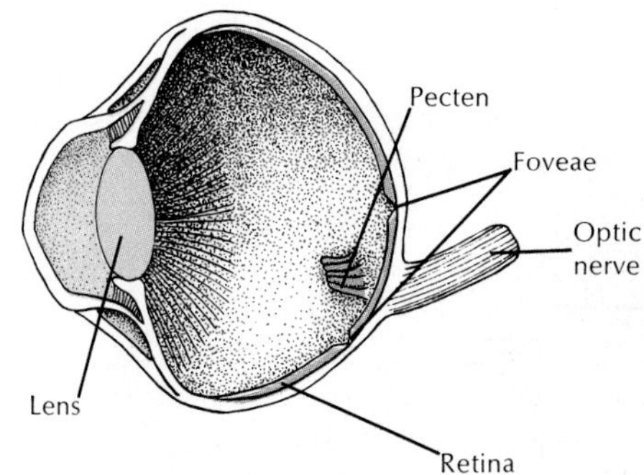

FIGURE 29-12

Hawk eye has all the structural components of the mammalian eye, plus a peculiar pleated structure, the pecten, believed to provide nourishment to the retina. The extraordinarily keen vision of the hawk is attributed to the extreme density of cone cells in the foveae: 1.5 million per fovea compared to 0.2 million for humans.

Flight

What prompted the evolution of flight in birds and the ability to rise free of earth-bound concerns, as almost every human being has dreamed of doing? The origin of flight was the result of complex adaptive pressures. The air was a relatively unexploited habitat stocked with flying insect food. Flight also offered escape from terrestrial predators and an opportunity to travel rapidly and widely to establish new breeding areas and to benefit from a year-round favorable climate and food supply by migrating north and south with the seasons.

How did bird flight begin? The fossil evidence is too meager to provide us with a recorded history of bird flight, but it must have happened in one of two ways: birds began to fly by climbing to a high place and gliding down, or by flapping their way into the air from the ground. The "ground-up" hypothesis holds that birds were ground-dwelling runners with primitive wings used to snare insects. With continued enlargement the proto-wings eventually enabled the running animal to flap its way into the air. The more favored "trees-down" hypothesis suggests that birds passed through an arboreal apprenticeship of tree climbing, leaping through trees, parachuting, gliding, and finally, fully powered flight.

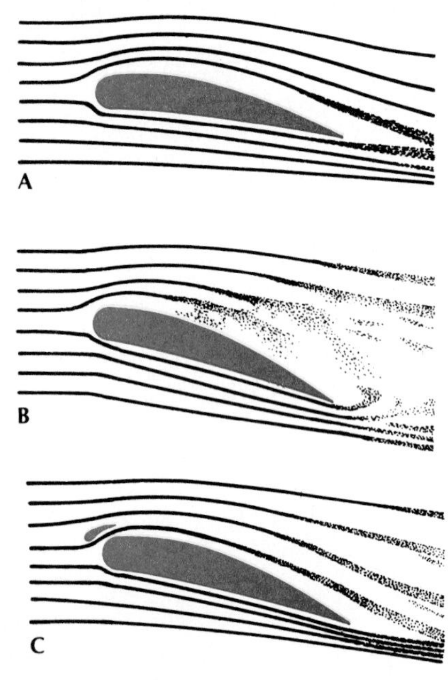

FIGURE 29-13

Air patterns formed by the airfoil, or wing, moving from right to left. **A,** Normal flight with a low angle of attack. As air moves smoothly over the wing, areas of negative pressure on the upper wing surface and of high pressure on the lower wing surface create lift. **B,** Appearance of lift-destroying turbulence on the upper wing surface when the angle of attack becomes too great. Stalling occurs. **C,** Prevention of stalling by directing a layer of rapidly moving air over upper surface with a wing-slot.

From Welty, J.C. 1962. The life of birds. Philadelphia, W.B. Saunders Co.

BIRD WING AS A LIFT DEVICE

The bird wing is an airfoil that is subject to recognized laws of aerodynamics. It is streamlined in cross section, with a slightly concave lower surface **(cambered)** and with small, tight-fitting feathers where the leading edge meets the air (Figure 29-13). Air slips efficiently over the wing, creating lift with minimum drag. Some lift is produced by positive pressure against the undersurface of the wing. But on the upper side, where the airstream must travel farther and faster over the convex surface, a negative pressure is created that provides more than two-thirds of the total lift.

The lift-to-drag ratio of an airfoil is determined by the angle of tilt (angle of attack) and the airspeed (Figure 29-13). A wing carrying a given load can pass through the air at high speed and small angle of attack or at low speed and larger angle of attack. But as speed decreases, a point is reached at which the angle of attack becomes too steep; turbulence appears on the upper surface, lift is destroyed, and stalling occurs. Stalling can be delayed or prevented by placing a **wing slot** along the leading edge so that a layer of rapidly moving air is directed across the upper wing surface. Wing slots were and still are used in aircraft traveling at low speed. In birds, two kinds of wing slots have developed: (1) the **alula,** or group of small feathers on the first digit (Figure 29-5), which provides a midwing slot, and (2) **slotting between primary feathers,** which provides a wing-tip slot. In many songbirds, these together provide stall-preventing slots for nearly the entire outer (and aerodynamically most important) half of the wing.

BASIC FORMS OF BIRD WINGS

Bird wings vary in size and form because the successful exploitation of different habitats has imposed special aerodynamic requirements. Four types of bird wings are easily recognized.*

Elliptical wings

Birds that must maneuver in forested habitats, such as sparrows, warblers, doves, woodpeckers, and magpies, have elliptical wings. This type has a **low aspect ratio** (ratio of length to width). The wings of the highly maneuverable British Spitfire fighter plane of World War II fame conformed closely to the outline of the sparrow wing. Elliptical wings are highly slotted between the primary feathers (Figure 29-14, *A*); this helps to prevent stalling during sharp turns, low-speed flight, and frequent landing and takeoff. Each separated primary feather behaves as a narrow wing with a high angle of attack, providing high lift at low speed. The high maneuverability of the elliptical wing is exemplified by the tiny chickadee, which, if frightened, can change course within 0.03 second.

High-speed wings

Birds that feed on the wing, such as swallows, hummingbirds, and swifts, or that make long migrations, such as plovers, sandpipers, terns, and gulls (Figure 29-14, *B*), have wings that sweep back and taper to a slender tip. They are rather flat in section, have a moderately high aspect ratio, and lack the wing-tip slotting characteristic of the preceding group. Sweepback and wide separation of the wing tips reduce "tip vortex," a drag-creating turbulence that tends to develop at wing tips. The fastest birds alive, such as sandpipers, clocked at 175 km per hour (109 miles per hour), belong to this group.

*Savile, D.B.O. 1957. Adaptive evolution in the avian wing. Evolution **11**:212-224.

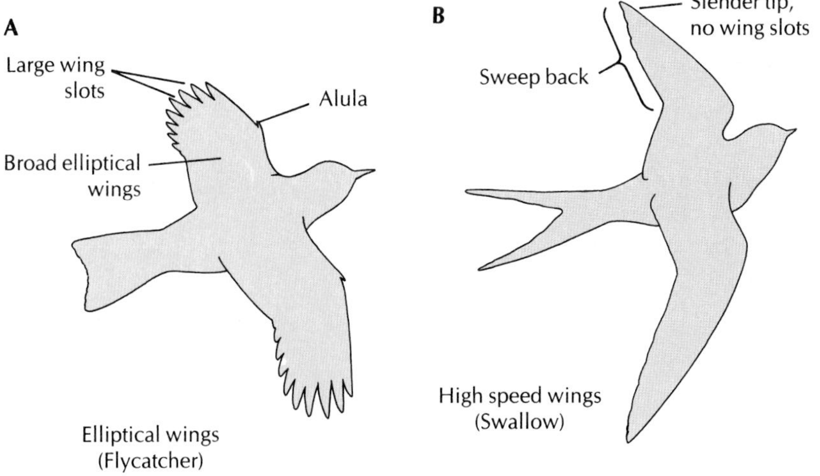

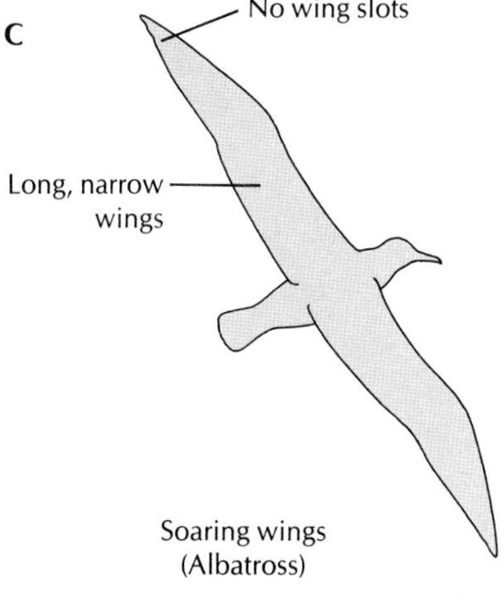

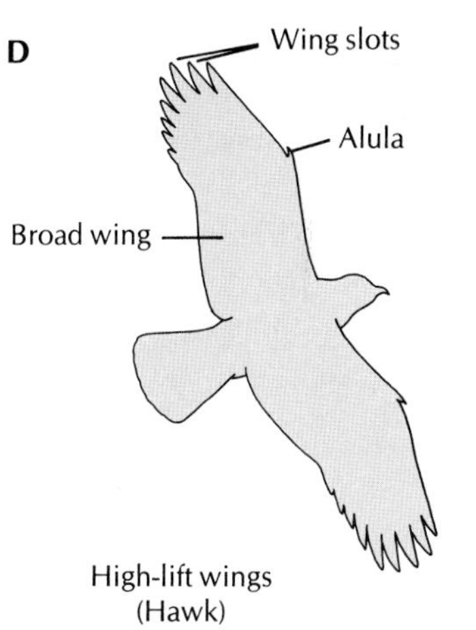

FIGURE 29-14

Four basic forms of bird wings.

Soaring wings

The oceanic soaring birds have **high-aspect ratio** wings resembling those of sailplanes. This group includes albatrosses (Figure 29-14, *C*), frigate birds, and gannets. Such long, narrow wings lack wing slots and are adapted for high speed, high lift, and dynamic soaring. They have the highest aerodynamic efficiency of all wings but are less maneuverable than the wide, slotted wings of land soarers. Dynamic soarers have learned how to exploit the highly reliable sea winds, using adjacent air currents of different velocities.

High-lift wings

Vultures, hawks, eagles, owls, and ospreys (Figures 29-14, *D*, and 29-15)—predators that carry heavy loads—have wings with slotting, alulas, and pronounced camber, all of which promote high lift at low speed. Many of these birds are land soarers, with broad, slotted wings that provide the sensitive response and maneuverability required for static soaring in the capricious air currents over land.

FIGURE 29-15

High-lift wings of the osprey, *Pandion haliaetus*, landing on nest. Note alulas *(top arrows)* and new primary feathers *(side arrows)*. Feathers are molted in sequence in exact pairs so that balance is maintained during flight.

Photograph by B. Tallmark.

FIGURE 29-16

In normal flapping flight of strong fliers like ducks, the wings sweep downward and forward fully extended. Thrust is provided by the primary feathers at the wing tips. To begin the upbeat, the wing is bent, bringing it upward and backward. The wing then extends, ready for the next downbeat.

FLAPPING FLIGHT

In flapping flight, the primary feathers at the wing tips generate thrust, while the secondary feathers of the inner wing, which do not move so far or so fast, act as an airfoil, providing lift. Greatest power is applied on the downstroke. The primary feathers are bent upward and twist to a steep angle of attack, biting into the air like a propeller (Figure 29-16). The entire wing (and the bird's body) is pulled forward. On the upstroke, the primary feathers bend in the opposite direction so that their upper surfaces twist into a positive angle of attack to produce thrust, just as the lower surfaces did on the downstroke. A powered upstroke is essential for hovering flight, as in hummingbirds, and is important for fast, steep takeoffs by small birds with elliptical wings.

Migration and Navigation

The advantages of migration were described in the prologue to this chapter. Not all birds migrate, of course, but the majority of North American and European species do, and the biannual journeys of some are truly extraordinary undertakings. Migration is both the greatest adventure and the greatest risk in the life of a migratory bird.

MIGRATION ROUTES

Most migratory birds have well-established routes trending north and south. Since most birds (and other animals) live in the northern hemisphere where most of the earth's land mass is concentrated, most birds are south-in-winter and north-in-summer migrants. Of the 4000 or more species of migrant birds (a little

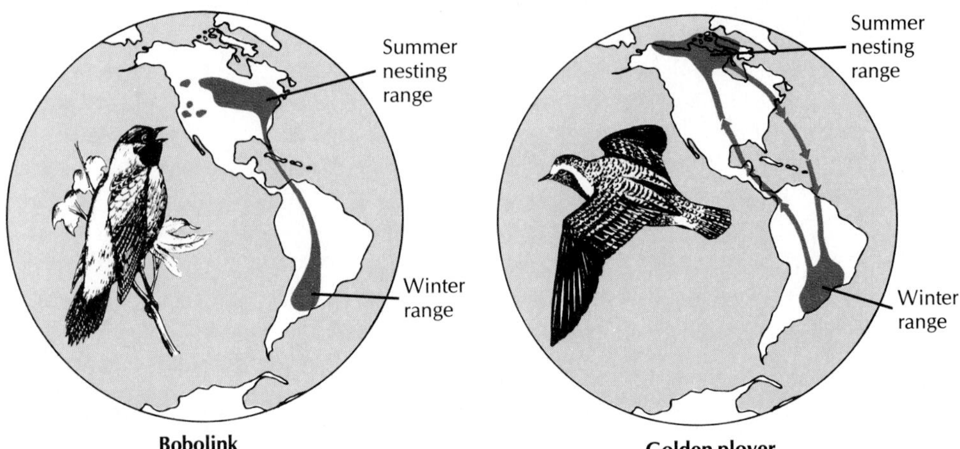

FIGURE 29-17

Migrations of the bobolink and golden plover. The bobolink commutes 22,500 km (14,000 miles) each year between nesting sites in North America and its wintering range in Argentina, a phenomenal feat for such a small bird. Although the breeding range has extended to colonies in western areas, these birds take no shortcuts but adhere to the ancestral seaboard route. The golden plover flies a loop migration, striking out across the Atlantic in its southward autumnal migration but returning in the spring by way of Central America and the Mississippi Valley because ecological conditions are more favorable at that time.

less than one half of the total bird species), most breed in the more northern latitudes of the hemisphere. Some use different routes in the fall and spring (Figure 29-17). Some, especially certain aquatic species, complete their migratory routes in a very short time. Others, however, make the trip in a leisurely manner, often stopping here and there to feed. Some of the warblers are known to take 50 or 60 days to migrate from their winter quarters in Central America to their summer breeding areas in Canada.

Some species are known for their long-distance migrations. The arctic tern, greatest globe spanner of all, breeds north of the Arctic Circle and in winter is found in the Antarctic regions, 17,500 km away. This species is also known to take a circuitous route in migrations from North America, passing over to the coastlines of Europe and Africa and thence to their winter quarters. Other birds that breed in Alaska follow a more direct line down the Pacific coast to North and South America.

Many small songbirds also make great migration treks (Figure 29-17). Africa is a favorite wintering ground for European birds, and many fly there from Central Asia as well.

In the early 1970s, W.T. Keeton showed that the flight bearings of homing pigeons were significantly disturbed by magnets attached to the bird's heads, or by minor fluctuations in the geomagnetic field. But until recently the nature and position of a magnetic receptor in pigeons remained a mystery. Deposits of a magnetic substance called magnetite (Fe_3O_4) have now been discovered in the neck musculature of pigeons and migratory white-crowned sparrows. If this material were coupled to sensitive muscle receptors, as has been proposed, the structure could serve as a magnetic compass that would enable birds to detect and orient their migrations to the earth's magnetic field.

STIMULUS FOR MIGRATION

Humans have known for centuries that the onset of the reproductive cycle of birds is closely related to season. Only within the last 50 years, however, has it been shown that the lengthening of days of late winter and early spring stimulate the development of the gonads and accumulation of fat—both important internal changes that predispose birds to migrate northward. There is evidence that increasing day length stimulates the anterior lobe of the pituitary into activity. The release of pituitary gonadotropic hormone in turn sets in motion a complex series of physiological and behavioral changes, resulting in gonadal growth, fat deposition, migration, courtship and mating behavior, and care of the young.

DIRECTION FINDING IN MIGRATION

Numerous experiments suggest that most birds navigate chiefly by sight. Birds recognize topographic landmarks and follow familiar migratory routes—a behavior assisted by flock migration, during which navigational resources and experience of older birds can be pooled. But in addition to visual navigation, birds make use of a variety of orientation cues at their disposal. Birds have an innate sense, a built-in clock of great accuracy; they have an innate sense of direction; and very recent work adds much credence to an old, much debated hypothesis that birds can detect and navigate by the earth's magnetic field. All of these resources are inborn and instinctive, although a bird's navigational abilities may improve with experience.

Recently experiments by German ornithologists G. Kramer and E. Sauer and American ornithologist S. Emlen have demonstrated convincingly that birds can navigate by celestial cues—the sun by day and the stars by night. Using special circular cages, Kramer concluded that birds possessed a built-in time sense that enabled them to maintain compass direction by referring to the sun, regardless of time of day (Figure 29-18). This is called **sun-azimuth orientation** (*azimuth*, compass-bearing of the sun). Sauer's and Emlen's ingenious planetarium experiments strongly suggest that some birds, probably many, are able to detect and navigate by the North Star axis around which the constellations appear to rotate.

Some of the remarkable feats of bird navigation still defy rational explanation. Most birds undoubtedly use a combination of environmental and innate cues to migrate. Migration is a rigorous undertaking; the target is often small, and natural selection relentlessly prunes off errors in migration, leaving only the best navigators to propagate the species.

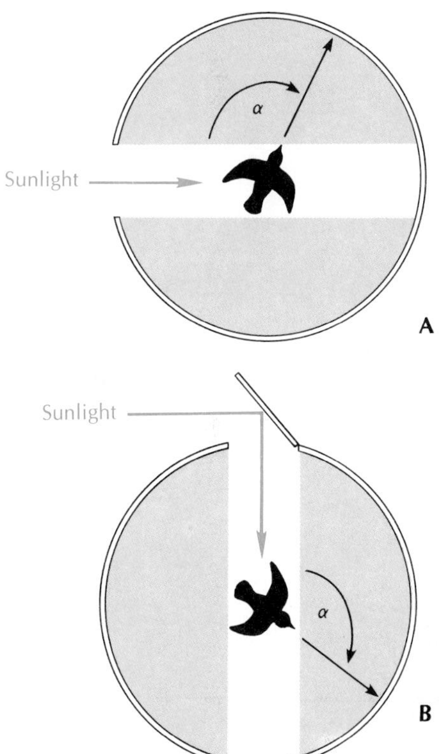

FIGURE 29-18

Gustav Kramer's experiments with sun-compass navigation in starlings. **A,** In a windowed, circular cage, the bird fluttered to align itself in the direction it would normally follow if it were free. **B,** When the true angle of the sun is deflected with a mirror, the bird maintains the same relative position to the sun. This shows that these birds use the sun as a compass. The bird navigates correctly throughout the day, changing its orientation to the sun as the sun moves across the sky.

Social Behavior and Reproduction

The adage says "birds of a feather flock together," and many birds are indeed highly social creatures. Especially during the breeding seasons, sea birds gather, often in enormous colonies, to nest and rear young. Land birds, with some conspicuous exceptions (such as swifts and blackbirds), tend to be less gregarious than sea birds during breeding and seek isolation for rearing their brood. But these same species that covet separation from their kind during breeding may aggregate for migration or feeding. Togetherness offers advantages — mutual protection from enemies, greater ease in finding mates, less opportunity for individual straying during migration, and mass huddling for protection against low night temperatures during migration. Certain species, such as pelicans, may use highly organized cooperative behavior to feed (Figure 29-19). At no time are the highly organized social interactions of birds more evident than during the breeding season, as they stake out territorial claims, select mates, build nests, incubate and hatch their eggs, and rear their young.

FIGURE 29-19

Cooperative feeding behavior by the white pelican, *Pelecanus onocrotalus*. **A,** Pelicans form a horseshoe to drive fish together. **B,** Then they plunge simultaneously to scoop up fish in their huge bills. These photographs were taken 2 seconds apart.
Photograph by B. Tallmark.

A

B

REPRODUCTIVE SYSTEM

The **testes** are tiny bean-shaped bodies during most of the year, which undergo a great enlargement at the breeding season, as much as 300 times larger than the nonbreeding size. Before discharge, the millions of sperm are stored in a **seminal vesicle,** which, like the testes, enlarges greatly during the breeding season. Since most birds lack a penis, copulation is a matter of bringing cloacal surfaces into contact, usually while the male stands on the back of the female (Figure 29-20). Some swifts copulate in flight.

In the female of most birds, only the **left ovary** and **oviduct** develop (Figure 29-21); those on the right dwindle to vestigial structures (the loss of one ovary is another adaptation of birds for reducing weight). Eggs discharged from the ovary are picked up by the oviduct, which runs posteriorly to the cloaca. While the eggs are passing down the oviduct, **albumin,** or egg white, from special glands is added to them; farther down the oviduct, the shell membrane, shell, and shell pigments are also secreted about the egg.

Fertilization takes place in the upper oviduct several hours before the layers of albumin, shell membranes, and shell are added. Sperm remains alive in the female oviduct for many days after a single mating.

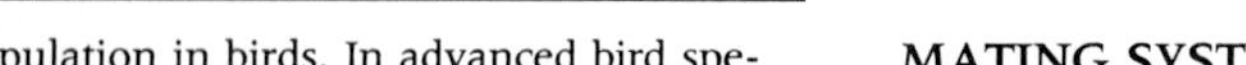

FIGURE 29-20

Copulation in birds. In advanced bird species, the male lacks a penis. The male copulates by standing on the back of the female, pressing his cloaca against that of the female, and passing sperm to the female.

MATING SYSTEMS

The two most common types of mating systems in animals are **monogamy** (meaning literally, "one marriage"), in which an individual mates with only one partner each breeding season, and **polygamy** ("many marriages"), in which an individual mates with two or more partners each breeding season. Although monogamy is rare among other animals, more than 90% of birds are monogamous. There are a few bird species, such as swans and geese, that choose partners for life. However, the great majority of birds are seasonally monogamous; that is, they pair up for the breeding season but lead independent lives the rest of the year.

One reason that monogamy is much more common among birds than among mammals is that female birds are not equipped, as mammals are, with a built-in food supply for the young. Thus the ability of the two sexes to provide parental care, especially food for the young, is more equal in birds than in mammals. A female bird will choose a male whose parental investment in their young is apt to be high and avoid a male that has mated with another female. If the male had mated with another female, he could at best divide his time between his two mates and might even devote most of his attention to the alternate mate. Thus females enforce monogamy.

Monogamy in birds is also encouraged by the need for the male to secure and defend a territory before he can attract a mate. The male may sing a great deal to announce his presence to females and to discourage rival males from entering his territory. The female wanders from one territory to another, seeking a male with foraging territory that offers the best chances for reproductive success. Usually a male is able to defend an area that provides just enough resources for one nesting female.

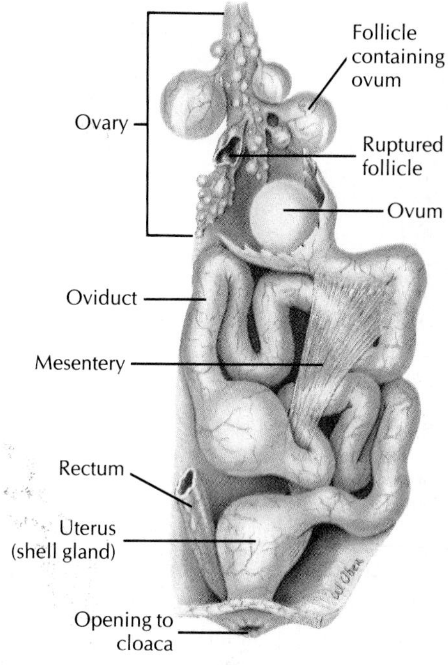

FIGURE 29-21

Reproductive system of a female bird.

The most common form of polygamy in birds, when it occurs, is **polygyny** ("many females"), in which a male mates with more than one female. In many species of grouse, the males gather in a collective display ground, the **lek,** which is divided into individual territories, each vigorously defended by the displaying male (Figure 29-22). There is nothing of value in the lek to the female except the male, and all he can offer are his genes, for only the females care for the young. Usually there are a dominant male and several subordinate males in the lek. Competition among males for females is intense, but the females appear to choose the dominant male for mating because, presumably, social rank correlates with genetic quality.

NESTING AND CARE OF YOUNG

To produce offspring, all birds lay eggs that must be incubated by one or both parents. The eggs of most songbirds require approximately 14 days for hatching; those of ducks or geese require at least twice that long. Most of the duties of

FIGURE 29-22

Dominant male sage grouse, *Centrocercus urophasianus,* surrounded by several hens that have been attracted by his "booming" display.

Photograph by L.L. Rue III.

FIGURE 29-23

Broad-tailed hummingbird, *Selasphorus platycercus,* feeding young in it's nest built of plant down and spider webs and decorated on the outside with lichens. The female builds the nest, incubates the two pea-sized eggs, and rears the young with no assistance from the male. These frail-looking but pugnacious little birds make arduous seasonal migrations between Canada and Mexico.

Photograph by L.L. Rue III.

incubation fall on the female, although in many instances both parents share the task, and occasionally only the male performs this work.

Most birds build some form of nest in which to rear their young. Some birds simply lay their eggs on the bare ground or rock. Others build elaborate nests, such as the pendent nests constructed by orioles, the neat lichen-covered nests of hummingbirds (Figure 29-23) and flycatchers, the chimney-shaped mud nests of cliff swallows, or the floating nest of the red-necked grebe. Most birds take considerable pains to conceal their nests from enemies. Woodpeckers, chickadees, bluebirds, and many others place their nests in tree hollows or other cavities; kingfishers often excavate tunnels in the banks of streams for their nests; and birds of prey build high in lofty trees or on inaccessible cliffs. A few birds, such as the American cowbird, the European cuckoo, and some ducks, build no nests at all but simply lay their eggs in the nests of other birds. When the eggs hatch, the young are taken care of by their foster parents. Most songbirds lay from 3 to 6 eggs, but the number of eggs laid in a clutch varies from 1 to 2 (some hawks and pigeons) to 18 or 20 (quail).

Newly hatched birds are of two types—**precocial** and **altricial** (Figure 29-24). The precocial young, such as quail, fowl, ducks, and most water birds, are covered with down when hatched and can run or swim as soon as their plumage is dry. The altricial ones, on the other hand, are naked and helpless at birth and remain in the nest for a week or more. The young of both types require care from the parents for some time after hatching. They must be fed, guarded, and protected against rain and sun. The parents of altricial species must carry food to their young almost constantly, since most young eat more than their weight each day. This food consumption explains the rapid growth of the young and their quick exit from the nest.

Bird Populations

Bird populations, like those of other animal groups, vary in size from year to year. Snowy owls, for example, are subject to population cycles that closely follow cycles in their food crop, mainly rodents. Voles, mice, and lemmings in the north have a fairly regular 4-year cycle of abundance; at population peaks, predator populations of foxes, weasels, and buzzards, as well as snowy owls, increase because there is abundant food for rearing their young. After a crash in the rodent population, snowy owls move south, seeking alternate food supplies. They occasionally appear in large numbers in southern Canada and northern United States, where their total absence of fear of humans makes them easy targets for thoughtless hunters.

Occasionally the activities of people bring about spectacular changes in bird distribution. Both starlings (Figure 29-25) and house sparrows have been accidentally or deliberately introduced to numerous countries, where they have become the two most abundant bird species on earth, with the exception of domestic fowl.

Altricial
One-day-old meadow lark

Precocial
One-day-old ruffed grouse

FIGURE 29-24

Comparison of 1-day-old altricial and precocial young. The altricial meadowlark *(left)* is born nearly naked, blind, and helpless. The precocial ruffed grouse *(right)* is covered with down, alert, strong legged, and able to feed itself.

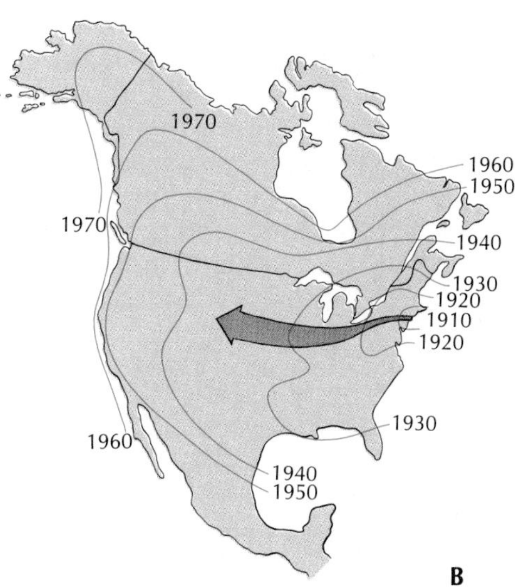

FIGURE 29-25

A, Starling, *Sturnus vulgaris.* Starlings are omnivorous, eating mostly insects in spring and summer and shifting to wild fruits in fall. **B,** Colonization of North America by starlings after the introduction of 120 birds into Central Park in New York City in 1890. There are now perhaps 100 million starlings in the United States, alone, testimony to the great reproductive potential of birds.

A, Photograph by L.L. Rue III; **B,** adapted from Johnson, S.R., and I. McT. Cowan. 1974. Can. J. Zool. **52:**1559-1576.

Humans also are responsible for the extinction of many bird species. More than 80 species of birds have, since 1695, followed the last dodo to extinction. Many died naturally, victims of changes in their habitat or competition with better-adapted species. But several have been hunted to extinction, among them the passenger pigeon, which only a century ago darkened the skies over North America in incredible numbers estimated in billions (Figure 29-26). Today, game bird hunting is a well-managed renewable resource in the United States and Canada, and while hunters kill millions of game birds each year, none of the 74 bird species legally hunted are endangered. Hunting interests, by acquiring large areas of wetlands for migratory bird refuges and sanctuaries, have contributed to the recovery of both game and nongame birds.

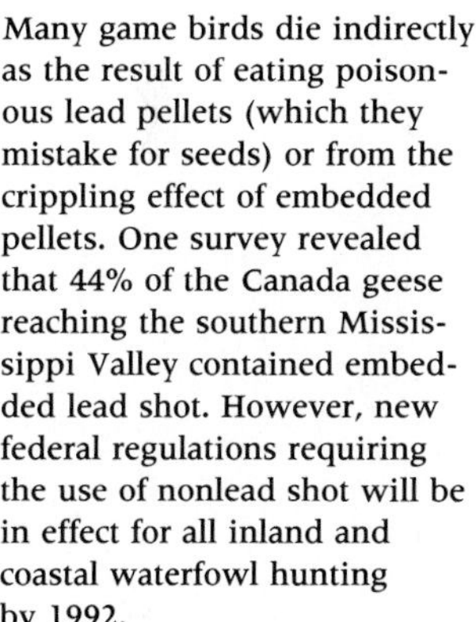

Many game birds die indirectly as the result of eating poisonous lead pellets (which they mistake for seeds) or from the crippling effect of embedded pellets. One survey revealed that 44% of the Canada geese reaching the southern Mississippi Valley contained embedded lead shot. However, new federal regulations requiring the use of nonlead shot will be in effect for all inland and coastal waterfowl hunting by 1992.

FIGURE 29-26

Sport-shooting passenger pigeons in Louisiana during the nineteenth century. Relentless sport and market hunting, before the establishment of state and federal hunting regulations, eventually dropped the population too low to sustain colonial breeding. The last passenger pigeon died in captivity in 1914.

Courtesy Culver Pictures

Our most destructive effects on birds are usually unintentional. The draining of marshes (more than 99% of Iowa's once extensive marshland is now farmland) has destroyed waterfowl nesting. Deforestation has likewise had great impact on tree-nesting species. The vertical appendages of civilization, such as television towers, monuments, tall buildings, and electric transmission towers and lines, take a fearful toll during bird migration in bad weather. Most birds, through their impressive reproductive potential, can replace in numbers those which become victims of human activities. Someone has calculated that a single pair of robins, producing two broods of four young a season, would leave 19,500,000 descendants in 10 years, should all survive at least that long. Although many birds have accommodated to the heavy-handed influence of humans on their environment, and some, like robins, house sparrows, and starlings, even thrive on it, most birds find the changes adverse, and to some species it is lethal.

Summary

The 8600 species of living birds are egg-laying, endothermic vertebrates covered with feathers and having the forelimbs modified as wings. Birds have evolved from archosaurian reptiles of the Mesozoic era and bear numerous reptilian morphological characters. The oldest known fossil bird, *Archaeopteryx*, from the Jurassic period of the Mesozoic era, was even more reptilelike than modern birds.

The adaptations of birds for flight are of two basic kinds — those reducing body weight and those promoting more power for flight. Feathers, the hallmark of birds, are complex derivatives of reptilian scales and combine lightness with strength, water repellency, and high insulative value. Body weight is further reduced by elimination of some bones, fusion of others (to provide rigidity for flight), and the presence of hollow, air-filled spaces in many bones. The light, horny bill, replacing the heavy jaws and teeth of reptiles, serves as both hand and mouth for all birds and is variously adapted for different feeding habits.

Kookaburra, the wood kingfisher of Australia, famous for its cry that resembles demented laughter.

Photograph by C.P. Hickman, Jr.

Adaptations that provide power for flight include a high metabolic rate and body temperature coupled with an energy-rich diet; a highly efficient respiratory system consisting of a system of air sacs arranged to pass air through the lungs during both inspiration and expiration; powerful flight and leg muscles arranged to place muscle weight near the bird's center of gravity; and an efficient, high-pressure circulation.

Birds have keen eyesight, good hearing, poorly developed sense of smell, and superb coordination for flight. The metanephric kidneys produce uric acid as the principal nitrogenous waste.

Birds fly by applying the same aerodynamic principles as an airplane and using similar equipment — wings for lift and support, a tail for steering and landing control, and wing slots for control at low flight speed. Flightlessness in birds is unusual but has evolved independently in several bird orders, usually on islands where terrestrial predators are absent; all are derived from flying ancestors.

Bird migration refers to regular movements between summer nesting places and wintering regions. Spring migration to the north where more food is available for nestlings enhances reproductive success. Many cues are used for direction finding in migration, including innate sense of direction and ability to navigate by the sun, the stars, or the earth's magnetic field.

The highly developed social behavior of birds is manifested in vivid courtship displays, mate selection, territorial behavior, and incubation of eggs and care of the young.

REVIEW QUESTIONS

1. Explain the significance of the discovery of *Archaeopteryx.* Why did this fossil prove beyond reasonable doubt that birds evolved from reptiles?
2. Birds are broadly divided into two groups—ratite and carinate. Explain what these terms mean and briefly discuss the appearance of flightlessness in birds.
3. The special adaptations of birds all contribute to two essentials for flight—more power and less weight. Explain how each of the following contributes to one or the other (or both) of these two essentials: feathers, skeleton, muscle distribution, digestive system, circulatory system, respiratory system, excretory system, reproductive system.
4. How do marine birds rid themselves of excess salt?
5. In what ways are the bird's ears and eyes specialized for their lifestyle?
6. Explain how the bird wing is designed to provide lift. What design features help to prevent stalling at low flight speeds?
7. Describe the four basic forms of bird wings.
8. What are the advantages of seasonal migration for birds?
9. Describe the different navigational resources birds may use in long-distance migration.
10. What are some of the advantages of social aggregation among birds?
11. More than 90% of all bird species are monogamous. Explain why monogamy is so much more common among birds than among mammals.
12. Briefly describe an example of polygyny among birds.
13. Define the terms precocial and altricial as they relate to birds.
14. Offer some examples of how human activities have been harmful to birds.

SELECTED REFERENCES

Baker, R. (ed.). 1981. The mystery of migration. New York, The Viking Press. *Handsomely illustrated, popularized account.*

Emlen, S.T. 1975. The stellar-orientation system of a migratory bird. Sci. Am. **233:**102-111 (Aug.). *Describes fascinating research with indigo buntings, revealing their ability to navigate by the center of celestial rotation at night.*

Feduccia, A. 1980. The age of birds. Cambridge, Mass., Harvard University Press. *Semipopular but authoritative account of bird evolution. Excellent text and illustrations.*

Terres, J.K. 1980. The Audubon Society encyclopedia of North American birds. New York, Alfred A. Knopf, Inc. *Comprehensive, authoritative, and richly illustrated.*

Welty, J.C., and L. Baptista. 1987. The life of birds, ed. 4. Philadelphia, W.B. Saunders Co. *Among the best of the ornithology texts; lucid style and excellent illustrations.*

THE TELL-TALE HAIR

If Fuzzy Wuzzy, the bear that had no hair (according to the children's rhyme), was truly hairless, it could not have been a mammal. For hair is as much an unmistakable characteristic of mammals as feathers are of birds. If an animal has hair it is a mammal; if it lacks hair it must be something else. It is true that many aquatic mammals are nearly hairless (whales, for example) but even they grow hair during development, and it can usually be found (with a bit of searching) at least in vestigal form somewhere on the body of the adult. Unlike feathers, which evolved from converted reptilian scales, mammalian hair is a completely new epidermal structure. Mammals use their hair for protection from the elements, for protective coloration and concealment, for waterproofing and buoyancy, and for behavioral signaling; they have turned it into sensitive vibrissae on their snouts and into prickly quills. Perhaps most important of all, mammals use their hair for thermal insulation, which allows them to enjoy the great advantages of homeothermy. Warm-blooded animals in most climates and at sunless times benefit from this natural and controllable protective insulation.

Hair, of course, is only one of several features that together define a mammal and help us to understand the mammalian evolutionary achievement. Among these are a highly developed placenta for feeding the embryo; mammary glands for nourishing the newborn; and an advanced nervous system that far exceeds in performance that of any other animal group. It is doubtful, however, that even with this winning combination of adaptations, the mammals could have triumphed as they have without their hair.▪

CHAPTER 30

Mammals

Mammals, with their highly developed nervous system and numerous ingenious adaptations, occupy almost every environment on earth that supports life. Although not a large group (some 4500 species as compared with 8600 species of birds, approximately 22,000 species of fishes, and nearly 1,000,000 species of insects), the class Mammalia (mam-may'lee-a) (L. *mamma*, breast) is arguably the most biologically successful group in the animal kingdom, with the possible exception of the insects. Many potentialities that dwell more or less latently in other vertebrates are highly developed in mammals. Mammals are exceedingly diverse in size, shape, form, and function. They range in size from the diminutive pigmy shrew, which has a body length of less than 4 cm and a weight of only a few grams, to the whales, some of which exceed 90,000 kg in weight.

An international moratorium on all commercial whaling took effect in 1986. However some countries that objected to the moratorium, notably Japan, are still killing hundreds of whales each year under the guise of "scientific" whaling.

Yet, despite their adaptability and in some instances because of it, mammals have been influenced by the heavy-handed presence of humans more than any other group of animals. We have domesticated numerous mammals for food and clothing, as beasts of burden, and as pets. We use millions of mammals each year in biomedical research. We have introduced alien mammals into new habitats, occasionally with benign results but more frequently with unexpected disaster. Although history provides us with numerous warnings, we continue to overcrop valuable wild stocks of mammals. The whale industry has threatened itself with total collapse by exterminating its own resource—a classic example of self-destruction in the modern world, in which competing segments of an industry are intent only on reaping all they can today as though tomorrow's supply were of no concern whatever. In some cases destruction of a valuable mammalian resource has been deliberate, such as the officially sanctioned (and tragically successful) policy during the Indian wars of exterminating the bison to drive the plains Indians into starvation. Although commercial hunting has declined, the ever-increasing human population with the accompanying destruction of wild habitats has harassed and disfigured the mammalian fauna in many parts of the world. We are becoming increasingly aware that our presence on this planet as the most powerful product of organic evolution makes us totally responsible for the character of our natural environment. Since our welfare has been and continues to be closely related to that

FIGURE 30-1

Evolution of the mammals. Mammals take their origin from the therapsid reptiles, which were probably warm-blooded. The most primitive mammals are the oviparous monotremes, which lay reptilelike eggs. They may have arisen from a group of therapsid reptiles distinct from those from which the viviparous therians, or "true" mammals, arose. The marsupials may have separated from the therian stock as early as the Jurassic period (160 million years BP), possibly in North America. The great radiation of modern mammals began in the Cretaceous period (135 million years BP).

of the other mammals, it is clearly in our interest to preserve the natural environment of which all mammals, ourselves included, are a part. We need to remember that nature can do without us, but we cannot exist without nature.

Origin and Relationships

In the early Mesozoic era, long before the great dinosaurs had reached the peak of their evolutionary success, a group of reptiles with mammal-like characteristics appeared. These were the **therapsids** (Figure 30-1). Their evolution and that of their descendants were accompanied by several structural changes that brought them ever closer to full mammalian status. The clumsy limbs of the reptile that stuck out laterally were replaced in therapsids by straight legs held close to the body, which provided speed and efficiency for hunting. Since reptilian stability was sacrificed by raising the animal from the ground, the muscular coordination center of the brain, the cerebellum, took on a greatly expanded role. Among the many changes in the bony structure of the head was the separation of air and food passages in the mouth. This enabled the animal to breathe while holding prey in its mouth. It also made possible prolonged chewing and some predigestion of food. The more advanced therapsids almost certainly had higher metabolic rates than the reptiles and some degree of endothermy. At some point the premammals acquired those two most characteristic of all mammalian identification tags: hair and mammary glands.

Most of the living mammals belong to the subclass Theria and have descended from a common ancestor of the Jurassic period, some 150 million years ago. However, monotremes (subclass Prototheria), the egg-laying mammals of Australia, Tasmania, and New Guinea (Figure 30-2), are so different from the others and possess so many reptilian characteristics that they are believed to have descended from an entirely different mammal-like reptile. The separation of Prototheria and Theria probably occurred approximately 50 million years earlier in the Triassic period. The geological record during the following Jurassic and Cretaceous periods is fragmentary, in large part because the mammals of these peri-

FIGURE 30-2

Spiny anteater, *Tachyglossus aculeatus*, the echidna. This egg-laying monotreme of Australia, Tasmania, and New Guinea is specialized for feeding on ants and termites.

Photograph by C.P. Hickman, Jr.

ods were creatures the size of a rat or smaller, with fragile bones that fossilized only under the most ideal circumstances.

When the dinosaurs vanished near the beginning of the Cenozoic era, the mammals suddenly expanded. This point, about 70 million years ago, marks the beginning of the Age of Mammals. This is partly attributable to the numerous ecological niches vacated by the reptiles, into which the mammals could move as their divergent adaptations fitted them. There were other reasons for their success. Mammals were agile, warm blooded, and insulated with hair; they had developed placental reproduction and suckled their young, thus dispensing with vulnerable eggs and nests; and they were more intelligent than any other animal alive. During the Eocene and Oligocene epochs of the Tertiary period (55 to 30 million years ago), the mammals flourished and reached their peak. In terms of numbers of species, this was the Golden Age of Mammals. They have declined somewhat in numbers since then, especially within the last million years, since humans have become formidable adversaries of wildlife. Nevertheless, mammals as a whole are a secure group, dominating the land environment now as they did 50 million years ago.

CHARACTERISTICS

1. **Body covered with hair,** but reduced in some
2. **Integument with sweat, scent, sebaceous,** and **mammary glands**
3. Mouth with **heterodont** teeth on both jaws
4. **Movable eyelids** and **fleshy external ears**
5. Four limbs (reduced or absent in some) adapted for many forms of locomotion
6. Circulatory system of a four-chambered heart, **persistent left aorta,** and **nonnucleated biconcave red blood corpuscles**
7. Respiratory system of lungs with alveoli, and voice box (larynx); **secondary (false) palate** separates air and food passages; **muscular diaphragm** separates thoracic and abdominal cavities.
8. **Muscular partition (diaphragm) between thorax** and **abdomen**
9. Excretory system of metanephric kidneys and ureters that usually open into a bladder.
10. Brain highly developed, especially **neocerebrum;** 12 pairs of cranial nerves
11. Endothermic and homeothermic
12. Separate sexes
13. Internal fertilization; **eggs developed in a uterus** with **placental attachment** (except in monotremes); **fetal membranes (amnion, chorion, allantois, yolk sac)**
14. Young nourished by **milk from mammary glands**

Since mammals and birds both evolved from reptiles, we can expect to find and do find many structural similarities among the three groups. It is, in fact, much easier to point to numerous resemblances between the mammals and the reptiles than to point to characteristics that are unique and diagnostic for mammals. **Hair,** discussed in the prologue to this chapter, is the most visible mammalian characteristic. A second unique characteristic of mammals is the method of nourishing their young with **milk-secreting glands;** reptiles have nothing remotely similar. Although less obvious, several important differences are present in cranial and jaw structure and jaw articulation. Placental mammals have **diphyodont teeth** (milk teeth replaced by a permanent set of teeth) rather than reptilian **polyphyodont teeth** (successive sets of teeth). Perhaps the single most important factor contributing to the success of mammals is the remarkable development of the **neocerebrum** permitting a level of adaptive behavior, learning, curiosity, and intellectual activity far beyond the capacity of any reptile.

CLASSIFICATION

Most mammalogists recognize 18 living orders of mammals, although some remove the aquatic carnivores (seals, sea lions, and walruses) from the order Carnivora and place them in a distinct order Pinnipedia. Fourteen extinct orders and five small living orders are not included in this classification.

Subclass Prototheria (pro'to-thir'e-a) (Gr. *prōtos,* first, + *thēr,* wild animal) — **egg-laying mammals.**

Order Monotremata (mon'o-tre'mah-tah) (Gr. *monos,* single, + *trēma,* hole) — **egg-laying mammals: duck-billed platypus** and **spiny anteater.** Only oviparous mammals; name of group referring to single opening from cloaca serving reproductive, digestive, and excretory systems; restricted to Australian region.

Subclass Theria (thir'e-a) (Gr. *thēr,* wild animal).

Infraclass Metatheria (met'a-thir'e-a) (Gr. *meta,* after, + *thēr,* wild animal) — **marsupial mammals.**

Order Marsupialia (mar-su'pe-ay'le-a) (Gr. *marsypion,* little pouch) — **pouched mammals,** for example, **opposums** (Figure 30-3), **kangaroos, koala.** Primitive mammals characterized in most by the abdominal pouch, the **marsupium,** in which young are reared; usually no typical placenta; mostly Australian with representatives in the Americas.

FIGURE 30-3

Opossum, *Didelphis marsupialis* (order Marsupialia), the only North American marsupial.

Photograph by Len Rue, Jr.

Infraclass Eutheria (yu-thir'e-a) (Gr. *eu,* true, + *thēr,* wild animal) — **placental mammals.**

Order Insectivora (in-sec-tiv'o-ra) (L. *insectum,* an insect, + *vorare,* to devour) — **insect-eating mammals,** for example, **shrews, hedgehogs, moles.** Most primitive of placental mammals; includes the smallest mammals (shrews).

Order Chiroptera (ky-rop'ter-a) (Gr. *cheir,* hand, + *pteron,* wing) — **bats.** All flying mammals with forelimbs modified into wings; use of echolocation by most bats; mostly nocturnal.

Order Carnivora (car-niv'o-ra) (L. *carn,* flesh, + *vorare,* to devour) — **flesh-eating mammals,** for example, **dogs, wolves, cats, bears, weasels.** Some of the most intelligent and strongest of animals; all with predatory habits; teeth especially adapted for tearing flesh; in most, canines used for killing their prey. Carnivora is divided into suborders Fissipedia, whose feet bear toes, and Pinnipedia, whose limbs are modified for aquatic life; **suborder Fissipedia,** worldwide except in the Australian and Antarctic regions and divided into certain familiar families, among which are the dog family, the cat family, the bear family, and the fur-bearing family; and **suborder Pinnipedia,** the sea lions, seals, sea elephants, and walruses.

Order Rodentia (ro-den'che-a) (L. *rodere,* to gnaw) — **gnawing mammals,** for example, **squirrels** (Figure 30-4), **rats, woodchucks.** Most numerous of all mammals both in numbers and species; dentition with two upper and two lower chisel-like incisors that grow continually and are adapted for gnawing.

FIGURE 30-4

Arctic ground squirrel, *Spermophilus parryi* (order Rodentia), one of the most northern in distribution of the large squirrel family.

Photograph by C.P. Hickman, Jr.

Order Lagomorpha (lag'o-mor'fa) (Gr. *lagōs,* hare, + *morphē,* form) — **rabbits, hares, pikas.** Dentition resembling that of rodents but with four upper incisors rather than two as in rodents.

Order Edentata (ee'den-ta'ta) (L. *edentatus,* toothless) — **toothless mammals,** for example, **sloths, anteaters, armadillos** (Figure 30-5). Either toothless or with degenerate peglike teeth; most representatives in South America; the nine-banded armadillo in the southern United States.

Order Cetacea (see-tay'she-a) (L. *cetus,* whale) — **fishlike mammals,** for example, **whales, dolphins, porpoises.** Anterior limbs modified into broad flippers; posterior limbs absent; nostrils represented by a single or

FIGURE 30-5

Young armadillo, *Dasypus novemcinctus* (order Edentata), digging for food. This is the only "armored" mammal of North America. With only degenerate teeth, the armadillo feeds on insects, other small invertebrates, berries, and other fruits.

Photograph by L.L. Rue III.

double blowhole on top of the head; when present, teeth all alike and lacking enamel.

Order Proboscidea (pro'ba-sid'e-a) (Gr. *proboskis,* elephant's trunk, from *pro,* before, + *boskein,* to feed)—**proboscis mammals; elephants.** Once contained the largest herbivores of the Cenozoic era; only two genera remaining: the Indian elephant, with relatively small ears, and the African elephant, with large ears.

Order Perissodactyla (pe-ris'so-dak'ti-la) (Gr. *perissos,* odd, + *dactylos,* toe)—**odd-toed hoofed mammals.** Mammals with an odd number (one or three) of toes and with well-developed hooves; includes **horses, zebras, tapirs,** and **rhinoceroses;** all herbivorous; both Perissodactyla and Artiodactyla often referred to as ungulates, or hoofed mammals, with teeth adapted for chewing.

Order Artiodactyla (ar'te-o-dak'ti-la) (Gr. *artios,* even, + *daktylos,* toe)—**even-toed hoofed mammals.** Even-toed ungulates, including **swine, camels, deer, hippopotamuses, giraffes, antelopes** (Figure 30-6), **cattle, pronghorn, sheep,** and **goats;** each toe sheathed in a cornified hoof; many, such as the cow, deer, and sheep, with horns; many ruminants, that is, herbivores with partitioned stomachs.

Order Primates (pry-may'teez) (L. *prima,* first)—**highest mammals,** for example, **lemurs, monkeys, apes, humans.** First in the animal kingdom in brain development with especially large cerebral hemispheres; five digits (usually provided with flat nails) on both forelimbs and hindlimbs; group singularly lacking in claws, scales, horns, and hooves; two suborders.

Suborder Prosimii (pro-sem'ee-i) (Gr. *pro,* before, + *simia,* ape)—**lemurs, tree shrews, tarsiers, lorises, pottos.** Primitive, arboreal, mostly nocturnal, primates restricted to the tropics of the Old World.

Suborder Anthropoidea (an'thro-poi'de-a) (Gr. *anthropos,* man). Consists of **monkeys, baboons, gibbons, apes,** and **humans.**

Structural and Functional Adaptations of Mammals

INTEGUMENT AND ITS DERIVATIVES

The mammalian skin and its modifications especially distinguish mammals as a group. As the interface between the animal and its environment, the skin is strongly molded by the animal's way of life. In general the skin is thicker in mammals than in other classes of vertebrates, although as in all vertebrates it is made up of **epidermis** and **dermis** (Figure 6-10, *B,* p. 157). Among the mammals the dermis becomes much thicker than the epidermis. The epidermis is relatively thin where it is well protected by hair, but in places subject to much contact and use, such as the palms or soles, its outer layers become thick and cornified with keratin.

FIGURE 30-6

Thompson gazelle, *Gazella thomsoni* (order Artiodactyla), one of the more than 70 species of antelope in Africa.

Photograph by C.P. Hickman, Jr.

Hair

Hair is especially characteristic of mammals, although humans are not very hairy creatures, and in whales hair is reduced to only a few sensory bristles on the snout. A hair grows out of a hair follicle, which, although an epidermal structure, is sunk into the dermis of the skin (see Figure 30-11). The hair grows continuously by rapid proliferation of cells in the follicle. As the hair shaft is pushed upward, new cells are carried away from their source of nourishment and die, turning into the same dense type of keratin that constitutes nails, claws, hooves, and feathers.

Mammals characteristically have two kinds of hair forming the **pelage** (fur coat): (1) dense and soft **underhair** for insulation and (2) coarse and longer

guard hair for protection against wear and to provide coloration. The underhair traps a layer of insulating air; in aquatic animals, such as the fur seal, otter, and beaver, it is so dense that it is almost impossible to wet it. In water the guard hairs wet and mat down over the underhair, forming a protective blanket (Figure 30-7). A quick shake when the animal emerges flings off the water and leaves the outer guard hair almost dry.

When a hair reaches a certain length, it stops growing. Normally it remains in the follicle until a new growth starts, whereupon it falls out. In humans, hair is shed and replaced throughout life. But in most mammals, there are periodic molts of the entire coat.

In the simplest cases, such as in foxes and seals, the coat is shed once each year during the summer months. Most mammals have two annual molts, one in the spring and one in the fall. The summer coat is always much thinner than the winter coat and is usually a different color. Several of the northern mustelid carnivores, such as the weasel, have white winter coats and brown-colored summer coats. It was once believed that the white inner pelage of arctic animals served to conserve body heat by reducing radiation loss, but recent research has shown that dark and white pelages radiate heat equally well. The winter white of arctic animals is simply camouflage in a land of snow. The varying hare of North America has three annual molts: the white winter coat is replaced by a brownish gray summer coat, and this is replaced in autumn by a grayer coat, which is soon shed to reveal the winter white coat beneath (Figure 30-8).

Outside the arctic, most mammals wear somber colors that are protective. Often the species is marked with "salt-and-pepper" coloration or a disruptive pattern that helps to make it inconspicuous in its natural surroundings. Examples are the leopard's spots, stripes of the tiger, and spots of fawns. Other mammals, for example, skunks, advertise their presence with conspicuous warning coloration.

The hair of mammals has become modified to serve many purposes. The bristles of hogs, vibrissae on the snouts of most mammals, and spines of porcupines and their kin are examples.

Vibrissae, commonly called "whiskers," are really sensory hairs that provide an additional special sense to many mammals. The bulb at the base of each follicle is provided with a large sensory nerve. The slightest movement of a vibrissa generates impulses in the nerve endings that travel to a special sensory area in the brain. The vibrissae are especially long in nocturnal and burrowing animals. In seals they apparently serve as a "distance touch" sensitive to pressure waves and turbulence in the water caused by objects or passing fishes. Vision is of little use to seals hunting in turbid water, where they are frequently found; investigators have noted that blind seals remain just as fat and healthy as normal seals.

Porcupines, hedgehogs, echidnas, and a few other mammals have developed an effective and dangerous spiny armor; the spines of the common North American porcupine break off at the bases when struck and, aided by backward-pointing hooks on the tops, work deeply into their victims (Figure 30-9). To assist slow learners, like dogs, in understanding what they are dealing with, porcupines rattle the spines and prominently display the white markings on the quills toward their antagonists.

Horns and antlers

Three kinds of horns or hornlike substances are found in mammals. **True horns,** found in **ruminants** (for example, cud chewers such as sheep and cattle), are hollow sheaths of keratinized epidermis that embrace a core of bone arising from the skull. Horns are not normally shed, are not branched (although they may be greatly curved), and are found in both sexes.

A hair is more than a strand of keratin. It consists of three layers: the medulla or pith in the center of the hair, the cortex with pigment granules next to the medulla, and the outer cuticle composed of imbricated scales. The hair of different mammals shows a considerable range of structure. It may be deficient in cortex, such as the brittle hair of deer, or it may be deficient in medulla, such as the hollow, air-filled hairs of the wolverine, so favored by northerners for trimming the hoods of parkas because it resists frost accumulation. The hairs of rabbits and some others are scaled to interlock when pressed together. Curly hair, such as that of sheep, grows from curved follicles.

FIGURE 30-7

American beaver, *Castor canadensis* (order Rodentia), cutting up a quaking aspen tree. This second largest rodent (the South American capybara is larger) has a heavy waterproof pelage consisting of long, tough guard hairs overlying the thick, silky underhair so valued in the fur trade.

Photograph by L.L. Rue III.

FIGURE 30-8

Snowshoe, or varying, hare, *Lepus americanus* (order Lagomorpha), in **A,** brown summer coat beginning to change to winter coat and, **B,** white winter coat. In winter, extra hair growth on the hind feet broadens the animal's support in snow. Snowshoe hares are common residents of the taiga (northern coniferous forests) and are an important food source for lynxes, foxes, and other carnivores. Population fluctuations of hares and their predators are closely related.

Photographs by L.L. Rue III.

FIGURE 30-9

Dogs are frequent victims of the porcupine's impressive armor. Unless removed (usually by a veterinarian) the quills will continue to work their way deeper in the flesh, causing great distress, and may lead to the victim's death.

Photograph by R.E. Treat.

Antlers of the deer family are entirely bone when mature. During their annual growth, antlers develop beneath a covering of highly vascular soft skin called **velvet** (Figure 30-10). Just before the breeding season, when the growth of the antlers is complete, the blood vessels constrict and the male removes the velvet by rubbing the antlers against trees. The antlers are dropped after the breeding season. New buds appear a few months later to herald the next set of antlers. For several years each new pair of antlers is larger and more elaborate than the previous set. The annual growth of antlers places a strain on the mineral metabolism, since during the growing season a large moose or elk must accumulate 50 or more pounds of calcium salts from its vegetable diet.

The **rhinoceros horn** is the third kind of horn. Hairlike horny fibers arise from the dermal papillae and are cemented together to form a single horn.

Glands

Of all vertebrates, mammals have the greatest variety of integumentary glands. Most fall into one of four classes: sweat, scent, sebaceous, and mammary. All are derivatives of the epidermis.

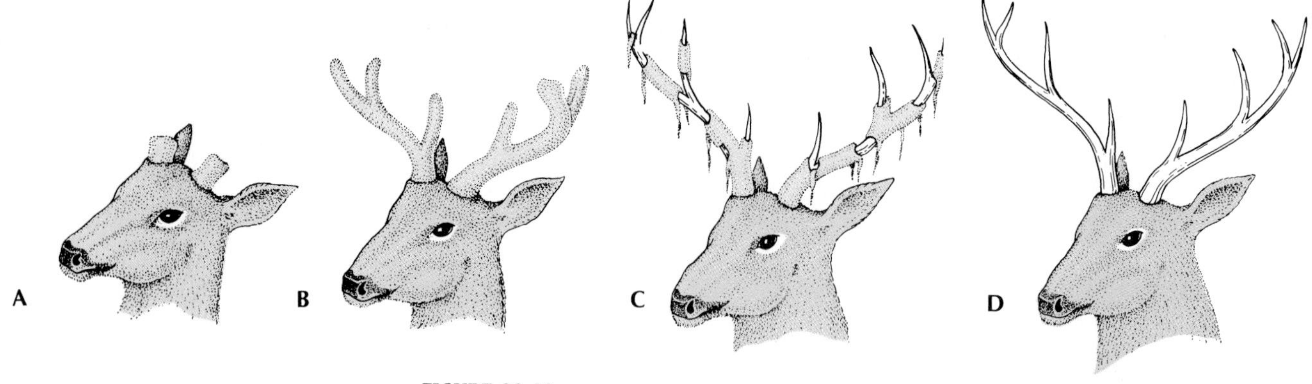

FIGURE 30-10

Annual growth of buck deer antlers. **A,** Antlers begin growth in late spring, stimulated by pituitary gonadotropins. **B,** The bone grows very quickly, until halted by a rapid rise in testosterone production by the testes. **C,** The skin (velvet) dies and sloughs off. **D,** Testosterone levels peak during the fall breeding season. The antlers are shed in January as testosterone levels subside.

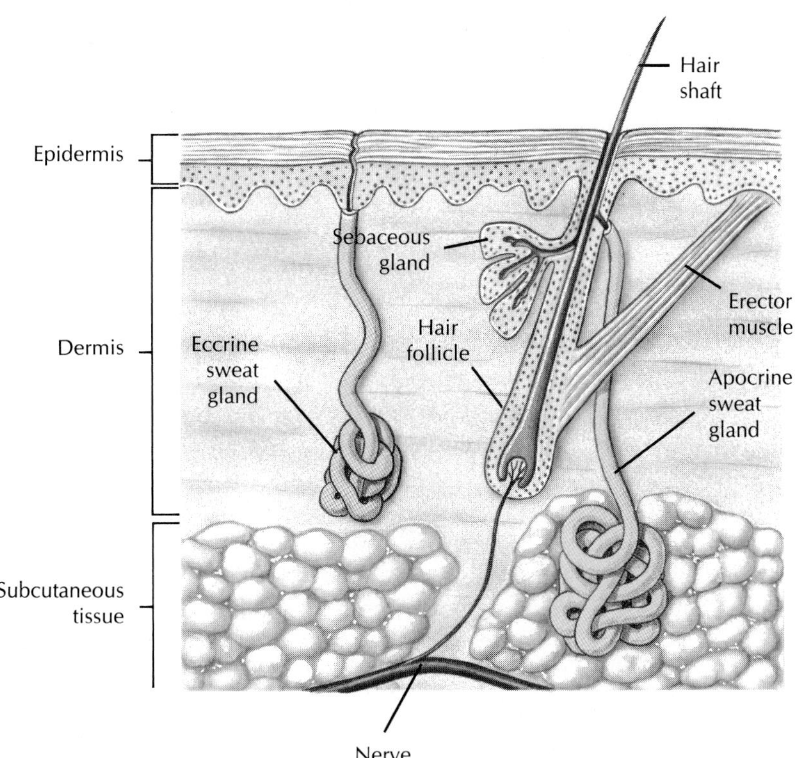

FIGURE 30-11

Glands in human skin. Sebaceous glands produce sebum which lubricates the hair and skin. Sweat glands are of two kinds. The more common eccrine glands secrete a watery sweat that cools the skin as it evaporates. Apocrine glands, more limited in body distribution, produce a milky secretion that is not involved in temperature regulation but may play a role in sexual attraction.

Sweat glands (Figure 30-11) are simple, tubular, highly coiled glands that occur over much of the body in most mammals. They are not present in other vertebrates. Two kinds of sweat glands may be distinguished: eccrine and apocrine. **Eccrine glands** secrete a watery sweat that functions mainly in temperature regulation (evaporative cooling). They occur in hairless regions (especially the foot pads) in most mammals, although in horses, some apes, and humans they are scattered all over the body. They are much reduced or absent in rodents, rabbits, whales, and others. Dogs are now known to have sweat glands all over the body. In human beings, racial differences are pronounced. Black people, who have more sweat glands than white people, can tolerate warmer temperatures.

Apocrine glands, the second type of sweat gland, are larger than eccrine glands and have longer and more winding ducts. Their secretory coil is in the subdermis. They always open into the follicle of a hair or where a hair has been. Black people have more apocrine glands than do white people, and women have twice as many as men. They develop approximately at sexual puberty and are restricted (in the human species) to the axillae (armpits), mons pubis, breasts, external auditory canals, prepuce, scrotum, and a few other places. Their secretion is not watery like ordinary sweat (eccrine gland) but rather a milky, whitish or yellow secretion that dries on the skin to form a plasticlike film. Apocrine glands are not involved in heat regulation, but their activity is known to be correlated with certain aspects of the sex cycle, among other possible functions.

Scent glands are present in nearly all mammals. Their location and functions vary greatly. They are used in communication with members of the same species, to mark territorial boundaries, for warning, or for defense. Scent-producing glands are located in orbital, metatarsal, and interdigital regions (deer); behind the eyes and on the cheek (pika and woodchuck); preputial regions on the penis (muskrats, beavers, and many canines); base of the tail (wolves and foxes); back of the head (dromedary); and anal region (skunks, minks, weasels). The latter

Milk varies in composition. Human milk is composed of 1% to 2% protein, 3% to 5% fat, 6% to 8% carbohydrate, 0.2% salts, and the remainder water. Cows' milk has more protein, about the same amount of fat, and less carbohydrate. In marine mammals (whales and seals) and arctic mammals (polar bears and caribou), where rapid growth of the young is important for species survival, the milk may contain 30% to 40% fat.

type, the most odoriferous of all glands, opens by ducts into the anus; the secretions of these glands can be discharged forcefully and propelled several feet. During the mating season many mammals give off strong scents for attracting the opposite sex. Humans also are endowed with scent glands. But civilization has taught us to dislike our own scent, a concern that has stimulated a lucrative deodorant industry to produce an endless output of soaps and odor-masking compounds.

Sebaceous glands are intimately associated with hair follicles (Figure 30-11), although some are free and open directly onto the surface. The cellular lining of the gland itself is discharged in the secretory process and must be renewed for further secretion. These gland cells become distended with a fatty accumulation, then die and are expelled as a greasy mixture called **sebum** into the hair follicle. Called a "polite fat" because it does not turn rancid, it serves as a dressing to keep the skin and hair pliable and glossy. Most mammals have sebaceous glands all over the body; in humans they are most numerous in the scalp and on the face.

Mammary glands, or **mammae,** which provide the name for mammals, are probably modified apocrine glands, although recent studies suggest that they may have been derived from sebaceous glands. Whatever their evolutionary origin, they occur on all female mammals and in a rudimentary form on all male mammals. They develop by the thickening of the epidermis to form a milk line along each side of the abdomen in the embryo. On certain parts of these lines the mammae appear, while the intervening parts of the ridge disappear.

In the human female, the mammary glands begin at puberty to increase in size because of fat accumulation and reach their maximum development in approximately the twentieth year. The mammae undergo additional development during pregnancy. In other mammals, the mammae are swollen only periodically when they are distended with milk during pregnancy and subsequent nursing of the young.

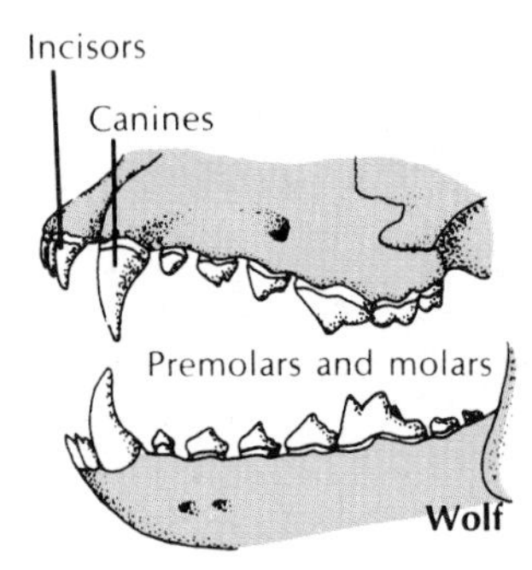

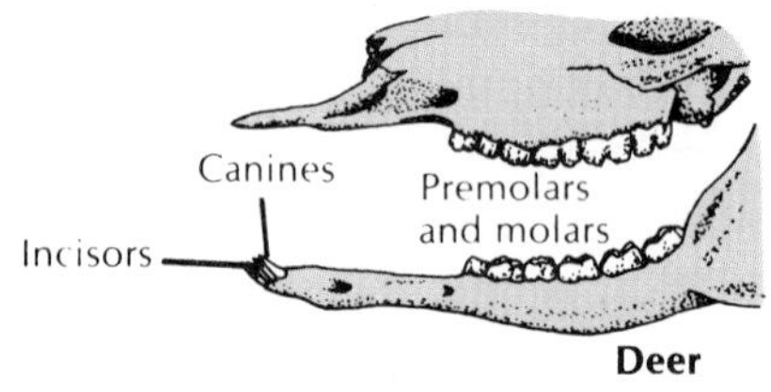

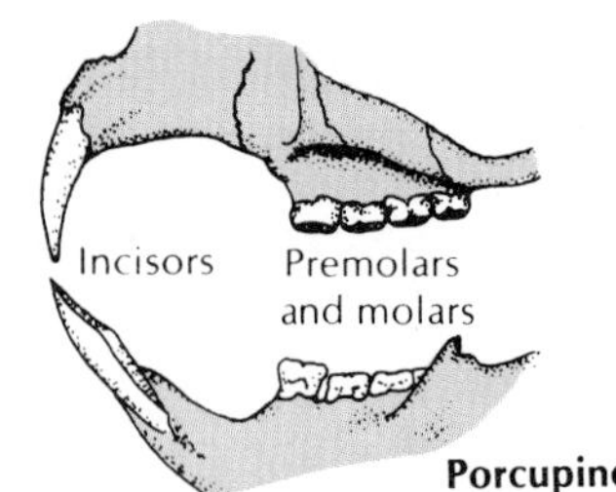

FIGURE 30-12

Adaptations of mammal tooth patterns for different diets. Sharp canines of the wolf are adapted for stabbing, and premolars and molars are for cutting rather than grinding. Browsing deer has predominantly grinding teeth; lower incisors and canines bite against a horny pad in the upper jaw. Porcupine has no canines; self-sharpening incisors are used for gnawing.

FOOD AND FEEDING

Mammals have exploited an enormous variety of food sources; some mammals require highly specialized diets, whereas others are opportunistic feeders that thrive on diversified diets. In all, food habits and physical structure are inextricably linked together. A mammal's adaptations for attack and defense and its specialization for finding, capturing, chewing or shearing, swallowing, and digesting food all determine a mammal's shape and habits.

Teeth, perhaps more than any other single physical characteristic, reveal the life-style of a mammal (Figure 30-12). It has been claimed that if all mammals except humans were extinct and represented only by fossil teeth, we could still construct a classification nearly as correct as the one we have now, which is based on all anatomical features. All mammals have teeth, except certain whales, monotremes, and anteaters, and their modifications are correlated with what the mammal eats.

Typically, mammals have a **diphyodont** dentition, that is, two sets of teeth: a temporary set, referred to as **deciduous** or **milk** teeth, that is replaced by a set of permanent teeth when the skull has grown large enough to accommodate a full set. In any given species, mammalian teeth are modified to perform specialized tasks such as cutting, nipping, gnawing, seizing, tearing, grinding, and chewing. Teeth differentiated in this manner in the individual are called **heterodont,** in contrast to the uniform, **homodont** dentition characteristics of lower vertebrates.

Usually four types of teeth are recognized. **Incisors,** with simple crowns and slightly sharp edges, are mainly for snipping or biting; **canines,** with long conical crowns, are specialized for piercing; **premolars,** with laterally compressed crowns and one or two cusps, are suitable for shearing and slicing; and **molars,** with large-surfaced crowns and variable cusp arrangement, are for crushing and mastication (chewing). Molars always belong to the permanent set; the others are deciduous.

Feeding specializations

On the basis of food habits, animals may be divided into herbivores, carnivores, omnivores, and insectivores.

Herbivorous animals that feed on grasses and other vegetation form two main groups: **browsers** and **grazers,** such as the ungulates (hooved mammals such as horses, deer, antelope, cattle, sheep, and goats), and the **gnawers,** such as the rodents, rabbits, and hares. In herbivores, the canines are smaller, or absent, whereas the molars, which are adapted for grinding, are broad and usually high-crowned. Rodents have chisel-sharp incisors that grow throughout life and must be worn away to keep pace with their continual growth.

Herbivorous mammals have a number of interesting adaptations for dealing with their massive diet of plant food. **Cellulose,** the structural carbohydrate of plants, is a potentially nutritious foodstuff, comprised of long chains of glucose units. However, the glucose molecules in cellulose are linked by a type of chemical bond that few enzymes can attack. Vertebrates do not synthesize cellulose-splitting enzymes. Instead, the herbivorous vertebrates harbor a **microflora** of anaerobic bacteria in huge fermentation chambers in the gut. These bacteria break down and metabolize the cellulose, releasing a variety of fatty acids, sugars, and starches that the host animal can absorb and use.

In some herbivores, such as horses and zebras, rabbits and hares, elephants, and many rodents, the gut has a spacious side-pocket, or diverticulum, called a **cecum,** which serves as a fermentation chamber and absorptive area. Hares, rabbits, and some rodents often eat their fecal pellets, giving the food a second pass through the fermenting action of the intestinal bacteria.

The **ruminants** (cattle, bison, buffalo, goats, antelopes, sheep, deer, giraffes, and okapis) have a huge **four-chambered stomach** (Figure 30-13). When a ruminant feeds, grass passes down the esophagus to the **rumen,** the first part of the stomach, where it is broken down by the rich microflora and then formed into small balls of cud. At its leisure the ruminant returns the cud to its mouth where it is deliberately chewed at length to crush the fiber. Swallowed again, the food returns to the rumen where it is digested by the cellulolytic bacteria. The pulp passes to the **reticulum** with its honeycombed epithelium where fermentation continues. Next the pulp passes into the **omasum,** where water, soluble food, and microbial products are absorbed. The smallest particles are separated out and allowed to proceed into the **abomasum** ("true" stomach). Here proteolytic enzymes are secreted and normal digestion takes place in an acid environment.

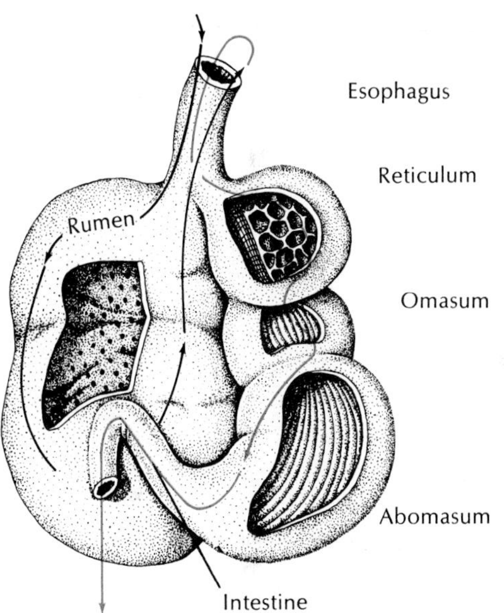

FIGURE 30-13

Ruminant's stomach. Food passes first to the rumen (sometimes through the reticulum) and then is returned to the mouth for chewing (chewing the "cud," or rumination) *(black arrow).* After reswallowing, food returns to the rumen or passes directly to reticulum, omasum, and abomasum for final digestion *(red arrow).*

Herbivores in general have large and long digestive tracts and must eat a large amount of plant food to survive. A large African elephant weighing 6 tons must consume 135 to 150 kg (300 to 400 pounds) of rough fodder each day to obtain sufficient nourishment for life.

Carnivorous mammals feed mainly on herbivores. This group includes foxes, weasels, cats, dogs, wolverines, fishers, lions, and tigers. Carnivores are well equipped with biting and piercing teeth and powerful clawed limbs for killing their prey. Since their protein diet is much more easily digested than is the woody food of herbivores, their digestive tract is shorter and the cecum small or absent. Carnivores eat separate meals and have much more leisure time for play and exploration.

In general, carnivores lead more active—and by human standards, more interesting—lives than do the herbivores. Since a carnivore must find and catch

FIGURE 30-14

Lionesses, *Panthera leo* (order Carnivora), eating a wildebeest. Lions stalk prey and then charge suddenly to surprise victim. They lack stamina for long chase. Lions gorge themselves with kill, then sleep and rest for periods as long as one week before eating again.

Photograph by L.L. Rue III.

its prey, there is a premium on intelligence; many carnivores, the cats, for example, are noted for their stealth and cunning in hunting prey (Figure 30-14). Although evolution seems to have favored the carnivores, their success has led to a selection of herbivores capable of either defending themselves or of detecting and escaping carnivores. Thus, for the herbivores, there has been a premium on keen senses and agility. Some herbivores, however, survive by virtue of their sheer size (for example, elephants) or by defensive group behavior (for example, musk-oxen).

Humans have changed the rules in the carnivore-herbivore contest. Carnivores, despite their intelligence, have suffered much from human presence and have been virtually exterminated in some areas. Herbivores, on the other hand, especially the rodents with their potent reproductive ability, have consistently defeated our most ingenious efforts to banish them from our environment. The problem of rodent pests in agriculture has been intensified; we have removed carnivores, which served as the herbivores' natural population control, but have not been able to devise a suitable substitute.

The renowned fecundity of meadow voles, and the effect of removing natural predators from rodent populations, is felicitously expressed in this excerpt from Thornton Burgess's "Portrait of a Meadow Mouse":

He's fecund to the nth degree
In fact this really seems to be
His one and only honest claim
To anything approaching fame.
In just twelve months, should all survive,
A million mice would be alive—
His progeny. And this, 'tis clear,
Is quite a record for a year.

Quite unsuspected, night and day
They eat the grass that would be hay.
On any meadow, in a year,
The loss is several tons, I fear.
Yet man, with prejudice for guide,
The checks that nature doth provide
Destroys. The meadow mouse survives
And on stupidity he thrives.

Omnivorous mammals live on both plant food and animals. Examples are pigs, raccoons, rats, bears (Figure 30-15), humans, and most other primates. Many carnivorous forms also eat fruits, berries, and grasses when hard pressed. The fox, which usually feeds on small rodents, insects, and birds, eats apples, beechnuts, berries, and corn when its normal food sources are scarce.

Insectivorous mammals are those that subsist chiefly on larval and adult insects. Examples are moles, shrews, anteaters, and most bats. The insectivorous category is not a sharply distinguished one, however, because many omnivores, carnivores, and even some herbivores eat insects on occasion.

For most mammals, searching for food and eating occupy most of their active life. Seasonal changes in food supplies are considerable in temperate zones. Living may be easy in the summer when food is abundant, but in winter many carnivores must range far and wide to eke out a narrow existence. Some migrate to regions where food is more abundant. Others hibernate and sleep the winter months away.

But there are many provident mammals that build up food stores during periods of plenty. This habit is most pronounced in rodents, such as squirrels, chipmunks, gophers, and certain mice. All tree squirrels—red, fox, and gray—collect nuts, conifer seeds, and fungi and bury these in caches for winter use.

Often each item is hidden in a different place (scatter hoarding) and scent marked to assist relocation in the future. Some of the caches of the chipmunks and squirrels may exceed a bushel.

FIGURE 30-15

Grizzly bear, *Ursus horribilis,* with a large salmon captured during the spawning run. All bears are opportunistic omnivores that take advantage of any seasonally abundant resource. They are highly dependent on vegetable food but also kill small mammals, scavenge, and rob all bees' nests they can find and enter. Bears are the least carnivorous of the Carnivora.

Photograph by L.L. Rue III.

MIGRATION

Migration is a much more difficult undertaking for mammals than for birds; not surprisingly, few mammals make regular seasonal migrations, preferring instead to center their activities in a defined and limited home range. Nevertheless, there are some striking examples of mammalian migrations. More migrators are found in North America than on any other continent.

An example is the barren-ground caribou of Canada and Alaska, which undertakes direct and purposeful mass migrations spanning 160 to 1100 km (100 to 700 miles) biannually (Figure 30-16). From winter ranges in the boreal forests (taiga) they migrate rapidly in late winter and spring to calving and foraging ranges on the barren grounds (tundra). The calves are born in mid-June. As the summer progresses, they are increasingly harassed by warble and nostril flies that bore into their flesh, by mosquitoes that drink their blood (estimated at a liter per caribou each week during the height of the mosquito season), and by wolves that prey on the calves. They move southward in July and August, feeding little along the way. In September they reach the forest and feed there almost continuously on low ground vegetation. Mating (rut) occurs in October.

The caribou have suffered a drastic decline in numbers. In primitive times there were several million of them; by 1958 less than 200,000 remained in Canada. The decline has been attributed to several factors, including habitat deterioration resulting from exploration and development in the North, but especially to excessive hunting. For example, the Western Arctic herd in Alaska exceeded 250,000 caribou in 1970. Following 5 years of heavy unregulated hunting, a 1976 census revealed only about 65,000 animals left. Hunting by people was then restricted and aerial hunting of wolves (a major predator of the calves) was opened. These changes produced a remarkable recovery: by 1980 the herd had increased to 140,000 and is expected to reach its original population of 250,000.

The plains bison, before its deliberate near extinction by white people, made huge circular migrations to separate summer and winter ranges.

The longest mammal migrations of all are made by the oceanic seals and whales. One of the most remarkable migrations is that of the fur seal (Figure 30-17), which breeds on the Pribilof Islands approximately 300 km off the coast of Alaska and north of the Aleutian Islands. From wintering grounds off southern California the females journey as much as 2800 km (1700 miles) across open ocean, arriving at the Pribilofs in early summer where they congregate in enormous numbers. The young are born within a few hours or days after the arrival of the cows. Then the bulls, having already arrived and established territories, collect harems of cows, which they guard with vigilance. After the calves have been nursed for approximately 3 months, cows and juveniles leave for their long migration southward. The bulls do not follow and remain in the Gulf of Alaska during the winter.

Although we might expect the only winged mammals, bats, to use their gift of flight to migrate, few of them do. Most spend the winter in hibernation. The four species of American bats that do migrate spend their summers in the northern or western states and their winters in the southern United States and Mexico.

FLIGHT AND ECHOLOCATION

Mammals have not exploited the skies to the same extent that they have the terrestrial and aquatic environments. However, many mammals scamper about in trees with amazing agility; some can glide from tree to tree, and one group, the

A

B

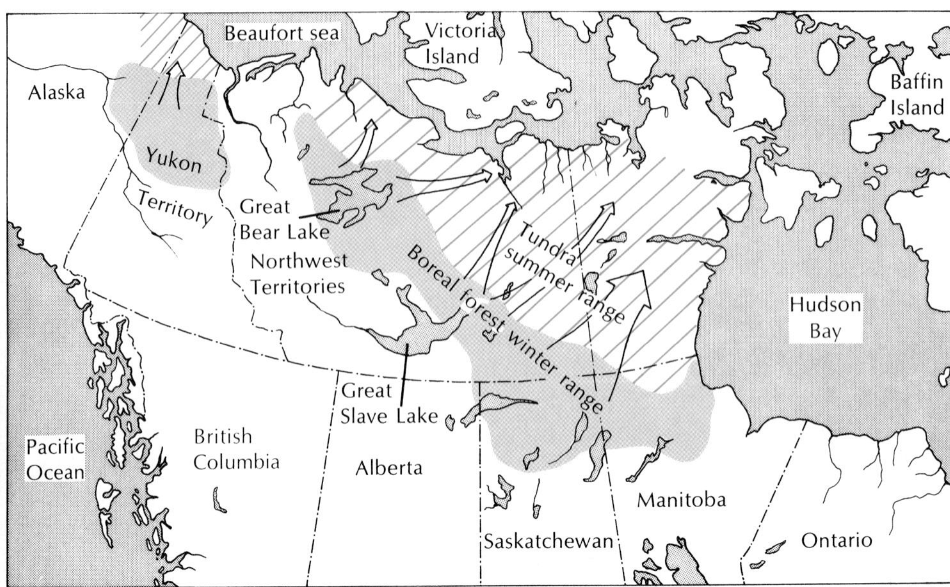

FIGURE 30-16

Barren-ground caribou, *Rangifer tarandus groenlandicus* (order Artiodactyla), of Canada and Alaska. **A,** Adult male caribou in autumn pelage and antlers in velvet. **B,** Summer and winter ranges of some major caribou herds in Canada and Alaska (other herds not shown occur on Baffin Island and in western and central Alaska). The principal spring migration routes are indicated by arrows; routes vary considerably from year to year.

Photograph by C.P. Hickman, Jr.

bats, is capable of full flight. Gliding and flying evolved independently in several groups of mammals, including the marsupials, rodents, flying lemurs, and bats. Anyone who has watched a gibbon perform in a zoo realizes there is something akin to flight in this primate, too. Among the arboreal squirrels, all of which are nimble acrobats, by far the most efficient is the flying squirrel. These forms actually glide rather than fly, using the gliding skin that extends from the sides of the body.

Bats, the only group of flying mammals, are nocturnal insectivores and thus occupy a niche left mostly vacant by birds (Figure 30-18). Their outstanding success is attributed to two things: flight and the capacity to navigate by echolocation. Together these adaptations enable bats to fly and avoid obstacles in absolute darkness, to locate and catch insects with precision, and to find their way deep into caves (another habitat largely ignored by other mammals and birds) where they sleep during the daytime hours.

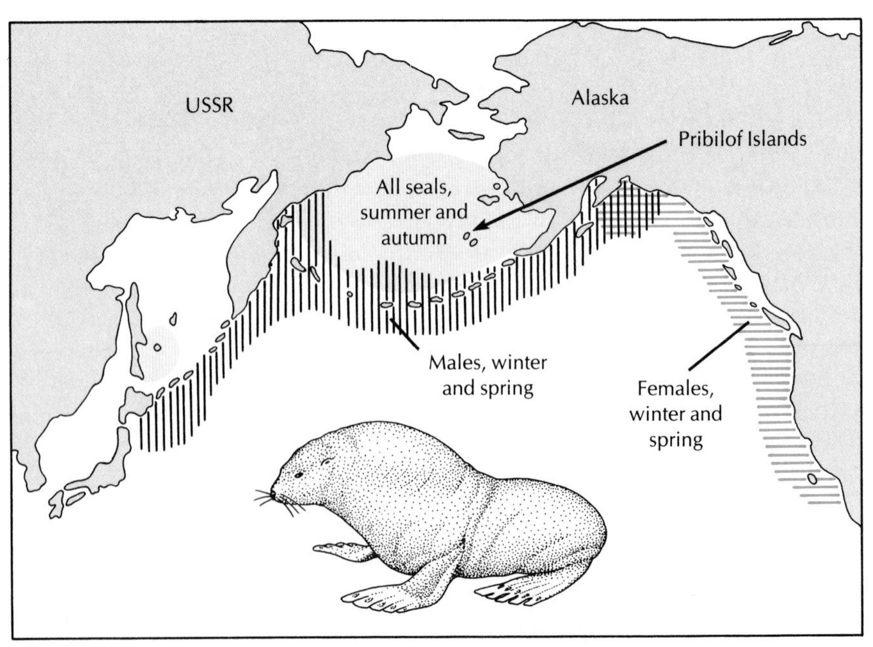

FIGURE 30-17

Annual migrations of the fur seal, showing the separate wintering grounds of males and females. Both males and females of the larger Pribilof population migrate in early summer to the Pribilof Islands, where the females give birth to their pups and then mate with the males.

When they are in flight, bats emit short pulses 5 to 10 milliseconds in duration in a narrow directed beam from the mouth (from the nose in some). Each pulse is frequency modulated; that is, it is highest at the beginning, up to 100,000 Hertz (Hz, cycles per second), and sweeps down to perhaps 30,000 Hz at the end. Sounds of this frequency are ultrasonic to the human ear, which has an upper limit of about 20,000 Hz. When the bat is searching for prey, it produces about 10 pulses per second. If a prey is detected, the pulse rate increases rapidly, up to 200 per second in the final phase of approach and capture. The pulses are spaced so that the echo of each is received before the next pulse is emitted, an adaptation that prevents jamming. Since the transmission-to-reception time decreases as the bat approaches an object, it can increase the pulse frequency to obtain more information about the object. The pulse length is also shortened as it nears the object.

FIGURE 30-18

The pallid bat, *Antozous pallidus,* in flight. This relatively large bat of western North America and Mexico is migratory in the northern part of its range.

Photograph by J.D. Cunningham/Visuals Unlimited.

The external ears of bats are large, like hearing trumpets, and shaped variously in different species. Less is known about the bat's inner ear, but it obviously is capable of receiving the ultrasonic sounds emitted. Bat navigation is so refined that biologists believe that the bat builds up a mental image of its surroundings from echo scanning that is virtually as complete as the visual image from eyes of diurnal animals.

For reasons not fully understood, all bats are nocturnal, even the fruit-eating bats that use vision and olfaction instead of sonar to find their food. The tropics have many kinds of bats, including the famed vampire bat. This species is provided with razor-sharp incisors used to shave away the epidermis to expose underlying capillaries. After infusing an anticoagulant to keep the blood flowing, it laps up its meal and stores it in a specially modified stomach.

REPRODUCTION

Most mammals have definite mating seasons, usually in the winter or spring and timed to coincide with the most favorable time of the year for rearing the young after birth. Many male mammals are capable of fertile copulation at any time, but the female mating function is restricted to a time during a periodic cycle, known as the **estrous cycle.** The female receives the male only during a relatively brief period known as **estrus,** or heat (Figure 30-19).

There are three different patterns of reproduction in mammals. One pattern is represented by the egg-laying mammals, the **monotremes.** The duck-billed platypus has one breeding season each year. The ovulated eggs, usually two, are fertilized in the oviduct. As they continue down the oviduct, various glands add albumin and then a thin, leathery shell to each egg. When laid, the eggs are about the size of a robin's egg. The platypus lays its eggs in a burrow nest where they are incubated for about 12 days. After hatching, the young are fed milk (which they obtain by licking, not suckling) for a prolonged period. Thus in monotremes there is no gestation (period of pregnancy) and the developing embryo draws on nutrients stored in the egg, much as do the embryos of reptiles and birds. But like all other mammals, the monotremes rear their young on milk.

The pouched mammals (**marsupials**) exhibit a second pattern of reproduction (Figure 30-20). The physiology of gestation and lactation may be complicated in members of this group, which have a brief gestation period. In red kangaroos the first pregnancy of the season is followed by a 33-day gestation, after which the young (joey) is born, crawls to the pouch without assistance from the mother, and attaches to a nipple. The mother immediately becomes pregnant again, but the presence of a suckling young in the pouch arrests development of the new embryo in the uterus at about the 100-cell stage. This period of arrest, called embryonic diapause, lasts approximately 235 days during which time the joey is growing in the pouch. When the joey leaves the pouch, the uterine embryo resumes development and is born about a month later. The mother again

Echolocation is believed to be used by many insectivores (e.g., shrews and tenrecs), but is crudely developed as compared with bats. The toothed whales, however, have a highly developed capacity to locate objects by echolocation. Totally blind sperm whales that are in perfect health have been captured with food in their stomachs. Although the mechanism of sound production and reception remains imperfectly understood, it is thought that low and high frequency clicks produced in the sinus passages are focused into a narrow beam by a lens-shaped body in the forehead (the "melon"). Returning echos are channeled through oil-filled sinuses in the lower jaw to the inner ear. Toothed whales can apparently determine the size, shape, speed, distance, direction, and density of objects in the water and know the position of every whale in the pod.

FIGURE 30-19

Mating elk, *Cervus canadensis* (order Artiodactyla). Male North American elk, more correctly called wapiti, begin bugling in August to herald the beginning of the rutting season. Elk are polygamous and males attempt to round up and mate with as many cows as they can defend against rival males. The rut ends in late October and calves are born the following June.

Photograph by Len Rue, Jr.

becomes pregnant, but because the second joey is suckling, once again development of the new embryo is arrested. Meanwhile, the first joey returns to the pouch from time to time to suckle. At this point the mother has three young of different ages dependent on her: a joey on foot, a joey in the pouch, and a diapause embryo in the uterus. There are variations on this remarkable sequence; not all marsupials have developmental delays like kangaroos, and some do not even have pouches. But in all marsupials the young are born at an extremely early stage of development and undergo prolonged development while dependent on a teat (Figure 30-21).

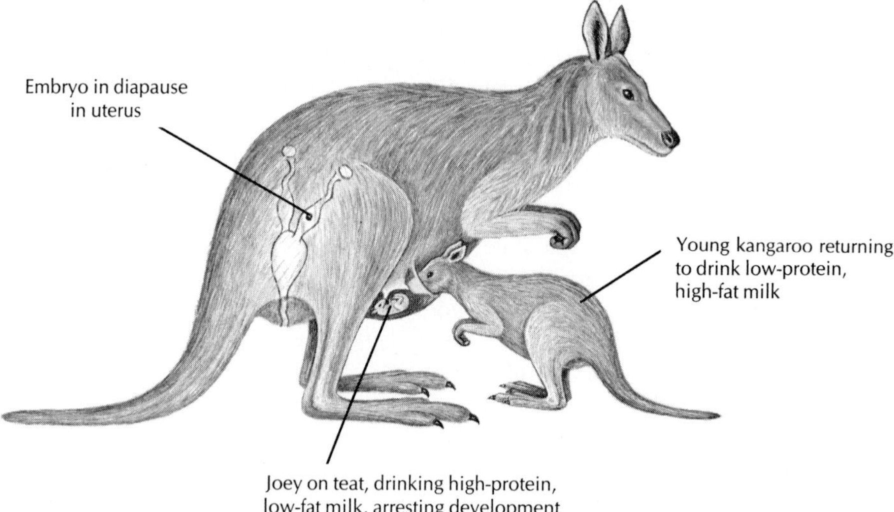

FIGURE 30-20

Kangaroos have a complicated reproductive pattern in which the mother may have three young in different stages of development dependent on her at once.

Adapted from Austin, C.R., and R.V. Short (eds.). 1972. Reproductive patterns, vol. 4. In Reproduction in mammals. New York, Cambridge University Press.

The third pattern of reproduction is that of the placental mammals, the **eutherians.** In this most successful of the mammal groups, the reproductive investment is in gestation. The embryo remains in the mother's uterus, nourished by food supplied through the placenta, an intimate connection between mother and young. The length of gestation varies greatly. In general, the larger the mammal, the longer the gestation time. For example, mice have a gestation period of 21 days; rabbits and hares, 30 to 36 days; cats and dogs, 60 days; cattle, 280 days; and elephants, 22 months. But there are important exceptions (nature seldom offers perfect correlations). Baleen whales, the largest mammals, carry their young for only 12 months. The condition of the young at birth also varies. An antelope bears its young well furred, eyes open, and able to run about. Newborn mice, however, are blind, naked, and helpless. We all know how long it takes a human baby to gain its footing. Human growth is in fact slower than that of any other mammal, and this is one of the distinctive attributes that sets us apart from other mammals.

A curious phenomenon that lengthens the gestation period of many mammals is **delayed implantation.** The blastocyst remains dormant while its implantation in the uterine wall is postponed for periods of a few weeks to several months. For many mammals (e.g., bears, seals, weasels, badgers, bats, many deer) delayed implantation is a device for extending gestation so that the young are born at the time of year which is best for their survival.

The number of young produced by mammals in a season depends on the mortality rate which, for some mammals such as mice, may be high at all age levels. Usually, the larger the animal, the smaller the number of young in a litter. Small rodents, which serve as prey for so many carnivores, produce as a rule more than one litter of several young each season. Meadow voles are known to produce as many as 17 litters of four to nine young in a year. Most carnivores have but one litter of three to five young per year. Large mammals, such as elephants and horses, give birth to a single young with each pregnancy. An elephant produces, on the average, four calves during her reproductive life of perhaps 50 years.

MAMMAL POPULATIONS

A population of animals includes all the members of a species that interbreed and share a particular space (Chapter 5). All mammals live in communities, each composed of numerous populations of different animal and plant species. Each species is affected by the activities of other species and by other changes (especially climatic changes) that occur. Thus populations are always changing in size. Populations of small mammals are lowest before the breeding season and greatest just after the addition of the new members. Beyond these expected changes in population size, animal populations may fluctuate from other causes.

Irregular fluctuations are commonly produced by variations in climate, such as unusually cold, hot, or dry weather, or by natural catastrophes, such as fires, hailstorms, and hurricanes. These are density-independent causes, since they affect a population whether it is crowded or dispersed. However, the most spectacular fluctuations are density dependent; that is, they are correlated with population crowding (pp. 142-143). Cycles of abundance are common among many rodent species.

The population peaks and mass migrations of the Scandinavian and arctic North American lemmings are well known. Lemmings breed all year round, although more in the summer than in winter. The gestation period is only 21 days; young born at the beginning of the summer are weaned in 14 days and are capable of reproducing by the end of the summer. At the peak of their population density, having devastated the vegetation by tunneling and grazing, they begin long, mass migrations to find new undamaged habitats for food and space. They swim across streams and small lakes as they go but cannot distinguish these from large lakes, rivers, and the sea, in which they drown. Since lemmings are the main diet of many carnivorous mammals and birds, any change in lemming population density affects all their predators as well.

The varying hare (snowshoe rabbit) of North America shows 10-year cycles in abundance. The well-known fecundity of rabbits enables them to produce litters

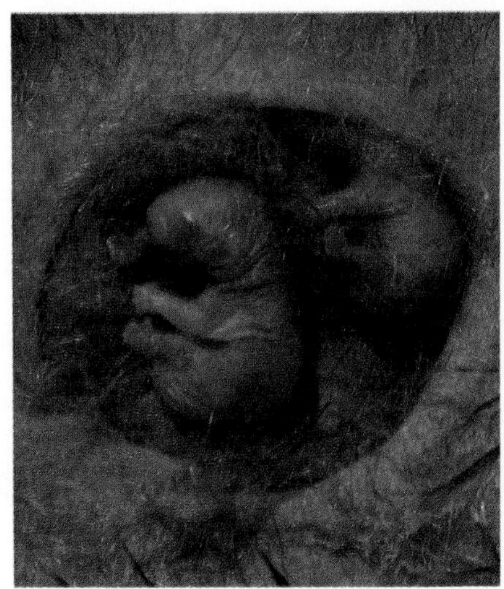

FIGURE 30-21

Fifteen-day-old opossums, *Didelphis marsupialis,* fastened to teats in mother's pouch. When born after a gestation period of only 12 days, they are the size of honeybees. They remain attached to the nipples for 50 to 60 days.

Photograph by L.L. Rue III.

FIGURE 30-22

Changes in population of varying hare and lynx in Canada as indicated by pelts received by the Hudson's Bay Company. The abundance of lynx (predator) follows that of the hare (prey).

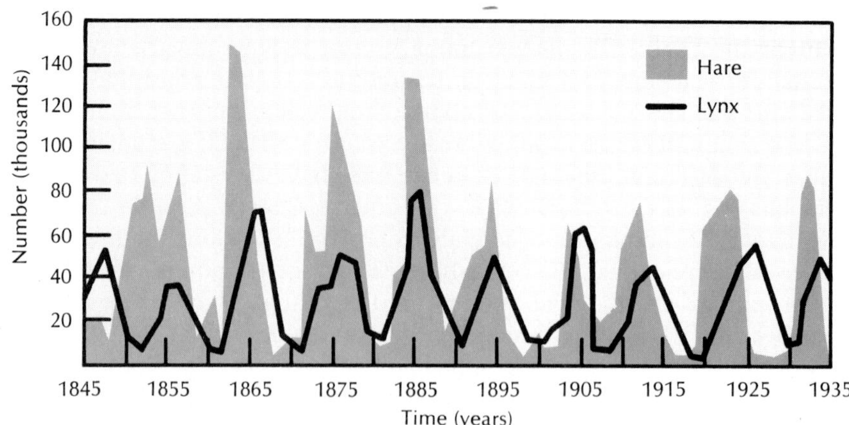

In his book *The Arctic,* Canadian naturalist Fred Bruemmer describes the growth of lemming populations in arctic Canada: "After a population crash one sees few signs of lemmings; there may be only one to every 10 acres. The next year, they are evidently numerous; their runways snake beneath the tundra vegetation, and frequent piles of rice-sized droppings indicate the lemmings fare well. The third year one sees them everywhere. The fourth year, usually the peak year of their cycle, the populations explode. Now more than 150 lemmings may inhabit each acre of land and they honeycomb it with as many as 4000 burrows. Males meet frequently and fight instantly. Males pursue females and mate after a brief but ardent courtship. Everywhere one hears the squeak and chitter of the excited, irritable, crowded animals. At such times they may spill over the land in manic migrations."

of three or four young as many as five times per year. The density may increase to 1500 hares competing for food in each square kilometer of northern forest. Predators (owls, minks, foxes, and especially lynxes) also increase (Figure 30-22). Then the population crashes precipitously, for reasons that have long been a puzzle to scientists. Rabbits die in great numbers, not from lack of food or from an epidemic disease (as was once believed) but evidently from some density-dependent psychogenic cause. As crowding increases, hares become more aggressive and stop breeding. The entire population reveals symptoms of pituitary–adrenal gland exhaustion, an endocrine imbalance called "shock disease," which results in death. There is much about these dramatic crashes that is not understood. Whatever the causes, population crashes that follow superabundance, although harsh, are clearly advantageous to the species, because the vegetation is allowed to recover, thus providing the survivors with a much better chance for successful breeding.

■ Summary

The 4500 living mammals are descended from a mammal-like reptile (therapsid) of the Jurassic period of the Mesozoic era. They diversified rapidly during the Tertiary period of the Cenozoic era to become the most intelligent and advanced animals on earth. Mammals are named for the glandular organs of the female (rudimentary in the male) that secrete milk for the nourishment of the young after their birth. An even more obvious mammalian characteristic is the presence of hair, an integumentary outgrowth that covers most mammals. Hair serves variously for mechanical protection, thermal insulation, protective coloration, and waterproofing. Mammalian skin is rich in glands: sweat glands that function in evaporative cooling, scent glands used in social interactions, and sebaceous glands that secrete lubricating skin oil. All placental mammals have deciduous teeth that are replaced by permanent teeth (diphyodont dentition). The four groups of teeth — incisors, canines, premolars, and molars — may be highly modified in different mammals for specialized feeding tasks, or they may be absent.

The food habits of mammals strongly influence their body form and physiology. Herbivorous mammals have special adaptations for harboring the intestinal microflora that break down cellulose of the woody diet, and they have developed adaptations for detecting and escaping predators. Carnivorous mammals feed mainly on herbivores, have a simple digestive tract, and have developed adaptations for a predatory life. Omnivores feed on both plant and animal foods. Insectivores feed mainly on insects.

Many marine, terrestrial, and aerial mammals migrate; some migrations, such

Yawning hippopotamus, *Hippopotamus amphibius* (order Artiodactyla).
Photograph by Len Rue, Jr.

as those of fur seals and caribou, are extensive. Migrations are usually made toward favorable climatic and optimal food and calving conditions, or to bring the sexes together for mating.

Mammals with true flight, the bats, are nocturnal and thus avoid direct competition with birds. Most employ ultrasonic echolocation to navigate and feed in darkness.

The most primitive living mammals are the egg-laying monotremes, a remnant group today. All other mammals are viviparous. Embryos of marsupials are born underdeveloped and complete their early growth in the mother's pouch, nourished by milk. The largest and most successful mammals are the placental mammals, in which the embryo undergoes an extensive development in the uterus and is nourished by the placenta, a specialization of the embryonic membranes. The unqualified success of mammals as a group cannot be attributed to greater organ system perfection, but rather to their impressive overall adaptability—the capacity to fit more perfectly in total organization to environmental conditions and thus exploit virtually every habitat on earth.

REVIEW QUESTIONS

1. In the early Mesozoic period a group of mammal-like reptiles, the therapsids, appeared that later gave rise to the true mammals. What new structural adaptations appeared in this group that made them more like mammals than reptiles?
2. The age of mammals began about 70 million years ago and was marked by a great radiation of mammal groups. Give some reasons why mammals were so successful.
3. What do the terms Prototheria, Theria, Metatheria, Eutheria, Monotremata, and Marsupalia literally mean, and what mammals are grouped under each term?
4. Hair is believed to have evolved in the therapsids in response to the need for insulation, but modern mammals have adapted hair for several other purposes. Describe some of these.
5. What is distinctive about each of the following: horns of ruminants, antlers of the deer family, and the horn of the rhinoceros? Briefly describe the growth cycle of antlers.
6. Describe the location and principal function(s) of each of the following skin glands: sweat glands (of two kinds, eccrine and apocrine); scent glands; sebaceous glands; mammary glands.
7. Define the terms diphyodont and heterodont and explain why both terms apply to mammalian dentition.
8. Describe the food habits of each of the following groups: herbivores, carnivores, omnivores, insectivores. Give the common names of several mammals belonging to each group.
9. Most herbivorous mammals depend on cellulose as their main energy source, yet no mammal synthesizes cellulose-splitting enzymes. How are the digestive tracts of herbivores specialized for symbiotic digestion of cellulose?

10. Describe the annual migrations of barren-ground caribou and fur seals.
11. Explain what is distinctive about the life-style and mode of navigation in bats.
12. Describe and distinguish the patterns of reproduction in monotremes, marsupials, and placental mammals. What aspects of mammalian reproduction are present in *all* mammals but in no other class of vertebrates?
13. What is the difference between density-dependent and density-independent causes of population fluctuations in mammals?
14. Describe the hare-lynx population cycle, considered a classic example of a prey-predator relationship (Figure 30-22). From your examination of the cycle, what do you think is causing the oscillations?

SELECTED REFERENCES

Eisenberg, J.F. 1981. The mammalian radiations: an analysis of trends in evolution, adaptation, and behavior. Chicago, University of Chicago Press. *Wide-ranging, authoritative synthesis of mammalian evolution and behavior.*

Halstead, L.B. 1978. The evolution of mammals. London, Eurobook, Ltd. *Richly illustrated, semipopular treatment.*

Kemp, T.S. 1982. Mammal-like reptiles and the origin of mammals. New York, Academic Press, Inc. *Comprehensive synthesis. The final chapter summarizes the earlier chapters and offers a model of evolutionary history of the mammal-like reptiles and primitive mammals.*

Macdonald, D. (ed.). 1984. The encyclopedia of mammals. New York, Facts on File, Inc. *Lucid coverage of all mammal orders and families, enhanced with the fine photographs and color artwork that have become a hallmark of this series.*

Young, J.Z., and M.J. Hobbs. 1975. The life of mammals: their anatomy and physiology, ed. 2. New York, Oxford University Press. *Standard source of information on anatomy, histology, physiology, and embryology of the placental mammals. Classification is not treated.*

GLOSSARY

This glossary lists definitions, pronunciations, and derivations of the most important recurrent technical terms, units, and names (excluding taxa) used in the text.

aboral (ab-o'rəl) (L. *ab*, from, + *os*, mouth) A region opposite the mouth.

acanthor (ə-kan'thor) (Gr. *akantha*, spine or thorn, + *or*) First larval form of acanthocephalans in the intermediate host.

aciculum (ə-sik'ū-ləm) (L. *acicula*, dim. of *acus*, a point) Supporting structure in notopodium and neuropodium of some polychaetes.

acclimatization (ə-klī'mə-də-zā'shən) (L. *ad*, to, + Gr. *klima*, climate) Gradual physiological adaptation in response to relatively long-lasting environmental changes.

acid (a'sid) (L. *acidus*, sour, tart) A compound that dissociates in water solution, one of the dissociation products being hydrogen ions (H^+).

acinus (as'ə-nəs), pl. **acini** (as'ə-ni) (L., grape) A small lobe of compound gland or a saclike cavity at the termination of a passage.

acoelomate (a-sēl'ə-māt') (Gr. *a*,, not, + *koilōma*, cavity) Without a coelom, as in flatworms and proboscis worms.

acontium (ə-kän'chē-əm), pl. **acontia** (Gr. *akontion*, dart) Threadlike structure bearing nematocysts located on mesentery of sea anemone.

acrocentric (ak'rō-sen'trək) (Gr. *akros*, tip, + *kentron*, center) Chromosome with centromere near the end.

actin (Gr. *aktis*, ray) A protein in the contractile tissue that forms the thin myofilaments of striated muscle.

active transport Mediated transport of substances into a cell against a concentration gradient; requires expenditure of energy.

adaptation (L. *adaptatus*, fitted) An anatomical structure, physiological process, or behavioral trait that improves an animal's fitness for existence.

adaptive value Degree to which a characteristic helps an organism to survive and reproduce; lends greater fitness in environment; selective advantage.

adductor (ə-duk'tər) (L. *ad*, to, + *ducere*, to lead) A muscle that draws a part toward a median axis, or a muscle that draws the two valves of a mollusc shell together.

adenine (ad'nēn, ad'ə-nēn) (Gr. *adēn*, gland, + *ine*, suffix) A purine base; component of nucleotides and nucleic acids.

adenosine (ə-den'ə-sen) **(di-, tri-) phosphate** (ADP and ATP) A nucleotide composed of adenine, ribose sugar, and two (ADP) or three (ATP) phosphate units; ATP is an energy-rich compound that, with ADP, serves as a phosphate bond-energy transfer system in cells.

adipose (ad'ə-pōs) (L. *adeps*, fat) Fatty tissue; fatty.

adrenaline (ə-dren'ə-lən) (L. *ad*, to, + *renalis*, pertaining to kidneys) A hormone produced by the adrenal, or suprarenal, gland; epinephrine.

adsorption (ad-sorp'shən) (L. *ad*, to, + *sorbere*, to suck in) The adhesion of molecules to solid bodies.

aerobic (a-rō'bik) (Gr. *aēr*, air, + *bios*, life) Oxygen-dependent form of respiration.

afferent (af'ə-rənt) (L. *ad*, to, + *ferre*, to bear) Adjective meaning leading or bearing toward some organ, for example, nerves conducting impulses toward the brain or blood vessels carrying blood toward an organ; opposed to efferent.

aggression (L. *aggredi*, attack) A primary instinct usually associated with emotional states; an offensive behavior action.

agonistic behavior (Gr. *agōnistēs*, combatant) An offensive action or threat directed toward another organism.

alate (ā'lāt) (L. *alatus*, wing) Winged.

allantois (ə-lan'tois) (Gr. *allas*, sausage, + *eidos*, form) One of the extraembryonic membranes of the amniotes that functions in respiration and excretion in birds and reptiles and plays an important role in the development of the placenta in most mammals.

allele (ə-lēl') (Gr. *allēlōn*, of one another) Alternative forms of genes coding for the same trait; situated at the same locus in homologous chromosomes.

allograft (a'lō-graft) (Gr. *allos*, other, + graft) A piece of tissue or an organ transferred from one individual to another individual of the same species, not identical twins; homograft.

allopatric (Gr. *allos*, other, + *patra*, native land) In separate and mutually exclusive geographical regions.

alpha helix (Gr. *alpha*, first, + L. *helix*, spiral) Literally the first spiral arrangement of the genetic DNA molecule; regular coiled arrangement of polypeptide chain in proteins; secondary structure of proteins.

altricial (al-tri'shəl) (L. *altrices*, nourishers) Referring to young animals (especially birds) having the young hatched in an immature dependent condition.

alula (al'yə-lə) (L. dim. of *ala*, wing) The first digit or thumb of a bird's wing, much reduced in size.

alveolus (al-vē'ə-ləs) (L. dim. of *alveus*, cavity, hollow) A small cavity or pit, such as a microscopic air sac of the lungs, terminal part of an alveolar gland, or bony socket of a tooth.

ambulacra (am'byə-lak'rə) (L. *ambulare*, to walk) In echinoderms, radiating grooves where podia of water-vascular system characteristically project to outside.

amebocyte (ə-mē'bə-sīt) (Gr. *amoibē*, change, + *kytos*, hollow vessel) Any free body cell capable of movement by pseudopodia; certain types of blood cells and tissue cells.

ameboid (ə-mē'boid) (Gr. *amoibē*, change, + *oid*, like) Ameba-like in putting forth pseudopodia.

amictic (ə-mik'tic) (Gr. *a*, without, + *miktos*, mixed or blended) Pertaining to female

bat / āpe / ärmadillo / herring / fēmale / finch / līce / crocodile / crōw / duck / ūnicorn / ə indicates unaccented vowel sound "uh" as in mammal, fishes, cardinal, heron, vulture / stress as in bi-ol'o-gy, bi'o-log'i-cal rotifers, which produce only diploid eggs that cannot be fertilized, or to the eggs produced by such females.

amino acid (ə-mē'nō) (amine, an organic compound) An organic acid with an amino group ($-NH_2$). Makes up the structure of proteins.

amitosis (ā'mī-tō'səs) (Gr. *a,* not, + *mitos,* thread) A form of cell division in which mitotic nuclear changes do not occur; cleavage without separation of daughter chromosomes.

amniocentesis (am'nē-ō-sin-te'səs) (Gr. *amnion,* membrane around the fetus, + *centes,* puncture) Procedure for withdrawing a sample of fluid around the developing embryo for examination of chromosomes in the embryonic cell and other tests.

amnion (am'nē-än) (Gr. *amnion,* membrane around the fetus) The innermost of the extraembryonic membranes forming a fluid-filled sac around the embryo in amniotes.

amniote (am'nē-ōt) Having an amnion; as a noun, an animal that develops an amnion in embryonic life, that is, reptiles, birds, and mammals.

amphid (am'fəd) (Gr. *amphidea,* anything that is bound around) One of a pair of anterior sense organs in certain nematodes.

amplexus (am-plek'səs) (L. embrace) The copulatory embrace of frogs or toads.

ampulla (am-pūl'ə) (L. flask) Membranous vesicle; dilation at one end of each semicircular canal containing sensory epithelium; muscular vesicle above the tube foot in water-vascular system of echinoderms.

amylase (am'ə-lās') (L. *amylum,* starch, + *ase,* suffix meaning enzyme) An enzyme that breaks down starch into smaller units.

anadromous (an-ad'rə-məs) (Gr. *anadromos,* running upward) Refers to fishes that migrate up streams from the sea to spawn.

anaerobic (an'ə-rō'bik) (Gr. *an,* not, + *aēr,* air, + *bios,* life) Not dependent on oxygen for respiration.

analogy (L. *analogus,* ratio) Similarity of function but not of origin.

anastomosis (ə-nas'tə-mō'səs) (Gr. *ana,* again, + *stoma,* mouth) A union of two or more blood vessels, fibers, or other structures to form a branching network.

androgen (an'drə-jən) (Gr. *anēr, andros,* man, + *genēs,* born) Any of a group of vertebrate male sex hormones.

androgenic gland (an'drō-jen'ək) (Gr. *anēr,* male, + *gennaein,* to produce) Gland in Crustacea that causes development of male characteristics.

aneuploidy (an'ū-ploid'ē) (Gr. *an,* without, not, + *eu,* good, well, + *ploid,* multiple of) Loss or gain of a chromosome; cells of the organism have one fewer than normal chromosome number, or one extra chromosome, for example, trisomy 21 (Down syndrome).

Angstrom (after Ångström, Swedish physicist) A unit of one ten millionth of a millimeter (one ten thousandth of a micrometer); it is represented by the symbol Å.

anhydrase (an-hī'drās) (Gr. *an,* not, + *hydōr,* water, + *ase,* enzyme suffix) An enzyme involved in the removal of water from a compound. Carbonic anhydrase promotes the conversion of carbonic acid into water and carbon dioxide.

anlage (än'lä-gə) (Ger. laying out, foundation) Rudimentary form; primordium.

annulus (an'yəl-əs) (L. ring) Any ringlike structure, such as superficial rings on leeches.

antenna (L. sail yard) A sensory appendage on the head of arthropods, or the second pair of the two such pairs of structures in crustaceans.

antennal gland Excretory gland of Crustacea located in the antennal metamere.

anterior (L. comparative of *ante,* before) The head end of an organism, or (as the adjective) toward that end.

anthropoid (an'thrə-poid) (Gr. *anthrōpos,* man + *eidos,* form) Resembling man; especially the great apes of the family Pongidae.

antibody (an'tē-bod'ē) A protein, usually circulating and dissolved in the blood, that is capable of combining with the antigen that stimulated its production.

anticodon (an'tī-kō'don) A sequence of three nucleotides in transfer RNA that is complementary to a codon in messenger RNA.

antigen (an'-ti-jən) Any substance capable of stimulating an immune response, most often a protein.

aperture (ap'ər-chər) (L. *apertura* from *aperire,* to uncover) An opening; the opening into the first whorl of a gastropod shell.

apical (ā'pə-kl) (L. *apex,* tip) Pertaining to the tip or apex.

apical complex A certain combination of organelles found in the protozoan phylum Apicomplexa.

apocrine (ap'ə-krən) (Gr. *apo,* away, + *krinein,* to separate) Applies to a type of mammalian sweat gland that produces a viscous secretion by breaking off a part of the cytoplasm of secreting cells.

apopyle (ap'-ə-pīl) (Gr. *apo,* away from, + *pylē,* gate) In sponges, opening of the radial canal into the spongocoel.

appendicular (L. *ada,* to, + *pendere,* to hang) Pertaining to appendages; pertaining to vermiform appendix.

arboreal (är-bōr'ē-al) (L. *arbor,* tree) Living in trees.

archenteron (ärk-en'tə-rän) (Gr. *archē,* beginning, + *enteron,* gut) The main cavity of an embryo in the gastrula stage; it is lined with endoderm and represents the future digestive cavity.

archeocyte (ar'kē-ə-sīt) (Gr. *archaios,* beginning, + *kytos,* hollow vessel) Ameboid cells of varied function in sponges.

archinephros (ärk'ē-nəf'rōs) (Gr. *archaois,* ancient, + *nephros,* kidney) Ancestral vertebrate kidney, existing today only in the embryos of hagfishes.

areolar (a-rē'ə-ler) (L. *areola,* small space) A small area, such as spaces between fibers of connective tissue.

Aristotle's lantern Masticating apparatus of some sea urchins.

artiodactyl (är'ti-o-dak'təl) (Gr. *artios,* even, + *daktylos,* toe) One of an order or suborder of mammals with two or four digits on each foot.

asconoid (Gr. *askos,* bladder) Simplest form of sponges, with canals leading directly from the outside to the interior.

asexual Without distinct sexual organs; not involving formation of gametes.

assimilation (L. *assimilatio,* bringing into conformity) Absorption and building up of digested nutriments into complex organic protoplasmic materials.

atoke (ā'tōk) (Gr. *a,* without, + *tokos,* offspring) Anterior, nonreproductive part of a marine polychaete, as distinct from the posterior, reproductive part (epitoke) during the breeding season.

atom (a'təm) (Gr. *atomos,* indivisible, uncut) A particle of matter indivisible by ordinary chemical means, consisting of a nucleus and one or more electrons.

ATP Adenosine triphosphate. In biochemistry, an ester of adenosine and triphosphoric acid.

atrium (ā'trē-əm) (L. *atrium*, vestibule) One of the chambers of the heart; also, the tympanic cavity of the ear; also, the large cavity containing the pharynx in tunicates and cephalochordates.

auricle (aw'ri-kl) (L. *auricula*, dim. of *auris*, ear) One of the less muscular chambers of the heart; atrium; the external ear, or pinna; any earlike lobe or process.

autogamy (aw-täg'ə-me) (Gr. *autos*, self, + *gamos*, marriage). Condition in which the gametic nuclei produced by meiosis fuse within the same organism that produced them to restore the diploid number.

autosome (aw'tō-sōm) (Gr. *autos*, self, + *sōma*, body) Any chromosome that is not a sex chromosome.

autotomy (aw-tät'ə-mē) (Gr. *autos*, self, + *tomos*, a cutting) The breaking off of a part of the body by the organism itself.

autotroph (aw'tō-trōf) (Gr. *autos*, self, + *trophos*, feeder) An organism that makes its organic nutrients from inorganic raw materials.

autotrophic nutrition (Gr. *autos*, self, + *trophia*, denoting nutrition) Nutrition characterized by the ability to use simple inorganic substances for the synthesis of more complex organic compounds, as in green plants and some bacteria.

axial (L. *axis*, axle) Relating to the axis, or stem; on or along the axis.

axolotl (ak'sə-lot'l) Nahuatl *atl*, water, + *xolotl*, doll, servant, spirit) Larval stage of any of several species of the genus *Ambystoma* (such as *Ambystoma tigrinum*) exhibiting neotenic reproduction.

axoneme (aks'ə-nēm) (L. *axis*, axle, + Gr. *nēma*, thread). The microtubules in a cilium or flagellum, usually arranged as a circlet of nine pairs enclosing one central pair; also, the microtubules of an axopodium.

B cell A type of lymphocyte that has antibodies in its cell membrane; important in humoral immune response; gives rise to plasma cells that secrete antibodies.

basal body Also known as kinetosome and blepharoplast, a cylinder of nine triplets of microtubules found basal to a flagellum or cilium; same structure as a centriole.

base (bās) (Gr. *basis*, base, bottom, foundation) A compound that dissociates in water solution, one of the dissociation products being hydroxyl ions (OH^-).

basis, basipodite (bā'səs, bā-si'pə-dīt) (Gr. *basis*, base, + *pous, podos*, foot) The distal or second joint of the protopod of a crustacean appendage.

benthos (ben'thäs) (Gr., depth of the sea) Organisms that live along the bottom of the seas and lakes; adj., **benthic**. Also, the bottom itself.

biogenesis (bī'ō-jen'ə-səs) (Gr. *bios*, life, + *genesis*, birth) The doctrine that life originates only from preexisting life.

bioluminescence Method of light production by living organisms in which usually certain proteins (luciferins), in the presence of oxygen and an exzyme (luciferase), are converted to oxyluciferins with the liberation of light.

biomass (Gr. *bios*, life, + *maza*, lump or mass) The weight of total living organisms or of a species population per unit of area.

biome (bī'ōm) (Gr. *bios*, life, + *ōma*, abstract group suffix) Complex of plant and animal communities characterized by climatic and soil conditions; the largest ecological unit.

biosphere (Gr. *bios*, life, + *sphaira*, globe) That part of earth containing living organisms.

biramous (bī-rām'əs) (L. *bi*, double, + *ramus*, a branch) Adjective describing appendages with two distinct branches.

bivalent (bī-vāl'ənt; biv'əl-ənt) (L. *bi*, double + *valen*, strength, worth) The pairs of homologous chromosomes of synapsis in the first meiotic division, a tetrad.

blastocoel (blas'tō-sēl) (Gr. *blastos*, germ, + *koilos*, hollow) Cavity of the blastula.

blastocyst (blast'o-sist) cnr. *blastos*, germ, + *kystis*, bladder) Mammalian embryo in the blastula stage.

blastomere (Gr. *blastos*, germ, + *meros*, part) An early cleavage cell.

blastopore (Gr. *blastos*, germ, + *poros*, passage, pore) External opening of the archenteron in the gastrula.

blastula (Gr. *blastos*, germ, + L. *ula*, dim) Early embryological stage of many animals; consists of a hollow mass of cells.

blepharoplast (blə-fä'rə-plast) (Gr. *blepharon*, eyelid, + *plastos*, formed) See basal body.

blood type Characteristic of human blood given by the particular antigens on the membranes of the erythrocytes, genetically determined, causing agglutination when incompatible groups are mixed; the blood types are designated A, B, O, AB, Rh negative, Rh positive, and others.

Bohr effect A characteristic of hemoglobin that causes it to dissociate from oxygen in greater degree at higher concentrations of carbon dioxide.

BP Before the present.

brachial (brak'ē-əl) (L. *brachium*, forearm) Referring to the arm.

branchial (brank'ē-ə) (Gr. *branchia*, gills) Referring to gills.

brown fat Mitochondria-rich, heat-generating adipose tissue of endothermic vertebrates.

buccal (buk'əl) (L. *bucca*, cheek) Referring to the mouth cavity.

buffer Any substance or chemical compound that tends to keep pH levels constant when acids or bases are added.

calorie (kal'ō-ry) (L. *calere*, to be warm) Unit of heat defined as the amount of heat required to heat 1 g of water from 14.5 to 15.5 C; 1 cal = 4.184 joules in the International System of Units.

cancellous (kan'səl-əs) (L. *cancelli*, latticework, + *osus*, full of) Having a spongy or porous structure.

capitulum (ka-pi'tə-ləm) (L. small head) Term applied to small, headlike structures of various organisms, including projection from body of ticks and mites carrying mouthparts.

carapace (kar'ə-pās) (F. from Sp. *carapacho*, shell) Shieldlike plate covering the cephalothorax of certain crustaceans; dorsal part of the shell of a turtle.

carbohydrate (L. *carbo*, charcoal, + Gr. *hydōr*, water) Compounds of carbon, hydrogen, and oxygen having the generalized formula $(CH_2O)_n$; aldehyde or ketone derivatives of polyhydric alcohols, with hydrogen and oxygen atoms attached in a 2:1 ratio.

carboxyl (kär-bäk'səl) (carbon + oxygen + yl, chemical radical suffix) The acid group of organic molecules —COOH.

carinate (kar'ə-nāt) (L. *carina*, keel) Having a keel, in particular the flying birds with a keeled sternum for the insertion of flight muscles.

carnivore (kar'nə-vōr') (L. *carnivorus*, flesh eating) One of the flesh-eating mammals of the order Carnivora. Also, any organism that eats animals. Adj., **carnivorous**.

carotene (kär'ə-tēn) (L. *carota*, carrot, + *ene*, unsaturated straight-chain hydrocarbons) A red, orange, or yellow pigment belonging to the group of carotenoids; precursor of vitamin A.

cartilage (L. *cartilago*; akin to L. *cratis*, wickerwork) A translucent elastic tissue that makes up most of the skeleton of embryos, very young vertebrates, and adult cartilaginous fishes, such as sharks and rays; in higher forms much of it is converted into bone.

caste (kast) (L. *castus*, pure, separated) One of the polymorphic forms within an insect society, each caste having its specific duties, as queen, worker, soldier, and so on.

catadromous (kə-tad'rə-məs) (Gr. *kata*, down, + *dromos*, a running) Refers to fishes that migrate from fresh water to the ocean to spawn.

catalyst (kad'ə-ləst) (Gr. *kata*, down, + *lysis*, a loosening) A substance that accelerates a chemical reaction but does not become a part of the end product.

cecum, caecum (sē'kəm) (L. *caecus*, blind) A blind pouch at the beginning of the large intestine; any similar pouch.

cellulose (sel'ūlōs) (L. *cella*, small room) Chief polysaccharide constituent of the cell wall of green plants and some fungi; an insoluble carbohydrate $(C_6H_{10}O_5)_n$ that is converted into glucose by hydrolysis.

centriole (sen'trē-ol) (Gr. *kentron*, center of a circle, + L. *ola*, small) A minute cytoplasmic organelle usually found in the centrosome and considered to be the active division center of the animal cell; organizes spindle fibers during mitosis and meiosis. Same structure as basal body or kinetosome.

centrolecithal (sen'tro-les'ə-thəl) (Gr. *kentron*, center, + *lekithos*, yolk, + Eng. *al*, adjective) Pertaining to an insect egg with the yolk concentrated in the center.

centromere (sen'trə-mir) (Gr. *kentron*, center, + *meros*, part) A small body or constriction on the chromosome to which a spindle fiber attaches during mitosis or meiosis.

centrosome (sen'trə-sōm) (Gr. *kentron*, center, + *sōma*, body) Minute body in cytoplasm of many plant and animal cells that contains one or two centrioles and is the center of dynamic activity in mitosis.

cephalization (sef'əli-zā-shən) (Gr. *kephalē*, head) The process by which specialization, particularly of the sensory organs and appendages, became localized in the head end of animals.

cephalothorax (sef'ə-lä-thō'raks) (Gr. *kephalē*, head, + thorax) A body division found in many Arachnida and higher Crustacea, in which the head is fused with some or all of the thoracic segments.

cerata (sə-ra'tə) (Gr. *keras*, a horn, bow) Dorsal processes on some nudibranchs for gaseous exchange.

cercaria (ser-kār'ē-ə) (Gr. *kerkos*, tail, + L. *aria*, like or connected with) Tadpolelike juvenile of trematodes (flukes).

chelicera (kə-lis'ə-rə), pl. **chelicerae** (Gr. *chēlē*, claw, + *keras*, horn) One of a pair of the most anterior head appendages on the members of the subphylum Chelicerata.

chelipeds (kēl'ə-peds) (Gr. *chēlē*, claw, + L. *pes*, foot) Pincerlike first pair of legs in most decapod crustaceans; specialized for seizing and crushing.

chemoautotroph (ke-mō-aw'tō-trōf) (Gr. *chemeia*, transmutation, + *autos*, self, + *trophos*, feeder) An organism utilizing inorganic compounds as a source of energy.

chemotaxis (kē'mō-tak'sis) (Gr. *chēmeia*, transmutation, + *taxis*, arrangement) An orientation movement of a (usually) simple organism in response to a chemical stimulus.

chiasma (kī-az'mə), pl. **chiasmata** (Gr., cross) An intersection or crossing, as of nerves; a connection point between homologous chromatids where crossing over has occurred at synapsis.

chitin (kī'tən) (Fr. *chitine*, from Gr. *chitōn*, tunic) A horny substance that forms part of the cuticle of arthropods and is found sparingly in certain other invertebrates; a nitrogenous polysaccharide insoluble in water, alcohol, dilute acids, and digestive juices of most animals.

chloragogue cells (klōr'ə-gog) (Gr. *chlōros*, light green, + *agōgos*, a leading, a guide) Modified peritoneal cells, greenish or brownish, clustered around the digestive tract of certain annelids; apparently they aid in elimination of nitrogenous wastes and in food transport.

chlorocruorin (klō'rō-kroo'ə-rən) (Gr. *chlōros*, light green, + L. *cruor*, blood) A greenish iron containing respiratory pigment dissolved in the blood plasma of certain marine polychaetes.

chlorophyll (klō'rō-fil) (Gr. *chlōros*, light green, + *phyllōn*, leaf) Green pigment found in plants and in some animals, necessary for photosynthesis.

chloroplast (Gr. *chlōros*, light green, + *plastos*, molded) A plastid containing chlorophyll and usually other pigments, found in cytoplasm of plant cells.

choanocyte (kō-an'ō-sīt) (Gr. *choanē*, funnel, + *kytos*, hollow vessel) One of the flagellate collar cells that line cavities and canals of sponges.

cholinergic (kōl'-i-nər'jik) (Gr. *chōle*, bile, + *ergon*, work) Type of nerve fiber that releases acetylcholine from axon terminal.

chorion (kō'rē-on) (Gr. *chorion*, skin) The outer of the double membrane that surrounds the embryo of reptiles, birds, and mammals; in mammals it contributes to the placenta.

choroid (kōr'oid) (Gr. *chorion*, skin, + *eidos*, form) Delicate, highly vascular membrane; in vertebrate eye, the layer between the retina and sclera.

choroid plexus (kōr'oid plek'sus) (Gr. *chorion*, skin, + *eidos*, form; L. *plexus*, interwoven) Vascular network in the brain ventricles that regulates secretion and absorption of cerebrospinal fluid.

chromatid (krō'mə-tid) (Gr. *chromato*, from *chrōma*, color, + L. *id*, feminine stem for particle of specified kind) A replicated chromosome joined to its sister chromatid by the centromere; separates and becomes daughter chromosome at anaphase of mitosis or anaphase of the second meiotic division.

chromatin (krō'mə-tin) (Gr. *chrōma*, color) The nucleoprotein material of a chromosome; the hereditary material containing DNA.

chromatophore (krō-mat'ə-fōr) (Gr. *chrōma*, color, + *pherein*, to bear) Pigment cell, usually in the dermis, in which usually the pigment can be dispersed or concentrated.

chromoplast (krō'mō-plast) (Gr. *chrōma*, color, + *plastos*, molded) A plastid containing pigment.

chromosome (krō'mə-sōm) (Gr. *chrōma*, color, + *sōma*, body) A complex body, spherical or rod shaped, that arises from the nuclear network during mitosis, splits longitudinally, and carries a part of the organism's genetic information as genes composed of DNA.

chrysalis (kris′ə-lis) (L. from Gr. *chrysos*, gold) The pupal stage of a butterfly.

cilium (sil′i-əm), pl. **cilia** (L. eyelid). A hairlike, vibratile organelle process found on many animal cells. Cilia may be used in moving particles along the cell surface or, in ciliate protozoans, for locomotion.

circadian (sər′kə-dē′-ən) (L. *circa*, around, + *dies*, day) Occurring at a period of approximately 24 hours.

cirrus (sir′əs) (L., curl) A hairlike tuft on an insect appendage; locomotor organelle of fused cilia; male copulatory organ of some invertebrates.

cisternae (sis-ter′nē) (L. *cista*, box) Space between membranes of the endoplasmic reticulum within cells.

cladistics (klad-is′-təks) (Gr. *cladus*, branch, sprout) A system of arranging taxa by analysis of primitive and derived characteristics so that the arrangement will reflect phylogenetic relationships.

cleavage (O.E. *cleofan*, to cut) Process of nuclear and cell division in animal zygote.

climax (klī′maks) (Gr. *klimax*, ladder) Stage of relative stability attained by a community of organisms, often the culminating development of a natural succession. Also, orgasm.

climax community (Gr. *klimax*, a ladder, staircase, climax) A self-perpetuating, more-or-less stable community of organisms that continues as long as environmental conditions under which it developed prevail.

cline (klīn) (Gr. *klinein*, slope, bend) A pattern of gradual genetic change in a population according to its geographical range.

clitellum (klī-tel′əm) (L. *clitellae*, packsaddle) Thickened saddelike portion of certain midbody segments of many oligochaetes and leeches.

cloaca (klō-ā′kə) (L., sewer). Posterior chamber of digestive tract in many vertebrates, receiving feces and urogenital products. In certain invertebrates, a terminal portion of digestive tract that serves also as respiratory, excretory, or reproductive duct.

clone (klōn) (Gr. *klōn*, twig) All descendants derived by asexual reproduction from a single individual.

cnidocil (nī′dō-sil) (Gr. *knidē*, nettle, + L. *cilium*, hair). Triggerlike spine on nematocyst.

cnidocyte (nī′dō-sīt) (Gr. *knidē*, nettle, + *kytos*, hollow vessel) Modified interstitial cell that holds the nematocyst; during development of the nematocyst, the cnidocyte is a cnidoblast.

coacervate (kō′ə-sər′vət) (L. *coacervatus*, to heap up) An aggregate of colloidal droplets held together by electrostatic forces.

cochlea (kōk′lēə) (L., snail, from Gr. *kochlos*, a shellfish) A tubular cavity of the inner ear containing the essential organs of hearing; occurs in crocodiles, birds, and mammals; spirally coiled in mammals.

cocoon (kə-kun′) (Fr. *cocon*, shell). Protective covering of a resting or developmental stage, sometimes used to refer to both the covering and its contents; for example, the cocoon of a moth or the protective covering for the developing embryos in some annelids.

codon (kō′dän) (L., code, + on) A sequence of three adjacent nucleotides that code for one amino acid in mRNA.

coelenteron (sē-len′tər-on) (Gr. *koilos*, hollow, + *enteron*, intestine) Internal cavity of a cnidarian; gastrovascular cavity; archenteron.

coelom (sē′lōm) (Gr. *koilōma*, cavity) The body cavity in triploblastic animals, lined with mesodermal peritoneum.

coelomocyte (sē′lō′mə-sīt) (Gr. *koilōma*, cavity, + *kytos*, hollow vessel) Another name for amebocyte; primitive or undifferentiated cell of the coelom and the water-vascular system.

coelomoduct (sē-lō′mə-dukt) (Gr. *koilos*, hollow, + L. *ductus*, a leading) A duct that carries gametes or excretory products (or both) from the coelom to the exterior.

coenzyme (kō-en′zīm) (L. prefix, *co*, with, + Gr. *enzymos*, leavened, from *en*, in, + *zymē*, leaven) A required substance in the activation of an enzyme; a prosthetic or nonprotein constituent of an enzyme.

collagen (käl′ə-jən) (Gr. *kolla*, glue, + *genos*, descent) A tough, fibrous protein occurring in vertebrates as the chief constituent of collagenous connective tissue; also occurs in invertebrates, for example, the cuticle of nematodes.

collencyte (käl′ən-sīt) (Gr. *kolla*, glue, + *kytos*, hollow vessel) A type of cell in sponges that secretes fibrillar collagen.

colloblast (käl′ə-blast) (Gr. *kolla*, glue, + *blastos*, germ) A glue-secreting cell on the tentacles of ctenophores.

comb plate One of the plates of fused cilia that are arranged in rows for ctenophore locomotion.

commensalism (kə-men′səl-iz′əm) (L. *com*, together with, + *mensa*, table) A relationship in which one individual lives close to or on another and benefits, and the host is unaffected; often symbiotic.

community (L. *communitas*, community, fellowship) An assemblage of organisms that are associated in a common environment and interact with each other in a self-sustaining and self-regulating relation.

competition Some degree of overlap in ecological niches of two populations in the same community, such that they both depend on the same food source, shelter, or other resources.

complement Collective name for a series of enzymes and activators in the blood, some of which may bind to antibodies and may lead to rupture of a foreign cell.

compound A chemical substance, the molecules of which are composed of two or more different elements.

conjugation (kon′jū-gā′shən) (L. *conjugare*, to yoke together) Temporary union of two ciliate protozoa while they are exchanging chromatin material and undergoing nuclear phenomena resulting in binary fission. Also, formation of cytoplasmic bridges between bacteria for transfer of plasmids.

conspecific (L. *com*, together, + *species*) A member of the same species.

contractile vacuole A clear fluid-filled cell vacuole in protozoa and a few lower metazoa; takes up water and releases it to the outside in a cyclical manner, for osmoregulation and some excretion.

control That part of a scientific experiment to which the experimental variable is not applied but which is similar to the experimental group in all other respects.

copulation (Fr., from L. *copulare*, to couple) Sexual union to facilitate the reception of sperm by the female.

corium (kō′re-um) (L. *corium*, leather) The deep layer of the skin; dermis.

cornea (kor′nē-ə) (L. *corneus*, horny) The outer transparent coat of the eye.

corneum (kor′nē-əm) (L. *corneus*, horny) Epithelial layer of dead, keratinized cells. Stratum corneum.

cornified (kor′nə-fīd) (L. *corneus*, horny) Adjective for conversion of epithelial cells into nonliving, keratinized cells.

corona (kə-rō'nə) (L., crown) Head or upper portion of a structure; ciliated disc on anterior end of rotifers.

corpora allata (kor'pə-rə əl-la'tə) (L. *corpus*, body + *allatum*, aided) Endocrine glands in insects that produce juvenile hormone.

cortex (kor'teks) (L., bark) The outer layer of a structure.

coxa, coxopodite (kox'ə, kəx-ä'pə-dīt) (L. *coxa*, hip, + Gr. *pous, podos*, foot) The proximal joint of an insect or arachnid leg; in crustaceans, the proximal joint of the protopod.

crista (kris'ta), pl. **cristae** (L. *crista*, crest) A crest or ridge on a body organ or organelle; a platelike projection formed by the inner membrane of mitochondria.

crossing-over Exchange of parts of nonsister chromatids at synapsis in the first meiotic division.

cryptobiotic (Gr. *kryptos*, hidden, + *biōticus*, pertaining to life) Living in concealment; refers to insects and other animals that live in secluded situations, such as underground or in wood; also tardigrades and some nematodes, rotifers, and others that survive harsh environmental conditions by assuming for a time a state of very low metabolism.

ctenoid scales (ten'oid) (Gr. *kteis, ktenos*, comb) Thin, overlapping dermal scales of the more advanced fishes; exposed posterior margins have fine, toothlike spines.

cuticle (kū'ti-kəl) (L. *cutis*, skin) A protective, noncellular, organic layer secreted by the external epithelium (hypodermis) of many invertebrates. In higher animals the term refers to the epidermis or outer skin.

cyanobacteria (sī-an-ō-bak-ter'ē-ə) (Gr. *kyanos*, a dark-blue substance, + *bakterion*, dim. of *baktron*, a staff) Photosynthetic prokaryotes, also called blue-green algae, cyanophytes.

cyanophyte (sī-an'ō-fīt) (Gr. *kyanos*, a dark blue substance, + , phyton, plant) A cyanobacterium, blue-green alga.

cycloid scales (sī'kloid) (Gr. *kyklos*, circle) Thin, overlapping dermal scales of the more primitive fishes; posterior margins are smooth.

cysticercus (sis'tə-ser'kəs) (Gr. *kystis*, bladder, + *kerkos*, tail) A type of juvenile tapeworm in which an invaginated and introverted scolex is contained in a fluid-filled bladder.

cystid (sis'tid) (Gr. *kystis*, bladder) In an ectoproct, the dead secreted outer parts plus the adherent underlying living layers.

cytochrome (sī'tə-krōm) (Gr. *kytos*, hollow vessel, + *chrōma*, color) Several iron-containing pigments that serve as electron carriers in aerobic respiration.

cytokinesis (sī'tə-kin-ē'sis) (Gr. *kytos*, hollow vessel, + *kinesis*, movement) Division of the cytoplasm of a cell.

cytopharynx (Gr. *kytos*, hollow vessel, + *pharynx*, throat) Short tubular gullet in ciliate protozoa.

cytoplasm (si'tə-plasm) (Gr. *kytos*, hollow vessel, + *plasma*, mold) The living matter of the cell, excluding the nucleus.

Darwinism Theory of evolution by natural selection.

data sing. **datum** (Gr. *dateomai*, to divide, cut in pieces) The results in a scientific experiment, or descriptive observations, upon which a conclusion is based.

deduction (L. *deductus*, led apart, split, separated) Reasoning from the general to the particular, that is, from given premises to their necessary conclusion.

definitive host The host in which sexual reproduction of a symbiont takes place; if no sexual reproduction, then the host in which the symbiont becomes mature and reproduces; contrast intermediate host.

demography (də-mäg'grə-fē) (Gr. *demos*, people, + *graphy*) The properties of the rate of growth and the age structure of populations.

deoxyribonucleic acid (DNA) The genetic material of all organisms, characteristically organized into linear sequences of genes.

deoxyribose (dē-ok'sē-rī'bōs) (L. *deoxy*, loss of oxygen, + ribose, a pentose sugar) A 5-carbon sugar having 1 oxygen atom less than ribose; a component of deoxyribonucleic acid (DNA).

dermal (Gr. *derma*, skin) Pertaining to the skin; cutaneous.

dermis The inner, sensitive mesodermal layer of skin; corium.

desmosome (dez'mə-sōm) (Gr. *desmos*, bond, + *sōma*, body) Buttonlike plaque serving as an intercellular connection.

determinate cleavage The type of cleavage, usually spiral, in which the fate of the blastomeres is determined very early in development; mosaic cleavage.

detritus (də-trī'tus) (L., that which is rubbed or worn away) Any fine particulate debris of organic or inorganic origin.

Deuterostomia (dū'də-rō-stō'mē-ə) (Gr. *deuteros*, second, secondary, + *stoma*, mouth) A group of higher phyla in which cleavage is indeterminate and primitively radial. The endomesoderm is entercoelous, and the mouth is derived away from the blastopore. Includes Echinodermata, Chordata, and a number of minor phyla. Compare with **Protostomia**.

diapause (dī'ə-pawz) (Gr. *diapausis*, pause) A period of arrested development in the life cycle of insects and certain other animals in which physiological activity is very low and the animal is highly resistant to unfavorable external conditions.

diffusion (L. *diffusus*, dispersion) The movement of particles or molecules from area of high concentration of the particles or molecules to area of lower concentration.

digitigrade (dij'ə-də-grād) (L. *digitus*, finger, toe, + *gradus*, step, degree) Walking on the digits with the posterior part of the foot raised; compare **plantigrade**.

dihybrid (dī-hī'brəd) (Gr. *dis*, twice, + L. *hibrida*, mixed offspring) A hybrid whose parents differ in two distinct characters; an offspring having diffuse alleles at two different loci, for example, *A/a B/b*.

dimorphism (dī-mor'fizm) (Gr. *di*, two, + *morphē*, form) Existence within a species of two distinct forms according to color, sex, size, organ structure, and so on. Occurrence of two kinds of zooids in a colonial organism.

dioecious (dī-ē'shəs) (Gr. *di*, two, + *oikos*, house) Having male and female organs in separate individuals.

diphycercal (dif'i-ser'kəl) (Gr. *diphyēs*, twofold, + *kerkos*, tail) A tail that tapers to a point, as in lungfishes; vertebral column extends to tip without upturning.

diphyodont (di'fi-ə-dänt) (Gr. *diphyēs*, twofold, + *odous*, tooth) Having deciduous and permanent sets of teeth successively.

diploblastic (di'plə-blas'tək) (Gr. *diploos*, double, + *blastos*, bud) Organism with two germ layers, endoderm and ectoderm.

diploid (dip'loid) (Gr. *diploos*, double, + *eidos*, form) Having the somatic (double, or 2n) number of chromosomes or twice the number characteristic of a gamete of a given species.

bat / āpe / ärmadillo / herring / fēmale / finch / līce / crocodile / crōw / duck / ūnicorn / ə indicates unaccented vowel sound "uh" as in mammal, fishes, cardinal, heron, vulture / stress as in bi-ol'o-gy, bi'o-log'i-cal

DNA see **deoxyribonucleic acid.**

dominance hierarchy A social ranking, formed through agonistic behavior, in which individuals are associated with each other so that some have greater access to resources than do others.

dorsal (dor'səl) (L. *dorsum*, back) Toward the back, or upper surface, of an animal.

Down syndrome A congenital syndrome including mental retardation, caused by the cells in a person's body having an extra chromosome 21; also called trisomy 21.

drive A state of activity directed toward satisfying a specific need.

dyad (dī'əd) (Gr. *dyas*, two) One of the groups of two chromosomes formed by the division of a tetrad during the first meiotic division.

eccrine (ek'rən) (Gr. *ek*, out of, + *krinein*, to separate) Applies to a type of mammalian sweat gland that produces a watery secretion.

ecdysis (ek'də-sis) (Gr. *ekdysis*, to strip off, escape) Shedding of outer cuticular layer; molting, as in insects or crustaceans.

ecdysone (ek-dī'sōn) (Gr. *ekdysis*, to strip off) Molting hormone of arthropods, stimulates growth and ecdysis, produced by prothoracic glands in insects and Y organs in crustaceans.

ecocline (ek'ō-klīn) (Gr. *oikos*, home, + *klino*, to slope, recline) The gradient between adjacent biomes; a gradient of environmental conditions.

ecology (Gr. *oikos*, home, + *logos*, discourse) Part of biology that deals with the relationship between organisms and their environment.

ecosystem (ek'ō-sis-təm) (eco[logy] from Gr. *oikos*, house, + system) An ecological unit consisting of both the biotic communities and the nonliving (abiotic) environment, which interact to produce a stable system.

ecotone (ek'ō-tōn) (eco[logy] from Gr. *oikos*, home, + *tonos*, stress) The transition zone between two adjacent communities.

ectoderm (ek'tō-derm) (Gr. *ektos*, outside, + *derma*, skin) Outer layer of cells of an early embryo (gastrula stage); one of the germ layers, also sometimes used to include tissues derived from ectoderm.

ectoplasm (ec'tō-plazm) (Gr. *ektos*, outside, + *plasma*, form) The cortex of a cell or that part of cytoplasm just under the cell surface; contrasts with **endoplasm**.

ectothermic (ek'tō-therm'ic) (Gr. *ektos*, outside, +*thermē*, heat) Having a variable body temperature derived from heat acquired from the environment; contrasts with **endothermic**.

effector (L. *efficere*, bring to pass) An organ, tissue, or cell that becomes active in response to stimulation.

efferent (ef'ə-rənt) (L. *ex*, out, + *ferre*, to bear) Leading or conveying away from some organ, for example, nerve impulses conducted away from the brain, or blood conveyed away from an organ; contrasts with **afferent**.

element (el'ə-mənt) (L. *elementum*, element) Matter represented by a particular kind of atom.

elephantiasis (el-ə-fən-tī'ə-səs) Disfiguring condition caused by chronic infection with filarial worms *Wuchereria bancrofti* and *Brugia malayi*.

embryogenesis (em'brē-ō-jen'ə-səs) (Gr. *embryon*, embryo, + genesis, origin) The origin and development of the embryo; embryogeny.

emigrate (L. *emigrare*, to move out) To move *from* one area to another to take up residence.

emulsion (ə-məl'shən) (L. *emulsus*, milked out) A colloidal system in which both phases are liquids.

endemic (en-dem'ik) (Gr. *en*, in, + *demos*, populace) Peculiar to a certain region or country; native to a restricted area; not introduced.

endergonic (en-dər-gän'ik) (Gr. *endon*, within, + *ergon*, work) Used in reference to a chemical reaction that requires energy; energy absorbing.

endochondral (en'dō-kän'drōl) (Gr. *endon*, within, + *chondros*, cartilage) Occurring within the substance of cartilage, especially bone formation.

endocrine (en'də-krən) (Gr. *endon*, within, + *krinein*, to separate) Refers to a gland that is without a duct and that releases its product directly into the blood or lymph.

endocytosis (en'dō-sī-tō'sis) (Gr. *endon*, within, + *kytos*, hollow vessel) The engulfment of matter by phagocytosis and of macromolecules by pinocytosis.

endoderm (en'də-dərm) (Gr. *endon*, within, + *derma*, skin) Innermost germ layer of an embryo, forming the primitive gut; also may refer to tissues derived from endoderm.

endolecithal (en'də-les'ə-thəl) (Gr. *endon*, within, + *lekithos*, yolk) Yolk for nutrition of the embryo incorporated into the egg cell itself.

endolymph (en'do-limf) (Gr. *endon*, within, + *lympha*, water) Fluid that fills most of the membranous labyrinth of the vertebrate ear.

endometrium (en'də-mē'trē-əm) (Gr. *endon*, within, + *mētra*, womb) The mucous membrane lining the uterus.

endoplasm (en'də-pla-zm) (Gr. *endon*, within, + *plasma*, mold or form) The portion of cytoplasm that immediately surrounds the nucleus.

endoplasmic reticulum A complex of membranes within a cell; may bear ribosomes (rough) or not (smooth).

endopod, endopodite (en'də-päd, endop'ə-dīt) (Gr. *endon*, within, + *pous*, *podos*, foot) Medial branch of a biramous crustacean appendage.

endopterygote (en'dəp-ter'i-gōt) (Gr. *endon*, within, + *pteron*, feather, wing) Insect in which the wing buds develop internally; has holometabolous metamorphosis.

endorphin (en-dor'fin) (contraction of endogenous morphine) Group of opiate-like brain neuropeptides that modulate pain perception and are implicated in many other functions.

endoskeleton (Gr. *endon*, within, + *skeletos*, hard) A skeleton or supporting framework within the living tissues of an organism; contrasts with **exoskeleton**.

endostyle (en'də-stīl) (Gr. *endon*, within, + *stylos*, a pillar) Ciliated groove(s) in the floor of the pharynx of tunicates, cephalochordates, and larval cyclostomes, useful for accumulating and moving food particles to the stomach.

endothermic (en'də-therm'ic) (Gr. *endon*, within, + *thermē*, heat) Having a body temperature determined by heat derived from the animal's own oxidative metabolism; contrasts with **ectothermic**.

enkephalin (en-kef'lin) (Gr. *endon*, within, + *kephale*, head) Group of small brain neuropeptides with opiate-like qualities.

enterocoel (en'tər-ō-sēl') (Gr. *enteron*, gut, + *koilos*, hollow) A type of coelom formed by the outpouching of a mesodermal sac from the endoderm of the primitive gut.

enterocoelic mesoderm formation Embryonic formation of mesoderm by a pouchlike outfolding from the archenteron, which then expands and obliterates the blastocoel, thus forming a large cavity, the coelom, lined with mesoderm.

enterocoelomate (en'ter-ō-sēl'ō-māte) (Gr. *enteron,* gut, + *koilōma,* cavity, + Engl. *ate,* state of) An animal having an enterocoel, such as an echinoderm or a vertebrate.

enteron (en'tə-rän) (Gr., intestine) The digestive cavity.

entropy (en'trə-pē) (Gr. *en,* in, on, + *tropos,* turn, change in manner) A quantity that is the measure of energy in a system not available for doing work.

enzyme (en'zīm) (Gr. *enzymos,* leavened, from *en,* in + *zyme,* leaven) A protein substance, produced by living cells, that is capable of speeding up specific chemical transformations, such as hydrolysis, oxidation, or reduction, but is unaltered itself in the process; a biological catalyst.

ephyra (ef'ə-rə) (Gr. *Ephyra,* Greek city) Refers to castlelike appearance. Medusa bud from a scyphozoan polyp.

epidermis (ep'ə-dər'məs) (Gr. *epi,* on, upon, + *derma,* skin) The outer, nonvascular layer of skin of ectodermal origin; in invertebrates, a single layer of ectodermal epithelium.

epididymis (ep'ə-did'ə-məs) (Gr. *epi,* on, upon, + *didymos,* testicle) Part of the sperm duct that is coiled and lying near the testis.

epipod, epipodite (ep'ə-päd, e-pip'ə-dīt) (Gr. *epi,* on, upon, + *pous, podos,* foot) A lateral process on the protopod of a crustacean appendage, often modified as a gill.

epithelium (ep'i-thē'lē-um) (Gr. *epi,* on, upon, + *thele,* nipple) A cellular tissue covering a free surface or lining a tube or cavity.

epitoke (ep'i-tōk) (Gr. *epitokos,* fruitful) Posterior part of a marine polychaete when swollen with developing gonads during the breeding season; contrast with **atoke**.

erythroblastosis fetalis (ə-rith'-rō-blas-tō'səs fə-tal'əs) (Gr. *erythros,* red, + *blastos,* germ, + *osis,* a disease; L. *fetalis,* relating to a fetus) A disease of newborn infants caused when Rh-negative mothers develop antibodies against the Rh-positive blood of the fetus. See **blood type**.

bat / āpe / ärmadillo / herring / fēmale / finch / līce / crocodile / crōw / duck / ūnicorn / ə indicates unaccented vowel sound "uh" as in mammal, fishes, cardinal, heron, vulture / stress as in bi-ol'o-gy, bi'o-log'i-cal

erythrocyte (ə-rith'rō-sīt) (Gr. *erythros,* red, + *kytos,* hollow vessel) Red blood cell; has hemoglobin to carry oxygen from lungs or gills to tissues; during formation in mammals, erythrocytes lose their nuclei, those of other vertebrates retain the nuclei.

estrus (es'trəs) (L. *oestrus,* gadfly, frenzy) The period of heat, or rut, especially of the female during ovulation of the egg. Associated with maximum sexual receptivity.

estuary (es'chə-we'rē) (L. *aestuarium,* estuary) An arm of the sea where the tide meets the current of a freshwater stream.

ethology (e-thäl'-ə-jē) (Gr. *ethos,* character, + *logos,* discourse) The study of animal behavior in natural environments.

euchromatin (ū'krō-mə-tən) (Gr. *eu,* good, well, + *chrōma,* color) Part of the chromatin that takes up stain less than does heterochromatin, contains active genes.

eukaryotic, eucaryotic (ū'ka-rē-ot'ik) (Gr. *eu,* good, true, + *karyon,* nut, kernel) Organisms whose cells characteristically contain a membrane-bound nucleus or nuclei; contrasts with **prokaryotic**.

euploidy (ū-ploid'ē) (Gr. *eu,* good, well, + *ploid,* multiple of) Change in chromosome number from one generation to the next in which there is an addition or deletion of a complete set of chromosomes in the progeny; the most common type is polyploidy.

euryhaline (ū'rə-hā'līn) (Gr. *eurys,* broad, + *hals,* salt) Able to tolerate wide ranges of saltwater concentrations.

evagination (ē-vaj'ə-nā'shən) (L. *e,* out, + *vagina,* sheath) An outpocketing from a hollow structure.

evolution (L. *evolvere,* to unfold) Organic evolution is any genetic change in organisms, or more strictly a change in gene frequency from generation to generation.

exergonic (ek'sər-gän'ik) (Gr. *exō,* outside of, + *ergon,* work) An energy-yielding reaction.

exocrine (ek'sə-krən) (Gr. *exō,* outside, + *krinein,* to separate) A type of gland that releases its secretion through a duct; contrasts with **endocrine**.

exon (ex'ən) (Gr. *exō,* outside) Part of the mRNA as transcribed from DNA that contains a portion of the information necessary for final gene product.

exopod, exopodite (ex'ə-päd, ex-äp'ə-dīt) (Gr. *exō,* outside, + *pous, podos,* foot) Lateral branch of a biramous crustacean appendage.

exopterygote (ek'səp-ter'i-gōt) (Gr. *exō,* without, + *pteron,* feather, wing) Insect in which the wing buds develop externally during nymphal instars; has hemimetabolous metamorphosis.

exoskeleton (ek'sō-skel'ə-tən) (Gr. *exō,* outside, + *skeletos,* hard) A supporting structure secreted by ectoderm or epidermis; external, not enveloped by living tissue, as opposed to endoskeleton.

experiment (L. *experiri,* to try) A trial made to support or disprove a hypothesis.

exteroceptor (ek'stər-ō-sep'tər) (L. *exter,* outward, + *capere,* to take) A sense organ excited by stimuli from the external world.

facilitated transport Mediated transport of a substance into a cell in the direction of a concentration gradient, contrast with **active transport**.

FAD Abbreviation for flavine adenine dinucleotide, an electron acceptor in the respiratory chain.

fatty acid Any of a series of saturated organic acids having the general formula $C_nH_{2n}O_2$, occurs in natural fats of animals and plants.

fermentation (L. *fermentum,* ferment) Enzymatic transformation, without oxygen, of organic substrates, especially carbohydrates, yielding products such as alcohols, acids, and carbon dioxide.

fiber, fibril (L. *fibra,* thread) These two terms are often confused. Fiber is a fiberlike cell or a strand of protoplasmic material produced or secreted by a cell and lying outside the cell. Fibril is a strandlike organelle produced by a cell and lying within the cell.

filter feeding Any feeding process by which particulate food is filtered from water in which it is suspended.

fission (L. *fissio,* a splitting) Asexual reproduction by a division of the body into two or more parts.

fitness Degree of adjustment and suitability for a particular environment. Genetic fitness is relative contribution of one genetically distinct organism to the next generation; organisms with high genetic fitness are naturally selected and become prevalent in a population.

flagellum (flə-jel'əm) (L., a whip) Whiplike organelle of locomotion.

flame bulb Specialized hollow excretory or osmoregulatory structure of one or several small cells containing a tuft of cilia (the "flame") and situated at the end of a minute tubule; connected tubules ultimately open to the outside. See **protonephridium**.

fluke (O.E. *flōc,* flatfish) A member of class Trematoda or class Monogenea. Also, certain of the flatfishes (order Pleuronectiformes).

FMN Abbreviation for flavin mononucleotide, the prosthetic group of a protein (flavoprotein) and a carrier in the electron transport chain in respiration.

food vacuole A digestive organelle in the cell.

fouling Contamination of feeding or respiratory areas of an organism by excrement, sediment, or other matter. Also, accumulation of sessile marine organisms on the hull of a boat or ship so as to impede its progress through the water.

fovea (fō'vē-ə) (L., a small pit) A small pit or depression; especially the fovea centralis, a small rodless pit in the retina of some vertebrates, a point of acute vision.

free energy The energy available for doing work in a chemical system.

gamete (ga'mēt, gə-mēt') (Gr. *gamos,* marriage) A mature haploid sex cell; usually, male and female gametes can be distinguished. An egg or a sperm.

gametic meiosis Meiosis that occurs during formation of the gametes, as in humans and other higher animals.

gametocyte (gə-mēt'ə-sīt) (Gr. *gametēs,* spouse, + *kytos,* hollow vessel) The mother cell of a gamete, that is, immature gamete.

ganoid scales (ga'noid) (Gr. *ganos,* brightness) Thick, bony, rhombic scales of some primitive bony fishes; not overlapping.

gap junction An area of tiny canals communicating the cytoplasm between two cells.

gastrodermis (gas'tro-dər'mis) (Gr. *gastēr,* stomach, + *derma,* skin) Lining of the digestive cavity of cnidarians.

gastrovascular cavity (Gr. *gastēr,* stomach, + L. *vasculum,* small vessel) Body cavity in certain lower invertebrates that functions in both digestion and circulation and has a single opening serving as both mouth and anus.

gastrula (gas'trə-lə) (Gr. *gastēr,* stomach, + L. *ula,* dim.) Embryonic stage, usually cap or sac shaped, with walls of two layers of cells surrounding a cavity (archenteron) with one opening (blastopore).

gastrulation (gas'trə-lā'shən) (Gr. *gastēr,* stomach) Process by which an early metazoan embryo becomes a gastrula, acquiring first two and then three layers of cells.

gemmule (je'mūl) (L. *gemma,* bud, + *ula,* dim.) Asexual, cystlike reproductive unit in freshwater sponges; formed in summer or autumn and capable of overwintering.

gene (Gr. *genos,* descent) The part of a chromosome that is the hereditary determiner and is transmitted from one generation to another. Specifically, a gene is a nucleic acid sequence (usually DNA) that encodes a functional polypeptide or RNA sequence.

gene pool A collection of all of the alleles of all of the genes in a population.

genetic drift Change in gene frequencies by chance processes in the evolutionary process of animals. In small populations, one allele may drift to fixation, becoming the only representative of that gene locus.

genome (jē'nōm) (Gr. *genos,* offspring, + L. *oma,* abstract group) All the genes in a haploid set of chromosomes.

genotype (jēn'ō-tīp) (Gr. *genos,* offspring, + *typos,* form) The genetic construction, expressed and latent, of an organism; the total set of genes present in the cells of an organism; contrasts with **phenotype.**

genus (jē'nus) pl. **genera** (L., race) A group of related species with taxonomic rank between family and species.

germ layer In the animal embryo, one of three basic layers (ectoderm, endoderm, mesoderm) from which the various organs and tissues arise in the multicellular animal.

germ plasm The germ cells of an organism, as distinct from the somatoplasm; the hereditary material (genes) of the germ cells.

gestation (je-stā'shən) (L. *gestare,* to bear) The period in which offspring are carried in the uterus.

glochidium (glō-kid'i-əm) (Gr. *glochis,* point, + *idion,* dim.) Bivalved larval stage of freshwater mussels.

glomerulus (glä-mer'u-ləs) (L. *glomus,* ball) A tuft of capillaries projecting into a renal corpuscle in a kidney. Also, a small spongy mass of tissue in the proboscis of hemichordates, presumed to have an excretory function. Also, a concentration of nerve fibers situated in the olfactory bulb.

glycogen (glī'kə-jən) (Gr. *glykys,* sweet, + *genes,* produced) A polysaccharide constituting the principal form in which carbohydrate is stored in animals; animal starch.

glycolysis (glī-kol'i-sis) (Gr. *glykus,* sweet, + *lyein,* to loosen) Enzymatic breakdown of glucose (especially) or glycogen into pyruvic acid with release of energy.

Golgi complex (gōl'jē) (after Camillo Golgi, 1844–1926, Italian histologist) An organelle in cells that serves as a collecting and packaging center for secretory products.

gonad (gō'nad) (N.L. *gonas,* a primary sex organ) An organ that produces gametes (ovary in the female and testis in the male).

gonangium (gō-nan'jē-əm) (N.L. *gonas,* primary sex organ + *angeion,* dim. of vessel) Reproductive zooid of hydroid colony (Cnidaria).

gonoduct (Gr. *gonos,* seed, progeny, + duct) Duct leading from a gonad to the exterior.

gonopore (gän'ə-pōr) (Gr. *gonos,* seed, progeny, + *poros,* an opening) A genital pore found in many invertebrates.

green gland Excretory gland of certain Crustacea; the antennal gland.

gregarious (L. *grex,* herd) Living in groups or flocks.

guanine (gwä'nēn) (Sp., from Quechua, *huanu,* dung) A white crystalline purine base, $C_5H_5N_5O$, occurring in various animal tissues and in guano and other animal excrements.

gynecophoric canal (gī'nə-kə-fōr'ik) (Gr. *gynē,* woman, + *pherein,* to carry). Groove in male schistosomes (certain trematodes) that carries the female.

habitat (L. *habitare,* to dwell) The place where an organism normally lives or where individuals of a population live.

habituation A kind of learning in which continued exposure to the same stimulus produces diminishing responses.

halter (hal'tər) pl. **halteres** (hal-ti'rēz) (Gr., leap) In Diptera, small club-shaped structure on each side of the metathorax representing the hind wings; believed to be sense organs for balancing; also called balancer.

haploid (Gr. *haploos,* single) The reduced, or n, number of chromosomes, typical of gametes, as opposed to the diploid, or 2n, number found in somatic cells. In certain phyla, some mature animals have a haploid number of chromosomes.

hectocotylus (hek-tə-kät'ə-ləs) (Gr. *hekaton,* hundred, + *kotylē,* cup) Specialized, and somtimes autonomous, arm that serves as a male copulatory organ in cephalopods.

hemal system (hē'məl) (Gr. *haima,* blood) System of small vessels in echinoderms; function unknown.

hemerythrin (hē'mə-rith'rin) (Gr. *haima,* blood, + *erythros,* red) A red, iron-containing respiratory pigment found in the blood of some polychaetes, sipunculans, priapulids, and brachiopods.

hemimetabolous (hē'mi-mə-ta'bə-ləs) (Gr. *hēmi,* half, + *metabolē,* change) Refers to gradual metamorphosis during development of insects, without a pupal stage.

hemocoel (hē'mə-sēl) (Gr. *haima,* blood, + *koilos,* hollow) Main body cavity of arthropods, may be subdivided into sinuses, through which the blood flows.

hemoglobin (Gr. *haima,* blood, +L. *globulus,* globule) An iron-containing respiratory pigment occurring in vertebrate red blood cells and in blood plasma of many invertebrates; a compound of an iron porphyrin heme and a protein globin.

hemolymph (hē'mə-limf) (Gr. *haima,* blood, + L. *lympha,* water) Fluid in the coelom or hemocoel of some invertebrates that represents the blood and lymph of higher forms.

hepatic (hə-pat'ic) (Gr. *hēpatikos,* of the liver) Pertaining to the liver.

herbivore ([h]erb'ə-vōr') (L. *herba,* green crop, + *vorare,* to devour) Any organism subsisting on plants. Adj., **herbivorous.**

hermaphrodite (hər-maf'rə-dīt) (Gr. *hermaphroditos,* containing both sexes; from Greek mythology, Hermaphroditos, son of Hermes and Aphrodite) An organism with both male and female functional reproductive organs. **Hermaphroditism** may refer to an aberration in unisexual animals; monoecism implies that this is the normal condition for the species.

heterocercal (het'ər-o-sər'kəl) (Gr. *heteros,* different, + *kerkos,* tail) In some fishes, a tail with the upper lobe larger than the lower, and the end of the vertebral column somewhat upturned in the upper lobe, as in sharks.

heterochromatin (he'tə-rō-krōm'ə-tən) (Gr. *heteros,* different, + *chrōma,* color) Chromatin that stains intensely and appears to represent inactive genetic areas.

heterodont (hed'ə-ro-dänt) (Gr. *heteros,* different + *odous,* tooth) Having teeth differentiated into incisors, canines, and molars for different purposes.

heterotroph (het'ə-rō-träf) (Gr. *heteros,* different, + *trophos,* feeder) An organism that obtains both organic and inorganic raw materials from the environment in order to live; includes most animals and those plants that do not carry on photosynthesis.

heterozygote (het'ə-rō-zī'gōt) (Gr. *heteros,* different, + *zygōtos,* yoked) An organism in which the pair of alleles for a trait is composed of different genes (usually dominant and recessive); derived from a zygote formed by the union of gametes of dissimilar genetic constitution.

hexamerous (hek-sam'ər-əs) (Gr. *hex,* six, + *meros,* part) Six parts, specifically, symmetry based on six or multiples thereof.

hibernation (L. *hibernus,* wintry) Condition, especially of mammals, of passing the winter in a torpid state in which the body temperature drops nearly to freezing and the metabolism drops close to zero.

histology (hi-stäl'-ə-jē) (Gr. *histos,* web, tissue, + *logos,* discourse) The study of the microscopic anatomy of tissues.

histone (hi'stōn) (Gr. *histos,* tissue) Any of several simple proteins found in cell nuclei and complexed at one time or another with DNA. Histones yield a high proportion of basic amino acids on hydrolysis; characteristic of eukaryotes.

holoblastic cleavage (Gr. *holo,* whole, + *blastos,* germ) Complete and approximately equal division of cells in early embryo. Found in mammals, amphioxus, and many aquatic invertebrates that have eggs with a small amount of yolk.

holometabolous (hō'lō-mə-ta'bə-ləs) (Gr. *holo,* complete, + *metabolē,* change) Complete metamorphosis during development.

holophytic nutrition (hōl'ō-fit'ik) (Gr. *holo,* whole, + *phyt,* plant) Occurs in green plants and certain protozoa and involves synthesis of carbohydrates from carbon dioxide and water in the presence of light, chlorophyll, and certain enzymes.

holozoic nutrition (hōl'ō-zō'ik) (Gr. *holo,* whole, + *zoikos,* of animals) Type of nutrition involving ingestion of liquid or solid organic food particles.

home range The area over which an animal ranges in its activities. Unlike territories, home ranges are not defended.

homeostasis (hō'mē-ō-stā'sis) (Gr. *homeo,* similar, + *stasis,* state or standing) Maintenance of an internal steady state by means of self-regulation.

homeothermic (hō'mē-ō-thər'mik) (Gr. *homeo,* alike, + *thermē,* heat) Having a nearly uniform body temperature, regulated independent of the environmental temperature; "warm blooded."

hominid (häm'ə-nid) (L. *homo, hominis,* man) A member of the family Hominidae, now represented by one living species, *Homo sapiens.*

homonoid (häm'ə-noid) Relating to the Hominoidea, a superfamily of primates to which the great apes and humans are assigned.

homocercal (hō'mə-ser'kal) (Gr. *homos,* same, common, + *kerkos,* tail) A tail with the upper and lower lobes symmetrical and the vertebral column ending near the middle of the base, as in most teleost fishes.

homodont (hō'mō-dänt) (Gr. *homos,* same + *odous,* tooth) Having all teeth similar in form.

homograft See **allograft.**

homology (hō-mäl'ə-jē) (Gr. *homologos,* agreeing) Similarity of parts or organs of different organisms caused by similar embryonic origin and evolutionary development from a corresponding part in some remote ancestor. Also, correspondence in structure of different parts of the same individual. May also refer to a matching pair of chromosomes. Adj., **homologous.**

homozygote (hō-mə-zī'gōt) (Gr. *homos,* same, + *zygotos,* yoked) An organism in which the pair of alleles for a trait is composed of the same genes (either dominant or recessive but not both). Adj., **homozygous.**

humoral (hū'mər-əl) (L. *humor,* a fluid) Pertaining to an endocrine or other secretion carried in the body fluids.

hyaline (hī'ə-lən) (Gr. *hyalos,* glass) Adj., glassy, translucent. Noun, a clear, glassy structureless material occurring, for example, in cartilage, vitreous body, mucin, and glycogen.

hybridoma (hī-brid-ō'mah) (contraction of hybrid + myeloma) Fused product of a normal and a myeloma (cancer) cell, which has some of the characteristics of the normal cell.

hydatid cyst (hī-da'təd) (Gr. *hydatis,* watery vesicle) A type of cyst formed by juveniles of certain tapeworms (*Echinococcus*) in their vertebrate hosts.

hydranth (hī'dranth) (Gr. *hydōr,* water, +

anthos, flower) Nutritive zooid of hydroid colony.

hydrocoral (Gr. *hydōr,* water, + *korallion,* coral) Certain members of the cnidarian class Hydrozoa that secrete calcium carbonate, resembling true corals.

hydroid The polyp form of a cnidarian as distinguished from the medusa form. Any cnidarian of the class Hydrozoa, order Hydroida.

hydrolysis (Gr. *hydōr,* water, + *lysis,* a loosening) The decomposition of a chemical compound by the addition of water; the splitting of a molecule into its groupings so that the split products acquire hydrogen and hydroxyl groups.

hydrostatic skeleton A mass of fluid or plastic parenchyma enclosed within a muscular wall to provide the support necessary for antagonistic muscle action; for example, parenchyma in acoelomates and perivisceral fluids in pseudocoelomates serve as hydrostatic skeletons.

hydroxyl (hydrogen + oxygen, + yl) Containing an OH^- group, a negatively charged ion formed by alkalies in water.

hyperosmotic (Gr. *hyper,* over, + *ōsmos,* impulse) Refers to a solution whose osmotic pressure is greater than that of another solution with which it is compared; contains a greater concentration of dissolved particles and gains water through a selectively permeable membrane from a solution containing fewer particles; contrasts with **hypoosmotic**.

hypodermis (hī'pə-dər'mis) (Gr. *hypo,* under, + L. *dermis,* skin) The cellular layer lying beneath and secreting the cuticle of annelids, arthropods, and certain other invertebrates.

hypoosmotic (Gr. *hypo,* under, + *ōsmos,* impulse) Refers to a solution whose osmotic pressure is less than that of another solution with which it is compared or taken as standard; contains a lesser concentration of dissolved particles and loses water during osmosis; contrasts with **hyperosmotic.**

hypophysis (hī-pof'ə-sis) (Gr. *hypo,* under, + *physis,* growth) Pituitary body.

hypostome (hī'pə-stōm) (Gr. *hypo,* under, + *stoma,* mouth) Name applied to structure in various invertebrates (such as mites and ticks), located at posterior or ventral area of mouth; elevation supporting mouth of hydrozoan.

hypothalamus (hī-pō-thal'ə-mis) (Gr. *hypo,* under, + *thalamos,* inner chamber) A ventral part of the forebrain beneath the thalamus; one of the centers of the autonomic nervous system.

hypothesis (hī-poth'ə-səs) (Gr. *hypothesis,* foundation, supposition) A statement or conclusion based on inductive reasoning about prior observations, which can be tested by an experiment.

immunoglobulin (im'yə-nə-glä'byə-lən) (L. *immunis,* free, + *globus,* globe) Any of a group of plasma proteins, produced by plasma cells, that participates in the immune response by combining with the antigen that stimulated its production. Antibody.

imprinting (im'print-ing) (L. *imprimere,* to impress, imprint) Rapid and usually stable learning pattern appearing early in the life of a member of a social species and involving recognition of its own species; may involve attraction to the first moving object seen.

indeterminate cleavage A type of embryonic development in which the fate of the blastomeres is not determined very early as to tissues or organs, for example, in echinoderms and vertebrates.

indigenous (in-dij'ə-nəs) (L. *indigena,* native) Pertains to organisms that are native to a particular region; not introduced.

induction (L. *inducere, inductum,* to lead) Reasoning from the particular to the general, that is, deriving a general statement (hypothesis) based on individual observations.

inflammation (in'fləm-mā'shən) (L. *inflammare,* from *flamma,* flame) The complicated physiological process in mobilization of body defenses against foreign substances and infectious agents and repair of damage from such agents.

infraciliature (in-frə-sil'e-ə-tər) (L. *infra,* below, + *cilia,* eyelashes) The organelles just below the cilia in ciliate protozoa.

infundibulum (in'fun-dib'u-ləm) (L., funnel) Stalk of the neurohypophysis linking the pituitary to the diencephalon.

innate (i-nāt') (L. *innatus,* inborn) A characteristic based partly or wholly on gene differences.

instar (inz'tär) (L., form) Stage in the life of an insect or other arthropod between molts.

instinct (L. *instinctus,* impelled) Stereotyped, predictable, genetically programmed behavior. Learning may or may not be involved.

integument (in-teg'ū-mənt) (L. *integumentum,* covering) An external covering or enveloping layer.

intermediate host A host in which some development of a symbiont occurs, but in which maturation and sexual reproduction do not take place (contrasts with **definitive host**).

interstitial (in-tər-sti'shəl) (L. *inter,* among, + *sistere,* to stand) Situated in the interstices or spaces between structures such as cells, organs, or grains of sand.

intron (in'trän) (L. *intra,* within) Portion of mRNA as transcribed from DNA that will not form part of mature mRNA, that is, that does not include part of message for gene product.

introvert (L. *intro,* inward, + *vertere,* to turn) The anterior narrow portion that can be withdrawn (introverted) into the trunk of a sipunculid worm.

invagination (in-vaj'ə-nā'shən) (L. *in,* in, + *vagina,* sheath) An infolding of a layer of tissue to form a saclike structure.

inversion (L. *invertere,* to turn upside down) A turning inward or inside out, as in embryogenesis of sponges; also, reversal in order of genes or reversal of a chromosome segment.

irritability (L. *irritare,* to provoke) A general property of all organisms involving the ability to respond to stimuli or changes in the environment.

isolecithal (ī'sə-les'ə-thəl) (Gr. *isos,* equal, + *lekithos,* yolk, + *al)* Pertaining to a zygote (or ovum) with yolk evenly distributed; homolecithal.

isotonic (Gr. *isos,* equal, + *tonikos,* tension) Said of solutions having the same or equal osmotic pressure; isosmotic.

isotope (ī'sə-tōp) (Gr. *isos,* equal, + *topos,* place) A form of an element with the same number of protons in the nucleus as an alternative form, but with a different number of neutrons.

juvenile hormone Hormone produced by the corpora allata of insects; among its effects are maintenance of larval or nymphal characteristics during development.

keratin (ker'ə-tən) (Gr. *kera,* horn, + *in,* suffix of proteins) A scleroprotein found in epidermal tissues and modified into hard structures such as horns, hair, and nails.

kinesis (kə-nē'səs) (Gr. *kinēsis,* movement) Movements by an organism in random directions in response to a stimulus.

kinetosome (kin-et'ə-sōm) (Gr. *kinētos,*

moving, + *sōma,* body) The self-duplicating granule at the base of the flagellum or cilium; similar to centriole, also called basal body or blepharoplast.

kinety (kə-nē'tē) (Gr. *kinein,* to move) All the kinetosomes and kinetodesmata of a row of cilia.

kinin (kī'nin) (Gr. *kinein,* to move, + *in,* suffix of hormones) A type of local hormone that is released near its site of origin; also called parahormone or tissue hormone.

K-selection (from the K term in the logistic equation) Natural selection under conditions that favor survival when populations are controlled primarily by density-dependent factors.

kwashiorkor (kwash-ē-or'kər) (from Ghana) Malnutrition caused by diet high in carbohydrate and extremely low in protein.

labium (lā'bē-əm) (L., a lip) The lower lip of the insect formed by fusion of the second pair of maxillae.

labrum (lā'brəm) (L., a lip) The upper lip of insects and crustaceans situated above or in front of the mandibles; also refers to the outer lip of a gastropod shell.

labyrinth (L. *labyrinthus,* labyrinth) Vertebrate internal ear, composed of a series of fluid-filled sacs and tubules (membranous labyrinth) suspended within bone cavities (osseous labyrinth).

labyrinthodont (lab'ə-rin'thə-dänt) (Gr. *labyrinthos,* labyrinth, + *odous, odontos,* tooth) A group of fossil-stem amphibians from which most amphibians later arose. They date from the late Paleozoic era.

lacteal (lak'te-əl) (L. *lacteus,* of milk) Noun, one of the lymph vessels in the villus of the intestine. Adj., relating to milk.

lacuna (lə-kū'nə), pl. **lacunae** (L., pit, cavity) A sinus; a space between cells; a cavity in cartilage or bone.

lagena (lə-jē'nə) (L., large flask) Portion of the primitive ear in which sound is translated into nerve impulses; evolutionary beginning of cochlea.

Lamarckism Hypothesis, as expounded by Jean Baptiste de Lamarck, of evolution by the acquisition during an organism's lifetime of characteristics that are transmitted directly to offspring.

lamella (lə-mel'ə) (L. dim. of *lamina,* plate) One of the two plates forming a gill in a bivalve mollusc. One of the thin layers of bone laid concentrically around an osteon (Haversian) canal. Any thin, platelike structure.

larva (lar'və) pl. **larvae** (L., a ghost) An immature stage that is quite different from the adult.

lek (lek) (Sw., play, game) An area where animals assemble for communal courtship display and mating.

lentic (len'tik) (L. *lentus,* slow) Of or relating to standing water such as swamp, pond, or lake.

leukocyte (lū'kə-sīt) (Gr. *leukos,* white, + *kytos,* hollow vessel) A blood cell with a nucleus and ameboid properties; a white blood cell. Has no hemoglobin.

lipase (lī'pās) (Gr. *lipos,* fat, + *ase,* enzyme suffix) An enzyme that accelerates the hydrolysis or synthesis of fats.

lipid, lipoid (li'pid) (Gr. *lipos,* fat) Certain fatlike substances, often containing other groups such as phosphoric acid; lipids combine with proteins and carbohydrates to form principal structured components of cells.

littoral (lit'ə-rəl) (L. *litoralis,* seashore) Adj., pertaining to the shore. Noun, that portion of the sea floor between the extent of high and low tides, intertidal; in lakes, the shallow part from the shore to the lakeward limit of aquatic plants.

locus (lō'kəs) pl. **loci** (lō'sī) (L., place) Position of a gene in a chromosome.

logistic equation A mathematical expression describing an idealized sigmoid curve of population growth.

lophophore (lōf'ə-fōr) (Gr. *lophos,* crest, + *phoros,* bearing) Tentacle-bearing ridge or arm within which is an extension of the coelomic cavity in lophophorate animals (ectoprocts, brachiopods, and phoronids).

lorica (lor'ə-kə) (L. *lorica,* corselet) A secreted, protective covering, as in phylum Loricifera.

lotic (lō'tik) (L. *lotus,* action of washing or bathing) Of or pertaining to running water, such as a brook or river.

lumen (lū'mən) (L., light) The cavity of a tube or organ.

lymphocyte (lim'fō-sīt) (L. *lympha,* water, goddess of water, + Gr. *kytos,* hollow vessel) Cell in blood and lymph that has central role in immune responses. See **T cells** and **B cells**.

lysosome (lī'sə-sōm) (Gr. *lysis,* loosing, + *sōma,* body) Intracellular organelle consisting of a membrane enclosing several digestive enzymes that are released when the lysosome ruptures.

macromolecule A very large molecule, such as a protein, polysaccharide, or nucleic acid.

macronucleus (ma'krō-nū'klē-əs) (Gr. *makros,* long, large, + *nucleus,* kernel) The larger of the two kinds of nuclei in ciliate protozoa; controls all cell functions except reproduction.

madreporite (ma'drə-pōr'īt) (Fr. *madrépore,* reef-building coral, + *ite,* suffix for some body parts) Sievelike structure that is the intake for the water-vascular system of echinoderms.

major histocompatibility complex (MHC) Genes which code for cell surface proteins involved in self-nonself recognition.

malacostracan (mal'ə-käs'trə-kən) (Gr. *malako,* soft, + *ostracon,* shell) Any member of the crustacean subclass Malacostraca, which includes both aquatic and terrestrial forms of crabs, lobsters, shrimps, pillbugs, sand fleas, and others.

Malpighian tubules (mal-pig'ē-ən) (Marcello Malpighi, Italian anatomist, 1628-1694) Blind tubules opening into the hindgut of nearly all insects and some myriapods and arachnids, and functioning primarily as excretory organs.

mantle Soft extension of the body wall in certain invertebrates, for example, brachiopods and molluscs, which usually secretes a shell; thin body wall of tunicates.

marasmus (mə-raz'məs) (Gr. *marasmos,* to waste away) Malnutrition, especially of infants, caused by a diet deficient in both calories and protein.

marsupial (mär-sū'pē-əl) (Gr. *marsypion,* little pouch) One of the pouched mammals of the subclass Metatheria.

mastax (mas'təx) (Gr., jaws) Pharyngeal mill of rotifers.

matrix (mā'triks) (L. *mater,* mother) The intercellular substance of a tissue, or that part of a tissue into which an organ or process is set.

maturation (L. *maturus,* ripe) The process of ripening; the final stages in the preparation of gametes for fertilization.

maxilla (mak-sil'ə) (L. dim. of *mala,* jaw) One of the upper jawbones in vertebrates; one of the head appendages in arthropods.

maxilliped (mak-sil'ə-ped) (L. *maxilla,* jaw, + *pes,* foot) One of the pairs of head appendages located just posterior to the maxilla in crustaceans; a thoracic appendage that has become incorporated into the feeding mouthparts.

mediated transport Transport of a substance across a cell membrane mediated by a carrier molecule in the membrane.

medulla (mə-dul'ə) (L., marrow) The inner portion of an organ in contrast to the cortex or outer portion. Also, hindbrain.

medusa (mə-dū-sə) (Gr. mythology, female monster with snake-entwined hair) A jellyfish, or the free-swimming stage in the life cycle of cnidarians.

Mehlis' gland (me'ləs) Glands of uncertain function surrounding the junction of yolk duct, oviduct, and uterus in trematodes and cestodes.

meiofauna (mī'ō-faə-nə) (Gr. *meion*, smaller, + L. *faunus*, god of the woods) Small invertebrates found in the interstices between sand grains.

meiosis (mī-ō'səs) (Gr. from *meioun*, to make small) The nuclear changes by means of which the chromosomes are reduced from the diploid to the haploid number; in animals, usually occurs in the last two divisions in the formation of the mature egg or sperm.

melanin (mel'ə-nin) (Gr. *melas*, black) Black or dark-brown pigment found in plant or animal structures.

menopause (men'ō-pawz) (Gr. *men*, month, + *pauein*, to cease) In the human female, that time of life when ovulation ceases; cessation of the menstrual cycle.

menstruation (men'stroo-ā'shən) (L. *menstrua*, the menses, from *mensis*, month) The discharge of blood and uterine tissue from the vagina at the end of a menstrual cycle.

meroblastic (mer-ə-blas'tik) (Gr. *meros*, part, + *blastos*, germ) Partial cleavage occurring in zygotes having a large amount of yolk at the vegetal pole; cleavage restricted to a small area on the surface of the egg.

merozoite (me'rə-zō'īt) (Gr. *meros*, part, + *zōon*, animal) A very small trophozoite at the stage just after cytokinesis has been completed in multiple fission of a protozoan.

mesenchyme (me'zn-kīm) (Gr. *mesos*, middle, + *enchyma*, infusion) Embryonic connective tissue; irregular or amebocytic cells often embedded in gelatinous matrix.

mesocoel (mez'ō-sēl) (Gr. *mesos*, middle, + *koilos*, hollow) Middle body coelomic compartment in some deuterostomes, anterior in lophophorates, corresponds to hydrocoel in echinoderms.

mesoderm (me'zə-dərm) (Gr. *mesos*, middle, + *derma*, skin) The third germ layer, formed in the gastrula between the ectoderm and endoderm; gives rise to connective tissues, muscle, urogenital and vascular systems, and the peritoneum.

mesoglea (mez'ō-glē'ə) (Gr. *mesos*, middle, + *glia*, glue) The layer of jellylike or cement material between the epidermis and gastrodermis in cnidarians and ctenophores; also may refer to jellylike matrix between epithelial layers in sponges.

mesohyl (me'sə-hil) (Gr. *mesos*, middle, + *hyle*, a wood) Gelatinous matrix surrounding sponge cells; mesoglea, mesenchyme.

mesonephros (me-zō-nef'rōs) (Gr. *mesos*, middle, + *nephros*, kidney) The middle of three pairs of embryonic renal organs in vertebrates. Functional kidney of fishes and amphibians; its collecting duct is a wolffian duct. Adj., **mesonephric**.

messenger RNA (mRNA) A form of ribonucleic acid that carries genetic information from the gene to the ribosome, where it determines the order of amino acids as a polypeptide is formed.

metabolism (Gr. *metabolē*, change) A group of processes that includes digestion, production of energy (respiration), and synthesis of molecules and structures by organisms; the sum of the constructive (anabolic) and destructive (catabolic) processes.

metacentric (me'tə-sen'trək) (Gr. *meta*, among, + *kentron*, center) Chromosome with centromere at or near the middle.

metacercaria (me'tə-sər-ka'rē-ə) (Gr. *meta*, after, + *kerkos*, tail, + L. *aria*, connected with) Fluke juvenile (cercaria) that has lost its tail and has become encysted.

metacoel (met'ə-sēl) (Gr. *meta*, after, + *koilos*, hollow) Posterior coelomic compartment in some deuterostomes and lophophorates; corresponds to somatocoel in echinoderms.

metamere (met'ə-mēr) (Gr. *meta*, after, + *meros*, part) A repeated body unit along the longitudinal axis of an animal; a somite, or segment.

metamerism (mə-ta'mə-ri'zəm) (Gr. *meta*, between, after, + *meros* part). Condition of being made up of serially repeated parts (metameres); serial segmentation.

metamorphosis (Gr. *meta*, after, + *morphē*, form, + *osis*, state of) Sharp change in form during postembryonic development, for example, tadpole to frog or larval insect to adult.

metanephridium (me'tə-nə-fri'di-əm) (Gr. *meta*, after, + *nephros*, kidney) A type of tubular nephridium with the inner open end draining the coelom and the outer open end discharging to the exterior.

metanephros (me'tə-ne'fräs) (Gr. *meta*, between, after, + *nephros*, kidney) Embryonic renal organs of vertebrates arising behind the mesonephros; the functional kidney of reptiles, birds, and mammals. It is drained from a ureter.

metazoa (met-ə-zō'ə) (Gr. *meta*, after, + *zōon*, animal) Multicellular animals; all animals above the Protozoa.

microfilament (mī'krō-fil'ə-mənt) (Gr. *mikros*, small, + *filum*, a thread) A thin, linear structure in cells; of actin in muscle cells and others.

micron (μ)(mī-krän) (Gr., neuter of *mikros*, small) One one thousandth of a millimeter; about 1/25,000 of an inch. Now largely replaced by micrometer (μm).

microneme (mī'krə-nēm) (Gr. *mikros*, small, + *nēma*, thread) One of the types of structures comprising the apical complex in the phylum Apicomplexa, slender and elongate, leading to the anterior and thought to function in host cell penetration.

micronucleus A small nucleus found in ciliate protozoa; controls the reproductive functions of these organisms.

microthrix (mī'krə-thrix) pl. **microtriches** (Gr. *mikros*, small, + *thrix*, hair) A small microvillus-like structure on the surface of a tapeworm's tegument.

microtubule (Gr. *mikros*, small, + L. *tubule*, pipe) A long, tubular cytoskeletal element with an outside diameter of 20 to 27 μm. Microtubules influence cell shape and play important roles during cell division.

microvillus (Gr. *mikros*, small, + *villus*, shaggy hair) Narrow, cylindrical cytoplasmic projection from epithelial cells; microvilli form the brush border of several types of epithelial cells.

mictic (mik'tik) (Gr. *miktos*, mixed or blended) Pertaining to haploid egg of rotifers or the females that lay such eggs.

miracidium (mīr'ə-sid'ē-əm) (Gr. *meirakidion*, youthful person) A minute ciliated larval stage in the life of flukes.

mitochondrion (mīt'ə-kän'drē-ən) (Gr. *mitos*, a thread, + *chondrion*, dim. of *chondros*, corn, grain) An organelle in the cell in which aerobic metabolism takes place.

mitosis (mī-tō′səs) (Gr. *mitos,* thread, + *osis,* state of) Nuclear division in which there is an equal qualitative and quantitative division of the chromosomal material between the two resulting nuclei; ordinary cell division (indirect).

molecule (mo′lə-kyūl) (L. *moles,* a mass). A group of atoms of the same or different elements that are bound together by chemical bonds.

monocyte (mon′ə-sīt) (Gr. *monos,* single, + *kytos,* hollow vessel) A type of leukocyte that becomes a phagocytic cell (macrophage) after moving into tissues.

monoecious (mə-nē′shəs) (Gr. *monos,* single, + *oikos,* house) Having both male and female gonads in the same organism; hermaphroditic.

monohybrid (Gr. *monos,* single, + L. *hybrida,* mongrel) A hybrid offspring of parents different in one specified character.

monomer (mä′nə-mər) (Gr. *monos,* single, + *meros,* part) A molecule of simple structure, but capable of linking with others to form polymers.

monophyletic (mä′nə-phī-le′tik) (Gr. *monos,* single, + *phyletikos,* pertaining to a phylum) Referring to a taxon whose units all evolved from a single parent stock; contrasts with **polyphyletic**.

monosaccharide (mä′nə-sa′kə-rīd) (Gr. *monos,* one, + *sakcharon,* sugar, from Sanskrit *sarkarā,* gravel, sugar) A simple sugar that cannot be decomposed into smaller sugar molecules; the most common are pentoses (such as ribose) and hexoses (such as glucose).

morphogenesis (mor′fə-je′nə-səs) (Gr. *morphē,* form, + *genesis,* origin) Development of the architectural features of organisms; formation and differentiation of tissues and organs.

morphology (Gr. *morphē,* form, + *logos,* discourse) The science of structure. Includes cytology, the study of cell structure, histology, the study of tissue structure; and anatomy, the study of gross structure.

morula (mär′u-lə) (L. *morum,* mulberry, + *ula,* dim.) Solid ball of cells in early stage of embryonic development.

mosaic cleavage Type characterized by independent differentiation of each part of the embryo; determinate cleavage.

mucin (mū′sən) (L. *mucus,* nasal mucus) Any of a group of glycoproteins secreted by certain cells, especially those of salivary glands.

bat / āpe / ärmadillo / herring / fēmale / finch / līce / crocodile / crōw / duck / ūnicorn / ə indicates unaccented vowel sound "uh" as in mammal, fishes, cardinal, heron, vulture / stress as in bi-ol′o-gy, bi′o-log′i-cal

mucus (mū′kəs) adj., **mucous** (L. *mucus,* nasal mucus) Viscid, slippery secretion rich in mucins produced by secretory cells such as those in mucous membranes.

mutation (mū-tā′shən) (L. *mutare,* to change) A stable and abrupt change of a gene; the heritable modification of a character.

mutualism (mü′chə-wə-li′zəm) (L. *mutuus,* lent, borrowed, reciprocal) A type of interaction in which two different species derive benefit from the association and in which the association is necessary to both; often symbiotic.

myocyte (mī′ə-sīt) (Gr. *mys,* muscle, + *kytos,* hollow vessel) Contractile cell (pinacocyte) in sponges.

myofibril (Gr. *mys,* muscle, + L. dim. of *fibra,* fiber) A contractile filament within muscle or muscle fiber.

myogenic (mī′o-jen′ik) (Gr. *mys,* muscle, + N.L. *genic,* giving rise to) Originating in muscle, such as heart beat arising in vertebrate cardiac muscle because of inherent rhythmical properties of muscle rather than because of neural stimuli.

myosin (mī′ə-sin) (Gr. *mys,* muscle, + *in,* suffix, belonging to) A large protein of contractile tissue that forms the thick myofilaments of striated muscle. During contraction it combines with actin to form actomyosin.

myotome (mī′ə-tōm) (Gr. *mys,* muscle + *tomos,* cutting) A voluntary muscle segment in cephalochordates and vertebrates; that part of a somite destined to form muscles; the muscle group innervated by a single spinal nerve.

nacre (nā′kər) (F., mother-of-pearl) Innermost lustrous layer of mollusc shell, secreted by mantle epithelium. Adj., **nacreous**.

NAD Abbreviation of nicotinamide adenine dinucleotide, an electron acceptor or donor in many metabolic reactions.

naiad (nā′əd) (Gr. *naias,* a water nymph) An aquatic, gill-breathing immature insect (nymph).

nares (na′rēz), sing. **naris** (L., nostrils) Openings into the nasal cavity, both internally and externally, in the head of a vertebrate.

natural selection A nonrandom reproduction of genotypes that results in the survival of those best adapted to their environment and elimination of those less well adapted; leads to evolutionary change.

nauplius (naw′plē-əs) (L., a kind of shellfish) A free-swimming microscopic larval stage of certain crustaceans, with three pairs of appendages (antennules, antennae, and mandibles) and median eye. Characteristic of ostracods, copepods, barnacles, and some others.

nekton (nek′tən) (Gr. neuter of *nēktos,* swimming) Term for actively swimming organisms, essentially independent of wave and current action. Compare with **plankton**.

nematocyst (ne-mad′ə-sist′) (Gr. *nēma,* thread, + *kystis,* bladder) Stinging organelle of cnidarians.

neoteny (nē′ə-tē′nē, nē-ot′ə-nē) (Gr. *neos,* new, + *teinein,* to extend) The attainment of sexual maturity in the larval condition. Also, the retention of larval characters into adulthood.

nephridium (nə-frid′ē-əm) (Gr. *nephridios,* of the kidney) One of the segmentally arranged, paired excretory tubules of many invertebrates, notably the annelids. In a broad sense, any tubule specialized for excretion and/or osmoregulation; with an external opening and with or without an internal opening.

nephron (ne′frän) (Gr. *nephros,* kidney) Functional unit of kidney structure of vertebrates, consisting of a Bowman's capsule, an enclosed glomerulus, and the attached uriniferous tubule.

nephrostome (nef′rə-stōm) (Gr. *nephros,* kidney, + *stoma,* mouth) Ciliated, funnel-shaped opening of a nephridium.

neritic (nə-rid′ik) (Gr. *nērites,* a mussel) Portion of the sea overlying the continental shelf, specifically from the subtidal zone to a depth of 200 m.

neurogenic (nū-rä-jen′ik) (Gr. *neuron,* nerve, + N.L. *genic,* give rise to) Originating in nervous tissue, as does the rhythmical beat of some arthropod hearts.

neuroglia (nū-räg′le-ə) (Gr. *neuron,* nerve, + *glia,* glue) Tissue supporting and filling the spaces between the nerve cells of the central nervous system.

neurolemma (nū-rə-lem′ə) (Gr. *neuron,* nerve, + *lemma,* skin) Delicate nucleated outer sheath of a nerve cell; sheath of Schwann.

neuromast (Gr., *neuron,* sinew, nerve, + *mastos,* knoll) Cluster of sense cells on or near the surface of a fish or amphibian that is sensitive to vibratory stimuli and water current.

neuron (Gr. nerve) A nerve cell.

neuropodium (nū'rə-pō'de-əm) (Gr. *neuron*, nerve, + *pous, podos*, foot) Lobe of parapodium nearer the ventral side in polychaete annelids.

neurosecretory cell (nu'rō-sə-krēd'ə-rē) Any cell (neuron) of the nervous system that produces a hormone.

niche The role of an organism in an ecological community; its unique way of life and its relationship to other biotic and abiotic factors.

nitrogen fixation (Gr. *nitron*, soda, + *gen*, producing) Reduction of molecular nitrogen to ammonia by some bacteria and cyanobacteria, often followed by **nitrification**, the oxidation of ammonia to nitrites and nitrates by other bacteria.

notochord (nōd'ə-kord') (Gr. *nōtos*, back + *chorda*, cord) An elongated cellular cord, enclosed in a sheath, that forms the primitive axial skeleton of chordate embryos and adult cephalochordates.

notopodium (nō'tə-pō'de-əm) (Gr. *nōtos*, back, + *pous, podos*, foot) Lobe of parapodium nearer the dorsal side in polychaete annelids.

nucleic acid (nu-klē'ik) (L. *nucleus*, kernel) One of a class of molecules composed of joined nucleotides; chief types are deoxyribonucleic acid (DNA), found in cell nuclei (chromosomes and mitochondria), and ribonucleic acid (RNA), found both in cell nuclei (chromosomes and nucleoli) and in cytoplasmic ribosomes.

nucleolus (nu-klē'ə-ləs) (dim. of L. *nucleus*, kernel) A deeply staining body within the nucleus of a cell and containing RNA; nucleoli are specialized portions of certain chromosomes that carry multiple copies of the information to synthesize ribosomal RNA.

nucleoplasm (nu'klē-ə-plazm') (L. *nucleus*, kernel, + Gr. *plasma*, mold) Protoplasm of nucleus, as distinguished from cytoplasm.

nucleoprotein A molecule composed of nucleic acid and protein; occurs in the nucleus and cytoplasm of all cells.

nucleosome (nu'klē-ə-som) (L. *nucleus*, kernel, + *sōma*, body) A repeating subunit of chromatin in which one and three quarters turns of the double-helical DNA are wound around eight molecules of histones.

nucleotide (nu'klē-ə-tīd) A molecule consisting of phosphate, 5-carbon sugar (ribose or deoxyribose), and a purine or a pyrimidine; the purines are adenine and guanine, and the pyrimidines are cytosine, thymine, and uracil.

nuptial flight (nup'shəl) The mating flight of insects, especially that of the queen with male or males.

nymph (L. *nympha*, nymph, bride) An immature stage (following hatching) of a hemimetabolous insect that lacks a pupal stage.

ocellus (ō-sel'əs) (L. dim. of *oculus*, eye) A simple eye or eyespot in many types of invertebrates.

octomerous (ok-tom'ər-əs) (Gr. *oct*, eight, + *meros*, part) Eight parts, specifically, symmetry based on eight.

ommatidium (ä'mə-tid'ē-əm) (Gr. *omma*, eye, + *idium*, small) One of the optical units of the compound eye of arthropods and molluscs.

oncogene (on'kə-gēn) (Gr. *onkos*, protuberance, tumor, + *genos*, descent) Any of a number of genes that, when activated, are associated with neoplastic growth (cancer). The gene in its benign, inactivated state is a **proto-oncogene**.

ontogeny (än-tä'jə-nē) (Gr. *ontos*, being, + *geneia*, act of being born, from *genēs*, born) The course of development of an individual from egg to senescence.

oocyst (ō'ə-sist) (Gr. *ōion*, egg, + *kystis*, bladder) Cyst formed around zygote of malaria and related organisms.

ooecium (ō-ēs'ē-əm) (Gr. *ōion*, egg, + *oikos*, house, + L. *ium*, from Ger. *ion*, dim.) Brood pouch; compartment for developing embryos in ectoprocts.

oogenesis (o'ə-jen'ə-səs) (Gr. *ōion*, egg, + *genesis*, origin) Formation and maturation of an egg.

oogonium (ō'əgōn'ē-əm) (Gr. *ōion*, egg + *gonos*, offspring) A cell that, by continued division, gives rise to oocytes; an ovum in a primary follicle immediately before the beginning of maturation.

ookinete (ō-ə-kī'nēt) (Gr. *ōion*, egg, + *kinein*, to move) The motile zygote of malaria organisms.

ootid (o'ə-tid) (Gr. *ōion*, egg, + *eidos*, form) A growth stage of an egg that arises from an oocyte and matures into an ovum.

operculum (ō-per'kū-ləm) (L., cover) The gill cover in bony fishes; horny plate in some snails.

operon (äp'ə-rän) A genetic unit consisting of a cluster of genes that are under the control of other genes, found in prokaryotes.

ophthalmic (äf-thal'mik) (Gr. *ophthalmos*, an eye) Pertaining to the eye.

opisthosoma (ō-pis'thə-sō'mə) (Gr. *opisthe*, behind, + *sōma*, body) Posterior body region in arachnids and pogonophorans.

opsonization (op'sən-i-zā'shən) (Gr. *opsonein*, to buy victuals, to cater) The facilitation of phagocytosis of foreign particles by phagocytes in the blood or tissues, mediated by antibody bound to the particles.

organelle (Gr. *organon*, tool, organ, + L. *ella*, dim.) Specialized part of a cell; literally, a small organ that performs functions analogous to organs of multicellular animals.

organic compound Any compound of carbon in which the principal bonds are between carbon atoms or carbon and hydrogen atoms.

organizer (or'gan-ī-zer) (Gr. *organos*, fashioning) Area of an embryo that directs subsequent development of other parts.

osmoregulation Maintenance of proper internal salt and water concentrations in a cell or in the body of a living organism; active regulation of internal osmotic pressure.

osmosis (oz-mō'sis) (Gr. *ōsmos*, act of pushing, impulse) The flow of solvent (usually water) through a selectively permeable membrane.

osmotic potential Osmotic pressure.

osphradium (äs-frā'dē-əm) (Gr. *osphradion*, small bouquet, dim. of *osphra*, smell) A sense organ in aquatic snails and bivalves that tests incoming water.

ossicles (L. *ossiculum*, small bone) Small separate pieces of echinoderm endoskeleton. Also, tiny bones of the middle ear of vertebrates.

osteoblast (os'tē-ō-blast) (Gr. *osteon*, bone, + *blastos*, bud) A bone-forming cell.

osteoclast (os'tē-ō-clast) (Gr. *osteon*, bone, + *klan*, to break) A large, multinucleate cell that functions in bone dissolution.

osteon (os'tē-än) (Gr., bone) Unit of bone structure; Haversian system.

ostium (L., door) Opening.

otolith (ōd'əl-ith') (Gr. *ous, otos*, ear, + *lithos*, stone) Calcareous concretions in the membranous labyrinth of the inner ear of lower vertebrates, or in the auditory organ of certain invertebrates.

oviger (ō'vi-jər) (L. *ovum*, egg, + *gerere*, to

bear) Leg that carries eggs in pycnogonids.

oviparity (ō′və-pa′rəd-ē) (L. *ovum*, egg, + *parere*, to bring forth) Reproduction in which eggs are released by the female; development of offspring occurs outside the maternal body. Adj., **oviparous** (ō-vip′ə-rəs).

ovipositor (ō′ve-päz′əd-ər) (L. *ovum*, egg, + *positor*, builder, placer, + *or*, suffix denoting agent or doer) In many female insects a structure at the posterior end of the abdomen for laying eggs.

ovoviviparity (ō′vo-vī-və-par′ə-dē) (L. *ovum*, egg, + *vivere*, to live, + *parere*, to bring forth) Reproduction in which eggs develop within the maternal body without additional nourishment from the parent and hatch within the parent or immediately after laying. Adj., **ovoviviparous** (ō′vo-vī-vip′ə-rəs).

ovum (L. *ovum*, egg) Mature female germ cell (egg).

oxidation (äk′sə-dā-shən) (Fr. *oxider*, to oxidize, from Gr. *oxys*, sharp, + *ation*) The loss of an electron by an atom or molecule; sometimes addition of oxygen chemically to a substance. Opposite of reduction, in which an electron is accepted by an atom or molecule.

oxidative phosphorylation (äk′sə-dād′iv fäs′fər-i-lā′shən) The conversion of inorganic phosphate to energy-rich phosphate of ATP, involving electron transport through a respiratory chain to molecular oxygen.

paedogenesis (pē-dō-jen′ə-sis) (Gr. *pais*, child, + *genēs*, born) Reproduction by immature or larval animals caused by acceleration of maturation. Progenesis.

paedomorphosis (pē-dō-mor′fə-səs) (Gr. *pais*, child, + *morphē*, form) Displacement of ancestral juvenile features to later stages of the ontogeny of descendants.

pair bond An affiliation between an adult male and an adult female for reproduction. Characteristic of monogamous species.

pallium (pal′e-əm) (L., mantle) Mantle of a mollusc or brachiopod.

pangenesis (pan-jen′ə-sis) (Gr. *pan*, all, + *genesis*, descent) Darwin's hypothesis that hereditary characteristics are carried by individual body cells that produce particles that collect in the germ cells.

papilla (pə-pil′ə) (L., nipple) A small nipplelike projection. A vascular process that nourishes the root of a hair, feather, or developing tooth.

papula (pa′pū-lə) (L., pimple) Respiratory processes on skin of sea stars; also, pustules on skin.

parapodium (pa′rə-pō′dē-əm) (Gr. *para*, beside, + *pous, podos*, foot) One of the paired lateral processes on each side of most segments in polychaete annelids; variously modified for locomotion, respiration, or feeding.

parasitism (par′ə-sit′iz-əm) (Gr. *parasitos*, from *para*, beside, + *sitos*, food) The condition of an organism living in or on another organism (host) at whose expense the parasite is maintained; destructive symbiosis.

parasympathetic (par′ə-sim-pə-thed′ik) (Gr. *para*, beside, + *sympathes*, sympathetic, from *syn*, with, + *pathos*, feeling) One of the subdivisions of the autonomic nervous system, whose fibers originate in the brain and in anterior and posterior parts of the spinal cord.

parenchyma (pə-ren′kə-mə) (Gr., anything poured in beside) In lower animals, a spongy mass of vacuolated mesenchyme cells filling spaces between viscera, muscles, or epithelia; in some, the cells are cell bodies of muscle cells. Also, the specialized tissue of an organ as distinguished from the supporting connective tissue.

parthenogenesis (pär′thə-nō-gen′ə-sis) (Gr. *parthenos*, virgin, + L. from Gr. *genesis*, origin) Unisexual reproduction involving the production of young by females not fertilized by males; common in rotifers, cladocerans, aphids, bees, ants, and wasps. A parthenogenetic egg may be diploid or haploid.

pathogenic (path′ə-jen′ik) (Gr. *pathos*, disease, + N.L. *genic*, giving rise to) Producing or capable of producing disease.

peck order A hierarchy of social privilege in a flock of birds.

pecten (L., comb) Any of several types of comblike structures on various organisms, for example, a pigmented, vascular, and comblike process that projects into the vitreous humor from the retina at a point of entrance of the optic nerve in the eyes of all birds and many reptiles.

pectoral (pek′tə-rəl) (L. *pectoralis*, from *pectus*, the breast) Of or pertaining to the breast or chest; to the pectoral girdle; or to a pair of horny shields of the plastron of certain turtles.

pedal laceration Asexual reproduction found in sea anemones, a form of fission.

pedalium (pə-dal′ē-əm) (Gr. *pedalion*, a prop, rudder) The flattened, bladelike base of a tentacle or group of tentacles in the cnidarian class Cubozoa.

pedicel (ped′ə-sel) (L. *pediculus*, little foot) A small or short stalk or stem. In insects, the second segment of an antenna or the waist of an ant.

pedicellaria (ped′ə-sə-lar′ē-ə) (L. *pediculus*, little foot, + *aria*, like or connected with) One of many minute pincerlike organs on the surface of certain echinoderms.

pedipalps (ped′ə-palps′) (L. *pes, pedis*, foot, + *palpus*, stroking, caress) Second pair of appendages of arachnids.

pedogenesis See **paedogenesis**.

peduncle (pē′dun-kəl) (L. *pedunculus*, dim. of *pes*, foot) A stalk. Also, a band of white matter joining different parts of the brain.

pelage (pel′ij) (Fr., fur) Hairy covering of mammals.

pelagic (pə-laj′ik) (Gr. *pelagos*, the open sea) Pertaining to the open ocean.

pellicle (pel′ə-kəl) (L. *pellicula*, dim. of *pellis*, skin) Thin, translucent, secreted envelope covering many protozoa.

pentadactyl (pen-tə-dak′təl) (Gr. *pente*, five, + *daktylos*, finger) With five digits, or five fingerlike parts, to the hand or foot.

peptidase (pep′tə-dās) (Gr. *peptein*, to digest, + *ase*, enzyme suffix) An enzyme that breaks down simple peptides, releasing amino acids.

peptide bond A bond that binds amino acids together into a polypeptide chain, formed by removing an OH from the carboxyl group of one amino acid and an H from the amino group of another to form an amide group —CO—NH—.

periostracum (pe-rē-äs′trə-kəm) (Gr. *peri*, around, + *ostrakon*, shell) Outer horny layer of a mollusc shell.

periproct (per′ə-präkt) (Gr. *peri*, around, + *prōktos*, anus) Region of aboral plates around the anus of echinoids.

perisarc (per′ə-särk) (Gr. *peri*, around, + *sarx*, flesh) Sheath covering the stalk and branches of a hydroid.

perissodactyl (pə-ris′ə-dak′təl) (Gr. *perissos*, odd, + *daktylos*, finger, toe) Pertaining to an order of ungulate mammals with an odd number of digits.

peristalsis (per′ə-stal′səs) (Gr. *peristalti-*

kos, compressing around) The series of alternate relaxations and contractions that force food through the alimentary canal.

peristomium (per'ə-stō'mē-əm) (Gr. *peri,* around, + *stoma,* mouth) Foremost true segment of an annelid; it bears the mouth.

peritoneum (per'ə-tə-nē-əm) (Gr. *peritonaios,* stretched around) The membrane that lines the coelom and covers the coelomic viscera.

pH (*p*otential of *h*ydrogen) A symbol referring to the relative concentration of hydrogen ions in a solution; pH values are from 0 to 14, and the lower the value, the more acid or hydrogen ions in the solution. Equal to the negative logarithm of the hydrogen ion concentration.

phagocyte (fag'ə-sīt) (Gr. *phagein,* to eat, + *kytos,* hollow vessel) Any cell that engulfs and devours microorganisms or other particles.

phagocytosis (fag'ə-sī-tō'səs) (Gr. *phagein,* to eat, + *kytos,* hollow vessel) The engulfment of a particle by a phagocyte or a protozoan.

phagosome (fa'gə-sōm) (Gr. *phagein,* to eat, + *sōma,* body) Membrane-bound vesicle in cytoplasm containing food material engulfed by phagocytosis.

pharynx (far'inks) pl. **pharynges** (Gr. *pharynx,* gullet) The part of the digestive tract between the mouth cavity and the esophagus that, in vertebrates, is common to both digestive and respiratory tracts. In cephalochordates the gill slits open from it.

phenetics (fen-et'iks) (Gr. *phainein,* to show) A system of arranging taxonomic groupings according to the statistically highest number of similarities without regard to phylogenetic significance.

phenotype (fē'nə-tīp) (Gr. *phainein,* to show) The visible or expressed characters of an organism, controlled by the genotype, but not all genes in the genotype are expressed.

pheromone (fer'ə-mōn) (Gr. *pherein,* to carry, + *hormōn,* exciting, stirring up) Chemical substance released by one organism that influences the behavior or physiological processes of another organism.

phosphagen (fäs'fə-jən) (phosphate + gen) A term for creatine phosphate and arginine phosphate, which store and may be sources of high-energy phosphate bonds.

phosphorylation (fäs'fə-rə-lā'shən) The addition of a phosphate group such as $—PO_3^{2-}$, to a compound.

photoautotroph (fōd-ə-aw'tō-trōf) (Gr. *phōtos,* light, + *autos,* self, + *trophos,* feeder) An organism requiring light as a source of energy for making organic nutrients from inorganic raw materials.

photosynthesis (fōd'ō-sin'thə-sis) (Gr. *phōs,* light, + *synthesis,* action or putting together) The synthesis of carbohydrates from carbon dioxide and water in chlorophyll-containing cells exposed to light.

phylogeny (fī-läj'ə-nē) (Gr. *phylon,* tribe, race, + *geneia,* origin) The origin and development of any taxon, or the evolutionary history of its development.

phylum (fī'ləm), pl. **phyla** (N.L. from Gr. *phylon,* race, tribe) A chief category, between kingdom and class, of taxonomic classifications into which are grouped organisms of common descent that share a fundamental pattern of organization.

pinacocyte (pin'ə-kō-sīt') (Gr. *pinax,* tablet, + *kytos,* hollow vessel) Flattened cells comprising dermal epithelium in sponges.

pinna (pin'ə) (L., feather, sharp point) The external ear. Also a feather, wing, or fin or similar part.

pinocytosis (pin'o-sī-tō'sis, pīn'o-sī-tō'sis) (Gr. *pinein,* to drink, + *kytos,* hollow vessel, + *osis,* condition) Taking up of fluid by endocytosis; cell drinking.

placenta (plə-sen'tə) (L., flat cake) The vascular structure, embryonic and maternal, through which the embryo and fetus are nourished while in the uterus.

placoid scale (pla'koid) (Gr. *plax, plakos,* tablet, plate) Type of scale found in cartilaginous fishes, with basal plate of dentin embedded in the skin and a backward-pointing spine tipped with enamel.

plankton (plank'tən) (Gr. neuter of *planktos,* wandering) The passively floating animal and plant life of a body of water; compares with **nekton.**

planula (plan'yə-lə) (N.L. dim from L. *planus,* flat) Free-swimming, ciliated larval type of cnidarians; usually flattened and ovoid, with an outer layer of ectodermal cells and an inner mass of endodermal cells.

planuloid ancestor (plan'yə-loid) (L. *planus,* flat, + Gr. *eidos,* form) Hypothetical form representing ancestor of Cnidaria and Platyhelminthes.

plasma cell (plaz'mə) (Gr. *plasma,* a form, mold) A descendant cell of a B cell, functions to secrete antibodies.

plasma membane (plaz'mə) (Gr. *plasma,* a form, mold) A living, external, limiting, protoplasmic structure that functions to regulate exchange of nutrients across the cell surface.

plasmid (plaz'məd) (Gr. *plasma,* a form, mold) A small circle of DNA carried by a bacterium in addition to its large chromosome.

plastid (plas'təd) (Gr. *plast,* formed, molded, + L. *id,* feminine stem for particle of specified kind) A membranous organelle in plant cells functioning in photosynthesis and/or nutrient storage, for example, chloroplast.

plastron (plas'trən) (Fr. *plastron,* breast plate) Ventral bony shield of turtles; or structure in corresponding position in certain arthropods; or, thin film of gas retained by epicuticle hairs of aquatic insects.

platelet (plāt'lət) (Gr. dim. of *plattus,* flat) A tiny, incomplete cell in the blood that releases substances initiating blood clotting.

pleiotropic (plī-ə-trō'pic) (Gr. *pleiōn,* more, + *tropos,* to turn) Pertaining to a gene producing more than one effect; affecting multiple phenotypic characteristics.

pleura (plu'rə) (Gr. side, rib) The membrane that lines each half of the thorax and covers the lungs.

plexus (plek'səs) (L. network, braid) A network, especially of nerves or blood vessels.

podium (pō'de-əm) (Gr. *pous, podos,* foot) A footlike structure, for example, the tube foot of echinoderms.

poikilothermic (poi-ki'lə-thər'mik) (Gr. *poikilos,* variable, + thermal) Pertaining to animals whose body temperature is variable and fluctuates with that of the environment; cold blooded; compares with ectothermic.

polarization (L. *polaris,* polar, + Gr. *iz,* make) The arrangement of positive electrical charges on one side of a surface membrane and negative electrical charges on the other side (in nerves and muscles).

polar molecule A molecule with an unequal distribution of electrical charge, one end being more positive and the other end more negative.

Polian vesicles (pō'le-ən) (from G.S. Poli, 1746–1825, Italian naturalist) Vesicles opening into ring canal in most asteroids and holothuroids.

polyandry (pol'y-an'drē) (Gr. *polys*, many, + *anēr*, man) Condition of having more than one male mate at one time.

polygamy (pə-lig'ə-mē) (Gr. *polys*, many, + *gamos*, marriage) Condition of having more than one mate at one time.

polygyny (pə-lij'ə-nē) (Gr. *polys*, many, + *gynē*, woman) Condition of having more than one female mate at one time.

polymer (pä'lə-mər) (Gr. *polys*, many, + *meros*, part) A chemical compound composed of repeated structural units called monomers.

polymerization (pə-lim'ər-ə-zā'shən) The process of forming a polymer or polymeric compound.

polymorphism (pä'lē-mor'fi-zəm) (Gr. *polys*, many, + *morphē*, form) The presence in a species of more than one structural type of individual.

polynucleotide (poly + nucleotide) A nucleotide of many mononucleotides combined.

polyp (päl'əp) (Fr. *polype*, octopus, from L. *polypus*, many footed) The sessile stage in the life cycle of cnidarians.

polypeptide (pä-lē-pep'tīd) (Gr. *polys*, many, + *peptein*, to digest) A molecule consisting of many joined amino acids, not as complex as a protein.

polyphyletic (pä'lē-fī-led'ik) (Gr. *polys*, many, + *phylon*, tribe) Derived from more than one ancestral source; opposed to monophyletic.

polyphyodont (pä'lē-fi'ə-dänt) (Gr. *polyphyes*, manifold, + *odous*, tooth) Having several sets of teeth in succession.

polypide (pä'li-pīd) (L. *polypus*, polyp) An individual or zooid in a colony, specifically in ectoprocts, which has a lophophore, digestive tract, muscles, and nerve centers.

polyploid (pä'lə-ploid') (Gr. *polys*, many, + *ploidy*, number of chromosomes) Characterized by a chromosome number that is greater than two full sets of homologous chromosomes.

polysaccharide (pä'lē-sak'ə-rid, -rīd) (Gr. *polys*, many, + *sakcharon*, sugar, from Sanskrit *sarkarā*, gravel, sugar) A carbohydrate composed of many monosaccharide units, for example, glycogen, starch, and cellulose.

polysome (**polyribosome**) (Gr. *polys*, many, + *sōma*, body) Two or more ribosomes connected by a molecule of messenger RNA.

pongid (pän'jəd) (L. *Pongo*, type genus of orangutan) Of or relating to the primate family Pongidae, comprising the anthropoid apes (gorillas, chimpanzees, gibbons, orangutans).

population (L. *populus*, people) A group of organisms of the same species inhabiting a specified geographical locality.

porocyte (pō'rə-sīt) (Gr. *poros*, passage, pore + *kytos*, hollow vessel) Type of cell found in asconoid sponges through which water enters the spongocoel.

portal system (L. *porta*, gate) System of large veins beginning and ending with a bed of capillaries; for example, hepatic portal and renal portal systems in vertebrates.

preadaptation The possession of a condition that is selected in an ancestral environment and that coincidentally predisposes an organism for survival in some other environment.

prebiotic synthesis The chemical synthesis that occurred before the emergence of life.

precocial (prē-kō'shəl) (L. *praecoquere*, to ripen beforehand) Referring (especially) to birds whose young are covered with down and are able to run about when newly hatched.

predaceous, predacious (prē-dā'shəs) (L. *praeda*, prey) Living by killing and consuming other animals; predatory.

predator (pred'ə-tər) (L. *praeda*, prey) An organism that preys on other organisms for its food.

prehensile (prē-hen'səl) (L. *prehendere*, to seize) Adapted for grasping.

primate (prī'māt) (L. *primus*, first) Any mammal of the order Primates, which includes the tarsiers, lemurs, marmosets, monkeys, apes, and humans.

primitive (L. *primus*, first) Primordial; ancient; little evolved; said of species closely approximating their early ancestral types.

proboscis (prō-bäs'əs) (Gr. *pro*, before, + *boskein*, feed) A snout or trunk. Also, tubular sucking or feeding organ with the mouth at the end as in planarians, leeches, and insects. Also the sensory and defensive organ at the anterior end of certain invertebrates.

producers (L. *producere*, to bring forth) Organisms, such as plants, able to produce their own food from inorganic substances.

progesterone (prō-jes'tə-rōn') (L. *pro*, before, + *gestare*, to carry) Hormone secreted by the corpus luteum and the placenta; prepares the uterus for the fertilized egg and maintains the capacity of the uterus to hold the embryo and fetus.

proglottid (prō-gläd'əd) (Gr. *proglōttis*, tongue tip, from *pro*, before, + *glōtta*, tongue, + *id*, suffix) Portion of a tapeworm containing a set of reproductive organs; usually corresponds to a segment.

prokaryotic, procaryotic (pro-kar'ē-ät'ik) (Gr. *pro*, before, + *karyon*, kernel, nut) Not having a membrane-bound nucleus or nuclei. Prokaryotic cells are more primitive than eukaryotic cells and persist today in the bacteria and cyanobacteria.

pronephros (prō-nef'rəs) (Gr. *pro*, before, + *nephros*, kidney) Most anterior of three pairs of embryonic renal organs of vertebrates; functional only in adult hagfishes and larval fishes and amphibians; vestigial in mammalian embryos. Adj., **pronephric**.

proprioceptor (prō'prē-ə-sep'tər) (L. *proprius*, own, particular, + receptor) Sensory receptor located deep within the tissues, especially muscles, tendons, and joints, that is responsive to changes in muscle stretch, body position, and movement.

prosimian (prō-sim'ē-ən) (Gr. *pro*, before, + L. *simia*, ape) Any member of a group of primitive, arboreal primates: lemurs, tarsiers, lorises, and so on.

prosopyle (präs'ə-pīl) (Gr. *prosō*, forward, + *pylē*, gate) Connections between the incurrent and radial canals in some sponges.

prostaglandins (präs'tə-glan'dəns) A family of fatty acid tissue hormones, originally discovered in semen, known to have powerful effects on smooth muscle, nerves, circulation, and reproductive organs.

prostomium (prō-stō'mē-əm) (Gr. *pro*, before, + *stoma*, mouth) In most annelids and some molluscs, that part of the head located in front of the mouth.

protease (prō'tē-ās) (Gr. *protein*, + *ase*, enzyme) An enzyme that digests proteins; includes proteinases and peptidases.

protein (prō'tēn, prō'tē-ən) (Gr. *protein*, from *proteios*, primary) A macromolecule of carbon, hydrogen, oxygen, and nitrogen and sometimes sulfur and phosphorus; composed of chains of amino acids joined by peptide bonds; present in all cells.

prothoracic glands Glands in the prothorax of insects that secrete the hormone ecdysone.

prothoracicotropic hormone Hormone secreted in brain of insects that stimulates the prothoracic glands to secrete ecdysone.

prothrombin (pro-thräm'bən) (Gr. *pro,* before, + *thrombos,* clot) A constituent of blood plasma that is changed to thrombin by a catalytic sequence that includes thromboplastin, calcium, and plasma globulins; involved in blood clotting.

protist (prō'tist) (Gr. *protos,* first) A member of the kingdom Protista, generally considered to include the unicellular eukaryotic organisms (protozoa and eukaryotic algae).

protocoel (prō'tə-sēl) (Gr. *protos,* first, + *kilos,* hollow) The anterior coelomic compartment in some deuterostomes, corresponds to the axocoel in echinoderms.

protocooperation A mutually beneficial interaction between organisms in which the interaction is not physiologically necessary to the survival of either.

protonephridium (prō'tō-nə-frid'ē-əm) (Gr. *protos,* first, + *nephros,* kidney) Primitive osmoregulatory or excretory organ consisting of a tubule terminating internally with flame bulb or solenocyte; the unit of a flame bulb system.

protoplasm (prō'tə-plazm) (Gr. *protos,* first, + *plasma,* form) Organized living substance; cytoplasm and nucleoplasm of the cell.

protopod, protopodite (prō'tə-päd, prō-top'ə-dīt) (Gr. *protos,* first, + *pous, podos,* foot) Basal portion of crustacean appendage, containing coxa and basis.

proto-oncogene see **oncogene**.

Protostomia (prō'tə-stō'mē-ə) (Gr. *protos,* first, + *stoma,* mouth) A group of phyla in which cleavage is determinate, the coelom (in coelomate forms) is formed by proliferation of mesodermal bands (schizocoelic formation), the mesoderm is formed from a particular blastomere (called 4d), and the mouth is derived from or near the blastopore. Includes the Annelida, Arthropoda, Mollusca, and a number of minor phyla. Compares with **Deuterostomia**.

proventriculus (pro'ven-trik'ū-ləs) (L. *pro,* before, + *ventriculm,* ventricle) In birds, the glandular stomach between the crop and gizzard. In insects, a muscular dilation of foregut armed internally with chitinous teeth.

proximal (L. *proximus,* nearest) Situated toward or near the point of attachment; opposite of distal, distant.

pseudocoel (sū'do-sēl) (Gr. *pseudēs,* false, + *koilōma,* cavity) A body cavity not lined with peritoneum and not a part of the blood or digestive systems, embryonically derived from the blastocoel.

pseudopodium (sū'də-pō'dē-əm) (Gr. *pseudēs,* false, + *podion,* small foot, + *eidos,* form) A temporary cytoplasmic protrusion extended out from a protozoan or ameboid cell, and serving for locomotion or for taking up food.

pupa (pū'pə) (L., girl, doll, puppet) Inactive quiescent stage of the holometabolous insects. It follows the larval stages and precedes the adult stage.

purine (pū'rēn) (L. *purus,* pure, + *urina,* urine) Organic base with carbon and nitrogen atoms in two interlocking rings. The parent substance of adenine, guanine, and other naturally occurring bases.

pyrimidine (pī-rim'ə-dēn) (alter. of pyridine, from Gr. *pyr,* fire, + *id,* adj. suffix, + *ine*) an organic base composed of a single ring of carbon and nitrogen atoms; parent substance of several bases found in nucleic acids.

queen In entomology, the single fully developed female in a colony of social insects such as bees, ants, and termites, distinguished from workers, nonreproductive females, and soldiers.

radial cleavage Type in which early cleavage planes are symmetrical to the polar axis, each blastomere of one tier lying directly above the corresponding blastomere of the next layer; indeterminate cleavage.

radial symmetry A morphological condition in which the parts of an animal are arranged concentrically around an oral-aboral axis, and more than one imaginary plane through this axis yields halves that are mirror images of each other.

radiole (rā'dē-ol) (L. *radiolus,* dim. of *radius,* ray, spoke of a wheel) Structure extending from head of some sedentary polychaetes used in feeding on suspended particles.

radula (ra'jə-lə) (L., scraper) Rasping tongue found in most molluscs.

ratite (ra'tīt) (L. *ratis,* raft) Having an unkeeled sternum; compares with **carinate**.

recapitulation Summing up or repeating; hypothesis that an individual repeats its phylogenetic history in its development.

recombinant DNA DNA from two different species, such as a virus and a mammal, combined into a single molecule.

redia (rē'dē-ə) pl. **rediae** (rē'dē-ē) (from Francesco Redi, 1626–1697, Italian biologist) A larval stage in the life cycle of flukes; it is produced by a sporocyst larva, and in turn gives rise to many cercariae.

regulative development Progressive determination and restriction of initially totipotent embryonic material.

releaser (L. *relaxare,* to unloose) Simple stimulus that elicits an innate behavior pattern.

renin (rē'nin) (L. *ren,* kidney) An enzyme produced by the kidney juxtaglomerular apparatus that initiates changes leading to increased blood pressure and increased sodium reabsorption.

replication (L. *replicatio,* a folding back) In genetics, the duplication of one or more DNA molecules from the preexisting molecule.

respiration (L. *respiratio,* breathing) Gaseous interchange between an organism and its surrounding medium. In the cell, the release of energy by the oxidation of food molecules.

restriction endonuclease An enzyme that cleaves a DNA molecule at a particular base sequence.

rete mirabile (rē'tē mə-rab'ə-lē) (L., wonderful net) A network of small blood vessels so arranged that the incoming blood runs countercurrent to the outgoing blood and thus makes possible efficient exchange between the two bloodstreams. Such a mechanism serves to maintain the high concentration of gases in the fish swim bladder.

reticular (rə-tīk'ū-lər) (L. *reticulum,* small net) Resembling a net in appearance or structure.

reticuloendothelial system (rə-tic'ū-lō-en-dō-thēl'i-əl) (L. *reticulum,* dim. of net, + Gr. *endon,* within, + *thele,* nipple) The fixed phagocytic cells in the tissues, especially the liver, lymph nodes, spleen, and others; also called RE system.

retrovirus (L. *retro,* backward, + *virus,* slime, poison) Any member of the Retroviridae, which are among the RNA viruses. RNA viruses must have their genetic material transcribed into DNA in the host cell by reverse transcriptase.

rhabdite (rab'dīt) (Gr. *rhabdos,* rod) Rodlike structures in the cells of the epidermis or underlying parenchyma in certain turbellarians. They are discharged in mucous secretions.

rheoreceptor (rē'ə-rē-cep'tər) (Gr. *rheos,*

a flowing, + receptor) A sensory organ of aquatic animals that responds to water current.

rhopalium (rō-pā'lē-əm) (N.L. from Gr. *rhopalon,* a club) One of the marginal, club-shaped sense organs of certain jellyfishes; tentaculocyst.

rhoptries (rōp'trēz) (Gr. *rhopalon,* club, + *tryō,* to rub, wear out) Club-shaped bodies in Apicomplexa comprising one of the structures of the apical complex; open at anterior and apparently functioning in penetration of host cell.

rhynchocoel (ring'kō-sēl) (Gr. *rhynchos,* snout, + *koilos,* hollow) In nemertines, the dorsal tubular cavity that contains the inverted proboscis. It has no opening to the outside.

ribosomal RNA (rRNA) Type of RNA found in ribosomes.

ribosome (rī'bə-sōm) A small organelle composed of protein and ribonucleic acid. May be free in the cytoplasm or attached to the membranes of the endoplasmic reticulum; functions in protein synthesis.

ritualization In ethology, the evolutionary modification, usually intensification, of a behavior pattern to serve communication.

RNA Ribonucleic acid, of which there are several different kinds, such as messenger RNA, ribosomal RNA, and transfer RNA (mRNA, rRNA, tRNA).

rostellum (räs-tel'ləm) (L., small beak) Projecting structure on scolex of tapeworm, often with hooks.

rostrum (räs'trəm) (L. ship's beak) A snoutlike projection on the head.

r-selection (from the r term in the logistic equation) Natural selection under conditions that favor survival when populations are controlled primarily by density-independent factors; contrast with **K-selection**.

ruminant (rūm'ə-nənt) (L. *ruminare,* to chew the cud) Cud-chewing artiodactyl mammals with a complex four-chambered stomach.

saccule (sa'kūl) (L. *sacculus,* small bag) Small chamber of the membranous labyrinth of the inner ear.

sagittal (saj'ə-dəl) (L. *sagitta,* arrow) Pertaining to the median anteroposterior plane that divides a bilaterally symmetrical organism into right and left halves.

bat / āpe / ärmadillo / herring / fēmale / finch / līce / crocodile / crōw / duck / ūnicorn / ə indicates unaccented vowel sound "uh" as in mammal, fishes, cardinal, heron, vulture / stress as in bi-ol'o-gy, bi'o-log'i-cal

salt (l. *sal,* salt) The reaction product of an acid and a base; dissociates in water solution to negative and positive ions, but not H^+ or OH^-.

saprophagous (sə-präf'ə-gəs) (Gr. *sapros,* rotten, + *phagos,* from *phagein,* to eat) Feeding on decaying matter; saprobic; saprozoic.

saprophyte (sap'rə-fīt) (Gr. *sapros,* rotten, + *phyton,* plant) A plant living on dead or decaying organic matter.

saprozoic nutrition (sap'rə-zō'ik) (Gr. *sapros,* rotten, + *zōon,* animal) Animal nutrition by absorption of dissolved salts and simple organic nutrients from surrounding medium; also refers to feeding on decaying matter.

sarcolemma (sär'kə-lem'ə) (Gr. *sarx,* flesh, + *lemma,* rind) The thin, noncellular sheath that encloses a striated muscle fiber.

sarcomere (sär'kə-mir) (Gr. *sarx,* flesh, + *meros,* part) Transverse segment of striated muscle believed to be the fundamental contractile unit.

sarcoplasm (sär'kə-plaz'-əm) (Gr. *sarx,* flesh, + *plasma,* mold) Clear, semifluid substance between fibrils of muscle tissue.

schizocoel (skiz'ə-sēl) (Gr. *schizo,* from *schizein,* to split, + *koilōma,* cavity) A coelom formed by the splitting of embryonic mesoderm. Noun, **schizocoelomate**, an animal with a schizocoel, such as an arthropod or mollusc. Adj., **schizocoelous**.

schizocoelous mesoderm formation (skiz'ō-sēl-ləs) Embryonic ectoderm of the mesoderm as cords of cells between ectoderm and endoderm; splitting of these cords results in the coelomic space.

schizogony (skə-zä'gə-nē) (Gr. *schizein,* to split, + *gonos,* seed) Multiple asexual fission.

sclerite (skle'rit) (Gr. *sklēros,* hard) A hard chitinous or calcareous plate or spicule; one of the plates making up the exoskeleton of arthropods, especially insects.

scleroblast (skler'ə-blast) (Gr. *sklēros,* hard, + *blastos,* germ) An amebocyte specialized to secrete a spicule, found in sponges.

sclerotic (skle-räd'ik) (Gr. *sklēros,* hard) Pertaining to the tough outer coat of the eyeball.

sclerotin (skler'ə-tən) (Gr. *sklērotēs,* hardness) Insoluble, tanned protein permeating the cuticle of arthropods.

sclerotization (skle'rə-tə-zā'shən) Process of hardening of the cuticle of arthropods by the formation of stabilizing cross linkages between peptide chains of adjacent protein molecules.

scolex (skō'leks) (Gr. *skōlēx,* worm, grub) The holdfast, or so-called head, of a tapeworm; bears suckers and, in some, hooks, and posterior to it new proglottids are differentiated.

scrotum (skrō'təm) (L., bag) The pouch that contains the testes in most mammals.

scyphistoma (sī-fis'tə-mə) (Gr. *skyphos,* cup, + *stoma,* mouth) A stage in the development of scyphozoan jellyfish just after the larva becomes attached; the polyp form of a scyphozoan.

sebaceous (sə-bāsh'əs) (L. *sebaceus,* made of tallow) A type of mammalian epidermal gland that produces a fatty substance.

sedentary (sed'ən-ter-ē) Stationary, sitting, inactive; staying in one place.

seminiferous (sem-ə-nif'rəs) (L. *semen,* semen, + *ferre,* to bear) Pertains to the tubules that produce or carry semen in the testes.

sensillum pl. **sensilla** (sin-sil'-əm) (L. *sensilus,* sensitive) Cuticular receptors in arthropods for stimuli other than light.

septum pl. **septa** (L., fence) A wall between two cavities.

serosa (sə-rō'sə) (N.L. from L. *serum,* serum) The outer embryonic membrane of birds and reptiles; chorion. Also, the peritoneal lining of the body cavity.

serotonin (sir'ə-tōn'ən) (L. *serum,* serum) A phenolic amine, found in the serum of clotted blood and in many other tissues, that possesses several poorly understood metabolic, vascular, and neural functions; 5-hydroxytryptamine.

serum (sir'əm) (L., whey, serum) The liquid that separates from the blood after coagulation; blood plasma from which fibrinogen has been removed. Also, the clear portion of a biological fluid separated from its particulate elements.

sessile (ses'əl) (L. *sessilis,* low, dwarf) Attached at the base; fixed to one spot, not able to move about.

seta (sēt'ə), pl. **setae** (sē'tē) (L., bristle) A needlelike chitinous structure of the integument of annelids, arthropods, and others.

siliceous (sə-li'shəs) (L. *silex,* flint) Containing silica.

sinus (sī'nəs) (L., curve) A cavity or space in tissues or in bones.

siphonoglyph (sī-fän'ə-glif') (Gr. *siphōn,*

reed, tube, siphon, + *glyphē*, carving) Ciliated furrow in the gullet of sea anemones.

siphuncle (sī'fun-kl) (L. *siphunculus*, small tube) Cord of tissue running through shell of nautiloid, connecting all chambers with body of animal.

soma (sō'mə) (Gr., body) The whole of an organism except the germ cells (germ plasm).

somatic (sō-mat'ik) (Gr. *sōma*, body) Refers to the body, for example, somatic cells in contrast to germ cells.

somatoplasm (sō'ma-də-pla'zm) (Gr. *sōma*, body, + *plasma*, anything formed) The living matter that makes up the mass of the body as distinguished from germ plasm, which makes up the reproductive cells. The protoplasm of body cells.

somite (sō'mīt) (Gr. *sōma*, body) One of the blocklike masses of mesoderm arranged segmentally (metamerically) in a longitudinal series beside the digestive tube of the embryo; metamere.

speciation (spē'sē-ā'shən) (L. *species*, kind) The evolutionary process by which new species arise; the process by which variations become fixed.

species (spē'shez, spē'sēz) sing. and pl. (L., particular kind) A group of interbreeding individuals of common ancestry that are reproductively isolated from all other such groups; a taxonomic unit ranking below a genus and designated by a binomen consisting of its genus and the species name.

spermatheca (spər'mə-thē'kə) (Gr. *sperma*, seed, + *thēkē*, case) A sac in the female reproductive organs for the reception and storage of sperm.

spermatid (spər'mə-təd) (Gr. *sperma*, seed, + *eidos*, form) A growth stage of a male reproductive cell arising by division of a secondary spermatocyte; gives rise to a spermatozoon.

spermatocyte (spər-mad'ə-sīt) (Gr. *sperma*, seed, + *kytos*, hollow vessel) A growth stage of a male reproductive cell; gives rise to a spermatid.

spermatogenesis (spər-mad'ə-jen'-ə-səs) (Gr. *sperma*, seed, + *genesis*, origin) Formation and maturation of spermatozoa.

spermatogonium (spər'mad-ə-gō'nē-əm) (Gr. *sperma*, seed, + *gonē*, offspring) Precursor of mature male reproductive cell; gives rise directly to a spermatocyte.

spermatophore (spər-mad'ə-fōr') (Gr. *sperma*, *spermatos*, seed, + *pherein*, to bear) Capsule or packet enclosing sperm, produced by males of several invertebrate groups and a few vertebrates.

sphincter (sfingk'tər) (Gr. *sphinkter*, band, sphincter, from *sphingein*, to bind tight) A ring-shaped muscle capable of closing a tubular opening by constriction.

spicule (spi'kyul) (L. dim. of *spica*, point) One of the minute calcareous or siliceous skeletal bodies found in sponges, radiolarians, soft corals, and sea cucumbers.

spiracle (spi'rə-kəl) (L. *spiraculum*, from *spirare*, to breathe) External opening of a trachea in arthropods. One of a pair of openings on the head of elasmobranchs for passage of water. Exhalent aperture of tadpole gill chamber.

spiral cleavage A type of early embryonic cleavage in which cleavage planes are diagonal to the polar axis and unequal cells are produced by the alternate clockwise and counterclockwise cleavage around the axis of polarity; determinate cleavage

spongin (spun'jin) (L. *spongia*, sponge) Fibrous, collagenous material making up the skeletal network of horny sponges.

spongioblast (spun'jeo-blast) (G. *spongos*, + *blastos*, bud) Cell in a sponge that secretes spongin, a protein.

spongocoel (spun'jō-sēl) (Gr. *spongos*, sponge, + *koilos*, hollow) Central cavity in sponges.

spongocyte (spun'jō-sīt) (Gr. *spongos*, sponge, + *kytos*, hollow vessel) A cell in sponges that secretes spongin.

sporocyst (spō'rə-sist) (Gr. *sporos*, seed, + *kystis*, pouch) A larval stage in the life cycle of flukes; it originates from a miracidium.

sporogony (spor-äg'ə-nē) (Gr. *sporos*, seed, + *gonos*, birth) Multiple fission to produce sporozoites after zygote formation.

sporozoite (spō'rə-zō'īt) (Gr. *sporos*, seed + *zōon*, animal, + *ite*, suffix for body part) A stage in the life history of many sporozoan protozoa; released from oocysts.

squamous epithelium (skwā'məs) (L. *squama*, scale, + *osus*, full of) Simple epithelium of flat, nucleated cells.

statoblast (stad'ə-blast) (Gr. *statos*, standing, fixed, + *blastos*, germ) Biconvex capsule containing germinative cells and produced by most freshwater ectoprocts by asexual budding. Under favorable conditions it germinates to give rise to new zooid.

statocyst (Gr. *statos*, standing, + *kystis*, bladder) Sense organs of equilibrium; a fluid-filled cellular cyst containing one or more granules (statoliths) used to sense direction of gravity.

statolith (Gr. *statos*, standing, + *lithos*, stone) Small calcareous body resting on tufts of cilia in the statocyst.

stenohaline (sten-ə-hā'līn, -lən) (Gr. *stenos*, narrow, + *hals*, salt) Pertaining to aquatic organisms that have restricted tolerance to changes in environmental saltwater concentration.

stereogastrula (ste'rē-ə-gas'trə-lə) (Gr. *stereos*, solid, + *gastēr*, stomach, + *L. ula*, dim.) A solid type of gastrula, such as the planula of cnidarians.

sternum (ster'nəm) (L., breastbone) Ventral plate of an arthropod body segment; breastbone of vertebrates.

sterol (ste'rōl) **steroid**(ste'roid) (Gr. *stereos*, solid + L. *ol* [from *oleum*, oil]) One of a class of organic compounds containing a molecular skeleton of four fused carbon rings; it includes cholesterol, sex hormones, adrenocortical hormones, and vitamin D.

stigma (Gr. *stigma*, mark, tattoo mark) Eyespot in certain protozoa. Spiracle of certain terrestrial arthropods.

stolon (stō'lən) (L. *stolō*, *stolonis*, a shoot, or sucker of a plant) A rootlike extension of the body wall that gives rise to buds that may develop into new zooids, thus forming a compound animal in which the zooids remain united by the stolon. Found in some colonial anthozoans, hydrozoans, ectoprocts and ascidians.

stoma (stō'mə) (Gr., mouth) A mouthlike opening.

stomochord (stō'mə-kord) (Gr. *stoma*, mouth, + *chordē*, cord) Anterior evagination of the dorsal wall of the buccal cavity into the proboscis of hemichordates; the buccal diverticulum.

strobila (strō'bə-lə) (Gr. *strobilos*, pine cone) A stage in the development of the scyphozoan jellyfish. Also, the chain of proglottids of a tapeworm.

stroma (strō'mə) (Gr. *strōma*, bedding) Supporting connective tissue framework of an animal organ; filmy framework of red blood corpuscles and certain cells.

structural gene A gene carrying the information to construct a protein.

substrate The substance upon which an enzyme acts; also a base or foundation (substratum); and the substance or base on which an organism grows.

succession Changes in species and structure of a community leading eventually to a climax community, either in previously unoccupied habitats (primary succession) or following disturbance (secondary succession).

sycon (sī'kon) (Gr. *sykon,* fig) A type of canal system in certain sponges. Sometimes called syconoid.

symbiosis (sim'bī-ōs'əs, sim'bē-ō'əs) (Gr. *syn,* with, + *bios,* life) The living together of two different species in an intimate relationship. Symbiont always benefits; host may benefit, may be unaffected, or may be harmed (mutualism, commensalism, and parasitism).

sympatric (sim-pa'trik) (Gr. *syn,* with, + *patra,* native land) Having the same or overlapping regions of distribution. Noun, **sympatry.**

synapse (si'naps, si-naps') (Gr. *synapsis,* contact, union) The place at which a nerve impulse passes between neuron processes, typically from an axon of one nerve cell to a dendrite of another nerve cell.

synapsis (si-nap'səs) (Gr. *synapsis,* contact, union) The time when the pairs of homologous chromosomes lie alongside each other in the first meiotic division.

syncytium (sin-sish'ē-əm) (Gr. *syn,* with, + *kytos,* hollow vessel) A mass of protoplasm containing many nuclei and not divided into cells.

syndrome (sin'drōm) (Gr. *syn,* with + *dramein,* to run) A group of symptoms characteristic of a particular disease or abnormality.

syngamy (sin'gə-mē) (Gr. *syn,* with, + *gamos,* marriage) Fertilization of one gamete with another individual gamete to form a zygote, found in most animals with sexual reproduction.

syrinx (sir'inks) (Gr., shepherd's pipe) The vocal organ of birds located at the base of the trachea.

systematics (sis-tə-mad'iks) Science of classification and evolutionary biology.

T cell A type of lymphocyte most important in cellular immune response and in regulation of most immune responses.

tactile (tak'til) (L. *tactilis,* able to be touched, from *tangere,* to touch) Pertaining to touch.

tagma pl. **tagmata** (Gr. *tagma,* arrangement, order, row) A compound body section of an arthropod resulting from embryonic fusion of two or more segments; for example, head, thorax, abdomen.

tagmatization, tagmosis Organization of the arthropod body into tagmata.

taiga (tī'gä) (Russ.) Habitat zone characterized by large tracts of coniferous forests, long, cold winters, and short summers; most typical in Canada and Siberia.

taxis (tak'sis) pl. **taxes** (Gr. *taxis,* order, arrangement) An orientation movement by a (usually) simple organism in response to an environmental stimulus.

taxon (tak'son) pl. **taxa** (Gr. *taxis,* order, arrangement) Any taxonomic group or entity.

taxonomy (tak-sän'ə-mi) (Gr. *taxis,* order, arrangement, + *nomos,* law) Study of the principles of scientific classification; systematic ordering and naming of organisms.

tectum (tek'təm) (L., roof) A rooflike structure, for example, dorsal part of capitulum in ticks and mites.

tegument (teg'ū-ment) (L., *tegumentum,* from *tegere,* to cover) An integument; specifically external covering in cestodes and trematodes, formerly believed to be a cuticle.

telencephalon (tel'en-sef'ə-lon) (Gr. *telos,* end, + *encephalon,* brain) The most anterior vesicle of the brain; the anteriormost subdivision of the prosencephalon that becomes the cerebrum and associated structures.

telocentric (tē'lō-sen'trək) (Gr. *telos,* end, + *kentron,* center) Chromosome with centromere at the end.

telolecithal (tel-lō-les'ə-thəl) (Gr. *telos,* end, + *lekithos,* yolk, + *al*) Having the yolk concentrated at one end of an egg.

telson (tel'sən) (Gr., extremity) Posterior projection of the last body segment in many crustaceans.

template (tem'plət) A pattern or mold guiding the formation of a duplicate; often used with reference to gene duplication.

tentaculocyst (ten-tak'u-lō-sist) (L. *tentaculum,* feeler, + Gr. *kystis,* pouch) One of the sense organs along the margin of medusae; a rhopalium.

tergum (ter'gəm) (L., back) Dorsal part of an arthropod body segment.

territory (L. *territorium,* from *terra,* earth) A restricted area preempted by an animal or pair of animals, usually for breeding purposes, and guarded from other individuals of the same species.

test (L. *testa,* shell) A shell or hardened outer covering.

tetrad (te'trad) (Gr. *tetras,* four) Group of two pairs of chromatids at synapsis and resulting from the replication of paired homologous chromosomes; the bivalent.

tetrapods (te'trə-päds) (Gr. *tetras,* four, + *pous, podos,* foot) Four-footed vertebrates; the group includes amphibians, reptiles, birds, and mammals.

therapsid (thə-rap'sid) (Gr. *theraps,* an attendant) Extinct Mesozoic mammal-like reptile from which true mammals evolved.

thermocline (thər'mō-klīn) (gr. *thermē* heat, + *klinein,* to swerve) Layer of water separating upper warmer and lighter water from lower colder and heavier water in a lake or sea; a stratum of abrupt change in water temperature.

Tiedemann's bodies (tēd'ə-mənz) (from F. Tiedemann, German anatomist) Four or five pairs of pouchlike bodies attached to the ring canal of sea stars, apparently functioning in production of coelomocytes.

tight junction Region of actual fusion of cell membranes between two adjacent cells.

tissue (ti'shu) (M.E. *tissu,* tissue) An aggregation of cells, usually of the same kind, organized to perform a common function.

torsion (L. *torquere,* to twist) A twisting phenomenon in gastropod development that alters the position of the visceral and pallial organs by 180 degrees.

toxicyst (tox'i-sist) (Gr. *toxikon,* poison, + *kystis,* bladder) Structures possessed by predatory ciliate protozoa, which on stimulation expel a poison to subdue the prey.

trachea (trā'kē-ə) (M.L., windpipe) The windpipe. Also, any of the air tubes of insects.

transcription Formation of messenger RNA from the coded DNA.

transduction Condition in which bacterial DNA (and the genetic characteristics it bears) is transferred from one bacterium to another by the agent of viral infection.

transfer RNA (tRNA) A form of RNA of about seventy nucleotides, which are adapter molecules in the synthesis of proteins. A specific amino acid molecule is carried by transfer RNA to a ribosome-messenger RNA complex for incorporation into a polypeptide.

transformation Condition in which DNA in the environment of bacteria somehow penetrates them and is incorporated into their genetic complement, so that their progeny inherit the genetic characters so acquired.

translation (L., a transferring) The process in which the genetic information present in messenger RNA is used to direct the order of specific amino acids during protein synthesis.

trichinosis (trik-ən-o'səs) Disease caused by infection with the nematode *Trichinella spiralis.*

trichocyst (trik'ə-sist) (Gr. *thrix,* hair, + *kystis,* bladder) Saclike protrusible organelle in the ectoplasm of ciliates, which discharges as a threadlike weapon of defense.

triglyceride (trī-glis'ə-rīd) (Gr. *tria,* three, + *glykys,* sweet, + *ide,* suffix denoting compound) A triester of glycerol with three acids.

triploblastic (trip'lō-blas'tik) (Gr. *triploos,* triple, + *blastos,* germ) Pertaining to metazoa in which the embryo has three primary germ layers—ectoderm, mesoderm, and endoderm.

trisomy 21 See **Down syndrome.**

trochophore (trōk'ə-fōr) (Gr. *trochos,* wheel, + *pherein,* to bear) A free-swimming ciliated marine larva characteristic of most molluscs and certain ectoprocts, brachiopods, and marine worms; an ovoid or pyriform body with preoral circlet of cilia and sometimes a secondary circlet behind the mouth.

trophallaxis (trōf'ə-lak'səs) (Gr. *trophē,* food, + *allaxis,* barter, exchange) Exchange of food between young and adults, especially certain social insects.

trophoblast (trōf'ə-blast) (Gr. *trephein,* to nourish, + *blastos,* germ) Outer ectodermal nutritive layer of blastodermic vesicle; in mammals it is part of the chorion and attaches to the uterine wall.

trophozoite (trōf'ə-zo'īt) (Gr. *trophē,* food, + *zōon,* animal) Adult stage in the life cycle of a protozoan in which it is actively absorbing nourishment.

tropomyosin (trōp'ə-mī'ə-sən) (Gr. *tropos,* turn, + *mys,* muscle) Low-molecular weight protein surrounding the actin filaments of striated muscle.

troponin (trə-pōn'in) Complex of globular proteins positioned at intervals along the actin filament of skeletal muscle; thought to serve as a calcium-dependent switch in muscle contraction.

tube feet (podia) Numerous small, muscular, fluid-filled tubes projecting from body of echinoderms; part of water-vascular system; used in locomotion, clinging, food handling, and respiration.

tubulin (tū'bū-lən) (L. *tubulus,* small tube, + *in,* belonging to) Globular protein forming the hollow cylinder of microtubules.

tundra (tun'drə) (Russ. from Lapp, *tundar,* hill) Terrestrial habitat zone, located between taiga and polar regions; characterized by absence of trees, short growing season, and mostly frozen soil during much of the year.

tunic (L. *tunica,* tunic, coat) In tunicates, a cuticular, cellulose-containing covering of the body secreted by the underlying body wall.

turbellarian (tər'bə-lar'ə-an) (L. *turbellae,* a stir or tumult) Flatworm of phylum Platyhelminthes, class Turbellaria.

typhlosole (tif'lə-sōl) (Gr. *typhlos,* blind, + *sōlēn,* channel, pipe) A longitudinal fold projecting into the intestine in certain invertebrates such as the earthworm.

umbilical (L. *umbilicus,* navel) Refers to the navel, or umbilical cord.

umbo (um'bō) pl. **umbones** (əm-bō'nēz) (L., boss, or knob, of a shield) One of the prominences on either side of the hinge region in a bivalve mollusc shell. Also, the "beak" of a brachiopod shell.

ungulate (un'gū-lət) (L. *ungula,* hoof) Hoofed. Noun, any hoofed mammal.

urethra (ū-rē'thrə) (Gr. *ourethra,* urethra) The tube from the urinary bladder to the exterior in both sexes.

utricle (ū'trə-kəl) (L. *utriculus,* little bag) The part of the inner ear containing the receptors for dynamic body balance; the semicircular canals lead from and to the utricle.

vacuole (vak'yə-wōl) (L. *vacuus,* empty, + Fr. *ole,* dim.) A membrane-bounded, fluid-filled space in a cell

valence (vā'ləns) (L. *valere,* to have power) Degree of combining power of an element as expressed by the number of atoms of hydrogen (or its equivalent) that that element can hold (if negative) or displace in a reaction (if positive). The oxidation state of an element in a compound. The number of electrons gained, shared, or lost by an atom when forming a bond with one or more other atoms.

valve (L. *valva,* leaf of a double door) One of the two shells of a typical bivalve mollusc or brachiopod.

vector (L., a bearer, carrier, from *vehere, vectum,* to carry) Any agent that carries and transmits pathogenic microorganisms from one host to another host.

veliger (vēl'ə-jər, vel-) (L. *velum,* veil, covering) Larval form of certain molluscs; develops from the trochophore and has the beginning of a foot, mantle, shell, and so on.

velum (vē'ləm) (L., veil, covering) A membrane on the subumbrella surface of jellyfish of class Hydrozoa. Also, a ciliated swimming organ of the veliger larva.

vestige (ves'tij) (L. *vestigium,* footprint) A rudimentary organ that may have been well developed in some ancestor or in the embryo.

villus (vil'əs) pl. **villi** (L., tuft of hair) A small fingerlike, vascular process on the wall of the small intestine. Also, one of the branching, vascular processes on the embryonic portion of the placenta.

virus (vī'rəs) (L., slimy liquid, poison) A submicroscopic noncellular particle composed of a nucleoprotein core and a protein shell; parasitic; will grow and reproduce in a host cell.

viscera (vis'ər-ə) (L., pl. of *viscus,* internal organ) Internal organs in the body cavity.

vitalism (L. *vita,* life) The view that natural processes are controlled by supernatural forces and cannot be explained through the laws of physics and chemistry alone, as opposed to mechanism.

vitamin (L. *vita,* life + *amine,* from former supposed chemical origin) An organic substance required in small amounts for normal metabolic function; must be supplied in the diet or by intestinal flora because the organism cannot synthesize it.

vitelline membrane (vi-tel'ən, vī'təl-ən) (L. *vitellus,* yolk of an egg) The noncellular membrane that encloses the egg cell.

viviparity (vī'və-par'ə-dē) (L. *vivus,* alive + *parere,* to bring forth) Reproduction in which eggs develop within the female body, with nutritional aid of maternal parent as in therian mammals, many reptiles, and some fishes; offspring are born as juveniles. Adj., **viviparous** (vī-vip'ə-rəs).

water-vascular system System of fluid-filled closed tubes and ducts peculiar to echinoderms; used to move tentacles and tube feet that serve variously for clinging, food handling, locomotion, and respiration.

X-organ Neurosecretory organ in eyestalk of crustaceans that secretes molt-inhibiting hormone.

Y-organ Gland in the antennal or maxillary segment of some crustaceans that secretes molting hormone.

zoecium, zooecium (zō-ē′shē-əm) (Gr. *zōon* animal, + *oikos,* house) Cuticular sheath or shell of Ectoprocta.

zoochlorella (zō′ə-klōr-el′ə) (Gr. *zōon,* animal, + *Chlorella*) Any of various minute green algae (usually *Chlorella*) that live symbiotically within the cytoplasm of some protozoa and other invertebrates.

bat / āpe / ärmadillo / herring / fēmale / finch / līce / crocodile / crōw / duck / ūnicorn / ə indicates unaccented vowel sound "uh" as in mammal, fishes, cardinal, heron, vulture / stress as in bi-ol′o-gy, bi′o-log′i-cal

zooid (zō-oid) (Gr. *zōon,* animal) An individual member of a colony of animals, such as colonial cnidarians and ectoprocts.

zooxanthella (zo′ə-zan-thəl′ə) (Gr. *zōon,* animal, + *xanthos,* yellow) A minute dinoflagellate alga living in the tissues of many types of marine invertebrates.

zygote (Gr. *zygōtos,* yoked) The fertilized egg.

zygotic meiosis Meiosis that takes place within the first few divisions after zygote formation; thus all stages in the life cycle other than the zygote are haploid.

INDEX

D

F

G

H

I

CREDITS

CHAPTER 1
The Andromeda Galaxy.
Photograph courtesy NASA (83-HC-315).

CHAPTER 2
California redwoods.
Photograph by C.P. Hickman, Jr.

CHAPTER 3
Black fly chromosomes.
Photograph by P. Adler, Clemson University.

CHAPTER 4
Charles Darwin, 1881.
Photograph courtesy Department of Library Services, American Museum of Natural History. (Neg. No. 32662.)

CHAPTER 5
Mountain Stream, Smoky Mountains.
Photograph by C.P. Hickman, Jr.

CHAPTER 6
Skeletal muscle.
TEM by W. Rosenberg, Iona College/BPS.

CHAPTER 7
Frontispiece from William Harvey's *An Anatomical Essay on the Movement of the Heart and Blood in Animals, 1628.*
Photograph by C.P. Hickman, Jr.

CHAPTER 8
Walter Cannon's *The way of an investigator.*
Photograph by C.P. Hickman, Jr.

CHAPTER 9
Zebra and topi, Kenya.
Photograph by C.P. Hickman, Jr.

CHAPTER 10
Tick on plant.
Photograph by D.H. Elles/Visuals Unlimited.

CHAPTER 11
Normal thyroid cells.
Photograph by D.M. Phillips/Visuals Unlimited.

CHAPTER 12
Trained circus elephants.
Photograph by J.D. Cunningham/Visuals Unlimited.

CHAPTER 13
Fertilization of hamster egg.
Photograph by D.M. Phillips/Visuals Unlimited.

CHAPTER 14
Nineteenth-century wood engraving.
From Animals; A Pictorial Archive from Nineteenth-Century Sources selected by Jim Harter. New York: Dover Publications, 1979.

CHAPTER 15
Ciliate protozoa, *Vorticella.*
Photograph by A.M. Siegelman/Visuals Unlimited.

CHAPTER 16
Giant tub sponge, *Xestospongia muta,* and diver.
Photograph by W.C. Ober.

CHAPTER 17
Crimson or snakelock anemone, *Cribrinopsis fernaldi.*
Photograph by R. Harbo/Marine Images.

CHAPTER 18
Planaria.
Photograph by J.D. Cunningham/Visuals Unlimited.

CHAPTER 19
Trichinella spiralis.
Photograph by A.M. Siegelman/Visuals Unlimited.

CHAPTER 20
Flamingo tongue, *Cyphoma gibbosum,* on gorgonian coral.
Photograph by L.S. Roberts.

CHAPTER 21
Clamworm, *Nereis virens.*
Photograph by R. DeGoursey/Visuals Unlimited.

CHAPTER 22
Original art by W. Ober.

CHAPTER 23
Ectoproct, *Reteporella graeffei,* on Australia's Great Barrier Reef.
Photograph by L.S. Roberts

CHAPTER 24
Mass of sea stars, *Pisaster ochraceous,* Pacific Coast, British Columbia.
Photograph by C.E. Mills, University of Washington/BPS.

CHAPTER 25
Branchiostoma (amphioxus) buried in shell gravel, anterior ends protruding.
Photograph by H. Angel.

CHAPTER 26
School of grunts, *Xenistius jessiae,* around coral reef.
Photograph by G.R. Robinson/Visuals Unlimited.

CHAPTER 27
Pickeral frog, *Rana palustris,* during metamorphosis.
Photograph by C.P. Hickman, Jr.

CHAPTER 28
Chicken turtle, *Dierochelys reticularia,* hatching.
Photograph by D. Jackson/Visuals Unlimited.

CHAPTER 29
Sandhill cranes migrating at night.
Photograph by W.J. Weber/Visuals Unlimited.

CHAPTER 30
Three-month-old grizzly bear cub.
Photograph by L.L. Rue, III.